D0407449

ANNUAL REVIEW OF BIOCHEMISTRY

EDITORIAL COMMITTEE (2003)

TANIA A. BAKER
ROBERT H. FILLINGAME
MICHAEL M. GOTTESMAN
ROGER D. KORNBERG
ROWENA G. MATTHEWS
GREGORY PETSKO
CHRISTIAN R.H. RAETZ
CHARLES C. RICHARDSON
JOANNE STUBBE
JEREMY W. THORNER
ROBERT TJIAN

RESPONSIBLE FOR THE ORGANIZATION OF VOLUME 72
(EDITORIAL COMMITTEE, 2001)

MICHAEL M. GOTTESMAN
DANIEL HERSCHLAG
ROWENA G. MATTHEWS
MICHAEL E. O'DONNELL
GREGORY PETSKO
CHRISTIAN R.H. RAETZ
CHARLES C. RICHARDSON
ROBERT D. SIMONI
THOMAS A. STEITZ
JEREMY W. THORNER
ROBERT TJIAN
CAROLYN BERTOZZI (GUEST)
JACK F. KIRSCH (GUEST)
ROGER D. KORNBERG (GUEST)
ERIN K. O'SHEA (GUEST)
WILLIAM T. WICKNER (GUEST)

Production Editor: JESSLYN S. HOLOMBO
Bibliographic Quality Control: MARY A. GLASS
Color Graphics Coordinator: EMÉ O. AKPABIO
Electronic Content Coordinator: SUZANNE K. MOSES
Subject Indexer: KYRA KITTS

ANNUAL REVIEW OF BIOCHEMISTRY

VOLUME 72, 2003

CHARLES C. RICHARDSON, *Editor*
Harvard Medical School

ROGER D. KORNBERG, *Associate Editor*
Stanford University School of Medicine

CHRISTIAN R.H. RAETZ, *Associate Editor*
Duke University Medical Center

JEREMY W. THORNER, *Associate Editor*
University of California, Berkeley

www.annualreviews.org science@annualreviews.org 650-493-4400

ANNUAL REVIEWS
4139 El Camino Way • P.O. Box 10139 • Palo Alto, California 94303-0139

ANNUAL REVIEWS
Palo Alto, California, USA

COPYRIGHT © 2003 BY ANNUAL REVIEWS, PALO ALTO, CALIFORNIA, USA. ALL RIGHTS RESERVED. The appearance of the code at the bottom of the first page of an article in this serial indicates the copyright owner's consent that copies of the article may be made for personal or internal use, or for the personal or internal use of specific clients. This consent is given on the condition that the copier pay the stated per-copy fee of $14.00 per article through the Copyright Clearance Center, Inc. (222 Rosewood Drive, Danvers, MA 01923) for copying beyond that permitted by Section 107 or 108 of the U.S. Copyright Law. The per-copy fee of $14.00 per article also applies to the copying, under the stated conditions, of articles published in any *Annual Reviews* serial before January 1, 1978. Individual readers, and nonprofit libraries acting for them, are permitted to make a single copy of an article without charge for use in research or teaching. This consent does not extend to other kinds of copying, such as copying for general distribution, for advertising or promotional purposes, for creating new collective works, or for resale. For such uses, written permission is required. Write to Permissions Dept., Annual Reviews, 4139 El Camino Way, P.O. Box 10139, Palo Alto, CA 94303-0139 USA.

International Standard Serial Number: 0066-4154
International Standard Book Number: 0-8243-0872-7
Library of Congress Catalog Card Number: 32-25093

All Annual Reviews and publication titles are registered trademarks of Annual Reviews.

♾ The paper used in this publication meets the minimum requirements of American National Standards for Information Sciences—Permanence of Paper for Printed Library Materials. ANSI Z39.48-1992.

Annual Reviews and the Editors of its publications assume no responsibility for the statements expressed by the contributors to this *Annual Review.*

TYPESET BY CADMUS PROFESSIONAL COMMUNICATIONS, LINTHICUM, MD
PRINTED AND BOUND BY QUEBECOR WORLD PRINTING, KINGSPORT, TN

Annual Review of Biochemistry
Volume 72, 2003

CONTENTS

ERRATA

An online log of corrections to *Annual Review of Biochemistry* chapters may be found at http://biochem.annualreviews.org/errata.shtml

ERRATA

Erratum: *Annu. Rev. Biochem.* 1999. 68:821–861

Adam J. Shaywitz and Michael E. Greenberg, **CREB: A STIMULUS-INDUCED TRANSCRIPTION FACTOR ACTIVATED BY A DIVERSE ARRAY OF EXTRACELLULAR SIGNALS**

The second sentence of the printed abstract contained errors. The abstract should be:

■ **Abstract** Extracellular stimuli elicit changes in gene expression in target cells by activating intracellular protein kinase cascades that phosphorylate transcription factors within the nucleus. One of the best characterized stimulus-induced transcription factors, cyclic AMP response element (CRE)-binding protein (CREB), activates transcription of target genes in response to a diverse array of stimuli, including peptide hormones, growth factors, and neuronal activity, that activate a variety of protein kinases including protein kinase A (PKA), pp90 ribosomal S6 kinase (pp90RSK), and Ca2+/calmodulin-dependent protein kinases (CaMKs). These kinases all phosphorylate CREB at a particular residue, serine 133 (Ser133), and phosphorylation of Ser133 is required for CREB-mediated transcription. Despite this common feature, the mechanism by which CREB activates transcription varies depending on the stimulus. In some cases, signaling pathways target additional sites on CREB or proteins associated with CREB, permitting CREB to regulate distinct programs of gene expression under different conditions of stimulation. This review will discuss the molecular mechanisms by which Ser133–phosphorylated CREB activates transcription, intracellular signaling pathways that lead to phosphorylation of CREB at Ser133, and features of each signaling pathway that impart specificity at the level of CREB activation.

Related Articles

From the ***Annual Review of Biomedical Engineering,*** Volume 4 (2002)

DNA Microarray Technology: Devices, Systems, and Applications, Michael J. Heller

Peptide Aggregation in Neurodegenerative Disease, Regina M. Murphy

Advances in In Vivo Bioluminescent Imaging of Gene Expression, Christopher H. Contag and Michael H. Bachmann

Advances in Proteomic Technologies, Martin L. Yarmush and Arul Jayaraman

From the ***Annual Review of Biophysics and Biomolecular Structure,*** Volume 32 (2003)

The Crystallographic Model of Rhodopsin and Its Use in Studies of Other G Protein–Coupled Receptors, Slawomir Filipek, David C. Teller, Krzysztof Palczewski, and Ronald Stenkamp

Nucleic Acid Recognition by OB-Fold Proteins, Douglas L. Theobald, Rachel M. Mitton-Fry, and Deborah S. Wuttke

Structure and Function of the Calcium Pump, David L. Stokes and N. Michael Green

Proteome Analysis by Mass Spectrometry, P. Lee Ferguson and Richard D. Smith

New Insight into Site-Specific Recombination from FLP Recombinase-DNA Structures, Yu Chen and Phoebe A. Rice

From the ***Annual Review of Cell and Developmental Biology,*** Volume 18 (2002)

A Cell Biological Perspective on Alzheimer's Disease, Wim Annaert and Bart De Strooper

Receptor Kinase Signaling in Plant Development, Philip Becraft

Chromosome-Microtubule Interactions During Mitosis, J. Richard McIntosh, Ekaterina L. Grishchuk, and Robert R. West

Proteolysis and Sterol Regulation, Randolph Hampton

Autoinhibitory Domains: Modular Effectors of Cellular Regulation, Miles A. Pufall and Barbara J. Graves

Molecular Mechanisms of Epithelial Morphogenesis, Frieder Schöck and Norbert Perrimon

Transcriptional and Translational Control in the Mammalian Unfolded Protein Response, Heather P. Harding, Marcella Calfon, Fumihiko Urano, Isabel Novoa, and David Ron

From the ***Annual Review of Genetics,*** Volume 36 (2002)

DNA Topology-Mediated Control of Global Gene Expression in Escherichia coli, G. Wesley Hatfield and Craig J. Benham

Origins of Spontaneous Mutations: Specificity and Directionality of Base-Substitution, Frameshift, and Sequence-Substitution Mutageneses, Hisaji Maki

Allosteric Cascade of Spliceosome Activation, David A. Brow

Chromosome Rearrangements and Transposable Elements, Wolf-Ekkehard Lönnig and Heinz Saedler

Toward Maintaining the Genome: DNA Damage and Replication Checkpoints, Kara A. Nyberg, Rhett J. Michelson, Charles W. Putnam, and Ted A. Weinert

From the ***Annual Review of Genomics and Human Genetics,*** Volume 3 (2002)

Hedgehog Signaling and Human Disease, Allen E. Bale

Deciphering the Genetic Basis of Alzheimer's Disease, Dennis J. Selkoe and Marcia B. Podlisny

Molecular Mechanisms for Genomic Disorders, Ken Inoue and James R. Lupski

Structuring the Universe of Proteins, Stephen K. Burley and Jeffrey B. Bonanno

Databases and Tools for Browsing Genomes, Ewan Birney, Michele Clamp, and Tim Hubbard

From the ***Annual Review of Immunology,*** Volume 21 (2003)

IgA Fc Receptors, Renato C. Monteiro and Jan G.J. van de Winkel

Toll-Like Receptors, Kiyoshi Takeda, Tsuneyasu Kaisho, and Shizuo Akira

Location is Everything: Lipid Rafts and Immune Cell Signaling, Michelle Dykstra, Anu Cherukuri, Hae Won Sohn, Shiang-Jong Tzeng, and Susan K. Pierce

Molecular Interactions Mediating T Cell Antigen Recognition, P. Anton van der Merwe and Simon J. Davis

Molecular Mechanisms Regulating Th1 Immune Responses, Susanne J. Szabo, Brandon M. Sullivan, Stanford L. Peng, and Laurie H. Glimcher

From the ***Annual Review of Medicine,*** Volume 54 (2003)

Molecular Genetics of Lung Cancer, Yoshitaka Sekido, Kwun M. Fong, and John D. Minna

Calorie Restriction, Aging, and Cancer Prevention: Mechanisms of Action and Applicability to Humans, Stephen D. Hursting, Jackie A. Lavigne, David Berrigan, Susan N. Perkins, and J. Carl Barrett

From the ***Annual Review of Microbiology,*** Volume 56 (2002)

Biofilms as Complex Differentiated Communities, P. Stoodley, K. Sauer, D.G. Davies, and J.W. Costerton

Transition Metal Transport in Yeast, Anthony Van Ho, Diane McVey Ward, and Jerry Kaplan

Inteins: Structure, Function, and Evolution, J. Peter Gogarten, Alireza G. Senejani, Olga Zhaxybayeva, Lorraine Olendzenski, and Elena Hilario

Metabolic Diversity in Aromatic Compound Utilization by Anaerobic Microbes, Jane Gibson and Caroline S. Harwood

The Molecular Biology of West Nile Virus: A New Invader of the Western Hemisphere, Margo A. Brinton

Bacterial Chromosome Segregation, Geoffrey C. Draper and James W. Gober

Impact of Genomic Technologies on Studies of Bacterial Gene Expression, Virgil Rhodius, Tina K. Van Dyk, Carol Gross, and Robert A. LaRossa

Prions as Protein-Based Genetic Elements, Susan M. Uptain and Susan Lindquist

From the ***Annual Review of Neuroscience,*** Volume 25 (2002)

Transcriptional Codes and the Control of Neuronal Identity, Ryuichi Shirasaki and Samuel L. Pfaff

A Decade of Molecular Studies of Fragile X Syndrome, William T. O'Donnell and Stephen T. Warren

From the ***Annual Review of Pharmacology and Toxicology,*** Volume 43 (2003)

Protein Flexibility and Computer-Aided Drug Design, Chung F. Wong and J. Andrew McCammon

ANNUAL REVIEWS is a nonprofit scientific publisher established to promote the advancement of the sciences. Beginning in 1932 with the *Annual Review of Biochemistry,* the Company has pursued as its principled function the publication of high-quality, reasonably priced *Annual Review* volumes. The volumes are organized by Editors and Editorial Committees who invite qualified authors to contribute critical articles reviewing significant developments within each major discipline. The Editor-in-Chief invites those interested in serving as future Editorial Committee members to communicate directly with him. Annual Reviews is administered by a Board of Directors, whose members serve without compensation.

2003 Board of Directors, Annual Reviews

Richard N. Zare, *Chairman of Annual Reviews*
Marguerite Blake Wilbur, Professor of Chemistry, Stanford University
John I. Brauman, *J. G. Jackson–C. J. Wood Professor of Chemistry, Stanford University*
Peter F. Carpenter, *Founder, Mission and Values Institute*
Sandra M. Faber, *Professor of Astronomy and Astronomer at Lick Observatory, University of California at Santa Cruz*
Susan T. Fiske, *Professor of Psychology, Princeton University*
Eugene Garfield, *Publisher,* The Scientist
Samuel Gubins, *President and Editor-in-Chief, Annual Reviews*
Daniel E. Koshland, Jr., *Professor of Biochemistry, University of California at Berkeley*
Joshua Lederberg, *University Professor, The Rockefeller University*
Sharon R. Long, *Professor of Biological Sciences, Stanford University*
J. Boyce Nute, *Palo Alto, California*
Michael E. Peskin, *Professor of Theoretical Physics, Stanford Linear Accelerator Ctr.*
Harriet A. Zuckerman, *Vice President, The Andrew W. Mellon Foundation*

Management of Annual Reviews

Samuel Gubins, President and Editor-in-Chief
Richard L. Burke, Director for Production
Paul J. Calvi, Jr., Director of Information Technology
Steven J. Castro, Chief Financial Officer

Annual Reviews of

Anthropology
Astronomy and Astrophysics
Biochemistry
Biomedical Engineering
Biophysics and Biomolecular Structure
Cell and Developmental Biology
Earth and Planetary Sciences
Ecology and Systematics
Entomology
Environment and Resources
Fluid Mechanics
Genetics
Genomics and Human Genetics
Immunology
Materials Research
Medicine
Microbiology
Neuroscience
Nuclear and Particle Science
Nutrition
Pharmacology and Toxicology
Physical Chemistry
Physiology
Phytopathology
Plant Biology
Political Science
Psychology
Public Health
Sociology

SPECIAL PUBLICATIONS
Excitement and Fascination of Science, Vols. 1, 2, 3, and 4

Annu. Rev. Biochem. 2003. 72:1–18
doi: 10.1146/annurev.biochem.72.081902.140918
Copyright © 2003 by Annual Reviews. All rights reserved
First published online as a Review in Advance on December 2, 2002

WITH THE HELP OF GIANTS

Irwin Fridovich
Department of Biochemistry, Duke University Medical Center, Box 3711, Durham, North Carolina 27710; email: fridovich@biochem.duke.edu

■ **Abstract** A childhood fascination with animals, plants, and insects was aided and abetted by many giants, beginning with my parents. The Bronx High School of Science and the City College of New York (CCNY) made a solid and priceless grounding in chemistry and biology available free of charge. Abe Mazur at CCNY revealed the wonders of biochemistry and illustrated that it was possible to pursue these wonders while being paid to do so. He also directed me to Duke University Medical School for PhD work under the tutelage of Phil Handler. With the exception of a sabbatical year at Harvard with Frank Westheimer, my entire career has been spent at Duke serving under three fine and supportive chairmen: Handler, Hill, and Raetz. The premier discoveries to emanate from my laboratory have been the sulfite oxidase, the several superoxide dismutases, the manganese catalase, and the catalase/peroxidase. Many other topics piqued my interest and resulted in ~ 400 publications. Herein I have recounted some of the circumstances surrounding that work and named a few of the people involved. The first 20 years I worked happily at the bench and the next 35 years just as happily facilitating the work of younger people. It has been so rewarding that I wish for nothing more than to be allowed to keep at it.

CONTENTS

INTRODUCTION

Now in my seventy-third year of life, I am keenly aware of several things, and the first is good fortune and privilege. Imagine being allowed to spend each day in the pursuit of new knowledge about the inner workings of living things, in the company of congenial coworkers and colleagues, and being paid to do so—all the while enjoying the feelings that the endeavor is worthwhile and that progress is being made! That is privilege indeed! The second is the amazing capacity of the human brain to store information. What is stored may not always be accessible at will, but it is there ready to be called up and displayed in the mind's eye by a chance sight, smell, sound, or thought. Thus while contemplating this writing obligation, I found myself recalling events that transpired six decades ago and that had not been consciously examined since. What is the molecular basis of such immense storage capacity? Someday, if the scientific enterprise is encouraged to proceed, we shall have the answer to that question and to much else besides. The third is that I have had lots of help along the way from parents, teachers, employers, coworkers, and others. They are the giants referred to in the title. Fourth and finally is family. The love and sense of continuity, provided first by children and now by grandchildren, are ultimately the finest things that life provides. The scientific work done in my laboratory is a matter of record—having been described in numerous papers published in several scientific journals. In contrast, the circumstances surrounding this work are known only to a small number of people whose sands of life have already or will soon have run. Hence, I devote much of the space allotted for this prefatory chapter to these heretofore unrecorded anecdotes. I hope some of this material will engage the interest of the reader.

Family History

My paternal grandfather was a furrier who produced coats, hats, and gloves for the Russian Army. One day an officer of the quartermaster corps of that Army, with whom he had frequent dealings, warned him that the government was planning to encourage attacks on Jews by the peasantry. A word to the wise is sufficient. Grandfather promptly sold his home and machinery and sailed to New York City with his family. The year was 1904, and my father was then four years old. Grandfather went back into his skilled trade, and as his sons grew, they were inducted into the fur trade. My father quit school at the end of the sixth grade so that he could contribute to the family's industry. Moreover he was, by his own admission, something of a young hell-raiser and much preferred work to study. My mother's history was similar in that she was also brought to the "new world," while still a toddler, by her parents who were fleeing Russian anti-Semitism.

My father's older brother, Sol, was a infantryman during World War I and came home in need of convalescence. Grandfather thought that he would recover best in a rural environment. So, once again sold his home and equipment and bought a 400 acre dairy farm in the Catskill region of New York State. To supplement the income from the dairy, he built several cabins that were rented to people seeking summer vacations in the country. My mother was then working as a seamstress in the garment district of Manhattan and was brought, by her older sister, for a vacation on grandfather's farm. There my parents met, romance blossomed, and they were married a few years later.

By this time Sol had regained good health. Indeed I remember him as big, strong, cigar smoking, whiskey drinking, and boisterous. Grandfather decided to return to the fur trade; so he sold the farm and used the proceeds to move the entire family back to the city and back to producing fur garments. I recall their factory as a loft in Manhattan with racks of 4 x 8 ft sheets of wood on which dampened furs were nailed to stretch them. There were several sewing machines and lots of naked incandescent bulbs hanging from the ceiling. All unused surfaces were covered with a thick layer of adherent hairs from the furs being unbaled, stretched, trimmed, and sewed.

The sewing machines had two counter rotating horizontal wheels that pulled in the edges of the furs being joined; while a stout needle oscillated just above the surfaces of those wheels. The operator had to guide the edges of the furs into those wheels while using his thumb to push the hairs out of the seam being sewed. The thumb was thus always close to and in danger of being caught by those wheels. Competition was fierce, and speed was essential. My father was a fast and skillful operator. Indeed he was so fast that he once failed to keep his finger out of harms way. He was treated at a Catholic hospital and for years afterward had only praise for the good sisters who took care of him. He was soon back at work, but for the remainder of his life had one finger that could not bend. My older brother also became a furrier, but although I tried my hand at it, I proved neither skillful nor fast.

Living things, whether animal or vegetable, fascinated me. Crotona Park, located close to our apartment, contained an area of several acres that was divided into 5 × 10 ft plots separated by footpaths. Interested youngsters could be assigned a plot in which a variety of vegetables could be grown. My two brothers and I spent every spare minute at the plots and managed to coax a surprising amount of Swiss chard, carrots, beans, and corn from our plots. The plots were overseen by two wonderful women, Mrs. Dayton and Mrs. Hickey; they provided tools, seeds, fertilizer, and instructions. They knew the names and life histories of the plants, birds, and insects we encountered, and we learned a great deal from them.

My parents were amazingly tolerant of our hobbies and enthusiasms even when they must have been a bother in our small apartment. Thus over the years we were allowed a variety of pets including tropical fish, snakes, lizards, a dog, and even a praying mantis. We became adept at catching houseflies to feed our insectivorous pets. Our most serious undertaking was the breeding of canaries. At one point we had more than two dozen canaries, and one room of the apartment that was given over to cages of all sizes, mostly of our own making. Our sources of information were a Mr. Rondoni, who bred canaries in his pet shop, and a book on the diseases of birds written by a man we referred to as the Birdman of Alcatraz—whose real name I cannot now recall.

Another major influence was the Bronx Young Men's Hebrew Association. They sponsored a Boy Scout troop (#242) and also had a woodworking shop and a photo guild. Woodworking and photography have remained lifelong interests.

The Boy Scouts of America then owned ~ 500 acres of woodland on Route 9W, directly across the Hudson River from Yonkers, NY. It could be reached by inexpensive public transportation. There were hiking trails, open lean-tos, and spring water at several locations. We spent many weekends camping at the Alpine scout property, and I have fond memories of splitting wood, making cooking fires, and walking the trails of that, to me, magical place. One of our camping trips coincided with an emergence of 17-year cicadas. The ground was perforated by thousands of cigarette-sized holes made by the emerging nymphs, whose cast off shells clung to every tree trunk, while the air positively vibrated with their shrill calls. It was then that I first appreciated the limited lifespan allotted to humans. How many times could one hope to witness this once-in-17-year spectacle?

Our scoutmaster was Murray Rabinowitz, and his assistant was Phil Kantrowitz. In addition to scout lore, in which both were well versed, Murray was a chemist and Phil a physicist, so we learned a great deal from our association with them. We lost touch with Murray, but Phil and his wife Alice remained family friends for decades after we had outgrown scouting. In 1984 Phil authored a text titled *Industrial Robots and Robotics* under the pen name Mark Stephans and dedicated it my father, mother, and their three sons.

THE BRONX HIGH SCHOOL OF SCIENCE When the time approached for high school, it seemed natural that I apply to the Bronx High School of Science. Attending that school solidified my interest in science. The principal was Morris Meister, father of the noted biochemist Alton Meister. The faculty was knowledgeable and dedicated, the facilities were excellent, and the students, drawn from all over New York City, were serious. Dr. Frankel taught a course in clinical chemistry. During the section on urine analysis, we were required to collect our own urine and bring it to class for both microscopic and chemical analyses. One day I was occupied at the lockers in the rear of the classroom and did not hear him order all desktops cleared. Seeing my briefcase still on my desk he grasped it by the handle and forcibly threw it down. The bottle of urine it contained did not survive. It took several weeks for the odor of stale urine to fade from my briefcase and from the books it contained.

Another notable member of the faculty was Miss Vodovich. She ran the biology preparation laboratory. I spent many happy hours helping her by loading carts with microscopes, dissecting kits, frogs, and reagents as needed by the different biology classes. Many of the graduates of Bronx Science went on to distinguished careers in chemistry, biochemistry, physics, and other sciences. One day, years later, while riding on a train to attend a meeting at Crans Sur Sierre in Switzerland, I noticed a fellow passenger staring at me. He finally approached me saying, "Didn't you go to the Bronx High School of Science?" We had been classmates, but his beard provided effective camouflage. He was going to a meeting of physicists.

FARMING Beginning at age 16, I spent five summers working as a farmhand. There was an employment agency in Manhattan through which such jobs could be arranged. My first job was on a farm devoted to raising vegetables. I liked the rural life well enough, but there were problems with that particular farm. George Ovsanikow drove himself, his son, and his three teenaged employees to ever greater efforts. One day, after we had loaded his truck with cucumbers, carrots, and cabbages, he directed us to dig potatoes while he went to sell the vegetables at the market in Menands. We worked at it all afternoon and had the potatoes lying in windrows on the field. George showed up, surveyed the field, and then began to berate us, in an angry tone, for not getting enough done. "Three men should have dug 40 bushels in the time you've been at it." We silently loaded the potatoes into bushels and found that we had 60 heaping bushels full. At the supper table, we so informed George. He was subdued for a while but could not bring himself to apologize. We boys agreed that we would quit the next time George unjustly berated us. That was an easy decision since we had suffered George's bad humor and his wife's poor cooking all summer and were close to the end of the summer in any case. Before a week had passed George blew up again, and we quit. He was willing to pay the wages we were due but thought he could keep us from leaving by refusing to drive us to the bus station. Fortunately we had helped a neighbor on our time off one Sunday. He owed us a return favor and was glad to repay his debt by providing the needed transportation.

The following summer I secured a job on a dairy farm, and it was better in every way. Herman Menke was a hardworking, good-natured fellow, and Elfrieda, his wife, was a good cook. Moreover, most of the work, i.e., haying, splitting firewood, cultivating, and even milking, could be done while standing or sitting—rather than the stooping and kneeling, which characterized truck farming. My typical day began with spreading ground grain in front of each milking position and then gathering the herd of cows, from the dew-wet pasture, into the barn at ~ 5:30 A.M. Then I started the vacuum pump and hung the milking machines on the first few cows. By this time, Herman showed up and stripped the milk that the milking machines did not get. Soon after we had finished with over 40 cows, the dairy truck showed up to haul away the ~ 1000 lbs of milk that represented the day's yield, and we went into the kitchen for breakfast. I recall that the truck driver, whose last name was Toliver, could carry two 110 lb cans of milk at the same time—one in each hand.

After fueling up on ham, eggs, milk, bread, and jam, I would shovel out the barn gutters, spread lime, and generally clean up. The next job entailed spreading the 3–4 tons of hay that had been dumped into the hay mow the previous afternoon. By the time that was done, I was soaked in sweat and had hay dust sticking to my wet hair, back, and trousers. One day some tourists were passing by just as I emerged from this chore, and I heard a woman exclaim, "Oh, look at that poor man!" I was at the moment feeling very fine. I had finished that job and was on my way to dunking my head in the spring and taking a long, cool drink. It struck me then that many things are worse to contemplate than to do. I returned to that farm for the following three summers and still keep in touch with Hilda, one of the Menke children.

At this time, I contemplated becoming a veterinarian and actually applied to the Cornell College of Veterinary Medicine. I was not accepted. I gathered that they thought that a New York City boy was likely to set up a small animal practice in the city rather than a rural large animal practice. They completely misread my intentions.

CITY COLLEGE The expenses of a private college were out of reach, so I applied to the City College (now the City University) of New York. Tuition was free, I could live at home, and the college could be reached by subway. Selection was on the basis of a competitive examination administered in an impressive auditorium called the Great Hall. I performed well enough to gain admission. The first semester I enrolled in a variety of science and nonscience courses. Among the latter was psychology.

After just two lectures, I decided that the instructor was trying to teach us things that my grandmother would have characterized as common sense. That sealed my commitment to the "real stuff," namely the sciences. Hence, I majored in chemistry with a minor in biology. Pocket money was obtained by way of part-time jobs. One of these evening jobs was in the college employment office. Working in an adjacent office was a very pretty girl who, one evening, asked me to help her close the very large windows in her office. I was glad to do so and in this way met Mollie, who was to become my wife.

The City College years were filled with study, riding the subway, and part-time work. In my senior year (1950), I was taking a course in advanced organic chemistry to complete my chemistry major and another in biochemistry, purely out of curiosity. Prof. Kuch, the chemist, was an excellent glassblower and, responding to my interest, showed me a number of useful glass working techniques. Dr. Mazur, the biochemist, was an inspiring teacher, and the subject matter was a revelation. When he taught us some intermediary metabolism, I recall thinking "this is what the chemistry is for—to understand how living things work."

CORNELL MEDICAL SCHOOL Both Kuch and Mazur had jobs in addition to teaching at CCNY. Kuch spent a few days a week working for the Cyanamid Corp in New Jersey, and Mazur similarly moonlighted at Cornell Medical School in Manhattan. Toward the end of the school year, each of them offered me a job—Kuch at Cyanamid and Mazur at Cornell. I was so taken by biochemistry and by Abe Mazur's informality and friendly personality that it was easy to decide. Immediately following the outdoor graduation, which took place in a driving rainstorm, I started to work in Dr. Mazur's Cornell laboratory. The facilities exceeded anything I had ever seen, and I got busy isolating proteins from rabbit muscle, hog kidneys, and an occasional human autopsy liver. Our elderly lab helper, Catherine, smoked constantly, and one day I told her that she should try to quit because it was bad for her health. She responded, "A short life but a merry one." She was like a second mother to me, and when I had a cold, she would bring hot soup to the lab for my lunch.

After working at Cornell for a year, Abe Mazur told me that I should go to graduate school. So great was my regard for him that I was prepared to act on any advice he might offer. To my query about which school he recommended, he said, "Duke." I did not recognize the name and did not know where it was located but responded, "OK, I'll write for an application." At which point, he informed me that I did not need an application because I had already been accepted. Phil Handler, the chairman of Biochemistry at Duke University Medical School, and Abe Mazur were friends and had agreed that every once in a while Phil would accept, sight unseen, anyone that Abe would recommend. So I rode the train from New York City to Durham, NC, and found the biochemistry department.

DUKE UNIVERSITY MEDICAL SCHOOL Phil Handler was easily the most impressive person I have ever met. He was blessed with a photographic memory that he, unlike most of us, could access at will. He had an unequaled command of language, a deep voice, and, with all that, a pleasant personality. I chose him as my thesis adviser, and he set me to work helping a senior graduate student, Murray Heimberg. The department was small, i.e., six faculty and a like number of graduate students. There was one UV visible spectrophotometer, and that was a Beckman DU, which was hellishly sensitive to humidity. Hence, it was difficult to use when the humidity was high. Because the labs were not air-conditioned, that meant at least half the year. As a reminder of the laboratory temperature during the long Durham summers, I recall that diethyl ether, when poured into graduated cylinder, boiled! The physical discomfits could not negate the benefits of a small, informal, and congenial department. Of course at the time, living quarters, businesses, and automobiles were also not air-conditioned, so we were somewhat inured to the heat and humidity.

The climate did lead to occasional frivolous behavior. Students working in the laboratory of Dr. George Schwert armed themselves with water pistols. During particularly uncomfortable afternoons, they would erupt from their laboratory and attack the people in other laboratories. This naturally led to a water pistol arms race and to running battles in the hallways.

When George accepted the chairmanship of Biochemistry at the University of Kentucky in Lexington (1961), we honored him with a plaque bearing a crossed pair of water pistols over the inscription "To George Schwert: the Fastest Gun in the West." He took that plaque home where it greatly impressed his six-year-old son, who was seen practicing quick draws with his toy pistol while muttering, "George Schwert Junior, the second fastest gun in the west."

SUPEROXIDE DISMUTASE

The road to superoxide dismutase (SOD) was long and tortuous. It began with sulfite oxidation, and that, in turn, derived from studies of sarcosine oxidase. Sulfite was an intermediate in the catabolism of cysteine, and how it got to the end product sulfate was of legitimate interest. The relevant enzyme, sulfite oxidase, was ultimately isolated by Harvey Cohen, a gifted MD/PhD student, and shown to contain both heme and molybdenum (1–3). K.V. Rajagopalan, my long time colleague and friend, ultimately discovered that this enzyme contained a pterin that he named molybdopterin. He and his coworkers subsequently deduced the structure and elucidated the biosynthesis of this very unstable dithiolene pterin that is found in virtually all molybdoenzymes (4).

A few stories about Raj come to mind. He came to Duke directly from Madras, in southern India, to do postdoctoral work with Phil Handler. Hence, we were both working in the Handler lab. At the time, I kept a rubber tube attached to the faucet in the sink for convenience in directing a stream of water into vessels needing rinsing or into my mouth when I was thirsty. During his first day in the lab, Raj saw me about to drink from that tube, and he shouted, "Don't drink that, you'll get sick!" When I asked for an explanation, he informed me that the potable water was at the drinking fountain in the hall. I then explained that there was only one water system in the building and that it was all treated and safe to drink. In a clear tone of disbelief, he said, "You mean to tell me that even the water flushed down the toilet is treated water?" To my affirmative answer, he commented, "What a rich country!"

Another story that involves Raj and that illustrates the narrow perspective that accompanies the extreme specialization of current students should be recounted. Raj and I were seated side by side during a student's "prelim" examination. The student was proposing a study of the gene coding for sulfite oxidase. There was so much topology and so little chemistry in the narrative that I whispered to Raj, "I wonder whether he knows what sulfite is." The time for rounds of questions by the committee finally came and Raj put the first question, "What is sulfite?" The student wanted some clarification of the question so Raj said, "Write the formula for sulfite." Not only did the student not know the formula for sulfite, but he took the position that he was interested in the gene and did not have to know what reaction the encoded enzyme catalyzed.

There were two quite different kinds of sulfite oxidation promoted by liver extracts. One was catalyzed by sulfite oxidase, while the other was a free radical chain oxidation initiated by superoxide that was produced by the action of xanthine oxidase on hypoxanthine. The role of hypoxanthine was exposed when dialysis of the extract was seen to eliminate the free radical oxidation of sulfite. A laborious isolation of the dialyzeable factor, during 1956, from two kilos of beef liver yielded hypoxanthine.

The great length of the reaction chains provided amplification of the initiating event. Thus illumination of a dilute solution of riboflavin served to initiate sulfite oxidation and provided the basis for manometric actinometry (5). So sensitive was this process to light that the experiments had to be done at night, in the dark, with a penlight used to make the periodic readings of the manometers. To get some idea of the lifetime of the flavin radical created by illumination, I constructed a jacketed flow machine that allowed illumination of the riboflavin stream at different distances and hence at different times from its confluence with the stream of sulfite. The glass blowing skills acquired in CCNY were useful in this endeavor. I recall that George Schwert was favorably impressed by my homemade flow machine.

A reasonably complete account of the several lines of evidence that led to the realization that the aerobic xanthine oxidase reaction released superoxide has already been given (6) and will not be recounted here. Suffice it to point out that it took $\sim$ 15 years to traverse the puzzling terrain from sulfite oxidation to superoxide dismutase and to emphasize that Joe M. McCord was there at the end and contributed mightily to the denouement (7, 8).

During the year 1970, Joe McCord and Bernard Keele, who were then both doing postdoctoral work in my laboratory, started asking me for reprints of the papers I had published. Without questioning their motives, I searched my files and found the requested reprints. Weeks later they presented me with a book containing those reprints. It was inscribed to "a good friend for the advice, encouragement, and scientific inspiration which you have given us." I treasured those sentiments but thought that binding one's own reprints in chronological order was just an affectation. I was wrong. I had many occasions to use that book of reprints when I needed a reminder about some experimental detail. Consequently I have continued the practice of binding all the reprints of our work and am indebted to McCord and Keele who started this practice and who chose as the color for the book jacket the color of a concentrated solution of the manganese-containing superoxide dismutase that Keele had isolated from *Escherichia coli*. We were informed by Ann Autor that the correct name for that color is puce. I take her word for it.

ACETOACETIC DECARBOXYLASE AND OTHER BYWAYS

It is not really true that our entire research efforts between 1953–1968 were devoted to the puzzles that led to superoxide dismutase. There were interesting byways that cried out for exploration. Thus during a sabbatical year at Harvard, I worked on a problem of interest to F.H. Westheimer, who was one of the giants who helped me along. The anaerobe that made this enzyme was *Clostridium acetobutylicum*, and my multicarboy cultures outgassed so much CO_2 and H_2 and had such a bad smell that I became notorious as the person responsible for stinking up the entire Mallinkrodt laboratories. After isolating the enzyme and devising a convenient spectrophotometric assay, I started some kinetic studies. While varying pH, I used additions of potassium chloride to keep the ionic strength constant and so found that monovalent anions inhibited. When I brought this fact to Frank suggesting that I would like to explore it, he asked how long I thought the project would take. I answered three weeks, and he responded that it would probably take nine months. He was right! He had a rule for judging how long a project would last–"take the first estimate, move up to the next units, and multiply by three." What I learned about the effect of anions on acetoacetic decarboxylase has been reported (9).

My habit at this time was to arrive at work very early so as to avoid the worst of the ferocious Boston traffic. One morning as I climbed the stairs, I saw a huge frog at the head of the stairs. It was so big and so immobile that I took it to be a rubber facsimile of a frog that someone had placed as a practical joke. However, when I nudged it with my shoe, it moved slightly. It was alive but badly dehydrated. I picked it up, needing both hands to encompass its girth, and carried it to my lab. After a few hours in water, it revived, and I was planning to carry it home to show to my two little daughters. Late in the day, someone came by looking for a "hot frog." This frog had been injected with a large dose of radioactive iodide and had then escaped. I was glad to hand it over.

The people responsible for the radioactive frog were careless in more ways than one. The next day I was walking down the hall carrying a Geiger counter. This counter, which was on, suddenly started to click ominously. Pointing it this way and that led me to the adjacent cold room. There was the source—a four-liter beaker precariously perched on a magnetic stirrer. There was no shielding and no warning that the contents were radioactive. I called Radiological Safety and then walked them down the hall with my Geiger counter running. They were furious and took appropriate action against the perpetrators.

Another side project, done back at Duke, dealt with the enzymatic oxidation of aldehydes. The textbook view then being that the gem diol hydrate of the aldehyde was the true substrate. A paper by Pocker & Meany (10), which showed that the hydration of acetaldehyde was conveniently slow and could be catalyzed by carbonic anhydrase, suggested that this textbook explanation could be conveniently tested. Thus anhydrous acetaldehyde could be added to a buffered solution of xanthine oxidase, which also acts as an aldehyde oxidase, in the presence of cytochrome *c*, and the rate of the reaction could be followed at 550 nm in terms of the reduction of the cytochrome. If anhydrous aldehyde was the true substrate, the rate would be fast initially and then slow down as the hydration equilibrium was established. This is precisely what happened, and carbonic anhydrase hastened the transition from a fast to slow reaction. Clearly the anhydrous aldehyde, rather than its hydrate, was the true substrate (11). John Edsall, the editor of the *Journal of Biological Chemistry* and another of my "giants," took the trouble to let me know how much he liked this work. That encouragement was very important to me.

Another interesting, but probably not very important, finding had to do with the polymerizaton of ferricytochrome *c*. Larry Howell was preparing anaerobic solutions of the cytochrome by sparging with N_2. We noticed that a copious, semi-stable foam formed above ferricytochrome *c* but not above ferrocytochrome *c*. Moreover, the sparged ferricytochrome *c* inhibited the sulfite/cytochrome *c* oxidoreductase. Gel exclusion chromatography revealed oligomers in the sparged ferricytochrome *c* but not in the similarly treated ferrocytochrome *c*. We surmised that the oxidized cytochrome was more prone to surface denaturation than was the reduced form and that refolding, when the bubbles collapsed, led to entanglement and hence oligomerization (12). Others have more recently noticed the comparative lability of the ferri form (13).

THE HABER/WEISS CONTROVERSY

S.F. Yang was interested in the mode of formation of the plant hormone ethylene. He reported that ethylene was formed when methional was incubated with an aerobic mixture of Mn (II), sulfite, a phenol, and horseradish peroxidase (14, 15). He proposed a free radical mechanism in which either O_2^- or HO• attacked methional. We knew that the xanthine oxidase reaction produced O_2^- and we had superoxide dismutase to remove O_2^-, so we could easily test his proposal. Charles (Chuck) Beauchamp exposed methional to the xanthine oxidase reaction and noted a production of ethylene that was inhibited by the dismutase. My immediate reaction was problem solved, it is O_2^-. But when the data were plotted as a function of time, a brief lag in the accumulation of ethylene was apparent. Something else essential for the conversion of methional to ethylene was accumulating at neutral pH; it could not have been O_2^- but could have been H_2O_2. Accordingly, Chuck tested the effect of added H_2O_2, and it eliminated the lag. Moreover catalase inhibited ethylene production as SOD had done. The unavoidable conclusion was that both O_2^- and H_2O_2 were needed to make ethylene from methional. A search of the literature led to a paper by Haber & Weiss (16) in which it was proposed that O_2^- could reduce H_2O_2 to OH- plus HO•. Hence, we supposed that HO• was the radical that converted methional to ethylene. That deduction was supported by our observation that hydroxyl radical scavengers blocked ethylene production (17).

The idea that an enzymatic reaction could lead to the production of a radical as reactive as HO• was anathema to some physical chemists who delighted in proving that the production of HO• from O_2^- plus H_2O_2, which we had referred to as the Haber/Weiss reaction, did not occur. What disappointed me was that not one of them bothered to repeat our observations. Had they done so they would have realized that O_2^- plus H_2O_2 did somehow contrive to produce HO•. Many biochemists did repeat our observations, and the gulf between the physical chemists and the biochemists was bridged by McCord & Day (18), who demonstrated that Fe (III) complexes could catalyze the Haber/Weiss reaction. Evidently the 50 mM phosphate buffer that we had used contained enough Fe (III) to mediate the reduction of H_2O_2 by O_2^-.

Chuck Beauchamp was hard working, knowledgeable, and productive. I thought that he had a bright future as a scientist. However, his father was worried about Chuck's ability to make a living from esoteric biochemical research and encouraged him to go to medical school following the completion of his PhD. That is what Chuck did, and due to his high level of intelligence and capacity for sustained effort, he has made a successful career in medicine—but he could have been a great biochemist. I conclude that well-meaning parents should not micromanage their children's lives.

More Controversy

We took lots of flak from those who doubted that O_2^- could possibly be a problem in vivo and who therefore doubted that the elimination of O_2^- was the function of the superoxide dismutases. They preferred a metal storage function for these metalloenzymes. This seems quaint now in view of the mass of accumulated evidence, but it was a source of aggravation to me. I recall attending an international symposium held in Malta during 1979. This had been organized by Joe and Willy Bannister with the help of a few others (19, 20). I presented work, which was done by Doug Malinowski, showing the essentiality of an arginine residue for the activity of the Cu, Zn SOD. There were many other interesting and informative talks, but one raised my blood pressure, and that was by Jim Fee. His message was that O_2^- was benign and that its dismutation was an incidental activity of certain metalloproteins. During the discussion, I arose and did my best to demolish each of his talking points. When I sat down, Joe McCord, who was seated beside me, whispered "You killed him."

At the close of that session we all went to lunch. Joe and I happened to take a table already occupied by two Spanish scientists. After mutual introductions, I asked whether I might put a question to them. They were agreeable so I asked, "Could you tell, from what you heard, whether I or Fee was right?" They looked uncomfortable, and then one finally said that he could see that there was a disagreement but could not know who was in the right. Unfortunately, that is often the case in scientific disputes. A very small number of people, who happen to be intimately involved in the subject matter, can make an informed judgment. The vast majority merely enjoy the spectacle of a good fight. This has serious consequences when the dispute causes poorly informed committees at granting agencies to deny funding. The people in my lab referred to Fee as the "prince of darkness."

Tetrazolium Oxidase

Chuck Beauchamp, who had demonstrated the production of HO• from O_2^- plus H_2O_2, also devised an activity stain for SOD that is still widely used (21). It is based on the photochemical production of O_2^- by a mixture of riboflavin plus a reductant such as methionine and the reduction of nitroblue tetrazolium (NBT) by that O_2^-. Hence, the gel electropherograms turned uniformly blue when illuminated, except at the positions where SOD intercepted the O_2^-. At the time this method was published, there were already reports of achromatic bands forming on starch gel electropherograms that had been exposed to ambient light after soaking in NBT. The workers who saw this did not recognize the role of light and ascribed the achromatic bands to a tetrazolium oxidase activity.

The Haber/Weiss Reaction, In Vitro and In Vivo

The iron-catalyzed interaction of H_2O_2 and O_2^-, which was seen to produce HO• in vitro, was assumed to occur in vivo as well. But Winterbourn (22) argued that Fe (III) available inside a cell would be kept reduced by ascorbate, so there was no need for O_2^- to do so. The findings that the [4Fe–4S] clusters of dehydratases, such as aconitase, were rapidly oxidized by O_2^- led to a new view of the deleterious interaction of O_2^- with H_2O_2 (23). Thus, the Fe released from the [4Fe–4S] cluster, after oxidation by O_2^-, would then be available to reduce H_2O_2 to HO• + OH-. The validity of this view was affirmed by Keyer & Imlay (24). Thus, the constant winnowing by the scientific enterprise gets us closer and closer to the kernal of truth, while the transient disagreements serve to energize the participants.

A MISSED OPPORTUNITY Skip Kessler, an MD/PhD student working with Rajagopalan, put forward a most interesting proposal for his prelim exam. He knew about O_2^- and SOD, and he had been reading about the respiratory burst of neutrophils. He proposed that the product of the respiratory burst was O_2^- made to discomfort infecting microorganisms. A short time after this prelim exam, I heard Manfred Karnovsky speak about his studies of neutrophils. He reported that the respiratory burst was different than ordinary respiration in that it was not inhibitable by cyanide. I was sorely tempted to reveal Kessler's proposal but did not for fear of jeopardizing his thesis project.

At the very next Federation of American Societies for Experimental Biology meeting, Bernard Babior requested a sample of SOD. He needed it to test his hypothesis that the respiratory burst of neutrophils was making O_2^-. Apparently this was an idea whose time had come. When I explained that I felt constrained not to help him (because his work competed with Rajagopalan's student), he was sympathetic. The end of this story is that Babior made SOD for himself and used it to prove O_2^- production by activated neutrophils (25), and Kessler never followed up on his idea.

NEVER DISCOUNT THE BUFFER Cu, Zn SOD was seen to be rapidly reduced and more slowly inactivated by H_2O_2. E. Kay Hodgson explored these processes. She routinely worked at pH 10.2 in a carbonate buffer because the rates increased with pH. Kay noticed that the enzyme could act as a peroxidase toward a range of substrates and that certain substrates such as azide, formate, and urate could protect against the inactivating effect of H_2O_2. These data were explained on the basis of a bound oxidant, possibly Cu (II) OH, produced at the active site by the reaction of the Cu (I) Zn SOD with H_2O_2. The oxidant was taken to be bound Cu (II) OH rather than free HO• because ethanol or benzoate, which could certainly intercept free HO•, did not protect (26, 27). We did not suspect that the carbonate buffer was an active participant in those reactions even though we should have been alerted to this possibility by a previous observation.

Thus another of Kay's projects was studying the weak luminescence emitted during the xanthine oxidase reaction. She was using a phosphate buffer and was measuring the light emitted during the oxidation of acetaldehyde. One day this luminescence could no longer be seen. When she told me about this sudden failure, I asked whether she had changed anything. She then stated that she had made a fresh phosphate buffer after the old stock had been exhausted. I immediately suspected carbonate because all neutral or alkaline solutions slowly accumulate carbonate derived from the CO_2 in the air. When carbonate was added to the new phosphate buffer, the light returned. Indeed the luminescence intensity proved proportional to the square of the carbonate concentration. Because SOD, or catalase, or scavengers of HO• all inhibited, we proposed the formation of HO• from O_2^- plus H_2O_2, followed by the oxidation of carbonate by HO•. Dimerization of the resultant carbonate radical was thought to yield a product that decomposed with production of light (28). Years later Kay was employed by the U.S. Navy in the design of propellers for atomic submarines. Her goal then was to avoid turbulence and cavitation in order to eliminate noise, so that sonar detection would be more difficult.

Carbonate was brought back to center stage by Sankarapandi & Zweier (29), who reported that carbonate was essential for the peroxidase activity of Cu, Zn SOD. They thought that carbonate increased the binding of H_2O_2 to the enzyme. This was easily shown not to be the case (30). The most likely explanation involves the oxidation of bicarbonate to the carbonate radical, which then diffuses from the active site to oxidize molecules in free solution (31). In pursuing this problem, Stefan Liochev again demonstrated his capacity for innovative thinking.

A STRANGER IN PARADISE Stefan first came to my lab on a fellowship provided by the International Atomic Energy Agency. They made all the arrangements, and I did not know when he was due to arrive. Stefan sent me this information in a telegram that was delivered two weeks after his arrival. We never found out who was responsible for holding up this telegram. In any case, Stefan arrived at the Raleigh/Durham Airport. I was not there to meet him, and he did not try to reach me by telephone due to unfamiliarity with our excellent telephone system. He waited for a long while and then wandered around the airport and finally encountered some Duke students who gave him a ride to the Medical Center. He had been allowed only $28 when he left Bulgaria and had neither a credit card nor a driver's license. Hence, the hotel adjacent to the medical center would not rent him a room. The desk clerk directed him to another cheaper motel to which Stefan dragged his heavy suitcase. It was, by then, very late, and he was exhausted and understandably discouraged. The next morning I received a call from a helpful secretary in Duke Hospital. She said that a man from Bulgaria was looking for me. I asked her to put him on the phone and then asked him where he was. His response was "the main building." The secretary then got back on and specified her location. I hurried over and so ended Stefan's problems. I had a large check for him and also a place to live. He was then taken shopping for groceries by Che Fu Kuo, who came back to the lab to report that Stefan had never seen anything like the cornucopia that is a modern supermarket. The discomfits surrounding his arrival were thus soon forgotten, and he has proven a very valuable coworker.

His first project was to investigate the reality of the vanadate-dependent NADH oxidase reported by Ramasarma. I expected that there was no such thing and that the more likely explanation was the production of O_2^- by a NADH oxidase amplified by a vanadate-catalyzed oxidation of NADH by O_2^-. When the experiment intended to probe this idea was explained to Stefan, he doubted our ability to do it because we would need an infusion pump and several reagents. He was used to working under severely restricted conditions in Bulgaria. The needed materials were available, and after a few days of work he said, "I think I am undertaking a criminal investigation." When asked for an explanation, he said, "Everything they claim is so, is not so." Actually Doug Darr had already shown that O_2^- plus vanadate would oxidize NADH (32), and Stefan elucidated the mechanism (33, 34).

O_2^- AND ANION CHANNELS R.E. Lynch came to the lab from Salt Lake City to spend a year doing research under funding by the Howard Hughes Medical Institute. His project was to see whether O_2^- could cross the erythrocyte stroma. His knowledge of experimental hematology was invaluable. He prepared vesicles of washed erythrocyte stroma and soon showed that O_2^- generated on one side would reduce cytochrome *c* on the other side. Moreover stilbene sulfonates, known to block anion channels, prevented this cross stromal action of O_2^- (35). When R.E. perceived the reluctance of other members of my lab group to donate blood, he simply drew blood from his own arm. One day I overheard snatches of a conversation between R.E. and someone out in the lab. From the words I heard I assumed they were talking about beautiful women. They were actually discussing wines. R.E. was a connoisseur of wines.

THE DANGERS OF THE AEROBIC LIFESTYLE Once when invited to give a seminar on O_2^- and SOD, I submitted the title "The Dangers of the Aerobic Lifestyle." There was then a bestseller extolling the benefits of aerobic exercise. The author of this book became aware of the title of my seminar and jumped to the conclusion that I intended to attack his thesis. He was a friend of Andy Wallace, the head of Cardiology at Duke, and complained. Andy called me and the misunderstanding was clarified.

AN INHIBITOR OF SOD

For an interval of 10 years, I was one of many consultants retained by Merck. During this period, I visited their Rahway or West Point labs and told them what I was up to and listened to their doings while making such helpful comments as

I could. It was an entirely pleasant association. After my consultancy had expired, I was called by one of the scientists at Merck who wanted my opinion of a specific inhibitor of the iron-containing SOD (FeSOD). I was enthusiastic but skeptical because the FeSOD is found in bacteria and plants, not in mammals. I was skeptical because an enzyme whose substrate is the small and simple O_2^- was not likely to specifically bind some larger organic compound. To my query of "How will you find such an inhibitor?" he responded, "By a random screen of ~ 100,000 compounds." I remonstrated that "You can't do research like that." He responded, "The hell we can't. We do it all the time, and we never fail." A year later he happily informed me that they had the inhibitor but would not disclose its identity since I was no longer consulting for them. He agreed to call back when he felt free to name the compound and did so six months later. By then the compound had been tested in infected mice and failed to provide benefit. The compound was pamoic acid, and I gave it to Hosni Hassan who found that it did not inhibit any SOD. What it did do was interfere with the assay (36). So much for random searches by technicians who do not understand the mechanism of the assay they are using.

We had occasion, more recently, to question a putative inhibitor of SOD, which was a catecholestrogen (37). This was another case of interference with the assay, and we submitted a letter to the editor correcting this error. However, *Nature* does not like to admit that it occasionally publishes nonsense and would not accept it. We therefore involved several collaborators and published results leaving no doubt about the issue (38).

Olen R. Brown and his students observed the bacteriostatic effect of hyperoxia and noted that it could be overcome by branched chain amino acids. Further work led them to conclude that the dihydroxy acid dehydratase, which catalyzes the penultimate step of branch chain biosynthesis in *E. coli,* was the oxygen-sensitive enzyme (39). It was almost a reflex for me to ascribe inactivation of this enzyme to O_2^-, rather than to O_2. Che Fu Kuo and Tadahiko Mashino did the work confirming that suspicion (40). Other members of the [4Fe−4S] containing dehydratase were subsequently found to be sensitive to oxidative inactivation by O_2^-.

ENDNOTE

This narrative has left much unsaid and the contributions of a number of "helpful giants" unacknowledged. To tell it all would take a book, not a chapter, and would surely strain the forbearance of most readers. All my papers and reviews were written out long hand and then converted to perfect typed copy by my secretary/bookkeeper Pattie Lewis. Her expert decades-long help must be acknowledged. I did not even feel the need for a personal computer until she decided to retire. I wish also to acknowledge the three departmental chairmen under whom I worked, Phil Handler, Bob Hill, and Chris Raetz. All were indispensable for their unfailing support and wise counsel.

My lab group is smaller now, and I am slower, but there is still the pleasure of solving nature's mysteries and the occasional excitement of scientific disagreements and the ensuing good fight.

The *Annual Review of Biochemistry* is online at http://biochem.annualreviews.org

LITERATURE CITED

1. Cohen HJ, Fridovich I. 1971. *J. Biol. Chem.* 246:359–66
2. Cohen HJ, Fridovich I. 1971. *J. Biol. Chem.* 246:367–73
3. Cohen HJ, Fridovich I, Rajagopalan KV. 1971. *J. Biol. Chem.* 246:374–82
4. Schindelin H, Kisker C, Rajagopalan KV. 2001. *Adv. Protein Chem.* 58:47–94
5. Fridovich I, Handler P. 1959. *Biochem. Biophys. Acta* 35:546–47
6. Fridovich I. 2001. *J. Biol. Chem.* 276: 28629–36
7. McCord JM, Fridovich I. 1968. *J. Biol. Chem.* 243:5753–60
8. McCord JM, Fridovich I. 1969. *J. Biol. Chem.* 243:6049–55
9. Fridovich I. 1963. *J. Biol. Chem.* 238: 592–98
10. Pocker Y, Meany JE. 1965. *J. Am. Chem. Soc.* 87:1809–11
11. Fridovich I. 1966. *J. Biol. Chem.* 241: 3126–28
12. Howell LG, Fridovich I. 1968. *J. Biol. Chem.* 243:5941–47
13. Hitgren-Willis S, Bowden EF, Pielak GJ. 1991. *J. Inorg. Chem.* 51:649–53
14. Yang SF. 1967. *Arch. Biochem. Biophys.* 122:481–87
15. Yang SF. 1969. *J. Biol. Chem.* 244: 4360–65
16. Haber F, Weiss J. 1934. *Proc. R. Soc. London Ser. A* 147:332–51
17. Beauchamp C, Fridovich I. 1970. *J. Biol. Chem.* 245:4641–46
18. McCord JM, Day ED Jr. 1978. *FEBS Lett.* 86:139–42
19. Bannister JV, Hill HAO. 1980. *Chemical and Biochemical Aspects of Superoxide and Superoxide Dismutase.* Netherlands: Elsevier
20. Bannister WH, Bannister JV. 1980. *Biological and Clinical Aspects of Superoxide and Superoxide Dismutase.* Netherlands: Elsevier
21. Beauchamp C, Fridovich I. 1971. *Anal. Biochem.* 44:276–87
22. Winterbourn CC. 1979. *Biochem. J.* 182: 625–28
23. Liochev SI, Fridovich I. 1994. *Free Radic. Biol. Med.* 16:29–33
24. Keyer K, Imlay JA. 1996. *Proc. Natl. Acad. Sci. USA.* 93:13635–40
25. Babior BM, Kipnes RS, Curnutte JT. 1973. *J. Clin. Investig.* 52:741–44
26. Hodgson EK, Fridovich I. 1975. *Biochemistry* 14:5294–99
27. Hodgson EK, Fridovich I. 1975. *Biochemistry* 14:5299–303
28. Hodgson EK, Fridovich I. 1976. *Arch. Biochem. Biophys.* 172:202–5
29. Sankarapandi S, Zweier JL. 1999. *J. Biol. Chem.* 274:1226–32
30. Liochev SI, Fridovich I. 1999. *Free Radic. Biol. Med.* 27:1444–47
31. Liochev SI, Fridovich I. 2002. *J. Biol. Chem.* 277:34674–78
32. Darr K, Fridovich I. 1984. *Arch. Biochem. Biophys.* 232:562–65
33. Liochev S, Fridovich I. 1985. *Free Radic. Biol. Med.* 1:287–92
34. Liochev S, Fridovich I. 1986. *Arch. Biochem. Biophys.* 250:139–45
35. Lynch S, Fridovich I. 1978. *J. Biol. Chem.* 253:1838–45

36. Hassan HM, Dougherty M, Fridovich I. 1980. *Arch. Biochem. Biophys.* 199:349–54
37. Huang P, Feng L, Oldham EA, Keating MJ, Plunkett W. 2000. *Nature* 407: 390–95
38. Kachadourian R, Liochev SI, Cabelli DE, Patel MN, Fridovich I, Day BJ. 2001. *Arch. Biochem. Biophys.* 392:349–53
39. Brown OR, Seither RL. 1983. *Fundam. Appl. Toxicol.* 3:209–14
40. Kuo CF, Mashino T, Fridovich I. 1987. *J. Biol. Chem.* 262:4724–27

Annu. Rev. Biochem. 2003. 72:19–54
doi: 10.1146/annurev.biochem.72.121801.161737
Copyright © 2003 by Annual Reviews. All rights reserved
First published online as a Review in Advance on December 11, 2002

THE ROTARY MOTOR OF BACTERIAL FLAGELLA

Howard C. Berg
Department of Molecular and Cellular Biology, Harvard University, 16 Divinity Avenue, Cambridge, Massachusetts 02138; email: hberg@biosun.harvard.edu

Key Words *Escherichia coli*, motility, chemotaxis, ion flux, torque generation

■ **Abstract** Flagellated bacteria, such as *Escherichia coli*, swim by rotating thin helical filaments, each driven at its base by a reversible rotary motor, powered by an ion flux. A motor is about 45 nm in diameter and is assembled from about 20 different kinds of parts. It develops maximum torque at stall but can spin several hundred Hz. Its direction of rotation is controlled by a sensory system that enables cells to accumulate in regions deemed more favorable. We know a great deal about motor structure, genetics, assembly, and function, but we do not really understand how it works. We need more crystal structures. All of this is reviewed, but the emphasis is on function.

CONTENTS

INTRODUCTION

The bacterial flagellar motor is a nanotechnological marvel, no more than 50 nm in diameter, built from about 20 different kinds of parts. It spins clockwise (CW) or counterclockwise (CCW) at speeds on the order of 100 Hz, driving long thin helical filaments that enable cells to swim. Peritrichously flagellated cells (*peri,* around; *trichos*, hair), such as *Escherichia coli*, execute a random search, moving steadily at about 30 diameters per second, now in one direction, now in another. Steady motion requires CCW rotation. Receptors near the surface of the cell count molecules of interest (sugars, amino acids, dipeptides) and control the direction of flagellar rotation. If a leg of the search is deemed favorable, it is extended, i.e., the motors spin CCW longer than they otherwise would. This bias enables cells to actively find regions in their environment where life is better. Thus, the flagellar motor is the output organelle of a remarkable sensory system, the components of which have been honed to perfection by billions of years of evolution.

We know a great deal about the structure of the flagellar motor but not very much at atomic resolution. We know a great deal about regulation of the genes that specify the motor's component parts and how those parts are assembled. We know a great deal about motor function: about the fuel that powers the motor, the torque that it can generate at different speeds, and what controls the likelihood that it changes direction. However, we do not know how the motor actually works, i.e., the details of what makes it go, or how it manages to shift abruptly from forward (CCW) to reverse (CW).

The work described here has been done primarily with the gram-negative organisms, *E. coli* and *Salmonella enterica* serovar Typhimurium (*Salmonella* for short), that are closely related. Their motors are driven by protons powered by a transmembrane electrochemical gradient, or protonmotive force. ATP plays no role, except in motor assembly and in chemotactic signaling (coupling of receptors and flagella). However, if cells are grown anaerobically, ATP generated by glycolysis powers a membrane H^+-ATPase that maintains the protonmotive force. A gram-positive *Streptococcus* has been useful for studies of membrane energetics. Gram-negative marine *Vibrio* spp. have motors that are driven by sodium ions rather than protons, and comparisons between different kinds of ion-driven machines are illuminating. Flagellar motors are important in a wide variety of other bacteria, from *Caulobacter* to *Leptospira*, not considered here.

In this chapter I review what is known about motor structure, genetics, and assembly, say more about function, and then touch upon motor mechanisms.

STRUCTURE

E. coli is rod-shaped, about 1 μm in diameter by 2 μm long. A cell is propelled by a set of four helical flagellar filaments (four, on average) that arise at random points on its sides and extend several body lengths out into the external medium. Each filament is driven at its base by a rotary motor embedded in the cell envelope. A cell swims steadily in a direction roughly parallel to its long axis for about a second—it is said to "run"—and then moves erratically in place for a small fraction of a second—it is said to "tumble"—and then swims steadily again in a new direction. When a cell runs at top speed, all of its flagellar filaments spin

CCW (as viewed along a filament from its distal end toward the cell body); the filaments form a bundle that pushes the cell steadily forward. When a cell tumbles, one or more filaments spin CW (1–3); these filaments leave the bundle, and the cell changes course (see below). Motors switch from CCW to CW and back again approximately at random. The likelihood of spinning CW is enhanced by a chemotactic signaling protein, CheY. When phosphorylated, CheY binds to the cytoplasmic face of the flagellar motor. The phosphorylation of CheY is catalyzed by a kinase, the activity of which is controlled by chemoreceptors. The activity of the kinase is depressed by the addition of attractant. (For recent reviews on bacterial chemotaxis, see References 4–11.)

A bacterial flagellum is shown schematically in Figure 1, along with a reconstruction of the motor core obtained by rotationally averaging images taken by electron microscopy. A gram-negative bacterium has a multilayered cell envelope. Note the positions of the inner (cytoplasmic) membrane, the peptidoglycan layer, and the outer membrane. The different components of the motor are named after the genes that encode them. A parts list is given in Table 1.

Originally, genes for which mutant cells lacked flagellar filaments were called *fla* (for flagellum), but after more than 26 had been found and the correspondence between genes in *E. coli* and *Salmonella* became clear, the nomenclature was simplified (12); *fla* genes are now called *flg, flh, fli,* or *flj,* depending upon their location on the genetic map. Genes for which mutant cells produce paralyzed flagella are called *mot* (for motility). Four of the 40 gene products listed in the table are involved in gene regulation (FlgM, FlhC, FlhD, FliA); see below. About half appear in the final structure (Figure 1). The hook (FlgE), the hook-associated proteins (FlgK, FlgL, and FliD), and the filament (FliC) are outside the cell; the MS-ring (FliF) and the P- and L-rings (FlgI and FlgH) are embedded in the cell wall; and the C-ring (FliM and FliN) is inside the cell. FliG is bound to the inner face of the MS-ring near its periphery. In some reports, it is treated as part of the C-ring. MotA and MotB, which are arranged in a circular array around the MS- and C-rings, span the inner membrane.

The hook and filament are polymers (crystals) of single polypeptides, hook protein (FlgE) and flagellin (FliC), respectively. They comprise 11 parallel rows of subunits on the surface of a cylinder, with the rows tilted (twisted) slightly relative to the local cylinder axis, as shown in Figure 2. The subunits pack in two different ways: The subunits in "short" protofilaments (R-type) are closer together than the subunits in "long" protofilaments (L-type). R and L refer to the direction of twist. If both types are present at the same time, the filament has curvature as well as twist and is helical, with the short protofilaments running along the inside of the helix (13). Mechanical strain energy is minimized when short protofilaments are next to short protofilaments and long protofilaments are next to long protofilaments, leading to 12 possible conformations, 2 straight (all short or all long) and 10 helical (14, 15).

The hook is flexible. As it rotates, its protofilaments continuously switch from short to long, so that short protofilaments always appear at the inside of the bend.

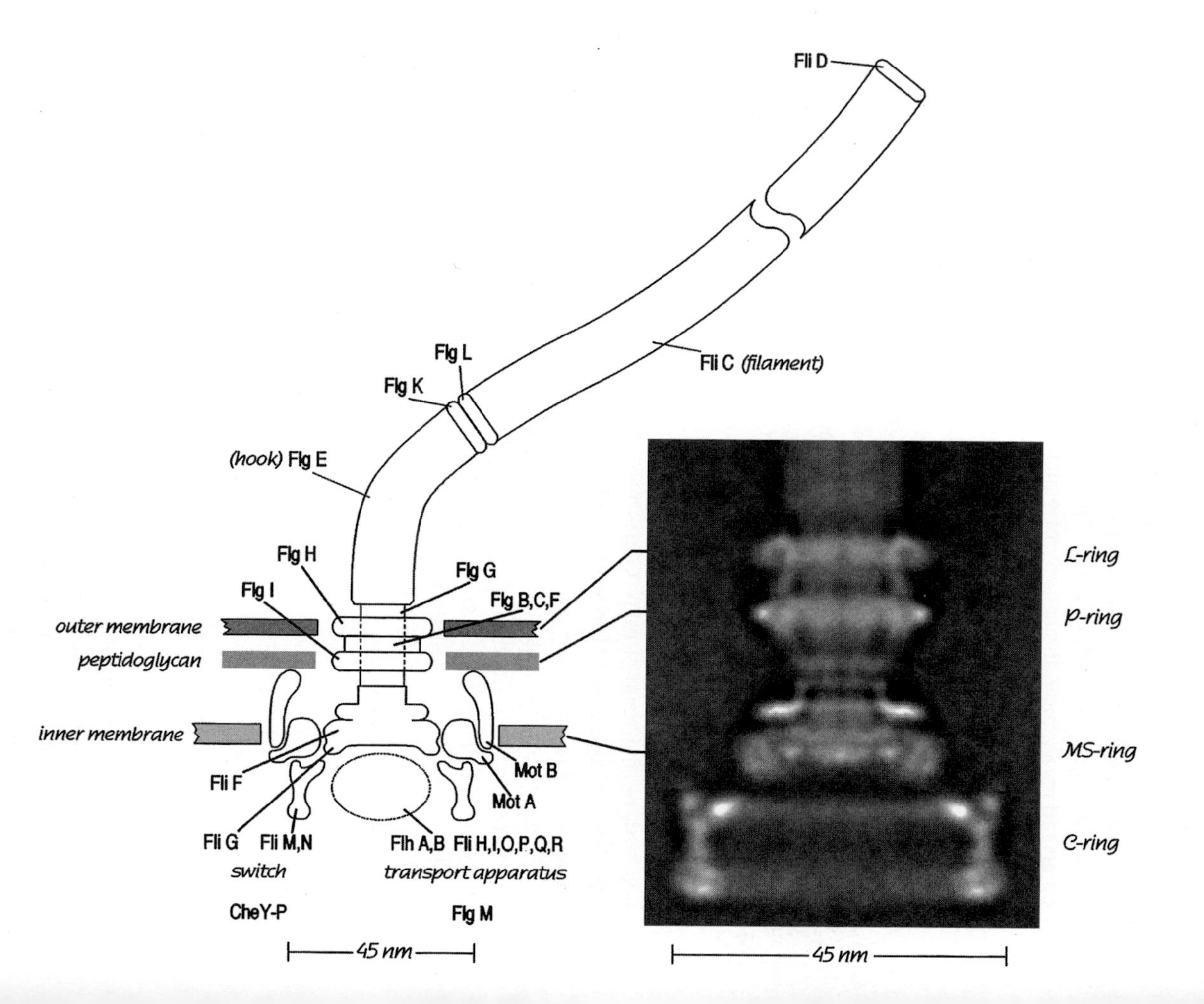

Fli D
Fli C (filament)
Flg L
Flg K
(hook) Flg E
Flg H
Flg G
Flg I
Flg B,C,F
outer membrane
peptidoglycan
inner membrane
Fli F
Mot B
Mot A
Fli G
Fli M,N
switch
Flh A,B
Fli H,I,O,P,Q,R
transport apparatus
CheY-P
Flg M
45 nm
L-ring
P-ring
MS-ring
C-ring
45 nm

The filament, on the other hand, is rigid, with a shape that depends on the amino acid sequence of the flagellin, the pH, ionic strength, and torsional load. When one or more motors on the cell switch from CCW to CW, their filaments go through a sequence of transformations, from normal to semicoiled to curly 1; when the motors switch back to CCW, they revert to normal, as shown in Figure 3. The normal filament, with 2 short protofilaments, is left-handed. The semicoiled filament, with 4, is right-handed with half the normal pitch. The curly 1 filament, with 5, is right-handed with half the normal pitch and half the normal amplitude. Evidently, the hook-associated proteins FlgK and FlgL (Figure 1) allow the protofilaments of the hook to switch from short to long but require that the protofilaments of the filament remain fixed. There are mutants of *flgL* that allow filaments to switch from normal to straight or from curly 1 to straight, depending upon the direction of rotation of the flagellar motor (16). These transformations are forbidden in the wild type. So, the response of the filament to torsion depends, in part, on how it is held at its base. (For recent discussions of filament structure, see References 17–19.) A flagellin truncated at both its N and C termini (to block filament formation) has been crystallized, and the transformation responsible for the switch from "short" to "long" has been identified by computer simulation (20).

Early on, it was thought that the basal body (the structure proximal to the hook) comprises four rings (M, S, P and L) and a rod, because these elements could be seen by electron microscopy when flagella were purified and negatively stained. In this procedure, cell walls were weakened by treatment with EDTA and lysozyme, cells were lysed with a nonionic detergent and treated with DNase I, and flagella were fractionated in detergent by differential sedimentation (21). The rings were named by DePamphilis & Adler (22), who found that the M-ring (for membrane) has affinity for inner-membrane fractions, the S-ring (for supramembranous) is seen just above the inner membrane, the P-ring (for

Figure 1 A schematic diagram of the flagellar motor, drawn to scale, compared to a rotationally averaged reconstruction of images of hook-basal bodies seen in an electron microscope. The different proteins are named for their genes and are listed in Table 1. CheY-P is the chemotaxis signaling molecule that binds to FliM, and FlgM is the anti-sigma factor pumped out of the cell by the transport apparatus; see the text. The general morphological features are C-ring, MS-ring, P-ring, L-ring, hook, hook-associated proteins (which include the distal cap), and filament. MotA, MotB, and components of the transport apparatus (dashed ellipse) do not survive extraction with detergent and, therefore, are not shown on the right. This reconstruction is derived from rotationally averaged images of about 100 hook–basal body complexes of *Salmonella* polyhook strain SJW880 embedded in vitreous ice (29). The radial densities have been projected from front to back along the line of view, so this is what would be seen if one were able to look through the spinning structure. Connections between the C-ring and the rest of the structure appear relatively tenuous. Digital print courtesy of D.J. DeRosier.

TABLE 1 Proteins of *E. coli* involved in flagellar motor assembly and function[a]

Gene product	Function or motor component	Size (kDa)	Copies per motor[b]	Operon class[c]
FlgA	Assembly of P-ring	24		2
FlgB	Proximal rod	15	6	2
FlgC	Proximal rod	14	6	2
FlgD	Assembly of hook	24		2
FlgE	Hook	42	130	2
FlgF	Proximal rod	26	6	2
FlgG	Distal rod	28	26	2
FlgH	L-ring	22	26	2
FlgI	P-ring	36	26	2
FlgJ	Muramidase	34		2
FlgK	Hook-filament junction; at hook	59	11	3a
FlgL	Hook-filament junction; at filament	34	11	3a
FlgM	Anti-sigma factor	11		3a
FlgN	FlgK, FlgL chaperone	16		3a
FlhA	Protein export	75		2
FlhB	Hook-length control	42		2
FlhC	Master regulator for class 2 operons	22		1
FlhD	Master regulator for class 2 operons	14		1
FlhE	?	12		2
FliA	Sigma factor for class 3 operons	27		2
FliC	Filament (flagellin)	55	5340	3b
FliD	Filament cap	50	10	3a
FliE	Rod MS-ring junction (?)	11	9?	2
FliF	MS-ring	61	26	2
FliG	Rotor component; binds MotA	37	26	2
FliH	Protein export	26		2
FliI	Protein export ATPase	49		2
FliJ	Rod, hook, filament chaperone	17		2
FliK	Hook-length control	39		2
FliL	?	17		2
FliM	Switch component; binds CheY-P	38	32?	2
FliN	Switch component	14	110	2
FliO	Protein export	11		2

TABLE 1 Continued

Gene product	Function or motor component	Size (kDa)	Copies per motor[b]	Operon class[c]
FliP	Protein export	27		2
FliQ	Protein export	10		2
FliR	Protein export	29		2
FliS	FliC chaperone	15		3a
FliT	FliD chaperone	14		3a
MotA	Force-generator	32	32?	3b
MotB	Force-generator	34	16?	3b

[a]Including proteins involved in gene regulation but not in signal processing. *flg* genes are in map region I (*E. coli* 24 min, *Salmonella* 23 min); *flh* and *mot* genes are in map region II (41 min, 40 min); and *fli* genes are in map region III (43 min, 40 min). For operons, additional gene products in *Salmonella*, and references to gene sequences, see Table 1 of Macnab (246).

[b]Approximate values. The figure given for FliC (flagellin) is subunits per turn of the normal helix (17).

[c]Class 3 operons that have some FliA-independent expression are designated 3a and those that do not, 3b (254, 258).

peptidoglycan) is at the right place to be embedded in the peptidoglycan, and the L-ring (for lipopolysaccharide) has affinity for outer-membrane fractions (23).

In the earliest models for the rotary motor (24), the M-ring was thought to rotate relative to the S-ring, which served as the stator. Later it was found that both the M- and S-rings (now called the MS-ring) comprise different domains of the same protein, FliF (25, 26). Therefore, they function as a unit. The C-ring (for cytoplasmic) was discovered much later when extracts were treated more gently, i.e., subjected to smaller extremes of pH and ionic strength (27–29). The image shown in Figure 1 was reconstructed from basal bodies prepared gently and examined in frozen-hydrated preparations in a cryoelectron microscope. Freeze-etch replicas made in situ show a knob in the center of each C-ring; the knob is thought to comprise the main body of the transport apparatus (30; see also Reference 31).

FliG, FliM, and FliN are also referred to as the "switch complex," since many mutations of *fliG, fliM,* and *fliN* lead to defects in switching (in control of the direction of rotation) (32, 33). Other mutations are nonmotile, and the null phenotypes are nonflagellate. The chemotactic signaling protein, CheY-P, binds to FliM (34–37). A variety of binding studies argue that FliG binds to FliF (38, 39) and FliG, FliM, and FliN bind to each other (36, 39, 40). Functional or partially functional in-frame fusions have been obtained between FliF and FliG (41, 42) and between FliM and FliN (43), but not between FliG and FliM (44). Fusions of the latter type block flagellar assembly. An electron micrographic analysis of basal-body structures found in nonmotile missense mutations of *fliG, fliM,* and *fliN* indicates loss of the C-ring, the components of which (FliM and

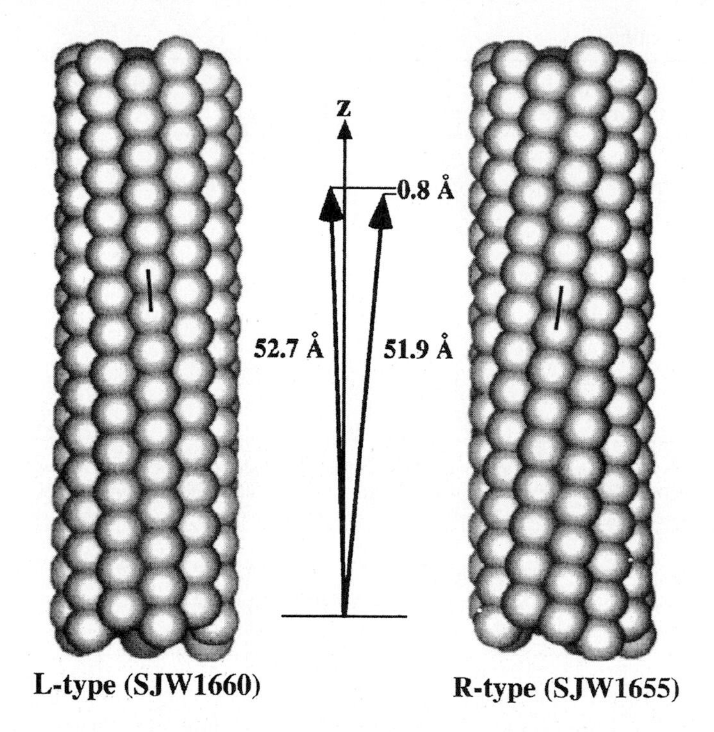

Figure 2 The surface lattice of L- and R-type straight flagellar filaments. The spacing between flagellin subunits along an 11-start helix (a protofilament) of the R-type is 0.07 nm less than between corresponding subunits of the L-type. L and R refer to the handedness of the filament twist. The SJW numbers designate particular bacterial strains. The distances are measured at a radius of 4.5 nm and are shown magnified in the middle of the drawing. (From Reference 19, Figure 19.)

FliN) can be recovered in the cytoplasm (45, 46). And nonmotile mutations of *fliM* and *fliN,* but not *fliG,* can be cured by overexpression (47). So attachment of the C-ring appears to be labile, as suggested by the region of low density between this structure and the rest of the basal body evident in the image reconstruction of Figure 1. Although it is conceivable that these structures rotate relative to one another (48), most workers assume that they rotate as a unit, i.e., that the rotor comprises both the MS- and C-rings.

The stator is thought to comprise the MotA and MotB proteins, which are membrane embedded and do not fractionate with the rest of the hook–basal body complex (49). However, they can be visualized as circular arrays of membrane particles ("studs") in freeze-fracture preparations of the inner membrane. Studs were seen first at the poles of *Aquaspirillum serpens* in sets of 14–16 (50), later in *Streptococcus* in similar numbers and in *E. coli* in sets of 10–12 (51), and finally in *Salmonella* and different species of *Bacillus* in sets of about 12 (52, 53).

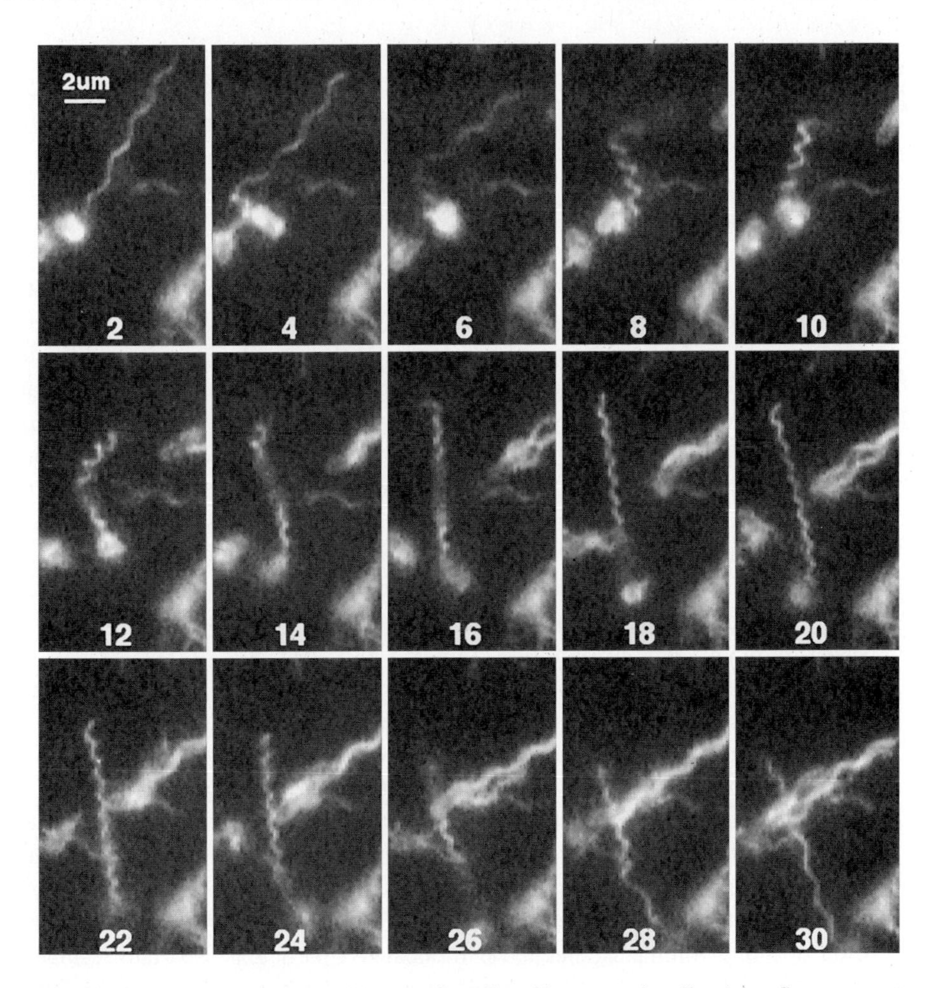

Figure 3 An *E. coli* cell with one flagellar filament, visualized by fluorescence microscopy. The recording was made at 60 Hz, but only every other field is shown. The numbers are in units of 1/60 s. When the motor switched from CCW to CW after field 2, the filament changed its shape from normal to semicoiled, 10, and then to curly 1, e.g., 20. When the motor switched back to CCW after field 26, the filament relaxed back to normal, 30. Initially, the cell swam toward 7 o'clock. After the normal to semicoiled transformation, it swam toward 5 o'clock. Flagellar filaments can also be visualized by dark-field or interference-contrast microscopy (259, 260), but fluorescence has the advantage that one can see the filaments all the way to the surface of the cell with reasonable depth of field. (From Reference 3, Figure 6.)

Both MotA and MotB span the cytoplasmic membrane. MotA has four membrane-spanning α-helical segments (54–56). The rest of the molecule (about two-thirds) is in the cytoplasm. MotB has one membrane-spanning α-helical segment near its N terminus, but most of the molecule is in the periplasmic space (57, 58). There is a peptidoglycan-binding domain near its C terminus (59) (although such binding has not been shown directly). MotB is thought to anchor MotA to the rigid framework of the cell wall. Elements of the stator must be anchored to this framework somewhere, or torque cannot be delivered to the flagellar filament (24). Evidently, MotA and MotB form a complex that acts as a torque-generating unit. The stoichiometry of each unit is not known for certain, but it is likely to be four MotA and two MotB. This arrangement is suggested from reconstitution of the *Vibrio* homologs, PomA and PomB (60). MotA and MotB can be co-isolated by an affinity tag on MotB, and thus these proteins bind to each other (40). Targeted disulfide cross-linking of the transmembrane segment of MotB indicates a symmetric arrangement of parallel α-helices (61), suggesting that each torque-generating unit contains at least two copies of MotB. Earlier, tryptophan-scanning mutagenesis had suggested a model in which the transmembrane segment of one MotB is bundled slantwise with the four transmembrane segments of one MotA to constitute a proton channel (62, 63). It appears likely, now, that there are two proton channels per complex, each comprising eight transmembrane segments from two copies of MotA and one transmembrane segment from one copy of MotB (61).

Studies of extragenic suppression of dominant missense mutations of *motA* (64) and *motB* (65, 66) suggest that MotA and MotB interact with FliG (a component at the cytoplasmic face of the MS-ring; see Figure 1) as well as with each other. Mutations near the putative peptidoglycan-binding region of MotB appear to misalign the stator and the rotor (66). Comparison of residues conserved in different bacterial species and site-directed mutagenesis have identified charged groups in the cytoplasmic domain of MotA that interact with other charged groups (primarily of opposite sign) in the C-terminal domain of FliG (67–69). Similar studies have implicated a particular aspartate residue of MotB (Asp32), located at the cytoplasmic end of the membrane channel, as a proton acceptor (70). Two proline residues in MotA (Pro173 and Pro222), also located at the cytoplasmic end of this channel, have been shown to be important for function (67, 71). Mutations in either Asp32 or Pro173 in membrane-bound complexes of MotA and MotB alter the susceptibility of MotA to proteolysis, providing additional evidence for changes in its conformation (72). Thus, it appears that torque is generated as protonation and deprotonation of Asp32 of MotB modulates the conformation of MotA, changing the interaction of a specific charged region in the cytoplasmic domain of MotA with a complementary charged region in the C-terminal domain of FliG. To see how this might happen, we need crystal structures of MotA, MotB, and FliG. Crystal structures of the C-terminal and middle domains of FliG have been obtained from the hyperthermophylic eubacterium, *Thermotoga maritima* (73, 74). These structures suggest that the charged groups implicated by site-directed mutagenesis might, indeed, be arrayed on the periphery of the rotor. To understand torque generation,

TABLE 2 Operons encoding the proteins of the chemotaxis system of *E. coli*[a]

Class 1	Class 2	Class 3
flhDC	*flgAMN*	*fliC*
	flgBCDEFGHIJKL	*motABcheAW*
	flhBAE	*tar tap cheRBYZ*
	fliAZY	*aer*
	fliDST	*trg*
	fliE	*tsr*
	fliFGHIJK	
	fliLMNOPQR	

[a]The underlined genes belong to the operons shown, activated by FlhDC, but they have additional promoters activated by FliA. Thus, they are expressed partially as class 2 genes and fully as class 3 genes. Class 3 genes not mentioned in the early part of the text encode receptors for aspartate (*tar*), dipeptides (*tap*), ribose and galactose (*trg*), and serine (*tsr*), a sensor for redox potential (*aer*), enzymes involved in sensory adaptation, a methyltransferase (*cheR*) and methylesterase (*cheB*), and an enzyme that accelerates the removal of phosphate from CheY-P (*cheZ*).

however, we need to know how the complementary groups on MotA interact with these sites and how these interactions change during proton translocation.

GENETICS

Genes expressing flagellar components are arranged in hierarchical order (75, 76) in three classes, as shown in Table 2. Class 1 contains the master operon, *flhDC*, the expression of which is required for transcription of class 2 and class 3 operons (77). Class 2 contains eight operons that encode components required for construction of the hook-basal body complex, and class 3 contains six more that encode components required for filament assembly and motor function. If nutrients are plentiful, motility and chemotaxis are considered luxuries, and cells dispense with them; for example, when *E. coli* is grown on glucose, flagellar synthesis is suppressed (78). *flhDC* is subject to activation by the catabolite repressor/activator protein (CAP) and cyclic AMP (79). It also is activated by the histone-like protein H-NS (80). Also, *flhDC* stimulates *fliA* expression; and *fliA*, in turn, further stimulates *flhDC* expression, which provides a self-reinforcing feedback loop to ensure expression of flagellar genes when needed (81) (see below). Connections between *flhDC* and other systems exist, e.g., those mediating response to heat shock (82), controlling cell division (83), or regulating synthesis of type 1 pili (84). These are not discussed further here (for a short review, see Reference 85). However, one dramatic example of flagellar upregulation that should be mentioned is swarming, in which cells lengthen, produce large numbers of flagella, and spread rapidly over the surface of hard agar (86, 87). The chemotaxis system itself, which requires expression of *flhDC*, appears to be involved in this process (88).

FliA, the gene product of a class 2 operon, is the sigma factor for transcription of class 3 operons (89). FlgM is an anti-sigma factor, i.e., an anti-FliA (90–92). FlgM is encoded by a class 3 gene, but it can also be expressed by readthrough from the class 2 *flgA* promoter (93, 94). Upon completion of the hook–basal body complex, just before the hook-associated proteins are added, the motor transport apparatus pumps FlgM out of the cell (94, 95). As a result, class 3 genes are activated. FlgM is a small protein (97 amino acids) that is largely unfolded, and its conformational plasticity is thought to expedite export (96). The removal of this protein allows cells to finish construction of the machinery needed for motility and chemotaxis. The economy here is that cells do not waste energy synthesizing the large amount of flagellin required for flagellar filaments unless rotary motors are assembled and ready to put these filaments to use. Nor do cells synthesize the torque-generating units MotA and Mot B, or components of the chemotaxis system, such as CheY unless hook and basal body assembly has been successful. The hook-associated proteins are encoded by class 3 operons, but they also are expressed at low levels in the absence of FliA, so the hook–basal body complex can be completed prior to the synthesis of flagellin (97, 98). In short, flagellar genes are expressed in the order in which their products are needed for assembly (99, 100).

ASSEMBLY

The motor is built from the inside out (for minireviews, see References 101, 102). This order was recognized by Suzuki et al. (103), who studied mutants of *Salmonella* defective for different *fla* genes and searched in pellets obtained from detergent extracts for incomplete flagellar structures. The simplest structure found was a "rivet," comprising the MS-ring and rod. Similar results were obtained with mutants of *E. coli* (104). A more recent study identified an even simpler initial structure, the MS-ring alone, and provided many details of the morphological pathway (105). FliG and the C-ring (FliM and FliN) are added to the MS-ring (FliF); see Figure 1. No other proteins are required for this construction (106). Then the transport apparatus is assembled (FlhA, FlhB, FliH, FliI, FliO, FliP, FliQ, FliR) (see References 107–113). This apparatus is used to pass components for other axial structures through a channel at the center of the MS-ring.

One of the key components of the transport apparatus, FliI, shows homology to both the β subunit of the F_0F_1-ATPase (107) and to components of bacterial type III secretory systems (114). Purified His-tagged FliI binds and hydrolyzes

ATP (115). FliI is inhibited by FliH, a component thought to ensure that hydrolysis is properly linked to transport of exported substrates (116).

In *Salmonella,* the type III secretion system injects virulence factors into epithelial cells of the small intestine, inducing them to engulf bacteria. This injection has been shown to involve needle structures, having components homologous to those of the transport apparatus of the flagellar motor (117, 118; for a general review, see Reference 119).

Cytoplasmic chaperones (Table 1) aid in the transport process, in part by preventing aggregation (120–125; for reviews, see References 126, 127).

The next components to be added include the proximal (FlgB, FlgC, FlgF) and distal rod (FlgG) (128). FliE is needed for this assembly and is thought to form a junction between the MS ring and the proximal part of the rod (129, 130). Construction of the hook (FlgE) begins, but it does not proceed very far until the P- and L-rings (FlgI and FlgH) are assembled. Components for these structures are secreted into the periplasmic space by the signal-peptide-dependent (Sec) secretory machinery (131, 132). Assembly of the P-ring also requires FlgA (133), as well as formation of disulfide bonds catalyzed by DsbA and DsbB (134). Formation of the L-ring requires the activity of a flagellum-specific muramidase, FlgJ, that also plays a role in rod formation (135). The L-ring constituent, FlgH, is a lipoprotein (136).

The hook is assembled with the aid of a cap at its distal end (FlgD), which is then discarded (137). Hook length, normally ~55 nm (138), is determined to a precision of about 10%, but the mechanism for this length control is not known. The key player is a cytoplasmic protein called FliK. If this protein is missing, cells form long hooks, called polyhooks (139, 140). However, some control remains, because the distribution of polyhook lengths still peaks at ~55 nm (141). If FliK were simply a molecular ruler, truncated FliK proteins should form shorter hooks, but all *fliK* mutants studied thus far produce longer hooks (142). FliK is exported during hook assembly, and export-deficient *fliK* mutants also produce long hooks (143). Normally, cellular levels of hook protein do not matter, but if FliK is missing, overproduction of hook protein produces superpolyhooks (144). One idea is that FliK functions with FlhB, a membrane protein of the transport apparatus, to switch the export substrate specificity from hook protein to hook-associated proteins and flagellin once the hook reaches its proper length (142, 145). What the signal might be that triggers this transition or why FliK export is required is not clear. Another idea, suggested by the fact that some mutations in genes encoding C-ring proteins produce short hooks, is that the C-ring has a set capacity for hook protein, which is exported en bloc (146). Somehow, this triggers secretion of FliK, which switches the export substrate specificity to flagellin. Again, how this might happen is not clear.

The hook-associated proteins (147) are added in the order HAP1 (FlgK), HAP3 (FlgL), and HAP2 (FliD). Finally, the FliC subunits (flagellin) required for growth of the filament are inserted under a cap (FliD) at its distal end (148, 149). Mutants that lack the cap simply dump flagellin into the external medium (150).

One can polymerize flagellin onto FlgL in such mutants by adding it exogenously (151, 152) or grow filaments in the normal fashion by supplementing such mutants, either with endogenous FliD (153) or with FliD preassembled into the cap structure (154). The cap promotes polymerization of flagellin; it has five legs that leave room for only one flagellin subunit at a time, and it counter rotates to accommodate insertion of additional subunits, one after another (155). Filament extension appears to be limited by the rate of flagellin export because filaments grow at a rate that decreases exponentially with length (156). For a recent review of the mechanism of filament assembly, see (157).

Presumably the torque-generating units, MotA and MotB, can be incorporated into the structure anytime after class 3 genes are expressed. Motors that are paralyzed because MotA or MotB is missing or defective (with defects induced genetically or mechanically) can be repaired by expression of functional copies; see below. When MotA and MotB are expressed together, they do not make membranes leaky to protons, as judged by the lack of any impairment of growth (158). No mutants have been found that implicate specific binding sites for MotB on other components of the flagellar motor; MotB simply has a peptidoglycan binding motif (59). These considerations have led to an ingenious model in which the periplasmic tails of MotB block proton channels in the MotA/MotB complex until the complex finds itself oriented properly at the periphery of the flagellar motor. Then the periplasmic tails bind to the peptidoglycan, thereby opening the proton channels (159).

FUNCTION

Power Source

Flagellar motors of *E. coli* and *S. typhimurium* are powered not by ATP (160) but rather by protons moving down an electrochemical gradient. Other cations and anions have been ruled out (161–163). The work per unit charge that a proton can do in crossing the cytoplasmic membrane is the protonmotive force, Δp. In general, it comprises two terms, one due to the transmembrane electrical potential difference, $\Delta\psi$, and the other to the transmembrane pH difference $(-2.3\ kT/e)\ \Delta\text{pH}$, where k is Boltzmann's constant, T the absolute temperature, and e the proton charge. At 24°C, $2.3\ kT/e = 59$ mV. By convention, $\Delta\psi$ is the internal potential less the external potential, and ΔpH is the internal pH less the external pH. *E. coli* maintains its internal pH in the range 7.6 to 7.8. For cells grown at pH 7, $\Delta p \approx -170$ mV, $\Delta\psi \approx -120$ mV, and $-59\ \Delta\text{pH} \approx -50$ mV. For cells grown at pH 7.6 to 7.8, $\Delta p \approx -140$ mV. For a general discussion of chemiosmotic energy coupling, see Harold & Maloney (164).

The dependence of speed on voltage has been measured in *E. coli* by wiring motors to an external voltage source. Filamentous cells were drawn roughly halfway into micropipettes, and the cytoplasmic membrane of the segment of the cell inside the pipette was made permeable to ions by exposure to gramicidin S. An inert marker was attached to a flagellar motor on the segment of the cell outside the pipette, and its motion was recorded on videotape. Application of an electrical potential between the external medium and the inside of the pipette (the latter negative) caused the marker to spin (165). The rotation speed was directly proportional to Δp over the full physiological range (up to -150 mV). These experiments were done with large markers (heavy loads) at speeds less than 10 Hz. They are being repeated in a different way with small markers (light loads) at speeds up to nearly 300 Hz. So far, the rotation speed still appears proportional to Δp, or to be more precise, to $\Delta\psi$ (C. Gabel & H.C. Berg, unpublished data).

Motors slow down at extremes of pH (usually external pH), below 6 or above 9. This effect is true both for cells tethered to glass by a single flagellum (166–168) and for swimming cells (161, 162, 167, 169). However, swimming cells show thresholds below which cells do not swim ($\Delta p \approx -30$ mV) and above which speed saturates ($\Delta p \approx -100$ mV), neither of which is evident with tethered cells. These thresholds might be due to problems with bundle formation and changes in filament shape, respectively.

The only measurement of proton flux that has been made is with motors of the motile *Streptococcus* sp. strain V4051 (170), a peritrichously flagellated, primarily fermentative, gram-positive organism that lacks an endogenous energy reserve and is sensitive to ionophores and uncouplers. Unlike *E. coli*, this organism can be starved and artificially energized, either with a potassium diffusion potential (by treating cells with valinomycin and shifting them to a medium with a lower concentration of potassium ion) or with a pH gradient (by shifting cells to a medium of lower pH); see Manson et al. (171, 172). If cells are energized in this way in a medium of low buffering capacity, one can follow proton uptake by the increase in external pH. The frequency of rotation of filaments in flagellar bundles can be determined by monitoring the swimming speed—the experiments were done with a smooth-swimming mutant—given the ratio of swimming speed to bundle frequency determined separately by videotaping cells under phase-contrast microscopy and measuring their vibration frequencies by power spectral analysis; see Lowe et al. (173). Finally, the data can be normalized to single motors by counting the number of cells and the number of flagellar filaments per cell. The total proton flux into the cell is much larger than the flux through its flagellar motors. However, the two can be distinguished by suddenly stopping the motors by adding an antifilament antibody—this cross-links adjacent filaments in the flagellar bundles (174)—and measuring the change in flux. This change was found to be directly proportional to the initial swimming speed, as would be expected if a fixed number of protons carries a motor through each revolution. This number is about 1200 (175). One might do better by patch-clamping motors, provided that one could devise a means for monitoring speed. For example, it should be possible to patch-clamp flagella from protoplasts obtained from gram-positive cells by treatment with a suitable muramidase [see, for example, Weibull (176)], but how would one follow the rotation of elements of the stator, now free of attachment to the rigid framework of the cell wall? Another problem is the small proton flux. The top speed encountered in the experiments just described (175) was 65 Hz, corresponding to a flux of 7.8×10^4 protons per motor per second, or a current of 1.2×10^{-2} pA. Currents flowing through single channels from excitable animal cell nerve membranes are typically 100 times larger.

Some bacteria, notably marine bacteria or bacteria that live at high pH, use sodium ions instead of protons (177, 178). Thus, flagellar motors can be ion driven, not only proton driven. Some Mot components of the motor of the marine bacterium *Vibrio alginolyticus* are homologous to MotA and MotB, namely PomA and PomB, but others are not, namely MotX and MotY (179). When flagella are driven with a large sodium gradient, their rotation speeds can be remarkably high, up to 1700 Hz (180, 181). And rotation can be blocked with specific inhibitors of sodium transport, such as amiloride (182) or phenamil (183). This property has made it possible to screen for sodium-channel mutants (184, 185). Also, functional chimeras have been constructed using components from proton- and sodium-ion-driven motors (see, for example, References 186, 187).

Torque-Generating Units

The flux through the flagellar motor is divided into as many as eight distinct proton channels (or pairs of proton channels), comprising one or more copies of the proteins MotA and MotB (currently thought to be 4 MotA and 2 MotB). It was shown by Stocker et al. (188) in the early days of bacterial genetics that phage grown on motile strains of *Salmonella* could transduce flagellar characters into nonmotile strains. Silverman et al. (189) utilized λ transducing phage to "resurrect" nonmotile mutants of *E. coli,* a process that occurred more rapidly when the basal body was already assembled and only *mot* genes needed to be transferred. Such activation was studied at the level of a single motor by Block & Berg (190), who tethered *motB* cells to a glass surface by a single flagellum (191) and expressed the wild-type gene from a plasmid under control of the *lac* promoter—λ phage was tried but did not work, because the phage heads adhered to the glass surface and prevented tethered cells from rotating (S.M. Block & H.C. Berg, unpublished data). This work was extended to *motA* cells in a more carefully controlled study by Blair & Berg (192). The speed of a tethered cell increased in a number of equally spaced steps, as shown in Figure 4, indicating that each additional torque-generating unit (comprising MotA and MotB) adds the same increment of torque (applies a similar force at the same distance from the axis of rotation). The main argument for a complement of 8 such torque-generating units is that resurrections of this kind have produced 8 equally spaced levels more than once, but never 9. As noted above in the section on structure, the number of studs seen in freeze-fracture experiments range from about 10 to 16. In particular, the number seen for *E. coli* is 10–12 (51). Blair & Berg (192), wondering whether this might represent an incomplete set, produced MotA and

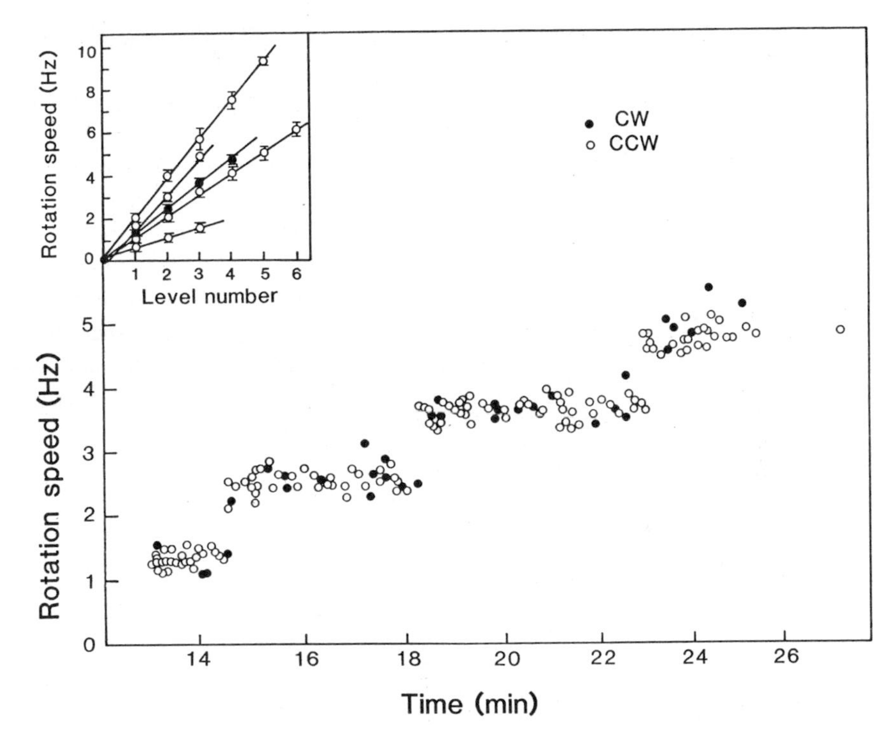

Figure 4 Rotation speed of a tethered *motA* cell, *E. coli* strain MS5037(pDFB36), following addition (at time 0) of the inducer IPTG (added in a minimal medium containing glycerol, glucose, and essential amino acids). Filled circles indicate CW rotation, open circles CCW rotation. The inset shows the mean rotation speed (plus or minus the standard error of the mean) at each level (step of the staircase) as a function of level number for this cell (closed circles) and for four additional cells (open circles). (From Reference 192, Figure 1, reprinted with permission from the American Association for Advancement of Science.)

MotB at a slight excess of wild-type levels and found that the torque increased by about 20%. They also found that the torque for wild-type motors was only about 5 times that for a one-generator motor, whereas following complete resurrection, this factor was about 8. So it is possible that the full complement of torque generators is 8, and the full complement of studs is 16, which yields 2 studs per torque generator.

Stepping

It is likely that the passage of each proton (or each proton pair) moves a torque generator (a MotA/MotB complex) one step (one binding site) along the periphery of the rotor, suddenly stretching the components that link that generator to the rigid framework of the cell wall. As this linkage relaxes, a tethered cell should rotate by a fixed increment. In other words, this molecular machine should behave like a stepping motor. Since proton passage is likely to occur at random times, the steps will occur with exponentially distributed waiting times. We have been looking for such steps since 1976 (193) but without success. The main reason, advanced then, is that the torque applied to the structure linking the rotor to the tethering surface (a series of elastic elements, comprising the rod, hook, and filament) causes that structure to twist (for measurements of the torsional compliance, see References 194, 195). When less torque is applied, these elements tend to untwist, carrying the cell body forward. Therefore, discontinuities in the relative motion of rotor and stator are smoothed out. To succeed, one probably needs to work at reduced torque, e.g., with a one-generator motor driving a small viscous load, perhaps just a hook. Such an object is expected to spin quite rapidly, so the technical problems are formidable.

One route around this difficulty is to examine variations in rotation period. If n steps occur at random over each revolution, then the ratio of the standard deviation to the mean should be $n^{-1/2}$ (196, p. 24; 197, appendix). An early analysis of this kind led to an estimate of $n \approx 400$ (198), which has been borne out by more recent work (197). The more recent analysis also showed that a tethered cell is restrained: It is not free to execute rotational Brownian motion. Thus, the rotor and stator are interconnected most of the time.

This stochastic analysis was repeated with tethered cells undergoing resurrection (as in Figure 4), and the number of steps per revolution was found to increase linearly with level number, increasing by about 50 steps per level (199). If torque generators interact with a fixed number of binding sites on the rotor, say 50, then why is the number of steps per revolution not just 50? If m torque generators are attached to the rotor and one steps, suddenly stretching its linkage to the rigid framework of the cell wall, then when that linkage relaxes and moves the rotor, it also must stretch the linkages of the $m - 1$ torque generators that have not stepped. If $m = 2$, the net movement of the rotor is half of what it would be at $m = 1$, so the apparent step number is 100 per revolution. If $m = 8$, the apparent step number is 400 per revolution. If, on the other hand, each torque generator is detached most of the time (for most of its duty cycle), then the apparent step number would remain 50. So, this experiment argues not only that each force generator steps independently of all the others, but that each remains connected to the rotor most of the time. In fact, the torque generators must be attached nearly all of the time; see below.

If steps occur at random, then the numbers 50, 100, . . . , 400 all are lower bounds. The smoother the rotation, the larger the estimate of the number of steps per revolution. Therefore, any noise in the system that adds to variation in rotation period reduces that estimate. If steps do not occur at random, i.e., if steps are clocked or successive steps are not independent of one another, then similar statistics could be generated with fewer steps; see Svoboda et al. (200).

Coarser fluctuations, probably associated with variations in the number of active torque-generating units, have been studied by Kara-Ivanov et al. (201).

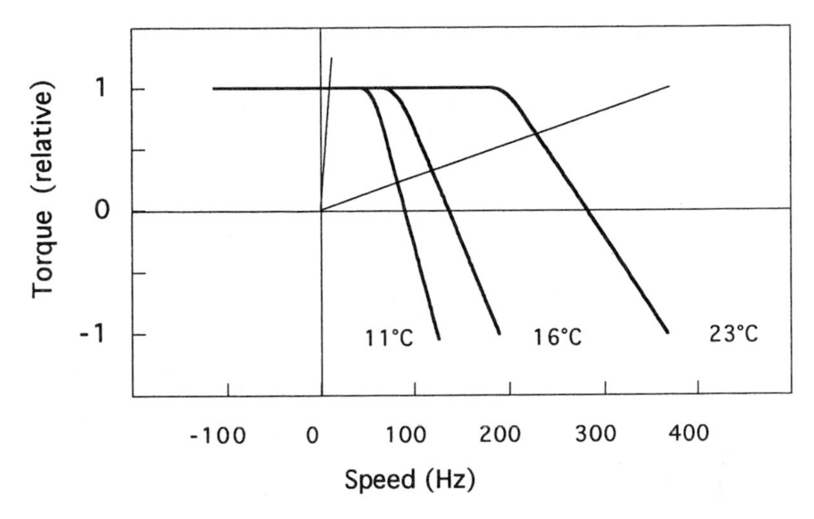

Figure 5 The torque-speed curve for the flagellar motor of *E. coli* shown at three temperatures (thick lines), together with two load lines (thin lines), one for an object the size of the cell body of wild-type *E. coli* (effective radius about 1 μm, left), the other for a minicell (effective radius about 0.3 μm, right). The strains used were derived from *E. coli* wild-type strain AW405 (219). Later work (203) showed that the torque declines somewhat in the low-speed regime, by about 10% between stall and the knee; see the text. (Adapted from Reference 202, Figure 16.)

Torque-Speed Dependence

A crucial test of any rotational motor model is its torque-speed dependence. Measurements of the torque generated by the flagellar motors of *E. coli* have been made over a wide range of speeds, including speeds in which the motor is driven backward, with the results shown in Figure 5. At 23°C, the torque exerted by the motor is approximately constant, all the way from negative speeds of at least -100 Hz to positive speeds of nearly 200 Hz. At higher speeds it declines approximately linearly, crossing the 0-torque line at about 300 Hz. At lower temperatures, the region of transition from constant torque to declining torque—we call this the "knee"—shifts to lower speeds, and the region of decline steepens (202, 203); the latter parts of the curves can be mapped onto one another with scaling of the speed axis.

Estimates of the torque generated in the low-speed regime range from about 2.7×10^{-11} dyn cm (2700 pN nm) to 4.6×10^{-11} dyn cm (4600 pN nm), the smaller value from estimates of the viscous drag on tethered cells of *Streptococcus* (173) and the larger value from the force exerted by tethered cells of *E. coli* on latex beads held in an optical trap (204).

A motor driving an inert object (a cell body, a latex bead, etc.) will spin at the speed at which the torque generated by the motor is balanced by the torque exerted on the object by viscous drag. The latter torque is defined by load lines, such as those shown in Figure 5 (thin lines), the one at the left for a large object and the one at the right for a small object. To appreciate this concept, note that the torque, *N*, required to rotate an object of fixed shape in a viscous medium is its rotational frictional drag coefficient, *f*, times its angular velocity, Ω (2π times its rotation speed, in Hz). In a torque versus speed plot, this function is a straight line passing through the origin, with slope *f*. Here, we assume that the medium is Newtonian, i.e., that the frictional drag coefficient does not depend on Ω, a condition satisfied in a dilute aqueous medium that does not contain long unbranched molecules, such as methylcellulose or polyvinylpyrrolidone (205). For such a medium, *f* is a geometrical factor times the bulk viscosity, η, where η is independent of Ω (independent of the rate of shear). For an isolated sphere of radius *a* spinning about an axis through its center, for example, this geometrical factor is $8\pi a^3$. For compact globular objects, the actual shape is not very critical; however, accurate values can be computed (206). The distance from the tethering surface does not really matter, either, provided that the gap between the object and the surface is at least 0.2 cell radii (193, 207).

At 23°C and for the load line shown at the left in Figure 5, the motor runs at 10 Hz; for the load line shown at the right, it runs at about 220 Hz. For a very shallow load line, e.g., one for a free hook, the speed would be close to the zero-torque speed, about 290 Hz. A motor free-running in this way always operates in the upper right quadrant of Figure 5. It cannot drive itself backward; however, it can redefine what is meant by forward by switching from CCW to CW or back again. Nor can it spin faster than its speed at zero load. To probe the upper left or lower right quadrants of Figure 5, one needs to subject the motor to torque applied externally.

One way to do this is by electrorotation (208). Cells were tethered and exposed to a high-frequency (2.25 MHz) rotating electric field (202). As explained in the cited reference, the external electric field polarizes the cell. The dipole field due to the polarization rotates at the same rate and in the same direction as the applied electric field. However, because of the finite time required for redistribution of charges, the polarization vector leads or lags the electric-field vector. The externally applied torque is the cross product of these vectors. The applied torque varies as the square of the magnitude of the electric field and changes sign with changes in the direction of rotation of that field. Therefore, it is possible to spin a tethered cell either forward or backward. Speeds of several hundred Hz are readily attainable (202). For reasons that we do not understand, the motor of a cell driven backward (CW if it is trying to spin CCW, or CCW if it is trying to spin CW) often breaks catastrophically: Motor torque suddenly drops to zero, the cell appears free to execute rotational Brownian motion, and the motor fails to recover. Our best guess is that the C-ring is sheared off the bottom of the rotor (Figure 1), disengaging all torque-generating units but leaving the bearings intact. If one were to break the rod, the cell would simply come off the tethering surface. We know this because certain mutations in the gene for the MS-ring weaken the rod MS-ring attachment, allowing rod, hook, and filament to pull out of the cell (209). In any event, once the motor has broken, one can compare the speed at which the cell body turns at a given value of externally applied torque with the speed at which it turned at the same value of externally applied torque before the break occurred. That difference is proportional to the torque generated by the motor at the speed at which it turned when intact. The data shown by the thick lines in Figure 5 were determined in this way. Difficulties encountered along the way are described elsewhere (210, 211). In particular, we had thought that there might be a barrier to backward rotation, but this proved to be an artifact due to ellipticity in the applied electric field. The possibility of a barrier was ruled out in experiments utilizing optical tweezers (204).

Additional work on the behavior of the motor in the upper right quadrant of Figure 5 was done by manipulating load lines. Flagella were shortened by viscous shear, and cells were adsorbed onto positively charged glass. Latex beads of various sizes were attached to the flagellar stubs, and the slopes of their load lines were increased by addition of the Ficoll (203). In the low-speed regime, torque was found to drop by about 10% from zero speed (stall) to the knee. In this regime, torque was independent of temperature, and solvent isotope effects (effects if shifts from H_2O to D_2O) were relatively small, as found earlier for artificially energized cells of *Streptococcus* (212). Evidently, at low speeds the motor operates near thermodynamic equilibrium, where rates of displacement of internal mechanical components or translocation of protons are not limiting. In the high-speed regime, torque was strongly temperature dependent, as seen in Figure 5, and solvent isotope effects were large (168), as found earlier for swimming cells of *E. coli, S. typhimurium,* and *Streptococcus* (173, 175, 213). This is what one would expect if the decline in torque at high-speed results from limits in rates of proton transfer (proton dissociation).

Slowly declining torque in the low-speed regime argues for a model in which the rate-limiting step depends strongly on torque and dissipates most of the available free energy, that is, for a powerstroke mechanism. The absence of a barrier to backward rotation rules out models (e.g., thermal ratchets) that contain a step that is effectively irreversible and insensitive to external torque (210). Eventually, we would like to understand why the low-speed regime is so broad, why the boundary between the low-speed and high-speed regimes is so narrow, and why the position of that boundary is sensitive to temperature.

The power output, the power dissipated when a torque N sustains rotation at angular velocity Ω, is $N\Omega$. For a torque of 4600 pN nm and a speed of 10 Hz, this is 2.9×10^5 pN nm s^{-1}. The power input, the rate at which protons can do work, is proton flux times proton charge times protonmotive force. Assuming 1200 protons per revolution and speed 10 Hz, the proton flux is $1.2 \times 10^4\ s^{-1}$. For *E. coli* at pH 7, $\Delta p \approx -170$ mV. Therefore, the power input is $(1.2 \times 10^4\ s^{-1})\ (e)\ (0.17\ V) = 2.0 \times 10^3$ eV s^{-1}, where e is the proton charge. Since 1 eV (one electron volt) $= 1.6 \times 10^{-12}$ erg $=$ 160 pN nm, the power input is 3.2×10^5 pN nm s^{-1}. Therefore, by this crude estimate, the efficiency of the motor, power output divided by power input, is about 90%. Within the uncertainty of the measurements—the proton flux has not been measured in *E. coli*—the efficiency could be 1, the value expected for a tightly coupled engine running slowly close to stall (175).

The power output, $N\Omega$, increases linearly with speed up to the boundary between the low-speed and high-speed regimes, and then it declines. If a fixed number of protons carries the motor through each revolution, the power input also increases linearly with speed. Therefore, the efficiency remains approximately constant up to the knee, and then it declines. There is no discontinuity in torque as one crosses the zero-speed axis (204). As the motor turns backward, it must pump protons, just as the F_0-ATPase pumps protons when driven backward by F_1.

The force exerted by each force-generating unit is substantial but not large on an absolute scale. If we take a ballpark figure for the stall torque of 4000 pN nm and assume that force-generating units act at the periphery of the rotor at a radius of about 20 nm, then 200 pN is applied. If there are 8 independent force-generating units, then each contributes 25 pN. This force is about equal in magnitude to that between two electrons 4.8 Å apart in a medium of dielectric constant 40 (midway between water, 80, and lipid, about 2). So, almost any kind of chemistry will do.

The energy available from one proton moving down the electrochemical gradient is $e\Delta p$. Given $\Delta p \approx -170$ mV, this is 0.17 eV, or 27 pN nm. At unit efficiency, this equals the work that the force-generator can do, Fd, where F is the force that it exerts, and d is the displacement generated by the transit of one proton. Assuming 52 steps per revolution (twice the number of FliG subunits) and a rotor radius of 20 nm, $d \approx 2.4$ nm. So $F \approx 11$ pN. If two protons are required per elementary step, the force is twice as large, and $F \approx 22$ pN. So the displacement of 2 protons per step is likely.

Angular Dependence of Torque

When optical tweezers were used to drive cells slowly backward or to allow them to turn slowly forward (204), torque did not vary appreciably with angle. When the motor is fully energized and has a full complement of torque-generating units, there is no discernible periodic fluctuation. On the other hand, the rotation rates of tethered cells often peak at some point in the cycle and pass through a minimum one-half revolution away. But this behavior is to be expected for cells tethered near one end when the axis of the tether is not normal to the plane of the glass. A very different result is obtained when one energizes and de-energizes tethered cells and asks where they stop or watches them spin when the proton-motive force is very low. When this manipulation was done with *Streptococcus*, periodicities were observed of order 5 or 6 (214). This probably reflects small periodic barriers to rotation intrinsic to the bearings.

Duty Ratio

In our stochastic analysis of steps (in the earlier section Stepping), we argued that the apparent number of steps per revolution would increase with the number of torque generators, as observed, if each torque generator remained attached to the rotor most of the time, i.e., if the torque-generating units had a high duty ratio. This issue was addressed directly in an experiment in which motors were resurrected at low viscous loads (215). If each unit remains attached to the rotor for most of its mechanochemical cycle, then near zero load, one generator can spin the rotor as fast as two or more. The speed is limited by the rate at which the first torque-generating unit can complete its mechanochemical cycle. The smallest load studied was that of a 0.3-μm-diameter latex sphere, and the best we could do was to conclude that the duty ratio was greater than 0.6.

In fact, the duty ratio must be close to 1; i.e., torque generators, like molecules of kinesin, are processive. The argument goes as follows: Consider a tethered cell driven by a single torque-generating unit, as in the first step of the resurrection shown in Figure 4. If a wild-type motor with 8 torque-generating units generates a torque of about 4×10^{-11} dyn cm (4000 pN nm), then the single-unit motor generates a torque of about 5×10^{-12} dyn cm. The torsional spring constant of the tether (the compliance is mostly in the hook) is about 5×10^{-12} dyn cm rad^{-1} (194), so the tether is twisted up about 1 radian, or 57°. Now, the viscous drag on the cell body is enormous compared to that on the rotor, so if the torque-generating unit lets go, the tether will unwind, driving the rotor backward. If the single unit steps 50 times per revolution, the displacement is 7.2° per step. If the cell is spinning ~1.2 Hz (Figure 4), the step interval is 1.6×10^{-2} s. If the duty ratio is 0.999, so that the torque-generating unit detaches for 1.6×10^{-5} s during each cycle, how far will the tether unwind? The tether unwinds exponentially: $\theta = \theta_0 \exp(-\alpha t)$, where θ_0 is the initial twist and α is the torsional spring constant divided by the rotational frictional drag coefficient. If we approximate the rotor as a sphere of radius $a = 20$ nm immersed in a medium of viscosity $\eta =$ 1 P (1 g cm^{-1} s^{-1}), which is about right for a lipid membrane, then the frictional drag coefficient, $8\pi\eta a^3$, is 2×10^{-16} dyn cm per rad s^{-1}, and $\alpha = 2.5 \times 10^4$ s^{-1}. So, in 1.6×10^{-5} s, the twist in the tether decreases from 57° to 57° $\exp(-2.5 \times 10^4 \text{ s}^{-1} \times 1.6 \times 10^{-5}\text{s}) = 38°$, or by 19°, i.e., by more than twice the step angle. Thus, the torque-generating unit would not be able to keep up. So the duty ratio must be close to 1. The interaction between the torque-generating unit and the rotor must be such that the rotor is not able to slip backward. If one imagines that a torque-generating unit binds to successive sites along the periphery of the rotor, then it has no unbound states. If each torque-generating unit has two proton channels, it is possible that a MotA associated with one channel remains attached to a FliG, while the MotA associated with the other channel takes the next step.

Switching

If one follows the direction of rotation of tethered cells and plots CW and CWW interval distributions, the plots are, to a first approximation, exponential. This relationship is true even during responses to constant chemotactic stimuli (216). Exceptions might apply to short events that are difficult to observe (217), and to long events that fall outside the time span of the usual measurements. Also, cells occasionally pause, particularly when the CW bias is high, for example, when responding to repellents (218). In our experiments done with wild-type *E. coli* strain AW405 (219) to a resolution of about 50 ms (220), pauses occur at a frequency of at most 3% of all events, including changes in direction, pauses, and events in which cells might, instead, stick to the tethering surface (S.M. Block & H.C. Berg, unpublished data). So, right or wrong, we have not considered pauses to be very important for chemotactic behavior. To deal properly with short events and pauses, one probably needs better time and spatial resolution than available with standard video. And then one needs to think hard about artifacts introduced by twisting of the tether, Brownian motion, and missed events.

When switching occurs, it appears to be all-or-none: One does not see motors step through an intermediate set of angular velocities, as would be expected if different force-generating units were to switch independently. With tethered cells, the time delay is no more than 10 ms, including the time required for the tether to untwist and then twist up again in the opposite direction (193). With single filaments observed by laser light scattering, reversals appear to be complete within 1 ms (221). If different torque-generating units are to switch synchronously, a global conformational change must occur that involves the arrays of sites with which MotA and MotB interact, probably through flexing of the MS- and/or C-rings (222). This conclusion is consistent with the biochemical and genetic evidence, discussed earlier, in which CheY-P binds to FliM and in which FliG, FliM, and FliN constitute a switch complex.

It was suggested by Khan & Macnab (161) and reaffirmed by Macnab (44) that switching is a thermal isomerization, in which the system sits in one of two potential wells and, with exponentially distributed waiting times, jumps from one to the other. Let the free energies of the CCW and CW states be G_{CCW} and G_{CW} and the free energy of the intervening transition state be G_T. The transition rate constants between the CCW and CW states, k^+ and k^-, are characterized by a factor that represents the frequency at which the system tries to jump, and a factor that represents the probability that it has enough energy to cross the activation barrier $G_T - G_{CCW}$ or $G_T - G_{CW}$. The ratio of the probabilities of the CW or CCW state, (CW bias)/(1 − CW bias) = $k^+/k^- = \exp(-\Delta G/kT)$, where $\Delta G = G_{CW} - G_{CCW}$ (defined in units of kT, the energy of thermal fluctuation for one particle, i.e., Boltzmann's constant times absolute temperature).

Strains that do not express the signaling molecule, CheY, or the kinase required for its phosphorylation, CheA, rotate exclusively CCW, so in the absence of CheY-P, G_{CCW} is much smaller than G_{CW}. However, the relative depths of these wells can be shifted by lowering the temperature (223). CW intervals appear at about 10°C and become as long as CCW intervals at about −1°C. ΔG changes linearly with temperature. An extrapolation back to room temperature (23°C) yields a value there of 14.4 kT. A similar effect on the free energies of the CW and CCW states has been found on varying the intracellular concentration of fumarate (224).

Several recent studies have addressed the question of how this state of affairs is perturbed by chemotactic signaling. In one study, CheY was replaced by the double-mutant CheY(D13K Y106W), abbreviated CheY**, a protein active without phosphorylation, in a strain lacking the kinase CheA and the phosphatase CheZ (225). In a second study, CheY was expressed in a strain in which all of CheY is phosphorylated, a strain that has the kinase but lacks the phosphatase and the receptor-demethylating enzyme CheB (226). In both cases, plots of CW bias versus CheY concentration were sigmoidal and could be characterized by Hill coefficients of 4.2 and 2.5, respectively. However, this nonlinearity arises not from the binding per se but from the effect of the binding of CheY** or CheY-P to FliM. Scharf et al. (225) assumed linear binding and found that ΔG decreased by about $r = 0.8\ kT$ for each molecule of CheY** bound, with the level of the CCW state rising and the level of the CW state falling by similar amounts, 0.4 kT. Alon et al. (226) used the allosteric model of Monod et al. (227), which could be fit with dissociation constants for binding in the CW (or tight) state, K_T, and in the CCW (or relaxed) state, K_R, that differed by a factor of about 2. The two results are equivalent, since $r = kT \ln(K_R/K_T)$. That is, both treatments assume that probabilities of switching are affected by stabilization of the CW state relative to the CCW state. Scharf et al. treat the flagellar motor as an open system and proceed phenomenologically, with r a free parameter. They do not specify a mechanism for the energy shift. Alon et al. treat the flagellar motor as a closed system, with r determined by the difference in binding affinities between tight and relaxed states. The models make similar predictions, because the latter differences are small. In more recent work, the energy shift, r, has been determined over a range of temperatures. It increases linearly from about 0.3 kT at 5°C to about 0.9 kT at 25°C (228).

A third study, done with individual cells rather than with cell populations, revealed a much steeper motor response, characterized by a Hill coefficient of 10.3 (229). CheY-GFP (green fluorescent protein) was expressed in the strain used by Alon et al., in which all CheY is phosphorylated. The CheY-GFP concentration was measured in single cells by fluorescence correlation spectroscopy, and the rotational behavior of a bead attached to a single flagellar filament was monitored. All of the data obtained from different cells expressing different levels of CheY-GFP fell on the same curve, shown in Figure 6 (open triangles, right ordinate). The dashed line is a fit to the allosteric model (227). Also shown in this figure is a binding curve of CheY-P to FliM, obtained from measurements of fluorescence resonance energy transfer (FRET) between CFP-FliM and CheY-YFP (230). The binding is approximately linear (closed circles, left ordinate, $K_D = 3.7 \pm 0.4\ \mu M$, Hill coefficient 1.7 ± 0.3) but also can be fit to the allosteric model (dashed line). The dissociation constants for the two fits are given in the

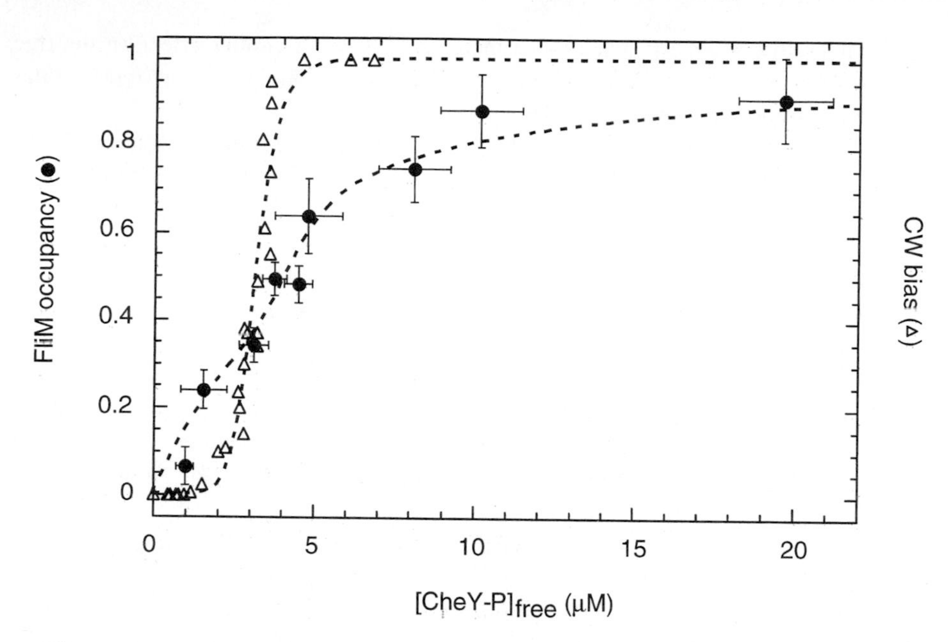

Figure 6 Comparison of dependence of motor bias (△) and FliM occupancy (●) on concentration of free cytoplasmic CheY-P. Dashed lines are fits to the allosteric model: for motor response, $K_T = 2.3 \pm 0.6\ \mu M$, $K_R = 6.6 \pm 3.0\ \mu M$; for binding, $K_T = 2.4 \pm 0.2\ \mu M$, $K_R = 5.7 \pm 1.0\ \mu M$. (Data for the motor bias are from Reference 229 and for the FliM binding from Reference 230. Adapted from Reference 230, Figure 2b.)

figure legend. As before, values for the CW and CCW states differ by a factor of about 2. (For other means of observing the binding of CheY-P to the base of the flagellar motor, see Reference 231.)

A more general stochastic model has been developed in which a ring of proteins (34 copies of FliM) display cooperative interactions (232). Each protein, or protomer, can adopt a CW or CCW configuration, and the direction of rotation depends upon how many protomers are in either state. Given a large enough coupling energy between adjacent protomers, the ensemble switches from a state in which nearly all of the protomers are in the CW configuration to one in which nearly all of the protomers are in the CCW configuration. As before, the motor response is more nonlinear than the binding. This model readily accommodates pauses, which are frequent when the coupling energy is small and the protomers tend to behave independently.

None of this tells us how the shapes of the C- or MS-rings differ in the CCW or CW states or how the binding of CheY-P to FliM stabilizes one conformation and destabilizes the other. We need to learn more about the structures of FliG, FliM, and FliN and the dynamics of their interactions.

MODELS

The fundamental question is how the flagellar motor generates torque, namely, how inward motion of one or more ions through a torque-generating unit causes it to advance circumferentially along the periphery of the rotor. Once that is understood, the nature of the conformational change required for switching, namely, how the direction of advance is distinguished from that of retreat, is likely to be self-evident. Criteria for success in modeling should be based on the behavior of individual motors, not groups of motors, as reflected in the way peritrichously flagellated bacteria swim. As is apparent in Figure 3, cell behavior is complicated by polymorphic transformations of the flagellar filaments.

Moving parts of the motor are submicroscopic and immersed in a viscous medium (water or lipid), so the Reynolds number is very small (233–236). And everything is overdamped (see Reference 237, pp. 41–45). So the designer does not have the benefit of flywheels or tuning forks. If, for example, the operator of the motor driving a tethered cell of *E. coli* 10 Hz were to disengage in the clutch, the cell body would coast no more than a millionth of a revolution (24). So, if there is a stage in the rotational cycle in which the torque changes sign, the motor will stop. Predicting net torque after averaging over a complete cycle is not sufficient. And mechanisms in which energy is stored in vibrational modes are not viable. However, one can use energy available from an electrochemical potential to stretch a spring and then use that spring to apply a steady force. As we have seen, the force required is modest, and almost any kind of chemistry will do.

Motion of the torque-generating units relative to the periphery of the rotor is driven by a proton (or sodium-ion) flux. Only one experiment has attempted to measure this flux (175), and flux and speed were found to be linearly related. Unless protons flow through the motor when it is stalled, this implies that a fixed number of protons carry the motor through each revolution. The running torque at low speeds is close to the stall torque (Figure 5). If the motor is stalled and no protons flow, no free energy is dissipated; therefore, the stalled motor is at thermodynamic equilibrium. For slow rotation near stall, the motor must operate reversibly at unit efficiency, with the free energy lost by protons traversing the motor equal to the mechanical work that it performs. This implies that the torque near stall should be proportional to the protonmotive force over its full physiological range, as observed. So the evidence is consistent with a model in which the motor is tightly coupled.

An important question is whether the ion that moves down the electrochemical gradient is directly involved in generating torque, i.e., participates in a powerstroke in which dissipation of energy available from the electrochemical gradient and rotational work occur synchronously, or whether the ion is indirectly involved in generating torque, e.g., by enabling a ratchet that is powered thermally. In the powerstroke case, protons can be driven out of the cell by backward rotation and steep barriers are not expected. In addition, if the rate-limiting step is strongly torque dependent, then the torque-speed curve (as plotted in Figure 5) can have a relatively flat plateau, because small changes in torque can generate large changes in speed (238) (see Reference 210, Figure 6c). In the ratchet case, with tight coupling, the likelihood of transit of ions against the electrochemical gradient is small, so the system must wait, even when large backward torques are applied, and barriers to backward rotation are expected. Also, the torque-speed curves are relatively steep (167) (see Reference 210, Figure 6b). So, the torque-speed curves of Figure 5 favor a powerstroke mechanism.

Tightly coupled models can be distinguished from one another only by their behavior at high speeds, far from the point of thermodynamic equilibrium. A seminal test for any such model is the torque-speed curve of Figure 5. The thermal-ratchet model that we proposed (222, 239), which (with improvements) has been applied successfully to F_0 (240), fails this test. At negative speeds, it predicts barriers to rotation, and at positive speeds, it predicts a torque that falls steadily toward the zero-torque speed. It does not predict a constant-torque plateau or an abrupt transition from a low-speed to a high-speed regime (see Reference 222, Figure 7).

As argued in the section on duty ratio, a torque generator must not have unbound states, i.e., states in which the rotor is free to spin backward. The ratchet model meets this criterion, even though its channel complex is not bound to the rotor in the usual biochemical sense. It is simply free to move forward or backward one step at a time, depending upon the occupancy of adjacent proton-accepting sites.

Finally, a successful model must be consistent with the general structural features outlined in Figure 1, where the filament, hook, rod, MS-ring, and FliG rotate relative to the rigid framework of the cell, defined by the peptidoglycan layer, and the Mot proteins do not. FliM and FliN are likely part of the rotor; however, the evidence for this is not airtight. Since MotA and B are embedded in the cytoplasmic membrane, they are not free to execute movements out of the plane of that membrane. Their movements are presumably cyclic. One can imagine a model in which a MotA/MotB complex rolls along the periphery of the rotor, but not if more than one MotB is bound to the peptidoglycan.

There appear to be essential electrostatic interactions between specific residues in the cytoplasmic domain of MotA and the C-terminal domain of FliC (69). Here, charge complementarity is more important than surface complementarity, i.e., long-range interactions appear to be more important than tight binding. Since some models for torque generation require transfer of protons from the stator to the rotor, it was expected that acidic residues on FliG might be more important than basic residues. However, replacement of the acidic residues deemed important for torque generation with alanine still allowed some rotation, whereas reversing their charge had a more severe effect (68). An extension of this study failed to identify any conserved basic residues critical for rotation in MotA, MotB, FliG, FliM, or FliN and only one conserved acidic residue critical for rotation, Asp32 of MotB (70). Other alternatives were considered and either ruled out or deemed unlikely. Therefore, the only strong candidate for a residue that functions directly in proton conduction is Asp32 of MotB.

Given this work, I would bet on a cross-bridge mechanism of the kind that Blair and colleagues propose (71). In such a scheme, proton transport drives the following cyclic sequence: First, a proton binds to an outward-facing binding site. Second, the protonmotive force drives a conformational change, a power-stroke that moves the rotor forward (or stretches a spring that moves it forward) and transforms the binding site to an inward-facing site. Finally, proton dissociation triggers detachment of the cross-bridge from the rotor, its relaxation to the original shape, and reattachment to an adjacent site. If the MotA/MotB complex is two headed, one head could remain attached while the other stepped, thus ensuring a high duty ratio. The cross-bridge mechanism was analyzed earlier by Läuger (Model II of Reference 241), but at a time when the torque-speed curve was not known to have a plateau, so this work should be revisited. To formulate models adequate for physiological testing, it would help to have more structural information.

REVIEWS

Models proposed for the flagellar motor are cataloged elsewhere (202, 242). Other reviews on the structure and function of proton-driven motors are available (211, 243–251). Also available are reviews on sodium motors (252, 253) and on flagellar genetics and assembly (101, 246, 254–256), as well as tutorials on the mathematical treatment of motor models (249, 257).

ACKNOWLEDGMENTS

I thank Lloyd Schoenbach and Linda Turner for help with the figures, and Bob Bourret and Bob Macnab for comments on sections of the manuscript. This review was adapted from a chapter "The Bacterial Rotary Motor" written for a volume of *The Enzymes*, first submitted in 1999 (242). Some of that material was presented in a discussion meeting of the Royal Society (211). Work in my laboratory has been supported by the National Institutes of Health, the National Science Foundation, and the Rowland Institute for Science.

The *Annual Review of Biochemistry* is online at http://biochem.annualreviews.org

LITERATURE CITED

1. Larsen SH, Reader RW, Kort EN, Tso W, Adler J. 1974. *Nature* 249:74–77
2. Macnab RM, Ornston MK. 1977. *J. Mol. Biol.* 112:1–30
3. Turner L, Ryu WS, Berg HC. 2000. *J. Bacteriol.* 182:2793–801
4. Blair DF. 1995. *Annu. Rev. Microbiol.* 49:489–522

5. Stock JB, Surette MG. 1996. See Ref. 261, pp. 1103–29
6. Falke JJ, Bass RB, Butler SL, Chervitz SA, Danielson MA. 1997. *Annu. Rev. Cell Dev. Biol.* 13:457–512
7. Stock AM, Robinson VL, Goudreau PN. 2000. *Annu. Rev. Biochem.* 69:183–215
8. Bren A, Eisenbach M. 2000. *J. Bacteriol.* 182:6865–73
9. Falke JJ, Hazelbauer GL. 2001. *Trends Biochem. Sci.* 26:257–65
10. Bray D. 2002. *Proc. Natl. Acad. Sci. USA* 99:7–9
11. Bourret RB, Stock AM. 2002. *J. Biol. Chem.* 277:9625–28
12. Iino T, Komeda Y, Kutsukake K, Macnab RM, Matsumura P, et al. 1988. *Microbiol. Rev.* 52:533–35
13. Asakura S. 1970. *Adv. Biophys.* 1:99–155
14. Calladine CR. 1978. *J. Mol. Biol.* 118: 457–79
15. Kamiya R, Asakura S, Wakabayashi K, Namba K. 1979. *J. Mol. Biol.* 131:725–42
16. Fahrner KA, Block SM, Krishnaswamy S, Parkinson JS, Berg HC. 1994. *J. Mol. Biol.* 238:173–86
17. Hasegawa K, Yamashita I, Namba K. 1998. *Biophys. J.* 74:569–75
18. Yamashita I, Hasegawa K, Suzuki H, Vonderviszt F, Mimori-Kiyosue Y, Namba K. 1998. *Nat. Struct. Biol.* 5:125–32
19. Namba K, Vonderviszt F. 1997. *Q. Rev. Biophys.* 30:1–65
20. Samatey FA, Imada K, Nagashima S, Vonderviszt F, Kumasaka T, et al. 2001. *Nature* 410:331–37
21. DePamphilis ML, Adler J. 1971. *J. Bacteriol.* 105:376–83
22. DePamphilis ML, Adler J. 1971. *J. Bacteriol.* 105:384–95
23. DePamphilis ML, Adler J. 1971. *J. Bacteriol.* 105:396–407
24. Berg HC. 1974. *Nature* 249:77–79
25. Ueno T, Oosawa K, Aizawa S-I. 1992. *J. Mol. Biol.* 227:672–77
26. Ueno T, Oosawa K, Aizawa S-I. 1994. *J. Mol. Biol.* 236:546–55
27. Driks A, DeRosier DJ. 1990. *J. Mol. Biol.* 211:669–72
28. Khan IH, Reese TS, Khan S. 1992. *Proc. Natl. Acad. Sci. USA* 89:5956–60
29. Francis NR, Sosinsky GE, Thomas D, DeRosier DJ. 1994. *J. Mol. Biol.* 235: 1261–70
30. Katayama E, Shiraishi T, Oosawa K, Baba N, Aizawa S-I. 1996. *J. Mol. Biol.* 255:458–75
31. Suzuki H, Yonekura K, Murata K, Hirai T, Oosawa K, Namba K. 1998. *J. Struct. Biol.* 124:104–14
32. Yamaguchi S, Aizawa S-I, Kihara M, Isomura M, Jones CJ, Macnab RM. 1986. *J. Bacteriol.* 168:1172–79
33. Yamaguchi S, Fujita H, Ishihara A, Aizawa S-I, Macnab RM. 1986. *J. Bacteriol.* 166:187–93
34. Welch M, Oosawa K, Aizawa S-I, Eisenbach M. 1993. *Proc. Natl. Acad. Sci. USA* 90:8787–91
35. Welch M, Oosawa K, Aizawa S-I, Eisenbach M. 1994. *Biochemistry* 33: 10470–76
36. Toker AS, Macnab RM. 1997. *J. Mol. Biol.* 273:623–34
37. Bren A, Eisenbach M. 1998. *J. Mol. Biol.* 278:507–14
38. Oosawa K, Ueno T, Aizawa S-I. 1994. *J. Bacteriol.* 176:3683–91
39. Marykwas DL, Schmidt SA, Berg HC. 1996. *J. Mol. Biol.* 256:564–76
40. Tang H, Braun TF, Blair DF. 1996. *J. Mol. Biol.* 261:209–21
41. Francis NR, Irikura VM, Yamaguchi S, DeRosier DJ, Macnab RM. 1992. *Proc. Natl. Acad. Sci. USA* 89:6304–8
42. Thomas DR, Morgan DG, DeRosier DJ. 2001. *J. Bacteriol.* 183:6404–12
43. Kihara M, Francis NR, DeRosier DJ, Macnab RM. 1996. *J. Bacteriol.* 178: 4582–89
44. Macnab RM. 1995. In *Two-Component Signal Transduction*, ed. JA Hoch, TJ

Silhavy, pp. 181–99. Washington, DC: Am. Soc. Microbiol.
45. Zhao R, Schuster SC, Khan S. 1995. *J. Mol. Biol.* 251:400–12
46. Zhao R, Pathak N, Jaffe H, Reese TS, Khan S. 1996. *J. Mol. Biol.* 261:195–208
47. Lloyd SA, Tang H, Wang X, Billings S, Blair DF. 1996. *J. Bacteriol.* 178:223–31
48. Thomas DR, Morgan DG, DeRosier DJ. 1999. *Proc. Natl. Acad. Sci. USA* 96:10134–39
49. Ridgway HF, Silverman M, Simon MI. 1977. *J. Bacteriol.* 132:657–65
50. Coulton JW, Murray RGE. 1978. *J. Bacteriol.* 136:1047–49
51. Khan S, Dapice M, Reese TS. 1988. *J. Mol. Biol.* 202:575–84
52. Khan S, Khan IH, Reese TS. 1991. *J. Bacteriol.* 173:2888–96
53. Khan S, Ivey DM, Krulwich TA. 1992. *J. Bacteriol.* 174:5123–26
54. Dean GE, Macnab RM, Stader J, Matsumura P, Burke C. 1984. *J. Bacteriol.* 159:991–99
55. Blair DF, Berg HC. 1991. *J. Mol. Biol.* 221:1433–42
56. Zhou J, Fazzio RT, Blair DF. 1995. *J. Mol. Biol.* 251:237–42
57. Stader J, Matsumura P, Vacante D, Dean GE, Macnab RM. 1986. *J. Bacteriol.* 166:244–52
58. Chun SY, Parkinson JS. 1988. *Science* 239:276–78
59. De Mot R, Vanderleyden J. 1994. *Mol. Microbiol.* 12:333–34
60. Sato K, Homma M. 2000. *J. Biol. Chem.* 275:5718–22
61. Braun TF, Blair DF. 2001. *Biochemistry* 40:13051–59
62. Sharp LL, Zhou J, Blair DF. 1995. *Biochemistry* 34:9166–71
63. Sharp LL, Zhou J, Blair DF. 1995. *Proc. Natl. Acad. Sci. USA* 92:7946–50
64. Garza AG, Bronstein PA, Valdez PA, Harris-Haller LW, Manson MD. 1996. *J. Bacteriol.* 178:6116–22
65. Garza AG, Harris-Haller LW, Stoebner RA, Manson MD. 1995. *Proc. Natl. Acad. Sci. USA* 92:1970–74
66. Garza AG, Biran R, Wohlschlegel JA, Manson MD. 1996. *J. Mol. Biol.* 258: 270–85
67. Zhou J, Blair DF. 1997. *J. Mol. Biol.* 273:428–39
68. Lloyd SA, Blair DF. 1997. *J. Mol. Biol.* 266:733–44
69. Zhou J, Lloyd SA, Blair DF. 1998. *Proc. Natl. Acad. Sci. USA* 95:6436–41
70. Zhou J, Sharp LL, Tang HL, Lloyd SA, Billings S, et al. 1998. *J. Bacteriol.* 180: 2729–35
71. Braun TF, Poulson S, Gully JB, Empey JC, Van Way S, et al. 1999. *J. Bacteriol.* 181:3542–51
72. Kojima S, Blair DF. 2001. *Biochemistry* 40:13041–50
73. Lloyd SA, Whitby FG, Blair DF, Hill CP. 1999. *Nature* 400:472–75
74. Brown PN, Hill CP, Blair DF. 2002. *EMBO J.* 21:3225–34
75. Komeda Y. 1982. *J. Bacteriol.* 150:16–26
76. Kutsukake K, Ohya Y, Iino T. 1990. *J. Bacteriol.* 172:741–47
77. Liu X, Matsumura P. 1994. *J. Bacteriol.* 176:7345–51
78. Adler J, Templeton B. 1967. *J. Gen. Microbiol.* 46:175–84
79. Komeda Y, Suzuki H, Ishidsu J-I, Iino T. 1975. *Mol. Gen. Genet.* 142:289–98
80. Bertin P, Terao E, Lee EH, Lejeune P, Colson C, et al. 1994. *J. Bacteriol.* 176: 5537–40
81. Kutsukake K. 1997. *Mol. Gen. Genet.* 254:440–48
82. Shi W, Zhou Y, Wild J, Adler J, Gross CA. 1992. *J. Bacteriol.* 174:6256–63
83. Pruss BM, Matsumura P. 1997. *J. Bacteriol.* 179:5602–4
84. Clegg S, Hughes KT. 2002. *J. Bacteriol.* 184:1209–13
85. Aizawa S-I, Kubori T. 1998. *Genes Cells* 3:625–34
86. Harshey RM. 1994. *Mol. Microbiol.* 13:389–94

87. Harshey RM, Matsuyama T. 1994. *Proc. Natl. Acad. Sci. USA* 91:8631–35
88. Burkart M, Toguchi A, Harshey RM. 1998. *Proc. Natl. Acad. Sci. USA* 95:2568–73
89. Ohnishi K, Kutsukake K, Suzuki H, Iino T. 1990. *Mol. Gen. Genet.* 221:139–47
90. Gillen KL, Hughes KT. 1991. *J. Bacteriol.* 173:2301–10
91. Gillen KL, Hughes KT. 1991. *J. Bacteriol.* 173:6453–59
92. Ohnishi K, Kutsukake K, Suzuki H, Iino T. 1992. *Mol. Microbiol.* 6:3149–57
93. Gillen KL, Hughes KT. 1993. *J. Bacteriol.* 175:7006–15
94. Kutsukake K. 1994. *Mol. Gen. Genet.* 243:605–12
95. Hughes KT, Gillen KL, Semon MJ, Karlinsey JE. 1993. *Science* 262:1277–80
96. Daughdrill GW, Chadsey MS, Karlinsey JE, Hughes KT, Dahlquist FW. 1997. *Nat. Struct. Biol.* 4:285–91
97. Homma M, Kutsukake K, Iino T, Yamaguchi S. 1984. *J. Bacteriol.* 157:100–8
98. Kutsukake K, Ide N. 1995. *Mol. Gen. Genet.* 247:275–81
99. Karlinsey JE, Tanaka S, Bettenworth V, Yamaguchi S, Boos W, et al. 2000. *Mol. Microbiol.* 37:1220–31
100. Kalir S, McClure J, Pabbaraju K, Southward C, Ronen M, et al. 2001. *Science* 292:2080–83
101. Aizawa S-I. 1996. *Mol. Microbiol.* 19:1–5
102. Macnab RM. 1999. *J. Bacteriol.* 181:7149–53
103. Suzuki T, Iino T, Horiguchi T, Yamaguchi S. 1978. *J. Bacteriol.* 133:904–15
104. Suzuki T, Komeda Y. 1981. *J. Bacteriol.* 145:1036–41
105. Kubori T, Shimamoto N, Yamaguchi S, Namba K, Aizawa S-I. 1992. *J. Mol. Biol.* 226:433–46
106. Kubori T, Yamaguchi S, Aizawa S-I. 1997. *J. Bacteriol.* 179:813–17
107. Vogler AP, Homma M, Irikura VM, Macnab RM. 1991. *J. Bacteriol.* 173:3564–72
108. Malakooti J, Ely B, Matsumura P. 1994. *J. Bacteriol.* 176:189–97
109. Fan F, Ohnishi K, Francis NR, Macnab RM. 1997. *Mol. Microbiol.* 26:1035–46
110. Ohnishi K, Fan F, Schoenhals GJ, Kihara M, Macnab RM. 1997. *J. Bacteriol.* 179:6092–99
111. Minamino T, Macnab RM. 1999. *J. Bacteriol.* 181:1388–94
112. Minamino T, Macnab RM. 2000. *J. Bacteriol.* 182:4906–14
113. Kihara M, Minamino T, Yamaguchi S, Macnab RM. 2001. *J. Bacteriol.* 183:1655–62
114. Dreyfus G, Williams AW, Kawagishi I, Macnab RM. 1993. *J. Bacteriol.* 175:3131–38
115. Fan F, Macnab RM. 1996. *J. Biol. Chem.* 271:31981–88
116. Minamino T, Macnab RM. 2000. *Mol. Microbiol.* 37:1494–503
117. Kubori T, Matsushima Y, Nakamura D, Uralil J, Lara-Tejero M, et al. 1998. *Science* 280:602–5
118. Kubori T, Sukhan A, Aizawa S-I, Galán JE. 2000. *Proc. Natl. Acad. Sci. USA* 97:10225–30
119. Hueck CJ. 1998. *Microbiol. Mol. Biol. Rev.* 62:379–433
120. Yokoseki T, Kutsukake K, Ohnishi K, Iino T. 1995. *Microbiology* 141:1715–22
121. Fraser GM, Bennett JCQ, Hughes C. 1999. *Mol. Microbiol.* 32:569–80
122. Minamino T, Macnab RM. 2000. *Mol. Microbiol.* 35:1052–64
123. Minamino T, Chu R, Yamaguchi S, Macnab RM. 2000. *J. Bacteriol.* 182:4207–15
124. Bennett JCQ, Thomas J, Fraser GM, Hughes C. 2001. *Mol. Microbiol.* 39:781–91
125. Auvray F, Thomas J, Fraser GM, Hughes C. 2001. *J. Mol. Biol.* 308:221–29
126. Bennett JCQ, Hughes C. 2000. *Trends Microbiol.* 8:202–4

127. Karlinsey JE, Lonner J, Brown KL, Hughes KT. 2000. *Cell* 102:487–97
128. Homma M, Kutsukake K, Hasebe M, Iino T, Macnab RM. 1990. *J. Mol. Biol.* 211:465–77
129. Müller V, Jones CJ, Kawagishi I, Aizawa S-I, Macnab RM. 1992. *J. Bacteriol.* 174:2298–304
130. Minamino T, Yamaguchi S, Macnab RM. 2000. *J. Bacteriol.* 182:3029–36
131. Homma M, Komeda Y, Iino T, Macnab RM. 1987. *J. Bacteriol.* 169:1493–98
132. Jones CJ, Homma M, Macnab RM. 1989. *J. Bacteriol.* 171:3890–900
133. Nambu T, Kutsukake K. 2000. *Microbiology* 146:1171–78
134. Dailey FE, Berg HC. 1993. *Proc. Natl. Acad. Sci. USA* 90:1043–47
135. Hirano T, Minamino T, Macnab RM. 2001. *J. Mol. Biol.* 312:359–69
136. Schoenhals GJ, Macnab RM. 1996. *J. Bacteriol.* 178:4200–7
137. Ohnishi K, Ohto Y, Aizawa S-I, Macnab RM, Iino T. 1994. *J. Bacteriol.* 176: 2272–81
138. Hirano T, Yamaguchi S, Oosawa K, Aizawa S-I. 1994. *J. Bacteriol.* 176:5439–49
139. Silverman M, Simon M. 1972. *J. Bacteriol.* 112:986–93
140. Patterson-Delafield J, Martinez RJ, Stocker BAD, Yamaguchi S. 1973. *Arch. Microbiol.* 90:107–20
141. Koroyasu S, Yamazato M, Hirano T, Aizawa S-I. 1998. *Biophys. J.* 74:436–43
142. Williams AW, Yamaguchi S, Togashi F, Aizawa S-I, Kawagishi I, Macnab RM. 1996. *J. Bacteriol.* 178:2960–70
143. Minamino T, González-Pedrajo B, Yamaguchi K, Aizawa S-I, Macnab RM. 1999. *Mol. Microbiol.* 34:295–304
144. Muramoto K, Makishima S, Aizawa S-I, Macnab RM. 1999. *J. Bacteriol.* 181: 5808–13
145. Kutsukake K, Minamino T, Yokoseki T. 1994. *J. Bacteriol.* 176:7625–29
146. Makishima S, Komoriya K, Yamaguchi S, Aizawa S-I. 2001. *Science* 291: 2411–13
147. Ikeda T, Homma M, Iino T, Asakura S, Kamiya R. 1987. *J. Bacteriol.* 169: 1168–73
148. Iino T. 1969. *J. Gen. Microbiol.* 56:227–39
149. Emerson SU, Tokuyasu K, Simon MI. 1970. *Science* 169:190–92
150. Homma M, Fujita H, Yamaguchi S, Iino T. 1984. *J. Bacteriol.* 159:1056–59
151. Kagawa H, Morisawa H, Enomoto M. 1981. *J. Mol. Biol.* 153:465–70
152. Kagawa H, Nishiyama T, Yamaguchi S. 1983. *J. Bacteriol.* 155:435–37
153. Ikeda T, Yamaguchi S, Hotani H. 1993. *J. Biochem.* 114:39–44
154. Ikeda T, Oosawa K, Hotani H. 1996. *J. Mol. Biol.* 259:679–86
155. Yonekura K, Maki S, Morgan DG, DeRosier DJ, Vonderviszt F, et al. 2000. *Science* 290:2148–52
156. Iino T. 1974. *J. Supramol. Struct.* 2:372–84
157. Yonekura K, Maki-Yonekura S, Namba K. 2002. *Res. Microbiol.* 153:191–97
158. Stolz B, Berg HC. 1991. *J. Bacteriol.* 173:7033–37
159. Van Way S, Hosking ER, Braun TF, Manson MD. 2000. *J. Mol. Biol.* 297: 7–24
160. Larsen SH, Adler J, Gargus JJ, Hogg RW. 1974. *Proc. Natl. Acad. Sci. USA* 71:1239–43
161. Khan S, Macnab RM. 1980. *J. Mol. Biol.* 138:563–97
162. Khan S, Macnab RM. 1980. *J. Mol. Biol.* 138:599–614
163. Ravid S, Eisenbach M. 1984. *J. Bacteriol.* 158:1208–10
164. Harold FM, Maloney PC. 1996. See Ref. 261, pp. 283–306
165. Fung DC, Berg HC. 1995. *Nature* 375: 809–12
166. Meister M, Berg HC. 1987. *Biophys. J.* 52:413–19
167. Khan S, Dapice M, Humayun I. 1990. *Biophys. J.* 57:779–96

168. Chen X, Berg HC. 2000. *Biophys. J.* 78:2280–84
169. Shioi J-I, Matsuura S, Imae Y. 1980. *J. Bacteriol.* 144:891–97
170. van der Drift C, Duiverman J, Bexkens H, Krijnen A. 1975. *J. Bacteriol.* 124: 1142–47
171. Manson MD, Tedesco P, Berg HC, Harold FM, van der Drift C. 1977. *Proc. Natl. Acad. Sci. USA* 74:3060–64
172. Manson MD, Tedesco PM, Berg HC. 1980. *J. Mol. Biol.* 138:541–61
173. Lowe G, Meister M, Berg HC. 1987. *Nature* 325:637–40
174. Berg HC, Anderson RA. 1973. *Nature* 245:380–82
175. Meister M, Lowe G, Berg HC. 1987. *Cell* 49:643–50
176. Weibull C. 1953. *J. Bacteriol.* 66:688–95
177. Imae Y, Atsumi T. 1989. *J. Bioenerg. Biomembr.* 21:705–16
178. Imae Y. 1991. In *New Era of Bioenergetics*, ed. Y Mukohata, pp. 197–221. Tokyo: Academic
179. Asai Y, Kojima S, Kato H, Nishioka N, Kawagishi I, Homma M. 1997. *J. Bacteriol.* 179:5104–10
180. Magariyama Y, Sugiyama S, Muramoto K, Maekawa Y, Kawagishi I, et al. 1994. *Nature* 371:752
181. Muramoto K, Kawagishi I, Kudo S, Magariyama Y, Imae Y, Homma M. 1995. *J. Mol. Biol.* 251:50–58
182. Sugiyama S, Cragoe EJ Jr, Imae Y. 1988. *J. Biol. Chem.* 263:8215–19
183. Atsumi T, Sugiyama S, Cragoe EJ Jr, Imae Y. 1990. *J. Bacteriol.* 172:1634–39
184. Kojima S, Atsumi T, Muramoto K, Kudo S, Kawagishi I, Homma M. 1997. *J. Mol. Biol.* 265:310–18
185. Kojima S, Asai Y, Atsumi T, Kawagishi I, Homma M. 1999. *J. Mol. Biol.* 285: 1537–47
186. Asai Y, Kawagishi I, Sockett RE, Homma M. 1999. *J. Bacteriol.* 181:6332–38
187. Asai Y, Kawagishi I, Sockett RE, Homma M. 2000. *EMBO J.* 19:3639–48
188. Stocker BAD, Zinder ND, Lederberg J. 1953. *J. Gen. Microbiol.* 9:410–33
189. Silverman M, Matsumura P, Simon M. 1976. *Proc. Natl. Acad. Sci. USA* 73:3126–30
190. Block SM, Berg HC. 1984. *Nature* 309: 470–72
191. Silverman M, Simon M. 1974. *Nature* 249:73–74
192. Blair DF, Berg HC. 1988. *Science* 242: 1678–81
193. Berg HC. 1976. In *Cell Motility, Cold Spring Harbor Conferences on Cell Proliferation*, ed. R Goldman, T Pollard, J Rosenbaum, pp. 47–56. Cold Spring Harbor, NY: Cold Spring Harbor Lab.
194. Block SM, Blair DF, Berg HC. 1989. *Nature* 338:514–17
195. Block SM, Blair DF, Berg HC. 1991. *Cytometry* 12:492–96
196. Cox DR, Lewis PAW. 1966. *The Statistical Analysis of Series of Events*. London: Methuen. 285 pp.
197. Samuel ADT, Berg HC. 1995. *Proc. Natl. Acad. Sci. USA* 92:3502–6
198. Berg HC, Manson MD, Conley MP. 1982. *Symp. Soc. Exp. Biol.* 35:1–31
199. Samuel ADT, Berg HC. 1996. *Biophys. J.* 71:918–23
200. Svoboda K, Mitra PP, Block SM. 1994. *Proc. Natl. Acad. Sci. USA* 91:11782–86
201. Kara-Ivanov M, Eisenbach M, Caplan SR. 1995. *Biophys. J.* 69:250–63
202. Berg HC, Turner L. 1993. *Biophys. J.* 65:2201–16
203. Chen X, Berg HC. 2000. *Biophys. J.* 78:1036–41
204. Berry RM, Berg HC. 1997. *Proc. Natl. Acad. Sci. USA* 94:14433–37
205. Berg HC, Turner L. 1979. *Nature* 278: 349–51
206. Garcia de la Torre J, Bloomfield VA. 1981. *Q. Rev. Biophys.* 14:81–139
207. Jeffery GB. 1915. *Proc. London Math. Soc.* 14:327–38

208. Washizu M, Kurahashi Y, Iochi H, Kurosawa O, Aizawa S-I, et al. 1993. *IEEE Trans. Ind. Appl.* 29:286–94
209. Okino H, Isomura M, Yamaguchi S, Magariyama Y, Kudo S, Aizawa S-I. 1989. *J. Bacteriol.* 171:2075–82
210. Berry RM, Berg HC. 1999. *Biophys. J.* 76:580–87
211. Berg HC. 2000. *Philos. Trans. R. Soc. London Ser. B* 355:491–501
212. Khan S, Berg HC. 1983. *Cell* 32:913–19
213. Blair DF, Berg HC. 1990. *Cell* 60:439–49
214. Khan S, Meister M, Berg HC. 1985. *J. Mol. Biol.* 184:645–56
215. Ryu WS, Berry RM, Berg HC. 2000. *Nature* 403:444–47
216. Block SM, Segall JE, Berg HC. 1983. *J. Bacteriol.* 154:312–23
217. Kuo SC, Koshland DE Jr. 1989. *J. Bacteriol.* 171:6279–87
218. Eisenbach M, Wolf A, Welch M, Caplan SR, Lapidus IR, et al. 1990. *J. Mol. Biol.* 211:551–63
219. Armstrong JB, Adler J, Dahl MM. 1967. *J. Bacteriol.* 93:390–98
220. Berg HC, Block SM, Conley MP, Nathan AR, Power JN, Wolfe AJ. 1987. *Rev. Sci. Instrum.* 58:418–23
221. Kudo S, Magariyama Y, Aizawa S-I. 1990. *Nature* 346:677–80
222. Meister M, Caplan SR, Berg HC. 1989. *Biophys. J.* 55:905–14
223. Turner L, Caplan SR, Berg HC. 1996. *Biophys. J.* 71:2227–33
224. Prasad K, Caplan SR, Eisenbach M. 1998. *J. Mol. Biol.* 280:821–28
225. Scharf BE, Fahrner KA, Turner L, Berg HC. 1998. *Proc. Natl. Acad. Sci. USA* 95:201–6
226. Alon U, Camarena L, Surette MG, Aguera y Arcas B, Liu Y, et al. 1998. *EMBO J.* 17:4238–48
227. Monod J, Wyman J, Changeux J-P. 1965. *J. Mol. Biol.* 12:88–118
228. Turner L, Samuel ADT, Stern AS, Berg HC. 1999. *Biophys. J.* 77:597–603
229. Cluzel P, Surette M, Leibler S. 2000. *Science* 287:1652–55
230. Sourjik V, Berg HC. 2002. *Proc. Natl. Acad. Sci. USA.* 99:12669–74
231. Khan S, Pierce D, Vale RD. 2000. *Curr. Biol.* 10:927–30
232. Duke TAJ, Le Novère N, Bray D. 2001. *J. Mol. Biol.* 308:541–53
233. Ludwig W. 1930. *Z. Vergl. Physiol.* 13:397–504
234. Taylor GI. 1952. *Proc. R. Soc. London Ser. A* 211:225–39
235. Purcell EM. 1977. *Am. J. Phys.* 45:3–11
236. Berg HC. 1996. *Proc. Natl. Acad. Sci. USA* 93:14225–28
237. Howard J. 2001. *Mechanics of Motor Proteins and the Cytoskeleton.* Sunderland, MA: Sinauer
238. Iwazawa J, Imae Y, Kobayasi S. 1993. *Biophys. J.* 64:925–33
239. Berg HC, Khan S. 1983. In *Mobility and Recognition in Cell Biology*, ed. H Sund, C Veeger, pp. 485–97. Berlin: de Gruyter
240. Elston TC, Oster G. 1997. *Biophys. J.* 73:703–21
241. Läuger P. 1988. *Biophys. J.* 53:53–65
242. Berg HC. 2003. In *The Enzymes,* Vol. XXIII, *Energy Coupling and Molecular Motors.* New York: Academic. In press
243. Läuger P, Kleutsch B. 1990. *Comments Theor. Biol.* 2:99–123
244. Caplan SR, Kara-Ivanov M. 1993. *Int. Rev. Cytol.* 147:97–164
245. Schuster SC, Khan S. 1994. *Annu. Rev. Biophys. Biomol. Struct.* 23:509–39
246. Macnab RM. 1996. See Ref. 261, pp. 123–45
247. Khan S. 1997. *Biochim. Biophys. Acta* 1322:86–105
248. Berry RB, Armitage JP. 1999. *Adv. Microb. Physiol.* 41:291–337
249. Berry RB. 2000. *Philos. Trans. R. Soc. London Ser. B* 355:503–9
250. Berg HC. 2000. *Phys. Today* 53:24–29
251. Walz D, Caplan SR. 2002. *Bioelectrochemistry* 55:89–92
252. McCarter LL. 2001. *Microbiol. Mol. Biol. Rev.* 65:445–62

253. Yorimitsu T, Homma M. 2001. *Biochim. Biophys. Acta* 1505:82–93
254. Macnab RM. 1992. *Annu. Rev. Genet.* 26:131–58
255. Chilcott GS, Hughes KT. 2000. *Microbiol. Mol. Biol. Rev.* 64:694–708
256. Aldridge P, Hughes KT. 2002. *Curr. Opin. Microbiol.* 5:160–65
257. Bustamante C, Keller D, Oster G. 2001. *Acc. Chem. Res.* 34:412–20
258. Iino T. 1985. In *Sensing and Response in Microorganisms*, ed. M Eisenbach, M Balaban, pp. 83–92. Amsterdam: Elsevier
259. Macnab RM. 1976. *J. Clin. Microbiol.* 4:258–65
260. Block SM, Fahrner K, Berg HC. 1991. *J. Bacteriol.* 173:933–36
261. Neidhardt FC, Curtiss R, Ingraham JL, Lin ECC, Low KB, et al., eds. 1996. *Escherichia coli and Salmonella: Cellular and Molecular Biology*. Washington, DC: Am. Soc. Microbiol.

Annu. Rev. Biochem. 2003. 72:55–76
doi: 10.1146/annurev.biochem.72.121801.161820
Copyright © 2003 by Annual Reviews. All rights reserved
First published online as a Review in Advance on January 8, 2003

ALIPHATIC EPOXIDE CARBOXYLATION

Scott A. Ensign[1] and Jeffrey R. Allen[2]
[1]*Department of Chemistry and Biochemistry, Utah State University, Logan, Utah 84322; email: ensigns@cc.usu.edu*
[2]*The Dow Chemical Company, 5501 Oberlin Drive, San Diego, California 92121; email: jrallen2@dow.com*

Key Words carboxylase, coenzyme M, dehydrogenase/reductase, disulfide oxidoreductase, epoxyalkane, short-chain

■ **Abstract** Aliphatic epoxides (epoxyalkanes) are highly reactive electrophilic molecules that are formed from the monooxygenase-catalyzed epoxidation of aliphatic alkenes. The bacterial metabolism of short-chain epoxyalkanes occurs by a three-step pathway resulting in net carboxylation to β-ketoacids. This pathway uses the atypical cofactor coenzyme M (CoM; 2-mercaptoethanesulfonic acid) as the nucleophile for the epoxide ring opening and as the carrier of 2-hydroxyalkyl- and 2-ketoalkyl-CoM intermediates. Four enzymes are involved in epoxide carboxylation: a zinc-dependent alkyltransferase, two short-chain dehydrogenases with specificities for the chiral products of the R- and S-1,2-epoxyalkane ring opening, and an NADPH:disulfide oxidoreductase/carboxylase that reduces the thioether bond of the 2-ketoalkyl-CoM conjugate and carboxylates the resulting carbanion. In this review, we summarize the biochemical, mechanistic, and structural features of the enzymes of epoxide carboxylation and show how these enzymes, together with CoM, work in concert to achieve this highly unusual carboxylation reaction.

CONTENTS

INTRODUCTION

Although diverse in terms of structure, substrate specificity, and cofactor usage, all carboxylases share an important common feature: the ability to generate a stabilized carbanion for attack on electrophilic CO_2 or an activated CO_2 species (e.g., carboxy-phosphate). In some enzymes, e.g., the biotin-dependent carboxylases, general acid-base chemistry alone is sufficient to generate the requisite carbanion due to the inherent acidity of protons alpha to a carboxylic acid or a thioester functional group (1). In the case of ribulose-1,5-diphoshate carboxylase (Rubisco), a combination of general acid-base and metal ion catalysis facilitates the formation of the *cis*-enediolate intermediate (2). In the case of phosphoenolpyruvate (PEP) carboxylase, the enolate tautomer of the pyruvate carbanion is formed when bicarbonate attacks and removes the phosphate group of PEP in a reaction facilitated by metal ion catalysis (3). In the case of the vitamin K-dependent carboxylases, the oxygenation of vitamin K is the driving force for the formation of a strong base that abstracts a proton from the γ-carbon of glutamate (4). For each of these reactions, the substrate itself is a potential nucleophile, which is conditional on enzymatic proton abstraction to generate the necessary carbanion.

In light of this general strategy for organic substrate carboxylation, it was a surprise to discover an apparently new type of carboxylation reaction that did not fit the established pattern: that of an aliphatic epoxide to form a β-keto acid in the pathway of bacterial alkene metabolism (Figure 1). Aliphatic epoxides are highly electrophilic molecules and are themselves subject to nucleophilic attack, as illustrated by the reactions catalyzed by glutathione-S-transferase and epoxide hydrolase (Figure 1*a* and *b*). Thus, it was apparent that a new strategy, which goes beyond general acid-base chemistry, must exist for activating an electro-philic epoxide to a suitable nucleophilic species. The elucidation of this strategy has revealed a novel sequence of transformations that uses an atypical cofactor and variations on themes for well-characterized families of enzymes. The elucidation of these reactions and the mechanism of epoxide activation are the topic of this review.

BIOLOGICAL REACTIVITY, FORMATION, AND METABOLISM OF EPOXIDES

Aliphatic epoxides are highly reactive molecules with mutagenic and in some cases carcinogenic properties (5). The reactivity of these compounds derives from their electrophilicity and resultant reaction with cellular nucleophiles

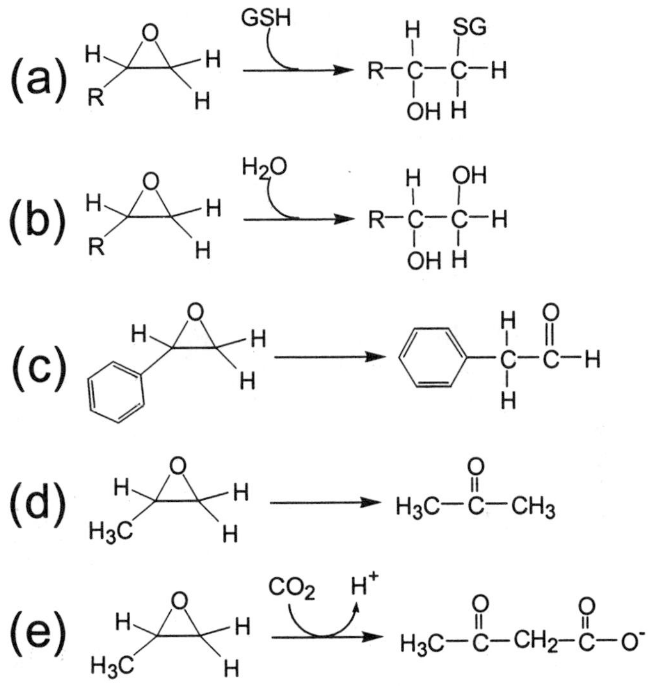

Figure 1 Enzymatic transformations of aliphatic and aromatic epoxides: (*a*) glutathione conjugation; (*b*) hydration; (*c*) isomerization to an aldehyde; (*d*) isomerization to a ketone; and (*e*) carboxylation to a β-keto acid.

(e.g., proteins and DNA). Epoxides are formed biogenically by the epoxidation of alkenes in reactions catalyzed by a number of enzymes with monooxygenase activity. These epoxidation reactions may be catalyzed fortuitously (i.e., in place of the natural oxygenase substrate), as the first step in detoxification of alkenes (as in the case of cytochrome P450) (6) or as the first step in the utilization of alkenes as primary carbon and energy sources for growth (7, 8).

Many organisms have detoxification enzymes that catalyze the transformation of epoxides to less toxic products, specifically, glutathione S-transferases, which form glutathione conjugates of epoxides (Figure 1*a*) (9), and epoxide hydrolases, which hydrate epoxides to the corresponding dihydrodiols (Figure 1*b*) (10). Epoxide hydrolases initiate epoxide catabolism in certain bacteria that grow using epoxypropane or epichlorohydrin (3-chloroepoxypropane) as carbon sources (11, 12). In the case of bacteria that grow using isoprene (2-methyl-1,3-butadiene) and styrene (phenylethene) as carbon sources, the epoxide products are further metabolized by a conventional strategy (glutathione conjugation in the case of isoprene monoxide metabolism) (13, 14) or a nonconventional one (isomerization to phenylacetaldehyde in the case of phenylepoxyethane metabolism, see Figure 1*c*) (15). The bacterial metabolism of the epoxides derived from propylene and some other linear short-chain aliphatic alkenes involves carboxylation as described below.

DISCOVERY OF THE PATHWAY, ENZYMES, AND COFACTORS OF ALIPHATIC EPOXIDE CARBOXYLATION

A CO_2-Dependent Pathway of Aliphatic Epoxide Metabolism

A number of bacteria have been isolated that are capable of aerobic growth using short-chain aliphatic alkenes (ethylene, propylene, and butylene) as the sole added source of carbon (7, 8). Initial metabolic studies suggested that a strategy other than conventional hydration or glutathione conjugation was involved in the metabolism of the epoxides formed from alkene epoxygenation (16–18). The first clue to the nature of the reaction was the observation that ketones were formed as products of epoxide degradation in resting cell suspensions (16) and cell extracts (19) of propylene-grown *Xanthobacter autotrophicus* (Figure 1*d*). Further analyses revealed that epoxide to ketone isomerization was a fortuitous reaction that occurred only in the absence of CO_2, a required cosubstrate for the productive metabolism of epoxides and for growth with aliphatic alkenes (Figure 1*e*) (18, 20). In vivo and in vitro studies of epoxide metabolism demonstrated that acetoacetate and 2-ketovalerate were, respectively, the products of epoxypropane and epoxybutane carboxylation (21). Studies of epoxypropane metabolism in a phylogenetically distinct bacterium, *Rhodococcus rhodochrous*, revealed an identical CO_2-dependent pathway (22, 23). Thus, aliphatic epoxide carboxylation appears to be a general strategy for epoxide metabolism in bacteria that grow using aliphatic alkenes as carbon sources.

In Vitro Requirements for Epoxide Carboxylation

Weijers, de Bont, and coworkers discovered that epoxyalkane degradation could be reconstituted in cell extracts of *X. autotrophicus* upon addition of NAD^+ and a dithiol (e.g., dithiothreitol) (19). They demonstrated further that epoxyalkane degradation was dependent on a homodimeric flavoprotein of 57 kDa subunits, identified as a member of the NADPH:disulfide oxidoreductase (DSOR) family, as well as unidentified additional protein components (24). NADPH was shown to substitute for dithiothreitol in the in vitro assay, consistent with a reductive role for this flavoprotein (24). Chion & Leak furthered these observations by identifying an additional protein, a homohexamer of 44 kDa subunits, that, together with the flavoprotein and additional fractions, restored epoxyalkane degradation (25). The importance of these proteins was reinforced by genetic studies wherein a fragment of DNA was identified that complemented mutants of *X. autotrophicus* unable to degrade epoxyalkanes (26). This complementing DNA encoded several hypothetical open reading frames (ORFs), the first two of which matched the apparent mol wt and N-terminal sequences of the flavoprotein and homohexameric protein (26). Together, these studies demonstrated that epoxyalkane metabolism was a complex process dependent on at least two proteins, NADPH and NAD^+, and other unidentified proteins and/or cofactors.

Purification and Characterization of the Proteins of Epoxide Carboxylation

In our laboratory, we expanded on the studies described above and purified to homogeneity four separate proteins that, when recombined with NADPH, NAD^+, CO_2, and epoxypropane, reconstituted epoxide carboxylase activity (27, 28). Two of the four proteins, designated components I and II, were the aforementioned homohexameric and flavoproteins. The two additional proteins, components III and IV, were dimeric proteins of 25–26 kDa subunits; the properties of which matched two additional ORFs in the complementing fragment of DNA characterized by Swaving et al. (26). Using purified components, the overall stoichiometry of epoxypropane carboxylation was found to occur as shown in Equation 1 (27):

$$Epoxypropane + CO_2 + NADPH + NAD^+ \rightarrow$$
$$acetoacetate + H^+ + NADP^+ + NADH. \qquad 1.$$

Biochemical Properties of Epoxide Carboxylase Components and Proposed Strategy of Epoxyalkane Carboxylation

Interestingly, the carboxylation of epoxypropane is coupled to the transhydrogenation of pyridine nucleotides. The biochemical properties of the four proteins involved in this reaction, together with functions predicted from the amino acid sequences of the proteins, were examined in order to determine how this process occurs (Table 1) (27). Purified component I contained zinc, and the analysis of its sequence indicated it is a homolog of cobalamin-independent methionine synthase (MetE) (29). MetE is a zinc metalloenzyme that, as discussed in more detail below, uses zinc to activate the thiol of homocysteine for nucleophilic attack on methyl-tetrahydrofolate (methyl-THF) (30). Thus, by analogy, a possible role for component I would be the activation and addition of a nucleophilic thiol to the electrophilic epoxide substrate, which results in epoxide ring opening and the formation of an alcohol-nucleophile adduct.

The amino acid sequences of components III and IV allowed their classification as members of the short-chain dehydrogenase/reductase (SDR) family of enzymes (23). The requirement of two short-chain dehydrogenases

TABLE 1 Properties of the four proteins required for epoxide carboxylation

Protein	Composition	Cofactor	Expression level[a]	Gene	Enzyme family	Abbreviated name[b] and Enzyme Commission (EC) number
Component I	α_6, 43 kDa subunits	1 Zn^{2+}/ subunit	4%	*xecA*	Zn-dependent alkyltransferase	EaCoMT EC 4.2.99.19
Component II	α_2, 57 kDa subunits	1 FAD/ subunit	18%	*xecC*	NADPH:disulfide oxidoreductase	2-KPCC EC 1.8.1.5
Component III	α_2, 26 kDa subunits	None	1%	*xecD*	Short-chain dehydrogenase	R-HPCDH EC 1.1.1.268
Component IV	α_2, 25 kDa subunits	None	1%	*xecE*	Short-chain dehydrogenase	S-HPCDH EC 1.1.1.269

[a] Percent of total protein in the clarified cell extract.

[b] As defined in the text.

in epoxypropane metabolism was explained by the need to deal with chiral alcohol intermediates presumed to be formed from the two enantiomers of epoxypropane. As shown in Figure 2, alkene monooxygenase, the enzyme catalyzing the epoxygenation of propylene, forms both enantiomers of epoxypropane, although the R-isomer is formed in an enantiomeric excess (23, 31, 32). For the assay of enzyme activity during the isolation of the proteins of epoxide carboxylation, a stock solution of epoxypropane that is a racemic mixture of R- and S-epoxypropane was used as the substrate (this is the common commercially available form). The protein requirements for epoxide carboxylation were reevaluated with the individual epoxypropane enantiomers as substrates with the following results: In the presence of NADPH and NAD^+, components I, II, and III alone catalyzed the carboxylation of R-epoxypropane, whereas components I, II, and IV alone catalyzed the carboxylation of S-epoxypropane (23). Thus, components III and IV catalyze a redundant oxidation reaction involving chiral substrates.

As mentioned above, component II is a member of the NADPH:disulfide oxidoreductase family based on its amino acid sequence. This enzyme was thus predicted to catalyze a reductive step involving the product of the component III/IV-catalyzed dehydrogenation (27). Together, the characterization of components I–IV suggested a strategy of epoxypropane activation involving nucleophilic epoxide ring opening, stereoselective alcohol oxidation, and reductive cleavage of the resultant nucleophile-ketone adduct. In this manner, the electrophilic epoxypropane would undergo net isomerization to the nucleophilic enolacetone, which generates a suitable substrate for attack on CO_2.

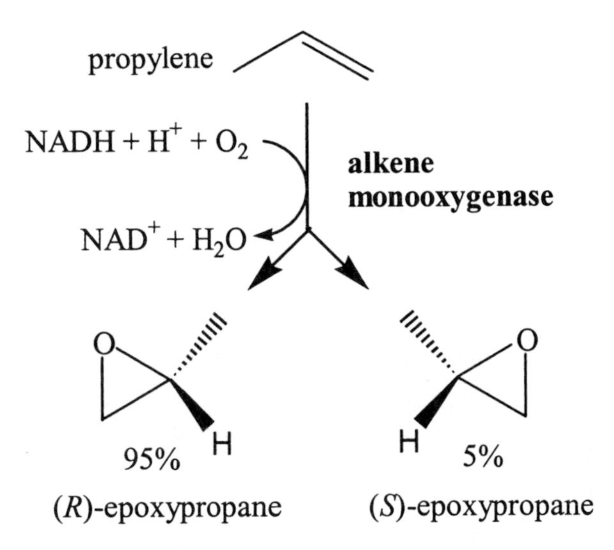

Figure 2 Alkene monooxygenase-catalyzed formation of epoxypropane isomers.

Identification of the Nucleophilic Cofactor of Epoxyalkane Carboxylation

An important unanswered question was the nature of the nucleophile, thought to be an organic thiol, involved in epoxyalkane transformations. Of particular interest was whether the nucleophile was a dissociable or bound cofactor or possibly a Cys amino acid side chain. Although the combination of purified protein components NAD^+ and NADPH were sufficient to support epoxide carboxylation, the specific rates were substantially lower than the corresponding in vivo rates, which suggested that some factor was limiting in the assays (27). Because component I was believed to catalyze the first reaction involving the putative nucleophile, a strategy was designed to determine whether the cofactor was associated with component I and in what form (Figure 3). Briefly, radiolabeled epoxypropane was incubated with component I, with the idea that component I would catalyze a single turnover of epoxypropane and the putative thiol cofactor. Component I was then resolved from small molecules by gel filtration chromatography. The component I thus treated retained none of the radiolabel and was completely inactive when used in place of untreated component I in epoxide carboxylase assays (33). A nonvolatile radiolabeled compound (i.e., not unreacted epoxypropane, which is volatile) was isolated from the salt fraction of the gel filtration eluant, and the addition of this compound to the inactivated component I restored epoxide carboxylase activity (33). Thus, component I contained a bound cofactor; the reaction of which with epoxypropane resulted in the formation of a dissociable product presumably involved in a subsequent step in the reaction pathway.

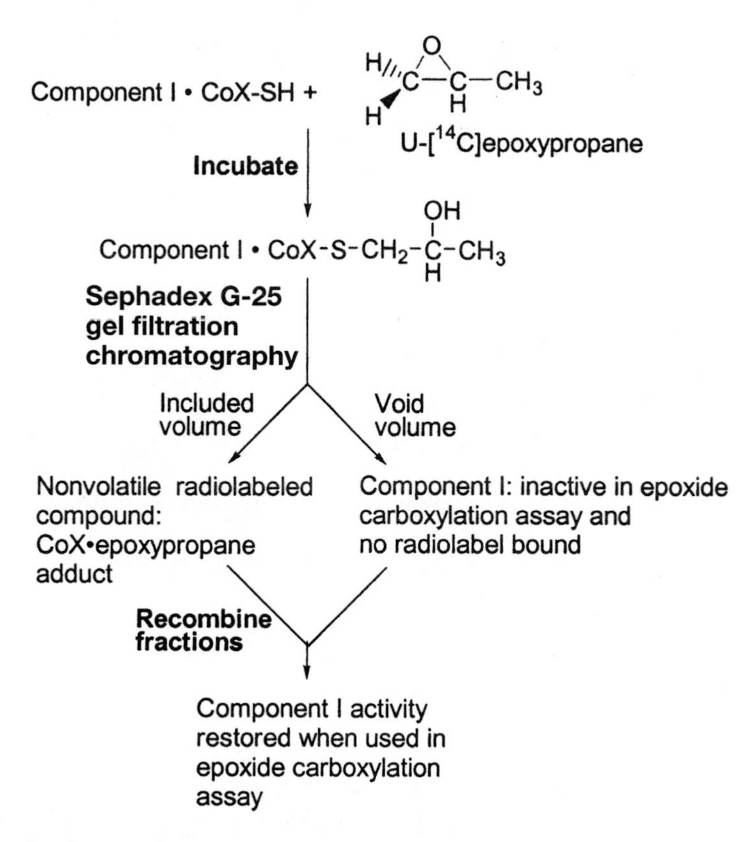

Figure 3 Strategy for identification and isolation of the thiol cofactor. The cofactor is designated CoX for cofactor from *Xanthobacter*.

The dissociable product formed upon incubation of epoxypropane and component I was structurally characterized by NMR. Unexpectedly, the product was identified as the thioether adduct of ring-opened epoxypropane and the organic thiol coenzyme M (CoM; 2-mercaptoethanesulfonate) (33). In 1974, the laboratory of Ralph Wolfe discovered that CoM is a central cofactor of archaebacterial methanogenesis (34). Specifically, CoM serves as a methyl group carrier in key reactions within the pathway of methane formation from C1 precursors (35, 36). Prior to our rediscovery of CoM as a cofactor of aliphatic epoxide carboxylation, CoM was believed to be restricted to the methanogenic archaea and to function solely in reductive methane formation.

The discovery of CoM as the thiol cofactor allowed the individual roles of the four components of the epoxide carboxylase system to be fully defined [(33) and Figure 4]. As predicted, epoxyalkane carboxylation involves nucleophilic ring opening, stereoselective dehydrogenation of chiral alcohol adducts, and reductive cleavage and carboxylation of the ketopropyl-CoM adduct. Importantly, each

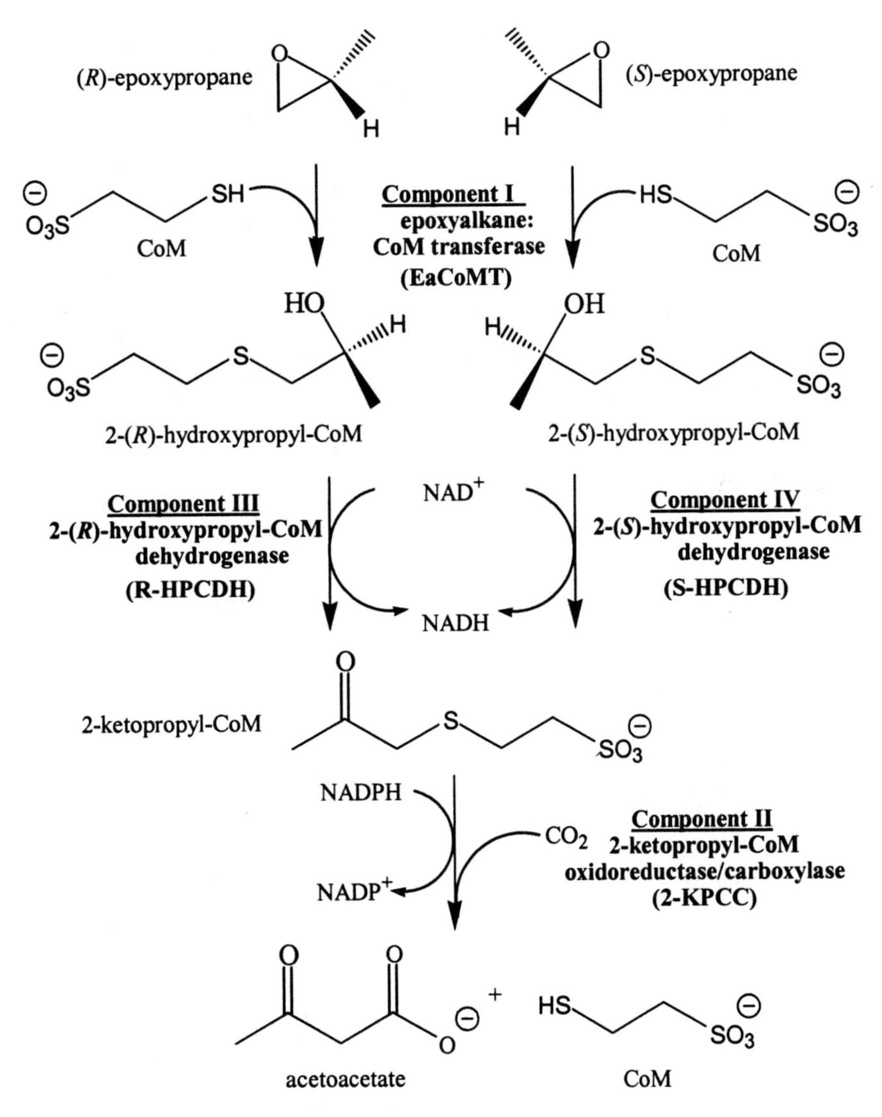

Figure 4 Pathway and enzymes of CoM-dependent epoxypropane carboxylation.

enzyme of the pathway catalyzes the transformation of its respective substrate independent of the other enzymes of the pathway. Thus, aliphatic epoxide carboxylation is more properly regarded as a three step linear pathway involving four separate enzymes than as a transformation catalyzed by a multiprotein enzyme complex. Previous assays of epoxypropane carboxylation relied on the recycling of CoM that had copurified with component I during enzyme isolation (27). These assays were grossly limiting for CoM, which explained the lower than expected activities for the purified enzyme system. CoM was found to be present in cell extracts of *X. autotrophicus* grown with propylene or epoxypropane as the carbon source, but it was not present in extracts prepared from cells grown on other carbon sources (e.g., succinate, glucose, or acetone) (37). Identical results have been obtained in comparative studies of *R. rhodochrous,* which was found to use an indistinguishable system for the metabolism of propylene and epoxypropane (22, 23, 37). It is not believed that the hydroxypropyl- and ketopropyl-adducts of CoM have any other function than as intermediates in the pathway of epoxypropane carboxylation. Likewise, in archaeabacterial methanogenesis, the only function of CoM and the methyl-CoM adduct is in methane production. Presumably, the four enzymes of the epoxide carboxylation pathway interact closely within the cell, although this has not to date been investigated.

CHARACTERIZATION AND MECHANISTIC FEATURES OF THE ENZYMES OF EPOXIDE CARBOXYLATION

Epoxyalkane:CoM Transferase, A Zinc-Dependent Epoxidase

Component I was renamed epoxyalkane:CoM transferase [EaCoMT; E.C 4.2.99.19; systematic name, 2-hydroxypropyl-CoM:2-mercaptoethanesulfonate lyase (epoxyalkane-ring-forming)] based on the nature of the reaction it catalyzes. As noted earlier, EaCoMT is a hexameric protein containing 1 Zn per subunit (Table 1). EaCoMT catalyzes the addition of CoM to either R- or S-epoxypropane, although the rate with R-epoxypropane is about twofold higher than that with S-epoxypropane (33). Kinetic analysis of EaCoMT with R-epoxypropane as the substrate showed the enzyme to follow a random sequential mechanism, with kinetic parameters $K_{m,\ epoxypropane} = 1.8\ \mu M$, $K_{m,\ CoM} = 34\ \mu M$, and $k_{cat} = 6.5\ s^{-1}$ (on the basis of one active site per subunit) (38). An analysis of the pH dependence of the reaction indicated that binding of CoM to the enzyme lowers the pKa of the thiol of CoM by 1.7 units, from a value of 9.1 to 7.4 (38). EaCoMT is highly specific for using CoM as the organic thiol substrate; when a range of other organic thiols was tested as substrates, only 3-mercaptopropionate, 2-mercaptoethanol, and cysteine were substrates, with very low affinities and specific activities at best 0.6% of the rate with CoM (38). Illustrative of the low affinities of these organic thiols, 2-mercaptoethanol was a competitive inhibitor versus CoM for EaCoMT with a K_I of 192 mM, a value 5600 times higher than the K_m for CoM (38).

The biochemical properties of EaCoMT and an analysis of its primary sequence suggested that it belongs to the family of alkyl transferases that use zinc to activate a thiol for nucleophilic attack. For example, the ADA protein (39, 40), protein farnesyl transferase (41), betaine-homocysteine methyltransferase (42), S-methylmethionine: homocysteine methyltransferase (43), cobalamin-depen-

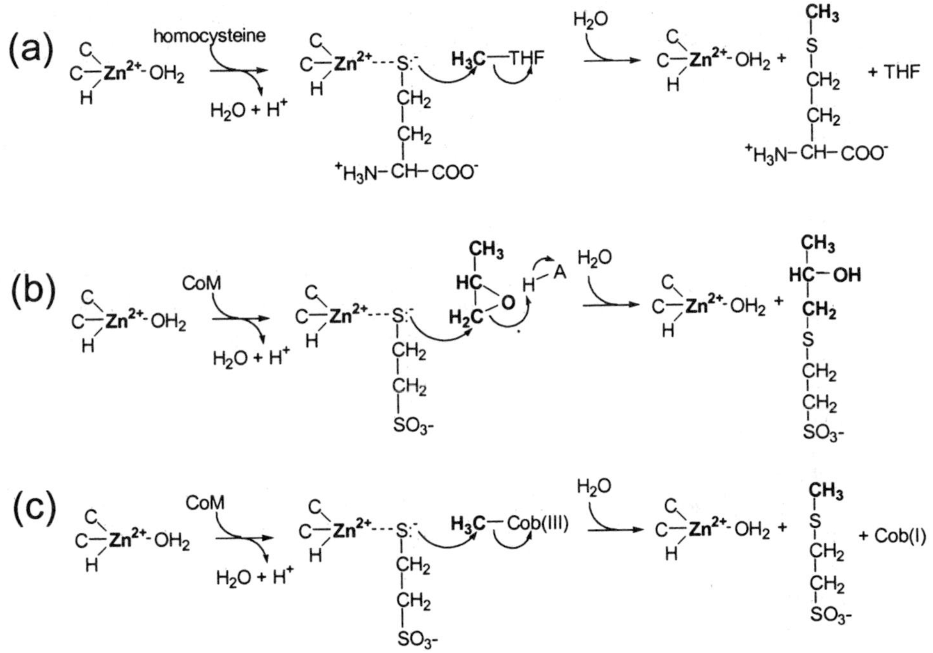

Figure 5 Zinc-mediated activation of thiol cofactors: (*a*) cobalamin-independent methionine synthase; (*b*) epoxyalkane:CoM transferase; and (*c*) methylcob(III)alamin:CoM methyltransferase.

dent (MetH) and cobalamin-independent methionine synthases (MetE) (44), and various methanogenic methyltransferases (45, 46) are all zinc metalloproteins.

Multiple sequence alignments revealed that EaCoMT has a putative zinc binding motif most similar to that of MetE (29). Matthews, Penner-Hahn, and coworkers have shown biochemically and spectroscopically that zinc in MetE reversibly binds homocysteine during catalysis via ligand exchange (29, 44, 47, 48). Two Cys and 1 His residues serve as permanent ligands to the zinc center while a fourth ligand, believed to be water, exchanges for the thiol of homocysteine. The identity of the three permanent ligands, in this case His641, Cys643, and Cys726, was confirmed by site-directed mutagenesis followed by biochemical and spectroscopic studies of the mutant enzymes (29, 47, 48). The coordination of homocysteine to zinc lowers the pKa of the thiol group such that it is activated for nucleophilic attack on the second substrate, methyl-THF. Once this occurs, methionine dissociates, and the dissociable ligand rebinds zinc as shown in Figure 5*a*.

Preparatory to characterizing the ligand environment of EaCoMT by biochemical and spectroscopic techniques, it was necessary to prepare the enzyme in a CoM-free form. Purified native EaCoMT was shown to contain on average 0.5 equivalents of bound CoM, a feature that was crucial to the identification of the epoxyalkane-CoM adduct (Figure 3) (33, 37). In order to prepare CoM-free EaCoMT, the protein was expressed in a CoM-free background, for which *E. coli* was used (37).

The importance of zinc in EaCoMT was established by removing zinc from the enzyme by chelation, which resulted in loss of catalytic activity that was partially restored by adding zinc back to the enzyme (38). Due to the harshness of the chelation conditions, an apoenzyme for biochemical studies was instead prepared by expressing the enzyme in zinc-deficient medium and purifying the enzyme in the presence of ethylenediaminctetraacetic acid (EDTA) (38). Even under these conditions the purified enzyme contained 0.2 mol zinc/mol subunit. The activity of this semiapoenzyme was 25% of the holoenzyme, which demonstrated a direct correlation between zinc content and activity.

Calorimetry was used to investigate the thermodynamics of CoM binding to EaCoMT and to further explore the role of zinc in this activity. Holo-EaCoMT had a high affinity for CoM; it bound 1 mol CoM/mol monomer with $K_D = 3.8$ μM (38). In the case of the semiapo EaCoMT, the stoichiometry of CoM binding was best fit by basing it on the zinc content of the enzyme: 0.21 mol CoM were bound with high affinity on the basis of monomer content, whereas 0.97 mol zinc were bound on the basis of zinc content. A site-directed mutant in one of the three zinc ligands of EaCoMT, C220A (H218, C220, and C341 are the predicted ligands), had a substantially lower affinity for and complement of zinc, and again, CoM binding was proportional to zinc content rather than monomer content (38). The binding of CoM to EaCoMT was further analyzed by comparing the thermodynamics of ethanethiol and ethanesulfonate binding to that of CoM binding. By this analysis, the interaction of the thiol with zinc was predicted to contribute 4.2 kcal/mol of the 7.5 kcal/mol of total binding energy in the EaCoMT•CoM binary complex (38). This translates to a 12-fold higher binding affinity for the thiol group than for the sulfonate group of CoM.

Extended X-ray absorption fine structure analysis of recombinant EaCoMT supports a mechanism of thiol group activation similar to that seen in MetE (8; K. Peariso, S.A. Ensign, and J.E. Penner-Hahn, unpublished data). Specifically, the addition of CoM to EaCoMT resulted in a change of the zinc environment from 2S + 2N/O to 3S + 1N/O, consistent with CoM replacing a dissociable water (or other N/O ligand). The addition of epoxypropane resulted in a return to a 2S + 2N/O environment, consistent with the mechanism shown in Figure 5*b*.

Recent biochemical and spectroscopic studies of CoM-dependent methyltransferases from a methanogenic archaebacterium support similar mechanisms of CoM activation. CoM-dependent methyltransferases catalyze the transfer of a methyl group from a donor molecule or corrinoid protein to CoM to form methyl-CoM, which is subsequently reduced to methane by the enzyme methyl-CoM reductase (36). Two such methyltransferases, designated MtaA and MtbA, contain the conserved $HXCX_nC$ motif first identified in MetE (45, 46). Penner-Hahn, Grahame, and coworkers showed that the binding of CoM to MtbA resulted in a change of zinc coordination from 2 S + 2 N/O to 3S + 1N/O (46).

Dau, Thauer, and coworkers also observed ligand exchange for MtaA, with the interesting difference that the starting environment was 1S + 3N/O, which changed to 2S + 2N/O upon CoM addition (49). The basis for the difference in the starting environment for these two enzymes is not known, but the net result is the same, i.e., replacement of a N/O ligand by S.

As summarized in Figure 5, the activation of CoM and homocysteine as nucleophiles is believed to be facilitated by coordination to a similar type of zinc center. EaCoMT differs from MetE and the methanogenic CoM transferases, as well as all other members of the zinc alkyl transferase family, in that it catalyzes a nucleophilic addition reaction rather than nucleophilic substitution (Figure 5). In this regard, the reaction is more similar to the epoxidase activity of glutathione-S-transferases (GSTs) (Figure 1*a*). Notably, GSTs are not zinc proteins and use a different strategy to activate glutathione as a nucleophile. The three-dimensional structures of a number of soluble GSTs indicate that the pKa of the thiol of glutathione is lowered by hydrogen bonding interactions at the active site of the enzyme (9, 50). Specifically, an interaction of the thiol with an active site tyrosine has been implicated in this stabilization (9, 50).

Stereoselective Dehydrogenation of 2-Hydroxypropyl-CoM Enantiomers

As shown in Figures 2 and 4, both enantiomers of epoxypropane are involved in microbial alkene oxidation and are substrates for EaCoMT. The dehydrogenases that catalyze the oxidation of 2-hydroxypropyl-CoM enantiomers (Table 1) are highly specific for their respective substrates; they exhibit only 0.5% to 1% activity with the opposing enantiomer (33). The amino acid sequences of the dehydrogenases reveal that they belong to the short-chain dehydrogenase/reductase (SDR) family of enzymes (23, 26). The SDR family is composed of numerous NAD^+ or $NADP^+$-dependent enzymes that are ~250 amino acids in length, active as dimers or tetramers, and carry out catalysis in the absence of a metal cofactor (51, 52). The enzymes have three distinct domains: a conserved N-terminal NAD^+- (or $NADP^+$) binding domain, a conserved central domain containing a catalytic triad of Ser, Tyr, and Lys residues, and a variable C-terminal domain that confers specificity for the specific enzyme substrate (51, 52). Extensive structural, kinetic, and mechanistic characterization of SDR enzymes has allowed the formulation of a conserved mechanism of action (51, 52). During catalysis, the Tyr residue of the catalytic triad is deprotonated and serves as a general base for abstraction of the proton from the hydroxyl group of the substrate. The Ser residue is believed to increase the acidity of the substrate hydroxyl through hydrogen bonding. The Lys residue hydrogen bonds to the ribose attached to the nicotinamide ring and stabilizes the deprotonated Tyr general base.

Among the SDR family, there are very few enzymes that have a partner catalyzing the same reaction but with opposite stereospecificity. To our knowledge, the only SDRs capable of this are a pair of plant tropinone reductases (53,

54) and the R- and S-specific hydroxypropyl-CoM dehydrogenases (HPCDHs) purified from *X. autotrophicus* and *R. rhodochrous.* Thus, these enzymes are excellent models for studying the molecular basis of stereospecificity in the SDR family.

R-HPCDH has been cloned and expressed in a fully active form, but unfortunately, S-HPCDH has not been successfully expressed due to its tendency to form inclusion bodies in heterologous expression systems (55). Therefore, studies of S-HPCDH are lagging behind those of R-HPCDH. Studies of native and mutant forms of R-HPCDH showed that it follows a mechanistic pattern consistent with other members of the SDR family. The enzyme binds and releases substrates in an ordered fashion and exhibits a k_{cat} of $26s^{-1}$ and K_m for 2-R-hydroxypropyl-CoM of 105 μM (55). Chemical modification and site-directed mutagenesis studies of R-HPCDH suggest that one or more arginine residues are important in substrate binding (55; D.D. Clark & S.A. Ensign, unpublished results). These results are intriguing because arginine would be an ideal candidate for imparting substrate specificity by specific interactions with the sulfonate of CoM. For the other two CoM-dependent enzymes whose structures are known, i.e., methyl-CoM reductase (56) and 2-KPCC (see below), arginine residues are key to the binding and orientation of substrate in this manner. Substrate specificity in the SDR family is conferred largely by differences within the C-terminal domain of the enzymes. In the case of R- and S-HPCDH, there are similarities as well as differences in this domain, which include differentially placed arginine residues (8). At present, we believe that differences in the spatial orientation of the sulfonate group and resultant interactions with key arginine residues(*s*) are key to the specificity of R-HPCDH and S-HPCDH, according to the model shown in Figure 6. R-HPCDH has been succesfully crystallized at resolutions higher than 2.0 Å (57), and its structure is currently being solved in the presence and absence of substrates and inhibitors. We anticipate that this structure will provide valuable insights into the molecular basis of stereoselectivity for both dehydrogenases.

NADPH:2-Ketopropyl-CoM Carboxylase/Oxidoreductase (2-KPCC), A CO_2-Fixing Member of the DSOR Family

The actual CO_2-fixing step in the pathway of aliphatic epoxide carboxylation, the reductant-dependent cleavage of 2-ketopropyl-CoM and carboxylation of the epoxide-derived fragment, is catalyzed by an unexpected enzyme: component II, the flavoprotein whose biochemical properties and primary sequence place it in the family of NADPH:disulfide oxidoreductases (24, 26, 28, 58) (Figure 4). The chemistry of this reaction is unprecedented for a DSOR enzyme. In all but one case, members of this family catalyze the reduction of a disulfide bond in an organic substrate, as, for example, in the case of glutathione reductase and dihydrolipoamide dehydrogenase (59). Mercuric reductase is an atypical member of this family, which catalyzes the reduction of mercuric ion to elemental

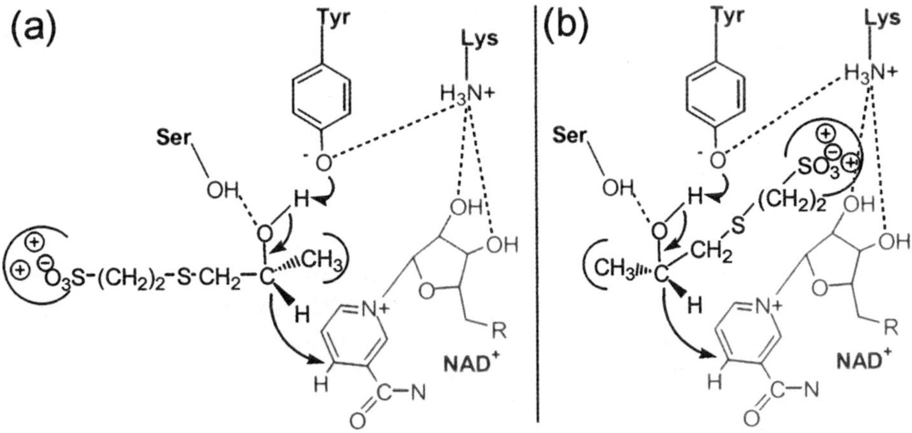

Figure 6 Model for the molecular basis for stereoselectivity by R- and S-2-hydroxypropyl-CoM dehydrogenases. The amino acid residues forming the catalytic triad are highlighted in blue. NAD^+ is highlighted in green. The putative alkyl- and ethanesulfonate-binding pockets are highlighted in red. Part (*a*) is the R-HPCDH active site; part (*b*) the S-HPCDH active site.

mercury (60). However, mercuric reductase shares the common feature of catalyzing a two-electron reduction of substrate.

All members of the DSOR family contain a highly conserved redox active cysteine pair, separated by four intervening residues, that participates in substrate reduction (59). A general mechanism for the initial reactions of the DSOR enzymes, which diverge into the specialized reactions of those enzymes that reduce organic substrates, is shown in Figure 7. The first step in the catalytic cycle is the binding of NADPH to the fully oxidized enzyme. NADPH reduces FAD to $FADH_2$ and is followed by nucleophilic attack of the C4A carbon of $FADH_2$ on the proximal cysteine thiol of the disulfide. Heterolytic cleavage of the C4A-thiol bond leaves FAD oxidized and the redox active disulfide in its two-electron reduced form. In the case of the disulfide-reducing DSOR enzymes, the distal cysteine thiol then attacks one of the S atoms of the disulfide substrate. This releases one organic thiol in its reduced form and forms a mixed disulfide with the other thiol (Figure 7*b*). Attack of the proximal cysteine thiol on the distal thiol promotes heterolytic cleavage of the mixed disulfide, which releases the second substrate to complete the catalytic cycle (Figure 7*b*).

Based on kinetic and mechanistic studies of the 2-KPCC-catalyzed reaction (61), together with information from kinetic and spectroscopic studies performed by Westphal and coworkers using 1,3-propanedithiol before the physiological substrate for 2-KPCC was identified (58), we proposed a catalytic mechanism similar to that described above, but wherein the distal thiol attacks the thioether linkage of 2-ketopropyl-CoM, which forms a mixed disulfide of Cys and CoM and with formation of the enolate (or enol) of acetone (Figure 7*c*). An early piece of evidence supporting the formation of enolacetone as a reaction intermediate was the observation that acetone was formed as the product of epoxide degradation when CO_2 was excluded from assays (18, 21). After the identification of CoM, we verified that acetone is indeed formed as the product of the 2-KPCC-catalyzed reaction in the absence of CO_2 (2-KPCC-catalyzed acetone formation occurs at about 40% of the maximal rate of acetoacetate formation) (61). Enolacetone is a strong nucleophile, and if formed in the absence of CO_2, it would be expected to abstract a proton from the bulk solvent to form acetone. This step would be essentially irreversible due to the high pK_a of the methyl group proton of acetone. As expected, 2-KPCC was unable to catalyze the reverse reaction with acetone, CO_2, $NADP^+$, and CoM as substrates. However, 2-KPCC did catalyze the reverse reaction when acetoacetate was used as the substrate, as well as the decarboxylation of acetoacetate and exchange of $^{14}CO_2$ into acetoacetate in the absence of other substrates. In all, five catalytic activities were shown to be catalyzed by 2-KPCC; each of which can be rationalized by the formation of enolacetone as a stabilized intermediate (61):

$$\textit{2-ketopropyl-CoM} + NADPH + CO_2 \rightarrow \textit{acetoacetate} + NADP^+ + CoM, \quad 2.$$

$$\textit{2-ketopropyl-CoM} + NADPH + H^+ \rightarrow \textit{acetone} + NADP^+ + CoM, \quad 3.$$

$$\textit{acetoacetate} + NADP^+ + CoM \rightarrow \textit{2-ketopropyl-CoM} + NADPH + CO_2, \quad 4.$$

$$\textit{acetoacetate} + H^+ \rightarrow \textit{acetone} + CO_2, \textit{ and} \quad 5.$$

$$\textit{acetoacetate} + [^{14}C]\, CO_2 \rightarrow 1\text{-}[^{14}C]\, \textit{acetoacetate} + CO_2. \quad 6.$$

The structure of 2-KPCC has recently been determined both in the presence and absence of the substrate 2-ketopropyl-CoM (62). The overall structure of the dimer, which contains the characteristic FAD binding, NADPH binding, and dimerization domains, is similar to the other members of the DSOR family for which structures have been determined (63–67). There are, however, some notable differences in the 2-KPCC structure. 2-KPCC has an extended N-terminus and a 13-amino-acid insertion within the interface domain that effectively eliminates the larger cleft that provides access to the active site for other DSOR enzymes (62). In the substrate-bound form, 2-ketopropyl-CoM is buried within the protein with no discernable substrate access channel. 2-ketopropyl-CoM is oriented in the active site in such a manner that there is a linear pathway for electron pair transfer between FAD, the redox active disulfide, and the thioether bond of the substrate (Figure 8) (62). The sulfonate of 2-ketopropyl-CoM forms a salt bridge with two Arg residues, whereas a Phe residue provides a backstop

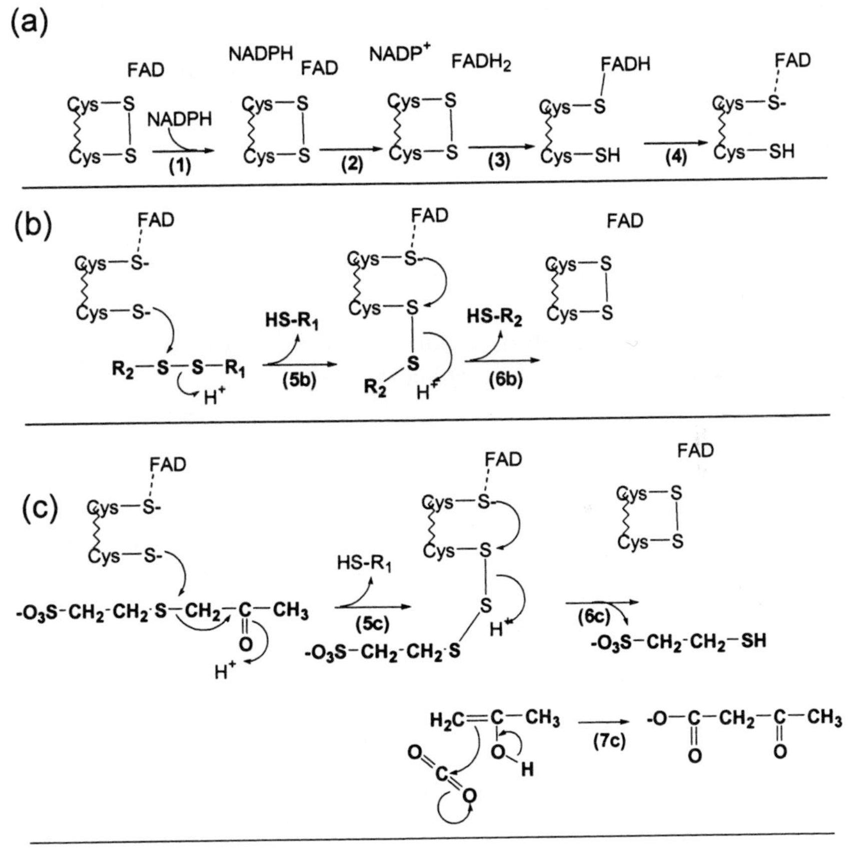

Figure 7 General mechanism of DSOR reactions and mechanism of 2-ketopropyl-CoM carboxylation: (*a*) general mechanism of disulfide bond reduction in DSOR enzymes; (*b*) disulfide bond reduction as in the case of glutathione reductase or trypanathione reductase; and (*c*) thioether bond reduction and carboxylation by 2-KPCC.

for proper alignment of the substrate (Figure 8). The keto group of the substrate is oriented within hydrogen bonding distance of an ordered water molecule that might donate a proton for enolacetone stabilization. Aside from this ordered water, the active site environment is largely hydrophobic, presumably a reflection of the need to exclude additional water that could serve as a source of protons for the undesirable protonation side reaction (Equation 3).

A putative substrate access channel, 8 Å long and 5 Å wide, leads to the active site disulfide and can be discerned in the structure of substrate-free 2-KPCC (62). The superimposition of the substrate-bound and substrate-free enzyme structures indicates that the binding of 2-ketopropyl-CoM induces a conformational change resulting in collapse of the channel and sequestration of substrate within the active site (62). This substrate sequestration is presumably intended to protect the reactive enolacetone from undesirable side reactions such as protonation. Notably, a similar sequestration is observed upon binding of ribulose-1,5-diphosphate to Rubisco (68). In summary, the structural characterization of 2-KPCC supports the general features of the mechanism shown in Figure 7 and provides additional details into the strategy of enolacetone stabilization and carboxylation.

WHY CARBOXYLATION AND WHY CoM?

It is interesting that alkene-oxidizing bacteria have evolved such a unique strategy and selected such an atypical cofactor for the metabolism of epoxyalkanes formed from alkene epoxidation. Epoxides are readily subject to the more ubiquitous reactions of hydrolysis and glutathione conjugation, and these strategies might be predicted to occur instead of the carboxylation strategy described herein. However, there are a number of features of the epoxide carboxylation reactions that make them attractive from metabolic and toxicological standpoints.

Epoxyalkanes are highly reactive molecules and, when formed as metabolic intermediates, should not be allowed to accumulate to toxic levels. As a reflection of this, EaCoMT has a very high affinity for epoxypropane ($K_{m,\ epoxypropane}$ = 1.8 μM) and is expressed at high levels (4% of soluble protein) in propylene-grown cell cultures (27, 38). In fact, epoxypropane does not accumulate to detectable levels in cells grown with propylene as the carbon source (33). In contrast to EaCoMT, epoxide-utilizing enzymes involved in detoxification generally have lower affinities and broader substrate specificities. Glutathione-S-transferases, for example, use a variety of electrophiles other than epoxides as substrates (9), although epoxide hydrolases generally have broader substrate specificities than EaCoMT (10).

The metabolic outcome of epoxide carboxylation is quite favorable and efficient, as β-ketoacids such as acetoacetate and 2-ketovalerate are easily converted to central metabolites (69). It is unclear how glutathione conjugates of epoxides would be converted to central metabolites, and this remains an unanswered question for the glutathione-dependent isoprene catabolic pathway (13). In order to convert diols formed from epoxide hydration to central metabolites, more reaction steps than those involved in epoxypropane carboxylation would be required. For example, hydrolysis of epoxypropane would form 1,2-propanediol, which, according to known metabolic pathways, would need to undergo subsequent dehydration, oxidation, thiolation, carboxylation, mutation, and isomerization to arrive at succinyl-CoA (12).

The selection of CoM as the nucleophile and carrier molecule appears well thought out as well. Composed simply of an ethyl spacer between sulfonate and thiol functional groups, CoM is the smallest known organic cofactor (35, 70). As opposed to the bulky glutathione-alcohol conjugates formed by glutathione-S-transferases, the 2-hydroxypropyl-CoM conjugate is of a manageable size for

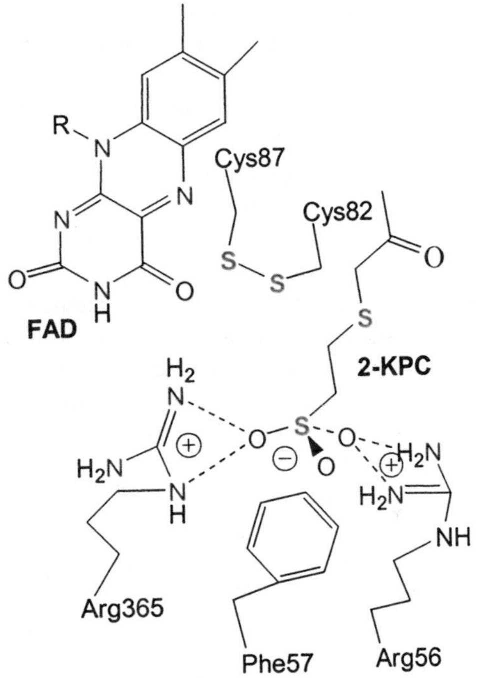

Figure 8 Cartoon model for the binding of 2-ketopropyl-CoM at the active site of 2-KPCC. The active site model is based on the crystallographic structure determined by Nocek et al. (60).

efficient and stereoselective dehydrogenation and reductive carboxylation in the subsequent steps (Figure 4). The sulfonate moiety of CoM provides a selective handle for recognition by the stereoselective dehydrogenases and 2-KPCC (Figures 6 and 8). These features of CoM also make it ideally suited to the reactions of methanogenesis, which to some degree mirror those of epoxide carboxylation [methanogenesis involves nucleophilic transfer of a methyl group to CoM, reduction of the methyl group to methane with formation of a mixed disulfide of CoM and another thiol cofactor, and subsequent reduction of the mixed disulfide by a heterodisulfide reductase (35, 36, 70)]. Thus, the epoxide carboxylation pathway appears to have evolved out of a specific metabolic need, based on the availability of a suitable cofactor, rather than by adaption of a broad specificity detoxification enzyme.

How alkene-oxidizing bacteria acquired the ability to synthesize CoM and what the original function of this cofactor was are intriguing and unanswered questions. Based on the ubiquity of CoM in the prevalent methanogenic archaea, which until 1999 were the only organisms known to have CoM, it seems likely that the cofactor originated in these microorganisms. Importantly, the genes of alkene metabolism are located on linear megaplasmids in *X. autotrophicus* and *R.*

rhodochrous, the two microorganisms that have been most intensively studied with regard to aliphatic alkene metabolism (71, 72). The genes for epoxide metabolism are located just downstream of the genes for alkene epoxygenation in the linear megaplasmid of *X. autotrophicus* (72; R. Larsen, W. Metcalf, and S. Ensign, unpublished results). The location of the epoxide carboxylation genes have not yet been determined for *R. rhodochrous* but are presumed to reside on the same plasmid. Just downstream of the genes for the four proteins of epoxide carboxylation on the megaplasmid of *X. autotrophicus* is the gene *xecG,* which encodes a hypothetical protein with homology to phosphosulfolactate synthase, recently shown to be a key enzyme of methanogenic CoM biosynthesis (72, 73). Thus, horizontal gene transfer to an extrachromosomal element could explain how two phylogenetically distinct eubacteria acquired the ability to synthesize a methanogenic cofactor for a highly specialized function.

In summary, the elucidation of the pathway of bacterial aliphatic alkene metabolism has revealed a new strategy for organic substrate carboxylation. The key to this strategy is the use of CoM, whose unique properties facilitate the net isomerization of an electrophilic epoxide to a nucleophilic enol tautomer. The gap between the discoveries of CoM in the pathways of methanogenesis and alkene catabolism was 25 years. We are hopeful that the new-found realization that CoM is more ubiquitous and versatile than previously thought will stimulate studies leading to its identification in other unique metabolic pathways and enzymes.

ACKNOWLEDGMENTS

Work in the laboratory of Scott Ensign is supported by National Institutes of Health grant GM51805. Special thanks to John Peters and Boguslaw Nocek, Montana State University, and Jim Penner-Hahn and Katrina Peariso, University of Michigan, for their contributions to these studies.

The *Annual Review of Biochemistry* is online at http://biochem.annualreviews.org

LITERATURE CITED

1. Knowles JR. 1989. *Annu. Rev. Biochem.* 58:195–221
2. Cleland WW, Andrews TJ, Gutteridge S, Hartman FC, Lorimer GH. 1998. *Chem. Rev.* 98:549–61
3. Janc JW, Oleary MH, Cleland WW. 1992. *Biochemistry* 31:6421–26
4. Dowd P, Hershline R, Ham SW, Naganathan S. 1995. *Science* 269: 1684–91
5. Wade DR, Airy SC, Sinsheimer JE. 1978. *Mutat. Res. Genet. Toxicol. Test.* 58: 217–23
6. Guengerich FP. 1991. *J. Biol. Chem.* 266: 10019–22
7. Hartmans S, de Bont JAM, Harder W. 1989. *FEMS Microbiol. Rev.* 63:235–64
8. Ensign SA. 2001. *Biochemistry* 40: 5845–53
9. Armstrong RN. 1997. *Chem. Res. Toxicol.* 10:2–18
10. Thomas H, Timms CW, Oesch F. 1990. In

Principles, Mechanisms and Biological Consequences of Induction, ed. K Ruckpaul, H Rein, pp. 280–337. New York: Taylor & Francis
11. Jacobs MHJ, van den Wijngaard AJ, Pentenga M, Janssen DB. 1991. *Eur. J. Biochem.* 202:1217–22
12. De Bont JAM, van Dijken JP, van Ginkel CG. 1982. *Biochim. Biophys. Acta* 714: 465–70
13. Vlieg J, Kingma J, Kruizinga W, Janssen DB. 1999. *J. Bacteriol.* 181:2094–101
14. Vlieg J, Kingma J, van den Wijngaard AJ, Janssen DB. 1998. *Appl. Environ. Microbiol.* 64:2800–5
15. Hartmans S, Smits JP, van der Werf MJ, Volkering F, de Bont JAM. 1989. *Appl. Environ. Microbiol.* 55:2850–55
16. Small FJ, Tilley JK, Ensign SA. 1995. *Appl. Environ. Microbiol.* 61:1507–13
17. Ensign SA, Hyman MR, Arp DJ. 1992. *Appl. Environ. Microbiol.* 58:3038–46
18. Small FJ, Ensign SA. 1995. *J. Bacteriol.* 177:6170–75
19. Weijers CAGM, Jongejan H, Franssen MCR, de Groot A, de Bont JAM. 1995. *Appl. Microbiol. Biotechnol.* 42:775–81
20. Sluis MK, Small FJ, Allen JR, Ensign SA. 1996. *J. Bacteriol.* 178:4020–26
21. Allen JR, Ensign SA. 1996. *J. Bacteriol.* 178:1469–72
22. Allen JR, Ensign SA. 1998. *J. Bacteriol.* 180:2072–78
23. Allen JR, Ensign SA. 1999. *Biochemistry* 38:247–56
24. Swaving J, de Bont JAM, Westphal A, de Kok A. 1996. *J. Bacteriol.* 178:6644–46
25. Chion CKNCK, Leak DJ. 1996. *Biochem. J.* 319:499–506
26. Swaving J, Weijers CA, van Ooyen AJ, de Bont JAM. 1995. *Microbiology* 141: 477–84
27. Allen JR, Ensign SA. 1997. *J. Biol. Chem.* 272:32121–28
28. Allen JR, Ensign SA. 1997. *J. Bacteriol.* 179:3110–15
29. Zhou ZHS, Peariso K, Penner-Hahn JE, Matthews RG. 1999. *Biochemistry* 38: 15915–26
30. Matthews RG, Goulding CW. 1997. *Curr. Opin. Chem. Biol.* 1:332–39
31. Gallagher SC, Cammack R, Dalton H. 1997. *Eur. J. Biochem.* 247:635–41
32. Small FJ, Ensign SA. 1997. *J. Biol. Chem.* 272:24913–20
33. Allen JR, Clark DD, Krum JG, Ensign SA. 1999. *Proc. Natl. Acad. Sci. USA* 96:8432–37
34. Taylor CD, McBride BC, Wolfe RS, Bryant MP. 1974. *J. Bacteriol.* 120:974–75
35. Wolfe RS. 1991. *Annu. Rev. Microbiol.* 45:1–35
36. Thauer RK. 1998. *Microbiology* 144: 2377–406
37. Krum JG, Ensign SA. 2000. *J. Bacteriol.* 182:2629–34
38. Krum JG, Ellsworth H, Sargeant RR, Rich G, Ensign SA. 2002. *Biochemistry* 41:5005–14
39. Myers LC, Terranova MP, Ferentz AE, Wagner G, Verdine GL. 1993. *Science* 261:1164–67
40. Myers LC, Verdine GL, Wagner G. 1993. *Biochemistry* 32:14089–94
41. Saderholm MJ, Hightower KE, Fierke CA. 2000. *Biochemistry* 39:12398–405
42. Breksa AP, Garrow TA. 1999. *Biochemistry* 38:13991–98
43. Thanbichler M, Neuhierl B, Bock A. 1999. *J. Bacteriol.* 181:662–65
44. Peariso K, Goulding CW, Huang S, Matthews RG, Penner-Hahn JE. 1998. *J. Am. Chem. Soc.* 120:8410–16
45. Sauer K, Thauer RK. 2000. *Eur. J. Biochem.* 267:2498–504
46. Gencic S, LeClerc GM, Gorlatova N, Peariso K, Penner-Hahn JE, Grahame DA. 2001. *Biochemistry* 40:13068–78
47. Peariso K, Zhou ZHS, Smith AE, Huang S, Matthews RG, Penner-Hahn JE. 1999. *J. Inorg. Biochem.* 74:262
48. Peariso K, Zhou ZHS, Smith AE, Matthews RG, Penner-Hahn JE. 2001. *Biochemistry* 40:987–93
49. Kruer M, Haumann M, Meyer-Klaucke

W, Thauer RK, Dau H. 2002. *Eur. J. Biochem.* 269:2117–23
50. Armstrong RN. 1998. *Curr. Opin. Chem. Biol.* 2:618–23
51. Jörnvall H, Persson B, Krook M, Atrian S, Gonzàlez-Duarte R, et al. 1995. *Biochemistry* 34:6003–13
52. Persson B, Krook M, Jörnvall H. 1991. *Eur. J. Biochem.* 200:537–43
53. Nakajima K, Hashimoto T, Yamada Y. 1993. *Proc. Natl. Acad. Sci. USA* 90:9591–95
54. Nakajima K, Hashimoto T, Yamada Y. 1994. *J. Biol. Chem.* 269:11695–98
55. Clark DD, Ensign SA. 2002. *Biochemistry* 41:2727–40
56. Ermler U, Grabarse W, Shima S, Goubeaud M, Thauer RK. 1997. *Science* 278:1457–62
57. Nocek B, Clark DD, Ensign SA, Peters JW. 2002. *Acta Crystallogr. D* 58: 1470–73
58. Westphal AH, Swaving J, Jacobs L, de Kok A. 1998. *Eur. J. Biochem.* 257:160–68
59. Pai EF. 1991. *Curr. Opin. Struct. Biol.* 1:796–803
60. Ghisla S, Massey V. 1989. *Eur. J. Biochem.* 181:1–17
61. Clark DD, Allen JR, Ensign SA. 2000. *Biochemistry* 39:1294–304
62. Nocek B, Jang SB, Jeong MS, Clark DD, Ensign SA, Peters JW. 2002. *Biochemistry* 41:12907–13
63. Kuriyan J, Kong XP, Krishna TSR, Sweet RM, Murgolo NJ, et al. 1991. *Proc. Natl. Acad. Sci. USA* 88:8764–68
64. Karplus PA, Schulz GE. 1989. *J. Mol. Biol.* 210:163–80
65. Schierbeek AJ, Swarte MBA, Dijkstra BW, Vriend G, Read RJ, et al. 1989. *J. Mol. Biol.* 206:365–79
66. Karplus PA, Pai EF, Schulz GE. 1989. *Eur. J. Biochem.* 178:693–703
67. Karplus PA, Schulz GE. 1987. *J. Mol. Biol.* 195:701–29
68. Taylor TC, Andersson I. 1997. *J. Mol. Biol.* 265:432–44
69. Ensign SA, Small FJ, Allen JR, Sluis MK. 1998. *Arch. Microbiol.* 169:179–87
70. DiMarco AA, Bobik TA, Wolfe RS. 1990. *Annu. Rev. Biochem.* 59:355–94
71. Saeki H, Akira M, Keizo F, Averhoff B, Gottschalk G. 1999. *Microbiology* 145: 1721–30
72. Krum JG, Ensign SA. 2001. *J. Bacteriol.* 83:2172–77
73. Graham DE, Xu HM, White RH. 2002. *J. Biol. Chem.* 277:13421–29

Annu. Rev. Biochem. 2003. 72:77–109
doi: 10.1146/annurev.biochem.72.121801.161700

FUNCTION AND STRUCTURE OF COMPLEX II OF THE RESPIRATORY CHAIN*

Gary Cecchini
Molecular Biology Division, Veterans Administration Medical Center, San Francisco, California 94121, and Department of Biochemistry and Biophysics, University of California, San Francisco, California 94143; email: ceccini@itsa.ucsf.edu

Key Words succinate dehydrogenase, fumarate reductase, quinone oxidoreductase, reactive oxygen species, respiratory chain

■ **Abstract** Complex II is the only membrane-bound component of the Krebs cycle and in addition functions as a member of the electron transport chain in mitochondria and in many bacteria. A recent X-ray structural solution of members of the complex II family of proteins has provided important insights into their function. One feature of the complex II structures is a linear electron transport chain that extends from the flavin and iron-sulfur redox cofactors in the membrane extrinsic domain to the quinone and *b* heme cofactors in the membrane domain. Exciting recent developments in relation to disease in humans and the formation of reactive oxygen species by complex II point to its overall importance in cellular physiology.

CONTENTS

*The U.S. Government has the right to retain a nonexclusive, royalty-free license in and to any copyright covering this paper.

OVERVIEW OF MEMBRANE-BOUND RESPIRATORY CHAIN

Cells oxidize a variety of substrates to generate the energy used for metabolism. Membrane-associated respiratory reactions energize vectorial proton translocation to generate this energy. In mitochondria and many aerobic bacteria this process occurs through the electron transport chain with oxygen serving as the terminal electron acceptor. In anaerobic and facultative anaerobic bacteria, however, organic and inorganic compounds other than oxygen can serve as the ultimate electron acceptor. Fumarate, for example, can act as the terminal electron acceptor for growth on glycerol, lactate, formate, or molecular hydrogen during anaerobic respiration. The protein components of the respiratory chain are oligomeric complexes located in the inner mitochondrial membrane in eukaryotes and the cytoplasmic membrane of prokaryotes. In the field of bioenergetics these protein complexes are often referred to as the multisubunit electron transport complexes I, II, III, and IV, and the mitochondrial oxidative phosphorylation system is composed of these complexes along with ATP synthase (complex V) (Figure 1) (1). During the past few years great strides have been made in our understanding of the three-dimensional structures of the membrane-bound enzyme complexes that interact to form aerobic and anaerobic electron transport chains (2). This review focuses on complex II [succinate: ubiquinone oxidoreductase (SQR) and menaquinol:fumarate oxidoreductase (QFR), its homolog utilized for anaerobic respiration]; however, a brief discussion of the components of the other respiratory complexes is necessary in order to place the function of complex II in context.

The entry point for electrons into the mitochondrial electron transport chain is through NADH:ubiquinone oxidoreductase (complex I). This is the largest respiratory complex, with a molecular mass greater than 900 kDa; the bovine enzyme contains at least 45 different subunits (3). Complex I catalyzes electron transfer from NADH to quinone through a series of redox centers that include a flavin mononucleotide (FMN) moiety, seven to nine iron-sulfur centers, and up to three detectable ubisemiquinone species (4–6). Coupled to electron transfer, protons are vectorially translocated across the mitochondrial inner membrane such that complex I is one of three respiratory chain complexes where energy is conserved. Electron microscopy shows that complex I exhibits an overall L shape with the membrane domain connected by a thin collar to the stalk domain in the matrix of the mitochondrion (or cytoplasm of bacteria) (7, 8). Recent electron microscopy studies of the *Escherichia coli* complex I suggest, however, that the native conformation exhibits a horseshoe-shaped structure and that the complex can convert its shape depending upon ionic strength of the buffer used for isolation (9). Mammalian complex I also exhibits an active/de-active transition that can be modulated by divalent cations and other factors in intact mitochondria, and it has been suggested that conformational changes in the enzyme are responsible for this transition (10). Whether the conformational change observed by electron cryomicroscopy is responsible for the active/de-

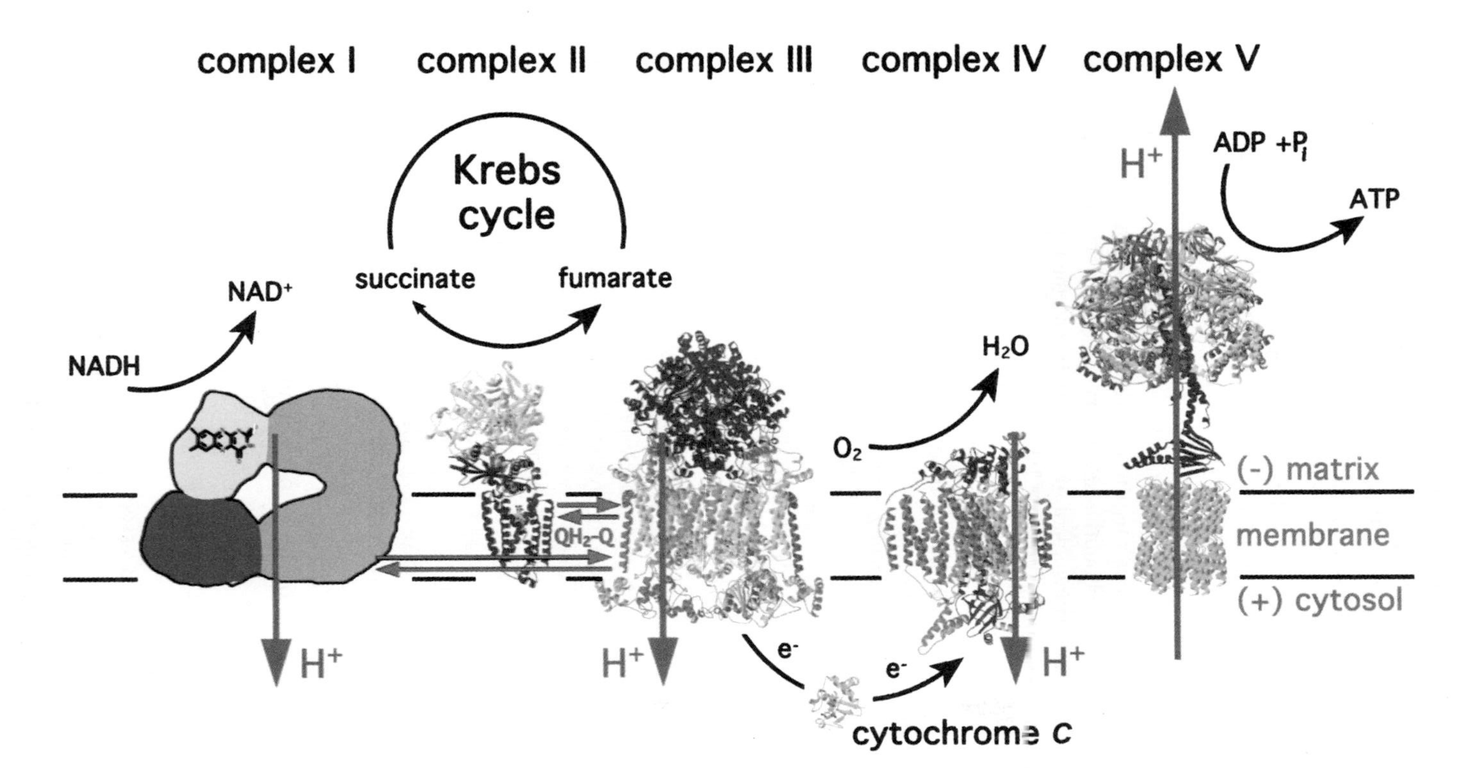

Figure 1 Diagram of respiratory chain from mitochondria. The complex II Protein Data Bank accession code is 1FUM; complex III, 1BGY; cytochrome *c*, 1CXA; complex IV, 1OCC; complex V, 1QO1. The complex I structure is a representation of the architecture of NADH:ubiquinol oxidoreductase determined in Reference 9.

active transition will require further experimentation. An X-ray structure for complex I is not yet available, and further understanding of its intricate architecture and regulation will necessitate much more experimental study.

Ubiquinol produced by the action of membrane-bound dehydrogenases such as complexes I, II, and electron transfer flavoprotein-ubiquinone oxidoreductase (ETF Q-reductase) is oxidized by complex III (ubiquinol-cytochrome *c* oxidoreductase or cytochrome bc_1 complex). Complex III in mammalian mitochondria contains 11 subunits, which include a membrane-bound diheme cytochrome *b*, and a membrane-anchored cytochrome c_1 and [2Fe-2S]-containing Rieske iron-sulfur protein. The electrons from ubiquinol are transferred to cytochrome *c* and this reaction develops the protonmotive Q cycle (11). Complex III is thus another of the mitochondrial respiratory complexes where energy is conserved. The transmembrane arrangement of complex III and the topographical orientation of the redox groups and mechanistic implications of this structure have been reviewed in detail in previous volumes of this series (12, 13). The final member of the mitochondrial electron transport chain that generates a transmembrane proton gradient is the terminal cytochrome oxidase (complex IV). Complex IV is a member of a superfamily of heme-copper oxidases found in many bacteria as well as the mitochondrion. The mammalian enzyme contains 13 different subunits, 3 of which are mitochondrially encoded (12, 14, 15). Complex IV has four redox metal centers, Cu_A, heme *a*, heme a_3, and Cu_B, that are part of a pathway from the substrate cytochrome *c*. Electrons are first transferred from cytochrome *c* to the mixed valence copper center (Cu_A) in subunit II. The electrons are subsequently transferred to cytochrome *a* in subunit I and then to the a_3 Cu_B binuclear active site, also in subunit I, where they reduce oxygen to two water molecules (2, 12, 14, 15).

The final component of the oxidative phosphorylation system of mitochondria is the ATP synthase (complex V or F_1F_0 ATPase). This enzyme is functionally reversible; it can use the proton gradient generated by the electron transport system described above to synthesize ATP and it can also hydrolyze ATP and pump protons against the electrochemical gradient. The crystal structure of the F_1 component of the bovine ATPase was the first reported for the members of the oxidative phosphorylation system described above (16) (Figure 1), and this structure supported the elegant binding exchange mechanism proposed for catalysis by the ATPase (17). The *E. coli* F_1F_0 ATPase contains 8 different subunits, whereas the bovine enzyme contains 16 different proteins (18). Both the bacterial and mammalian enzymes have a proton channel in the F_0 portion, which is linked to the catalytic F_1 portion by a stalk that is necessary for the structural rotation of the F_1 portion during catalysis (19).

COMPLEX II

Complex II has been associated with many seminal discoveries involving the structure and function of the bioenergetic complexes over the past 50 years. An excellent description of the part that complex II played in the discovery of covalently bound flavin cofactors, labile sulfide and properties of iron-sulfur clusters, and utilization of protein-stabilized ubiquinones by mitochondria for electron transfer can be found in a review by Helmut Beinert (20). The enzyme studied was succinate dehydrogenase (SDH), which catalyzes the oxidation of succinate to fumarate as part of the Krebs cycle. Subsequent studies showed an additional role for SDH besides its part in the Krebs cycle: Following succinate oxidation, the enzyme transfers electrons directly to the quinone pool; hence complex II is more precisely termed succinate:quinone oxidoreductase (SQR). Fumarate reductase (FRD), or menaquinol:fumarate oxidoreductase (QFR), is found in anaerobic and facultative organisms such as bacteria and parasitic helminths, where membrane-bound forms of the complex catalyze the oxidation of reduced quinones coupled to the reduction of fumarate. It was first shown in *E. coli* that separate enzymes for fumarate reduction and succinate oxidation are present depending upon growth conditions (21). In organisms that contain both a FRD and a SDH the synthesis of SDH is repressed and FRD is induced by anaerobic conditions (22–24). Nevertheless, it has been shown using genetic manipulation of *E. coli* that in vivo QFR and SQR can functionally replace each other if appropriate conditions are used to allow expression of their respective gene products (25, 26). As more information on the structure and genetic organization of SQR and QFR has become available the fact that they can functionally replace each other seems less surprising. It should be emphasized that QFR and SQR can catalyze the same reactions in vivo and in vitro, attesting to their high degree of sequence and structural similarity.

In eubacteria the genes for complex II are usually encoded as part of a compact operon. For example, *E. coli* SQR genes are in the order *sdhCDAB*, whereas those for QFR are ordered *frdABCD*. In both cases the *C* and *D* genes encode hydrophobic membrane anchor proteins that interact with quinones and are necessary to anchor the catalytic domain to the membrane surface. The gene order for archaeal complex IIs, however, can vary; for many enzymes termed SQRs the gene order is *sdhABCD*, like the QFR sequence (27). Similarly, the *Wolinella succinogenes* QFR is ordered *frdCAB* (28), as is the SQR from the gram-positive organism *Bacillus subtilis* (*SdhCAB*) (29). These latter organisms are examples of the class of complex II in which the hydrophobic C and D polypeptides have apparently fused into a single membrane anchor subunit. In all cases the polypeptide containing the dicarboxylate binding site and the flavin cofactor is encoded by the *A* gene, and the iron-sulfur-containing subunit of complex II is encoded by the *B* gene. In the case of eukaryotic organisms, mitochondrial DNA encodes a number of the protein components of the other electron transport and oxidative phosphorylation complexes (complexes I, III, IV, and V); however, with only a few exceptions all the genes for complex II are nuclear encoded. The exceptions include mitochondrial genomes from red algae such as *Porphyra purpurea* and heterotrophic zooflagellates such as *Reclinomonas americana,* in which the *sdhB, sdhC,* and *sdhD* genes are mitochondrially encoded (30). The high degree of sequence similarity across species for the hydrophilic subunit genes (*A* and *B*) and for the membrane anchor *C* subunit, as well as their nuclear and mitochondrial locations, has been used to support the idea that mitochondria and citric acid cycle evolution originated from within the α-proteobacterial branch of eubacteria (30, 31).

It had been predicted, based upon nucleotide and amino acid sequence comparison of QFR and SQR as well as biochemical analysis of their proposed structures, that they have a common evolutionary precursor (32–36). A class of soluble fumarate reductases is found in many anaerobic or microaerophilic bacteria such as from the genus *Shewanella* (37), yeast (38), and unicellular parasites like *Trypanosoma brucei* (39). Although these enzymes do not exhibit many of the properties of the classical complex II, the flavoprotein domain is homologous to the flavoprotein subunit of SDH and FRD. Two characteristics of this soluble class of FRD differ from classical complex II: (*a*) the flavin is noncovalently bound to the enzyme; and (*b*) the enzymes are essentially unable to oxidize succinate (37–40). Similar to what is found in nature, when site-directed mutants of *E. coli* QFR (41) or *Saccharomyces cerevisiae* SQR (42) were constructed that produce enzymes containing noncovalently bound flavin, the enzymes had lost the ability to oxidize succinate, although they retained the ability to reduce fumarate. This led to the suggestion that during evolution of complex II the primordial form of the enzyme contained a noncovalently linked flavin cofactor and that this protein was able only to reduce fumarate (34, 36). Upon acquisition of the covalent flavin linkage the enzyme would become able to catalyze succinate oxidation, which is the physiological reaction for SDH. It would also have been necessary for the enzymes to acquire other intermediate electron carriers of suitable redox potential, such as iron-sulfur clusters or cytochromes as in the *Shewanella* fumarate reductases (37). In order for the enzyme to become a complex II, the capacity to bind to the membrane domain and interact with electron carriers such as quinones would have to be acquired. The high degree of sequence conservation of the flavin and iron-sulfur domain of complex II is in agreement with suggestions that FRD and SDH evolved from a common evolutionary ancestor. The membrane domain of complex II has less sequence conservation across species; nevertheless there is a common structural motif of a four-helix bundle for the transmembrane domain. The sites of quinone reduction/oxidation are found in the membrane domain. It is presumed that once the soluble forms of SDH and FRD bound to the membrane domain over the course of evolution, subtle changes in the potential of the redox cofactors allowed the enzyme to interact with different quinone species in the membrane. As ubiquinone has a higher redox potential than menaquinone, the more positive redox potential of the iron-sulfur clusters of SQR (compared to QFR) allows a thermodynamically more favorable reaction for reduction of ubiquinone compared to menaquinol oxidation by QFR.

OVERALL STRUCTURE OF COMPLEX II

After more than 50 years of biochemical study on complex II, within a span of six months in 1999 a number of high-resolution X-ray structures for QFR and soluble forms of fumarate reductase became available (43–48). These structures came from several different laboratories, and the enzymes were isolated from different organisms, which has resulted in the most significant advances in our understanding of the overall architecture required for fumarate reduction and succinate oxidation. A gratifying aspect of the structural analysis is that to a large extent it has confirmed many of the speculations about the probable structure of the enzymes based on biochemical and biophysical analysis (20, 34–36, 49, 50). Of the available structures, however, only two are from membrane-bound forms of QFR, those from *E. coli* (43) and *W. succinogenes* (45). The other structures are for soluble forms of fumarate reductase from bacteria of the genus *Shewanella* (46–48) or L-aspartate oxidase from *E. coli* (44). L-Aspartate oxidase is a member of this family, as it can function as an L-aspartate/fumarate oxidoreductase generating iminoaspartate and succinate (51), in addition to its role in the biosynthesis of NAD^+. Although this review focuses on the properties of the membrane-bound forms, important mechanistic information about the catalytic process of fumarate reduction has been derived from studying the soluble fumarate reductases and they are discussed in that context.

Various classification schemes have been proposed for complex II based upon in vivo function, particularly quinone substrate used by the enzyme, differences in *b* heme composition, and number of membrane domain polypeptides (35, 52–55). In these schemes the *E. coli* and *W. succinogenes* QFR complexes would fall into two separate classes. The *E. coli* QFR enzyme, like mammalian and *E. coli* SQR, contains two hydrophobic membrane anchor subunits; however, it lacks the *b* heme moiety. The *W. succinogenes* QFR is of the class that contains two *b* type hemes but only a single hydrophobic membrane anchor subunit. Mammalian complex II and *E. coli* SQR would be part of the same class in which the enzymes are poised to reduce ubiquinone, contain a single *b* heme, and are anchored to the membrane by two hydrophobic subunits. Although no X-ray structure is available for any SQR, one is anticipated soon (56).

In addition to the reports of the X-ray structures (43–48), several insightful reviews have discussed the details of the fumarate reductase structures (44, 50, 52, 57, 58). As predicted from analysis of the sequences of the subunits and biochemical data, complex II is essentially a modular protein complex. The structure of the *E. coli* QFR in Figure 2 clearly shows a demarcation between the hydrophilic and hydrophobic subunits. In all of the crystal forms analyzed to date for either the *E. coli* or *W. succinogenes* QFR, the complexes in the asymmetric unit are associated in a fashion that forms a dimer. In the *E. coli* QFR the crystal contact buries 325 Å^2 of surface area and this is mediated by molecules of the detergent Thesit ($C_{12}E_9$) (43), whereas in the *W. succinogenes* QFR structure 3665 Å^2 is buried upon formation of the dimer (45). Although this dimer has also

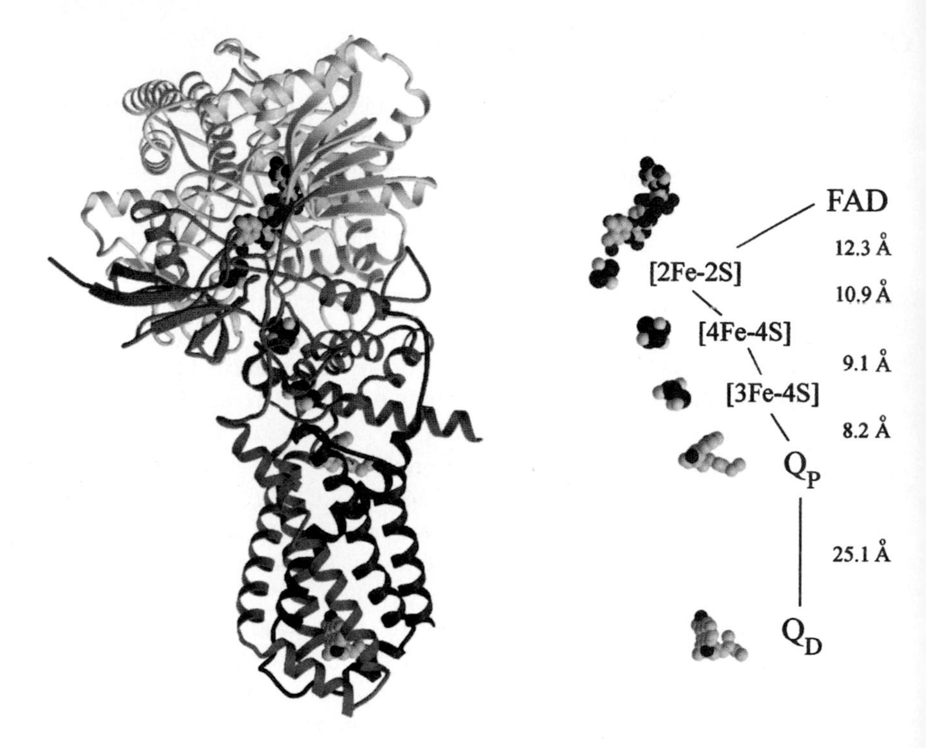

Figure 2 Ribbon diagram of the *E. coli* QFR structure (1FUM) is shown on the *left*. The FrdA flavoprotein subunit is shown in yellow, the FrdB iron-sulfur protein in brown, the FrdC subunit in blue, and the FrdD subunit in purple. Cofactors are shown as space-filling models. On the *right* of the figure the distances between the redox cofactors (edge-to-edge) in *E. coli* QFR are shown.

been reported in the detergent-solubilized *W. succinogenes* enzyme, based upon gel filtration experiments (58), no compelling evidence suggests that the dimer is a necessary prerequisite for function of QFR. In contrast to the dimers found in the QFR crystals, the *E. coli* SQR crystals available show that SQR is packed as a trimer with the monomers related by a crystallographic threefold symmetry axis (56); however, as for QFR there is no indication that this arrangement has any functional significance.

Both QFR structures have an overall similar length. They are oriented perpendicular to the membrane; *W. succinogenes* QFR is 120 Å long and *E. coli* QFR is 110 Å long (57, 58). Parallel to the membrane, the maximum width (FrdAB catalytic domain) of the monomer of both QFRs is 70 Å. The enzymes are attached to the membrane domain by interactions between the iron-sulfur subunit (FrdB) and their respective hydrophobic membrane anchor subunits. It has been suggested that an intact [3Fe-4S] cluster is necessary in order for succinate dehydrogenase and fumarate reductase hydrophilic domains (SdhAB and FrdAB) to bind to the membrane and thus be active in quinone reduction (34, 60–62). The QFR structures show a close interaction between the [3Fe-4S] iron-sulfur cluster and the membrane anchor subunit(s). The *E. coli* QFR membrane domain is composed of two subunits (FrdC and FrdD), each of which contains three membrane-spanning segments with helical secondary structures. These helices have been termed helices I–VI (I–III in FrdC, and IV–VI in FrdD). The single subunit FrdC of *Wolinella* QFR contains five transmembrane helices; it lacks the one corresponding to helix III in *E. coli* FrdC. This may indicate that a gene fusion between a corresponding *frdD* gene and the 5′ end of an ancestral *frdC* gene is responsible for the five transmembrane helices and single FrdC subunit forms of QFR and SQR. As mentioned, the structures of the hydrophobic anchor subunits from the two organisms can be aligned to a significant degree; however, this requires a rotation around the membrane normal and involves a difference in the relative orientation of the soluble subunits for the *E. coli* and *Wolinella* QFRs (45). As pointed out by Lancaster and coworkers, in the membrane domain of its QFR the *E. coli* enzyme, which lacks heme, has the four transmembrane helices forming the helical core bundle packed closer together than in the *Wolinella* QFR, which contains two *b* hemes (45). Thus, in *E. coli* QFR there does not seem to be room for heme insertion. Also, the *E. coli* enzyme contains additional bulky amino acid residues in this core as compared to *Wolinella* QFR; both factors probably contribute to the lack of a *b* heme moiety (45). For example, whereas the distances between the Cα atoms of the *W. succinogenes* heme axial b_P ligands are 12.7 Å (HisC93-HisC182), the corresponding residues in *E. coli* (FrdC His82 and FrdD Cys77) are only 11.2 Å apart (45). Mutagenesis to change FrdD Cys77 to His in *E. coli* QFR, in order to provide a second heme axial ligand, fails to allow heme *b* insertion in the *E. coli* enzyme (D.A. Berthold & G. Cecchini, unpublished data) in agreement with the idea that the tight packing of the helices in the *E. coli* QFR membrane domain precludes heme insertion.

In *E. coli* QFR the interaction between the membrane domain and the iron-sulfur subunit is through amino acid residues near the N terminus of the FrdC and FrdD subunits on the cytoplasmic side of the membrane. The first 21 amino acid residues of the FrdC subunit extend away from the membrane into the cytoplasm and appear to help stabilize the binding of the hydrophilic domain to the membrane. The first eight residues of FrdD may also stabilize this interaction. In the *Wolinella* QFR there is also a short helix in the cytoplasm between helices II–IV (FrdC residues 105–118) that may stabilize the binding of the FrdAB subunits to FrdC. In *E. coli* QFR there is an equivalent loop in FrdC between helices II and III (FrdC residues 91–104) and in FrdD between helices V and VI (FrdD residues 89–97) that also may help bind FrdB to the membrane domain. It should also be noted that the quinone binding site termed Q_P (proximal to the [3Fe-4S] cluster) is part of this binding region. No one factor by itself appears to be responsible for the binding of the hydrophilic subunits to the membrane domain. Reconstitution of the catalytic subunits with the membrane anchor subunits is sensitive to ionic strength, certain types of anions, and pH, indicative of a strong electrostatic attraction between the membrane subunits and the iron-sulfur subunit of complex II (34). It has been suggested that amino groups on the surface of the catalytic subunits of mammalian complex II are important for the stability of the complex, based on chemical modification studies (63, 64), and some of these groups were predicted to be near the [3Fe-4S] cluster of the enzyme (34). Several conserved lysine residues are near the C terminus of the iron-sulfur subunit (in *E. coli* FrdB Lys216, Lys228, Lys241), and it is probable that modification of these residues could be responsible for the dissociation of the catalytic subunits from the membrane domain (64). For example, *E. coli* FrdB Lys228 seems to be in hydrogen bond contact with the quinone, indicative of close association between the membrane subunits and the C terminus of FrdB.

Clearly the presence of redox cofactors such as iron-sulfur clusters and *b* heme in the case of mammalian and *E. coli* SQR (61, 65, 66), *Wolinella* QFR (67), and *Bacillus subtilis* SQR (35, 54, 68) are necessary for proper assembly of an intact complex II. In addition to these factors, other protein components have been suggested as requirements for proper assembly of complex II in eukaryotic systems. Two chaperone-like proteins, ABC1 (69) and TCM62 (70), have been implicated in assembly of complex II in yeast. Deletion of the *ABC1* gene leads to deficiencies in complex II and IV, suggesting that its role is not specific to complex II (69). The role of TCM62 seems more specific to complex II and it has been suggested to play a role in the assembly of the iron-sulfur clusters of the enzyme (70). Recent results, however, suggest that TCM62 is part of a large protein complex in the mitochondrial matrix with a mass similar to chaperonins and is required for essential mitochondrial functions at high temperature (71), not just for stability or assembly of SQR. The factors required for the assembly of iron-sulfur clusters into proteins is a complex process requiring numerous factors (72). One of these factors shown to affect SQR assembly is a cysteine desulfurase (Nfs1p), which plays a central role in iron-sulfur cluster synthesis (73). In addition to these specific factors it is known that nutritional iron deficiency affects complex I and II levels in the mitochondrial membrane, although the respiratory complexes that do assemble are fully functional (74). In the case of complex II this has been shown to be the result of a fully functional iron-responsive element (IRE) in the mRNA for the iron-sulfur protein subunit SdhB. This IRE functions to mediate translational repression by iron regulatory proteins and thus affects overall levels of complex II in the mitochondrial membrane (75). The overall picture that emerges for assembly of complex II is that although it contains only four subunits and is the smallest of the electron transport complexes, the numerous cofactors it contains (flavin, iron-sulfur clusters, heme) mean that interference with their insertion into the complex affects assembly. Thus, like other members of the mitochondrial respiratory chain, complex II may turn out to require additional assembly factors (76).

FLAVOPROTEIN SUBUNIT AND FORMATION OF THE COVALENT FAD LINKAGE

Although there were indications that SDH was a flavoprotein as far back as 1939, it was 15 years later before direct spectral evidence was obtained that flavin is present in the enzyme [reviewed in (77)]. It became apparent that SDH contained flavin adenine dinucleotide (FAD) attached to the protein through a covalent flavin linkage (78), the first covalent flavin linkage reported for any protein. A decade later, the covalent flavin was identified as an 8α-*N*3-histidyl-FAD linkage (79) and the primary amino acid sequence of the flavin peptide was determined (80). Since this early work on succinate dehydrogenase, a number of proteins have been identified that contain covalent flavin linkages and a variety of modes of linkage to the peptide backbone have become apparent (81). Despite this vast increase in our knowledge of the types and number of covalent flavin proteins in nature, the reason for the presence of a covalent flavin linkage in a particular enzyme remains enigmatic.

All SQR and QFR complexes from both prokaryotes and eukaryotes that have been examined in detail contain a histidyl-FAD covalent linkage. By contrast, soluble fumarate reductase homologs of the flavoprotein subunit of complex II such as those from yeast (38), bacteria from the genus *Shewanella* for example (37), and unicellular parasites (39) contain noncovalently bound FAD. It has been speculated that the covalent linkage prevents the loss of the flavin cofactor from proteins in a membrane or periplasmic environment where concentrations of flavin mononucleotide (FMN) and FAD would be low and unable to replace the lost flavin (81). Now that structures are available for both membrane-bound QFRs (43, 45) and soluble fumarate reductase homologs (46–48), however, there is not an obvious structural difference as to how the covalent flavin linkage is acquired. The X-ray structures of the membrane-bound QFRs and the noncovalent flavin-containing structures from *Shewanella* show that the overall topology of the flavin and capping domains and the site of flavin binding are conserved. The evolution and maintenance of the covalent flavin link suggest that it may serve mechanistic requirements of these enzymes (81).

Whether covalent flavin attachment is protein mediated or self-catalytic has been discussed in detail (81). A general consensus is that the process is self-catalyzed and not mediated by another enzyme, although a mitochondrial chaperone, heat shock protein 60 (hsp60), has been shown to assist in the covalent flavinylation process in yeast SDH (82). Formation of the covalent flavin linkage is nevertheless a complex process that appears to require the flavoprotein to fold into the proper scaffold in order for the flavin cofactor and the amino acid to which it will covalently link to be in the proper steric orientation for the self-catalysis to occur. Insightful work on covalent flavinylation has come from the laboratories of Edmondson (83) and McIntire (84, 85) and their colleagues. The latter group has studied this process in *p*-cresol methylhydroxylase (PCMH) from *Pseudomonas putida* (81, 84, 85). Although PCMH contains an 8α-*N*3-tyrosyl FAD linkage rather than the 8α-*N*3-histidyl FAD linkage found in SQR and QFR, the proposed mechanism for covalent flavin linkage most likely has remarkable similarity to the process occurring in SQR/QFR. PCMH is an $\alpha_2\beta_2$ flavocytochrome composed of a flavoprotein (α_2) and cytochrome *c* (β_2) subunits. In PCMH it was found that it was necessary for the cytochrome *c* subunit to bind to the flavoprotein subunit before covalent flavinylation could occur (86). This suggests that structural alterations in the flavoprotein subunit induced by cytochrome *c* binding are necessary before covalent flavinylation can occur (81, 84, 86). Recently these authors have also shown that the redox potential of the FAD cofactor is significantly increased when the FAD binds to the apoflavoprotein and then increases again upon formation of the covalent flavin bond (85). This increase in reduction potential of the noncovalently bound FAD would make the isoalloxazine moiety of FAD a better electrophile for attack at its 8α-carbon by the nucleophile tyrosine O^- (85). The authors suggest that this would increase the equilibrium concentration of the iminoquinone methide intermediate required for nucleophilic attack at the 8α-carbon. These results are consistent with the quinone-methide mechanism of covalent flavin linkage that has been proposed from several different laboratories (83, 87, 88). In this mechanism a quinone-methide tautomer is formed by loss of a proton from the 8α-position of the isoalloxazine ring, followed by an attack by a nucleophilic amino acid residue at the 8α-methide. This forms a reduced 8α-flavin adduct, which it is important to note must then be oxidized to produce the active covalently bound flavin (83).

Although the formation of the covalent flavin linkage described above has not been experimentally demonstrated for SQR or QFR, evidence suggests that the same mechanism applies to complex II. An interesting series of studies has used the yeast SQR to investigate when in the process of assembly of complex II the flavin is inserted covalently. In *S. cerevisiae,* His90 is the attachment site in the SdhA subunit for the covalent FAD. When this residue is mutated, it was found that the SdhA subunit could still be translocated to the mitochondrion and assembled into the enzyme complex; however, the flavin was attached noncovalently (42). Further studies showed that cofactor attachment occurs within the mitochondrial matrix following cleavage of the mitochondrial targeting presequence (82). In a situation reminiscent of PCMH discussed above, FAD attachment was stimulated by the presence of the iron-sulfur subunit (SdhB). Flavinylation was also stimulated by the dicarboxylic acid intermediates such as succinate, malate, or fumarate, all of which would bind at the active site of the enzyme. An additional observation was that C-terminal-truncated SdhA subunit imported by the mitochondrion was not able to incorporate covalent flavin, suggesting that the imported protein must fold properly for covalent FAD insertion to take place (82). Although the mitochondrial chaperonin hsp60 apparently does interact with the SdhA subunit, it was not absolutely required for FAD attachment (89). Complex II in the mutated protein assembled in *S. cerevisiae* contained tightly bound noncovalent FAD, and was also shown to be catalytically competent for fumarate reduction but inactive in its normal physiological reaction, succinate oxidation (42).

Similar results were found for *E. coli* QFR. When His44, which is the site of the covalent FAD linkage, was mutated to a Ser, Cys, or Tyr residue, QFR retained tightly bound FAD but in noncovalent form (41), and the enzyme complexes assembled normally within the membrane. These mutant enzyme forms also essentially lost the ability to oxidize succinate, although they retained significant fumarate reductase activity. The tightly bound noncovalent flavin cofactor could be removed by dialysis against potassium bromide, which resulted in inactive enzyme. Fumarate reductase activity, however, could be restored by reconstitution of the enzyme with FAD (41). In both the yeast and *E. coli* studies it was suggested that the reason the noncovalent-FAD-containing QFR or SQR was unable to oxidize succinate was that the redox potential of the FAD cofactor had lowered sufficiently to preclude this reaction (41, 42). Unfortunately, no direct measurement of the redox potential of the flavin in the mutant enzymes has been done in order to address this hypothesis.

Recently additional site-directed mutations of *E. coli* QFR have been constructed that allow normal enzyme assembly; however, the enzymes contain noncovalent FAD (I. Schröder, E. Maklashina, Y. Sher, G. Cecchini, unpublished data). This was accomplished by mutating conserved residues thought to be part of the catalytic site in the FrdA subunit. In Figure 3 the spatial location of the residues that cause noncovalent flavin incorporation is shown. It is noteworthy that the residues which cause noncovalent flavin assembly are also those implicated as being involved in substrate binding and catalysis of the enzyme. As shown in Figure 3, FrdA Glu245 and Arg287 are two residues from the capping domain that when mutated result in noncovalent flavin assembly in QFR. In addition, FrdA His355 and Arg390 from the flavin domain also produce noncovalent flavins. Although not shown in Figure 3, when alanine insertions are placed in the hinge region connecting the capping and flavin domains, the enzyme contains noncovalent FAD. These results are consistent with previous observations that suggest that proper folding and alignment of residues around the site of flavin attachment are required for self-catalysis to be initiated in flavoproteins containing covalent flavin linkages. FrdA His355 and Arg390 are hydrogen bonded to dicarboxylate oxygens at the C4 position of fumarate, and in the closed conformation Arg287 also would bind this region of the substrate/inhibitor. FrdA His232, by contrast, is hydrogen bonded to the other end of the dicarboxylate at the C1 position, and mutation of His232 still allows the covalent FAD linkage to form (90). FrdA Glu245 is part of the capping domain and may be part of a proton pathway to Arg287, so it may have a similar effect by affecting the protonation state of the arginine. The alanine insertions would be expected to affect the movement of the hinge region connecting the capping and flavin domains. They might be expected to affect the molecular architecture of the active site such that there would be a misalignment of amino acid residues around the flavin and substrate binding site that could affect the covalent flavin

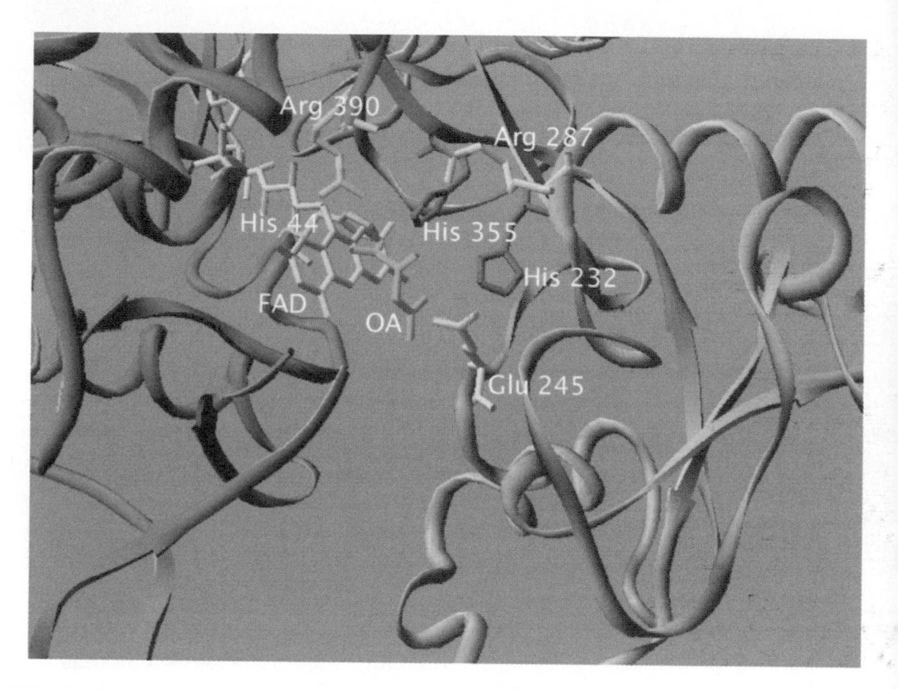

Figure 3 View of the substrate and FAD binding site region in *E. coli* QFR. The flavoprotein domain of the FrdA subunit is shown in blue and the capping domain is shown in yellow-gold. Inhibitory oxaloacetate (OA), which binds tightly at the substrate binding site, is shown in green. FrdA His232, which binds to the C1 carboxylate of fumarate, is shown in purple. FrdA Arg390 and His44 are shown in cyan. FrdA His355 is in purple, and Arg287 and Glu245 in beige. Mutation of His232 allows covalent FAD incorporation into the enzyme, whereas for all other residues shown, mutations cause the flavin to bind noncovalently.

binding. An additional point is that when a covalent FAD linkage forms based on the quinone-methide mechanism, a positively charged amino acid residue in the vicinity of the N1 and C2 positions of the flavin ring would facilitate this reaction. The positively charged amino acid would help stabilize the negative charge in the N1/C2 region of the flavin ring system to make the 8α-position more electrophilic and thus more reactive with the nucleophilic His44. The most likely candidate for this amino acid is FrdA Arg390, since it is the only positively charged amino acid near the N1/C2 position (3.4 Å away in *E. coli* QFR). This amino acid residue thus is critical to covalent bond formation, and in humans the equivalent residue (an Arg408 to Cys mutation in human SdhA) has been shown to be responsible for late-onset neurodegenerative disease (91). In total these data suggest that the precise orientation of the flavin ring system, amino acids at the active site, and overall topology of the flavin and capping domains must be exquisitely maintained in order for covalent bond formation to occur similar to the situation in vanillyl-alchohol oxidase, another covalent flavoprotein (92). The presence of a substrate or dicarboxylic acid inhibitor at the active site region also appears to be necessary in order to maintain the proper conformation by causing closure of the capping domain. The stimulation of covalent flavinylation shown by the presence of the SdhB subunit (82) may also be explained in two ways. One, it may contribute to the conformational changes necessary for proper alignment of the amino acid residues around the flavin; second, it may act as an electron acceptor for the reduced histidyl-FAD product that would be formed upon nucleophilic attack on the quinone-methide (83).

The noncovalent mutant proteins described above maintain fumarate reductase activity but lack succinate oxidase activity. Recently Heffron, Armstrong, and colleagues have been able to determine by protein film voltammetry the redox potential of the noncovalent flavin in the FrdA Arg287 mutation. The direct measurement of the FAD redox potential in this noncovalent mutation showed the redox potential of the flavin to be some 100 mV lower than wild-type covalent FAD in fumarate reductase (93) [wild-type QFR, FAD $E_{m,7} = -50$ mV (94)]. This lowered redox potential in *E. coli* fumarate reductase noncovalent FAD mutants is consistent with the redox potential reported for those fumarate reductase homologs that naturally contain a noncovalent flavin. For example, the redox potential of the noncovalent FAD from L-aspartate oxidase of *E. coli* is -216 mV (95) and that for *Shewanella frigidimarina* flavocytochrome c_3 is -152 mV (96). Both of these enzymes, although they reduce fumarate proficiently, are essentially unable to oxidize succinate. The significant rise in the redox potential upon formation of a covalent flavin bond is a consequence of both the covalent linkage and the protein environment around the flavin and is consistent with that for other enzymes containing this linkage (81, 83, 85). Thus, the higher FAD redox potential is also a prerequisite for succinate oxidation.

CATALYSIS IN COMPLEX II

The availability of high-resolution X-ray structures of QFR has contributed significantly to our understanding of the mechanism of fumarate reduction by the succinate dehydrogenase/fumarate reductase family of flavoproteins. In particular the structures of the soluble fumarate reductases, which have been solved to 1.8 Å (47), have provided important insights into the mechanism of fumarate reduction. The residues shown to be part of the substrate binding/active site and involved in catalysis are absolutely conserved throughout the family of fumarate reductases and succinate dehydrogenases whether they are membrane bound or soluble forms. The mechanism originally proposed for the *S. frigidimarina* fumarate reductase (47) has been supported by elegant studies using site-directed mutagenesis and X-ray crystallography of mutant enzyme forms (96–99). The basic mechanism proposed by these workers is shown in Figure 4 with the amino acids from the *E. coli* FrdA subunit used for comparison. The structures of all

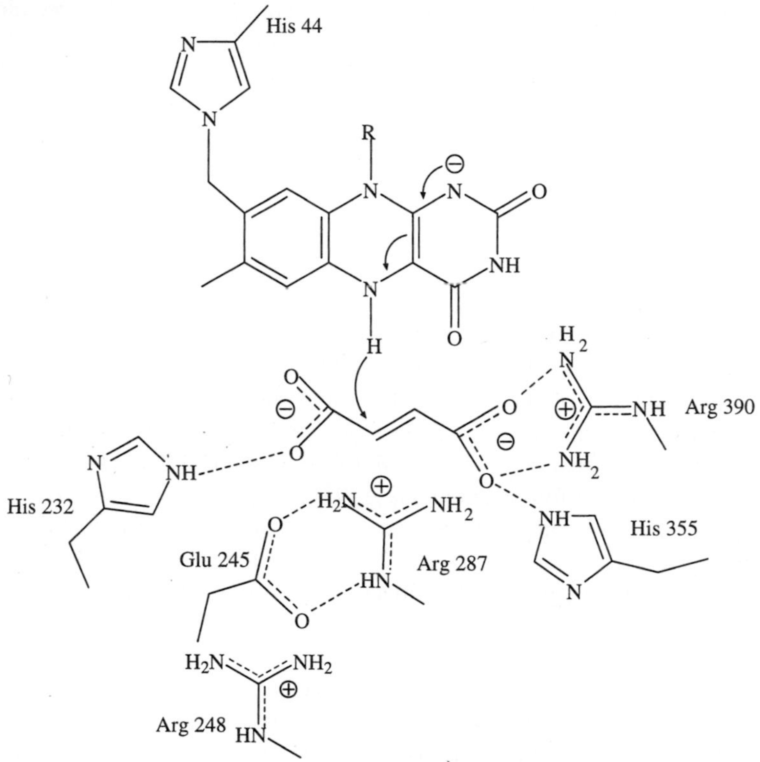

Figure 4 Catalytic mechanism proposed for fumarate reduction. Fumarate is polarized by interactions with His232, Arg390, and His355, which facilitates hydride transfer from the N5 position of the isoalloxazine ring of the reduced FAD cofactor. Arg287 is then positioned to donate a proton to fumarate, which results in succinate formation. Reprotonation of Arg287 is accomplished via a proton pathway that includes Arg248 and Glu245. (Adapted from References 97, 98.)

fumarate reductases in the closed conformation are similar, suggesting that this is the catalytically competent state of the enzyme (100). In this conformation, as shown in Figure 4, the C4 carboxylate of fumarate is in a highly polar environment and bound by hydrogen bonds and electrostatic attraction to FrdA Arg287 and Arg390 and by a hydrogen bond to FrdA His355 (101). The C1 carboxylate of fumarate is in a less polar environment and hydrogen bonded to FrdA His232 (47, 48, 57, 58, 96–101). The bound substrate is also distorted during closure of the capping domain, which induces polarization of the substrate (45, 48, 101). The polar nature of the hydrogen bonding environment around the C4 carboxylate group has been suggested to polarize the fumarate (101). The combined effect of twisting of the substrate and electronic effects generates a positive charge at the C2 position, making it a candidate for nucleophilic attack from the N5 position of the flavin ring. Hydride transfer from the N5 of the flavin to C2 is followed by protonation at the C3 position by FrdA Arg287 (96, 101). Reprotonation of FrdA Arg287 is accomplished by a proton pathway involving FrdA Glu245 and FrdA Arg248 (96). The product of this reaction succinate is then released by movement of the capping domain, which also would allow entrance of another molecule of the substrate. In principle, the reverse reaction (succinate oxidation), which is catalyzed efficiently by membrane-bound QFRs containing covalently bound flavin, would proceed by the reverse of the mechanism described above. Mutation of the amino acid residues indicated in Figure 4 has been done in *E. coli* QFR, and results are consistent with the mechanism proposed using the *S. frigidimarina* enzyme [(47, 90, 96–99); I. Schröder, E. Maklashina, & G. Cecchini, unpublished data].

Another interesting aspect of catalysis that is restricted to succinate dehydrogenase and not shown by fumarate reductase is the "diode effect" (102). This effect shows that catalytic fumarate reduction by succinate dehydrogenase abruptly slows to a diminished catalytic rate below a redox potential of approximately −60 to −80 mV despite an increase in the driving force of the reaction (103, 104). At redox potentials above −60 mV, however, succinate dehydrogenase will reduce fumarate very rapidly, and in that sense the enzyme is fully reversible, like fumarate reductase containing covalently bound flavin. The diode effect has been used to classify fumarate reductase and succinate dehydrogenase as to their normal physiological function (105). It is suggested that at pH values below 7.64, SDH is energetically poised to catalyze fumarate reduction under conditions of low driving force (106). It was concluded that the reduction of FAD is the factor responsible for the diode effect and that a conformational change may occur upon formation of $FADH_2$ (106). This hypothesis seems reasonable in light of the domain movements that obviously occur in QFR, and presumably SQR, around the FAD binding site. Nevertheless there must be differences between succinate dehydrogenase and fumarate reductase to account for their different catalytic behavior even though their flavoprotein subunits are likely to have similar structures. It is pertinent therefore that for *E. coli* FrdAB the two-electron oxidation of the anionic FAD hydroquinone is associated with the loss of one proton, whereas the reoxidation of the hydroquinone in SdhAB appears to involve two protons (94, 103, 104). Additionally, in fumarate reductase the flavin semiquinone appears to be neutral and most likely protonated at the N5 position (94), whereas succinate dehydrogenase is thought to contain an anionic flavin semiquinone (34).

The physiological significance of the diode effect is unclear. It has been suggested, however, that it could provide a means for control of the Krebs cycle during periods of hypoxia, in which the quinone pool would become reduced (106), by shutting down succinate dehydrogenase so that it would not reduce fumarate and thus interfere with the Krebs cycle. This effect might be more relevant in bacteria that vary their ratio of low- and high-potential quinones in response to environmental stimuli (107) but also could be useful in mammalian cells where a hypoxic response is initiated by mutations in SQR genes, for example.

IRON-SULFUR CLUSTERS OF COMPLEX II AND THE ELECTRON TRANSFER PATHWAY

After years of controversy the fact that three distinct types of iron-sulfur clusters are present in prototypical SQR/QFRs finally became apparent during the 1980s. The history of these controversies and discovery of the numbers and types of iron-sulfur clusters have been extensively reviewed (20, 34, 49). It was determined that the cluster composition of complex II is a $[2Fe\text{-}2S]^{2+,1+}$ cluster, often termed Center 1; a $[4Fe\text{-}4S]^{2+,1+}$ cluster, Center 2; and a $[3Fe\text{-}4S]^{1+,0}$ cluster, Center 3. The application of biophysical technologies such as electron paramagnetic resonance (EPR), Mössbauer, low-temperature magnetic circular dichroism (MCD), and other spectroscopies (20) allowed these answers about the numbers and types of iron-sulfur clusters. It is thus quite appropriate that the new knowledge obtained from the X-ray crystal structures of complex II (43, 45) has helped to answer a question about the role of the [4Fe-4S] cluster of the enzyme. The midpoint potential (pH 7) of the [4Fe-4S] cluster has been reported to range from −320 mV for *E. coli* QFR to −175 to −260 mV for *E. coli* and mammalian SQR, respectively (108). The low midpoint potential of the [4Fe-4S] cluster had led to speculation that this cluster was not part of the electron transfer pathway but rather might play a structural or regulatory role in the enzyme (34, 49, 109). The X-ray structures show, however, that the [4Fe-4S] cluster is part of a linear electron transport chain between the FAD cofactor and the quinone or *b* heme(s) located in the membrane domain, as indicated in Figure 2. The relatively close physical association of the iron-sulfur clusters (~11 Å, edge-to-edge) suggests that they all participate in electron transfer and that the low potential of the [4Fe-4S] cluster does not present a thermodynamic barrier (52). Studies of *E. coli* FrdAB subunits using protein film voltammetry show a boost in catalytic current becoming apparent at the redox potential of the [4Fe-4S] cluster. This is also consistent with the [4Fe-4S] cluster relaying electrons to FAD as part of the electron transport chain within the enzyme (110).

The reported reduction potentials of all three iron-sulfur clusters in QFR are lower than the respective counterparts in SQR in accord with their physiological donors/acceptors. In *E. coli* QFR the reduction potentials are −74, −67, −310, −35, −50 mV, respectively, for menaquinol, [3Fe-4S], [4Fe-4S], [2Fe-2S], FAD (94), whereas for *E. coli* SQR the reduction potentials are −60 to −79 for FAD, and +10, −175, +65, +36, +90 for [2Fe-2S], [4Fe-4S], [3Fe-4S], heme *b*, and ubiquinone, respectively (34, 66, 111–113). The reduction potential for FAD has not actually been measured for *E. coli* SQR; however, a potential of −79 mV has been determined for beef SQR (111). The value of −60 mV listed above is derived from the potential at which the diode effect makes itself apparent in *E.*

coli SQR (104) versus a potential of −80 mV for the beef enzyme (103). The different value for the reduction potentials of the iron-sulfur clusters in QFR versus SQR most likely reflects differences in the protein environment surrounding the clusters. The arrangement of the iron-sulfur clusters in SQR is without a doubt the same as in QFR, considering the high degree of sequence similarity of the FrdB and SdhB subunits from different species, including the absolute arrangement of conserved cysteinyl ligands for the Fe-S clusters (34, 35, 108). *E. coli* SQR is unusual in one respect in that the third cysteine of the [2Fe-2S] signature sequence (**C**xxxx**C**xx**C**. . . . . **C**) is replaced by an aspartate residue (114); however, site-directed mutagenesis of the equivalent residue in *E. coli* QFR has shown that the [2Fe-2S] cluster is retained albeit with a slightly higher redox potential (115). Based on sequence and EPR studies it had been predicted that the iron-sulfur protein of complex II would fold into two domains (34, 35), which is as found in the QFR structures (43, 45). The N-terminal domain containing the [2Fe-2S] cluster folds similarly to plant-type ferredoxins, whereas the C-terminal domain containing the [4Fe-4S] and [3Fe-4S] clusters is similar in topology to bacterial ferredoxins (43, 45, 57, 58). The C-terminal domain of FrdB also contains several helices that associate with the membrane anchor subunits and thus are required to hold the hydrophilic subunits to the membrane domain.

MEMBRANE DOMAIN OF COMPLEX II

Although the hydrophilic flavoprotein and iron-sulfur protein subunits seem highly conserved among eukaryotic and prokaryotic organisms, there is a much greater variation in the primary amino acid sequences of the membrane-spanning subunits. The single-subunit membrane anchor complexes such as QFR from *W. succinogenes* (45) or *B. subtilis* SQR (29) have five membrane-spanning helices rather than the six found in the two-subunit membrane anchor forms. The suggestion that the single membrane anchor subunits evolved by a fusion event where the third helix of the C-subunit has been deleted (35, 36) is supported by the two available QFR structures. The complexes that contain a single membrane-spanning polypeptide also contain two heme groups while the mammalian, *S. cerevisiae,* and *E. coli* SQR contain a single heme group, and *E. coli* QFR lacks heme.

There is no evidence in the single *b* heme–containing complex II or *E. coli* QFR that the enzymes can act as proton pumps. It has been suggested that this is because the reactions catalyzed are not sufficiently exergonic to promote proton translocation (116). Nevertheless, in the single-subunit diheme containing *B. subtilis* SQR the uphill electron transfer reaction from menaquinol to succinate has shown to be sensitive to uncouplers (53, 117), and this process is thought to be driven by the $\Delta\mu_H{}^+$ produced by the aerobic respiratory chain as discussed by Ohnishi and coworkers (52). It is also possible that the *W. succinogenes* QFR may produce $\Delta\mu_H^+$, which may occur by a Mitchellian type Q loop [reviewed in (52, 58, 67)]. This contention awaits experimental verification, as the available evidence is not entirely consistent with this hypothesis (118). Therefore, it is quite pertinent that a recent report shows that in *W. succinogenes* QFR mutation of FrdC Glu180 almost completely abolishes the ability of the enzyme to interact with quinones (119). This glutamate residue is conserved in all diheme QFRs but not in diheme SQRs and lies in the middle of the membrane domain in FrdC in a helical region parallel to the *b* heme electron transfer pathway. Fourier transform infrared (FTIR) spectroscopy data support the idea that this glutamate is involved in proton exchange, and it is suggested that FrdC Glu180 is part of a proton translocation pathway that would take protons from the periplasm back across the membrane to the cytoplasm, a process termed the E-pathway hypothesis (119). The net effect would be a coupled proton/electron transfer system so that *W. succinogenes* QFR is not electrogenic, which would agree with previous data showing that *W. succinogenes* QFR is not a classical proton pump (58, 118, 119). This result seems to suggest that diheme QFRs and diheme SQRs have a different mechanism with regard to the process of proton translocation, in agreement with the different thermodynamics of fumarate reduction and succinate oxidation in the two enzymes.

The interactions of quinones with complex II is an area of active interest, and the structural data from *E. coli* QFR in particular has shed light on the amino acid residues that line the binding site(s). In the *E. coli* 3.3-Å structure, two menaquinone binding sites were observed (43). That two binding sites were found for quinones in the structure for complex II was not unexpected based on previous site-directed mutagenesis studies in *E. coli* QFR (120) and *S. cerevisiae* SQR (121) and by photoaffinity labeling studies of mitochondrial and *E. coli* SQR (122, 123). It had also been known that a stabilized semiquinone pair exists in complex II from beef heart (124), and EPR studies had suggested that this interacting pair were approximately 8–9 Å apart. The two quinones seen in the *E. coli* QFR structure were, however, some 25 Å apart (Figure 2) and localized on opposite sides of the membrane (43), suggesting that the quinone at Q_D was not part of the semiquinone pair. In addition the large spatial separation between the two quinones would suggest that they were not part of an electron transfer chain unless some other redox active species was placed between them to shorten the electron transfer distances (52). A single stabilized semiquinone species has been found in an *E. coli* QFR mutant (FrdC Glu29Leu) with a signal of g = 2.005, a midpoint potential of -57 mV, and a stability constant of $\sim 1.2 \times 10^{-2}$ at pH 7.2 (125). This stability constant is some eight orders of magnitude greater than that found for free quinone within the membrane. These same studies also showed the presence of a stabilized semiquinone species in wild-type QFR, however, with a stability constant four orders of magnitude lower than in the mutant enzyme (125). The results indicate that the FrdC Glu29 residue destabilizes the semiquinone produced during electron transfer at the proximal Q_P site in *E. coli* QFR (125). The EPR studies also indicated that the [3Fe-4S] cluster was the dominant spin relaxation enhancer (125), which is consistent with the close spatial separation (8 Å) between the two entities. The quinone site inhibitor 2-heptyl-4-hydroxyquinoline *N*-oxide (HQNO) was shown to affect the EPR spectrum of the [3Fe-4S] cluster, suggesting that its binding site was also at or near the Q_P site. This suggestion was confirmed with the recent 2.7-Å resolution of the X-ray structure of *E. coli* QFR containing HQNO that binds at the Q_P site (126). HQNO is nearly isosteric with menaquinol and it was found that two hydrogen bond donors, FrdB Lys228 and FrdD Trp14, were positioned within hydrogen bonding distance of the negatively charged *N*-oxide (126). The hydroxyl group on the other side of the HQNO ring, like that for menaquinone, is within hydrogen bonding distance of FrdC Glu29 and FrdD Arg81. Thus, as noted above, replacement of FrdC Glu29 with a neutral amino acid such as leucine apparently disrupts this hydrogen bonding interaction and stabilizes the semiquinone species present at Q_P. Another inhibitor of quinone reactions in *E. coli* QFR, DNP-19 (2-[1-(*p*-chlorophenyl)ethyl]4,6-dinitrophenol) (127), was also found to bind at the Q_P site (126). The finding was interpreted to mean that this inhibitor bound at the Q_P site such that it would sterically prevent quinol binding in agreement with inhibitor studies (126, 127). Importantly, in both the HQNO and DNP-19 structures, quinone was absent at the Q_D site even though electron density had been very strong in the original 3.3-Å-resolution structure determined in the absence of inhibitors. This is reminiscent of the binding of inhibitors at the Q_o site of the cytochrome bc_1 complex where their binding reduces the affinity for quinol at the second Q_o site in the dimer (128).

The structure for the *W. succinogenes* QFR did not contain bound quinones (45), although they are necessary for function with this enzyme. Further study using site-directed mutagenesis approaches has identified a glutamate residue (Glu-C66) that is essential for menaquinol oxidation (100). This residue was chosen for mutagenesis, as it was positioned between two cavities on the periplasmic side of the FrdC subunit and modeling suggested that a hydrogen bond could form from one of the hydroxyl groups of menaquinol to the carboxylate oxygen of Glu-C66. Substitution of Glu-C66 with a Gln residue did not significantly alter the structure or change the redox potential of the two *b* hemes of the *Wolinella* QFR, but it did dramatically alter the ability of the enzyme complex to oxidize menaquinol (100). Thus, it was concluded that this residue probably accepts a proton from the menaquinol during electron transfer to the distal heme b_D (100), a role analogous to that previously proposed for FrdC Glu29 in *E. coli* QFR (120, 129). Mechanistically, however, the different spatial location of these two residues has important implications. In *E. coli* QFR the glutamate residue is localized on the cytoplasmic side, and thus protons liberated during menaquinol oxidation would stay in the cytoplasm. By contrast, in *Wolinella* QFR the glutamate residue is located near the periplasm and the protons released by menaquinol oxidation would likely be delivered there. Thus, it was noted by Lancaster and coworkers (100) that since protons resulting from the reduction of fumarate are taken up from the cytoplasmic side of the membrane, the oxidation of quinol by fumarate should be coupled to the generation of an electrochemical proton gradient across the membrane (100). However, as noted above it is possible that protons are transported back across the membrane so that the net process is not electrogenic (119). In *B. subtilis* SQR a similar residue (SdhC Asp52) has been suggested to be part of the menaquinone reduction site (130). This site would likely be located on the periplasmic side of the membrane and sensitive to the inhibitor HQNO based on EPR studies and their effect on the potential of heme b_L (131). Overall all these studies suggest that the sites for menaquinol oxidation are on different sides of the membrane in *E. coli* and *W. succinogenes* QFR and that HQNO also inhibits *B. subtilis* SQR on the periplasmic side of the membrane (125). As noted above, the *B. subtilis* SQR oxidation of succinate by menaquinone is sensitive to uncouplers and may be driven by the electrochemical proton gradient (53, 117). The topology of the sites for quinone oxidation/reduction in *B. subtilis* SQR and *W. succinogenes* QFR and the presence of two *b* hemes spanning the membrane in these complexes suggest that they both carry out analogous reactions but in different directions (52, 100). Thus, for membrane-spanning, diheme-containing complex IIs, a major difference between those with a single subunit and those containing two subunits may be their coupling to a proton potential across the membrane.

The presence of an acidic amino acid residue (like the Glu and Asp residues discussed above) as part of a quinone binding site is prevalent in respiratory proteins. There is a conserved aspartate in the cytochrome bc_1 complex that is hydrogen bonded to ubiquinone, and quinol oxidase also contains a conserved aspartate (132). Particularly relevant to complex II is that in *Paracoccus denitrificans* SQR a mutation of SdhD Asp88 confers resistance to carboxins, which are specific inhibitors of the reaction of SQR with quinones (133). In eukaryotic complex II there are also conserved aspartate residues that have been localized as part of the quinone binding sites either by site-directed mutagenesis or by chemical labeling with azido-quinones (123, 134).

The role of the quinone found at the Q_D site in *E. coli* QFR remains enigmatic. It does not appear to play an essential structural role, since it is absent in the enzyme containing HQNO (126), which otherwise shows a normal structure. Nevertheless, mutagenesis data on both *E. coli* QFR (120) and *S. cerevisiae* SQR (121) suggest that residues in the region of the hydrophobic Q_D binding site have effects on enzymatic quinone activity. Yeast SQR has recently been reported to contain a *b* heme (135), and it is not clear if the reported mutations (121) affect the properties of the heme. In the case of *E. coli* QFR, the large spatial separation between the Q_P and Q_D quinones (25 Å) makes electron transfer seem unlikely over such a long distance (52). The *b* heme in yeast SQR by contrast might play a role as an intervening redox cofactor to allow electron transfer across the membrane, similar to the case in *B. subtilis* and *W. succinogenes* complex II. It should be stated, however, that no data suggest that yeast SQR is involved in producing a transmembrane proton gradient, nor does it respond to the electrical potential of the membrane. In the case of *E. coli* QFR, available data suggest that electron transfer occurs near the cytoplasmic side of the membrane (108), making it seem even less likely that Q_D plays a role in catalysis with quinones. This question requires further study and it is relevant that an unassigned density in the membrane-spanning domain of *E. coli* QFR has been reported (126). This density, which became more apparent at the higher resolution (2.7 Å) structure, coincides with the only major cavity found in the *E. coli* QFR structure. This factor, termed M, appears about equidistant from the Q_P and Q_D binding sites, and the amino acids lining the cavity where it apparently resides are polar, giving the cavity characteristics of a quinone binding site (126). It remains to be proven what the M factor is and whether it participates in electron transfer in *E. coli* QFR.

ROLE OF THE *b* HEME IN COMPLEX II

It is apparent that the *b* hemes in *W. succinogenes* QFR and *B. subtilis* SQR participate in electron transfer to/from quinone during menaquinol oxidation/reduction, respectively. In *B. subtilis* SQR there is direct evidence that the high-potential heme *b* is reduced by succinate at the same rate as enzyme turnover (130). In *W. succinogenes* QFR the high-potential heme *b* is partially reduced by succinate, whereas it is fully reduced by low-potential menaquinol analogues and reoxidized by fumarate at the same rate as enzyme turnover (136). The axial ligands of the heme seem to be the same in all SQRs so far examined. Measurements using near infrared MCD spectroscopy and EPR have suggested that all of the hemes have a bis-histidine axial ligation (137–139). This is also consistent with what is observed in the *W. succinogenes* QFR structure (45).

The role of the heme in the complex IIs containing only a single *b* heme is, however, not understood. As this is the type found in mammalian mitochondria, it is important to our understanding of how complex II participates in the disease process to understand the function of the single heme in SQR. One well-established fact is that there is a structural role for the *b* heme in those complex IIs containing heme (29, 35). The structural role for the *b* hemes in *B. subtilis* SQR is well studied (29, 35), and using heme biosynthesis mutants in both *B. subtilis* (68) and *E. coli* SQR (65), it was found that the hydrophilic subunits failed to assemble with the membrane subunits unless heme was present. The crystal structure of *W. succinogenes* QFR also shows significant interaction between four of the transmembrane helices and the heme(s), indicating an important role in assembly of that complex (45). The histidyl ligands for the *b* hemes are conserved throughout complex II whether they contain one or two hemes (35, 54). In the *E. coli* SQR containing a single *b* heme, SdhC His84 and SdhD His71 have been identified as the axial heme ligands (66, 140). Mutation of the SdhC His84 residue results in formation of a hexacoordinated low spin heme, but since this mutant is able to bind carbon monoxide, the data suggest that carbon monoxide is able to displace an alternative ligand that presumably replaces the imidazole nitrogen of SdhC His84 (66). Mutation of SdhD His71 has, however, a more severe effect on the enzyme in that the heme becomes high spin and pentacoordinate and its redox potential is lowered by ~100 mV (66). The mutant enzymes are also less stable and the catalytic subunits more easily dissociate from the membrane domain, in agreement with results that heme is important for assembly (35, 54, 65, 68).

Therefore, one reason the role of the *b* heme in single-heme-containing SQR remains enigmatic is that in the bovine SQR the measured redox potential is too low for the heme to be fully reduced by succinate. The *b* heme redox potential for bovine SQR was found to be −185 mV, although this potential was raised somewhat (−144 mV) in the isolated SdhCD peptide fraction (141). The isolated SdhCD-containing fraction, however, was reactive with carbon monoxide, whereas SQR itself is not reactive toward carbon monoxide. This suggests that the isolated heme *b*–containing peptides may have altered their axial ligands such that the heme becomes pentacoordinate and thus reactive with carbon monoxide (141). The heme *b* in bovine SQR when fully reduced is, however, rapidly oxidized by fumarate, suggesting a possible role for the heme in fumarate reduction (139). This implies that the heme may play a role in fumarate reduction but not succinate oxidation in the mammalian enzyme. In contrast to the bovine enzyme, the *E. coli* SQR b_{556} heme, which has a much higher potential (+36 mV), is fully reducible by succinate (142). The rate of the *b* heme reduction is, however, significantly affected by the presence of UQ/UQH_2; reduction is slower than turnover at low concentration of quinone and accelerated by anaerobic conditions and saturating levels of ubiquinone (143). Like the bovine enzyme, the heme in *E. coli* SQR, once it is reduced, can be oxidized by fumarate at the same rate as the turnover of the enzyme. One interpretation of these data is that the [3Fe-4S] cluster directly reduces quinone without participation of the *b* heme, but in reverse electron transfer the *b* heme, which must be in close spatial proximity to the quinone and [3Fe-4S] cluster, becomes part of the electron transfer pathway (108). These data could also be interpreted to suggest that in SQR containing a single *b* heme, the heme is in redox equilibrium with the quinone pool in the membrane. The understanding of the role of the *b* heme in SQR requires further study, as there are intriguing suggestions that it plays more than a structural role in the single *b* heme class of SQR.

RELATION OF COMPLEX II TO DISEASE AND REACTIVE OXYGEN SPECIES

The genes for complex II in mammals are all encoded in the nucleus. The *SDHA* gene has been localized in the human genome to chromosome 5p15; the gene contains 15 exons over 38 kb (144). The *SDHB* gene has been mapped to chromosome 1p35–36.1 and contains eight exons over 40 kb (145). The genes for the small hydrophobic subunits that contain the cytochrome *b* component were mapped to chromosome 1q21 (*SDHC*) and 11q23 (*SDHD*) (146). The *SDHC* gene is composed of six exons covering 35 kb (147), and the *SDHD* gene contains four exons over 18 kb (148).

The availability of human genomic sequence data has generated considerable interest in human complex II gene defects. Although complex II–related diseases are relatively rare, they have been associated with a wide spectrum of clinical phenotypes (149, 150). In fact the first reported nuclear mutation that causes a respiratory chain defect was a mutation in the *SDHA* gene associated with Leigh syndrome (151). This mutation changed human Arg544 to a tryptophan residue and significantly lowered SDH activity. Based on the *E. coli* QFR structure (43), the equivalent arginine in the bacterial enzyme is near the mouth of the substrate entrance channel. This suggests, as one possibility, that the channel is altered or that closure of the capping domain is restricted, affecting substrate entry/exit. Another mutation recently described is alteration of human Arg408 to a cysteine residue, which causes late-onset neurodegenerative disease (91). This mutation results in the loss of the covalent FAD linkage in the flavoprotein subunit of SQR. The loss of covalent FAD linkage results in enzyme that is no longer able to oxidize succinate. This is what is seen in *E. coli* QFR when the equivalent arginine (FrdA Arg390, Figure 3) is mutated.

Mutations in human *SDHB, SDHC*, and *SDHD* genes appear to cause a different clinical phenotype than that found with the *SDHA* mutations (149). Mutations of the iron-sulfur protein or the cytochrome *b* subunits cause a clinical syndrome termed familial paraganglioma and pheochromocytoma (149, 150). This condition is characterized by neural crest–derived tumors of the paraganglia, mostly in the head or neck. Frequently they are localized in highly vascular organs such as the carotid body. A more complete description of the clinical effects in patients carrying these mutations has been described (149, 150, 152). A mutation in human *SDHB* that causes a pheochromocytoma is substitution of human Pro197 by arginine (153). This proline residue is conserved in all QFRs and SQRs and is next to one of the cysteinyl residues involved in ligation of the [3Fe-4S] cluster. In *E. coli* QFR, substitution of this proline by either a histidine or glutamine causes the enzyme to lose succinate oxidase activity in an oxygen-sensitive manner (154). Although the [3Fe-4S] cluster apparently remains intact, the enzyme in the presence of quinones appears to generate reactive oxygen species, which in turn rapidly inactivate the enzyme. It is quite likely that a similar phenomenon would take place in subjects carrying this mutation in SQR.

Although fewer mutations have been located in the human *SDHC* gene, mutations that disrupt its start codon cause autosomal dominant hereditary paraganglioma (155). Analysis suggested that *SDHC* functions as a tumor-suppressor gene, since tumors have lost the wild-type gene (155). Although not reported for humans, an interesting mutation termed *mev-1* has been found in the *SdhC* gene of the nematode *Caenorhabditis elegans* (156). The *mev-1* mutation is a missense mutation of SdhC Gly69 (*C. elegans* numbering) to a glutamic acid residue (156). Nematodes harboring this mutation show a dramatic decrease in life span and also accumulate markers of aging such as protein carbonyls. Further study in the Ishii laboratory has shown that this defect causes increased superoxide production and that complex II containing this mutation is potentially a potent producer of reactive oxygen species (157). It was suggested that the mutated glycine residue was near a quinone binding site (156), and the *E. coli* QFR structure (43) indicates that the Q_P site would be very near this glycine residue, based on amino acid sequence alignment. As an acidic residue at this site seems to be involved in protonation reactions in *E. coli* QFR (125, 126) and may help destabilize the quinone, it seems reasonable that alteration of this site with such a residue could cause alteration of the ability of complex II to produce reactive oxygen species. It is thus quite pertinent that studies from the Imlay laboratory have shown that *E. coli* QFR is a very potent producer of the superoxide anion (158) and recently that the more fully reduced enzyme also produces H_2O_2 (159). These studies also suggested that SQR produces only superoxide (159) and at a much lower level than QFR, although in the case of mutations like that found in *C. elegans* (156) SQR might also produce H_2O_2 like QFR by redistributing electrons more to the FAD cofactor than found in wild-type enzyme. These studies indicated that the primary site of production of superoxide and H_2O_2 was at the FAD cofactor of wild-type QFR. They also showed that wild-type SQR produced much less superoxide and virtually no H_2O_2 (159), in agreement with the known propensity of flavins to react with molecular oxygen (160). Nevertheless, the studies are also consistent with the interpretation that mutant forms of both enzymes could produce superoxide and/or H_2O_2 if the iron-sulfur or quinone reaction sites should become exposed and available to oxygen. It is reasonable to assume that both the flavin and site of mutation would contribute to reactive oxygen species formation in mutant forms of complex II, although it is likely that the primary site is at the FAD cofactor as proposed (159).

The largest number of human complex II mutations linked to disease have been found in *SDHD;* these all contribute to hereditary paraganglioma or pheochromocytoma (149). Some are in the leader sequence of SdhD or are frameshift mutations, so they likely cause assembly defects in complex II. Several of the mutations are interesting for their similarity to mutations made in the bacterial complex II, where their biochemistry has been studied in greater detail (108). For example, the human His102Leu mutation, which contributes to hereditary paraganglioma (161), is equivalent by sequence alignment to SdhD His71 in *E. coli* SQR. This residue is one of the axial heme ligands of SQR, and mutation of the residue lowers the redox potential of the *b* heme by 100 mV and makes the enzyme less stable (66). The enzyme did retain significant activity; however, the lower potential of the heme might contribute to increasing the electron distribution toward the FAD cofactor, which as noted above could be deleterious. Examination of the structure of *E. coli* QFR, although it lacks the *b* heme, suggests that other missense mutations may also affect the environment of the *b* heme. For example, human SdhD Arg70, Pro81, Asp92, Leu95, all reported

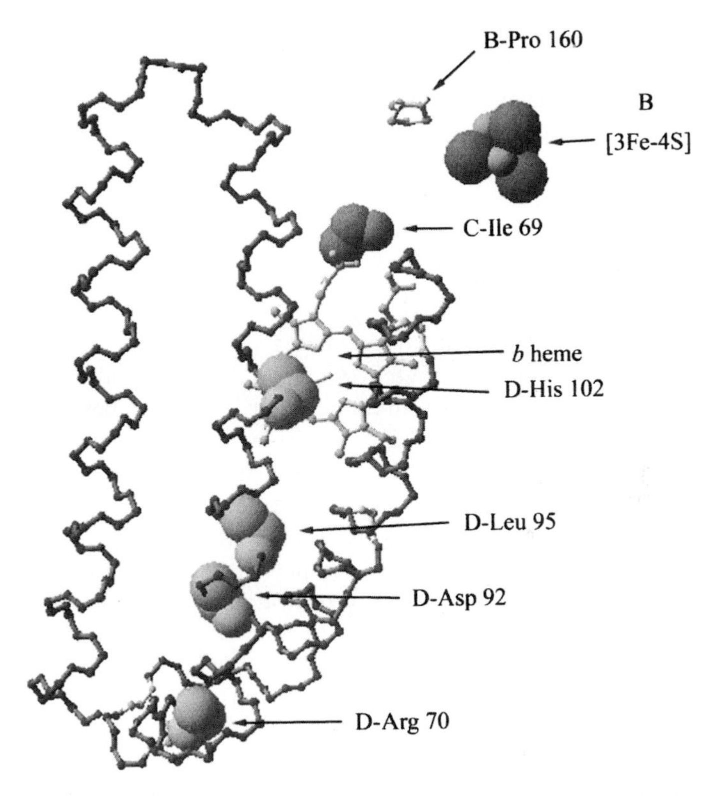

Figure 5 Cartoon of the human SDHD subunit representing hypothetical locations of mutations that contribute to hereditary paraganglioma. The numbered residues are the equivalent human SDHD amino acids known to contribute to hereditary paraganglioma (149, 150). Exceptions are residue C-Ile69, which is the human residue equivalent to the *mev-1* mutation from *C. elegans* that causes premature aging (156), and *E. coli* B-Pro160, which is equivalent to human B-Pro197 associated with hereditary paraganglioma (149). Also shown is the location of the [3Fe-4S] cluster. The diagram was drawn using the *E. coli* FrdD subunit as a template with FrdD Cys77 represented as equivalent to human His102, one of the *b* heme ligands.

human mutations (149), may also reside within the second transmembrane helix of SdhD (equivalent to helix V, in QFR; see Figure 5). Thus mutation might be expected to disrupt the heme environment, which may affect the redox potential of the *b* heme and stability of the enzyme complex (66).

Overall it has been suggested that *SDHB, SDHC,* and *SDHD* act as tumor suppressor genes (150), and it is noted that their mutation mimics chronic hypoxia (161). This led to the suggestion that the gene products could be important components of the oxygen-sensing system and their malfunction would lead to hypoxic stimulation, cellular proliferation, and vascularization (161). There is no direct evidence, however, suggesting that wild-type SQR can respond to oxygen, although it is possible that mutations in the *b* heme could provide an oxygen binding site. Alternatively, the membrane domain could respond to the state of reduction of the quinone pool, which might affect succinate oxidation and quinone reduction by SQR. This is an important question that needs further study.

CONCLUDING REMARKS

Significant advances in our understanding of complex II have come from the available crystal structures of QFR (43, 45). Clearly it would be advantageous to obtain the three-dimensional structure of an SQR member of the complex II family of respiratory proteins, particularly one containing a single *b* heme. Such a structure may help shed light on the fascinating observations of a tunnel diode effect in the enzyme (102) and help us understand the biochemical regulation of complex II. A structure may also provide information on the possible role of the *b* heme cofactor in the function of complex II and on whether it plays a role in regulating the hypoxic response noted in mammals carrying complex II mutations. Like complexes I and III of the respiratory chain, complex II contains quinone binding sites in the membrane domain. Elucidation of the ubiquinone binding site in SQR should aid in defining amino acid residues that may participate in the quinone binding sites in the other respiratory complexes.

As the only membrane-bound enzyme in the Krebs cycle and as part of the mitochondrial electron transport chain, complex II is uniquely positioned to act as a regulator of both metabolic pathways. Although it is the simplest respiratory complex in terms of number of subunits, it does contain multiple redox cofactors, such as complexes I, III, and IV, and it is unique because it is not a proton pump. Nevertheless, although simpler in overall structure, the way electron flow is regulated through the complex is incompletely understood. As with other respiratory complexes, understanding how the redox cofactors and individual subunits are assembled into the holoenzyme may require the discovery of additional protein assembly factors. The mitochondrial respiratory chain, particularly complexes I and III, has long been suggested as a primary source of reactive oxygen species. However, the role of complex II in this activity has been less appreciated. Recent findings in eukaryotic systems have shown that complex II can be a significant source of reactive oxygen species (157, 159). How specific complex II mutations generate reactive oxygen species and how this effects mitochondrial-related diseases are likely to receive increased attention as a result of its documented importance in such processes (149, 159). Bacterial systems will probably continue to be seminal in addressing important questions on the assembly and function of complex II because of the relative ease of biochemical, genetic, and structural studies using them. Complex II continues to fascinate researchers, and its unique physiological role in both the respiratory chain and Krebs cycle makes it an important area to study.

ACKNOWLEDGMENTS

The author gratefully acknowledges funding from the Department of Veterans Affairs and National Institutes of Health (NIH) grant GM61606. The author is also appreciative of past support from the National Science Foundation and the NIH Heart, Lung, and Blood Institute. The author would also like to thank his collaborators for many fruitful discussions and help over the years and Roy Lancaster for communicating results prior to publication. He is indebted to Tina Iverson, Elena Maklashina, and Victoria Yankovskaya for preparing the figures and helpful discussions.

The *Annual Review of Biochemistry* is online at http://biochem.annualreviews.org

LITERATURE CITED

1. Hatefi Y. 1985. *Annu. Rev. Biochem.* 54:1015–69
2. Saraste M. 1999. *Science* 283:1488–93
3. Carroll J, Shannon RJ, Fearnley IM, Runswick MJ, Walker JE, Hirst J. 2002. *J. Biol. Chem.* 277:50311–17
4. Ohnishi T. 1998. *Biochim. Biophys. Acta* 1364:186–206
5. Friedrich T, Scheide D. 2000. *FEBS Lett.* 479:1–5
6. Yano T, Magnitsky S, Ohnishi T. 2000. *Biochim. Biophys. Acta* 1459:299–304
7. Grigorieff N. 1998. *J. Mol. Biol.* 277: 1033–46
8. Guénebaut V, Schlitt A, Weiss H, Friedrich T. 1998. *J. Mol. Biol.* 276: 105–12
9. Böttcher B, Scheide D, Hesterberg M, Nagel-Steger L, Freidrich T. 2002. *J. Biol. Chem.* 277:17970–77
10. Grivennikova VG, Kapustin AN, Vinogradov AD. 2001. *J. Biol. Chem.* 276: 9038–44
11. Mitchell P. 1976. *J. Theor. Biol.* 62: 327–67
12. Trumpower BL, Gennis RB. 1994. *Annu. Rev. Biochem.* 63:675–716
13. Berry EA, Guergova-Kurus M, Huang L, Crofts AR. 2000. *Annu. Rev. Biochem.* 69:1005–75
14. Yoshikawa S, Shinzawa-Itoh K, Tsukihara T. 2000. *J. Inorg. Biochem.* 82:1–7
15. Ludwig B, Bender E, Arnold S, Hüttemann M, Lee I, Kadenbach B. 2001. *Chembiochem.* 2:392–403
16. Abrahams JP, Leslie AGW, Lutter R, Walker JE. 1994. *Nature* 370:621–28
17. Boyer PD. 1997. *Annu. Rev. Biochem.* 66:717–49
18. Lutter R. 1993. *Biochem. J.* 295:799–806
19. Junge W, Pänke O, Cherepanov DA, Gumbiowski K, Müller M, Engelbrecht S. 2001. *FEBS Lett.* 504:152–60
20. Beinert H. 2002. *Biochim. Biophys. Acta* 1553:7–22
21. Hirsch CA, Rasminsky M, Davis BD, Lin ECC. 1963. *J. Biol. Chem.* 238:3770–74
22. Spencer ME, Guest JR. 1973. *J. Gen. Microbiol.* 114:563–70
23. Van Hellemond JJ, Tielens AGM. 1994. *Biochem. J.* 304:321–31
24. Kita K, Hirawake H, Miyadera H, Amino H, Takeo S. 2002. *Biochim. Biophys. Acta* 1553:123–39
25. Guest JR. 1981. *J. Gen. Microbiol.* 122: 171–79
26. Maklashina E, Berthold DA, Cecchini G. 1998. *J. Bacteriol.* 180:5989–96
27. Schäfer G, Anemüller S, Moll R. 2002. *Biochim. Biophys. Acta* 1553:57–73
28. Lauterbach F, Körtner C, Albracht SP,

Unden G, Kröger A. 1990. *Arch. Microbiol.* 154:386–93
29. Hederstedt L. 2002. *Biochim. Biophys. Acta* 1553:74–83
30. Burger G, Lang BF, Reith M, Gray MW. 1996. *Proc. Natl. Acad. Sci. USA* 93:2328–32
31. Schnarrenberger C, Martin W. 2002. *Eur. J. Biochem.* 269:868–83
32. Gest H. 1980. *FEMS Microbiol. Lett.* 7:73–77
33. Wood D, Darlison MG, Wilde RJ, Guest JR. 1984. *Biochem. J.* 222:519–34
34. Ackrell BAC, Johnson MK, Gunsalus RP, Cecchini G. 1992. In *Chemistry and Biochemistry of Flavoenzymes*, ed. F Müller, Vol. 3, pp. 229–97. Boca Raton: CRC
35. Hägerhäll C. 1997. *Biochim. Biophys. Acta* 1320:107–41
36. Hederstedt L. 1999. *Science* 284: 1941–42
37. Morris CJ, Black AC, Pealing SL, Manson FD, Chapman SK, et al. 1994. *Biochem. J.* 302:587–93
38. Muratsubaki M, Katsume T. 1985. *J. Biochem.* 97:1201–9
39. Besteiro S, Biran M, Biteau N, Coustou V, Baltz T, et al. 2002. *J. Biol. Chem.* 277:38001–12
40. Tisdale H, Hauber J, Prager G, Turini P, Singer TP. 1968. *Eur. J. Biochem.* 4:472–77
41. Blaut M, Whittaker K, Valdovinos A, Ackrell BAC, Gunsalus RP, Cecchini G. 1989. *J. Biol. Chem.* 264:13599–604
42. Robinson KM, Rothery RA, Weiner JH, Lemire BD. 1994. *Eur. J. Biochem.* 222: 983–90
43. Iverson T, Luna-Chavez C, Cecchini G, Rees DC. 1999. *Science* 284:1961–66
44. Mattevi A, Tedeschi G, Bacchella L, Coda A, Negri A, Ronchi S. 1999. *Struct. Fold. Des.* 7:745–56
45. Lancaster CRD, Kröger A, Auer M, Michel H. 1999. *Nature* 402:377–85
46. Bamford V, Dobbin PS, Richardson DJ, Hemming AM. 1999. *Nat. Struct. Biol.* 6:1104–7
47. Taylor P, Pealing SL, Reid GA, Chapman SK, Walkinshaw MD. 1999. *Nat. Struct. Biol.* 6:1108–12
48. Leys D, Tsapin AS, Nealson KH, Meyer TE, Cusanovich MA, Van Beeumn JJ. 1999. *Nat. Struct. Biol.* 6:1113–17
49. Singer TP, Johnson MK. 1985. *FEBS Lett.* 190:189–98
50. Ackrell BAC. 2000. *FEBS Lett.* 466:1–5
51. Tedeschi G, Negri A, Mortarino M, Ceciliani F, Simonic T, et al. 1996. *Eur. J. Biochem.* 239:427–33
52. Ohnishi T, Moser CC, Page CC, Dutton PL, Yano T. 2000. *Struct. Fold. Des.* 8:23–32
53. Schirawski J, Unden G. 1998. *Eur. J. Biochem.* 257:210–15
54. Hägerhäll C, Hederstedt L. 1996. *FEBS Lett.* 389:25–31
55. Lancaster CRD. 2002. *Biochim. Biophys. Acta* 1553:1–6
56. Törnroth S, Yankovskaya V, Cecchini G, Iwata S. 2002. *Biochim. Biophys. Acta* 1553:171–76
57. Iverson TM, Luna-Chavez C, Schröder I, Cecchini G, Rees DC. 2000. *Curr. Opin. Struct. Biol.* 10:448–55
58. Lancaster CRD, Kröger A. 2000. *Biochim. Biophys Acta* 1459:422–31
59. Deleted in proof
60. Ackrell BAC, Kearney EB, Singer TP. 1977. In *Structure and Function of Energy-Transducing Membranes,* ed. K van Dam, BF Van Gelder, pp. 37–48. New York: Elsevier
61. Ackrell BAC, Ball MB, Kearney EB. 1980. *J. Biol. Chem.* 255:2761–69
62. Cecchini G, Ackrell BAC, Deshler JO, Gunsalus RP. 1986. *J. Biol. Chem.* 261: 1808–14
63. Yu L, Yu C-A. 1981. *Biochim. Biophys. Acta* 637:383–86
64. Choudhry ZM, Gavrikova EV, Kotlyar AB, Tushurashvili PR, Vinogradov AD. 1985. *FEBS Lett.* 182:171–75
65. Nakamura K, Yamaki M, Sarada M,

Natayama S, Vibat CRT, et al. 1996. *J. Biol. Chem.* 271:521–27
66. Maklashina E, Rothery RA, Weiner JH, Cecchini G. 2001. *J. Biol. Chem.* 276: 18968–76
67. Kröger A, Biel S, Simon J, Gross R, Unden G, Lancaster CRD. 2002. *Biochim. Biophys. Acta* 1553:23–38
68. Hederstedt L, Rutberg L. 1980. *J. Bacteriol.* 144:941–51
69. Brasseur G, Tron G, Dujardin G, Slonimski PP, Brivet-Chevillotte P. 1997. *Eur. J. Biochem.* 246:103–11
70. Dibrov E, Fu S, Lemire BD. 1998. *J. Biol. Chem.* 273:32042–48
71. Klanner C, Neupert W, Langer T. 2000. *FEBS Lett.* 470:365–69
72. Lill R, Kispal G. 2000. *Trends Biochem. Sci.* 25:352–56
73. Li J, Kogan M, Knight SA, Pain D, Dancis A. 1999. *J. Biol. Chem.* 274: 33025–34
74. Ackrell BAC, Maguire JJ, Dallman P, Kearney EB. 1984. *J. Biol. Chem.* 259: 10053–59
75. Gray NK, Pantopoulos K, Dandekar T, Ackrell BAC, Hentze MW. 1996. *Proc. Natl. Acad. Sci. USA* 93:4925–30
76. Tzagoloff A, Dieckmann CL. 1990. *Microbiol. Rev.* 54:211–25
77. Kearney EB. 1960. *J. Biol. Chem.* 235: 865–77
78. Singer TP, Kearney EB, Massey V. 1956. *Arch. Biochem. Biophys.* 60: 255–57
79. Walker WH, Singer TP. 1970. *J. Biol. Chem.* 245:4424–25
80. Kenney WC, Walker WH, Singer TP. 1972. *J. Biol. Chem.* 247:4510–13
81. Mewies M, McIntire WS, Scrutton NS. 1998. *Protein Sci.* 7:7–20
82. Robinson KM, Lemire BD. 1996. *J. Biol. Chem.* 271:4055–60
83. Edmondson DE, Newton-Vinson P. 2001. *Antioxid. Redox Signal.* 3:789–806
84. Engst S, Kuusk V, Efimov I, Cronin CN, McIntire WS. 1999. *Biochemistry* 38:16620–28
85. Efimov I, Cronin CN, McIntire WS. 2001. *Biochemistry* 40:2155–66
86. Kim J, Fuller JH, Kuusk V, Cunane L, Chen Z, Mathews FS, McIntire WS. 1995. *J. Biol. Chem.* 270:31202–9
87. Frost JW, Rastetter WH. 1980. *J. Am. Chem. Soc.* 102:7157–59
88. Walsh C. 1980. *Acc. Chem. Res.* 13:148–55
89. Robinson KM, Lemire BD. 1996. *J. Biol. Chem.* 271:4061–67
90. Schröder I, Gunsalus RP, Ackrell BAC, Cochran B, Cecchini G. 1991. *J. Biol. Chem.* 266:13572–79
91. Birch-Machin MA, Taylor RW, Cochran B, Ackrell BAC, Turnbull DM. 2000. *Ann. Neurol.* 48:330–35
92. Fraaije MW, van den Heuvel RH, van Berkel WJ, Mattevi A. 2000. *J. Biol. Chem.* 275:38654–58
93. Heffron K. 2001. *Studies of the redox and catalytic properties of the anaerobic respiratory enzymes of Escherichia coli.* PhD thesis. Oxford Univ., Oxford, UK. 256 pp.
94. Léger C, Heffron K, Pershad HR, Maklashina E, Luna-Chavez C, et al. 2001. *Biochemistry* 40:11234–45
95. Bossi RT, Negri A, Tedeschi G, Mattevi A. 2002. *Biochemistry* 41:3018–24
96. Turner KL, Doherty MK, Heering HA, Armstrong FA, Reid GA, Chapman SK. 1999. *Biochemistry* 38:3302–9
97. Doherty MK, Pealing SL, Miles CS, Moysey R, Taylor P, et al. 2000. *Biochemistry* 39:10695–701
98. Pankhurst KL, Mowat CG, Miles CS, Leys D, Walkinshaw MD, et al. 2002. *Biochemistry* 41:8551–56
99. Mowat CG, Pankhurst KL, Miles CS, Leys D, Walkinshaw MD, et al. 2002. *Biochemistry* 41:11990–96
100. Lancaster CRD, Groß R, Haas A, Ritter M, Mäntele W, et al. 2000. *Proc. Natl. Acad. Sci. USA* 97:13051–56
101. Reid GA, Miles CS, Moysey RK, Pan-

khurst KL, Chapman SK. 2000. *Biochim. Biophys. Acta* 1459:310–15
102. Sucheta A, Ackrell BAC, Cochran B, Armstrong FA. 1992. *Nature* 356:361–62
103. Hirst J, Ackrell BAC, Armstrong FA. 1997. *J. Am. Chem. Soc.* 119:7434–38
104. Pershad HR, Hirst J, Cochran B, Ackrell BAC, Armstrong FA. 1999. *Biochim. Biophys. Acta* 1412:262–72
105. Ackrell BAC, Armstrong FA, Cochran B, Sucheta A, Yu T. 1993. *FEBS Lett.* 326:92–94
106. Hirst J, Sucheta A, Ackrell BAC, Armstrong FA. 1996. *J. Am. Chem. Soc.* 118:5031–38
107. Eliot SJ, Léger C, Pershad HR, Hirst J, Heffron K, et al. 2002. *Biochim. Biophys. Acta* 1555:54–59
108. Cecchini G, Schröder I, Gunsalus RP, Maklashina E. 2002. *Biochim. Biophys. Acta* 1553:140–57
109. Kowal AT, Werth MT, Manodori A, Cecchini G, Schröder I, et al. 1995. *Biochemistry* 34:12284–93
110. Sucheta A, Cammack R, Weiner JH, Armstrong FA. 1993. *Biochemistry* 32:5455–65
111. Ohnishi T, King TE, Salerno JC, Blum H, Bowyer JR, Maida T. 1981. *J. Biol. Chem.* 256:5577–82
112. Thauer RK, Jungermann K, Decker K. 1977. *Microbiol. Rev.* 41:100–80
113. Condon C, Cammack R, Patil DS, Owen P. 1985. *J. Biol. Chem.* 260:9427–34
114. Darlison MG, Guest JR. 1984. *Biochem. J.* 223:507–17
115. Werth MT, Sices H, Cecchini G, Schröder I, Lasage S, et al. 1992. *FEBS Lett.* 299:1–4
116. Schultz BE, Chan SI. 2001. *Annu. Rev. Biophys. Biomol. Struct.* 30:23–65
117. Schnorpfeil M, Janausch IG, Biel S, Kröger A, Unden G. 2001. *Eur. J. Biochem.* 268:3069–74
118. Lancaster CRD, Simon J. 2002. *Biochim. Biophys. Acta* 1553:84–101
119. Lancaster CRD. 2002. *Biochim. Biophys. Acta.* 1565:215–31
120. Westenberg DJ, Gunsalus RP, Ackrell BAC, Sices H, Cecchini G. 1993. *J. Biol. Chem.* 268:815–22
121. Oyedotun KS, Lemire BD. 2001. *J. Biol. Chem.* 276:16936–43
122. Lee GY, He DY, Yu L, Yu C-A. 1995. *J. Biol. Chem.* 270:6193–98
123. Shenoy SK, Yu L, Yu C-A. 1997. *J. Biol. Chem.* 272:17867–72
124. Ruzicka FJ, Beinert H, Schepler KL, Dunham WR, Sands RH. 1975. *Proc. Natl. Acad. Sci. USA* 72:2886–90
125. Hägerhäll C, Magnitsky S, Sled VD, Schröder I, Gunsalus RP, et al. 1999. *J. Biol. Chem.* 274:26157–64
126. Iverson TM, Luna-Chavez C, Croal LR, Cecchini G, Rees DC. 2002. *J. Biol. Chem.* 277:16124–30
127. Yankovskaya V, Sablin SO, Ramsay RR, Singer TP, Ackrell BAC, et al. 1996. *J. Biol. Chem.* 271:21020–24
128. Gutierrez-Cirlos EB, Trumpower BL. 2002. *J. Biol. Chem.* 277:1195–202
129. Westenberg DJ, Gunsalus RP, Ackrell BAC, Cecchini G. 1990. *J. Biol. Chem.* 265:19560–67
130. Matsson M, Tolstoy D, Aasa R, Hederstedt L. 2000. *Biochemistry* 39:8617–24
131. Smirnova I, Hägerhäll C, Konstantinov A, Hederstedt L. 1995. *FEBS Lett.* 359: 23–26
132. Abramson J, Riistama S, Larsson G, Jasaitis A, Svensson-Ek M, et al. 2000. *Nat. Struct. Biol.* 7:910–17
133. Matsson M, Hederstedt L. 2001. *J. Bioenerg. Biomembr.* 33:99–105
134. Shenoy SK, Yu L, Yu C-A. 1999. *J. Biol. Chem.* 274:8717–22
135. Oyedotun KS, Lemire BD. 1999. *FEBS Lett.* 442:203–7
136. Unden G, Albracht SPJ, Kröger A. 1984. *Biochim. Biophys. Acta* 767:460–69
137. Friden H, Cheesman MR, Hederstedt L, Anderson KK, Thomson AJ. 1990. *Biochim. Biophys. Acta* 1041:207–15
138. Crouse BR, Yu C-A, Yu L, Johnson MK. 1995. *FEBS Lett.* 367:1–4

139. Peterson J, Vibat CRT, Gennis RB. 1994. *FEBS Lett.* 355:155–56
140. Vibat CRT, Cecchini G, Nakamura K, Kita K, Gennis RB. 1998. *Biochemistry* 37:4148–59
141. Yu L, Xu J-X, Haley PE, Yu C-A. 1987. *J. Biol. Chem.* 262:1137–43
142. Kita K, Vibat CRT, Meinhardt S, Guest JR, Gennis RB. 1989. *J. Biol. Chem.* 264:2672–77
143. Maklashina E, Cecchini G. 1998. *Biochim. Biophys. Acta EBEC Short Rep.* 10:216
144. Parfait B, Chretien D, Rötig A, Marsac C, Munnich A, Rustin P. 2000. *Hum. Genet.* 106:236–43
145. Au HC, Ream-Robinson D, Bellew LA, Broomfield PL, Saghbini M, Scheffler IE. 1995. *Gene* 159:249–53
146. Hirawake H, Taniwaki M, Tamura A, Kojima S, Kita K. 1997. *Cytogenet. Cell Genet.* 79:132–38
147. Elbehti-Green A, Au HC, Mascarello JT, Ream-Robinson D, Scheffler IE. 1998. *Gene* 213:133–40
148. Hirawake H, Taniwaki M, Tamura A, Amino H, Tomitsuka E, Kita K. 1999. *Biochim. Biophys. Acta* 1412:295–300
149. Baysal BE, Rubinstein WS, Taschner PEM. 2001. *J. Mol. Med.* 79:495–503
150. Rustin P, Rötig A. 2002. *Biochim. Biophys. Acta* 1553:117–22
151. Bourgeron T, Rustin P, Chretien D, Birch-Machin M, Bourgeois M, et al. 1995. *Nat. Genet.* 11:144–49
152. Dluhy RG. 2002. *N. Engl. J. Med.* 346: 1486–88
153. Astuti D, Latif F, Dallol A, Dahia PL, Douglas F, et al. 2001. *Am. J. Hum. Genet.* 69:49–54
154. Cecchini G, Sices H, Schröder I, Gunsalus RP. 1995. *J. Bacteriol.* 177:4587–92
155. Niemann S, Müller U. 2000. *Nat. Genet.* 26:268–70
156. Ishii N, Fujii M, Hartman PS, Tsuda M, Yasuda K, et al. 1998. *Nature* 394: 694–97
157. Senoo-Matsuda N, Yasuda K, Tsuda M, Ohkubo T, Yoshimura S, et al. 2001. *J. Biol. Chem.* 276:41553–58
158. Imlay JA. 1995. *J. Biol. Chem.* 270: 19767–77
159. Messner KR, Imlay JA. 2002. *J. Biol. Chem.* 277:42563–71
160. Massey V. 1994. *J. Biol. Chem.* 269: 22459–62
161. Baysal BE, Ferrell RE, Willett-Brozick JE, Lawrence EC, Myssiorek D, et al. 2000. *Science* 287:848–51

Annu. Rev. Biochem. 2003. 72:111–135
doi: 10.1146/annurev.biochem.72.121801.161459
Copyright © 2003 by Annual Reviews. All rights reserved
First published online as a Review in Advance on January 9, 2003

PROTEIN DISULFIDE BOND FORMATION IN PROKARYOTES

Hiroshi Kadokura, Federico Katzen, and Jon Beckwith
Department of Microbiology and Molecular Genetics, Harvard Medical School, Boston, Massachusetts 02115, email: hiroshi_kadokura@hms.harvard.edu; fkatzen@hms.harvard.edu; jbeckwith@hms.harvard.edu

Key Words protein folding, *Escherichia coli*, periplasm, thioredoxin, electron transfer

■ **Abstract** Disulfide bonds formed between pairs of cysteines are important features of the structure of many proteins. Elaborate electron transfer pathways have evolved *Escherichia coli* to promote the formation of these covalent bonds and to ensure that the correct pairs of cysteines are joined in the final folded protein. These transfers of electrons consist, in the main, of cascades of disulfide bond formation or reduction steps between a series of proteins (DsbA, DsbB, DsbC, and DsbD). A surprising variety of mechanisms and protein structures are involved in carrying out these steps.

CONTENTS

INTRODUCTION

Covalent linkages of amino acids in proteins are largely limited to the peptide bond. The most common exception to this rule is the disulfide bond, a sulfur-sulfur chemical bond that results from an oxidative process that links nonadjacent (in most cases) cysteines of a protein. Proteins that contain disulfide bonds can be divided into two classes: those in which the cysteine-cysteine linkage is a stable part of their final folded structure and those in which pairs of cysteines alternate between the reduced and oxidized states. For the first class, the disulfide bond may contribute to the folding pathway of the protein and to the stability of its native state. For the second, the oxidative-reductive cycling of the disulfide bond may be central to a protein's activity as an enzyme (e.g., certain ribonucleotide reductases) or may be involved in a protein's activation and deactivation (see, e.g., OxyR).

It has long been noted that proteins containing stable disulfide bonds are rarely found in the cytoplasmic compartments of any organism. Instead, in bacteria, they are usually located in extracytoplasmic compartments or secreted into the media and, in eukaryotic cells, in compartments such as the endoplasmic reticulum and the plasma membrane or secreted into the external milieu. The restriction of stably disulfide bonded proteins to noncytoplasmic environments has been attributed to the contrasting reductive nature of the cytoplasm and the oxidative nature of certain other compartments and the external milieu. A recent exception to this rule is the discovery of large numbers of disulfide-bonded proteins in the cytoplasm of certain archaea (1).

Based on in vitro studies on protein folding, it appeared that simply the presence of oxygen or another strong oxidant (e.g., oxidized glutathione) was sufficient to promote disulfide bond formation in proteins in vivo. However, this is not the case. Genetic studies in bacteria and yeast revealed that enzymatic systems in extracytoplasmic compartments are necessary for the efficient formation of disulfide bonds in proteins. In the absence of such systems, the formation of cysteine-cysteine linkages is extremely slow. Similarly, the absence of disulfide-bonded proteins in the cytoplasm cannot be attributed solely to the generally reducing environment of this compartment. It is possible to observe protein disulfide bond formation in the cytoplasm of *E. coli* without changing the overall reductive nature of this compartment by mutationally altering certain electron transfer pathways (2). These alterations lead to the accumulation of oxidized thioredoxins, which then act as catalysts of disulfide bond formation.

Proteins that are capable of catalyzing protein disulfide bond formation are members of a large collection of thiol-disulfide oxidoreductases found in all living cells. Many of these enzymes belong to the thioredoxin superfamily, which is defined by an active site containing a CXXC motif (cysteines separated by two amino acids) and by a thioredoxin fold seen in the three-dimensional structure of the prototypical thioredoxin 1 of *E. coli.* While in extracytoplasmic compartments these proteins can act as oxidants, those located in the cytoplasm perform mainly reductive steps. One of the most important of the cytoplasmic activities for many bacteria is the reduction of ribonucleotide reductase (an essential enzyme that converts ribonucleotides to deoxyribonucleotides) by thioredoxins and glutaredoxins.

Increasingly, enzymes are being identified that are not members of the thioredoxin superfamily but that use redox active cysteines in transferring electrons in oxidative and reductive pathways. Typically, these show entirely different three-dimensional structures from thioredoxin and use redox active cysteine pairs that are separated by more than two amino acids. In addition, these enzymes may use small molecule electron donor and receptor cofactors, such as FAD, NADPH, NADH, quinones, and lipoic acid (3). Unlike the thioredoxin family, the structures of this latter group of proteins do not fall into a single or even a few classes. Many of these nonthioredoxin-like enzymes themselves receive from or donate electrons to proteins belonging to the thioredoxin-like class. For example, the protein DsbB does not contain a thioredoxin fold, but it transfers electrons via two pairs of redox active cysteines from the thioredoxin-like oxidant DsbA to quinones (see below). So far, those enzymes whose major role is to form disulfide bonds in extracytoplasmic proteins are members of the thioredoxin superfamily and not of this other class. Similarly, most of those cytoplasmic reductants that carry out the last step in a pathway of electron transfer, such as the reduction of ribonucleotide reductase, are also members of the thioredoxin superfamily.

In this review, we focus on the enzymes of prokaryotes that are responsible for the formation of disulfide bonds in proteins and for assuring that the disulfide bonds formed appear in the final folded protein product. The need for and the very existence of the enzymes for disulfide bond formation were not recognized until a little more than ten years ago. It was then that genetic studies in a number of laboratories began to reveal the complex pathways of electron transfer that result in these important steps in protein folding. The studies in prokaryotes have, in turn, catalyzed renewed investigations of disulfide bond formation in eukaryotes from yeast to mammals. The studies in higher organisms have led to the discovery of systems comparable to those found in bacteria.

THE PATHWAY FOR OXIDATIVE PROTEIN FOLDING

DsbA, A Primary Catalyst of Disulfide Bond Formation in Bacteria

In the 1960s, Anfinsen and his coworkers showed that disulfide bond formation in proteins can occur spontaneously in vitro. Using reduced and denatured ribonuclease A as a model protein, they found that molecular oxygen was capable of promoting disulfide bond formation (4). These studies led to the long-held assumption that disulfide bond formation was a spontaneous process in vivo. However, efficient disulfide bond formation in vitro of proteins that contain disulfide bonds requires hours or even days of incubation, while disulfide bond formation in the cell is a much more rapid process, which occurs within seconds or minutes after synthesis of proteins. It was not until thirty years later that discovery of mutations in the gene *dsbA* of *E. coli* and *Vibrio cholerae* revealed that disulfide bond formation is, in fact, a catalyzed process in vivo (5–7).

DsbA is a primary oxidant of proteins exported to the cell envelope of *E. coli*. Consistent with its crucial role in the folding of cell envelope proteins, *dsbA* mutants exhibit pleiotropic phenotypes. In these mutants, many cell envelope proteins such as alkaline phosphatase and the outer membrane protein OmpA are rapidly degraded due to the lack of disulfide bonds necessary for their stability (5). They do not exhibit motility because of the lack of proper assembly of the flagellar motor (8) and are hypersensitive to the reductant dithiothreitol, benzylpenicillin, and metals (9, 10). In addition, *dsbA* mutants in many pathogenic bacteria are avirulent since virulence components such as pili and toxins contain disulfide bonds (7).

DsbA is a monomeric, 21-kDa periplasmic enzyme containing a redox active sequence, Cys30-Pro31-His32-Cys33, embedded in thioredoxin-like fold (11). These two cysteines are found in the disulfide-bonded form in vivo (12). DsbA rapidly oxidizes proteins secreted into the periplasm by donating its disulfide to pairs of cysteines in substrate proteins by a thiol-disulfide exchange reaction (see Figure 1 oxido-reduction reaction and also Figure 2a) (13). Inserted into the thioredoxin domain of DsbA is an α-helical domain (11). Based on structural modeling, it is suggested that this α-helical domain may act to prevent the unwanted interaction between DsbA and DsbD (see below DsbD section) (14).

Among all known thiol-disulfide oxidoreductases, DsbA, with a standard redox potential of ~ -120 mV (15–17), is the strongest thiol oxidant, capable of promoting disulfide bond formation extremely rapidly in vivo and in vitro. The ability of DsbA to act as a strong oxidant derives in part from its biophysical properties. The first cysteine residue of the active site (Cys30) has an unusually low pKa value of about 3.5 (18). Cys30 is thus entirely in the thiolate anion state at physiological pH. This characteristic renders the oxidized form of this protein less stable and more reactive than its reduced form (15, 17, 19), which provides a thermodynamic drive to promote the disulfide transfer from DsbA to folding protein. The three-dimensional structures of the reduced and oxidized DsbA explain how the thiolate ion of Cys30 of DsbA is stabilized (20, 21). First, this residue is located at the N terminus of the active site helix $\alpha 1$. Thus, the partial positive charge from the dipole of the helix $\alpha 1$ can stabilize the thiolate. Second, the second residue in the dipeptide between the active-site cysteines in the CXXC motif, His32, is hydrogen-bonded to Cys30 in the reduced but not in the oxidized DsbA (20, 21).

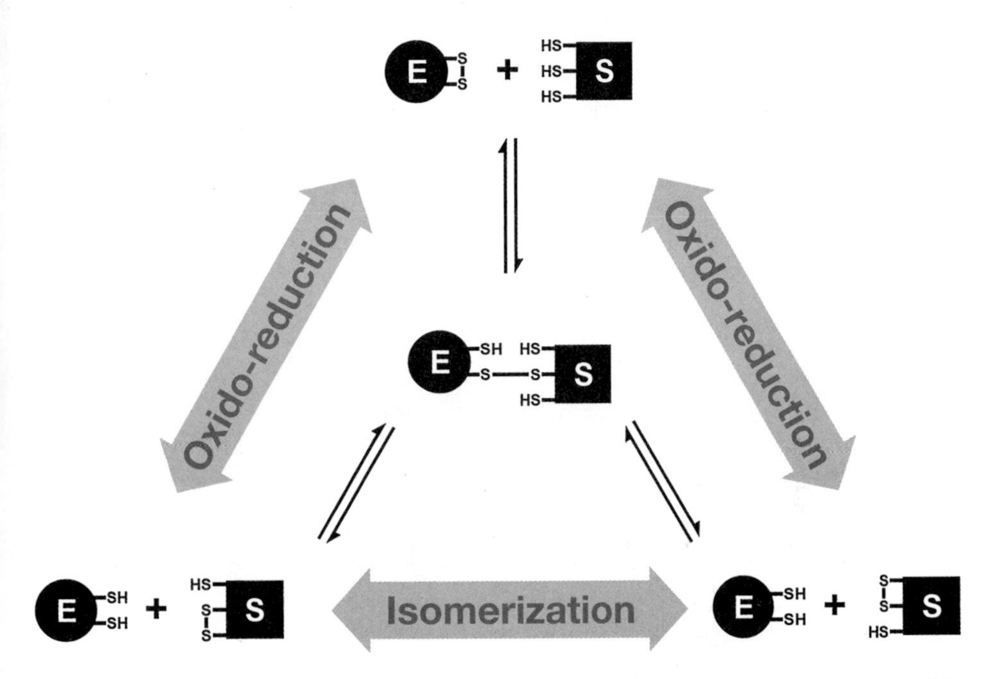

Figure 1 Thiol-redox reactions. The figure shows the three possible thiol-disulfide interconversions between an enzyme (E) with two active cysteine residues and a substrate (S) with three cysteine residues. The central complex represents one of the six possible mixed disulfide intermediates. This figure illustrates the commonality of mechanism for the three series of reactions. Because the mixed disulfide intemediate is formed by the attack of the disulfide by a thiolate anion, the thiol of the attacking cysteine must be depronated (not shown).

DsbA catalyzes disulfide bond formation in a wide variety of proteins that are localized to the bacterial cell envelope. How is it possible for this enzyme to recognize so many different substrates? Recent studies probing for the protease accessible sites on the surface of DsbA revealed that the oxidized form is more flexible than the reduced form. This finding has led to the proposal that greater flexibility of the oxidized form might facilitate the accommodation of its various substrates, while the rigid nature of the reduced form might facilitate the release of the oxidized products (22).

Despite the broadness of its substrate specificity, DsbA apparently avoids interaction with a specific set of proteins containing potentially oxidizable cysteines (e.g., DsbC and DsbD; see below). It is suggested that a relatively deep hydrophobic groove running alongside the active site of DsbA might participate in the binding of peptides (23). Additionally, NMR analysis of interactions between DsbA and a model substrate suggests that DsbA binds peptides via hydrophobic interactions (24). Thus, only unfolded or partially folded proteins might be able to serve as substrates of DsbA. This may provide a partial

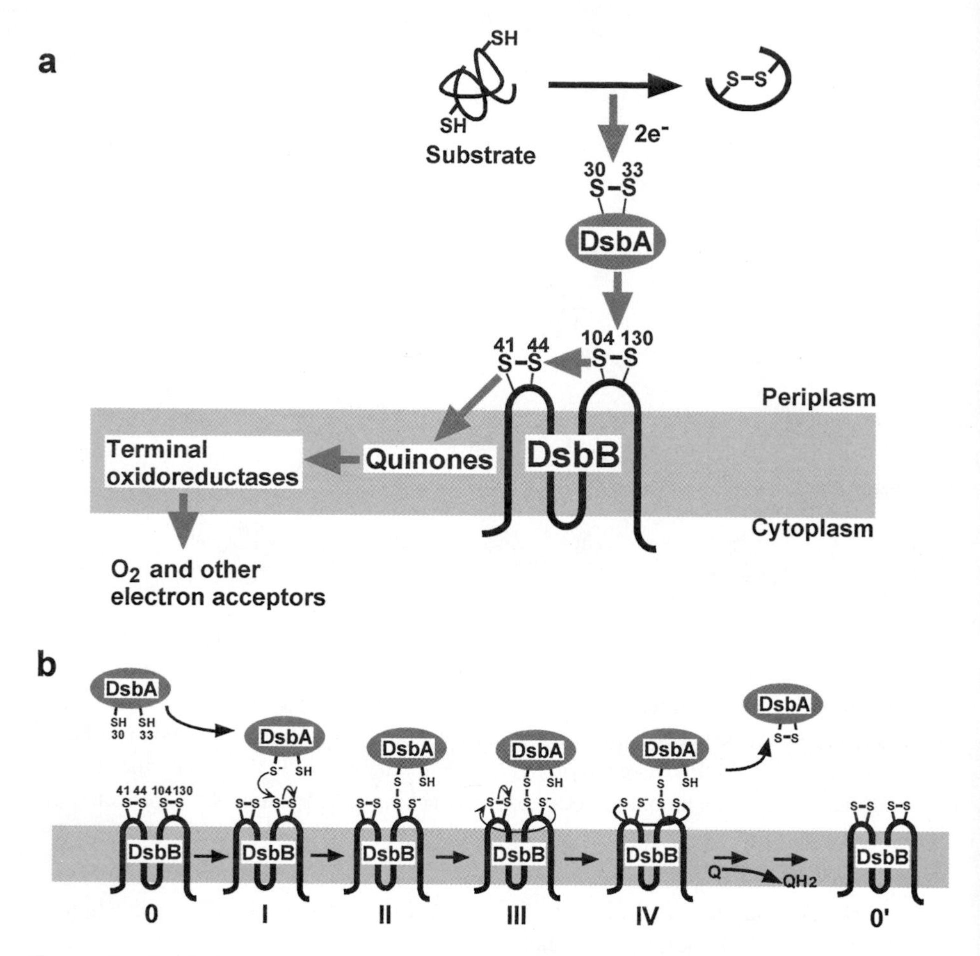

Figure 2 Oxidative pathway for protein disulfide bond formation. (*a*) Electron transfer from substrate to the final electron acceptors via DsbA and DsbB. Direction of electron flow is indicated by gray arrows. (*b*) A model for early steps in the reoxidation of DsbA by DsbB (39). Deprotonated Cys30 of DsbA attacks the Cys104-Cys130 (stage I) to form the DsbA-DsbB complex (stage II), which transfers one electron from DsbA to the second periplasmic domain of DsbB. This is immediately followed by the attack on Cys41-Cys44 by the reduced Cys130 (stage III), which transfers one electron from the second periplasmic domain to the first periplasmic domain of DsbB. This interdomain electron transfer occurs prior to the movement of the second electron from DsbA to DsbB and causes the formation of the ternary complex (stage IV). In the next step, the reduced Cys44 in the ternary complex might further attack quinone so as to transfer one electron from DsbB to quinone (not shown). This reaction may then trigger the release of fully oxidized DsbA from the DsbA-DsbB complex (39).

explanation for the very slow electron transfer in vitro between oxidized DsbA and reduced periplasmic proteins such as DsbC (25) and DsbD (26) (see below for further explanation).

When a protein is being translocated across the cytoplasmic membrane, how does DsbA, on the other side of that membrane, determine with which cysteines it will interact? It is generally assumed that DsbA acts on each cysteine of a substrate rather randomly, based on the finding that incorrect disulfide bonds are often formed and must be corrected (see below). However, recent studies suggest that this might not always be the case. Elastase of *Pseudomonas aeroginosa* is synthesized as a preproenzyme. Disulfide bond formation in this enzyme in vivo appears to occur in three steps (27). First, DsbA catalyzes the formation of a C-terminal disulfide bond in the proenzyme. This step is essential for the subsequent autocatalytic processing of its propeptide. Finally, DsbA catalyzes N-terminal disulfide bond formation before translocation of the mature enzyme across the outer membrane. Thus, disulfide bond formation by DsbA in vivo could be an ordered process (27), guided by structural changes of the folding protein.

DsbB, The Enzyme Responsible for Maintaining DsbA Oxidized

As the result of the transfer of the disulfide bond of DsbA to a substrate protein, the two cysteines in the active site of DsbA become reduced. The fact that DsbA is found exclusively in the oxidized form in normal cells indicates that a mechanism exists for its reoxidation (12). This oxidation step is carried out by a cytoplasmic membrane protein DsbB (Figure 2a). DsbB was discovered by three groups independently through different genetic screens and selections (8, 9, 28). Mutations of the *dsbB* gene cause the accumulation of DsbA in its reduced form and exhibit the same pleiotropic phenotype as *dsbA* mutants (8, 9, 28).

DsbB restores the disulfide bond to DsbA by using the oxidizing power of the electron transport chain (29, 30) (Figure 2a). Aerobically, DsbB transfers two electrons from DsbA onto oxidized ubiquinone (30, 31). The reduced ubiquinone then is reoxidized by the terminal cytochrome oxidases *bd* and *bo*, which finally transfer electrons onto oxygen. Anaerobically, DsbB passes electrons from DsbA onto menaquinone (the major anaerobic quinone) (30). Electrons are then transferred from menaquinone to final electron acceptors other than oxygen. That quinones are the electron acceptors for DsbB explains why the oxidative process of disulfide bond formation still occurs under anaerobic conditions.

DsbB (20 kDa) spans the cytoplasmic membrane four times with both N and C termini facing the cytoplasm (32). Each periplasmic domain contains one pair of essential cysteines: Cys41 and Cys44 in the first periplasmic domain and Cys104 and Cys130 in the second periplasmic domain (32). These pairs of cysteines are disulfide bonded in vivo and redox active (33). Even though the first pair of cysteines is arranged in a CXXC sequence, DsbB probably does not have a thioredoxin-like fold because the periplasmic domains of DsbB are very short and DsbB shares no other sequence homology with thioredoxins.

As the oxidative power leading to disulfide formation in all cell envelope proteins appears to pass through DsbB (28), uncovering the mechanism of this protein is important for understanding disulfide bond formation in vivo. Genetic and biochemical evidence has provided insights into the mechanism of electron transfer from the reduced DsbA to the oxidized quinone through the active-site cysteines of DsbB. First, the finding of a mixed disulfide complex involving Cys30 of DsbA and Cys104 of DsbB suggests that the reduced DsbA probably interacts directly with the Cys104-Cys130 disulfide of DsbB (34, 35). This reaction will result in transfer of one electron from DsbA to DsbB, which renders Cys130 reduced. Second, the Cys41-Cys44 disulfide probably acts as the acceptor of electrons from the Cys104-Cys130 cysteine pair. This direction of electron transfer has been proposed because both in vivo and in vitro the oxidation of the Cys104-Cys130 pair requires the Cys41-Cys44 pair, whereas the Cys41-Cys44 pair is still oxidized in the absence of the Cys104-Cys130 pair in vivo (33, 36). Finally, quinones are the direct recipient of electrons from the Cys41-Cys44 pair; the oxidation of this cysteine pair requires the function of the respiratory components in vivo (33) and only ubiquinone in vitro (36). Thus, electrons from DsbA are presumed to flow first to Cys104-Cys130, then to Cys41-Cys44, and finally to quinones in the respiratory chain (Figure 2a).

Previously, it had been presumed that DsbA was oxidized by a simple thiol-disulfide exchange reaction that occurred between the Cys30-Cys33 pair of DsbA and the Cys104-Cys130 disulfide of DsbB (37, 38). However, recent results indicate that this may not be the case. Kadokura and Beckwith (39) reconstituted DsbB activity by coexpressing the two halves of DsbB, each containing one of the disulfide bonds. They detected a ternary complex that has the properties of an intermediate in the transfer of electron between DsbB and DsbA. This complex consists of the two halves of DsbB and the intact DsbA, linked together by two disulfide bonds: an inter-protein disulfide between Cys30 of DsbA and Cys104 of DsbB and an inter-domain disulfide between Cys130 and Cys41 of DsbB. Such a complex is not predicted by the simple model of disulfide exchange. In addition, the simpler model predicts that the Cys104-Cys130 pair would become reduced after releasing oxidized DsbA. However, in cells in which DsbB is actively oxidizing DsbA, such reduced DsbB is not detected. Instead, DsbB is found either in its oxidized state or in a complex with DsbA (39, 40). These results argue for a model in which DsbB coordinates the action of all four of its essential cysteines before releasing the oxidized DsbA from the DsbA-DsbB complex (Figure 2b).

Surprisingly, recent results indicate that both cysteine pairs of DsbB are significantly less oxidizing than the cysteine pair of DsbA with redox potentials of -240mV for Cys41-Cys44 and -267mV for Cys104-Cys130. The reported potential for the DsbA Cys30-Cys33 pair is -120mV. [The value for each of the DsbB cysteine pairs is the average of potentials reported in two studies (36, 41).] These findings present a paradox, as it is generally assumed that in such pathways electrons flow cleanly along a sequence of proteins (and/or small molecules) that form a gradient of redox potentials. The coordinated electron transfer described above (Figure 2b) in the DsbA-DsbB pathway may be necessary to compensate for the relative redox potentials that appear to oppose the direction of the actual electron transfer process. For example, after the formation of the DsbB-DsbA complex, the domain that includes the reduced Cys130 might be structurally arranged so that this cysteine is poised to interact with the Cys41-Cys44 disulfide bond (39). The resulting Cys130-Cys41 disulfide might then act to prevent the back reaction, because Cys130 has to be in its free form in order for this residue to attack the inter-protein disulfide bond to release reduced DsbA. In addition, the high redox potential of the quinones (e.g.,+113mV for ubiquinone) (42) in conjunction with the complex described above may be sufficient to overcome the apparently unfavorable redox potentials of the DsbB cysteine pairs (39, 41).

Recently, Regeimbal and Bardwell (36) proposed a model in which DsbA is oxidized directly by a quinone on DsbB. These authors suggest the need for this alternative model based on the apparent paradox of the redox potential differences between DsbB and DsbA. However this proposal fails to explain the ternary complex that forms in cells when DsbB is actively oxidizing DsbA (39). Furthermore, examples of in vivo physiological roles of such proteins that appear to be inconsistent with their apparent redox potentials may be more common than thought (43), as is also indicated in the discussion of the functioning of the DsbC protein presented below.

Thus far, the molecular mechanism of electron transfer between quinones and the Cys41-Cys44 pair of DsbB is unknown. However, recent genetic and biochemical studies revealed two regions on DsbB that appear to be involved in this process. First, the Arg48 residue and its distance from the CXXC sequence in the first periplasmic domain of DsbB play a role in the interaction of DsbB with quinones. Conversion of the Arg48 residue to histidine or cysteine results in an enzyme that can function aerobically but not anaerobically. The properties of the purified mutant Arg48His DsbB protein provide an explanation for this phenotype. The mutant enzyme can no longer utilize menaquinone for electron transfer and exhibits reduced affinity for ubiquinone (40). In addition, insertion of extra alanine residues at any position between Cys44 and Arg48 abolishes the respiratory chain-dependent oxidation of DsbB, which results in its serious dysfunction (44). Second, the segment from His91 to Glu112 in the second periplasmic domain of DsbB may also be involved in ubiquinone binding, because a photo-activatable ubiquinone analogue is cross-linked to this region when DsbB is illuminated with UV light in the presence of the reagent (45).

THE PATHWAY FOR PROTEIN DISULFIDE BOND ISOMERIZATION

DsbC, The Protein Disulfide Bond Isomerase

Anfinsen's pioneering studies on protein folding in the 1960s not only showed that disulfide bonds could form spontaneously, but they also revealed that oxidation of proteins with more than two cysteines often resulted in the formation of incorrect disulfide bonds, which produced a randomly scrambled set of protein isomers. The slowness of the spontaneous unscrambling process led Anfinsen and coworkers to propose and eventually isolate a protein disulfide bond isomerase that efficiently promotes rearrangements of disulfide bonds (46).

More recently it has become clear that in vivo the disulfide bond formation catalysts in both bacteria and eukaryotes, far from being perfect, regularly introduce nonnative disulfides into proteins with more than two cysteines. The cell thus relies on a parallel safekeeping system that rescues misfolded proteins from a dead-end pathway. In *E. coli*, this system is composed of the periplasmic disulfide bond isomerase DsbC and its paralogue DsbG (47–50). Because the major difference between these two proteins seems to be just their substrate specificity (50), in this review we focus exclusively on DsbC.

DsbC is a V-shaped homodimeric molecule (23.4-kDa for each monomer), where each arm is a monomer consisting of a C-terminal thioredoxin fold with an active CXXC motif and an N-terminal dimerization domain (51) (Figure 3*a*). The dimeric arrangement of the molecule is fundamental for its proposed roles in the periplasm as a thiol disulfide bond isomerase (25, 47, 52) and as a chaperone (53). Although no physiological substrates are known for DsbC and *E. coli* null mutants do not exhibit any obvious phenotype, the role of DsbC is indicated by its ability to refold, in vivo and in vitro, a number of heterologous proteins.

As a thiol-disulfide bond isomerase/reductase, DsbC promotes the rearrangements of nonnative disulfide bonds in vivo and in vitro in a number of proteins with multiple cysteine residues. Incorrect disulfide bonds in proteins, such as RNase A, bovine pancreatic trypsin inhibitor, tissue plasminogen activator, and urokinase among others, (54–56) can be successfully corrected by DsbC. As a chaperone, DsbC assists in the reactivation in vitro of denatured proteins that do not contain cysteine residues (53). The active site cysteines of DsbC are not required for this latter activity (57). This dual role presumably allows DsbC synergistically to identify (chaperone activity) and unscramble nonnative disulfide-bonded intermediates (isomerase activity) in which the formation of incorrect disulfide bonds likely triggers the appearance of aggregation-prone surfaces.

The recognition of unfolded structures probably resides in the uncharged cleft embodied by the two arms of the molecule (Figure 3*a*). This cleft can switch between open and closed conformations, depending on the redox state of their active thioredoxin folds (14). This flexibility allows DsbC to adjust the cleft according to the size and shape of the binding partner with the potential to accommodate a variety of unfolded structures.

Dimerization mutants provided important information about the mechanism of action of DsbC. First, it has been shown that a monomeric variant of DsbC, which lacks the N-terminal dimerization interface, can still work as a thiol-oxidoreductase, but it is devoid of both isomerase and chaperone activities (55). Second, mutants selected by their ability to complement a DsbA deficiency were unable to dimerize and promoted disulfide bond formation. In contrast to wild-type DsbC, they can be readily reoxidized in vitro by DsbB (58).

These findings provide an explanation of why DsbA and DsbC, with similar structures and nearly identical redox potentials (-120 and -130 mV, respectively) (15–17, 25), behave so differently. The dimeric structure of DsbC serves as a shield against oxidation by DsbB, leaving its active site reduced, which is a prerequisite for function as a thiol disulfide reductase or isomerase (see Figure 1). Reduction of the active site of DsbC is necessary so that one of its active cysteines, Cys-98, can make a nucleophilic attack on the disulfide bond in the substrate protein. This reaction results in an unstable mixed disulfide between DsbC and its substrate, which can be resolved in two ways as depicted in Figure 1. On the one hand, a third cysteine of the substrate could attack this mixed disulfide releasing DsbC with its active site cysteines restored to the reduced state and leaving the substrate with rearrangements of its disulfide bonds. On the other hand, the action of the most C-terminal cysteine of the DsbC active site, Cys-101, may attack the mixed disulfide and generate an oxidized DsbC and a reduced substrate. In this latter case DsbC is acting as a reductase, which leaves the substrate protein another chance to be oxidized by DsbA. Oxidized DsbC must be reduced to regain activity (59, 60).

DsbD, The Membrane-Bound Enzyme Responsible for Maintaining DsbC in the Reduced State

The maintenance of DsbC in the reduced state in *E. coli* depends on the cytoplasmic membrane protein DsbD (61, 62), which utilizes the thioredoxin/thioredoxin reductase system in the cytosol as a source of reducing equivalents (52, 54). As electrons are passed from NADPH to thioredoxin via thioredoxin reductase, periplasmic DsbC is thus kept reduced by DsbD at the expense of NADPH oxidation in the cytoplasm.

DsbD is in fact a branching point for the electrons coming from the cytoplasm. Not only does it reduce DsbC (and its homologue DsbG) (50, 54), but it also transfers electrons to CcmG, a membrane-tethered thioredoxin-like protein involved in maintaining cysteines in the heme-attachment site of *c*-type apocytochromes in the reduced state. This activity may be required because DsbA may act to promote disulfide bond formation among these cysteines [for a review of *c*-type cytochrome biogenesis see (63)]. In this manner, DsbD connects the thioredoxin system in the cytoplasm with two seemingly unrelated periplasmic

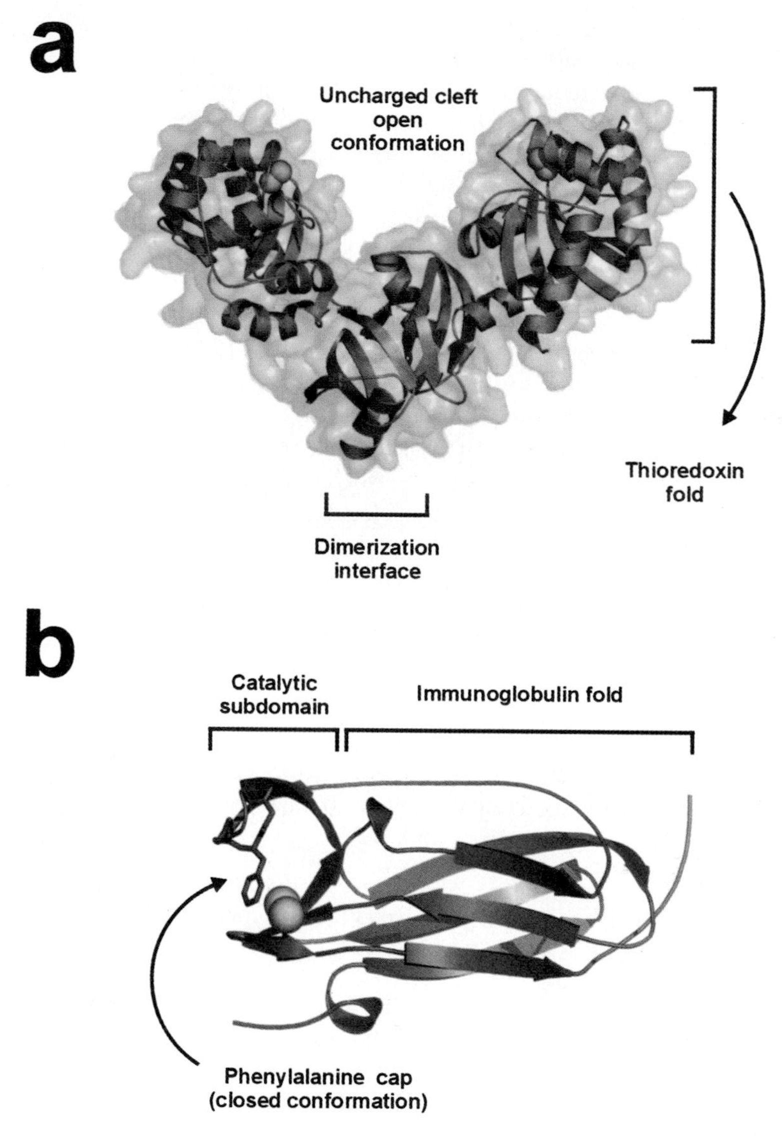

Figure 3 Crystal structures of (*a*) oxidized DsbC, (*b*) oxidized DsbD(α), and (*c*) the DsbD(α)-DsbC complex. Spheres represent cysteine residues.

machineries, the thiol-disulfide bond isomerization pathway and the cytochrome *c* biogenesis. In this section we focus mainly on the former pathway.

DsbD is a 546 amino-acid protein with a three-domain structure consisting of an amino-terminal periplasmic domain (DsbDα) with a cleavable signal peptide, a central hydrophobic core with 8 transmembrane segments (DsbDβ), and a carboxy-terminal thioredoxin-like domain (DsbDγ) (64–66) (Figure 4). Each of

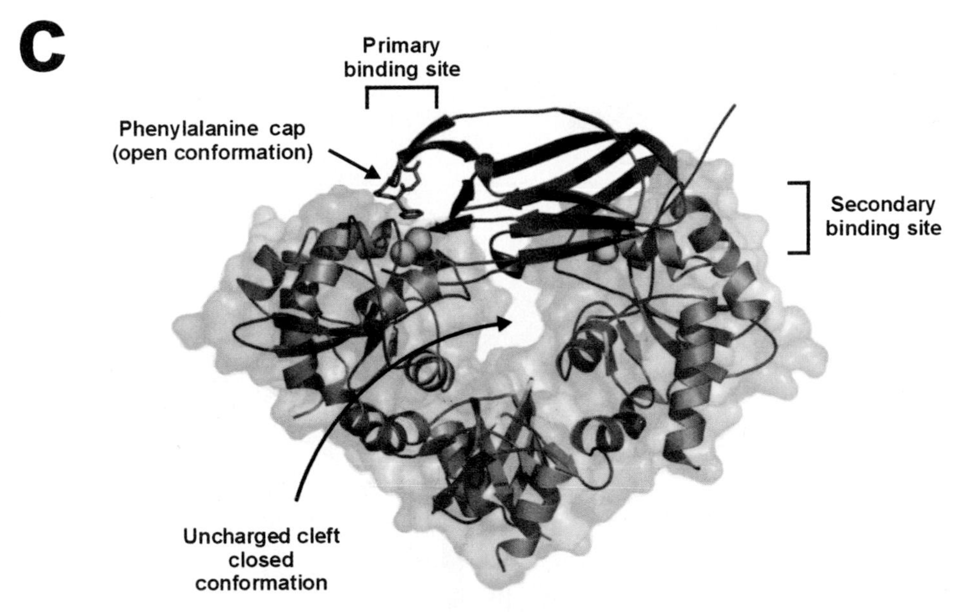

Figure 3 (*Continued*).

these domains includes a pair of cysteines that is essential for the protein's activity (64).

The three separated domains of DsbD can be expressed from separate plasmids and can reconstitute a functional unit when coexpressed in vivo (67). The preferential accumulation of oxidized isoforms of these domains when the immediate upstream electron donor is missing suggests that electron transfer is based on a sequential reduction and oxidation of the 3 structural domains. Thereby, reducing potential is transferred first from thioredoxin to DsbDβ and then successively to DsbDγ, DsbDα, and DsbC (67) (Figure 4).

Topological studies indicate that the central pair of cysteines are embedded in the membrane, probably one close to the cytoplasm and the other close to the periplasm (64–66) (Figure 4). This topology is consistent with the finding that Cys163 (proposed to face the cytoplasm) can be trapped in a disulfide bond with Cys33 of cytoplasmic thioredoxin (67, 68). This feature of the hydrophobic β domain of DsbD raises questions about the mechanism of electron transfer via this part of the protein. In particular, how do Cys163 and Cys285, which must interact with proteins on opposite sides of the membrane, interact with each other?

The model in which a cascade of electron transfer reactions via disulfide bond reduction steps takes place with DsbD (67) has been validated through several approaches. First, it is possible to trap predicted mixed disulfide intermediates such as DsbDα-DsbC and DsbDβ-thioredoxin in vivo (67, 68). Second, the interaction between DsbDα and DsbC was characterized in vitro, and the

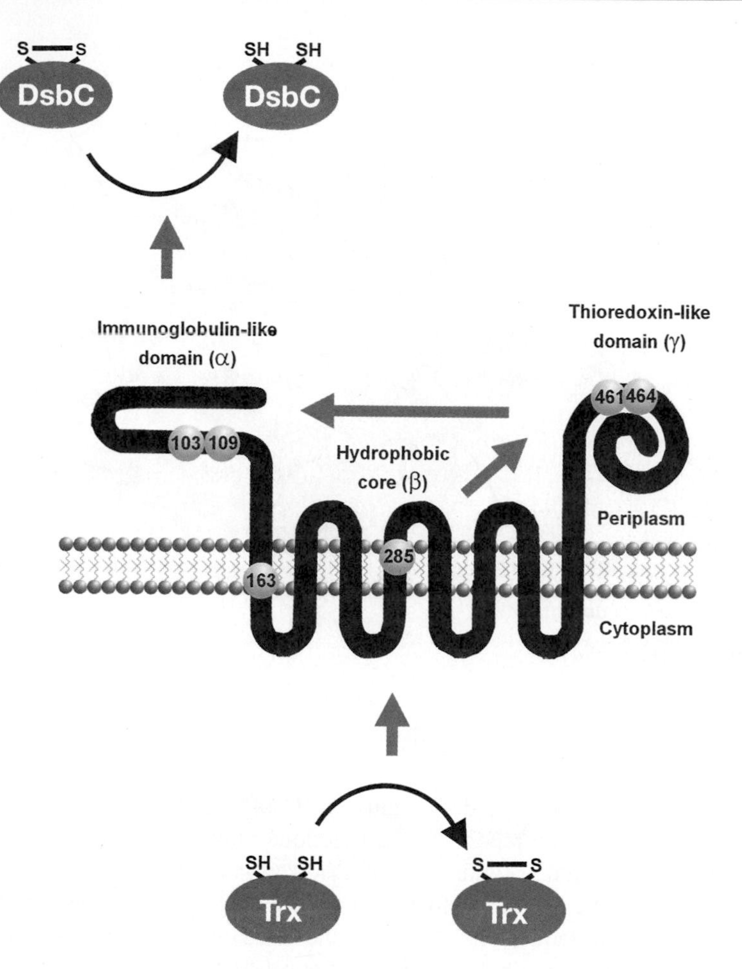

Figure 4 Membrane topology, structural domains, and electron flow in DsbD. Numbered spheres represent the essential cysteines. Gray arrows indicate the direction of the electron flow. Trx, thioredoxin. For simplicity, the only periplasmic substrate of DsbD shown is DsbC.

structure of this intermediate has been solved (14, 69). And, finally, the entire DsbD system has been reconstituted employing purified components (26). Together, this set of results explains several features of the mechanism of electron transfer described below.

In vitro studies on the reduction of DsbC by DsbDα show that electron transfer can proceed with dimeric and not monomeric DsbC (69). Thus, DsbC dimerization not only protects the enzyme from oxidation by DsbB, but it is also a prerequisite for reduction by DsbD.

DsbC reduction by DsbD is thermodynamically favored. The redox potentials of DsbDα and DsbDγ are -229 mV and -241 mV, respectively (26), which, as predicted, are lower than that of DsbC (-130 mV) but higher than that of thioredoxin (-270 mV) (70). These data show that, in contrast to the paradox observed with the redox potentials of DsbB domains and DsbA, the last two stages of electron transfer by DsbD are energetically driven and can be explained by two conventional thiol-disulfide exchange steps. Although thermodynamics would also allow electron transfer from DsbDα to DsbA, this reaction does not proceed in vitro (26), which suggests that DsbD rather than being a general thioredoxin reductase has a specific role in the periplasm and is kinetically isolated from the thiol oxidizing branch.

This specificity might be imposed by the three-dimensional structure itself. DsbDα assumes an immunoglobulin-like fold, with the active site located in the antigen binding end (14, 71) (Figure 3*b*). When complexed with DsbC, DsbDα is found within the central cleft and contacts both arms of the dimer (14) (Figure 3*c*). The existence of two binding sites explains why DsbDα selectively forms a complex with dimeric but does not with monomeric DsbC nor with DsbA. In addition, structural modeling suggests that the affinity between DsbDα and DsbA might be further reduced by steric hindrance of an extended helical domain of DsbA (14). Finally, the active site of DsbDα is shielded from the environment by a flexible cap (71), which lifts up to move away from this site in the mixed disulfide intermediate (14) (compare Figure 3*b* and 3*c*). This feature might protect the protein from promiscuous interactions. Overall, the accumulated data show that the specificity and efficiency achieved by DsbD are kinetically, thermodynamically, and structurally driven.

This close relationship between DsbD and DsbC might be the result of a relatively recent coevolutionary process. These two proteins are restricted to a small subset of proteobacteria. The rest of the eubacteria, together with archaea, appear to lack DsbC and modular DsbD homologues. Instead, they do have a DsbDβ-like protein, called CcdA, which (rather than diverting the reducing equivalents into multiple electron acceptors) appears to be involved only in cytochrome *c* biogenesis (72). These findings suggest that DsbD evolutionarily expanded its substrate range by the acquisition of the thiol redox active domains DsbDα and DsbDγ.

It is intriguing then to ask how the rest of the prokaryotes and archaea ensure accurate disulfide bond formation of extracytoplasmic proteins. It might well be that in these other bacteria either small thiol-active molecules are acting as reductants, or other not well-characterized proteins are responsible for this proofreading task.

THE ELECTRON JOURNEY: FROM CYTOPLASMIC NADPH TO TERMINAL ACCEPTORS

The periplasmic disulfide bond formation and isomerization pathways constitute two branches of the same machinery that are kept functionally separated. However, a number of misfolded proteins with three or more cysteine residues are substrates of both of these branches. Thus, these substrates can act as an exceptional link between the two otherwise isolated branches, which allows electrons originating from cytoplasmic NADPH to jump across this barrier and eventually contribute to the proton motive force through cytochrome oxidases (Figure 5).

DETERMINANTS OF THE DIRECTION OF ELECTRON TRANSFER BY THIOL-DISULFIDE OXIDOREDUCTASES: STRUCTURAL FEATURES WITHIN THE PROTEINS AND ENVIRONMENTAL INFLUENCES

Members of the family of thiol-disulfide oxidoreductases perform functions inside living cells mostly as reductants or oxidants. Ordinarily, as far as we know, each of these proteins is restricted physiologically in its activity to one or the other role. What are the factors that determine that one protein will act as a reductant in vivo and another as an oxidant?

First of all, not surprisingly, the redox potential of the proteins plays an important role in determining the function (or property) of an enzyme. It should be remembered that redox potentials are determined experimentally by measuring the relative amounts of oxidized and reduced protein species in redox equilibrium with a compound of known redox potential such as glutathione. Because these enzymes are usually acting on protein substrates in vivo, redox potentials measured in this way in vitro may not always indicate the function of the enzymes (reductant or oxidant) inside the cell. Nevertheless, the physiological role of a protein in this family is usually correlated with the redox properties of its active site. Studies supporting this conclusion are available for those proteins containing the CXXC motif embedded in a thioredoxin-like fold. For example, DsbA, which acts as a protein-thiol oxidant in the periplasm, has a highly oxidizing potential of -120 mV; thioredoxin, which acts as a disulfide reductant in the cytoplasm, has a more reducing redox potential of -270 mV (15–17).

Figure 5 A pathway of electron flow during periplasmic protein disulfide bond formation of *E. coli*. Thiol-active branches are designated with respect to their redox effect on an hypothetical periplasmic substrate (white spheres). Black arrows denote the direction of the electron flow. The following abbreviations are used: Q, quinone; Trx, thioredoxin; TrxR, thioredoxin reductase. Although DsbC may act both as a reductant and an isomerase of incorrectly formed disulfide bonds, in this Figure we have only chosen the flow of electrons that results from its activity as a reductant. Only a pathway under aerobic growth conditions is shown.

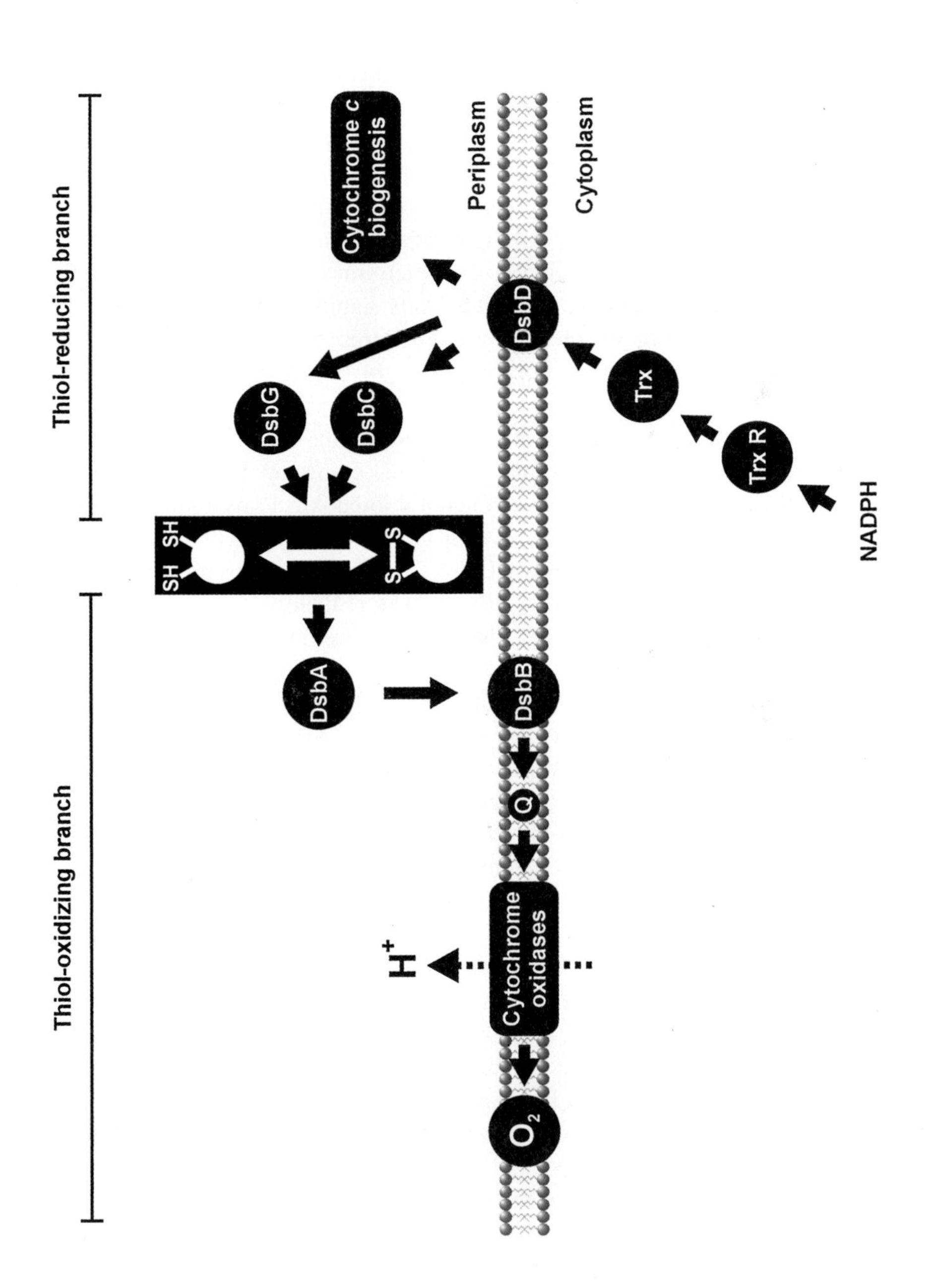
Thiol-reducing branch
Thiol-oxidizing branch
Cytochrome c biogenesis
Periplasm
Cytoplasm
DsbD
DsbG
DsbC
Trx
Trx R
NADPH
SH SH
S—S
DsbA
DsbB
Q
Cytochrome oxidases
H+
O2

Importantly, the in vitro redox properties of the thioredoxin family of proteins are strongly influenced by the amino acid residues between the two active site cysteine residues. When the intervening dipeptide of one member of this family is changed to that of another, the new enzyme exhibits a redox potential that is consistently shifted in the direction of the redox potential of the latter protein. In the case of DsbA, purified variants that contain CXXC sequences from more reducing proteins, such as thioredoxin 1, now exhibit lower redox potentials (17, 73). Similarly, when the active site of thioredoxin 1 is replaced with that of DsbA or from other more oxidizing proteins, its redox potential is higher (74). Another feature of DsbA mentioned earlier, which influences the redox property of its CXXC sequence, is its location at the N terminus of helix α1 of DsbA. In this position, the thiolate anion of Cys30 is likely stabilized by the dipole effect of α1 helix. Recent random mutagenesis studies address the importance of the helix α1 region in determining the redox properties of this enzyme including its redox potential (75).

A second important factor that can influence the function of a thiol-disulfide oxidoreductase is the nature of the subcellular environment in which it is acting. When the CXXC replacement versions of thioredoxin 1 are studied in vivo, their physiological functioning reflects well their redox potentials (70, 74). That is, when thioredoxin 1 containing the DsbA CXXC active site replaces the wild-type thioredoxin 1, it is much less effective at reducing its cytoplasmic protein substrates (e.g., ribonucleotide reductase) and results in observable phenotypic differences. However, in the case of DsbA, the redox potential itself might be less critical for satisfying its in vivo function; active-site variants, which are more than three magnitudes weaker oxidants than the wild-type protein in vitro, can still fulfill the function of DsbA under normal growth conditions (73). Consistent with these results, when the highly reducing cytoplasmic thioredoxin 1 is exported to the *E. coli* periplasm, it can act, albeit weakly, as an oxidant of cell envelope proteins. If the exported thioredoxin active CXXC sequence is replaced with that of DsbA, it then behaves as a highly efficient oxidant. Both exported thioredoxin 1 and its DsbA variant are able to act as oxidants because they are oxidized by DsbB (76–78). These results show that the redox environment (more precisely the nature of the redox partners) plays an important role in determining the redox properties of a protein of this family. In the cytoplasm where thioredoxin reductase reduces thioredoxin, this protein acts as a protein reductant (2). In the periplasm where DsbB acts to oxidize the exported thioredoxin, this protein now behaves as a protein oxidant. These findings are not a surprise as a number of studies have shown that a protein of this family can perform electron transfer—oxidation or reduction—in either direction.

One of the most striking examples of a redox potential paradox is that of the protein DsbC. We have described above the structural features of this protein that maintain it as a reductant in the periplasm, even though its redox potential is little different from that of the very strong oxidant, DsbA.

Then what is the role of the redox potential? The redox potential most often appears to be important for determining the efficiency of the reaction. Even though the exported thioredoxin can act as a protein oxidant in periplasm, the active-site variants with higher redox potentials can much more effectively transfer disulfide bonds to protein substrates both in vivo and in vitro (17, 77, 78).

HOMOLOGUES AND ANALOGUES OF DSB PROTEINS IN OTHER ORGANISMS

Homologues in Gram-Negative Bacteria

Homologues of *E. coli* DsbA and DsbB are found in many gram-negative bacteria. In contrast, DsbC and DsbD are mostly restricted to the β- and γ-subdivisions of eubacteria (72).

The DsbB-DsbA pathway plays an important role in the virulence of many bacteria (79). For example, lack of DsbA causes a deficiency in the folding of exotoxins such as cholera toxin (7) and *Bordetella pertussis* toxin (6 subunits with 11 intramolecular disulfide bonds) (80) and in the biogenesis of cell surface structures such as enteropathogenic pili that are required for adhesion to cells (7, 81). *Shigella flexneri* DsbA oxidizes a component of the type III secretion apparatus necessary for the secretion of effector molecules, which allows the cell-to-cell spread of this pathogen (82).

Interestingly, *Salmonella typhymurium* has two *E. coli* DsbA homologues: the virulence plasmid-encoded SrgA (83) and chromosomally encoded DsbA (84). Like *E. coli* DsbA, the active site of *S. typhymurium* DsbA is CPHC, while that of SrgA is CPPC. When introduced into an *E. coli dsbA*-null strain, *S. typhymurium* DsbA acts as a more efficient oxidant than SrgA for substrates such as alkaline phosphatase. Because alternating the active-site dipeptides of DsbA and DsbC can greatly affect their in vivo abilities to promote the folding of multidisulfide proteins (85), it could be that these two *Salmonella* DsbAs have different substrate specificities or exhibit different kinetics that are suited to the nature of their substrates.

An *E. coli dsbC* mutant does not show any obvious phenotype except for the defect in expression of heterologous proteins with multiple disulfide bonds (see above). However, interestingly, *B. pertussis dsbC* is reported to be required for the motility of this organism and for the secretion of pertussis toxin (80).

Are There Homologues in Gram-Positive Bacteria?

Very few extracytoplasmic proteins with disulfide bonds have been identified in gram-positive bacteria; this raises the question of whether this class of bacteria has catalytic systems for disulfide bond formation (86, 87). However, recent genetic studies suggest the existence of such systems in *Bacillus subtilis*.

B. subtilis contains two putative extracytoplasmic proteins, BdbA and BdbD, that are homologous to thioredoxins and two putative membrane proteins, BdbB and BdbC, that show sequence similarity to *E. coli* DsbB (88–91). Genetic evidence indicates that at least three of these genes play some role in disulfide bond formation. In the case of *bdbD* or *bdbC*, which comprise an operon, mutations in these genes lead to a severe reduction in the production of the disulfide-bonded *E. coli* PhoA when expressed in *B. subtilis* and the disulfide bond-containing competence protein ComG. Other competence proteins, none of which contain disulfide bonds, are unaffected in these mutants (90, 91).

The *bdbA* and *bdbB* genes are also together in an operon. This operon is involved in the synthesis of the secreted lantibiotic, sublancin 168 (87). Sublancin 168 contains two disulfide bonds. BdbB, but not BdbA, is required for active sublancin production and also for efficient *E. coli* PhoA secretion, but not for competence development by *B. subtilis* (89). BdbC can partially replace BdbB for sublancin production, which suggests that these two DsbB homologues have different but overlapping roles in oxidative folding of *B. subtilis* proteins. Thus far, the function of BdbA is not clear because inactivation of the *bdbA* gene does not result in any of the detectable phenotypic changes described for the other three genes (89).

Consistent with the roles of BdbB and BdbC in oxidative protein folding, mutations in the genes for these homologues of the *E. coli* DsbB suppress cytochrome *c* deficiency caused by the lack of CcdA (91). As we have described above, the function of CcdA, like DsbD, is presumed to be maintenance of reduced cysteines in apocytochrome *c* in the face of the oxidative activity of proteins such as DsbA. The *bdbB* and *bdbC* mutations may reduce the extracellular oxidative activity, thus allowing maintenance of free cysteines, even in the absence of CcdA.

In a related gram-positive organism, *Bacillus brevis*, an extracellular thioredoxin homologue named Bdb has been identified that restores motility and alkaline phosphatase activity to an *E. coli dsbA* mutant. However, its physiological role in *B. brevis* is unknown because of the lack of a mutant strain (92). Nevertheless, homologues of DsbA and DsbB have also been identified by BLAST searches of protein databases of gram-positive bacteria in addition to the *Bacilli*; this strengthens the evidence that oxidative pathways similar to the *E. coli* system are present in these bacteria.

Disulfide-Bonded Proteins in Archaea

The archaea include inhabitants of quite extreme environments: Some live in water at temperatures over 100°C, while others live in extremely salty, alkaline, or acid environments. One way in which these organisms have overcome the harshness of their surroundings is to evolve their proteins to be resistant to inactivation by these various extreme conditions. It now appears that, at least for some Archaea, this evolution has resulted in the unusual situation in which large numbers of cytoplasmic proteins are disulfide bonded.

Stable disulfide bonds are usually very rare in cytoplasmic proteins. However, recent studies by Mallick et al. (1) reveal that a high fraction of the intracellular proteins contain disulfide bonds in the case of several thermophilic archaea that grow at over 100°C. They combined genomic protein sequence database information and available three-dimensional structures of proteins to estimate the number of disulfide bonds in intracellular proteins. Four of the analyzed organisms were predicted to have 30%–43% of their cysteines in disulfide bonds. Consistent with their estimates, a large fraction of the cysteines of intracellular proteins of *Pyrobaculum aerophilium* was refractory to modification by reagents that react with free sulfhydryl groups (1). These results raise interesting research questions: Do the disulfide bonds, in fact, contribute to the stability of these proteins? What kind of cytoplasmic system catalyzes the formation of disulfide bonds in the cytoplasm of these archaea? Do archaea also make exported disulfide bond-containing proteins? If so, do the systems involved resemble the prokaryotic or eukaryotic ones?

Analogous Systems in Eukaryotic Cells

In eukaryotes, it had been thought that oxidized glutathione, abundant in the endoplasmic reticulum (ER), promotes disulfide bond formation in proteins in this compartment. However, following the discovery of the DsbA-DsbB system in *E. coli*, genetic studies in yeast *Saccharomyces cerevisiae* revealed the existence of Ero1p (93, 94), a protein required for disulfide bond formation in the ER. These and further studies established the necessity of an oxidative enzyme catalyst in the yeast ER. Such studies showed that an ER protein disulifde isomerase, Pdi1p, which contains thioredoxin domains, is used by yeast as a primary catalyst for oxidation of substrate proteins (94). Pdi1p, in turn, is maintained in the oxidized state by the membrane-associated Ero1p, which contains four cysteines essential for its function (95–97). The recipient of electrons from Ero1p is unclear. However, because, like DsbB, Ero1p can function both aerobically and anaerobically (98), Ero1p should be able to use sources other than molecular oxygen. Analogous oxidative pathways appear to exist also in higher eukaryotes (99–101).

Whether a protein reduction or isomerization pathway analogous to the bacterial DsbC-DsbD system exists in eukaryotes is uncertain. In yeast, inactivation of the glutathione synthesis pathway can suppress the phenotype of the *ero1–1* mutant, which indicates that gluthathione acts as a net reductant in ER counteracting the oxidizing activity of the Ero1p pathway (102). This finding led to the proposal that the reducing equivalents from glutathione are used to reduce either the improperly paired cysteine residues of substrates or to reduce Pdi1p to allow it to act as an isomerase. In higher eukaryotes, the ER resident PDI-homologue, ERp57, is thought to reduce the misfolded proteins (103), allowing its retrotranslocation from the ER to the cytoplasm for degradation (104). In addition, the reduced form of PDI is proposed to act as a chaperone to unfold cholera toxin in ER for its retrograde protein transport into the cytoplasm (105). These results suggest the necessity of some reductive pathways in these organisms.

CONCLUSION AND PERSPECTIVES

The presence of disulfide bonds in proteins was originally thought to result from the simplest of processes—spontaneous formation in an oxidizing environment. However, beginning with the discovery of protein disulfide bond isomerase and continuing into studies of recent years, pathways that transfer electrons for this process have been revealed that exhibit unimagined complexity. These pathways involve cascades of disulfide bond formation and reduction steps that result from communication between cytoplasm, membrane, and periplasmic space. There are still further complexities to be unraveled. Do cysteines in a substrate protein differ in their reactivity towards DsbA? How does an apparent disulfide bond form in the intramembraneous portion of DsbD and how is it transferred from periplasm to cytoplasm? Do enzymes of this pathway, DsbA and DsbC, exhibit any substrate specificity? How does DsbB use quinones for generation of disulfide bonds? How do other bacteria with different redox environments and thiol reactive agents carry out similar steps?

ACKNOWLEDGMENTS

We thank David Goldstone for help with the scripts for Figure 3. Federico Katzen is a Charles A. King trust fellow. Jon Beckwith is an American Cancer Society Professor. This work was supported by grants #41883 and #55090 from the National Institute of General Medical Sciences.

LITERATURE CITED

1. Mallick P, Boutz DR, Eisenberg D, Yeates TO. 2002. *Proc. Natl. Acad. Sci. USA* 99:9679–84
2. Stewart EJ, Åslund F, Beckwith J. 1998. *EMBO J.* 17:5543–50
3. Bryk R, Lima CD, Erdjument-Bromage H, Tempst P, Nathan C. 2002. *Science* 295:1073–77
4. Anfinsen CB, Haber E, Sela M, White FH. 1961. *Proc. Natl. Acad. Sci. USA* 47:1309–14
5. Bardwell JCA, McGovern K, Beckwith J. 1991. *Cell* 67:581–89
6. Kamitani S, Akiyama Y, Ito K. 1992. *EMBO J.* 11:57–62
7. Peek JA, Taylor RK. 1992. *Proc. Natl. Acad. Sci. USA* 89:6210–14
8. Dailey FE, Berg HC. 1993. *Proc. Natl. Acad. Sci. USA* 90:1043–47
9. Missiakas D, Georgopoulos C, Raina S. 1993. *Proc. Natl. Acad. Sci. USA* 90: 7084–88
10. Stafford SJ, Humphreys DP, Lund PA. 1999. *FEMS Microbiol. Lett.* 174:179–84
11. Martin JL, Bardwell JCA, Kuriyan J. 1993. *Nature* 365:464–68
12. Kishigami S, Akiyama Y, Ito K. 1995. *FEBS Lett.* 364:55–58

13. Darby NJ, Creighton TE. 1995. *Biochemistry* 34:3576–87
14. Haebel PW, Goldstone D, Katzen F, Beckwith J, Metcalf P. 2002. *EMBO J.* 21:4774–84
15. Zapun A, Bardwell JCA, Creighton TE. 1993. *Biochemistry* 32:5083–92
16. Åslund F, Berndt KD, Holmgren A. 1997. *J. Biol. Chem.* 272:30780–86
17. Huber-Wunderlich M, Glockshuber R. 1998. *Fold. Des.* 3:161–71
18. Nelson JW, Creighton TE. 1994. *Biochemistry* 33:5974–83
19. Wunderlich M, Jaenicke R, Glockshuber R. 1993. *J. Mol. Biol.* 233:559–66
20. Schirra HJ, Renner C, Czisch M, Huber-Wunderlich M, Holak TA, Glockshuber R. 1998. *Biochemistry* 37:6263–76
21. Guddat LW, Bardwell JCA, Martin JL. 1998. *Structure* 6:757–67
22. Vinci F, Couprie J, Pucci P, Quéméneur E, Moutiez M. 2002. *Protein Sci.* 11:1600–12
23. Guddat LW, Bardwell JCA, Zander T, Martin JL. 1997. *Protein Sci.* 6:1148–56
24. Couprie J, Vinci F, Dugave C, Quéméneur E, Moutiez M. 2000. *Biochemistry* 39:6732–42
25. Zapun A, Missiakas D, Raina S, Creighton TE. 1995. *Biochemistry* 34:5075–89
26. Collet JF, Riemer J, Bader MW, Bardwell JCA. 2002. *J. Biol. Chem.* 277: 26886–92
27. Braun P, Ockhuijsen C, Eppens E, Koster M, Bitter W, Tommassen J. 2001. *J. Biol. Chem.* 276:26030–35
28. Bardwell JCA, Lee JO, Jander G, Martin N, Belin D, Beckwith J. 1993. *Proc. Natl. Acad. Sci. USA* 90:1038–42
29. Kobayashi T, Kishigami S, Sone M, Inokuchi H, Mogi T, Ito K. 1997. *Proc. Natl. Acad. Sci. USA* 94:11857–62
30. Bader M, Muse W, Ballou DP, Gassner C, Bardwell JCA. 1999. *Cell* 98:217–27
31. Bader MW, Xie T, Yu CA, Bardwell JCA. 2000. *J. Biol. Chem.* 275: 26082–88
32. Jander G, Martin NL, Beckwith J. 1994. *EMBO J.* 13:5121–27
33. Kobayashi T, Ito K. 1999. *EMBO J.* 18:1192–98
34. Guilhot C, Jander G, Martin NL, Beckwith J. 1995. *Proc. Natl. Acad. Sci. USA* 92:9895–99
35. Kishigami S, Kanaya E, Kikuchi M, Ito K. 1995. *J. Biol. Chem.* 270:17072–74
36. Regeimbal J, Bardwell JCA. 2002. *J. Biol. Chem.* 277:32706–13
37. Kishigami S, Ito K. 1996. *Genes Cells* 1:201–8
38. Debarbieux L, Beckwith J. 1999. *Cell* 99:117–19
39. Kadokura H, Beckwith J. 2002. *EMBO J.* 21:2354–63
40. Kadokura H, Bader M, Tian H, Bardwell JCA, Beckwith J. 2000. *Proc. Natl. Acad. Sci. USA* 97:10884–89
41. Inaba K, Ito K. 2002. *EMBO J.* 21:2646–54
42. Gennis RB, Stewart V. 1996. In *Escherichia coli* and *Salmonella-Cellular and Molecular Biology*, ed. FC Neidhardt, pp. 217–61. Washington, DC: ASM Press
43. Gon S, Giudici-Orticoni MT, Méjean V, Iobbi-Nivol C. 2001. *J. Biol. Chem.* 276: 11545–51
44. Kobayashi T, Takahashi Y, Ito K. 2001. *Mol. Microbiol.* 39:158–65
45. Xie T, Yu L, Bader MW, Bardwell JCA, Yu CA. 2002. *J. Biol. Chem.* 277: 1649–52
46. De Lorenzo F, Goldberger RF, Steers E Jr, Givol D, Anfinsen CB. 1966. *J. Biol. Chem.* 241:1562–67
47. Shevchik VE, Condemine G, Robert-Baudouy J. 1994. *EMBO J.* 13:2007–12
48. Missiakas D, Georgopoulos C, Raina S. 1994. *EMBO J.* 13:2013–20
49. Andersen CL, Matthey-Dupraz A, Missiakas D, Raina S. 1997. *Mol. Microbiol.* 26:121–32
50. Bessette PH, Cotto JJ, Gilbert HF, Georgiou G. 1999. *J. Biol. Chem.* 274: 7784–92

51. McCarthy AA, Haebel PW, Törönen A, Rybin V, Baker EN, Metcalf P. 2000. *Nat. Struct. Biol.* 7:196–99
52. Rietsch A, Belin D, Martin N, Beckwith J. 1996. *Proc. Natl. Acad. Sci. USA* 93:13048–53
53. Chen J, Song JL, Zhang S, Wang Y, Cui DF, Wang CC. 1999. *J. Biol. Chem.* 274:19601–5
54. Rietsch A, Bessette P, Georgiou G, Beckwith J. 1997. *J. Bacteriol.* 179: 6602–8
55. Sun XX, Wang CC. 2000. *J. Biol. Chem.* 275:22743–49
56. Joly JC, Swartz JR. 1997. *Biochemistry* 36:10067–72
57. Liu X, Wang CC. 2001. *J. Biol. Chem.* 276:1146–51
58. Bader MW, Hiniker A, Regeimbal J, Goldstone D, Haebel PW, et al. 2001. *EMBO J.* 20:1555–62
59. Walker KW, Gilbert HF. 1997. *J. Biol. Chem.* 272:8845–48
60. Darby NJ, Raina S, Creighton TE. 1998. *Biochemistry* 37:783–91
61. Crooke H, Cole J. 1995. *Mol. Microbiol.*15:1139–50
62. Missiakas D, Schwager F, Raina S. 1995. *EMBO J.* 14:3415–24
63. Thöny-Meyer L. 2002. *Biochem. Soc. Trans.* 30:633–38
64. Stewart EJ, Katzen F, Beckwith J. 1999. *EMBO J.* 18:5963–71
65. Chung J, Chen T, Missiakas D. 2000. *Mol. Microbiol.* 35:1099–109
66. Gordon EH, Page MD, Willis AC, Ferguson SJ. 2000. *Mol. Microbiol.* 35:1360–74
67. Katzen F, Beckwith J. 2000. *Cell* 103: 769–79
68. Krupp R, Chan C, Missiakas D. 2001. *J. Biol. Chem.* 276:3696–701
69. Goldstone D, Haebel PW, Katzen F, Bader MW, Bardwell JC, et al. 2001. *Proc. Natl. Acad. Sci. USA* 98:9551–56
70. Mössner E, Huber-Wunderlich M, Rietsch A, Beckwith J, Glockshuber R, Åslund F. 1999. *J. Biol. Chem.* 274: 25254–59
71. Goulding CW, Sawaya MR, Parseghian A, Lim V, Eisenberg D, Missiakas D. 2002. *Biochemistry* 41:6920–27
72. Katzen F, Deshmukh M, Daldal F, Beckwith J. 2002. *EMBO J.* 21:3960–69
73. Grauschopf U, Winther JR, Korber P, Zander T, Dallinger P, Bardwell JCA. 1995. *Cell* 83:947–55
74. Mössner E, Huber-Wunderlich M, Glockshuber R. 1998. *Protein Sci.* 7:1233–44
75. Philipps B, Glockshuber R. 2002. *J. Biol. Chem.* 277:43050–57
76. Debarbieux L, Beckwith J. 1998. *Proc. Natl. Acad. Sci. USA* 95:10751–56
77. Debarbieux L, Beckwith J. 2000. *J. Bacteriol.* 182:723–27
78. Jonda S, Huber-Wunderlich M, Glockshuber R, Mössner E. 1999. *EMBO J.* 18:3271–81
79. Yu J, Kroll JS. 1999. *Microbes Infect.* 1:1221–28
80. Stenson TH, Weiss AA. 2002. *Infect. Immunol.* 70:2297–303
81. Zhang HZ, Donnenberg MS. 1996. *Mol. Microbiol.* 21:787–97
82. Yu J, Edwards-Jones B, Neyrolles O, Kroll JS. 2000. *Infect. Immunol.* 68: 6449–56
83. Friedrich MJ, Kinsey NE, Vila J, Kadner RJ. 1993. *Mol. Microbiol.* 8:543–58
84. Turcot I, Ponnampalam TV, Bouwman CW, Martin NL. 2001. *Can. J. Microbiol.* 47:711–21
85. Bessette PH, Qiu J, Bardwell JCA, Swartz JR, Georgiou G. 2001. *J. Bacteriol.*183:980–88
86. Chung YS, Breidt F, Dubnau D. 1998. *Mol. Microbiol.* 29:905–13
87. Paik SH, Chakicherla A, Hansen JN. 1998. *J. Biol. Chem.* 273:23134–42
88. Bolhuis A, Venema G, Quax WJ, Bron S, van Dijl JM. 1999. *J. Biol. Chem.* 274:24531–38
89. Dorenbos R, Stein T, Kabel J, Bruand C,

Bolhuis A, et al. 2002. *J. Biol. Chem.* 277:16682–88
90. Meima R, Eschevins C, Fillinger S, Bolhuis A, Hamoen LW, et al. 2002. *J. Biol. Chem.* 277:6994–7001
91. Erlendsson LS, Hederstedt L. 2002. *J. Bacteriol.* 184:1423–29
92. Ishihara T, Tomita H, Hasegawa Y, Tsukagoshi N, Yamagata H, Udaka S. 1995. *J. Bacteriol.* 177:745–49
93. Frand AR, Kaiser CA. 1998. *Mol. Cell* 1:161–70
94. Pollard MG, Travers KJ, Weissman JS. 1998. *Mol. Cell* 1:171–82
95. Frand AR, Kaiser CA. 1999. *Mol. Cell* 4:469–77
96. Frand AR, Kaiser CA. 2000. *Mol. Biol. Cell* 11:2833–43
97. Tu BP, Ho-Schleyer SC, Travers KJ, Weissman JS. 2000. *Science* 290: 1571–74
98. Sevier CS, Cuozzo JW, Vala A, Åslund F, Kaiser CA. 2001. *Nat. Cell Biol.* 3:874–82
99. Cabibbo A, Pagani M, Fabbri M, Rocchi M, Farmery MR, et al. 2000. *J. Biol. Chem.* 275:4827–33
100. Pagani M, Fabbri M, Benedetti C, Fassio A, Pilati S, et al. 2000. *J. Biol. Chem.* 275:23685–92
101. Mezghrani A, Fassio A, Benham A, Simmen T, Braakman I, Sitia R. 2001. *EMBO J.* 20:6288–96
102. Cuozzo JW, Kaiser CA. 1999. *Nat. Cell Biol.* 1:130–35
103. Antoniou AN, Ford S, Alphey M, Osborne A, Elliott T, Powis SJ. 2002. *EMBO J.* 21:2655–63
104. Fagioli C, Mezghrani A, Sitia R. 2001. *J. Biol. Chem.* 276:40962–67
105. Tsai B, Rodighiero C, Lencer WI, Rapoport TA. 2001. *Cell* 104:937–48

Annu. Rev. Biochem. 2003. 72:137–174
doi: 10.1146/annurev.biochem.72.121801.161712
Copyright © 2003 by Annual Reviews. All rights reserved
First published online as a Review in Advance on January 16, 2003

THE ENZYMES, REGULATION, AND GENETICS OF BILE ACID SYNTHESIS

David W. Russell
Department of Molecular Genetics, University of Texas Southwestern Medical Center, 5323 Harry Hines Blvd., Dallas, Texas 75390-9046; email: david.russell@utsouthwestern.edu

Key Words cholesterol, catabolism, cytochrome P450, nuclear receptors, genetic disease

■ **Abstract** The synthesis and excretion of bile acids comprise the major pathway of cholesterol catabolism in mammals. Synthesis provides a direct means of converting cholesterol, which is both hydrophobic and insoluble, into a water-soluble and readily excreted molecule, the bile acid. The biosynthetic steps that accomplish this transformation also confer detergent properties to the bile acid, which are exploited by the body to facilitate the secretion of cholesterol from the liver. This role in the elimination of cholesterol is counterbalanced by the ability of bile acids to solubilize dietary cholesterol and essential nutrients and to promote their delivery to the liver. The synthesis of a full complement of bile acids requires 17 enzymes. The expression of selected enzymes in the pathway is tightly regulated by nuclear hormone receptors and other transcription factors, which ensure a constant supply of bile acids in an ever changing metabolic environment. Inherited mutations that impair bile acid synthesis cause a spectrum of human disease; this ranges from liver failure in early childhood to progressive neuropathy in adults.

CONTENTS

INTRODUCTION

Approximately 500 mg of cholesterol is converted into bile acids each day in the adult human liver. Newly synthesized bile acids are secreted into the bile and delivered to the lumen of the small intestine where they act as emulsifiers of dietary lipids, cholesterol, and fat-soluble vitamins. The solubilized nutrients are incorporated into lipoproteins, which are delivered to the liver and metabolized. Bile acids are transported from the intestine to the liver via the portal circulation and then resecreted into the bile (1). About 95% of bile acids are recovered in the gut during each cycle of the enterohepatic circulation, and the 5% that are lost are replaced by new synthesis in the liver. This production (~500 mg/day) accounts for about 90% of the cholesterol that is actively metabolized in the body, and steroid hormone biosynthesis accounts for the remainder.

Bile acid synthesis is tightly regulated to ensure that sufficient amounts of cholesterol are catabolized to maintain homeostasis and to provide adequate emulsification in the intestine. When an organism is replete, excess bile acids repress further synthesis, and conversely when bile acids are in short supply, synthesis is increased. In some species, including rats and mice, output from the bile acid biosynthetic pathway is increased in response to the accumulation of cholesterol.

We last reviewed the enzymes that participate in bile acid synthesis and their regulation in 1992 (2). Since that time, new pathways of bile acid synthesis have been defined, the genes encoding the biosynthetic enzymes of these pathways have been isolated, and the contributions of individual enzymes and pathways to cholesterol metabolism have been elucidated in genetically engineered mice and genetically deficient humans. In addition, the transcription factors that regulate output from the pathways have been identified, and the mechanisms by which these proteins increase and decrease bile acid synthesis have been elucidated. These advances are reviewed here.

PATHWAYS OF BILE ACID SYNTHESIS

Cholesterol is converted into bile acids by pathways that involve 17 different enzymes, many of which are preferentially expressed in the liver. The immediate products of these pathways are referred to as primary bile acids; the structures of which vary widely between different vertebrate species. For example, in humans and rats, cholic acid and chenodeoxycholic acid are the primary bile acids, whereas in mice cholic acid and β–muricholic acid predominate. The chemical diversity of the bile acid pool is further expanded by the actions of anaerobic bacteria in the gut, which convert primary bile acids into dozens of secondary and tertiary bile acids (3). The plethora of different bile acids in the enterohepatic circulation ensures complete solubilization of hydrophobic nutrients in the small intestine. In addition, individual bile acids differ in their abilities to regulate the synthesis of bile acids, which allows an organism to fine-tune output from the biosynthetic pathways.

The steps leading to synthesis of a primary bile acid include: (*a*) the initiation of synthesis by 7α-hydroxylation of sterol precursors (Figure 1), (*b*) further modifications to the ring structures (Figure 2), (*c*) oxidation and shortening of the side chain (Figure 3), and (*d*) conjugation of the bile acid with an amino acid (Figure 4). The 17 enzymes that participate in these four steps are listed in Table 1. The exact order of steps in the biosynthetic pathway remains unclear because many of the intermediates serve as substrates for more than one biosynthetic enzyme.

Initiation

The synthesis of bile acids begins by one of several routes. In the classic pathway, cholesterol is converted into 7α-hydroxycholesterol by cholesterol 7α-hydroxylase (Figure 1, reaction 1), a microsomal cytochrome P450 enzyme (CYP7A1) expressed only in the liver (Table 1). P450 enzymes are mixed function mono-oxidases located in the microsomal and mitochondrial compartments of cells. Six members of the P450 superfamily participate in bile acid synthesis. Genes and cDNAs encoding cholesterol 7α-hydroxylase have been isolated from many species (4), and sequence comparisons between these reveal a hydrophobic enzyme of ~500 amino acids. When purified from rat liver (5–7), or expressed in cultured cells (7), *Escherichia coli* (8–10), or transgenic mice (11), the cholesterol 7α-hydroxylase enzyme has a low turnover number and a preference for cholesterol as substrate. Expression of the cholesterol 7α-hydroxylase gene and possibly the activity of the enzyme are highly regulated (see Regulation of Bile Acid Synthesis).

Mice deficient in the cholesterol 7α-hydroxylase gene (*Cyp7a1*) have a high incidence of postnatal lethality due to liver failure, vitamin deficiencies, and lipid malabsorption (12–14). The bile acid pool size in these animals is reduced by 75% (15), and the reduction in bile acid synthesis is not compensated for by increased expression of other bile acid biosynthetic enzymes (16). Intestinal

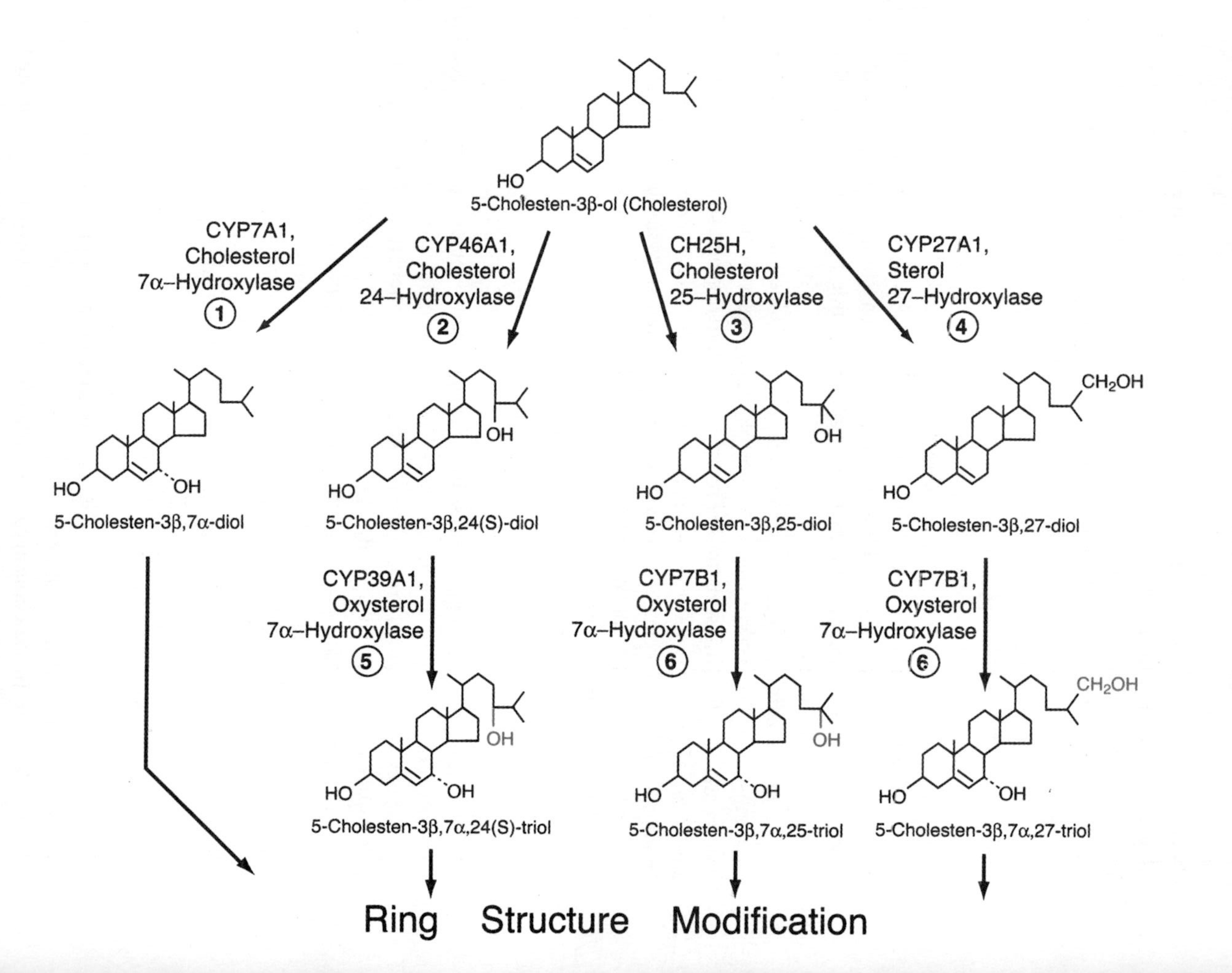

HO
5-Cholesten-3β-ol (Cholesterol)
CYP7A1, Cholesterol 7α–Hydroxylase 1
CYP46A1, Cholesterol 24–Hydroxylase 2
CH25H, Cholesterol 25–Hydroxylase 3
CYP27A1, Sterol 27–Hydroxylase 4
HO
OH
CH2OH
5-Cholesten-3β,7α-diol
5-Cholesten-3β,24(S)-diol
5-Cholesten-3β,25-diol
5-Cholesten-3β,27-diol
CYP39A1, Oxysterol 7α–Hydroxylase 5
CYP7B1, Oxysterol 7α–Hydroxylase 6
CYP7B1, Oxysterol 7α–Hydroxylase 6
5-Cholesten-3β,7α,24(S)-triol
5-Cholesten-3β,7α,25-triol
5-Cholesten-3β,7α,27-triol
Ring Structure Modification

cholesterol absorption is reduced to <5% of normal in the mutant mice, which results in a 200% increase in hepatic cholesterol synthesis. These alterations maintain cholesterol homeostasis (15). About 90% of the cholesterol 7α-hydroxylase deficient mice die within the first three weeks of birth. These short-lived mice produce only small amounts of hepatotoxic monohydroxy bile acids (14). Animals that survive this period begin synthesizing normal bile acids. Although the bile acid pool size never exceeds 25% of normal, this amount is sufficient to reverse the pathologic phenotype (13). The mice begin to synthesize bile acids via a different pathway in which oxysterols rather than cholesterol serve as substrates for 7α-hydroxylation (Figure 1).

Hydroxylation of cholesterol at three different positions on the side chain produces 24-hydroxycholesterol, 25-hydroxycholesterol, and 27-hydroxycholesterol. That these oxysterols could serve as substrates for bile acid synthesis was discovered in the 1960s (17). The finding that cholesterol 7α-hydroxylase deficient mice produced bile acids from an alternate pathway led to the identification of the three enzymes required to synthesize the various oxysterols and two enzymes that hydroxylate these intermediates at the 7α-position (Figure 1).

The first of these enzymes is cholesterol 24-hydroxylase, a microsomal cytochrome P450 (CYP46A1, Table 1) that synthesizes the oxysterol 24(*S*)-hydroxycholesterol (Figure 1, reaction 2). This enzyme is expressed in neurons of the mouse brain and at much lower levels in the liver (18). The 24-hydroxylase mRNA and protein are detectable in nerve cells of the normal human brain (18), and they are induced in glial cells of the central nervous system in subjects with Alzheimer's disease (18, 19).

Mice lacking the cholesterol 24-hydroxylase gene (*Cyp46a1*) have a 40% reduction in de novo cholesterol synthesis in the brain but no detectable alterations in bile acid metabolism (E.G. Lund et al., unpublished information). These findings are consistent with the observations of Björkhem et al. that 24-hydroxylation represents an important pathway by which cholesterol is secreted from the brain (21), and they indicate that a decrease in cholesterol catabolism in the central nervous system is compensated for by a reduction in synthesis. The rate of 24(*S*)-hydroxycholesterol synthesis in mouse brain is estimated to be 0.85 mg per day per 100 kg body weight, which pales in comparison to the hepatic synthesis of bile acids (~100 mg per day per 100 kg body weight) (E.G. Lund et al., unpublished information). Humans produce 6–7 mg of 24(*S*)-hydroxycholesterol per day (22), of which ~3.5 mg is catabolized

Figure 1 Biochemical steps involved in the initiation of bile acid synthesis. Circled numbers designate individual reactions, and the enzymes that catalyze each step are indicated next to the arrows. The chemical modification introduced by a particular enzyme is indicated in red on the product of the reaction. Cumulative changes to sterol intermediates are indicated in green. The products of the sterol 7α-hydroxylase enzymes are substrates for ring structure modification.

TABLE 1 Enzymes of bile acid synthesis

Reaction[a]	Enzyme	Mass[b] (Da)	Subcellular Localization	Comments	cDNA Accession Numbers[c]
1	Cholesterol 7α-hydroxylase	57,660	Endoplasmic reticulum	P450 (CYP7A1)[d], regulated, liver specific	(h) M93133 (m) L23754
2	Cholesterol 24-hydroxylase	56,821	Endoplasmic reticulum	P450 (CYP46A1), brain selective	(h) AF094480 (m) AF094479
3	Cholesterol 25-hydroxylase	31,700	Endoplasmic reticulum	Diiron cofactor, many tissues	(h) AF059214 (m) AF059213
4	Sterol 27-hydroxylase	56,900	Mitochondria	P450 (CYP27A1), many tissues	(h) M62401 (m) AK004977
5	Oxysterol 7α-hydroxylase	54,129	Endoplasmic reticulum	P450 (CYP39A1), many tissues	(h) AF237982 (m) AF237981
6	Oxysterol 7α-hydroxylase	58,255	Endoplasmic reticulum	P450 (CYP7B1), many tissues	(h) AF029403 (m) U36993
7	3β-Hydroxy-Δ^5-C_{27} steroid oxidoreductase	40,929	Endoplasmic reticulum	Selective for C_{27} substrates, many tissues	(h) AF277719 (m) AF277718
8	Sterol 12α-hydroxylase	58,078	Endoplasmic reticulum	P450 (CYP8B1), regulated, liver specific	(h) AF090320 (m) AF090317
9	Δ^4-3-Oxosteroid 5β-reductase	37,377	Cytoplasm	Aldo-keto reductase (AKR1D1), many tissues	(h) Z28339 (m) BF234820/AI931261
10	3α-Hydroxysteroid dehydrogenase	37,095	Cytoplasm	Aldo-keto reductase (AKR1C4), many tissues, many isozymes	(h) S68287 (m) NM_030611

11	Sterol 27-hydroxylase	56,900	Mitochondria	P450 (CYP27A1), many tissues	See above
12	Bile acid CoA ligase	70,312	Endoplasmic reticulum, peroxisome	Also uses very long chain fatty acids as substrates	(h) AF096290 (m) AF033031
13	2-Methylacyl-CoA racemase	42,359	Peroxisome, mitochondria	Also acts on 2-methyl fatty acids, liver and kidney	(h) AF158378 (m) U89906
14	Branched-chain acyl-CoA oxidase	76,826	Peroxisome	ACOX2, also acts on 2-methyl fatty acids, many tissues	(h) X95190 (m) AJ238492
15	D-Bifunctional protein	79,686	Peroxisome	Multiple enzyme activities, many tissues	(h) X87176 (m) X89998
16	Peroxisomal thiolase 2	58,993	Peroxisome	Multiple isoforms (SCP2, SCPx), liver enriched	(h) U11313 (m) M91458
17	Bile acid CoA:amino acid N-acyltransferase	46,296	Peroxisome	Liver selective	(h) L34081 (m) U95215

[a] Reaction numbers refer to Figures 1, 2, 3, and 4.

[b] Exact molecular weights are derived from cDNA sequences and are for the mature human enzymes.

[c] (h), human; (m), mouse GenBank accession numbers (3a).

[d] Cytochrome P450 enzymes are named and numbered according to the convention described in (199).

to bile acids (23). This amount is again much smaller than the mass of bile acids synthesized each day (~500 mg). Together these findings indicate that the 24-hydroxylase enzyme contributes little to overall bile acid synthesis but is important in the turnover of cholesterol in the brain.

The microsomal enzyme cholesterol 25-hydroxylase (Figure 1, reaction 3) is not a P450 but rather is a member of a family of lipid metabolizing enzymes that utilize oxygen and a diiron-oxygen cofactor to hydroxylate, desaturate, epoxidate, or acetylinate substrates (24). The mouse cholesterol 25-hydroxylase cDNA was isolated by expression cloning (25). The mRNA is present at low levels in most tissues and at a higher level in the lung. Expression of recombinant cholesterol 25-hydroxylase in cultured cells results in the synthesis of 25-hydroxycholesterol (25), and as expected (26), the concomitant repression of cholesterol biosynthetic enzymes. Cholesterol 25-hydroxylase knockout mice have no alterations in bile acid synthesis or cholesterol metabolism, and to date there is no in vivo evidence that this enzyme catalyzes the formation of 25-hydroxycholesterol in the whole animal (G. Liang, J. Li-Hawkins, D. W. Russell, unpublished observations). Cholesterol 25-hydroxylase may play a role in cholesterol catabolism in a tissue-specific fashion, as does the cholesterol 24-hydroxylase.

27-Hydroxycholesterol is the most abundant oxysterol in the plasma of the mouse (27) and human (28), and it is synthesized from cholesterol by sterol 27-hydroxylase (Figure 1, reaction 4), a mitochondrial cytochrome P450 (CYP27A1, Table 1). This enzyme can also hydroxylate cholesterol at carbons 24 and 25 to form 24-hydroxycholesterol and 25-hydroxycholesterol, respectively (29). Unlike the 24-hydroxylase and 25-hydroxylase enzymes, which contribute only modestly to bile acid synthesis in the mouse, about 25% of the bile acid pool originates from oxysterols produced by sterol 27-hydroxylase. This percentage is derived from the observations that the 24- and 25-hydroxylase knockout mice have no discernible alterations in bile acid synthesis and that the cholesterol 7α-hydroxylase deficient mice, which cannot convert cholesterol directly into bile acids, have a bile acid pool size that is 25% of normal (15, 16). The relative contributions of the oxysterol biosynthetic enzymes to bile acid synthesis in humans are difficult to assess. Analysis of the bile acids in normal individuals (22, 30, 31) and in a subject with cholesterol 7α-hydroxylase deficiency (32) suggest that the oxysterol pathways together are responsible for only 5% to 10% of bile acid production. The consequences for loss of the sterol 27-hydroxylase gene in mice and in humans are discussed below in Side Chain Oxidation.

To be converted into bile acids, oxysterols must undergo 7α-hydroxylation. Two microsomal cytochrome P450 enzymes catalyze this step. The CYP39A1 oxysterol 7α-hydroxylase acts on 24(*S*)-hydroxycholesterol (Figure 1, reaction 5). The mRNA encoding this enzyme is abundantly and constitutively expressed in mouse and human liver as well as in the nonpigmented epithelium of the eye (33–35). The CYP39A1 oxysterol 7α-hydroxylase contains ~470 amino acids, and when expressed in cultured cells, it demonstrates a rank order preference for oxysterol substrates of 24(*S*)-hydroxycholesterol (1.0) > 24,25-epoxycholesterol (0.16) > 25-hydroxycholesterol (0.08) > 27-hydroxycholesterol (0.06) > 22-hydroxycholesterol (0.05) (33). The contributions of this enzyme to bile acid synthesis in the mouse, or any other species, have not been determined.

The conversion of 25-hydroxycholesterol and 27-hydroxycholesterol to bile acid intermediates is catalyzed by the CYP7B1 oxysterol 7α-hydroxylase (Figure 1, reaction 6). The cDNA for CYP7B1 was isolated originally from brain by differential hybridization (36), but it was only later shown to encode an enzyme with oxysterol 7α-hydroxylase activity (37–39). The deduced sequence of the protein is ~40% identical to that of cholesterol 7α-hydroxylase and ~30% identical to the CYP39A1 oxysterol 7α-hydroxylase (33). The CYP7B1 oxysterol 7α-hydroxylase is expressed at high levels in the adult liver and at lower levels in the kidney, brain, and prostate. In mice, hepatic expression of this enzyme is induced during the third week of life and thereafter exhibits a sexually dimorphic expression pattern in this and other tissues in which expression is higher in the male (13, 27). The CYP7B1 oxysterol 7α-hydroxylase utilizes 27-carbon oxysterols and 19-carbon steroids, such as dehydroepiandrosterone (37), as substrates. Mice that lack the encoding gene (*Cyp7b1*) have elevated plasma levels of 25-hydroxycholesterol and 27-hydroxycholesterol but not 24-hydroxycholesterol (27, 39a). These animals also have increased levels of cholesterol 7α-hydroxylase, presumably to compensate for the reduced bile acid biosynthetic capacity (27). This increase (~30%) is roughly equal in size to that of the residual bile acid pool in cholesterol 7α-hydroxylase-deficient mice (15), which confirms that the CYP7B1 oxysterol 7α-hydroxylase pathway synthesizes 25% to 30% of all bile acids in this species. As noted previously, 5% to 10% of the bile acid pool in humans originates from oxysterols, but the fraction that derives from the CYP7B1 versus the CYP39A1 oxysterol 7α-hydroxylase is unknown.

Ring Structure Modification

The 7α-hydroxylated intermediates derived from cholesterol and the oxysterols are next converted into their 3-oxo, Δ^4 forms (Figure 2, reaction 7) by a microsomal 3β-hydroxy-Δ^5-C_{27}-steroid oxidoreductase (C_{27} 3β-HSD, Table 1). There is only one C_{27} 3β-HSD, and loss of this enzyme blocks the synthesis of all bile acids. The human and mouse C_{27} 3β-HSDs are membrane bound enzymes of 369 amino acids that share ~34% sequence identity with the C_{19} and C_{21}3β-HSDs involved in steroid hormone biosynthesis (40). The reaction catalyzed by these enzymes is complex and involves isomerization of the double bond from the 5 to the 4 position and the oxidation of the 3β-hydroxyl to a 3-oxo group (Figure 2, reaction 7). The C_{27} 3β-HSD enzyme will act only on sterols with a 7α-hydroxyl group (40–42), and thus this step lies downstream of the cholesterol and oxysterol 7α-hydroxylases.

The products of the C_{27} 3β-HSD enzyme take one of two routes in subsequent steps of bile acid synthesis. If the intermediate is acted upon by sterol 12α-

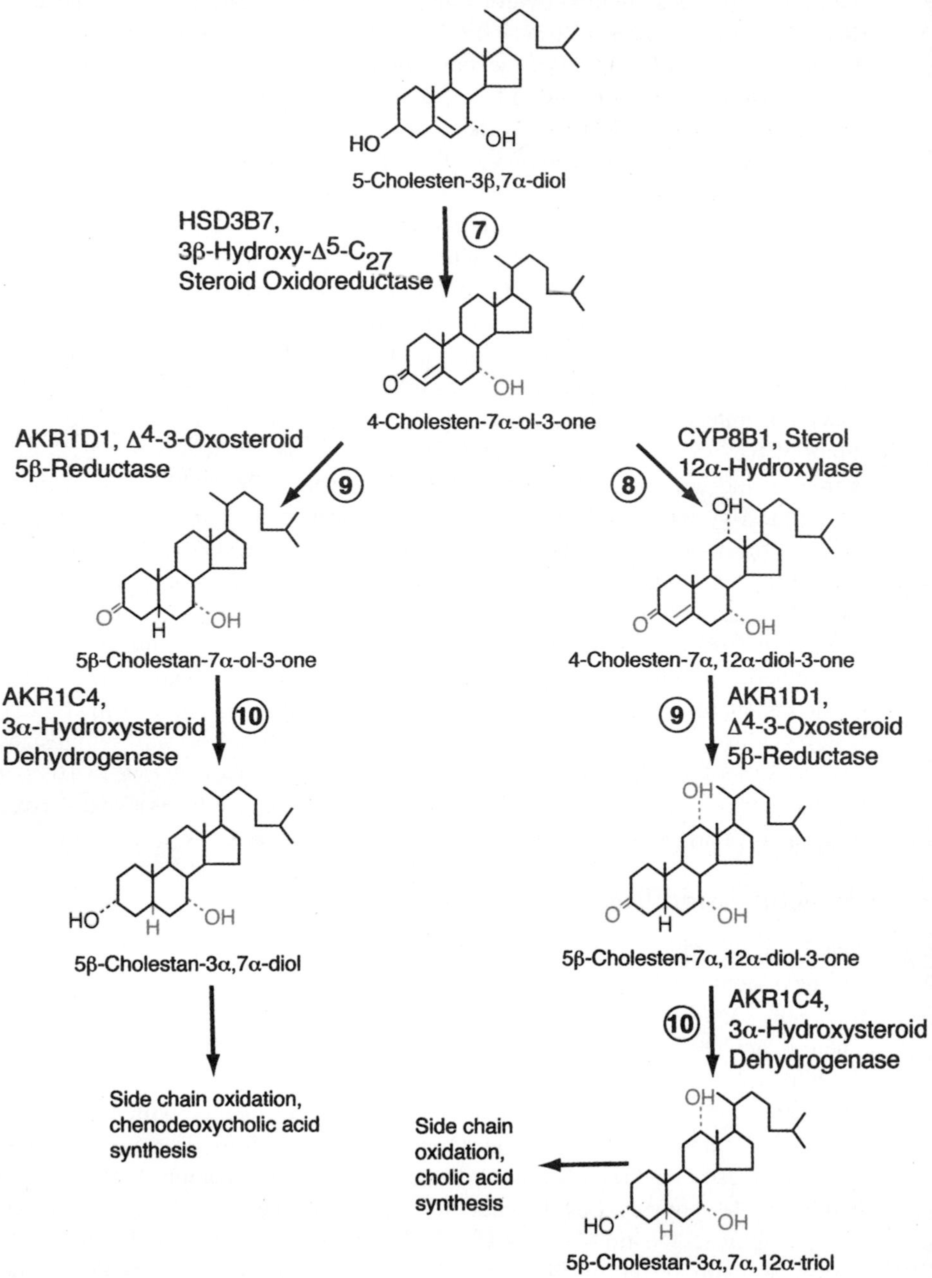

5-Cholesten-3β,7α-diol

HSD3B7,
3β-Hydroxy-Δ^5-C_{27}
Steroid Oxidoreductase

(7)

4-Cholesten-7α-ol-3-one

AKR1D1, Δ^4-3-Oxosteroid
5β-Reductase

(9)

CYP8B1, Sterol
12α-Hydroxylase

(8)

5β-Cholestan-7α-ol-3-one

4-Cholesten-7α,12α-diol-3-one

AKR1C4,
3α-Hydroxysteroid
Dehydrogenase

(10)

AKR1D1,
Δ^4-3-Oxosteroid
5β-Reductase

(9)

5β-Cholestan-3α,7α-diol

5β-Cholesten-7α,12α-diol-3-one

Side chain oxidation,
chenodeoxycholic acid
synthesis

AKR1C4,
3α-Hydroxysteroid
Dehydrogenase

(10)

Side chain
oxidation,
cholic acid
synthesis

5β-Cholestan-3α,7α,12α-triol

hydroxylase (Figure 2, reaction 8), a microsomal cytochrome P450 (CYP8B1, Table 1), then the resulting product will be converted ultimately into cholic acid. In the absence of 12α-hydroxylation, a metabolic fate of conversion to chenodeoxycholic acid or another bile acid is met. Two primary bile acids are produced by most vertebrate species; one of which is usually cholic acid and the other a bile acid like chenodeoxycholic acid (rat, human, and hamster), muricholic acid (mouse), ursodeoxycholic acid (bear), or hyodeoxycholic acid (pig). The level of sterol 12α-hydroxylase in the liver determines the relative amounts of the two primary bile acids, and in some species like the mouse, the ratio of one bile acid to another controls output from the biosynthetic pathway because cholic acid mediates feedback regulation of the pathway.

Complementary DNAs and genes encoding sterol 12α-hydroxylase have been isolated from the rabbit (43), human (44), and mouse (45). Transcription of the gene in rodents is subject to many of the same regulatory inputs as the cholesterol 7α-hydroxylase gene (46, 47). Loss of sterol 12α-hydroxylase in mice eliminates cholic acid from the bile acid pool and leads to an increase in the synthesis of muricholates (45). The altered composition of the pool affects the absorption of dietary sterols and the regulation of bile acid synthesis. With no cholic acid and an excess of muricholates, the overall hydrophilicity of the bile acid pool is increased, which causes a decrease in the absorption of dietary cholesterol and a compensating increase in hepatic cholesterol synthesis. In normal mice, cholic acid mediates feedback regulation of bile acid synthesis, and its loss from the sterol 12α-hydroxylase knockout mice causes a derepression of cholesterol 7α-hydroxylase and a corresponding increase in the synthesis of bile acids (see Regulation of Bile Acid Synthesis).

The 12α-hydroxylated intermediates, and those produced by the C_{27}-3β-HSD enzyme that escape 12α-hydroxylation, are subject to reduction of the double bond in the A-ring by the enzyme Δ^4-3-oxosteroid 5β-reductase (Figure 2, reaction 9). This cytosolic enzyme of 326 amino acids utilizes NADH as a cofactor and is a member (AKR1D1, Table 1) of a large family of proteins, the aldo-keto reductases. These enzymes are present in all species and catalyze oxidation-reduction reactions on endogenous as well as xenobiotic substrates

←

Figure 2 Modifications to the sterol ring structure in bile acid synthesis. Circled numbers designate individual reactions, and the enzymes that catalyze each step are indicated next to the arrows. The chemical modification introduced by a particular enzyme is indicated in red on the product of the reaction. Cumulative changes to sterol intermediates are indicated in green. The products of the 3α-hydroxysteroid dehydrogenase enzyme are substrates for side chain oxidation. Intermediates not acted upon by sterol 12α-hydroxylase (left arm of pathway scheme) are converted ultimately into chenodeoxycholic acid as indicated or into another primary bile acid depending on the species (see text for further details). Intermediates that are acted upon by sterol 12α-hydroxylase (right arm of pathway scheme) are converted ultimately into cholic acid.

(48). The enzymes prior to 5β-reductase in the bile acid biosynthetic pathway are located in microsomal or mitochondrial membranes, and thus the cell faces the challenge at this step of transporting intermediates from a hydrophobic environment (the membrane) to a hydrophilic environment (the cytosol). It is not known whether specific transport proteins or 5β-reductase itself mediate this transfer or whether the actions of the prior membrane-bound enzymes increase the hydrophilicity of intermediates to the point that they spontaneously enter the cytosol (see Subcellular Itinerary of Bile Acid Synthesis).

The final step of ring modification involves reduction of the 3-oxo group to an alcohol in the alpha stereochemical configuration and is catalyzed by 3α-hydroxysteroid dehydrogenase (Figure 2, reaction 10). Rat cDNAs encoding an enzyme with this activity specify a soluble protein that is a member (AKR1C4, Table 1) of the aldo-keto reductase family (49, 50). While the rat AKR1C4 enzyme will utilize bile acid intermediates as substrates in vitro and is abundant in the liver, other enzymes also may have this activity in vivo. The aldo-keto reductase family is large and includes multiple members with sequence identity to AKR1C4 that differ in number and enzymatic properties depending on the species (49). The rat AKR1C4 enzyme also reduces C_{19} and C_{21} steroid hormones, and thus it is possible that a 3α-hydroxysteroid dehydrogenase with specificity for C_{27} intermediates in the bile acid pathways remains to be identified.

Side Chain Oxidation

The products of ring modification next undergo progressive oxidation and shortening of the sterol side chain. The first few steps in this portion of the pathway are performed by sterol 27-hydroxylase (Figure 3, reaction 11), the same mitochondrial cytochrome P450 (CYP27A1) that initiates bile acid synthesis through the formation of 27-hydroxycholesterol (Figure 1, reaction 4). The enzyme introduces a hydroxyl group at carbon 27 and then oxidizes this group to an aldehyde and then to a carboxylic acid. At one time, it was thought that separate aldehyde and alcohol dehydrogenases catalyze the oxidation of the initial hydroxyl group (2), but studies in transfected cells and with purified sterol 27-hydroxylase are consistent with one enzyme performing all three reactions (51–53).

The participation of sterol 27-hydroxylase in the initiation and side chain oxidation steps of the biosynthetic pathways means that all intermediates,

Figure 3 Side chain oxidation in bile acid synthesis. Circled numbers designate individual reactions, and the enzymes that catalyze each step are indicated next to the arrows. The chemical modification introduced by a particular enzyme is indicated in red on the product of the reaction. Cumulative changes to sterol intermediates are indicated in green. The products of side chain oxidation are substrates for conjugation. Sterol 27-hydroxylase catalyzes three sequential oxidation steps in this portion of the pathway.

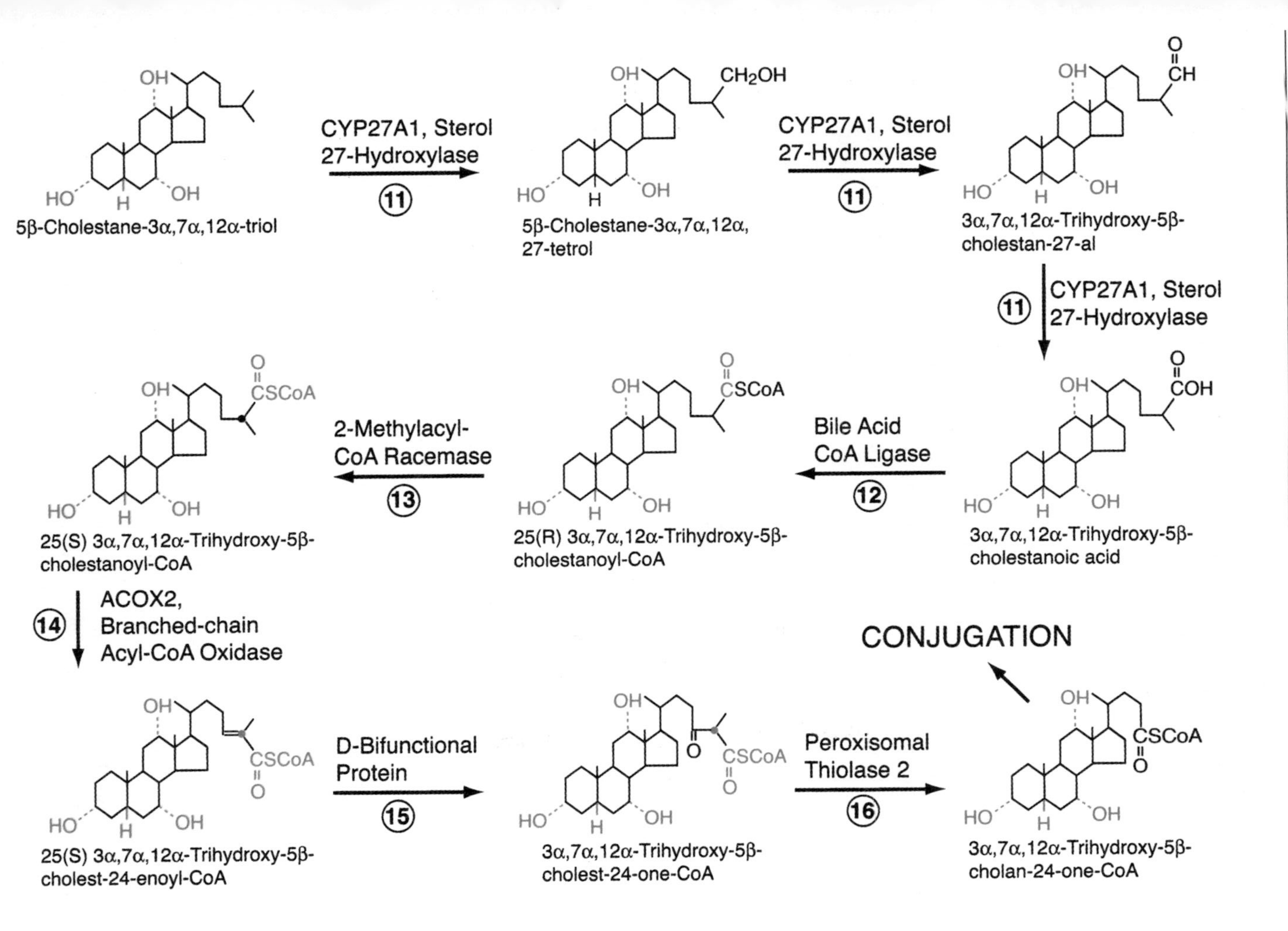
5β-Cholestane-3α,7α,12α-triol
CYP27A1, Sterol 27-Hydroxylase
11
5β-Cholestane-3α,7α,12α, 27-tetrol
CH₂OH
CYP27A1, Sterol 27-Hydroxylase
11
3α,7α,12α-Trihydroxy-5β-cholestan-27-al
CH
CYP27A1, Sterol 27-Hydroxylase
11
3α,7α,12α-Trihydroxy-5β-cholestanoic acid
COH
Bile Acid CoA Ligase
12
25(R) 3α,7α,12α-Trihydroxy-5β-cholestanoyl-CoA
CSCoA
2-Methylacyl-CoA Racemase
13
25(S) 3α,7α,12α-Trihydroxy-5β-cholestanoyl-CoA
ACOX2, Branched-chain Acyl-CoA Oxidase
14
25(S) 3α,7α,12α-Trihydroxy-5β-cholest-24-enoyl-CoA
D-Bifunctional Protein
15
3α,7α,12α-Trihydroxy-5β-cholest-24-one-CoA
Peroxisomal Thiolase 2
16
3α,7α,12α-Trihydroxy-5β-cholan-24-one-CoA
CONJUGATION
HO
OH
H
O

regardless of their origin, must be acted upon by the enzyme prior to formation of a bile acid. As a consequence of the multiple reactions catalyzed by sterol 27-hydroxylase, inactivation of the gene in mice has severe effects on bile acid synthesis (54–57). The bile acid pool size is 25% of normal in the mutant mice; this results in decreased intestinal cholesterol absorption and increased hepatic and peripheral tissue synthesis of cholesterol. Cholesterol 7α-hydroxylase mRNA and activity are greatly increased in these mice (54, 55). The composition of the bile acid pool is not altered (55).

Unexpectedly, plasma cholesterol and triglyceride concentrations are elevated in sterol 27-hydroxylase knockout mice, and both the liver and adrenal are enlarged (55). The dislipidemia appears to have two causes. First, the expression of sterol regulatory element binding proteins 1 and 2 (SREBP-1 and SREBP–2) is elevated in these mice. These transcription factors increase the expression of genes involved in lipid biosynthesis leading to a greater rate of triglyceride synthesis. Second, the reduced bile acid pool causes a decrease in the activity of the farnesoid X receptor (FXR) and a corresponding decline in the expression of apolipoprotein C-II, which is required for the hydrolysis of triglycerides (58). Dietary bile acids normalize plasma lipids and liver size in the mutant mice but have no effects on the enlarged adrenal gland (55).

Humans lacking sterol 27-hydroxylase accumulate cholestanol, a 5α-reduced derivative of cholesterol that is neurotoxic (59). Cholestanol is synthesized by an alternate catabolic pathway that is active when bile acid synthesis is disrupted. In contrast, mice with an induced mutation in the sterol 27-hydroxylase gene do not accumulate cholestanol and consequently do not develop the severe neuropathy characteristic of the human disease (54). Protection in the mouse is due in part to the induction of another cytochrome P450, CYP3A11, which hydroxylates sterol intermediates on carbon 25. This induction reduces the buildup of sterols and compensates for the loss of the sterol 27-hydroxylase in the mutant mice (56, 57). The corresponding enzyme in humans [CYP3A4, (60)] is not induced in patients with sterol 27-hydroxylase deficiency, which may explain the phenotypic difference between mice and humans.

The oxidized bile acid intermediates arising from sterol 27-hydroxylase exit the mitochondria and are next subject to shortening of the side chain. The terminal three carbon atoms are removed in peroxisomes by a series of reactions analogous to those involved in the β–oxidation of fatty acids (61). The first reaction is catalyzed by bile acid coenzyme A ligase (Figure 3, reaction 12), which activates the sterol intermediate by conjugation with coenzyme A. Two enzymes have been identified with this activity, very long-chain coenzyme A synthetase [a 620 amino acid protein of the endoplasmic reticulum and peroxisome (62–64)] and very long-chain acyl-coenzyme A synthetase homolog 2 (65) [alternatively bile acyl-coenzyme A synthetase (66), a related (~45% identical) 690 amino acid protein of the endoplasmic reticulum]. These enzymes also activate very long chain fatty acids containing 18 or more carbons. The very long-chain coenzyme A synthetase is found in liver and kidney and appears to be largely responsible for the activation of C_{27} intermediates of bile acid biosynthesis (Figure 3, reaction 12). The very long-chain acyl-coenzyme A synthetase homolog 2 is present only in liver and is involved in the activation of C_{24} bile acids that are deconjugated in the small intestine and returned to the liver via the enterohepatic circulation (66).

The carbon atom at position 25 of intermediates in the bile acid biosynthetic pathways is prochiral, having two apparently equivalent methyl groups (C26 and C27) as substituents. Sterol 27-hydroxylase, like most enzymes, recognizes the nonequivalence of the two methyl groups and catalyzes a stereospecific hydroxylation event, producing almost exclusively the 25(*R*) isomer (67, 68). After activation of intermediates with coenzyme A by the bile acid ligase, the 25(*R*) isomers must be converted into 25(*S*) isomers before subsequent steps in side chain shortening can take place. This reaction is catalyzed by 2-methylacyl-coenzyme A racemase (Figure 3, reaction 13), an enzyme located in both mitochondria and peroxisomes (69, 70). This protein can be purified in the absence of detergents (71), which suggests that it is not membrane bound. The 2-methylacyl-coenzyme A racemase also acts on branched chain fatty acids like phytanic acid derived from the catabolism of isoprenoids. Consequently, mutations in the encoding gene result in the accumulation of both bile acid intermediates and phytanic acid (see Genetics of Bile Acid Synthesis).

The sterol products of the racemase enzyme are next subject to dehydrogenation catalyzed by the FAD-containing peroxisomal enzyme branched chain acyl-coenzyme A oxidase to yield 24,25-*trans*-unsaturated derivatives (Figure 3, reaction 14). During the course of this reaction, the enzyme transfers electrons to molecular oxygen and produces hydrogen peroxide as a byproduct. There are two related (45% sequence identity) acyl-coenzyme A oxidase enzymes in humans and mice, abbreviated as ACOX1 and ACOX2. ACOX1 is a peroxisomal proliferator activated receptor-α (PPARα) target gene that dehydrogenates straight chain fatty acids and eicosanoids (72, 73). Mice that lack the ACOX1 gene accumulate long chain fatty acids, have reduced numbers of peroxisomes, and develop fatty livers but do not appear to have defects in bile acid synthesis (74, 75). A similar phenotype is observed in humans with ACOX1 deficiency (76). ACOX2 also may be regulated by PPARα and acts on intermediates in the bile acid pathway and 2-methyl-branched chain fatty acids such as pristanic acid (77, 78). The consequences of an ACOX2 knockout in mice have not yet been reported. Three ACOX enzymes exist in the rat, and these have substrate specificities that are different from those of the orthologous human and mouse enzymes (73).

The next step in the biosynthetic pathway involves hydration and oxidation at the Δ^{24} bond and is catalyzed by the D-bifunctional protein (Figure 3, reaction 15). This remarkably complex peroxisomal enzyme of 736 amino acids catalyzes both a hydration step in which a molecule of water is added across the double bond to form a C_{24}-alcohol intermediate and the subsequent oxidation of the alcohol to form the C_{24}-oxo product shown in Figure 3.

Different domains in the amino-terminal half of the enzyme catalyze these two reactions, while the carboxyl-terminal half of the enzyme contains a domain postulated to be involved in sterol transport (76, 79). Mice deficient in the D-bifunctional protein gene accumulate unsaturated C_{27} bile acid intermediates and very long chain fatty acids such as pristanic and phytanic acid (77). They continue to synthesize some C_{24} bile acids; this indicates that another enzyme can also perform this step in the pathway. The compensating enzyme is thought to be the related L-bifunctional protein, which normally utilizes long chain fatty acids as substrates (77).

The last step in the oxidation of the side chain of bile acid intermediates is catalyzed by peroxisomal thiolase 2 (Figure 3, reaction 16), which cleaves the C_{24}-C_{25} bond to form propionyl-coenzyme A and a C_{24}-coenzyme A bile acid intermediate. This enzyme is also referred to as sterol carrier protein-chi (SCPχ) in the literature (80) and is encoded by a gene with two promoters (81). Transcription from sequences flanking exon 1 of the gene produces a mRNA that encodes a 547 amino acid peroxisomal thiolase 2 precursor protein, which following import into the peroxisome is cleaved after amino acid 424 to produce the mature active enzyme. The carboxyl-terminal 123 amino acid product of this cleavage event is referred to as sterol carrier protein 2. X-ray crystallography studies show that this fragment is capable of binding a variety of lipids (82). The second promoter is located in intron 11 and produces a mRNA that encodes only sterol carrier protein 2 (81). Mice deficient in the peroxisomal thiolase 2 gene accumulate bile acid intermediates, C_{23} bile acids, and branched chain fatty acids (83, 84), which indicates that both the full-length thiolase enzyme and the truncated sterol carrier protein 2 are involved in bile acid synthesis and the metabolism of branched chain fatty acids (85). The knockout mice contain some normal C_{24} bile acids; this suggests the presence of another thiolase with activity towards sterols (84).

Conjugation

The terminal step in bile acid synthesis involves the addition of an amino acid, usually glycine or taurine, in amide linkage to carbon 24 (Figure 4, reaction 17). The reaction is catalyzed by the bile acid coenzyme A:amino acid N-acyltransferase enzyme (Table 1). The N-acyltransferase is a remarkably efficient enzyme as more than 98% of bile acids excreted from the liver are amidated. This enzyme, like the preceding four enzymes in the pathway, is located in the peroxisomes (86). The substrates of the N-acyltransferase are a bile acid coenzyme A thioester and either taurine [mice (87)] or glycine and taurine [humans (88)]. The ratio of glycine to taurine conjugated bile acids in humans is solely dependent on the relative abundance of the two amino acids and appears to have no functional or regulatory consequences. The ratio of free to conjugated bile acids may be regulated by PPARα, which activates a peroxisomal enzyme, coenzyme A thioesterase 2, that catalyzes the hydro-

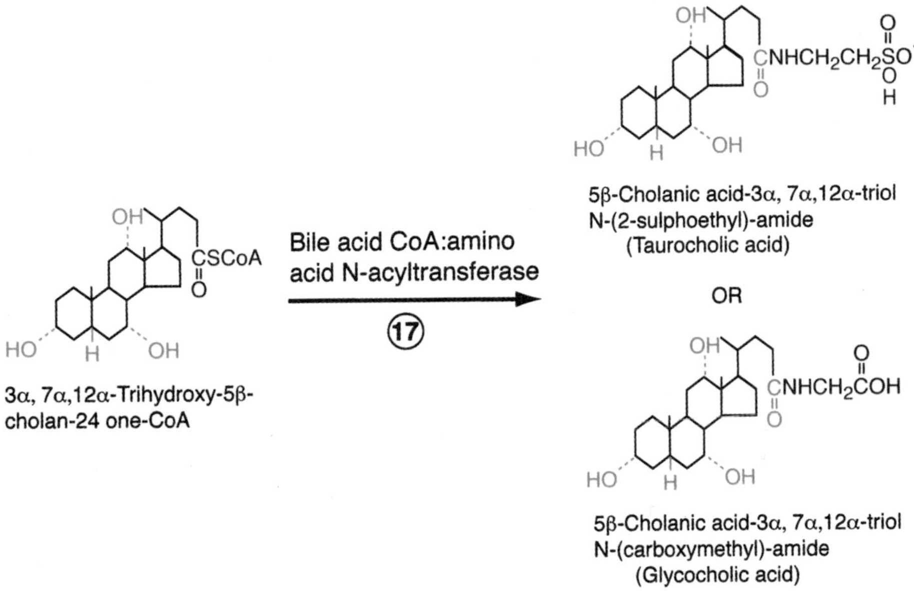

Bile acid CoA:amino acid N-acyltransferase

(17)

3α, 7α,12α-Trihydroxy-5β-cholan-24 one-CoA

5β-Cholanic acid-3α, 7α,12α-triol N-(2-sulphoethyl)-amide (Taurocholic acid)

OR

5β-Cholanic acid-3α, 7α,12α-triol N-(carboxymethyl)-amide (Glycocholic acid)

Figure 4 Conjugation of bile acids. A circled number designates the reaction, and the enzyme that catalyzes this step is indicated next to the arrow. The chemical modifications introduced by the enzyme are indicated in red on the products of the reaction. Cumulative changes to sterol intermediates are indicated in green. Depending on the species of origin, the bile acid CoA:amino acid N-acyltransferase will utilize either glycine or taurine or both amino acids as substrates for conjugation. Conjugated bile acids are secreted from the liver into the bile by ABC transporters and other proteins located in the canalicular membrane.

lysis of bile acid coenzyme A thioesters into free bile acids and coenzyme A (89). The biological consequences arising from changes in this ratio are not known.

Conjugation of bile acids increases the amphipathicity and enhances the solubility of the molecules, which makes them impermeable to cell membranes. Oxygens on the sulfur of taurocholic acid and the terminal carbon of glycocholic acid are ionized at physiological pH, which together with the planar structure of the bile acid and the hydroxyl groups on the rings renders the bile salt (the ionized form of the bile acid) very amphipathic. Conjugation of cholic acid with glycine reduces the pK from 6.4 to 4.4 ensuring that the bile acid is completely ionized and highly soluble. Before conjugation, the monohydrate form of cholic acid is soluble to a concentration of 0.28 g/l in water at 15°C, whereas the sodium salt of free cholic acid is soluble to >569 g/l (90). The sodium salt of glycocholic acid is somewhat less soluble at 274 g/l, which may protect the gut and liver from the exceptionally strong detergent properties of the free bile acid. Because conjugated and free bile acids do not cross cell membranes, dedicated transport systems such as the ileal bile acid transporter and members of the ABC family of transporters are required to move bile acids into and out of cells (91, 92). The need for a transport system increases the half-lives of bile acids in the enterohepatic circulation and protects cells that are otherwise ill equipped to handle the detergent properties of these molecules.

SUBCELLULAR ITINERARY OF BILE ACID SYNTHESIS

As intermediates progress down the bile acid biosynthetic pathways they gradually increase in hydrophilicity to the point that the final conjugated bile acids are exceptionally soluble in water. When these changes in chemical properties are considered together with the distribution of the biosynthetic enzymes in the endoplasmic reticulum, cytosol, mitochondria, and peroxisome (Table 1) and the mass of cholesterol converted into bile acids each day (~500 mg), it is clear that the hepatocyte faces an enormous challenge in transporting intermediates and products of the pathway throughout its interior, and finally, to the exterior of the cell.

How intermediates are moved from one compartment to another remains uncharacterized. In the case of steroid hormone biosynthesis, the rate-limiting step is the movement of cholesterol to the first enzyme in the pathway, the CYP11A1 side chain cleavage enzyme. This transport is accomplished by the steroidogenic acute regulatory (StAR) protein, whose expression is restricted to tissues that synthesize large quantities of steroids (93). Although the expression of StAR in cultured cells stimulates the formation of 27-hydroxycholesterol by sterol 27-hydroxylase (Figure 1, reaction 4) (94, 94a), the StAR transporter is not detectable in the liver. StAR is but one member of a family of proteins referred to as STARTS that are thought to be involved in intracellular cholesterol movement (95). It is possible that one or more START family members is involved in the intracellular transport of bile acid intermediates.

Many steps in the oxidation and shortening of the side chain of bile acid intermediates resemble similar reactions in the fatty acid oxidation pathway. Moreover, the enzymes that catalyze these steps in bile acid synthesis also utilize branched chain fatty acids as substrates (Table 1). This analogy suggests that a transport system similar to the carnitine/carnitine palmitoyl transferase machinery used to import fatty acids into the mitochondria may exist to shuttle bile acid intermediates between intracellular compartments; however, no covalently modified derivatives of bile acid intermediates that could represent a transport form have been reported in mass spectrometry studies.

Adrenoleukodystrophy, a human disease, is caused by mutations in the *ABCD1* gene, which encodes a member of the ABC transporter family that is implicated in the import of very long chain fatty acids into peroxisomes (96). Similarly, genetic studies in yeast (97–99) and *Arabidopsis* (100) indicate that several different ABC transporters are required for the metabolism of fatty acids in these species. With respect to sterol movement, mutations in yeast genes encoding ABC transporters influence the uptake of sterols (100a) and the ability of glucocorticoids to modulate the activity of an ectopically expressed glucocorticoid receptor in this species (101, 102). The ABC transporters active in the latter regard (*LEM* gene products) pump glucocorticoids out of the cell, thereby reducing the activity of the glucocorticoid receptor. By analogy, ABC transporters may export bile acid intermediates out of the various subcellular compartments and into the cytosol en route to another organelle.

The availability of cDNAs encoding all of the enzymes in the bile acid pathway will enable the development of in vitro systems that reconstitute the pathway and allow the biochemical purification of proteins involved in transport. Alternatively, expression cloning approaches in cells that do not normally synthesize bile acids may identify candidate importers and exporters of bile acid intermediates.

REGULATION OF BILE ACID SYNTHESIS

Role of Nuclear Receptors

It has been known since the late 1960s that the amount of bile acid synthesized by the liver is regulated precisely (103, 104). When bile acids accumulate, synthesis is reduced by a negative feedback mechanism that decreases the expression of two enzymes in the biosynthetic pathway, cholesterol 7α-hydroxylase (Figure 1, reaction 1) and sterol 12α-hydroxylase (Figure 2, reaction 8). Conversely, cholesterol accumulation induces bile acid synthesis by activating cholesterol 7α-hydroxylase in some but not all species. The regulatory responses of cholesterol 7α-hydroxylase were shown to be mediated at the transcriptional level in the late 1980s (2). Since that time, transcription factors that mediate negative and positive feedback regulation of bile acid synthesis have been identified, and the mechanisms by which these proteins act have been elucidated.

A striking finding from this body of work is that many of the transcription factors regulating the expression of the cholesterol 7α-hydroxylase and sterol 12α-hydroxylase genes are nuclear receptors (Figure 5). Suppression is triggered by the binding of bile acids (when in excess) to the farnesoid X receptor (FXR, NR1H4), which then activates transcription of the short heterodimeric partner (SHP, NR0B2) gene, a second nuclear receptor (105, 106). SHP binds to and inhibits a third receptor, the liver receptor homologue-1 (LRH-1, NR5A2), which normally activates the genes encoding cholesterol 7α-hydroxylase and sterol 12α-hydroxylase (47, 107, 108). In the absence of LRH-1 activity, transcription from these two genes decreases, and consequently, the synthesis of bile acids declines. The increase in bile acid synthesis that occurs when cholesterol accumulates in rodents is mediated by the liver X receptor α (LXRα, NR1H3),

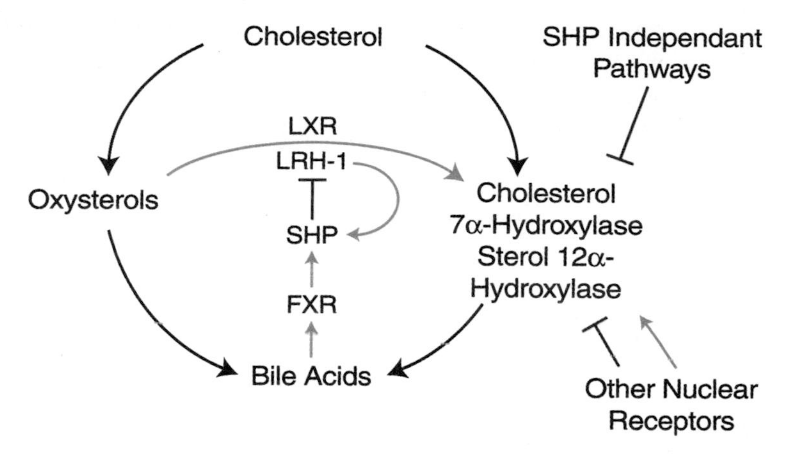

Figure 5 Regulation of bile acid synthesis by nuclear hormone receptors and other inputs. Black arrows indicate the different pathways by which cholesterol is converted into bile acids. Green arrows indicate positive regulatory inputs, and red brakes indicate negative inputs. The nuclear receptor acronyms used are: LXR, liver X receptor α; LRH-1, liver receptor homologue-1; SHP, short heterodimer partner; FXR, farnesoid X receptor.

a nuclear receptor that is activated on binding oxysterols (109–111). In conjunction with LRH-1, LXRα stimulates transcription from the cholesterol 7α-hydroxylase gene (105, 106). Three retinoid X receptors (RXRα, β, and γ; NR2B1–3) serve as coreceptors for FXR and LXR, which brings the basal number of receptors that activate and suppress bile acid synthesis to seven.

The individual contributions of nuclear receptors to the regulation of bile acid synthesis are described below.

Farnesoid X Receptor

Suppression of bile acid synthesis is mediated by FXR, which binds bile acids and activates the transcription of genes involved in bile acid and lipid metabolism (112–114). Target genes include those encoding the ileal bile acid binding protein (112), SHP (105, 106), the phospholipid transfer protein (115), several ABC transporters (116, 117), the organic anion transporting polypeptide 8 (118), and apolipoprotein C-II (58). Mice deficient in FXR express high levels of cholesterol 7α-hydroxylase and sterol 12α-hydroxylase mRNAs; this results in increased bile acid synthesis (119). The mutant mice accumulate bile acids in plasma due to markedly decreased levels of ABCB11, which normally transfers bile acids from the hepatocyte into the bile (92). FXR knockout mice also are hypertryglyceridemic, due in part to decreased expression of apolipoprotein C-II, which is required for the metabolism of triglycerides (58). High concentrations of dietary bile acids cause death in these mice, most likely due to liver failure associated with the loss of ABCB11 expression (119).

FXR agonists that mimic the ability of bile acids to activate the receptor have been identified. The best characterized of these is GW4064, an isoxazole derivative with high selectivity and nanomolar affinity for the receptor (120). When administered to rats, the compound decreases hepatic cholesterol 7α-hydroxylase mRNA levels and reduces serum triglyceride levels (106). A naturally occurring FXR antagonist, guggulsterone, also has been identified. Guggulsterone inhibits transcription from FXR-responsive genes in cultured cells (121, 122) and, when administered to cholesterol-fed mice, prevents the accumulation of cholesterol in the liver (121). The mechanism responsible for the reduction in hepatic cholesterol associated with the administration of guggulsterone remains to be defined.

The bile acid pool of an organism contains twenty or more different bile acids that vary in their abilities to activate FXR, in their susceptibilities to metabolism by the gut flora and in their half-lives in the enterohepatic circulation. In vitro studies indicate that some bile acids, such as chenodeoxycholic acid, are potent FXR ligands, whereas others such as cholic acid, ursodeoxycholic acid, and β–muricholic acid are less active (112–114). Differences in potency can be due to the failure of hydrophilic bile acids to gain entry into cells lacking appropriate transporters (114) or due to differences in the relative affinities of the various bile acids for the nuclear receptor. An example of the latter phenomenon is observed in knockout mice lacking sterol 12α-hydroxylase, which cannot make cholic acid and have unregulated synthesis of bile acids. This derepression is attributable to a failure of the bile acids made in the mutant mice, chiefly muricholic acid derivatives, to activate FXR and hence to mediate feedback regulation (45). Structural differences between receptors isolated from different species also determine the effective concentration of a given bile acid (123); this suggests that a bile acid active in one species may be inactive in another.

Ligand potency is a function of many variables that include the relative affinities of the ligand for the nuclear receptor, of the nuclear receptor for the DNA response element, and of the receptor for auxiliary proteins, such as chaperones, coactivators, and corepressors. The half-life of the ligand also affects its potency, and in the case of bile acids, this parameter is determined by how long the bile acid persists in the enterohepatic circulation and its susceptibility to metabolism by bacteria in the intestinal tract. The relative effects of each of these mechanisms on the regulation of bile acid synthesis remains to be determined, but clearly the possible means by which output from the biosynthetic pathways can be regulated are numerous.

Short Heterodimer Partner

Most nuclear receptor response elements are positive; this means transcription of the target gene is increased when it is bound by a ligand-activated receptor. Although negative response elements are known, the mechanisms by which these act often involve the participation of a second transcription factor rather than the receptor acting as a classic repressor like those defined in bacteria. In agreement with this general rule, FXR suppresses the expression of bile acid synthesis indirectly by activating the transcription of SHP, an unusual nuclear receptor that lacks a DNA binding domain (124). The promoter for the SHP gene contains an FXR response element composed of inverted repeats separated by a single base pair (105, 106), which explains the ability of FXR to stimulate SHP transcription. SHP in turn suppresses bile acid syntheis by binding to and inhibiting LRH-1, which is required for expression of the cholesterol 7α-hydroxylase and sterol 12α-hydroxylase genes (46, 47, 105, 106).

In addition to the FXR binding site in the SHP promoter, there is a functional LRH-1 response element (105, 106, 125). This arrangement provides an autoregulatory mechanism by which the SHP-dependent inhibition of bile acid synthesis can be attenuated (Figure 5). In this loop, the accumulation of SHP protein results in a gradual titration of LRH-1 activity, which in turn decreases SHP transcription and allows the system to return to baseline.

Chow-fed mice deficient in SHP have elevated levels of cholesterol 7α-hydroxylase and sterol 12α-hydroxylase mRNAs, increased synthesis of bile acids, and a larger bile acid pool size (126, 127), which are phenotypic traits consistent with SHP repressing bile acid synthesis. Nevertheless, cholesterol 7α-hydroxylase and sterol 12α-hydroxylase mRNA levels fall when SHP knockout mice are fed bile acids but not when a specific FXR agonist (GW4064) is administered to the mice. These results indicate that additional, SHP-independent, pathways exist that are capable of suppressing bile acid synthesis (see Additional Regulatory Responses).

SHP also inhibits the activity of several other nuclear receptors in transfected cells via direct protein-protein interactions (124). One of these nuclear receptor targets may be hepatic nuclear factor-4α (HNF-4α, NR2A1). In vitro studies indicate that HNF-4α activates transcription from the sterol 12α-hydroxylase promoter (128), and consistent with these findings, HNF-4α deficient mice express low levels of the 12α-hydroxylase mRNA. The increased expression of sterol 12α-hydroxylase in the SHP deficient mice may thus result from derepression of either of the positive transcription factors, HNF-4α or LRH-1.

Liver Receptor Homologue-1

Cholesterol 7α-hydroxylase is expressed exclusively in the liver (7). This tissue specific pattern of expression is mediated by *cis*-acting regulatory sequences in the gene (129, 130), which bind the transcription factor LRH-1 (107). LRH-1 has been isolated several times and each time given a different name. The initial description appeared as a submission (M813985) to GenBank in 1991 with the name liver receptor homologue-1 (LRH-1). Orthologous cDNAs were isolated thereafter from several sources and given various names, including from human (PHR-1) (131), *Xenopus* (xFF1rA) (132), rat (FTF) (133), and human again (CPF) (107). The term LRH-1 is used here to denote the protein's christened name.

LRH-1 is expressed in the liver, ovary, small intestine, pancreas, and colon of the adult mouse (134) and shares sequence identity with steroidogenic factor-1 (SF-1, NR5A1), a nuclear receptor that directs the tissue-specific expression of genes involved in gonadal and adrenal steroidogenesis (135). SF-1 was shown previously to be antagonized by another member of the nuclear receptor family termed DAX-1 (136), which prompted the discovery that antagonism of LRH-1 by SHP was the mechanism by which bile acids mediated feedback regulation (105). The finding that hepatocytes treated with bile acids or the FXR agonist GW4064 induced the expression of SHP was the impetus for this discovery in another laboratory (106). Inhibition involves the binding of SHP to the carboxy-terminal *trans*-activation domain of LRH-1, which prevents the latter from interacting with coactivator proteins (108).

Conventional elimination of LRH-1 from mice causes early embryonic lethality (T.A. Kerr and D.W. Russell, unpublished observations), and thus the expectations for bile acid synthesis arising from loss of this factor cannot be tested in this manner. The application of conditional knockout strategies is underway and should provide both confirmation and further insight into the biological roles of LRH-1.

Liver X Receptor α

That cholesterol could induce the expression of cholesterol 7α-hydroxylase and hence bile acid synthesis was initially controversial; however, the isolation of cDNAs encoding this enzyme allowed an unambiguous demonstration that this is the case in rats (7) and mice (137). Substrate mediated induction occurs via LXRα, which binds oxysterol ligands and induces transcription of the cholesterol 7α-hydroxylase gene (109–111). An LXRα response element is located in the 5′-flanking region of the mouse and rat 7α-hydroxylase genes (110), but this element is either missing in species that do not respond to cholesterol (e.g., human and rabbit) or present but nonfunctional in other nonresponding species (e.g., hamster) (138).

Mice deficient in LXRα appear phenotypically normal until challenged with diets high in cholesterol, which cause dramatic accumulation of the sterol in the liver and eventually death (139). The mutant mice fail to induce cholesterol 7α-hydroxylase and thus are unable to convert excess cholesterol into bile acids. These results explain in part the resistance of plasma cholesterol levels to dietary cholesterol in rodents, and conversely, the sensitivity of other species such as rabbits and humans that fail to induce cholesterol 7α-hydroxylase when a high cholesterol diet is consumed (140, 141). There are two LXR genes in mammals, one encoding LXRα and the other encoding LXRβ (NR1H2) (142). LXRα is more highly expressed in the liver than LXRβ (134), and in agreement with this expression pattern, mice lacking LXRβ do not exhibit dislipidemia when fed cholesterol (143). Mice that lack both LXRα and -β have a more severe phenotype than those without LXRα, which suggests that LXRβ is able to compensate partially for LXRα (144).

LXR activates expression from the cholesterol 7α-hydroxylase gene but does not induce transcription from the sterol 12α-hydroxylase gene. In contrast to 7α-hydroxylase, sterol 12α-hydroxylase is suppressed by cholesterol feeding in murine species (145). As a consequence, the bile acid profile changes so proportionally less cholic acid is synthesized. Cholic acid is a potent ligand for FXR, which means that a decrease in the production of this bile acid attenuates the negative feedback regulation of bile acid synthesis and facilitates cholesterol disposal (45). Moreover, cholic acid enhances the solubilization and absorption of cholesterol in the small intestine, and a reduction in its level decreases the delivery of dietary cholesterol to the liver in the face of a cholesterol challenge.

LXR plays a crucial role in integrating the pathways of cholesterol supply and catabolism by regulating the expression of the sterol response element binding protein-1c (SREBP-1c) gene. SREBP-1c is a transcription factor that activates many genes involved in cholesterol and lipid biosynthesis (146). As cholesterol accumulates, LXR induces the expression of SREBP-1c, which in turn activates the synthesis of enzymes that produce fatty acids (144). These fatty acids are then utilized in the formation of cholesteryl esters, the intracellular storage form of cholesterol. LXR also stimulates the transcription of genes encoding cholesterol efflux proteins [the ABCA1, ABCG5, and ABCG8 transporters (147, 148)], and in this manner enhances the direct secretion of the sterol from the liver. Numerous other genes involved in lipid metabolism are targets of LXR (149); these include the cholesterol ester transport protein, ABCG1, apolipoprotein E, lipoprotein lipase, and the phospholipid transfer protein (150), and the activation of these contributes to the integration of the supply and catabolism pathways.

Synthetic agonists that activate LXR have been developed, initially as potential therapeutic agents to treat hyperlipidemias. These compounds activate LXR in vitro and induce the expected transcriptional responses of the receptor in vivo (147, 151, 152). Treatment of mice and hamsters with one of these agonists (T0901317), however, is reported to increase plasma cholesterol and triglyceride levels (151). Although the mechanism responsible for these increases is not known, it may be that the effects of stimulation of SREBP-1c by LXR and the ensuing increase in lipid synthesis predominate over the stimulation of cholesterol 7α-hydroxylase and bile acid synthesis. An added challenge in the development of LXR therapeutics is that activation of this receptor in human hepatocytes causes a decrease in cholesterol 7α-hydroxylase expression (153). Thus, species-specific regulatory responses must be taken into account. LXR agonists may find application as antiatherogenic agents in macrophages where they promote the excretion of cholesterol from the vessel wall (154, 155); alternatively, receptor antagonists may be useful in lowering plasma lipid levels.

Retinoid X Receptor

LXR and FXR form obligate heterodimers with RXR proteins. The activities of the heterodimeric LXR and FXR receptors with respect to the regulation of bile acid synthesis are modulated by ligands that interact with the RXR subunit (147).

Mice deficient in hepatic RXRα express high levels of cholesterol 7α-hydroxylase, which is consistent with the loss of FXR-mediated feedback inhibition, and they fail to induce expression of the 7α-hydroxylase gene in response to dietary cholesterol, which is consistent with reduced LXRα input (156). Two additional RXR isoforms, β and γ, are expressed in the liver (157), but the above results suggest that RXRα is the predominant heterodimerization partner for nuclear receptors in this tissue.

Other Nuclear Receptors

The expression of cholesterol 7α-hydroxylase is increased while that of sterol 12α-hydroxylase is decreased in rats administered thyroid hormone (145, 158–160). Thyroid hormone response elements in the rat cholesterol 7α-hydroxylase promoter have not yet been defined, but a negative element has been identified in the human gene (161), which suggests that induction of this gene may result from the alleviation of inhibition rather than ligand-dependent stimulation of transcription. The mechanism by which thyroid hormone represses 12-hydroxylase expression is not known (160). Analyses in receptor knockout mice suggest that thyroid hormone receptor β (NR1A2) is responsible for the activation of the cholesterol 7α-hydroxylase gene (162).

Mice with an induced hepatic deficiency of the nuclear receptor HNF-4α (NR2A1) have markedly reduced expression of cholesterol 7α-hydroxylase (163). Many genes involved in lipid metabolism have altered expression patterns in these animals, and thus it is not clear whether the response of cholesterol 7α-hydroxylase is direct or indirect. The expression of the CYP7B1 oxysterol 7α-hydroxylase and sterol 12α-hydroxylase genes is reduced to very low levels in the HNF-4α-deficient mice, and in vitro studies indicate that these two genes are direct targets of the receptor. The effects of these changes on bile acid metabolism in the mutant mice remain to be determined, but the prediction based on results from other knockout animals is that oxysterols will accumulate and that cholic acid levels will be very low in the HNF-4α-deficient mice (27, 45). Certain cytokines decrease the expression of cholesterol 7α-hydroxylase (see Additional Regulatory Responses), and in vitro studies indicate that these act by decreasing the activity of HNF-4α (164).

Agonists of PPARα (NR1C1) such as fibrates have multiple effects on bile acid metabolism (165). In rats and mice, dietary fibrates decrease the expression of cholesterol 7α-hydroxylase resulting in a reduction in bile acid secretion (166). In vitro studies indicate that PPARα antagonizes the positive effects of HNF4-α on the promoter of the cholesterol 7α-hydroxylase gene (167, 168). PPARα deficient mice express normal levels of cholesterol 7α-hydroxylase suggesting that the input of this receptor is modest under physiological conditions (168). However, these animals are unresponsive to dietary fibrates (166, 168), and this indicates the response of bile acid synthesis to these agents is receptor mediated. The expression of sterol 12α-hydroxylase is increased by fibrates in the mouse (169), which would have the net effect of elevating the level of cholic acid in the bile acid pool and of reducing the expression of cholesterol 7α-hydroxylase via activation of the FXR-SHP pathway (45).

When certain bile acids, such as lithocholic acid, accumulate in an organism, they induce a number of responses that reduce bile acid synthesis in an effort to protect the liver. These responses are mediated *inter alia* by three nuclear receptors, the pregnane X receptor (PXR, NR1I2), the constitutive androstane receptor (CAR, NRI3), and the vitamin D receptor (VDR, NR1I1). Lithocholic acid is an agonist for PXR and VDR, and through these receptors stimulates the transcription of cytochrome P450 and sulfotransferase genes whose products detoxify the bile acid (170–172). In addition, activated PXR reduces the expression of cholesterol 7α-hydroxylase by an as yet undefined mechanism (170, 173). The regulatory contributions of CAR to bile acid synthesis remain to be determined, but the ligand specificity of this receptor overlaps with that of PXR (174), which suggests an important role in this regard.

Additional Regulatory Reponses

The actions of nuclear receptors underlie many of the responses of bile acid synthesis to various dietary and physiological inputs. There are clearly other regulatory pathways that impinge on synthesis and for which the identities of the participating proteins are more or less well defined. Underlying transcription factors have been identified for pathways that involve hepatic nuclear factor-1α (HNF-1α) and two proteins involved in circadian rhythm, the D site binding protein (DBP) and the adenovirus E4 promoter ATF site binding protein (E4BP4). Mice that lack HNF-1α have elevated levels of cholesterol 7α-hydroxylase and increased bile acid synthesis (175). The loss of feedback regulation in these animals is due in part to decreased expression of FXR and a corresponding decline in SHP expression. Cholesterol 7α-hydroxylase is subject to diurnal regulation mediated by DBP and E4BP4 (176, 177), and the expression of DBP itself is regulated by circadian rhythm in the liver as well as regions of the brain involved in an organism's response to light (177, 178). This level of control integrates the expression of cholesterol 7α-hydroxylase and bile acid synthesis with food intake and exposure to light, which are postulated to be at the ancestral heart of metabolic gene regulation (179).

Evidence for additional regulatory pathways with as yet undefined mediators comes from the analysis of mice that lack SHP, the central repressor in the scheme shown in Figure 5. Based on this model, the removal of SHP should result in the constitutive expression of cholesterol 7α-hydroxylase and sterol 12α-hydroxylase, and it should eliminate the response of these genes to dietary bile acids. Experiments in two independent lines of knockout mice reveal that SHP deficient mice exhibit only modest elevations in hydroxylase gene expression (~twofold) and respond to dietary bile acids by decreasing expression of cholesterol 7α-hydroxylase and sterol 12α-hydroxylase (126, 127). The mutant animals do not respond to an FXR agonist (GW4064), which indicates that FXR mediates suppression via activation of SHP as predicted by the model.

These data indicate the existence of one or more SHP-independent pathways that negatively regulate bile acid synthesis. The modest elevation in hydroxylase gene expression in the mutant mice is ascribed to a tonic suppression of synthesis mediated by the bile acid pool size. Two observations support this interpretation. First, the pool size is increased by $\sim$30% in these mice and, as a consequence of elevated sterol 12α-hydroxylase expression, has more cholic acid in it. These alterations cause suppression of bile acid synthesis in wild-type mice and in the knockout mice serve to limit derepression of the biosynthetic enzymes. Second, the SHP knockout mice induce cholesterol 7α-hydroxylase and sterol 12α-hydroxylase in response to cholestyramine, a drug whose only known mechanism of action is to reduce the bile acid pool size (126, 127).

A second SHP-independent pathway comes into play when the SHP knockout animals consume bile acids. This damages the liver and induces a stress-response pathway that decreases the expression of bile acid biosynthetic enzymes (126). The c-Jun N-terminal kinase (JNK) is thought to mediate this repression (127, 180).

Additional lines of evidence support the notion that kinases and other intracellular signaling molecules regulate bile acid synthesis. For example, certain strains of mice fail to repress cholesterol 7α-hydroxylase in response to bile acid feeding, and this outcome is correlated with alterations in cytokine secretion, in particular with the levels of tumor necrosis factor and interleukin-1, which repress 7α-hydroxylase expression (181). In addition, mice deficient in the fibroblast growth factor 4 receptor (FGFR4), a cell surface tyrosine kinase, express high levels of cholesterol 7α-hydroxylase and have markedly elevated bile acid pool sizes and fecal excretion rates (182). When fed bile acids and cholesterol, they develop massive liver enlargement, which is correlated with the induction of a corepressor of nuclear hormone receptors termed receptor interacting peptide 140 (RIP140). This induction in turn is postulated to antagonize the positive actions of LXR on the cholesterol 7α-hydroxylase promoter and thus to mimic the phenotype associated with loss of LXR (182).

The involvement of cytokines and a tyrosine kinase receptor in regulating bile acid synthesis means that any number of intracellular signaling molecules may affect the system, which include intracellular kinases, phosphatases, and adaptor proteins. The identification of these regulatory proteins is an active area of current research.

GENETICS OF BILE ACID SYNTHESIS

Inherited genetic diseases that affect bile acid synthesis provide a unique window on the biological roles of bile acids and of individual enzymes in the pathways. Seven defects involving enzymes that catalyze both early and late steps in the pathways are known (Table 2), and several more have been reported in abstract form. In general, those that affect early biosynthetic steps cause disease in

TABLE 2 Diseases caused by mutations in bile acid synthesis genes

Reaction[a]	Enzyme	Gene	Chromosome	Phenotype	OMIM (182a)
1	Cholesterol 7α-hydroxylase	*CYP7A1*	8q11-q12	Hypercholesterolemia	118455
4	Sterol 27-hydroxylase	*CYP27A1*	2p23.3-p24.1	Cerebrotendinous xanthomatosis, progressive CNS neuropathy, cholestanol, and bile alcohol accumulation	606530
6	Oxysterol 7α-hydroxylase	*CYP7B1*	8q21.3	Hyperoxysterolemia, neonatal liver failure	603711
7	3β-hydroxy-Δ^5-C_{27}-steroid oxidoreductase	*HSD3B7*	16p11.2-p12	Neonatal liver failure, hepatotoxic bile acid intermediate accumulation	231100
9	Δ^4-3-Oxosteroid 5β-reductase	*AKR1C4*	7q32-q33	Neonatal liver failure, hepatotoxic bile acid intermediate accumulation	604741
13	2-Methylacyl-CoA racemase	*AMACR*	5p13.2-q11.1	Adult onset sensory motor neuropathy, pristanic acid accumulation, neonatal liver disease	604489
15	D-Bifunctional protein	*EHHADH*	3q27	Hypotonia, liver enlargement, developmental defects, pristanic acid, and C_{27} bile acid accumulation	261515

[a] Reaction numbers refer to Figures 1, 2, and 3.

newborns, while the consequences of those affecting later steps are varied. Several disorders, including Zellweger syndrome, neonatal adrenoleukodystrophy, and infantile Refsum disease, affect peroxisome assembly and thereby impair bile acid synthesis. These diseases are not considered further here, as they do not involve mutations in genes encoding biosynthetic enzymes.

Cholesterol 7α-Hydroxylase Deficiency

A prismatic pedigree of cholesterol 7α-hydroxylase deficiency is known in which affected individuals present with elevated plasma cholesterol levels, decreased bile acid excretion, and accumulation of cholesterol in the liver (32). A two base pair deletion in exon 6 of the gene (*CYP7A1*) in these subjects causes a shift in the translational reading frame and the synthesis of a truncated protein with no enzymatic activity. Individuals heterozygous for the mutant allele have modest elevations in serum cholesterol, which suggests that the disorder is inherited in a codominant fashion. The hyperlipidemic phenotype associated with cholesterol 7α-hydroxylase deficiency in humans contrasts with the normolipidemic phenotype of mice that lack this enzyme (12). This difference may reflect alterations in how bile acid synthesis is regulated in the two species, for example, via LXR (153). The hyperlipidemia of human cholesterol 7α-hydroxylase deficiency is resistant to treatment with hydroxymethylglutaryl CoA reductase inhibitors (statins) (32). The effects of bile acid therapy have not yet been tested in these individuals.

Sterol 27-Hydroxylase Deficiency

Loss of sterol 27-hydroxylase (Figure 1, reaction 4 and Figure 3, reaction 11) causes the neuropathological disorder cerebrotendinous xanthomatosis (CTX) (59). CTX is characterized by the synthesis of abnormal bile alcohols, a reduced synthesis of normal bile acids, and the accumulation of cholesterol and the 5α-reduced derivative of cholesterol, cholestanol, in the blood and tissues of affected individuals. The buildup of the latter sterols in the myelin sheaths surrounding neurons in the brain gradually disrupts the ordered structure of this tissue and causes progressive neurological dysfunction and eventual death. Since an initial description of the molecular basis of this disease in 1991 (183), more than 43 mutations have been described in the *CYP27A1* gene that cause CTX (184), and the consequences of an induced mutation in the mouse are well characterized (54–56). There are marked differences between the phenotypes of the mutant mice and human subjects with this disorder, which, as described above, appear to be due to alterations in transcriptional regulatory mechanisms and biosynthetic enzyme expression. CTX is treated effectively with oral bile acid therapy if diagnosed by mass spectrometry or molecular methods at an early age (59).

Oxysterol 7α-Hydroxylase Deficiency

As with cholesterol 7α-hydroxylase deficiency, one family with CYP7B1 oxysterol 7α-hydroxylase deficiency is known (185). The affected individual presented with liver failure as a newborn, a marked elevation of oxysterols in the serum, and the hyperexcretion of 3β-hydroxy-5-cholestenoic and 3β-hydroxy-5-cholenoic acids in the urine. 3β-Hydroxy-5-cholestenoic acid is the expected product arising from repeated side chain oxidations of cholesterol by sterol 27-hydroxylase (Figure 3, reaction 11), and 3β-hydroxy-5-cholenoic acid is derived from cholesterol after undergoing all of the side chain oxidation steps shown in Figure 3. Molecular analysis of this subject's genomic DNA revealed a nonsense mutation in exon 5 of the CYP7B1 oxysterol 7α-hydroxylase gene (*CYP7B1*) that caused premature truncation of the protein and eliminated enzymatic activity (185).

The chemical phenotype in this individual is consistent with a generalized failure to 7α-hydroxylate oxysterols produced by the alternate pathways of bile acid synthesis (Figure 1, reactions 2, 3, and 4), while the liver failure is consistent with the known hepatotoxic properties of oxysterols and the unsaturated monohydroxy bile acids listed above. Cholesterol 7α-hydroxylase protein was present in the liver of this subject as determined by immunoblotting; however, no corresponding enzyme activity was detectable (185). Presumably, the accumulated oxysterols inhibited this enzyme and effected a complete loss of bile acid synthesis. This subject did not respond to oral bile acid therapy and was treated subsequently by liver transplantation. The phenotype of the CYP7B1 oxysterol 7α-hydroxylase deficient individual, together with data showing that the contributions of the alternate pathways to bile acid synthesis are modest in humans (22, 30, 31), suggest that the most important function of this enzyme is to inactivate oxysterols by facilitating their conversion to bile acids.

3β-Hydroxy-Δ^5-C_{27}-Steroid Oxidoreductase Deficiency

Patients with a defect in the C_{27} 3β-HSD enzyme were first described in 1987 (186). They present in the clinic with neonatal jaundice, liver enlargement, fat-soluble vitamin deficiency, and lipid malabsorption. Mass spectrometry reveals an accumulation in blood and urine of hepatotoxic C_{24} and C_{27} steroids with 3β-hydroxy-Δ^5 structures. Oral bile acid therapy is curative and is thought to cause suppression of cholesterol 7α-hydroxylase and thus prevent further synthesis of the cholestatic intermediates. The encoding C_{27} 3β-HSD gene (*HSD3B7*) was isolated in 2000 (40), and since then 12 different mutations that inactivate the enzyme have been identified in affected individuals (J. B. Cheng et al., unpublished observations). Very little, if any, cholic acid and chenodeoxycholic acid are detected by sensitive chemical methods in these subjects, which indicates that the C_{27}3β-HSD enzyme is unique in the bile acid biosynthetic pathway.

Δ^4-3-Oxosteroid 5β-Reductase Deficiency

Loss of this enzyme activity (Figure 2, reaction 9) causes an accumulation of C_{24} bile acids that retain the Δ^4-3-oxo structure in the A-ring and C_{24} bile acids that have a 5α-reduced configuration (*allo*-bile acids) (187). These sterols are hepatotoxic and over time cause liver failure. The Δ^4-3-oxosteroid 5β-reductase enzyme has been purified to homogeneity, and a putative cDNA cloned (188). Antibodies raised against the protein failed to detect a signal in liver samples from affected individuals (2); however, no formal description of the molecular basis of this disease has appeared to date. With respect to bile acid synthesis, the chemical phenotype of these individuals indicates that the 5β-reductase acts preferentially on intermediates in the pathway, but when these accumulate in the absence of the enzyme, the structurally unrelated, membrane-bound steroid 5α-reductase enzymes (189) are capable of reducing the Δ^4 bond to produce *allo*-bile acids. Steroid 5β-reductase deficiency is effectively treated with oral bile acid therapy.

2-Methylacyl-CoA Racemase Deficiency

Ferdinandusse et al. reported two patients with adult-onset sensory motor neuropathy who had elevated plasma levels of pristanic acid, a polyisoprenoid fatty acid, and of two intermediates in the bile acid biosynthetic pathway, 25(*R*)3α,7α,12α-trihydroxy-5β-cholestanoic acid and 25(*R*)3α,7α-dihydroxy-5β-cholestanoic acid (190). The accumulation of these compounds was consistent with defects in the 2-methylacyl-CoA racemase (Figure 3, reaction 13), and two point mutations were identified in the encoding gene (*AMACR*) that inactivated the enzyme. Cultured fibroblasts from these subjects exhibited a reduced ability to oxidize pristanic acid, but the activities of downstream enzymes involved in β–oxidation, which included the D-bifunctional protein and peroxisomal thiolase 2, were normal. Liver function also was apparently normal in these adult individuals. This suggested that the above C_{27} intermediates were efficient as bile acids. In agreement with this outcome, several species, including alligators and frogs, synthesize 3α,7α,12α-trihydroxy-5β-cholestanoic acid and 3α,7α-dihydroxy-5β-cholestanoic acid as primary bile acids (191). A similar phenotype of progressive peripheral neuropathy is characteristic of infantile Refsum disease in which branched chain fatty acids alone accumulate (192). These observations together suggest that the buildup of C_{27} bile acid intermediates in the racemase deficient subjects does not contribute to the symptoms of the disease.

An infant with 2-methylacyl-CoA racemase deficiency presented a very different clinical picture (192a). In contrast to the adult subjects, this child had coagulopathy, vitamin D and E deficiencies, and mild liver impairment. No neurologic disease was noted. Chemical analysis revealed an accumulation of C_{27} racemase substrates and pristanic acid. A point mutation specifying a missense substitution (S52P) was present in exon 1 of the *AMACR* gene. Treatment with cholic acid restored liver function and reduced levels of bile acid intermediates. Together, these findings suggest that racemase deficiency causes liver failure in children that resolves prior to adulthood, at which time the progressive accumulation of phytanic acid leads to neuropathy. The ontological changes in the phenotype of affected individuals suggests that cholic acid therapy is appropriate in childhood and that diets low in branched chain fatty acids may be palliative in adulthood.

D-Bifunctional Protein Deficiency

Loss of this enzyme activity is associated with the accumulation of C_{27} bile acid intermediates and pristanic acid, an enlarged liver, developmental defects, hypotonia, and seizures (193–195). Early studies ascribed the biochemical defect in these patients to the absence of the L-bifunctional protein (193); however, later studies showed that mutations in the structurally related D-bifunctional protein were the cause of this genetic disease (196–198). Some, but not all, patients manifest symptoms of liver failure although all show the associated neurological deficiencies (76). This presentation makes it difficult to determine whether the accumulation of bile acid intermediates or pristanic acid is the major contributor to the observed phenotype. An effective treatment for D-bifunctional protein deficiency is not reported.

FUTURE DIRECTIONS

In conclusion, interest in the synthesis and metabolism of bile acids has continued to grow over the past decade due in large part to advances in the biochemistry, regulation, and genetics of the enzymes involved. Much remains to be learned about the transport of intermediates between subcellular organelles, the roles of nuclear receptors and their various spliced forms and ligands, the identities of other transcription factors and intracellular signaling enzymes that regulate bile acid synthesis, and the physiological importance of tissue specific cholesterol catabolism. The next several years should be equally exciting.

ACKNOWLEDGMENTS

I thank Trevor Penning for insight into the aldo-keto reductase family of enzymes, and Mike Brown, Joe Goldstein, Helen Hobbs, Steve Kliewer, and David Mangelsdorf for critical review of the manuscript. Research described here from the author's lab is supported by grants from the National Institutes of Health (HL20948, HD38127).

The *Annual Review of Biochemistry* is online at http://biochem.annualreviews.org

LITERATURE CITED

1. Carey MC. 1982. *The Liver, Biology and Pathobiology*, pp. 429–65. New York: Raven
2. Russell DW, Setchell KDR. 1992. *Biochemistry* 31:4737–49
3. Hylemon PB, Harder J. 1998. *FEMS Microbiol. Rev.* 22:475–88

3a. Natl. Cent. Biotechnol. Inf. 2002. *GenBank.* http://www.ncbi.nlm.nih.gov/

4. Chiang JYL. 1998. *Front. Biosci.* 3:D176–93
5. Ogishima T, Deguchi S, Okuda K. 1987. *J. Biol. Chem.* 262:7646–50
6. Li YC, Wang DP, Chiang JYL. 1990. *J. Biol. Chem.* 265:12012–19
7. Jelinek DF, Andersson S, Slaughter CA, Russell DW. 1990. *J. Biol. Chem.* 265:8190–97
8. Li YC, Chiang JYL. 1991. *J. Biol. Chem.* 266:19186–91
9. Karam WG, Chiang JYL. 1994. *J. Lipid Res.* 35:1222–31
10. Nakayama K, Puchkaev A, Pikuleva IA. 2001. *J. Biol. Chem.* 276:31459–65
11. Miyake JH, Doung X-DT, Strauss W, Moore GL, Castellani LW, et al. 2001. *J. Biol. Chem.* 276:23304–11
12. Ishibashi S, Schwarz M, Frykman PK, Herz J, Russell DW. 1996. *J. Biol. Chem.* 271:18017–23
13. Schwarz M, Lund EG, Setchell KDR, Kayden HJ, Zerwekh JE, et al. 1996. *J. Biol. Chem.* 271:18024–31
14. Arnon R, Yoshimura T, Reiss A, Budai K, Lefkowitch JH, Javitt NB. 1998. *Gastroenterology* 115:1223–28
15. Schwarz M, Russell DW, Dietschy JM, Turley SD. 1998. *J. Lipid Res.* 39:1833–43
16. Schwarz M, Russell DW, Dietschy JM, Turley SD. 2001. *J. Lipid Res.* 42:1594–603
17. Javitt NB. 2000. *Biochim. Biophys. Acta* 1529:136–41
18. Lund EG, Guileyardo JM, Russell DW. 1999. *Proc. Natl. Acad. Sci. USA* 96:7238–43
19. Bogdanovic N, Bretillon L, Lund EG, Diczfalusy U, Lannfelt L, et al. 2001. *Neurosci. Lett.* 314:45–48
20. Deleted in proof
21. Björkhem I, Diczfalusy U, Lutjohann D. 1999. *Curr. Opin. Lipidol.* 10:161–65
22. Björkhem I, Lutjohann D, Diczfalusy U, Stahle L, Ahlborg G, Wahren J. 1998. *J. Lipid Res.* 39:1594–600
23. Björkhem I, Andersson U, Ellis E, Alvelius G, Ellegard L, et al. 2001. *J. Biol. Chem.* 276:37004–10
24. Shanklin J, Cahoon EB. 1998. *Annu. Rev. Plant Physiol. Plant Mol. Biol.* 49:611–41
25. Lund EG, Kerr TA, Sakai J, Li W-P, Russell DW. 1998. *J. Biol. Chem.* 273: 34316–48
26. Schroepfer GJ. 2000. *Physiol. Rev.* 80:361–554
27. Li-Hawkins J, Lund EG, Turley SD, Russell DW. 2000. *J. Biol. Chem.* 275: 16536–42
28. Dzeletovic S, Breuer O, Lund E, Diczfalusy U. 1995. *Anal. Biochem.* 225: 73–80
29. Lund E, Björkhem I, Furster C, Wikvall K. 1993. *Biochim. Biophys. Acta* 1166: 177–82
30. Babiker A, Andersson O, Lindblom D, van der Linden J, Wiklund B, et al. 1999. *J. Lipid Res.* 40:1417–25
31. Duane WC, Javitt NB. 1999. *J. Lipid Res.* 40:1194–99
32. Pullinger CR, Eng C, Salen G, Shefer S, Batta AK, et al. 2002. *J. Clin. Investig.* 110:109–17
33. Li-Hawkins J, Lund EG, Bronson AD, Russell DW. 2000. *J. Biol. Chem.* 275: 16543–49
34. Li-Hawkins J, Meunch C, Russell DW. 2000. *Pathways of bile acid synthesis.* Presented at Biol. Bile Acids Health Dis. XVI Int. Bile Acid Meet., ed. GP van Berge Henegouwen, D Keppler, U Leuschner, G Paumgartner, A Stiehl., Dordrecht, Neth.: Kluwer Acad.
35. Ikeda H, Ueda M, Ikeda M, Kobayashi H, Honda Y. 2003. *Lab. Investig.* 83:349–55
36. Stapleton G, Steel M, Richardson M, Mason JO, Rose KA, et al. 1995. *J. Biol. Chem.* 270:29739–45
37. Rose KA, Stapleton G, Dott K, Kieny MP, Best R, et al. 1997. *Proc. Natl. Acad. Sci. USA* 94:4925–30
38. Schwarz M, Lund EG, Lathe R,

Björkhem I, Russell DW. 1997. *J. Biol. Chem.* 272:23995–24001
39. Martin KO, Reiss AB, Lathe R, Javitt NB. 1997. *J. Lipid Res.* 38:1053–58
39a. Rose K, Allan A, Gauldie S, Stapleton G, Dobbie L, et al. 2001. *J. Biol. Chem.* 276:23937–44
40. Schwarz M, Wright AC, Davis DL, Nazer H, Björkhem I, Russell DW. 2000. *J. Clin. Investig.* 106:1175–84
41. Wikvall K. 1981. *J. Biol. Chem.* 256:3376–80
42. Furster C, Zhang J, Toll A. 1996. *J. Biol. Chem.* 271:20903–7
43. Eggertsen G, Olin M, Andersson U, Ishida H, Kubota S, et al. 1996. *J. Biol. Chem.* 271:32269–75
44. Gafvels M, Olin M, Chowdhary BP, Raudsepp T, Andersson U, et al. 1999. *Genomics* 56:184–96
45. Li-Hawkins J, Gafvels M, Olin M, Lund EG, Andersson U, et al. 2002. *J. Clin. Investig.* 110:1191–200
46. Castillo-Olivares A, Gil G. 2000. *J. Biol. Chem.* 275:17793–99
47. Castillo-Olivares A, Gil G. 2001. *Nucleic Acids Res.* 29:4035–42
48. Jez JM, Penning TM. 2001. *Chem.-Biol. Interact.* 130(1–3):499–525
49. Penning TM. 1997. *Endocr. Rev.* 18:281–305
50. Usui E, Okuda K, Kato Y, Noshiro M. 1994. *J. Biochem.* 115:230–37
51. Andersson S, Davis DL, Dahlback H, Jornvall H, Russell DW. 1989. *J. Biol. Chem.* 264:8222–29
52. Dahlback H, Holmberg I. 1990. *Biochem. Biophys. Res. Commun.* 167:391–95
53. Pikuleva IA, Babiker A, Waterman MR, Björkhem I. 1998. *J. Biol. Chem.* 273:18153–60
54. Rosen H, Reshef A, Maeda N, Lippoldt A, Shpizen S, et al. 1998. *J. Biol. Chem.* 273:14805–12
55. Repa JJ, Lund EG, Horton JD, Leitersdorf E, Russell DW, et al. 2000. *J. Biol. Chem.* 275:39685–92
56. Honda A, Salen G, Matsuzaki Y, Batta AK, Xu G, et al. 2001. *J. Biol. Chem.* 276:34579–85
57. Honda A, Salen G, Matsuzaki Y, Batta AK, Xu G, et al. 2001. *J. Lipid Res.* 42:291–300
58. Kast HR, Nguyen CM, Sinal CJ, Jones SA, Laffitte BA, et al. 2001. *Mol. Endocrinol.* 15:1720–28
59. Björkhem I, Boberg KM, Leitersdorf E. 2001. In *The Metabolic and Molecular Bases of Inherited Disease*, ed. CR Scriver, AL Beaudet, WS Sly, D Valle, B Childs, et al., 2:2961–88. New York: McGraw-Hill. 8th ed.
60. Furster C, Wikvall K. 1999. *Biochim. Biophys. Acta* 1437:46–52
61. Reddy JK, Hashimoto T. 2001. *Annu. Rev. Nutr.* 21:193–230
62. Uchiyama A, Aoyama T, Kamijo K, Uchida Y, Kondo N, et al. 1996. *J. Biol. Chem.* 271:30360–65
63. Berger J, Truppe C, Neumann H, Forss-Petter S. 1998. *FEBS Lett.* 425:305–9
64. Steinberg SJ, Wang SJ, Kim DG, Mihalik SJ, Watkins PA. 1999. *Biochem. Biophys. Res. Commun.* 257:615–21
65. Steinberg SJ, Wang SJ, McGuinness MC, Watkins PA. 1999. *Mol. Genet. Metab.* 68:32–42
66. Mihalik SJ, Steinberg SJ, Pei Z, Park J, Kim DG, et al. 2002. *J. Biol. Chem.* 277:24771–79
67. Gustafsson J, Sjostedt S. 1978. *J. Biol. Chem.* 253:199–201
68. Shefer S, Cheng FW, Batta AK, Dayal B, Tint GS, et al. 1978. *J. Biol. Chem.* 253:6386–92
69. Kotti TJ, Savolainen K, Helander HM, Yagi A, Novikov DK, et al. 2000. *J. Biol. Chem.* 275:20887–95
70. Amery L, Fransen M, De Nys K, Mannaerts GP, van Veldhoven PP. 2000. *J. Lipid Res.* 41:1752–59
71. Schmitz W, Albers C, Fingerhut R, Conzelmann E. 1995. *Eur. J. Biochem.* 231:815–22

72. Lee SS, Pineau T, Drago J, Lee EJ, Owens JW, et al. 1995. *Mol. Cell. Biol.* 15:3012–22
73. VanVeldhoven PP, Vanhove G, Asssel-berghs S, Eyssen HJ, Mannaerts GP. 1992. *J. Biol. Chem.* 267:20065–74
74. Fan C-Y, Pan J, Chu R, Lee D, Kluckman KD, et al. 1996. *J. Biol. Chem.* 271:24698–710
75. Infante JP, Tschanz CL, Shaw N, Michaud AL, Lawrence P, Brenna JT. 2002. *Mol. Genet. Metab.* 75:108–19
76. Wanders RJA, Barth PG, Heymans HSA. 2001. See Ref. 59, Vol. II, pp. 3219–56
77. Baes M, Huyghe S, Carmeliet P, Declercq PE, Collen D, et al. 2000. *J. Biol. Chem.* 275:16329–36
78. Vanhove GF, Van Veldhoven PP, Fransen M, Denis S, Eyssen HJ, et al. 1993. *J. Biol. Chem.* 268:10335–44
79. Haapalainen AM, van Aalten DMF, Merilainen G, Jalonen JE, Pirila P, et al. 2001. *J. Mol. Biol.* 313:1127–38
80. Seedorf U, Assmann G. 1991. *J. Biol. Chem.* 266:630–36
81. Ohba T, Holt JA, Billheimer JT, Strauss JF III. 1995. *Biochemistry* 34: 10660–68
82. Choinowski T, Hauser H, Piontek K. 2000. *Biochemistry* 39:1897–902
83. Seedorf U, Rabbe M, Ellinghaus P, Kannenberg F, Fobker M, et al. 1998. *Genes Dev.* 12:1189–201
84. Kannenberg F, Ellinghaus P, Assmann G, Seedorf U. 1999. *J. Biol. Chem.* 274:35455–60
85. Seedorf U, Ellinghaus P, Nofer JR. 2000. *Biochim Biophys Acta* 1486: 45–54
86. Solaas K, Ulvestad A, Soreide O, Kase BF. 2000. *J. Lipid Res.* 41:1154–62
87. Falany CN, Fortinberry H, Leiter EH, Barnes S. 1997. *J. Lipid Res.* 38: 1139–48
88. Falany CN, Johnson MR, Barnes S, Diasio RB. 1994. *J. Biol. Chem.* 269: 19375–79
89. Hunt MC, Solaas K, Kase BF, Alexson SEH. 2002. *J. Biol. Chem.* 277: 1128–38
90. Budavari S, ed. 1989. *The Merck Index.* Rahway, NJ: Merck & Co. 11th ed.
91. Love MW, Dawson PA. 1998. *Curr. Opin. Lipidol.* 9:225–29
92. Borst P, Elferink RO. 2002. *Annu. Rev. Biochem.* 71:537–92
93. Stocco DM. 2001. *Annu. Rev. Physiol.* 63:193–213
94. Sugawara T, Lin D, Holt JA, Martin KO, Javitt NB, et al. 1995. *Biochemistry* 34:12506–12
94a. Pandak WM, Ren S, Marques D, Hall E, Redford K, et al. 2002. *J. Biol. Chem.* 277:48,158–64. 10.1074/jbc. M205244200
95. Soccio RE, Adams RM, Romanowski MJ, Sehayek E, Burley SK, Breslow JL. 2002. *Proc. Natl. Acad. Sci. USA* 99:6943–48
96. Moser HW, Smith KD, Watkins PA, Powers J, Moser AB. 2001. See Ref. 59, Vol. II, pp. 3257–3301
97. Shani N, Watkins PA, Valle D. 1995. *Proc. Natl. Acad. Sci. USA* 92:6012–16
98. Shani N, Valle D. 1996. *Proc. Natl. Acad. Sci. USA* 93:11901–6
99. Swartzmann EE, Viswanathan MN, Thorner J. 1996. *J. Cell Biol.* 132: 549–63
100. Zolman BK, Silva ID, Bartel B. 2001. *Plant Physiol.* 127:1266–78
100a. Wilcox LJ, Balderes DA, Wharton B, Tinkelenberg AH, Rao G, et al. 2002. *J. Biol. Chem.* 277:32466–72
101. Kralli A, Bohen SP, Yamamoto KR. 1995. *Proc. Natl. Acad. Sci. USA* 92:4701–5
102. Sitcheran R, Emter R, Kralli A, Yamamoto KR. 2000. *Genetics* 156: 963–72
103. Danielsson H, Einarsson K, Johansson G. 1967. *Eur. J. Biochem.* 2:44–49
104. Shefer S, Hauser S, Mosbach EH. 1968. *J. Lipid Res.* 9:328–33
105. Lu TT, Makishima M, Repa JJ,

Schoonjans K, Kerr TA, et al. 2000. *Mol. Cell.* 6:507–15
106. Goodwin B, Jones SA, Price RR, Watson MA, McKee DD, et al. 2000. *Mol. Cell.* 6:517–26
107. Nitta M, Ku S, Brown C, Okamoto AY, Shan B. 1999. *Proc. Natl. Acad. Sci. USA* 96:6660–65
108. Lee YK, Moore DD. 2001. *J. Biol. Chem.* 277:2463–67
109. Janowski BA, Willy PJ, Devi TR, Falck JR, Mangelsdorf DJ. 1996. *Nature* 383: 728–31
110. Lehmann JM, Kliewer SA, Moore LB, Smith-Oliver TA, Oliver BB, et al. 1997. *J. Biol. Chem.* 272:3137–40
111. Forman BM, Ruan B, Chen J, Schroepfer GJ Jr, Evans RM. 1997. *Proc. Natl. Acad. Sci. USA* 94:10588–93
112. Makishima M, Okamoto AY, Repa JJ, Tu H, Learned RM, et al. 1999. *Science* 284:1362–65
113. Parks DJ, Blanchard SG, Bledsoe RK, Chandra G, Consler TG, et al. 1999. *Science* 284:1365–68
114. Wang HB, Chen J, Hollister K, Sowers LC, Forman BM. 1999. *Mol. Cell.* 3:543–53
115. Urizar NL, Dowhan DH, Moore DD. 2000. *J. Biol. Chem.* 275:39313–17
116. Ananthanarayanan M, Balasubramanian N, Makishima M, Mangelsdorf DJ, Suchy FJ. 2001. *J. Biol. Chem.* 276: 28857–65
117. Kast HR, Goodwin B, Tarr PT, Jones SA, Anisfeld AM, et al. 2002. *J. Biol. Chem.* 277:2908–15
118. Jung D, Podvinec M, Meyer UA, Mangelsdorf DJ, Fried M, et al. 2002. *Gastroenterology* 122:1954–66
119. Sinal CJ, Tohkin M, Miyata M, Ward JM, Lambert G, Gonzalez FJ. 2000. *Cell* 102:731–44
120. Maloney PR, Parks DJ, Haffner CD, Fivush AM, Chandra G, et al. 2000. *J. Med. Chem.* 43:2971–74
121. Urizar NL, Liverman AB, Dodds DT, Silva FV, Ordentlich P, et al. 2002. *Science* 296:1703–6
122. Wu J, Xia CS, Meier J, Li SZ, Hu X, Lala DS. 2002. *Mol. Endocrinol.* 16:1590–97
123. Cui J, Heard TS, Yu JH, Lo JL, Huang L, et al. 2002. *J. Biol. Chem.* 277: 25963–69
124. Seol W, Choi HS, Moore DD. 1996. *Science* 272:1336–39
125. Lee YK, Parker KL, Choi HS, Moore DD. 1999. *J. Biol. Chem.* 274: 20869–73
126. Kerr TA, Saeki S, Schneider M, Schaefer K, Berdy S, et al. 2002. *Dev. Cell* 2:713–20
127. Wang L, Lee Y-K, Bundman D, Han YQ, Thevananther S, et al. 2002. *Dev. Cell* 2:721–31
128. Zhang M, Chiang JY. 2001. *J. Biol. Chem.* 276:41690–99
129. Goodhart SA, Huynh C, Chen W, Coooper AD, Levy-Wilson B. 1999. *Biochem. Biophys. Res. Commun.* 266: 454–59
130. Agellon LB, Drover VAB, Cheema SK, Gbaguidi GF, Walsh A. 2002. *J. Biol. Chem.* 277:20131–34
131. Becker-Andre M, Andre E, DeLamarter JF. 1993. *Biochem. Biophys. Res. Commun.* 194:1371–79
132. Ellinger-Ziegelbauer H, Hihi AK, Laudet V, Keller H, Wahli W, Dreyer C. 1994. *Mol. Cell. Biol.* 14:2786–97
133. Galarneau L, Pare JF, Allard D, Hamel D, Levesque L, et al. 1996. *Mol. Cell. Biol.* 16:3853–65
134. Repa JJ, Mangelsdorf DJ. 1999. *Curr. Opin. Biotechnol.* 10:557–63
135. Parker KL, Rice DA, Lala DS, Ikeda Y, Luo X, et al. 2002. *Recent Prog. Horm. Res.* 57:19–36
136. Ito M, Yu R, Jameson JL. 1997. *Mol. Cell. Biol.* 17:1476–83
137. Duel S, Drisko J, Graf L, Machleder D, Lusis AJ, Davis RA. 1993. *J. Lipid Res.* 34:923–31

138. Chiang JY, Kimmel R, Stroup D. 2001. *Gene* 262:257–65
139. Peet DJ, Turley SD, Ma W, Janowski BA, Lobaccaro J-MA, et al. 1998. *Cell* 93:693–704
140. Rudel L, Deckelman C, Wilson M, Scobey M, Anderson R. 1994. *J. Clin. Investig.* 93:2463–72
141. Xu G, Salen G, Shefer S, Ness GC, Nguyen LB, et al. 1995. *J. Clin. Investig.* 95:1497–504
142. Alberti S, Steffensen KR, Gustafsson J-A. 2000. *Gene* 243:93–103
143. Alberti S, Schuster G, Parini P, Feltkamp D, Diczfalusy U, et al. 2001. *J. Clin. Investig.*107:565–73
144. Repa JJ, Liang G, Ou J, Bashmakov Y, Lobaccaro J-MA, et al. 2000. *Genes Dev.* 14:2819–30
145. Vlahcevic ZR, Eggertsen G, Björkhem I, Hylemon PB, Redford K, Pandak WM. 2000. *Gastroenterology* 118:599–607
146. Horton JD, Goldstein JL, Brown MS. 2002. *J. Clin. Investig.* 109:1125–31
147. Repa JJ, Turley SD, Lobaccaro J-MA, Medina J, Li L, et al. 2000. *Science* 289:1524–29
148. Repa JJ, Berge KE, Pomajzl C, Richardson JA, Hobbs H, Mangelsdorf DJ. 2002. *J. Biol. Chem.* 277:18793–800
149. Edwards PA, Kast HR, Anisfeld AM. 2002. *J. Lipid Res.* 43:2–12
150. Cao G, Beyer TP, Yang XP, Schmidt RJ, Zhang Y, et al. 2002. *J. Biol. Chem.* 277:39561–65
151. Schultz JR, Tu H, Luk A, Repa JJ, Medina JC, et al. 2000. *Genes Dev.* 14:2831–38
152. Menke JG, Macnaul KL, Hayes NS, Baffic J, Chao Y-S, et al. 2002. *Endocrinology* 143:2548–58
153. Goodwin B, Watson MA, Kim H., Miao J, Kemper JK, Kliewer SA. 2002. *Mol. Endocrinol.* 17:386–94
154. Claudel T, Leibowitz MD, Fievet C, Tailleux A, Wagner B, et al. 2001. *Proc. Natl. Acad. Sci. USA* 98:2610–15
155. Tangirala RK, Bischoff ED, Joseph SB, Wagner BL, Walczak R, et al. 2002. *Proc. Natl. Acad. Sci. USA.* 99:11896–901
156. Wan Y-JY, An D, Cai Y, Repa JJ, Hung-Po Chen T, et al. 2000. *Mol. Cell. Biol.* 20:4436–44
157. Mangelsdorf DJ, Borgmeyer U, Heyman RA, Zhou JY, Ong ES, et al. 1992. *Genes Dev.* 6:329–44
158. Ness GC, Pendelton LC, Li YC, Chiang JY. 1990. *Biochem. Biophys. Res. Commun.* 172:1150–56
159. Pandak WM, Heuman DM, Redford K, Stravitz RT, Chiang JY, et al. 1997. *J. Lipid Res.* 38:2483–91
160. Andersson U, Yang YZ, Björkhem I, Einarsson C, Eggertsen G, Gafvels M. 1999. *Biochim. Biophys. Acta* 1438: 167–74
161. Drover VA, Wong NC, Agellon LB. 2002. *Mol. Endocrinol.* 16:14–23
162. Gullberg H, Rudling M, Forrest D, Angelin B, Vennstrom B. 2000. *Mol. Endocrinol.* 14:1739–49
163. Hayhurst GP, Lee Y-H, Lambert G, Ward JM, Gonzalez FJ. 2001. *Mol. Cell. Biol.* 21:1393–403
164. de Fabiani E, Mitro N, Anzulovich AC, Pinelli A, Galli G, Crestani M. 2001. *J. Biol. Chem.* 276:30708–16
165. Grundy SM, Vega GL. 1987. *Am. J. Med.* 83:9–20
166. Post SM, Duez H, Gervois PP, Staels B, Kuipers F, Princen HM. 2001. *Arterioscler. Thromb. Vasc. Biol.* 21:1840–45
167. Marrapodi M, Chiang JY. 2000. *J. Lipid Res.* 41:514–20
168. Patel DD, Knight BL, Soutar AK, Gibbons GF, Wade DP. 2000. *Biochem. J.* 351:747–53
169. Hunt MC, Yang Y-Z, Eggertsen G, Carneheim CM, Gafvels M, et al. 2000. *J. Biol. Chem.* 275:28947–53
170. Staudinger JL, Goodwin B, Jones SA, Hawkins-Brown D, MacKenzie KI, et al. 2001. *Proc. Natl. Acad. Sci. USA* 98:3369–74
171. Xie W, Radominska-Pandya A, Shi Y,

Simon CM, Nelson MC, et al. 2001. *Proc. Natl. Acad. Sci. USA* 98:3375–80
172. Makishima M, Lu TT, Xie W, Whitfield GK, Domoto H, et al. 2002. *Science* 296:1313–16
173. Staudinger J, Liu Y, Madan A, Habeebu S, Klaasen CD. 2001. *Drug Metab. Dispos.* 29:1467–72
174. Moore LB, Maglich JM, McKee DD, Wisely B, Willson TM, et al. 2002. *Mol. Endocrinol.* 16:977–86
175. Shih DQ, Bussen M, Sehayek E, Ananthanarayanan M, Shneider BL, et al. 2001. *Nat. Genet.* 27:375–82
176. Lavery DJ, Schibler U. 1993. *Genes Dev.* 7:1871–84
177. Mitsui S, Yamaguchi S, Matsuo T, Ishida Y, Okamura H. 2001. *Genes Dev.* 15:995–1006
178. Ripperger JA, Shearman LP, Reppert SM, Schibler U. 2000. *Genes Dev.* 14:679–89
179. Rutter J, Reick M, McKnight SL. 2002. *Annu. Rev. Biochem.* 71:307–31
180. Gupta S, Stravitz RT, Dent P, Hylemon PB. 2001. *J. Biol. Chem.* 276:15816–22
181. Miyake JH, Wang S-L, Davis RA. 2000. *J. Biol. Chem.* 75:21805–8
182. Yu C, Wang F, Kan M, Jin C, Jones RB, et al. 2000. *J. Biol. Chem.* 275:15482–89
182a. Natl. Cent. Biotechnol. Inf. 2002. *Online Mendelian Inheritance in Man.* http://www.ncbi.nlm.nih.gov/omim/
183. Cali JJ, Hsieh C, Francke U, Russell DW. 1991. *J. Biol. Chem.* 266:7779–83
184. Lee M-H, Hazard S, Carpten JD, Yi S, Cohen J, et al. 2001. *J. Lipid Res.* 42:159–69
185. Setchell KDR, Schwarz M, O'Connell NC, Lund EG, Davis DL, et al. 1998. *J. Clin. Investig.* 102:1690–703
186. Clayton PT, Leonard JV, Lawson AM, Setchell KD, Andersson S, et al. 1987. *J. Clin. Investig.* 79:1031–38
187. Setchell KD, Suchy FJ, Welsh MB, Zimmer-Nechemias L, Heubi J, Balistreri WF. 1988. *J. Clin. Investig.* 82:2148–57
188. Onishi Y, Noshiro M, Shimosato T, Okuda K. 1991. *FEBS Lett.* 283:215–18
189. Russell DW, Wilson JD. 1994. *Annu. Rev. Biochem.* 63:25–61
190. Ferdinandusse S, Denis S, Clayton PT, Graham A, Rees JE, et al. 2000. *Nat. Genet.* 24:188–91
191. Haslewood GAD. 1967. *J. Lipid Res.* 8:535–50
192. Wanders RJA, Jakobs C, Skjeldal OH. 2000. See Ref. 59, II:3303–21
192a. Setchell KDR, Heubi JE, Bove KE, O'Connell NC, Brewsaugh T, et al. 2003. *Gastroenterology* 124:217–32
193. Watkins PA, Chen WW, Harris CJ, Hoefler G, Hoefler S, et al. 1989. *J. Clin. Investig.* 83:771–77
194. Suzuki Y, Jiang LL, Souri M, Miyazawa S, Fukuda S, et al. 1997. *Am. J. Hum. Genet.* 61:1153–62
195. Une M, Konishi M, Suzuki Y, Akaboshi S, Yoshii M, et al. 1997. *J. Biochem.* 122:655–58
196. Dieuaide-Noubhani M, Novikov D, Baumgart E, Vanhooren JC, Fransen M, et al. 1996. *Eur. J. Biochem.* 240:660–66
197. van Grunsven EG, van Berkel E, Ijlst L, Vreken P, de Klerk JBC, et al. 1998. *Proc. Natl. Acad. Sci. USA* 95:2128–33
198. van Grunsven EG, van Berkel E, Mooijer PAW, Watkins PA, Moser HW, et al. 1999. *Am. J. Hum. Genet.* 64:99–107
199. Nelson DR, Koymans L, Kamataki T, Stegeman JJ, Feyereisen R, et al. 1996. *Pharmacogenetics* 6:1–42

Annu. Rev. Biochem. 2003. 72:175–207
doi: 10.1146/annurev.biochem.72.121801.161504

PROTEIN-LIPID INTERPLAY IN FUSION AND FISSION OF BIOLOGICAL MEMBRANES*

Leonid V. Chernomordik[1] and Michael M. Kozlov[2]
[1]Section on Membrane Biology, Laboratory of Cellular and Molecular Biophysics, NICHD, National Institutes of Health, 10 Center Drive, Bethesda, Maryland 20892-1855; email: lchern@helix.nih.gov
[2]Department of Physiology and Pharmacology, Sackler Faculty of Medicine, Tel Aviv University, Tel Aviv 69978, Israel; email: michk@post.tau.ac.il

Key Words hemagglutinin, hemifusion, stalk, fusion pore, membrane curvature

■ **Abstract** Disparate biological processes involve fusion of two membranes into one and fission of one membrane into two. To formulate the possible job description for the proteins that mediate remodeling of biological membranes, we analyze the energy price of disruption and bending of membrane lipid bilayers at the different stages of bilayer fusion. The phenomenology and the pathways of the well-characterized reactions of biological remodeling, such as fusion mediated by influenza hemagglutinin, are compared with those studied for protein-free bilayers. We briefly consider some proteins involved in fusion and fission, and the dependence of remodeling on the lipid composition of the membranes. The specific hypothetical mechanisms by which the proteins can lower the energy price of the bilayer rearrangement are discussed in light of the experimental data and the requirements imposed by the elastic properties of the bilayer.

CONTENTS

*The US Government has the right to retain a nonexclusive, royalty-free license in and to any copyright covering this paper.

1. INTRODUCTION

A complex choreography of membrane transformations, in which two membranes merge (fusion) and divide (fission), underlies the fundamental biological processes in normal and pathological conditions. Fusion and fission in exo- and endocytosis, in intracellular trafficking, in enveloped virus infection, and in many other reactions are all tightly controlled by protein machinery but also dependent on the lipid composition of the membranes. Whereas each protein has its own individual personality, membrane lipid bilayers have rather general properties manifested by their shapes, their elastic behavior in the course of deformations, and their resistance to structural changes.

Most of the current research on membrane remodeling is concentrated on the identification of proteins involved in fusion and fission and analysis of their conformational changes, which are supposed to induce membrane rearrangements. In this review we discuss membrane fusion and fission from a broader view. The starting point is a consideration of the physical factors that determine the tendency of the membrane bilayers to change their topology. Specifically, our analysis focuses on the elastic forces that drive membranes toward fusion or fission and on the intermediate membrane structures that constitute the pathways of these processes. The possible roles of proteins and lipids are addressed in the context of requirements imposed by these physical factors. In addition, a unifying consideration of fusion and fission helps us to identify the similar features of the two oppositely directed processes.

We discuss the emerging pathways of biological fusion and fission with special emphasis on the best-characterized examples of viral fusion reactions in which the proteins involved are reliably identified. Because a number of excellent recent reviews have summarized current understanding of the structure of identified and suspected fusion and budding-fission proteins, we only briefly touch upon some of these issues, those most relevant for our analysis. We also discuss the dependence of membrane remodeling on the lipid composition of the membranes. Finally, we address the hypothetical mechanisms by which lipids and proteins can cooperate in lowering and paying the energy price of the required membrane transformation.

2. REMODELING OF LIPID BILAYERS: A JOB DESCRIPTION FOR PROTEINS

Membrane remodeling in biological systems necessarily includes the rearrangement of membrane lipid matrices. The lipid bilayer is stabilized against any structural changes by a powerful hydrophobic effect (1). Hence, remodeling requires energy, which either comes from the thermal fluctuations exerted by the membrane or is delivered to the membrane by specialized proteins.

The energy provided by thermal fluctuations within an experimentally reasonable timescale of 1 s has been estimated on the basis of electroporation studies (2) as $F_{therm} \approx 40k_BT$ (where $k_BT \approx 4 \cdot 10^{-21}$ J, or $\sim 10^{-21}$ cal, is the product of the Boltzmann constant, k_B, and the absolute temperature, T) (3).

If the energy required for membrane rearrangements exceeds F_{therm}, it has to be delivered by the proteins. In this section we formulate a possible job description for the proteins suggested by theoretical analysis of the energy consumed at each step of lipid bilayer remodeling. This discussion focuses mainly on the intermediate structures of the fusion reaction, which have been explored much more thoroughly than those of fission.

2.1 How Do Membranes Resist Remodeling?

The complicated membrane rearrangements consist of elementary constituents: ruptures and deformations. Two major theoretical approaches are currently used to describe membranes. The first is based on the modeling of membranes by powerful numerical methods (4, 5), and the second uses continuous description of membranes in terms of their average effective properties [reviewed in (6)].

The goal of the numerical approaches, such as molecular dynamics, is the simulation of the fusion reaction on the basis of the dynamics of each atom in the system (6). These methods are promising for the future, when our knowledge about the atom-atom interactions within the membrane and in the surrounding aqueous solution will reach a new level. Currently, these interactions are described by a number of fitting parameters, which cannot be measured directly.

A continuous approach, in contrast, does not account for the atomic details of the membranes. However, this method gives an effective description based on just a few measurable parameters and therefore provides a good background for qualitative and quantitative predictions. Thanks to these features, for the last three decades the continuous approach has been successful in understanding membrane behavior. Our analysis below is based on this approach.

2.1.1 RUPTURES Membrane rupture (Figure 1) exposes the hydrophobic interior of the bilayer to the aqueous solution. The energy of such exposure can be estimated from the hydrocarbon-water surface tension, $\gamma = 50$ mN/m (1). The energy of a circular monolayer rupture of radius r (Figure 1A) comes from the areas of the hydrophobic bottom and the hydrophobic wall,

$$F_{mr} = (\pi r^2 + 2\pi r d_m) \cdot \gamma,$$

where $d_m \approx 1.5$ nm is the monolayer thickness. For $r = d_m = 1.5$ nm, the energy of the rupture is $\sim 160 k_B T$. A circular rupture of the whole bilayer (Figure 1*B*) has the energy $F_{br} = 4\pi r d_m \cdot \gamma$, which for the same radius $r = 1.5$ nm equals $\sim 350 k_B T$. This hydrophobic energy is so large that lipid molecules at the edge of the pore reorient to cover the edge with the polar heads and to form the hydrophilic pore (Figure 1*C*).

2.1.2 DEFORMATIONS Among deformations, the most important for our purpose are bending of the membrane and tilt of the hydrocarbon chains (Figure 2*A–C*). It is convenient to characterize *bending of a monolayer* by the curvatures of a plane lying inside the monolayer close to the interface between the polar heads and the hydrocarbon chains of the lipids (Figure 2*A*, dashed lines) (7). *Bending of a bilayer* is described by the curvatures of the plane between the monolayers (Figure 2*B*, dashed lines). Although a surface bending is characterized by two principal curvatures (Figure 2*D*), in practice one uses their combinations, called the total curvature, $J = c_1 + c_2$, and the Gaussian curvature, $K = c_1 \cdot c_2$, which have a profound geometrical meaning. For mathematical reasons, the Gaussian curvature is only rarely relevant for the description of membrane deformations. For a monolayer, the curvature is defined as positive, $J > 0$, if the surface bends toward the polar heads (Figure 2*A*). For the bilayer of a closed vesicle or a cell, positive curvature corresponds to bending toward the outside medium (Figure 2*B*).

If there are no forces acting on its surface, a lipid monolayer adopts a curvature J_s called the spontaneous curvature (8, 9). Lipid molecules are then characterized by their effective shapes (10). The spontaneous curvatures of the monolayers of the most important lipids have been measured by Rand and collaborators (7, 11–13). The one-chain oleoyl lysophosphatidylcholine (LPC) has the effective shape of an inverted cone described by a positive spontaneous curvature of 1/(5.8 nm) (12). Regular two-chain lipids are almost cylindrical and have a slightly negative spontaneous curvature such as −1/(87.3 nm) for dioleoylphosphatidylcholine (DOPC). Finally, the two-chain lipids having relatively small polar heads have strongly negative spontaneous curvatures such as −1/(2.8 nm) for dioleoylphosphatidylethanolamine (DOPE) and even −1/(1.01 nm) for dioleoylglycerol (13). Lipids of pronounced positive or negative spontaneous curvature are often referred to as nonbilayer lipids.

The spontaneous curvature of bilayers, J_s^B, is generated by the difference between their monolayers. According to the bilayer-couple hypothesis (14), the bilayer spontaneous bending is produced by a difference in the areas of the outer, A_{out}, and inner, A_{in}, lipid monolayers, resulting in J_s^B proportional to $A_{out} - A_{in}$.

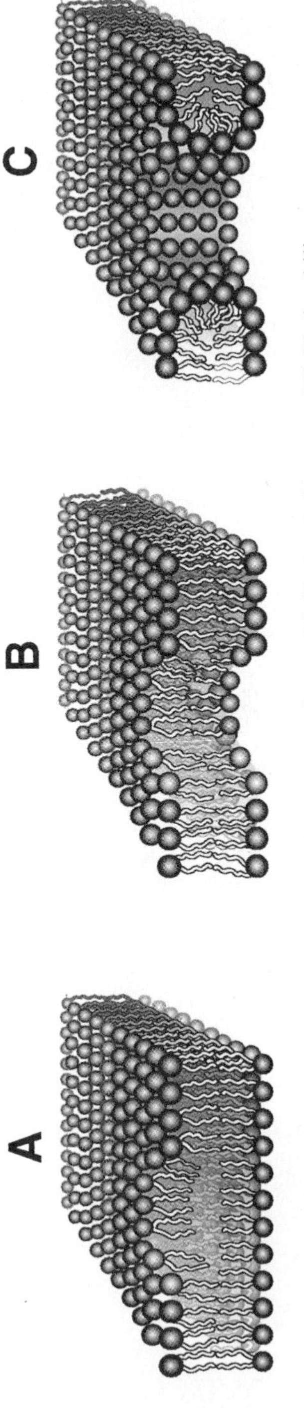

Figure 1 Bilayer rupture. (*A*) Hydrophobic defect in one monolayer. (*B*) Hydrophobic pore. (*C*) Hydrophilic pore.

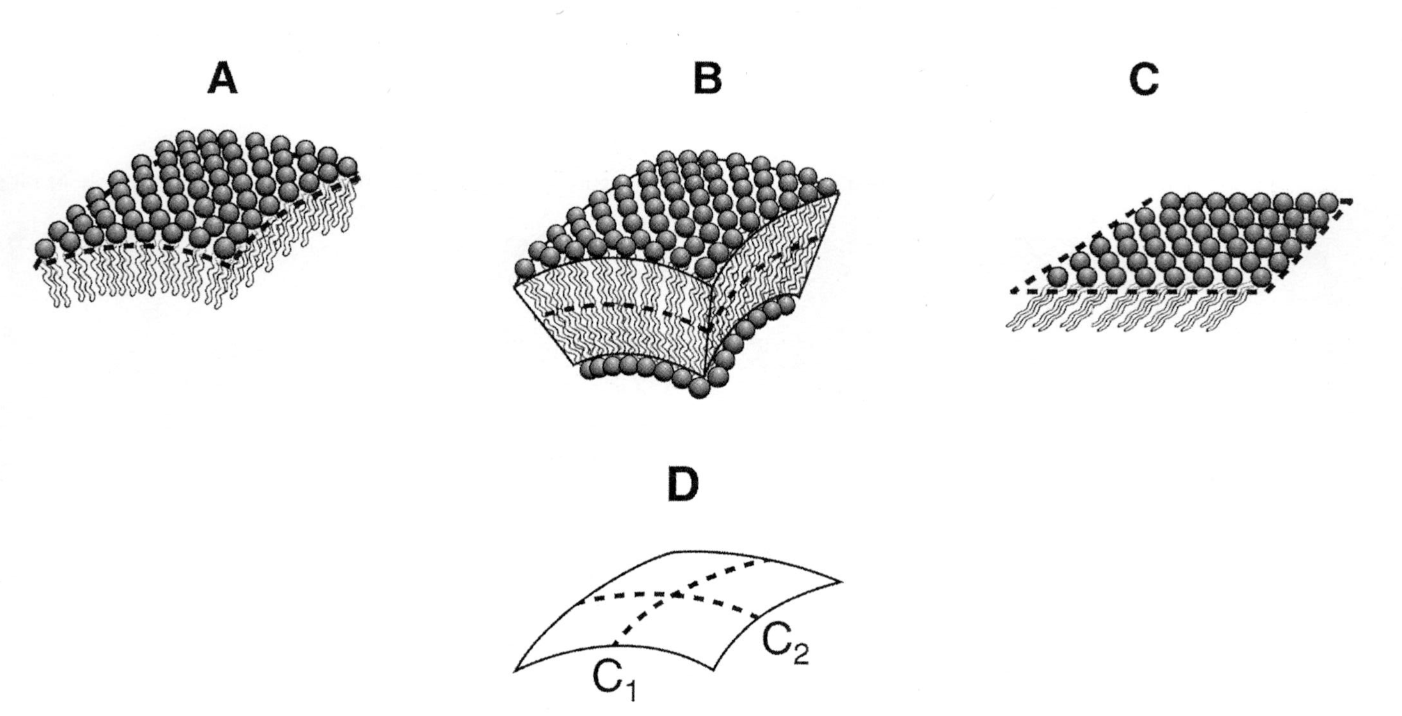

Figure 2 Membrane deformation. (*A*) Bending of lipid monolayer. (*B*) Bending of lipid bilayer. (*C*) Tilt of hydrocarbon chains. (*D*) Curvature of the surface, which represents the membrane.

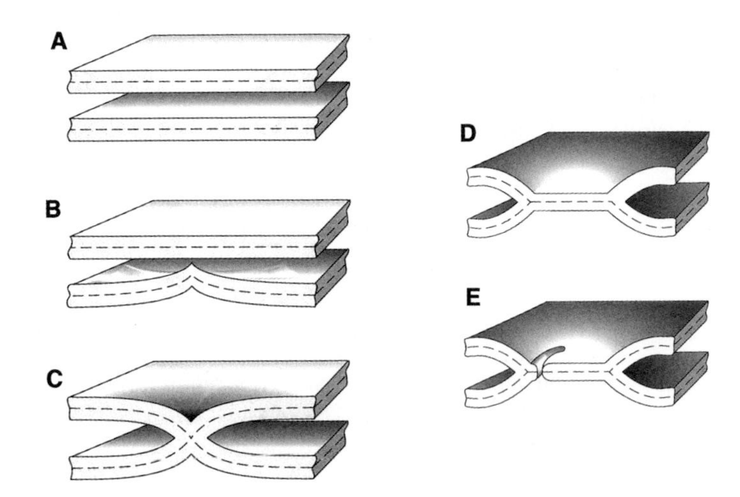

Figure 3 Fusion of lipid bilayers. (*A*) Establishment of membrane contact. (*B*) Pointlike protrusion at the prefusion stage. (*C*) Fusion stalk. (*D*) Hemifusion diaphragm. (*E*) Cracklike fusion pore.

Another reason is a difference in the spontaneous curvatures of the outer, J_s^{out}, and inner, J_s^{in}, monolayers, leading to J_s^B proportional to $J_s^{\text{out}} - J_s^{\text{in}}$.

Under external forces, a membrane bends with respect to its spontaneous curvature. The bending energy F_{bend} as a function of deformation is given by: $F_{\text{bend}} = (1/2)A \cdot \kappa \cdot (J - J_s)^2$, where A is the membrane area and κ is the bending modulus or bending rigidity (8). The rigidity of a lipid monolayer is $\kappa_m \approx 10k_BT$. For a bilayer it is twice as large, $\kappa_b \approx 20k_BT$. Using the concept of bending energy, one can calculate the elastic energy of a spherical vesicle of radius R whose membrane has zero spontaneous curvature, $J_s = 0$, area $A = 4\pi R^2$, and total curvature $J = 2/R$. This energy does not depend on the vesicle radius and is $F_{\text{bend}} = 8\pi\kappa_b \approx 500k_bT$.

One can also estimate the energy of the hydrophilic pore (Figure 1*C*) as the bending energy of the monolayer covering the pore edge. We obtain the energy per unit length of the edge as $f_{\text{pore}} = (1/2)\kappa_m(1/d_m) \approx 3k_bT/\text{nm} \approx 10^{-11}$ J/m. This energy, also referred to as the line tension, is close to those measured experimentally (15) and estimated in a more sophisticated way (16). Based on this value the edge energy of a pore of 1-nm radius can be estimated as $\sim 20k_bT$.

Treatment of the tilt of hydrocarbon chains (Figure 2*C*) requires a more complicated modeling and is presented in detail elsewhere [(17) and references therein].

2.2 Job 1: Bringing the Membranes Together

Before remodeling starts, the membranes have to establish a contact (Figure 3*A*). The bilayers, which do not carry an electric charge, tend to approach each other spontaneously up to the equilibrium distance of 2–3 nm (18). The initial distances between biological membranes are much larger, 10–20 nm, because of the electrostatic repulsion between the bilayers and the steric interaction of membrane proteins. Thus, fusion/fission proteins have to bring membrane lipid bilayers into reasonably close contact (a distance of a few nanometers), allowing the downstream fusion/fission stages to proceed. This aim might require the proteins to either pull or push the membranes toward each other and to bend them locally to minimize the area of the strongest intermembrane repulsion (19).

2.3 Job 2: Prefusion/Prefission Stage

Merger of the contacting membrane surfaces requires the formation of some transient membrane discontinuities. The energy price of this stage includes the local membrane approach to almost zero distance against strong intermembrane repulsion, the energy of rupture of the merging monolayers, and the deformation energy of the monolayers accompanying their local approach and rupture.

Earlier attempts to analyze this stage of fusion (19) predicted large contributions to the energy from intermembrane repulsion and hydrophobic monolayer discontinuities. This result was related to the assumption that membranes approach each other at rather large areas on the tops of smooth membrane bulges where strong hydration repulsion (18) prevents them from establishing a dehydrated contact. We suggest a different configuration of the prefusion intermediate, which is illustrated in Figure 3*B*. Such pointlike protrusions must exhibit a minimal repulsion from the apposing membrane, thus producing a minimal contribution to the energy. Formation of a pointlike dehydrated contact between the membranes will decrease the hydrophobic energy of the monolayer rupture. Pointlike protrusions have not been considered before because of the seemingly high energy of sharp bending at the top of the protrusion. However, recent developments of the elastic model of membrane rearrangements, including tilt deformation (17), have demonstrated the feasibility of structures of this kind. A similar sharp ridgelike fluctuation of the membrane neck might be considered as a possible prefission intermediate. Theoretical analysis of these prefusion and prefission intermediates is a matter for future work.

Proteins can facilitate this stage by inducing local dehydration of the membrane contact and elastic stresses supporting membrane fluctuations.

2.4 Job 3: Fusion Stalk and Hemifission Intermediate

The junction of the transient discontinuities of the contacting monolayers of two membranes yields a first intermembrane lipid connection (Figure 3*C*) called the fusion stalk (20). In the case of fission, this stage corresponds to fission of only the internal monolayer of the membrane neck.

The evolution of ideas on the fusion stalk, beginning from the original model (21), ran into an important roadblock in 1993 when Siegel formulated the energy crisis of the stalk model. When the energy of the hydrophobic interstice, the structural defect unavoidably emerging in the stalk middle, was considered, the energy of a stalk was found to be prohibitively high (22).

This energy crisis was solved by optimized packing of the hydrocarbon chains within interstices for the case of strongly bent bilayers in a very close (~1 nm) contact (23) and then for the canonical case of flat membranes at arbitrary intermembrane distances (17), along with finding the stalk shape that minimizes its bending energy (17, 24). The stalk structure (Figure 3*C*) suggested in (17) combines deformation of bending, tilt, and splay of the monolayers with optimized monolayer shape and no vacuum voids inside the interstices. This model predicts feasible energies for stalk without any requirements on prefusion membrane deformation and at all experimentally relevant intermembrane distances. For the membranes of DOPC the predicted stalk energy is $\sim 45k_BT$ (17), which exceeds by $\sim 5k_BT$ the characteristic thermal energy $F_{\text{therm}} \approx 40k_BT$ (see above). The more negative the spontaneous curvature of the contacting monolayers is, the lower the energy of the stalk becomes. For DOPE, which is known to undergo spontaneous fusion in lipid mesophases, the stalk energy is negative, $\sim -30k_BT$, meaning that the stalk is energetically favorable for this lipid and its formation does not limit the fusion process. Stalklike structures have recently been detected by X-ray diffraction for a dehydrated lipid of the negative spontaneous curvature (25).

The task of the fusion proteins might be to deliver the energy required for stalk formation in a bilayer of biological lipid composition. Protein might generate bilayer stresses that relax during stalk formation and release the energy needed.

Analysis of the hemifission intermediate is currently being undertaken in the spirit of the updated stalk model (Y. Kozlovsky and M.M. Kozlov, in preparation).

2.5 Job 4: Hemifusion Diaphragm

Three different scenarios have been suggested for evolution of the stalk into a fusion pore, which connects the aqueous volumes bounded by the fusing membranes. In the stalk-pore hypothesis (15, 26), radial expansion of the stalk results in a dimpling of the distal monolayers toward each other and the formation of a single bilayer (21, 27), referred to as a hemifusion diaphragm (HD) (Figure 3*D*). Formation of a pore in the HD (26) completes fusion. To allow formation of the pore edge, the HD radius has to reach at least a few nanometers. An alternative model proposes that the fusion pore forms directly from the stalk, avoiding the stage of stalk expansion (22, 23). The third approach, based on numerical simulations of bilayer fusion, suggests that the initial fusion stalk extends anisotropically into an elongated stalk, which induces the formation of holes in the stalk vicinity in both fusing membranes (5). Fusion of the rims of the two apposing holes yields a fusion pore.

Very recently, stalk evolution has been analyzed using the elastic model of tilt and bending deformations (28). We have found that a circular HD is always more favorable energetically than an elongated stalk. The intrinsic tendency of a fusion stalk to expand into an HD is controlled by the spontaneous curvature, J_s, of the contacting monolayers of the fusing membranes. For positive or moderately negative J_s, such as that of DOPC, the HD does not form spontaneously. However, even in these conditions an HD can form provided that its rim is subject to an external force pulling the HD apart. An obvious job for proteins is to induce this pulling force.

2.6 Job 5: Fusion Pore Formation

Formation of the edge of the hydrophilic pore is facilitated by the positive spontaneous curvature of the HD monolayers (15). Opening of a fusion pore is also promoted by the lateral tension, σ, developed in the diaphragm. The characteristic time of pore formation depends on both the lateral tension and the area of the stressed membrane. The larger the area, the higher the probability of pore formation within a given time span and under a given tension (15). The usual tension leading to pore formation within seconds in the membranes of unilamellar vesicles of a diameter of ~40 nm is $\sigma \approx 10$ mN/m (29). What mechanism can produce this tension in an HD?

The recent HD model predicts a very large tension, $\sigma \approx 20$ mN/m, generated by the elastic stresses of tilt and bending in the narrow region along the diaphragm rim (28), which may lead to formation of a cracklike fusion pore expanding along the diaphragm rim (Figure 3*E*). This is different from the usual circular pores formed in homogeneously stressed membranes.

A task for proteins is to pull the HD rim apart, thus increasing the probability of pore nucleation. The analogues of the HD and fusion pore in the case of fission are not obvious at this stage and are a matter for future work.

The importance of the bending deformations in membrane remodeling, and in particular the role of stalk and pore intermediates in fusion, have been verified by the experimental studies on protein-free bilayers.

2.7 Lipid Bilayer Fusion Observed Experimentally

Bringing two protein-free bilayers into close contact is not sufficient for fusion. Bilayers formed from phosphatidylcholine (PC), a major constituent of the membranes of mammalian cells, do not fuse when kept for hours and even days in very close contact (~3 nm). Only application of tension or dehydration of the contact zones with high concentrations of polyethylene glycol fuses bilayers of this composition (15, 30, 31). However, under certain conditions bilayers do fuse [reviewed in (15, 31)]. In addition, reliable evidence is provided for the existence of the hemifusion state. For instance, for two fusing planar bilayers, the specific electrical capacitance, and thus the thickness of the contact zone, coincides with that of a single bilayer (15). In liposome-planar bilayer and liposome-liposome fusion, mixing of the lipids of the contacting monolayers of the membranes proceeds when neither lipid mixing between distal monolayers nor a fusion pore is detected (32–35).

The hemifusion phenotype can represent the end-state of the fusion reaction (15, 33, 36, 37). Alternatively, the reaction proceeds to the opening of a fusion pore (32, 33). Hemifusion depends on the lipid composition of the contacting monolayers of the membranes (15, 31, 35, 36, 38). Lipids of positive and negative spontaneous curvature (e.g., LPC versus phosphatidylethanolamine, PE, or oleic acid, OA), respectively, inhibit and promote hemifusion. In contrast to hemifusion, the subsequent opening of a fusion pore depends on the lipid composition of the distal membrane monolayers. Lipids that promote hemifusion (OA and PE) inhibit pore formation, and vice versa: LPC, which inhibits hemifusion, promotes pore formation (15, 36). Thus, fusion dependence on lipids in the contacting membrane monolayers supports the hypothesis that fusion proceeds through stalk intermediates. The effects of the lipids in the distal monolayers are consistent with pore formation in HDs as predicted by the stalk-pore model [see above and (39)].

The dependence of bilayer fusion on the lipid composition has also been discussed in terms of the correlation between fusion promotion and increase in the hydrophobicity of the bilayer surface, increase in bilayer tension, and the presence of lipid packing defects, including formation of so-called extended conformations of the lipids (40–45).

2.8 Lipid Bilayer Fission Observed Experimentally

Even a narrow lipid bilayer tube resists fission. Lipid tethers of a few nanometers in diameter can be very stable (46). However, under certain conditions bilayers do divide. One of the experimental models for studying fission of protein-free bilayers is based on a membrane tube formed by fusion of two black lipid membranes (15). Fission of such tubes is driven by surface tension. Shape transformation and fission can also be driven by the domain boundary in heterogeneous membranes (47), as has been observed for vesicles composed of dimyristoyl phosphatidylcholine and cholesterol and for sphingomyelin vesicles at the temperature of the lipid phase transition that generates inhomogeneities (48). Finally, enzymatic formation of ceramide-enriched microdomains in fluid giant vesicles composed of phosphatidylcholine and sphingomyelin causes budding and release of the smaller vesicles (49). In this case, vesiculation is driven by the spontaneous curvature and area difference between the monolayers generated by the asymmetric production of the ceramide. The underlying effects of the spontaneous curvature of the bilayer on the shapes of the protein-free vesicles are discussed in (50).

3. REMODELING OF BIOLOGICAL MEMBRANES

The intermediate structures in the fusion and fission pathways for biological membranes are better characterized for membrane fusion and, in particular, for fusion mediated by viral fusion proteins (51, 52). In this section we compare the intermediates of the prototype biological fusion reaction mediated by influenza hemagglutinin (HA) with those identified for protein-free lipid bilayers. We then briefly discuss the much less characterized membrane pathway of protein-mediated fission.

3.1 Fusion: Experimental Approaches

Numerous enveloped viruses enter the host cells either by endocytosis or by direct fusion of the viral envelope to a plasma membrane (53). In the former mechanism, fusion proteins such as HA of influenza virus A are activated by low pH within the endosome. In the latter mechanism, fusion proteins such as HIV gp120/41 are activated at neutral pH by complex interactions with cellular receptors.

Fusion is usually studied in experimental systems, which are much simpler than actual virus entry into a host cell. These in vitro systems include low-pH-triggered or receptor-triggered fusion of virus to liposomes or target cells. Alternatively, viral fusion proteins are expressed in cells. Fusion of these cells with target cells or with lipid bilayers is assayed as redistribution of membrane probes and aqueous content probes using fluorescence microscopy, spectrofluorimetry, and electrophysiology.

3.2 Fusion Phenotypes and Hypothetical Pathways

Under optimal conditions, fusion is usually too fast to be dissected into distinct stages. Thus, current understanding of the pathway of viral fusion (Figure 4*A*– *E*) is based mostly on experiments in which fusion was slowed down by lowering temperature, modifying fusogenic proteins, decreasing their numbers, and/or altering lipid composition to that unsuitable for fusion. Blocking, or at least slowing down, of fusion progression beyond a certain stage allows characterization of that stage.

Some of the phenotypes of biological fusion resemble those identified for protein-free lipid bilayers. Early fusion pores (Figure 4*D*) in HA-mediated fusion

---

Figure 4 Two hypothetical pathways in protein-mediated membrane fusion. (*A*) Membrane expressing fusion protein (for instance, HA) in contact with the target membrane. (*B*) Conformational change in the activated fusion proteins. (*C*) Fusion stalk. (*D*) Opening of a lipidic fusion pore. (*E*) Fusion pore expansion. (*F, G*) Alternative pathway involves formation of a proteinaceous pore (*F*), which then expands to allow membranes to establish lipidic connection (*G*). Stalk formation (*C*) is facilitated by cone-shaped lipids (e.g., OA), and hindered by inverted-cone-shaped lipids (e.g., LPC). Lipid dependence of the pore (*D*) is opposite to that in the stalk. Dashed lines show the boundaries of the hydrophobic surfaces of monolayers. The diagrams of intermediates *F* and *G* include a view from below with large and small circles to represent proteins and lipid polar heads, respectively.

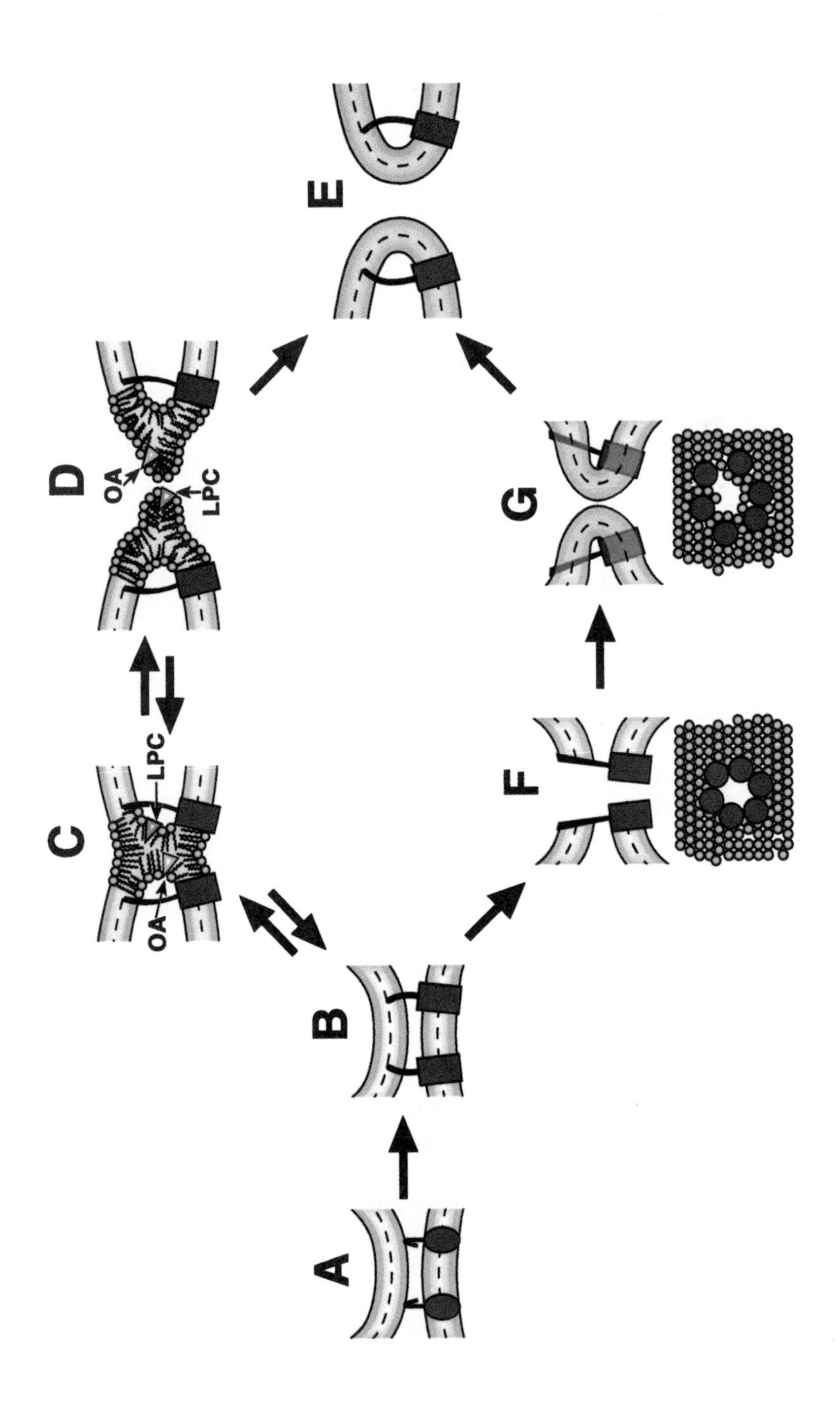
A
B
C
OA
LPC
D
OA
LPC
E
F
G

(54, 55) open abruptly, fluctuate, and then either close or expand irreversibly with characteristic electrophysiological features similar to those in bilayer fusion (33). As in fusion of bilayers, protein-mediated fusion yields a hemifusion phenotype (56) (Figure 4*C*). The unambiguous evidence for HA-mediated hemifusion, lipid mixing in the absence of a fusion pore, has come from combining the most sensitive electrophysiological assay for fusion pores with fluorescence microscopy used to assay lipid probe redistribution (57, 58). Additional fusion intermediates, characteristic only of protein-mediated fusion, have also been identified. In contrast to lipid bilayer fusion, the opening of a small fusion pore in HA-mediated fusion precedes the onset of detectable lipid mixing (54, 59). Lipid flow between membranes is also restricted in another type of HA-formed fusion intermediate, which can be transformed into complete fusion (lipid- and content-mixing) only with treatments known to destabilize the HD (58, 60). This fusion phenotype apparently represents local hemifusion (58, 61, 62) and is referred to as restricted hemifusion because lipid flow through these structures is restricted by activated HAs. Formation of the restricted and reversible (61) hemifusion connections happens as fast as the opening of a fusion pore but requires less HA. Therefore, the opening of a pore is likely preceded by formation of multiple restricted hemifusion sites (61, 63).

Even if the fusion pathway under biologically relevant conditions involves membrane structures, which are arrested and identified as distinct phenotypes, the specific fusion pathway is always a matter of conjecture rather than a direct experimental finding. For instance, it remains to be verified that pores are formed in the hemifusion structures and that large fusion pores are formed by expansion of initial small pores. However, the set of available pieces of the puzzle is consistent with the hypothesis that fusion starts with local hemifusion, which then breaks into a small pore that subsequently expands (Figure 4*A*– *E*). Note that the prediction of a cracklike pore propagating along the HD rim (Section 2.6) (28) is supported by the observation in HA-mediated fusion of the lack of movement of aqueous dyes while total fusion pore conductance increases (54). Elastic stresses generated along the perimeter of HD might also explain the intriguing recent finding that homotypic fusion of yeast vacuoles proceeds along the rim of the membrane contact area (63a).

3.3 Budding-Fission: Intermediate Stages

In contrast to fusion, the pathway of budding-fission reactions has been well studied in only one in vitro system, receptor-mediated endocytosis of transferrin (Tfn) in semi-intact or perforated cells (64–66). Narrowing of the cross section of the neck was assayed as changes in the accessibility of Tfn in the budding membrane compartments for different probes. The size of the probes varied from many nanometers in the case of antibodies and avidin to a few angstroms in the case of 2-mercaptoethanesulfonic acid. Three sequential stages of budding fission have been identified: (*a*) the budding initiation, where receptor-bound Tfn is accessible for all probes; (*b*) the constricted bud, where Tfn is inaccessible to the bulky probes while still accessible to the small probe, 2-mercaptoethanesulfonic acid; and finally, (*c*) the sealed vesicle, where Tfn is inaccessible for all probes. Note that at the sealed vesicle stage, this experimental approach does not distinguish between a bud that has undergone hemifission but is still connected to the initial membrane and a completely separated vesicle.

4. PROTEINS THAT DRIVE MEMBRANE REMODELING

4.1 Fusion Proteins

Do different fusion proteins share any structural motifs reflecting their common functionality? A definitive answer to this question is hindered by the paucity of available information about these proteins along with the fact that most (if not all) reliably identified fusion proteins are viral envelope proteins. The structures and properties of the best-characterized viral fusion proteins have recently been reviewed in depth [for instance, HA (67), gp120/gp41 (67–69), E protein of flaviviruses (70), and G protein of rhabdoviruses (71)]. Below, we briefly discuss only those properties of the fusion proteins that appear to be most relevant for lipid bilayer remodeling.

4.1.1 OVERALL STRUCTURE OF THE FUSION PROTEIN AND ITS REFOLDING UPON ACTIVATION All characterized viral fusion proteins are anchored in the envelope by transmembrane domains (TM). Although for some proteins there is significant latitude in the TM sequence that supports fusion (72, 73), the specific anchor by which protein is attached to a membrane is of importance for its fusogenic activity. Replacing the TM of HIV gp41 (74, 75), HA (56, 57), and parainfluenza virus fusion protein (76) with a lipid anchor blocks the ability of these proteins to form expanding fusion pores. Modifying the TM either of HA (73, 77, 78) or of the fusion protein of vesicular stomatitis virus (79, 80) also inhibits fusion.

On the basis of the overall structure of the ectodomain, the majority of the viral fusion proteins can be divided into two classes (81). In class 1 (for instance, HA and gp120/gp41), the fusion peptide (see below) is located at or near the NH2 terminus of the fusion protein created by the proteolytic cleavage upon protein maturation. Activated proteins of this class share the common 6-helix bundle, a hairpin arrangement with a central α-helical coiled-coil domain. In this very stable structure, the fusion peptide and TM are located at the same end of the rodlike molecule.

Class 2 proteins (for instance, E protein of flaviviruses and E1 protein of alphaviruses such as Semliki Forest virus) have the fusion peptide in an "internal" location rather than at the NH2 terminus of the protein. In this case, proteolytic cleavage leading to mature conformation occurs in the accessory protein rather than in the fusion protein itself. Class 2 proteins are not predicted to form coiled-coils and contain predominantly β-strand secondary structures.

As a rule, conformational change of the activated proteins of both classes is profound and likely releases significant conformational energy. For most viral fusion proteins, with the intriguing exception of G protein of rhabdoviruses (71), activation in the absence of a target membrane results in the irreversible functional inactivation of the proteins. Although fusion can, in principle, be mediated by some transient protein conformations that are closer to the initial than to the stable inactivated form of the protein (82–84), at least some features of the major restructuring are important for fusion. In particular, HA mutations expected to block the extension of the central coiled-coil (85) inhibit fusion.

4.1.2 FUSION PEPTIDE The early work on the HA-mediated fusion identified the functional importance of the fusion peptide, i.e., the NH2 terminus of the HA2 subunit of HA (67) that inserts into the membranes upon activation (86). Mutations in this region of ~20 amino acid residues inhibit fusion (87, 88). NH2-terminal or internal peptides similar to the fusion peptide of HA have been found in many other viruses.

The structures of the different fusion peptides and the effects of these peptides on lipid bilayers have been examined extensively (88–92). These studies indicate that the functional role of the membrane insertion of the fusion peptide is not limited to just establishing an additional membrane anchor. It is also clear that the functionality of the fusion peptide in the context of the entire protein cannot be fully explained by studying the synthetic peptide/bilayer interactions. For instance, changes in the HA fusion peptide, which at the level of corresponding synthetic peptides significantly alter peptide ability to modulate the spontaneous curvature of lipid monolayers, do not affect the fusogenic properties of the whole protein (92).

The existence of a distinct fusion peptide—usually a stretch of 10–30 amino acid residues that is amphiphilic, highly conserved, critical for fusion, hidden in the initial conformation of the protein, exposed in the activated conformation and then inserted in the membrane—is widely assumed to be one of the most important prerequisites for the generic fusion protein. However, the influx of additional structural information and the broadening of the collection of characterized fusion proteins somewhat blur the exact meaning of the term fusion peptide. A conserved uncharged sequence in the fusion proteins of rhabdoviruses, recognized as a fusion domain and found to insert into the target membranes, contains some polar amino acids and is much less hydrophobic than other fusion peptides (71). Ectodomains of several fusion proteins have multiple regions with the properties expected for fusion peptides (93–95).

It remains possible that the functional role of some of the protein regions identified as fusion peptides is not only limited to direct interaction with the membranes but also, or even instead, involves supporting or destabilizing certain protein structures. In particular, fusion peptides can mediate some functionally important interactions between adjacent proteins (96), or within the same protein. For instance, interactions between the fusion peptide and the TM of a fusion protein might be instrumental in transition from hemifusion to an opening of an expanding fusion pore (87, 88, 97, 98).

4.1.3 POSTACTIVATION OLIGOMERIZATION Activated fusion proteins form high-order oligomers. HA aggregation is detected as a loss of HA mobility along the membrane surface (99). Morphologically, HA aggregation is seen as the transformation of the well-defined spikes of the neutral-pH form of HA into a "thick layer of entangled thin threads" (100). Low-pH-induced aggregation at the surface of the membrane has also been reported for the large fragment of the HA2 subunit, FHA2 (101). Two specific HA domains exposed in activated HA, the fusion peptides (102, 103) and the kinked regions of HA2 (residues 106–112) responsible for the aggregation of FHA2 (101, 104), have been implicated in interactions of HA trimers.

Fusion proteins other than HA also assemble into high-order oligomers. Multimeric aggregates are reported for HIV gp41 (105) and dengue virus E protein (106) released from the interactions stabilizing the initial form of fusion proteins. Under conditions of fusion, baculovirus gp64 forms aggregates containing up to 10 trimers (107). Stability of the gp64 trimers is a prerequisite for formation of these aggregates, indicating that trimers serve as a fundamental building block. Destabilization of the gp64 trimers by reducing the intermonomer S–S bonds completely abolishes both aggregation and fusogenic activity of this protein (107).

Not only does activation promote interactions between adjacent fusion proteins, but vice versa: Interactions between adjacent fusion proteins promote activation. For the same conditions of activation (pH and temperature), the percentage of low-pH-activated HA increases with higher surface density of cleaved HA trimers capable of undergoing such conformational changes (96). There is an early reversible form of low-pH-activated HA from which HA can still revert to the initial conformation (108). Intertrimer interaction shifts HA restructuring toward irreversible stages. Reversible stages prior to a major restructuring have been discussed for fusion proteins of some other viruses [e.g., flavivirus (109), rabies virus (71), and HIV (110, 111)] and might be a common feature of different fusion proteins.

Interactions between fusion proteins, a distinct reversible stage of refolding, and positive cooperativity of the activation might be crucial for the synchronized discharge of the conformational energy of the multiple adjacent proteins in a hypothetical fusion machine. The notion that fusion is mediated by a concerted action of multiple fusion proteins is supported by numerous functional studies (58, 61, 99, 107, 112–117).

4.2 Budding-Fission Proteins

4.2.1 INTRACELLULAR BUDDING The protein machinery involved in budding and fission has been reviewed recently (66, 118, 119). In brief, bending of the membranes into spherical buds in different pathways of intracellular trafficking and virus budding-off is driven by the layer of proteins (coat) assembled at the membrane surface. Intracellular trafficking utilizes clathrin, COPI, and COPII coats.

In addition, trafficking requires some cylinder-forming proteins, which constrict membranes into tubular shapes and are apparently involved in the formation and constriction of membrane necks connecting the buds to the nearly flat initial membrane.

The first discovered cylinder-forming protein, dynamin, a 100-kDa GTPase, self-assembles into helices on the membrane surface and constricts flat membranes into collared tubes with an external diameter of ~50 nm (120–122). Dynamin can bend membranes in concert with clathrin and on its own (120, 121, 123). The budding pathway driven by dynamin without involvement of sphere-forming coats is referred to as fission-only and is implicated in a kiss-and-run mechanism of exocytosis (124).

Recently, additional proteins, amphiphysin 1 and 2 (121) and endophilin (125), have been found to tubulate the membranes both on their own and in concert with dynamin. Tubes formed by either amphiphysin or endophilin have diameters similar to those formed by dynamin. Another protein, epsin, binds to phosphatidylinositol-4,5-bisphosphate–containing liposomes and bends them into tubes with the outer diameter of approximately one third of that of the tubes formed by the dynamin (125a). Epsin-induced membrane invagination may go hand in hand with clathrin polymerization and AP2 complex recruitment.

4.2.2 INTRACELLULAR FISSION Most sphere-forming protein coats are capable of severing the membrane necks and separating the coated vesicles from the initial membranes (126, 127). The exception is the clathrin coat, which does not pinch off the coated buds without the assistance of its cylinder-forming partners (128). Fission of membrane tubes mediated by dynamin correlates with its GTPase activity (120, 121). Amphiphysin supports dynamin-induced tube fragmentation, and endophilin inhibits it (125).

Although dynamin can induce fission on its own, in biologically relevant situations its function in fission can include, or even be limited to, recruitment of additional proteins. A proposed candidate is endophilin, the protein involved in synaptic vesicle formation, which, in addition to its cylinder-forming properties, exhibits acyltransferase activity (129, 130). Strikingly, another acyltransferase, CtBP/BARS, is reported to induce constriction and fission of Golgi tubes (131).

Tightly controlled reactions of intracellular budding fission involve many other important proteins, including phosphatidylinositol (PI)-transfer proteins (132) and protein kinase D (133). The yet unresolved challenge is to reliably distinguish proteins that control the reactions from those that actually mediate them.

4.2.3 VIRAL BUDDING AND FISSION Budding of the enveloped viruses releases them into the extracellular space (118). For influenza virus, cell expression of just one viral matrix protein, M1, which assembles under the membrane, is sufficient to bend cell membrane into buds and release spherical viruslike particles (134).

Similarly, just two flavivirus proteins, envelope Glycoproteins E and prM, assemble at the membrane and drive the formation and release of small subviral particles (135).

4.3 Lipids in Biological Fusion and Fission

4.3.1 FUSION AND NONBILAYER LIPIDS Protein-mediated fusion is sensitive to membrane lipid composition. Some lipids are of importance for specific fusion reactions. For instance, PE is important for fusion of Golgi membranes (136). Sphingolipid and cholesterol are strictly required for fusion mediated by Semliki Forest virus (137, 138). Polyphosphoinositides play a signaling role in membrane trafficking (139).

Although lipids can undoubtedly control the local concentration and activity of specific proteins, some lipid effects on fusion apparently reflect changes in the lipid bilayer propensity to fuse. The latter effects are strikingly universal and follow the predictions of the stalk-pore hypothesis. For instance, LPC, a lipid of positive spontaneous curvature, which inhibits fusion of protein-free bilayers, also inhibits disparate biological fusion reactions if added to the contacting membrane monolayers (Figure 4*C*). This lipid reversibly blocks fusion mediated by fusion proteins of baculovirus (140), Sendai virus (141), rabies virus (142), HIV (98), paramyxovirus SV5 (143), and Semliki Forest and avian leukosis viruses (personal communication with J. Wilschut and with G. Melikyan). LPC also inhibits a number of nonviral fusion reactions triggered by Ca^{2+}, GTP-γ-S, and GTP [reviewed in (144)].

Although inhibition of biological fusion by these amphiphiles is consistent with their expected effects on the energy of the stalk and HD, alternative interpretations have been also discussed (145). The most important alternative suggested is that LPC affects fusion by direct interaction with the exposed amphiphilic regions of the fusion protein (140, 146). However, a recent report showing very weak interaction between LPC and activated HA (147), along with an earlier analysis (145), argues against this interpretation.

In contrast to lipids of positive spontaneous curvature, lipids of negative curvature such as PE, diacylglycerol (DAG), and *cis*-unsaturated fatty acids (e.g., OA) promote biological fusion (142, 144, 148) (Figure 4*C*).

No specific lipid moiety is responsible for this dependence of fusion on nonbilayer lipids. Different naturally occurring nonbilayer lipids and synthetic surfactants that are quite dissimilar in the structures of their hydrocarbon chains and polar heads reversibly inhibit or promote fusion in specific correlation with their spontaneous curvature. The lipid-dependent fusion stage is prior to lipid mixing and fusion pore opening but subsequent to the activation of HA (148), gp120/gp41 (98), and F protein of paramyxovirus SV5 (143).

As in fusion of protein-free lipid bilayers, a stage of biological fusion that depends on the composition of the contacting membrane monolayers is followed by the stage dependent on the distal membrane monolayers. Amphiphiles of positive spontaneous curvature (LPC and chlorpromazine) added to the distal monolayer of the fusing membranes (the inner monolayer of the cell membrane) promote the opening of a fusion pore in HA-mediated fusion (58, 60, 149) (Figure 4*D*). Similar promotion has been reported for fusion of rabies virus with liposomes (142). Likewise, secretion from SNARE (soluble *N*-ethylmaleimide-sensitive factor attachment protein receptor proteins) mutant cells is rescued by LPC added to the outer monolayer of the plasma membrane, which in this case represents the distal monolayer of the fusing membranes (150). Thus, the lipids that facilitate lipid monolayer bending into the edge of a hydrophilic pore (15, 60) promote fusion pore formation in protein-mediated fusion.

To conclude, the effects of nonbilayer lipids on biological fusion suggest that proteins catalyze this process through the same stalk-pore sequence of intermediates as that identified for fusion of protein-free lipid bilayers.

Importantly, lipid inhibitors of fusion such as LPC present a unique and powerful tool for dissecting complex and multistep pathways of biological fusion reactions (98, 143, 151, 152). An LPC block was used in (152) to uncouple the formation of the SNARE complexes in sea urchin egg cortical exocytosis from the membrane fusion event. Similarly, reversible inhibition of membrane merger by LPC, in combination with peptide inhibitors blocking restructuring of fusion proteins into 6-helix bundles, has been elegantly used to demonstrate that fusion mediated by HIV (98) and parainfluenza virus (143) is a result of the transition of the fusion proteins into the bundle configuration and did not occur after this protein restructuring.

4.3.2 LIPIDS IN FISSION Some effects of the lipids on the budding-fission reaction are linked to the recruitment of the proteins involved. For instance, AP-2, part of the clathrin coat machinery, binds to phosphoinositides [reviewed in (52, 132)]. Dynamin binds to acidic lipids (132). DAG binds protein kinase D regulating fission from the *trans*-Golgi network (133). Cholesterol is needed for clathrin coat–mediated budding (153). PE is required for disassembly of contractile ring in cytokinesis (154).

In addition to these specific effects, lipids can directly modulate budding-fission reactions by generating the membrane spontaneous curvature, J_s^B (Section 2.1). Indeed, stimulation of endocytosis by the flippase-dependent translocation of aminolipids from the outer to inner monolayer of the plasma membrane (50, 155) is explained by the developing difference between the areas of the membrane monolayers that, according to the bilayer-couple mechanism (Section 2.1), promotes the spontaneous bending toward the cytoplasm.

The asymmetry in the spontaneous curvatures of the two monolayers is a possible result of action of two acyltransferases implicated in intracellular fission, endophilin (130) and CtBP/BARS (131), which can transform lysophosphatidic acid (LPA) into phosphatidic acid (PA) in the necks of budding vesicles and Golgi tubes. The resulting J_s^B has been suggested to be crucial for fission.

5. MECHANISMS OF MEMBRANE REMODELING

5.1 Hypothetical Intermediates of Biological Fusion

Two radically different pathways of protein-mediated fusion have been suggested. The first proceeds through the entirely proteinaceous fusion pore (Figure 4*F,G*), while the second starts with membrane hemifusion (Figure 4*C*). Below we discuss both pathways and consider the specific mechanisms by which proteins can mediate fusion and fission.

5.1.1 PROTEINACEOUS PORE The hypothetical proteinaceous fusion pore establishes an aqueous connection between fusing volumes before any lipid monolayer merger (156, 157). This concept has been strengthened by recent work indicating that the terminal stage in vacuolar fusion in yeast is mediated by the transmembrane complex of two proteolipid hexamers formed by the V0 subunit of vacuolar H^+-ATPase (157). It has been suggested that these V0 *trans*-complexes establish a proteolipid-lined channel at the fusion site. Ca^{2+}-triggered conformational change in proteolipids opens a pore by a radial expansion of the proteolipid oligomeric ring (Figure 4*G*). Lipid polar heads migrate into the amphiphilic clefts in the lumen of the pores, and hydrocarbon chains of the lipids fill the space between the adjacent proteolipids. The argument for this mechanism boils down to the ability of V0 to form membrane *trans*-complexes and channels that expand in a Ca^{2+}/calmodulin-dependent fashion.

This mechanism readily accounts for the earliest fusion stages, in which membranes get into close contact and break their continuity. However, the mechanism addresses neither the force driving the lateral separation of the subunits nor the specific pathway by which a proteinaceous pore transforms into a lipid-involving structure. One might suggest that subunit separation yields two apposing pores, one in each membrane, with edges covered partially by the polar interface of the proteolipid and partially by lipid polar heads. Then the lipidic components of the edges of the adjacent pores merge to establish the continuity of two lipid bilayers. In energy terms, this scenario means that the energy released during separation of the V0 subunits should be sufficient for the formation of a lipid pore edge with energy per unit length of 10^{-11} J/m (Section 2.1). Assuming that each of the V0 subunits has a cross-section radius of ~1 nm, the energy of repulsion between the subunits required for their separation to a distance of 1 nm (the cross-section diameter of a lipid) is $\sim 3k_BT$. Protein refolding can provide such energies in the form of steric repulsion only as long as the proteins are adjacent rather than separated. On the other hand, it is hard to identify strong enough long-range repulsive interactions between the subunits required to allow lipids into the pore edge. For instance, electrostatic repulsion between two elementary charges located at the centers of the subunits separated by a 1-nm gap gives only $\sim 0.35k_BT$. Even stronger energy requirements have to be fulfilled to ensure the further expansion of the fusion pore. Hence, the mechanism of fusion through a proteinaceous pore still awaits clarification.

5.1.2 HEMIFUSION INTERMEDIATE The phenomenology of biological fusion is strikingly similar to that of fusion of protein-free bilayers including the lipid dependence, properties of early fusion pores (see above), and the activation energies for membrane rearrangements (31, 158). This suggests that protein-mediated fusion, in analogy to lipid bilayer fusion, proceeds through a hemifusion intermediate (Figure 4*C*). Although it is still not established that hemifusion is in fact a true intermediate of any biological fusion, the indirect evidence in favor of this hypothesis versus the proteinaceous pore hypothesis is extensive:

1. As discussed above, fusion proteins are indeed able to create hemifusion structures, as has been shown for wild-type and mutant HA and some other fusion proteins [e.g., parainfluenza (76), coronavirus (159), Sendai virus (160), and HIV (161)].
2. HA-expressing cells fusing with planar lipid bilayer demonstrate spreading of lipid dye prior to the opening of an expanding fusion pore (162). Similarly, in an elegant recent study of insulin granule exocytosis, lipid mixing was shown to precede fusion pore opening by ~0.3 s (163). Thus, both for viral and for exocytotic fusion, one might observe the sequence of events expected if hemifusion indeed is an intermediate in fusion pore formation. It remains possible, however, that the pore forms outside of the hemifusion structure and independently of it.
3. Proteins that form ionic channels have strict sequence requirements to support both a polar interface with water filling the channel lumen and the hydrophobic interface with a bilayer interior. Thus, the fusion pore formation observed for lipid-anchored HA (63, 164, 165), and the wide range of TM sequences supporting the formation of expanding fusion pores (73) are difficult to reconcile with the hypothesis that HA forms proteinaceous fusion pores and indirectly favor the alternative hypothesis that fusion starts with hemifusion intermediates.
4. LPC in the contacting membrane monolayers blocks fusion at a stage prior to the opening of a fusion pore, showing that the earliest stages of fusion are already dependent on the lipids (98, 148). The dependence of pore formation on the lipids in the distal membrane monolayers also indicates that complete fusion proceeds through lipidic rather than proteinaceous intermediates (58).

To conclude, in our view, there is now a rather solid case for the hypothesis that at least some of the better-characterized reactions of biological fusion, and in particular viral fusion, proceed through hemifusion intermediates similarly to the fusion of protein-free bilayers. These intermediates, if they indeed exist, are most likely either too small or too labile to be identified by electron microscopy (63).

5.1.3 FUSION PORE AS THE MOST DEMANDING STAGE Assuming that biological fusion, as fusion of protein-free bilayers, starts from local hemifusion and proceeds to the opening of a fusion pore, we can now ask which of the fusion stages is the most energy consuming; in other words, what is the most demanding job for the fusion protein? Experimental results indicate that early fusion stages including the establishment of the close contact and the local merger of the contacting monolayers into a fusion stalk are less demanding than the subsequent opening of a fusion pore and, especially, its expansion. Only at the highest number of activated HAs does fusion reach the stage of an expanding fusion pore (58, 61). Less HA is needed to open a flickering pore, and still less to establish the hemifusion phenotype. Many mutant fusion proteins (56, 76, 87, 159) and fragments of proteins including even short peptides (166, 167) mediate hemifusion but are incapable of forming expanding pores.

These results are consistent with the theoretical analysis above according to which a stalk might require just a few k_BT beyond the energy provided by the thermal fluctuations. The expansion of the HD is already fairly expensive and likely requires the concerted action of multiple fusion proteins (28). Fusion pores are even more expensive. To conclude, the main challenge for proteins likely involves the generation of nonlocal maintainable force that forms an expanding fusion pore.

5.2 Models for Local Rearrangements

Several mechanisms have been suggested to specify the way in which fusion and fission proteins might perform their jobs.

5.2.1 LOCAL MODELS FOR FISSION Most fission mechanisms suggested for the best-characterized dynamin-mediated reaction [reviewed and classified in (168)] can be subdivided into two groups. The first group considers dynamin as a mechanochemical enzyme that changes its conformation upon GTP hydrolysis, resulting in an additional constriction [reviewed in (122)] and/or stretching (169, 170) of the collared tube. Alternative models suggest that dynamin plays the role of a molecular switch (171) that triggers a cascade of downstream reactions mediated by endophilin and leading to fission (129).

A recent mechanism of dynamin-mediated fission reconciles, at least partially, the previous models (172). It proposes that the dynamin helix plays the role of a rigid external skeleton in the membrane neck, imposing constrictions on the neck shape. In these conditions the neck membrane loses its stability and collapses if the dynamin helix stretches and/or tightens as a result of GTP hydrolysis and if the spontaneous curvature of the neck membrane becomes more negative as a result of the endophilin-mediated reaction (Section 4.3) of conversion of the lipid of positive spontaneous curvature, LPA, to a lipid of almost zero spontaneous curvature, PA. The conformational change in dynamin and the change in lipid composition can act concertedly, making the fission phenomenon more robust.

This model, as well as earlier ideas [reviewed in (173)], describes only the first stages of the fission process, namely constriction of the membrane neck. None of the models address the nucleation of the membrane discontinuity, hemifission (if it exists), and the completion of membrane separation, which are the subject of future work.

5.2.2 LOCAL MODELS FOR FUSION Many suggested fusion mechanisms are based on the assumption that fusion protein restructuring exerts a pulling force on a membrane. The hypothetical fusion machines comprise several protein molecules assembled around the future fusion site. In some models (68, 98, 143, 174, 175), at the first stage of fusion protein restructuring the proteins establish holds on both membranes. In the case of viral fusion proteins such as HA or gp120/gp41 that are anchored in the viral membrane by the TM, this stage involves insertion of the fusion peptide into the target membrane. It has been estimated that binding of the HA fusion peptide to the membrane is strong enough to serve as a reliable hold (84). In the SNARE-pin hypothesis (175), suggested for intracellular fusion proteins, a similar configuration develops upon binding of two integral proteins expressed in different membranes. Further conformational change pulls two membrane anchors of the same protein (or both TMs of the *trans*-complex of the proteins) into close proximity. This stage, best characterized both at the structural and at the functional levels for gp120/gp41 (68, 98) and parainfluenza virus (143), consists of the formation of the outer layer of the 6-helix bundle. This bundle formation directly induces fusion (98). At this stage TMs of the fusion protein can become structural elements of the otherwise lipidic edge of the initial pore (165).

Another hypothetical fusion mechanism suggests that fusion is mediated by activated HAs with their fusion peptides inserted into the viral membrane (84). The subsequent coiled-coil transition in the HA2 ectodomain results in a force pulling the fusion peptides and deforming the viral membrane in the vicinity of the HA trimer. As a result, the trimers self-assemble into ringlike clusters and dimple the viral membrane toward the target membrane. The top of the dimple accumulates the bending stress, which is released when the membranes hemifuse.

A similar mechanism has been suggested for paramyxovirus-induced membrane fusion (95). In this case, fusion peptides are first inserted into the target membrane, and then the fusion protein coiled-coil extends toward the viral rather than target membrane (in the direction opposite that described for HA and gp120/gp41), which dimples the target membrane toward the viral envelope. This kind of scenario was suggested earlier for exocytosis (176). In the scaffold model (176), fusion proteins induce dimpling of the plasma membrane toward the granule membrane, stress the tip of the dimple, and promote its fusion. Both exocytosis (177) and HA-mediated fusion (178) are inhibited by inflating cells and, thus, by damping of dimple formation by membrane tension. These findings confirm the functional importance of the membrane dimples for fusion.

A recent model suggests that the specific mechanism by which activated HA might form a hydrophobic defect at the top of the dimple is the extraction of the previously inserted fusion peptides (179).

All these models suggest specific protein-driven mechanisms of establishing local intermembrane contact and generating local membrane stresses. Yet none of them address the membrane tension necessary for the most demanding stage—expansion of the fusion pore—because a local protein machine is unable to generate tension in a membrane section larger than the initial fusion site.

5.3 Models for Global Rearrangements

To generate a membrane stress that can drive fusion pore expansion, the fusion machine must act on a larger area of the membrane. Fusion proceeds via stages that are analogous to those of membrane budding-off and fission but in the opposite direction (Figure 5). Therefore, it is tempting to speculate about the force that drives fusion pore expansion on the basis of what is known about the force that drives the constriction of the membrane neck in the budding reaction (180).

Lipid bilayers of membrane buds (Figure 5) in intracellular vesicles and viral particles are often very strongly bent to a radius of curvature as small as 15–20 nm (181, 182). How can proteins bend membranes to the radii, which in protein-free lipid dispersions can be achieved only by using the harshest ultrasound treatments (183)? What drives unbending of these vesicles during fusion?

5.3.1 MODELS BASED ON THE BILAYER-COUPLE HYPOTHESIS A global force driving membrane remodeling may come from interplay of the areas of two membrane monolayers (Section 2.1), which generates the tendency of the membrane to bend for budding or to unbend for fusion. For instance, epsin-induced asymmetry in the monolayer areas apparently drives tubulation of liposomes (125a). However, the bilayer-couple mechanism might work for large membranes only if the absolute number of the protein or lipid molecules inserted into or extracted from one of the membrane monolayers is very large because the action is averaged over the whole monolayer area. Alternatively, the site of fusion or budding/fission represents a small membrane patch mechanically isolated from the remaining membrane by a rigid wall. The stiffness of this wall must greatly exceed the stretching-compression rigidity of a lipid monolayer, so that the tensions generated within the membrane patch do not propagate into the surrounding membrane. To the best of our knowledge, the existence of such rigid walls around fusion or fission sites has never been reported.

In contrast to the large membranes, a small vesicle can be primed to unbending, and hence fusion, by depletion of lipids or proteins from its external

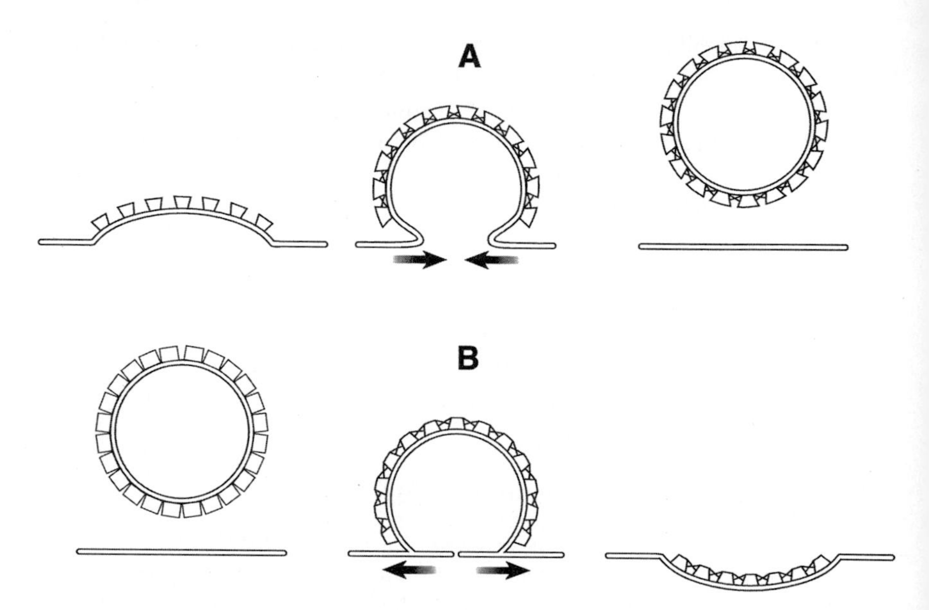

Figure 5 Coat mechanism for membrane remodeling. (*A*) Budding and fission. (*B*) Fusion and expansion of the fusion pore.

monolayer. Currently, there is no experimental support for this kind of remodeling mechanism in cells. Budding of protein-free membranes driven by the monolayer area asymmetry results in vesicles, which are much larger than the intracellular vesicles and often remain connected to the parent vesicles (50). In our opinion, the global driving force for membrane remodeling is most likely generated by interconnected protein coats.

5.3.2 PROTEIN COAT MODEL FOR BUDDING To induce budding and fission (Figure 5*A*), a protein coat has to satisfy two major conditions. First, the attractive interaction between the coat-forming proteins has to be sufficiently strong. The interaction energy has to exceed the bending energy of a membrane fragment of area a covered by one protein, $f_{\text{int}} \geq (1/2)\kappa \cdot a \cdot J^2$. For a bud radius of 15 nm (i.e., a curvature of $J = 2/15\ \text{nm}^{-1}$), an area per protein molecule of $a \approx 100\ \text{nm}^2$, and a bilayer bending rigidity of $\kappa_b \approx 20k_BT$, we obtain $f_{\text{int}} \geq 20k_BT$. This energy is ~100 times larger than the energy of the physical interaction between the protein molecules, such as electrostatic attraction and the estimated energy of interaction of clathrin triskelions (184), but it can be provided by a chemical bond.

The second condition relates to the rigidity, κ_p, of the protein coat, which has to exceed the rigidity of the lipid bilayer, κ_b. The bending rigidity of protein layers has never been measured experimentally but was estimated as $\kappa_p = 2000\ k_BT$ (180) based on the elasticity of dilatation measured for a number of protein layers. Accordingly, the dense coat of fusion proteins is likely to be a hundred times more rigid than the underlying lipid bilayer, and the protein coat can, indeed, control the shape of lipid bilayer.

In virus release and in some intracellular reactions, sphere-forming coat proteins such as influenza M1 and COPs might provide the global driving force for membrane budding-fission and also determine the specific pathway of the actual membrane fission. In other reactions, these two functions can involve different proteins. For instance, coats consisting of clathrin, adaptor proteins, and their partners might provide the global force for budding, while dynamin, endophilin, and other cylinder-forming proteins mediate fission.

5.3.3 PROTEIN COAT MODEL FOR FUSION The protein coat model of membrane fusion (180) (Figure 5*B*) assumes that (*a*) activated fusion proteins form a dense interconnected protein coat surrounding the zone of intermembrane contact; (*b*) this coat has a spontaneous curvature opposite to that of the budding coat, and to adopt its preferable configuration, the fusion coat bends out of the initial shape of the membrane surface; and (*c*) the bending of the protein coat deforms the underlying lipid bilayer and produces tension that drives fusion and expands the fusion pore.

In contrast to the local models, the protein coat model accounts for the force driving the fusion pore expansion until it reaches the dimension of the coat itself. For a virus, whose surface is completely covered by the coat, this means a complete insertion of the viral membrane into the target one.

Some of the features of viral fusion reactions, discussed above, are consistent with this hypothesis. Aggregation of activated fusion proteins (Section 4.1) is crucial for coat formation. The larger the coat, the larger the fusion intermediates that would remain subject to the tension, explaining the observed dependence of the fusion phenotype on the numbers of available fusion proteins (Section 3.2).

In contrast to the local models, the coat hypothesis accounts for the extremely high surface densities of fusion proteins characteristic of many viruses as a factor that drives, rather than sterically hinders, fusion. Furthermore, the protein coat hypothesis is independent of any specific features of fusion proteins, such as the 6-helix bundle motif shared by many but not all viral fusion proteins (67, 70, 71, 185), and can be fairly general.

To the best of our knowledge, the coat hypothesis is the only one suggesting the driving force for all fusion stages from the beginning through the end. However, it remains possible that two functions of the fusion protein, generation of the sustained driving force and local action initiating the reaction, may be separate. For instance, the local action can proceed according to one of the local mechanisms discussed above, while coat assembly drives the later stages. In intracellular fusion, local fusion and pore expansion can be controlled by different proteins. It has also been suggested that the global force that drives the pore expansion is of an osmotic nature [(163), but see (186)].

6. CONCLUDING REMARKS

Does a universal mechanism underlie all reactions of membrane remodeling? Are the remodeling proteins structurally involved in the earliest local intermediates of fusion and fission? Alternatively, do these proteins promote remodeling by priming the contacting bilayers for fusion or fission and then allowing them to reorganize through the sequence of lipid-involving intermediates? Currently, we can answer neither of these fundamental questions unambiguously. However, the available data are consistent with a tempting hypothesis that a universal element in diverse remodeling reactions is rearrangement of lipid bilayer patches involving formation of nonbilayer intermediates. Fusion and fission are then driven by similar but oppositely directed elastic forces, and knowledge accumulated for one of these reactions can be used to better understand another.

If this hypothesis is correct, the direct and local action of the proteins at the contacting monolayers of the merging membranes can be less crucial than commonly assumed. Indeed, in contrast to all known fusion reactions, budding and fission are often mediated by proteins, which are located at the distal membrane monolayers and thus cannot directly interact with the contacting monolayers. Therefore, any similarity between the mechanisms by which proteins merge bilayers in fusion and fission has to be related to the way in which proteins, present on either of the two sides of the membrane, shape and stress it.

This hypothesis is clearly too simple to fully describe biological reactions. Although it gives some idea of the possible forces involved in membrane rearrangements and the related requirements for protein action, it leaves open the question of the structure of the specific protein machinery that releases the conformational energy of the proteins and delivers it to the intermediates of membrane fusion and fission. Future studies will, hopefully, bring together detailed characterizations of different protein machines, functional studies describing the pathways of membrane rearrangements, and theoretical analysis of configurations and energies of the intermediate structures of membrane remodeling to uncover the answers to the fascinating question of how proteins merge and divide lipid bilayers.

ACKNOWLEDGMENTS

We are very grateful to Eugenia Leikina, Gregory Melikyan, Aditya Mittal, Corinne Ramos, and Joshua Zimmerberg for critically reading the manuscript. We apologize to authors of relevant research for the impossibility of citing all the references relevant to this review because of severe space constraints. The work of M.K. is supported by the Human Frontier Science Program Organization.

The *Annual Review of Biochemistry* is online at http://biochem.annualreviews.org

LITERATURE CITED

1. Tanford C. 1973. *The Hydrophobic Effect: Formation of Micelles and Biological Membranes*. New York: Wiley. 200 pp.
2. Abidor IG, Arakelyan VB, Chernomordik LV, Chizmadzhev YA, Pastushenko VF, Tarasevich MR. 1979. *Bioelectrochem. Bioenerg.* 6:37–52
3. Koslov MM, Markin VS. 1984. *J. Theor. Biol.* 109:17–39
4. Noguchi H, Takasu M. 2001. *J. Chem. Phys.* 115:9547–51
5. Muller M, Katsov K, Schick M. 2002. *J. Chem. Phys.* 116:2342–45
6. Jahn R, Grubmuller H. 2002. *Curr. Opin. Cell Biol.* 14:488–95
7. Leikin S, Kozlov MM, Fuller NL, Rand RP. 1996. *Biophys. J.* 71:2623–32
8. Helfrich W. 1973. *Z. Naturforsch. Teil C* 28:693–703
9. Gruner S. 1989. *J. Phys. Chem.* 93: 7562–70
10. Israelachvili JN, Mitchell DJ, Ninham BW. 1976. *J. Chem. Soc. Faraday Trans. 2* 72:1525–68
11. Kozlov MM, Leikin S, Rand RP. 1994. *Biophys. J.* 67:1603–11
12. Fuller N, Rand RP. 2001. *Biophys. J.* 81:243–54
13. Szule JA, Fuller NL, Rand RP. 2002. *Biophys. J.* 83:977–84
14. Sheetz MP, Singer SJ. 1974. *Proc. Natl. Acad. Sci. USA* 71:4457–61
15. Chernomordik LV, Melikyan GB, Chizmadzhev YA. 1987. *Biochim. Biophys. Acta* 906:309–52
16. May S. 2000. *Eur. Phys. J. E* 3:37–44
17. Kozlovsky Y, Kozlov MM. 2002. *Biophys. J.* 88:882–95
18. Rand RP, Parsegian VA. 1989. *Biochim. Biophys. Acta* 988:351–76
19. Leikin SL, Kozlov MM, Chernomordik LV, Markin VS, Chizmadzhev YA. 1987. *J. Theor. Biol.* 129:411–25
20. Gingell D, Ginsberg I. 1978. In *Membrane Fusion*, ed. G Poste, GL Nicholson, pp. 791–833. Amsterdam: Elsevier
21. Kozlov MM, Markin VS. 1983. *Biofizika* 28:255–61
22. Siegel DP. 1993. *Biophys. J.* 65: 2124–40
23. Kuzmin PI, Zimmerberg J, Chizmadzhev YA, Cohen FS. 2001. *Proc. Natl. Acad. Sci. USA* 98:7235–40
24. Markin VS, Albanesi JP. 2002. *Biophys. J.* 82:693–712
25. Yang L, Huang HW. 2002. *Science* 297:1877–79
26. Kozlov MM, Leikin SL, Chernomordik LV, Markin VS, Chizmadzhev YA. 1989. *Eur. Biophys. J.* 17:121–29
27. Markin VS, Kozlov MM, Borovjagin VL. 1984. *Gen. Physiol. Biophys.* 3:361–77
28. Kozlovsky Y, Chernomordik LV, Kozlov MM. 2002. *Biophys. J.* 83:2634–51
29. Sandre O, Moreaux L, Brochard-Wyart F. 1999. *Proc. Natl. Acad. Sci. USA* 96:10591–96
30. Zimmerberg J, Cohen FS, Finkelstein A. 1980. *J. Gen. Physiol.* 75:241–50
31. Lentz BR, Lee JK. 1999. *Mol. Membr. Biol.* 16:279–96
32. Lee J, Lentz BR. 1997. *Biochemistry* 36:6251–59
33. Chanturiya A, Chernomordik LV, Zimmerberg J. 1997. *Proc. Natl. Acad. Sci. USA* 94:14423–28
34. Wenk MR, Seelig J. 1998. *Biochim. Biophys. Acta* 1372:227–36
35. Meers P, Ali S, Erukulla R, Janoff AS. 2000. *Biochim. Biophys. Acta* 1467: 227–43
36. Chernomordik L, Chanturiya A, Green J, Zimmerberg J. 1995. *Biophys. J.* 69: 922–29
37. Garcia RA, Pantazatos SP, Pantazatos DP, MacDonald RC. 2001. *Biochim. Biophys. Acta* 1511:264–70
38. Basanez G, Goni FM, Alonso A. 1998. *Biochemistry* 37:3901–8

39. Chernomordik L, Kozlov MM, Zimmerberg J. 1995. *J. Membr. Biol.* 146:1–14
40. Duzgunes N, Wilschut J, Fraley R, Papahadjopoulos D. 1981. *Biochim. Biophys. Acta* 642:182–95
41. Aldwinckle TJ, Ahkong QF, Bangham AD, Fisher D, Lucy JA. 1982. *Biochim. Biophys. Acta* 689:548–60
42. Wilschut J, Hoekstra D. 1986. *Chem. Phys. Lipids* 40:145–66
43. Ohki S. 1988. In *Molecular Mechanisms of Membrane Fusion,* ed. S Ohki, D Doyle, TD Flanagan, SW Hui, E Mayhew, pp. 123–39. New York: Plenum
44. Lentz BR. 1994. *Chem. Phys. Lipids* 73:91–106
45. Kinnunen PK, Holopainen JM. 2000. *Biosci. Rep.* 20:465–82
46. Waugh RE, Song J, Svetina S, Zeks B. 1992. *Biophys. J.* 61:974–82
47. Lipowsky R. 1995. *Curr. Opin. Struct. Biol.* 5:531–40
48. Dobereiner HG, Kas J, Noppl D, Sprenger I, Sackmann E. 1993. *Biophys. J.* 65:1396–403
49. Holopainen JM, Angelova MI, Kinnunen PK. 2000. *Biophys. J.* 78:830–38
50. Devaux PF. 2000. *Biochimie* 82:497–509
51. Jahn R, Sudhof TC. 1999. *Annu. Rev. Biochem.* 68:863–911
52. Burger KNJ. 2000. *Traffic* 1:605–13
53. Hernandez LD, Hoffman LR, Wolfberg TG, White JM. 1996. *Annu. Rev. Cell Dev. Biol.* 12:627–61
54. Zimmerberg J, Blumenthal R, Sarkar DP, Curran M, Morris SJ. 1994. *J. Cell Biol.* 127:1885–94
55. Melikyan GB, Niles WD, Ratinov VA, Karhanek M, Zimmerberg J, Cohen FS. 1995. *J. Gen. Physiol.* 106:803–19
56. Kemble GW, Danieli T, White JM. 1994. *Cell* 76:383–91
57. Melikyan GB, White JM, Cohen FS. 1995. *J. Cell Biol.* 131:679–91
58. Chernomordik LV, Frolov VA, Leikina E, Bronk P, Zimmerberg J. 1998. *J. Cell Biol.* 140:1369–82
59. Tse FW, Iwata A, Almers W. 1993. *J. Cell Biol.* 121:543–52
60. Melikyan GB, Brener SA, Ok DC, Cohen FS. 1997. *J. Cell Biol.* 136:995–1005
61. Leikina E, Chernomordik LV. 2000. *Mol. Biol. Cell* 11:2359–71
62. Markosyan RM, Melikyan GB, Cohen FS. 2001. *Biophys. J.* 80:812–21
63. Frolov V, Cho M-S, Bronk P, Reese T, Zimmerberg J. 2000. *Traffic* 1:622–30
63a. Wang L, Seeley ES, Wickner W, Merz AJ. 2002. *Cell* 108:357–69
64. Schmid SL, Smythe E. 1991. *J. Cell Biol.* 114:869–80
65. Schmid SL. 1993. *Trends Cell Biol.* 3:145–48
66. Schmid SL. 1997. *Annu. Rev. Biochem.* 66:511–48
67. Skehel JJ, Wiley DC. 2000. *Annu. Rev. Biochem.* 69:531–69
68. Eckert DM, Kim PS. 2001. *Annu. Rev. Biochem.* 70:777–810
69. Wyatt R, Sodroski J. 1998. *Science* 280:1884–88
70. Heinz FX, Allison SL. 2001. *Curr. Opin. Microbiol.* 4:450–55
71. Gaudin Y. 2000. *Subcell. Biochem.* 34:379–408
72. Schroth-Diez B, Ponimaskin E, Reverey H, Schmidt MF, Herrmann A. 1998. *J. Virol.* 72:133–41
73. Melikyan GB, Lin S, Roth MG, Cohen FS. 1999. *Mol. Biol. Cell* 10:1821–36
74. Weiss CD, White JM. 1993. *J. Virol.* 67:7060–66
75. Salzwedel K, Johnston PB, Roberts SJ, Dubay JW, Hunter E. 1993. *J. Virol.* 67:5279–88
76. Tong S, Compans RW. 2000. *Virology* 270:368–76
77. Kozerski C, Ponimaskin E, Schroth-Diez B, Schmidt MF, Herrmann A. 2000. *J. Virol.* 74:7529–37
78. Armstrong RT, Kushnir AS, White JM. 2000. *J. Cell Biol.* 151:425–38

79. Cleverley DZ, Lenard J. 1998. *Proc. Natl. Acad. Sci. USA* 95:3425–30
80. Cleverley DZ, Geller HM, Lenard J. 1997. *Exp. Cell Res.* 233:288–96
81. Lescar J, Roussel A, Wien MW, Navaza J, Fuller SD, et al. 2001. *Cell* 105:137–48
82. Stegmann T, White JM, Helenius A. 1990. *EMBO J.* 9:4231–41
83. Shangguan T, Siegel DP, Lear JD, Axelsen PH, Alford D, Bentz J. 1998. *Biophys. J.* 74:54–62
84. Kozlov MM, Chernomordik LV. 1998. *Biophys. J.* 75:1384–96
85. Gruenke JA, Armstrong RT, Newcomb WW, Brown JC, White JM. 2002. *J. Virol.* 76:4456–66
86. Durrer P, Galli C, Hoenke S, Corti C, Gluck R, et al. 1996. *J. Biol. Chem.* 271:13417–21
87. Qiao H, Armstrong RT, Melikyan GB, Cohen FS, White JM. 1999. *Mol. Biol. Cell* 10:2759–69
88. Cross KJ, Wharton SA, Skehel JJ, Wiley DC, Steinhauer DA. 2001. *EMBO J.* 20:4432–42
89. Durell SR, Martin I, Ruysschaert JM, Shai Y, Blumenthal R. 1997. *Mol. Membr. Biol.* 14:97–112
90. Pecheur EI, Sainte-Marie J, Bienvenue A, Hoekstra D. 1999. *J. Membr. Biol.* 167:1–17
91. Han X, Bushweller JH, Cafiso DS, Tamm LK. 2001. *Nat. Struct. Biol.* 8:715–20
92. Korte T, Epand RF, Epand RM, Blumenthal R. 2001. *Virology* 289:353–61
93. Ghosh JK, Peisajovich SG, Shai Y. 2000. *Biochemistry* 39:11581–92
94. Nieva JL, Suarez T. 2000. *Biosci. Rep.* 20:519–33
95. Peisajovich SG, Shai Y. 2002. *Trends Biochem. Sci.* 27:183–90
96. Markovic I, Leikina E, Zhukovsky M, Zimmerberg J, Chernomordik LV. 2001. *J. Cell Biol.* 155:833–44
97. Schoch C, Blumenthal R. 1993. *J. Biol. Chem.* 268:9267–74
98. Melikyan GB, Markosyan RM, Hemmati H, Delmedico MK, Lambert DM, Cohen FS. 2000. *J. Cell Biol.* 151:413–23
99. Gutman O, Danieli T, White JM, Henis YI. 1993. *Biochemistry* 32:101–6
100. Ruigrok RW, Wrigley NG, Calder LJ, Cusack S, Wharton SA, et al. 1986. *EMBO J.* 5:41–49
101. Epand RF, Yip CM, Chernomordik LV, LeDuc DL, Shin YK, Epand RM. 2001. *Biochim. Biophys. Acta* 1513:167–75
102. Ruigrok RW, Aitken A, Calder LJ, Martin SR, Skehel JJ, et al. 1988. *J. Gen. Virol.* 69:2785–95
103. Han X, Tamm LK. 2000. *J. Mol. Biol.* 304:953–65
104. Kim CH, Macosko JC, Shin YK. 1998. *Biochemistry* 37:137–44
105. Caffrey M, Braddock DT, Louis JM, Abu-Asab MA, Kingma D, et al. 2000. *J. Biol. Chem.* 275:19877–82
106. Kelly EP, Greene JJ, King AD, Innis BL. 2000. *Vaccine* 18:2549–59
107. Markovic I, Pulyaeva H, Sokoloff A, Chernomordik LV. 1998. *J. Cell Biol.* 143:1155–66
108. Leikina E, Ramos C, Markovic I, Zimmerberg J, Chernomordik LV. 2002. *EMBO J.* 21:5701–10
109. Stiasny K, Allison SL, Marchler-Bauer A, Kunz C, Heinz FX. 1996. *J. Virol.* 70:8142–47
110. Doranz BJ, Baik SS, Doms RW. 1999. *J. Virol.* 73:10346–58
111. Kliger Y, Peisajovich SG, Blumenthal R, Shai Y. 2000. *J. Mol. Biol.* 301:905–14
112. Bentz J. 2000. *Biophys. J.* 78:227–45
113. Blumenthal R, Sarkar DP, Durell S, Howard DE, Morris SJ. 1996. *J. Cell Biol.* 135:63–71
114. Danieli T, Pelletier SL, Henis YI, White JM. 1996. *J. Cell Biol.* 133:559–69
115. Ellens H, Bentz J, Mason D, Zhang F, White JM. 1990. *Biochemistry* 29:9697–707

116. Roche S, Gaudin Y. 2002. *Virology* 297:128–35
117. Plonsky I, Zimmerberg J. 1996. *J. Cell Biol.* 135:1831–39
118. Garoff H, Hewson R, Opstelten DJ. 1998. *Microbiol. Mol. Biol. Rev.* 62:1171–90
119. Brodsky FM, Chen CY, Knuehl C, Towler MC, Wakeham DE. 2001. *Annu. Rev. Cell Dev. Biol.* 17:517–68
120. Sweitzer SM, Hinshaw JE. 1998. *Cell* 93:1021–29
121. Takei K, Haucke V, Slepnev V, Farsad K, Salazar M, et al. 1998. *Cell* 94:131–41
122. Hinshaw JE. 2000. *Annu. Rev. Cell Dev. Biol.* 16:483–519
123. Koenig JH, Ikeda K. 1989. *J. Neurosci.* 9:3844–60
124. Hannah MJ, Schmidt AA, Huttner WB. 1999. *Annu. Rev. Cell Dev. Biol.* 15:733–98
125. Farsad K, Ringstad N, Takei K, Floyd SR, Rose K, De Camilli P. 2001. *J. Cell Biol.* 155:193–200
125a. Ford MG, Mills IG, Peter BJ, Vallis Y, Praefcke GJ, et al. 2002. *Nature* 419: 361–66
126. Spang A, Matsuoka K, Hamamoto S, Schekman R, Orci L. 1998. *Proc. Natl. Acad. Sci. USA* 95:11199–204
127. Matsuoka K, Orci L, Amherdt M, Bednarek SY, Hamamoto S, et al. 1998. *Cell* 93:263–75
128. Schmid SL, McNiven MA, De Camilli P. 1998. *Curr. Opin. Cell Biol.* 10:504–12
129. Schmidt A, Wolde M, Thiele C, Fest W, Kratzin H, et al. 1999. *Nature* 401: 133–41
130. Huttner WB, Schmidt A. 2000. *Curr. Opin. Neurobiol.* 10:543–51
131. Weigert R, Silletta MG, Spano S, Turacchio G, Cericola C, et al. 1999. *Nature* 402:429–33
132. Huijbregts RP, Topalof L, Bankaitis VA. 2000. *Traffic* 1:195–202
133. Baron CL, Malhotra V. 2002. *Science* 295:325–28
134. Gomez-Puertas P, Albo C, Perez-Pastrana E, Vivo A, Portela A. 2000. *J. Virol.* 74:11538–47
135. Schalich J, Allison SL, Stiasny K, Mandl CW, Kunz C, Heinz FX. 1996. *J. Virol.* 70:4549–57
136. Pecheur EI, Martin I, Maier O, Bakowsky U, Ruysschaert JM, Hoekstra D. 2002. *Biochemistry* 41:9813–23
137. Nieva JL, Bron R, Corver J, Wilschut J. 1994. *EMBO J.* 13:2797–804
138. Ahn A, Gibbons DL, Kielian M. 2002. *J. Virol.* 76:3267–75
139. Martin TF. 1998. *Annu. Rev. Cell Dev. Biol.* 14:231–64
140. Chernomordik LV, Vogel SS, Sokoloff A, Onaran HO, Leikina EA, Zimmerberg J. 1993. *FEBS Lett.* 318:71–76
141. Yeagle PL, Smith FT, Young JE, Flanagan TD. 1994. *Biochemistry* 33:1820–27
142. Gaudin Y. 2000. *J. Cell Biol.* 150:601–12
143. Russell CJ, Jardetzky TS, Lamb RA. 2001. *EMBO J.* 20:4024–34
144. Chernomordik L. 1996. *Chem. Phys. Lipids* 81:203–13
145. Chernomordik LV, Leikina E, Kozlov MM, Frolov VA, Zimmerberg J. 1999. *Mol. Membr. Biol.* 16:33–42
146. Gunther-Ausborn S, Praetor A, Stegmann T. 1995. *J. Biol. Chem.* 270:29279–85
147. Baljinnyam B, Schroth-Diez B, Korte T, Herrmann A. 2002. *J. Biol. Chem.* 277:20461–67
148. Chernomordik LV, Leikina E, Frolov V, Bronk P, Zimmerberg J. 1997. *J. Cell Biol.* 136:81–94
149. Razinkov VI, Melikyan GB, Epand RM, Epand RF, Cohen FS. 1998. *J. Gen. Physiol.* 112:409–22
150. Grote E, Baba M, Ohsumi Y, Novick PJ. 2000. *J. Cell Biol.* 151:453–66
151. Vogel SS, Leikina EA, Chernomordik LV. 1993. *J. Biol. Chem.* 268: 25764–68
152. Tahara M, Coorssen JR, Timmers K,

Blank PS, Whalley T, et al. 1998. *J. Biol. Chem.* 273:33667–73
153. Subtil A, Gaidarov I, Kobylarz K, Lampson MA, Keen JH, McGraw TE. 1999. *Proc. Natl. Acad. Sci. USA* 96:6775–80
154. Emoto K, Umeda M. 2000. *J. Cell Biol.* 149:1215–24
155. Farge E, Ojcius DM, Subtil A, Dautry-Varsat A. 1999. *Am. J. Physiol. Cell Physiol.* 276:C725–33
156. Lindau M, Almers W. 1995. *Curr. Opin. Cell Biol.* 7:509–17
157. Peters C, Bayer MJ, Buhler S, Andersen JS, Mann M, Mayer A. 2001. *Nature* 409:581–88
158. Lee J, Lentz BR. 1998. *Proc. Natl. Acad. Sci. USA* 95:9274–79
159. Chang KW, Sheng Y, Gombold JL. 2000. *Virology* 269:212–24
160. Rocheleau JV, Petersen NO. 2001. *Eur. J. Biochem.* 268:2924–30
161. Munoz-Barroso I, Durell S, Sakaguchi K, Appella E, Blumenthal R. 1998. *J. Cell Biol.* 140:315–23
162. Razinkov VI, Melikyan GB, Cohen FS. 1999. *Biophys. J.* 77:3144–51
163. Takahashi N, Kishimoto T, Nemoto T, Kadowaki T, Kasai H. 2002. *Science* 297:1349–52
164. Nussler F, Clague MJ, Herrmann A. 1997. *Biophys. J.* 73:2280–91
165. Markosyan RM, Cohen FS, Melikyan GB. 2000. *Mol. Biol. Cell* 11:1143–52
166. Leikina E, LeDuc DL, Macosko JC, Epand R, Shin YK, Chernomordik LV. 2001. *Biochemistry* 40:8378–86
167. Langosch D, Crane JM, Brosig B, Hellwig A, Tamm LK, Reed J. 2001. *J. Mol. Biol.* 311:709–21
168. Sever S, Damke H, Schmid SL. 2000. *Traffic* 1:385–92
169. Stowell MH, Marks B, Wigge P, McMahon HT. 1999. *Nat. Cell Biol.* 1:27–32
170. Higgins MK, McMahon HT. 2002. *Trends Biochem Sci.* 27:257–63
171. Sever S, Damke H, Schmid SL. 2000. *J. Cell Biol.* 150:1137–48
172. Kozlov MM. 2001. *Traffic* 2:51–65
173. Huttner WB, Zimmerberg J. 2001. *Curr. Opin. Cell Biol.* 13:478–84
174. Weissenhorn W, Dessen A, Harrison SC, Skehel JJ, Wiley DC. 1997. *Nature* 387:426–30
175. Weber T, Zemelman BV, McNew JA, Westermann B, Gmachl M, et al. 1998. *Cell* 92:759–72
176. Monck JR, Fernandez JM. 1992. *J. Cell Biol.* 119:1395–404
177. Solsona C, Innocenti B, Fernandez JM. 1998. *Biophys. J.* 74:1061–73
178. Markosyan RM, Melikyan GB, Cohen FS. 1999. *Biophys. J.* 77:943–52
179. Bentz J. 2000. *Biophys. J.* 78:886–900
180. Kozlov MM, Chernomordik LV. 2002. *Traffic* 3:256–67
181. Matsuoka K, Schekman R, Orci L, Heuser JE. 2001. *Proc. Natl. Acad. Sci. USA* 98:13705–9
182. Kuhn RJ, Zhang W, Rossmann MG, Pletnev SV, Corver J, et al. 2002. *Cell* 108:717–25
183. Lichtenberg D, Barenholz Y. 1988. In *Methods of Biochemical Analysis*, ed. D Glick, pp. 337–461. New York: Wiley Intersci.
184. Nossal R. 2001. *Traffic* 2:138–47
185. Chen L, Gorman JJ, McKimm-Breschkin J, Lawrence LJ, Tulloch PA, et al. 2001. *Structure* 9:255–66
186. Monck JR, Oberhauser AF, Alvarez de Toledo G, Fernandez JM. 1991. *Biophys. J.* 59:39–47

Annu. Rev. Biochem. 2003. 72:209–247
doi: 10.1146/annurev.biochem.72.121801.161828
Copyright © 2003 by Annual Reviews. All rights reserved

THE MANY FACES OF VITAMIN B_{12}: CATALYSIS BY COBALAMIN-DEPENDENT ENZYMES[1]

Ruma Banerjee and Stephen W. Ragsdale
Department of Biochemistry, University of Nebraska, Lincoln, Nebraska 68588-0664; email:rbanerjee1@unl.edu; sragsdale1@unl.edu

Key Words cobalamin, methyltransferase, dehalogenase, isomerase, adenosylcobalamin, methylcobalamin, methanogenesis, acetogenesis, homocysteine, methionine, adenosylmethionine

■ **Abstract** Vitamin B_{12} is a complex organometallic cofactor associated with three subfamilies of enzymes: the adenosylcobalamin-dependent isomerases, the methylcobalamin-dependent methyltransferases, and the dehalogenases. Different chemical aspects of the cofactor are exploited during catalysis by the isomerases and the methyltransferases. Thus, the cobalt-carbon bond ruptures homolytically in the isomerases, whereas it is cleaved heterolytically in the methyltransferases. The reaction mechanism of the dehalogenases, the most recently discovered class of B_{12} enzymes, is poorly understood. Over the past decade our understanding of the reaction mechanisms of B_{12} enzymes has been greatly enhanced by the availability of large amounts of enzyme that have afforded detailed structure-function studies, and these recent advances are the subject of this review.

CONTENTS

[1]Abbreviations used: MeCbl, methylcobalamin; AdoCbl, 5′ deoxyadenosyl cobalamin; CH_3-H_4folate, methyltetrahydrofolate; CH_3-H_4MPT: methyltetrahydromethanopterin; CoM, Coenzyme M or mercaptoethanesulfonate; CODH, carbon monoxide dehydrogenase; ACS, acetyl-CoA synthase; AdoMet, S-adenosylmethionine; CFeSP, corrinoid iron-sulfur protein; PCB, polychlorinated biphenyl.

INTRODUCTION

The history of vitamin B_{12}, originating in 1925 with the descriptions by Whipple and Minot & Murphy of the antipernicious anemia factor, is rich and replete with scientific milestones (1, 2). A hunt spanning two decades led independently to the isolation of the cofactor by Smith & Folkers in 1948 and dramatically changed the therapeutic solution for pernicious anemia from several grams of uncooked liver to a few micrograms of a red crystalline compound (3, 4)! Solution of the crystal structure of vitamin B_{12} by Hodgkin in 1956 (5) and an intense transatlantic collaboration culminating in the total synthesis of the cofactor involving the Woodward and Eschenmoser laboratories were some of the other highlights of the ensuing decades (5a).

Vitamin B_{12} is a tetrapyrrolic cofactor in which the central cobalt atom is coordinated by four equatorial nitrogen ligands donated by pyrroles A-D of the corrin ring. B_{12} is portly and, unlike other tetrapyrroles, has a built-in axial ligand appended from the periphery of ring D of the corrin macrocycle (Figure 1). The identity of the base at the terminus of the propanolamine tether varies in different organisms and is the unusual ribonucleoside, dimethylbenzimidazole, found in cobalamins. Diversity is also present at the upper axial ligand where cyano-, methyl-, and deoxyadenosyl-groups are seen in vitamin B_{12}, methylcobalamin (MeCbl[1]), and AdoCbl or coenzyme B_{12} respectively.

As major strides were being made on elucidating the identity and structure of the antipernicious anemia factor, a biological role for B_{12} was unraveled in Barker's laboratory where an AdoCbl derivative was found to be the cofactor for glutamate mutase (6). This was followed by the discovery several years later of the other biologically active alkylcobalamin, MeCbl (7). Presently, three classes of B_{12} enzymes are recognized, the isomerases, the methyltransferases, and the reductive dehalogenases. Recent advances in our understanding of their reaction mechanisms are the subject of this review.

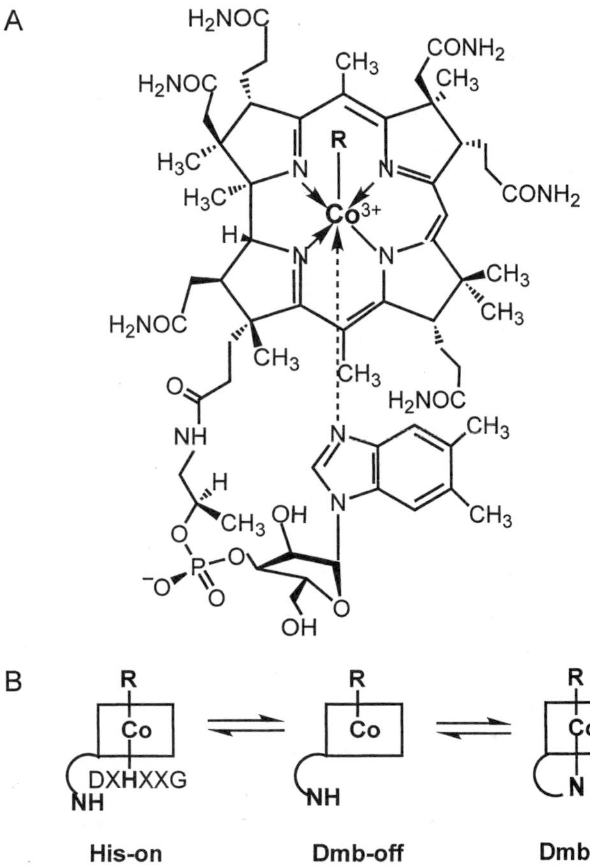

Figure 1 Structure and alternative conformations of cobalamin found in B_{12}-dependent enzymes. (*A*) Structure of cobalamin, where R is deoxyadenosine in AdoCbl, methyl in MeCbl, -OH in hydroxocobalamin, and -CN in vitamin B_{12}. (*B*) The free cofactor can exist in the base-on (or Dmb-on) or base-off (or Dmb-off) conformations, with the former predominating at physiological pH. The His-on conformation in which the endogenous ligand, dimethylbenzimidazole, is replaced by an active site histidine is seen is some B_{12}-dependent enzymes. In the corrinoid iron-sulfur (CFeSP) protein, the cofactor is in the base-off conformation, and a protein ligand does not appear to occupy the lower axial position.

B_{12}-DEPENDENT ISOMERASES

The isomerases are the largest subfamily of B_{12}-dependent enzymes found in bacteria, where they play important roles in fermentation pathways (8–10). The only exception is methylmalonyl-CoA mutase, which is found in both bacteria and in man. In some organisms, a B_{12}-dependent ribonucleotide reductase

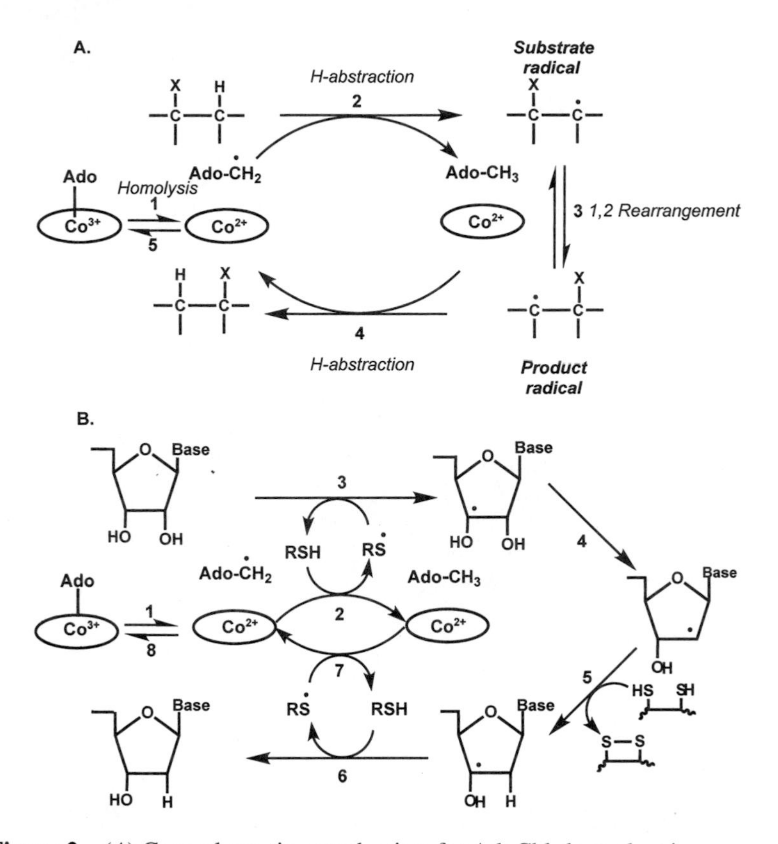

Figure 2 (*A*) General reaction mechanism for AdoCbl-dependent isomerases. (*B*) In class II ribonucleotide reductases, a thiyl radical rather than the initially formed 5′-deoxyadenosyl radical abstracts a hydrogen atom from the substrate and reduction occurs at C2′ of the substrate.

catalyzes the conversion of ribonucleotides to deoxyribonucleotides, which is fundamentally important for DNA replication and repair. B_{12}-dependent isomerases catalyze the 1,2 interchange between a variable substituent and a hydrogen atom on vicinal carbons (Figure 2). Considerable diversity exists in the nature of the migrating group ranging from carbon (in methylmalonyl-CoA mutase, isobutyryl-CoA mutase, glutamate mutase, and methyleneglutarate mutase) to nitrogen (in ethanolamine ammonia lyase, D-ornithine aminomutase, and β-lysine mutase) and oxygen (in diol dehydrase).

Subclasses of Isomerases

Recent crystal structures and spectroscopic studies have led to the recognition of two subclasses of B_{12}-dependent enzymes that differ with respect to their mode of cofactor binding (Figure 1). In solution and at physiological pH, the lower ligand position is occupied by the endogenous base, dimethylbenzimidazole, in a conformation that has been referred to in the literature as base-on. Protonation of the lower base, which has a pKa of 3.7 in AdoCbl (11), leads to the base-off conformer. Crystal structures of the Class I or His-on family of B_{12}-dependent isomerases have revealed the presence of yet another cofactor conformation in which the lower axial ligand, dimethylbenzimidazole, is replaced by a histidine residue donated by the protein (12, 13). The histidine is embedded in a DX**H**XXG sequence, which is the only primary sequence motif that appears to be conserved in both the B_{12}-dependent isomerase and methyltransferase family members that use the same cofactor-binding mode (14). In this mode, the nucleotide tail is bound in an extended conformation that results in dimethylbenzimidazole being >10Å removed from the cobalt to which it is coordinated in solution. The isomerases in this subfamily include methylmalonyl-CoA mutase, glutamate mutase, methyleneglutarate mutase, isobutyryl-CoA mutase, and lysine 5,6 aminomutase.

In the Class II or Dmb-on family of B_{12}-dependent isomerases, the solution conformation of the cofactor is retained upon binding to the active site. Thus, dimethylbenzimidazole is coordinated to the cobalt in the lower axial position. This mode of B_{12} binding has been seen in the crystal structures of diol dehydratase (15) and ribonucleotide reductase (16) and is expected for ethanolamine ammonia lyase based on the electron paramagnetic resonance (EPR) spectra (17, 18).

Despite the differences in the conformations of the bound cofactor, the B_{12}-dependent isomerases, with the exception of ribonucleotide reductase, exhibit similarities in the structural motifs they employ (Figure 3). A TIM barrel is juxtaposed on the upper face of cobalamin and encases the active site where the substrate binds. Numerous hydrogen bonds extend between the peripheral side chains on the corrin ring and the protein and likely contribute to the tight binding of the cofactor. In methylmalonyl-CoA mutase (12) and in glutamate mutase (13), the lower face of cobalamin is perched on the carboxy-terminal end of a Rossman (α/β)-fold, and a histidine residue extends from a loop to coordinate to the cobalt. In diol dehydratase, a Rossman-fold like structure exists in the central part of the β-subunit, which cups the lower face of the cofactor and may be important in making contact with the nucleotide (15). In contrast, ribonucleotide reductase exhibits a novel fold for binding B_{12}, which is more similar to the related structural elements used in the diiron-tyrosyl radical-dependent ribonucleotide reductases than it is to other B_{12} enzymes (16).

Cofactor Is a Latent Radical Reservoir

The organometallic bond in AdoCbl is water stable but is relatively weak with a bond dissociation energy estimated to be 31.5 kcal mol^{-1} for the base-on conformer and 34.5 kcal mol^{-1} for the base-off conformer (19–21). The inherent lability of the Co-carbon bond is exploited by AdoCbl-dependent isomerases to

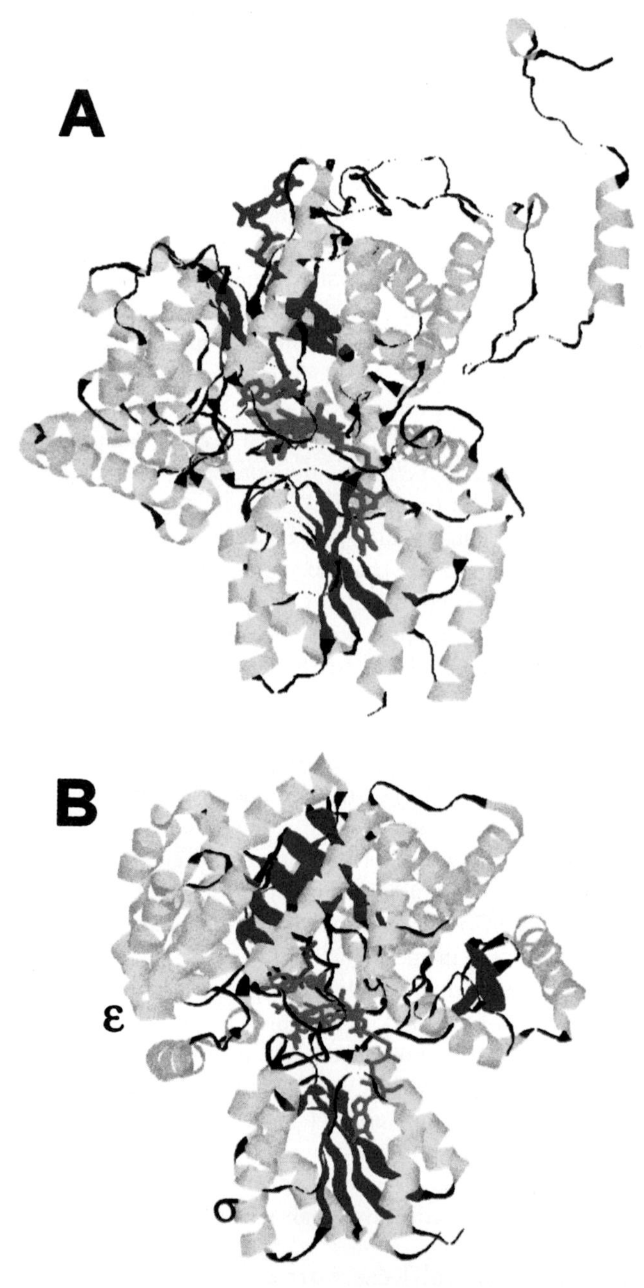

Figure 3 Comparison of the B_{12}-binding domains of AdoCbl-dependent isomerases: *A*, methylmalonyl-CoA mutase; *B*, glutamate mutase; *C*, dioldehydratase; and *D*, ribonucleotide reductase.

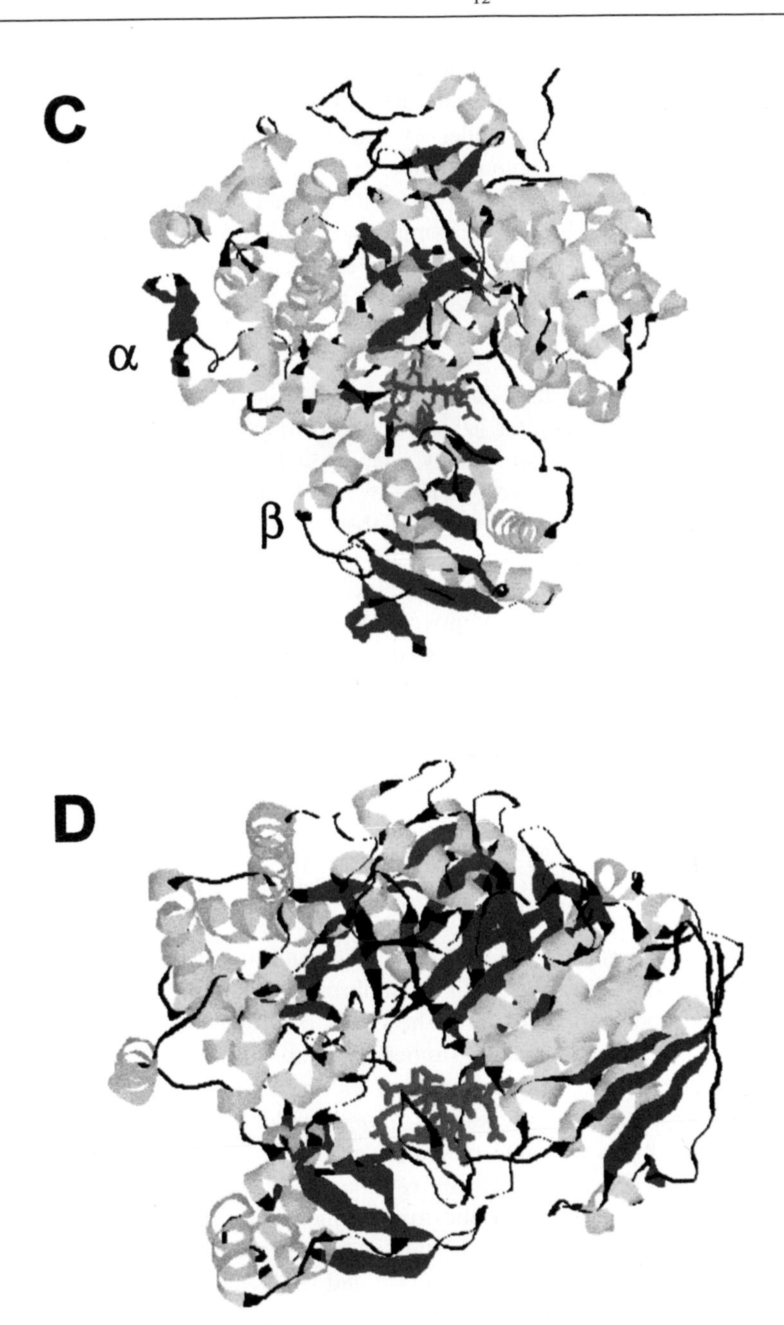

Figure 3 *Continued.*

effect radical based rearrangements that are initiated by homolysis of the Co-carbon bond (22). In the absence of substrate, the homolysis products, cob(II)alamin and the deoxyadenosyl radical, are not observed, indicating that the equilibrium for the homolysis reaction greatly favors geminate recombination. However, in the presence of substrate, the radical reservoir is mobilized into action as evidenced by an approximately trillionfold acceleration of the homolytic cleavage rate compared to the uncatalyzed rate in solution (20). Thus, utilization of substrate binding energy is key to controlling the homolysis equilibrium and protecting the latent radical pool from dissipation in side reactions.

Although ground state destabilization of the Co-carbon bond had been considered as a strategy for achieving the trillionfold homolysis rate enhancement (23–26), it is mechanistically unappealing. Stabilization of the deoxyadenosyl radical in the absence of substrate would lead to its rapid extinction and accumulation of inactive enzymes in the aerobic milieu where most of these enzymes operate. The extent of ground state stabilization has been experimentally evaluated by resonance Raman studies on methylmalonyl-CoA mutase in which isotope editing was employed to identify the Co-carbon stretching frequency for the free and enzyme-bound cofactor (27). The difference in their positions (6 cm^{-1}) was estimated to correspond to ~0.5 kcal mol^{-1} weakening of the Co-carbon bond, which is miniscule compared to the ~17 kcal mol^{-1} that occurs during the catalytic cycle (27). Resonance Raman spectra of glutamate mutase have also been reported (28); however, in the absence of isotope editing, conclusions regarding the identity and location of the Co-carbon stretching frequency can not be made.

Homolysis Timing Control

A key control issue for AdoCbl-dependent radical enzymes is containing the reactivity of the radical reservoir to minimize inactivating side reactions of the homolysis products. Insights into how this control is achieved have emerged from kinetic and mutagenesis studies on these enzymes. Propagation of the organic radical from deoxyadenosine to substrate is believed to occur directly in these enzymes with the exception of ribonucleotide reductase where an intermediate protein based cysteinyl radical is generated and operates as the working radical (Figure 2). The homolysis rate has been estimated in several of these enzymes by stopped-flow kinetic measurements and in each case found to be rapid and not rate limiting for catalytic turnover. A curious observation, first made with methylmalonyl-CoA mutase and later with other enzymes, was that the rate of cofactor homolysis is sensitive to isotopic substitution in the substrate (29). Thus, substitution of protiated methylmalonyl-CoA with [CD_3]-methylmalonyl-CoA decelerated homolysis ~20-fold at 25°C (29, 30). Similarly, use of the corresponding deuterated substrates led to substantial slowing of the homolysis rates in glutatmate mutase (31) and in ethanolamine ammonia lyase (32). These observations were curious not only because of the sensitivity of bond cleavage in

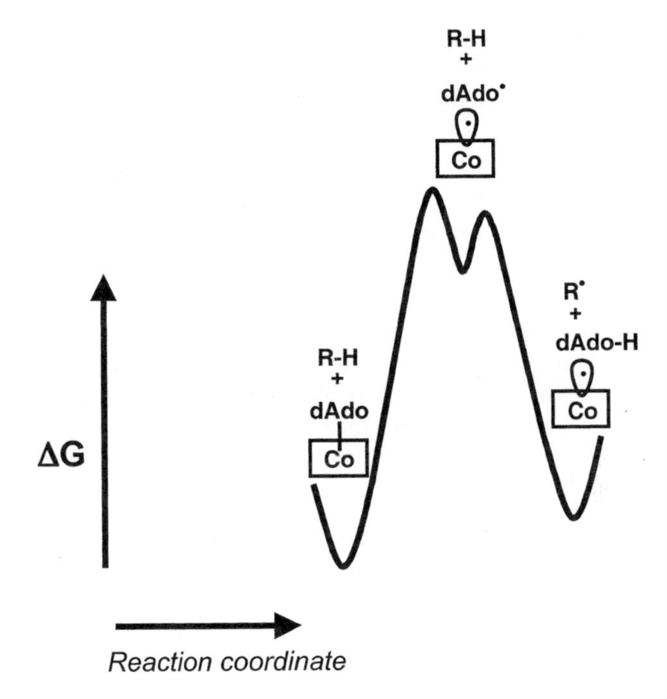

Figure 4 Qualitative free energy profile demonstrating kinetic coupling as a strategy for controlling the timing of the homolysis reaction.

the cofactor to isotopic substitution in the substrate, but also because the magnitude of the isotopic discrimination was anomalous and outside the range predicted for semiclassical behavior.

These results were interpreted as evidence for kinetic coupling between the homolytic cleavage and substrate radical generation steps as shown in Figure 4. According to this model, the homolysis equilibrium is unfavorable and lies in the direction of geminate recombination. However, in the presence of substrate, the high-energy dAdo•intermediate abstracts a hydrogen atom from the substrate to generate a more stable substrate-centered radical. The net effect of this coupled reaction is a shift in the overall homolysis equilibrium from recombination to radical propagation. This model provides an explanation for the sensitivity of the homolysis step to isotopic substitution in the substrate. Because detectable accumulation of the homolysis product, cob(II)alamin, is dependent on the generation of the substrate radical, its rate of formation is sensitive to whether a hydrogen or deuterium atom is being abstracted.

In ribonucleotide reductase, kinetic coupling involves generation of a thiyl radical rather than a substrate radical by the deoxyadenosyl radical (33). A working thiyl radical is common to both the diiron-tryosyl radical-dependent and B_{12}-dependent ribonucleotide reductases, and it is responsible in turn for generating the substrate radical. Mutation of C408, the site of the thiyl radical in the

Lactobacillus leichmannii B_{12}-dependent ribonucleotide reductase, leads to failure of the mutant enzyme to catalyze Co-carbon bond homolysis (33).

The basis of the anomalously large isotope effects has been characterized by examining the temperature dependence of the isotope effect in methylmalonyl-CoA mutase (30). The deuterium isotope effect increases from 35.6 to 49.9 as the temperature decreases from 20°C to 5°C. The magnitude of the isotope effect on the Arrhenius preexponentials (0.078) and the difference in the activation energies (Ea_D-Ea_H = 3.41 kcal mol^{-1}) deviate significantly from the values predicted for semiclassical behavior and provide strong evidence for quantum mechanical tunneling. Exalted tritium isotope effects of the order of 80–110 have been reported for diol dehydratase (34) and ethanolamine ammonia lyase (35, 36), and it is likely that nonclassical transfer pathways are involved in hydrogen atom abstraction by these enzymes as well.

Control of Radical Trajectories

One of the challenges for enzymes that deploy radical chemistry is the danger of extinction of reactive intermediates lurking in active sites lined with potential hydrogen atom donors. This problem is magnified in B_{12}-dependent isomerases where significant distances are bridged as the organic radical propagates intermolecularly between deoxyadenosine and substrate. EPR spectroscopy reveals the presence of biradical intermediates, which are either strongly or weakly coupled depending on whether the interradical distance is moderate (5–7Å) or large (10–11Å) (37). In most cases, the organic radical has been assigned as being either substrate- or product-derived, based on isotopic perturbation of the EPR spectra. The deoxyadenosyl radical has eluded detection in AdoCbl-dependent enzymes consistent with the kinetic coupling model in which it is not predicted to accumulate.

Depending on the interradical distance, the interaction between cob(II)alamin and the organic radical is either strong or weak as evidenced by the very different EPR spectral morphologies seen in B_{12}-dependent isomerases. In diol dehydratase and in ethanolamine ammonia lyase, a doublet signal is seen in spectra that consist of a broad resonance near $g = 2.3$ [due to cob(II)alamin] and a doublet of lines split by 70–140 Gauss near $g = 2.0$. The doublet signal arises from the organic radical split by weak isotropic exchange and dipolar coupling interactions with cob(II)alamin (38, 39). Simulations yield an electron-electron distance of ~10Å, which is consistent with the crystal structure of diol dehydratase (15).

In contrast, ribonucleotide reductase (40), glutamate mutase (41), methyleneglutarate mutase (42, 43), and methylmalonyl-CoA mutase (44, 45) exhibit EPR spectra with very different characteristics and have an effective g value of ~2.1 and resolved cobalt hyperfine splittings with ~55 G spacing. This class of EPR spectra arise from strong interactions between cob(II)alamin and the radical species via exchange and dipolar interactions (37). Simulations yield an electron-electron distance of 5–7Å for the spectra associated with ribonucleotide

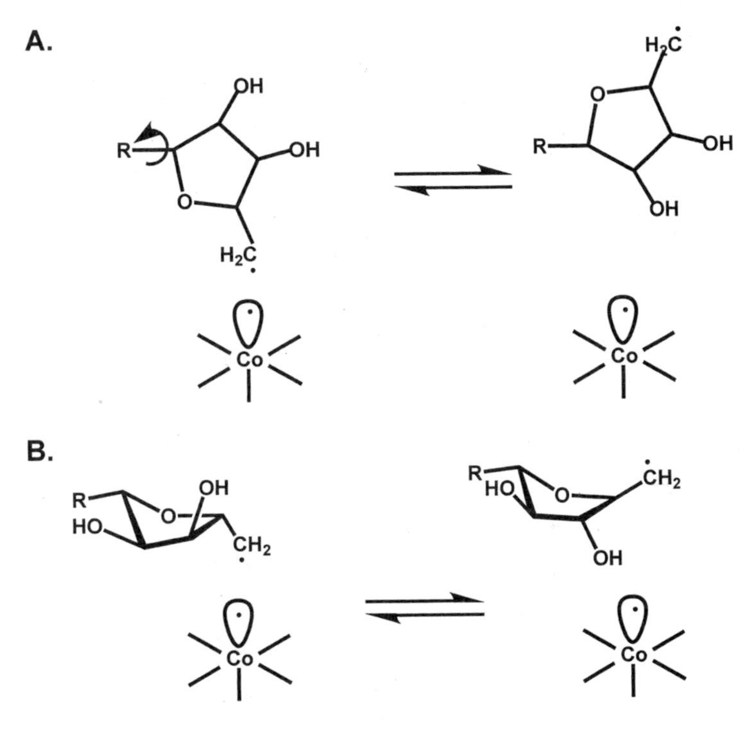

Figure 5 Ribose conformations control radical trajectories during propagation steps in diol dehydratase (*A*) and in glutamate mutase (*B*).

reductase (46) and glutamate mutase (41), which is consistent with the crystal structures of these enzymes (13, 16). In ribonucleotide reductase, a thiyl radical coupled to cob(II)alamin has been observed and was identified by substitution of the protons on the β-carbon of cysteine, which resulted in spectral narrowing (40).

The EPR spectra reveal that significant distances are traversed as the organic radical propagates between deoxyadenosine and substrate/product during each catalytic cycle, and this raises the issue of how side reactions are suppressed. Insights into how these enzymes control radical trajectories have emerged from structures of diol dehydratase (47) and glutamate mutase (48). In both cases, conformational toggling of the deoxyadenosine radical results in its reorientation from one favoring recombination to one abutting the substrate hydrogen atom destined for abstraction. In glutamate mutase, the interradical distance between the substrate and cob(II)alamin is estimated to be ~6.6Å by simulations of the EPR spectra (41). This distance is bridged by a pseudorotation of the ribose ring that moves the C5′ carbon (where the radical is localized) from an axial to an equatorial location and thereby guides the radical away from geminate recombination and toward substrate radical generation (Figure 5). The crystal structure of glutamate mutase shows a mixed population of conformers, C2′-*endo* (C5′ is positioned above the cobalt) and C3′-*endo* (C5′ is directed toward the substrate) (48).

A different movement, i.e., rotation around the N-glycosidic bond, swings the C5′ radical in diol dehydratase away from cob(II)alamin and toward the substrate (Figure 5) (47). This rather large movement accomplishes the transfer of the organic radical over an ~7Å distance from its original position in the spin-correlated radical pair to within van der Waal's contact distance of the substrate hydrogen atom destined for abstraction has been monitored in ethanolamine ammonia lyase by electron nuclear double resonance (49) and electron spin echo envelope modulation spectroscopy (50). Together, these studies provide fascinating glimpses into strategies adopted by B_{12}-dependent enzymes for controlling radical trajectories and stereospecificity of hydrogen atom abstraction.

Rearrangement Mechanisms

The differences in the chemical reactivities of the groups that are rearranged by B_{12}-dependent isomerases have necessitated the reliance on different strategies for achieving their rearrangements (Figure 6). Crystal structure determinations combined with computational studies have played a synergistic role in guiding mutational and kinetic studies aimed at understanding the mechanisms of rearrangement. In this section, isomerizations catalyzed by diol dehydratase, methylmalonyl-CoA mutase, and glutamate mutase are discussed in detail as prototypes for rearrangement mechanisms for B_{12}-dependent isomerases.

In diol dehydratase, an active site potassium coordinates to the two hydroxyl groups of the substrate that replace two water molecules in the resting enzyme (Figure 6). It is seven-coordinate with the remaining ligands being supplied by the protein (15). Experiments with ^{18}O-labeled propane 1,2 diols have demonstrated that the oxygen atom that is eliminated is derived from the O1 or O2 hydroxyl depending on the chirality at C2 of the substrate (51). These results necessitate the formation of a 1,1 gem-diol as an obligatory intermediate in the dehydration of propanediol and, in analogy, of ethanediol catalyzed by diol dehydratase.

Computational studies have been conducted to examine the energetics for alternative pathways for rearrangement (52–54). A transition state for migration was found only for the concerted mechanism but not for the stepwise pathway in which the hydroxyl group dissociates from C2 before adding to C1 (52). Ab initio studies have led to the proposal of a retro-push-pull mechanism in which the migration barrier is reduced by a combination of partial protonation and deprotonation of the migrating and spectator hydroxyl groups respectively (53). Candidate active site residues that could participate in such a mechanism are E170, positioned for pull catalysis with respect to the spectator hydroxyl, and H143, poised for retro-push catalysis via hydrogen bonding to the migrating hydroxyl (Figure 6). Potassium, whose Lewis acidity is expected to be diminished by the seven oxygen atoms surrounding it in the active site, appears to be chiefly important in binding and orienting the substrate rather than in lowering the migration barrier (53, 54).

Polar radical pathways are also predicted for the carbon skeleton rearrangements catalyzed by methylmalonyl-CoA mutase and glutamate mutase. However, a key difference between the two reactions is the migration of an sp^2-hybridized carbon in methylmalonyl-CoA mutase and an sp^3 center in glutamate mutase. For the methylmalonyl-CoA mutase-catalyzed reaction, both fragmentation/recombination and intramolecular addition/elimination pathways can be entertained. However, computational studies indicate that the energetic barrier for the intramolecular pathway is lower than for the dissociative pathway and reveal that partial proton transfer to the migrating group would facilitate rearrangement (55, 56). H244 is positioned in the active site of methylmalonyl-CoA mutase to serve as a general acid and engage in a hydrogen bonding interaction with the substrate carbonyl group (Figure 6). Mutation of H244 (to glycine, alanine, or glutamine) results in a 10^2-10^3-fold decrease in kcat and loss of one of the two kinetic pKa's seen in the wild-type enzyme (57, 58). The overall integrity of the mutant enzyme is confirmed in the crystal structure of the H244A mutant, which is superimposable on that of the wild-type enzyme (58). The impact of the H244 mutation on rearrangement was evaluated by monitoring the ratio of tritium partitioned from the 5′ position of AdoCbl to substrate and product as a function of which was used to initiate this reversible isomerization reaction (58). The partition ratio was altered with respect to wild-type enzyme and was sensitive to the direction of the reaction in contrast to the reaction catalyzed by the wild-type enzyme. These results indicate that a catalytic penalty incurred by mutation of H244 is an increase in the rearrangement barrier. Interestingly mutation of this residue greatly enhances the oxidative sensitivity and leads to frequent interception of cob(II)alamin during the reaction indicating that H244 plays a key role in protecting radical intermediates (57, 58).

Glutamate mutase is the only known carbon skeleton isomerase in which a saturated carbon moiety undergoes migration. A cyclic intramolecular pathway is chemically implausible for this reaction, while a fragmentation/recombination route is feasible. Rapid quench flow studies have provided evidence for the formation of a glycyl radical and acrylate at kinetically competent rates, although the amount of these species detected was very low (59). Ab initio molecular orbital calculations predict that the fragmentation barrier for a neutral substrate is significantly lower than for one in which the glutamyl radical is either deprotonated or protonated (60). Three arginine residues interact with the two carboxylate groups while E171 is within hydrogen bonding distance to the α-amino group of the substrate in the active site of glutamate mutase (Figure 6). Mutation of E171 to glutamine decreases kcat 50-fold and alters the pH profile for Vmax (61). In addition, the overall deuterium isotope effect decreases from $^{D}V = 4.6$ in wild type to 2.1 in the mutant, which indicates that some other step contributes more significantly to limiting the reaction rate in the mutant. Together, these data are consistent with a role for the glutamate residue as a

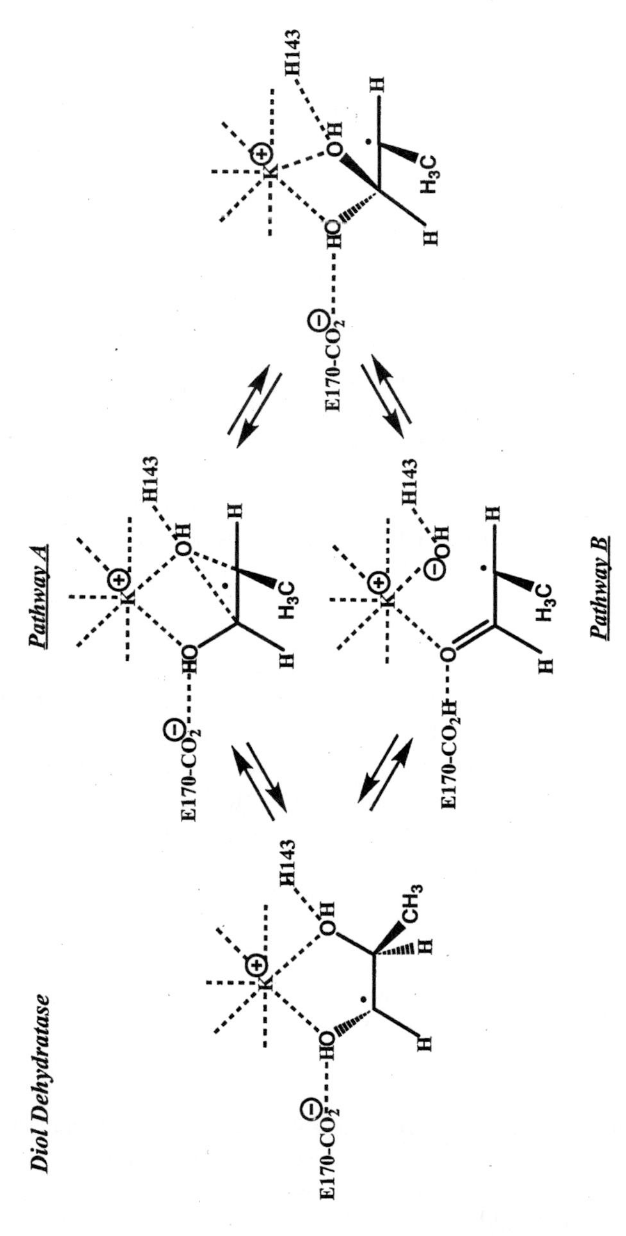

Figure 6 Alternative pathways for the rearrangement reactions in representative members of AdoCbl-dependent isomerases.

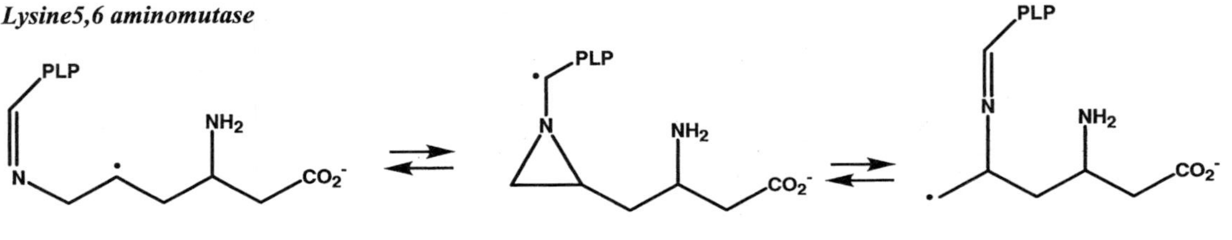

Figure 6 *Continued.*

general base that deprotonates the α-amino group and facilitates the rearrangement reaction.

In the subclass of aminomutases encompassing D-lysine 5,6 aminomutase (62) and D-ornithine aminomutase (63), an entirely different strategy is employed for stabilization of the migrating group. These enzymes are dependent on pyridoxal phosphate, and the amino group of the substrate forms a Schiff base with this cofactor. The resulting sp^2 hybridized amino group can then undergo rearrangement via an intramolecular addition/elimination pathway (64). Computational studies predict that protonation of the pyridoxal phosphate ring would enhance the reaction via captodative stabilization of the cyclic intermediate (65).

Although the strategies deployed by B_{12}-dependent enzymes to orchestrate the approximately trillion fold rate acceleration in cleavage of the Co-carbon bond has been the focus of much attention, an equally amazing facet of these reactions is the magnitude of stabilization afforded for reactive carbon radical intermediates. Mutagenesis studies guided by crystal structures, in concert with computational studies, are beginning to illuminate this issue and reveal the roles of active site residues in stabilizing intermediates and facilitating the rearrangement reaction.

Mechanism-Based Inactivation and Reactivation of Isomerases

B_{12}-dependent isomerases exhibit both substrate-induced and oxygen-dependent lability, which leads to the gradual accumulation of inactive enzyme during turnover. The relative susceptibilities of these enzymes to inactivation vary over at least 8 orders of magnitude, from 0.7 min^{-1} in lysine 5,6 aminomutase (66) to $\sim 4 \times 10^{-8}$ min^{-1} in methylmalonyl-CoA mutase [estimated from data in (58)]. In the latter enzyme, inactivation is dependent on the presence of oxygen and, although the mechanism has not been investigated, presumably results from interception of cob(II)alamin and/or the organic radical by oxygen during catalytic turnover.

Substrate-dependent inactivation of lysine 5,6 aminomutase does not require the presence of oxygen. Interestingly in this enzyme, radical extinction occurs by intermolecular electron transfer from cob(II)alamin to either the substrate or product radical. This leads to the formation of oxidized, cob(III)alamin and a carbanion that is rapidly quenched by protonation. Similar mechanism-based inactivation has been observed with substrate and cofactor analogs in other systems (67, 68). Substrate-dependent inactivation has also been observed in diol dehydratase and glycerol dehydratase, but the mechanism has not been elucidated (69–71).

Because glycerol dehydratase and its isofunctional enzyme, diol dehydrate, play an essential role in glycerol fermentation in some bacteria, their propensity to undergo substrate-induced inactivation poses a conundrum. A clue to the existence of a repair system came from the demonstration that glycerol-inactivated enzymes in permeabilized bacterial cells are rapidly reactivated by the

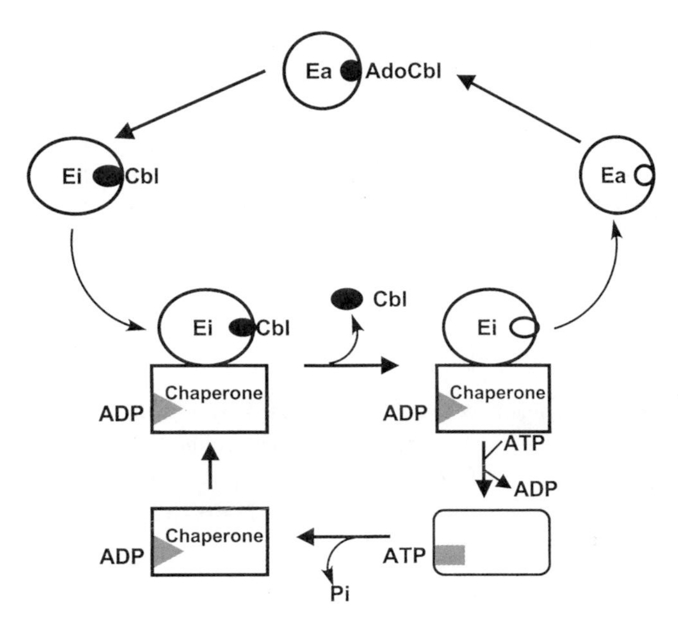

Figure 7 Postulated mechanism of reactivation of diol dehydratase by an ATP-dependent chaperone.

exchange of modified cofactor for intact AdoCbl in the presence of ATP and Mg^{2+} (72, 73). This in turn led to the discovery of a class of B_{12} chaperones that function to release inactive cofactor from their target isomerases and regenerate apoenzyme that can be reloaded with AdoCbl (Figure 7) (74, 75).

The chaperone for diol dehydratase has been best characterized and exists as an $\alpha_2\beta_2$ complex (76). It exhibits low intrinsic ATPase activity and forms a tight complex with diol dehydratase and ADP. It is in this complex that inactive cofactor is eliminated from the isomerase and binding of ATP results in dissociation of the chaperone-target enzyme pair. The resulting apoenzyme can bind AdoCbl to form active holoenzyme.

B_{12}-DEPENDENT METHYLTRANSFERASES

The B_{12}-dependent methyltransferases play an important role in amino acid metabolism in many organisms (including humans) as well as in one-carbon metabolism and CO_2 fixation in anaerobic microbes. In this section, we describe the general characteristics of methyltransferase reactions, then discuss specific methyltransferases and their metabolic roles, and conclude with a discussion of their reaction mechanism.

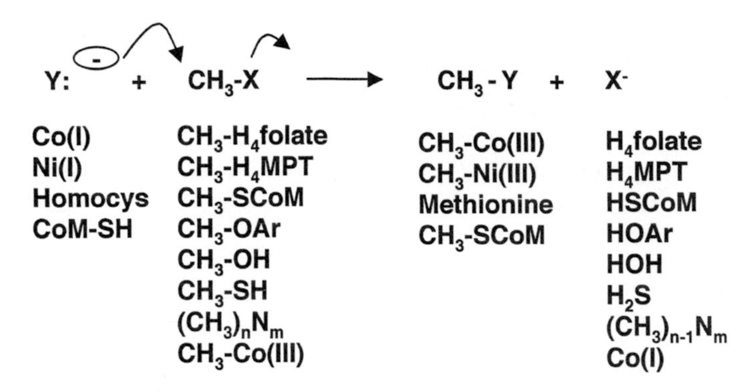

Figure 8 Methyl donors and acceptors used by the B_{12}-dependent methyltransferases. B_{12}-dependent methyltransferases catalyze the movement of a methyl group from a methyl donor (X-CH_3) to a methyl group acceptor (Nuc), where X and Nuc are the leaving group and the nucleophile, respectively.

Classes of B_{12}-Dependent Methyltransferases

As a subclass of the group transfer enzymes, B_{12}-dependent methyltransferases catalyze the movement of a methyl group from a methyl donor to a methyl group acceptor (Figure 8), where X and Nuc are the leaving group and the nucleophile, respectively. Three components are required for the methyl transfer reaction, and each component is found on a different polypeptide or domain (Figure 9). A specific three-component system is utilized for each methyl donor. A B_{12}-containing protein alternatively undergoes methylation by an MT1-component, which binds the methyl donor (CH_3-X), and demethylation by an MT2 component, which binds the final methyl group acceptor (Y).

There are two MT1 classes: one subclass binds simple substrates like methanol (MtaB), methylated amines (MttB, MtbB, MtmB), methylated thiols (MtsB), methoxylated aromatics (MtvB), and methylated heavy metals, while the

Figure 9 Three components of the B_{12}-dependent methyltransferases. Three components are required for the methyl transfer reaction; each component is found on a different polypeptide or domain. Mt designates methyltransferase, while the third letter denotes the methyl donor (a, methanol; v, vanillate; m, methylamine; t, trimethylamine; and s, dimethylsulfide). Acs designates that the protein is a component of the acetyl-CoA synthase (ACS) operon, with AB denoting the CODH/ACS, CD the two subunits of the CFeSP, and E the methyltetrahydrofolate:CFeSP MeTr. The last letter designates the specific component: *B* methylates the corrinoid protein, *C* is the corrinoid protein, and *A* methylates CoM. Thus, B_{12}-containing protein (*Mt_C*) alternatively undergoes methylation by a *MT1*- or *Mt_B*- component, which binds and activates the methyl donor, and demethylation by a *MT2* or *Mt_A* homolog that binds the final methyl group acceptor.

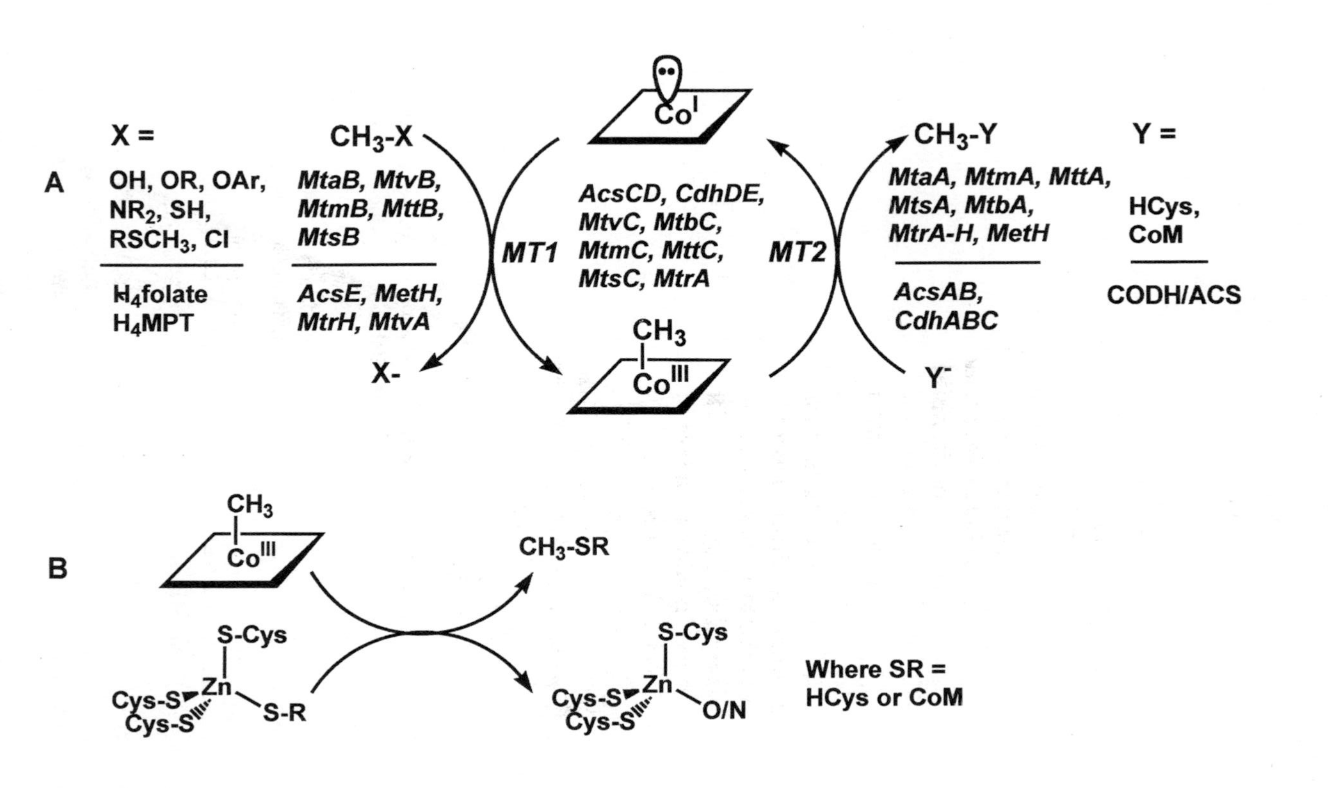
A
X =
OH, OR, OAr,
NR₂, SH,
RSCH₃, Cl
H₄folate
H₄MPT
CH₃-X
MtaB, MtvB,
MtmB, MttB,
MtsB
AcsE, MetH,
MtrH, MtvA
X-
MT1
AcsCD, CdhDE,
MtvC, MtbC,
MtmC, MttC,
MtsC, MtrA
Co^I
CH₃
Co^III
MT2
CH₃-Y
MtaA, MtmA, MttA,
MtsA, MtbA,
MtrA-H, MetH
AcsAB,
CdhABC
Y⁻
Y =
HCys,
CoM
CODH/ACS
B
CH₃
Co^III
S-Cys
Zn
Cys-S
Cys-S
S-R
CH₃-SR
S-Cys
Zn
Cys-S
Cys-S
O/N
Where SR =
HCys or CoM

other catalyzes methyl transfer from CH_3-H_4folate (AcsE-MeTr) and the methanogenic analog, methyltetrahydromethanopterin (CH_3-H_4MPT). Transfer of a methyl group from MT1 to the second B_{12}-containing component leads to formation of an organometallic methylcobalt intermediate. The third component (MT2) catalyzes transfer of the Co-bound methyl group to an acceptor (e.g., CoM, acetyl-CoA synthase, and H_4MPT). All the MT2-type enzymes isolated so far contain Zn, which coordinates to and thereby activates the thiolate methyl acceptor (Figure 9*B*). The methyl transfer to CODH/ACS involves a different type of reaction in which the methyl group appears to be transferred from cobalt to nickel.

Methionine Synthase

Methionine synthase from *Escherichia coli* (MetH) is the most extensively studied B_{12}-dependent methyltransferase. It catalyzes transfer of the methyl group from CH_3-H_4folate to homocysteine to form methionine and H_4folate (Figure 10) (77–78a). Compared to model reactions, methyl transfer from enzyme-bound CH_3-H_4folate to Co(I) is accelerated 35-million-fold and, from CH_3-Co(III) to bound homocysteine, 6-million-fold (77). Methionine synthase is a modular enzyme containing separate binding domains for homocysteine, CH_3-H_4folate, B_{12}, and AdoMet (79) (Figure 11). The B_{12} domain in its different oxidation states must interact punctually and specifically with each of the other three domains: The Co(I) form with the CH_3-H_4folate binding domain, the Co(II) form with the AdoMet binding domain, and the CH_3-Co(III) form with the homocysteine binding domain. The various conformational changes required to accomplish this molecular juggling act apparently are rate limiting during steady-state turnover (77).

The independently expressed modules of methionine synthase retain most of the functional properties of the native protein (79). A serendipitous proteolytic cleavage of methionine synthase in the crystallization process excised the B_{12} binding domain, which led to the first crystal structure of a protein-bound B_{12} (80). Surprisingly, cobalamin binding is accompanied by replacement of the lower axial benzimidazole ligand by a histidine residue to generate the His-on conformation (Figure 1*B*). The histidine residue is part of a catalytic triad consisting of H759, D757, and S810 that controls the coordination state of cobalt, i.e., His-on or His-off, by modulating the protonation state of the histidine.

Methyltransferases Involved in Anaerobic CO_2 Fixation and Energy Metabolism

Derivatives of B_{12} play key roles in energy metabolism and in cell carbon synthesis in anaerobic microbes such as methanogenic archaea (81) and acetogenic bacteria (82).

ACETOGENESIS AND B_{12} Acetogenic bacteria are important in the carbon cycle for degrading a wide variety of compounds derived from lignin and complex

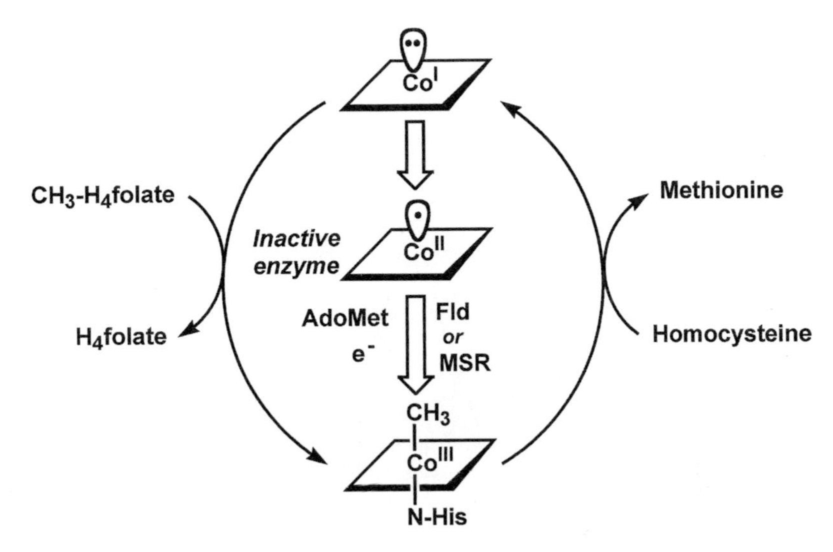

Figure 10 Reactions catalyzed by cobalamin-dependent methionine synthase. During the catalytic cycle, B_{12} cycles between CH_3-Co(III) and Co(I). Occasionally, Co(I) undergoes oxidative inactivation to Co(II), which requires reductive activation. During reactivation, the methyl donor is AdoMet, and the electron donor is reduced flavodoxin in *E. coli* (78a) and methionine synthase reductase in humans (138).

carbohydrates and for fixing CO_2 and the toxic gas, CO. The key methyl donors are CH_3-H_4folate, methoxylated aromatics (CH_3-OAr), methanol, and CH_3-Co(III), although the important methyl acceptors are Co(I), and Ni(I) (Figure 8). The B_{12}-dependent reactions are catalyzed by CODH/ACS (AcsAB), vanillate O-demethylase (MtvABC), and the corrinoid iron-sulfur protein (AcsCD) (CFeSP) and its methyltransferase (AcsE) (Figure 9). These methyltransferase reactions feed into the central CO/CO_2 fixing and energy yielding metabolic pathway, which is called the Wood-Ljungdahl pathway (83).

CH_3-H_4folate is formed by a six- or four-electron reduction from CO_2 or CO, and the methyl group becomes the methyl of acetic acid (Figure 12). The CFeSP interfaces with two completely different types of proteins: a methyltransferase and CODH/ACS. The AcsE-methyltransferase catalyzes transfer of a methyl group from CH_3-H_4folate to the CFeSP, whereas the reaction with CODH/ACS involves methyl migration from the CFeSP presumably to Ni(I) in ACS. MeTr is a small dimeric protein containing two equivalent 33 kDa subunits and is a member of the TIM barrel protein family, which contains 8 beta strands and 8 alpha helices (84) (Figure 13*A*). AcsE-MeTr shares significant sequence homology with residues 340–580 of the *E. coli* methionine synthase where CH_3-H_4folate is presumed to bind (84). In AcsE-MeTr, CH_3-H_4folate binds within a negatively charged cleft in the alpha-beta barrel (85) (Figure 13*B*). Residues

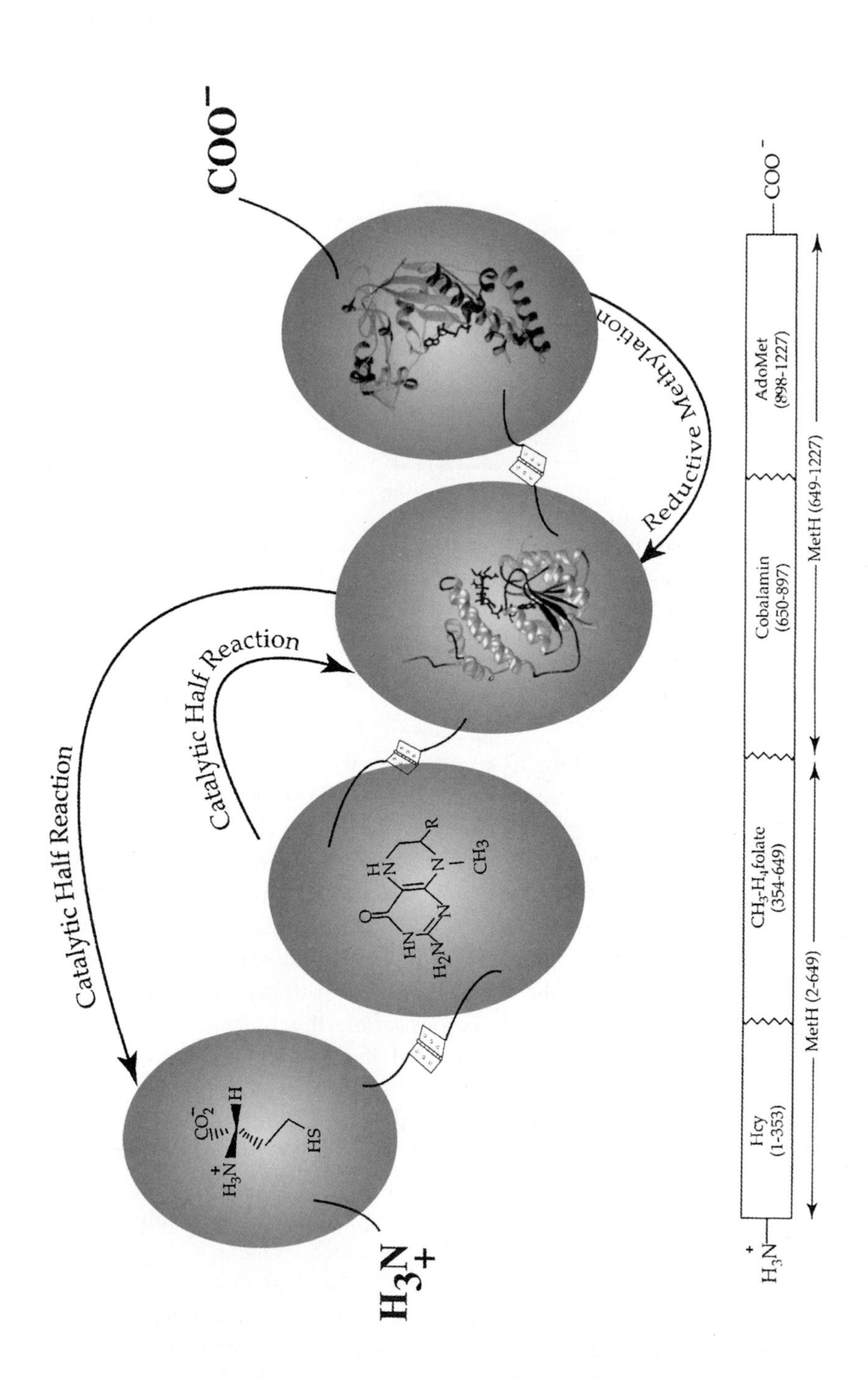
COO⁻
H₃N⁺
Reductive Methylation
Catalytic Half Reaction
Catalytic Half Reaction
Hcy
(1-353)
CH₃-H₄folate
(354-649)
Cobalamin
(650-897)
AdoMet
(898-1227)
MetH (2-649)
MetH (649-1227)
H₃N⁺
COO⁻

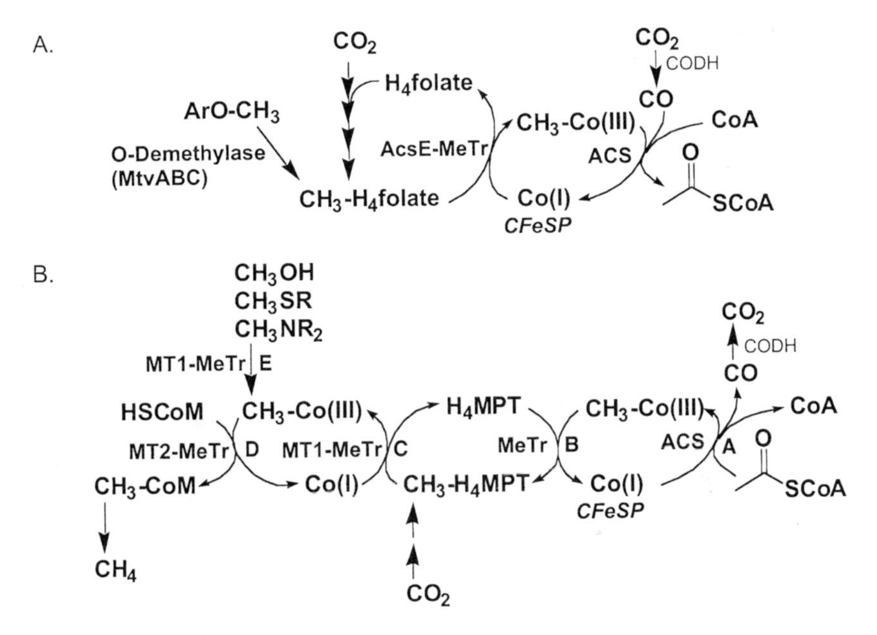

Figure 12 (*A*) The Wood-Ljungdahl Pathway. ACS, acetyl-CoA synthase; CODH, CO dehydrogenase; CFeSP, corrinoid iron-sulfur protein; and MeTr, AcsE-methyltransferase. (*B*) B_{12}-dependent methyltransferases in methanogenesis. The letters on the individual steps designate MeTr reactions that are discussed in the text.

D75, N96, and D160 interact with the pterin substrate in AcsE-MeTr and are strictly conserved in methionine synthase.

After a role for cobamides was discovered in acetogenesis (86, 87), twenty years elapsed before the physiological methyl carrier, the CFeSP, was isolated (88). When this protein was purified to homogeneity and characterized, a modified form of B_{12}, 5′-methoxybenzimidazolylcobamide, and a $[4Fe\text{-}4S]^{2+/1+}$ cluster were found; thus the name corrinoid iron-sulfur protein was proposed (89). A distinguishing feature of the CFeSP is its base-off His-off conformation, i.e., neither the intramolecular benzimidazole base nor a protein-derived histidine residue coordinates to the cobalt in any of its redox states (89–91). The absence of a lower axial ligand is believed to be important for promoting heterolytic cleavage of the Co-C bond and disfavoring homolytic cleavage (89, 92).

←

Figure 11 Modular structure of methionine synthase. The four modules are connected by flexible linker regions, shown as hinges, that facilitate a molecular juggling act, which allows the CH_3-H_4folate-, AdoMet-, or homocysteine-binding domains to alternatively access the B_{12}-binding module. The region of the amino acid sequence comprising each module is shown below the diagram. From Figure 3 of (77).

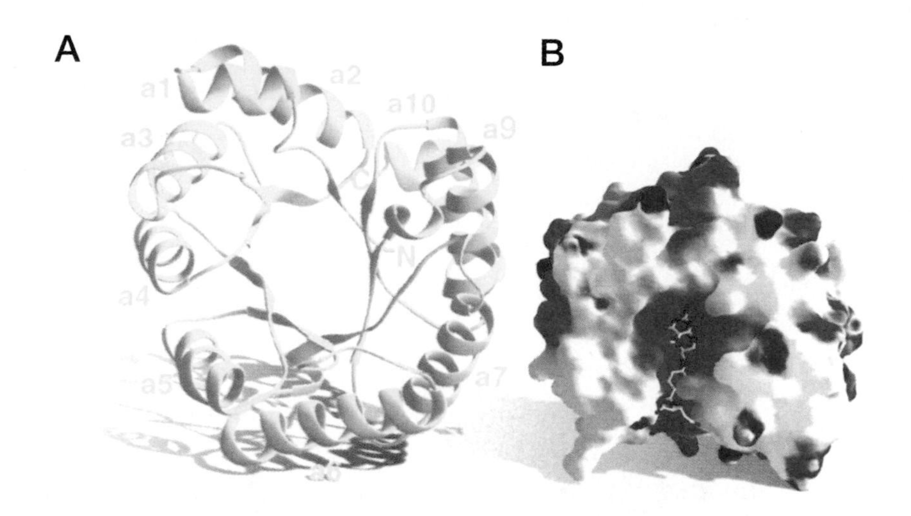

Figure 13 Structure of the AcsE-MeTr. (*A*) TIM barrel fold of the dimeric MeTr. (*B*) Electrostatic diagram showing CH_3-H_4folate binding within a negatively charged cleft in the TIM barrel.

Furthermore, removal of the lower axial ligand is correlated with an increase in the midpoint potential of the Co(II)/(I) couple by ca. 150 mV, which would favor the active Co(I) state within the cell (92).

O-DEMETHYLASES Another B_{12}-dependent methyltransferase that is important to acetogenic microbes is the O-demethylase, which couples the demethylation of an aromatic methyl ether to the formation of CH_3-H_4folate (Figure 12) (93). Anaerobic cleavage of the methyl group in an O-demethylation reaction leaves the ether oxygen in the phenolic product (94), which is excreted. The methyl group is further metabolized by the Wood-Ljungdahl pathway (Figure 12*A*).

The proteins involved in the O-demethylation reaction are MtvA, -B, and -C, which are induced when *Moorella thermoacetica* is exposed to the phenylmethylether (95). In this three-component system, MtvB binds the phenylmethylether and catalyzes methyl transfer to form a CH_3-Co species bound to MtvC. MtvA binds H_4folate and catalyzes transmethylation from CH_3-Co(III) to form CH_3-H_4folate (96–99).

METHANOGENESIS AND B_{12} In methanogens, the key methyl donors are CH_3-H_4MPT, methanol, protein-bound CH_3-Co(III), methylthiols, methylamines, and CH_3-SCoM, while the methyl acceptors are CoM, homocysteine, Co(I), and Ni(I) (Figure 8). The cobalamin-dependent reactions involve MtaABC, MtmABC, MttABC, MtsAB, MtrH, and CODH/ACS (CdhABC). The methanogenic methionine synthase at least in one methanogen appears to be a B_{12}-dependent enzyme (100).

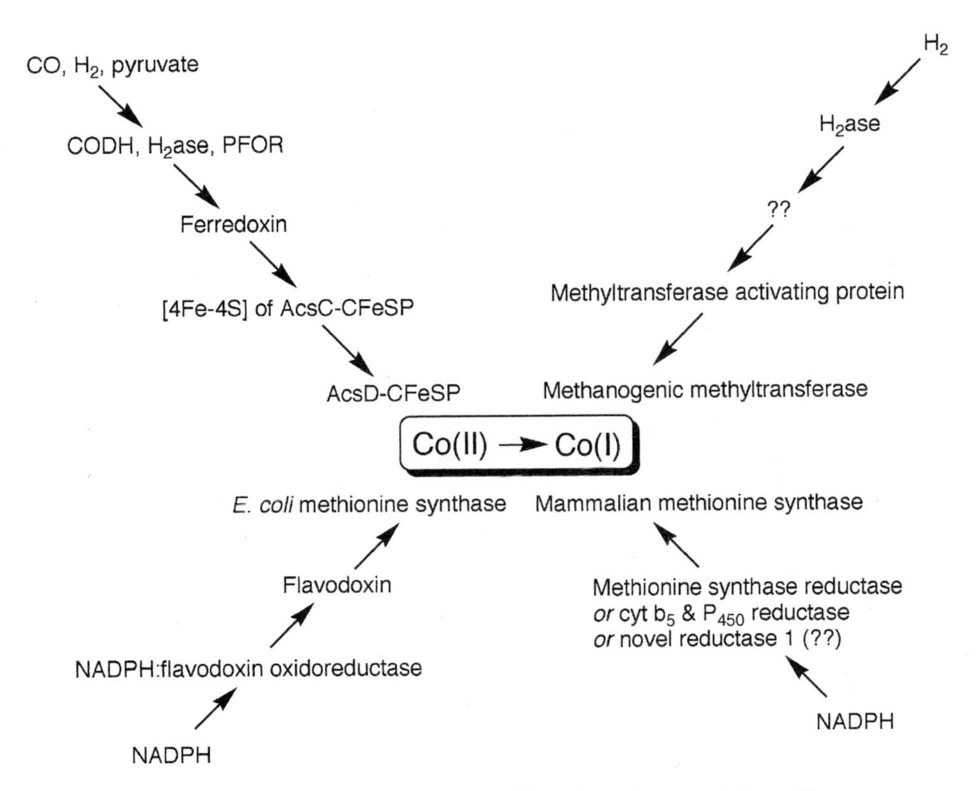

Figure 14 Different systems for activation of B_{12}-dependent methyltransferases.

Conversion of the methyl group of acetyl-CoA to methane by acetoclastic methanogens involves 4 methyl transfer steps (Figure 12*B*, Steps A–D). This reaction sequence is essentially the reverse of acetyl-CoA synthesis by acetogens (above). In methanogens, acetyl-CoA is disassembled by the bifunctional CODH/ACS (CdhABC) (Step A) (101) to form enzyme-bound intermediates (E-CO, E-SCoA, E-CH_3). CO undergoes oxidation and is released as CO_2, CoA is released into solution, and the methyl group is transferred first to a CFeSP and then to H_4MPT en route to methane (102, 103).

Methylotrophic methanogenesis denotes the generation of methane from simple methyl group donors like methanol, tri-, di-, and mono-methylamines and methylsulfides. Methylamines are important methane precursors in marine environments, where they arise from the breakdown of common osmolytes (104). Dimethylsulfide is responsible for up to 90% of the sea to air biogenic sulfur flux and is produced by marine phytoplankton as the main volatile sulfur compound emitted from the ocean (105). The methylotrophic MT1 enzymes catalyze transfer of the methyl group from methanol (MtaB), methylamines (MtbB and MttB), and methylthioethers (MtsB) to their specific cognate corrinoid proteins (106, 107) (Figure 9, reaction E in Figure 12*B*, and Figure 14). The MtmB structure (108) reveals a TIM-barrel fold, like the AcsE-MeTr. An unusual feature of the methylamine methyltransferases is a novel amino acid, pyrolysine, which is encoded by a UAG codon within the coding regions of the *mtm*B, *mtb*B, and *mtt*B genes (109). MtsB of the methylthiol:CoM methyl transferase contains both the methanethiol-binding protein and its corrinoid protein on a single subunit (110), which, like all the corrinoid proteins involved in methylotrophic methanogenesis, is homologous to the B_{12}-binding domain of methionine synthase (106, 107, 110). The MT2-dependent methylation of CoM generates CH_3-SCoM, the direct methane precursor from all methanogenic substrates (112).

A novel bioenergetic mechanism is exhibited by the MT2-like component of the methyltetrahydromethanopterin:CoM methyltransferase (Mtr) (113). This membrane-bound enzyme contains 8-subunits (MtrA-H) and generates a sodium ion gradient, which is used to produce ATP (114, 115). The reaction involves the methylation of the Co(I) center in MtrA by MtrH [MtrH is homologous to the AcsE-MeTr (116)] followed by transfer of the methyl group from CH_3-cobalt to CoM by one of the components of the MtrA-H complex. The latter step is the energy-yielding component of the reaction (114).

Mechanism of Methyl Transfer

The key steps in the methyltransferase mechanism are substrate binding, activation of the methyl group to enhance its reactivity toward nucleophilic attack, activation/generation of the Co(I) nucleophile, the methyl group transfer reaction itself, and product release.

BINDING AND ACTIVATION OF THE METHYL GROUP AND METHYL GROUP DONOR

CH_3-H_4folate binds tightly to the AcsE-MeTr from *M. thermoacetica* within a negatively charged region in the crevice (Figure 13*B*) of a TIM barrel structure (84). The *Methanosarcina barkeri* monomethylamine methyltransferase (MtmB) shares this TIM barrel fold (108) and, based on sequence homology with AcsE-MeTr (84), the CH_3-H_4folate binding domain of methionine synthase is predicted to also adopt this fold.

Methyl transfer from a tertiary amine requires activation of the methyl donor. Protonation at N5 is a plausible mechanism, which is consistent with the observed proton uptake during the reaction (117). This would lower the activation barrier to nucleophilic displacement of the methyl group by the Co(I) nucleophile. Proton transfer could occur either in the binary complex between enzyme and CH_3-H_4folate (117, 118) or later along the reaction coordinate (119). The second-order rate constant for the AcsE-MeTr-catalyzed reaction of CH_3-H_4folate and the CFeSP (120) or of methionine synthase (2-649) with exogenous cobalamin (77) increases as the pH is lowered; this exhibits an apparent pKa of 5.6–6.0. The reverse reactions exhibit the opposite pH-rate profile, with the second-order rate constant decreasing as the pH is lowered. Binding of CH_3-H_4folate to AcsE-MeTr results in proton uptake from solution (117), which suggests protonation in the binary complex. However, the MetH (2-649) domain does not bind protonated CH_3-H_4folate, and proton release rather than proton uptake accompanies formation of the binary complex (119). Thus, while it is clear that general acid catalysis is involved in the methyl transfer reaction, whether this occurs in the binary complex, in the ternary complex (with the methyl acceptor), or in the transition state for the methyl transfer reaction remains open. Methionine synthase and AcsE-MeTr appear to differ in this aspect of catalysis.

Activation of the methyl group of other methyl donors (methanol, methyl amines, and methane thiols) is less well understood. Although one possibility is protonation, another is coordination of the heteroatom of the methyl donor substrate to a Lewis acid. Activation of methanol in the methanol methyltransferase system requires Zn (121), and ligation of methanol to the Zn^{2+} ion in the methanol methyltransferase system (81) is presumed to be important. Model studies support a role for Zn^{2+} in catalysis of methyl transfer to Co(I) (122).

In the MT2-type reactions involving transfer of the methyl group from CH_3-Co(III) (Figure 9), activation of the methyl acceptor is required. The substrates for methionine synthase and the CoM methyltransferases (homocysteine or CoM) are not very nucleophilic in their protonated state, which is the predominant form at physiological pH. For homocysteine, the thiol exhibits a microscopic pKa of 9.0 when the amino group is protonated (123). The pH-activity profile of the reaction of 2-mercaptoethanol with MeCbl indicates participation of the thiolate as a nucleophile (124). In the enzymatic system, the thiolate appears to be generated by ligation to an active-site Zn^{2+} (125). The homocysteine-binding module of methionine synthase contains a zinc ion, which is coordinated to three cysteine residues and one O/N ligand in its resting state (126). Homocysteine replaces the O/N ligand to form a tetrahedral Zn-S_4 site (127). Similarly, Zn-dependent activation appears to be important in MT2-type reactions involving transfer of the methyl group of CH_3-Co(III) to CoM (121, 128).

ACTIVATION OF COBALAMIN Co(I) in B_{12} is a supernucleophile, with a Pearson constant of 14 (129), and is weakly basic, with a p*K*a below 1 for the Co(I)-H complex (130). Thus, Co(I), with a free lone pair of electrons, is poised for nucleophilic attack on an activated methyl group. Co(I) is also highly reducing, with the redox potential for the Co(II)/(I) couple below −500 mV (92, 131, 132). These properties make Co(I) fairly unstable and easy to oxidize. In the anaerobic microbial system (AcsA-E), the cobalt center is oxidized to the inactive Co(II) state once in every 100 turnovers (133). In methionine synthase, oxidative inactivation is estimated to occur once every 2000 turnovers (134). Reductive reactivation is necessary for reentry to the catalytic cycle, a feature common to all the B_{12}-dependent methyltransferases. A number of different reactivation systems have evolved to accomplish this (Figure 14).

An important aspect of reactivation, which was first shown for the acetogenic CFeSP (89) and is probably shared among all B_{12}-dependent methyltransferases, is that the cofactor adopts a base-off conformation. The base-off His-on conformation is found in many B_{12}-dependent methyltransferases (135), whose sequences contain a consensus motif (DX**H**XXG), where H represents the lower axial histidine ligand. In the acetogenic system, the Co(II) state of the CFeSP appears to already be four-coordinate, which facilitates electron transfer to form a four-coordinate Co(I) state (89–92). Electron transfer occurs from an external electron donor (CO, H_2, pyruvate) to an oxidoreductase (CODH, hydrogenase, pyruvate ferredoxin oxidoreductase), to a low-potential ferredoxin, to the [4Fe-4S] cluster in the AcsC subunit of the CFeSP, which transfers an electron to Co(II) (133).

Another pathway for reductive reactivation is exemplified by methionine synthase, where an unfavorable one-electron reduction of the inactive Co(II) to the Co(I) state is coupled to the highly exergonic demethylation of AdoMet to form MeCbl (136, 137). This is a striking example of how coupling of an unfavorable redox reaction to a highly favorable chemical reaction can markedly affect the apparent Co(II)/(I) redox potential (136). In *E. coli*, flavodoxin is the proximal electron donor, while in humans, the electron donor is methionine synthase reductase (138). Besides the requirement for a low potential electron donor, the lower axial histidine ligand to the cobalt of methionine synthase must be removed to facilitate reduction of the inactive Co(II) enzyme. A catalytic triad consisting of H759, D757, and S810 appears to facilitate removal of the axial histidine ligand by its protonation. Proton uptake is associated with reduction of Co(II) to Co(I) (139). In fact, binding of flavodoxin is sufficient to induce the conversion of 5- to 4-coordinate Co(II) (139), the coordination state naturally found in the Co(II) form of the acetogenic CFeSP. Generation of a 4-coordinate Co(II) complex also is an intermediate step in the electrochemical reduction to the Co(I) state of B_{12} in solution (139a).

A third ATP requiring activation system is exhibited by the MT2-type methanol- and methylamine-methyltransferases (140). A protein, which is called RamM, appears to contain two [4Fe-4S clusters] that are involved in catalyzing the ATP-dependent reductive activation of any of the methylamine MT2-type methyltransferases (Joseph Krzycki, personal communication). ATP-dependent activation also appears to be required for the aromatic O-demethylase from some acetogens (141).

THE METHYL TRANSFER: HETEROLYSIS VERSUS HOMOLYSIS, S_N2 VERSUS OXIDATIVE ADDITION Theoretically, a methyl group can be transferred as a cation, radical, or anion (Equations 1–3). Although, of the reactions shown in Figure 8, only the methionine synthase and Wood-Ljungdahl pathway reactions have been studied in sufficient detail to assign the oxidation states of the reaction intermediates, it seems reasonable to conclude that all the methyltransferases share a common mechanism that involves transfer of the methyl group formally as a carbocation and cycling of the cobamide cofactor between Co(I) and CH_3-Co(III) (Equation 1). Transfer of a radical would involve a Co(II) intermediate, either in the activation of the methyl group (Equation 2*B*) or after homolytic cleavage (Equation 2*A*). Mutagenesis, rapid kinetics, and single-turnover studies of the methyl transfer reactions in the Wood-Ljungdahl pathway indicate the absence of either Co(II) intermediates or electron transfer reactions during the catalytic cycle (120, 133, 142, 143). Additionally, there is no evidence for a methyl anion transfer, which would result in a Co(III) product or intermediate after Co-C bond cleavage (Equation 3).

$$\text{Heterolysis/S}_N\text{2: CH}_3\text{-Co(III)} \rightarrow [\text{CH}_3^{\,+}] + \text{Co(I)} \qquad 1.$$

$$\text{Homolytic cleavage: CH}_3\text{-Co(III)} \rightarrow [\text{CH}_3\bullet] + \text{Co(II)} \qquad 2a.$$

$$\text{Electron transfer/homolysis: CH}_3\text{-Co(III)}$$

$$+ 1\,e^- \rightarrow \text{CH}_3\text{-Co(II)} \rightarrow [\text{CH}_3\bullet] + \text{Co(I)} \qquad 2b.$$

$$\text{Heterolysis: CH}_3\text{-Co(III)} \rightarrow [\text{CH}_3^{\,-}] + \text{Co(III)} \qquad 3.$$

The heterolytic cleavage of the Co-C bond could occur by an S_N2-type reaction or by an oxidative addition reaction (77). An S_N2 reaction (Figure 15*A*) involves back-side attack on the methyl group, which leads to inversion of configuration at carbon. This mechanism requires in-line geometry of the methyl group donor and the methyl group acceptor. The oxidative addition reaction (Figure 15*B*) is a ligand substitution or ligand interchange reaction in which the dz^2 orbital of the Co(I) initially bonds with the methyl carbon and with the heteroatom to which the CH_3 group is attached and would lead to retention of configuration at the transferred carbon. As pointed out by Matthews (77), the critical issue for an oxidative addition mechanism is whether CH_3-H_4folate can be positioned close enough to the corrin ring to afford the appropriate steric and symmetry interactions necessary for orbital overlap. The empty 4s cobalt orbital would hybridize with the symmetry-matched σ orbital of the C-N bond and the filled $3d_{xz}$ or $3d_{yz}$ cobalt orbital of the metal would hybridize with the empty σ^* orbital of the C-N bond to weaken the N5-methyl bond of CH_3-H_4folate.

Stereochemical analyses of the overall reactions catalyzed by the Wood-Ljungdahl pathway (144) and by methionine synthase (145) are consistent with two consecutive S_N2-type reactions, i.e., transfer of the methyl group first to Co(I) and subsequently to the A-Cluster of ACS or to homocysteine respectively. However, the individual methyl transfer reactions, e.g., CH_3-H_4folate to acetyl-CoA (or methionine), have not been interrogated. Therefore, the stereochemical results for the overall reaction require that the MT1- and MT2-catalyzed half reactions occur by the same reaction mechanism, i.e., either two S_N2 or two ligand interchange reactions, because either would yield net retention of the stereochemical configuration. Crystallization of a binary complex of CH_3-H_4folate with intact methionine synthase or a ternary complex of CH_3-H_4folate with AcsE-MeTr and the CFeSP would illuminate the mechanism of methyl

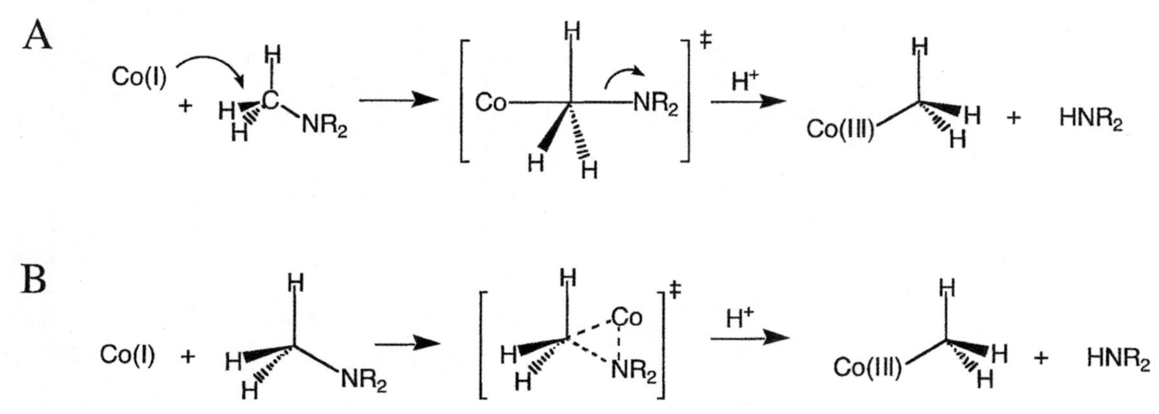

Figure 15 Comparison of S_N2 (*A*) and oxidative addition (*B*) mechanisms for methyltransferases.

transfer by revealing the relative orientations of cobalt, the methyl group, and the heteroatom attached to the methyl group.

B_{12}-DEPENDENT REDUCTIVE DEHALOGENASES

Classes of Dehalogenases and Their Environmental Role

Anaerobic microbes containing B_{12}-dependent reductive dehalogenases play an important role in the detoxification of aromatic and aliphatic chlorinated organics, which include compounds on the EPA priority pollutant list such as chlorinated phenols, chlorinated ethenes, and PCBs (146, 147). Exposure to these compounds induces a host of genes involved in removing the chloride ion. Both aerobic and anaerobic microbes can perform dehalogenation; however, anaerobic bacteria are more efficient in removing halogen atoms from polyhalogenated compounds (148). *Desulfomonile tiedjei* was the first dehalogenating anaerobe to be isolated and characterized (149), and ~20 strains of anaerobic organisms capable of reductive dehalogenation have so far been isolated, including *Desulfomonile*, *Desulfitobacterium*, and *Dehalobacter*.

Substrate Specificity and Bioremediation

Organisms have distinct preference for the chlorine substituent they can remove; for example, *D. tiedjei* removes the *meta* chlorine group of chlorophenols and chlorobenzoates (149), while *Desulfitobacterium dehalogenans* (150), *Desulfitobacterium hafniense*, and *Desulfitobacterium chlororespirans* (151–153) dehalogenate at the position *ortho* to a hydroxy group. *D. dehalogenans* can also dehalogenate hydroxy-PCBs (154). On the other hand, *Desulfitobacterium frappieri* PCP-1 catalyzes dehalogenation of chlorophenols (155) or anilines with chlorine groups at the ortho, meta, and para positions. *D. tiedjei* also can dehalogenate trichloroethylene (156). The purified *D. chlororespirans* enzyme can catalyze the dechlorination of a hydroxy-PCB (3,3′,5,5′-tetrachloro-4,4′-biphenyldiol) (153),

$$\text{R-Cl} + 2\,e^- + 2\,H^+ \rightarrow \text{RH} + \text{HCl}. \qquad 4.$$

There are two major classes of anaerobic dehalogenases: the heme containing 3-chlorobenzoate reductive dehalogenase from *D. tiedjei* (157) and the vitamin B_{12}-dependent enzymes. The B_{12}-dependent reductive dehalogenases, most of which also contain iron-sulfur clusters, are either embedded in or attached to the membrane by a small anchoring protein. This membrane association is important in linking the dehalogenation reaction to the electron transport pathway of anaerobic respiration.

The gene encoding the reductive dehalogenase (CprA) is embedded in a cluster of genes that encodes a membrane anchor (CprB), a membrane-associated regulatory protein (CprC), a DNA binding protein (CprK), and several chaperones (CprT, CprD, CprB) (158). Several B_{12}-dependent dehalogenases have been purified, including the 3-chloro-4-hydroxy-phenylacetate reductive dehalogenases from *D. hafniense* (159) and *D. dehalogenans* (160), the 3-chloro-4-hydroxybenzoate dehalogenase from *D. chlororespirans* (153), and the haloalkane (perchloroethylene and trichloroethene) dehalogenases from *Desulfitobacterium* strain PCE-S (161), *Dehalobacter restrictus* (162), and *Dehalobacter multivorans* (163, 164). These proteins contain a corrinoid, a [4Fe-4S] cluster, and a [3Fe-4S] cluster per monomeric unit. In some anaerobes, an unusual corrinoid cofactor appears to be present, because a recently isolated strain of *D. multivorans* that contains the standard corrinoids, but lacks the corrinoid cofactor that has been identified in the dehalogenase, is unable to dehalogenate TCE (165). Multiple dehalogenases are evident in the genome sequence of *D. hafniense* strain DCB-2 (166). Seventeen genes homologous to reductive dehalogenases are present in *Dehalococcoides ethenogenes* strain 195 (Steve Zinder, personal communication). Thus, the ability of certain dehalorespiring microbes to metabolize various types of chlorinated organic compounds (e.g., chloroalkanes plus chlorobenzoates) is conferred by separate (but homologous) enzymes (167).

Role of Cobalamin: Organometallic Adduct or Electron Transfer?

The role of B_{12} in the reductive dehalogenases appears to be significantly different from those of the AdoCbl-dependent isomerases and the MeCbl-dependent methyltransferases. Mechanistic studies on the anaerobic dehalogenases have lagged behind studies on the aerobic enzymes. Two working models for the mechanisms of dehalorespiring reductive dehalogenases are shown in Figure 16. In Path A, an organocobalt adduct is formed as in the methyltransferases. In Path B, the corrinoid serves as an electron donor. In Path A, the halide is eliminated as the aryl-Co(III) intermediate is formed. Path B resembles the Birch reduction of hydroxylated aromatics in which a radical anion is formed. Both mechanisms assume that the corrinoid is the site of dehalogenation based on the ability of B_{12} to catalyze dehalogenation in solution (168) and light-reversible inhibition by propyl iodide, a characteristic of corrinoid-dependent reactions (162, 169). Because only low potential reductants can drive this reaction, Co(I) is likely to be the active species; however, none of the dehalogenases-bound intermediates shown in Figure 16 have been identified. When B_{12} is reacted with perchloroethylene, the Co(I) spectrum disappears as Co(II) is formed, and it is consistent with pathway B, which involves radical intermediates and Co(I) as an electron donor without an organocobalt intermediate. (170). Radical trap experiments provide further evidence for pathway B.

The k_{cat} for dehalogenation of 3-chloro-4-hydroxybenzoate by the *D. chlororespirans* dehalogenase is 2.3-fold higher in 100% H_2O than in 100% ^2H_2O, and the proton inventory plot is linear, which indicates that a single proton is transferred in a partially rate limiting reaction in the dehalogenase mechanism (153). These results suggest deprotonation of an active-site water molecule,

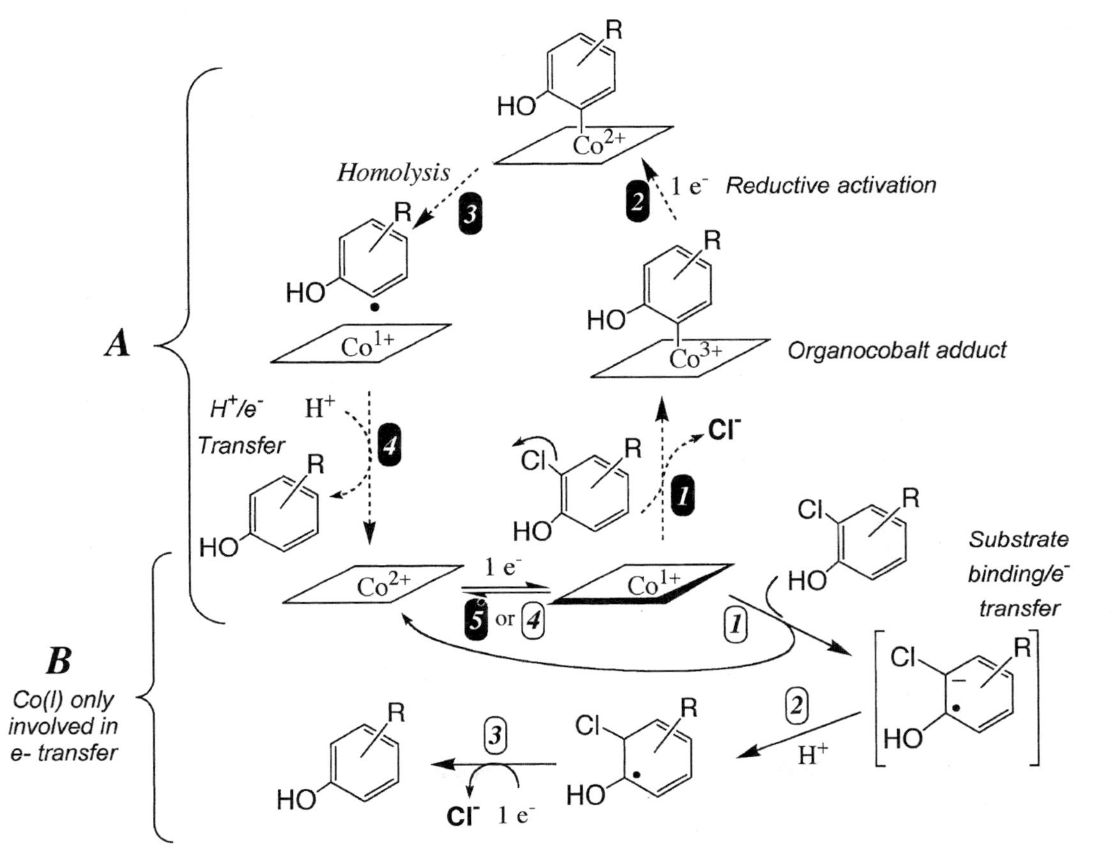

Figure 16 Alternative mechanisms for reductive dehalogenation. Adapted from (153).

because deuterium from solvent is incorporated specifically into the carbon of 2,5-dichlorobenzoate that undergoes dehalogenation (171).

Energy Conservation by Dehalorespiration

Bacteria derive energy by dehalorespiration, which couples the dehalogenation reaction to an electron transfer chain. The energetics are favorable because low potential reductants like pyruvate, formate, or hydrogen, with redox potentials below −400 mV, donate electrons to fairly oxidizing electron acceptors (the RCl/RH couple are between +250 and +600 mV) (148, 172). For example, for perchloroethylene dechlorination, 189 kJ per mol of H_2, is available, equivalent to ~2.5 ATP. However, with pyruvate as the electron donor, only 1 mol of ATP or less per mol of chloride was removed (167, 173). This poor efficiency may reflect the recent evolution of the ability to metabolize these man-made compounds, most of which are not naturally produced.

Inhibition of dehalorespiration by uncouplers indicates a chemiosmotic ATP synthesis mechanism (174) and implies that the dehalogenase and the electron donating enzymes are membrane associated. The reductive dehalogenases appear to face the inside and the hydrogenases to face the outside of the cytoplasmic membrane (162, 175). A b-type cytochrome(*s*) (176) and menaquinone are believed to be involved in the electron transport chain (177).

Many fundamental questions regarding the reaction mechanism of dehalogenases remain. Are organocobalt adducts analogous to those in B_{12}-dependent methyltransferases formed? How is the chloride replaced by a hydride equivalent? Does the reaction involve radical intermediates? A hurdle in mechanistic studies has been the very poor growth of the organisms, which results in very low amounts of protein to investigate. Development of better strategies for growth of the organisms and/or heterologous expression of the genes will be a boon to this field.

ACKNOWLEDGMENTS

This work was supported by grants from the National Institutes of Health (DK45776 to R. Banerjee and GM49451 to S.W. Ragsdale), National Science Foundation (MCB0211730 to S.W. Ragsdale), and Department of Energy (ER20297 to S.W. Ragsdale). R. Banerjee is an established investigator of the American Heart Association.

The *Annual Review of Biochemistry* is online at http://biochem.annualreviews.org

LITERATURE CITED

1. Whipple GH, Robscheit-Robbins FS. 1926. *Am. J. Physiol.* 72:408–18
2. Minot GR, Murphy WP. 1926. *JAMA* 87:470–76
3. Smith EL. 1948. *Nature* 161:638
4. Rickes EL, Brink NG, Konivszy FR, Wood TR, Folkers K. 1948. *Science* 107:396–97
5. Hodgkin DC, Kamper J, Mackay M, Pickworth JW, Trueblood KN, White JG. 1956. *Nature* 178:64

5a. Stevens R. 1982. In *B_{12}*, Vol. 1, ed. D.

Dolphin, pp. 169–200. New York: Wiley
6. Barker HA, Weissbach H, Smyth RD. 1958. *Proc. Natl. Acad. Sci. USA* 44:1093–97
7. Guest JR, Friedman S, Woods DD, Smith EL. 1962. *Nature* 195:340–42
8. Banerjee R. 1997. *Chem. Biol.* 4:175–86
9. Banerjee R. 2001. *Biochemistry* 40: 6191–98
10. Marsh EN, Drennan CL. 2001. *Curr. Opin. Chem. Biol.* 5:499–505
11. Brown KL, Hakimi JM. 1984. *J. Am. Chem. Soc.* 106:7894–99
12. Mancia F, Keep NH, Nakagawa A, Leadlay PF, McSweeney S, et al. 1996. *Structure* 4:339–50
13. Reitzer R, Gruber K, Jogl G, Wagner UG, Bothe H, et al. 1999. *Struct. Fold. Des.* 7:891–902
14. Marsh ENG, Holloway DE. 1992. *FEBS Lett.* 10:167–70
15. Shibata N, Masuda J, Tobimatsu T, Toraya T, Suto K, et al. 1999. *Struct. Fold. Des.* 7:997–1008
16. Sintchak MD, Arjara G, Kellogg BA, Stubbe J, Drennan CL. 2002. *Nat. Struct. Biol.* 9:293–300
17. Ke SC, Torrent M, Museav DG, Morokuma K, Warncke K. 1999. *Biochemistry* 38:12681–89
18. Abend A, Bandarian V, Nitsche R, Stupperich E, Retey J, Reed GH. 1999. *Arch. Biochem. Biophys.* 370:138–41
19. Finke RG, Hay BP. 1984. *Inorg. Chem.* 23:3041–43
20. Hay BP, Finke RG. 1987. *J. Am. Chem. Soc.* 109:8012–18
21. Brown KL, Zou X. 1999. *J. Inorg. Biochem.* 77:185–95
22. Halpern J. 1985. *Science* 227:869–75
23. Grate JH, Schrauzer GN. 1979. *J. Am. Chem. Soc.* 101:4601–11
24. Marzilli LG, Toscano J, Randaccio L, Bresciani-Pahor N, Calligaris M. 1979. *J. Am. Chem. Soc.* 101:6754–56
25. Krouwer JS, Holmquist B, Kipnes RS, Babior BM. 1980. *Biochim. Biophys. Acta* 612:153–59
26. Glusker JP. 1982. See Ref. 5a, pp. 23–106
27. Dong S, Padmakumar R, Maiti N, Banerjee R, Spiro TG. 1998. *J. Am. Chem. Soc.* 120:9947–48
28. Huhta MS, Chen HP, Hemann C, Hille CR, Marsh EN. 2001. *Biochem. J.* 355: 131–37
29. Padmakumar R, Padmakumar R, Banerjee R. 1997. *Biochemistry* 36: 3713–18
30. Chowdhury S, Banerjee R. 2000. *J. Am. Chem. Soc.* 122:5417–18
31. Marsh ENG, Ballou DP. 1998. *Biochemistry* 37:11864–72
32. Bandarian V, Reed GH. 2000. *Biochemistry* 39:12069–75
33. Booker S, Licht S, Broderick J, Stubbe J. 1994. *Biochemistry* 33:12676–85
34. Essenberg MK, Frey PA, Abeles RH. 1971. *J. Am. Chem. Soc.* 93:1242–51
35. Weisblat DA, Babior BM. 1971. *J. Biol. Chem.* 246:6064–71
36. Chih HW, Marsh EN. 2001. *Biochemistry* 40:13060–67
37. Gerfen GJ. 1999. In *Chemistry and Biochemistry of B_{12}*, ed. R. Banerjee, pp. 165–95. New York: Wiley
38. Buettner GR, Coffman RE. 1977. *Biochim. Biophys. Acta* 480:495–505
39. Schepler KL, Dunham WR, Sands RH, Fee JA, Abeles RH. 1975. *Biochim. Biophys. Acta* 97:510–18
40. Licht S, Gerfen GJ, Stubbe J. 1996. *Science* 271:477–81
41. Bothe H, Darley DJ, Albracht SP, Gerfen GJ, Golding BT, Buckel W. 1998. *Biochemistry* 37:4105–13
42. Michel C, Albracht SP, Buckel W. 1992. *Eur. J. Biochem.* 205:767–73
43. Zelder O, Buckel W. 1993. *Biol. Chem. Hoppe-Seyler* 374:85–90
44. Zhao Y, Abend A, Kunz M, Such P, Retey J. 1994. *Eur. J. Biochem.* 225: 891–96

45. Padmakumar R, Banerjee R. 1995. *J. Biol. Chem.* 270:9295–300
46. Gerfen GJ, Licht S, Willems J-P, Hoffman BM, Stubbe J. 1996. *J. Am. Chem. Soc.* 118:8192–97
47. Masuda J, Shibata N, Morimoto Y, Toraya T, Yasuoka N. 2000. *Struct. Fold. Des.* 8:775–88
48. Gruber K, Reitzer R, Kratky C. 2001. *Angew. Chem. Int. Ed. Engl.* 40: 3377–80
49. LoBrutto R, Bandarian V, Magnusson OT, Chen X, Schramm VL, Reed GH. 2001. *Biochemistry* 40:9–14
50. Warncke K, Utada AS. 2001. *J. Am. Chem. Soc.* 123:8564–72
51. Retey J, Arigoni D. 1966. *Experientia* 22:783–4
52. Toraya T, Yoshizawa K, Eda M, Yamabe T. 1999. *J. Biochem.* 126:650–54
53. Smith DM, Golding BT, Radom L. 2001. *J. Am. Chem. Soc.* 123:1664–75
54. Toraya T, Eda M, Kamachi T, Yoshizawa K. 2001. *J. Biochem.* 130: 865–72
55. Smith DM, Golding BT, Radom L. 1999. *J. Am. Chem. Soc.* 121:1383–84
56. Smith DM, Golding BT, Radom L. 1999. *J. Am. Chem. Soc.* 121:9388–99
57. Maiti N, Widjaja L, Banerjee R. 1999. *J. Biol. Chem.* 274:32733–37
58. Thoma NH, Evans PR, Leadlay PF. 2000. *Biochemistry* 39:9213–21
59. Chih H-W, Marsh ENG. 2000. *J. Am. Chem. Soc.* 122:10732–33
60. Wetmore SD, Smith DM, Golding BT, Radom L. 2001. *J. Am. Chem. Soc.* 123:7963–72
61. Madhavapeddi P, Marsh EN. 2001. *Chem. Biol.* 8:1143–49
62. Chang CH, Frey PA. 2000. *J. Biol. Chem.* 275:106–14
63. Chen HP, Wu SH, Lin YL, Chen CM, Tsay SS. 2001. *J. Biol. Chem.* 276: 44744–50
64. Frey PA, Chang CH. 1999. See Ref. 37, pp. 835–57
65. Wetmore SD, Smith DM, Radom L. 2000. *J. Am. Chem. Soc.* 122:10208–9
66. Tang KH, Chang CH, Frey PA. 2001. *Biochemistry* 40:5190–99
67. Magnusson OT, Frey PA. 2002. *Biochemistry* 41:1695–702
68. Huhta MS, Ciceri D, Golding BT, Marsh EN. 2002. *Biochemistry* 41: 3200–6
69. Pawelkiewicz J, Zagalak B. 1965. *Acta Biochim. Pol.* 12:207–18
70. Toraya T, Shirakashi T, Kosuga T, Fukui S. 1976. *Biochem. Biophys. Res. Commun.* 69:475–80
71. Poznanskaya AA, Yakusheva MI, Yakovlev VA. 1977. *Biochim. Biophys. Acta* 484:236–43
72. Honda S, Toraya T, Fukui S. 1980. *J. Bacteriol.* 143:1458–65
73. Ushio K, Honda S, Toraya T, Fukui S. 1982. *J. Nutr. Sci. Vitaminol.* 28:225–36
74. Mori K, Tobimatsu T, Hara T, Toraya T. 1997. *J. Biol. Chem* 272:32034–41
75. Tobimatsu T, Kajiura H, Yunoki M, Azuma M, Toraya T. 1999. *J. Bacteriol.* 181:4110–13
76. Mori K, Toraya T. 1999. *Biochemistry* 38:13170–78
77. Matthews RG. 2001. *Acc. Chem. Res.* 34:681–89
78. Matthews RG. 1999. See Ref. 37, pp. 681–706
78a. Fujii K, Huennekens FM. 1974. *J. Biol. Chem.* 249:6745–53
79. Goulding CW, Postigo D, Matthews RG. 1997. *Biochemistry* 36:8082–91
80. Drennan CL, Huang S, Drummond JT, Matthews RG, Ludwig ML. 1994. *Science* 266:1669–74
81. Sauer K, Thauer RK. 1999. See Ref. 37, pp. 655–80
82. Ragsdale SW. 1999. See Ref. 37, pp. 633–54
83. Ragsdale SW, Kumar M, Zhao S, Menon S, Seravalli J, Doukov T. 1998. In *Vitamin B_{12} and B_{12}-Proteins*, ed. B

Krautler, pp. 167–77. Weinheim, Ger.: Wiley-VCH
84. Doukov T, Seravalli J, Stezowski J, Ragsdale SW. 2000. *Structure* 8: 817–30
85. Doukov T, Seravalli J, Ragsdale SW, Drennan CL. 2003. *Science* 298:567–72
86. Ljungdahl L, Irion E, Wood HG. 1966. *Fed. Proc.* 25:1642–48
87. Poston JM, Kuratomi K, Stadtman ER. 1964. *Ann. NY Acad. Sci.* 112:804–6
88. Hu S-I, Pezacka E, Wood HG. 1984. *J. Biol. Chem.* 259:8892–97
89. Ragsdale SW, Lindahl PA, Münck E. 1987. *J. Biol. Chem.* 262:14289–97
90. Wirt MD, Kumar M, Ragsdale SW, Chance MR. 1993. *J. Am. Chem. Soc.* 115:2146–50
91. Wirt MD, Wu J-J, Scheuring EM, Kumar M, Ragsdale SW, Chance MR. 1995. *Biochemistry* 34:5269–73
92. Harder SA, Lu W-P, Feinberg BF, Ragsdale SW. 1989. *Biochemistry* 28: 9080–87
93. Frazer AC. 1994. In *Acetogenesis*, ed. HL Drake, pp. 445–83. New York: Chapman & Hall
94. DeWeerd KA, Saxena A, Nagle DP Jr, Suflita JM. 1988. *Appl. Environ. Microbiol.* 54:1237–42
95. Naidu D, Ragsdale SW. 2001. *J. Bacteriol.* 183:3276–81
96. Berman MH, Frazer AC. 1992. *Appl. Environ. Microbiol.* 58:925–31
97. Meßmer M, Reinhardt S, Wohlfarth G, Diekert G. 1996. *Arch. Microbiol.* 165: 18–25
98. El Kasmi A, Rajasekharan S, Ragsdale SW. 1994. *Biochemistry* 33:11217–24
99. Kaufmann F, Wohlfarth G, Diekert G. 1997. *Arch. Microbiol.* 168:136–42
100. Schroder I, Thauer RK. 1999. *Eur. J. Biochem.* 263:789–96
101. Maupin-Furlow JA, Ferry JG. 1996. *J. Bacteriol.* 178:6849–56
102. Jablonski PE, Lu WP, Ragsdale SW, Ferry JG. 1993. *J. Biol. Chem.* 268: 325–29
103. Abbanat DR, Ferry JG. 1991. *Proc. Natl. Acad. Sci. USA* 88:3272–76
104. King GM. 1984. *Geomicrobiol. J.* 3:276–301
105. Charlson RJ, Schwartz SE, Hales JM, Cess RD, Coakely JA, et al. 1992. *Science* 255:423–30
106. Burke SA, Lo SL, Krzycki JA. 1998. *J. Bacteriol.*180:3432–40
107. Sauer K, Harms U, Thauer RK. 1997. *Eur. J. Biochem.* 243:670–77
108. Hao B, Gong W, Ferguson TK, James CM, Krzycki JA, Chan MK. 2002. *Science* 296:1462–66
109. Paul L, Ferguson DJ, Krzycki JA. 2000. *J. Bacteriol.* 182:2520–29
110. Tallant TC, Krzycki JA. 1997. *J. Bacteriol.*179:6902–11
111. Deleted in proof
112. Thauer RK. 1998. *Microbiology* 144: 2377–406
113. Gottschalk G, Thauer RK. 2001. *Biochim. Biophys. Acta* 1505:28–36
114. Weiss DS, Gartner P, Thauer RK. 1994. *Eur. J. Biochem.* 226:799–809
115. Harms U, Weiss DS, Gartner P, Linder D, Thauer RK. 1995. *Eur. J. Biochem.* 228:640–48
116. Hippler B, Thauer RK. 1999. *FEBS Lett.* 449:165–68
117. Seravalli J, Shoemaker RK, Sudbeck MJ, Ragsdale SW. 1999. *Biochemistry* 38: 5736–45
118. Seravalli J, Zhao S, Ragsdale SW. 1999. *Biochemistry* 38:5728–35
119. Smith AE, Matthews RG. 2000. *Biochemistry* 39:13880–90
120. Zhao S, Roberts DL, Ragsdale SW. 1995. *Biochemistry* 34:15075–83
121. Sauer K, Thauer RK. 1997. *Eur. J. Biochem.* 249:280–85
122. Schnyder A, Darbre T, Keese R. 1998. *Angew. Chem. Int. Ed. Engl.* 37: 1283–85
123. Reuben DME, Bruice TC. 1976. *J. Am. Chem. Soc.* 98:114–21
124. Hogenkamp HP, Bratt GT, Sun SZ. 1985. *Biochemistry* 24:6428–32

125. Matthews RG, Goulding CW. 1997. *Curr. Opin. Chem. Biol.* 1:332–39
126. Goulding CW, Matthews RG. 1997. *Biochemistry* 36:15749–57
127. Peariso K, Zhou ZS, Smith AE, Matthews RG, Penner-Hahn JE. 2001. *Biochemistry* 40:987–93
128. Leclerc GM, Grahame DA. 1996. *J. Biol. Chem.* 271:18725–31
129. Schrauzer GN, Deutsch E, Windgassen RJ. 1968. *J. Am. Chem. Soc.* 90: 2441–42
130. Tackett SL, Collat JW, Abbott JC. 1963. *Biochemistry* 2:919–23
131. Lexa D, Savéant J-M, Zickler J. 1977. *J. Am. Chem. Soc.* 99:2786–90
132. Banerjee RV, Frasca V, Ballou DP, Matthews RG. 1990. *Biochemistry* 29:11101–9
133. Menon S, Ragsdale SW. 1999. *J. Biol. Chem.* 274:11513–18
134. Drummond JT, Huang S, Blumenthal RM, Matthews RG. 1993. *Biochemistry* 32:9290–95
135. Drennan CL, Matthews RG, Ludwig ML. 1994. *Curr. Opin. Struct. Biol.* 4:919–29
136. Banerjee RV, Harder SR, Ragsdale SW, Matthews RG. 1990. *Biochemistry* 29: 1129–35
137. Jarrett JT, Hoover DM, Ludwig ML, Matthews RG. 1998. *Biochemistry* 37: 12649–58
138. Olteanu H, Banerjee R. 2001. *J. Biol. Chem.* 276:35558–63
139. Hoover DM, Jarrett JT, Sands RH, Dunham WR, Ludwig ML, Matthews RG. 1997. *Biochemistry* 36:127–38
139a. Lexa D, Saveant JM. 1983. *Acc. Chem. Res.* 16:235–43
140. Wassenaar RW, Keltjens JT, van der Drift C. 1998. *Eur. J. Biochem.* 258: 597–602
141. Kaufmann F, Wohlfarth G, Diekert G. 1998. *Eur. J. Biochem.* 253:706–11
142. Seravalli J, Kumar M, Ragsdale SW. 2002. *Biochemistry* 41:1807–19
143. Menon S, Ragsdale SW. 1998. *Biochemistry* 37:5689–98
144. Lebertz H, Simon H, Courtney LF, Benkovic SJ, Zydowsky LD, et al. 1987. *J. Am. Chem. Soc.* 109:3173–74
145. Zydowsky LD, Courtney LF, Frasca V, Kobayashi K, Shimizu H, et al. 1986. *J. Am. Chem. Soc.* 108:3152–53
146. Copley SD. 1998. *Curr. Opin. Chem. Biol.* 2:613–17
147. Janssen DB, Oppentocht JE, Poelarends GJ. 2001. *Curr. Opin. Biotechnol.* 12:254–58
148. El Fantroussi S, Naveau H, Agathos SN. 1998. *Biotechnol. Prog.* 14:167–88
149. Linkfield TG, Tiedje JM. 1990. *J. Ind. Microbiol.* 5:9–15
150. Utkin I, Dalton DD, Wiegel J. 1995. *Appl. Environ. Microbiol.* 61:346–51
151. Christiansen N, Ahring BK. 1996. *Int. J. Syst. Bacteriol.* 46:442–48
152. Sanford R, Cole J, Loffler F, Tiedje J. 1996. *Appl. Environ. Microbiol.* 62: 3800–8
153. Krasotkina J, Walters T, Maruya KA, Ragsdale SW. 2001. *J. Biol. Chem.* 276: 40991–97
154. Wiegel J, Zhang X, Wu Q. 1999. *Appl. Environ. Microbiol.* 65:2217–21
155. Dennie D, Gladu I, Lepine F, Villemur R, Bisaillon J, Beaudet R. 1998. *Appl. Environ. Microbiol.* 64:4603–6
156. Cole JR, Fathepure BZ, Tiedje JM. 1995. *Biodegradation* 6:167–72
157. Ni S, Fredrickson JK, Xun L. 1995. *J. Bacteriol.* 177:5135–39
158. Smidt H, van Leest M, van der Oost J, de Vos WM. 2000. *J. Bacteriol.* 182: 5683–91
159. Christiansen N, Ahring BK, Wohlfarth G, Diekert G. 1998. *FEBS Lett.* 436: 159–62
160. van de Pas BA, Smidt H, Hagen WR, van der Oost J, Schraa G, et al. 1999. *J. Biol. Chem.* 274:20287–92
161. Miller E, Wohlfarth G, Diekert G. 1998. *Arch. Microbiol.* 169:497–502

162. Miller E, Wohlfarth G, Diekert G. 1997. *Arch. Microbiol.* 168:513–19
163. Neumann A, Wohlfarth G, Diekert G. 1996. *J. Biol. Chem.* 271:16515–19
164. Neumann A, Wohlfarth G, Diekert G. 1998. *J. Bacteriol.* 180:4140–45
165. Siebert A, Neumann A, Schubert T, Diekert G. 2002. *Arch. Microbiol.* 178: 443–49
166. genome_hafniense. *http://www.jgi.doe.gov/JGI_microbial/html/desulfito/desulf_homepage.html*
167. van de Pas BA, Gerritse J, de Vos WM, Schraa G, Stams AJ. 2001. *Arch. Microbiol.* 176:165–69
168. Krone UE, Thauer RK, Hogenkamp HPC. 1989. *Biochemistry* 28:4908–14
169. Neumann A, Wohlfarth G, Diekert G. 1995. *Arch. Microbiol.* 163:276–81
170. Shey J, van der Donk WA. 2000. *J. Am. Chem. Soc.* 122:12403–4
171. Griffith GD, Cole JR, Quensen JF, Tiedje JM. 1992. *Appl. Environ. Microbiol.* 58:409–11
172. Dolfing J, Harrison BK. 1992. *Environ. Sci. Technol.* 26:2213–18
173. Holliger C, Hahn D, Harmsen H, Ludwig W, Schumacher W, et al. 1998. *Arch. Microbiol.* 169:313–21
174. Mohn WW, Tiedje JM. 1991. *Arch. Microbiol.* 157:1–6
175. Miller E, Wohlfarth G, Diekert G. 1997. *Arch. Microbiol.* 166:379–87
176. Holliger C, Wohlfarth G, Diekert G. 1999. *FEMS Rev.* 22:383–98
177. Schumacher W, Holliger C. 1996. *J. Bacteriol.* 178:2328–33

Annu. Rev. Biochem. 2003. 72:249–289
doi: 10.1146/annurev.biochem.72.121801.161900
Copyright © 2003 by Annual Reviews. All rights reserved
First published online as a Review in Advance on February 27, 2003

SEMISYNTHESIS OF PROTEINS BY EXPRESSED PROTEIN LIGATION

Tom W. Muir
Laboratory of Synthetic Protein Chemistry, The Rockefeller University, 1230 York Avenue, New York, New York 10021; email: muirt@mail.rockefeller.edu

Key Words chemical ligation, inteins, semisynthesis, protein engineering

■ **Abstract** Expressed protein ligation (EPL) is a protein engineering approach that allows recombinant and synthetic polypeptides to be chemoselectively and regioselectively joined together. The approach makes the primary structure of most proteins accessible to the tools of synthetic organic chemistry, enabling the covalent structure of proteins to be modified in an unprecedented fashion. The ability to incorporate noncoded amino acids, biophysical probes, and stable isotopes into specific locations within proteins provides research tools to peer into the inner workings of these molecules. In this review I discuss the development of this technology, its broad application to biological systems, and its possible role in the area of proteomics.

CONTENTS

0066-4154/03/0707-0249$14.00

INTRODUCTION

The recent unveiling of draft versions of the human genome revealed between 30,000 and 40,000 genes in our genetic makeup (1, 2). The number of gene products must be much larger, in part because of alternative splicing of pre-mRNAs; 70–90% of the transcripts encoded in our genome contain two or more exons. If one also superimposes the myriad ways in which proteins can be posttranslationally modified—for example, about one-third of mammalian proteins are thought to be phosphorylated (3)—then one quickly arrives at numbers usually associated with the subject of astronomy. It will be incredibly difficult to determine an accurate catalog of the human proteome because such an inventory would have to consider tissue and cell-type specific expression (and modification) patterns, as well as how those patterns vary over biological timescales that range from a few minutes (e.g., signal transduction) to years (e.g., development). It is fair to say, however, that a staggeringly large number (perhaps hundreds of thousands) of distinct protein molecules are present in our bodies over the course of a lifetime.

Understanding protein function is at the heart of experimental biology. This endeavor requires a full description of the posttranslational modifications of a protein and how they affect intrinsic function, stability, localization, and three-dimensional structure, as well as interactions with other molecules. To achieve this goal at the level of a proteome is a heart-stopping proposition, yet the assembly of this encyclopedia must be one of the long-term objectives of biology. The magnitude of this task is fueling considerable interest in the development of new technologies that can accelerate the acquisition of protein function information. The emergence of dynamic new fields such as proteomics, structural genomics, bioinformatics, and chemical biology illustrates how this problem is attracting the attention of researchers from a broad spectrum of disciplines.

Chemistry has always played an important role in studying biological processes. As might be expected, proteins have been a major focus of biological chemistry research, both from the perspective of fully understanding their innate biological function and from the perspective of harnessing that function for nonbiological applications, e.g., nonphysiological chemical reaction catalysis. Many useful approaches require access to protein molecules impossible to prepare using standard ribosomal synthesis. These strategies include generating proteins possessing unnatural or modified amino acids, but also proteins possessing most natural posttranslational modifications. It is extremely difficult to obtain, by standard purification, homogeneous preparations of posttranslationally modified proteins for biochemical or structural studies. This deficiency is highlighted by the paucity of high-resolution structures for modified proteins in the protein structure databank (http://www.rcsb.org/pdb/). For example, of the 18,200 entries present at the time this review was written, only ~150 contain a phosphorylated amino acid within the protein. The number is even less if one looks at glycosylated or lipidated proteins.

The demand for specifically modified proteins has spurred development of a variety of different protein engineering approaches over the years. These techniques range from classical chemical labeling methods to the more recent technologies, nonsense suppression mutagenesis (4, 5) and expressed protein ligation (EPL) (6–8). The latter two approaches overcome most, if not all, of the shortcomings of chemical labeling methods and are truly interdisciplinary in their execution, largely because they combine the strengths of organic synthesis and recombinant DNA technology. In nonsense suppression mutagenesis, the ribosomal machinery is essentially tricked into incorporating the desired unnatural amino acid into the protein [reviewed in (9)]. This tactic is achieved by mutating the codon of interest with an amber codon. The mutant gene is then added to an in vitro translation mix that includes semisynthetic suppressor tRNA charged with the unnatural residue. Alternatively, it is possible to microinject the gene and the tRNA directly into a living cell, typically a *Xenopus* oocyte, leading to in vivo generation of the mutant protein (10). A general limitation of nonsense suppression mutagenesis has been the difficulty associated with the preparation and delivery of the tRNA acylated with the nonnatural amino acid of interest. However, spectacular progress has been made recently in engineering strains of Escherichia coli in which the native translation system can enzymatically generate tRNA acylated with an unnatural amino acid (11). This important breakthrough is likely to have a broad impact on the area of protein science.

Expressed protein ligation employs a different tactic to incorporate an unnatural residue at the desired position within a protein. In essence, the approach involves linking together premade synthetic and recombinant peptide building blocks to give the final protein product. This semisynthesis is achieved by incorporating "molecular Velcro" at appropriate ends of the fragments, allowing their assembly to take place with high regioselectivity in water at physiological pH. Although EPL involves more chemical steps than the in vivo version of nonsense suppression mutagenesis, it has two important advantages: multiple unnatural amino acids can be introduced into the protein, and a much broader range of modifications is possible. For these reasons, since its introduction in 1998, EPL has been applied to a wide variety of protein engineering problems. This technology and its applications are the subject of this review.

ORIGINS OF EXPRESSED PROTEIN LIGATION

Expressed protein ligation has its conceptual and technical roots in established areas of peptide and protein chemistry. The technology also has direct input from a recently discovered biological process, protein splicing, the characterization of which resulted in some cases from studies on proteins from rather obscure microorganisms. To some extent, EPL came into existence because of a convergence of chemical and biological principles. To see how this conjunction happened, it is worthwhile reviewing the relevant areas of protein chemistry.

Protein Semisynthesis

The term protein semisynthesis originally referred to processes in which proteolytic or chemical cleavage fragments of natural proteins were used as the building blocks for the resynthesis of the protein [reviewed in (12)]. For example, it was noted nearly 30 years ago that CNBr fragments of pancreatic trypsin inhibitor spontaneously condense to re-form the native peptide bond between them (13), an effect that is also observed with cytochrome *c* (14). In both cases, the two fragments produced after CNBr cleavage cooperatively refold, bringing the homoserine lactone at the carboxy terminus of one fragment in close proximity to the free amino terminus of the other. This autocatalytic fragment religation approach has been used to incorporate natural and unnatural amino acids into cytochrome *c* (14, 15).

The definition of protein semisynthesis may be broadened slightly to include any process that results in the site-specific modification of a natural protein. Although standard chemical labeling or bioconjugation approaches would not meet this criterion, several protein labeling strategies exist that are moderately to completely selective. The simplest of these involves the introduction, by standard site-directed mutagenesis, of a unique cysteine into a given position in the protein of interest, permitting selective derivatization of the sulfhydryl group with a thiol-reactive chemical probe. For example, this approach has been used to incorporate photoactivatable-cross-linkers [e.g. (16)] and fluorophores [e.g. (17)] into proteins and also as a method to prepare photocaged enzymes (18). The lower pK_a of the α-amino group of proteins as compared to the ϵ-amino groups of internal lysines can also be exploited for semisynthesis. Wetzel and coworkers demonstrated that moderately selective acylation reactions could be performed on the N terminus of polypeptides under slightly acidic conditions (19). Another example of site-specific protein labeling has been described by the Bayley group (20). This strategy involves a chemo-enzymatic process. In the first step, a thiophosphate group is introduced enzymatically into the protein by employing [ATP]-γ-S as the substrate for a kinase. The second step involves S-alkylation of the reactive thiophosphate with a chemical probe. This clever approach was used recently to prepare a photocaged version of protein kinase A (21). The technique is limited, however, by the modest efficiency of the initial thiophosphorylation step.

Another general approach to protein semisynthesis involves the use of proteolytic enzymes to facilitate the regioselective ligation of peptide fragments. Classically, this reverse-proteolysis strategy involved altering the reaction conditions such that aminolysis of an acyl-enzyme intermediate is favored over hydrolysis. This is typically achieved by including high concentrations of organic solvents such as glycerol, dimethylformamide (DMF), or acetonitrile in the reaction medium. Under these conditions a peptidyl-enzyme intermediate will undergo aminolysis with the second peptide fragment, leading to the formation of an amide-linked product (22, 23). Useful results have been obtained with a variety of proteases including subtilisin, trypsin, and papain. The power of reverse-proteolysis in protein synthesis was recently illustrated by the generation of novel glycoforms of ribonuclease B (24). Using a series of enzyme-mediated oligosaccharide and polypeptide ligation steps, Witte et al. were able to site-specifically introduce a branched oligosaccharide, sialyl Lewis X, at a single site in the protein molecule. In another recent example, a vinyl-sulfone probe was attached to the C terminus of the small protein, ubiquitin, using a chemo-enzymatic approach employing a trypsin-mediated esterification reaction (25). This semisynthetic protein was then used to characterize a series of deubiquitinating enzymes.

Significant progress in the area of enzyme-mediated peptide ligation has been achieved by engineering the active site of proteolytic enzymes. This approach was pioneered in the mid-1980s in the Kaiser laboratory. This group chemically converted the active site serine in the protease subtilisin into a cysteine. The resulting thiolsubtilisin analog possessed improved acylation activity relative to the wild-type enzyme (26). In an elegant extension of this work, Wells and coworkers further engineered the active site of subtilisin, affording an enzyme capable of efficiently catalyzing the ligation of peptide fragments (27). Their double mutant form of the enzyme, termed subtiligase, functions as an effective acyl transferase and importantly has vastly reduced proteolytic activity compared to the wild-type enzyme. Mutation of the active site serine to a cysteine allows the enzyme to be acylated with peptides esterified at their C terminus with a glycolate phenylalanyl amide group. A second Pro→Ala mutation within the active site relieves steric crowding, allowing efficient aminolysis of the enzyme-peptide thioester intermediate by the amino terminus of a second peptide fragment. Subtiligase has been used in the total synthesis of proteins (27), the preparation of cyclic peptides (28), and of relevance to the present discussion, semisynthetic proteins (29).

Given the individual utility of chemical synthesis and enzymatic semisynthesis, strategies that combine the two can be of great value. The protein semisynthesis approaches developed by Offord & Rose do just that. They pioneered the use of hydrazone- and oxime-forming reactions for chemically ligating synthetic and recombinant peptide fragments together (30–32). In one example, ligation through hydrazone formation was used to investigate the biological activity of a series of granulocyte colony stimulating factor (G-CSF) analogs (30, 31). Recombinant fragments of the protein were first generated through specific proteolysis at an engineered Lys-Ser sequence. N-terminal aldehyde groups or C-terminal hydrazides were then incorporated into the appropriate fragments through periodate oxidation (30, 33) and reverse-proteolysis, respectively. Backbone engineered derivatives of the 174-residue protein were thus obtained by ligating the fragments back together, and importantly these were shown to retain full biological activity (30). There followed a sophisticated extension of this work in which a central fragment of G-CSF was replaced by a synthetic peptide, essentially a kind of synthetic cassette mutagenesis (31). The appropriate reactive functionalities were introduced into the synthetic peptide insert that allowed it to be regioselectively reacted with the flanking recombinant pieces of the protein.

The related oxime-based strategy (32) has also been used to prepare semi-synthetic proteins with noncoded amino acids. In a beautiful series of studies, the Proudfoot & Offord groups used this kind of chemistry to identify potent inhibitors of HIV-1 infection based on using the β-chemokine RANTES (34). A natural ligand for the HIV-1 coreceptor CCR5, RANTES acts as a competitive inhibitor of HIV infection in vitro, but its proinflammatory properties render it of little value as an antiviral agent in patients. These researchers were able generate an N-terminally modified version of the protein, dubbed AOP-RANTES, which blocks viral infection in a variety of immune cells but importantly has no proinflammatory effects. This molecule was prepared by reacting an aminooxy-pentane molecule with the N terminus of RANTES, which had previously been converted into a reactive glyoxyl group by selective periodate oxidation.

The strength of the hydrazone and oxime chemistries used in the above examples is that they can be performed against the background of all the various amino acid side chain functionalities commonly found in proteins. That is to say, they are chemoselective. As discussed below, this concept has revolutionized the field of total chemical synthesis of proteins and provides the chemical foundation of expressed protein ligation.

Chemical Ligation of Unprotected Peptides

During the last decade chemical ligation has emerged as an approach to total synthesis of proteins. Chemical ligation allows unprotected peptides to be chemoselectively joined together. Chemical ligation was developed as a solution to the intrinsic problems associated with classical approaches for fragment condensation, which are often handicapped by the insolubility of the large protected peptide building blocks. Several schemes have been developed over the years to solve these problems. These include the thiol-capture procedure of Kemp (35, 36) that relies on a proximity-based entropic activation step and the minimal protected fragment condensation approach of Blake and Yamashiro (37, 38) that utilizes peptide α-thioacid derivatives. However, it was not until the early 1990s that the idea of using a chemoselective coupling reaction with fully unprotected synthetic peptide building blocks was devised in the Kent laboratory (39). In a seminal piece of work, the 99-residue HIV-1 protease was assembled from two ~50-residue unprotected peptides using a thioester bond–forming chemical reaction performed in aqueous solution at pH 7.

Given the elegance and practicality of chemical ligation, it is not surprising that a large amount of research effort has gone into expanding and refining the chemical ligation approach over the last several years. A full discussion of this now burgeoning field is beyond the scope of this article [the topic has been reviewed extensively elsewhere (40, 41)]; however, it is worth noting that the

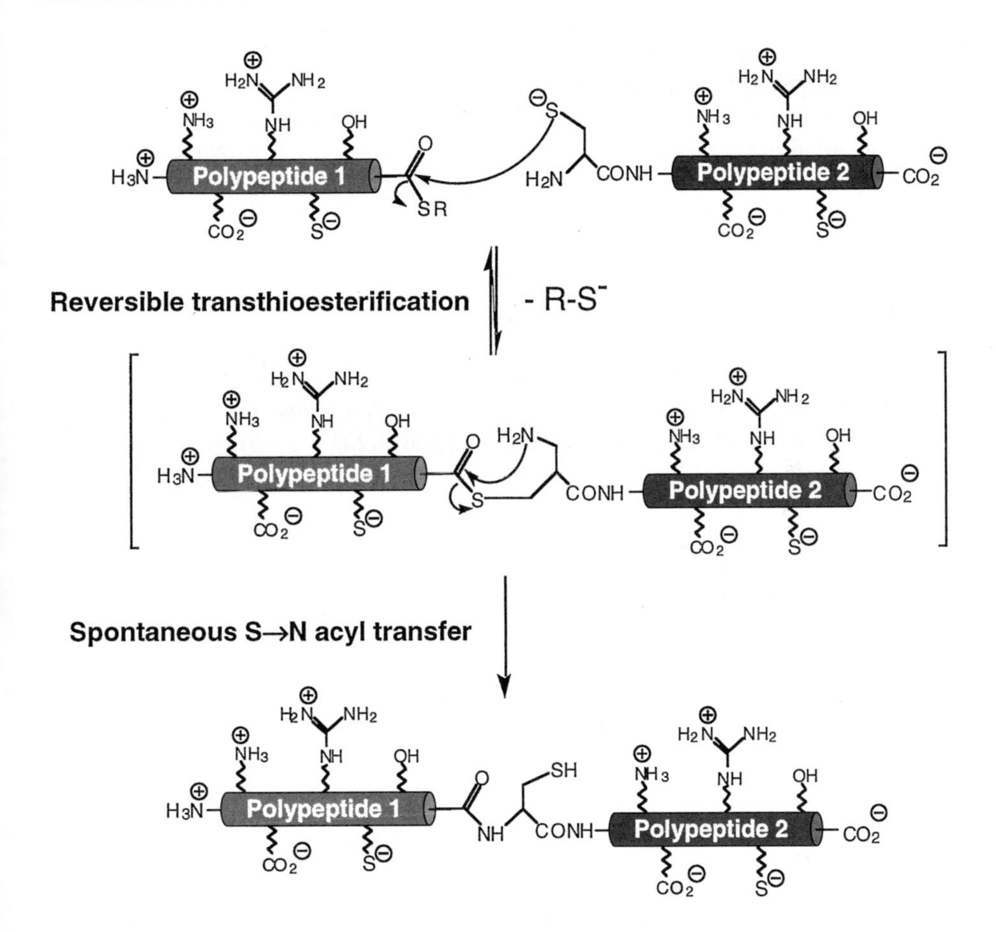

Figure 1 Principle of native chemical ligation. Both peptide reactants are fully unprotected and the reaction proceeds in water at or around neutral pH. In principle, either polypeptide building block could be synthetic or natural in origin provided they contain the requisite reactive groups, namely α-Cys or α-thioester.

original thioester ligation strategy has now been joined by ligation approaches based on thioether (42), oxime/hydrazone (30–32) (as noted above), thiazolidine (43), and amide (44–47) derivatives.

Native Chemical Ligation

A defining point in establishing chemical ligation as a general synthetic route to proteins was the development of native chemical ligation in the Kent laboratory in 1994 (44). In this technique, two fully unprotected synthetic peptide fragments are reacted under neutral aqueous conditions with the formation of a normal (native) peptide bond at the ligation site. As shown in Figure 1, the first step in native chemical ligation involves the chemoselective reaction that occurs at physiological pH between a peptide fragment containing an N-terminal cysteine (α-Cys) residue and a second peptide fragment containing an α-thioester group. The initial transthioesterification reaction is followed by a spontaneous intramolecular S→N acyl shift to generate an amide bond at the ligation junction. This type of acyl rearrangement chemistry was pioneered in the 1950s by Wieland et al. using amino acid derivatives (48), and is also reminiscent of the thiol capture method of Kemp alluded to earlier (35). However, the thiol capture strategy requires regiospecific protection of cysteine sulfhydryls prior to the ligation step. In contrast, native chemical ligation is compatible with all naturally occurring side chain groups including the sulfhydryl group of cysteine. Indeed, it is worth emphasizing that internal cysteine residues are permitted in both peptide segments because of the reversible nature of the initial transthioesterification step (44).

Native chemical ligation has undergone a number of important refinements since it was introduced. These include the development of auxiliary groups that in favorable cases overcome the requirement for an N-terminal cysteine (49–55), the identification of catalytic thiol cofactors (56), and the development of solution-phase (57–59) and solid-phase (60–62) sequential ligation strategies that allow several peptides to be linked together in series. These developments, as well as the numerous applications of native chemical ligation to the total synthesis of proteins, have been reviewed recently (41, 63).

As noted in the initial description of the technique (44), native chemical ligation, because of its compatibility with all naturally occurring amino acids, is ideally suited to protein semisynthesis. The only requirement is that the natural protein fragment contains one or the other of the necessary reactive groups, either α-Cys or α-thioester. This premise was put to the test first in the Verdine laboratory, employing a generally useful mutagenesis/proteolysis strategy (see later for details) to introduce an α-Cys into a recombinant version of the transcription factor protein, AP-1 (64, 65). This work confirmed that native chemical ligation could be used in a semisynthetic context, and provided the strategy by which to attach a synthetic molecule to the N terminus of a recombinant protein. This left the question of how to prepare recombinant polypeptide α-thioesters, which are required if synthetic peptides are to be ligated to the C terminus, or inserted into the middle of recombinant proteins. The solution to this problem fell directly out of studies of the naturally occurring process known as protein splicing.

Protein Splicing and *Trans*-Splicing

Protein splicing is a posttranslational process in which a precursor protein undergoes a series of intramolecular rearrangements and internal reactions that result in precise removal of an internal segment, referred to as an intein, and ligation of the two flanking portions, termed exteins [for excellent reviews, see (66, 67)]. With one exception (noted below), splicing has no sequence requirements in either of the exteins. In contrast, inteins are characterized by several conserved sequence motifs and well over a hundred members of this protein domain family have now been identified (for a comprehensive and up-to-date listing see www.neb.com/neb/inteins.html). As shown in Figure 2, the first step in the standard protein splicing mechanism involves an N→S (or N→O) acyl shift in which the N-extein unit is transferred to the side chain SH or OH group of a Cys/Ser residue, located at the immediate N terminus of the intein. [Note that although the overwhelming majority of inteins contain an N-terminal Cys/Ser, a few examples contain Ala at this position; these inteins are thought to use the +1 nucleophile within the C-extein in the first step (68).] On first analysis, this rearrangement is thermodynamically unfavorable; however, high-resolution structural analysis indicates that the intein structure may catalyze this step in part, by twisting the scissile amide-bond into a higher energy conformation (69), thereby helping to push the equilibrium to the (thio)ester side. In the next step in the process, the entire N-extein unit is transferred to a second conserved Cys/Ser/Thr residue at the intein-C-extein boundary (or +1 position) in a transesterification step. The resulting branched intermediate is then resolved through a cyclization reaction involving a conserved asparagine residue at the C terminus of the intein. The intein is thus excised as a C-terminal succinimide derivative. In the final (and possible concerted) step, an amide bond is formed between the two exteins as a result of an S→N (or O→N) acyl shift. The final step of protein splicing resembles the second step of native chemical ligation. Indeed, native chemical ligation provided a mechanistic framework for understanding the last step in the protein splicing mechanism (70).

Inteins have been found in proteins from eubacteria, archaea, and eucarya, which suggests protein splicing has an ancient evolutionary origin (71). A defined biological role for protein splicing has yet to be found, although the biological functions of a series of autoprocessing domains (which share structural homology with inteins and in some cases may be evolutionarily linked) in higher eukaryotes are known. The fascinating subject of intein distribution and evolution is beyond the scope of this review, but is discussed at some length in Giriat et al. (72).

Although the biological role(s) of protein splicing remains an open question, the process has been exploited extensively in the area of biotechnology. These applications fall into two general categories; those that employ intact inteins bearing various beneficial mutations, and those that utilize artificially or naturally split inteins. The former specifically exploit our understanding of the protein splicing mechanism. A number of mutant inteins (many contain a C-terminal Asn→Ala mutation) have been designed that can promote only the first step of protein splicing (70, 73–75). Proteins expressed as in-frame N-terminal fusions to one of these engineered inteins can be cleaved by thiols via an intermolecular transthioesterification reaction (Figure 3). In effect, this system provides a traceless chemical protease that can be exploited for purifying recombinant

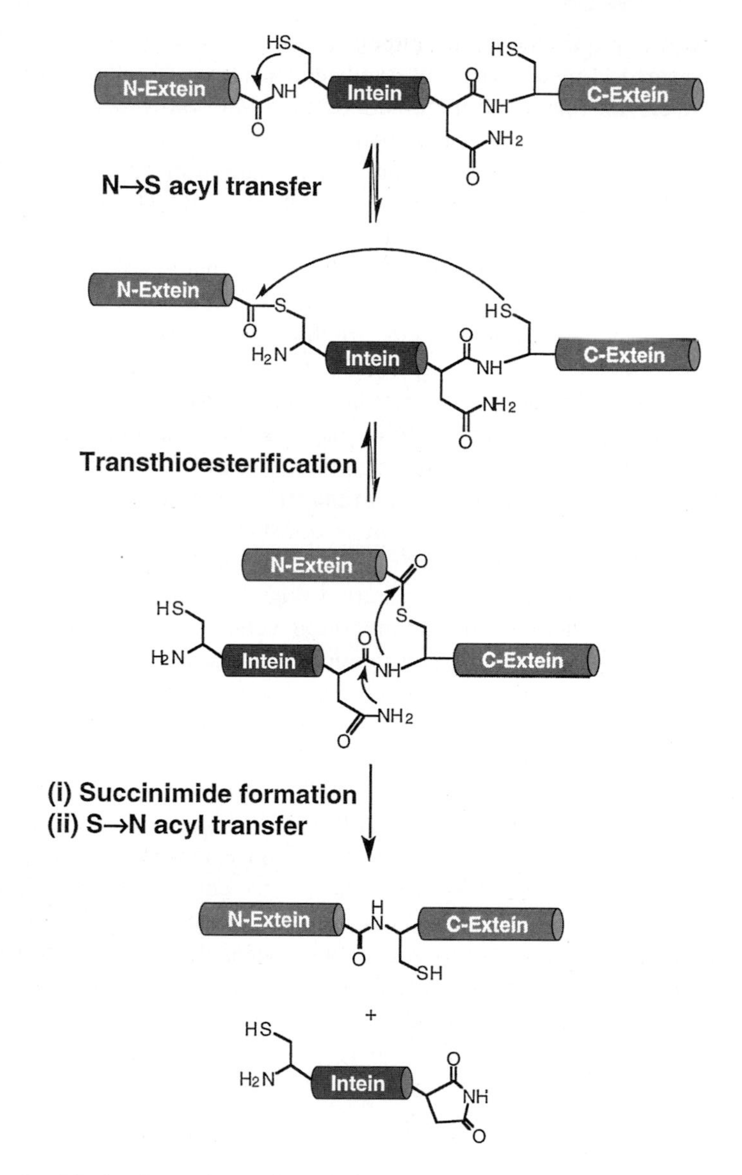

Figure 2 Mechanism of protein splicing. The process involves a series of acyl rearrangements and internal reactions catalyzed by the central intein protein domain. This process results in linkage of the two flanking polypeptides, the N- and C-exteins, via a normal peptide bond.

proteins that lack any sort of residual tag or appendage (73). Importantly, one of the key ingredients of native chemical ligation, protein α-thioester derivatives, can also be prepared by this route.

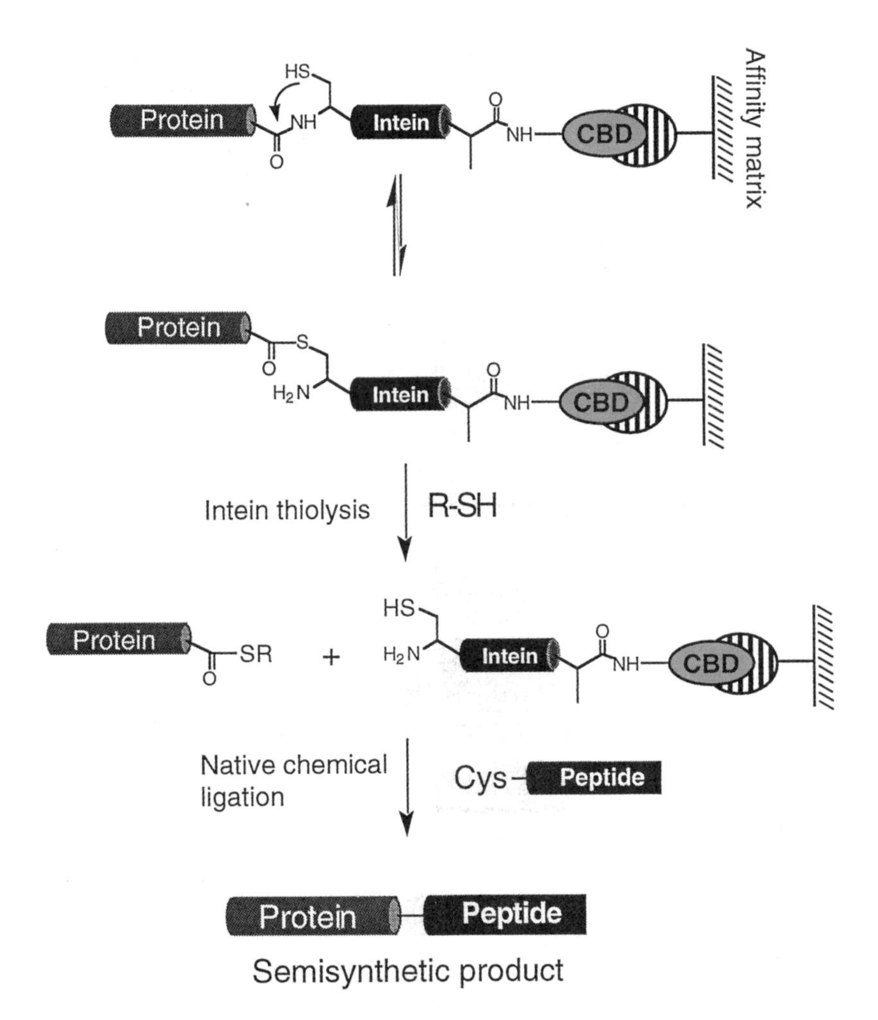

Figure 3 Recombinant protein α-thioesters and expressed protein ligation. Mutation of the C-terminal asparagine residue within the intein to an alanine residue blocks the final step in protein splicing (see Figure 2). Proteins expressed as in-frame N-terminal fusions to such mutant inteins can be cleaved by thiols to give the corresponding protein α-thioester derivatives. Commercial expression vectors are available that allow the generation of recombinant intein fusion proteins possessing chitin binding domain (CBD) affinity purification handles. The recombinant protein α-thioester can then react with an α-Cys-containing peptide (either synthetic or recombinant) via native chemical ligation to give a semisynthetic product. Expressed protein ligation refers to the combination of the intein thiolysis and native chemical ligation steps.

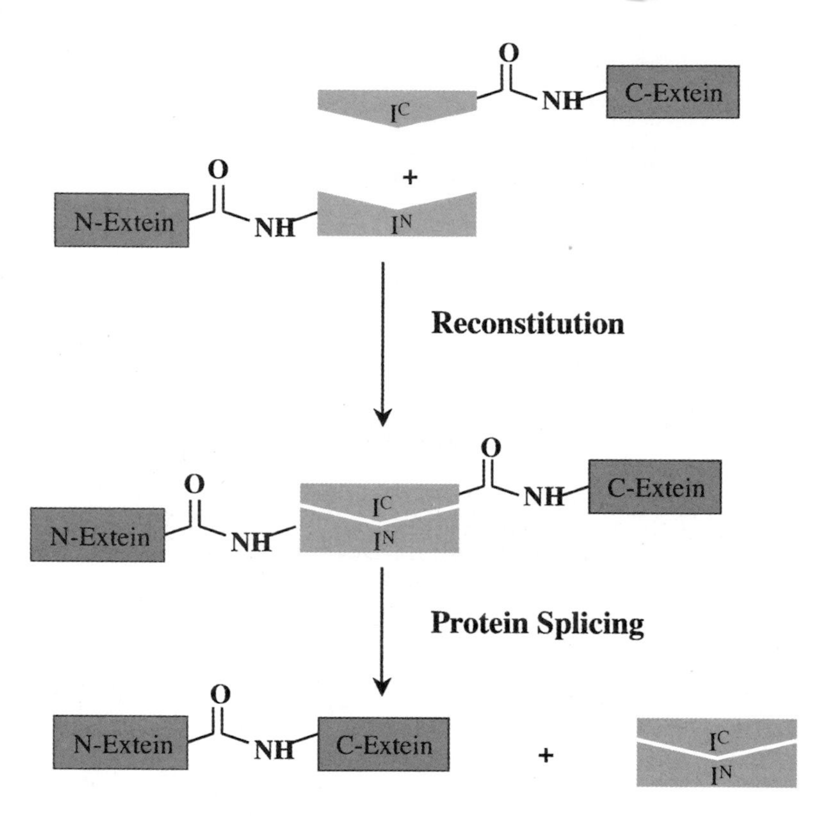

Figure 4 Protein *trans*-splicing by split inteins. Inteins can be cut into N- and C-terminal pieces (referred to in the figure as I^N and I^C) that individually have no activity but that when combined, associate noncovalently to give a functional protein splicing element.

Several groups independently demonstrated that inteins can be cut into two pieces that individually have no activity but that when combined will associate noncovalently to give a functional protein (Figure 4) (76–79). Affinity tags can be fused to the split site for purification of precursor fragments (76) and the reconstituted split intein then mediates a normal protein splicing reaction. Protein *trans*-splicing, as this process is generally known, can occur spontaneously in vivo when both fragments are expressed in the same cell (80). However, in vitro *trans*-splicing usually requires denaturation and renaturation of the isolated precursor fragments (76–79). Most recently, the protein *trans*-splicing area has received a considerable boost through the discovery of the naturally occurring split *Synechocytis* (*Ssp*) DnaE intein (78). Significantly, this split intein does not require reconstitution from denaturants to initiate splicing, with obvious implications for applications of the approach to protein engineering. Protein *trans*-splicing, like expressed protein ligation, provides a way of selectively ligating two different polypeptides together. The approach has been applied to the generation of a semisynthetic protein [where one of the split intein-extein units was synthetic (81)] and proteins labeled with isotopes in specific segments (79). As discussed later in this article, the approach has also been exploited for the generation of head-to-tail cyclic peptides and proteins (82), for detecting protein-protein interactions (83), and for controlling protein function (84).

Expressed Protein Ligation

As noted above, recombinant protein α-thioesters can be obtained by thiolysis of the corresponding protein-intein fusion [where the intein has been suitably mutated (73)]. In principle, this means that synthetic and recombinant building blocks can be intermixed in a semisynthetic version of native chemical ligation. Such an approach was first reported in 1998 by two groups and was dubbed expressed protein ligation (EPL) (6, 7) or intein-mediated protein ligation (IPL) (8). Although EPL and IPL refer to exactly the same process, namely that illustrated in Figure 3, the latter term is slightly misleading since, strictly speaking, the actual ligation step involves a chemical reaction (i.e., native chemical ligation) and does not involve the intein. Indeed, IPL better describes the protein *trans*-splicing technique illustrated in Figure 4. Thus, EPL is used in this review to refer to a semisynthetic version of native chemical ligation in which one or both of the reactants is a recombinant (expressed) protein.

SCOPE OF EXPRESSED PROTEIN LIGATION

Since its introduction just a few years ago, expressed protein ligation has been applied to a large number of different protein engineering problems. More than 40 different proteins have been studied using EPL to date, with multiple semisynthetic analogues of the same protein having been prepared in several instances. The number of applications of EPL is increasing rapidly each year, from half a dozen or so reports in 1998 to ~30 from 2001 through 2002. These studies have provided a number of important insights into the scope and limitations of the approach.

Protein Engineering Using EPL

The underlying premise of EPL, and indeed any semisynthetic approach, is that the modifications one might wish to make to the protein can be restricted to a small region of the primary structure, for example a single domain within a multidomain protein. Thus, one can use EPL to replace that domain with a corresponding synthetic cassette containing whatever modification(s) is of interest. This leads to the great advantage that EPL has over protein total synthesis; namely, most of the protein sequence can be made via ribosomal synthesis, which means that extremely large proteins are accessible to the approach. For example, EPL has been used to prepare analogs of the β' subunit of *E. coli* RNA polymerase, which contains 1407 amino acids (17). Proteins of this size are well beyond the range of total synthesis, even using sophisticated sequential native chemical ligation approaches (57–62). The modular nature of EPL is, however, a limitation if one wishes to make multiple modifications that are peppered throughout the sequence of a protein. To achieve this end, it would be necessary to ligate several synthetic and recombinant pieces, a technically demanding undertaking (86). In contrast, total chemical synthesis is perfectly suited to this problem since it makes little practical difference whether one or all of the residues are unnatural. This advantage of total synthesis is nicely illustrated by the construction of HIV-1 protease containing all D-amino acids (87).

In the most straightforward application of EPL, a small chemical probe is appended to the natural end of a protein, either the N or the C terminus. Studies of this type typically involve attachment of affinity handles (86, 88) or fluorophores (17, 86) to the appropriate end of the protein. The majority of EPL studies have involved modifying a native residue within the protein. From a protein engineering perspective, this type of application is more challenging since it requires an optimal ligation junction to be identified, followed by synthesis of the appropriate recombinant and synthetic fragments of the protein. Ideally, one would want to generate the semisynthetic protein from two segments, one synthetic and one recombinant, since this would involve a single ligation reaction. This situation obviously requires that the region of interest be within synthetic striking distance of one or the other end of the protein. Optimized solid-phase peptide synthesis (SPPS) approaches allow peptides of up to ~50 residues to be reliably prepared (89). Thus, unless the region of interest is within 50 residues of the N or C terminus of the protein, a three-piece ligation strategy must be used (86, 90). Finally, EPL has also been used to link together purely recombinant protein fragments. This tactic has proved useful for the in vitro assembly of cytotoxic proteins (8) and for segmental isotopic labeling of proteins (91).

Reactive Recombinant Proteins

EPL requires that the synthetic and recombinant building blocks contain either an α-thioester or an α-Cys group. A large amount of work has gone into the development of solid-phase methodologies for the incorporation of these groups into synthetic peptides. In particular, synthetic peptide α-thioesters can be easily prepared by using *tert*-butoxycarbonyl (Boc) SPPS and, in less straightforward fashion, by using 9-fluorenylmethoxycarbonyl (Fmoc) SPPS [reviewed in (41, 63)]. Importantly for EPL, robust approaches are now available for introducing these reactive handles into recombinant proteins.

PROTEIN α-THIOESTERS As noted above, α-thioester derivatives of recombinant proteins can be prepared by thiolysis of mutated intein fusions (Figure 3). *E. coli* expression vectors are now commercially available that allow the generation of protein-fusions to two different engineered inteins, the *Saccharomyces cerevisiae* vacuolar ATPase (*Sce* VMA1) intein, and the *Mycobacterium xenopi* DNA gyrase A (*Mxe* GyrA) intein. In principle, many other inteins could be used for this purpose [e.g. (74, 92)]. It is worth noting that the *Sce* VMA1 intein contains an endonuclease domain [not required for splicing activity (93, 94)] inserted between two conserved regions in the sequence. The presence of the additional catalytic unit accounts for the large difference in size between the *Sce* VMA1 and *Mxe* GyrA inteins, 455 versus 198 residues, respectively. As one might expect, these two inteins have different characteristics. For example, the overall expression levels of fusion proteins have been found to differ depending on which intein they are fused to (95, 96). In addition, the *Mxe* GyrA intein can be efficiently refolded from chemical denaturants (97, 98), whereas the *Sce* VMA1 refolds less efficiently (99, 100). Consequently, fusion proteins to the former intein can be reconstituted from bacterial inclusion bodies, whereas those to the latter cannot. This property of the *Mxe* GyrA intein has been used to express highly cytotoxic sequences for use in EPL (97).

A variety of molecules have been used to induce thiolysis of intein fusion proteins (6–8, 73, 100, 101). All of these thiols are small enough to enter into the catalytic pocket of the intein, thereby inducing the cleavage of the thioester bond connecting the −1 residue in the protein to the side chain thiol group of the first residue in the intein. The choice of thiol very much depends on whether the protein α-thioester is to be isolated prior to ligation. Ethyl α-thioester derivatives (produced by thiolysis with ethanethiol) have been shown to be quite stable toward hydrolysis, allowing the ethyl α-thioester proteins to be purified and stored until use (100). However, simple alkyl thioesters are not very reactive in EPL. This unreactivity can, however, be overcome by adding thiol cofactors such as thiophenol (6, 56) or 2-mercaptoethanesulfonic acid (MESNA) (8) to the reaction mixture. Both these additives generate more reactive α-thioesters in situ through transthioesterification. Thiophenol and MESNA can also be used in the initial protein-intein cleavage step if the resulting protein α-thioester derivative does not need to be isolated—that is to say, if thiolysis and ligation can be performed in the same pot (6–8).

α-Cys PROTEINS Methods for generating recombinant proteins possessing amino-terminal cysteine residues (α-Cys) all rely on cleavage of an appropriate precursor protein (Figure 5). In the simplest scenario, a cysteine is introduced next to the initiating methionine in the protein sequence. Endogenous methionyl aminopeptidases (MetAPs) then remove the Met, thereby generating the α-Cys protein (102). This approach has met with some success (103–105); however, the in vivo processing event is often inefficient [e.g., (64)], resulting in lower yields of the desired material. The MetAP strategy may be utilized more frequently in light of recent work from the Rose group, which developed an elegant "covalent capture" approach that allows α-Cys proteins to be fished out of crude cell lysates using an aldehyde-bearing resin (106).

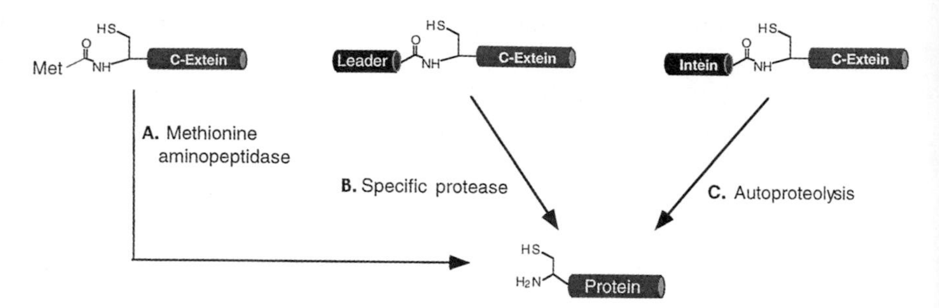

Figure 5 Recombinant α-Cys proteins. An N-terminal cysteine (α-Cys) can be introduced by specific proteolysis of recombinant fusion proteins containing a cryptic α-Cys, C-terminal to a protease cleavage site. In one strategy, the leader motif can simply be methionine, which is removed in vivo by specific Met aminopeptidases (*A*), or it can be a short peptide recognition sequence removed in vitro by cleavage with heterologous proteases such as factor Xa or TEV (*B*). (*C*) In an alternative strategy, the leader-sequence consists of a mutated intein that undergoes specific cleavage at the intein-C-extein junction (autoproteolysis) in response to changes in pH or temperature.

Most approaches currently in use for preparing α-Cys proteins involve the in vitro use of exogenous proteases. In the approach developed in the Verdine laboratory, a factor Xa recognition sequence is appended immediately in front of the cryptic N-terminal cysteine in the protein of interest (64). Treatment of such a recombinant fusion protein with the protease gives the requisite α-Cys protein directly, which can then be used in subsequent ligation reactions (64, 65, 91). Tolbert & Wong recently showed that the cysteine protease from tobacco etch virus (TEV) can also be used to release α-Cys proteins from suitable precursors (107). This group demonstrated that the TEV protease would accept Cys in the P1′ site rather than the usual Ser or Gly. The advantages of this approach are the high specificity of the protease and the ability to overexpress the protease in *E. coli.* In principle, other proteases that cleave on the C-terminal side of their recognition site, such as enterokinase or ubiquitin C-terminal hydrolase, should also allow the generation of α-Cys proteins.

Protein splicing can also be exploited to prepare α-Cys proteins. The *Methanobacterium thermoautotrophicum* (*Mth)* RIR1 (74), *Ssp* DnaB (75), and *Mxe* GyrA (109) inteins have all been mutated such that cleavage at the C-terminal splice junction, that is, between the intein and the C-extein, is induced in a pH- and temperature-dependent fashion (Figure 5). Release of these inteins allows the production of an α-Cys containing protein without the need to resort to exogenous protease treatment (which can often result in cleavage at nonspecific sites). Several expression vectors, based on this system, are commercially available. A potential drawback of this intein-based approach is that spontaneous cleavage can occur to varying degrees, resulting in premature loss of the intein and, more importantly, the associated affinity purification tag.

Ligation Strategies

EPL, like native chemical ligation, uses a Cys residue at the site of peptide bond formation. This means that a Cys residue must be located relatively close (in the primary sequence of the protein) to the region where the unnatural residue(s) will be introduced. As noted earlier, it is now possible to synthesize 50 residue peptides in a reliable fashion. Thus, in order for the covalent structure of a protein to be completely accessible to EPL, or native chemical ligation, there must be a native Cys residue every 50 residues or less in the primary sequence. Thousands of proteins meet this requirement and are thus ideal targets for EPL. In cases where there is not a native Cys at a suitable position, the simplest solution is to introduce one through mutation. This tactic has been applied to the semisyntheses of a number of fully active proteins [e.g., (7, 110–112)]. In fact the vast majority of proteins prepared by EPL contain a nonnative cysteine. In most cases, the Cys residue was not arbitrarily introduced; such a mutation could affect the structure and function of the protein. Rather, the mutation was designed to be as conservative as possible in terms of sequence [Ala→Cys or Ser→Cys (91)] and structure [e.g., in a linker or loop region (7)]. Moreover, pains were usually taken to avoid mutating highly conserved residues. In some cases, the effect of a Cys mutation on protein function has been evaluated by making the site-mutant prior to beginning a semisynthesis (97, 113). The availability of straightforward site-directed mutagenesis strategies makes this quasi-empirical approach quite attractive in many systems. As noted earlier, auxiliary groups are now available that overcome the requirement for an N-terminal cysteine in some cases (e.g., Gly-Gly) (49–55). In principle, these groups could be used in the context of EPL, although this is yet to be reported.

Another factor in choosing where to introduce a Cys residue for ligation is the identity of the preceding amino acid; this residue will be at the C terminus in the α-thioester fragment. The Dawson group exhaustively explored the effect of varying this residue on the kinetics of native chemical ligation (59). Not surprisingly, they found that increasing the steric bulk of the side chain (particularly at the β-position) slowed down the reaction. Thus, one would try to avoid using Thr-Cys, Ile-Cys, and Val-Cys ligation junctions. A related issue here is the effect of the -1 residue on the efficiency of the protein-intein thiolysis step. The commercial *Sce* VMA1 and *Mxe* GyrA inteins each have their own set of preferences at this position (114): certain residues are associated with increased levels of premature cleavage in vivo (e.g., aspartic acid), and other residues are associated with no cleavage at all (e.g., proline).

EPL reactions can be performed in two ways; either the thiolysis and native chemical ligation reactions are carried out in one pot, or the recombinant protein α-thioester is isolated first. The former approach limits, to some extent, the type of additives that can be present in the reaction mixture, since the intein must remain folded during the ligation reaction. Nonetheless, these one-pot EPL reactions have been successfully performed in the presence of detergents (86, 115), guanidinium chloride [GdmCl, up to 2 M (100)], urea [up to 4 M (98)], mixed aqueous/organic solvent systems (100), and a host of different thiols (6–8, 73, 100, 101). If the protein α-thioester is first isolated, then the subsequent native chemical ligation step can be performed with a much broader range of additives, including agents that fully denature the proteins (41). The advantage of additives such as chaotropes or detergents that increase solubility is that they allow high concentrations (millimolar) of the reactant polypeptides to be achieved, thereby improving the ligation yields. Note that excellent ligation yields have been observed in a few cases using quite low reactant concentrations. In one striking example involving a bacterial sigma factor, the two peptide fragments were found to interact noncovalently, making the ligation reaction essentially a unimolecular process (115). Such conformationally assisted native chemical ligations have also been found to obviate the need for the α-Cys residue (116).

Most applications of EPL to date have involved just two building blocks and thus a single ligation reaction. As discussed earlier, this requires that the region of interest in the protein be relatively close (within 50 residues) to the native N or C terminus. To address this limitation, Cotton et al. developed a sequential ligation strategy that allows multiple (three or more) building blocks to be linked together in series (90). Note that the ability to ligate just three peptide fragments together in a spatially controlled manner opens up the entire primary sequence of a protein to chemical modification. In the sequential EPL method, the synthetic peptide insert contains an α-thioester group and a cryptic α-Cys residue masked by a factor Xa–cleavable pro-sequence. The reversible cysteine protection is necessary to prevent the peptide from reacting with itself in either an intra- or intermolecular fashion. As shown in Figure 6, removal of the protecting group sequence following the first ligation reaction allows the second ligation to be performed in a controlled and directed fashion. Other cysteine protection strategies have been used in sequential ligation reactions (57–62); however, the advantage of the proteolytic approach is that the protecting group sequence can be genetically encoded. This advantage has allowed recombinant protein inserts to be used in sequential EPL reactions (86, 96, 117).

APPLICATIONS OF EXPRESSED PROTEIN LIGATION

EPL has been applied to many different classes of protein from both prokaryotic and eukaryotic organisms. These include kinases, phosphatases, transcription factors, polymerases, ion channels, cytoplasmic signaling proteins, and antibodies. A variety of modifications have been introduced into these proteins, allowing functional questions to be addressed that would be difficult to explore by other means. These applications are discussed in the following sections.

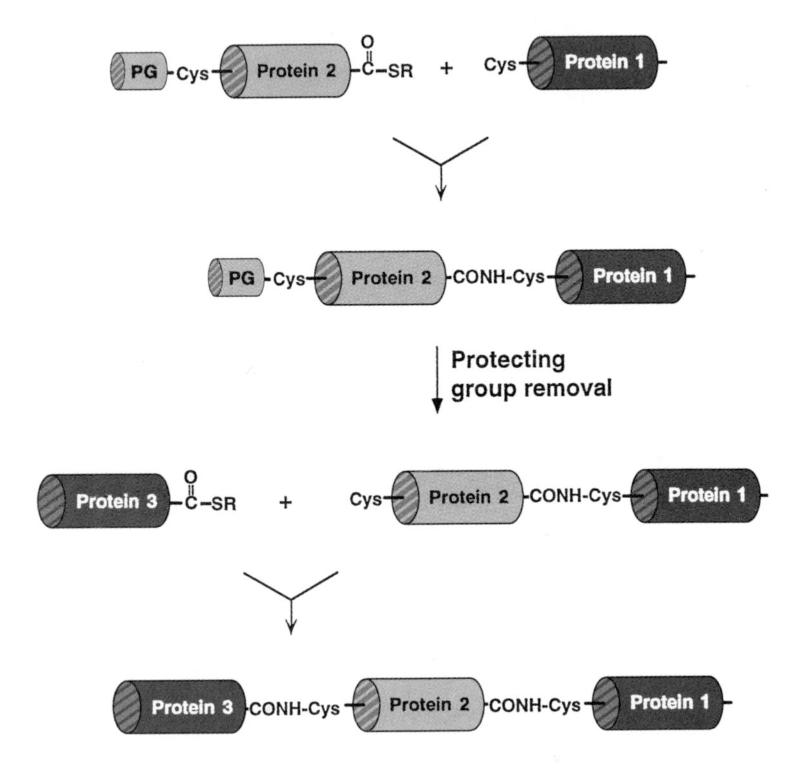

Figure 6 Sequential expressed protein ligation. Three protein fragments are linked together in a tandem ligation procedure. Key to the process is the semipermanent protection of the α-Cys group in the second α-thioester fragment. This tactic allows the two flanking polypeptides, 1 and 3, to be ligated to the appropriate end of the central polypeptide. The protecting group (PG) can be a short peptide, removable with a protease such as factor Xa. Thus, each of the three fragments can be prepared by either synthetic or recombinant methods.

Introduction of Fluorescent Probes

It has long been recognized that fluorescence spectroscopy, because of its extremely high sensitivity, is a powerful method for studying protein structure and function. Nature provides us with two fluorescent amino acids, tyrosine and tryptophan, and these have been used to great effect by protein biophysicists, particularly for studying the thermodynamic stability of proteins and the way they fold (118). For many applications, however, it is desirable to employ fluorescent probes that have spectroscopic properties quite different from those of the two intrinsic fluorophores. Site-specific attachment of chemical probes to a unique cysteine residue offers one route to these fluorescent proteins. In addition, the discovery of the green fluorescent protein (GFP) from the jellyfish *Aequorea victoria* has provided biologists with a straightforward genetic approach for tagging proteins with a fluorescent probe (119). Despite their widespread application, neither of these strategies is without its limitations; the former often requires that the protein be heavily mutated in order to have a unique cysteine, and the latter involves attachment of a 30-kDa fluorescent protein. The use of EPL to incorporate fluorescent probes into proteins overcomes both these problems.

A variety of fluorescent probes have been introduced into proteins using EPL (6, 17, 86, 88, 90, 98, 100, 104, 107, 112, 117, 120, 121). These probes are typically attached to an amino acid side chain within the synthetic peptide building block (e.g., the ϵ-amino group of lysine). Several orthogonal protection strategies are available that allow fluorophores, such as fluorescein or tetramethylrhodamine, to be introduced during SPPS (6, 100). Simple derivatives of several fluorophores have also been synthesized for use in EPL reactions (88, 107, 121).

The ability to site-specifically introduce small fluorescent probes into proteins opens up many avenues for assaying function. Perhaps the simplest of these involves monitoring the intrinsic fluorescence of the probe during the biological process under investigation. Several fluorophores are known to be environmentally sensitive, that is, their quantum yields and/or Stokes shifts are sensitive to changes in the dielectric constant in their immediate environment. Thus, the fluorescence of such a probe will change if it is located close to residues that undergo a structural change in the protein or to a site of ligand binding. The environmentally sensitive 5-(dimethylamino)-naphthalene-sulfonamide (dansyl) probe has been introduced into several proteins by EPL. For example, a sequential EPL procedure was used to incorporate this probe between the Src homology 3 (SH3) and Src homology 2 (SH2) domains of the Abelson nonreceptor protein tyrosine kinase (c-Abl) (90). The fluorescence properties of semisynthetic c-Abl were found to be sensitive to changes in interdomain orientation, allowing this construct to act as a biosensor for high-affinity bidentate ligand interactions. In another example, Alexandrov and coworkers incorporated a dansyl probe into a semisynthetic version of Rab7, a small GTPase (120). The fluorophore was incorporated close to a region at the C terminus of the protein that is known to be posttranslationally prenylated by the enzyme Rab geranylgeranyl transferase (RabGGTase). Both steady-state and time-resolved fluorescence measurements were then used to determine the affinity of Rab7 for RabGGTase and for the escort protein REP-1. EPL-compatible probes such as dansyl and fluorescein are also suitable for fluorescence anisotropy measurements. This is particularly powerful for determining protein-ligand affinities and has the big attraction that the fluorophore does not need to be close to the ligand binding site in the protein. Indeed, proteins labeled at their native N or C terminus, which are easily prepared by EPL, are well suited to anisotropy-based binding assays. This application of EPL is likely to be common in the future.

Fluorescence resonance energy transfer (FRET) represents yet another way in which structural and functional information can be obtained using fluorescent proteins. FRET is a physical phenomenon that is dependent upon the distance between two fluorophores, a donor and an acceptor, specifically located within the molecule of interest (122). Cotton & Muir utilized a solid-phase sequential EPL approach to incorporate the FRET pair, tetramethylrhodamine and fluorescein, into the c-Crk-II signaling protein, which is a substrate of the c-Abl protein tyrosine kinase. By careful placement of the fluorophores within c-Crk-II, it was possible to monitor phosphorylation of this protein using FRET measurements (86). In subsequent studies, it was possible to design an extremely sensitive dual labeled c-Crk-II analog that enabled real-time monitoring of c-Abl kinase activity and provided a rapid nonradioactive approach to screening potential kinase inhibitors (117). The Ebright group has used a combination of EPL-labeling and cysteinyl-labeling to introduce FRET pairs at various locations within *E. coli* RNA polymerase (RNAP) holoenzyme (17, 121). In one lovely set of experiments, the researchers were able to study the mechanisms of transcription initiation and RNAP translocation by using combinations of fluorescent-labeled promoter DNA, σ^{70}, and RNAP core enzyme, the latter prepared using EPL (17). In a follow-up study (121), the group created a structural model for the RNAP holoenzyme complex by using numerous FRET-derived distance constraints to dock promoter DNA and σ^{70} onto the previously determined crystal structure (123) of the core enzyme.

Introduction of Posttranslational Modifications

It is becoming increasingly clear that the functional state of many proteins is regulated by posttranslational modification (124). Indeed, it is useful to think about this as nature's way of conferring functional diversity on the same translated sequence (40). One of the goals of proteomics is to catalog all of the posttranslational modifications within the proteome of an organism. Indeed, some progress has already been made in characterizing phosphorylation at the proteome level (125). However, knowing that a protein is modified is only half the story. The next stage is to determine what that modification does to function. Some information can be obtained by mutating the modified residue(s) in the protein to something that either cannot be modified or that mimics the modified amino acid (e.g., Asp/Glu are often used to mimic phosphoserine/phosphothreonine). Spectacular progress has also been made recently in developing specific inhibitors of enzymes that carry out the modifications [reviewed in (126)]. In most cases, however, a full molecular understanding of how a modification affects function can be achieved only through biochemical and structural studies on the modified protein. This necessity is complicated by the difficulty associated with preparing sufficient amounts of chemically defined, posttranslationally modified proteins using standard biotechnology approaches. EPL provides a solution to this problem since chemistry is now available to allow the preparation of appropriately modified peptides that can be ligated to recombinant protein fragments.

Nowhere is the utility of EPL in this area better illustrated than in the preparation of phosphoproteins. Indeed, one of the first applications of the approach involved the incorporation of phosphotyrosine residue into the kinase Csk (6). The standard method for preparing phosphorylated proteins is through enzymatic treatment of the protein substrate with a kinase. These reactions do not always proceed in good yield, and the positions of phosphorylation cannot always be controlled. EPL can eliminate both of these problems, as illustrated through the semisynthesis of a hyperphosphorylated version of the type 1 transforming growth factor β receptor (TβR-I) (113, 127). In this example, a tetraphosphorylated synthetic peptide α-thioester was ligated to an α-Cys fragment of TβR-I, comprising the cytoplasmic serine/threonine kinase domain. Access to a homogeneous preparation of hyperphosphorylated TβR-I allowed the molecular mechanism of receptor activation to be studied for the first time. Accordingly, phosphorylation was shown to increase the binding affinity of TβR-I for its natural substrate, the Smad2 transcription factor, and decrease its affinity for the inhibitory protein FKBP12 (128). These observations led to a new model of receptor activation in which phosphorylation of the receptor switches the protein from an inhibited state into an activated form capable of binding substrate. In a related series of experiments, EPL was used to study the effect of phosphorylation on the Smad2 protein. Biochemical studies on semisynthetic Smad2 indicated that phosphorylation induces trimerization of the protein (129). As shown in Figure 7, this conclusion was confirmed when the crystal structure of the semisynthetic protein was determined. These investigations revealed how phosphorylation of Smad2 allows dissociation from the activated TβR-I receptor and simultaneously induces hetero-oligomerization with a key regulatory protein, Smad4 (128, 129).

Our understanding of how glycosylation regulates protein function has also been hampered by the difficulties associated with producing homogeneous samples. Indeed, the complex structures of oligosaccharide modifications make the production of chemically defined glycoproteins a considerably more daunting challenge than the synthesis of most phosphoproteins. Nonetheless, rapid progress continues to be made on the chemical and enzymatic synthesis of large carbohydrates and glycopeptides [reviewed in (130)] and these advances are beginning to extend into the realm of proteins using chemical ligation approaches. The Bertozzi group has pioneered the use of oxime and hydrazone ligation chemistry to prepare neoglycoproteins [reviewed in (131)]. This same group was also the first to apply native chemical ligation (132) and most recently EPL (133) to the preparation of O-linked glycoproteins. The work of Tolbert & Wong indicates that N-linked glycoproteins should also be accessible using EPL (88).

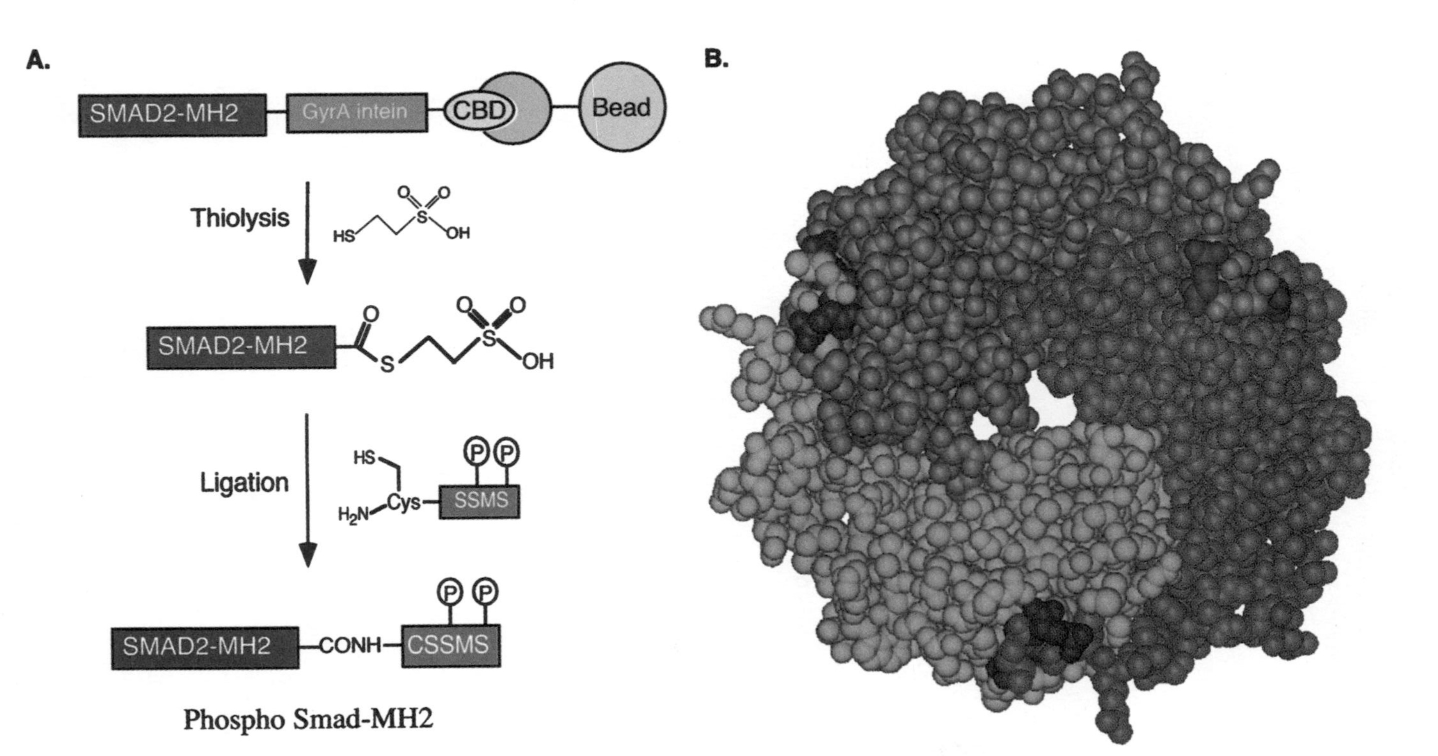

Figure 7 Semisynthesis of phosphorylated Smad2. (*A*) The MH2 domain (residues 241–462) of Smad2, lacking the C-terminal phosphopeptide tail, was expressed as a GyrA intein fusion and then ligated to a synthetic phosphopeptide (residues 463–467) corresponding to the missing tail sequence. (*B*) CPK rendering of the crystal structure (129) (Protein Data Bank accession number 1KHX) of semisynthetic Smad2-MH2. The three monomers in the homotrimeric complex are shown in green, blue, and red. The phosphoserine residues in the tail of each monomer interact with a positively charged surface on the adjacent monomer.

Lipidation represents an important class of protein modification whose role in targeting proteins to cellular membranes is well documented (134). As with phosphorylation and glycosylation, the preparation of homogeneously lipidated proteins has been problematic. A recent paper by Alexandrov and coworkers indicates that, once again, EPL may provide a practical solution to the problem (112). In an impressive piece of work, EPL was used to prepare a monoprenylated version of the small GTPase Rab7. Specifically, a geranylgeranyl group was attached to one of the two known prenylation sites in the protein. A dansyl probe was also introduced into this protein and allowed the mechanism of the second prenylation step, by RabGGTase, to be studied using stopped-flow fluorescence approaches. On the basis of their measurements, the authors conclude that the RabGGTase may attach the two prenyl groups to Rab7 through a processive mechanism.

Introduction of Stable Isotopes

Expressed protein ligation has been used to develop a segmental isotopic labeling strategy designed to overcome the practical size limit for protein structure determination using nuclear magnetic resonance (NMR) spectroscopy. This limit is attributable to the loss of spectral resolution occurring both from increased line widths at longer rotational correlation times, and from the increased number of signals of similar chemical shifts—both effects are proportional to the number of amino acids in the protein. The former of these problems has to a large extent been addressed with the development of new NMR experiments such as transverse relaxation optimized spectroscopy (TROSY) (135) and approaches for measuring residual dipolar coupling constants (136). However, standard isotopic labeling strategies involving uniform incorporation of ^{13}C, ^{15}N, and selective deuteration of amino acid side chains cannot avoid the signal overlap problem for larger systems. Segmental isotopic labeling resolves this issue by allowing selected portions of a protein to be isotopically labeled with NMR active isotopes; unlabeled regions of the protein can then be filtered out of the NMR spectrum using suitable heteronuclear correlation experiments. Thus, segmental labeling significantly reduces the spectral complexity for large proteins and allows a variety of solution-based NMR experiments to be applied.

Segmental isotopic labeling has been accomplished using both protein *trans*-splicing and expressed protein ligation. In a pioneering study, Yamazaki and coworkers used a protein *trans*-splicing system based on a split PI-*PfuI* intein to selectively label with ^{15}N the C-terminal domain of the *E. coli* RNA polymerase α subunit (79). EPL was first applied to this area in the Muir and Cowburn laboratories, using the technique to selectively ^{15}N-label a single domain within the Src-homology domain 3 and 2 segment derived from the Abl protein tyrosine kinase (91). In both the *trans*-splicing and EPL strategies, one half of the protein of interest was bacterially expressed using a growth medium enriched with the ^{15}N isotope. Subsequent ligation of this labeled fragment with the second protein fragment, in this case unlabeled, yielded the desired segmentally labeled product.

Thus, segmental isotopic labeling fundamentally exploits the ability of EPL and protein *trans*-splicing to join together two recombinant proteins.

The real attraction of the segmental isotopic labeling approach is that it allows the structure and function of a protein domain to be determined in its native context. The value of this approach is illustrated by some recent work on an ~50-kDa bacterial sigma factor (115). In this example, segmental isotopic labeling and NMR were used to study the molecular mechanism of sigma autoregulation. The promoter DNA binding function of many sigma factors is thought to be autoinhibited by a direct physical interaction between the C-terminal DNA binding domain (region 4.2) and the N-terminal autoregulatory domain (region 1.1). To test this model, EPL was used to prepare two labeled versions of a sigma factor from *Thermotoga maritima.* Each protein was selectively labeled in region 4.2 with ^{13}C, ^{15}N, and ^{2}H; however, one sample lacked the autoregulatory domain (Figure 8). The $^{1}H\{^{15}N\}$ and $^{1}H\{^{15}C\}$ heteronuclear NMR spectra obtained for these two samples were found to be almost identical, which indicates that region 1.1 and region 4.2 do not directly interact in this sigma factor. On the basis of these data, the researchers suggest that autoinhibition occurs through an indirect electrostatic mechanism.

In many cases it may be desirable to label internal regions (endosegmental labeling) of a protein with NMR probe nuclei. Protein *trans*-splicing and EPL have both been used to generate proteins in which a central domain was isotopically labeled. The Yamazaki group developed a clever tandem *trans*-splicing technique to ^{15}N-label an internal region of maltose binding protein (137). This approach utilized two different split inteins, PI-*pfuI* and PI-*pfuII*; the locations of the two split sites were carefully selected to maintain the fidelity of intein reconstitution, allowing the two splicing reactions to occur simultaneously. As noted earlier, sequential EPL strategies have been developed that allow multiple recombinant protein fragments to be ligated together. Such an approach has been used to assemble the adaptor protein c-Crk-II from its three constitutive Src homology domains (96). We have recently extended this study to generate multi-milligram amounts of endosegmentally labeled c-Crk-II in which the central SH3 domain was labeled with ^{15}N (J. Ottesen, U. Blaschke, and T.W. Muir, unpublished information).

Although NMR has been the focus of segmental labeling studies to date, it is worth noting that other forms of spectroscopy could utilize selective isotopic labeling patterns. In particular, isotope-edited vibrational spectroscopy (138), which requires access to specifically ^{13}C-labeled samples, is an approach that would benefit tremendously from the availability of site-specifically labeled or segmentally-labeled proteins.

Introduction of Unnatural Amino Acids

Fundamentally, EPL is an exceedingly useful tool for introducing unnatural amino acids into proteins. Shown in Figure 9 are some of the noncoded amino acids that have been introduced into proteins using the approach. Typically, these

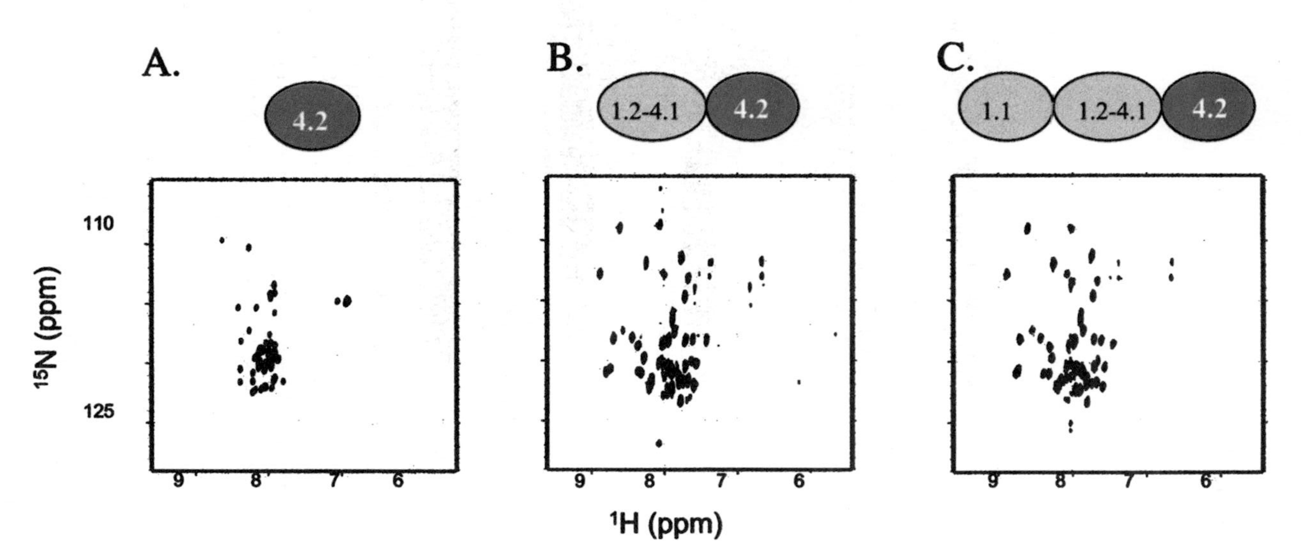

Figure 8 Segmental isotopic labeling of σ^A from *Thermatoga maritima* showing the effect of context on the structure of the promoter binding domain (region 4.2) of the 50-kDa protein. (*A*) $^1H\{^{15}N\}$ HSQC-TROSY spectra of the isolated domain, (*B*) the labeled domain in σ^A lacking regulatory region 1.1, (*C*) the labeled domain in full-length σ^A. Spectra *B* and *C* are nearly superimposable, indicating that region 1.1 is not in direct contact with region 4.2.

unnatural residues were used to probe some aspect of protein function that was difficult or impossible to study by other means. For example, the Cole group used EPL to incorporate a tyrosine phosphonate, a nonhydrolyzable analog of phosphotyrosine, at the C terminus of the SHP-2 phosphatase (139). Introduction of this residue allowed the effect of tyrosine phosphorylation on the structure and function of SHP-2 to be studied, and ultimately allowed the authors to propose a novel mechanism for how tyrosine phosphoryation of SHP-2 regulates the MAP kinase signal transduction pathway. In another example from this same group, a series of fluorinated tyrosine analogs were introduced into the protein tyrosine kinase, Src, which itself is the substrate of the protein tyrosine kinase Csk (110). Kinase assays carried out using these semisynthetic protein substrates provided detailed insights into the mechanism of Src phosphorylation by Csk. In a third example, the Verdine group prepared semisynthetic versions of the AP-1 transcription activator complex (composed of c-Fos and c-Jun subunits) in order to study how it interacts with DNA. An affinity cleavage reagent was prepared by chemically ligating a synthetic thioester derivative of EDTA to a series of recombinant AP-1 complexes, each possessing an N-terminal cysteine residue generated by proteolytic cleavage (64). These semisynthetic proteins were then used in an elegant series of studies to elucidate what structural elements dictated protein orientation within an NFAT•AP-1•DNA ternary complex (64), and to study how the orientation of these proteins affects transcription activation (65).

Unnatural amino acids can also be introduced at the ligation site within semisynthetic proteins by changing the chemical apparatus used in the ligation process. Tam & Yu were the first to demonstrate that an N-terminal homocysteine can substitute for cysteine in native chemical ligation reactions (140), although the ligation reactions are somewhat slower than cysteine-mediated ligation reactions as a result of the less favorable 1–6 S→N acyl transfer step. The Walsh group extended the use of homocysteine to EPL through the preparation of an analog of the precursor to the peptide antibiotic microcin N17 (141). Maturation of this precursor peptide by the enzyme MccB17 synthetase results in cyclization of several serine and cysteine residues to give the oxazole and thiazole heterocycles, respectively. Replacement of cysteine with homocysteine did not result in the generation of a six-membered heterocycle, thus revealing the high degree of specificity of the synthetase.

Several groups have shown that selenocysteine (Sec) can be used in native chemical ligation and expressed protein ligation reactions (111, 142, 143). No general method yet exists for the generation of recombinantly expressed selenocysteine-containing proteins, so the ability to incorporate this residue using native chemical ligation and EPL is an exciting advance for the study of proteins in which selenium is an essential catalytic group. In addition, replacement of native cysteines with Sec could have much utility for probing protein structure and function. Selenium is easier to ionize than sulfur; thus a selenol (RSeH) has a lower pKa than a thiol and the corresponding selenolate (Rse^-) is more nucleophilic than a thiolate. One practical implication is that Sec-mediated

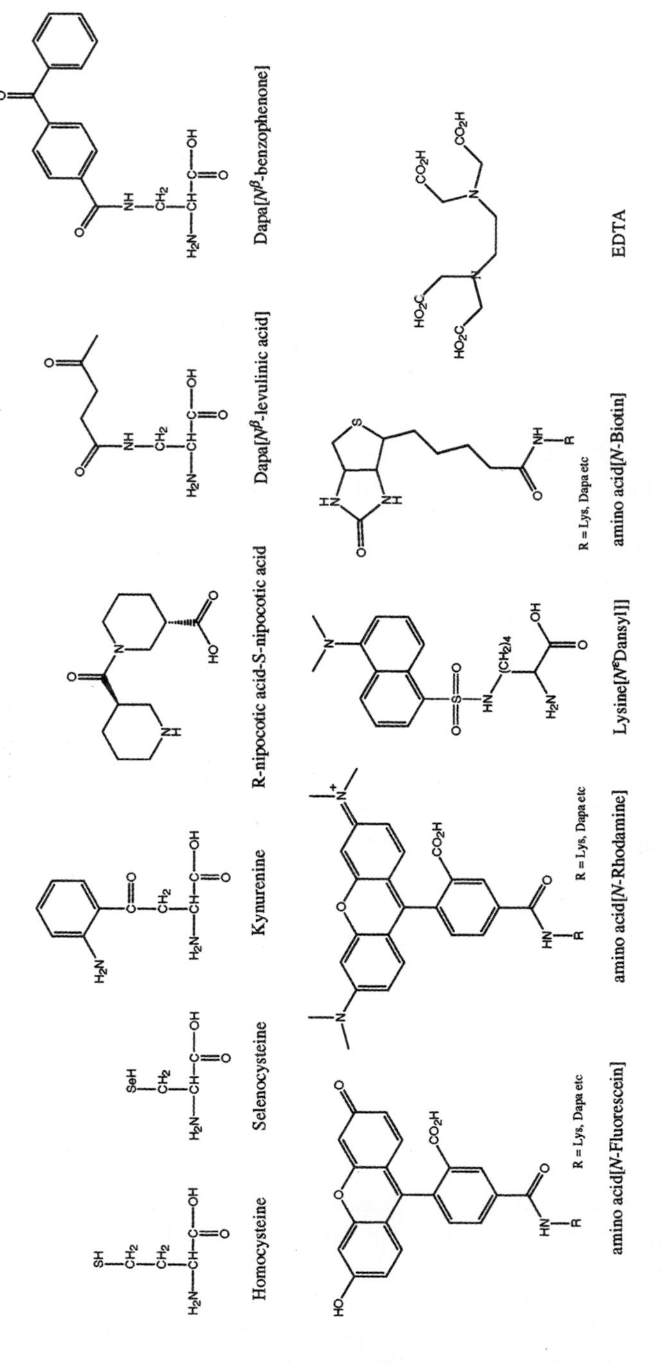

Figure 9 Examples of unnatural amino acids and probes incorporated into proteins by EPL.

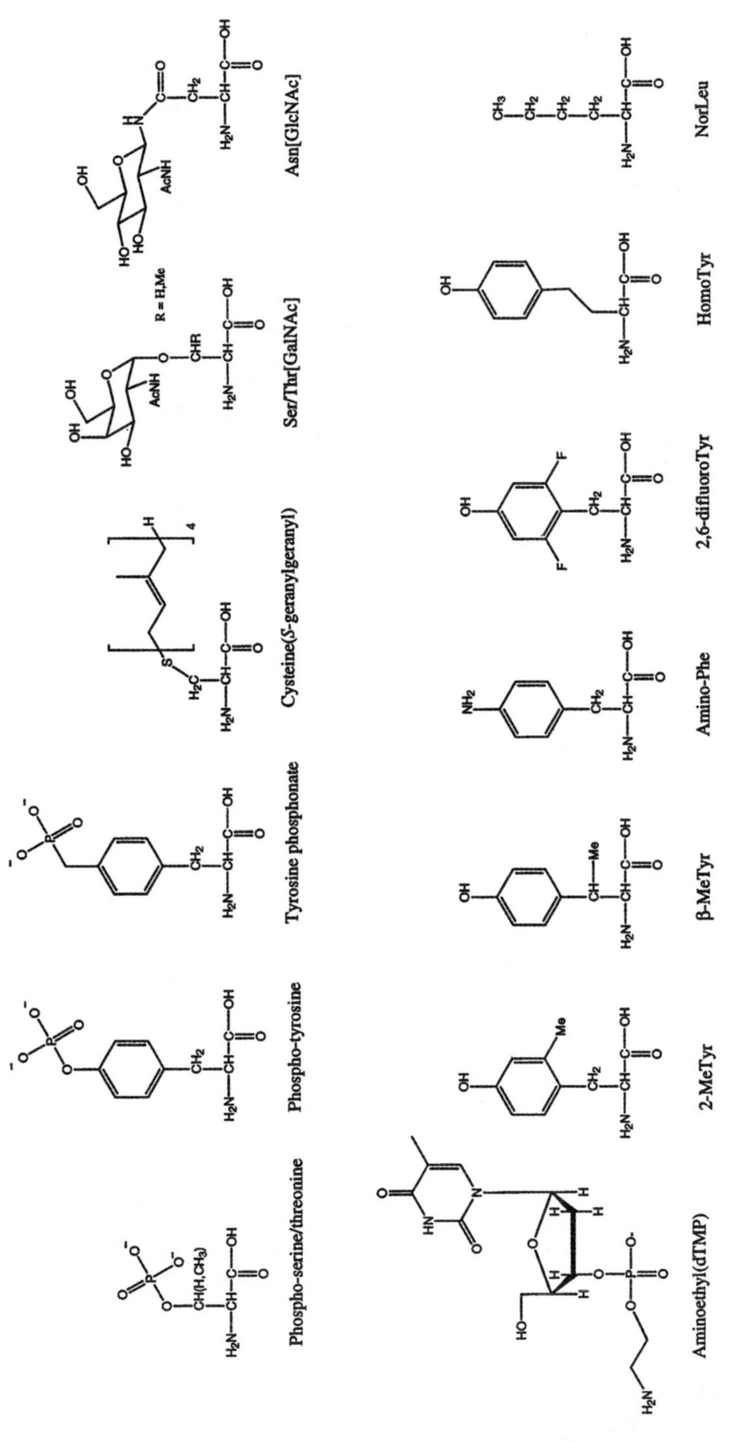

Phospho-serine/threonine

Phospho-tyrosine

Tyrosine phosphonate

Cysteine(*S*-geranylgeranyl)

Ser/Thr[GalNAc]

R = H,Me

Asn[GlcNAc]

Aminoethyl(dTMP)

2-MeTyr

β-MeTyr

Amino-Phe

2,6-difluoroTyr

HomoTyr

NorLeu

Figure 9 *Continued.*

ligation reactions can be performed at a much lower pH than the corresponding Cys-mediated ligations [the reaction with Sec is 1000-fold faster than with Cys at pH 5 (111)]. The softer nature of selenolate compared to thiolate was recently used to fine-tune the copper binding site within the metalloprotein azurin (144). Sec-mediated EPL was used to replace a cysteine copper ligand in azurin with Sec. The resulting analog exhibited altered spectroscopic properties compared to the wild type, perhaps reflecting a subtle change in metal coordination geometry; however, the reduction potential of the variant was similar to that of the wild type. The lower reduction potential of selenium compared to sulfur means that diselenides are more stable to reducing agents than disulfides and thus could be used to stabilize protein structures. In addition, Zhu & van der Donk recently showed that Sec residues in peptides could undergo chemoselective oxidative elimination to give dehydroalanines, which can then undergo Michael-type addition reactions with thiolate nucleophiles such as farnesylthiolates or thioglycosides (145). In principle, this useful transformation allows chemical probes to be introduced site-specifically into protein templates prepared by semisynthesis.

The backbone structure of a protein, which includes the native N and C termini, accounts for ~50% of the mass of a typical protein. Despite the importance of the polypeptide backbone in maintaining native structure and function, the effect of modifying this region of a protein has been essentially unexplored, at least in comparison to changing amino acid side chains. Native chemical ligation has already been used to modify the main chain of several small proteins [e.g., (146)], so a natural extension would be to use EPL for this purpose. Several groups have reported altering the native N and C termini of proteins using EPL. The ability to prepare recombinant protein α-thioesters has been particularly useful in this regard; α-thioesters are versatile synthetic intermediates that can be transformed into a wide variety of other functionalities under mild conditions (147). Accordingly, recombinant proteins possessing a range of C-terminal modifications including thioacids (148), esters (149), and carboxyamides (150) have been prepared from the corresponding protein α-thioesters. The Raines group recently reported the first use of EPL to alter backbone structure in the interior of a protein (151). In this work, an entire element of secondary structure, a β-turn, within the enzyme ribonuclease A (RNase A) was replaced with a reverse-turn mimetic, *R*-nipecotic acid-*S*-nipecotic acid. The resulting semisynthetic protein displayed wild-type enzymatic activity and was slightly more resistant to thermal denaturation. This important piece of work raises the tantalizing possibility that EPL might be used to pepper proteins with prefabricated structural units (what the authors refer to as protein prosthesis) for the purposes of improving thermodynamic or physiological stability.

Topology Engineering of Proteins

Protein engineering can be defined as the alteration of the natural covalent structure of a protein. Typically an amino acid side chain is changed, either by site-directed mutagenesis or, as discussed in the preceding sections, using chemical synthesis or semisynthesis. A less explored branch of protein engineering involves changing the normal linear backbone architecture into an entirely different topology. Examples include cyclic and branched polypeptides. The last few years have seen the application of EPL and protein *trans*-splicing to the area of topology engineering.

Total chemical synthesis has been used to prepare a variety of polypeptides possessing unusual topology (152). In particular, it is now well established that head-to-tail (backbone) circular proteins can be prepared using an intramolecular version of native chemical ligation (153–156); circularization occurs when the polypeptide precursor contains both an α-Cys and an α-thioester. It is therefore not surprising that EPL would eventually be adapted for this purpose (see Figure 10*A*). Accordingly, several EPL-based strategies have now been developed that allow the generation of circular recombinant proteins in vitro (103, 157, 158). These approaches differ from each other only in the way in which the α-Cys is introduced into the precursor. Exposing an α-Cys can be achieved by using an exogenous protease like factor Xa (157), using an engineered intein that undergoes C-terminal cleavage (158), or by removal of the initiating Met by an endogenous methionyl amino-peptidase (103). Interestingly, these intramolecular EPL reactions do not seem to require the addition of thiol cofactors, as is typically the case for standard intermolecular ligation reactions. This is most likely because of conformational assistance in the reactions; the linear precursor protein is chosen such that its N and C termini are juxtaposed, thereby promoting circularization. The spontaneous nature of these reactions has been exploited for the in vivo preparation of a circular protein inside bacteria (159).

Two independent groups reported the use of split inteins for the production of backbone cyclized peptides and proteins (82, 160). In this clever system, the peptide to be cyclized is nested between the two halves of the naturally occurring *Ssp* DnaE split intein such that the C-terminal intein fragment is fused to the N terminus of the peptide and vice versa. As shown in Figure 10*B*, protein splicing leads to the generation of the desired backbone cyclized peptide in vivo, although the approach has also been used in vitro. More recently, two artificially split inteins, the PI-*pfuI* (161) intein and the *Mycobacterium tuberculosis* RecA intein (162), have been used in a similar circularization approach.

Protein circularization is of considerable interest to the protein engineering and protein folding communities (163). Basic polymer theory predicts that circularization will lead to a net thermodynamic stabilization of a protein fold owing to reduced conformational entropy in the denatured state. Indeed, several circular proteins prepared by EPL and protein *trans*-splicing have proven more stable than their linear counterparts. The Plückthun group has prepared circular versions of β-lactamase (103) and GFP (161) and found that the former is more stable to heat-precipitation and proteolytic degradation than its linear counterpart,

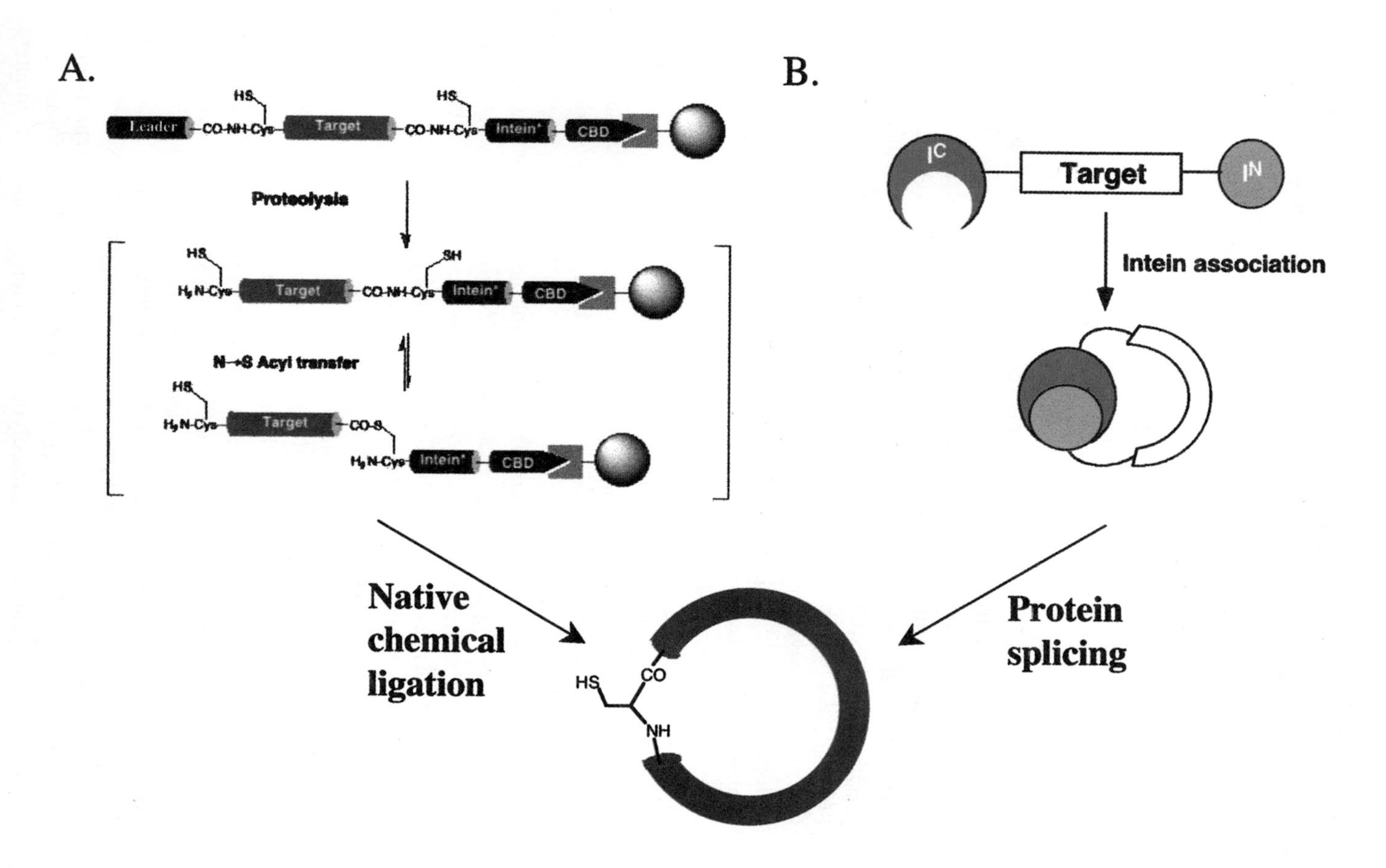
A.
Leader
CO-NH-Cys
HS
Target
Intein*
CBD
Proteolysis
H2N-Cys
SH
N→S Acyl transfer
CO-S
B.
IC
Target
IN
Intein association
Native chemical ligation
Protein splicing
HS
CO
NH

and the latter has a much slower rate of unfolding at high concentrations of guanidinium chloride compared to the wild type. Likewise, backbone cyclized dihydrofolate reductase (DHFR) was found to be more thermostable (82) than the linear protein, and a circular version of the IIA subunit of the *E. coli* glucose transporter was found to be slightly more resistant to thermal and chemical denaturation (162). Not all circular proteins have been found to be more stable; circularization of the c-Crk-II SH3 domain was found to have little effect on the thermodynamic stability of the protein, although the folding and unfolding rates were both faster than in the wild type (164). Similarly, Goldenberg & Creighton observed no significant improvement in stability upon circularizing bovine pancreatic trypsin inhibitor (165). In both these examples, it is likely that unfavorable enthalpic effects (e.g., strain) offset the beneficial entropic effect resulting from circularization.

Many pharmaceutically important natural products, including several antibiotics and immunosuppressants, are based on cyclic peptides. Indeed, peptide cyclization is commonly used in medicinal chemistry to improve the properties of bioactive peptides. Thus, the ability to biosynthesize backbone cyclic peptides using EPL or split inteins has important implications for drug development, particularly since the cyclic peptide can be made in vivo. Work in the Benkovic laboratory extended the possibilities by demonstrating that the split intein approach can be used to generate large combinatorial libraries of cyclic peptides in *E. coli* (166). In this case the peptide sequence between the two intein fragments was randomized at the genetic level. A recent report from Payan and coworkers suggests that this split intein library approach will be a powerful method for rapidly identifying bioactive peptides (167). This group generated a random library of cyclic pentapeptides in human B cells using a retroviral delivery system. A sophisticated cell-based screen was then used to identify a series of cyclic peptides that inhibit the IL-4 signaling pathway. These compounds are potentially therapeutic as anti-inflammatory compounds or may serve as leads for the development of even more efficacious compounds.

Figure 10 Generation of circular proteins. Head-to-tail (backbone) cyclized versions of recombinant peptides and proteins can be prepared (*A*) by EPL or (*B*) by a split-intein approach. In the EPL strategy, circularization occurs spontaneously following proteolytic removal of a leader peptide. The resulting α-Cys attacks the α-thioester at the intein junction in an intramolecular ligation process. Note that unlike intermolecular EPL reactions, this process does not require the addition of thiols. In the split-intein approach the N- and C-terminal intein halves (referred to in the figure as I^N and I^C) are fused to the C and N termini, respectively, of the target sequence. Association of I^N and I^C in *cis* leads to protein splicing and generation of the circular peptide.

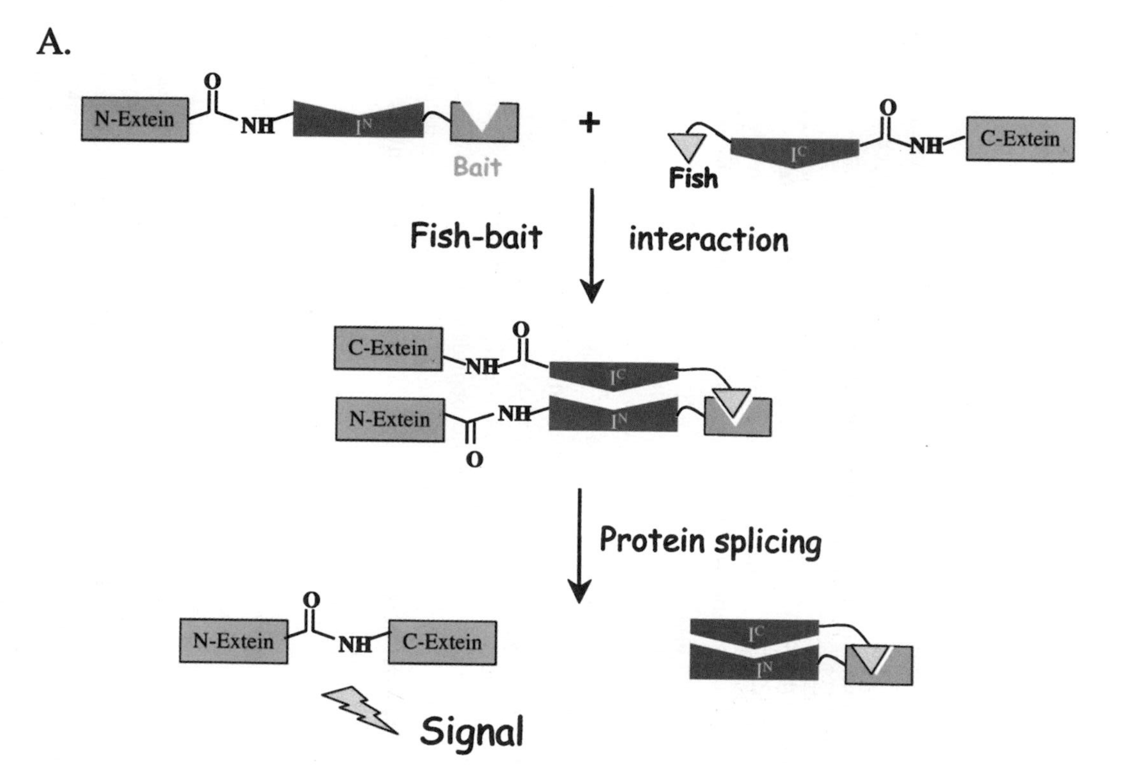

Figure 11 Application of protein *trans*-splicing to proteomics and chemical genetics. (*A*) Principle of the two-hybrid split-intein approach for detecting protein-protein interactions. (*B*) Principle of the conditional *trans*-splicing approach for triggering the posttranslational synthesis of proteins.

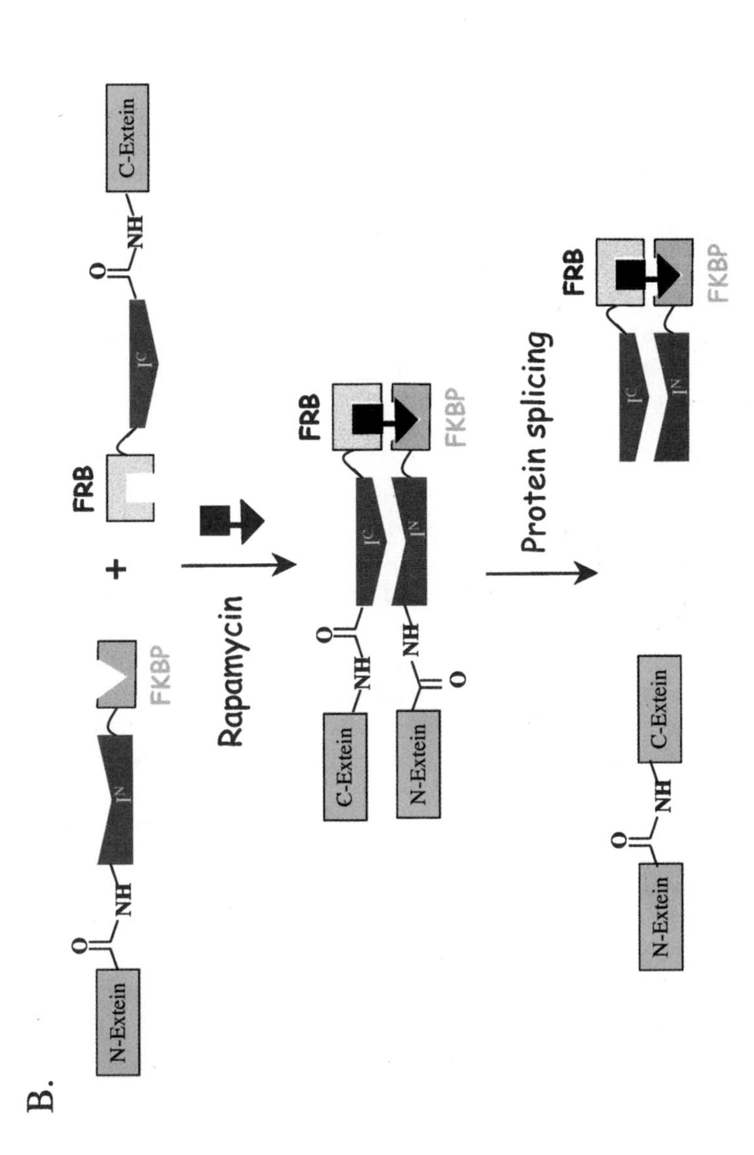

Figure 11 *Continued.*

FUTURE OUTLOOK AND SUMMARY

As noted at the outset of this review, the success of genome sequencing efforts has fueled considerable interest in the area of proteomics, that is, the characterization of the protein complement of a cell or organism. The challenges associated with the acquisition, interrogation, and interpretation of proteomic data are substantial, and a variety of technologies, both experimental and computational, will have to be brought to bear if significant progress is to be made in this endeavor. Expressed protein ligation provides researchers with a versatile tool for the study of protein function by allowing the preparation of proteins containing both natural and artificial modifications. The technology is certainly powerful for the biochemical and biophysical analysis of individual proteins, as discussed in the preceding sections. EPL may also be a valuable tool for more global analyses of protein function, i.e., proteomic studies. For example, the Yao group recently prepared a protein microarray by first biotinylating proteins using EPL and then spatially arraying these on an avidin-coated slide (168). Importantly, the use of EPL ensured that all the proteins were homogeneous with respect to the site of immobilization, in this case the C terminus of the proteins. In principle, EPL could be used to attach other reactive handles to the N or C terminus of proteins that would allow their covalent immobilization on suitable surfaces.

Approaches are also beginning to emerge that use protein *trans*-splicing to analyze and control protein function in ways suitable for proteomics. The Umezawa group has developed a clever two-hybrid approach that allows protein-protein interactions to be detected in the cytosol of prokaryotic (83) and eukaryotic (169) cells. As shown in Figure 11*A*, the strategy involves fusing each half of a reporter protein (GFP or luciferase) to the appropriate end of a split intein. The intein fragments are also fused to either a receptor protein (bait) or to a library of potential ligands (fish). Interaction between fish and bait leads to protein *trans*-splicing and hence generation of the reporter protein. Mootz & Muir recently described another use of protein *trans*-splicing, in this case as a way to trigger a change in the primary structure of a protein of interest (84). The conditional protein *trans*-splicing approach illustrated in Figure 11*B* works by fusing a split intein to the FKBP12/rapamycin/FRB three-hybrid system. Protein *trans*-splicing occurs only in the presence of the small molecule rapamycin owing to the low intrinsic affinity of the intein fragments. The technique effectively provides a way to trigger the posttranslational synthesis of a target protein from two fragments and thus may be useful in the emerging area of chemical genetics (126).

In the future, expressed protein ligation should facilitate important advances in biomedicine through the generation of novel protein therapeutic drugs and diagnostic tools. In one important step toward this goal, Sydor et al. recently established conditions that allow single-chain antibodies to be utilized in EPL reactions (98). Using these procedures, it should now be possible to specifically attach almost any synthetic molecule to the C terminus of an antibody. Such semisynthetic antibodies could have applications in many areas of biomedicine, including diagnostics, vaccine development, and targeted-drug delivery. Molecules of this type could also have value in the area of medical imaging. Lastly, and as noted earlier, EPL allows the preparation of fluorescent proteins suitable for addressing basic biological questions. By extension, it should be possible to prepare fluorescent protein biosensors for use in diagnostic and high-throughput screening assays; indeed, an example of the latter application has been reported (117).

Expressed protein ligation, introduced in 1998, is still a relatively new technique. Nonetheless, it has already established a firm foothold in the in vitro biochemical and biophysical analysis of protein function. In the future, we can certainly anticipate seeing the approach adapted for the in vivo analysis of protein function; indeed, EPL has very recently been used to ligate a cell-permeating peptide to a recombinant protein that could then enter cultured cells (170). Approaches of this type, when coupled with fluorescence-reporter or photocaging strategies, could have great utility for studying protein function in a cellular context.

In conclusion, EPL is a wonderfully malleable platform for the chemical engineering of proteins. To date, the technique has been applied only in the biological arena; however, there appears to be no technical reason to prohibit the preparation of protein-conjugates for use in nonbiological fields such as nano-technology or material science.

ACKNOWLEDGMENTS

Some of the work discussed in this article was performed in the author's laboratory at the Rockefeller University and was supported by the National Institutes of Health, the Pew Charitable Trusts, the Alfred P. Sloan Foundation, and the Rockefeller University.

The *Annual Review of Biochemistry* is online at http://biochem.annualreviews.org

LITERATURE CITED

1. Venter JC, Adams MD, Myers EW, Li PW, Mural RJ, et al. 2001. *Science* 291: 1304–51
2. Lander ES, Linton LM, Birren B, Nusbaum C, Zody MC, et al. 2001. *Nature* 409:860–921
3. Cohen P. 1999. *Curr. Opin. Chem. Biol.* 3:459–65
4. Noren CJ, Anthony-Cahill SJ, Griffith MC, Schultz PG. 1989. *Science* 244:182–88
5. Bain JD, Glabe CG, Dix TA, Chamberlin AR, Diala ES. 1989. *J. Am. Chem. Soc.* 111:8013–14
6. Muir TW, Sondhi D, Cole PA. 1998. *Proc. Natl. Acad. Sci. USA* 95:6705–10
7. Severinov K, Muir TW. 1998. *J. Biol. Chem.* 273:16205–9
8. Evans TC Jr, Benner J, Xu M-Q. 1998. *Protein Sci.* 7:2256–64
9. Cornish VW, Mendel D, Schultz PG. 1995. *Angew. Chem. Int. Ed. Engl.* 34:621–33

10. Nowack MW, Gallivan JP, Silverman SK, Labarca CG, Dougherty DA, Lester HA. 1998. *Methods Enzymol.* 293:504–29
11. Wang L, Brock A, Herberich B, Schultz PG. 2001. *Science* 292:498–500
12. Wallace CJA. 1995. *Curr. Opin. Biotech.* 6:403–10
13. Dyckes DF, Creighton T, Sheppard RC. 1974. *Nature* 247:202–4
14. Wallace CJA, Clark-Lewis I. 1992. *J. Biol. Chem.* 267:3852–61
15. Wallace CJA, Clark-Lewis I. 1997. *Biochemistry* 36:14733–14740
16. Chen Y, Ebright YW, Ebright RH. 1994. *Science* 265:90–92
17. Mukhopadhyay J, Kapanidis AN, Mekler V, Kortkhonjia E, Ebright YW, Ebright RH. 2001. *Cell* 106:453–63
18. Curley K, Lawrence DS. 1998. *J. Am. Chem. Soc.* 120:8573–74
19. Wetzel R, Halualani R, Stults JT, Quan C. 1990. *Bioconjugate Chem.* 1:114–22
20. Zou K, Miller WT, Givens RS, Bayley H. 2001. *Angew. Chem. Int. Ed.* 40:3049–51
21. Zou K, Cheley S, Givens RS, Bayley H. 2002. *J. Am. Chem. Soc.* 124:8220–29
22. Homandberg GA, Laskowski M. 1979. *Biochemistry* 18:586–92
23. Vogel K, Chmielewski J. 1994. *J. Am. Chem. Soc.* 116:11163–64
24. Witte K, Sears P, Martin R, Wong CH. 1997. *J. Am. Chem. Soc.* 119:2114–18
25. Borodovsky A, Kessler BM, Casagrande R, Overkleeft HS, Wilkinson KD, Ploegh HL. 2001. *EMBO J.* 20:5187–96
26. Nakatsuka T, Sasaki T, Kaiser ET. 1987. *J. Am. Chem. Soc.* 109:3808–10
27. Jackson DY, Burnier J, Quan C, Stanley M, Tom J, Wells JA. 1994. *Science* 266: 243–47
28. Jackson DY, Burnier JP, Wells JA. 1995. *J. Am. Chem. Soc.* 117:819–20
29. Chang TK, Jackson DY, Burnier JP, Wells JA. 1994. *Proc. Natl. Acad. Sci. USA* 91:12544–48
30. Gaertner HF, Rose K, Cotton R, Timms D, Camble R, Offord RE. 1992. *Bioconjugate Chem.* 3:262–68
31. Gaertner HF, Offord RE, Cotton R, Timms D, Camble R, Rose K. 1994. *J. Biol. Chem.* 269:7224–30
32. Rose K. 1994. *J. Am. Chem. Soc.* 116: 30–33
33. Geoghegan KF, Stroh JG. 1992. *Bioconjugate Chem.* 3:138–46
34. Simmons G, Clapham PR, Picard L, Offord RE, Rosenkilde MM, et al. 1997. *Science* 276:276–79
35. Kemp DS, Leung S-L, Kerkman DJ. 1981. *Tetrahedron Lett.* 22:181–84
36. Kemp DS, Carey RI. 1993. *J. Org. Chem.* 58:2216–22
37. Blake J, Li CH.1981. *Proc. Natl. Acad. Sci. USA* 78:4055–58
38. Yamashiro D, Li CH. 1988. *Int. J. Pept. Protein Res.* 31:322–34
39. Schnölzer M, Kent SBH. 1992. *Science* 256:221–25
40. Cotton GJ, Muir TW. 1999. *Chem. Biol.* 6:R247–56
41. Dawson PE, Kent SBH. 2000. *Annu. Rev. Biochem.* 69:923–60
42. Muir TW, Williams MJ, Ginsberg MH, Kent SBH. 1994. *Biochemistry* 33: 7701–8
43. Liu C-F, Tam JP. 1994. *Proc. Natl. Acad. Sci. USA* 91:6584–88
44. Dawson PE, Muir TW, Clark-Lewis I, Kent SBH. 1994. *Science* 266:776–79
45. Tam JP, Lu Y-A, Shao J. 1995. *Proc. Natl. Acad. Sci. USA* 92:12485
46. Nilsson BL, Kiessling LL, Raines RT. 2000. *Org. Lett.* 2:1939–41
47. Saxon E, Armstrong JI, Bertozzi CR. 2000. *Org. Lett.* 2:2141–43
48. Wieland T, Bokelmann E, Bauer L, Lang HU, Lau H. 1953. *Liebigs Ann. Chem.* 583:129
49. Canne L, Bark S, Kent SBH. 1996. *J. Am. Chem. Soc.* 118:5891–96
50. Bark SJ, Kent SBH. 1999. *FEBS Lett.* 460:67–76
51. Meutermans WDF, Golding SW, Bourne

GT, Miranda LP, Dooley MJ, et al. 1999. *J. Am. Chem. Soc.* 121:9790–96
52. Botti P, Carrasco MR, Kent SBH. 2001. *Tetrahedron Lett.* 42:1831–33
53. Marinzi C, Bark SJ, Offer J, Dawson PE. 2001. *Bioorg. Med. Chem.* 9:2323–28
54. Kawakami T, Akaji K, Aimoto S. 2001. *Org. Lett.* 3:1403–5
55. Offer J, Boddy CNC, Dawson PE. 2002. *J. Am. Chem. Soc.* 124:4642–46
56. Dawson PE, Churchill MJ, Ghadiri MR, Kent SBH. 1997. *J. Am. Chem. Soc.* 119:4325–29
57. Muir TW, Dawson PE, Kent SBH. 1997. *Methods Enzymol.* 289:266–98
58. Hennard C, Tam JP. 1997. In Peptides, frontiers in peptide science. *Proc. Am. Pept. Symp., 15th, Nashville, Tenn,* ed. JP Tam, PTP Kaumaya, pp. 247–48. Dordrecht, Neth.: Kluwer Acad.
59. Hackeng TM, Griffin JH, Dawson PE. 1999. *Proc. Natl. Acad. Sci. USA* 96:10068–73
60. Camarero JA, Cotton GJ, Adeva A, Muir TW. 1998. *J. Pept. Res.* 51:303–16
61. Canne LE, Botti P, Simon RJ, Chen Y, Dennis EA, Kent SBH. 1999. *J. Am. Chem. Soc.* 121:8720–27
62. Brik A, Keinan E, Dawson PE. 2000. *J. Org. Chem.* 65:3829–35
63. Hofmann RM, Muir TW. 2002. *Curr. Opin. Biotech.* 13:297–303
64. Erlanson DA, Chytil M, Verdine GL. 1996. *Chem. Biol.* 3:981–91
65. Chytil M, Peterson BR, Erlanson DA, Verdine GL. 1998. *Proc. Natl. Acad. Sci. USA* 95:14076–81
66. Noren CJ, Wang JM, Perler FB. 2000. *Angew. Chem. Int. Ed. Engl.* 39:450–66
67. Paulus H. 2000. *Annu. Rev. Biochem.* 69:447–96
68. Southworth MW, Benner J, Perler FB. 2000. *EMBO J.* 19:5019–26
69. Klabunde T, Sharma S, Telenti A, Jacobs WR Jr, Sacchettini JC. 1998. *Nat. Struct. Biol.* 5:31–36
70. Xu MQ, Perler FB. 1996. *EMBO J.* 15:5146–53
71. Perler FB. 1998. *Cell* 92:1–4
72. Giriat I, Muir TW, Perler FB. 2001. *Genet. Eng.* 23:171–99
73. Chong S, Mersha FB, Comb DG, Scott ME, Landry D, et al. 1997. *Gene* 192: 271–81
74. Evans TC, Benner J, Xu MQ. 1999. *J. Biol. Chem.* 274:3923–26
75. Mathys S, Evans TC Jr, Chute CI, Wu H, Chong SR, et al. 1999. *Gene* 231:1–13
76. Southworth MW, Adam E, Panne D, Byer R, Kautz R, Perler FB. 1998. *EMBO J.* 17:918–26
77. Mills KV, Lew BM, Jiang S-Q, Paulus H. 1998. *Proc. Natl. Acad. Sci. USA* 95:3543–48
78. Wu H, Hu Z, Liu XQ. 1998. *Proc. Natl. Acad. Sci. USA* 95:9226–31
79. Yamazaki T, Otomo T, Oda N, Kyogoku Y, Uegaki K, et al. 1998. *J. Am. Chem. Soc.* 120:5591–92
80. Shingledecker K, Jiang S-Q, Paulus H. 1998. *Gene* 207:187–95
81. Lew BM, Mills KV, Paulus H. 1998. *J. Biol. Chem.* 273:15887–90
82. Scott CP, Abel-Santos E, Wall M, Wahnon DC, Benkovic SJ. 1999. *Proc. Natl. Acad. Sci. USA* 96:13638–43
83. Ozawa T, Nogami S, Sato M, Ohya Y, Umezawa Y. 2000. *Anal. Chem.* 72:5151–57
84. Mootz HD, Muir TW. 2002. *J. Am. Chem. Soc.* 124:9044–45
85. Deleted in proof
86. Cotton GJ, Muir TW. 2000. *Chem. Biol.* 7:253–61
87. Milton RC, Milton SC, Kent SBH. 1992. *Science* 256:1445–48
88. Tolbert TJ, Wong C-H. 2000. *J. Am. Chem. Soc.* 122:5421–28
89. Schnölzer M, Alewood P, Jones A, Alewood D, Kent SBH. 1992. *Int. J. Pept. Protein Res.* 40:180–93
90. Cotton GJ, Ayers B, Xu R, Muir TW. 1999. *J. Am. Chem. Soc.* 121:1100–1
91. Xu R, Ayers B, Cowburn D, Muir TW. 1999. *Proc. Natl. Acad. Sci. USA* 96:388–93

92. Wood DW, Wu W, Belfort G, Derbyshire V, Belfort M. 1999. *Nat. Biotechnol.* 17:889–92
93. Chong SR, Xu MQ. 1997. *J. Biol. Chem.* 272:15587–90
94. Chong SR, Montello GE, Zhang AH, Cantor EJ, Liao W, et al. 1998. *Nucleic Acids Res.* 26:5109–15
95. Blaschke UK, Silberstein J, Muir TW. 2000. *Methods Enzymol.* 328:478–96
96. Blaschke UK, Cotton GJ, Muir TW. 2000. *Tetrahedron* 56:9461–70
97. Valiyaveetil FI, MacKinnon R, Muir TW. 2002. *J. Am. Chem. Soc.* 124:9113–20
98. Sydor JR, Mariano M, Sideris S, Nock S. 2002. *Bioconjugate Chem.* 13:707–12
99. Kawasaki M, Makino S, Matsuzawa H, Satow Y, Ohya Y, Anraku Y. 1996. *Biochem. Biophys. Res. Commun.* 222: 827–32
100. Ayers B, Blaschke UK, Camarero JA, Cotton GJ, Holford M, Muir TW. 2000. *Biopolymers* 51:343–54
101. Welker E, Scheraga HA. 1999. *Biochem. Biophys. Res. Commun.* 254:147–51
102. Hirel PH, Schmitter JM, Dessen P, Fayat G, Blanquet S. 1989. *Proc. Natl. Acad. Sci. USA* 86:8247–51
103. Iwai H, Plückthun A. 1999. *FEBS Lett.* 459:166–72
104. Dwyer MA, Lu W, Dwyer JJ, Kossiakoff AA. 2000. *Chem. Biol.* 7:263–74
105. Camarero JA, Muir TW. 1999. In *Current Protocols in Protein Science,* Unit 18.4.1–18.4.21. New York: Wiley
106. Villain M, Vizzavona J, Rose K. 2001. *Chem. Biol.* 8:673–79
107. Tolbert TJ, Wong C-H. 2002. *Angew. Chem. Int. Ed.* 41:2171–74
108. Deleted in proof
109. Southworth MW, Amaya K, Evans TC, Xu M-Q, Perler FB. 1999. *BioTechniques* 27:110–20
110. Wang D, Cole PA. 2001. *J. Am. Chem. Soc.* 123:8883–86
111. Hondal RJ, Nilsson BL, Raines RT. 2001. *J. Am. Chem. Soc.* 123:5140–41
112. Alexandrov K, Heinemann I, Durek T, Sidorovitch V, Goody RS, Waldmann H. 2002. *J. Am. Chem. Soc.* 124:5648–49
113. Huse M, Holford MN, Kuriyan J, Muir TW. 2000. *J. Am. Chem. Soc.* 122:8337–38
114. Chong SR, Williams KS, Wotkowicz C, Xu M-Q. 1998. *J. Biol. Chem.* 273:10567–77
115. Camarero JA, Shekhman A, Campbell E, Chlenov M, Gruber TM, et al. 2002. *Proc. Natl. Acad. Sci. USA* 99:8536–41
116. Beligere GS, Dawson PE. 1999. *J. Am. Chem. Soc.* 121:6332–33
117. Hofmann RM, Cotton GJ, Chang EJ, Vidal E, Veach D, et al. 2001. *Bioorg. Med. Chem. Lett.* 11:3091–94
118. Tanford C. 1970. *Adv. Protein Chem.* 24:1–95
119. Tsien RY. 1998. *Annu. Rev. Biochem.* 67:509–44
120. Iakovenko A, Rostkova E, Merzlyak E, Hillebrand AM, Thoma NH, et al. 2000. *FEBS Lett.* 468:155–58
121. Mekler V, Kortkhonjia E, Mukhopadhyay J, Knight J, Revyankin A, et al. 2002. *Cell* 108:599–614
122. Selvin PR. 1995. *Methods Enzymol.* 246: 300–34
123. Zhang G, Campbell E, Minakhin L, Richter C, Severinov K, Darst S. 1999. *Cell* 98:811–24
124. Pawson T, Nash P. 2000. *Genes Dev.* 14:1027–47
125. McLachlin DT, Chait BT. 2001. *Curr. Opin. Chem. Biol.* 5:591–602
126. Shogren-Knaak MA, Alaimo PJ, Shokat KM. 2001. *Annu. Rev. Cell. Dev. Biol.* 17:405–33
127. Flavell RR, Huse M, Goger M, Trester-Zedlitz M, Kuriyan J, Muir TW. 2002. *Org. Lett.* 4:165–68
128. Huse M, Muir TW, Xu L, Chen Y-G., Kuriyan J, Massague J. 2001. *Mol. Cell* 6:671–82
129. Wu J, Hu M, Chai J, Seoane J, Huse M, et al. 2001. *Mol. Cell* 8:1277–89
130. Hecht SM, ed. 1999. *Bioorganic Chem-*

istry: Carbohydrates. New York: Oxford Univ. Press
131. Lemieux GA, Bertozzi CR. 1998. *Trends Biotech.* 16:506–13
132. Shin Y, Winans KA, Backes BJ, Kent SBH, Ellman JA, Bertozzi CR. 1999. *J. Am. Chem. Soc.* 121:11684–89
133. Macmillan D, Bertozzi CR. 2000. *Tetrahedron* 56:9515–25
134. Casey PJ. 1995. *Science* 268:221–25
135. Pervushin K, Riek R, Wider G, Wuthrich K. 1997. *Proc. Natl. Acad. Sci. USA* 94:12366–71
136. Tjandra N, Bax A. 1997. *Science* 278: 1111–14
137. Otomo T, Ito N, Kyogoku Y, Yamazaki T. 1999. *Biochemistry* 38:16040–44
138. Cheng H, Sukal S, Deng H, Leyh TS, Callender R. 2001. *Biochemistry* 40: 4035–43
139. Lu W, Gong D, Bar-Sagi D, Cole PA. 2001. *Mol. Cell* 8:759–69
140. Tam JP, Yu QT. 1998. *Biopolymers* 46:319–27
141. Roy RS, Allen O, Walsh CT. 1999. *Chem. Biol.* 6:789–99
142. Gieselman MD, Xie L, van Der Donk WA. 2001. *Org. Lett.* 3:1331–34
143. Quaderer R, Sewing A, Hilvert D. 2001. *Helv. Chim. Acta* 84:1197–206
144. Berry SM, Gieselman MD, Nilges MJ, van der Donk WA, Lu Y. 2002. *J. Am. Chem. Soc.*124:2084–85
145. Zhu Y, van der Donk WA. 2001. *Org. Lett.* 3:1189–92
146. Lu W, Randal M, Kossiakoff A, Kent SBH. 1999. *Chem. Biol.* 6:419–27
147. Camarero JA, Adeva A, Muir TW. 2000. *Lett. Pept. Sci.* 7:17–21
148. Kinsland C, Taylor SV, Kelleher NL, McLafferty FW, Begley TP. 1998. *Protein Sci.* 7:1839–42
149. Lykke-Andersen J, Christiansen J. 1998. *Nucleic Acids Res.* 26:5631–35
150. Cottingham IR, Millar A, Emslie E, Colman A, Schnieke AE, McKee C. 2001. *Nat. Biotechnol.* 19:974–77
151. Arnold U, Kinderaker MP, Nilsson BL, Huck BR, Gellman SH, Raines RT. 2002. *J. Am. Chem. Soc.* 124:8522–23
152. Muir TW. 1995. *Structure* 3:649–52
153. Zhang L, Tam JP. 1997. *J. Am. Chem. Soc.* 119:2363–70
154. Camarero JA, Muir TW. 1997. *J. Chem. Soc. Chem. Commun.* pp. 1369–70
155. Camarero JA, Pavel J, Muir TW. 1998. *Angew. Chem. Int. Ed. Eng.* 37:347–49
156. Tam JP, Lu YA. 1998. *Protein Sci.* 7:1583–92
157. Camarero JA, Muir TW. 1999. *J. Am. Chem. Soc.* 121:5597–98
158. Evans TC, Benner J, Xu M-Q. 1999. *J. Biol. Chem.* 274:18359–81
159. Camarero JA, Fushman D, Cowburn D, Muir TW. 2001. *Bioorg. Med. Chem.* 9:2479–84
160. Evans TC Jr, Martin D, Kolly R, Panne D, Sun L, et al. 2000. *J. Biol. Chem.* 275:9091–94
161. Iwai H, Lingel A, Plückthun A. 2001. *J. Biol. Chem.* 276:16548–54
162. Siebold C, Erni B. 2002. *Biophys. Chem.* 96:163–71
163. Trabi M, Craik DJ. 2002. *Trends Biochem. Sci.* 27:132–38
164. Camarero JA, Fushman D, Sato S, Giriat I, Cowburn D, et al. 2001. *J. Mol. Biol.* 308:1045–62
165. Goldenberg DP, Creighton TE. 1984. *J. Mol. Biol.* 179:527–45
166. Scott CP, Abel-Santos E, Jones AD, Benkovic SJ. 2001. *Chem. Biol.* 8:801–15
167. Kinsella TM, Ohashi CT, Harder AG, Yam GC, Li W, et al. 2002. *J. Biol. Chem.* 277:37512–18
168. Lesaicherre ML, Lue RYP, Chen GYJ, Zhi Q, Yao SQ. 2002. *J. Am. Chem. Soc.* 124:8768–69
169. Ozawa T, Kaihara A, Sato M, Tachihara K, Umezawa Y. 2001. *Anal. Chem.* 73:2516–21
170. Machova Z, Muhle C, Krauss U, Trehin R, Koch A, et al. 2002. *ChemBioChem* 3:672–77

Annu. Rev. Biochem. 2003. 72:291–336
doi: 10.1146/annurev.biochem.72.121801.161720
Copyright © 2003 by Annual Reviews. All rights reserved
First published online as a Review in Advance on February 27, 2003

MECHANISMS OF ALTERNATIVE PRE-MESSENGER RNA SPLICING

Douglas L. Black
Department of Microbiology, Immunology, and Molecular Genetics, Howard Hughes Medical Institute, University of California-Los Angeles, 6-762 MacDonald Building, Box 951662, 675 Charles Young Drive South, Los Angeles, California 90095-1662; email: dougb@microbio.ucla.edu

Key Words alternative splicing, regulatory mechanisms, protein diversity, RNA binding proteins, spliceosome

■ **Abstract** Alternative pre-mRNA splicing is a central mode of genetic regulation in higher eukaryotes. Variability in splicing patterns is a major source of protein diversity from the genome. In this review, I describe what is currently known of the molecular mechanisms that control changes in splice site choice. I start with the best-characterized systems from the *Drosophila* sex determination pathway, and then describe the regulators of other systems about whose mechanisms there is some data. How these regulators are combined into complex systems of tissue-specific splicing is discussed. In conclusion, very recent studies are presented that point to new directions for understanding alternative splicing and its mechanisms.

CONTENTS

INTRODUCTION

The splicing reaction that assembles eukaryotic mRNAs from their much longer precursors provides a uniquely versatile means of genetic regulation. Alterations in splice site choice can have many different effects on the mRNA and protein products of a gene. Commonly, alternative splicing patterns determine the inclusion of a portion of coding sequence in the mRNA, giving rise to protein isoforms that differ in their peptide sequence and hence chemical and biological activity (1). Alternative splicing is a major contributor to protein diversity in metazoan organisms. Estimates of the minimum number of human gene products that undergo alternative splicing are as high as 60% (2). Moreover, many gene transcripts have multiple splicing patterns and some have thousands (3, 4). To understand this complexity of gene expression, we must study how changes in splice site choice come about.

In a typical multiexon mRNA, the splicing pattern can be altered in many ways (Figure 1). Most exons are constitutive; they are always spliced or included in the final mRNA. A regulated exon that is sometimes included and sometimes excluded from the mRNA is called a cassette exon. In certain cases, multiple cassette exons are mutually exclusive—producing mRNAs that always include one of several possible exon choices but no more. In these systems, special mechanisms must enforce the exclusive choice (5, 6). Exons can also be lengthened or shortened by altering the position of one of their splice sites. One sees both alternative 5′ and alternative 3′ splice sites. The 5′-terminal exons of an mRNA can be switched through the use of alternative promoters and alternative splicing. Similarly, the 3′-terminal exons can be switched by combining alternative splicing with alternative polyadenylation sites. Alternative promoters are primarily an issue of transcriptional control. Control of polyadenylation appears mechanistically similar to control of splicing, although it is not discussed here (7). Finally, some important regulatory events are controlled by the failure to remove an intron, a splicing pattern called intron retention. Particular pre-mRNAs often have multiple positions of alternative splicing, giving rise to a family of related proteins from a single gene (Figure 1*H*).

Changes in splice site choice can have all manner of effects on the encoded protein. Small changes in peptide sequence can alter ligand binding, enzymatic activity, allosteric regulation, or protein localization. In other genes, the synthesis of a whole polypeptide, or a large domain within it, can depend on a particular splicing pattern. Genetic switches based on alternative splicing are important in

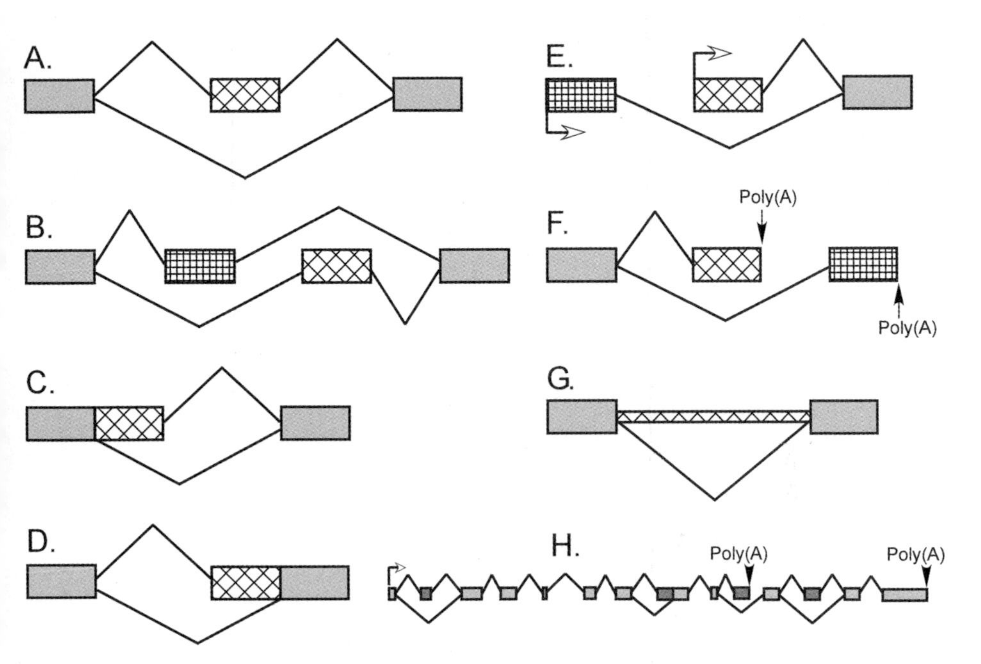

Figure 1 Patterns of alternative splicing. Constitutive sequences present in all final mRNAs are gray boxes. Alternative RNA segments that may or may not be included in the mRNA are hatched boxes. (*A*) A cassette exon can be either included in the mRNA or excluded. (*B*) Mutually exclusive exons occur when two or more adjacent cassette exons are spliced such that only one exon in the group is included at a time. (*C, D*) Alternative 5′ and 3′ splice sites allow the lengthening or shortening of a particular exon. (*E, F*) Alternative promoters and alternative poly(A) sites switch the 5′- or 3′-most exons of a transcript. (*G*) A retained intron can be excised from the pre-mRNA or can be retained in the translated mRNA. (*H*) A single pre-mRNA can exhibit multiple sites of alternative splicing using different patterns of inclusion. These are often used in a combinatorial manner to produce many different final mRNAs.

many cellular and developmental processes, including sex determination, apoptosis, axon guidance, cell excitation and contraction, and many others. Errors in splicing regulation have been implicated in a number of different disease states. The roles played by alternative splicing in particular cellular processes and in disease have been reviewed extensively (1, 8–20). Here we focus on the common mechanistic features of a number of well-studied model systems.

The excision of the introns from a pre-mRNA and the joining of the exons is directed by special sequences at the intron/exon junctions called splice sites (21). The 5′ splice site marks the exon/intron junction at the 5′ end of the intron (Figure 2*A*). This includes a GU dinucleotide at the intron end encompassed within a larger, less conserved consensus sequence (21). At the other end of the intron, the 3′ splice site region has three conserved sequence elements: the

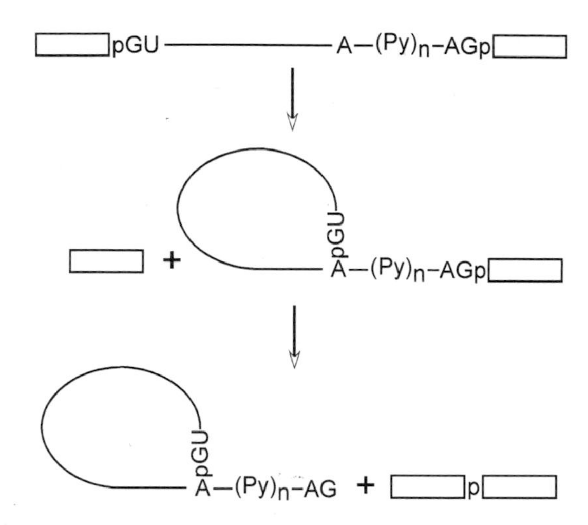

Figure 2*A* Splicing takes place in two transesterification steps. The first step results in two reaction intermediates: the detached 5′ exon and an intron/3′-exon fragment in a lariat structure. The second step ligates the two exons and releases the intron lariat. See text for details.

branch point, followed by a polypyrimidine tract, followed by a terminal AG at the extreme 3′ end of the intron. Splicing is carried out by the spliceosome, a large macromolecular complex that assembles onto these sequences and catalyzes the two transesterification steps of the splicing reaction (Figure 2*A*, Figure 2*B*). In the first step, the 2′-hydroxyl group of a special A residue at the branch point attacks the phosphate at the 5′ splice site. This leads to cleavage of the 5′ exon from the intron and the concerted ligation of the intron 5′ end to the branch-point 2′-hydroxyl. This step produces two reaction intermediates, a detached 5′ exon and an intron/3′-exon fragment in a lariat configuration containing a branched A nucleotide at the branch point. The second transesterification step is the attack on the phosphate at the 3′ end of the intron by the 3′-hydroxyl of the detached exon. This ligates the two exons and releases the intron, still in the form of a lariat.

The spliceosome assembles onto each intron from a set of five small nuclear ribonucleoproteins (snRNPs) and numerous accessory proteins (Figure 2*B*) (21–23, 23a). During assembly, the U1 snRNP binds to the 5′ splice site via base pairing between the splice site and the U1 snRNA. The 3′ splice site elements are bound by a special set of proteins. SF1 is a branch-point binding protein (also called BBP in yeast). The 65-kDa subunit of the dimeric U2 auxiliary factor (U2AF) binds to the polypyrimidine tract. In at least some cases, the 35-kDa subunit of U2AF binds to the AG at the intron/exon junction. The earliest defined complex in spliceosome assembly, called the E (early) or commitment complex, contains U1 and U2AF bound at the two intron ends (24). The E complex is

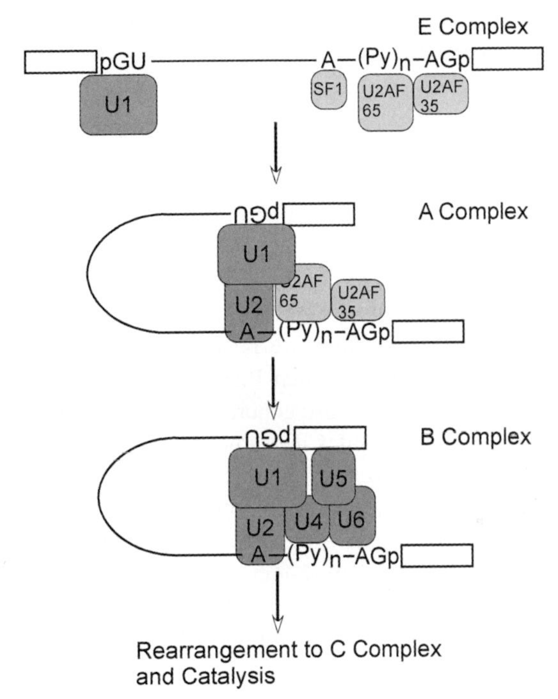

Figure 2*B* The spliceosome contains five small nuclear ribonucleoproteins that assemble onto the intron. The Early (E) complex contains the U1 snRNP bound to the 5′ splice site. Each element of the 3′ splice site is bound by a specific protein, the branch point by SF1 (BBP), the polypyrimidine tract by U2AF 65, and the AG dinucleotide by U2AF 35. This complex also apparently contains the U2 snRNP not yet bound to the branch point (24). The A complex forms when U2 engages the branch point via RNA/RNA base-pairing. This complex is joined by the U4/5/6 Tri-snRNP to form the B complex. The B complex is then extensively rearranged to form the catalytic C complex. During this rearrangement the interactions of the U1 and U4 snRNPs are lost and the U6 snRNP is brought into contact with the 5′ splice site. The spliceosome contains many additional proteins (359–363). Moreover, other pathways of assembly may be possible (23a, 364, 365).

joined by the U2 snRNP, whose snRNA base-pairs at the branch point, to form the A complex. The A complex is joined by the U4/U5/U6 tri-snRNP to form the B complex. The B complex undergoes a complicated rearrangement to form the C complex, in which the U1 snRNP interaction at the 5′ splice site is replaced with the U6 snRNP and the U1 and U4 snRNPs are lost from the complex. It is the C complex that catalyzes the two chemical steps of splicing (Figure 2*B*). There is also a minor class of spliceosome that excises a small family of introns that use different consensus sequences (25). Only the major class of spliceosome is discussed here.

Changes in splice site choice arise from changes in the assembly of the spliceosome. In most systems, splice site choice is thought to be regulated by altering the binding of the initial factors to the pre-mRNA and the formation of early spliceosome complexes. By the time the E complex is formed, it appears that the splice sites are paired in a functional sense and the defined intron is committed to being spliced. However, this need not always be the case and there is evidence in one system for the regulation of 3′ splice site choice after the first catalytic step of splicing [i.e., after branch formation (see below) (26)].

The splice site consensus sequences are generally not sufficient information to determine whether a site will assemble a spliceosome and function in splicing. Other information and interactions are necessary to activate their use (27, 28). Introns can range in size from less than 100 nucleotides to hundreds of thousands of nucleotides. In contrast, exons are generally short and have a fairly narrow size distribution of 50–300 nucleotides. Commonly, spliceosomal components binding on opposite sides of an exon can interact to stimulate excision of the flanking introns (29). This process is called exon definition and apparently occurs in most internal exons (30). On top of this process, there are many non–splice site regulatory sequences that strongly affect spliceosome assembly. RNA elements that act positively to stimulate spliceosome assembly are called splicing enhancers. Exonic splicing enhancers are commonly found even in constitutive exons. Intronic enhancers also occur and appear to differ from exonic enhancers. Conversely, other RNA sequences act as splicing silencers or repressors to block spliceosome assembly and certain splicing choices. Again, these silencers have both exonic and intronic varieties. Some regulatory sequences create an RNA secondary structure that affects splice site recognition (31–33), but most seem to be protein binding sites. This review focuses on the nonspliceosomal pre-mRNA binding proteins that act through these splicing regulatory sequences.

THE SEX DETERMINATION GENES OF *DROSOPHILA* AS MODELS FOR SPLICING REGULATION

Sex Lethal Protein Is a Splicing Repressor

By far the best-understood systems of splicing regulation come from the pathway of somatic sex determination in *Drosophila melanogaster.* A series of remarkable genetic studies identified the key regulators of the sex determination pathway as RNA binding proteins that alter the splicing of particular transcripts (8, 18, 34). These studies provided an essential starting point for the biochemical analysis of splicing regulation (Figure 3).

The master regulatory gene at the top of the sex determination pathway encodes the RNA binding protein Sex lethal (Sxl). Sxl protein is expressed specifically in female flies, where it represses splicing patterns that would lead to male development. In the presence of Sxl, female splicing patterns are expressed, leading to gene products needed for female development. Sex lethal contains two RNA binding domains (RBD) of the RNP consensus type (RNP-cs, also called an RRM) (Figure 4) (35, 36). In the crystal structure of the Sxl protein bound to RNA, the RNA-interacting surfaces of these domains face each other to form a cleft that specifically interacts with a nine-nucleotide U-rich element found in the Sxl-target RNAs (37, 38). An additional N-terminal glycine-rich domain in Sxl influences the cooperative assembly of the protein onto multiple binding sites (39).

The downstream targets of the Sxl protein include transcripts from the *Transformer* (Tra) and *Male-specific lethal 2* (Msl2) genes (Figure 3*A*, Figure 3*B*). The Tra gene also encodes a splicing regulator. In the absence of Sxl, a splicing pattern is used that produces a truncated and inactive Tra protein. In female flies, where Sxl is present, it binds to its recognition element within the 3′ splice site of Tra exon 2, blocking recognition by U2AF (Figure 3*A*) (40–43). This causes a shift in the 3′ splice site to a position downstream, thus deleting a stop codon from the Tra mRNA and allowing translation of active Tra protein. This appears to be the simplest mechanism for altering a splicing pattern: the Sxl protein directly competes with an essential splicing factor (U2AF) for its RNA binding site. However, Sxl regulation of Tra requires two other *Drosophila* genes, *virilizer* and *Female-specific lethal 2D*, so it is likely that there is more to this mechanism than is currently understood (44, 45).

A second target of Sxl is the gene Msl2, which regulates X chromosome dosage compensation in male flies (Figure 3*B*) (46–48). In females, Sxl binds to two sites in the first intron of Msl2 pre-mRNA: one in the polypyrimidine tract and one adjacent to the 5′ splice site. Sxl binding to the polypyrimidine tract again blocks U2AF binding (49). Sxl binding near the 5′ splice site blocks a regulatory factor called TIA-1 (see below) and binding of the U1 snRNP to the 5′ splice site (50, 51). Both Sxl binding sites are needed for the inhibition of splicing and retention of the first Msl2 intron in the final mRNA. Interestingly, the retained Msl2 intron is in the 5′ UTR of the transcript and does not affect its open reading frame. It has been shown that the Sxl bound within this region blocks translation of the transcript in the cytoplasm (52–54). Thus, Sxl is affecting both the splicing of the transcript in the nucleus and its translation in the cytoplasm. There appear to be other examples in animal cells of predominantly nuclear splicing regulators having additional cytoplasmic functions (55–58).

In addition to regulating transcripts downstream in the sex determination pathway, Sxl also autoregulates its own splicing to maintain the female splicing phenotype (8, 18, 34, 59) (Figure 3*C*). In male flies, the Sxl gene is transcribed but the inclusion of Sxl exon 3 introduces a premature stop codon to produce a truncated and inactive protein. In female flies, the same Sxl promoter is active, producing the same RNA precursor. However, in the early female embryo an additional Sxl promoter is briefly active. This early promoter produces a transcript missing exon 3 that is translated into active Sxl protein (60, 61). This embryonic Sxl protein initiates the whole splicing cascade by repressing the

A. Transformer

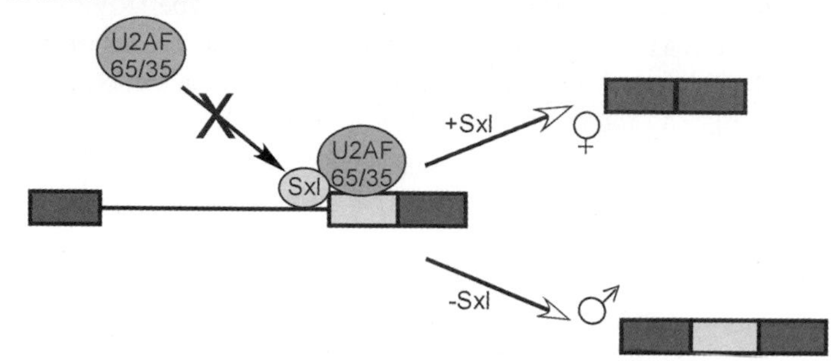

B. Male-specific lethal 2

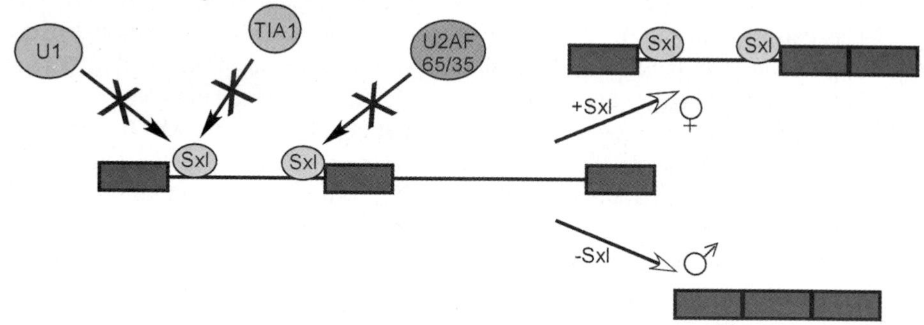

C. Sex lethal autoregulation

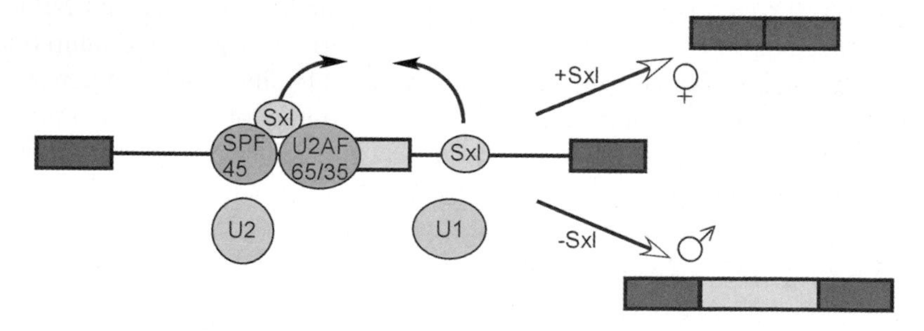

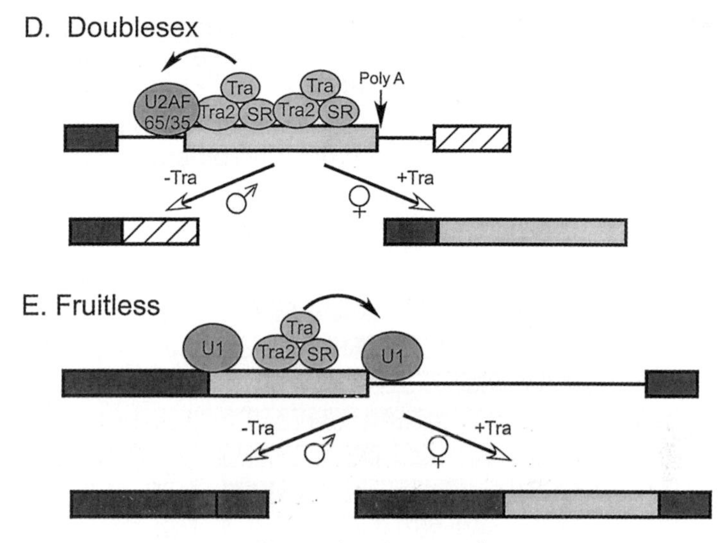

Figure 3 *Continued.*

Figure 3 Splicing regulation in the *Drosophila* sex determination cascade. (*A*) Sex lethal represses a splicing pattern in the Transformer RNA. In male flies, the absence of Sxl protein allows U2AF binding to an upstream 3′ splice site, producing the male spliced product. In female flies, Sxl protein binds to the upstream 3′ splice site, causing splicing in the female-specific pattern and producing an mRNA encoding active Transformer protein. (*B*) Sxl also binds to the Msl2 transcript. Here, Sxl binds both in the 3′ splice site, blocking U2AF binding; and near the 5′ splice site, blocking binding by the TIA-1 protein and the U1 snRNP. This causes retention of the intron in the final Msl2 mRNA in female flies. In male flies, the absence of Sxl allows the mRNA to be fully spliced. (*C*) In the Sxl transcript itself, the 3′ splice site of the third exon contains two AG dinucleotides. In male flies, the downstream AG is bound by U2AF and is required for activation of the exon for splicing. The upstream AG is bound by the SPF45 protein and is the site of exon ligation. Thus, in male flies, exon 3 is included in an mRNA that encodes an inactive Sxl protein. In female flies, Sxl protein derived from the activation of an early embryonic promoter is present. Sxl binds on both sides of the exon to induce exon skipping and the production of additional Sxl protein from the constitutive promoter. Under these conditions, it appears that lariat formation may still occur at the exon 3′ splice site. [See (26, 67)] (*D*) The Tra protein produced in female flies is a positive regulator of Doublesex splicing. Doublesex encodes two different transcription factors. Exon 4 of dsx contains a poor polypyrimidine tract causing exon 4 skipping in male flies. Exon 4 also contains a Tra-dependent splicing enhancer that binds Tra as well as the Tra2 protein and two different SR proteins, in females. This enhancer complex stimulates U2AF binding and splicing of exon 4, producing the female product of the dsx gene. (*E*) The Tra protein also regulates the splicing of Fruitless mRNAs. In male flies, the upstream-most of a pair of alternative 5′ splice sites is used. In female flies, Tra, Tra2, and the SR protein RBP1 bind to an exonic enhancer to activate splicing of the downstream 5′ splice site.

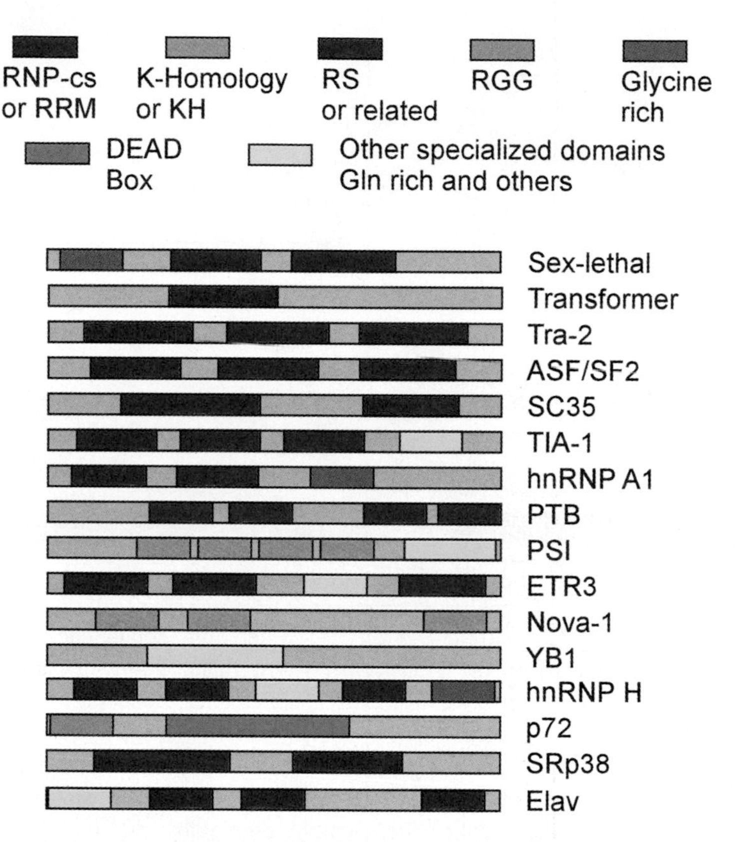

Figure 4 Domain structures of splicing regulatory proteins. Splicing regulators have common domains but combine them in various ways. Nearly all have RNA binding domains of either the RNP-cs (RRM) or KH type, repeated in different numbers and combined with different auxiliary domains such as RS or RS-like domains. Although structures of individual or pairs of domains have been solved, little is known about the overall structure of these multidomain proteins. Shown here are some of the proteins discussed in the text, normalized to equal length. The positions and number of their known domains are indicated by color-coded boxes.

splicing of Sxl exon 3 from the constitutive promoter. The mechanism of Sxl autoregulation is quite different from its activity on Tra and Msl2 (39, 62–64). Multiple Sxl binding sites flank exon 3 and are required for its repression. Sxl protein assembles cooperatively onto these sites and its glycine-rich amino terminus is required for this cooperative binding. In this case, rather than blocking the binding of a particular factor, Sxl is creating an RNP structure that encompasses the whole repressed exon. How this structure actually represses splicing is the subject of some interest. The repression requires the Sans fille protein, the *Drosophila* homolog of the U1A and U2 B″ proteins, which may indicate interactions of Sxl with the snRNPs assembling on both sides of the exon (65, 66).

A recent study has given new insight into the mechanism of Sxl autoregulation (26, 67). Sxl exon 3 has an unusual 3′ splice site containing two AG dinucleotides that flank the polypyrimidine tract (Figure 3*C*). The branch point is upstream from both of these AG dinucleotides. In the absence of Sxl, the polypyrimidine tract is bound by U2AF65, and the downstream AG by U2AF35. Interestingly, this downstream AG is required for the splicing of the exon but is not the site of exon ligation, which occurs at the upstream AG (63). The upstream AG is bound by a factor called SPF45, and in the absence of SPF45, splicing occurs at the downstream AG (26). Thus, for this exon the role of U2AF in promoting spliceosome assembly at the 3′ splice site has been separated from a role in defining the AG for exon ligation. Both in vivo and in vitro, the repression of the exon by Sxl requires SPF45. Most interestingly, in a model pre-mRNA carrying the Sxl 3′ splice site, the first step of splicing (5′ exon cleavage and lariat formation) occurs in vitro in the presence of Sxl, but the second step (3′ splice site cleavage and exon ligation) is blocked. Thus, at least in this in vitro system, Sxl inhibits the second step of splicing but not the first step. Although it needs to be confirmed in vivo on a standard Sxl RNA, this result seems to imply that a spliceosome can be drastically rearranged after the first step of splicing and redirected to a distant 3′ splice site. It will be interesting to find out if 3′ splice site choice can commonly occur so late in the spliceosome assembly pathway, and if such late choice always requires SPF45 and a similar arrangement of AG dinucleotides.

Transformer Is a Splicing Activator

Sex lethal regulates the splicing of the Transformer transcript to produce the female-specific Tra protein that in turn regulates the splicing of the Doublesex (dsx) and Fruitless (fru) mRNAs in the sex determination pathway (8, 18, 34). Unlike Sxl, Tra is a positive regulator of splicing that activates female-specific splicing patterns in its targets (Figure 3*D*, Figure 3*E*).

Tra contains an extended RS domain, rich in arginine/serine dipeptides (Figure 4). These domains are a common feature of splicing regulatory proteins (see below). The Tra protein does not have an RNA binding domain and must cooperatively assemble with other proteins onto its target RNA sequences (68–70). In male flies, Tra is absent and mRNAs of the dsx gene are spliced from exon 3 to exon 5 to encode a transcription factor that determines male differentiation (71, 72). Conversely, the presence of Tra in females activates the splicing of exon 3 to exon 4 and results in an mRNA that encodes a transcriptional regulator leading to female differentiation. Exon 4 of dsx carries an unusually weak 3′ splice site that is not normally recognized by the U2AF protein. Within this exon is a splicing enhancer sequence whose activity is controlled by the presence of Tra protein (73–77). This enhancer has a series of six 13-nucleotide repeat elements and an additional purine-rich element. The repeat element is bound by a trio of proteins including Tra, an SR protein RBP1, and an SR-related protein called Tra2 (76). SR proteins compose an important family of splicing regulators discussed in detail below. Tra2 is not sex specific but is required for Tra-dependent splicing regulation. Unlike Tra, both RBP1 and Tra2 contain RNA binding domains and directly interact with the repeat sequence. Tra is required for Tra2 binding to the repeat and the cooperative assembly of the complex. Interestingly, the purine-rich element also binds Tra and Tra2, but in this case they cooperate with a different SR protein: dSRp30 (76). Via its interactions with Tra and two different SR proteins, Tra2 is induced to bind to two different regulatory sequences. The assembly of these complexes onto the dsx splicing enhancer is thought to stabilize U2AF binding to the upstream 3′ splice site, leading to spliceosome assembly and splicing at exon 4 [(78); also see below and (79, 80) for discussion]. Tra and its assembly with SR proteins and Tra2 is the major paradigm for how exonic splicing enhancers work. However, the nature of the interactions that lead to U2AF binding is not completely clear and there is much to be learned about how splicing enhancement actually takes place.

Tra also positively regulates the splicing of the Fruitless gene transcript (Figure 3*E*) (81– 83). An internal exon of fru has a pair of alternative 5′ splice sites. In male flies, the upstream 5′ splice site is used. This gene product, encoding a BTB-ZF transcription factor, goes on to regulate male courtship behavior and other aspects of male sexual development. In females, Tra and Tra2 activate the use of a downstream 5′ splice site, leading to a different mRNA. Like dsx, this activation of the female 5′ splice site also appears to involve RBP1 and requires several copies of the same repeated element found in the dsx enhancer. However, here the repeats are just upstream of the activated 5′ splice site. Thus, Tra and its cofactors Tra2 and RBP1 can act through exonic splicing enhancers to stimulate splicing at either a 3′ splice site or a 5′ splice site.

COMPLEX SYSTEMS OF TISSUE-SPECIFIC SPLICING

The *Drosophila* sex determination pathway has provided the central examples of how a choice between two possible splicing patterns is regulated. In addition to alterations by sex, metazoan organisms regulate the splicing of thousands of other transcripts depending on cell type, developmental state, or external stimulus. The analysis of these systems, mostly in mammalian cells, has identified many of the same kinds of proteins seen in the fly sex determination pathway (34, 69, 70). However, these proteins are often combined in complex ways into multiple layers of regulation. Many of these proteins can act either positively or negatively depending on their binding context. We discuss individual proteins and regulatory elements according to their location in introns or exons and then describe the challenges to understanding how they are combined to give a precise pattern of regulation.

Exonic Regulatory Elements and Proteins That Bind to Them

POSITIVE REGULATION FROM EXONS Exons often contain enhancer or silencer elements that affect their ability to be spliced. There are many exonic splicing enhancers (ESEs), similar to the dsx enhancer. These RNA regulatory elements are diverse in sequence and often embedded within nucleotides that also code for protein. Such enhancers have been identified through exon mutations that block splicing, through computational comparisons of exon sequences, and through the selection of sequences that activate splicing or that bind to splicing regulatory proteins—most notably the SR proteins (12, 20, 68, 84, 85).

The SR proteins constitute the best-studied family of splicing regulators (80, 86, 87). The SR proteins have a common domain structure of one or two RNP-cs RNA binding domains followed by what is called an RS domain containing repeated arginine/serine dipeptides (Figure 4). The serines in an RS domain can be highly phosphorylated. The family includes the proteins SRp20, SRp30c, 9G8, SRp40, SRp55, SRp70, and the best-studied members ASF/SF2 and SC35 (80, 88). The SR family is also defined by particular properties in splicing (87). Additional proteins in the splicing reaction have RS domains but serve different roles. These include U2AF, U1 70K, SRm160/300, Tra2, and numerous others. Some of these SR-related proteins are general splicing factors and some are apparently specific inhibitors of splicing (89–92).

The true SR proteins have a wide range of activities in the splicing reaction (80). Splicing in cellular extracts requires the presence of at least one member of the family, and the different members are generally interchangeable in their ability to fulfill this requirement. However, the proteins are not always interchangeable in vivo (93–96). A separate activity of these proteins that we focus on here is the activation of splicing through exonic splicing enhancers. This activity was first seen in the dsx enhancer as described above. It has become clear that many if not all exons, constitutive or regulated, contain ESE elements that bind to specific members of the SR family (12). The presence of these elements seems to be a general mechanism for defining exons.

Most naturally occurring ESEs have been shown to bind specific SR proteins (12, 20). The most commonly studied are purine-rich sequences, sometimes given the consensus sequence $(GAR)_n$, that are bound by the proteins ASF/SF2 and Tra2 (97, 98). In addition to naturally occurring enhancers, in vitro selection has identified many optimal binding sequences for different SR family members. The binding sites for a given family member can be fairly degenerate. For example, only some of the selected sequences for ASF/SF2 are purine rich and show similarity to the GAR element (97). When these optimal sites are introduced into model splicing substrates, they act as SR protein–dependent splicing enhancers. The families of SR protein recognition sequences generated through in vitro selection experiments have proven to be an effective means of predicting the location of ESEs in natural exons (12, 97, 99).

It is not clear whether the activity of all ESEs is SR protein dependent. A recent computational analysis identified a number of elements with ESE activity (85). Similarly, functional selection of sequences that activate splicing has identified additional elements (84, 100–103). Some are known to be SR protein dependent; it will be interesting to see if any are not. An AC-rich enhancer element is apparently mediated by two non-SR proteins, YB-1 and p72. p72 is a member of the DEAD Box RNA helicase family, raising interesting questions about the mechanisms of splicing activation (104, 105).

The two domains of a SR protein are modular in function. The RBD targets the protein to a particular exonic element, with the different RBDs targeting different sequences (106–109). This RNP-cs domain can be replaced with the RNA binding MS2 coat protein from the MS2 bacteriophage. When the MS2 coat protein binding site was placed in an enhancer-dependent exon, an MS2/RS domain fusion protein activated splicing in vitro (110). Thus, just tethering the RS domain to the exon can activate splicing. The RS domains themselves are largely interchangeable (106, 111). One can switch the RS domains of different proteins and they will maintain their activity, both in vitro and in vivo. Moreover, a natural RS domain can be replaced with a synthetic sequence of 10 RS dipeptides and the protein will maintain its activity in at least some assays (112). RS domains have several proposed functions (80). The unphosphorylated domain is highly positively charged and may act as a counter ion to enhance protein affinity for RNA or RNA/RNA hybridization. However, phosphorylation is required for activity in splicing (113–118), and there is clear evidence for phosphorylated RS domains engaging in protein/protein interactions (119). In biochemical experiments, ASF/SF2 was shown to interact with the RS domain containing U1 snRNP protein, U1 70K (120). Yeast two-hybrid experiments showed that the RS domain could act as a protein/protein interaction domain for binding to other SR proteins (121). It was further shown that this interaction in yeast required the presence of the SR protein kinase and presumably phosphorylation (122). Since RS domains are largely interchangeable in these assays, it may be that any RS domain can bind to any other. However, it has not been shown that two RS domains directly contact each other in these interactions rather than contacting other protein domains. Indeed, other tests of SR protein function both in vivo and in vitro indicate that there is specificity to these protein/protein interactions that may be determined by contacts outside of the RS domain (111, 123, 124).

There are many questions about the mechanisms of SR protein action. As described above, an exonic enhancer can stimulate U2AF or U2 binding to weak 3′ splice sites, and U1 snRNP binding to 5′ splice sites (78, 79, 120, 125–129) (Figure 5). However, in naturally occurring enhancers, SR proteins apparently bind as a component of a large, multiprotein complex, as seen with dsx. The components of these exon complexes are not all identified but can include such known splicing factors as the large RS domain proteins SRm160/300, the mammalian homolog of Tra2, the U1 snRNP, and the heterogeneous nuclear (hn)

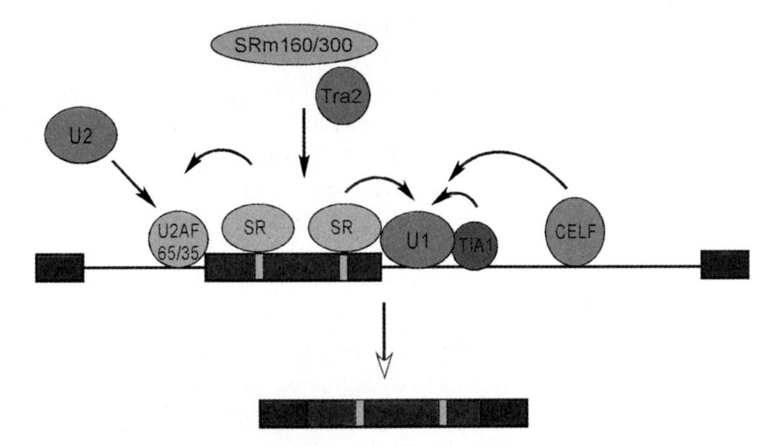

Figure 5 Mechanisms of splicing activation. Splicing activators are generally thought to interact with components of the spliceosome to stabilize their binding to adjacent splice sites. SR proteins bind to exonic splicing enhancer elements (green boxes) to stimulate U2AF binding to the upstream 3′ splice site, or U1 snRNP binding to the downstream 5′ splice site. SR proteins often need other factors to function, such as the SRm160/300 proteins or Tra2. Proteins that bind to intronic splicing enhancers include TIA-1 and the CELF proteins. TIA-1 binds immediately downstream from the 5′ splice site to stimulate U1 binding. CELF protein binding sites can be more distant from the regulated exon and it is not known how they might interact with the spliceosome.

RNPs A1 and H (see below). Thus, the particular interactions of the SR protein that activate splicing are not entirely clear. Splicing activation by the dsx enhancer in vitro requires the RS domain on U2AF 35 (78). However, this domain is not required for splicing regulation in vivo, indicating that additional interactions take place between U2AF and the enhancer complex (79, 130). Splicing enhancement by at least some ESEs requires SRm160 and 300, two large splicing factors that also contain RS domains (131–133). The enhancer complex thus contains multiple RS domains, and which factors directly interact is not fully known. Careful kinetic analyses of the rate of splicing activated by repeated ESEs indicate that only one SR protein complex at a time can interact with the spliceosome at the 3′ splice site (134). More structural data regarding the protein/protein interactions in these complexes is needed to develop more precise models of ESE function. Finally, in some systems, enhancer-bound SR proteins may stimulate splicing by counteracting repressor molecules rather than enhancing U2AF binding (135, 136).

One of the most interesting aspects of SR protein function is their phosphorylation. Both hyper- and hypophosphorylation of SR proteins seems to inhibit their activity in vitro (113–115). In several biological contexts, the phosphorylation of SR proteins correlates with their activity (137, 138). Several kinases have been identified that phosphorylate SR proteins. The SR protein kinases

(SRPK) 1 and 2 are conserved from humans to yeast, and the activity of these kinases alters SR protein localization and protein/protein interactions (122, 140–145). The Clk/Sty group of kinases also phosphorylate SR proteins. They show a different pattern of preferred phosphorylation sites than the SRPKs and affect in vitro splicing activity, protein localization, and U1 70K binding by ASF/SF2 (113, 119, 146, 147). Genetic studies of a *Drosophila* homolog of Clk/Sty called Doa clearly implicate the kinase in splicing regulation (148). Doa phosphorylates the *Drosophila* SR protein RBP1 as well as Tra and Tra2. Doa mutations block enhancer-dependent splicing of dsx exon 4. Interestingly, the splicing of Fruitless is not affected by Doa mutations. Thus, splicing events that use common regulators have different phosphorylation dependence for those regulators. As with other responses to signaling pathways, a widely expressed protein like RBP1 or Tra2 might control very precisely regulated splicing through its own precise modifications.

Negative Regulation in Exons

In opposition to the positive effects of exonic enhancers, exonic silencer or repressor elements have been identified. The best characterized of these are bound by particular hnRNP proteins. The hnRNP proteins are a large group of molecules identified by their association with unspliced mRNA precursors (hnRNA), and are not a single family of related proteins (149).

The most studied of these proteins, hnRNP A1, has been implicated in several processes including splicing and, oddly, the maintenance of telomere length (150–152). hnRNP A1 contains two RNP-cs RNA binding domains and a glycine-rich auxiliary domain (Figure 4). It belongs to a family of related proteins arising from both multiple genes and alternative splicing (153–155). A crystal structure of the two RBDs of A1 bound to the telomeric repeat DNA $d(TTAGGG)_2$ shows that an A1 protein dimer forms a four-RNP-domain surface, which binds two DNAs or four copies of the element (156). The topology of each DNA strand extends from the N-terminal RBD of one dimer subunit to the C-terminal RBD of the other subunit. If similar interactions occur with RNA elements in the pre-mRNA, this structure would allow cooperative binding to multiple splicing silencer elements and looping of the RNA between binding elements. This A1 structure is also remarkably different from the double-RNP domain structures of Sxl and nucleolin (38, 157). It seems that the many multi-RNP domain proteins vary considerably in their RNA binding, oligomerization, and orientation of RNP-cs domains.

A1 was originally implicated in splicing as a factor that counteracted SR proteins in an in vitro assay of splice site shifting (158, 159). Several transcripts have splicing patterns that are sensitive to the relative ratio of A1 to ASF/SF2, which changes between different tissues (160, 161). A1 has since been shown to bind to exonic splicing silencers (ESSs) in the HIV, FGFR2, and other transcripts (162, 163). The protein is required for the silencing effects of these sequences in vitro. The silencing can be recapitulated by tethering just the glycine-rich domain

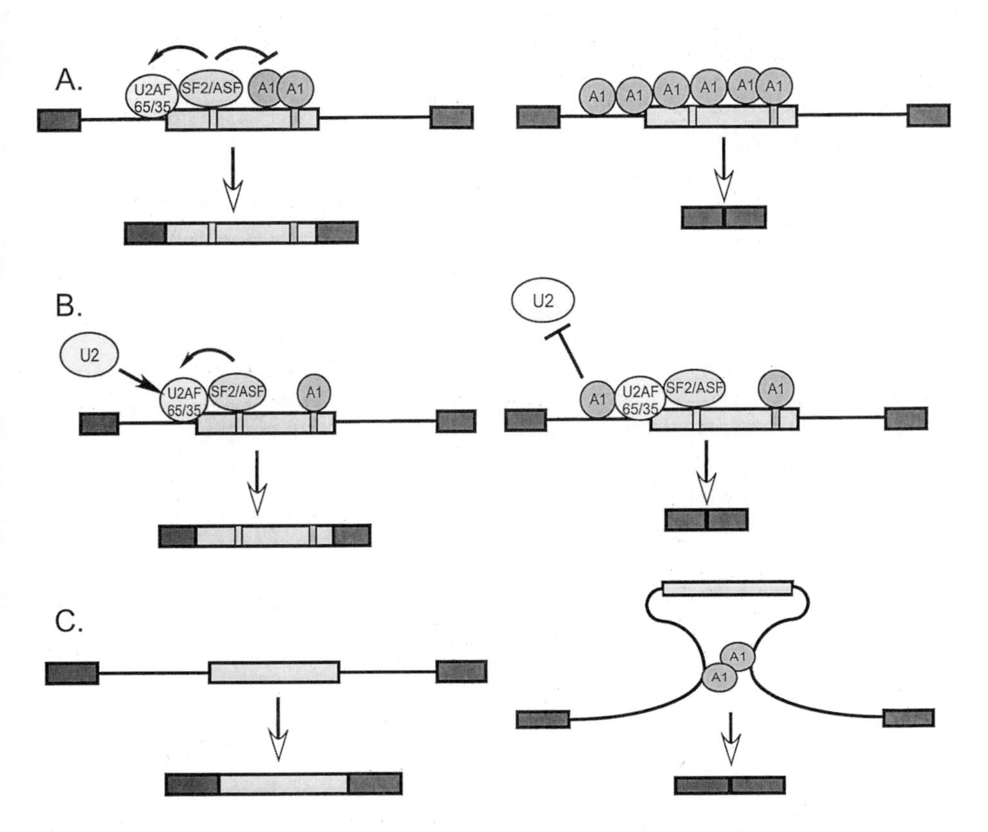

Figure 6 Models for splicing repression by hnRNP A1. (*A*) In HIV Tat exon 3, specific A1 binding to an ESS is thought to nucleate the assembly of additional A1 molecules along the RNA, creating a zone of RNA where spliceosome assembly is repressed. The A1 repression can be blocked by the strong binding of SR proteins to ESEs, which presumably also stimulates spliceosome assembly at the upstream 3′ splice site, and allows exon inclusion. (*B*) There is an additional A1 binding site adjacent to the branch point for Tat exon 3 that blocks splicing in conjunction with the exonic A1 sites. A1 binding to this intronic element does not block U2AF binding to the 3′ splice site but does block U2 snRNP binding to the branch point. (*C*) hnRNP A1 also represses an exon in its own transcript using intronic binding sites. A1 bound to these sites can multimerize, thus looping out the exon and causing exon skipping.

to the exon via an MS2 fusion (163). Several mechanisms have been proposed for A1-mediated splicing repression, and its mechanism may differ between different transcripts (Figure 6). A1 could interfere directly with the assembly of spliceosomal components, it could block the exon bridging interactions that occur during exon definition, or it could block splicing activation by SR proteins binding to adjacent ESEs. There is evidence for all of these activities. A1-dependent silencer elements have also been found in introns (see below).

Several groups have studied the role of A1 in repressing exons in the HIV precursor RNA. The mechanistic analysis is most advanced for Tat exon 3. This exon contains several enhancer elements that bind the SR proteins SF2/ASF and SC35 (136, 164–167). There are also A1 binding sites both within the exon and adjacent to a site of branch formation upstream. In studies of this exon, the Krainer lab focused on the exonic elements that bind A1, SC35, and SF2/ASF (136). They used an S100 extract that is depleted for all the SR proteins and further depleted it for A1. In the absence of A1, either SC35 or SF2/ASF can activate Tat exon 3 splicing. The addition of A1 specifically inhibited SC35- but not SF2/ASF-activated splicing. It was found that in addition to binding the ESS, A1 crosslinked to exonic RNA distal to the ESS in the region of the SR protein binding sites. This crosslinking occurred only if the ESS was present. It was proposed that specific A1 binding at the ESS nucleates the cooperative nonspecific binding of A1 upstream. This higher order complex of multiple A1s would create a zone of inhibition along the RNA. SF2/ASF is proposed to block the propagation of this complex, whereas SC35 does not bind tightly enough to do this (Figure 6*A*) (136). This model is appealing because it provides a rationale for two long-known modes of RNA binding by A1: specific binding to particular short RNA elements and non-sequence-specific binding to longer RNAs. On the other hand, this crosslinking experiment could also be interpreted to support cooperative binding to separated specific sites, as has been proposed for A1 autoregulation (see below). There are additional effects of A1 in inhibiting HIV Tat splicing through blocking U2 assembly [(168); see below]. Additional experiments are needed to create a model that incorporates exonic A1 effects on SR protein activity and effects from other binding sites.

In addition to A1, other exonic splicing silencers have been shown to bind hnRNP H and its close relative hnRNP F (169, 170). H and F are a different structural family from A1 and contain three RNP-cs-type RNA binding domains that recognize G-rich elements (171, 172). Interestingly, hnRNP H acts as a splicing repressor when bound to an ESS in β-tropomyosin, but as an activator when bound to a similar element in HIV Tat exon 2 (170, 173). It also binds to splicing regulatory elements in introns.

Intronic Regulatory Elements

Many splicing regulatory sequences are present in introns rather than exons. Binding sites for regulators are often found within the polypyrimidine tract or immediately adjacent to the branch point or 5′ splice site. However, splicing regulatory elements can also act from a distance, being found hundreds of nucleotides away from the regulated exon. As in exonic regulation, positive- and negative-acting sequences compose intronic splicing enhancers and silencers respectively (ISEs and ISSs). Also like the exonic elements, groups of elements are often found clustered to make composite regulatory sequences that mediate both positive and negative regulation. These regulatory regions can be highly conserved between species, and are often identified by sequence alignments.

Positive Regulation from Introns

Several elements are known to act as ISEs, but the proteins that mediate their effects are less well characterized than for ESEs. Some ISEs do appear to be SR protein dependent (174, 175). In other cases, SR proteins do not appear to directly bind to the regulatory element, and at least some ISEs appear to be mechanistically different from ESE sequences.

Some 5′ splice sites are activated for splicing by a uridine-rich sequence immediately downstream. The regulated intron of *Drosophila* Msl2 requires this element for splicing in heterologous HeLa cell extracts (50). Similarly, the K-SAM exon in the FGFR 2 transcript is activated by a U-rich element, called IAS1, immediately adjacent to the K-SAM 5′ splice site (176). In both cases, this element was found to bind the protein TIA-1 (Figure 4, Figure 5). Extracts depleted for TIA-1 are inhibited for Msl2 splicing. Overexpression of TIA-1 induces K-SAM splicing in transfected cells. Significantly, TIA-1 stimulates U1 snRNP binding to 5′ splice sites that are dependent on the U-rich element for function (50). So far, this is the only intronic enhancer protein shown to directly affect spliceosome assembly. TIA-1 and its relative TIAR contain three RNP-cs domains and a C-terminal prion-like aggregation domain (Figure 4). Interestingly, the yeast homolog of TIA-1, NAM8, is required for the function of certain regulated 5′ splice sites (177, 178). However, yeast NAM8 is a component protein of the U1 snRNP, whereas in animal cells TIA-1 appears to mainly exist as a separate factor. TIA-1 and TIAR were discovered as regulators of apoptosis, and as components of the cytoplasmic mRNA granules formed in response to cellular stress (179). It would be interesting to find a relationship between these various functions.

The CUGBP and ETR-like factors (CELF) are a large family of proteins generated from both multiple genes and alternative splicing (180, 181). These proteins have been implicated in many different aspects of RNA metabolism and given many different names, including the Bruno-like factors in *Drosophila* and the *Drosophila* splicing factor Elav (182–185) (Figure 4). At least some members of this family activate splicing through intronic enhancer elements. One target of these proteins is chicken cardiac troponin T (TnT) (180, 186). TnT exon 5 is spliced in embryonic muscle but excluded from the TnT mRNA in the adult. This exon has been extensively analyzed and is controlled by a complex set of regulatory elements in both the exon and the flanking introns. Some of the positive intronic elements, called muscle-specific enhancers (MSEs), contain UG elements that bind to CELF family members, including ETR3 (180) (Figure 5). The overexpression of ETR3 in transfected cells, or the addition of ETR3 to in vitro splicing extracts, increases TnT exon 5 splicing in an MSE-dependent manner. Interestingly, the loss of exon 5 inclusion during muscle development coincides with a change in the expressed isoforms of ETR3, which may be a key factor in determining exon 5 splicing during development (180).

There are a large number of CELF family proteins, and their effects are not limited to muscle cells. The NAPOR1 protein is a splice variant of ETR3 that is enriched in portions of the nervous system (187, 188). Overexpression of NAPOR in cells has divergent effects on two *N*-methyl-D-aspartate (NMDA) receptor 1 (NR1) exons (188). NR1 exon 5 splicing is decreased by NAPOR, whereas exon 21 splicing is increased. This effect also requires particular intronic elements. Interestingly, the relative inclusion of exons 5 and 21 strongly correlates in vivo with expression of NAPOR. In the forebrain, where NAPOR is highly expressed, NR1 exon 5 is mostly excluded from the mRNA and exon 21 is included. The reverse occurs in the cerebellum, where NAPOR expression is low, exon 5 is included, and exon 21 is excluded. Thus, whether this protein is a positive-acting factor or a negative one depends on the exon and cellular context. The CELF proteins as a family are widely expressed and apparently take part in a wide variety of cellular activities. Models for the function of these proteins in splicing and their effects on spliceosome assembly will await better biochemical assays.

Another common ISE element is the hexanucleotide UGCAUG. This element, particularly when duplicated, strongly enhances splicing (189–191). The element is known to enhance the splicing of exons in the c-src, fibronectin, nonmuscle myosin heavy chain (NMHC), and calcitonin transcripts (189, 190, 192–195). For several enhancers, mutagenesis analyses indicate this to be the key element in their activity. It is usually found repeated several times downstream of the activated exon with some elements being found at some distance from the exon (>500 nt) (191). UGCAUG was also shown to be the most common hexanucleotide downstream of a set of neuronally regulated exons (196). Although associated with tissue-specific exons, the enhancement effect of the element is not necessarily tissue specific (189, 190). The protein that mediates the effects of the UGCAUG hexanucleotide has not been identified definitively.

Intronic enhancers are often made up of intricate combinations of positive and negative elements that assemble into large RNP complexes. The enhancer downstream of the neural-specific exon N1 in the *c-src* transcript is an example of such a structure (Figure 7) (190, 192, 197). The most conserved portion of the enhancer, called the downstream control sequence (DCS), contains a GGGGG element needed for full enhancer activity, a CUCUCU element required for splicing repression, and the UGCAUG essential for enhancer activity (190, 198). In its native context, the DCS needs additional surrounding elements for enhancer activity. Alternatively, two copies of the DCS by itself, or three copies of the UGCAUG hexanucleotide, are sufficient for strong splicing enhancement. The DCS assembles a large RNP complex that has been analyzed in some detail (172, 199–202). The GGGGG element binds to hnRNPs H and F. The CUCUCU element binds to the polypyrimidine tract binding protein or its neuronal homolog (PTB/nPTB; see below). The UGCAUG binds to the KH-type splicing regulatory protein (KSRP) and at least one unidentified factor. RNA competition and immunodepletion or inhibition experiments indicate positive roles for

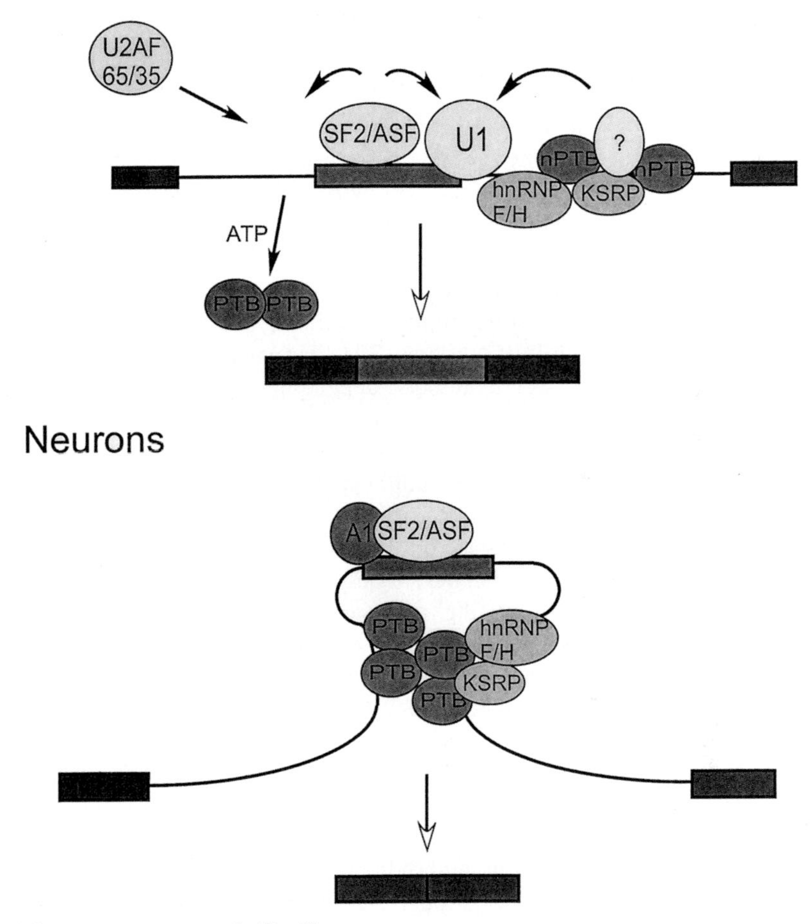

Figure 7 Combinatorial control of splicing in the c-src N1 exon. Tissue-specific exons use a combination of positive and negative inputs to maintain their regulation. The N1 exon of the c-src gene is repressed by the PTB protein in non-neuronal cells where it binds to silencer elements on both sides of the exon and is thought to form an RNA looping complex. PTB binds in the downstream intron within a large complex, containing the KSRP and hnRNP F/H proteins. In neurons, the PTB is replaced with the related nPTB protein, and PTB binding to the upstream silencer elements is destabilized by an ATP-dependent activity. Under these conditions, the downstream regulatory region can act as a splicing enhancer. The stimulatory proteins for this enhancer are not yet known, but require the UGCAUG element. The N1 exon itself also contains regulatory elements that bind to the ASF/SF2 and hnRNP A1 proteins. Similar models have been developed for most other tissue-specific exons that have been examined.

hnRNPs F/H and KSRP in N1 exon splicing. However, it is not clear that these proteins are responsible for the strong enhancing effect of the UGCAUG.

Negative Regulation from Introns

There are also intronic silencer elements. Some are bound by proteins that we have already discussed. SR proteins have been shown to bind to an intronic sequence near a branch point in the adenovirus L1 mRNA (203). The binding of SR proteins to this sequence early in infection blocks the use of this 3′ splice site and shifts splicing to an adjacent site. Late in infection, the cellular SR proteins become dephosphorylated and inactive in splicing repression, thus allowing the site to be used (114). Similarly, an intronic binding site for SRp30c is inhibitory for the splicing of an hnRNP A1 exon (204). It appears that SR proteins can be either splicing activators or repressors, depending on where in the pre-mRNA they bind.

hnRNP A1 also binds to intronic elements as well as exonic (Figure 6*B*, *C*). All of the A1 binding sites are inhibitory for splicing, unlike those for SR proteins. The Kjems group examined an A1 binding site adjacent to the branch point for HIV Tat exon 3, the same HIV exon analyzed for its A1-binding exonic silencer (see above) (168). As in the studies of the exonic elements, it was found that A1 is required for splicing repression. However, in this system, full repression required both the exonic and the branch-point elements. Interestingly, U2AF still binds to the polypyrimidine tract in an A1-repressed transcript, but splicing is blocked at the assembly of U2 snRNP at the branch point. This may imply, at least for transcripts containing the branch-point A1 binding site, that A1 is also acting at a later step than SR protein–mediated splicing enhancement.

A1 also autoregulates the splicing of its own transcript, but this appears to involve a different mechanism than that for HIV Tat (205) (Figure 6*C*). A cassette exon, 7b, is skipped in the production of A1, but included in the mRNA for A1b. This exon is regulated through A1 binding sites in the introns surrounding it. Using a truncated RNA to model the splicing shift in vitro, it was shown that competing away A1 protein would activate 7b splicing, and that addition of recombinant A1 could reinhibit 7b splicing. A1 binding sites on both sides of the repressed exon are needed for full inhibition. This led to a model in which A1 uses its dimerization ability to interact with the two elements simultaneously and loop out the region containing the repressed exon (Figure 6*C*). It was shown that A1 immobilized on an RNA affinity column could indeed interact with two RNA binding sites simultaneously. This simultaneous binding of two sites required the glycine-rich C-terminal domain of the protein. In the crystal structure, a dimeric A1 protein is missing the Gly domain (206). Thus the Gly domain may do more than form the dimerization interface, and the stoichiometry of the A1 in the RNA looping complex is not clear. This looping model is interesting for several reasons. First, other splicing regulators may also use a looping-out mechanism to sequester exons or splice sites away from the splicing apparatus. Second, such a loop provides an appealing model for the splicing of long introns. Introns can be extremely long and it is difficult to understand how the ends will find each other during spliceosome assembly. The presence of intronic A1 sites would allow large portions of intron to be looped out and kept away from the decision of where to splice. It will be interesting to develop other assays for this intron bridging effect, and to examine its role in the splicing of other transcripts.

Besides A1, the other splicing repressor commonly found associated with regulated exons is the polypyrimidine tract binding protein, PTB (also discovered as hnRNP I) (207–210). PTB contains four RNP-cs domains, as well as conserved N-terminal and linker peptides (Figure 4). PTB has been extensively studied for its roles in splicing regulation and in other processes such as viral translation (211, 212). In vitro selection experiments indicate that PTB optimally binds UUCU elements placed within a larger pyrimidine-rich region, although other combinations of cytosines and uridines also bind well (213, 214). PTB forms a dimer in solution containing eight RBDs (215, 216). It is not clear how many CU-rich elements are engaged by a protein monomer. How these many domains combine to interact with the RNA is an interesting unanswered question (215–218).

PTB has been implicated in the repression of a wide range of vertebrate tissue-specific exons (69, 211, 212). The depletion of PTB from in vitro splicing extracts can activate the splicing of such exons (202, 219). Added-back PTB re-represses splicing, and this repression is dependent on the presence of PTB binding sites in the pre-mRNA. Depletion of PTB in vivo by RNA interference (RNAi) also leads to increased exon inclusion for several exons (219a). PTB binding sites are often found in the polypyrimidine tract of regulated 3′ splice sites, but essential PTB repressor sites are also present elsewhere. The location of these binding sites has given rise to models for PTB-mediated splicing repression that are reminiscent of those for Sex lethal and hnRNP A1 (1, 212). In some transcripts, the high affinity of PTB for the polypyrimidine tracts is thought to allow PTB to outcompete U2AF for binding and thus inhibit splicing. Other transcripts have additional required PTB binding sites that are often in the intron downstream of the regulated exon. These additional PTB elements are frequently clustered with other, sometimes positive-acting, splicing regulatory elements (212). For these transcripts, PTB is thought to assemble into a higher order complex that bridges the two sets of sites (Figure 7) (212). Similar to the A1 model above, this will loop out the repressed exon and presumably sequester it from the splicing machinery. However, the simultaneous binding of two sites by a PTB monomer or dimer, as seen with A1, has not been demonstrated with PTB. Thus, the actual contacts that form the bridge are not clear and additional proteins could be involved. Moreover, the clustering of PTB sites with enhancer elements indicates that the PTB may also prevent enhancer activity.

Though not ubiquitous, PTB is a widely expressed protein. Interestingly, PTB is most often implicated in the repression of a highly tissue-specific exon in all cells outside of one particular tissue (69, 211, 212). For example, the rat β-tropomyosin exon 7 is apparently repressed by PTB in all tissues except skeletal muscle (220). A smooth muscle–specific exon in α-actinin is repressed by PTB outside of smooth muscle cells (219). The neuron-specific N1 exon of c-src is apparently repressed by PTB everywhere except neurons (221). An exception to this is the third exon of rat α-tropomyosin, which is repressed by PTB specifically in smooth muscle, and active elsewhere (222, 223). Thus, PTB most often appears to be a general repressor whose effects are specifically blocked for particular transcripts in particular tissues.

Mechanisms that allow release of PTB repression are likely to play an important role in the tissue specificity of splicing. Several mechanisms for the release of PTB repression are proposed. PTB concentration could simply be reduced in specific tissues; for example, PTB levels in neurons are low (224, 225). Alternatively, splicing activators such as CELF proteins that compete with PTB could be specifically expressed (186). This is similar to splicing regulation by the ratio of SF2/ASF to hnRNP A1 that affects a number of transcripts (161). Finally, proteins that block PTB action can be expressed. These could include inactive or dominant-negative splice variants of PTB or tissue-specific PTB homologs that bind similar RNA elements (172, 186, 224, 226, 228, 229). These mechanisms are not mutually exclusive and all may occur in different systems. In the case of the c-src N1 exon, the loss of PTB binding to repressor elements seems to be an active process that requires ATP. There are at least four PTB binding elements, two in the N1 3′ splice site and two within a splicing enhancer downstream (Figure 7). In neural extracts, the neural-specific homolog of PTB (called nPTB or brPTB) replaces the standard PTB in binding to the downstream repressor sites (172, 224, 228). Interestingly, in these extracts nPTB/PTB still crosslinks to the upstream sites within the active 3′ splice site (202). However, when ATP is added to the extract to initiate splicing, this PTB is stripped off. It is thought that the nPTB changes the PTB interaction with the repressor sites, allowing some general ATP-requiring activity to mediate its removal.

A better mechanistic understanding of PTB-mediated splicing repression will require better structural information about the nature of the repressed complex, better biochemical assays for splicing repression, and genetic data on the loss of PTB activity in specific tissues.

Multifactorial Systems of Splicing Control

One feature of the *Drosophila* sex determination transcripts appears to differ from tissue-specific splicing systems. For Sxl, Tra, and dsx splicing, an apparently default splicing pattern (male) is chosen in the absence of a regulatory protein. The splicing pattern can be shifted to the female pattern by the introduction of a single specific factor, Tra or Sxl. In contrast, many systems of tissue-specific splicing appear to be without a true default choice, or single determinative factor, but instead are under a combination of positive and negative control by factors that are fairly widely expressed (69). This combinatorial control is apparent for both mammalian and *Drosophila* transcripts, and may reflect a condition in which the ratio of splicing patterns must be adjusted under many different conditions or in multiple cell types.

To exemplify the many factors that contribute to a single splicing choice, we describe just two systems, but similar results have been obtained in several others (69, 176, 219, 230, 232, 233). Both chicken cardiac troponin T and mouse c-src mRNAs contain exons whose inclusion is limited to a specific tissue. Chicken cTnT exon 5 is spliced in embryonic muscle but skipped in adult muscle (234). c-src exon N1 is spliced in neurons but skipped in non-neuronal cells (Figure 7) (235, 236). Both exons are short (30 and 18 nucleotides respectively), which can lead to exon skipping, perhaps resulting from a loss of exon definition interactions between the flanking splice sites (237, 238). Improving the splice sites to better match the consensus, or increasing the exon length, can increase the unregulated inclusion of a short exon (237–240). Thus, both exons have features that limit their recognition by constitutive splicing factors. Both cTnT exon 5 and c-src N1 have ESEs that bind to SR proteins, including SF2/ASF, SRp40, SRp55, and SRp75 for TnT exon 5 and SF2/ASF for N1, and that stimulate the splicing of these exons (239, 241, 241a). The N1 exonic enhancer also binds hnRNPs A1 and H (241a). Both cTnT exon 5 and c-src N1 contain numerous positive and negative regulatory elements in their flanking introns (Figure 7) (190, 192, 197, 242–244). For both exons, a limited combination of these elements is sufficient to determine proper tissue specific splicing of a heterologous exon. For TnT exon 5, members of the CELF protein family bind to some of the intronic enhancer elements and activate splicing of the exon (180, 186). For c-src N1, CELF family members do not seem to activate splicing and the enhancer activity is dependent on a UGCAUG element, whose mediating factor is unknown (190). In both exons, PTB binding sites in the flanking introns repress splicing, and PTB is proposed to be a general repressor that turns off splicing in most tissues (186, 202, 221).

In the face of all these general positive and negative influences, how the precise tissue specificity of these exons is produced is an interesting question. For cTnT exon 5, an embryonic muscle variant of ETR3 is thought to counteract PTB and activate splicing specifically in embryonic muscle (180, 186). An ETR3 isoform is seen in immunoblot experiments whose developmental expression correlates beautifully with splicing of the exon. However, the identity of this isoform is not known and it is not clear why other CELF proteins, which activate the exon in transfection assays, do not show activity on the endogenous transcript. The enhancer for the N1 exon is active in more than just neuronal cells and is thus not the only source of tissue specificity (197). Instead, part of the specificity of splicing is thought to derive from the specific loss of PTB-mediated repression in neurons (Figure 7) (202). This may be due to the substitution of PTB with its neuronal homolog nPTB, which does not repress N1 splicing in vitro (172). However, nPTB has not yet been shown to affect N1 splicing in vivo. Thus, for both cTnT exon 5 and src N1, the decision to splice results from a balance of multiple positive and negative inputs. Protein overexpression experiments with wild-type and dominant-negative mutant proteins indicate that exon inclusion can be increased by the loss of repression activity, the gain of enhancement activity, or both (186). Moreover, splicing activation may occur for different reasons in different cells.

Highly Tissue-Specific Regulatory Proteins

Although a single critical factor has not been shown to determine the tissue specificity of splicing in any system, the expression of some splicing regulatory proteins is restricted to certain cells. The neuronal PTB protein and the variants of ETR3, such as NAPOR, are examples of such factors (172, 180, 186, 188, 228). The best-studied splicing regulators that show precise tissue-specific expression are probably the neuronal proteins Nova-1 and Nova-2 (245, 246). These two related proteins were first identified as autoantigens in paraneoplastic disorders. They each contain three RNA binding domains of the KH type, similar to hnRNP K (Figure 4). The Nova proteins are expressed almost exclusively in neurons of the central nervous system. In vitro selection experiments identified preferred RNA binding sites for the two proteins (246, 247). A Nova-1 binding site was found adjacent to the regulated 3A exon in the glycine receptor $\alpha 2$ (GlyR$\alpha 2$) and cotransfection experiments showed that Nova-1 overexpression did increase the inclusion of this exon. Nova-1 was also found to interact with neuronal PTB in yeast two-hybrid assays [called brain (br) PTB in this case] (228). A brPTB binding site upstream of the Nova-1 site for GlyR$\alpha 2$ exon 3A also affected exon 3A splicing. Most significantly, Nova-1 knockout mice were generated (248). These mice die postnatally from a loss of spinal cord and brain stem neurons. The Nova-1 null mice show a twofold decrease in GlyR$\alpha 2$ exon 3A inclusion and a threefold decrease in γ-aminobutyric acid A ($GABA_A$) $\gamma 2$ alternative exon splicing. From the partial effect of the knockout, Nova-1 is likely not the single controlling factor for these exons. However, the lethality makes clear that the protein is controlling some essential function in the neurons where it is expressed. It will be interesting to find out whether the lethal phenotype is due to multiple partial changes in splicing, as seen for GlyR$\alpha 2$ and $GABA_A$ $\gamma 2$, or instead is the result of a drastic change in one or a few transcripts.

Other cell-type-specific regulators of splicing are found in other species. The Elav protein is expressed in all *Drosophila* neurons and has been shown to regulate transcripts from the Neuroglian, Erectwing, and Armadillo genes (182–184). Also in *Drosophila*, the Halfpint protein controls ovary-specific splicing of transcripts from the Ovarian Tumor gene (249). In *Caenorhabditis elegans*, the protein Mec8 controls Unc52 splicing in body wall muscle (250). The availability of genetic analysis in these systems makes them ripe for more detailed mechanistic studies.

Genetic Dissection of Complex Systems of Regulation

It seems clear that the ratio of one splicing pattern to another for a typical alternatively spliced transcript is determined by the combination of factors. Most experiments using protein overexpression or mutation of specific protein binding sites show only partial effects on the use of a particular splicing pattern. It is thus difficult to assess the relative importance of one factor over another.

The biological roles of putative splicing regulatory proteins and their many targets are perhaps best approached through genetic studies. Genetic analysis and RNA interference methods in *Drosophila* and *C. elegans* have already provided important information on the function of U2AF, SR proteins, and other factors in particular splicing events (93, 94,130, 251). In mammals, this kind of data will come from genetic knockouts (such as for Nova-1 protein), from RNA interference experiments, or from somatic cell knockout methods. These have only begun to be generated for the many known splicing regulatory proteins.

The in vivo role of ASF/SF2 was examined in the chicken lymphocyte cell line DT40, which exhibits high levels of homologous recombination (96). Transfection of insertion mutant alleles and selection of homologous recombinants allowed the recovery of heterozygous but not homozygous ASF/SF2 knockout cells. That the homozygous knockout was lethal was demonstrated by complementation with a transgene under tetracycline regulation. This gene when active allowed the recovery of cells in which both endogenous genes were mutant. The repression of the transgene by tetracycline in the homozygous knockout cells was again lethal, demonstrating that ASF/SF2 was indeed essential in these cells. Through complementation experiments with various mutant transgenes, it was demonstrated that the RNP domains of the protein were essential, while the RS domain could be replaced with the RS domain of several other SR proteins without the loss of viability (111). This demonstrated that the ASF/SF2 RS domain is a generic interaction domain that is not targeted differently in different SR proteins. These DT40 studies are a potentially fruitful approach to analyzing the function of many other splicing regulatory proteins expressed in these cells.

The DT40 studies indicate that ASF/SF2 is an essential gene, at least in this cell line. This could pose a problem for generating traditional mouse knockouts of SR proteins. However, the potential embryonic lethal phenotype of the double knockout can be circumvented using the Cre/LoxP method for generating knockouts in specific cell lineages. This has been used very effectively for the SR protein SC35 (95). For SC35, the homozygous null mutation did indeed have an embryonic lethal phenotype in mice. However, by gencrating the "floxed" allele with flanking LoxP sites and breeding these mice with mice expressing Cre recombinase specifically in the thymus, the Fu group was able examine SC35 function specifically in the T cell lineage. The SC35 mutant shows a specific T cell defect and changes in the splicing of CD45 transcripts. This method shows great promise for determining the roles of SC35 and other factors in any particular splicing event in vivo.

MECHANISMS OF INTRON RETENTION

Most alternative splicing patterns involve a choice of one set of splice sites competing against another. Because the change involves two competing splicing pathways, any mechanism that alters the relative rates of spliceosome assembly for the two splicing patterns is a potential means of regulation. One alternative splicing pattern in which this may not be the case is intron retention. Here, the choice is between splicing with intron excision and no splicing with the retention of an intron in the final mRNA. This partially spliced RNA product must then be exported to the cytoplasm. Thus for intron retention, the competition may be between splicing and mRNA transport rather than between two splicing patterns. Studying these systems may yield important information about how the cell determines when an mRNA is complete and ready for movement to the cytoplasm.

In fact, many factors that affect intron retention are the same as those affecting other alternative splicing patterns. For the *Drosophila* Msl2 transcript, an intron is retained in the final mRNA in female flies (54). This intron retention requires Sxl protein binding to both the 5′ and 3′ splice sites. The blockage of both sites may be necessary to prevent any recognition of the intron as a substrate for the splicing reaction, thus allowing its transport. In this regard, it may be significant that the Sxl protein accompanies the transcript to the cytoplasm.

Drosophila P Element Splicing

Another well-characterized system of intron retention is the P element transcript in *Drosophila* (252) (Figure 8). The P element is a transposon whose movement in the genome is restricted to the *Drosophila* germ line. The transposase enzyme is encoded on the fully spliced P element mRNA. In germ cells, all of the P element introns are excised and transposase is expressed. In somatic cells, the P element third intron is retained in the mRNA, preventing translation of the transposase enzyme and hence blocking somatic transposition. A series of genetic and biochemical analyses by the Rio lab have identified components of this regulatory system. The exon upstream from the third intron contains a regulatory sequence that inhibits its splicing in somatic cells. Interestingly, this sequence contains elements called pseudo 5′ splice sites that are similar to 5′ splice sites and bind the U1 snRNP, but which are not used for splicing (253). The regulatory region also contains binding sites for proteins, including the P element Somatic Inhibitor (PSI) and hrp48 (254–256). hrp48 is a *Drosophila* homolog of hnRNP A and B. PSI has an interesting domain structure with four central KH-type RNA binding domains, flanked by large N- and C-terminal

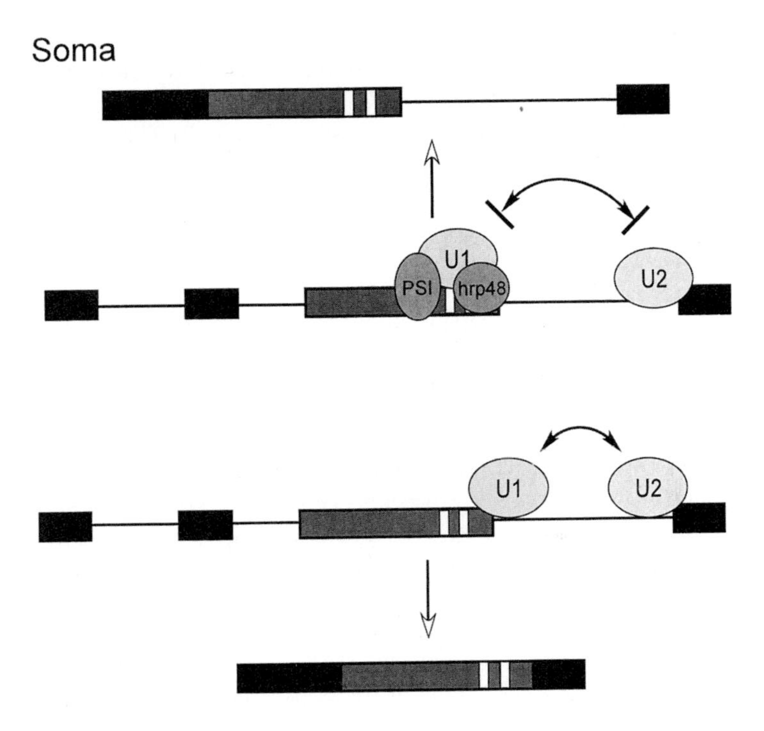

Figure 8 Regulation of *Drosophila* P element splicing. In the germ line the P element transcript is fully spliced to make an mRNA encoding the transposase. In somatic cells, the third intron of the P element transcript is retained in the mRNA. This intron is repressed by the P element Somatic Inhibitor (PSI). PSI binds in the exon upstream from intron 3 in conjunction with U1 snRNP binding to two pseudo 5′ splice sites. The protein hrp48 is also in this complex. The binding of U1 to the pseudo 5′ splice sites is thought to prevent its binding to the correct site downstream and somehow block splicing. In the germ line, the absence of PSI allows normal splicing of the intron.

regions (Figure 4). The C-terminal domain contains a novel repeated motif. PSI is homologous to several vertebrate proteins, including the KSRP that binds to the intronic enhancer for c-src N1 exon (200). In addition to splicing, these vertebrate proteins have been implicated in a variety of processes, including transcription, RNA localization, and RNA degradation (57, 257, 258). How all of these functions are reconciled with each other is not yet clear. The conserved C-terminal repeats in PSI bind to the U1 70K protein (259). PSI expression is excluded from the female germ line, and both biochemistry and genetics experiments indicate that this restriction to somatic cells is an important determinant of the splicing inhibition (255, 256, 260). PSI null mutations are lethal. However, hypomorphic alleles that lack the U1 70K interaction sequence show misregulation of both P element splicing and the splicing of Squid gene transcripts (260). It is thought that in somatic cells, PSI binds to the repressor element and through its interaction with the U1 70K protein induces the binding of the U1 snRNP to the pseudo 5′ splice sites (Figure 8). This assembly in some way prevents use of the normal 5′ splice site in the splicing reaction. Perhaps the purpose of the pseudo 5′ splice sites is to interact with the 3′ splice site in a nonfunctional manner and thus prevent its use with either the natural 5′ splice site or any possible cryptic sites nearby. If such a nonfunctional spliceosome does assemble, it will be interesting to determine what allows its disassembly and hence the transport of the mRNA retaining the intron to the cytoplasm.

Retroviral Splicing and the Balance of Splicing and Transport

The most familiar systems of intron retention are the retroviruses. In the simplest retroviruses the viral envelope protein is encoded on a spliced mRNA (261–265). When this intron is retained, the unspliced transcript serves both as mRNA for the Gag and polymerase proteins and as genomic RNA for packaging into virions. More complex retroviruses such as HIV create additional transcripts using alternative splice sites and cassette exons to produce mRNAs for additional viral proteins, most notably the regulatory proteins Tat and Rev.

Retroviruses make use of many of the same splicing regulatory elements and factors described above. In addition to splicing enhancers and repressors, there are also RNA sequence elements that allow transport of unspliced RNAs to the cytoplasm (261–265). Some of these transport elements are constitutive, interacting directly with cellular factors (266–271). Other elements interact with viral proteins, such as HIV Rev, to allow viral control of the nuclear export process (264). It is thought that partially spliced RNAs are retained in the nucleus by the binding of particular spliceosomal components. The viral transport elements are able to override this retention and allow transport of any RNA that bears them. Retroviruses ensure that some unspliced RNA is available for export by making the viral introns relatively inefficient. The rate of viral splicing is carefully balanced to produce enough of both spliced mRNA and transportable unspliced RNA (272–274). Many features make retroviral splicing inefficient, including poorly recognized splice sites, and ESS elements as discussed above [for example, see (275)]. More unusual features are also seen. Rous sarcoma virus (RSV) has an interesting negative regulator of splicing (NRS) downstream from the viral 5′ splice site (276, 277). This multipartite element has binding sites for numerous factors, including SR proteins, the U1 snRNP, and the minor class snRNP U11. The sequence assembles into a large complex in cellular extracts (278–282). Both the SR proteins and U1 are important for the splicing inhibition. It is thought that this complex interacts with the downstream viral 3′ splice sites to inhibit splicing. This is similar to the inhibition of P element splicing and again it will be interesting to examine how and when this complex is removed from the transported RNA.

Table 1 lists features of the splicing regulatory proteins described above as well as several others.

LONG-STANDING QUESTIONS AND EMERGING ISSUES

Many of the most interesting questions regarding the mechanisms of alternative splicing are still unanswered. Recent results have identified important links between splicing and the events that precede and follow it in the gene expression pathway. Moreover, the biological roles played by alternative splicing in cellular differentiation and adaptation, as well as in genetic variation between species, are only beginning to be examined.

How Does Pre-mRNP Structure Determine Spliceosome Assembly?

Of the identified splicing regulators, relatively few are known to interact directly with spliceosomal components such as the U1 snRNP (TIA-1, PSI, SR proteins) or the U2AF protein (SR proteins, Tra2). These proteins all assemble into large pre-mRNP complexes, and splicing commitment assays make clear that some pre-mRNP complexes determine subsequent spliceosome assembly. However, in nuclear extracts, a heterogeneous H complex containing many of these proteins forms on almost any RNA, and a majority of this material does not proceed through the splicing reaction (283). Because of their size and complexity, these complexes are difficult to define and isolate as homogenous assemblies. Thus, the actual structure of what is called the hnRNP or pre-mRNP complex is very unclear. At a more local level, how a multiple RBD protein interacts with RNA and cooperates with other proteins in assembling into a large pre-mRNP is also not understood. If we are to understand how proteins such as ASF/SF2 or PTB actually promote or block spliceosome assembly, the structures of complete multidomain proteins complexed with RNA will be essential.

There are also surely many splicing regulatory molecules yet to be discovered. Most known splicing regulators directly contact their target RNA. As in transcription, there are likely to be other splicing regulators that interact with other proteins but do not contact nucleic acid. Moreover, the regulators need not be proteins. Some of the recently discovered noncoding micro-RNAs are appealing candidates for regulators of the splicing reaction (284–287), although there is little evidence for this yet.

Coupling Splicing to Upstream and Downstream Processes

Although studied in vitro using presynthesized RNAs, spliceosomes are thought to assemble in vivo onto the pre-mRNA as it is being synthesized (288, 289). Despite assembly concurrent with transcription, the kinetics of intron removal is

TABLE 1 Known splicing regulators

Protein	Other names	Homologs	Species[a]	Target transcripts	Spliceosomal target	Other interactors	Activator or repressor	Domains	References
Sxl			Dm	Sxl, Tra, Msl2	U1/U2 via Snf	Itself	Repressor	RNP-cs	(34)
Tra			Dm	dsx, Fru	U2AF, U1	Tra2, SR proteins	Activator	RS	(34)
Tra2			Multiple	Multiple	U2AF	SR proteins	Activator	RNP-cs RS	(34)
ASF/SF2		SR proteins	Multiple	Many	U2AF, U1 snRNP	Tra2	Activator	RNP-cs RS	(80)
TIA-1		TIAR-1 NAM8	Multiple	Msl2, FGFR2	U1		Activator	RNP-cs	(50, 176)
hnRNP A1		hnRNP A2, B hrp48	Multiple	Many	SR proteins, U2AF	Itself	Repressor	RNP-cs	(136, 168, 205)
PTB	hnRNP I	nPTB	Multiple	Multiple		Napor1 Nova-1 hrp48	Repressor	RNP-cs	(212)
PSI		FBP KSRP	Multiple	P element, Squid	U1		Repressor	KH	(260)
ETR3	CELF2 Napor 1	Multiple	Vertebrate	cTnT, NR1, Cl channel 1, insulin receptor		PTB	Activator	RNP-cs	(180, 366, 367)
Nova 1		Nova-2	Vertebrate	GlyRα2 GABA$_A$Rγ2		nPTB	Activator	KH	(248)
YB 1			Vertebrate	CD44		p72	Activator	Cold shock	(105)
hnRNP H		hnRNP F hnRNP H′	Vertebrate	HIV, src, βTm[b]		hnRNP F	Both	RNP-cs	(169, 170, 173, 201, 280)
p72		p68	Multiple	CD44		YB1	Activator	DEAD Box	(104)
SRp38	SRrp35, 40 TASR		Vertebrate	Many?			Repressor	RNP-cs RS	(90, 92, 347a)
SRrp86			Vertebrate		SR proteins		Both	RNP-cs RS	(89)
RSF1			Dm		SR proteins		Repressor	RNP-cs RS	(91)
Elav		Hu proteins	Multiple	Neuroglian, Erectwing, Armadillo			Repressor	RNP-cs	(182)
Mec8			*C. elegans*	Unc52				RNP-cs	(250)
Fox1			*C. elegans*	Xol-1				RNP-cs	(369)
Quaking		SAM68, SLM2	Vertebrate	MAG, PLP, MBP			Repressor	KH	(370, 371)
SWAP			Dm	Itself				RS	(372)
Halfpint		FIR, PUF60	Multiple	Otu, eIF4E, APP	U2AF	Enc	Activator	RNP-cs	(249)
p32			Vertebrate	Adenovirus	SR proteins		Repressor		(373)

[a] Dm, *Drosophila melanogaster*

[b] βTm, β Tropomyosin

variable and the actual excision of some introns may be completed after the polymerase reaches the end of the transcription unit (290). The cotranscriptional assembly of the spliceosome has profound implications for the regulation of splice site choice. Recent studies have made clear the intimate relationship between pre-mRNA synthesis and processing (291–295). Simply changing the promoter on a gene can have a large effect on the rate of inclusion of a regulated exon (296). This is thought to result partially from changes in the transcription elongation rate through the gene (297, 298). Given equal rates of spliceosome assembly at each of two competing splice sites, by slowing the rate of the synthesis of the downstream choice (through a transcriptional pause site, for example), the upstream site can be made to predominate. Although not proven to occur in an endogenous gene, such transcriptional regulation seems likely to be an important component of many splicing choices [see for example (299–301)].

Genes where splicing seems particularly likely to occur during transcription are those with very long introns. It is a long-standing puzzle how a 5′ splice site at the beginning of a 100-kb intron can be accurately joined to the correct 3′ splice site so far downstream, rather than to an intervening cryptic site. This correct site not only is far away in the sequence, but is synthesized more than an hour after the 5′ splice site (302). This problem is neatly resolved by the idea of recursive splice sites (303). The initial 5′ splice site may splice to intermediate 3′ splice sites along the intron as they are synthesized. These recursive intermediate sites are special in that they regenerate a 5′ splice site as they are joined to the original site. Thus, the 5′ exon may hop along the long intron, being respliced several times at these ratcheting points, before being joined to the final correct site at the end of the long intron. This last site would presumably not regenerate a 5′ splice site and thus terminate the resplicing process. There is evidence for the use of recursive splice sites in several long introns (303). Interestingly, a resplicing mechanism offers another point of control for altering splice site choice. In the Ubx gene of *Drosophila,* a cassette exon containing a recursive splice site can be removed from the RNA even after joining to the upstream exon (303).

Besides indirect effects of the transcription reaction on splicing, there is also evidence for a more active role. This has given rise to the mRNA factory idea: that transcription and processing all take place in the same very large coordinated complex. Such mRNA factories have many appealing features for controlling gene expression. For example, they would ensure that the order in which introns assemble into spliceosomes is the order in which they are transcribed.

The C-terminal domain (CTD) of RNA polymerase II has been shown to directly interact with a number of RNA processing factors (304–306). The CTD consists of a long series of heptapeptide repeats and is thus capable of interacting with multiple factors simultaneously. Deletion of the CTD disrupts the constitutive splicing reaction (307). The CTD is phosphorylated and this phosphorylation is altered as the polymerase progresses through a gene, presumably altering its interactions with processing factors (308, 309). Phosphorylated CTD can stimulate splicing in vitro (310, 311). Conversely, splicing factors directly affect elongation in vitro (312). Only with further study will a real picture emerge of how spliceosomes might be loaded onto the polymerase, and how they are deposited onto the nascent RNA.

It seems likely that some factors bind directly onto the pre-mRNA rather than getting there via the CTD, and that some introns are removed after transcription is complete (290). However, proteins with RS domains have also been shown to interact with transcription factors that bind at promoters (313, 314). The WT1 protein seems to have a dual function, one splice variant acting on transcription and a very similar variant interacting with spliceosomes (315–317). The development of coupled in vitro transcription/splicing systems may enable us to look at the activity of such dual function proteins.

As discussed for retroviruses, splicing must also be controlled relative to downstream processes, most notably nuclear export (295, 318). How does the cell determine when the mRNA is finished and ready for export? In addition to the viral signals that allow export of unspliced RNA, there are also proteins that get deposited on the RNA during the splicing process (319–322). The exon junction complex (EJC) is a non-sequence-specific assembly that is deposited upstream from an exon/exon junction after splicing has occurred. The EJC contains a number of proteins that interact with the nuclear export pathway and the nonsense-mediated RNA decay (NMD) pathway (295, 318). Interestingly, one of its components, UAP56, has also been implicated in the splicing reaction, providing a link with the process that is apparently required for its deposition (323). The connection created by the EJC between splicing and the downstream fate of the mRNA allows for interesting speculation about additional roles for alternative splicing patterns. It is possible that multiple kinds of EJCs differentially affect the later export, translation, localization, or stability of the mRNA [see for example (324)]. If altering the splicing pattern could alter the components of the EJC, the significance of a particular change in splicing could go well beyond a change in its sequence.

Splicing Regulation and Cell Physiology

For practical reasons, most systems of alternative splicing analyzed at the molecular level exhibit stable differences in splicing between two different cell types. However, splicing can also be regulated within a given cell by external stimuli and growth conditions [(325–332); for a review see (333)]. Very little is known about how signal transduction pathways impinge on the splicing reaction. To observe a change in the expressed ratio of two spliced forms, the RNA must turn over. Thus, it is often difficult to separate changes in splicing from changes in mRNA stability. In a few systems, results reflecting clear differences in splicing have been obtained. The cell surface molecule CD44 exhibits a large number of splice variants owing to the differential inclusion of 10 variable exons (334). Several of these variable exons are heavily studied because their inclusion has been associated with the progression of certain tumors to an invasive phenotype (335). The inclusion of variable exon 5 is induced by activation of the PKC/Ras pathway and involves the downstream effector ERK (336, 337). Transcripts for the protein tyrosine phosphatase CD45, an important lymphocyte marker, contain several exons that are included in B cells but variably excluded in T cells [(338) and references therein]. During T cell activation by antigen, the splicing of CD45 is altered such that these alternative exons are repressed. This splicing switch also involves the PKC/Ras pathway. In both these systems, exonic elements have been identified that are required for the inducible change in splicing. For CD44 exon 5, this element binds hnRNP A1 (339).

In the nervous system, a number of gene transcripts are known to alter their splicing in response to cell activity (333, 340–342). Some of these RNAs contain exons that are repressed in response to cell depolarization and to activation of calmodulin dependent kinase IV (CaM K IV). In the stress inducible exon (STREX) of BK potassium channels and in exon 5 of NMDA receptor 1, the CaM kinase–dependent repression requires a special sequence in the 3′ splice site called a calcium-responsive RNA element (CaRRE) (343). This element is transferable and can confer CaM kinase repression on an otherwise constitutive exon. The CaRRE is thus thought to be the actual target of the repressive effect, although the factors that bind to it are not known. The CaRRE element and those in CD44 and CD45 provide an essential starting point for following the splicing response back to the signaling system. It is not clear in these examples whether the signaling pathway is directly altering splicing regulatory proteins or inducing the expression of a splicing regulator. The time course of the effect (or our ability to measure it) is certainly long enough to allow for the latter possibility. On the other hand, the CD44 switch is not blocked by protein synthesis inhibitors (336).

Besides the direct modification or the induced expression of splicing factors, splicing may also be regulated by alterations in the nucleo/cytoplasmic distribution of certain proteins. Changes in SR protein phosphorylation and localization are known to occur during early nematode development (138). hnRNP A1 is partially shifted to the cytoplasm in response to MKK/p38 signaling (344). In primary neuronal culture, Tra2β1 shows cytoplasmic accumulation in response to increasing the intracellular calcium concentration with the drug thapsigargin (345). A similar shift in the localization of both Tra2β1 and SR proteins was induced in response to ischemia in mice. For both A1 and Tra2β1, shifting of the proteins to the cytoplasm was accompanied by moderate changes in the spliced isoform ratio of particular transcripts (344, 345). It is not clear that these splicing changes are the direct result of splicing factor relocalization. However, such relocalization is an appealing mechanism for inducible regulation and is common in systems of transcriptional regulation (346, 347).

Splicing is also likely to be regulated in response to cell proliferation. During mitosis, the nuclear envelope breaks down and the nuclear contents mix with the cytoplasm. Very little is known about what happens to the splicing reaction during this period, but most other features of gene expression are shut down. A recent study shed the first light on this process (347a). Splicing extracts derived from mitotic cells were shown to have a specific inhibitor of splicing. This inhibitor was identified as the dephosphorylated form of SRp38, an SR-related protein. Thus, an attractive model is that a mitosis-specific phosphatase dephosphorylates p38 to inhibit splicing during this phase of the cell cycle. Several important cell cycle regulators are alternatively spliced (348, 349). It will be interesting to test the dependence of these splicing patterns on SRp38.

Alternative Splicing and the Genome

The prevalence of alternative splicing and our limited understanding of its mechanisms present a challenge for identifying all the proteins available to an organism (3, 4). It is currently difficult to use genomic sequence to predict splicing patterns. Nevertheless, new genomic approaches will have a broad impact on our understanding of both the mechanisms and the biological roles of alternative splicing (350, 350a).

The most studied systems of alternative splicing were chosen in part because of their conservation between species. This conservation provides some assurance of the physiological importance of the alternative transcripts. However, not all splicing patterns are conserved between different mammals or between different Dipteran insects. Even within a given species, there appears to be significant variation in the relative use of particular splice variants (351–353). One explanation for this is that alternative splicing provides an advantageous mechanism for testing new protein sequences during evolution. A single point mutation can extend an exon or create a new exon. Such a transcript, encoding a new protein, may comprise only a few percent of the product mRNA. Thus, mutations that alter splicing can allow production of new proteins without significant loss of the wild-type protein. Although this might be advantageous for protein evolution, the high degree of variability in splicing makes it difficult to prove the significance of a splice variant that is not conserved across species.

The large number of splice variants also makes it difficult to study splicing regulation across the whole genome. One would like to examine whether a particular exon is coregulated with others, and to test how a whole ensemble of splice variants is altered by a particular condition. There is consequently a great deal of interest in the use of DNA microarrays to study splicing regulation (350, 350a).

Several microarray approaches to splicing have been described, although all have technical limitations. The most straightforward system examined splicing across the yeast genome, using an array of deposited oligonucleotides that hybridized in exons, introns, or across splice junctions (354). This system was able to clearly categorize intron groups based on their different splicing factor dependence. In another approach, high-density oligonucleotide arrays were generated where probes were tiled across the entire human genome or more densely across chromosome 22 (355). Such probes do not give exon junction information; they only indicate the presence of a particular exon region in the mRNA population. This approach will not yield sufficient information to describe complicated splicing patterns and identify small exons. Nevertheless, it will give a clear first identification of many new exons and their tissue specificity.

Oligonucleotide arrays depend on the hybridization properties of the probes being well matched across the array for denaturation temperature (T_m) and other features. Some exons may have complementary oligonucleotides that will not work in parallel with others on the same array. In a novel fiber-optic-array system, the differences in hybridization properties were minimized by using the ligation of two oligos to measure each spliced junction (356, 357). It is not clear whether this system can be scaled up to measure thousands of splice variants simultaneously, as needed for a whole-genome analysis. However, it has potential advantages for the simultaneous measurement of particular exon groups.

Even if it is not possible to put all exons on a single array, these methods hold promise for analyzing the combinatorial mechanisms controlling splicing. The covariation of exon groups can be related to studies of splicing factor expression throughout an organism. This can be combined with genetic knockouts of individual splicing regulators and with the immunoprecipitation of specific transcripts with antibodies to regulatory proteins (260, 358). Microarrays promise to take studies of splicing regulation to the level of the whole genome. By studying alternative splicing at both the genomic and mechanistic levels, our understanding of splicing regulation can be integrated into the biology of cells and organisms.

ACKNOWLEDGMENTS

I thank Benoit Chabot, Thomas Cooper, Xiang-Dong Fu, Brenton Graveley, Kristen Lynch, Don Rio, and Juan Valcarcel for advice on the manuscript, and Kim Simmonds for help in preparing it. My own work has been supported by the National Institutes of Health and the Packard Foundation and as an Investigator of the Howard Hughes Medical Institute.

The *Annual Review of Biochemistry* is online at http://biochem.annualreviews.org

LITERATURE CITED

1. Grabowski PJ, Black DL. 2001. *Prog. Neurobiol.* 65:289–308
2. Modrek B, Lee C. 2002. *Nat. Genet.* 30:13–19
3. Black DL. 2000. *Cell* 103:367–70
4. Graveley BR. 2001. *Trends Genet.* 17:100–7
5. Smith CW, Nadal-Ginard B. 1989. *Cell* 56:749–58
6. Schmucker D, Clemens JC, Shu H, Worby CA, Xiao J, et al. 2000. *Cell* 101:671–84
7. Colgan DF, Manley JL. 1997. *Genes Dev.* 11:2755–66
8. Baker BS. 1989. *Nature* 340:521–24
9. Black DL. 1998. *Neuron* 20:165–68
10. Burgess RW, Nguyen QT, Son YJ, Lichtman JW, Sanes JR. 1999. *Neuron* 23:33–44
11. Burke JF, Bright KE, Kellett E, Ben-

jamin PR, Saunders SE. 1992. *Prog. Brain Res.* 92:115–25
12. Cartegni L, Chew SL, Krainer AR. 2002. *Nat. Rev. Genet.* 3:285–98
13. Cooper TA, Mattox W. 1997. *Am. J. Hum. Genet.* 61:259–66
14. Eckardt NA. 2002. *Plant Cell* 14:743–47
15. Jiang ZH, Wu JY. 1999. *Proc. Soc. Exp. Biol. Med.* 220:64–72
16. Matera AG. 1999. *Curr. Biol.* 9:R140–42
17. Philips AV, Cooper TA. 2000. *Cell. Mol. Life Sci.* 57:235–49
18. Schutt C, Nothiger R. 2000. *Development* 127:667–77
19. Caceres JF, Kornblihtt AR. 2002. *Trends Genet.* 18:186–93
20. Blencowe BJ. 2000. *Trends Biochem. Sci.* 25:106–10
21. Burge CB, Tuschl T, Sharp PA. 1999. In *The RNA World*, ed. RF Gesteland, TR Cech, JF Atkins, pp. 525–60. Cold Spring Harbor, NY: Cold Spring Harb. Lab. 2nd ed.
22. Nilsen TW. 2002. *Mol. Cell.* 9:8–9
23. Staley JP, Guthrie C. 1998. *Cell* 92:315–26
23a. Brow DA. 2002. *Annu. Rev. Genet.* 36: 333–60
24. Das R, Zhou Z, Reed R. 2000. *Mol. Cell.* 5:779–87
25. Nilsen TW. 1996. *Science* 273:1813
26. Lallena MJ, Chalmers KJ, Llamazares S, Lamond AI, Valcarcel J. 2002. *Cell* 109:285–96
27. Berget SM. 1995. *J. Biol. Chem.* 270: 2411–14
28. Black DL. 1995. *RNA* 1:763–71
29. Hoffman BE, Grabowski PJ. 1992. *Genes Dev.* 6:2554–68
30. Robberson BL, Cote GJ, Berget SM. 1990. *Mol. Cell. Biol.* 10:84–94
31. Libri D, Piseri A, Fiszman MY. 1991. *Science* 252:1842–45
32. Libri D, Stutz F, McCarthy T, Rosbash M. 1995. *RNA* 1:425–36
33. Jacquenet S, Ropers D, Bilodeau PS, Damier L, Mougin A, et al. 2001. *Nucleic Acids Res.* 29:464–78
34. Lopez AJ. 1998. *Annu. Rev. Genet.* 32:279–305
35. Varani G, Nagai K. 1998. *Annu. Rev. Biophys. Biomol. Struct.* 27:407–45
36. Antson AA. 2000. *Curr. Opin. Struct. Biol.* 10:87–94
37. Crowder SM, Kanaar R, Rio DC, Alber T. 1999. *Proc. Natl. Acad. Sci. USA* 96:4892–97
38. Handa N, Nureki O, Kurimoto K, Kim I, Sakamoto H, et al. 1999. *Nature* 398: 579–85
39. Wang J, Bell LR. 1994. *Genes Dev.* 8:2072–85
40. Sosnowski BA, Belote JM, McKeown M. 1989. *Cell* 58:449–59
41. Inoue K, Hoshijima K, Sakamoto H, Shimura Y. 1990. *Nature* 344:461–63
42. Valcarcel J, Singh R, Zamore PD, Green MR. 1993. *Nature* 362:171–75
43. Granadino B, Penalva LO, Green MR, Valcarcel J, Sanchez L. 1997. *Proc. Natl. Acad. Sci. USA* 94:7343–48
44. Granadino B, Penalva LO, Sanchez L. 1996. *Mol. Gen. Genet.* 253:26–31
45. Hilfiker A, Amrein H, Dubendorfer A, Schneiter R, Nothiger R. 1995. *Development* 121:4017–26
46. Kelley RL, Solovyeva I, Lyman LM, Richman R, Solovyev V, Kuroda MI. 1995. *Cell* 81:867–77
47. Zhou S, Yang Y, Scott MJ, Pannuti A, Fehr KC, et al. 1995. *EMBO J.* 14:2884–95
48. Bashaw GJ, Baker BS. 1995. *Development* 121:3245–58
49. Merendino L, Guth S, Bilbao D, Martinez C, Valcarcel J. 1999. *Nature* 402: 838–41
50. Forch P, Puig O, Kedersha N, Martinez C, Granneman S, et al. 2000. *Mol. Cell* 6:1089–98
51. Forch P, Merendino L, Martinez C, Valcarcel J. 2001. *RNA* 7:1185–91
52. Bashaw GJ, Baker BS. 1997. *Cell* 89:789–98

53. Kelley RL, Wang J, Bell L, Kuroda MI. 1997. *Nature* 387:195–99
54. Gebauer F, Merendino L, Hentze MW, Valcarcel J. 1998. *RNA* 4:142–50
55. Gebauer F, Merendino L, Hentze MW, Valcarcel J. 1997. *Semin. Cell Dev. Biol.* 8:561–66
56. Cote CA, Gautreau D, Denegre JM, Kress TL, Terry NA, Mowry KL. 1999. *Mol. Cell.* 4:431–37
57. Gu W, Pan F, Zhang H, Bassell GJ, Singer RH. 2002. *J. Cell Biol.* 156: 41–51
58. Kaminski A, Jackson RJ. 1998. *RNA* 4:626–38
59. Hughson FM, Schedl P. 1999. *Nat. Struct. Biol.* 6:499–502
60. Horabin JI, Schedl P. 1996. *RNA* 2:1–10
61. Zhu C, Urano J, Bell LR. 1997. *Mol. Cell. Biol.* 17:1674–81
62. Horabin JI, Schedl P. 1993. *Mol. Cell. Biol.* 13:7734–46
63. Penalva LO, Lallena MJ, Valcarcel J. 2001. *Mol. Cell. Biol.* 21:1986–96
64. Sakamoto H, Inoue K, Higuchi I, Ono Y, Shimura Y. 1992. *Nucleic Acids Res.* 20:5533–40
65. Salz HK, Flickinger TW. 1996. *Genetics* 144:95–108
66. Salz HK. 1992. *Genetics* 130:547–54
67. Graveley BR. 2002. *Cell* 109:409–12
68. Hertel KJ, Lynch KW, Maniatis T. 1997. *Curr. Opin. Cell Biol.* 9:350–57
69. Smith CW, Valcarcel J. 2000. *Trends Biochem. Sci.* 25:381–88
70. Wang J, Manley JL. 1997. *Curr. Opin. Genet. Dev.* 7:205–11
71. Coschigano KT, Wensink PC. 1993. *Genes Dev.* 7:42–54
72. Burtis KC. 2002. *Science* 297:1135–36
73. Tian M, Maniatis T. 1994. *Genes Dev.* 8:1703–12
74. Tian M, Maniatis T. 1993. *Cell* 74:105–14
75. Tian M, Maniatis T. 1992. *Science* 256: 237–40
76. Lynch KW, Maniatis T. 1996. *Genes Dev.* 10:2089–101
77. Hertel KJ, Lynch KW, Hsiao EC, Liu EH, Maniatis T. 1996. *RNA* 2:969–81
78. Zuo P, Maniatis T. 1996. *Genes Dev.* 10:1356–68
79. Graveley BR, Hertel KJ, Maniatis T. 2001. *RNA* 7:806–18
80. Graveley BR. 2000. *RNA* 6:1197–211
81. Ryner LC, Goodwin SF, Castrillon DH, Anand A, Villella A, et al. 1996. *Cell* 87:1079–89
82. Heinrichs V, Ryner LC, Baker BS. 1998. *Mol. Cell. Biol.* 18:450–58
83. Anand A, Villella A, Ryner LC, Carlo T, Goodwin SF, et al. 2001. *Genetics* 158:1569–95
84. Schaal TD, Maniatis T. 1999. *Mol. Cell. Biol.* 19:261–73
85. Fairbrother WG, Yeh RF, Sharp PA, Burge CB. 2002. *Science* 297:1007–13
86. Caceres JF, Krainer AR. 1997. *Eukaryotic mRNA Processing*, pp. 174–217. Oxford, UK: Oxford Univ. Press
87. Manley JL, Tacke R. 1996. *Genes Dev.* 10:1569–79
88. Zahler AM, Lane WS, Stolk JA, Roth MB. 1992. *Genes Dev.* 6:837–47
89. Barnard DC, Li J, Peng R, Patton JG. 2002. *RNA* 8:526–33
90. Cowper AE, Caceres JF, Mayeda A, Screaton GR. 2001. *J. Biol. Chem.* 276: 48908–14
91. Labourier E, Bourbon HM, Gallouzi IE, Fostier M, Allemand E, Tazi J. 1999. *Genes Dev.* 13:740–53
92. Yang L, Embree LJ, Tsai S, Hickstein DD. 1998. *J. Biol. Chem.* 273:27761–64
93. Longman D, Johnstone IL, Caceres JF. 2000. *EMBO J.* 19:1625–37
94. Ring HZ, Lis JT. 1994. *Mol. Cell. Biol.* 14:7499–506
95. Wang HY, Xu X, Ding JH, Bermingham JR Jr, Fu XD. 2001. *Mol. Cell* 7:331–42
96. Wang J, Takagaki Y, Manley JL. 1996. *Genes Dev.* 10:2588–99

97. Liu HX, Zhang M, Krainer AR. 1998. *Genes Dev.* 12:1998–2012
98. Tacke R, Manley JL. 1999. *Curr. Opin. Cell. Biol.* 11:358–62
99. Cartegni L, Krainer AR. 2002. *Nat. Genet.* 30:377–84
100. Tian H, Kole R. 1995. *Mol. Cell. Biol.* 15:6291–98
101. Tian H, Kole R. 2001. *J. Biol. Chem.* 276:33833–39
102. Woerfel G, Bindereif A. 2001. *Nucleic Acids Res.* 29:3204–11
103. Coulter LR, Landree MA, Cooper TA. 1997. *Mol. Cell. Biol.* 17:2143–50
104. Honig A, Auboeuf D, Parker MM, O'Malley BW, Berget SM. 2002. *Mol. Cell. Biol.* 22:5698–707
105. Stickeler E, Fraser SD, Honig A, Chen AL, Berget SM, Cooper TA. 2001. *EMBO J.* 20:3821–30
106. Chandler SD, Mayeda A, Yeakley JM, Krainer AR, Fu XD. 1997. *Proc. Natl. Acad. Sci. USA* 94:3596–601
107. Mayeda A, Screaton GR, Chandler SD, Fu XD, Krainer AR. 1999. *Mol. Cell. Biol.* 19:1853–63
108. Tacke R, Manley JL. 1995. *EMBO J.* 14:3540–51
109. Zuo P, Manley JL. 1993. *EMBO J.* 12:4727–37
110. Graveley BR, Maniatis T. 1998. *Mol. Cell* 1:765–71
111. Wang J, Xiao SH, Manley JL. 1998. *Genes Dev.* 12:2222–33
112. Zhu J, Krainer AR. 2000. *Genes Dev.* 14:3166–78
113. Prasad J, Colwill K, Pawson T, Manley JL. 1999. *Mol. Cell. Biol.* 19:6991–7000
114. Kanopka A, Muhlemann O, Petersen-Mahrt S, Estmer C, Ohrmalm C, Akusjarvi G. 1998. *Nature* 393:185–87
115. Sanford JR, Bruzik JP. 1999. *Genes Dev.* 13:1513–18
116. Mermoud JE, Cohen PT, Lamond AI. 1994. *EMBO J.* 13:5679–88
117. Mermoud JE, Cohen P, Lamond AI. 1992. *Nucleic Acids Res.* 20:5263–69
118. Roscigno RF, Garcia-Blanco MA. 1995. *RNA* 1:692–706
119. Xiao SH, Manley JL. 1997. *Genes Dev.* 11:334–44
120. Kohtz JD, Jamison SF, Will CL, Zuo P, Luhrmann R, et al. 1994. *Nature* 368:119–24
121. Wu JY, Maniatis T. 1993. *Cell* 75:1061–70
122. Yeakley JM, Tronchere H, Olesen J, Dyck JA, Wang HY, Fu XD. 1999. *J. Cell Biol.* 145:447–55
123. Dauwalder B, Mattox W. 1998. *EMBO J.* 17:6049–60
124. Xiao SH, Manley JL. 1998. *EMBO J.* 17:6359–67
125. Wang Z, Hoffmann HM, Grabowski PJ. 1995. *RNA* 1:21–35
126. Lavigueur A, La Branche H, Kornblihtt AR, Chabot B. 1993. *Genes Dev.* 7:2405–17
127. Bourgeois CF, Popielarz M, Hildwein G, Stevenin J. 1999. *Mol. Cell. Biol.* 19:7347–56
128. Cote J, Simard MJ, Chabot B. 1999. *Nucleic Acids Res.* 27:2529–37
129. Selvakumar M, Helfman DM. 1999. *RNA* 5:378–94
130. Rudner DZ, Breger KS, Rio DC. 1998. *Genes Dev.* 12:1010–21
131. Blencowe BJ, Issner R, Nickerson JA, Sharp PA. 1998. *Genes Dev.* 12:996–1009
132. Li Y, Blencowe BJ. 1999. *J. Biol. Chem.* 274:35074–79
133. Eldridge AG, Li Y, Sharp PA, Blencowe BJ. 1999. *Proc. Natl. Acad. Sci. USA* 96:6125–30
134. Hertel KJ, Maniatis T. 1998. *Mol. Cell* 1:449–55
135. Kan JL, Green MR. 1999. *Genes Dev.* 13:462–71
136. Zhu J, Mayeda A, Krainer AR. 2001. *Mol. Cell* 8:1351–61
137. Sacco-Bubulya P, Spector DL. 2002. *J. Cell Biol.* 156:425–36
138. Sanford JR, Bruzik JP. 2001. *Proc. Natl. Acad. Sci. USA* 98:10184–89

139. Deleted in proof
140. Siebel CW, Feng L, Guthrie C, Fu XD. 1999. *Proc. Natl. Acad. Sci. USA* 96:5440–45
141. Yun CY, Fu XD. 2000. *J. Cell Biol.* 150:707–18
142. Wang HY, Arden KC, Bermingham JR Jr, Viars CS, Lin W, et al. 1999. *Genomics* 57:310–15
143. Wang HY, Lin W, Dyck JA, Yeakley JM, Songyang Z, et al. 1998. *J. Cell Biol.* 140:737–50
144. Gui JF, Tronchere H, Chandler SD, Fu XD. 1994. *Proc. Natl. Acad. Sci. USA* 91:10824–28
145. Gui JF, Lane WS, Fu XD. 1994. *Nature* 369:678–82
146. Colwill K, Pawson T, Andrews B, Prasad J, Manley JL, et al. 1996. *EMBO J.* 15:265–75
147. Tacke R, Chen Y, Manley JL. 1997. *Proc. Natl. Acad. Sci. USA* 94:1148–53
148. Du C, McGuffin ME, Dauwalder B, Rabinow L, Mattox W. 1998. *Mol. Cell* 2:741–50
149. Krecic AM, Swanson MS. 1999. *Curr. Opin. Cell Biol.* 11:363–71
150. Ford LP, Wright WE, Shay JW. 2002. *Oncogene* 21:580–83
151. LaBranche H, Dupuis S, Ben-David Y, Bani MR, Wellinger RJ, Chabot B. 1998. *Nat. Genet.* 19:199–202
152. Dallaire F, Dupuis S, Fiset S, Chabot B. 2000. *J. Biol. Chem.* 275:14509–16
153. Buvoli M, Cobianchi F, Bestagno MG, Mangiarotti A, Bassi MT, et al. 1990. *EMBO J.* 9:1229–35
154. Burd CG, Dreyfuss G. 1994. *EMBO J.* 13:1197–204
155. Mayeda A, Munroe SH, Caceres JF, Krainer AR. 1994. *EMBO J.* 13:5483–95
156. Ding J, Hayashi MK, Zhang Y, Manche L, Krainer AR, Xu RM. 1999. *Genes Dev.* 13:1102–15
157. Allain FH, Bouvet P, Dieckmann T, Feigon J. 2000. *EMBO J.* 19:6870–81
158. Mayeda A, Helfman DM, Krainer AR. 1993. *Mol. Cell. Biol.* 13:2993–3001
159. Mayeda A, Krainer AR. 1992. *Cell* 68:365–75
160. Caceres JF, Stamm S, Helfman DM, Krainer AR. 1994. *Science* 265:1706–9
161. Hanamura A, Caceres JF, Mayeda A, Franza BR Jr, Krainer AR. 1998. *RNA* 4:430–44
162. Caputi M, Mayeda A, Krainer AR, Zahler AM. 1999. *EMBO J.* 18:4060–67
163. Del Gatto-Konczak F, Olive M, Gesnel MC, Breathnach R. 1999. *Mol. Cell. Biol.* 19:251–60
164. Tange TO, Kjems J. 2001. *J. Mol. Biol.* 312:649–62
165. Staffa A, Acheson NH, Cochrane A. 1997. *J. Biol. Chem.* 272:33394–401
166. Amendt BA, Si ZH, Stoltzfus CM. 1995. *Mol. Cell. Biol.* 15:6480
167. Amendt BA, Hesslein D, Chang LJ, Stoltzfus CM. 1994. *Mol. Cell. Biol.* 14:3960–70
168. Tange TO, Damgaard CK, Guth S, Valcarcel J, Kjems J. 2001. *EMBO J.* 20:5748–58
169. Jacquenet S, Mereau A, Bilodeau PS, Damier L, Stoltzfus CM, Branlant C. 2001. *J. Biol. Chem.* 276:40464–75
170. Chen CD, Kobayashi R, Helfman DM. 1999. *Genes Dev.* 13:593–606
171. Caputi M, Zahler AM. 2001. *J. Biol. Chem.* 276:43850–59
172. Markovtsov V, Nikolic JM, Goldman JA, Turck CW, Chou MY, Black DL. 2000. *Mol. Cell. Biol.* 20:7463–79
173. Caputi M, Zahler AM. 2002. *EMBO J.* 21:845–55
174. Hastings ML, Wilson CM, Munroe SH. 2001. *RNA* 7:859–74
175. Gallego ME, Gattoni R, Stevenin J, Marie J, Expert-Bezancon A. 1997. *EMBO J.* 16:1772–84
176. Del Gatto-Konczak F, Bourgeois CF, Le Guiner C, Kister L, Gesnel MC, et al. 2000. *Mol. Cell. Biol.* 20:6287–99
177. Puig O, Gottschalk A, Fabrizio P,

Seraphin B. 1999. *Genes Dev.* 13:569–80
178. Spingola M, Ares M Jr. 2000. *Mol. Cell* 6:329–38
179. Kedersha NL, Gupta M, Li W, Miller I, Anderson P. 1999. *J. Cell Biol.* 147: 1431–42
180. Ladd AN, Charlet-B N, Cooper TA. 2001. *Mol. Cell. Biol.* 21:1285–96
181. Timchenko LT, Miller JW, Timchenko NA, DeVore DR, Datar KV, et al. 1996. *Nucleic Acids Res.* 24:4407–14
182. Lisbin MJ, Qiu J, White K. 2001. *Genes Dev.* 15:2546–61
183. Koushika SP, Soller M, White K. 2000. *Mol. Cell. Biol.* 20:1836–45
184. Koushika SP, Lisbin MJ, White K. 1996. *Curr. Biol.* 6:1634–41
185. Good PJ, Chen Q, Warner SJ, Herring DC. 2000. *J. Biol. Chem.* 275:28583–92
186. Charlet-B N, Logan P, Singh G, Cooper TA. 2002. *Mol. Cell* 9:649–58
187. Choi DK, Ito T, Tsukahara F, Hirai M, Sakaki Y. 1999. *Gene* 237:135–42
188. Zhang W, Liu H, Han K, Grabowski PJ. 2002. *RNA* 8:671–85
189. Huh GS, Hynes RO. 1994. *Genes Dev.* 8:1561–74
190. Modafferi EF, Black DL. 1997. *Mol. Cell. Biol.* 17:6537–45
191. Lim LP, Sharp PA. 1998. *Mol. Cell. Biol.* 18:3900–6
192. Black DL. 1992. *Cell* 69:795–807
193. Hedjran F, Yeakley JM, Huh GS, Hynes RO, Rosenfeld MG. 1997. *Proc. Natl. Acad. Sci. USA* 94:12343–47
194. Huh GS, Hynes RO. 1993. *Mol. Cell. Biol.* 13:5301–14
195. Guo N, Kawamoto S. 2000. *J. Biol. Chem.* 275:33641–49
196. Brudno M, Gelfand MS, Spengler S, Zorn M, Dubchak I, Conboy JG. 2001. *Nucleic Acids Res.* 29:2338–48
197. Modafferi EF, Black DL. 1999. *RNA* 5:687–706
198. Chan RC, Black DL. 1997. *Mol. Cell. Biol.* 17:2970
199. Min H, Chan RC, Black DL. 1995. *Genes Dev.* 9:2659–71
200. Min H, Turck CW, Nikolic JM, Black DL. 1997. *Genes Dev.* 11:1023–36
201. Chou MY, Rooke N, Turck CW, Black DL. 1999. *Mol. Cell. Biol.* 19:69–77
202. Chou MY, Underwood JG, Nikolic J, Luu MH, Black DL. 2000. *Mol. Cell* 5:949–57
203. Kanopka A, Muhlemann O, Akusjarvi G. 1996. *Nature* 381:535–38
204. Simard MJ, Chabot B. 2002. *Mol. Cell. Biol.* 22:4001–10
205. Blanchette M, Chabot B. 1999. *EMBO J.* 18:1939–52
206. Xu RM, Jokhan L, Cheng X, Mayeda A, Krainer AR. 1997. *Structure* 5:559–70
207. Patton JG, Mayer SA, Tempst P, Nadal-Ginard B. 1991. *Genes Dev.* 5:1237–51
208. Garcia-Blanco MA, Jamison SF, Sharp PA. 1989. *Genes Dev.* 3:1874–86
209. Gil A, Sharp PA, Jamison SF, Garcia-Blanco MA. 1991. *Genes Dev.* 5:1224–36
210. Ghetti A, Pinol-Roma S, Michael WM, Morandi C, Dreyfuss G. 1992. *Nucleic Acids Res.* 20:3671–78
211. Valcarcel J, Gebauer F. 1997. *Curr. Biol.* 7:R705–8
212. Wagner EJ, Garcia-Blanco MA. 2001. *Mol. Cell. Biol.* 21:3281–88
213. Singh R, Valcarcel J, Green MR. 1995. *Science* 268:1173–76
214. Pérez I, Lin CH, McAfee JG, Patton JG. 1997. *RNA* 3:764–78
215. Pérez I, McAfee JG, Patton JG. 1997. *Biochemistry* 36:11881–90
216. Oh YL, Hahm B, Kim YK, Lee HK, Lee JW, et al. 1998. *Biochem. J.* 331: 169–75
217. Conte MR, Grune T, Ghuman J, Kelly G, Ladas A, et al. 2000. *EMBO J.* 19:3132–41
218. Liu H, Zhang W, Reed RB, Liu W, Grabowski PJ. 2002. *RNA* 8:137–49
219. Southby J, Gooding C, Smith CWJ. 1999. *Mol. Cell. Biol.* 19:2699–711

219a. Wagner EJ, Garcia-Blanco MA. 2002. *Mol. Cell* 10:943–49
220. Mulligan GJ, Guo W, Wormsley S, Helfman DM. 1992. *J. Biol. Chem.* 267: 25480–87
221. Chan RC, Black DL. 1997. *Mol. Cell. Biol.* 17:4667–76
222. Gooding C, Roberts GC, Moreau G, Nadal-Ginard B, Smith CW. 1994. *EMBO J.* 13:3861–72
223. Gooding C, Roberts GC, Smith CW. 1998. *RNA* 4:85–100
224. Zhang L, Liu W, Grabowski PJ. 1999. *RNA* 5:117–30
225. Lillevali K, Kulla A, Ord T. 2001. *Mech. Dev.* 101:217–20
226. Wagner EJ, Carstens RP, Garcia-Blanco MA. 1999. *Electrophoresis* 20:1082–86
227. Deleted in proof
228. Polydorides AD, Okano HJ, Yang YY, Stefani G, Darnell RB. 2000. *Proc. Natl. Acad. Sci. USA* 97:6350–55
229. Wollerton MC, Gooding C, Robinson F, Brown EC, Jackson RJ, Smith CW. 2001. *RNA* 7:819–32
230. Ashiya M, Grabowski PJ. 1997. *RNA* 3:996–1015
231. Deleted in proof
232. Carstens RP, Wagner EJ, Garcia-Blanco MA. 2000. *Mol. Cell. Biol.* 20:7388–400
233. Expert-Bezancon A, Le Caer JP, Marie J. 2002. *J. Biol. Chem.* 277:16614–23
234. Cooper TA, Ordahl CP. 1985. *J. Biol. Chem.* 260:11140–48
235. Levy JB, Dorai T, Wang LH, Brugge JS. 1987. *Mol. Cell. Biol.* 7:4142–45
236. Martinez R, Mathey-Prevot B, Bernards A, Baltimore D. 1987. *Science* 237:411–15
237. Black DL. 1991. *Genes Dev.* 5:389–402
238. Dominski Z, Kole R. 1991. *Mol. Cell. Biol.* 11:6075–83
239. Xu R, Teng J, Cooper TA. 1993. *Mol. Cell. Biol.* 13:3660–74
240. Dominski Z, Kole R. 1992. *Mol. Cell. Biol.* 12:2108–14
241. Ramchatesingh J, Zahler AM, Neugebauer KM, Roth MB, Cooper TA. 1995. *Mol. Cell. Biol.* 15:4898–907
241a. Rooke N, Markovtsov V, Cagavi E, Black DL. 2003. *Mol. Cell. Biol.* 23(6): In press
242. Ryan KJ, Cooper TA. 1996. *Mol. Cell. Biol.* 16:4014–23
243. Cooper TA. 1998. *Mol. Cell. Biol.* 18:4519–25
244. Chan RC, Black DL. 1995. *Mol. Cell. Biol.* 15:6377–85
245. Buckanovich RJ, Posner JB, Darnell RB. 1993. *Neuron* 11:657–72
246. Yang YY, Yin GL, Darnell RB. 1998. *Proc. Natl. Acad. Sci. USA* 95:13254–59
247. Buckanovich RJ, Darnell RB. 1997. *Mol. Cell. Biol.* 17:3194–201
248. Jensen KB, Dredge BK, Stefani G, Zhong R, Buckanovich RJ, et al. 2000. *Neuron* 25:359–71
249. Van Buskirk C, Schupbach T. 2002. *Dev. Cell* 2:343–53
250. Lundquist EA, Herman RK, Rogalski TM, Mullen GP, Moerman DG, Shaw JE. 1996. *Development* 122:1601–10
251. Zorio DA, Blumenthal T. 1999. *RNA* 5:487–94
252. Rio DC. 1991. *Trends Genet.* 7:282–87
253. Siebel CW, Fresco LD, Rio DC. 1992. *Genes Dev.* 6:1386–401
254. Siebel CW, Kanaar R, Rio DC. 1994. *Genes Dev.* 8:1713–25
255. Siebel CW, Admon A, Rio DC. 1995. *Genes Dev.* 9:269–83
256. Adams MD, Tarng RS, Rio DC. 1997. *Genes Dev.* 11:129–38
257. Duncan R, Bazar L, Michelotti G, Tomonaga T, Krutzsch H, et al. 1994. *Genes Dev.* 8:465–80
258. Chen CY, Gherzi R, Ong SE, Chan EL, Raijmakers R, et al. 2001. *Cell* 107: 451–64
259. Labourier E, Adams MD, Rio DC. 2001. *Mol. Cell* 8:363–73

260. Labourier E, Blanchette M, Feiger JW, Adams MD, Rio DC. 2002. *Genes Dev.* 16:72–84
261. Kiss-Laszlo Z, Hohn T. 1996. *Trends Microbiol.* 4:480–85
262. Boris-Lawrie K, Roberts TM, Hull S. 2001. *Life Sci.* 69:2697–709
263. Fischer U, Pollard VW, Luhrmann R, Teufel M, Michael MW, et al. 1999. *Nucleic Acids Res.* 27:4128–34
264. Pollard VW, Malim MH. 1998. *Annu. Rev. Microbiol.* 52:491–532
265. Cullen BR. 1998. *Virology* 249:203–10
266. Yang J, Bogerd HP, Wang PJ, Page DC, Cullen BR. 2001. *Mol. Cell* 8:397–406
267. Yang J, Cullen BR. 1999. *RNA* 5:1645–55
268. Pasquinelli AE, Ernst RK, Lund E, Grimm C, Zapp ML, et al. 1997. *EMBO J.* 16:7500–10
269. Ernst RK, Bray M, Rekosh D, Hammarskjold ML. 1997. *Mol. Cell. Biol.* 17:135–44
270. Gruter P, Tabernero C, von Kobbe C, Schmitt C, Saavedra C, et al. 1998. *Mol. Cell* 1:649–59
271. Saavedra C, Felber B, Izaurralde E. 1997. *Curr. Biol.* 7:619–28
272. Katz RA, Skalka AM. 1990. *Mol. Cell. Biol.* 10:696–704
273. Fu XD, Katz RA, Skalka AM, Maniatis T. 1991. *Genes Dev.* 5:211–20
274. Bouck J, Fu XD, Skalka AM, Katz RA. 1998. *J. Biol. Chem.* 273:15169–76
275. Pongoski J, Asai K, Cochrane A. 2002. *J. Virol.* 76:5108–20
276. McNally MT, Beemon K. 1992. *J. Virol.* 66:6–11
277. Ogert RA, Lee LH, Beemon KL. 1996. *J. Virol.* 70:3834–43
278. McNally LM, McNally MT. 1998. *Mol. Cell. Biol.* 18:3103–11
279. McNally LM, McNally MT. 1999. *J. Virol.* 73:2385–93
280. Fogel BL, McNally MT. 2000. *J. Biol. Chem.* 275:32371–78
281. Gontarek RR, McNally MT, Beemon K. 1993. *Genes Dev.* 7:1926–36
282. Hibbert CS, Gontarek RR, Beemon KL. 1999. *RNA* 5:333–43
283. Bennett M, Piñol-Roma S, Staknis D, Dreyfuss G, Reed R. 1992. *Mol. Cell. Biol.* 12:3165–75
284. Mourelatos Z, Dostie J, Paushkin S, Sharma A, Charroux B, et al. 2002. *Genes Dev.* 16:720–28
285. Ambros V. 2001. *Cell* 107:823–26
286. Lagos-Quintana M, Rauhut R, Lendeckel W, Tuschl T. 2001. *Science* 294:853–58
287. Lau NC, Lim LP, Weinstein EG, Bartel DP. 2001. *Science* 294:858–62
288. Osheim YN, Miller OL Jr, Beyer AL. 1985. *Cell* 43:143–51
289. Misteli T, Spector DL. 1999. *Mol. Cell* 3:697–705
290. Wetterberg I, Bauren G, Wieslander L. 1996. *RNA* 2:641–51
291. Hirose Y, Manley JL. 2000. *Genes Dev.* 14:1415–29
292. Proudfoot NJ, Furger A, Dye MJ. 2002. *Cell* 108:501–12
293. Bentley D. 1999. *Curr. Opin. Cell Biol.* 11:347–51
294. Bentley D. 2002. *Curr. Opin. Cell Biol.* 14:336–42
295. Maniatis T, Reed R. 2002. *Nature* 416:499–506
296. Cramer P, Caceres JF, Cazalla D, Kadener S, Muro AF, et al. 1999. *Mol. Cell* 4:251–58
297. Kadener S, Fededa JP, Rosbash M, Kornblihtt AR. 2002. *Proc. Natl. Acad. Sci. USA* 99:8185–90
298. Kadener S, Cramer P, Nogues G, Cazalla D, de la Mata M, et al. 2001. *EMBO J.* 20:5759–68
299. Eperon LP, Graham IR, Griffiths AD, Eperon IC. 1988. *Cell* 54:393–401
300. Roberts GC, Gooding C, Mak HY, Proudfoot NJ, Smith CW. 1998. *Nucleic Acids Res.* 26:5568–72
301. Ghosh S, Garcia-Blanco MA. 2000. *RNA* 6:1325–34

302. Thummel CS. 1992. *Science* 255: 39–40
303. Hatton AR, Subramaniam V, Lopez AJ. 1998. *Mol. Cell* 2:787–96
304. Corden JL, Patturajan M. 1997. *Trends Biochem. Sci.* 22:413–16
305. Morris DP, Greenleaf AL. 2000. *J. Biol. Chem.* 275:39935–43
306. Robert F, Blanchette M, Maes O, Chabot B, Coulombe B. 2002. *J. Biol. Chem.* 277:9302–6
307. McCracken S, Fong N, Yankulov K, Ballantyne S, Pan G, et al. 1997. *Nature* 385:357–61
308. Fong N, Bentley DL. 2001. *Genes Dev.* 15:1783–95
309. Komarnitsky P, Cho EJ, Buratowski S. 2000. *Genes Dev.* 14:2452–60
310. Hirose Y, Tacke R, Manley JL. 1999. *Genes Dev.* 13:1234–39
311. Zeng C, Berget SM. 2000. *Mol. Cell. Biol.* 20:8290–301
312. Fong YW, Zhou Q. 2001. *Nature* 414: 929–33
313. Monsalve M, Wu Z, Adelmant G, Puigserver P, Fan M, Spiegelman BM. 2000. *Mol. Cell* 6:307–16
314. Lai MC, Teh BH, Tarn WY. 1999. *J. Biol. Chem.* 274:11832–41
315. Bickmore WA, Oghene K, Little MH, Seawright A, van Heyningen V, Hastie ND. 1992. *Science* 257:235–37
316. Davies RC, Calvio C, Bratt E, Larsson SH, Lamond AI, Hastie ND. 1998. *Genes Dev.* 12:3217–25
317. Larsson SH, Charlieu JP, Miyagawa K, Engelkamp D, Rassoulzadegan M, et al. 1995. *Cell* 81:391–401
318. Reed R, Hurt E. 2002. *Cell* 108:523–31
319. Luo MJ, Reed R. 1999. *Proc. Natl. Acad. Sci. USA* 96:14937–42
320. Le Hir H, Gatfield D, Izaurralde E, Moore MJ. 2001. *EMBO J.* 20:4987–97
321. Le Hir H, Moore MJ, Maquat LE. 2000. *Genes Dev.* 14:1098–108
322. Le Hir H, Izaurralde E, Maquat LE, Moore MJ. 2000. *EMBO J.* 19:6860–69
323. Luo ML, Zhou Z, Magni K, Christoforides C, Rappsilber J, et al. 2001. *Nature* 413:644–47
324. Palacios IM. 2002. *Curr. Biol.* 12: R50–52
325. Berke JD, Sgambato V, Zhu P, Lavoie B, Vincent M, et al. 2001. *Neuron* 32:277–87
326. Chalfant CE, Mischak H, Watson JE, Winkler BC, Goodnight J, et al. 1995. *J. Biol. Chem.* 270:13326–32
327. Collett JW, Steele RE. 1993. *Dev. Biol.* 158:487–95
328. Rodger J, Davis S, Laroche S, Mallet J, Hicks A. 1998. *J. Neurochem.* 71: 666–75
329. Shifrin VI, Neel BG. 1993. *J. Biol. Chem.* 268:25376–84
330. Smith MA, Fanger GR, O'Connor LT, Bridle P, Maue RA. 1997. *J. Biol. Chem.* 272:15675–81
331. Wang A, Cohen DS, Palmer E, Sheppard D. 1991. *J. Biol. Chem.* 266:15598–601
332. Zacharias DA, Strehler EE. 1996. *Curr. Biol.* 6:1642–52
333. Stamm S. 2002. *Hum. Mol. Genet.* 11:2409–16
334. Herrlich P, Morrison H, Sleeman J, Orian-Rousseau V, Konig H, et al. 2000. *Ann. NY Acad. Sci.* 910:106–18; discussion 118–20
335. Stickeler E, Kittrell F, Medina D, Berget SM. 1999. *Oncogene* 18:3574–82
336. Konig H, Ponta H, Herrlich P. 1998. *EMBO J.* 17:2904–13
337. Weg-Remers S, Ponta H, Herrlich P, Konig H. 2001. *EMBO J.* 20:4194–203
338. Lynch KW, Weiss A. 2000. *Mol. Cell. Biol.* 20:70–80
339. Matter N, Marx M, Weg-Remers S, Ponta H, Herrlich P, Konig H. 2000. *J. Biol. Chem.* 275:35353–60
340. Daoud R, Berzaghi MD, Siedler F, Hubener M, Stamm S. 1999. *Eur. J. Neurosci.* 11:788–802
341. Vallano ML, Lambolez B, Audinat E, Rossier J. 1996. *J. Neurosci.* 16:631–39

342. Vallano ML, Beaman-Hall CM, Benmansour S. 1999. *NeuroReport* 10: 3659–64
343. Xie J, Black DL. 2001. *Nature* 410:936–39
344. van Oordt WV, Diaz-Meco MT, Lozano J, Krainer AR, Moscat J, Caceres JF. 2000. *J. Cell Biol.* 149:307–16
345. Daoud R, Mies G, Smialowska A, Olah L, Hossmann KA, Stamm S. 2002. *J. Neurosci.* 22:5889–99
346. Cyert MS. 2001. *J. Biol. Chem.* 276: 20805–8
347. Vandromme M, Gauthier-Rouviere C, Lamb N, Fernandez A. 1996. *Trends Biochem. Sci.* 21:59–64
347a. Shin C, Manley JL. 2002. *Cell* 111: 407–17
348. Robertson KD, Jones PA. 1999. *Oncogene* 18:3810–20
349. Sawa H, Ohshima TA, Ukita H, Murakami H, Chiba Y, et al. 1998. *Oncogene* 16:1701–12
350. Roberts GC, Smith CW. 2002. *Curr. Opin. Chem. Biol.* 6:375–83
350a. Woodley L, Valcárcel J. 2002. *Brief. Funct. Genomics Proteomics* 1:266–77
351. Marden JH, Fitzhugh GH, Girgenrath M, Wolf MR, Girgenrath S. 2001. *J. Exp. Biol.* 204:3457–70
352. Gavrilov DK, Shi X, Das K, Gilliam TC, Wang CH. 1998. *Nat. Genet.* 20:230–31
353. Nissim-Rafinia M, Kerem B. 2002. *Trends Genet.* 18:123–27
354. Clark TA, Sugnet CW, Ares M Jr. 2002. *Science* 296:907–10
355. Shoemaker DD, Schadt EE, Armour CD, He YD, Garrett-Engele P, McDonagh PD. 2001. *Nature* 409:922–27
356. Yeakley JM, Fan JB, Doucet D, Luo L, Wickham E, et al. 2002. *Nat. Biotechnol.* 20:353–58
357. Grabowski P. 2002. *Nat. Biotechnol.* 20:346–47
358. Brown V, Jin P, Ceman S, Darnell JC, O'Donnell WT, et al. 2001. *Cell* 107: 477–87
359. Rappsilber J, Ryder U, Lamond AI, Mann M. 2002. *Genome Res.* 12:1231–45
360. Neubauer G, King A, Rappsilber J, Calvio C, Watson M, et al. 1998. *Nat. Genet.* 20:46–50
361. Will CL, Luhrmann R. 1997. *Curr. Opin. Cell Biol.* 9:320–28
362. Gottschalk A, Neubauer G, Banroques J, Mann M, Luhrmann R, Fabrizio P. 1999. *EMBO J.* 18:4535–48
363. Zhou Z, Licklider LJ, Gygi SP, Reed R. 2002. *Nature* 419:182–85
364. Stevens SW, Ryan DE, Ge HY, Moore RE, Young MK, et al. 2002. *Mol. Cell* 9:31–44
365. Maroney PA, Romfo CM, Nilsen TW. 2000. *Mol. Cell* 6:317–28
366. Charlet N, Savkur RS, Singh G, Philips AV, Grice EA, Cooper TA. 2002. *Mol. Cell* 10:45–53
367. Savkur RS, Philips AV, Cooper TA. 2001. *Nat. Genet.* 29:40–47
368. Deleted in proof
369. Nicoll M, Akerib CC, Meyer BJ. 1997. *Nature* 388:200–4
370. Wu JI, Reed RB, Grabowski PJ, Artzt K. 2002. *Proc. Natl. Acad. Sci. USA* 99:4233–38
371. Stoss O, Olbrich M, Hartmann AM, Konig H, Memmott J, et al. 2001. *J. Biol. Chem.* 276:8665–73
372. Zachar Z, Chou TB, Kramer J, Mims IP, Bingham PM. 1994. *Genetics* 137: 139–50
373. Petersen-Mahrt SK, Estmer C, Ohrmalm C, Matthews DA, Russell WC, Akusjarvi G. 1999. *EMBO J.* 18:1014–24

Annu. Rev. Biochem. 2003. 72:337–366
doi: 10.1146/annurev.biochem.72.121801.161447
Copyright © 2003 by Annual Reviews. All rights reserved

COVALENT TRAPPING OF PROTEIN-DNA COMPLEXES

Gregory L. Verdine and Derek P.G. Norman
Department of Chemistry and Chemical Biology, Harvard University, Cambridge, Massachusetts 02138; email: verdine@chemistry.harvard.edu, dnorman@fas.harvard.edu

Key Words methyltransferase, glycosylase, recombinase, topoisomerase, reverse transcriptase, protein-DNA cross-linking, disulfide cross-linking, trapped complex

■ **Abstract** High-resolution structural studies of protein-DNA complexes have proven to be an invaluable means of understanding the diverse functions of proteins that manage the genome. Most of the structures determined to date represent proteins bound noncovalently to various DNA sequences or structures. Although noncovalent complexation is often adequate to study the structures of proteins that have robust, specific interactions with DNA, it is poorly suited to the study of transient intermediates in enzyme-catalyzed DNA processing reactions or of complexes that exist in multiple equilibrating forms. In recent years, strategies developed for the covalent trapping of protein-DNA complexes have begun to show promise as a window into an otherwise inaccessible world of structure.

CONTENTS

INTRODUCTION

One of the crowning achievements of modern biology has been to put a molecular face on practically every major aspect of genome function. High-resolution structures have illuminated, inter alia, how DNA is packaged into chromatin, bound by regulatory proteins, replicated, transcribed, recombined, relaxed, and repaired. The ironic yet inescapable conclusion of these studies, viewed collectively, is that DNA would be devoid of biologic activity were it

0066-4154/03/0707-0337$14.00

constrained to adopt only the canonical double-helical structure of Watson and Crick.

Despite such impressive progress, capturing structural snapshots of ordinarily fleeting intermediates in enzyme-mediated DNA processing events or of noncovalent protein-DNA complexes having multiple recognition modes remains a formidable challenge. For this reason, among others, the structural record is fragmentary, with certain areas of genome biology being broadly represented, whereas others are covered only sparsely in terms of available structures. Nearly every structural motif known to be involved in sequence-specific DNA recognition has been structurally characterized, often in multiple related forms. This wealth of structural information is a testament to the robustness and specificity that are typical of noncovalent complexes formed between regulatory proteins and their cognate sequences. On the other hand, structures of proteins that catalyze reactions on DNA have proven more difficult to obtain; for instance, not a single co-complex structure is available for a DNA helicase, Type II topoisomerase, or telomerase bound to its DNA substrate. These considerations illustrate the need for new ways of preparing protein-DNA complexes for structural studies.

This review is concerned with the use of covalent bond formation between proteins and DNA as a way of trapping the resulting complexes in discrete structural states. We focus specifically on examples in which covalent trapping has figured prominently in the determination of high-resolution structures.

Trapping strategies fall into two broad categories. Mechanism-based trapping relies on a detailed picture of the enzyme mechanism to subvert the normal course of the reaction through chemical manipulation, thereby resulting in the accumulation of an ordinarily transient intermediate or a close analog. Structure-based trapping takes advantage of known points of noncovalent contact between a protein and DNA as defined sites for engineering covalent cross-links. Examples of each are detailed in the pages that follow, with an emphasis on the trapping strategy and what has been learned from the resulting structures. Although the number of examples in which trapping strategies have been employed is still quite small, the following examples illustrate the unique advantages of such approaches for capturing proteins in action on DNA.

MECHANISM-BASED TRAPPING OF PROTEIN-DNA COMPLEXES

DNA (Cytosine-5)-Methyltransferases

Many organisms increase the information content of their genome by expanding the genetic alphabet from four bases (A, C, G, and T) to five, six, or even seven bases. The noncanonical bases are incorporated into the genome not by replication, but instead by covalent modification of canonical bases in DNA. The most common mode of such epigenetic covalent modification is the enzyme-catalyzed transfer of a methyl group from the cofactor S-adenosyl methionine (AdoMet) to either A or C, thereby producing N^6-methyl-A, N^4-methyl-C, or 5-methyl-C. All three of these bases are found in bacteria as components of the restriction/modification system that enables the host to distinguish its own DNA from that of foreign invaders, but only 5-methyl-C has been found thus far in eukaryotic genomes (1). In mammals, 5-methyl-C plays a key role in epigenetic silencing of gene expression and consequently is indispensable for genomic stability, embryonic development, genomic imprinting, and X-chromosome inactivation (2–5).

Just as the biology of DNA methylation has commanded much attention, so also has the chemistry of the methyl transfer process. On the one hand, methylation of the exocyclic nitrogen atoms in A and C does not seem to pose a tremendous challenge, as these nitrogens possess lone pairs of electrons with some degree of nucleophilic character. Methylation of the 5-carbon atom of cytosine is an altogether different proposition, appearing to require the abstraction of a proton having astronomically low acidity ($pK_a > 35$) at physiologic pH. The first inroad into the mechanism by which DNA cytosine-5 methyltransferases (MTases) accomplish this seemingly impossible feat was made by Santi and coworkers (6, 7), who proposed that MTases employ covalent catalysis involving a conserved Cys residue on the enzyme to activate C5 for methylation and proton abstraction. A subsequent refinement of the mechanism (Figure 1) (8) posited that proton transfers at N3, presumably facilitated by a general acid/base on the enzyme, lower the activation barrier for the operations at C6 and C5. Although the proposed mechanism seemed to solve the chemical questions surrounding methylation at C5 of cytosine, when viewed in the context of the structure of DNA, it raised two serious issues: (*a*) N3 of C is normally involved in Watson-Crick base pairing with the complementary G residue, and (*b*) the attack trajectories required for bond formation and breakage at C5 and C6 lie perpendicular to the plane of the base and are therefore blocked in duplex DNA by the stacked neighboring base pairs. Both of these issues raised the disturbing suggestion that significant disruption of duplex DNA structure might be a necessary prerequisite for catalysis by MTases (9). Testing this hypothesis through high-resolution structural studies would require some means of freezing the enzyme-substrate complex in the midst of catalysis.

A compelling piece of evidence in favor of the covalent catalysis scheme was the observation that DNA containing 5-fluoro-C, a mechanism-based inhibitor, became irreversibly attached to the bacterial MTase *Hha* I, but only in the presence of the cofactor AdoMet (Figure 1). The presumption was that the enzyme bound 5-fluoro-C and catalyzed methyl transfer to it just as with the normal substrate cytosine, but was then blocked from further progression since abstraction of F^+ at C5 is prohibitively high in energy. Were this indeed the case, the trapped covalent intermediate formed with 5-fluoro-C would differ from the corresponding catalytic intermediate (Figure 1, compare the two intermediate structures) only by the sterically conservative replacement of an H atom by F.

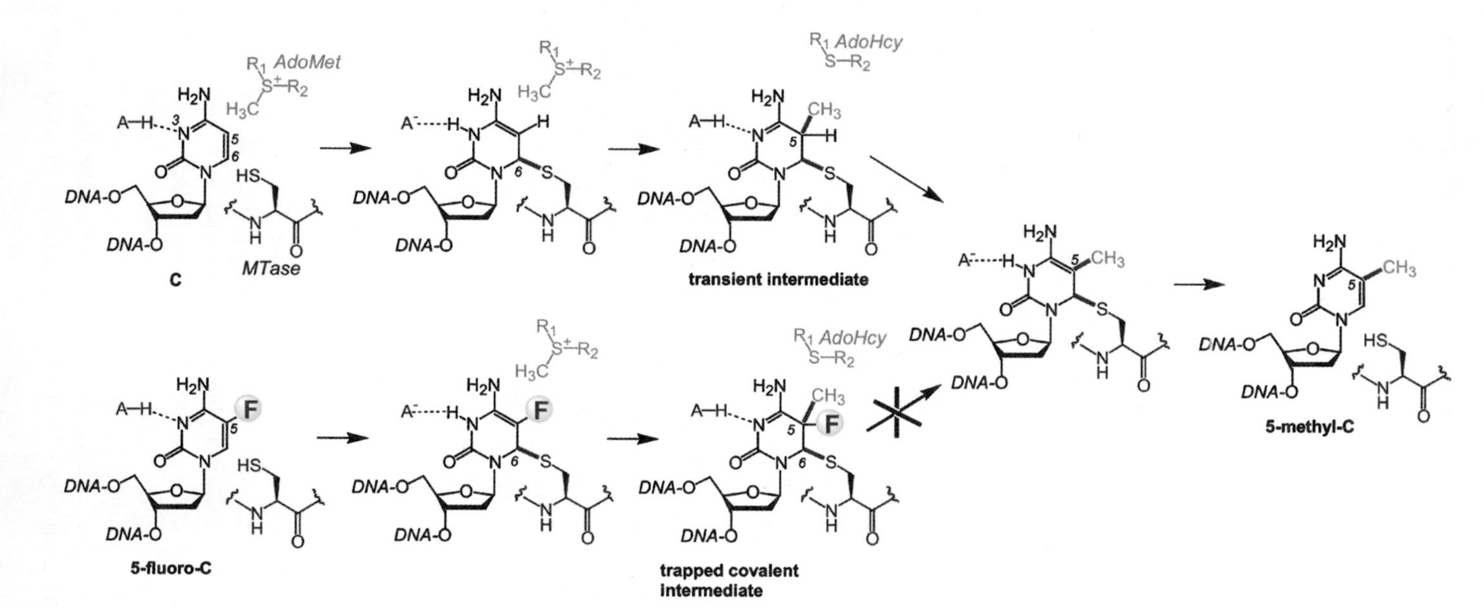

Figure 1 Catalytic mechanism of DNA (cytosine-5)-methyltransferases (*upper pathway*) and mode of inhibition by 5-fluoro-C in DNA (*lower pathway*). Following extrusion of the substrate cytosine from the DNA helix and insertion into the enzyme active site, a nucleophilic Cys residue (Cys 71 in *Hae* III methyltransferase) attacks C6 of the substrate. The resulting enamine undergoes alkylation at C5, then proton abstraction at C5, and conjugate elimination to provide the 5-methyl-C product and liberate the enzyme. With 5-fluoro-C, the enzyme and substrate become trapped in the midst of the catalytic cycle because of the inability of the enzyme to abstract F^+ from C5. Structures of the trapped covalent intermediate have been determined for *Hha* I and *Hae* III methyltransferases.

The availability of a reagent for site-specific incorporation of 5-fluoro-C into synthetic oligonucleotides (10) enabled the preparation of homogeneous trapped complexes of bacterial MTases, a prerequisite for structural studies, and led to the direct determination that the catalytic nucleophile is a Cys residue that is absolutely conserved in every member of this protein family, including all known mammalian MTases (DNMT1, DNMT3A, and DNMT3B).

The first structure of a trapped MTase-DNA structure was that of the *Hha* I enzyme, which transfers a methyl group to a cytosine (underlined) in the sequence 5′-G<u>C</u>GC (11). Not long thereafter, the structure of the trapped complex of *Hae* III methyltransferase was elucidated (12); this enzyme catalyzes methylation at 5′-GG<u>C</u>C. The essential features of both structures are quite similar, especially in the active site region, but the *Hae* III structure has an additional striking and completely unexpected rearrangment of DNA basc pairing, so we have chosen it for display here (Figure 2). The most dramatic feature of both MTase structures is the extrusion of the substrate cytosine from the DNA helix and its insertion into a deep concave pocket on the enzyme (Figure 2*A*, *B*). This "base-flipping" mechanism, more accurately described as nucleotide-flipping (Figure 2*C*), enables the enzyme to gain access to the substrate without interference from the surrounding duplex; equally important, the concave active site pocket provides a desolvated microenvironment within which to perform catalysis. These MTases are composed of two domains. The larger domain, which contains the active site, has the Rossman fold common to many nucleotide-binding proteins. The smaller domain, which makes extensive sequence-specific contacts in the major groove of the DNA, is highly unusual in the sense that it contains little if any repetitive secondary structure (α-helix or β-sheet). Even more intriguing is the fact that while the sequence-specific DNA recognition domains of both *Hae* III and *Hha* I MTases lack repetitive structure, their three-dimensional folds are distinct and nonhomologous.

The features of the MTase active sites are strikingly similar to each other and to those anticipated on the basis of biochemical considerations (Figure 2*B*). The sulfhydryl group of Cys 71 is clearly covalently bonded to C6, and a methyl group has been added at C5. The addition reaction to the 5,6-double bond takes place with *anti* stereochemistry, which dictates that the substituents involved in the subsequent elimination reaction have a *syn* relationship (Figure 2*B*, F = H in the natural substrate); this observation supports the notion that the elimination is not concerted but rather takes place in two steps (Figure 1). Curiously missing in both the *Hae* III and *Hha* I MTase structures is a candidate for the catalytic base that abstracts the C5-H in the elimination step, and indeed the identity of this base yet remains a mystery. Especially satisfying from a mechanistic standpoint is the positioning of a general acid/base, Glu 109 (Glu 119 in *Hha* I MTase), which is in intimate contact with both N3 and the 4-NH_2 of the substrate; the close proximity of these interacting partners essentially requires that either N3 or Glu 109 be protonated (compare Figure 2*B* and Figure 1).

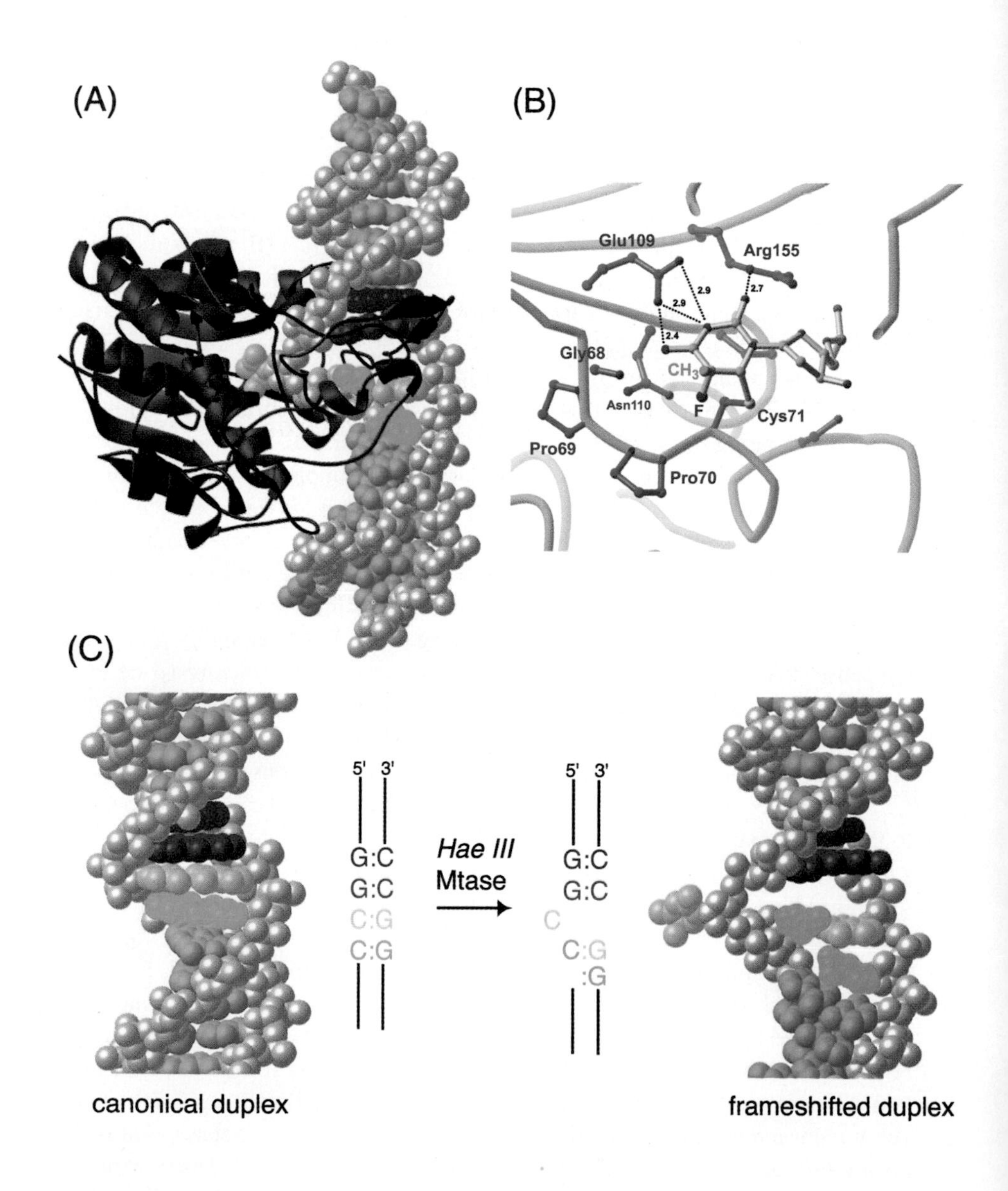
(A)
(B)
Glu109
Arg155
2.9
2.7
2.9
2.4
Gly68
CH_3
Asn110
F
Cys71
Pro69
Pro70
(C)
5' 3'
G:C
G:C
C:G
C:G
Hae III
Mtase
5' 3'
G:C
G:C
C
C:G
:G
canonical duplex
frameshifted duplex

Apart from its extrahelical dC nucleoside, the remainder of the DNA duplex has a rather normal B-form conformation when bound to *Hha* I MTase. This highly localized perturbation is in contrast to the drastic rearrangement of DNA structure seen in the complex with *Hae* III MTase (Figure 2). In this case, extrusion of the substrate nucleoside is accompanied by frameshifting of base pairing on the 3′-side of the recognition site. Thus, the guanine residue that was "estranged" by extrusion of the substrate dC from the helix is not left unpaired in the *Hae* III structure as it is in the *Hha* I structure, but instead forms a frameshifted pair with the cytosine base on the 3′-side of the substrate dC. This in turn leaves the guanine residue at the 3′-end of the recognition site without any pairing partner. This frameshifting appears to result from the intercalation of an amino acid residue, Ile 221, into the helical stack of the nonsubstrate strand, forcing it to dislocate. The substrate strand, on the other hand, cannot dislocate because it is held firmly in place through extensive interactions with the MTase.

Base-Excision Repair DNA Glycosylases

Inside the cell, DNA is subject to attack by a wide variety of reactive species, including electrophilic oxidants, ionizing radiation, ultraviolet light, methylating agents, and even water (13). Nearly all the resulting adducts interfere with the normal template function of the genome in one way or another, and many give rise to mutations by inducing illegitimate recombination or miscoding during DNA replication (14). All free-standing organisms dedicate a considerable fraction of their genomes to encoding proteins that search for and repair DNA

←

Figure 2 The structure of the *Hae* III MTase trapped on an oligonucleotide containing 5-fluoro-C. (*A*) Overall structure of the trapped protein-DNA complex. The protein is depicted as a ribbon trace, and the DNA is in space-filling model, with the bases of the recognition site color-coded; bases outside the recognition site are blue. Note that substrate cytosine (yellow) is extrahelical and inserted into an active site pocket on the protein. (*B*) Close-up view of the *Hae* III MTase active site. The protein backbone is depicted as a tube model in gray, with particular side chains shown in blue. The substrate cytosine is shown as a gold framework, with the 5-methyl group in green and 5-fluoro group in magenta. Note the covalent addition of Cys 71 to C6 and the key interaction of Glu 109 with N3 of the substrate (see also Figure 1). The precise stereochemistry at C6 could not be deduced from the experimental electron density, but it was inferred from the location of the AdoMet binding pocket (cofactor not shown). (*C*) Space-filling model of canonical B-form DNA (*left*) and the DNA in the crystal structure of the trapped *Hae* III-DNA complex (*right*). Bases of the recognition site are color-coded according to their pairing relationships in canonical DNA. In addition to the extrahelical orientation of the substrate cytosine, note the frameshifted base pair in the *Hae* III MTase structure (*right*, green guanine paired with yellow cytosine). Note also the large gap in the duplex above the frameshifted pair; Ile 221 is intercalated into the helix at this location (Ile 221 not shown).

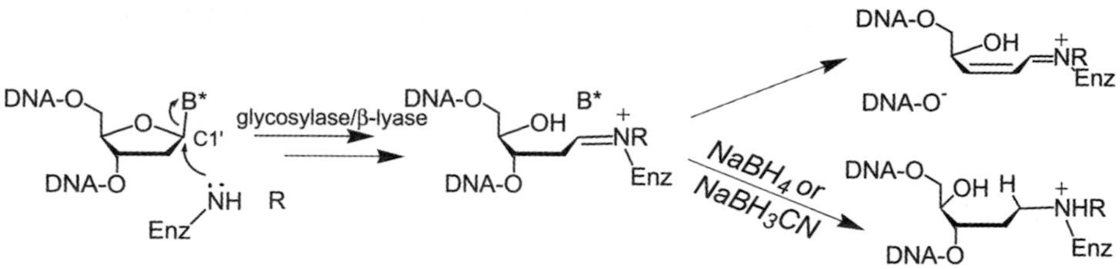

Figure 3 Catalytic mechanism of glycosylase/β-lyases and mode of trapping by borohydride. This class of glycosylase utilizes an enzyme-borne nitrogen nucleophile to displace the lesion base (B*) by attack at C1′. Rearrangement of the initial covalent intermediate leads to the formation of a Schiff base intermediate, which ordinarily proceeds further to yield a strand-cleaved product; the final step, hydrolysis to liberate the enzyme, is not shown. Alternatively, the Schiff base can be intercepted by addition of the borohydride reducing agents $NaBH_4$ or $NaBH_3CN$, leading to irreversible trapping of the glycosylase on its DNA substrate. Note that the product of borohydride trapping differs from the Schiff base intermediate only by the bond order at the site of attachment and the addition of hydrogen atoms to C1′ and the nucleophilic N atom. The R = H in the case of hOgg1, and R = CH_2- in the cases of MutM and Nei, which utilize an N-terminal proline as the nucleophile (see text for further explanation).

damage. A major challenge in this area has been to understand how DNA repair proteins distinguish such lesions from the vast excess of normal DNA.

Most small single-base lesions in DNA, including uracil, 8-oxoguanine, N7-methylguanine, N3-methyladenine, and thymine glycol, are repaired by the base-excision repair (BER) pathway (15, 16). BER is initiated by lesion-specific DNA *N*-glycosylases (17–21), which catalyze excision of the damaged base through nucleophilic substitution at C1′. Following removal of the damaged base, the resulting abasic site is cleaved from DNA and the strand is restored through an orchestrated series of processing events involving multiple enzymes of the BER pathway.

DNA glycosylases fall into two mechanistic classes that utilize distinct yet related mechanisms. (*a*) Monofunctional DNA glycosylases catalyze a single chemical transformation, namely, displacement of a damaged base by an activated water molecule to produce an abasic deoxyribose moiety. Numerous X-ray structures are available of monofunctional DNA glycosylases, both alone and in complex with DNA (UDG, MUG, AlkA, Aag, MutY). (*b*) DNA glycosylase/β-lyases also catalyze displacement of the base, but they use the nitrogen atom of an amine moiety on the protein as the nucleophile (Figure 3). The initially formed covalent protein-DNA adduct, an aminal species rearranges to a Schiff base, which subsequently undergoes β-elimination to cleave the DNA backbone at the 3′-carbon of the lesion. Some glycosylase/β-lyases proceed further to catalyze δ-elimination, resulting in backbone cleavage at the 5′-carbon and complete removal of the lesion nucleoside from DNA.

The cascade of at least five distinct chemical transformations catalyzed by glycosylase/β-lyases occurs within a single active site, a marvel of catalytic versatility otherwise unprecedented in the biochemistry of protein-nucleic acid complexes. Mutational inactivation of catalytic activity and synthesis of uncleavable substrates have enabled the co-crystallization of stable recognition complexes having glycosylase/lyases bound to DNA (19, 22–24). The structures of these complexes have shed light on the overall architecture of the protein-DNA interaction and in at least one case have provided insight into lesion recognition (23), but have been less informative with respect to catalysis. The recent application of mechanism-based trapping to this problem has led to significant new insights into catalysis of the β-lyase cascade.

The Schiff-base intermediate through which glycosylase/β-lyases proceed can be intercepted in situ by the reducing agents sodium borohydride ($NaBH_4$) and sodium cyanoborohydride ($NaBH_3CN$) (Figure 3) (25–28). Whereas the Schiff base (imine) intermediate is a transient species, the amine linkage resulting from borohydride reduction is so stable that it is incapable of being further processed by the enzyme. Hence, the glycosylase and its DNA substrate become irreversibly trapped at an intermediate stage of the catalytic cycle. Borohydride trapping has emerged as an important tool for in vitro studies of glycosylase/lyases; it has been used among other things to identify unknown glycosylases (29–31), to pinpoint their active site nucleophiles (26, 28, 32, 33), and to probe their catalytic mechanisms (26, 28, 34, 35). Recently, the scope of borohydride trapping has been expanded to include the preparation of complexes suitable for X-ray crystallographic structure determination. Thus far, high-resolution structures of three trapped glycosylase/β-lyases have been reported: Nei (36), MutM (37, 38), and hOgg1 (38a). Nei, also known as endonuclease VIII, is a bacterial DNA glycosylase that excises aberrant pyrimidines, including thymine glycol, uracil glycol, dihydrothymine, dihydrouracil, 5-hydroxycytosine, and 5-hydroxyuracil (39, 40). MutM, also known as Fpg, is a bacterial glycosylase that removes 8-oxoguanine (oxoG), formamidopyrimidines (FaPy) (41–43), and 5-hydroxycytosine (44) from DNA. Both Nei and MutM utilize the amine group of their N-terminal proline residue (Pro 2) as the catalytic nucleophile that displaces lesion bases (32, 33). hOgg1, a member of the HhH-GPD superfamily of DNA glycosylases (29, 45, 46), is responsible for the excision of 8-oxoguanine (oxoG) and FaPy lesions from DNA (47). The active site nucleophile of hOgg1 is a lysine residue located at position 249 (Lys 249) within the HhH-GPD motif; this Lys residue is absolutely conserved in all members of the superfamily that are glycosylase/β-lyases, but it is not conserved in those members that are monofunctional glycosylases.

The borohydride-trapped amine linkage is expected to represent a reasonably close mimic of the actual Schiff base intermediate (see Figure 3). In terms of chemical constitution, the former differs from the latter by only the addition of two H atoms, one on the C1′ of the substrate and the other on the N atom of the catalytic nucleophile (either N^{α} of Pro 2 or N^{ϵ} of Lys 249). This change in bond order is accompanied by slight elongation of the bond from ~1.3 Å (C = N) to ~1.5 Å (C-N). Although this difference in bond length is probably negligible, the change in hybridization from sp^2(C = N) to sp^3 (C-N) could result in some local conformational adjustments. Importantly, both the Schiff base and the amine are expected to bear formal positive charge on the N atom covalently attached to DNA (N^α of Pro 2 or N^ϵ atom of Lys 249).

COVALENT TRAPPING OF MutM Crystal structures of MutM from *Escherichia coli* (37) and *Bacillus stearothermophilus* (38) borohydride-trapped to oxoG-containing DNA have been determined independently. Apart from minor differences resulting from nonidentity of certain amino acid residues, the two structures are nearly indistinguishable (overall backbone RMSD = 1.05 Å). Both structures show continuous electron density between Pro 2 N^α and C1′ of the deoxyribose moiety, with an interatomic distance that is consistent with the presence of a covalent cross-link at this site (Figure 4*A*). In neither structure is electron density visible for the expelled oxoG base, nor does the segment of the protein likely to comprise the base-binding pocket possess a well-ordered structure. Most likely, this segment of protein structure becomes ordered in the presence of the lesion base but does not bind free oxoG strongly enough to retain the base following detachment from the DNA backbone (38).

The active sites of the MutM structures reveal a molecular architecture well suited to carry out the consecutive transformations of base removal and β- and δ-elimination. Comparison of the borohydride-trapped structures with other MutM structures lacking the covalent enzyme-substrate linkage, including an end product complex in which the lesion sugar is entirely absent (38), suggest that progression through the steps of the β-lyase cascade occurs with little if any change in the structure of the protein-DNA interface (24, 38). Curiously absent from the structures is a candidate catalytic acid/base to assist in catalysis of the β-lyase cascade. Even though the conserved residue Glu 3 is hydrogen bonded to $O^{4'}$ of the ring-opened deoxyribose moiety (Figure 4*A*), a role for Glu 3 in catalysis of the β-lyase cascade has been ruled out on biochemical grounds (36). Glu 3 does play an important structural role as an N-terminal helix cap for helix α-E, which points directly toward the active site; furthermore, the negative charge on Glu 3 is also believed to provide transition state stabilization for the positive charge developed on $O^{4'}$ in the glycosyl transfer step. Intriguingly, the borohydride-trapped *B. stearothermophilus* structure contains a well-ordered water molecule located 3.5 Å from the pro-*S* hydrogen of C2′, suggesting the possibility that solvent molecules may participate as specific acid/base catalysts in the reaction.

COVALENT TRAPPING OF Nei The only structure of Nei available thus far (36) was obtained by crystallization of a borohydride-trapped *E. coli* Nei-DNA complex. Although Nei has only 29% sequence identity (44% similarity) with MutM in *B. stearothermophilus* and acts on a different set of substrates, the two

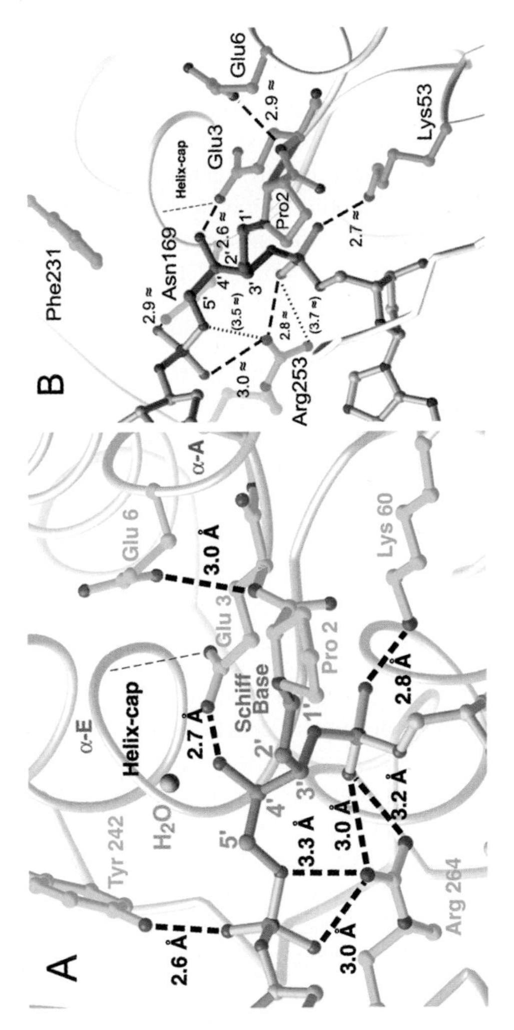

Figure 4 Active sites of the borohydride-trapped structures of (*A*) *B. stearothermophilus* MutM and (*B*) *E. coli* Nei (endonuclease VIII). Note the covalent bond between the N-terminal Pro (Pro 2) and the C1′ of the substrate sugar moiety. Note also the extensive interactions of the enzyme with the DNA backbone phosphates flanking the lesion. Reproduced with permission from Nature Publishing Group, http://www.nature.com/molcellbio/.

proteins have the same overall fold (main chain RMSD = 1.8 Å) and similar active sites (Figure 4*B*, compare with 4*A*). In the Nei structure, electron density corresponding to the lesion base (thymine glycol) is absent, just as with MutM. However, the segment of Nei most likely to be involved in base recognition appears to be well-ordered, whereas that of MutM is disordered. The stretch of protein sequence comprising the base-recognition pocket is almost devoid of conserved residues in Nei versus MutM, perhaps reflecting the fact that the two proteins recognize structurally dissimilar lesions. Efforts to soak thymine glycol into the crystal using a 20 mM solution of the lesion base were unsuccessful, suggesting that the free base has a low affinity for the protein. Elucidation of the structural basis for lesion recognition by both Nei and MutM will have to wait for co-crystal structures having an intact lesion nucleoside. The identity of the general base that facilitates β- and δ-elimination by deprotonating C2′ and C4′, respectively, is not apparent from the Nei structure; the authors propose that water molecules activated by Glu 3 (numbered Glu 2 in the original paper) perform this role (36).

COVALENT TRAPPING OF hOgg1 Unlike Nei and MutM, hOgg1 retains its lesion base for at least a few hours following borohydride trapping and much longer when the complex is crystallized. Thus, in the X-ray structure of hOgg1 borohydride-trapped in the presence of oxoG-containing DNA, electron density for oxoG is clearly evident within the base-recognition pocket (38a). Indeed, the position of the oxoG base is nearly superimposable on that of a catalytically inactive recognition complex having an intact oxoG 2′-deoxynucleoside bound within the active site (23). In the trapped complex, experimental electron density for all the atoms throughout the extended chain formed by cross-linking the substrate sugar to Lys 249 are readily discernible, leaving no doubt that the two are engaged in a covalent bond between C1′ and N^{ϵ} (Figure 5).

One of the most interesting biochemical questions with hOgg1 concerns the mechanism by which the protein catalyzes an entire cascade of reactions using a single active site. Most of the elementary steps in the mechanism involve transfer of protons among three atoms, N^{ϵ}, C2′, and $O^{4'}$, suggesting there should be residues positioned in the active site to carry out general acid/base chemistry at these three atoms. Curiously, however, the only obvious candidate acid/base, Asp 268, has been ruled out on the basis of biochemical and structural studies (22). The structure of the borohydride-trapped complex (Figure 5) (38a) has suggested a surprising answer to the question of what performs the proton transfer steps. The oxoG base itself appears to act as an acid/base cofactor. The structure reveals that not only is oxoG retained in the active site, but its N9 atom is hydrogen bonded to two of the three atoms that undergo proton transfers ($O^{4'}$ and N^{ϵ}) and is also close to the third (C2′). Such a hydrogen bonding arrangement is most consistent with the oxoG base being deprotonated at N9 (i.e., the oxoG is an anion), suggesting that the enzyme lowers the pK_a of the base below its apparent value in solution, ~10.8 (D.P.G.N., J.C. Fromme & G.L.V., unpublished observa-

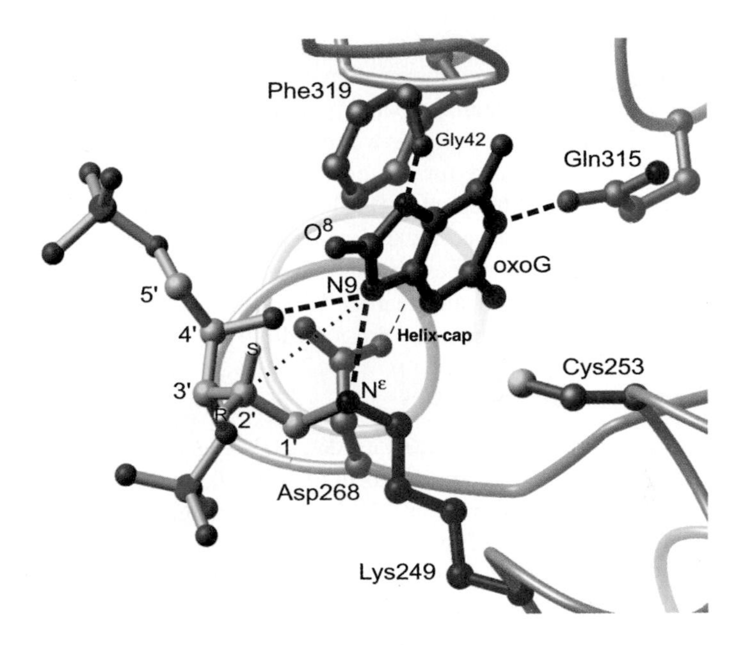

Figure 5 Active site of hOgg1 borohydride-trapped to oxoG-containing DNA. The N^{ϵ} atom of the active-site lysine is covalently linked to C1′ of the ring-opened ribose (see Figure 3). The oxoG base remains bound in the active site, and its N9 atom appears to make favorable hydrogen bonding interactions with O4′ (2.6 Å) and N^{ϵ} of Lys 249 (2.6 Å), suggesting that N9 is probably negatively charged. Specificity for oxoG versus G is conferred by a hydrogen bonding contact between the N7-H of oxoG and the backbone carbonyl of Gly 42.

tions). Such a dramatic pK_a lowering has been observed in another base-excision DNA repair system: The pK_a of uracil is lowered by 3.4 units relative to the solution value when bound to the active site of uracil-DNA glycosylase (48). Karplus and coworkers have proposed, based on calculations, that uracil DNA glycosylase excises uracil from DNA through a transition state in which the base is anionic (49), and a similar situation may pertain to hOgg1.

The placement of the putatively anionic oxoG base in intimate association with N^{ϵ}, C2′, and $O^{4'}$ suggests that the excised oxoG free base may act as an acid/base cofactor to catalyze the multiple transformations leading to DNA strand scission. This proposition has found independent biochemical support through findings that addition of the purine analogs 8-bromoguanine and 8-aminoguanine (oxoG is too insoluble to be used in these experiments) greatly accelerated the rate of DNA strand cleavage catalyzed by hOgg1 on an abasic DNA substrate (38a). The fact that the maximal rate of acceleration provided by 8-bromoguanine is different from that of 8-aminoguanine, yet the X-ray structures of borohydride-trapped complexes containing the two analogs are virtually identical, strongly suggests that the analogs, and by extension oxoG, participate directly in at least the rate-limiting step of the mechanism. Using a substrate in which tritium had been stereoselectively introduced to the pro-*R* position of C2′, the presence of either of these two base analogs profoundly influenced the stereochemical preference for proton abstraction at C2′ (38a). Although the pro-*R* selectivity exhibited by hOgg1 in the presence of purine analogs is reversed from that observed for endonuclease III (50), it is consistent with the results of molecular dynamics simulations conducted on the Schiff base intermediate (38a), which indicates that the lowest-energy state of the system places N9 of the oxoG base nearest the pro-*R* hydrogen.

DNA Strand-Invading Enzymes

COVALENT TRAPPING OF Cre RECOMBINASE DNA recombination plays a key role in a variety of crucial genomic processing events, including the generation of genetic diversity during meiosis, DNA repair through homologous recombination, and the integration of the bacteriophage and virus genomes into and out of their host genomes (51). A subset of these enzymes show exquisite sequence specificity in their selection of recombination targets; these sequence-specific recombinases have proven thus far the most amenable to high-resolution structural studies in complex with DNA. Covalent trapping has provided a valuable tool in this area for capturing recombination intermediates.

There are two families of DNA recombinases, the λ integrase family, also known as tyrosine recombinases, and the resolvase-invertase family, also known as serine recombinases. The defining characteristic of the λ integrase family is that its members utilize an active site tyrosine as a nucleophile to attack the phosphodiester backbone of the substrate DNA, creating a nick in which the enzyme is linked to a 3′-phosphate and leaving a free 5′-hydroxyl group (Figure 6*A*). Many members of the λ integrase family cannot perform recombination by themselves, but only do so in the obligatory presence of additional protein factors; however, Cre recombinase, a 38-kDa protein from the bacteriophage P1, carries out recombination at specific 34 base-pair elements known as *loxP* sites without the need for accessory proteins (52). The high sequence specificity and biochemical simplicity of the Cre-*lox* system have made it the nearly exclusive option for carrying out targeted recombination in genetically engineered mice (53, 54). These same properties have enabled Van Duyne and coworkers to acquire a wealth of structural information on the system, such that the entire recombination event can now be described at high resolution (55–57). As the structures of Cre recombinase have been the subject of a comprehensive review (58), discussion here is limited to the mechanism by which its DNA complex was covalently trapped.

Recombination is a topologically complex process that proceeds through a discrete sequence of molecular events (Figure 6*B*). Prior to the determination of the structures, it was difficult to envisage how a single enzyme could choreograph such a complex and demanding series of events. In Cre-mediated recombination, a tetramer of Cre recognizes two pseudopalindromic *loxP* sites, one on the donor (phage) DNA duplex and the other on the recipient (host) duplex. A set of Cre protamers located opposite one another deliver their active site tyrosine residues (Tyr 324) to symmetry-related phosphodiesters on each of the two strands, forming two nicks, each having a tyrosine 3′-phosphodiester bond to the enzyme and a free 5′-hydroxyl group. The DNA strands containing the freed 5′-hydroxyl groups become locally unpaired and reach across to attack the opposite tyrosine 3′-phosphodiester intermediate, displacing the tyrosine side chain to form the Holliday junction, a four-way branched structure. Then the two previously unreacted Tyr 324 side chains carry out a second round of strand invasion, which is once again followed by 5′-hydroxyl attack on the opposite tyrosine 3′-phosphodiester, completing the sequence of strand-exchange events.

An important breakthrough in this area was the development of a novel mechanism-based strategy to trap the 3′-phosphotyrosine recombination intermediate of λ integrase (59). Specifically, when λ integrase processes a DNA strand synthesized to contain a nick one base pair 3′ to the site of strand invasion, the nucleoside located between the successive nicks floats away into solution; hence its 5′-hydroxyl group is unavailable to participate in displacement of the 3′-phosphotyrosine intermediate (see below). Application of essentially this same strategy to the Cre-*lox* system (Figure 6*C*) made it possible to obtain the first co-crystals of a trapped covalent intermediate in DNA recombination by Cre (56). To simplify the system, a perfectly symmetrized version of *loxP*, dubbed *loxA*, was used. A *loxA* substrate was constructed from two single-stranded oligonucleotides, which annealed to form a duplex DNA containing two symmetric nicks (Figure 6*C*) (56). These nicks were introduced at a location one base pair 3′ to the site of tyrosine invasion. Attack on this substrate by Cre was expected to result in loss of the cytidine nucleoside between the enzyme-induced and synthetic nicks, producing a gapped duplex unsuitable for further reaction. The complementary strand also contains a nick, but this is bound to a Cre subunit that is inactive at this stage of the reaction sequence, therefore no reaction takes place at this site.

The structure of the Cre-*loxA* complex reveals a dimer of dimers containing two types of active sites, one is bound to the DNA via a tyrosine 3′-phosphodiester linkage, and the other contains an unreacted Tyr 324. Interestingly, in the unreacted active site, the interactions of the scissile phosphate are similar to those seen in the reacted active site, with several contacts between positively charged side chains and the nonbridging phosphate oxygens; however, the tyrosine nucleophile (Tyr 324) is retracted from the scissile phosphate (5.8 Å Tyr-O/phosphate-P distance), thereby preventing attack by tyrosine on the phosphate. The structure of the trapped complex thus explains why the Cre tetramer, despite being simultaneously bound to DNA at all four active sites, activates strand cleavage at only two. Interestingly, the pair of active subunits must convert to inactive ones, and vice versa, following formation of the Holliday junction. Subsequent structural elucidations of an unreacted Cre-*loxP*

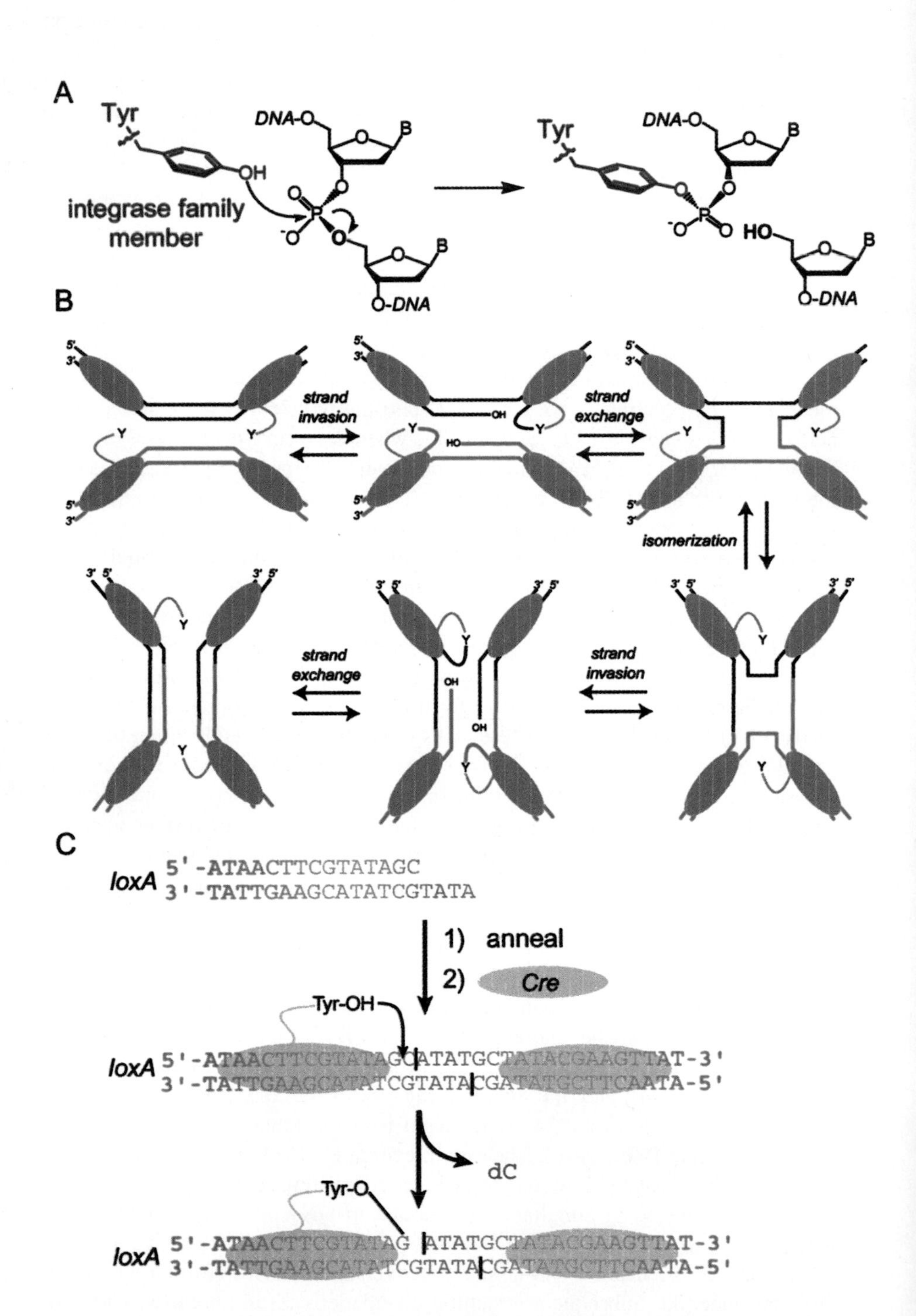
A
Tyr
DNA-O
B
OH
integrase family
member
O
B
O-DNA
Tyr
DNA-O
B
HO
B
O-DNA
B
strand
invasion
strand
exchange
isomerization
strand
exchange
strand
invasion
C
loxA 5'-ATAACTTCGTATAGC
3'-TATTGAAGCATATCGTATA
1) anneal
2) Cre
Tyr-OH
loxA 5'-ATAACTTCGTATAGCATATGCTATACGAAGTTAT-3'
3'-TATTGAAGCATATCGTATACGATATGCTTCAATA-5'
dC
Tyr-O
loxA 5'-ATAACTTCGTATAG ATATGCTATACGAAGTTAT-3'
3'-TATTGAAGCATATCGTATACGATATGCTTCAATA-5'

complex (57) and of a Cre-*loxP* Holliday junction intermediate (55) have indicated that this switch is accomplished through relatively modest movements at the intersubunit interfaces. In the trapped complex, the DNA strand that ordinarily migrates across the junction to attack the opposite duplex is missing an entire nucleoside at its 5′-end, making the next step of the reaction sequence unclear from this structure. However, it is noteworthy that the two other nucleotides on the 5′-end of the freed strand are not paired to their original partners but are seen reaching out into the solvent-filled cavity between the duplexes, as if trying to home in on the distal tyrosine 3′-phosphate target; the three base pairs on the 5′-end of the liberated strand are known to be swapped during recombination.

COVALENT CAPTURE OF TOPOISOMERASE I Topoisomerases perform a variety of functions essential for cell viability and proliferation, such as modulating the superhelical state of chromatin during replication and transcription and decatenating sister chromatids during mitosis (60). Two families of topoisomerases are known. Type II topoisomerases (topo II) catalyze ATP-dependent DNA decatenation and modulation of DNA supercoils by nicking both DNA strands to form a transient double-stranded break through which another duplex DNA can pass. Type I topoisomerases, which do not utilize ATP, can be divided into two subfamilies, Type IA and Type IB. Type IA topoisomerases (topo IA) introduce a nick in a single-stranded region of DNA, allowing strand passage to occur through the nick. Type IB topoisomerases (topo IB) nick only one of the two DNA strands in double-stranded DNA, thereby allowing the DNA to relax by swiveling about the backbone of the uncleaved strand; because topo I cannot utilize ATP as a source of energy, it is incapable of increasing the superhelical stress of DNA.

Figure 6 Chemical and biochemical mechanisms of recombination/integration. (*A*) Integrase family members use a tyrosine residue to attack the scissile phosphate, displacing the 5′-hydroxyl. (*B*) The overall mechanism of Cre-mediated recombination. After initial strand invasion by two symmetry-related Tyr residues (*Y*) on the Cre tetramer (purple) (see also part *A*), the free 5′-hydroxyls (OH) attack the opposite tyrosine phosphodiester to effect strand exchange. The resulting Holliday junction isomerizes to activate the opposite two Cre protamers, which then use their active site tyrosine residues to perform a second round of strand invasion. The 5′-hydroxyls resulting from strand invasion then attack the opposite tyrosine phosphodiester, completing the second strand exchange event. (*C*) Trapping strategy. A duplex oligonucleotide containing a Cre half-site was designed to self-assemble by pairing to provide a symmetric recognition site, *loxA*, containing two nicks (*vertical bars*). Upon incubation with Cre, Tyr 324 invades the strand one base pair upstream of the nick, displacing a 2′-deoxycytidine (dC) nucleotide, which diffuses away. Loss of the dC prevents religation of the DNA, hence the Cre protein becomes irreversibly trapped to the DNA.

Like members of the λ integrase family, topoisomerases nick DNA via attack of a nucleophilic tyrosine residue on a phosphodiester linkage (61–63). Topo IB produces the same type of nick as λ integrase, with a 3′-phosphotyrosine link to the enzyme and a free 5′-hydroxyl. Curiously, topo II and topo IA form a 5′-phosphotyrosine linkage and leave a 3′-hydroxyl. The lack of any significant degree of sequence specificity has been a major impediment to high-resolution structural studies on topo-DNA complexes. Thus, although numerous reports of various forms of topo protein domains have appeared in the literature (64–70), only two papers have reported structural studies on a topoisomerase in complex with duplex DNA (71, 71a). In the first work (71), Hol and coworkers took advantage of a clever covalent trapping strategy first employed with λ integrase and calf thymus topo I (72). Replacement of the bridging 5′-oxygen in a phosphodiester linkage with sulfur yields the 5′-bridging phosphorothioate analog (Figure 7*A*, *left*). This analog turns out to be an excellent substrate for cleavage (nicking) by enzymes like λ integrase and topo IB, perhaps because the P-S bond is weak relative to P-O and the 5′-thiolate is an excellent leaving group. Analysis of the single-turnover rate of cleavage of the natural and phosphorothioate substrates shows that they are cleaved at roughly the same rate, although the natural substrate's rate of cleavage was slightly higher (73). However, in the reverse direction, the ligation reaction is impeded by both the unfavorable energetics of replacing a P-O bond with P-S, and by the poor nucleophilicity of a thiolate with respect to attack on phosphorus (74). Thus, substitution by the bridging phosphorothioate shifts the equilibrium position of the DNA cleavage reaction ($K_{cleavage}$) completely toward the cleaved species, predominantly by decreasing the rate constant in the reverse direction for strand ligation ($k_{ligation}$).

Hol and coworkers also took advantage of the fact that human topoisomerase I exhibits a modest preference to nick at a defined position within a DNA sequence of *Tetrahymena thermophilus* encoding ribosomal RNA (75). By incorporation of the 5′-bridging phosphorothioate into a postion that showed preferential nicking on native DNA, the combined enhancements of strand nicking were sufficient to yield a single trapped product that could be crystallized and used for X-ray structure determination (Figure 7*B*, *C*). In a parallel effort, a catalytically inactive mutant of topo I (Y723F) yielded crystals with the active site bound over the same position in an intact DNA duplex.

The structures of the noncovalent and covalently trapped structures of human topo I proved remarkably similar, with an rms deviation of 0.45 Å for the C_α atoms. Unexpectedly, residues 440–614 of human topo I, as well as Tyr 723, were found to be structurally homologous to bacteriophage HP1 integrase (rms deviation of 3.8 Å), despite the two having just 12% sequence identity. Significantly, many of the features of the active site are conserved between the two, including a catalytically critical histidine side chain that in integrase is positioned to stabilize the nonbridging oxygens of the scissile phosphate. This underlying

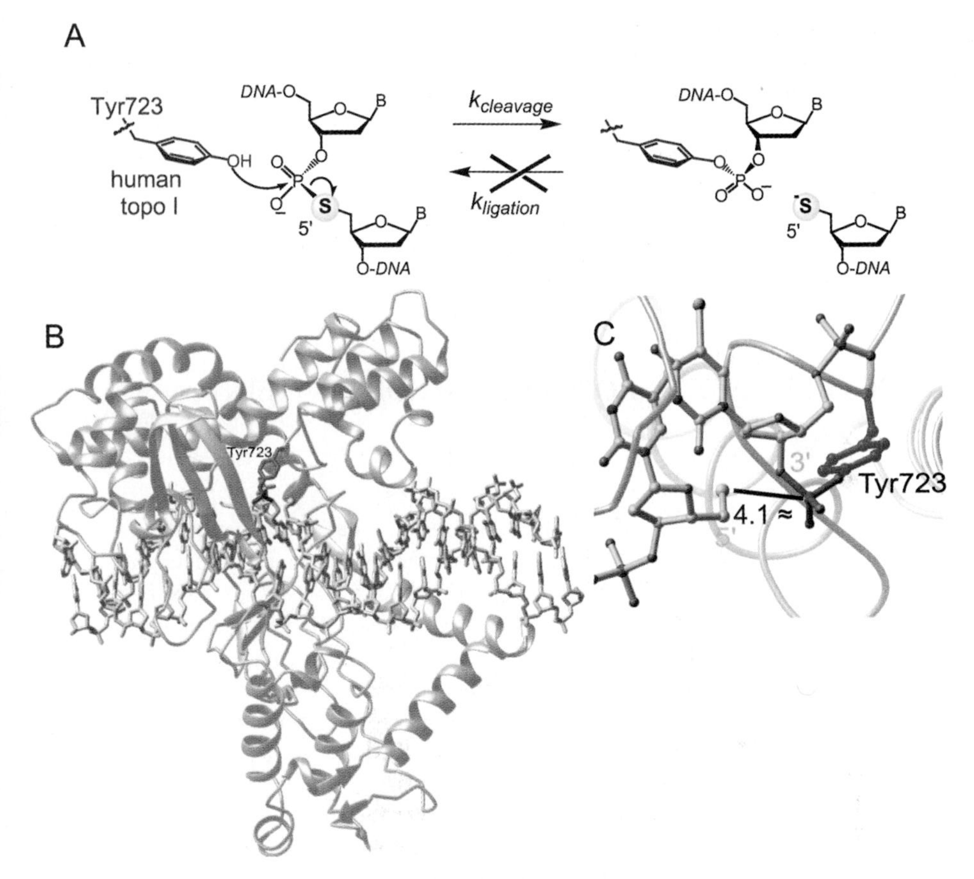

Figure 7 Topoisomerase I trapping mechanism and structure. (*A*) Trapping mechanism. The active site tyrosine of human topo I (Tyr 723) is depicted attacking a bridging 5′-phosphorothioate at the scissile position of the recognition sequence to yield a 5′-thiolate product, which is unable to perform religation. (*B*) The overall structure of the trapped complex. Human topoisomerase I is shown as a gray ribbon trace, and the DNA is depicted as a blue-green framework model. The active site Tyr (navy) is attached to the scissile phosphate (magenta); the two residues flanking the nick are in yellow. (*C*) Close-up view of the active site, with the same color scheme as in (*B*); the distance from the 5′-thiol(ate) to the scissile phosphate is shown explicitly.

structural similarity and active site architecture serve to highlight the similarities in fundamental reaction chemistry performed by the two enzymes, although the deployment of this chemisty toward a biological end differs so strikingly between the two.

Stewart and coworkers have recently reported an elaboration of this mechanism-based trapping methodology to enable capture of a topo I-DNA complex bound to Topotecan, a topoisomerase poison clinically used in the treatment of cancer (71a). The drug was found to intercalate into DNA at the site of incision, essentially substituting for a base pair. In addition to spreading apart the neighboring base pairs, the twist of the DNA was reduced, moving the 5′-SH 11.5 Å from the phosphorus atom of the phosphotyrosine linkage (compare to Figure 7*C*).

A topo II-duplex DNA complex awaits structural elucidation. This challenge very well may be overcome with the use of a strategy similar to that used for Cre recombinase or human topo I. Alternatively, the method outlined below, disulfide cross-linking, may prove amenable to use with topo II.

STRUCTURE-BASED CROSS-LINKING OF PROTEIN-DNA COMPLEXES

As mentioned above, elucidating the structures of protein-DNA complexes having modest affinity or specificity presents a major challenge. It has proven difficult to crystallize sequence- or structure-specific complexes having an equilibrium dissociation constant (K_d) greater than 10^{-8} M, perhaps because the lack of preorganization in such systems imposes a further entropic penalty on crystallization. Weakly bound complexes are likewise difficult to study by NMR because their dynamical properties typically yield line-broadened spectra. Protein-DNA complexes having less than tenfold higher affinity for a specific versus nonspecific oligonucleotide are also difficult to study by X ray or NMR because they are inherently inhomogeneous, even when tightly associated. Covalent trapping represents an attractive option to stabilize otherwise weakly bound complexes and to enforce structural homogeneity upon otherwise inhomogeneous ones. With certain DNA-modifying enzymes, exemplified by the cases above, enough information is known about the mode of enzyme action to devise a mechanism-based trapping strategy. In numerous other cases, including all DNA-binding proteins devoid of catalytic activity, no such opportunity for mechanism-based trapping presents itself, and hence an alternative trapping strategy is required. For such applications, it has proven valuable to engineer site-specific covalent cross-links between the protein and DNA.

The most widely used technique for covalent trapping of proteins on DNA is photo-cross-linking. Typically, photoactive nucleosides such as 5-bromouracil, 5-iodouracil (76), or 8-azido-dA (77) are incorporated site-specifically into DNA at or near positions that are contacted by the protein. These oligonucleotides are incubated with the protein, and light of an appropriate wavelength is shined on them to initiate the cross-linking reaction. Although photo-cross-linking has been of tremendous value in identifying points of contact between proteins and DNA (78), several factors make it unsuitable for use in structural studies. First, photo-cross-linking reactions rarely furnish yields of cross-linked product above 10% to 20%, rendering the technique impractical to prepare trapped complexes in the quantity required for crystal screening and structure determination.

Second, photo-cross-linking reactions involve high-energy intermediates that undergo kinetically controlled, irreversible reactions; the resulting structures may thus be strained or otherwise unrepresentative of the actual physiologic structure in the absence of the cross-link.

A recent development combines the use of protein-engineering and site-specific DNA-modification techniques to introduce disulfide cross-links into protein-DNA interfaces [(79); P. Zhou & G.L.V., unpublished observations]. The resulting disulfide-trapped complexes have proven amenable to structure determination by X ray and NMR, whereas the corresponding uncross-linked complexes had proven refractory to structure determination by standard methods. In disulfide trapping, a Cys residue is introduced by site-directed mutagenesis into the DNA-binding surface of a protein, and an alkanethiol tether is incorporated at a nearby site in the protein-binding surface of the DNA through synthetic chemistry. Application of this technology obviously requires some knowledge about likely contact points in the protein and DNA. Several convenient methods exist to tether thiol functionality to various positions in DNA. Convertible nucleosides (80–82), several of which are commercially available (Glen Research), permit attachment of tethers to the exocyclic amines of A, C, and G (N^6-thioalkyl-A, N^4-thioalkyl-C, and N^2-thioalkyl-G, respectively). The tethers in N^6-thioalkyl-A and N^4-thioalkyl-C project into the major groove from the "floor," whereas the tether in N^2-thioalkyl-G projects from the floor of the minor groove. All three of these modifications preserve Watson-Crick hydrogen bonding and have a negligible effect on the structure of the duplex. Reagents are also available (Glen Research and Link Technologies) to tether alkanethiols to the 5′- and 3′-phosphates in DNA (83, 84); these are introduced at DNA ends or at internal nicked positions in the duplex.

Several attributes of disulfide chemistry make it especially attractive for performing protein-DNA cross-linking. Although disulfide bonds are thermodynamically stable (65 kcal/mol), they are kinetically labile toward exchange in the presence of a free thiol at pH 8 and above. Therefore, disulfide bond formation can be carried out under conditions of thiol/disulfide equilibration, so as to avoid the formation of strained cross-links and prevent adventitious oxidation of the protein. Finally, the cross-linked product can usually be obtained in yields greater than 80%, and the product can be readily separated from unreacted DNA and protein by anion-exchange chromatography.

The following examples illustrate the use of disulfide cross-linking in structural studies of protein-DNA complexes.

DISULFIDE-CROSS-LINKING OF AN HIV-REVERSE TRANSCRIPTASE:PRIMER: TEMPLATE-NUCLEOTIDE COMPLEX HIV reverse transcriptase (RT) performs the essential task of converting the single-stranded RNA genome packaged in the HIV virion into a double-stranded DNA copy competent for insertion into the host DNA [for a detailed review, see (85)]. RT is a heterodimer of p66 and p51 subunits (86). The two subunits are generated by alternative proteolytic processing of p160$^{gag\text{-}pol}$; hence p66 comprises a full-length copy of p51 plus a C-terminal RNase H domain. Despite their extensive sequence identity, p51 and p66 interact asymmetrically in the functional heterodimer, with the smaller subunit acting as a scaffold for the larger one, which is entirely responsible for processing of the substrate in its RT and RNase active sites (87). Evident in the architecture of p66 are subdomains reminiscent of, but not strictly homologous to, those found in other polymerases: fingers, thumb, palm, and connector. The first insights into the interaction mode of RT with a nucleic acid substrate were gained through the X-ray structure of RT bound to a duplex DNA oligonucleotide and an antibody in the absence of any nucleotide (87). The DNA duplex was found to lie along a broad concave groove in the palm domain, with one end of the duplex projecting toward the active site located at the base of the fingers domain. This structure had many tantalizing features, including a gaping hole between the fingers and thumb, which seemed to provide a perfect entryway for the template strand; however, comprehensive interpretation of the structure was held back by its modest resolution (3.0 Å) and by the fact that it did not contain a bound nucleotide or the Mg^{2+} ions essential for assembly of a catalytic complex. Efforts in several labs to obtain the structure of such a ternary RT-primer:template-nucleotide complex through co-crystallization procedures standard in the polymerase field did not bear fruit.

The reasons for the failure to crystallize a ternary RT complex were suggested by biochemical experiments showing that in the presence of a terminating dideoxy-NTP (ddNTP), the polymerizable (structure-specific) complex was only six- to tenfold more stable than a nonpolymerizable (nonspecific) complex formed with duplex DNA (79). This narrow energetic window between specific and nonspecific binding modes precludes the formation of a homogeneous complex by simple mixing of the protein, DNA, and nucleotide components.

A disulfide cross-linking strategy was devised to capture HIV-RT only when it was bound to the proper end of the primer:template oligonucleotides (88). A modeling study (89) based on the X-ray structure of RT bound to blunt-ended DNA (87) hypothesized that amino acids presented on one face of α-helix H in the p66 subunit tracked in the minor groove of the template:primer (Figure 8*A*). A single Cys residue was engineered into each of three positions lying along this face at successive turns of helix-H (positions 258, 262, and 266; Figure 8*A*). To screen for the formation of only stable cross-links (Figure 8*B*), a combinatorial trapping strategy was employed (Figure 8*C*, *D*). A template:primer containing an alkanethiol tethered to a G residue on the template strand was designed such that addition of different mixtures of dNTPs could ratchet RT forward by single base-pair steps along the DNA duplex (Figure 8*C*). In this way, the precise position of RT relative to the template:primer could be controlled by the addition of different mixtures of nonterminating and terminating nucleotides.

All three of the engineered Cys mutant RT proteins were found to undergo positionally selective cross-linking to DNA (Figure 8*D*), with the Q258C mutant being the most selective (Figure 8*D*). The cross-linking reaction was found to

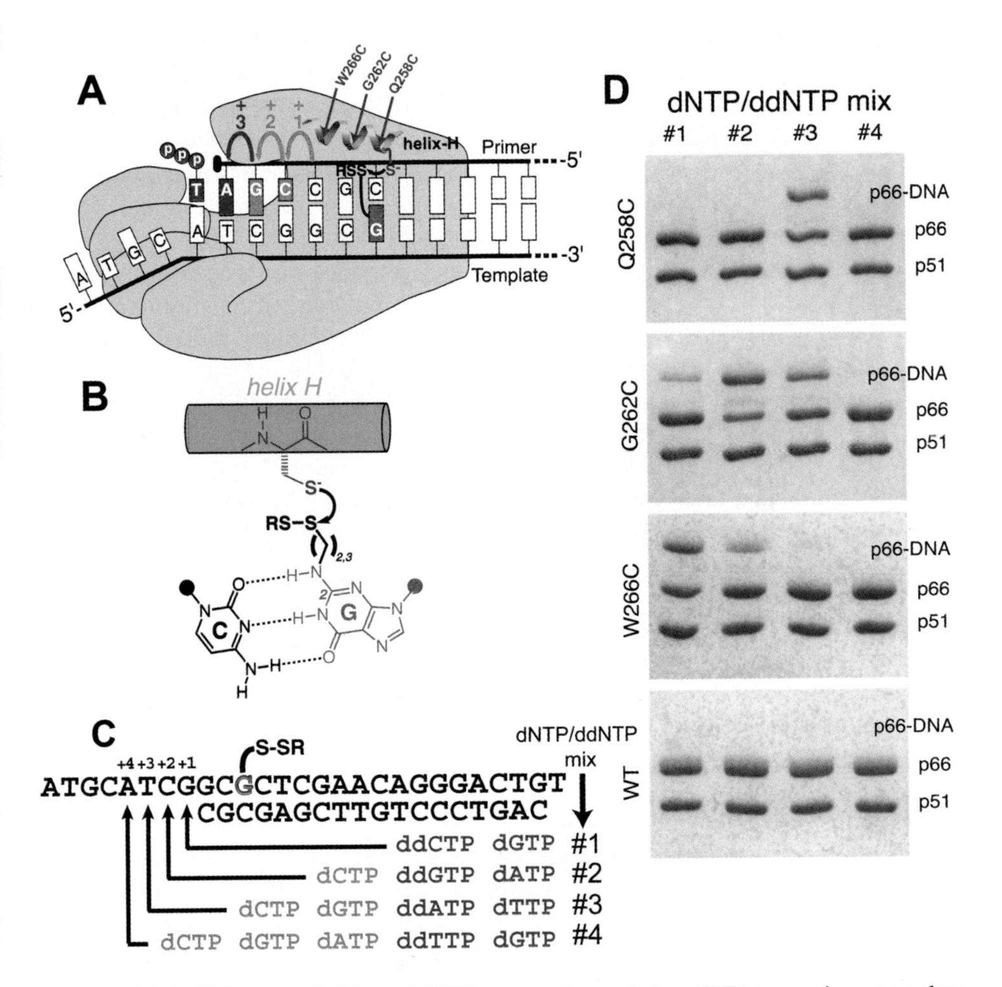

Figure 8 Disulfide cross-linking of HIV reverse transcriptase (RT) to a primer:template duplex. (*A*) Combinatorial cross-linking strategy. Three positions along helix-H were chosen as sites for the engineering of Cys residues to be screened for cross-link formation. *Curved arrows* depict the enzyme ratcheting forward on DNA through successive cycles of extension and termination. (*B*) Cross-linking chemistry. A Cys residue engineered into helix-H of RT attacks a disulfide-bearing tether attached to the N^2-position of a guanine residue in DNA, remote from the active site. (*C*) Mixtures of 2′-deoxynucleotides (dNTP, cyan) and 2′,3′-dideoxynucleotides (ddNTP, red) used to ratchet RT along the primer: template (mixtures are represented horizontally). (*D*) SDS gel analysis of the reactions between the three Cys-engineered RT mutants (Q258C, G262C, and W266C) and the four ratcheting conditions described in (*C*). The products were analyzed by nonreducing SDS-PAGE. Note the disappearance of band intensity at the position of p66 and the corresponding appearance of a band having retarded mobility (p66-DNA), indicative of cross-link formation. The Q258C mutant was chosen for X-ray analysis because it displayed the highest level of positional selectivity in cross-linking.

proceed efficiently even in the presence of 2 mM β-mercaptoethanol, representing a 200-fold excess over the concentration of protein. Surprisingly, the disulfide-trapped complex was stable in up to 10-mM DTT (1000-fold excess), conditions that ordinarily reduce disulfide bonds in proteins. The cross-linking reaction proceeded severalfold faster with a three-carbon thiol tether in the DNA than with a two-carbon tether and was greater than tenfold faster with a DNA:RNA primer:template than with a DNA:DNA duplex (88).

Whereas RT failed to yield diffraction-quality crystals when noncovalently bound to a primer:template and incoming nucleotide, the cross-linked complex formed with the Q258C mutant protein readily afforded crystals. Interestingly, the trapped crystals were unable to seed the formation of a non-cross-linked crystal with the same DNA sequence and incoming nucleotide (H. Huang & G.L.V., unpublished observations). Diffraction data on the trapped complex yielded the first ternary complex structure containing RT, a DNA:DNA primer:template, a bound nucleotide, and two metal ions (presumed to be Mg^{2+}) (88). In this structure, the previously observed gap between the fingers and thumb is closed, so as to envelop the bound nucleotide into the active site, and the template strand brushes along the surface of the fingers domain. The DNA adopts a nearly identical orientation in the nucleotide-bound and nucleotide-free states, suggesting that the open and closed structures represent different states of the catalytic cycle. Similar motions of the fingers domain have been observed for other polymerases lacking any homology to RT and may in fact be a general feature of these structurally diverse enzymes (90–92). Another interesting feature of the overall complex is the positioning of the RNase H domain active site over the phosphodiester backbone, as if poised to perform cleavage of the RNA genome in concert with polymerization, but frustrated by the presence of a DNA template in this complex. The features of the RT active site are completely consistent with mechanistic expectations. The triphosphate moiety of the incoming nucleotide is wrapped around a Mg^{2+} ion, and its α-phosphate is also coordinated to a second Mg^{2+} ion, which is perfectly placed to activate the 3′-OH of the primer strand (3′-H in the crystal structure) for in-line displacement. The Mg^{2+} ions are held in place through coordination to three conserved Asp residues in the active site (D110, D185, and D186).

RT is a target for two important classes of AIDS drugs. Nucleoside RT inhibitors (NRTIs), exemplified by AZT and ddC, are prodrugs that undergo intracellular elaboration to the corresponding 5′-triphosphates, which are then taken up by RT and incorporated into the viral DNA, whereupon they cause DNA chain termination. Non-nucleoside RT inhibitors (NNRTIs), exemplified by nevirapine, bind to a hydrophobic pocket in RT adjacent to the active site and abrogate the polymerase activity of the enzyme (93). The structure of the ternary complex helps in a general way to understand the development of resistance to nucleoside RT inhibitors, as most of the known mutations map to residues that directly contact the incoming nucleotide or immediately adjacent parts of the primer or template; certain other mutations map to residues that buttress the active site (88). A more detailed description of the effects of the mutations is not

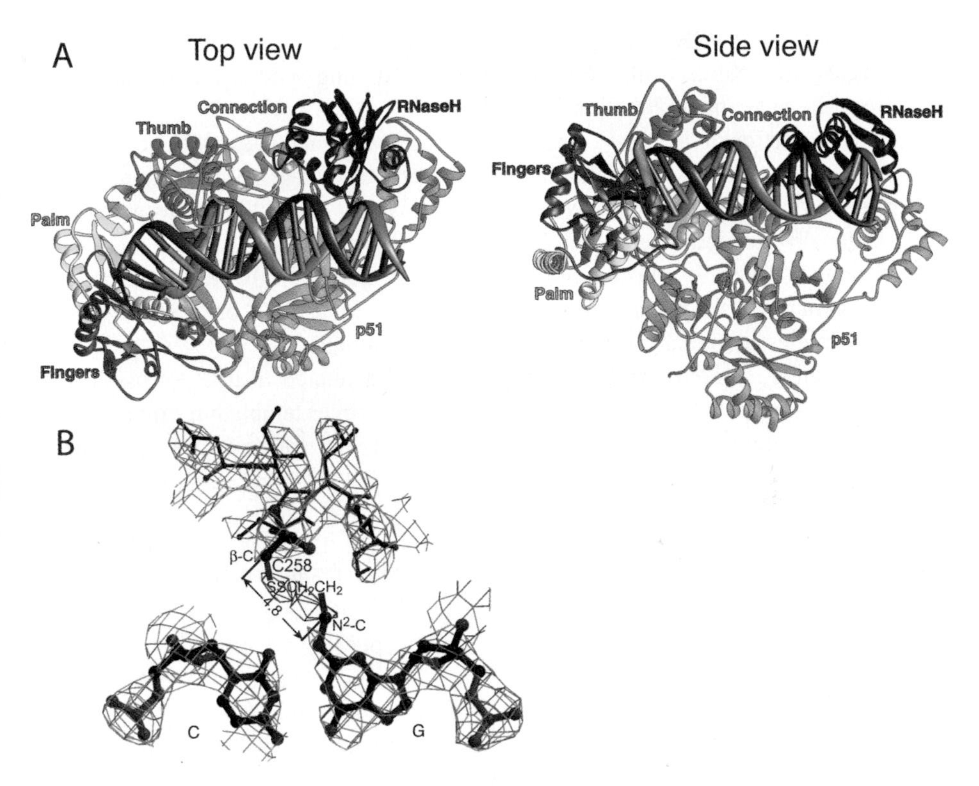

Figure 9 The structure of HIV-RT cross-linked to DNA primer:template. (*A*) Two views of the RT:DNA:dTTP complex, with the polymerase active site on the left and the RNase H domain on the right. The domains of p66 are in color: fingers (red), palm (yellow), thumb (orange), connection (cyan), and RNase H (blue); p51 is in gray. The DNA template strand (light green) contains 25 nucleotides, and the primer strand (dark green) contains 21 nucleotides. The dTTP noncovalently bound in the active site is in gold. Helix-H can be seen inserting into the major groove one-half turn behind (*rightward from*) the active site; the cross-link is not shown explicitly. (*B*) $2F_o$-F_c (cyan) and F_o-F_c (purple) electron density maps at 3.2 Å resolution of the region surrounding the tethered G:C pair in the template:primer and the Cys 258 residue in helix-H of RT. Oxygen, red; nitrogen, blue; phosphorus, purple. The terminal atoms of the disulfide tether (Cys β-C and guanine N^2-C) and the distance separating them are shown explicitly; N^2 was not included explicitly in the structure refinement, but it was modeled here by reference to the positions of other atoms in the G base. The diffuse electron density for most of the atoms of the disulfide-cross-linked tether indicates that it is conformationally mobile.

possible at present, owing to two factors: (*a*) many of the mutations have subtle or even unmeasurable effects on the kinetic properties of RT (94–98); and (*b*) the resolution of the crystal structure (3.1–3.2 Å) is not sufficiently high to allow discernment of modest structural changes. Indeed, the structure of a highly AZT-resistant quintuple mutant of RT (M41L, D67N, K70R, T215Y, K219Q) bound to AZTTP (H. Huang, S.C. Harrison & G.L.V., unpublished observations) reveals only subtle differences to the corresponding wild-type structure, apart from the side chain alteration itself.

Examination of the $2F_o$-F_c electron density showed an absence of electron density for the engineered disulfide bond, indicating that even though the surrounding protein and DNA atoms were statically positioned, the tether and sulfur atoms are conformationally mobile in the crystal, leading to diffuse electron density too weak to be observed (Figure 9). This observation reinforces the notion that the disulfide bond acts to maintain the physical proximity of the protein and DNA but does not enforce any obvious structural perturbation on either macromolecule.

Recently, Arnold and coworkers have used a related disulfide cross-linking strategy to capture structures of RT bound to a primer:template terminating in a 3′-azidothymidine (AZT) nucleotide (98a). These structures provide further insight into the mechanism of resistance to AZT.

CROSS-LINKING OF THE GCN4 BASIC REGION TO DNA GCN4 is a yeast transcription activator with a basic leucine zipper (bZip) DNA-binding domain. The protein homodimerizes via coiled-coil interactions involving the leucine zippers, thus creating a Y-shaped structure in which the helices splay to receive the DNA duplex between them. The ~20 amino acid segment in direct contact with DNA is rich in positively charged amino acids, hence its designation as the "basic region" (Figure 10*A*, yellow). In the absence of DNA, the basic regions of GCN4 are unstructured, but their folding into a continuous α-helix is rapidly induced upon interaction with DNA. A consensus DNA recognition site for GCN4, known as GCRE, is a pseudopalindrome that has two half-sites in a tandem repeat orientation (Figure 10*B*). Peptides representing the basic region of GCN4 alone failed to bind DNA with any measurable specificity. However, a 24-mer peptide comprising the GCN4 basic region plus a C-terminal Gly-Gly-Cys linker, when disulfide cross-linked to an adenine residue flanking the consensus half-site, was found to bind the DNA with high affinity and sequence specificity (99).

In a recent study, the structure of the disulfide-trapped 24-mer-DNA complex was determined by solution NMR spectroscopy (Figure 10*D*) (P. Zhou & G.L.V., unpublished observations). The GCN4 peptide-DNA interface has a repertoire of contacts essentially identical to those of the corresponding protein-DNA complex determined crystallographically (100). Remarkably, the monomeric peptide appears to cause bending of the DNA duplex distinct from that observed with the intact GCN4 homodimer (Figure 10*D*). Based on these findings, it has been suggested that both direct contacts and dimerization-"induced torque" contribute to the overall structure of the dimeric protein bound to DNA.

Presumably, the reduction of entropy loss from converting an otherwise intermolecular association into an intramolecular one is what enables the tethered peptide to bind sequence specifically, while the untethered peptide fails to do so. This case shows that disulfide cross-linking can be used to strip down, or

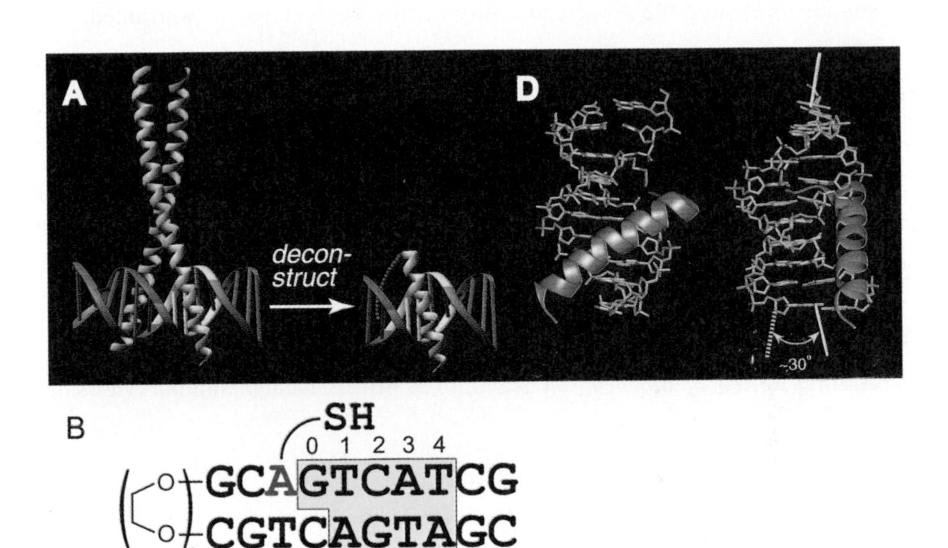

Figure 10 GCN4 peptide-DNA cross-link. (*A*) Structure of the full-length GCN4 leucine zipper motif bound to DNA. The basic region and DNA of the consensus half-site are colored yellow. (*B*) Sequence of the duplex DNA used to determine the structure of a basic region peptide disulfide trapped to DNA. The GNC4 consensus half-site is in yellow, and the thiol-tethered adenine is in red. At the left of the sequence is schematically depicted a hexa-ethyleneglycol linker connecting the 3′-end of the bottom strand to the 5′-end of the top strand; this was added to stabilize the duplex and ensure a 1:1 ratio of the two strands. (*C*) Chemistry used for introduction of the thiol-tethered *A* residue. A convertible nucleoside, O^6-phenyl inosine, was modified to generate an adenine containing a major groove presented disulfide. (*D*) The NMR structure of the peptide DNA cross-link shows that the duplex DNA is bent more significantly than in the full-length GCN4-DNA complex. The disulfide cross-link is visible above the peptide.

deconstruct, protein-DNA complexes to their minimal contact region. Future studies to explore the scope and utility of this method appear warranted.

ACKNOWLEDGMENTS

D.P.G.N. is the recipient of an NSF Predoctoral Fellowship. Work on covalent trapping in the Verdine laboratory is supported by NIH grants GM 51330, GM 44853, and GM 39589. The authors thank J.C. Fromme, S.D. Bruner, W. Yang, M. Karplus, and P. Zhou for making available unpublished results. The authors also thank K.A. Plummer, M.C. Spong, J.M. Finkelstein, and J.C. Fromme for a critical reading of the manuscript.

The *Annual Review of Biochemistry* is online at http://biochem.annualreviews.org

LITERATURE CITED

1. Jeltsch A. 2002. *ChemBioChem* 3: 382–93
2. Jones PA, Baylin SB. 2002. *Nat. Rev. Genet.* 3:415–28
3. Robertson KD. 2002. *Oncogene* 21: 5361–79
4. Li E. 2002. *Nat. Rev. Genet.* 3:662–73
5. Mann MRW, Bartolomei MS. 2002. *Genome Biol.* 3(2):1003.1–3.4
6. Santi DV, Norment A, Garrett CE. 1984. *Proc. Natl. Acad. Sci. USA* 81:6993–97
7. Wu JC, Santi DV. 1987. *J. Biol. Chem.* 262:4778–86
8. Chen L, MacMillan AM, Verdine GL. 1993. *J. Am. Chem. Soc.* 115:5318–19
9. Erlanson DA, Chen L, Verdine GL. 1993. *J. Am. Chem. Soc.* 115:12583–84
10. MacMillan AM, Chen L, Verdine GL. 1992. *J. Org. Chem.* 57:2989–91
11. Klimasauskas S, Kumar S, Roberts RJ, Cheng X. 1994. *Cell* 76:357–69
12. Reinisch KM, Chen L, Verdine GL, Lipscomb WN. 1995. *Cell* 82:143–53
13. Lindahl T. 1993. *Nature* 362:709–15
14. Friedberg EC, Walker GC, Siede W, eds. 1995. *DNA Repair and Mutagenesis.* Washington, DC: ASM
15. Dogliotti E, Fortini P, Pascucci B, Parlanti E. 2001. *Prog. Nucleic Acid Res. Mol. Biol.* 68:3–27
16. Memisoglu A, Samson L. 2000. *Mutat. Res.* 451:39–51
17. Krokan HE, Nilsen H, Skorpen F, Otterlei M, Slupphaug G. 2000. *FEBS Lett.* 476:73–77
18. David SS, Williams SD. 1998. *Chem. Rev.* 98:1221–62
19. Scharer OD, Jiricny J. 2001. *BioEssays* 23:270–81
20. Dodson ML, Lloyd RS. 2002. *Free Radic. Biol. Med.* 32:678–82
21. Hollis T, Lau A, Ellenberger T. 2001. *Prog. Nucleic Acid Res. Mol. Biol.* 68:305–14
22. Norman DPG, Bruner SD, Verdine GL. 2001. *J. Am. Chem Soc.* 123:359–60
23. Bruner SD, Norman DPG, Verdine GL. 2000. *Nature* 403:859–66
24. Serre L, de Jesus KP, Boiteux S, Zelwer C, Castaing B. 2002. *EMBO J.* 21:2854–65
25. Tchou J, Grollman AP. 1995. *J. Biol. Chem.* 270:11671–77
26. Dodson ML, Schrock RD 3rd, Lloyd RS. 1993. *Biochemistry* 32:8284–90
27. Schrock RD 3rd, Lloyd RS. 1991. *J. Biol. Chem.* 266:17631–39
28. Nash HM, Lu R, Lane WS, Verdine GL. 1997. *Chem. Biol.* 4:693–702
29. Nash HM, Bruner SD, Scharer OD,

Kawate T, Addona TA, et al. 1996. *Curr. Biol.* 6:968–80
30. Hilbert TP, Chaung WR, Boorstein RJ, Cunningham RP, Teebor GW. 1997. *J. Biol. Chem.* 272:6733–40
31. Hilbert TP, Boorstein RJ, Kung HC, Bolton PH, Xing D, et al. 1996. *Biochemistry* 35:2505–11
32. Zharkov DO, Rieger RA, Iden CR, Grollman AP. 1997. *J. Biol. Chem.* 272: 5335–41
33. Rieger RA, McTigue MM, Kycia JH, Gerchman SE, Grollman AP, Iden CR. 2000. *J Am. Soc. Mass Spectrom.* 11:505–15
34. Sun B, Latham KA, Dodson ML, Lloyd RS. 1995. *J. Biol. Chem.* 270:19501–8
35. McCullough AK, Sanchez A, Dodson ML, Marapaka P, Taylor JS, Lloyd RS. 2001. *Biochemistry* 40:561–68
36. Zharkov DO, Golan G, Gilboa R, Fernandes AS, Gerchman SE, et al. 2002. *EMBO J.* 21:789–800
37. Gilboa R, Zharkov DO, Golan G, Fernandes AS, Gerchman SE, et al. 2002. *J. Biol. Chem.* 277:19811–16
38. Fromme JC, Verdine GL. 2002. *Nat. Struct. Biol.* 9:544–52
38a. Fromme JC, Bruner SD, Yang W, Karplus M, Verdine GL. 2003. *Nat. Struct. Biol.* 3:204–11
38b. Norman DPG, Chung SJ, Verdine GL. 2003. *Biochemistry* 42:1564–72
39. Melamede RJ, Hatahet Z, Kow YW, Ide H, Wallace SS. 1994. *Biochemistry* 33:1255–64
40. Jiang D, Hatahet Z, Melamede RJ, Kow YW, Wallace SS. 1997. *J. Biol. Chem.* 272:32230–39
41. Boiteux S, O'Connor TR, Lederer F, Gouyette A, Laval J. 1990. *J. Biol. Chem.* 265:3916–22
42. O'Connor TR, Boiteux S, Laval J. 1989. *Ann. Ist Super Sanita* 25:27–31
43. Tchou J, Kasai H, Shibutani S, Chung MH, Laval J, et al. 1991. *Proc. Natl. Acad. Sci. USA* 88:4690–94
44. Hatahet Z, Kow YW, Purmal AA, Cunningham RP, Wallace SS. 1994. *J. Biol. Chem.* 269:18814–20
45. Labahn J, Scharer OD, Long A, Ezaz-Nikpay K, Verdine GL, Ellenberger TE. 1996. *Cell* 86:321–29
46. Thayer MM, Ahern H, Xing D, Cunningham RP, Tainer JA. 1995. *EMBO J.* 14:4108–20
47. Boiteux S, Radicella JP. 2000. *Arch. Biochem. Biophys.* 377:1–8
48. Drohat AC, Stivers JT. 2000. *J. Am. Chem. Soc.* 122:1840–41
49. Dinner AR, Blackburn GM, Karplus M. 2001. *Nature* 413:752–55
50. Mazumder A, Gerlt JA, Absalon MJ, Stubbe J, Cunningham RP, et al. 1991. *Biochemistry* 30:1119–26
51. Keeney S. 2001. *Curr. Top. Dev. Biol.* 52:1–53
52. Hoess RH, Abremski K. 1985. *J. Mol. Biol.* 181:351–62
53. Orban PC, Chui D, Marth JD. 1992. *Proc. Natl. Acad. Sci. USA* 89:6861–65
54. Gu H, Marth JD, Orban PC, Mossmann H, Rajewsky K. 1994. *Science* 265:103–6
55. Gopaul DN, Guo F, Van Duyne GD. 1998. *EMBO J.* 17:4175–87
56. Guo F, Gopaul DN, Van Duyne GD. 1997. *Nature* 389:40–46
57. Guo F, Gopaul DN, Van Duyne GD. 1999. *Proc. Natl. Acad. Sci. USA* 96:7143–48
58. Van Duyne GD. 2001. *Annu. Rev. Biophys. Biomol. Struct.* 30:87–104
59. Pargellis CA, Nunes-Duby SE, de Vargas LM, Landy A. 1988. *J. Biol. Chem.* 263:7678–85
60. Wang JC. 2002. *Nat. Rev. Mol. Cell Biol.* 3:430–40
61. Champoux JJ. 1994. *Adv. Pharmacol.* 29A:71–82
62. Champoux JJ. 1981. *J. Biol. Chem.* 256: 4805–9
63. Champoux JJ. 1977. *Proc. Natl. Acad. Sci. USA* 74:3800–4
64. Berger JM, Gamblin SJ, Harrison SC, Wang JC. 1996. *Nature* 379:225–32
65. Cabral JHM, Jackson AP, Smith CV,

Shikotra N, Maxwell A, Liddington RC. 1997. *Nature* 388:903–6
66. Cheng CH, Kussie P, Pavletich N, Shuman S. 1998. *Cell* 92:841–50
67. Holdgate GA, Tunnicliffe A, Ward WH, Weston SA, Rosenbrock G, et al. 1997. *Biochemistry* 36:9663–73
68. Feinberg H, Changela A, Mondragon A. 1999. *Nat. Struct. Biol.* 6:961–68
69. Mondragon A, DiGate R. 1999. *Struct. Fold. Des.* 7:1373–83
70. Nichols MD, DeAngelis K, Keck JL, Berger JM. 1999. *EMBO J.* 18:6177–88
71. Redinbo MR, Stewart L, Kuhn P, Champoux JJ, Hol WG. 1998. *Science* 279:1504–13
71a. Staker BL, Hjerrild K, Feese MD, Behnke CA, Burgin AB Jr, Stewart L. 2002. *Proc. Natl. Acad. Sci. USA* 99:15387–92
72. Burgin AB Jr, Huizenga BN, Nash HA. 1995. *Nucleic Acids Res.* 23:2973–79
73. Krogh BO, Shuman S. 2000. *Mol. Cell* 5:1035–41
74. Pearson RG. 1966. *Science* 151:172–77
75. Andersen AH, Gocke E, Bonven BJ, Nielsen OF, Westergaard O. 1985. *Nucleic Acids Res.* 13:1543–57
76. Willis MC, Hicke BJ, Uhlenbeck OC, Cech TR, Koch TH. 1993. *Science* 262:1255–57
77. Liu J, Fan QR, Sodeoka M, Lane WS, Verdine GL. 1994. *Chem. Biol.* 1:47–55
78. Christ F, Steuer S, Thole H, Wende W, Pingoud A, Pingoud V. 2000. *J. Mol. Biol.* 300:867–75
79. Huang H, Harrison SC, Verdine GL. 2000. *Chem. Biol.* 7:355–64
80. Xu YZ, Zheng Q, Swann PF. 1992. *J. Org. Chem.* 57:3841
81. MacMillan AM, Verdine GL. 1991. *Tetrahedron* 14:2603–16
82. Ferentz AE, Verdine GL. 1994. *Nucleic Acids Mol. Biol.* 8:14–40
83. Sinha ND, Cook RM. 1988. *Nucleic Acids Res.* 16:2659–69
84. Connolly BA, Rider P. 1985. *Nucleic Acids Res.* 13:4485–502
85. Swanstrom R, Wills J. 1997. In *Retroviruses*, ed. HE Varmus, J Coffin, S Hughes, pp. 263–334. Cold Spring Harbor, NY: Cold Spring Harbor Press
86. di Marzo Veronese F, Copeland TD, DeVico AL, Rahman R, Oroszlan S, et al. 1986. *Science* 231:1289–91
87. Jacobo-Molina A, Ding J, Nanni RG, Clark AD Jr, Lu X, et al. 1993. *Proc. Natl. Acad. Sci. USA* 90:6320–24
88. Huang H, Chopra R, Verdine GL, Harrison SC. 1998. *Science* 282:1669–75
89. Bebenek K, Beard WA, Darden TA, Li L, Prasad R, et al. 1997. *Nat. Struct. Biol.* 4:194–97
90. Doublie S, Tabor S, Long AM, Richardson CC, Ellenberger T. 1998. *Nature* 391:251–58
91. Sawaya MR, Prasad R, Wilson SH, Kraut J, Pelletier H. 1997. *Biochemistry* 36:11205–15
92. Brautigam CA, Steitz TA. 1998. *Curr. Opin. Struct. Biol.* 8:54–63
93. Kohlstaedt LA, Wang J, Friedman JM, Rice PA, Steitz TA. 1992. *Science* 256:1783–90
94. Anderson KS. 2002. *Biochim. Biophys. Acta* 1587:296–99
95. Kerr SG, Anderson KS. 1997. *Biochemistry* 36:14064–70
96. Ray AS, Anderson KS. 2001. *Nucleosides Nucleotides Nucleic Acids* 20:1247–50
97. Ray AS, Yang Z, Shi J, Hobbs A, Schinazi RF, et al. 2002. *Biochemistry* 41:5150–62
98. Vaccaro JA, Anderson KS. 1998. *Biochemistry* 37:14189–94
98a. Sarafianos SG, Clark AD Jr, Das K, Tuske S, Birktoft JJ, et al. 2002. *EMBO J.* 21:6614–24
99. Stanojevic D, Verdine GL. 1995. *Nat. Struct. Biol.* 2:450–57
100. Ellenberger TE, Brandl CJ, Struhl K, Harrison SC. 1992. *Cell* 71:1223–37

Annu. Rev. Biochem. 2003. 72:367–394
doi: 10.1146/annurev.biochem.72.121801.161824
Copyright © 2003 by Annual Reviews. All rights reserved
First published online as a Review in Advance on March 19, 2003

TEMPORAL AND SPATIAL REGULATION IN PROKARYOTIC CELL CYCLE PROGRESSION AND DEVELOPMENT

Kathleen R. Ryan and Lucy Shapiro
Department of Developmental Biology, Beckman Center, Stanford University School of Medicine, Stanford, California 94305-5329; email: kryan@cmgm.stanford.edu, shapiro@cmgm.stanford.edu

Key Words *Caulobacter*, *Bacillus subtilis*, sporulation, FtsZ, chromosome segregation

■ **Abstract** Bacteria exhibit a high degree of intracellular organization, both in the timing of essential processes and in the placement of the chromosome, the division site, and individual structural and regulatory proteins. We examine the temporal and spatial regulation of the *Caulobacter* cell cycle, bacterial chromosome segregation and cytokinesis, and *Bacillus subtilis* sporulation. Mechanisms that control timing of cell cycle and developmental events include transcriptional cascades, regulated phosphorylation and proteolysis of signal transduction proteins, transient genetic asymmetry, and intercellular communication. Surprisingly, many signal transduction proteins are dynamically localized to specific subcellular addresses during the cell division cycle and sporulation, and proper localization is essential for their function. The Min proteins that govern division site selection in *Escherichia coli* may be the first example of a system that generates positional information de novo.

CONTENTS

INTRODUCTION

The last decade has witnessed a renaissance in the study of the bacterial cell. The extensive focus on *Escherichia coli* over multiple decades provided the foundations of molecular biology, largely based on the power of genetics and the in vitro dissection of individual biochemical pathways common to all living entities. More recently, these fundamental mechanisms are being approached in the context of the living cell. Bacterial cells that were once thought of largely as unorganized containers of enzymes are now known to be highly organized with complex behavior patterns. Attention to a variety of bacterial species has revealed new insights into the cell cycle and interior organization of individual bacterial cells, their differentiation into biofilms and spores, their ability to grow in extreme environments, and their interactions with eukaryotes in pathogenesis.

Bacteria exert an unanticipated degree of temporal and spatial control over their internal processes. The primary task of the bacteria cell is self-duplication, which entails DNA replication, chromosome segregation, and cell division at the correct place and time to produce two genetically identical progeny. In addition, many bacteria elaborate on this basic program by creating asymmetric predivisional cells that yield progeny with distinct cell fates, either constitutively or as a response to environmental conditions. Even the basic program of cell division requires temporal and spatial precision to ensure that division occurs only after DNA replication and chromosome segregation have been completed and that the division plane is situated to yield daughter cells each containing a complete genome. In organisms undergoing asymmetric cell division, there are added layers of complexity, such as the dynamic localization of individual proteins to different sites within the predivisional cell and spatially restricted construction of appendages such as stalks, flagella, and pili, which differentiate the progeny morphologically. In this review, we address two fundamental questions: What processes are limited to specific times and locations in the life of the bacterial cell, and what mechanisms do bacteria use to achieve temporal and spatial regulation?

CELL CYCLE CONTROL IN *CAULOBACTER CRESCENTUS*

Temporal Regulation

The gram-negative, aquatic bacterium *Caulobacter crescentus* is particularly amenable to the study of prokaryotic cell cycle control because each cell replicates its chromosome only once per cell division cycle. Different stages of the cell cycle are associated with distinct morphologies, and cell cultures can be synchronized easily. Asymmetric cell divison yields two different progeny: a flagellated swarmer cell and a stalked cell (Figure 1). Motile swarmer cells are

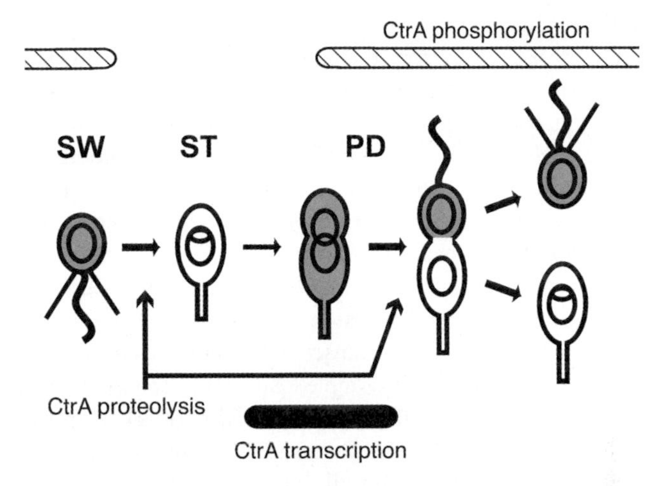

Figure 1 Diagram of the *Caulobacter crescentus* cell cycle. The motile swarmer cell (SW) has a polar flagellum (*curved* line) and pili (*straight lines*), contains the CtrA protein (gray shading), and cannot initiate DNA replication (circular chromosome). During differentiation into a stalked cell (ST), the flagellum is shed and a stalk is built at the same pole, CtrA is proteolyzed (*first arrow*), and the cell becomes able to initiate DNA replication (theta structure). CtrA is transcribed and reaccumulates in the predivisional cell (PD), and a flagellum is built at the pole opposite the stalk. CtrA is cleared from the stalked portion of the late PD cell (*second arrow*). Asymmetric division yields a motile swarmer cell and a replication-competent stalked cell. CtrA is transcribed during the time indicated by the black bar and phosphorylated during the times indicated by the striped bars.

incapable of initiating DNA replication and cell division until they differentiate into stalked cells. Swarmer-to-stalked cell differentiation, equivalent to the G1-S transition of the cell cycle, is marked by shedding of the polar flagellum and pili and construction of a stalk at the site previously occupied by the flagellum. As DNA replication progresses, the stalked cell elongates into an asymmetric predivisional cell and builds a flagellum at the pole opposite the stalk. At the conclusion of DNA replication and chromosome segregation, the cell divides into progeny with different cell cycle fates: The progeny stalked cell immediately enters a new period of chromosome replication and cell division, while the progeny swarmer cell must first differentiate into a stalked cell.

Analysis of the global transcription pattern during the cell cycle (1) using microarrays of the entire *Caulobacter* genome (2) showed that differential transcriptional regulation is an important mechanism for temporal control of cell cycle processes in prokaryotes. At least 19% of the genes in the *Caulobacter* genome have cell cycle-regulated expression patterns; the transcription of 553 genes varies by a factor of two or more in a cyclic pattern during the 150-min *Caulobacter* cell cycle, which includes 79 genes shown previously by *lacZ* transcriptional fusions to exhibit cell cycle-regulated expression profiles (1). The characteristics of cell cycle-regulated gene expression in this bacterial cell reveal several organizing principles. First, genes are transcribed just before their products are needed to perform specific cellular functions. For example, genes required for DNA replication, which include genes encoding initiation factors, polymerase subunits, topoisomerases, and nucleotide biosynthetic enzymes, are transcribed just before and during S phase. The genes required for DNA methylation are transcribed after DNA replication is completed, when the newly replicated, hemi-methylated chromosome is rapidly brought to a state of full methylation just before cell division. Second, the genes encoding subunits of large protein complexes are cotranscribed; ribosomal subunits and components of the NADH dehydrogenase complex are expressed at the swarmer-to-stalked cell transition, when the cell enters a period of marked growth and increased protein synthesis. Finally, genes required to construct complex macromolecular appendages, such as flagella and pili, are transcribed in cascades that reflect the temporal order of assembly of the protein products. The 41 genes of the flagellar expression hierarchy are organized so that (*a*) downstream genes are dependent upon upstream genes for their expression, and (*b*) transcription of genes encoding cell-proximal proteins of the flagellar basal body and motor precedes transcription of genes encoding cell-distal proteins of the hook, rod, and flagellar filament.

Whole-genome transcriptional profiling of the *Caulobacter* cell cycle also revealed that at least 25% of the genes with cell cycle-dependent expression patterns are regulated, directly or indirectly, by the two-component signal transduction protein CtrA (1). CtrA was discovered in a genetic screen for essential genes that are required for the regulation of flagellar gene expression and was then found to play a key role in cell cycle progression (3, 4). CtrA is a DNA-binding response regulator that directly induces or represses the transcription of at least 55 operons in *Caulobacter* (5). CtrA's direct transcriptional targets include genes involved in cell division (6), DNA methylation (7), flagellum (3) and pili (8) biosynthesis, chemotaxis (5), and metabolism (5). Importantly, CtrA binds to and silences the *Caulobacter* origin of replication, so that cells containing active CtrA cannot initiate DNA replication (4, 9).

Because CtrA must be present to activate transcriptional targets in the predivisional cell but must also be absent at the onset of DNA replication, CtrA itself is regulated by at least three redundant mechanisms (Figure 1). First, a positive transcriptional feedback loop causes a burst of CtrA synthesis in late stalked and predivisional cells (10). The *ctrA* gene is transcribed from two promoters. The weaker P1 promoter fires just after replication initiation, when the cells lack activated CtrA. As CtrA accumulates and is activated by phosphorylation, it represses P1 and activates the strong P2 promoter. Second, CtrA phosphorylation is under cell cycle control, so that CtrA is only phosphorylated and active in swarmer and predivisional cells (4). Two essential histidine kinases have been implicated in CtrA phosphorylation, CckA (11) and DivL (12), but it is not known how these kinases respond to cell cycle cues and cause a cell cycle-regulated pattern of CtrA phosphorylation. Finally, CtrA is rapidly proteolyzed at the swarmer-to-stalked cell transition, when the cell is preparing to initiate DNA replication (4). The mechanisms that activate CtrA proteolysis at the G1-S transition and then turn it off just after replication initiation are crucial for ensuring that the *Caulobacter* chromosome is replicated once and only once per cell division cycle. The ClpXP protease is required for CtrA proteolysis at the G1-S transition (13), but the factors that link CtrA proteolysis to a specific stage of the cell cycle remain to be determined. A mutant CtrA protein that mimics the active, phosphorylated state and cannot be proteolyzed causes a dominant cell cycle block in the G1 phase because DNA replication cannot be initiated (4).

Analysis of the genetic regulatory network of the *Caulobacter* cell cycle has been greatly advanced by genome-wide expression profiling and by the finding that so many cell cycle processes are integrated by the action of the CtrA master regulator. However, several pressing questions remain. First, what transcriptional regulators control the expression of the other 75% of cell cycle-regulated genes that are not affected by a *ctrA* loss-of-function mutation? Among these genes are those encoding critical factors required for DNA replication such as *dnaA*, *dnaX, dnaN, gyrB,* and *dnaK*. These genes are coordinately expressed prior to DNA replication (14, 15), and their promoters share a common sequence motif thought to be a binding site for a transcriptional repressor (16), but the identity of the repressor and its connection to the cell cycle are as yet unknown. Second, how does CtrA activate distinct sets of genes at different times in the cell cycle? Even before the *Caulobacter* genome was completed, there were several examples of genes clearly regulated by CtrA that had different peak expression times in the cell cycle. For example, the *fliQR* flagellar genes, *ccrM* encoding an essential DNA methyltransferase and *pilA* encoding the structual subunit of pili, are reproducibly transcribed at different times in late stalked and predivisional cells (7, 8). One study has shown that the precise timing of CtrA-regulated genes can be attributed in part to the affinity of CtrA for specific promoters, so that higher-affinity promoters such as P*fliQR* are activated first, when active CtrA is still accumulating in the predivisional cell, while lower-affinity promoters such as P*ccrM* are activated later, when the predivisional cell has achieved a high concentration of CtrA (7). Whole-genome microarray experiments further revealed that when a G1 cell cycle arrest is induced by the constitutively active *ctrA* mutant, the expression of 125 genes is significantly increased compared to a wild-type control (1). Interestingly, 85% of these genes are normally expressed in the G1 swarmer stage of the cell cycle, rather than being evenly distributed throughout the period during which CtrA is known to activate target genes. These results suggest a mechanism of combinatorial transcriptional control, where CtrA interacts with other factors that are active during specific stages of the cell cycle, to influence the transcription of subsets of genes at specific times. A major goal is to identify these factors and determine their roles in the *Caulobacter* cell cycle genetic network.

Although gene expression in *Caulobacter* is clearly linked to cell cycle progression, it is not yet known how important it is that each of the cell cycle-regulated genes be transcribed in the exact pattern seen in wild-type populations during logarithmic growth. In the case of pili biogenesis, the time of transcription of the *pilA* gene encoding the pilin monomer is critical for assembling the pilus only at the pole of the new swarmer progeny, and not in the predivisional cell (8). Also, the normal progression of cell cycle events is altered if the gene encoding the essential DNA methyltransferase CcrM is expressed constitutively during the cell cycle (17). Conversely, temporally controlled transcription does not always result in a cell cycle-regulated pattern of protein abundance. For example, the DivK response regulator that signals proteolytic events during differentiation (18) is present throughout the cell cycle (19), but the *divK* gene is transcribed in a burst in late predivisional cells (1). In some cases, genes with striking cell cycle transcription patterns are also regulated at other levels, such as (*a*) stability of flagellin mRNAs (20, 21), (*b*) cell cycle-regulated phosphorylation of CtrA (4), and (*c*) cell cycle-regulated proteolysis of CtrA (4), the chemoreceptor McpA (22), and the flagellar protein FliF (23), so that transcription contributes to, but does not fully determine, the activity pattern of the gene product. An important recent discovery, discussed below, is that many bacterial proteins are regulated at the level of subcellular localization. While the abundance of the CckA histidine kinase does not change during the cell cycle, its location does: CckA is transiently localized to the cell poles in the predivisional cell, coincident with its autophosphorylation and transfer of phosphate to CtrA (11).

Several studies have indicated that temporal control of the cell cycle is maintained in part by coupling specific processes to the completion of a prior cell cycle or morphological event. The transcription of essential cell division and DNA methylation genes late in the cell cycle is dependent upon DNA replication (24, 25). In addition, defects in genes involved in chromosome structure and partitioning prevent normal cell division (26, 27). The pathways connecting DNA replication and chromosome segregation to cell division may both be mediated by CtrA because inhibition of DNA replication using hydroxyurea or a deletion of the *smc* (structural maintenance of chromosomes) gene prevents the accumulation of CtrA in the predivisional cell at the completion of S-phase (27, 25). Without CtrA, predivisional cells are unable to express the cell division genes *ftsQA* (25), the DNA methyltransferase *ccrM* (7), or genes in class II of the flagellar transcriptional hierarchy (3). Wortinger et al. (25) found that inhibiting DNA replication greatly reduces *ctrA* transcription, but the mechanism connecting these two processes remains to be discovered.

Like *Salmonella typhimurium* and other enteric bacteria, *Caulobacter* cells possess a morphological checkpoint that senses assembly of the flagellar basal body and hook structures (28). *Caulobacter* is unusual, however, in that the completion of early flagellar structures is required not only for the accumulation of subunits to build the flagellar filament (20) but also for cell division (29). The *fliX* flagellar mutant is nonmotile and filamentous, yet FliX is not a structural component of the flagellum, but a trans-acting factor that senses flagellar assembly and transmits this information to the transcription factor FlbD. FlbD regulates transcription of late flagellar genes and is predicted to activate a key cell division protein as well (30). Gain-of-function mutations in FlbD can bypass *fliX* mutants to produce motile cells and can also bypass mutations in structural genes of the flagellum to allow normal cell division in the absence of flagellar assembly. The mechanisms used by FliX to sense flagellar assembly and communicate with FlbD are areas of ongoing investigation. Perhaps the flagellar assembly checkpoint is particularly important for *Caulobacter* because these cells typically inhabit very dilute aquatic environments, where nutrients are scarce (2). Coupling flagellar assembly to cell division ensures that one of the progeny of each division is motile and may increase the likelihood of its locating a new source of food.

Spatial Regulation

Many bacterial species have long been known to display spatial asymmetry of appendages such as flagella, pili, and stalks. Recent studies using immunogold-electron microscopy, immunofluorescence microscopy, and GFP-tagged proteins have yielded the surprising result that the interior of the bacterial cell is also highly organized; this organization includes the localization of signal transduction proteins that were not expected to have specific intracellular addresses and dynamics. The first examples of localized signal transduction proteins in bacteria were the methyl-accepting chemoreceptor proteins (MCPs) of *E. coli* and *Caulobacter* (31, 32). MCPs sense the presence of specific molecules, such as serine and aspartate, via a ligand-binding domain in the periplasm. The MCP cytoplasmic domain communicates this information to a two-component signal transduction pathway consisting of the histidine kinase CheA and the response regulator CheY. CheY~P then interacts with FliM, a switch protein on the flagellar motor, to change the direction of rotation and allow the bacterium to move up gradients of attractants. The MCPs form clusters at one or both poles of the cell, and maximal polar clustering requires a ternary complex of chemoreceptor CheA and the linker protein CheW, which is required for MCPs to activate or inhibit CheA phosphorylation (32, 33). In strains expressing the four MCPs of *E. coli* individually, each MCP is primarily located at the poles, but each one has a greater or lesser tendency to aggregate into clusters (34). Thus, targeting to the pole and clustering into higher-order assemblies are separable properties.

A series of studies using live cells and in vitro systems support a model in which (*a*) the input from several chemoreceptors is integrated at the level of CheA activation, and (*b*) different types of chemoreceptors can influence each other's signaling properties. In reconstituted in vitro systems, maximal CheA phosphorylation occurs in a complex containing ~6 MCP homodimers (33). Chemical crosslinking shows that different MCPs assemble into complexes together in vivo (35). Furthermore, a synthetic, multivalent ligand for one chemoreceptor amplifies the cellular response to a ligand sensed by a different chemoreceptor, which provides evidence that different MCPs communicate with each other in chemotaxis signaling (36).

The crystal structure of an MCP cytoplasmic signaling domain revealed dimers further organized into trimers (37); this structural arrangement suggests that homodimers of individual MCPs assemble into heterotrimers in the polar clusters. In addition to explaining the signaling interactions between different chemoreceptors, the assembly of chemosensory proteins into a two-dimensional lattice may underlie the striking sensitivity of bacterial chemotaxis; bacteria can respond to fractional changes in MCP receptor occupancy of less than 1% (38). A recent study modeling the structure of the MCP lattice (39) suggested that each MCP homodimer contacts two others in the cytoplasm via the signaling domains (37), but each also contacts two different dimers in the periplasm via the ligand-binding domains. This arrangement would allow ligand-binding by one homodimer to be communicated to other trimers in the lattice and would allow a small number of ligand-binding events to affect the activity of many CheA proteins, which enhances sensitivity. Though the mechanism is not yet defined, it is clear that the subcellular localization and clustering of MCP proteins is a critical determinant of their function in chemotactic signal transduction.

The discovery of MCP localization in both *E. coli* and *Caulobacter* prompted investigators to examine the subcellular distributions of other signal transduction proteins. The membrane-bound histidine kinases CckA (11), PleC (40), and DivJ (40) and the single-domain response regulator DivK (19), which regulate the *Caulobacter* cell cycle and polar organelle development, all display precise, cell cycle-dependent localization patterns (Figure 2). For example, CckA, which phosphorylates the CtrA response regulator, is diffuse in swarmer and stalked cells but becomes localized to the cell poles in predivisional cells. CckA localization is coincident with the time of CtrA activation (4). DivK, which is required for the proteolysis of CtrA and other proteins at the G1-S transition (18), is diffuse in swarmer cells and localizes first to the stalked pole, then to the nascent flagellar pole of the predivisional cell. At division, DivK remains at the pole of the stalked cell and diffuses in the new swarmer cell (19). Dynamic DivK localization depends on the activity of both DivJ and PleC. In a *divJ* null mutant, DivK fails to localize to the poles, while in the *pleC* null mutant, DivK is not released from the pole of the swarmer cell after division (19). The current challenges are (*a*) to determine how these two-component regulators are targeted and maintained at specific locations and (*b*) to understand the benefits of localizing signal transduction proteins in a dynamic fashion during cell cycle progression.

Two recent studies of *Caulobacter* pilus assembly have made important strides toward understanding the mechanisms and consequences of localizing signal transduction proteins. An early step in pilus assembly is the construction of a secretion channel, which the pilus eventually traverses to reach the exterior

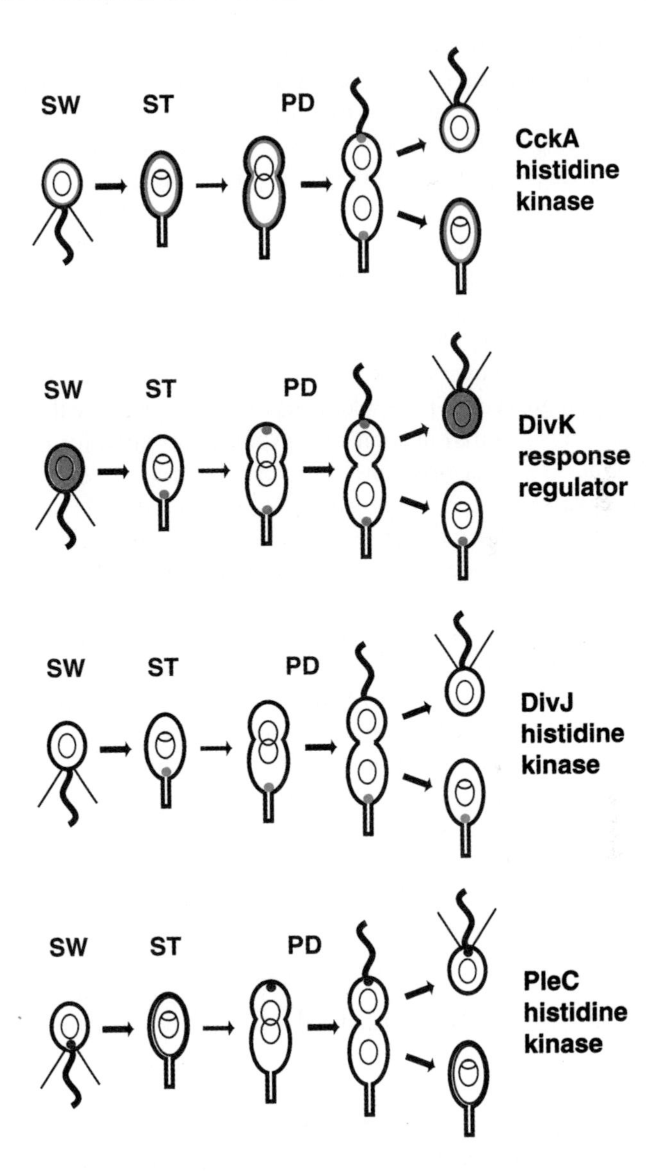

Figure 2 Dynamic localization of signal transduction proteins during the *Caulobacter* cell cycle. See text and legend to Figure 1 for details.

of the cell. In the predivisional cell, CpaE, an assembly factor for the secretion channel, and CpaC, which likely forms the secretion channel, are directed to the new flagellar pole of the predivisional cell, dependent upon the action of PodJ (41, 42). However, PodJ's role as a localizing factor is not restricted to structural components of the pilus; PodJ is also required to localize the histidine kinase PleC to the flagellar pole of the predivisional cell (42). Around the time of cell division, pilin subunits are accumulated in a process dependent upon PleC. In a *podJ* mutant, where PleC is active but delocalized, pilin subunits fail to accumulate (41). This result indicates that the transient localization of active PleC to the pole where pili will be assembled is necessary for it to function in the signal transduction pathway that culminates in pili biogenesis. In contrast, the PleC-H610A protein, which is incapable of autophosphorylation, is correctly localized to the flagellar pole of the predivisional cell but is not released upon differentiation into a stalked cell, as is the wild-type protein. *pleC610A* mutant cells, lacking PleC activity entirely, never release their flagella or build stalks and, consequently, lose the characteristic asymmetry of the wild-type *Caulobacter* cell. These studies show that localization of a signal transduction protein can be an important regulator of its cellular activity and conversely that the biochemical activation of a signal transduction protein can be required for it to be delocalized or cleared at the proper time during the cell cycle.

A recent dissection of the DivJ protein revealed that its polar localization is mediated by a ~60 amino acid cytoplasmic region between the transmembrane region and the conserved histidine kinase domain (43). Polar localization of the PilS histidine kinase, which regulates type IV pilus expression in *Pseudomonas aeruginosa*, also requires the linker region between the transmembrane domains and the histidine kinase domain (44). DivJ localization was previously found to require PleC activity (40). Sciochetti et al. (43) determined further that one cellular function of PleC, the gain of flagellar motility, is necessary for DivJ localization, but stalk formation, which is also downstream of PleC, is not required. Finally, some partial deletion mutants of *divJ* encoded proteins that failed to localize to the stalked pole, yet they still complemented the growth defects of the *divJ* null mutant, which suggests that DivJ localization may not be required for all of its functions.

DNA REPLICATION AND CHROMOSOME SEGREGATION

How do bacteria coordinate chromosome replication and segregation with cell division to ensure that each daughter cell receives a complete genome? Both the temporal and spatial aspects of this question have received intense study in the past decade, aided by new techniques that have enabled researchers to surmount the long-standing obstacle of visualizing chromosome dynamics in bacterial cells. These studies reveal that the bacterial chromosome is organized within the cell, that DNA replication, chromosome condensation, and transcription may all contribute to chromosome partitioning, and that there appears to be a dedicated mechanism for the rapid movement of replication origins to opposite poles of the cell just after they are duplicated. Surprisingly, proteins first implicated in the movement of replication origins are now believed to operate checkpoints that prevent subsequent cell division or sporulation if replication origins are not correctly positioned in the cell during S phase.

Two techniques have been used to uncover the organization of the bacterial chromosome within the cell. Fluorescent in situ hybridization (FISH) utilizes a labeled DNA probe to hybridize to a specific region of the chromosome in fixed cells (27, 45). Specific chromosome locations have also been visualized in living cells by inserting a tandem array of *lac* operator sequences into the chromosome and expressing in the same strain a GFP-lacI fusion protein, which binds to the operator array (46). Bacterial replication origins are usually located at or near the cell poles, though different species and different growth conditions of the same species vary somewhat. In slowly growing *E. coli* (45) and in *Caulobacter crescentus* (27) cells, the origin and terminus are located at opposite poles of the cell (for *Caulobacter*, the origin is always located at the stalked pole preceding S phase). Newly replicated origins migrate to opposite poles of the cell, and the terminus moves toward the midcell, near the site of future cell division. After replication and cell division are complete, two cells are born with origins and termini located at opposite poles. Chromosomal organization is slightly different in rapidly growing *E. coli* (47) and in *Bacillus subtilis* (46) cells in which replication origins are located at the ¼ and ¾ positions and termini are at the center. Just prior to DNA replication initiation, the origins move toward the midcell. One newly replicated origin from each pair then migrates back toward the cell pole, while the other origin remains at the center of the cell, where division will occur. In this case, cell division yields two progeny with origins near each pole.

The consistency of chromosome organization within the cell and the abrupt migration of chromosomal origins to opposite poles after they are replicated imply the existence of an active mechanism for nucleoid positioning and separation in bacteria. An early model of chromosome segregation posited that daughter chromosomes are attached to the cell wall and are moved apart passively, by zonal growth of the cell. However, rapid origin movement to the poles of *B. subtilis* occurs in the presence of vancomycin, which prevents cell wall growth (46). Attempts to identify the bacterial mitotic apparatus have focused on the *parA/parB* genes of *C. crescentus* and the *soj/spo0J* genes of *B. subtilis*, which are chromosomally encoded homologs of genes required for accurate partitioning of the *E. coli* F-factor and bacteriophage P1 plasmids [reviewed in (48)].

spo0J (the *parB* homolog) is required for an early stage of the *B. subtilis* sporulation pathway, specifically the transcription of Spo0A-dependent genes (49). Mutations in *spo0J* lead to the formation of 1% to 2% anucleate cells in a vegetatively growing population, which is ~100 times more frequent than in wild-type *B. subtilis*; this suggests an active role for *spo0J* in chromosome partitioning (49). Mutations in the *soj* gene suppress the *spo0J* sporulation defect by restoring Spo0A-dependent transcription (49). The Spo0J protein, like its ParB homolog in plasmid partitioning, binds to several conserved DNA sequences (called *parS* sites), which lie near the *soj spo0J* operon and the origin of replication (50). Because Spo0J foci colocalize with the replication origins in *B. subtilis* (51), and because insertion of a *parS* site confers stable partitioning on an unstable plasmid (50), Spo0J was hypothesized to be a component of the bacterial mitotic machinery. However, the *B. subtilis* chromosomal origins migrate and are positioned normally in *spo0J* mutants (46).

Studies of *soj*, the *B. subtilis parA* homolog, have helped to clarify the role of *spo0J*. Soj acts as a repressor of sporulation by physically disrupting transcription complexes of Spo0A, RNA polymerase, and certain sporulation promoters under the control of Spo0A (52). Soj also displays an unusual localization pattern, which is dependent upon Spo0J: In ~30% of cells, the Soj protein oscillates from pole to pole on a time scale of minutes (53). In the absence of Spo0J, Soj is stably associated with nucleoids, specifically with sporulation stage II promoters. As a member of the ParA family, Soj is predicted to have ATPase activity. Mutations in residues that likely prevent nucleotide binding cause Soj to be distributed throughout the cell, while a mutation predicted to interfere with nucleotide hydrolysis causes Soj to be stably localized at the cell poles (53). Furthermore, one *spo0J* mutant was isolated which increased the incidence of Soj oscillation (54). These data can be combined into a checkpoint model where the proper migration of chromosomal origins and Spo0J to the cell poles releases the repression of sporulation by Soj. The nucleoid-bound form of Soj, seen in the absence of Spo0J, is probably the repressive form, while the oscillating form of Soj would allow sporulation to occur. The nucleotide binding state of Soj is linked to its localization, and Spo0J may communicate the status of chromosome segregation to Soj by affecting its nucleotide binding state or ATPase activity.

The idea that *parA/parB* homologs participate in developmental or cell cycle checkpoints has been confirmed and extended in studies of the Par proteins in *C. crescentus*. The *Caulobacter parA* and *parB* genes are essential, and depletion of the ParB protein or an increase in the level of ParA results in the same phenotype: long, unpinched filamentous cells that continue to replicate DNA (26, 55). This phenotype is indicative of a cell division block, and in fact the depletion of ParB prevents the formation of FtsZ rings, the earliest visible step in bacterial cytokinesis (55). Movement of ParB foci to the cell poles precedes Z-ring formation by 20 min in the *Caulobacter* cell cycle; this suggests that correct ParB localization could be a signal for Z-ring formation when chromosome segregation is proceeding properly (55). In elegant in vitro experiments, ParB was revealed to act as a nucleotide exchange factor for ParA. The ATP-bound form of ParA binds to single stranded DNA and prevents ParB-*parS* complex formation (56). Further studies will be needed to integrate these molecular details into a model of how correct chromosome segregation is sensed by ParA and ParB and then communicated to the cell division apparatus.

If the *parA/parB* and *soj/spo0J* proteins do not segregate and position the chromosomal origins, what system is responsible? In *B. subtilis*, the DivIVA protein, which participates in division site selection (see below), was also shown to have a role in positioning the chromosomal origin inside the forespore of sporulating cells (57). Another study using chromosomal inversions found that the portion of the *B. subtilis* chromosome targeted first into the forespore is not determined by the origin itself or by *spo0J*, but by a diffuse region of the chromosome 150 kbp-300 kbp away from the origin (58). The authors speculate that DivIVA may bind to this centromeric region and mediate its early migration into the forespore.

Two other mechanisms proposed to provide the force and directionality for nucleoid separation and positioning are DNA replication itself and chromosome condensation by the structural maintenance of chromosomes (SMC) and MukB proteins. Localization of several GFP-tagged DNA polymerase subunits in living cells revealed that, in *B. subtilis*, the replisome is located at midcell during S phase (59), and in *Caulobacter*, the replisome assembles at the polar origin and moves slowly to the midcell during S phase (60). In both species, active DNA replication is required for the formation of replisome foci. When replication is stalled at a specific chromosomal location via the *B. subtilis* stringent response, this site colocalizes with the replisome (61). When *Caulobacter* cells are pulse-labeled with bromodeoxyuridine at different times during S phase to label newly replicated DNA, the foci are first located at the extreme cell poles, then are observed closer and closer to the division plane (60). The implications of these findings are that DNA is spooled through a replication factory that is either stationary or passively moving, and daughter chromosomes are extruded to opposite halves of the cell. The force expended by DNA polymerase to move the template may be used to power chromosome segregation (60, 61).

As DNA is released from the replisome, it is thought to be captured and organized by chromosome condensation proteins, SMC in some species and MukB in others [reviewed in (62)]. SMC and MukB are not homologous at the sequence level but share a common organization of N- and C-terminal globular domains separated by an α-helical coiled-coil rod with a central hinge region. The N-terminal domain contains a nucleotide binding site conferring ATPase activity, while the C-terminal domain binds DNA nonspecifically. SMC and MukB were originally thought to form antiparallel homodimers that bend at the central hinge region and bring distant DNA sequences together by a scissor-like motion (63). However, a recent structural study showed that SMC monomers interact exclusively through their hinge regions and that each arm of an SMC dimer is composed of an intramolecular coiled coil. Based on this geometry, a new model of SMC action was proposed in which the two arms of an SMC dimer encircle DNA strands and hold or release them to effect proper chromosome condensation and segregation (63a). *mukB* mutants in *E. coli* (64) and *smc* mutants in *B. subtilis* (65, 66) produce 20% to 30% anucleate cells. In *C. crescentus*, *smc* mutants do not produce anucleate cells but instead arrest at the predivisional stage; this suggests the presence of a chromosome segregation checkpoint in the cell division cycle (27).

In addition to the general effect of producing anucleate cells, *mukB* mutants cause aberrant localization of the SeqA protein (67), which is thought to bind to newly replicated, hemimethylated DNA as it emerges from the replisome at the center of the *E. coli* cell (68). *smc* mutants of *B. subtilis* show mislocalization of the origin-associated Spo0J protein (66), and 10% to 15% of *C. crescentus smc* mutant cells mislocalize the replication origin or terminus, as assayed by FISH (27). The effects of mutants upon the general disposition of the nucleoid and upon specific DNA sequences or DNA-binding proteins strongly suggest that MukB and SMC participate in chromosome segregation by compacting and organizing daughter chromosomes as they are extruded from the replisome into the two halves of the predivisional cell.

A recent report describing the effects of a transcriptional inhibitor on the separation of newly replicated DNA sequences indicated that the force exerted by RNA polymerase on the DNA template may be necessary for chromosome segregation (69). This model relies on the observation that the majority of genes in bacterial chromosomes are oriented away from the replication origin (70), so that when duplicated origins are oriented toward opposite poles, the net effect of transcription would be to move the DNA template toward the poles and away from the midcell. Because RNA polymerase is a powerful molecular motor, which exerts a force ~5 times stronger than myosin (71), and a highly abundant cellular protein, it could theoretically provide the force for nucleoid separation. This model requires additionally that the DNA template be translocated by stationary RNA polymerase complexes (69). Precedence for such transcription factories comes from the stationary DNA replication factory of *B. subtilis* and the nucleolar transcription factories of eukaryotic cells. *B. subtilis* RNA polymerase is located in the nucleoid (72), but the mobility of active transcription complexes remains to be addressed experimentally.

The significant strides made in understanding nucleoid organization and chromosome partitioning have raised new, fundamental questions. Despite much attention to replication origins, it is still unclear how origin regions are rapidly segregated to opposite poles of the predivisional or sporulating cell and whether polar anchoring proteins exist. In *B. subtilis*, it is also unknown whether a membrane anchor is required to keep the replisome stationary during S phase. Finally, the discovery of an actin-like protein, ParM, which forms dynamic filaments required for partitioning of the *E. coli* R1 plasmid (73), suggests that the chromosomally encoded prokaryotic actins, MreB and Mbl, may be involved in chromosome partitioning in some species, although chromosome segregation defects have not yet been attributed to *mreB* or *mbl* mutants (74).

REGULATION OF BACTERIAL CELL DIVISION

Timing and Force Generation

Segregation of newly replicated chromosomes to opposite ends of the bacterial cell will not yield genetically equivalent progeny unless cell division occurs after chromosome segregation and between the separated nucleoids. Temporal and spatial control of bacterial cell division is accomplished largely by regulating the behavior of FtsZ, a homolog of eukaryotic tubulin found in virtually all bacteria, which forms a cytokinetic ring (Z-ring) at the midcell that is absolutely required for cell division (75). Z-ring formation is the earliest known step in bacterial cell division; it occurrs well before cell invagination and nucleoid separation in *E. coli* (76) and before the conclusion of DNA replication in *Caulobacter* (77). The fact that Z-rings are assembled long before they constrict suggests that assembly and constriction are regulated separately by the cell.

Temporal control of Z-ring assembly has been the subject of several studies, but the cellular signal is still unknown. The availability of FtsZ protein appears not to control Z-ring assembly. Overproduction of FtsZ in *C. crescentus* or *E. coli* does not lead to premature midcell division, but instead it causes division at aberrant sites near the cell poles (6, 78). In *Caulobacter*, rapid proteolysis of FtsZ occurs at cell division, so that new FtsZ synthesis must occur before the next round of division in each daughter cell. FtsZ is normally synthesized during swarmer-to-stalked cell differentiation, and Z-rings form in the stalked cells (79). However, if swarmer cells are engineered to produce a high level of FtsZ protein prior to differentiation, Z-rings are not assembled prematurely (77); this finding indicates that other cellular factors regulate the timing of Z-ring assembly at the future division site. In *Caulobacter*, the initiation of DNA replication is not required for assembly of the Z-ring but is required for its proper placement near the future division site (80). Several studies have attempted and failed to uncover a direct link between the onset or completion of DNA replication and Z-ring formation. It appears, therefore, that the bacterial DNA replication and division cycles overlap and interact via checkpoints but are not directly linked.

Because FtsZ is a homolog of the cytoskeletal GTPase tubulin, many groups have examined its biochemical properties in vitro [reviewed in (81)] to uncover clues as to how FtsZ mediates cytokinesis in vivo. In the presence of GTP, FtsZ polymerizes into single linear polymers called protofilaments (82). Polymerization assays performed in the presence of ZipA, which stabilizes Z-rings in vivo (83), reveal networked bundles of ~10–20 protofilaments (84), but the type of FtsZ polymer that comprises the Z-ring in living cells is unknown. GTP hydrolysis is responsible for the dynamic behavior of FtsZ polymers: Kinetic experiments show that FtsZ polymerizes until the available GTP is exhausted, then depolymerizes (85). Polymers formed from mutants with impaired GTPase activity display increased stability (86), and nonhydrolyzable GTP analogs stabilize FtsZ assemblies (87).

Several models have emerged from in vitro and in vivo studies of FtsZ to describe how constriction of the Z-ring causes cell invagination and division. By analogy with tubulin, motive force could be generated in two ways: by regulated assembly and disassembly of the FtsZ polymer or by use of the Z-ring as a track for an unknown motor protein. To attempt to distinguish between these models, investigators have asked how important FtsZ disassembly dynamics are for cell division. The *ftsZ2* and *ftsZ84* alleles, which encode proteins with 10% or less of the wild-type GTPase activity, can both support cell division in *E. coli* (86, 88). In vitro, the FtsZ2 protein copolymerizes with wild-type FtsZ into filaments that display increased stability, which suggests that Z-rings composed entirely of FtsZ2 in vivo would be overly stable as well (86). These results seemed to indicate that cell division is mediated by an external motor acting on a relatively stable Z-ring track because a wild-type disassembly rate is not required. A recent study examining Z-ring remodeling in living cells using FtsZ-GFP and fluorescence recovery after photobleaching revealed that Z-rings are extremely dynamic structures that undergo complete remodeling with a half-time of ~30 s (88). Moreover, the remodeling of the Z-ring in dividing cells was just as rapid as in cells that had not yet begun to invaginate. Cells containing only the FtsZ84 protein divided normally, but recovery of Z-ring fluorescence after photobleaching took ~ninefold longer than in wild-type cells (88). These results imply that any mechanism of Z-ring contraction and force generation must be able to operate on the framework of a dynamic FtsZ polymer, and it must still be able to function when the rate of Z-ring remodeling is slowed by an order of magnitude. Furthermore, Z-ring remodeling could contribute to cell division: If depolymerized FtsZ monomers or short filaments were sequestered in the cytoplasm during cell division, then FtsZ would be removed from the ring faster than it is replaced, which could cause constriction of the Z-ring (88).

A different model for Z-ring constriction stems from the observation that FtsZ protofilaments formed with GDP in vitro are more curved than those formed with GTP (89, 90). In this model, Z-ring contraction is triggered by a burst of GTP hydrolysis, which increases the curvature of the Z-ring filaments and exerts contractile force on the membrane. However, because GTP hydrolysis occurs immediately upon polymerization of FtsZ subunits in vitro (91), it may be that phosphate release is the regulated step that triggers shape change in the FtsZ polymer (92). This mechanism would be similar to the association of phosphate release with the power stroke in eukaryotic myosin motors (93).

In order to cause cell invagination, Z-ring constriction must exert force on the cell membrane. Two essential cell division proteins bind directly to FtsZ and also interact with the cell membrane: FtsA is a structural homolog of actin (94) found in both the membrane and cytosolic fractions of *E. coli*, and ZipA contains a single membrane-spanning domain at its N-terminus and a FtsZ-binding domain at its C-terminus (95). FtsA and ZipA both stabilize Z-rings in vivo and recruit downstream components of the division apparatus to the division plane [(83, 96, 97) and references within]. A dissection of the ZipA protein revealed that a deletion mutant lacking the N-terminal membrane spanning domain was sufficient to bind to the Z-ring in vivo and bundle FtsZ filaments in vitro, but it was unable to support division as the only copy of *zipA* in the cell (95). Furthermore, when the *zipA* membrane anchor was replaced with a membrane anchor of the same topology from another protein, this chimera also failed to support cell division. These results suggest that the membrane anchor of ZipA may mediate an important interaction with other components of the division machinery, in addition to tethering the Z-ring to membrane (95). Further study of the FtsA protein is required to determine the nature and role of its interaction with the cell membrane.

Division Site Selection

Two forces have been proposed to mediate division site selection and account for the very low prevalence of anucleate cells in wild-type bacterial populations: nucleoid occlusion and the Min system [reviewed in (98)]. By manipulating the nucleoid position and DNA replication in *E. coli* and observing the effects upon cell division, it has been inferred that formation of Z rings is prevented in the region of the membrane overlying an unreplicating nucleoid (99, 100). The molecular interactions responsible for this effect are currently unknown, but one study found that mutations in *mukB* cause an increased proportion of cells to form Z-rings in the region of the nucleoid (101) suggesting that the MukB protein may participate in the nucleoid occlusion mechanism.

The *minCDE* genes, so named because mutations in them increase the frequency of minicells lacking DNA, constitute a more clearly defined system for placement of the Z ring. MinC interacts directly with FtsZ and prevents its polymerization (102). MinD is an ATPase that interacts with MinC and phospholipids, which brings MinC to the inner membrane (103–105). MinE is a topological regulator of MinCD whose function is to limit MinCD activity to the cell poles, allowing division at the midcell (106). By affecting the ATPase activity of MinD and thus its affinity for the membrane (105, 107), MinE sets the system in a surprising oscillatory motion that determines where the Z ring will be placed. MinC and MinD form a test-tube shaped sheath around the inside of one end of the cell (Figure 3). On a time scale of tens of seconds, the MinCD sheath breaks down at one end of the cell and reforms at the other end (104, 108). MinE forms a ring-shaped structure at the medial edge of the MinCD zone, and MinE molecules not present in the ring are found in a polar zone that overlaps the MinCD zone. The MinE ring moves poleward as the MinCD sheath breaks down, which gives it the appearance of a windshield wiper clearing out one end of the cell at a time (109, 110). The net result of the Min oscillations is to yield a time-averaged minimum occupancy of MinC at the midcell, making this the site where FtsZ polymerization is allowed to occur.

Reaction-diffusion models of the Min proteins based on their known biochemical properties suggest that, if a cell begins with a homogeneous distribution of all three Min proteins, they will spontaneously begin to oscillate in the manner seen in wild-type *E. coli* (111, 112). This result suggests that the Min system does not rely on any previously placed markers at the pole or the midcell but instead generates this information via intrinsic biochemical and biophysical properties; these include the average size of the cell, diffusion rates within the cytoplasm, binding affinties of the Min proteins for each other, the rate of MinD ATPase activity and the control of this rate by MinE, and finally the effect of the MinD

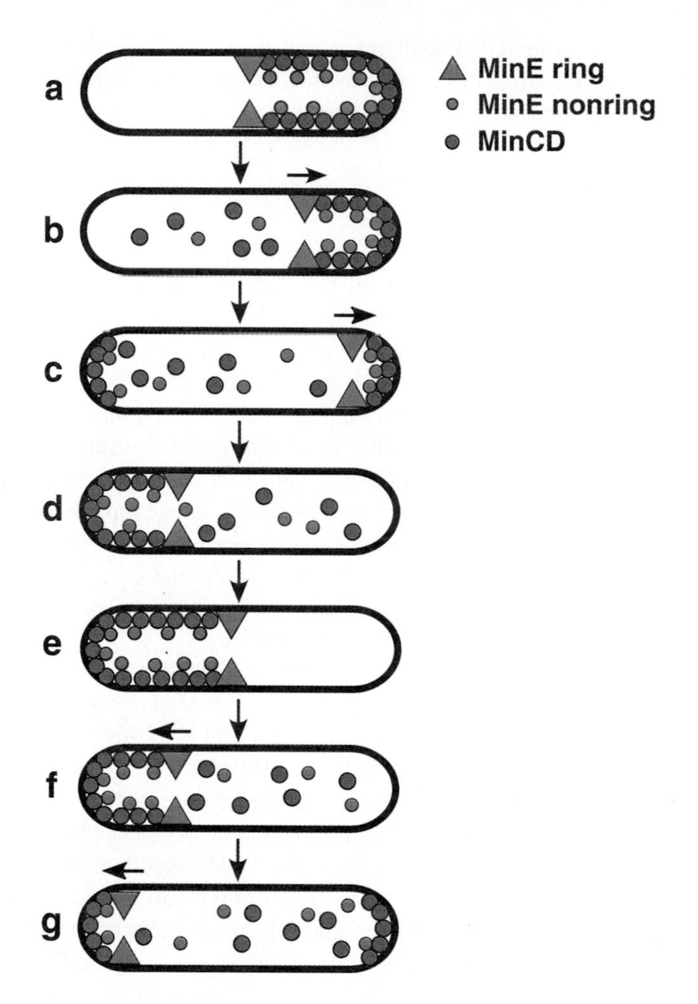

Figure 3 Oscillation of the Min system for division site selection. (*a*) The MinCD complex assembles in the shape of a test tube in the right half of the cell (*orange circles*). MinE forms both a polar zone (*blue circles*) coincident with MinCD and a ring (*blue triangles*) at the medial edge of the MinCD zone. (*b–c*) The MinCD sheath and MinE polar zone disassemble, and the MinE ring moves toward the right-hand pole. (*d–e*) The MinCD sheath reassembles in the left half of the cell, accompanied by a new MinE polar zone and MinE ring. The MinE ring appears before the MinCD zone is fully assembled. (*f–g*) MinCD disassembles and the MinE ring migrates toward the left-hand pole. Red arrows indicate the direction of MinE ring migration during disassembly of the polar MinCD sheath.

nucleotide binding state on its affinity for the membrane. Although these models do not exclude the possibility that the Min proteins interact with other topological markers, they present the exciting possibility that the Min system is a primary generator of positional information within the cell.

Despite the fact that the behavior of the Min system can be accurately modeled, several unexplained observations remain. In fluorescence micrographs, the MinD protein appears to coat the inside of roughly half of the cell at any one time, yet the number of MinD molecules in the cell is not sufficient to uniformly cover the area observed (113). In addition, the MinD protein reshapes phospholipid vesicles into tubes by forming an ordered polymer on the surface suggesting that its cooperative assembly and disassembly on a portion of the cell membrane may contribute to its oscillatory behavior (105). The structure of the MinD sheath and the MinE ring in vivo, which include the properties that cause MinE to localize preferentially at the medial edge of the MinCD zone, remain to be determined.

Most gram-positive bacterial species and subfamilies of the gram-negative bacteria lack some or all of the Min proteins and use different mechanisms to place the division site. *B. subtilis* contains homologs of *minC* and *minD*, but it has no *minE* homolog (114, 115). MinC and MinD are stationary and are located at both ends of the cell, as well as at the midcell late in the division cycle (116, 117). The localization of MinC and MinD is mediated by DivIVA, a coiled-coil protein with weak homology to tropomyosins. DivIVA arrives late at the division site, dependent upon the Z-ring and other division proteins (116). DivIVA recruits the MinCD complex to these committed division sites, and all three proteins remain at the new poles after division (116, 117). The MinCD complex then acts to inhibit Z-ring formation near the poles in the newly born cells (118, 119). It is remarkable that DivIVA and MinE are unrelated but have evolved to regulate the same protein complex that inhibits Z-ring formation. *Caulobacter* contains no homologues of the Min proteins or of DivIVA (2). Furthermore, *Caulobacter* divides not at the midcell, but consistently at 0.6 cell lengths away from the stalked pole. These departures from the characteristics of cell division in *E. coli* and *B. subtilis* suggest that division site selection in *Caulobacter* may rely exclusively upon inhibitory action of the nucleoid (77).

GENERATION OF ASYMMETRY IN *BACILLUS SUBTILIS* SPORULATION

In response to starvation, *B. subtilis* cells initiate a developmental program that culminates in the production of a dormant, environmentally resistant spore. The decision to sporulate is mediated by a complex phosphorelay composed of histidine kinases, response regulators, and phosphatases that integrate cell density and nutritional signals and lead to activation of Spo0A by phosphorylation [reviewed in (120)]. Growing *B. subtilis* cells divide centrally and produce two identical progeny. Activation of Spo0A leads instead to polar septation creating a larger mother cell compartment and a smaller forespore compartment with different fates. The mother cell is eventually lysed, while the forespore remains dormant until nutritional signals trigger germination. Mother cell- and forespore-specific transcriptional programs geared to these different cell fates are controlled largely by the activation of compartment-specific sigma factors that direct RNA polymerase to specific sets of genes (Figure 4) [reviewed in (121)]. This section addresses (*a*) the initial establishment of a polar division site, (*b*) the activation of early mother cell and forespore sigma factors just after formation of the asymmetric septum, and (*c*) the activation of late mother cell and forespore sigma factors just after engulfment of the prespore by the mother cell membranes.

The earliest morphological milestone in sporulation is formation of the asymmetric septum, which begins with the redeployment of FtsZ away from the midcell and toward polar division sites. If an activated form of Spo0A is produced during vegetative growth, it is sufficient to induce FtsZ relocalization and subsequent septation at a polar site (122). Until recently, it was hypothesized that Z-ring assembly was inhibited at the midcell and promoted at polar division sites during sporulation. It now appears, however, that FtsZ assembles at the midcell early in sporulation and is redeployed to potential polar division sites via a spiral intermediate (123). Two factors are necessary and sufficient to induce FtsZ spirals and polar Z-rings (123): an increase in FtsZ concentration, normally mediated by a σ^H-dependent burst of *ftsZ* transcription at the onset of sporulation (124), and the SpoIIE protein, which is produced before septation under the control of Spo0A (125). SpoIIE is an integral membrane protein that localizes to the polar septum (126) and binds directly to FtsZ and to itself (127, 128), but its precise function in the switch from midcell to polar division sites remains unclear.

After polar septation, SpoIIE plays a critical role in linking cell morphology to cell fate decisions by participating in the activation of σ^F, a forespore-specific sigma factor at the top of a hierarchy of compartmentalized transcriptional regulators. σ^F is transcribed prior to asymmetric septation along with *spoIIAA* and *spoIIAB* (129). σ^F is held in an inactive complex by the antisigma factor SpoIIAB until after septation, when σ^F is activated specifically in the forespore compartment (130–132). SpoIIAB also phosphorylates SpoIIAA, an anti-antisigma factor, which keeps SpoIIAA in an inactive state (132, 133). SpoIIE is a phosphatase that dephosphorylates SpoIIAA; dephosphorylated SpoIIAA binds SpoIIAB and displaces σ^F from the complex (134, 135). σ^F is then free to activate forespore-specific genes.

Dephosphorylation of SpoIIAA by SpoIIE is necessary, but not sufficient, for σ^F activation. Late-acting cell division mutants that prevent asymmetric septation produce abundant unphosphorylated SpoIIAA but still do not activate σ^F (136, 137). Furthermore, the *spoIIE48* mutant cannot activate σ^F but contains wild-type levels of unphosphorylated SpoIIAA (136). These studies indicate that SpoIIE itself participates in two steps to activate σ^F, dephosphorylation of

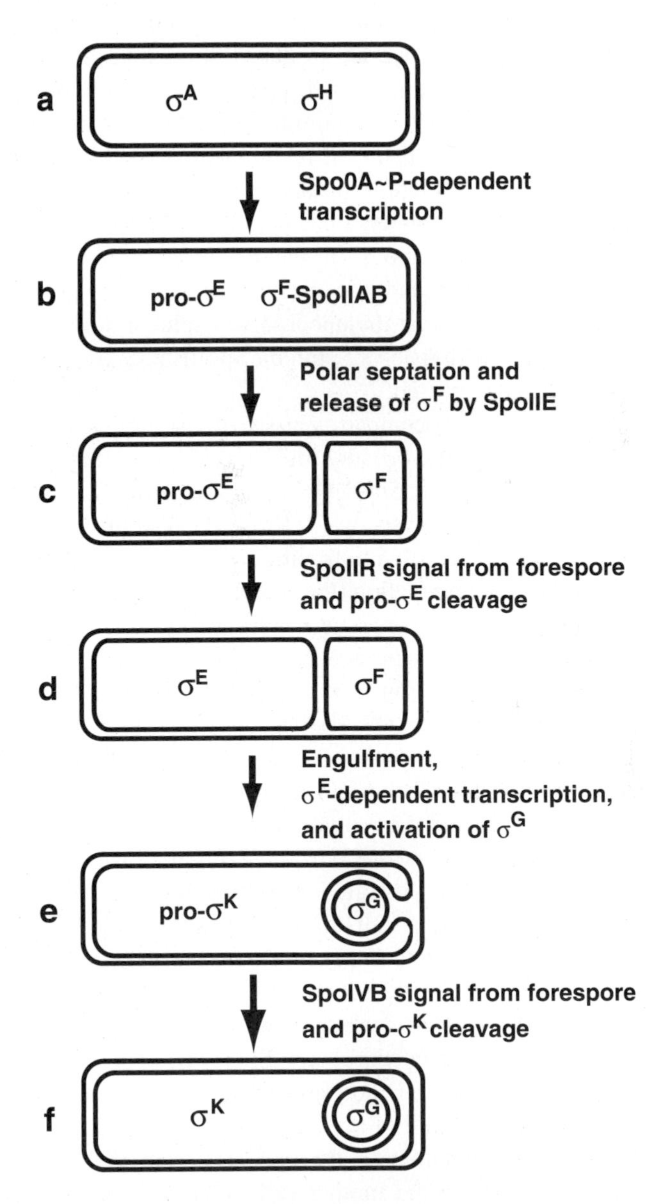

Figure 4 Activation of compartment-specific sigma factors during *B. subtilis* sporulation. See text for details. (*a–b*) The cell before polar septation. (*c–d*) The larger compartment is the mother cell, and the smaller compartment is the forespore. (*e–f*) Engulfment of the forespore by the mother cell membranes.

SpoIIAA and another event that is triggered by formation of the polar septum. It remains unclear how σ^F activation is restricted to the forespore compartment, because σ^F and its regulators are all produced throughout the cell before polar septation. The hypothesis that the SpoIIE protein is somehow sequestered on the forespore side of the asymmetric septum has been tested twice, with contradictory results (136, 138). Several *spoIIE* mutations bypass the requirement for an asymmetric septum and activate σ^F prematurely. Some bypass mutations delocalize SpoIIE so that it is no longer targeted to the asymmetric septum (136, 137), while others are point mutations in the unconserved domain between the membrane-spanning region and the phosphatase domain (139, 140). These mutant phenotypes suggest that the unconserved soluble domain of SpoIIE and proper localization to the division site inhibit SpoIIE activity until a morphological checkpoint has been passed.

The activation of the first compartment-specific sigma factor already appears very complex, yet there are two additional mechanisms that contribute to the regulation of σ^F activity. First, SpoIIAB, when it is not in a complex with σ^F or SpoIIAA, is subject to proteolysis by ClpCP (141). The proteolysis of SpoIIAB upon σ^F release establishes a self-reinforcing cycle that locks σ^F activity on in the forespore (141). During sporulation, the region of the chromosome comprising the origin-proximal 30% of the genome is consistently targeted into the forespore first (142, 143). Asymmetric septation precedes nucleoid separation, so that the remainder of the chromosome must be translocated through the septum into the forespore by the SpoIIIE protein (144). This mechanism of chromosome segregation leads to a situation in which, early in sporulation, the forespore contains only origin-proximal genes. The *spoIIA* locus encoding SpoIIAA, SpoIIAB, and σ^F is distal to the origin and normally enters the forespore later in sporulation. Moving the gene encoding σ^F to a site near the origin allows ~10% of wild-type sporulation in a *spoIIAA* mutant that alone prevents sporulation completely (145). Conversely, movement of the *spoIIAB* gene to a site near the origin inhibits sporulation (146). These results imply that transient genetic asymmetry of the *spoIIA* operon helps to generate the correct ratio of σ^F and SpoIIAB in the forespore. SpoIIAB proteolysis synergizes with transient genetic asymmetry so that during an early window in sporulation SpoIIAB, which is released from σ^F in the forespore and degraded, is not replenished because the *spoIIAB* gene has not yet arrived in the forespore (146).

The next step in differentiating the prespore from the mother cell is the specific activation of σ^E in the mother cell. σ^E is synthesized as an inactive precursor under the direction of Spo0A before polar septation (147, 148). However, pro-σ^E cleavage in the mother cell does not occur until after the asymmetric septum has been built. The protease SpoIIGA cleaves pro-σ^E specifically in the mother cell compartment (149, 150) under the direction of SpoIIR, a secreted signaling protein transcribed in the forespore by σ^F-RNAP (151). Although it is clear that pro-σ^E cleavage in the mother cell requires intercellular signaling, the molecular details of the SpoIIR-SpoIIGA interaction are unknown.

Because σ^E and its protease SpoIIGA are both made before septation, it is not clear how SpoIIR activates cleavage only in the mother cell compartment. Pro-σ^E is sequestered to the cytoplasmic membrane by its prosequence; when cleavage occurs, σ^E becomes free to interact with RNA polymerase in the nucleoid (152). One study has suggested further that pro-σ^E is targeted to the mother cell side of the asymmetric septum (153), which would provide a mechanism for release of cleaved σ^E only into the mother cell compartment. Another study has challenged these findings and supports a model in which preferential *spoIIG* transcription in the mother cell and selective σ^E degradation in the forespore combine to yield active σ^E only in the mother cell (154).

After the sporangium has been divided into forespore and mother cell compartments by the asymmetric septum, the forespore is engulfed by the mother cell and undergoes physiological and morphological changes that prepare it to withstand a hostile environment and to germinate when nutrients become available. Like polar septation and differentiation early in sporulation, these later physiological changes require compartment-specific transcriptional programs. Activation of the late sporulation sigma factors occurs through similar pathways as σ^F and σ^E activation, yet with important mechanistic differences. Like σ^F, σ^G appears to be held in an inactive complex in the forespore by the antisigma factor SpoIIAB (155). The release of σ^G, however, may be mediated by the proteolysis of SpoIIAB, which is concurrent with σ^G activation after engulfment (156).

σ^K is activated in the mother cell late in sporulation, also by a process of prosigma factor cleavage. Like pro-σ^E, the N-terminal extension of pro-σ^K promotes its association with the cell membrane (157), so cleavage causes release from the membrane and redeployment of σ^K to the nucleoid. Pro-σ^K is synthesized specifically in the mother cell, and its cleavage is dependent upon a signal from the forespore (158). The signaling protein is SpoIVB, a serine protease believed to be secreted by the forespore into the space between the surrounding membranes (159). SpoIVB is not the protease that cleaves pro-σ^K; instead it functions in a morphological checkpoint that restricts σ^K activation to the time after engulfment. *spoIVB* is under the transcriptional control of σ^G, and mutants that express *spoIVB* early activate σ^K prematurely and are impaired in sporulation (158). Furthermore, mutations in other genes can bypass the need for SpoIVB in pro-σ^K processing (160, 161).

What is the nature of the pro-σ^K processing machinery, and how does it respond to the forespore signal? The integral membrane protein SpoIVFB (not to be confused with SpoIVB) is the protease that cleaves pro-σ^K (162). Mutations that bypass the requirement for the forespore signal lie in the genes encoding SpoIVFA and BofA, suggesting that these proteins inhibit pro-σ^K processing until the forespore signal is received (160, 161). SpoIVFA, SpoIVFB, and BofA are all synthesized in the mother cell under the direction of σ^E (160, 161), and they colocalize in the membranes surrounding the forespore where pro-σ^K processing occurs (163). SpoIVFB, SpoIVFA, and BofA form a stable integral membrane complex, and SpoIVFA both localizes the complex to the forespore membranes and functions as a scaffold to bring together SpoIVFB and its inhibitor BofA (163). SpoIVFA levels decrease concurrent with pro-σ^K processing, but it is not clear if this step is sufficient to release the inhibition of SpoIVFB, or whether the loss of SpoIVFA is mediated directly by the SpoIVB signaling protease (164).

The SpoIVFB protease is a member of a newly-defined class of Zn^{2+} metalloproteases that cleave their substrates within transmembrane domains (165). The founding member of this family is the site-2 protease that cleaves a eukaryotic transcription factor, the sterol response element binding protein (SREBP), and releases it from membranes of the endoplasmic reticulum and nuclear envelope. Like pro-σ^K processing, regulated SREBP cleavage creates a soluble protein that activates transcription upon receipt of an intercellular signal. These results indicate that an important mechanism of regulated proteolysis has been conserved from bacteria to humans, and they suggest that the other bacterial homologs of SpoIVFB may play critical roles in prokaryotic development or cellular responses to environmental signals.

The *Annual Review of Biochemistry* is online at http://biochem.annualreviews.org

LITERATURE CITED

1. Laub MT, McAdams HH, Feldblyum T, Fraser CM, Shapiro L. 2000. *Science* 290:2144–48
2. Nierman WC, Feldblyum T, Laub MT, Paulsen IT, Nelson KE, et al. 2001. *Proc. Natl. Acad. Sci. USA* 98:4136–41
3. Quon KC, Marczynski GT, Shapiro L. 1996. *Cell* 84:83–93
4. Domian IJ, Quon KC, Shapiro L. 1997. *Cell* 90:415–24
5. Laub MT, Chen SL, Shapiro L, McAdams HH. 2002. *Proc. Natl. Acad. Sci. USA* 99:4632–37
6. Kelly AJ, Sackett MJ, Din N, Quardokus E, Brun YV. 1998. *Genes Dev.* 12: 880–93
7. Reisenauer A, Quon KC, Shapiro L. 1999. *J. Bacteriol.* 181:2430–39
8. Skerker JM, Shapiro L. 2000. *EMBO J.* 19:3223–34
9. Quon KC, Yang B, Domian IJ, Shapiro L, Marczynski GT. 1998. *Proc. Natl. Acad. Sci. USA* 95:120–25
10. Domian IJ, Reisenauer A, Shapiro L. 1999. *Proc. Natl. Acad. Sci. USA* 96:6648–53
11. Jacobs C, Domian IJ, Maddock JR, Shapiro L. 1999. *Cell* 97:111–20
12. Wu J, Ohta N, Zhao Y, Newton A. 1999. *Proc. Natl. Acad. Sci. USA* 96:13068–73
13. Jenal U, Fuchs T. 1998. *EMBO J.* 17:5658–69
14. Roberts RC, Shapiro L. 1997. *J. Bacteriol.* 179:2319–30
15. Zweiger G, Shapiro L. 1994. *J. Bacteriol.* 176:401–8
16. Keiler KC, Shapiro L. 2001. *J. Bacteriol.* 183:4860–65
17. Zweiger G, Marczynski G, Shapiro L. 1994. *J. Mol. Biol.* 235:472–85
18. Hung DY, Shapiro L. 2002. *Proc. Natl. Acad. Sci. USA* 99:13160–65
19. Jacobs C, Hung DY, Shapiro L. 2001. *Proc. Natl. Acad. Sci. USA* 98: 4095–100
20. Mangan EK, Malakooti J, Caballero A, Anderson P, Ely B, et al. 1999. *J. Bacteriol.* 181:6160–70

21. Anderson PE, Gober JW. 2000. *Mol. Microbiol.* 38:41–52
22. Alley MRK, Maddock JR, Shapiro L. 1993. *Science* 259:1754–57
23. Jenal U, Shapiro L. 1996. *EMBO J.* 15:2393–406
24. Stephens CM, Shapiro L. 1993. *Mol. Microbiol.* 9:1169–79
25. Wortinger M, Sackett MJ, Brun YV. 2000. *EMBO J.* 19:4503–12
26. Mohl DA, Gober JW. 1997. *Cell* 88:675–84
27. Jensen RB, Shapiro L. 1999. *Proc. Natl. Acad. Sci. USA* 96:10661–66
28. Hughes KT, Gillen KL, Semon MJ, Karlinsey JE. 1993. *Science* 262:1277–80
29. Muir RE, O'Brien TM, Gober JW. 2001. *Mol. Microbiol.* 39:1623–37
30. Muir RE, Gober JW. 2001. *Mol. Microbiol.* 41:117–30
31. Alley MRK, Maddock JR, Shapiro L. 1992. *Genes Dev.* 6:825–36
32. Maddock JR, Shapiro L. 1993. *Science* 259:1717–23
33. Levit MN, Stock JB. 2002. *J. Biol. Chem.* 277:36760–65
34. Lybarger SR, Maddock JR. 2000. *Proc. Natl. Acad. Sci. USA* 97:8057–62
35. Ames P, Studdert CA, Reiser RH, Parkinson JS. 2002. *Proc. Natl. Acad. Sci. USA* 99:7060–65
36. Gestwicki JE, Kiessling LL. 2002. *Nature* 415:81–84
37. Kim KK, Yokota H, Kim SH. 1999. *Nature* 400:787–92
38. Sourjik V, Berg HC. 2002. *Proc. Natl. Acad. Sci. USA* 99:123–27
39. Kim SH, Wang W, Kim KK. 2002. *Proc. Natl. Acad. Sci. USA* 99:11611–15
40. Wheeler RT, Shapiro L. 1999. *Mol. Cell* 4:683–94
41. Viollier PH, Sternheim N, Shapiro L. 2002. *EMBO J.* 21:4420–28
42. Viollier PH, Sternheim N, Shapiro L. 2002. *Proc. Natl. Acad. Sci. USA* 99:13831–36
43. Sciochetti SA, Lane T, Ohta N, Newton A. 2002. *J. Bacteriol.* 184:6037–49
44. Boyd JM. 2000. *Mol. Microbiol.* 36:153–62
45. Niki H, Hiraga S. 1998. *Genes Dev.* 12:1036–48
46. Webb CD, Graumann PL, Kahana JA, Teleman AA, Silver PA, et al. 1998. *Mol. Microbiol.* 28:883–92
47. Gordon GS, Sitnikov D, Webb CD, Teleman A, Straight A, et al. 1997. *Cell* 90:1113–21
48. Moller-Jensen J, Jensen RB, Gerdes K. 2000. *Trends Microbiol.* 8:313–20
49. Ireton K, Grossman AD. 1994. *EMBO J.* 13:1566–73
50. Lin DC, Grossman AD. 1998. *Cell* 92:675–85
51. Lewis PJ, Errington J. 1997. *Mol. Microbiol.* 25:945–54
52. Cervin MA, Spiegelman GB, Raether B, Ohlsen K, Perego M, et al. 1998. *Mol. Microbiol.* 29:85–95
53. Quisel JD, Lin DC, Grossman AD. 1999. *Mol. Cell* 4:665–72
54. Autret S, Nair R, Errington J. 2001. *Mol. Microbiol.* 41:743–55
55. Mohl DA, Easter JJ, Gober JW. 2001. *Mol. Microbiol.* 42:741–55
56. Easter JJ, Gober JW. 2002. *Mol. Cell* 10:427–34
57. Thomaides HB, Freeman M, El Karoui M, Errington J. 2001. *Genes Dev.* 15:1662–73
58. Wu LJ, Errington J. 2002. *EMBO J.* 21:4001–11
59. Lemon KP, Grossman AD. 1998. *Science* 282:1516–19
60. Jensen RB, Wang SC, Shapiro L. 2001. *EMBO J.* 20:4952–63
61. Lemon KP, Grossman AD. 2000. *Mol. Cell* 6:1321–30
62. Hiraga S. 2000. *Annu. Rev. Genet.* 34:21–59
63. Niki H, Imamura R, Kitaoka M, Yamanaka K, Ogura T, et al. 1992. *EMBO J.* 11:5101–9
63a. Haering CH, Löwe J, Hochwagen A, Nasmyth K. 2002. *Mol. Cell* 9:773–88

64. Niki H, Jaffe A, Imamura R, Ogura T, Hiraga S. 1991. *EMBO J.* 10:183–93
65. Britton RA, Lin DC, Grossman AD. 1998. *Genes Dev.* 12:1254–59
66. Moriya S, Tsujikawa E, Hassan AK, Asai K, Kodama T, et al. 1998. *Mol. Microbiol.* 29:179–87
67. Onogi T, Niki H, Yamazoe M, Hiraga S. 1999. *Mol. Microbiol.* 31:1775–82
68. Hiraga S, Ichinose C, Niki H, Yamazoe M. 1998. *Mol. Cell* 1:381–87
69. Dworkin J, Losick R. 2002. *Proc. Natl. Acad. Sci. USA* 99:14089–94
70. Lopez P, Philippe H. 2001. *C. R. Acad. Sci. Ser. III* 324:201–8
71. Wang MD, Schnitzer MJ, Yin H, Landick R, Gelles J, et al. 1998. *Science* 282:902–7
72. Lewis PJ, Thaker SD, Errington J. 2000. *EMBO J.* 19:710–18
73. Moller-Jensen J, Jensen RB, Lowe J, Gerdes K. 2002. *EMBO J.* 21:3119–27
74. Jones LJ, Carballido-Lopez R, Errington J. 2001. *Cell* 104:913–22
75. Bi E, Lutkenhaus J. 1991. *Nature* 354:161–64
76. Den Blaauwen T, Buddelmeijer N, Aarsman ME, Hameete CM, Nanninga N. 1999. *J. Bacteriol.* 181:5167–75
77. Quardokus EM, Din N, Brun YV. 2001. *Mol. Microbiol.* 39:949–59
78. Ward JEJ, Lutkenhaus J. 1985. *Cell* 42:941–49
79. Quardokus E, Din N, Brun YV. 1996. *Proc. Natl. Acad. Sci. USA* 93:6314–19
80. Quardokus EM, Brun YV. 2002. *Mol. Microbiol.* 45:605–16
81. Addinall SG, Holland B. 2002. *J. Mol. Biol.* 318:219–36
82. Romberg L, Simon M, Erickson HP. 2001. *J. Biol. Chem.* 276:11743–53
83. Pichoff S, Lutkenhaus J. 2002. *EMBO J.* 21:685–93
84. RayChaudhuri D. 1999. *EMBO J.* 18:2372–83
85. Mukherjee A, Lutkenhaus J. 1999. *J. Bacteriol.* 181:823–32
86. Mukherjee A, Saez C, Lutkenhaus J. 2001. *J. Bacteriol.* 183:7190–97
87. Scheffers DJ, den Blaauwen T, Driessen AJ. 2000. *Mol. Microbiol.* 35:1211–19
88. Stricker J, Maddox P, Salmon ED, Erickson HP. 2002. *Proc. Natl. Acad. Sci. USA* 99:3171–75
89. Lu C, Reedy M, Erickson HP. 2000. *J. Bacteriol.* 182:164–70
90. Rivas G, Lopez A, Mingorance J, Ferrandiz MJ, Zorrilla S, et al. 2000. *J. Biol. Chem.* 275:11740–49
91. Scheffers DJ, Driessen AJ. 2002. *Mol. Microbiol.* 43:1517–21
92. Scheffers DJ, Driessen AJ. 2001. *FEBS Lett.* 506:6–10
93. Tyska MJ, Warshaw DM. 2002. *Cell Motil. Cytoskelet.* 51:1–15
94. van den Ent F, Lowe J. 2000. *EMBO J.* 19:5300–7
95. Hale CA, Rhee AC, de Boer PA. 2000. *J. Bacteriol.* 182:5153–66
96. Hale CA, de Boer PA. 2002. *J. Bacteriol.* 184:2552–56
97. Chen JC, Beckwith J. 2001. *Mol. Microbiol.* 42:395–413
98. Harry EJ. 2001. *Mol. Microbiol.* 40:795–803
99. Yu XC, Margolin W. 1999. *Mol. Microbiol.* 32:315–26
100. Sun Q, Margolin W. 2001. *J. Bacteriol.* 183:1413–22
101. Yu XC, Sun Q, Margolin W. 2001. *Biochimie* 83:125–29
102. Pichoff S, Lutkenhaus J. 2001. *J. Bacteriol.* 183:6630–35
103. de Boer PA, Crossley RE, Hand AR, Rothfield LI. 1991. *EMBO J.* 10:4371–80
104. Hu Z, Lutkenhaus J. 1999. *Mol. Microbiol.* 34:82–90
105. Hu Z, Gogol EP, Lutkenhaus J. 2002. *Proc. Natl. Acad. Sci. USA* 99:6761–66
106. de Boer PAJ, Crossley RE, Rothfield LI. 1989. *Cell* 56:641–49
107. Hu Z, Lutkenhaus J. 2001. *Mol. Cell* 7:1337–43

108. Raskin DM, de Boer PA. 1999. *J. Bacteriol.* 181:6419–24
109. Fu X, Shih YL, Zhang Y, Rothfield LI. 2001. *Proc. Natl. Acad. Sci. USA* 98:980–85
110. Hale CA, Meinhardt H, de Boer PA. 2001. *EMBO J.* 20:1563–72
111. Meinhardt H, de Boer PA. 2001. *Proc. Natl. Acad. Sci. USA* 98:14202–7
112. Howard M, Rutenberg AD, de Vet S. 2001. *Phys. Rev. Lett.* 87(27):278102
113. Rothfield LI, Shih YL, King G. 2001. *Cell* 106:13–16
114. Levin PA, Margolis PS, Setlow P, Losick R, Sun D. 1992. *J. Bacteriol.* 174:6717–28
115. Varley AW, Stewart GC. 1992. *J. Bacteriol.* 174:6729–42
116. Marston AL, Thomaides HB, Edwards DH, Sharpe ME, Errington J. 1998. *Genes Dev.* 12:3419–30
117. Marston AL, Errington J. 1999. *Mol. Cell* 4:673–82
118. Cha JH, Stewart GC. 1997. *J. Bacteriol.* 179:1671–83
119. Edwards DH, Errington J. 1997. *Mol. Microbiol.* 24:905–15
120. Sonenshein AL. 2000. *Curr. Opin. Microbiol.* 3:561–66
121. Kroos L, Zhang B, Ichikawa H, Yu YT. 1999. *Mol. Microbiol.* 31:1285–94
122. Levin PA, Losick R. 1996. *Genes Dev.* 10:478–88
123. Ben-Yehuda S, Losick R. 2002. *Cell* 109:257–66
124. Gonzy-Treboul G, Karmazyn-Campelli C, Stragier P. 1992. *J. Mol. Biol.* 224:967–79
125. York K, Kenney TJ, Satola S, Moran C Jr, Poth H, et al. 1992. *J. Bacteriol.* 174:2648–58
126. Arigoni F, Pogliano K, Webb CD, Stragier P, Losick R. 1995. *Science* 270:637–40
127. Levin PA, Losick R, Stragier P, Arigoni F. 1997. *Mol. Microbiol.* 25:839–46
128. Lucet I, Feucht A, Yudkin MD, Errington J. 2000. *EMBO J.* 19:1467–75
129. Gholamhoseinian A, Piggot P. 1989. *J. Bacteriol.* 171:5747–49
130. Margolis P, Driks A, Losick R. 1991. *Science* 254:562–65
131. Duncan L, Losick R. 1993. *Proc. Natl. Acad. Sci. USA* 90:2325–29
132. Min KT, Hilditch CM, Diederich B, Errington J, Yudkin MD. 1993. *Cell* 74:735–42
133. Duncan L, Alper S, Losick R. 1996. *J. Mol. Biol.* 260:147–64
134. Duncan L, Alper S, Arigoni F, Losick R, Stragier P. 1995. *Science* 270:641–44
135. Feucht A, Magnin T, Yudkin MD, Errington J. 1996. *Genes Dev.* 10:794–804
136. King N, Dreesen O, Stragier P, Pogliano K, Losick R. 1999. *Genes Dev.* 13:1156–67
137. Feucht A, Daniel RA, Errington J. 1999. *Mol. Microbiol.* 33:1015–26
138. Wu LJ, Feucht A, Errington J. 1998. *Genes Dev.* 12:1371–80
139. Barak I, Youngman P. 1996. *J. Bacteriol.* 178:4984–89
140. Feucht A, Abbotts L, Errington J. 2002. *Mol. Microbiol.* 45:1119–30
141. Pan Q, Garsin DA, Losick R. 2001. *Mol. Cell* 8:873–83
142. Wu LJ, Errington J. 1994. *Science* 264:572–75
143. Sharpe ME, Errington J. 1995. *Proc. Natl. Acad. Sci. USA* 92:8630–34
144. Wu LJ, Lewis PJ, Allmansberger R, Hauser PM, Errington J. 1995. *Genes Dev.* 9:1316–26
145. Frandsen N, Barak I, Karmazyn-Campelli C, Stragier P. 1999. *Genes Dev.* 13:394–99
146. Dworkin J, Losick R. 2001. *Cell* 107:339–46
147. LaBell TL, Trempy JE, Haldenwang WG. 1987. *Proc. Natl. Acad. Sci. USA* 84:1784–88
148. Satola SW, Baldus JM, Moran C Jr. 1992. *J. Bacteriol.* 174:1448–53
149. Driks A, Losick R. 1991. *Proc. Natl. Acad. Sci. USA* 88:9934–38

150. Peters HK 3rd, Haldenwang WG. 1994. *J. Bacteriol.* 176:7763–66
151. Hofmeister AE, Londono-Vallejo A, Harry E, Stragier P, Losick R. 1995. *Cell* 83:219–26
152. Hofmeister AE. 1998. *J. Bacteriol.* 180: 2426–33
153. Ju J, Haldenwang WG. 1999. *J. Bacteriol.* 181:6171–75
154. Fujita M, Losick R. 2002. *Mol. Microbiol.* 43:27–38
155. Kellner EM, Decatur A, Moran CPJ. 1996. *Mol. Microbiol.* 21:913–24
156. Kirchman PA, DeGrazia H, Kellner EM, Moran C Jr. 1993. *Mol. Microbiol.* 8:663–71
157. Zhang B, Hofmeister A, Kroos L. 1998. *J. Bacteriol.* 180:2434–41
158. Cutting S, Oke V, Driks A, Losick R, Lu S, et al. 1990. *Cell* 62:239–50
159. Wakeley PR, Dorazi R, Hoa NT, Bowyer JR, Cutting SM. 2000. *Mol. Microbiol.* 36:1336–48
160. Cutting S, Driks A, Schmidt R, Kunkel B, Losick R. 1991. *Genes Dev.* 5:456–66
161. Ricca E, Cutting S, Losick R. 1992. *J. Bacteriol.* 174:3177–84
162. Resnekov O, Losick R. 1998. *Proc. Natl. Acad. Sci. USA* 95:3162–67
163. Rudner DZ, Losick R. 2002. *Genes Dev.* 16:1007–18
164. Kroos L, Yu YT, Mills D, Ferguson-Miller S. 2002. *J. Bacteriol.* 184:5393–401
165. Rudner DZ, Fawcett P, Losick R. 1999. *Proc. Natl. Acad. Sci. USA* 96:14765–70

Annu. Rev. Biochem. 2003. 72:395–447
doi: 10.1146/annurev.biochem.72.121801.161800
First published online as a Review in Advance on March 6, 2003

SIGNALS FOR SORTING OF TRANSMEMBRANE PROTEINS TO ENDOSOMES AND LYSOSOMES*

Juan S. Bonifacino[1] and Linton M. Traub[2]
[1]Cell Biology and Metabolism Branch, National Institute of Child Health and Human Development, National Institutes of Health, Bethesda, Maryland 20892; email: juan@helix.nih.gov
[2]Department of Cell Biology and Physiology, University of Pittsburgh School of Medicine, Pittsburgh, Pennsylvania 15261; email: traub+@pitt.edu

Key Words endocytosis, plasma membrane, *trans*-Golgi network, receptor downregulation, multivesicular bodies, vacuole, phosphoinositides

■ **Abstract** Sorting of transmembrane proteins to endosomes and lysosomes is mediated by signals present within the cytosolic domains of the proteins. Most signals consist of short, linear sequences of amino acid residues. Some signals are referred to as tyrosine-based sorting signals and conform to the NPXY or YXXØ consensus motifs. Other signals known as dileucine-based signals fit [DE]XXXL[LI] or DXXLL consensus motifs. All of these signals are recognized by components of protein coats peripherally associated with the cytosolic face of membranes. YXXØ and [DE]XXX-L[LI] signals are recognized with characteristic fine specificity by the adaptor protein (AP) complexes AP-1, AP-2, AP-3, and AP-4, whereas DXXLL signals are recognized by another family of adaptors known as GGAs. Several proteins, including clathrin, AP-2, and Dab2, have been proposed to function as recognition proteins for NPXY signals. YXXØ and DXXLL signals bind in an extended conformation to the $\mu 2$ subunit of AP-2 and the VHS domain of the GGAs, respectively. Phosphorylation events regulate signal recognition. In addition to peptide motifs, ubiquitination of cytosolic lysine residues also serves as a signal for sorting at various stages of the endosomal-lysosomal system. Conjugated ubiquitin is recognized by UIM, UBA, or UBC domains present within many components of the internalization and lysosomal targeting machinery. This complex array of signals and recognition proteins ensures the dynamic but accurate distribution of transmembrane proteins to different compartments of the endosomal-lysosomal system.

*The U.S. Government has the right to retain a nonexclusive, royalty-free license in and to any copyright covering this paper.

CONTENTS

INTRODUCTION

Lysosomes are the final destination for many macromolecules taken up by endocytosis from the extracellular space and from the cell surface (Figure 1). Nutrient carrier proteins such as cholesterol-laden low density lipoprotein (LDL) particles, for example, bind to specific receptors on the surface of cells and are rapidly internalized into endosomes. The acidic pH of endosomes induces dissociation of the LDL-receptor complexes, after which the LDL is delivered to lysosomes for degradation while the receptor returns to the cell surface for additional rounds of endocytosis. Binding of hormones, growth factors, and other signaling molecules to their cognate receptors can also trigger downregulation of the receptors by inducing their internalization and delivery to lysosomes. Finally, some endosomal and lysosomal proteins traffic via the plasma membrane en route to their main sites of residence at steady state.

Lysosomes can also be accessed via the biosynthetic pathway (Figure 1). In a process that mirrors receptor-mediated endocytosis, newly synthesized acidic hydrolases modified with mannose 6-phosphate groups bind to mannose 6-phosphate receptors (MPRs) in the *trans*-Golgi network (TGN), after which they are transported to endosomes following an intracellular route. The hydrolase-MPR

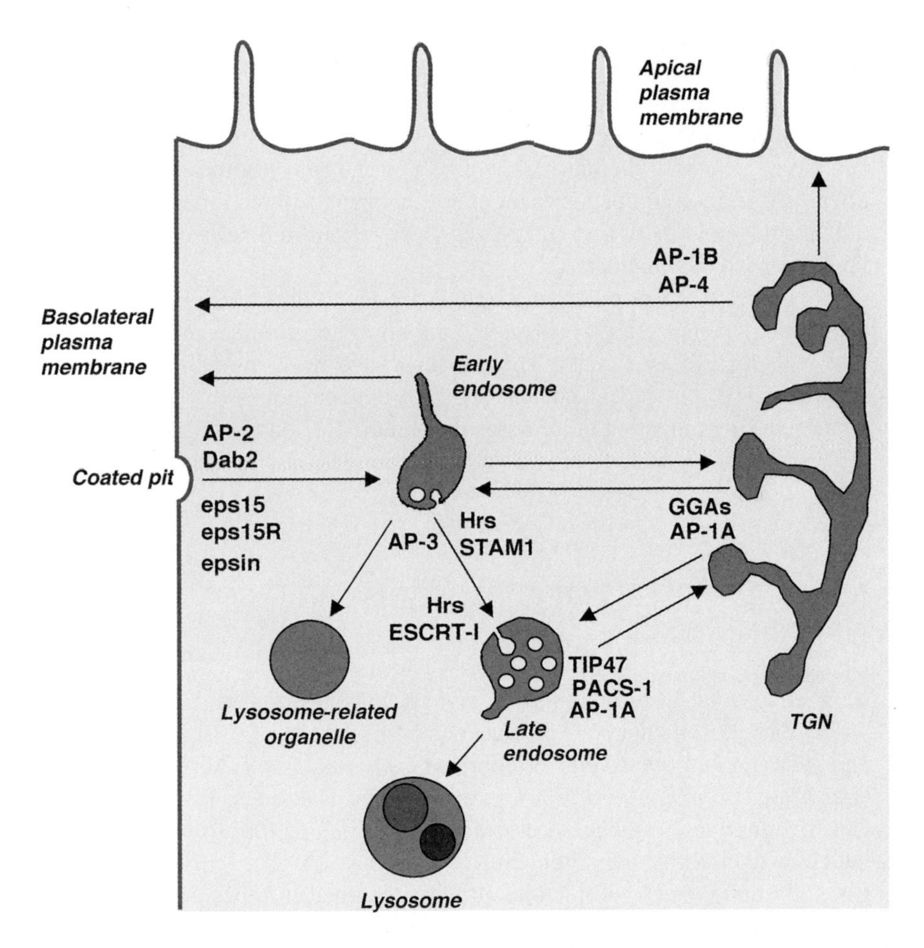

Figure 1 Sorting pathways leading to endosomes, lysosomes, and related organelles. The presumed location and site of action of different coat proteins are indicated by their placement next to an organelle or arrow origin. These should be considered tentative, as in most cases they have not been definitively established. Lysosome-related organelles include melanosomes, platelet-dense bodies, antigen-processing compartments, lytic granules, and other organelles that share some biogenetic pathways with endosomes and lysosomes.

complexes dissociate in the acidic pH of the endosomes, and then the hydrolases are transported to lysosomes with the fluid phase while the MPRs return to the TGN for further rounds of hydrolase sorting. Abnormal or excess proteins targeted for degradation, as well as many lysosomal resident proteins, also travel from the TGN to endosomes and lysosomes intracellularly.

Both the endocytic and biosynthetic routes are critical for the biogenesis and function of endosomes and lysosomes. The proper operation of these transport routes requires that several important sorting decisions be made along the way. At the plasma membrane, proteins can either remain at the cell surface or be rapidly internalized into endosomes. At the TGN, the choice is between going to the plasma membrane or being diverted to endosomes. In endosomes, proteins can either recycle to the plasma membrane or go to lysosomes. These decisions are governed by a complex system of sorting signals in the itinerant proteins and a molecular machinery that recognizes those signals and delivers the proteins to their intended destinations.

This article focuses on sorting signals present in transmembrane proteins that are targeted to endosomes, lysosomes, and related organelles, and on the proteins that recognize these signals. These issues have been the subject of previous reviews (1–3), but recent advances in the understanding of the nature of the signals, the identification of new signals, and the elucidation of their mechanisms of recognition allow us to provide a more complete and updated account of this topic.

GENERAL PROPERTIES OF ENDOSOMAL-LYSOSOMAL SORTING SIGNALS

Most endosomal-lysosomal sorting signals characterized to date are contained within the cytosolic domains of transmembrane proteins. A list of known signals and their recognition proteins or domains is presented in Table 1. In general, the signals consist of short, linear arrays of amino acid residues. These arrays are not exactly conserved sequences but degenerate motifs of four to seven residues of which two or three are often critical for function. The critical residues are generally bulky and hydrophobic, although charged residues are also important determinants of specificity for some signals. Two major classes of endosomal-lysosomal sorting signals are referred to as "tyrosine-based" and "dileucine-based" owing to the identity of their most critical residues. Sorting mediated by these signals is saturable in vivo, indicating that it relies on recognition of the signals by a limited number of "receptor-like" molecules. In cases in which these recognition molecules have been identified and characterized, interactions have been found to occur with low affinity relative to other protein-protein interactions, such as those between polypeptide hormones and their receptors. Sorting signals can be considered a subset of a larger group of peptide-motif recognition-domain interactions (including those listed in Table 2) that are a characteristic feature of the endocytic and lysosomal sorting machineries. The defining property that sets sorting signals apart from other motifs is their occurrence within the cytosolic domains of transmembrane proteins.

Not all signals are short peptide motifs, however. In some cases, the sorting determinants appear to be folded structures in which the critical residues are not necessarily colinear (4). A striking example of this type of conformational determinant is the protein ubiquitin, which can function as an endosomal-

TABLE 1 Endosomal/lysosomal sorting signals

Signal type	Proposed recognition protein or domain	Functions
NPXY	Clathrin terminal domain, μ2 subunit of AP-2, PTB domain of Dab2	Internalization
YXXØ	μ subunits of AP complexes	Internalization, lysosomal targeting, basolateral targeting
[DE]XXXL[LI]	μ and/or β subunits of AP complexes	Internalization, lysosomal targeting, basolateral targeting
DXXLL	VHS domain of the GGAs	TGN-to-endosomes sorting
Acidic cluster	PACS-1	Endosomes-to-TGN sorting
FW- or P-rich	TIP47	Endosomes-to-TGN sorting
NPFX(1,2)D	SHD1 domain of *Saccharomyces cerevisiae* Sla1p	Internalization
Ubiquitin	UIM, UBA, and UBC domains	Internalization, lysosomal/vacuolar targeting

Motifs in this and other tables are denoted using the PROSITE syntax (*http://www.expasy.ch/prosite/*). Amino acid residues are designated according to the single letter code as follows: A, alanine; C, cysteine; D, aspartic acid; E, glutamic acid; F, phenylalanine; G, glycine; H, histidine; I, isoleucine; K, lysine; L, leucine; M, methionine; N, asparagine; P, proline; Q, glutamine; R, arginine; S, serine; T, threonine; V, valine; W, tryptophan, and Y, tyrosine. X stands for any amino acid and Ø stands for an amino acid residue with a bulky hydrophobic side chain. Abbreviations: PTB, phosphotyrosine-binding; Dab2, disabled-2; AP, adaptor protein; VHS domain present in Vps27p, Hrs, Stam; GGAs, Golgi-localized, γ-ear-containing, ARF-binding proteins; PACS-1, phosphofurin acidic cluster sorting protein 1; TIP47, tail-interacting protein of 47 kDa; SHD1, Sla1p homology domain 1; UBA, ubiquitin associated; UBC, ubiquitin conjugating; UIM, ubiquitin interaction motif.

lysosomal sorting signal upon covalent conjugation onto the cytosolic domain of some transmembrane proteins (5).

SIGNAL-MEDIATED SORTING OCCURS WITHIN COATED AREAS OF MEMBRANES

Long before the identification of sorting signals and their recognition proteins, it had become widely accepted that sorting in the endosomal-lysosomal system occurs through coated areas of membranes. Receptor-mediated endocytosis of lipoproteins, for instance, was found to involve concentration of the lipoprotein-receptor complexes within plasma membrane "coated pits" that become deeply invaginated and eventually give rise to "coated vesicles" (6, 7). A similar process was shown to occur at the TGN, where acidic hydrolases bound to the MPRs gather within coated membrane domains before budding in a different type of coated vesicle (8, 9). Recent evidence has revealed the existence of endosomal coated domains where protein sorting also takes place (10–12). In light of these observations, it was logical to hypothesize that sorting involved interactions

TABLE 2 Endocytic machinery motifs and their recognition domains

Motif	Proposed recognition protein or domain
DP[FW]	Ear domain of the α subunit of AP-2
FXDXF	Ear domain of the α subunit of AP-2
NPF	EH domains
LØXØ[DE]	Clathrin terminal domain
LLDLL	Clathrin terminal domain
PWDLW	Clathrin terminal domain

Motifs are denoted as indicated in the legend to Table 1. EH, eps15 homology.

between signals present within the cytosolic domains of the receptors and components of the protein coats (13). The formulation of this hypothesis provided the impetus for the isolation and biochemical characterization of coat proteins (Figure 2). We now know of the existence of several coats that function in the endosomal-lysosomal system. Clathrin coats are composed of the structural protein clathrin (14), the heterotetrameric adaptor protein (AP) complexes, AP-1, AP-2, or AP-3 (15), and various accessory factors (16). Plasma membrane clathrin coats contain AP-2 and accessory factors such as AP180, epsin 1, eps15, eps15R, intersectin 1, disabled-2 (Dab2), and others. TGN and endosomal clathrin coats contain AP-1 and/or the monomeric adaptors, GGA1, GGA2, and GGA3. AP-3 is also found on clathrin coats associated with endosomes, although this complex may be able to function in the absence of clathrin. Another type of clathrin coat associated with endosomes contains the protein Hrs but no AP complexes. Finally, a fourth AP complex, AP-4, appears to be part of a nonclathrin coat associated with the TGN. Because of their localization to sites of signal-mediated protein sorting (Figure 1), these coat proteins are considered prime candidates to function in signal recognition.

Figure 2 Clathrin and adaptors. The basic unit of clathrin is a "triskelion" composed of three heavy chains (CHC) and three light chains (CLC). At the amino terminus of each CHC there is globular domain known as the terminal domain (TD), which serves as a binding site for many adaptor proteins, as indicated in the figure. The structure of the adaptor-protein (AP) complexes AP-1/4 has been modeled after that of AP-2 (69) (Figure 4). This structure consists of a brick-like core comprising the trunk domains of the two large subunits plus the μ and σ subunits, with two hinge-like sequences that connect the core to two ear domains. AP-1/3 interact with clathrin whereas AP-4 does not. Dab2, the Golgi-localized, γ-ear-containing, ADP-ribosylation factor-binding proteins (GGAs), epsin, eps15, Hrs, and STAM1 are modular clathrin-associated proteins that may also function as adaptors. Their domain structure is indicated, but proteins are not drawn to scale. All of the above proteins are part of coats associated with cytosolic faces of membranes. The sorting signals proposed to interact with each of these proteins are indicated in gray letters.

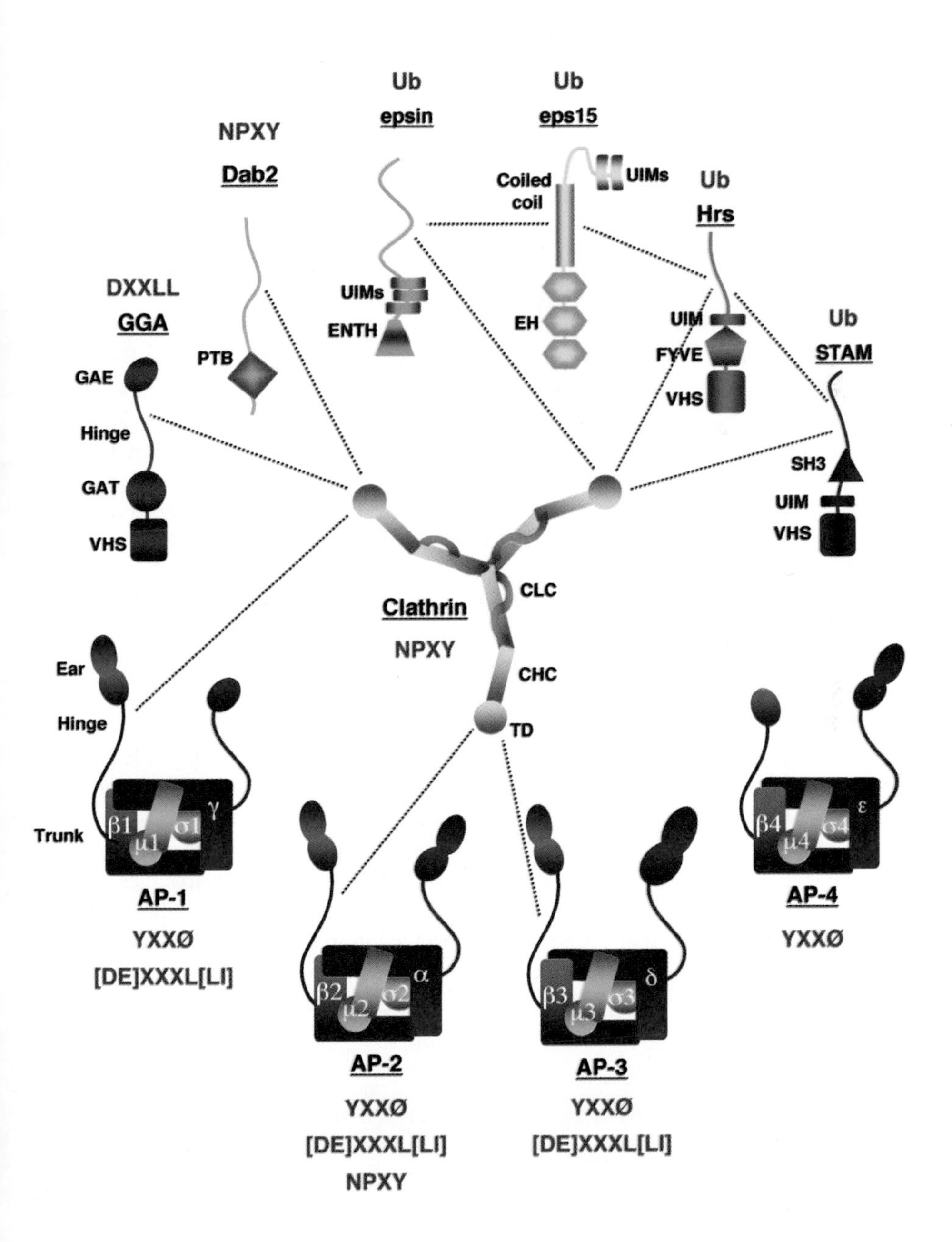

Ub
epsin
Ub
eps15
NPXY
Dab2
Coiled coil
UIMs
Ub
Hrs
DXXLL
GGA
UIMs
ENTH
EH
UIM
FYVE
VHS
Ub
STAM
GAE
PTB
Hinge
GAT
VHS
SH3
UIM
VHS
CLC
Clathrin
NPXY
CHC
TD
Ear
Hinge
Trunk
AP-1
YXXØ
[DE]XXXL[LI]
AP-2
YXXØ
[DE]XXXL[LI]
NPXY
AP-3
YXXØ
[DE]XXXL[LI]
AP-4
YXXØ

TYROSINE-BASED SORTING SIGNALS

Discovery of Tyrosine-Based Sorting Signals

The cloning of several genes encoding endocytic receptors in the early 1980s failed to reveal the presence of conserved sequences that could correspond to the hypothetical endocytic signals. This perplexing finding began to be explained with the discovery by the group of Brown & Goldstein that substitution of a cysteine codon for a tyrosine codon in the cytosolic domain of the LDL receptor, detected in a patient with familial hypercholesterolemia, abrogated the rapid internalization of the receptor (17). Later studies demonstrated that the critical tyrosine residue was part of the sequence motif NPXY (18), found not only in the LDL receptor but also in other cell surface proteins such as the LDL receptor-related protein 1 (LRP1), megalin, the β subunits of integrins, and the β-amyloid precursor protein (Table 3). Many other receptors known at the time, such as the transferrin receptor, the asialoglycoprotein receptor, and the cation-dependent and cation-independent mannose 6-phosphate receptors (CD- and CI-MPRs, respectively), however, lacked NPXY motifs but were rapidly internalized, suggesting the existence of other types of endocytic signals. An observation by Lazarovits & Roth suggested that tyrosine residues could nonetheless be key elements of these other signals (19). These investigators found that artificial replacement of a tyrosine residue for a cysteine residue in the cytosolic domain of influenza hemagglutinin (HA) enabled the protein to undergo rapid internalization via clathrin-coated pits. Several groups subsequently showed that substitution of tyrosine residues in the cytosolic domains of various endocytic receptors devoid of NPXY motifs impaired internalization (20–24). Systematic mutational analyses performed by Kornfeld and colleagues soon led to the definition of another tyrosine-based motif, YXXØ, as the major determinant of endocytosis of the MPRs (25, 26) as well as many other transmembrane proteins (Table 4). This motif is now known to mediate not only rapid internalization from the plasma membrane, but also targeting of certain proteins to lysosomes (27, 28) or to the basolateral plasma membrane domain of polarized epithelial cells (29, 30), sorting events that are thought to occur within clathrin-coated areas of the TGN or endosomes. The concept thus emerged of two types of tyrosine-based sorting signals represented by the NPXY and YXXØ consensus motifs.

NPXY-Type Signals

CHARACTERISTICS OF NPXY SIGNALS To date, NPXY signals have been shown to mediate only rapid internalization of a subset of type I integral membrane proteins and not other intracellular sorting events. These signals tend to occur in families such as members of the LDL receptor, integrin β, and β-amyloid precursor protein families (Table 3). They also seem to be conserved among different metazoans, including *Caenorhabditis elegans, Drosophila*, and mammals (Table 3), suggesting evolutionary conservation of the sorting mechanism

TABLE 3 NPXY-type signals

Protein	Species	Sequence
LDL receptor	Human	Tm-10-INFD**NP**V**Y**QKTT-29
LRP1 (1)	Human	Tm-21-VEIG**NP**T**Y**KMYE-64
LRP1 (2)	Human	Tm-55-TNFT**NP**V**Y**ATLY-33
LRP1	*Drosophila*	Tm-43-GNFA**NP**V**Y**ESMY-38
LRP1 (1)	*C. elegans*	Tm-54-TTFT**NP**V**Y**ELED-91
LRP1 (2)	*C. elegans*	Tm-140-LRVD**NP**L**Y**DPDS-4
Megalin (1)	Human	Tm-70-IIFE**NP**M**Y**SARD-125
Megalin (2)	Human	Tm-144-TNFE**NP**I**Y**AQME-53
Integrin β-1 (1)	Human	Tm-18-DTGE**NP**I**Y**KSAV-11
Integrin β-1 (2)	Human	Tm-30-TTVV**NP**K**Y**EGK
Integrin β (1)	*Drosophila*	Tm-26-WDTE**NP**I**Y**KQAT-11
Integrin β (2)	*Drosophila*	Tm-35-STFK**NP**M**Y**AGK
APLP1	Human	Tm-33-HGYE**NP**T**Y**RFLE-3
APP	Human	Tm-32-NGYE**NP**T**Y**KFFE-4
APP-like	*Drosophila*	Tm-38-NGYE**NP**T**Y**KYFE-3
Insulin receptor	Human	Tm-36-YASS**NP**E**Y**LSAS-379
EGR receptor (1)	Human	Tm-434-GSVQ**NP**V**Y**HNQP-96
EGR receptor (2)	Human	Tm-462-TAVG**NP**E**Y**LNTV-68
EGR receptor (3)	Human	Tm-496-ISLD**NP**D**Y**QQDF-34

Numbers in parentheses indicate motifs that are present in more than one copy within the same protein. The signals in this and other tables should be considered examples. Not all of these sequences have been shown to be active in sorting; some are included because of their conservation in members of the same protein family. Key residues are indicated in bold type. Numbers of amino acids before (i.e., amino-terminal) and after (i.e., carboxy-terminal) the signals are indicated. Abbreviations: Tm, transmembrane; LDL, low density lipoprotein; LRP1, LDL receptor related protein 1; APP, β-amyloid precursor protein; APLP1, APP-like protein 1.

in which they participate. Substitution of alanine for the N, P, or Y residues in the LDL receptor signal largely abolishes rapid endocytosis, whereas substitution of phenylalanine for tyrosine does not (18). Although NPXY is the minimal motif shared by all proteins with this type of signal (Table 3), alone it may not be sufficient for internalization. In the LDL receptor, a phenylalanine residue at two positions amino-terminal to the critical asparagine is also required for optimal internalization, indicating that the complete signal for this receptor is the sequence FDNPVY (18). Moreover, transplantation of only NPVY into the transferrin receptor fails to direct rapid endocytosis, but insertion of FDNPVY does (31). Similarly, internalization of the β-amyloid precursor protein is mediated by the longer GYENPTY sequence, in which the first of the two tyrosines is the most critical for function (32). Other proteins also have either a

TABLE 4 YXXØ-type signals

Protein	Species	Sequence
LAMP-1	Human	Tm-RKRSHAG**Y**QT**I**
LAMP-2a	Human	Tm-KHHHAG**Y**EQ**F**
LAMP-2a	Chicken	Tm-KKHHNTG**Y**EQ**F**
LAMP-2b	Chicken	Tm-RRKSRTG**Y**QS**V**
LAMP-2c	Chicken	Tm-RRKSYAG**Y**QT**L**
LAMP	*Drosophila*	Tm-RRRSTSRG**Y**MS**F**
LAMP	Earthworm	Tm-RKRSRRG**Y**ES**V**
CD63	Human	Tm-KSIRSG**Y**EV**M**
GMP-17	Human	Tm-HCGGPRPG**Y**ET**L**
GMP-17	Mouse	Tm-HCRTRRAE**Y**ET**L**
CD68	Human	Tm-RRRPSA**Y**QA**L**
CD1b	Human	Tm-RRRS**Y**QN**I**P
CD1c	Human	Tm-KKHCS**Y**QD**I**L
CD1d	Mouse	Tm-RRRSA**Y**QD**I**R
CD1	Rat	Tm-RKRRRS**Y**QD**I**M
Endolyn	Rat	Tm-KFCKSKERN**Y**HT**L**
Endolyn	*Drosophila*	Tm-KFYKARNERN**Y**HT**L**
TSC403	Human	Tm-KIRLRCQSSG**Y**QR**I**
TSC403	Mouse	Tm-KIRQRHQSSA**Y**QR**I**
Cystinosin	Human	Tm-HFCLYRKRPG**Y**DQ**L**N
Putative solute carrier	Human	Tm-12-SLSRGSG**Y**KE**I**
TRP-2	Human	Tm-RRLRKG**Y**TP**L**MET-11
HLA-DM β	Human	Tm-RRAGHSS**Y**TP**L**PGS-9
LmpA	Dictyostelium	Tm-KKLRQQKQQG**Y**QA**I**INNE
Putative lysosomal protein	Dictyostelium	Tm-RSKSNQNQS**Y**NL**I**QL
LIMP-II	Dictyostelium	Tm-RKTFYNNNQYNG**Y**NI**I**N
Transferrin receptor	Human	16-PLS**Y**TR**F**SLA-35-Tm
Asialoglycoprotein receptor H1	Human	MTKE**Y**QD**L**QHL-29-Tm
CI-MPR	Human	Tm-22-SYK**Y**SK**V**NKE-132
CD-MPR	Human	Tm-40-PAA**Y**RG**V**GDD-16
CTLA-4	Human	Tm-10-TGV**Y**VK**M**PPT-16
Furin	Human	Tm-17-LIS**Y**KG**L**PPE-29
TGN38	Rat	Tm-23-ASD**Y**QR**L**NLKL
gp41	HIV-1	Tm-13-RQG**Y**SP**L**SFQT-144
Acid phosphatase	Human	Tm-RMQAQPPG**Y**RH**V**ADGEDHA

See legends to Tables 1–3 for explanation of signal format.

phenylalanine or a tyrosine residue at this same position relative to the NPXY motif, although their functional importance has not been assessed (Table 3). These observations suggest that the most potent among these signals conform to the hexapeptide motif [FY]XNPXY.

NPXY signals are most often found within medium-length cytosolic domains ranging from ~40 to ~200 amino acid residues, although some are present within the larger cytosolic domains of signaling receptors such as the insulin receptor and the epidermal growth factor (EGF) receptor (Table 3). The distance of these signals from the transmembrane domains is variable, but none is closer than 10 residues (Table 3). They are also never found exactly at the carboxy-terminus of the proteins. Some proteins, most notably LRP-1 and megalin, contain two copies of this motif, suggesting the possibility of bivalent binding to their recognition proteins.

RECOGNITION PROTEINS FOR NPXY SIGNALS Ever since the discovery of NPXY signals, recognition proteins for these signals have been sought. As is often the case with this type of weak interaction, affinity purification approaches met with little success. The LDL receptor and other proteins having NPXY-type signals are internalized via clathrin-coated pits (6, 7), so clathrin-coat components are likely candidates for recognition proteins. Indeed, nuclear magnetic resonance (NMR) spectroscopy studies have shown that peptides containing the FDNPVY sequence from the LDL receptor bind directly to the globular, amino-terminal domain (TD) of the clathrin heavy chain (33) (Figure 2). These interactions exhibit the expected requirement for the critical tyrosine residue and occur with an equilibrium dissociation constant (K_D) of ~0.1 mM (33). The FDNPVY peptide also binds to clathrin cages, in which context the NPVY residues adopt a type 1 β-turn structure (33). A problem with the idea that clathrin functions as the primary recognition protein for NPXY signals is that clathrin is associated not only with the plasma membrane but also with the TGN and endosomes. Since NPXY signals only mediate internalization from the plasma membrane, it is unclear what would keep NPXY signals from engaging clathrin at other intracellular locations. Another caveat pointed out by Boll et al. (34) is that, because of its position within the cytosolic domain, the FDNPVY signal of the LDL receptor could extend at most 30–40Å from the transmembrane domain, whereas the terminal domain of clathrin lies at about 100Å from the membrane (35).

AP-2 and other putative adaptors (Figure 2) are situated between the clathrin lattice and the membrane and may therefore be in a better position to interact with endocytic signals. Early experiments by Pearse detected a weak interaction of AP-2 with a fusion protein containing the cytosolic domain of the LDL receptor, but the dependence of this interaction on the FDNPVY signal was not tested (36). More recently, Boll et al. (34) have reported the use of surface plasmon resonance (SPR) spectroscopy and photoaffinity labeling to demonstrate binding of FDNPVY peptides to purified AP-2. This binding is dependent on the phenylalanine and tyrosine residues of the signal. The very low affinity of these interactions and the insolubility of the peptides at high concentrations have precluded an estimation of the K_D of the binding reaction. Interestingly, competition and mutational analyses suggest that NPXY and YXXØ signals bind to distinct sites on the μ2 subunit of AP-2 (34) (see below). This observation is in line with the finding that inducible overexpression of the FXNPXY-containing LDL receptor or the YXXØ-containing transferrin receptor does not slow the endocytic uptake of the other, even though each is able to compete with itself for incorporation into endocytic structures (37). These two receptors also saturate the endocytic machinery at different receptor densities, suggesting distinct recognition components (37).

Recent work has led to the alternative hypothesis that NPXY signals are not directly recognized by AP-2 but rather by proteins containing a domain known as phosphotyrosine-binding (PTB) or phosphotyrosine-interacting domain (PID). Although one biologic use of the PTB domain is to drive recruitment of signaling adaptors such as IRS-1 or Shc onto NPXpY motifs (pY stands for phosphotyrosine) on activated receptor tyrosine kinases, this domain demonstrates a marked plasticity in ligand recognition. Some PTB domain-containing proteins actually display a substantially higher selectivity for NPXY than for NPXpY (38, 39), and others bind sequences unrelated to this consensus motif. There is mounting evidence that PTB domain-containing proteins do function in LDL receptor internalization. Dab2 (Figure 2) is one good candidate, as this protein colocalizes extensively with clathrin coat components at the cell surface at steady state (39, 40). The PTB domain of Dab2 is ~65% identical to that of Dab1, a protein that participates in neural migration and cortical lamination in the brain (41). Dab1 operates in a signaling pathway directly downstream of the very low density lipoprotein (VLDL) receptor and apoE receptor 2 by engaging FXNPXY sequences in these receptors via the PTB domain (41). Similarly, the Dab2 PTB domain binds directly FXNPXY sequences from various LDL receptor family members in the nonphosphorylated form (39, 40, 42), and overexpression of either the Dab1 (43) or Dab2 (40) PTB domain causes the LDL receptor to back up at the cell surface, as its endocytosis is selectively impeded. Conditional disruption of *Dab2* in mice results in viable animals that are apparently healthy but display a proteinuria (44) that is similar to, but less severe than, that seen in megalin$^{-/-}$ animals (45). As the LDL receptor family member megalin plays a pivotal role in protein reabsorption in the renal proximal tubule (46), these in vivo data tie Dab2 to LDL receptor family endocytosis.

Targeting of Dab2 to the clathrin bud site is due to several endocytic interaction motifs, including DPF/FXDXF, LVDLN, and NPF sequences that engage AP-2, clathrin, and EH domain-containing proteins, respectively (Table 2) (39, 40). The PTB domain of Dab2 also binds phosphatidylinositol 4,5-bisphosphate (PtdIns(4,5)P_2) via a surface separate from the FXNPXY binding site (40). These functional attributes are consistent with Dab2 being an intermediate sorting adaptor, but the mild phenotype of Dab2 nullizygous mice argues against it being singularly responsible for recognition of all NPXY-type internalization signals. In fact, there are other PTB domain proteins that could be functionally redundant with Dab2. ARH, for example, is a PTB domain protein that is defective in patients with an autosomal recessive form of hypercholesterolemia (47, 48). In cultured lymphoblasts from these patients, the LDL receptor stagnates at the plasma membrane, but the transferrin receptor continues to be internalized normally (49). This suggests that ARH is specifically involved in FXNPXY-driven endocytosis. The ARH PTB domain is most closely related to the PTB domains of Dab1/2, Numb, and GULP/Ced-6, all PTB domains that bind nonphosphorylated FXNPXY sequences preferentially. In *Drosophila*, Numb is involved in cell fate determination by segregating asymmetrically during cell division. Numb antagonizes Notch receptor signaling in the daughter cell into which it segregates by facilitating receptor endocytosis within clathrin-coated vesicles (50). The mammalian Numb orthologue, like Dab2, is located within clathrin-coated structures at steady state (51) and contains endocytic interaction motifs that facilitate interactions with AP-2 and eps15 (51).

YXXØ-Type Signals

CHARACTERISTICS OF YXXØ SIGNALS Although discovered after NPXY signals, YXXØ signals are now known to be much more widely involved in protein sorting. They are found in endocytic receptors such as the transferrin receptor and the asialoglycoprotein receptor, intracellular sorting receptors such as the CI- and CD-MPRs, lysosomal membrane proteins such as LAMP-1 and LAMP-2, and TGN proteins such as TGN38 and furin, as well as in proteins localized to specialized endosomal-lysosomal organelles such as antigen-processing compartments (e.g., HLA-DM) and cytotoxic granules (e.g., GMP-17) (Table 4). Each of these protein types has a distinct steady-state distribution, but they all share the property of trafficking via the plasma membrane to some extent. The YXXØ motifs are essential for the rapid internalization of these proteins from the plasma membrane. However, their function is not limited to endocytosis, since the same motifs have been implicated in the targeting of transmembrane proteins to lysosomes and lysosome-related organelles (27, 28, 52, 53), as well as the sorting of a subset of proteins to the basolateral plasma membrane of polarized epithelial cells (29, 30). The multiple roles of YXXØ signals suggest that they must be recognized not only at the plasma membrane but also at other sorting stations such as the TGN and endosomes. YXXØ signals are found in organisms as distantly related as protists and mammals (Table 4), suggesting that they participate in evolutionarily conserved mechanisms of sorting. As a matter of probability, sequences conforming to the YXXØ motif are common in the large cytosolic domains of signaling receptors and in cytosolic proteins. However, most of these sequences are not likely to be active as sorting signals because they are folded within the structure of the proteins and therefore are not accessible for interactions with components of the sorting machinery.

The YXXØ tetrapeptide is the minimal motif capable of conferring sorting information onto transmembrane proteins. However, the X residues and other residues flanking the motif also contribute to the strength and fine specificity of the signals. The Y residue is essential for function and in most cases cannot even be substituted by other aromatic amino acid residues, suggesting that the phenolic hydroxyl group of the tyrosine is a critical recognition element. In this regard, YXXØ signals differ from NPXY signals, which tolerate phenylalanine in place of tyrosine [reviewed in (1, 17)]. The Ø position can accommodate several residues with bulky hydrophobic side chains, although the exact identity of this residue can specify the properties of the signal (53, 54). The X residues are highly variable but tend to be hydrophilic. A salient feature of YXXØ signals involved in lysosomal targeting is that most have a glycine residue preceding the critical tyrosine residue (Table 4). Mutation of this glycine to alanine decreases lysosomal targeting but not endocytosis (28), indicating that this residue is an important determinant of sorting to lysosomes. Lysosomal-sorting YXXØ motifs tend to have acidic residues at the X positions, and these may also contribute to the efficiency of lysosomal targeting (54).

Just as important as the actual amino acid sequence of YXXØ signals is their position within the cytosolic domains. These signals can be found in all types of transmembrane proteins, including Type I (e.g., LAMP-1 and LAMP-2), Type II (e.g., the transferrin and asialoglycoprotein receptors) and multispanning integral membrane proteins (e.g., CD63 and cystinosin). Purely endocytic YXXØ signals are most often situated at 10–40 residues from the transmembrane domains, but not at the carboxy-termini of the proteins (Table 4). This means that the carboxy-terminus of the Ø residue need not be free for function in endocytosis. In contrast, lysosomal-targeting YXXØ signals are conspicuous for their presence at 6–9 residues from the transmembrane domain and at the carboxy-terminus of the proteins (Table 4). In some proteins targeted to late endosomal or lysosomal compartments such as CD1b or cystinosin, the Ø residue is followed by only one more residue (Table 4). The importance of the distance from the transmembrane domain has been emphasized by a study showing that changing the spacing of the GYQTI signal from LAMP-1 impairs targeting to lysosomes (55). The mutant proteins continue to be internalized at the same rates but recycle to the plasma membrane, a behavior typical of endocytic receptors (55). These observations indicate that the placement of YXXØ signals at 6–9 residues from the transmembrane domain allows their recognition as lysosomal targeting signals at the TGN and/or endosomes.

RECOGNITION OF YXXØ SIGNALS BY THE μ SUBUNITS OF AP COMPLEXES As expected from morphological studies, YXXØ signals have been shown to interact with AP-1 and AP-2 (56–58). Recognition of YXXØ signals was originally ascribed to the large subunits of AP-2 complexes. Subsequent studies using the yeast two-hybrid system and in vitro binding assays revealed that YXXØ-signal recognition is instead a property of the μ subunits of the four AP complexes (56, 57, 59, 60). Binding to the μ subunits is now known to occur with apparent affinities in the 0.05–100 μM range (56, 61, 62) and to be strictly dependent on the Y and Ø residues (56, 57, 59). Each μ subunit recognizes a distinct but overlapping set of YXXØ signals, as defined by the identity of residues other than the critical tyrosine (63, 64). The μ2 subunit of AP-2 exhibits the highest avidity and broadest specificity. These properties may allow sufficient leeway for μ2 to interact productively with YXXØ signals even when the placement of the signals or the identity of the X residues are suboptimal. This interpretation is supported by the fact that most YXXØ signals characterized to date function as endocytic signals. In addition, it has been shown that mutations in the X residues that substantially decrease interactions with μ2 have little effect on internalization (59). Moreover, mutations in or around the LAMP-1 signal that impair lysosomal targeting have minimal effects on internalization (28, 55). These observations suggest that the binding properties of μ2 endow the endocytic machinery with the ability to function efficiently with a wide variety of YXXØ signals.

Relative to μ2, μ1 (A and B isoforms), μ3 (A and B isoforms), and μ4 bind more weakly and display a narrower set of preferences for residues surrounding the critical tyrosine (63–65). μ1A preferences are relatively nondescript, whereas μ3A and μ3B exhibit clear preferences for acidic residues before and after the critical tyrosine (63). The most characteristic feature of μ4-binding signals is the presence of aromatic residues at various positions near the critical tyrosine residue (64). These properties may make interactions with μ1, μ3, and μ4 subunits more sensitive to variations in signal placement and amino acid sequence. Thus, these subunits, and by extension the AP complexes of which they are part, may be responsible for intracellular sorting events mediated by a subset of YXXØ signals. Exactly which sorting events are mediated by μ1, μ3, and μ4 subunits, however, cannot be inferred solely from their signal preferences. The μ3A and μ3B preferences hint at a role for AP-3 in protein sorting to lysosomes and lysosome-related organelles, since the signals from several proteins targeted to these organelles (i.e., LAMP-2a, CD63, and GMP-17) have acidic residues at the Y+1 position. This presumption is supported by genetic evidence linking AP-3 to the biogenesis of lysosome-related organelles in various organisms [reviewed in (66)]. A confounding fact is that, for the most part, none of the μ subunits analyzed shows a strong preference for a glycine preceding the critical tyrosine, a characteristic of lysosomal targeting signals. This suggests either that GYXXØ signals are recognized more specifically by another type of protein, or that the glycine is present for reasons other than recognition specificity. In this regard, the absence of a side chain in the glycine residue could allow more flexible or unhindered recognition of the YXXØ residues in close proximity to the membrane.

STRUCTURAL BASES FOR THE RECOGNITION OF YXXØ SIGNALS Yeast two-hybrid, proteolytic digestion, and crystallographic analyses have shown that the μ subunits are organized into an amino-terminal domain comprising one third of the protein and a carboxy-terminal domain comprising the remaining two thirds

(67–69). The amino-terminal domain mediates assembly with the β subunits of the AP complexes, whereas the carboxy-terminal domain harbors the binding site for YXXØ signals. This latter domain exhibits homology to the carboxy-terminal domain of *Drosophila* Stoned-B and the mammalian stonin 1 and stonin 2 proteins (70–72); in these proteins, however, this domain does not bind YXXØ signals (71). The crystal structure of the amino-terminal domain of μ2 in the context of the AP-2 core consists of a five-stranded β-sheet flanked on either side by α-helices (69). This domain makes extensive hydrophobic contacts with the trunk domain of β2 (69). The crystal structure of the carboxy-terminal domain of μ2 complexed with various YXXØ signal peptides has also been solved. This domain exhibits a banana-shaped, 16-strand β-sheet structure organized into two β-sandwich subdomains (*A* and *B*) (68) (Figure 3*A*). The YXXØ signal peptide binds in an extended conformation to β-sheet strands 1 and 16 of subdomain *A* and forms an extra β-strand paired with β-strand 16 of μ2. The critical Y and Ø residues fit into two hydrophobic pockets on the surface of the protein (Figure 3*B*). The Y residue interacts with residues lining the interior of the pocket via both its aromatic ring and phenolic hydroxyl group. This latter group participates in a network of hydrogen bonds with various μ2 residues, including ^{176}Asp located at the bottom of the pocket. The absence of this phenolic hydroxyl group in phenylalanine explains why this residue cannot substitute for tyrosine in YXXØ signals. The pocket for the Ø residue is lined with flexible aliphatic side chains that allow accommodation of structurally diverse bulky hydrophobic residues (68). The X residues can also be involved in backbone interactions with μ2, thus contributing to the fine specificity of binding (68). Residues amino-terminal to the critical tyrosine can provide additional points of attachment to μ2. For example, a leucine residue at position Y-3 from a YXXØ signal in P-selectin binds to a separate hydrophobic pocket on the μ2 surface (73) (Figure 3*C*). On the basis of these observations, interactions of YXXØ signals with μ2 have been likened to either two-pinned (i.e., most signals, Figure 3*B*) or three-pinned (i.e., P-selectin, Figure 3*C*) plugs fitting into complementary sockets (68, 73). The three-pin configuration may allow distribution of interaction forces over a larger surface area, such that internalization activity is less dependent on any one particular residue. Because of their homology and the conservation of residues involved in YXXØ recognition (3), other μ subunits are expected to have a similar structure and mechanism of binding.

The physiological role of μ2 in YXXØ signal recognition and the identity of the residues involved in recognition have been corroborated using a dominant-negative interference approach (74). Mutant μ2 constructs with substitutions of ^{176}Asp and ^{421}Trp are unable to bind YXXØ signals. Upon transfection into cells, these constructs are incorporated into AP-2 and impair the rapid internalization of the transferrin receptor (74).

The recent resolution of the crystal structure of the AP-2 core (69) has uncovered an unexpected complexity in the mechanism of YXXØ-signal recognition. This structure shows that the carboxy-terminal domain of μ2 rests on a

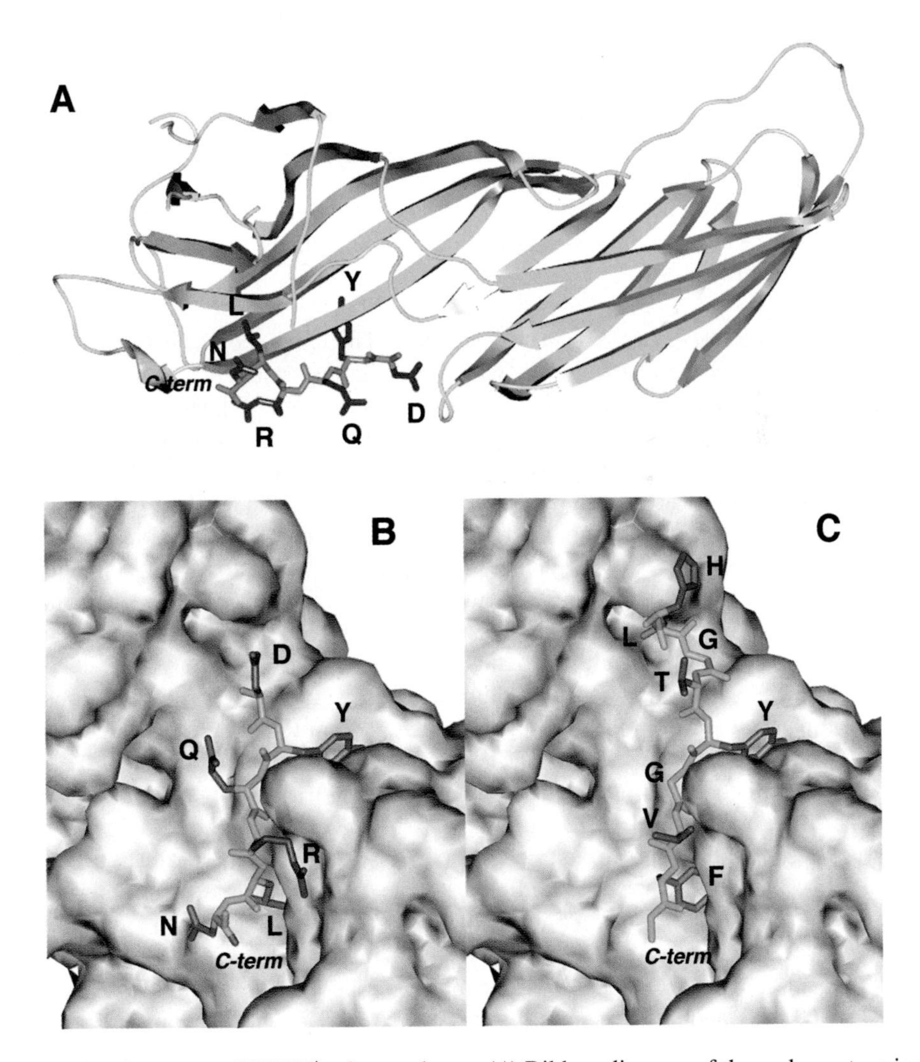

Figure 3 Structure of YXXØ-$\mu2$ complexes. (*A*) Ribbon diagram of the carboxy-terminal domain of $\mu2$ in a complex with the DYQRLN peptide from TGN38. Notice the orientation of the Y and L residues toward $\mu2$. (*B*) Binding of the DYQRLN peptide from TGN38 to the surface of $\mu2$. The peptide binds in an extended conformation with the Y and L residues fitting into two hydrophobic pockets (i.e., the "two-pinned" plug model). (*C*) Binding of the HLGTYGVF peptide from P-selectin to the surface of $\mu2$. This peptide binds also in an extended conformation with the L, Y, and F residues fitting into three hydrophobic pockets (i.e., the "three-pinned" plug model). The peptide backbone is indicated in green, the key side chains contacting $\mu2$ in red, and other side chains in purple. Data are from (68, 73).

furrow formed by the trunk domains of the α and $\beta2$ subunits (Figure 4, *A* and *B*). In this structure, the YXXØ-binding site on $\mu2$ is partially occluded by an interaction with the $\beta2$-trunk domain. The solved structure thus corresponds to an inactive form of AP-2, which must be activated by a conformational change. This change may involve displacement of the carboxy-terminal domain of $\mu2$ from the AP-2 core and its attachment to the membrane via a phosphoinositide-binding site on subdomain *B* (69, 75) (Figure 4*C*). Such a change would place the YXXØ-binding site in a favorable position to interact with sorting signals, especially with those from lysosomal membrane proteins, which are located close to the membrane.

REGULATION OF YXXØ-$\mu2$ INTERACTIONS The large conformational change required to activate AP-2 for YXXØ-signal binding is likely a major target for regulation. Indeed, a threonine residue (^{156}Thr) in the linker sequence that connects the amino-terminal and carboxy-terminal domains of $\mu2$ is phosphorylated (76, 77) specifically by a recently discovered serine/threonine kinase, AAK1 (78, 79). This kinase binds to the ear domain of the α subunit of AP-2 and largely colocalizes with this complex in clathrin-coated pits and vesicles (78). Phosphorylation of ^{156}Thr enhances the affinity of AP-2 for YXXØ signals by about one order of magnitude (79) and is required for normal receptor-mediated endocytosis of transferrin (77, 78). Thus, cycles of phosphorylation/dephosphorylation may allow for multiple rounds of cargo recruitment into forming clathrin-coated pits. How this phosphorylation could trigger release of the $\mu2$ carboxy-terminal domain from the core is not apparent from the structure. In addition to phosphorylation, other factors such as electrostatic attraction between the positively charged $\mu2$ surface and the negatively charged membrane surface could contribute to the release of $\mu2$ required for YXXØ-signal binding.

Localized changes in the levels of phosphoinositides [PtdIns(4,5)P_2 and PtdIns(3,4,5)P_3] by specific lipid kinases (80) and phosphatases (81) might also modulate YXXØ-$\mu2$ interactions. Studies have in fact shown that the interaction of AP-2 with phosphoinositides enhances the recognition of YXXØ signals (62). This modulation likely involves phosphoinositide-binding sites present on both $\mu2$ (69, 75) and the amino-terminal portion of the α-trunk domain (69, 82).

→

Figure 4 Structure of the AP-2 core. (*A*) Surface representation of the AP-2 core with its four subunits indicated in different colors. Two sites for polyphosphatidylinositol (PIP2/PIP3) binding, one on $\mu2$ and the other on α, are indicated by the squares. The partially occluded YXXØ-binding site on $\mu2$ is indicated by the circle. (*B*) Rotated view of the structure shown in *A*. (*C*) Hypothetical structure of AP-2 activated for YXXØ binding. The carboxy-terminal domain of $\mu2$ is shown displaced from its original position in the core. This conformational change would position the polyphosphatidylinositol-binding and YXXØ-binding sites on $\mu2$ close to the membrane, where they can interact with their corresponding ligands. Data are from (69).

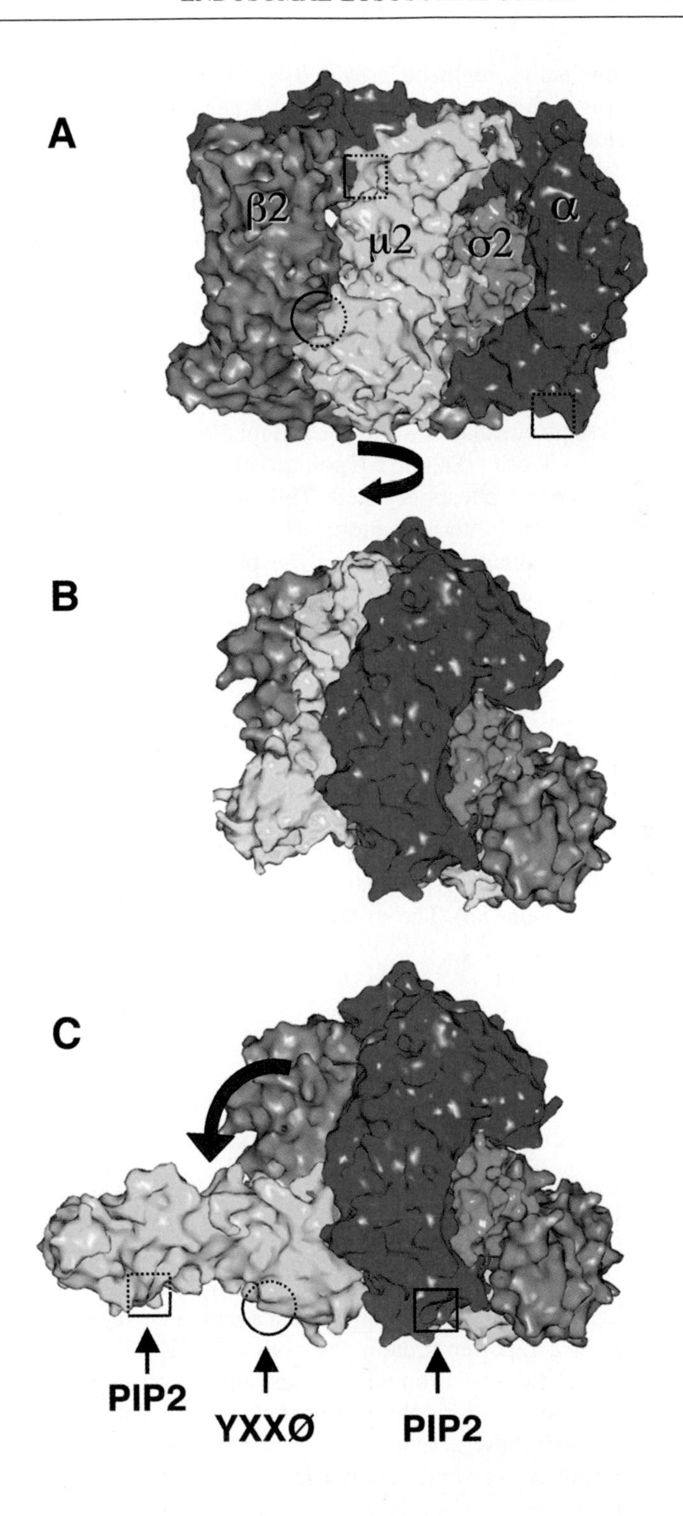
A
β2
μ2
σ2
α
B
C
PIP2
YXXØ
PIP2

Additional regulation might be provided by interaction of AP-2 with putative docking factors such as synaptotagmin (83). A general model incorporating all of these regulatory factors could posit that AP-2 exists in equilibrium between two conformers defined by the position of the μ2 carboxy-terminal domain: an unphosphorylated, inactive form and a phosphorylated, active form capable of binding YXXØ signals, phosphoinositides, and synaptotagmin via the μ2 carboxy-terminal domain. Phosphorylation and/or binding of any of these ligands would shift the equilibrium toward the open state of AP-2, resulting in a highly cooperative activation of the complex.

Phosphorylation of the YXXØ signals or adjacent residues can also modulate their recognition by μ2. The T cell costimulatory receptor CTLA-4, for example, has a YVKM sequence that binds to μ2 and mediates rapid internalization of the protein in resting T cells (84–86). Upon activation of the T cells, the Y residue of the signal becomes phosphorylated. This phosphorylation blocks interaction with μ2 and inhibits internalization, while at the same time allowing the recruitment of signaling molecules to the phosphotyrosine residue (84–86). Phosphorylation of a single tyrosine residue thus serves as a regulatory switch that determines whether the protein is removed from the plasma membrane in resting T cells or remains at the cell surface to transduce signals in activated T cells. A similar regulatory process has been proposed to operate for the neural cell adhesion protein L1 (87). The negative effect of tyrosine phosphorylation on interactions with μ2 can be easily explained by the impossibility of accommodating a large and negatively charged phosphate group into the Y-binding pocket. In contrast to this negative regulation, phosphorylation by casein kinase II of a serine residue immediately preceding the critical tyrosine residue in the YXXØ signal of aquaporin 4 (AQP4) enhances interactions with μ3A and lysosomal targeting of the protein (88). The structural bases for this effect, however, remain to be elucidated.

DILEUCINE-BASED SORTING SIGNALS

Discovery of Dileucine-Based Sorting Signals

At a time when the field of intracellular protein sorting was focused on the study of tyrosine-based sorting signals in endocytosis and lysosomal targeting came the surprising discovery of dileucine-based sorting signals. In the course of a deletion analysis of the cytosolic domain of the CD3-γ chain of the T-cell antigen receptor, Letourneur & Klausner noticed that a particular segment of this domain could confer on a reporter protein the ability to be rapidly internalized and delivered to lysosomes, even though it lacked any tyrosine residues (89). Further deletion analyses revealed that this segment contained a DKQTLL sequence that was responsible for those activities. An alanine scan mutagenesis showed that both leucine residues were required, whereas the other residues were dispensable for function (89). Shortly after the publication of these observations, Johnson & Kornfeld (90, 91) reported that deletion of the LLHV and HLLPM sequences from the carboxy-termini of the CI- and CD-MPRs, respectively, impaired sorting of the MPRs from the TGN to the endosomal system. Both leucines as well as a cluster of acidic amino acid residues preceding the leucines were found to be important for sorting (90–92). Subsequent studies uncovered the presence of dileucine-based sorting signals in many other transmembrane proteins (Tables 5 and 6) and demonstrated that these signals have as broad a range of functions as that of tyrosine-based sorting signals. All of these signals had come to be thought of as variants of the same family. Recent investigations into the nature of their recognition proteins, however, have revealed that the prototypical CD3-γ and MPR signals correspond to two distinct classes represented by the motifs, [DE]XXXL[LI] and DXXLL, respectively.

[DE]XXXL[LI]-Type Signals

CHARACTERISTICS OF [DE]XXXL[LI] SIGNALS [DE]XXXL[LI] signals play critical roles in the sorting of many type I, type II, and multispanning transmembrane proteins (Table 5). The DKQTLL sequence of the CD3-γ chain is now known to be part of a regulatable SDKQTLL signal that participates in serine phosphorylation-dependent downregulation of the T-cell antigen receptor from the cell surface, a process that involves rapid internalization and lysosomal degradation of the receptor (93, 94). The CD4 coreceptor protein undergoes a similar downregulation upon phosphorylation of a serine residue within its SQIKRLL signal (95). Other transmembrane proteins that contain constitutively active forms of this signal are mainly localized to late endosomes and lysosomes (e.g., NPC1, LIMP-II), as well as to specialized endosomal-lysosomal compartments such as endocytic antigen-processing compartments (e.g., Ii), synaptic dense-core granules (VMAT1, VMAT2), stimulus-responsive storage vesicles (e.g., GLUT4, AQP4), and premelanosomes and melanosomes (e.g., tyrosinase, TRP-1, Pmel17, QNR-71). The Nef gene product of human immunodeficiency virus (HIV-1) has also been shown to contain a [DE]XXXL[LI] signal that participates in downregulation of CD4 (96, 97). This type of signal appears to be conserved throughout the animal and protist kingdoms since it is found not only in mammals but also in birds, fish, *C. elegans*, and yeast (Table 5). For *Saccharomyces cerevisiae* alkaline phosphatase, a [DE]XXXL[LI]-type signal mediates sorting of alkaline phosphatase to the vacuole, the yeast counterpart of metazoan lysosomes (98). Like YXXØ signals, the [DE]XXXL[LI] signals in mammalian proteins mediate rapid internalization and targeting to endosomal-lysosomal compartments, suggesting that they can be recognized both at the plasma membrane and intracellular locations. Sequences conforming to this motif have also been implicated in basolateral targeting in polarized epithelial cells (99, 100).

Substitution of either of the critical leucines by alanine abrogates all activities of [DE]XXXL[LI] signals (89, 101). The [DE]XXXL[LI] motif highlights the

TABLE 5 [DE]XXX[LI]-type signals

Protein	Species	Signal
CD3-γ	Human	Tm-8-S**D**KQT**LL**PN-26
LIMP-II	Rat	Tm-11-D**E**RAP**LI**RT
Nmb	Human	Tm-37-Q**E**KDP**LL**KN-7
QNR-71	Quail	Tm-37-T**E**RNP**LL**KS-5
Pmel17	Human	Tm-33-G**E**NSP**LL**SG-3
Tyrosinase	Human	Tm-8-E**E**KQP**LL**ME-12
Tyrosinase	Medaka fish	Tm-16-G**E**RQP**LL**QS-13
Tyrosinase	Chicken	Tm-8-P**E**IQP**LL**TE-13
TRP-1	Goldfish	Tm-7-E**G**RQP**LL**GD-15
TRP-1	Human	Tm-7-E**A**NQP**LL**TD-20
TRP-1	Chicken	Tm-7-E**L**HQP**LL**TD-20
TRP-2	Zebrafish	Tm-5-R**E**FEP**LL**NA-11
VMAT2	Human	Tm-6-E**E**KMA**IL**MD-29
VMAT1	Human	Tm-6-E**E**KLA**IL**SQ-32
VAchT	Mouse	Tm-10-S**E**RDV**LL**DE-42
VAMP4	Human	19-S**E**RRN**LL**ED-88-Tm
Neonatal FcR	Rat	Tm-16-D**D**SGD**LL**PG-19
CD4	Human	Tm-12-S**Q**IKR**LL**SE-17
CD4	Cat	Tm-12-S**H**IKR**LL**SE-17
GLUT4	Mouse	Tm-17-R**R**TPS**LL**EQ-17
GLUT4	Human	Tm-17-H**R**TPS**LL**EQ-17
IRAP	Rat	46-EP**R**GSR**LL**VR-53-Tm
Ii	Human	MD**D**QRD**LI**SNNEQLP**ML**GR-11-Tm
Ii	Mouse	MD**D**QRD**LI**SNHEQLP**IL**GN-10-Tm
Ii	Chicken	MAE**E**QRD**LI**SSDGSSG**VL**PI-12-Tm
Ii-1	Zebrafish	MEPDH**Q**NES**LI**QRVPSAET**IL**GR-12-Tm
Ii-2	Zebrafish	MSSEG**N**ETP**LI**SDQSSVNMGPQP-8-Tm
Lamp	Trypanosome	Tm-RPRRRT**E**EDE**LL**PE**E**AEG**LI**DPQN
Menkes protein	Human	Tm-74-P**D**KHS**LL**VGDFREDDDTAL
NPC1	Human	Tm-13-T**E**RER**LL**NF
AQP4	Human	Tm-32-V**E**TDD**LI**L-29
RME-2	*C. elegans*	Tm-104-F**E**NDS**LL**
Vam3p	*S. cerevisiae*	153-N**E**QSP**LL**HN-121-Tm
ALP	*S. cerevisiae*	7-S**E**QTR**LV**P-18-Tm
Gap1p	*S. cerevisiae*	Tm-23-E**V**DLD**LL**K-24

See legends to Tables 1–3 for explanation of signal format.

residues found in the most active among these signals. An acidic residue at position −4 from the first leucine appears to be important for targeting to late endosomes or lysosomes, though not for internalization (101, 102). The first of the two leucines is generally invariant, probably because substitution by other amino acids, including isoleucine, greatly decreases the potency of the signal (89). The second leucine, in contrast, can be replaced by isoleucine without loss of activity (89). In some cases, another acidic residue or a phosphoacceptor serine further amino-terminal to the [DE]XXXL[LI] motif adds to the strength of the signals. Some [DE]XXXL[LI] signals have arginine residues in place of the acidic residues, as is the case for signals from GLUT4 and IRAP, two proteins localized to insulin-regulated storage compartments (Table 5). This difference may bear physiological significance, since replacement of two glutamates for the two arginines in the RRTPSLL signal of GLUT4 impairs its internalization and sorting to storage compartments (102). Conversely, substitution of two arginines for the aspartate-glutamate pair in the DERAPLI signal of LIMP-II impairs internalization and lysosomal targeting (102). Hence, various [DE]XXXL[LI] signals may be recognized differently at different intracellular locales.

As is the case for YXXØ signals, the position of [DE]XXXL[LI] signals relative to the transmembrane domains and to the carboxy or amino termini also appears to influence the function of the signals. In proteins targeted to late endosomes or lysosomes (e.g., NPC1, LIMP-II), synaptic dense-core granules (e.g., VMAT1 and VMAT2), and premelanosomes or melanosomes (e.g., tyrosinase, TRP-1), the signals are very close to the transmembrane domain (i.e., 6–11 residues away). Late endosomal or lysosomal proteins also tend to display their signals near their carboxy (e.g., NPC1, LIMP-II) or amino termini (e.g., Ii). These properties are similar to those of lysosomal YXXØ signals, suggesting that both are subject to the same positional requirements for lysosomal targeting. Also in this case, proximity to the transmembrane domain and to the carboxy or amino terminus may enable optimal binding to specific recognition proteins. Indeed, a minimum of 6–7 amino acids from the transmembrane domains has been experimentally demonstrated to be optimal for downregulation of CD3-γ chimeras (103). The presence of a few residues carboxy-terminal to the second leucine or isoleucine, though common in naturally occurring signals, does not appear to be essential for function, since a chimeric protein bearing the DKQTLL sequence from CD3-γ at the carboxy terminus can undergo rapid internalization and lysosomal targeting (89).

RECOGNITION OF [DE]XXXL[LI] SIGNALS BY AP COMPLEXES Given the functional similarities of YXXØ- and [DE]XXXL[LI] signals, it is not surprising that [DE]XXXL[LI] signals have also been found to bind AP complexes in various in vitro assays (96, 104–109). This binding is dependent on the LL or LI pairs and, in some cases, on the acidic residues at positions −4 and −5 from the first leucine, thus paralleling the sequence requirements for function of the signals. Each [DE]XXXL[LI] signal exhibits distinct preferences for different AP complexes. For example, the DDQRDLI and NEQLPML signals of Ii bind to AP-1 and AP-2, but not detectably to AP-3 (108). In contrast, the DERAPLI signal of LIMP-II and the EEKQPLL signal of tyrosinase bind to AP-3, but not to AP-1 or AP-2 (106). This binding specificity agrees with the observation that LIMP-II (110) and tyrosinase (111), but not class II MHC-associated Ii (112), are missorted in AP-3-deficient cells. Similarly to YXXØ signals, the fine specificity of interactions of [DE]XXXL[LI] signals may be dictated by the X residues or other contextual factors (113).

In vivo overexpression of transmembrane proteins bearing YXXØ- or [DE]XXXL[LI] signals has been shown to saturate the corresponding sorting machineries, causing missorting of proteins that have the same type of signal (114). However, YXXØ signals do not compete with [DE]XXXL[LI] signals and vice versa (114). This indicates that [DE]XXXL[LI] signals do not bind to the same site as YXXØ signals on μ2. Indeed, in vitro binding and yeast two-hybrid analyses have failed to demonstrate interactions of various [DE]XXXL[LI] signals with the carboxy-terminal domain of μ2 (56, 57, 59). Similar analyses, however, have documented interactions of [DE]XXXL[LI] signals from Ii and HIV-1 Nef with full-length μ1, μ2, and μ3A (115–117), and a phage display screen has identified the 119–123 segment of μ2 as a binding site for the Ii signals (118). Photoaffinity labeling analyses, in contrast, have demonstrated an interaction of various [DE]XXXL[LI] signals with the trunk domains of the β1 and β2 subunits of AP-1 and AP-2, respectively (119). Locating the [DE]XXX-L[LI]-binding site on the AP complexes will ultimately require structural and mutational analyses of the kind that have been performed for YXXØ-μ2 interactions. Curiously, the purified AP-2 core could not be cocrystallized with any of several [DE]XXXL[LI] peptides (69), suggesting that the binding site may inaccessible. In any event, given the apparent functional diversity of [DE]XXX-L[LI] signals, it cannot be ruled out that proteins other then AP complexes are involved in the recognition of certain subsets of signals. The recent discovery of a different kind of recognition protein for DXXLL signals suggests that this is a definite possibility.

DXXLL-Type Signals

CHARACTERISTICS OF DXXLL SIGNALS It has only recently become evident that DXXLL signals constitute a distinct type of dileucine-based sorting signals. These signals are present in several transmembrane receptors and other proteins that cycle between the TGN and endosomes, such as the CI- and CD-MPRs, sortilin, the LDL-receptor-related proteins LRP3 and LRP10, and β-secretase (Table 6). They also seem to be conserved in all metazoans. For the CI- and CD-MPRs, DXXLL signals appear to mediate incorporation into clathrin-coated vesicles that bud from the TGN for transport to the endosomal system (91, 120–122). The requirement for the D and LL residues in these signals is quite

TABLE 6 DXXLL-type dileucine-based signals

Protein	Species	Sequence
CI-MPR	Human	Tm-151-SFHDDSDEDLLHI
CI-MPR	Bovine	Tm-150-TFHDDSDEDLLHV
CI-MPR	Rabbit	Tm-151-SFHDDSDEDLLNI
CI-MPR	Chicken	Tm-148-SFHDDSDEDLLNV
CD-MPR	Human	Tm-54-EESEERDDHLLPM
CD-MPR	Chicken	Tm-54-DESEERDDHLLPM
Sortilin	Human	Tm-41-GYHDDSDEDLLE
SorLA	Human	Tm-41-ITGFSDDVPMVIA
Head-activator BP	Hydra	Tm-41-INRFSDDEPLVVA
LRP3	Human	Tm-237-MLEASDDEALLVC
ST7	Human	Tm-330-KNETSDDEALLLC
LRP10	Mouse	Tm-235-WVVEAEDEPLLA
LRP10	Human	Tm-237-WVAEAEDEPLLT
Beta-secretase	Human	Tm-9-HDDFADDISLLK
Mucolipin-1	Mouse	Tm-43-GRDSPEDHSLLVN
Nonclassical MHC-I	Deer mouse	Tm-6-VRCHPEDDRLLG
FLJ30532	Human	Tm-83-HRVSQDDLDLLTS
GGA1	Human	350-ASVSLLDDELMSL-275
GGA1	Human	415-ASSGLDDLDLLGK-211
GGA2	Human	408-VQNPSADRNLLDL-192
GGA3	Human	384-NALSWLDEELLCL-326
GGA	*Drosophila*	447-TVDSIDDVPLLSD-116

See legends to Tables 1–3 for explanation of signal format. Serine and threonine residues are underlined.

strict since mutation of any of these residues to alanine inactivates the signals and results in increased expression of the transmembrane proteins at the cell surface (91, 92). The D position does not even tolerate isoelectric or isosteric substitutions without drastic loss of activity (92). In contrast, the X residues or other residues amino-terminal to the critical D are less important for function (92). Given these distinct requirements, it is noteworthy that the D residue is generally found in the context of a cluster of acidic residues (Table 6). Because of this, these signals are also referred to as acidic cluster-dileucine signals. Another feature of these signals is the presence of one or more serine residues upstream of the acidic cluster (Table 6). Also of note is the fact that most DXXLL signals are separated by one or two variable residues from the carboxy-termini of the proteins. The distance of the signals from the transmembrane domain, on the other hand, is longer and more variable. In the case of the CD-MPR, this distance may be shortened in the cell by palmitoylation of two cysteine residues in the cytosolic domain (123). The palmitoyl chains likely get inserted into the lipid bilayer, pulling the rest of the cytosolic domain closer to the membrane (123).

RECOGNITION OF DXXLL SIGNALS BY THE VHS DOMAIN OF THE GGAs Unlike [DE]XXXL[LI] signals, DXXLL signals do not detectably bind to AP complexes (120, 124, 125). Instead, DXXLL signals from sortilin, the CI- and CD-MPRs, LRP3, SorLa, and β-secretase bind to the mammalian GGAs (120, 125–130), a recently described family of ADP-ribosylation factor (ARF)-dependent clathrin adaptors localized to the TGN and endosomes (131–134) (Figure 2). Molecular dissection of the GGAs has revealed that the DXXLL-binding activity resides within their amino-terminal VHS domains (120, 125–127). The VHS domains of the three human GGAs (i.e., GGA1, GGA2 and GGA3) bind DXXLL signals from various proteins with affinities ranging from 5 to 100 μM (135, 136). Despite differences in affinity, no sequence preferences for particular subsets of DXXLL signals are apparent among the three human GGAs. The VHS domains of other proteins, such as Hrs (Figure 2), STAM1, TOM1, and TOM1L1 do not bind DXXLL signals (120). Furthermore, the VHS domains of the GGAs do not bind [DE]XXXL[LI] or YXXØ signals (120). These observations emphasize the high degree of specificity of GGA-DXXLL signal interactions. In line with the sequence requirements for function in vivo, the D and LL residues, but not the X residues, of the signals are essential for interactions (120, 125, 127, 136). The physiological significance of these interactions is underscored by the findings that the CD-MPR and the GGAs exit the TGN on the same vesicular intermediates (120) and that a dominant-negative GGA construct causes retention of the CI- and CD-MPRs at the TGN (120). The mammalian GGAs are thus likely involved in the sorting of transmembrane proteins having DXXLL signals from the TGN to endosomes. In the case of the MPRs, this sorting is critical for the efficient delivery of acidic hydrolases to lysosomes. In line with this notion, the *S. cerevisiae* GGAs, Gga1p and Gga2p, have been functionally implicated in sorting hydrolases to the vacuole (132, 133, 137–139), although to date the signals that bind to the yeast GGAs have not been identified.

STRUCTURAL BASES FOR DXXLL-VHS DOMAIN INTERACTIONS Recent studies have solved the crystal structures of the VHS domains of GGA3 and GGA1 in complexes with DXXLL signal peptides from the CI- and CD-MPRs (135, 140) (Figure 5). The VHS domains of both proteins were found to be right-handed superhelices of eight helices (Figure 5*A*). Both the CI-MPR and CD-MPR peptides bind in an extended conformation to a groove formed by helices 6 and 8 (135, 140) (Figure 5*A*). The critical D and LL residues bind to an electropositive and two shallow hydrophobic pockets, respectively, whereas the X residues and flanking residues either point away from the VHS domain or are disordered (Figure 5*B*). The terminal carboxyl group appears to interact weakly with residues on the VHS domain (135, 140), which might explain why most DXXLL

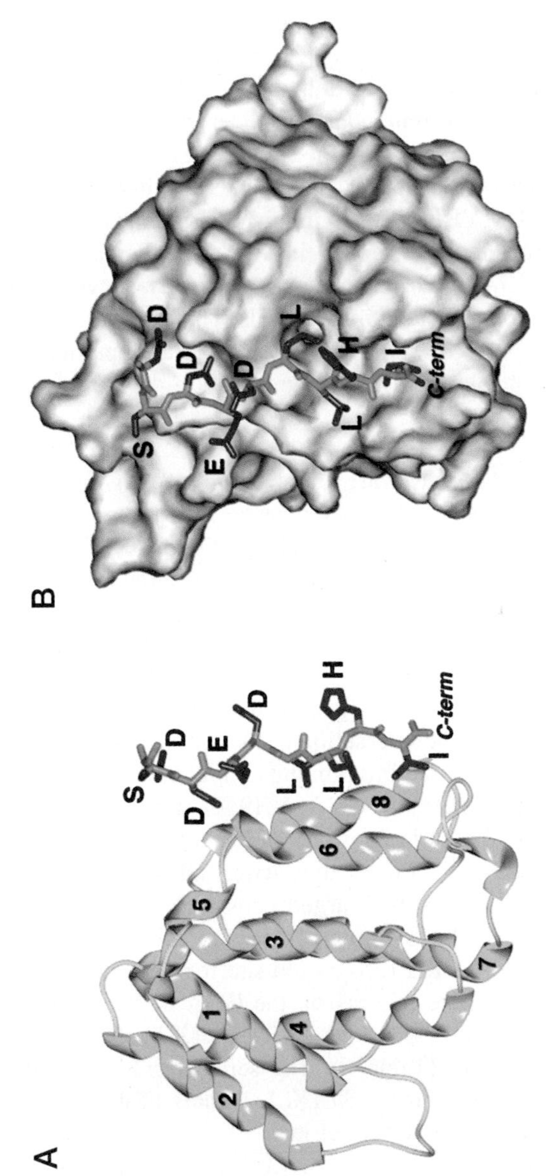

Figure 5 Structure of the complex between the GGA3 VHS domain and the SDEDLLHI signal from the CI-MPR. (*A*) Ribbon diagram showing the peptide binding in an extended conformation to helices 6 and 8 of the VHS domain. Note the position of the key D and LL residues facing the VHS domain. (*B*) Surface representation showing the key D and LL residues binding to pockets on the VHS domain. The peptide backbone is indicated in green, the key side chains contacting the VHS domain in red, and other side chains in purple. Graphics courtesy of Saurav Misra and James Hurley, NIDDK.

signals are positioned one or two residues away from the carboxy-terminus (Table 6). The identity of the DXXLL residues that contact the VHS domain is entirely consistent with the sequence requirements for the function of these signals. The structure of these complexes reveals that the failure of [DE]XXX-L[LI] signals to bind to the VHS domains is due, at least in part, to the placement of their acidic residue at position −4, instead of −3, from the first leucine. In addition, the structures of the VHS domains show that the pocket for the aspartate residue is too small to accommodate a bulkier glutamate residue, in agreement with the functional requirements of the signals (92). Thus, these structural studies provide compelling evidence for the distinct nature of [DE]XXXL[LI] and DXXLL signals. The VHS residues involved in interactions with the D and LL residues are completely conserved in the three human GGAs and mostly conserved in the GGAs in *Drosophila* and *C. elegans*, suggesting that all of these proteins may be able to bind similar signals. This conservation does not extend to the two GGAs in *S. cerevisiae*, indicating that their participation in vacuolar protein sorting probably involves recognition of a different type of signal.

REGULATION OF THE RECOGNITION OF DXXLL SIGNALS BY THE GGAs As mentioned above, serine residues are often found one to three positions amino-terminal to DXXLL signals. These serines are in a sequence context that fits the [ST]XX[DE] consensus motif for phosphorylation by casein kinase II (CKII), in which the X residues are generally acidic. For the CI- and CD-MPR, these serines have been shown to be phosphorylated both in vivo and in vitro by CKII or a CKII-like kinase (141–144). Although the physiological relevance of these phosphorylation events is still unclear (91, 92, 145), mutation of the serine residue upstream of the CI-MPR signal to alanine decreases the sorting efficiency of the receptor (92) and the interaction of its DXXLL signal with VHS domains (127, 136). Moreover, the in vitro binding affinity of a DXXLL signal containing peptide from the CI-MPR to GGA VHS domain increases ~threefold by phosphorylation of the serine residue (136). Crystallographic analyses have revealed that phosphorylation of this serine allows electrostatic interactions between one of the negatively charged phosphate oxygens and two positively charged residues on the VHS domain (136). The acidic cluster may therefore be conserved to provide a CKII recognition site for this type of regulation, rather than to interact with basic regions on the GGA VHS domain. An identically positioned serine residue next to a DXXLL signal is likely to afford the same type of regulation for sortilin (Table 6). The serine residues in other proteins are further amino-terminal from the DXXLL signals (Table 6); if and how they regulate signal recognition is not yet known.

Another form of regulation of DXXLL-GGA interactions has recently been uncovered. GGA1 and GGA3 have DXXLL motifs within their hinge domains (Table 6), which are able to bind to their own VHS domains (146). This binding is thought to be auto-inhibitory and to account for the poor binding activity of the full-length GGA1 and GGA3 relative to their isolated VHS domains (146).

Inhibition depends on CKII-mediated phosphorylation of a serine three residues upstream of the critical aspartate (146). These observations suggest that GGA1 and GGA3 may become activated by a dephosphorylation event that displaces its own DXXLL ligand and frees up the VHS domain for interaction with signals in the cytosolic domains of receptors. These interactions may in turn be enhanced by CKII-mediated phosphorylation of serine residues upstream of the receptors' signals. A recent study has presented evidence for cooperation between the GGAs and AP-1 in the sorting of MPRs (121). The GGAs were found to bind, via their hinge domains, to the ear domain of the γ subunit of AP-1. An AP-1-associated CKII then phosphorylates the GGAs, resulting in autoinhibition and possible transfer of the MPRs from the GGAs to AP-1 (121). This observation explains the presence of MPRs in TGN-derived, clathrin-coated vesicles that contain both GGAs and AP-1 (121).

OTHER SIGNALS

Acidic Clusters

Another family of sorting motifs consists of clusters of acidic residues containing sites for phosphorylation by CKII. This type of motif is often found in transmembrane proteins that are localized to the TGN at steady state, including the prohormone-processing enzymes furin, PC6B, PC7, CPD, and PAM, and the glycoprotein E of herpes virus 3 (Table 7). Several of these proteins cycle between the TGN and endosomes, and it is currently thought that the acidic clusters play a role in retrieval from endosomes to the TGN. This retrieval depends on phosphorylation of the CKII sites (147). Recent studies have identified a monomeric protein named PACS-1 (phosphofurin acidic cluster sorting protein 1), which binds to acidic clusters in a CKII phosphorylation-dependent manner (148). Interestingly, PACS-1 appears to function as a connector that links the phosphorylated acidic clusters to the clathrin-dependent sorting machinery. This connection is likely mediated by interaction of PACS-1 with AP-1 and AP-3 (149). Antisense RNA and dominant-negative interference studies support the notion that PACS-1 is required for transport of furin from endosomes to the TGN (148, 149).

Lysosomal Avoidance Signals

Discharge into the endosomal lumen of lysosomal hydrolases transported by the CD-MPR requires, in part, protonation of ^{133}Glu of the receptor (150). As the pK_a of glutamic acid is 4.1, this necessitates that the unloading occur within an endosomal compartment with a pH <6.0 (151). Indeed, MPRs can be visualized within the intralumenal vesicles of prelysosomal structures (152). Yet the half-life of the CD-MPR is in excess of 40 h (153); thus each receptor performs multiple delivery cycles. For the CD-MPR, endosomal retrieval

TABLE 7 Acidic cluster signals

Protein	Species	Sequence
Furin	Mouse	Tm-31-Q**EE**CPS**D**S**EEDE**G-14
PC6B (1)[a]	Mouse	Tm-39-R**D**R**D**Y**DEDDEDD**I-36
PC6B (2)	Mouse	Tm-69-L**DE**T**EDDE**L**E**Y**DDE**S-4
PC7	Human	Tm-38-K**D**P**DE**V**E**T**E**S-47
CPD	Human	Tm-36-H**E**FQ**DE**T**D**T**EEE**T-6
PAM	Human	Tm-59-Q**E**K**EDD**GS**E**S**EEE**Y-12
VMAT2	Human	Tm-35-G**EDEE**S**E**S**D**
VMAT1	Human	Tm-35-G**ED**S**DEE**P**D**H**EE**
VAMP4	Human	25-L**EDD**S**DEEED**F-81-Tm
Glycoprotein B	HCMV	Tm-125-K**D**S**DEEE**NV
Glycoprotein E	Herpes virus 3	Tm-28-F**ED**S**E**ST**D**T**EEE**F-21
Nef	HIV-1 (AAL65476)	55-L**E**AQ**EEEE**V-139
Kex1p (1)	*S. cerevisiae*	Tm-29-A**DD**L**E**SGLGA**EDD**L**E**Q**DE**QL**E**G-40
Kex1p (2)	*S. cerevisiae*	Tm-79-T**E**I**DE**SF**E**MT**D**F
Kex2p	*S. cerevisiae*	Tm-36-T**E**P**EE**V**ED**F**D**F**D**LS**DED**H-61
Vps10p	*S. cerevisiae*	Tm-112-F**E**I**EEDD**VPTL**EEE**H-37

See legends to Tables 1–3 for explanation of signal format. Serine and threonine residues are underlined.
[a]The number in parentheses is the motif number.

depends upon both cytosolic palmitoylation (123) and an ^{18}FW di-aromatic sorting signal; an FW→AA mutation causes a ~tenfold increase in CD-MPR recovered within dense lysosomes (154). A functionally analogous ^{18}YF di-aromatic signal is present in the cytosolic domain of the mannose receptor (155). The CD-MPR sequence is recognized with micromolar affinity by the carboxy-terminal segment of TIP47 (tail-interacting protein of 47 kDa) (156), a protein that also binds the cytosolic domain of CI-MPR despite the absence of an FW sequence (4, 157). The binding site on the CI-MPR is, in part, localized to the sequence ^{49}PPAPRP of the cytosolic domain, adjacent to the ^{26}YSKV-internalization signal, and appears to associate with TIP47 via a hydrophobic-based interaction (4). Still, the MPR interactions are selective as TIP47 does not bind appreciably in vitro to the cytosolic tails of furin, TGN38, or the LDL receptor, or to other trafficking signals within the CI-MPR (4, 156, 157). The intracellular distribution of TIP47 is compatible with a role in retrieval of MPRs from late endosomes, and decreasing TIP47 abundance with antisense oligonucleotides reduces substantially the half-life of the CI-MPR (156).

In the presence of the nonhydrolyzable GTP analogue, GTPγS, a fourfold increase in TIP47 association with endosome membranes occurs (156). The GTP requirement appears to reflect the involvement of Rab9, as TIP47 cooperatively binds Rab9•GTP and MPRs via separate sites (158). The improved affinity of the ternary complex for MPRs (K_d = 0.3 μM) is proposed to direct TIP47 to the appropriate intracellular site (158). Curiously, TIP47 is a member of the perilipin family of proteins associated with the cytosolic surface of intracellular lipid droplets (159), but the significance of this homology to the sorting of MPRs is not currently known.

NPFX(1,2)D-Type Signals

Another internalization sequence in *S. cerevisiae*, NPFSD, was uncovered within the cytosolic domain of the furin-like endoprotease, Kex2p, when this region was switched for the carboxy-terminal region of Ste2p (160). Kex2p does not usually traffic to the cell surface, but the NPFSD sequence, when appended to a noninternalizing reporter, independently directs efficient uptake (160–162). A related sequence, NPFSTD, near the carboxy-terminus of Ste3p is able to drive **a**-factor-dependent internalization of a truncated, nonubiquitinatable form of this receptor (160). Ste2p contains a lower-potency variant, GPFAD, which seems to function redundantly with ubiquitination (see below) in directing optimal receptor internalization (162). Although this type of internalization motif has an embedded NPF triplet, and the proline and phenylalanine are both essential side chains, the sequence does not engage EH domains (162). Instead, activity of the NPFX(1,2)D sequence is dependent upon Sla1p—specifically, the first *SLA1* homology domain (SHD1), to which it binds directly (162). Part of a known endocytic complex together with Pan1p and End3p (163), Sla1p couples cargo recognition and internalization with actin cytoskeletal dynamics, which plays a major role in endocytosis in yeast (164).

UBIQUITIN AS A SORTING SIGNAL

Discovery of Ubiquitin-Based Sorting

Ubiquitin is a 76-amino acid polypeptide virtually invariant in sequence throughout eukaryotes. The carboxy-terminal residue, ^{76}Gly, can be conjugated by an isopeptide linkage to lysine side chains on target proteins. Conjugation is the result of a sequential series of reactions involving the generation and transfer of a thioester-bound ubiquitin intermediate by the E1, E2, and E3 conjugation machinery proteins (165). Several endogenous lysines (e.g., ^{29}Lys, ^{48}Lys, ^{63}Lys) within the protein-linked ubiquitin molecule can, subsequently, be self-conjugated to additional ubiquitin molecules, generating polyubiquitin chains. Post-translational ubiquitination is at the core of regulated intracellular protein turnover, and ^{48}Lys-linked polyubiquitin chains of tetraubiquitin or greater serve as a signal for protein degradation via the 26S proteasome (166). Appreciation that ubiquitin also functions as an authentic trafficking signal came from the discovery that, in *S. cerevisiae*, internalization of the G protein-coupled α-factor

(Ste2p) and **a**-factor (Ste3p) pheromone receptors is accompanied by ubiquitination of these proteins (167, 168). Genetic background mutations that prevent endocytosis cause ubiquitinated species to accumulate, whereas compromising the cellular ubiquitination machinery retards surface uptake and prolongs receptor half-life (167, 168). It was quickly established that a multitude of yeast plasma membrane receptors and permeases/transporters (e.g., Ste6p, Gal2p, Gap1p, Fur4p, Pdr5p, Zrt1p,) use ubiquitination as an endocytic signal (169–171).

Apparently the major endocytic signal in budding yeast, in higher organisms ubiquitin addition also regulates the internalization of certain transmembrane proteins. One of the first indications came from studies on the growth hormone (GH) receptor in the CHO-ts20 cell line that exhibits a temperature-sensitive E1 ubiquitin-activating enzyme. At the nonpermissive temperature, no uptake or lysosomal degradation of transfected GH receptor occurs (172). Mutation of ^{327}Phe in the cytosolic domain of the receptor blocks endocytosis and also inhibits receptor ubiquitination (173). Specific inhibitors of the proteasome, which cause depletion of free ubiquitin within the cell, abrogate ligand-stimulated destruction of the GH receptor, but transferrin receptor endocytosis is unaffected (174, 175). Also, blocking endocytosis by cyclodextran-mediated cholesterol depletion, or with a dominant-negative dynamin mutant, causes ubiquitinated GH receptor to accumulate on the plasma membrane (176). These experiments firmly establish that internalization of the GH receptor requires the ubiquitin-conjugation machinery and that the receptor is ubiquitinated directly.

There is now good evidence for ubiquitination regulating the internalization of the EGF (177–179), MET (180), and CSF-1 (181, 182) receptors, the epithelial sodium channel (ENaC) (171), the aggregated IgG-bound Fc receptor FcRγIIA (183), and the transmembrane Notch ligand Delta (184–186). The α1 subunit of the glycine receptor is modified with one to three ubiquitin molecules at the plasma membrane as a prelude to uptake (187), and ubiquitination of E-cadherin precedes internalization (188). There is also evidence that endocytosis of the pre-T cell receptor in thymocytes is dependent upon ubiquitination (189). In *C. elegans*, targeted neural overexpression of ubiquitin reduces the surface density of glutamate receptors (190) in a manner that is dependent on clathrin-mediated endocytosis, as mutant *unc-11* (AP180) counteracts the effect of the excess ubiquitin. Direct ubiquitination of the glutamate receptor can be demonstrated biochemically (190). Final corroboration of the role of ubiquitination in endocytic mobilization comes from the demonstration that the Karposi's sarcoma-associated herpes virus protein K3 can route surface MHC class I for lysosomal degradation by possessing inherent E3 ubiquitin ligase activity that ubiquitinates a single MHC class I lysine (191, 192).

Multiple Sorting Steps Involving Ubiquitin

Ubiquitination has now been found to effect protein sorting at other intracellular stations as well. Ubiquitin conjugation is a cyclical process; because ubiquitin is a long-lived protein, the bulk of activated ubiquitin generated by the E1 enzyme comes from ubiquitin recovered by deubiquitinating enzymes (193). Inactivation of one deubiquitinating enzyme in yeast, Doa4p, results in pleiomorphic effects, including defects in surface uptake (194–197). Bypass suppressors of a *doa4* mutation turn out to be six class E *vps* (vacuolar protein sorting) mutants (194). Class E *vps* mutants are defective in multivesicular body formation, and as these suppressors essentially negate the role of Doa4p, the data indicate that the late endosome/prevacuolar compartment is a major sink for ubiquitin (198). However, even though overexpression of ubiquitin overcomes the *doa4* phenotype (195, 199), this does not indicate that Doa4p simply retrieves and recycles ubiquitin from tagged proteins only originating from the cell surface (198). Biosynthetic delivery of newly synthesized vacuolar proteins carboxypeptidase S (CPS) and the polyphosphatase Phm5p into the vacuolar lumen is also ubiquitin dependent (196, 200). These proteins are ubiquitinated directly; a single major lysine acceptor has been mapped in both Phm5p (196) and CPS (200), but other potential acceptors are also utilized in Phm5p (196). Biosynthetic ubiquitination of CPS occurs post Golgi but prior to delivery to late endosomes (200). If ubiquitination of CPS is blocked, the unprocessed form accumulates on the limiting membrane of the vacuole instead of inside (200). The same is true for Phm5p (196). In-frame addition of either a single unextendable ubiquitin (196, 201) or a ubiquitination-directing sequence (200) routes into the vacuole lumen a variety of proteins that normally traffic through the late endosome/prevacuolar compartment without entering intralumenal structures. This does not involve passage through the plasma membrane, showing that ubiquitin-directed sorting of proteins not originating from cell surface certainly occurs (Figure 6).

In mammalian cells, signaling receptors, like the EGF receptor, move into intralumenal membranes of multivesicular endosomes following ligand stimulation and internalization (202). TGFα, an alternate ligand for the EGF receptor, does not induce this translocation, nor substantial receptor degradation, because the ligand dissociates from the receptor within early endosomes (203, 204). TGFα also does not cause protracted ubiquitination of the EGF receptor (204). Proteasome inhibitors prevent both the sequestration of EGF-activated receptors into the interior of multivesicular bodies (204) and degradation of internalized receptors (204, 205). Also, lysosomal turnover of the HIV coreceptor CXCR4 requires a degradation motif (206). Mutation of the three lysine residues within this motif abolishes degradation of the G protein–coupled receptor while monoubiquitination of the receptor correlates with lysosomal delivery (206). Interestingly, proteasome inhibitors selectively disrupt delivery of membrane proteins into the lysosome; soluble protein traffic is normal (175, 204). Together, these studies reveal that transmembrane protein relocation into the interior elements of the multivesicular late endosome is a regulated step in lysosomal delivery that is under the control of ubiquitin.

Sorting decisions are, of course, made before reaching the late endosomal compartment. Minutes after endocytosis, EGF and GH receptors are actively

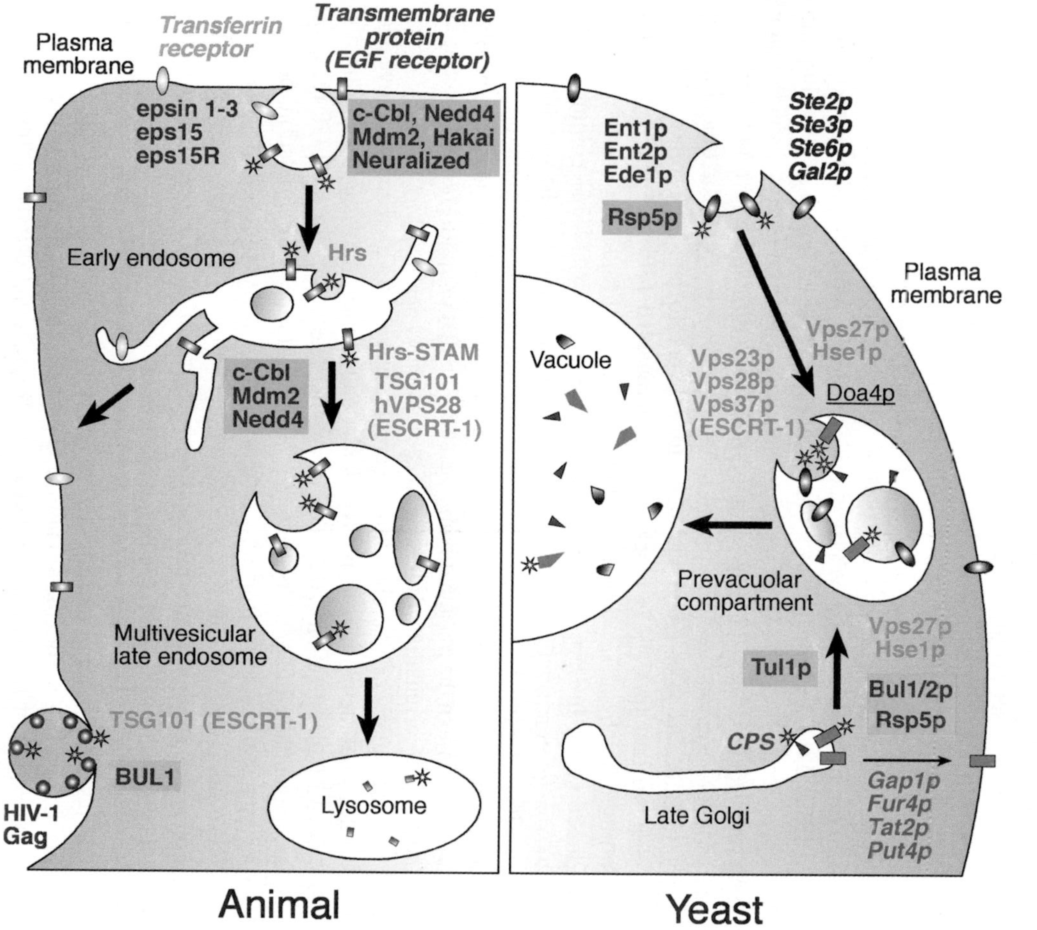
Plasma membrane
Transferrin receptor
Transmembrane protein (EGF receptor)
epsin 1-3
eps15
eps15R
c-Cbl, Nedd4
Mdm2, Hakai
Neuralized
Early endosome
Hrs
Hrs-STAM
TSG101
hVPS28
(ESCRT-1)
c-Cbl
Mdm2
Nedd4
Multivesicular late endosome
TSG101 (ESCRT-1)
BUL1
HIV-1 Gag
Lysosome
Animal
Ent1p
Ent2p
Ede1p
Rsp5p
Ste2p
Ste3p
Ste6p
Gal2p
Plasma membrane
Vps27p
Hse1p
Vacuole
Vps23p
Vps28p
Vps37p
(ESCRT-1)
Doa4p
Prevacuolar compartment
Vps27p
Hse1p
Tul1p
Bul1/2p
Rsp5p
CPS
Late Golgi
Gap1p
Fur4p
Tat2p
Put4p
Yeast

segregated away from components that recycle back to the cell surface (11) (Figure 6). Transferrin receptors with an in-frame ubiquitin at the amino terminus are not recycled to the surface as efficiently as the wild-type receptor (12). Cells treated with the proteasome inhibitor lactacystin (177, 204) or expressing a ubiquitination-defective receptor (179) display elevated EGF receptor recycling at the expense of receptor degradation. As proteasome inhibitors deplete cellular ubiquitin, this again suggests that ubiquitin participates in partitioning molecules away from recycling cargo. Lysosomal destruction of IL-2 (207, 208) and glycine (187) receptors is also prevented by proteasome inhibitors. Because proteins return to the plasma membrane instead of being delivered to the limiting membrane of the lysosome, this work reveals that ubiquitin participates in additional sorting decisions at the early endosome as well (Figure 6). In fact, the discovery that ubiquitin is also involved in sorting within the endosomal compartment makes definitive interpretation of the many studies making use of proteasome inhibitors difficult as all steps are likely to be perturbed under these conditions.

Finally, the trafficking of certain amino acid permeases in yeast is governed by nutritional status. Under appropriate environmental conditions, newly synthesized general amino acid permease, Gap1p, proline permease, Put4p, uracil permease, Fur4p, and tryptophan permease, Tat2p, are diverted to the vacuole for destruction (198, 209–212). This change in permease routing is ubiquitin dependent, a decision that appears to occur around Golgi exit (211), before delivery to the prevacuolar compartment (212) (Figure 6). Thus, a ubiquitin sorting tag can clearly function at multiple branchpoints en route to the lysosome/vacuole.

Characteristics of the Ubiquitin Signal

In yeast, addition of only one or two ubiquitin moieties to Ste3p is sufficient for constitutive endocytosis (213), whereas uptake of the galactose transporter, Gal2p, can occur within cells expressing an unextendable ubiquitin, devoid of any lysines (199). This demonstrates that the extent of ubiquitin derivation required to generate a functional ubiquitination signal is low. In fact, simply

←

Figure 6 Ubiquitin-directed endosomal trafficking. Schematic depiction of the likely site of action of ubiquitin-conjugating machinery (boxed in yellow), endocytic recognition/sorting components (bold type), and deubiquitinating enzymes (underlined) displayed in relation to the relevant subcellular organelles in either in animal (*left*) or yeast (*right*) cells. With the exception of the transferrin receptor, the trafficking itinerary of the various cargo proteins (italic type) shown is regulated by enzymatic ubiquitin addition (✫). In yeast, the activity of Rsp5p and Bul1/2p at the Golgi complex diverts the Gap1p permease from delivery to the cell surface by driving polyubiquitination and trafficking to multivesicular bodies. See text for complete details.

appending a single ubiquitin molecule to the carboxyl end of Ste2p is sufficient to drive endocytosis of this receptor (197). Similarly, fusion of ubiquitin to Pma1p, the resident plasma membrane H^+-ATPase, promotes internalization (214). In mammalian cells, addition of a single ubiquitin molecule essentially in place of the amino terminus of invariant chain (Ii) also changes the steady-state distribution from the plasma membrane to internal endocytic elements (215). In these systems, uptake is apparently not dependent upon extension of the fused ubiquitin as no difference in trafficking behavior of K48R mutants is seen (197, 215). A conjugation-incompetent, fully arginine-substituted form of ubiquitin expressed in a *doa4* background, which has defective Ste2p uptake, rescues α-factor internalization (197). Together, these studies indicate that a single molecule is the minimal ubiquitin tag required for endocytosis, and it is suggested that this distinguishes the endocytic ubiquitin signal from the proteasome degradation signal (216).

However, in vivo, the extent of ubiquitination and the type of linkage vary in both yeast and mammals. Even Ste2p is multiubiquitinated in vivo (197). Both Gap1p and Fur1p permeases are polyubiquitinated on two lysine residues within the cytosolic amino terminus, ^{9}Lys and ^{16}Lys in Gap1p (212), and ^{38}Lys and ^{41}Lys in Fur4p (217). In these proteins, simultaneous mutation of both acceptor lysines traps the permease at the surface (212, 217). A single utilizable lysine is sufficient to drive endocytosis, but the rate is slowed (212), as it is when only monoubiquitination occurs (218), suggesting that multiubiquitin is a more effective signal. Further, using Lys→Arg substituted ubiquitins, it has been established that chain extension in both permeases is via a ubiquitin ^{63}Lys linkage (212, 218). Similarly, a single ubiquitin fused to the transferrin receptor retards recycling of only ~25% of internalized transferrin receptors; most still recycle efficiently (12), suggesting that multiubiquitin might be the optimal signal form.

In many instances, it has not been established whether the multiple ubiquitin-positive species detected biochemically represent limited assembly of multiubiquitin chains or monoubiquitination of several distinct lysine residues. In vertebrates, a range from limited to extensive ubiquitination can be seen. Deubiquitinating enzymes, the largest family of ubiquitin-modifying enzymes, rapidly disassemble/salvage conjugated ubiquitin (193), possibly causing underestimation of the extent of ubiquitination in vivo. The efficiency of anti-ubiquitin antibodies is also poor, and consequently, in many studies ubiquitination is detected using overexpressed, epitope-tagged ubiquitin. Thus, the precise molecular details of the ubiquitin signal remain to be determined. Because fusion of only residues 36–44 of ubiquitin to a cytosolically truncated form of the α chain of the IL-2 receptor facilitates its internalization, it was suggested that ^{43}Leu-^{44}Ile are part of a [DE]XXXL[LI] signal within ubiquitin (215). However, in yeast, a fusion of this dileucine-bearing segment of ubiquitin to a truncated Ste2p does not promote internalization whereas whole ubiquitin does (214). Although this might be due to failure of this putative [DE]XXXL[LI] signal to interface with yeast AP-2, accessibility of the dileucine within folded ubiquitin might be limited.

Substitution of aliphatic hydrophobic side chains buried within the internal core of ubiquitin disrupts folding and abrogates the sorting capacity of a Ste2p-ubiquitin chimera (214). This observation suggests that in *S. cerevisiae*, the ubiquitin tag is detected as a folded interaction surface rather than as a linear sequence in an extended conformation. Systematic mutation of a ubiquitin appended to Ste2p as the only internalization information uncovered two roughly contiguous surface patches containing important side chains. Most important are ^{4}Phe and ^{44}Ile, with ^{8}Leu and ^{70}Val, in proximity to ^{44}Ile, also participating in the generation of the internalization signal (214). Interestingly, ^{8}Leu, ^{44}Ile, and ^{70}Val are also important elements of the polyubiquitin determinant necessary for proteasomal degradation (219), which suggests that cellular identification of these ubiquitin signals might employ related components (see below).

Finally, although ubiquitination of the GH receptor occurs at the plasma membrane, and ubiquitinated GH receptor is found in clathrin lattices (176), receptor-bound ubiquitin is not necessary for internalization (220). Simultaneous Lys→Arg substitution of all 16 potential ubiquitin acceptors in the cytosolic domain of the GH receptor is still compatible with internalization (220). Intriguingly, this internalization is still dependent upon both a competent cellular ubiquitination machinery and a ubiquitin-dependent endocytosis (UbE) motif within the GH receptor cytosolic domain (220). This suggests that, in this system, either ubiquitination of another component serves as an internalization signal for the receptor, or that ubiquitination of an inhibitory molecule promotes exposure of a ubiquitin-independent sorting signal, or that the docked ubiquitination machinery tags the complex for uptake (220). An example of the last possibility is the recent discovery that CIN85 bridges a receptor-bound E3 ligase and endophilin, a known endocytic protein, to promote internalization (221, 222).

Generation of the Ubiquitin Sorting Signal

Selection of proteins for ubiquitination is mediated by E3 ubiquitin ligases, which contain substrate-recognition modules. Two general classes of E3s can be distinguished on the basis of sequence homology and mode of operation with respect to the ubiquitin-carrying E2 component (165, 223). HECT (homologous to *E*6-AP carboxy terminus) domain E3s are often large proteins with an ~350-residue HECT domain, commonly positioned at the carboxy terminus. The HECT domain transfers activated ubiquitin from a designated E2 onto an invariant HECT cysteine prior to conjugation onto the substrate; thus, HECT E3s conjugate ubiquitin directly. RING (Really Interesting New Gene) finger E3 ligases all contain an ~70-amino acid sequence with conserved cysteine and histidine residues positioned to coordinate two zinc atoms (165, 223). The folded finger provides an E2 interaction surface and this class of E3 provides target specificity while working in catalytic conjunction with an E2, as there is no thiol ester-linked ubiquitin directly associated with these E3s.

In many instances, ubiquitination of proteins at the plasma membrane is preceded by phosphorylation that generates a docking site for a particular E3 ubiquitin ligase. In *S. cerevisiae*, the major and best-characterized E3 ligase operative in endocytic trafficking is Rsp5p (171) (Figure 6). In various *rsp5* mutants, plasma membrane proteins fail to be ubiquitinated and, consequently, internalized (171). Rsp5p is a HECT-type E3 that recognizes its targets via three amino-terminal, tandem WW domains. The ~40-amino acid WW fold, bearing two conserved tryptophan residues, is a protein–protein interaction domain that recognizes several proline-rich ligand motifs, including the PPXY and PPLP motifs, polyproline stretches with included glycine and methionine residues, as well as phosphoserine- and phosphothreonine-containing sequences (224). The three Rsp5p WW domains appear to function redundantly, although differential contributions to fluid-phase uptake and protein endocytosis can be mapped (225, 226). For Ste2p ubiquitination, Rsp5p recognizes a hyperphosphorylated ^{331}SINNDAKSS sequence and ubiquitinates the included ^{337}Lys as well as other lysines within the carboxy-terminal segment (197). A related sequence, DAKTI, which controls Ste6p endocytosis (227), is likely also an Rsp5p target. Unexpectedly, internalization of a Ste2 p-ubiquitin chimera is defective in an *rsp5* background (161), indicating that a monoubiquitin tag in the absence of functional endocytic conjugation machinery is insufficient to promote pheromone receptor internalization. These results are in general accord with the GH receptor data (220) and suggest that additional ubiquitination of core endocytic machinery might be a prerequisite for efficient endocytosis (see below).

G protein-coupled receptor internalization in vertebrates is also triggered by phosphorylation, but in contrast to the pheromone receptors, the intermediate adaptor β-arrestin provides critical internalization information and interfaces with the clathrin-coat machinery (228). Still, β-arrestin and the β2-adrenergic receptor itself are ubiquitinated in a stimulus-dependent fashion (229). The E3 ligase in this instance is Mdm2, a RING E3 better known for promoting the proteasomal degradation of the cell-cycle regulator p53 (230). Mdm2-null cells show that Mdm2-catalyzed ubiquitination of arrestin is required for β2-adrenergic receptor internalization, whereas Mdm2-driven polyubiquitination of the receptor is required for lysosomal degradation because a lysine-free receptor form is turned over very slowly (229).

Nedd4 is an Rsp5p orthologue in vertebrates that is linked to internalization of the $\alpha_2\beta\gamma$-ENaC heterotetramer, as overexpression of catalytically defective Nedd4 results in increased channel number and activity at the cell surface (171, 231). Human Nedd4 has four WW domains that bind, primarily via WW domain 2 and/or 3 (232, 233), to PPPXY (PY) motifs found in the cytosolic segment of each ENaC subunit (171). The physiologically relevant PY motifs are contained within the β and γ subunits (234, 235), and inherited mutations of the PY motifs in either the β- or γ-ENaC subunits cause Liddle's syndrome (236, 237), a form of severe hypertension. Disease appears to be due to elevated channel activity because ubiquitin-dependent endocytosis is disrupted and direct, lysine-dependent ubiquitination of α- and γ-ENaC subunits by Nedd4 has been demonstrated (231). Additional regulation of the E3-substrate interaction also occurs; phosphorylation of a threonine residue flanking the PY motif within the carboxy-terminal segment of either β- or γ-ENaC subunits increases the affinity for Nedd4 WW domains by threefold (238). *Drosophila* Nedd4 is involved in internalization of Commissureless (Comm), a regulatory transmembrane protein operative in axonal migration (239). DNedd4 binds, via PY—WW-domain interactions, to Comm to induce ubiquitination and facilitate Comm-dependent downregulation of the Roundabout (Robo) receptor from the cell surface (239). A Nedd4-like E3 also appears likely to regulate internalization of the ClC-5 chloride channel by interacting with a PPYTPP internalization sequence (240), and Itch is yet another Nedd4-like HECT-domain E3 that recognizes Notch via WW-dependent recognition to prompt ubiquitin addition (241).

c-Cbl, a RING finger E3 ligase, is recruited onto the EGF receptor via an SH2-like domain adjacent to the E2-binding RING finger domain (177, 178) (Figure 6). Phosphorylation of ^{1045}Tyr in the cytosolic domain of the receptor generates a major docking site for c-Cbl. A Y1045F substitution diminishes internalization of the receptor while potentiating recycling and mitogenic signaling (179). Dominant-negative Cbl diminishes EGF receptor ubiquitination and reduces lysosomal degradation (204), whereas H_2O_2 inhibition of EGF receptor uptake correlates with inhibition of receptor ubiquitination (242). This system also illustrates nicely how ongoing ubiquitination can guide a receptor along its trajectory toward the interior of the lysosome. Ubiquitination begins at the cell surface; the Y1045A mutant (179), or wild-type EGF receptor in dominant-negative dynamin-expressing cells (243), arrests at the cell surface. Yet maximal ubiquitination of the receptor occurs as it penetrates the endosomal compartment (204). Internalized TGFα-activated EGF receptors recycle back to the cell surface because the ligand dissociates from the receptor within early endosomes and ubiquitination is terminated (203, 204). As kinase-deficient (202) or Y1045F-substituted (179) EGF receptors do not translocate into the multivesicular body interior but instead recycle to the cell surface, sustained activation of the receptor, to maintain the ubiquitinated state, is required for directed passage to the lysosome. In fact, c-Cbl and the EGF receptor traffic toward the lysosome bound together (204, 244, 245); after 60 min both the EGF receptor and c-Cbl are found within the intralumenal vesicles of multivesicular bodies (245) (Figure 6). Thus, progressive ubiquitination generates sorting signals to ensure downregulation of the receptor.

c-Cbl is also involved in negative regulation of other receptor tyrosine kinases including the platelet-derived growth factor (PDGF) receptor (246). CSF-1 receptor internalization is slow in c-Cbl mutant cells (181), and c-Cbl also translocates onto tyrosine-phosphorylated Notch1, mediating lysosomal degradation of the uncleaved transmembrane form of the receptor (247). The immunological phenotypes of c-Cbl$^{-/-}$ mice suggest that this E3 is also involved in negative regulation of antigen receptors (189). Neuralized, a RING E3 ligase in flies ubiquitinates the Notch ligand Delta in vitro and using the RING domain, facilitates endocytosis and lysosome degradation of Delta (184–186). Another RING E3, Hakai, drives internalization of E-cadherin (188). Tyrosine phosphorylation facilitates Hakai recruitment, and overexpression of Hakai in epithelial cells perturbs cell-cell adhesion by inducing E-cadherin endocytosis (188).

An interesting variation on the generation of a ubiquitin sorting signal is seen in the vacuole-directed diversion of the Gap1p permease. Here, Rsp5p is the responsible E3 (211), in accord with localization of the protein to both cell surface and perinuclear structures (248). Bul1p/Bul2p are additionally required at the Golgi complex to deflect Gap1p from surface delivery, as a *bul1Δbul2Δ* strain delivers excess permease to the surface (211) (Figure 6). Bul1p is ~50% identical to Bul2p, and they act redundantly by binding, via a PY motif, to the WW-domain region of Rsp5p and driving extensive polyubiquitination of Gap1p (211) as a signal for intracellular vacuolar delivery (211). A subsequent study (212) revealed that Rsp5p, Bul1/2p, Doa4p, and ^{9}Lys and ^{16}Lys of Gap1p are also all required for internalization. As ^{63}Lys-ubiquitin linkages are generated (212), perhaps Bul1/2p are required to modulate Rsp5p activity specifically to generate the poly-^{63}Lys-linked chains.

Tul1p is an atypical ligase that ubiquitinates biosynthetic cargo for intracellular delivery to the vacuole. Unusual in that the RING domain is located at the carboxyl end of a membrane protein with seven predicted transmembrane domains, Tul1p localizes to the Golgi complex at steady state (249) (Figure 6). Tul1p appears to be the E3 that directs both CPS and Phm5p into the interior of the vacuole as, in a *tul1Δ* background, GFP-tagged forms of these two vacuolar proteins accumulate on the limiting vacuolar membrane instead of internally (249). Missorting correlates with a fivefold decrease in CPS ubiquitination, and Tul1p-ubiquitinated proteins proceed to the vacuole without passing through the cell surface, showing that sorting occurs at the Golgi (249). The mode of substrate recognition is also unusual; transmembrane domains harboring polar side chains mark proteins for Tul1p modification (249), explaining how a single L→D substitution within the transmembrane domain of Pep12p causes ubiquitin-dependent translocation of this endosomal SNARE to the vacuole interior (250). This mode of substrate recognition has led to the suggestion that, in addition to routing normal vacuolar components, Tul1p acts as a quality control device to send abnormal proteins that reach the Golgi complex for destruction (249). The dichotomy between Tul1p and Rsp5p-Bul1/2p in directing different transmembrane cargo along a similar path to the vacuole probably reflects the nutritionally inducible switching required for Gap1p, Fur4p, and other permeases.

Recognition of Ubiquitinated Transmembrane Proteins

Since added ubiquitin serves as a sorting determinant at the cell surface, at endosomes, and at the TGN (Figure 6), ubiquitinated transmembrane cargo must interface with several different sorting components. At present, it appears that decoding the ubiquitin sorting tag might be performed, primarily, by a limited group of ubiquitin-binding modules embedded within compartment-specific components. The 26S proteasome degrades polyubiquitinated targets, and S5a (Rpn10p in *S. cerevisiae*) is a ubiquitin-binding subunit of the 19S regulatory component of the proteasome. Using a bivalent ubiquitin-binding sequence from S5a (251) as a search model, the endocytic proteins Ent1p/Ent2p/epsin 1–3, eps15/eps15R, Vps27/Hrs, Hse1p/STAM1, and STAM2 (also termed Hrs binding protein, Hbp) were each revealed to contain one to three copies of a degenerate consensus sequence termed the ubiquitin-interacting motif (UIM) (252). The UIMs in the epsins, eps15/eps15R, Hrs/Vps27p, and Hse1p all bind directly to ubiquitin, to monoubiquitin relatively weakly, and to polyubiquitin chains with better apparent affinity (12, 205, 253–256). Point mutation of conserved side chains in the Vps27p and Hrs UIMs, or UIM deletion, blocks ubiquitin association (12, 205, 253, 256), and interaction of yeast epsin, Ent1p, with ubiquitin is sensitive to a ubiquitin mutation (I44A) that disrupts Ste2p internalization (256). UIM binding to ubiquitin chains with alternate linkages has not yet been addressed.

The overall domain architecture of these UIM-bearing endocytic proteins (Figure 2) suggests that they could act as cargo-dedicated intermediate adaptors for sorting ubiquitinated cargo at discrete intracellular sites. The epsins superficially resemble Dab2; a PtdIns(4,5)P_2-binding ENTH domain is followed by a largely unstructured carboxy-terminal segment (257, 258) containing interaction motifs for engaging AP-2, clathrin, and eps15, respectively (Table 2). The UIMs are positioned directly following the ENTH domain and thus proximal to the membrane. eps15, on the other hand, has a central coiled-coil domain that can hetero-oligomerize with eps15R (or intersectin), flanked by amino-terminal tandem EH domains and multiple AP-2 binding triplets at the carboxy-terminal end. Two UIMs are located at the extreme carboxyl terminus of eps15/eps15R. As epsin and eps15 are binding partners, and epsin contributes to clathrin lattice assembly, they could cooperatively capture ubiquitinated cargo within clathrin buds at the cell surface. This idea is supported by UIM deletions in the yeast epsin, Ent1p (Figure 6). Although an *ent1Δent2Δ* strain expressing a UIM-deleted Ent1p from a plasmid exhibits minimally impeded Ste2p endocytosis (256), this is apparently due to functional redundancy with Ede1p, the orthologue of mammalian eps15, in which the UIM is replaced with a ubiquitin-associated (UBA) domain. The UBA domain is a ~40-amino acid region found in some E2, E3, and several other proteins and binds multiubiquitin chains (259). In *ent1Δent2Δede1Δ* yeast, production of the UIM-deleted Ent1p does not support α-factor internalization, whereas wild-type Ent1p does. As Ste2p is normally ubiquitinated in the UIM-deleted strain, the data reveal the importance of the UIM and functionally analogous UBA domains in endocytic sorting.

Analogously to Ent1p and Ede1p, Vps27p hetero-oligomerizes with Hse1p (253), forming a complex similar to the Hrs-STAM complex in mammals. Both CPS and Ste2p delivery to the vacuolar lumen is blocked in either *vps27Δ* or *vps27* UIM-mutated strains (256), which confirms that this sorting complex gathers both biosynthetic and endocytic cargo for transport to the vacuole (Figure 6). That a complex of two UIM-harboring proteins works redundantly is shown by the correct delivery of ubiquitinated Ste3p to the vacuole interior in *vps27-ΔUIM* or in *hse1-ΔUIM* strains but not in a *vps27-ΔUIMhse1-ΔUIM* double mutant, where the pheromone receptor ends up missorted on the vacuolar outer membrane (253). Indeed, vacuoles isolated from the double mutant do not contain ubiquitinated polypeptides as do normal vacuoles (253). A vacuolar protein that does not require ubiquitin for delivery, Sna3p (196), is still appropriately delivered to the lumen, which also contains internal membranes, in a *vps27/hse1* UIM-deleted strain (253). This again highlights the sorting function of the UIM.

In animal cells, localization of an Hrs-STAM complex to clathrin-scaffold regions upon early endosomes places the UIMs in an appropriate position to retain ubiquitinated cargo at this site (Figure 6). The importance of the Hrs lipid-binding FYVE domain (Figure 2) is revealed by the disappearance of the endosomal clathrin coats in the presence of the PtdIns3-kinase inhibitor wortmannin (11, 12), a drug that prevents sequestration of EGF receptors within multivesicular bodies (260). Overexpression of Hrs causes redistribution of clathrin onto early endosomes (261). Coexpression of Hrs and a ubiquitin-tagged transferrin receptor markedly increases retention of this receptor within endosomal structures, a phenomenon overcome by inactivating the Hrs UIM by mutagenesis (12). Overexpressed Hrs alone causes relocation of endogenous ubiquitinated proteins to endosomes and slows degradation of EGF (205, 261, 262). Thus, increasing the density of the UIM-containing proteins on endosomal structures concentrates ubiquitinated cargo at the site. These recent studies all show a strong correlation between normal function of UIM-containing traffic proteins and appropriate sorting of ubiquitin.

A severely truncated form of Hrs expressed in mutant *Drosophila* causes strikingly enlarged endosomes and persistent activation of several receptor tyrosine kinases that are not appropriately downregulated (254). Targeted disruption of Hrs in mice is embryonically lethal, but primary embryonic cultures again show defects in early, but not late, endosome morphology (263). The cultured Hrs$^{-/-}$ cells contain numerous large transferrin receptor-positive vacuoles, mirroring wortmannin-treated cells (263). Although this is all in accord with the clathrin/Hrs sorting scaffold preferentially operating at the early endosomes, where Hrs has consistently been localized, Hrs-mutant flies have a pronounced defect in multivesicular body formation (254), and Vps27p, the homologue in *S. cerevisiae*, is a class E vps protein involved in multivesicular body formation (256). This suggests either that, in the absence of sorted ubiquitinated cargo, no multivesicular body formation occurs, or that Hrs is also involved in the involution process directly. Support for the latter idea comes unexpectedly from recent work on HIV budding from the surface of infected cells where, with respect to the cytosol, virus budding is topologically analogous to endosome involution. The HIV-1 Gag viral polyprotein is alone sufficient to drive virion budding. The p6 late domain of Gag carries a PTAP peptide sequence that is required for budding, and this sequence binds directly to a cellular component, tumor-susceptibility gene 101 (TSG101) (264). Remarkably, TSG101 is the orthologue of Vps23p, which, in a 350-kDa functional complex with Vps28p and Vps37p termed ESCRT-I (endosomal sorting complex required for transport), is involved in sorting ubiquitinated cargo into multivesicular endosomes (200, 265). RNAi-mediated knockdown of TSG101 protein level arrests virus separation and the release of infectious particles (264). The same cellular machinery that retroviruses usurp to promote efficient viral particle release normally facilitates lysosomal delivery. After 60 min in A-431 cells, which express very high levels of the EGF receptor, internalized EGF colocalizes with early endosome autoantigen 1 (EEA1), ubiquitinated proteins, TSG101, and hVPS28 at large scattered intracellular structures (205). *tsg101* mutant cells recycle to the surface, instead of degrading, ligand-bound EGF receptors and also missort the CI-MPR and lysosomal hydrolases (265). Antibodies against hVPS28 or depletion of TSG101 with RNAi arrests EGF receptor traffic in a similar fashion to overexpression of Hrs (205). TSG101 RNAi also prevents viral E3-directed degradation of MHC class I (191).

Budding of several retroviruses, including HIV, is inhibited by proteasome inhibitors, and spherical viral particles, still tethered to the plasma membrane, accumulate (266, 267). The effect can be bypassed by overexpression of ubiquitin, and Gag itself is ubiquitinated (266). In fact, a number of viral Gag proteins display PY motifs, and a Nedd4-like HECT E3 with two WW domains, termed BUL1, appears to be involved in retrovirus budding (268). For virus release, the PTAP sequence does not function independently in the absence of ubiquitination (269) because the interaction of Gag with TSG101 is bipartite, utilizing both the PTAP sequence and conjugated ubiquitin (264). TSG101/Vps23p contains a ubiquitin recognition module termed the ubiquitin conjugating (UBC)-variant, or UEV domain. E2 conjugating enzymes, which receive activated ubiquitin from E1, exhibit a conserved ubiquitin-conjugation domain, termed UBC, containing an invariant catalytic cysteine for thioester linkage to ubiquitin. Both Vps23p and TSG101 bear sequence homology to the UBC domain but lack the active site cysteine. Like the UIM and UBA, the UBC domain imparts to both proteins the capacity to bind ubiquitin (200, 205, 270). The Gag PTAP sequence binds directly to the UEV domain to replace an α helix normally found in the canonical UBC fold, but which is absent in TSG101 (270), and although TSG101 binds the Gag PTAP sequence directly, the affinity increases roughly tenfold if ubiquitin is conjugated to the Gag p6 (264). These studies indicate that the viral Gag polyprotein uses both PTAP and ubiquitin to redirect the endosomal ESCRT-I complex to the plasma membrane (Figure 6). If pirating a regular cellular function, what contributes the PTAP sequence to TSG101/ESCRT-I normally at the endosomal sorting site?

Intriguingly, Hrs has a PSAP sequence, and in fact, direct interaction between Hrs and ESCRT-I has been reported (12). Thus, Hrs could be part of the machinery that recruits ESCRT-I to the surface of the nascent multivesicular body. Hrs would play a sequential, but coupled role, beginning at the early endosome by segregating ubiquitinated cargo away from the recycling population and then driving ESCRT-I translocation to initiate intralumenal budding (Figure 6). This is in full accord with the limited number of intralumenal membrane elements seen in early endosomes containing Hrs-clathrin bilayered coats (11). This model would account for the observed phenotype of $Hrs^{-/-}$ flies and mice, and it integrates ubiquitin-dependent lysosomal sorting with the morphological transition that accompanies early to late endosome maturation.

An unexpected observation is that the UIM-sequence also directs monoubiquitination of proteins containing this sequence, or of extraneous proteins fused to a UIM region, as the acceptor lysine is not contained within the UIM (255, 271–273). The E3 ligase involved appears to be Nedd4 (255, 273), but the significance of this modification has not been precisely defined. It has been suggested that the UIM could participate directly in E3 binding by associating with the ligase-bound ubiquitin, or bind to the newly conjugated ubiquitin, in either case possibly favoring only monoubiquitination (255). Although a substantial amount of eps15/eps15R is monoubiquitinated upon addition of EGF, the fraction of ubiquitinated epsin and Hrs is, however, low (255, 271). Further, the effect of monoubiquitination on the capacity of UIM-containing proteins to bind ubiquitin has not yet been reported. The observed monoubiquitin modification could be linked to the requirement for a functional ubiquitination machinery to facilitate endocytosis of GH receptor (220) or Ste2p-ubiquitin (161). Ubiquitination of the endocytic machinery could amplify, or create, via UIM-ubiquitin interactions, a specifically required protein-interaction web at the appropriate sorting site. One argument for this is that the amount of ubiquitinated protein amassed on endosomes in Hrs-overexpressing cells stimulated with EGF cannot be accounted for by ubiquitinated EGF receptor alone (205). In this model, mirroring HIV-1 Gag, monoubiquitinated Hrs and internal PSAP sequence, and not ubiquitinated cargo itself, might be the physiologic binding partner of the ESCRT-I complex. The Hrs UIM could synchronously bind a ubiquitin tag attached to cargo. That ESCRT-I might not necessarily bind ubiquitinated cargo directly is indicated by the ability of overexpressed ESCRT-II complex, which functions downstream of ESCRT-I in multivesicular body formation, to rescue sorting in an ESCRT-I-mutant (*vps23*Δ) strain (274). As ESCRT-II is not known to interact with ubiquitin, perhaps the Vps27p and Hse1p UIMs mediate the critical cargo sorting (253).

Some of the ubiquitin signal on early endosomes could also be derived from eps15, as this protein binds to Hrs2 (275) which, in neurons, is enriched focally on multivesicular endosomes (275), highly reminiscent of the bilayered clathrin scaffold sorting sites (11). A physiological role for direct ubiquitination of the endocytic machinery is also supported by functional studies of the *Drosophila* epsin homologue, Liquid facets (Lqf) (276). Fat facets, a deubiquitinating enzyme, interacts genetically with Lqf and is specifically required for Lqf function, possibly by recycling the ubiquitinated epsin for further rounds of endocytic sorting (276, 277). In this regard, it has also been proposed that ubiquitination of UIM-bearing endocytic proteins causes functional inactivation by the conjugated ubiquitin engaging the UIM (273).

Finally, Doa4p appears to deubiquitinate cargo proteins just before final packaging into the involuting buds (200) (Figure 6), accounting for the strong sorting phenotype of *doa4* mutants; in fact, Doa4p accumulates in the expanded class E structure in class E mutants (194). However, deubiquitination is not essential for progression into intralumenal vesicles. Ubiquitinated forms of yeast and mammalian proteins are detectable in multivesicular bodies and vacuoles (204, 205, 253).

CONCLUDING REMARKS

The information reviewed in this article illustrates how far we have come in the understanding of the molecular mechanisms of signal-mediated protein trafficking to endosomes and lysosomes. Many issues remain unresolved, but with the powerful tools now available we can expect these to be elucidated in the not too distant future. Among the outstanding questions are the identification of the complete repertoire of sorting signals and their binding partners, the explanation of the structural bases for the recognition of all signals, the demonstration of the exact sorting events mediated by particular signals and recognition proteins, the detailed molecular description of complex pathways of signal addition and detection (as is the case for ubiquitin), the regulation of signal recognition, and the contribution of luminal/extracellular and transmembrane domains to sorting. The current knowledge of the mechanisms of signal-mediated protein sorting has already contributed to the elucidation of the pathogenesis of genetic diseases such as familial hypercholesterolemia, Hermansky-Pudlak syndrome type 2, and Liddle's syndrome. A more complete understanding of protein sorting is likely to illuminate the participation of the endosomal-lysosomal system in more common physiological and pathological processes as well as to provide novel avenues for therapeutic intervention.

ACKNOWLEDGMENTS

We thank Mickey Marks, Markus Boehm, and James Hurley for helpful discussions and critical reading of the manuscript, and Matthew Wood and Saurav Misra for help with the crystallography figures.

The *Annual Review of Biochemistry* is online at http://biochem.annualreviews.org

LITERATURE CITED

1. Trowbridge IS, Collawn JF, Hopkins CR. 1993. *Annu. Rev. Cell Biol.* 9:29–61
2. Sandoval IV, Bakke O. 1994. *Trends Cell Biol.* 4:292–97
3. Bonifacino JS, Dell'Angelica EC. 1999. *J. Cell Biol.* 145:923–26
4. Orsel JG, Sincock PM, Krise JP, Pfeffer SR. 2000. *Proc. Natl. Acad. Sci. USA* 97:9047–51
5. Hicke L. 2001. *Cell* 106:527–30
6. Roth TF, Cutting JA, Atlas SB. 1976. *J. Supramol. Struct.* 4:527–48
7. Anderson RG, Brown MS, Goldstein JL. 1977. *Cell* 10:351–64
8. Schulze-Lohoff E, Hasilik A, von Figura K. 1985. *J. Cell Biol.* 101:824–29
9. Klumperman J, Hille A, Veenendaal T, Oorschot V, Stoorvogel W, et al. 1993. *J. Cell Biol.* 121:997–1010
10. Futter CE, Gibson A, Allchin EH, Maxwell S, Ruddock LJ, et al. 1998. *J. Cell Biol.* 141:611–23
11. Sachse M, Urbe S, Oorschot V, Strous GJ, Klumperman J. 2002. *Mol. Biol. Cell* 13:1313–28
12. Raiborg C, Bache KG, Gillooly DJ, Madshus IH, Stang E, Stenmark H. 2002. *Nat. Cell Biol.* 4:394–98
13. Pearse BM, Robinson MS. 1990. *Annu. Rev. Cell Biol.* 6:151–71
14. Kirchhausen T. 2000. *Annu. Rev. Biochem.* 69:699–727
15. Kirchhausen T. 1999. *Annu. Rev. Cell Dev. Biol.* 15:705–32
16. Slepnev VI, De Camilli P. 2000. *Nat. Rev. Neurosci.* 1:161–72
17. Davis CG, Lehrman MA, Russell DW, Anderson RG, Brown MS, Goldstein JL. 1986. *Cell* 45:15–24
18. Chen J-J, Goldstein JL, Brown MS. 1990. *J. Biol. Chem.* 265:3116–23
19. Lazarovits J, Roth M. 1988. *Cell* 53: 743–52
20. Lobel P, Fujimoto K, Ye RD, Griffiths G, Kornfeld S. 1989. *Cell* 57:787–96
21. McGraw TE, Maxfield FR. 1990. *Cell Regul.* 1:369–77
22. Collawn JF, Stangel M, Kuhn LA, Esekogwu V, Jing SQ, et al. 1990. *Cell* 63:1061–72
23. Alvarez E, Girones N, Davis RJ. 1990. *Biochem. J.* 267:31–35
24. Breitfeld PP, Casanova JE, McKinnan WC, Mostov KE. 1990. *J. Biol. Chem.* 265:13750–57
25. Canfield WM, Johnson KF, Ye RD, Gregory W, Kornfeld S. 1991. *J. Biol. Chem.* 266:5682–88
26. Jadot M, Canfield WM, Gregory W, Kornfeld S. 1992. *J. Biol. Chem.* 267: 11069–77
27. Williams MA, Fukuda M. 1990. *J. Cell Biol.* 111:955–66
28. Harter C, Mellman I. 1992. *J. Cell Biol.* 117:311–25
29. Hunziker W, Harter C, Matter K, Mellman I. 1991. *Cell* 66:907–20
30. Rajasekaran AK, Humphrey JS, Wagner M, Miesenböck G, Le Bivic A, et al. 1994. *Mol. Biol. Cell* 5:1093–103
31. Collawn JF, Kuhn LA, Liu L-FS, Tainer JA, Trowbridge IS. 1991. *EMBO J.* 10:3247–53
32. Perez RG, Soriano S, Hayes JD, Ostaszewski B, Xia W, et al. 1999. *J. Biol. Chem.* 274:18851–56
33. Kibbey RG, Rizo J, Gierasch LM, Anderson RG. 1998. *J. Cell Biol.* 142: 59–67
34. Boll W, Rapoport I, Brunner C, Modis Y, Prehn S, Kirchhausen T. 2002. *Traffic* 3:590–600
35. Musacchio A, Smith CJ, Roseman AM, Harrison SC, Kirchhausen T, Pearse BM. 1999. *Mol. Cell* 3:761–70
36. Pearse BM. 1988. *EMBO J.* 7:3331–36
37. Warren RA, Green FA, Stenberg PE, Enns CA. 1998. *J. Biol. Chem.* 273: 17056–63
38. Howell BW, Lanier LM, Frank R,

Gertler FB, Cooper JA. 1999. *Mol. Cell. Biol.* 19:5179–88
39. Morris SM, Cooper JA. 2001. *Traffic* 2:111–23
40. Mishra SK, Keyel PA, Hawryluk MJ, Agostinelli NR, Watkins SC, Traub LM. 2002. *EMBO J.* 21:4915–26
41. Herz J, Bock HH. 2002. *Annu. Rev. Biochem.* 71:405–34
42. Oleinikov AV, Zhao J, Makker SP. 2000. *Biochem. J.* 347(Part 3):613–21
43. Gotthardt M, Trommsdorff M, Nevitt MF, Shelton J, Richardson JA, et al. 2000. *J. Biol. Chem.* 275:25616–24
44. Morris SM, Tallquist MD, Rock CO, Cooper JA. 2002. *EMBO J.* 21:1555–64
45. Nykjaer A, Dragun D, Walther D, Vorum H, Jacobsen C, et al. 1999. *Cell* 96:507–15
46. Verroust PJ, Kozyraki R. 2001. *Curr. Opin. Nephrol. Hypertens.* 10:33–38
47. Garcia CK, Wilund K, Arca M, Zuliani G, Fellin R, et al. 2001. *Science* 292: 1394–98
48. Al-Kateb H, Bahring S, Hoffmann K, Strauch K, Busjahn A, et al. 2002. *Circ. Res.* 90:951–58
49. Norman D, Sun XM, Bourbon M, Knight BL, Naoumova RP, Soutar AK. 1999. *J. Clin. Invest.* 104:619–28
50. Berdnik D, Torok T, Gonzalez-Gaitan M, Knoblich J. 2002. *Dev. Cell* 3:221–31
51. Santolini E, Puri C, Salcini AE, Gagliani MC, Pelicci PG, et al. 2000. *J. Cell Biol.* 151:1345–52
52. Marks MS, Roche PA, van Donselaar E, Woodruff L, Peters PJ, Bonifacino JS. 1995. *J. Cell Biol.* 131:351–69
53. Gough NR, Zweifel ME, Martinez-Augustin O, Aguilar RC, Bonifacino JS, Fambrough DM. 1999. *J. Cell Sci.* 112: 4257–69
54. Rous BA, Reaves BJ, Ihrke G, Briggs JA, Gray SR, et al. 2002. *Mol. Biol. Cell* 13:1071–82
55. Rohrer J, Schweizer A, Russell D, Kornfeld S. 1996. *J. Cell Biol.* 132:565–76
56. Ohno H, Stewart J, Fournier MC, Bosshart H, Rhee I, et al. 1995. *Science* 269:1872–75
57. Boll W, Ohno H, Zhou SY, Rapoport I, Cantley LC, et al. 1996. *EMBO J.* 15:5789–95
58. Höning S, Griffith J, Geuze HJ, Hunziker W. 1996. *EMBO J.* 15:5230–39
59. Ohno H, Fournier MC, Poy G, Bonifacino JS. 1996. *J. Biol. Chem.* 271: 29009–15
60. Hirst J, Bright NA, Rous B, Robinson MS. 1999. *Mol. Biol. Cell* 10:2787–802
61. Stephens DJ, Crump CM, Clarke AR, Banting G. 1997. *J. Biol. Chem.* 272: 14104–9
62. Rapoport I, Miyazaki M, Boll W, Duckworth B, Cantley LC, et al. 1997. *EMBO J.* 16:2240–50
63. Ohno H, Aguilar RC, Yeh D, Taura D, Saito T, Bonifacino JS. 1998. *J. Biol. Chem.* 273:25915–21
64. Aguilar RC, Boehm M, Gorshkova I, Crouch RJ, Tomita K, et al. 2001. *J. Biol. Chem.* 276:13145–52
65. Ohno H, Tomemori T, Nakatsu F, Okazaki Y, Aguilar RC, et al. 1999. *FEBS Lett.* 449:215–20
66. Boehm M, Bonifacino JS. 2002. *Gene* 286:175–86
67. Aguilar RC, Ohno H, Roche KC, Bonifacino JS. 1997. *J. Biol. Chem.* 272: 2760–66
68. Owen DJ, Evans PR. 1998. *Science* 282: 1327–32
69. Collins BM, McCoy AJ, Kent HM, Evans PR, Owen DJ. 2002. *Cell* 109: 523–35
70. Andrews J, Smith M, Merakovsky J, Coulson M, Hannan F, Kelly LE. 1996. *Genetics* 143:1699–711
71. Martina JA, Bonangelino CJ, Aguilar RC, Bonifacino JS. 2001. *J. Cell Biol.* 28:1111–20
72. Walther K, Krauss M, Diril MK, Lemke S, Ricotta D, et al. 2001. *EMBO Rep.* 2:634–40
73. Owen DJ, Setiadi H, Evans PR, McEver RP, Green SA. 2001. *Traffic* 2:105–10

74. Nesterov A, Carter RE, Sorkina T, Gill GN, Sorkin A. 1999. *EMBO J.* 18:2489–99
75. Rohde G, Wenzel D, Haucke V. 2002. *J. Cell Biol.* 158:209–14
76. Pauloin A, Thurieau C. 1993. *Biochem. J.* 296:409–15
77. Olusanya O, Andrews PD, Swedlow JR, Smythe E. 2001. *Curr. Biol.* 11:896–900
78. Conner SD, Schmid SL. 2002. *J. Cell Biol.* 156:921–29
79. Ricotta D, Conner SD, Schmid SL, von Figura K, Höning S. 2002. *J. Cell Biol.* 156:791–95
80. Gaidarov I, Smith ME, Domin J, Keen JH. 2001. *Mol. Cell.* 7:443–49
81. Haffner C, Di Paolo G, Rosenthal JA, de Camilli P. 2000. *Curr. Biol.* 10:471–74
82. Gaidarov I, Chen Q, Falck JR, Reddy KK, Keen JH. 1996. *J. Biol. Chem.* 271: 20922–29
83. Haucke V, De Camilli P. 1999. *Science* 285:1268–71
84. Shiratori T, Miyatake S, Ohno H, Nakaseko C, Isono K, et al. 1997. *Immunity* 6:583–89
85. Zhang Y, Allison JP. 1997. *Proc. Natl. Acad. Sci. USA* 94:9273–78
86. Chuang E, Alegre ML, Duckett CS, Noel PJ, Vander Heiden MG, Thompson CB. 1997. *J. Immunol.* 159:144–51
87. Schaefer AW, Kamei Y, Kamiguchi H, Wong EV, Rapoport I, et al. 2002. *J. Cell Biol.* 157:1223–32
88. Madrid R, Le Maout S, Barrault MB, Janvier K, Benichou S, Merot J. 2001. *EMBO J.* 20:7008–21
89. Letourneur F, Klausner RD. 1992. *Cell* 69:1143–57
90. Johnson KF, Kornfeld S. 1992. *J. Biol. Chem.* 267:17110–15
91. Johnson KF, Kornfeld S. 1992. *J. Cell Biol.* 119:249–57
92. Chen HJ, Yuan J, Lobel P. 1997. *J. Biol. Chem.* 272:7003–12
93. Dietrich J, Hou X, Wegener AM, Geisler C. 1994. *EMBO J.* 13:2156–66
94. von Essen M, Menne C, Nielsen BL, Lauritsen JP, Dietrich J, et al. 2002. *J. Immunol.* 168:4519–23
95. Pitcher C, Höning S, Fingerhut A, Bowers K, Marsh M. 1999. *Mol. Biol. Cell* 10:677–91
96. Bresnahan PA, Yonemoto W, Ferrell S, Williams-Herman D, Geleziunas R, Greene WC. 1998. *Curr. Biol.* 8:1235–38
97. Greenberg M, DeTulleo L, Rapoport I, Skowronski J, Kirchhausen T. 1998. *Curr. Biol.* 8:1239–42
98. Vowels JJ, Payne GS. 1998. *EMBO J.* 17:2482–93
99. Matter K, Yamamoto EM, Mellman I. 1994. *J. Cell Biol.* 126:991–1004
100. Miranda KC, Khromykh T, Christy P, Le TL, Gottardi CJ, et al. 2001. *J. Biol. Chem.* 276:22565–72
101. Pond L, Kuhn LA, Teyton L, Schutze MP, Tainer JA, et al. 1995. *J. Biol. Chem.* 270:19989–97
102. Sandoval IV, Martinez-Arca S, Valdueza J, Palacios S, Holman GD. 2000. *J. Biol. Chem.* 275:39874–85
103. Geisler C, Dietrich J, Nielsen BL, Kastrup J, Lauritsen JP, et al. 1998. *J. Biol. Chem.* 273:21316–23
104. Heilker R, Manning-Krieg U, Zuber JF, Spiess M. 1996. *EMBO J.* 15:2893–99
105. Dietrich J, Kastrup J, Nielsen BL, Odum N, Geisler C. 1997. *J. Cell Biol.* 138: 271–81
106. Höning S, Sandoval IV, von Figura K. 1998. *EMBO J.* 17:1304–14
107. Fujita H, Saeki M, Yasunaga K, Ueda T, Imoto T, Himeno M. 1999. *Biochem. Biophys. Res. Commun.* 255:54–58
108. Hofmann MW, Höning S, Rodionov D, Dobberstein B, von Figura K, Bakke O. 1999. *J. Biol. Chem.* 274:36153–58
109. Peden AA, Park GY, Scheller RH. 2001. *J. Biol. Chem.* 276:49183–87
110. Le Borgne R, Alconada A, Bauer U, Hoflack B. 1998. *J. Biol. Chem.* 273: 29451–61
111. Huizing M, Sarangarajan R, Strovel E,

Zhao Y, Gahl WA, Boissy RE. 2001. *Mol. Biol. Cell* 12:2075–85
112. Caplan S, Dell'Angelica EC, Gahl WA, Bonifacino JS. 2000. *Immunol. Lett.* 72:113–17
113. Kongsvik TL, Höning S, Bakke O, Rodionov DG. 2002. *J. Biol. Chem.* 277: 16484–88
114. Marks MS, Woodruff L, Ohno H, Bonifacino JS. 1996. *J. Cell Biol.* 135:341–54
115. Rodionov DG, Bakke O. 1998. *J. Biol. Chem.* 273:6005–8
116. Deleted in proof
117. Craig HM, Reddy TR, Riggs NL, Dao PP, Guatelli JC. 2000. *Virology* 271:9–17
118. Bremnes T, Lauvrak V, Lindqvist B, Bakke O. 1998. *J. Biol. Chem.* 273: 8638–45
119. Rapoport I, Chen YC, Cupers P, Shoelson SE, Kirchhausen T. 1998. *EMBO J.* 17:2148–55
120. Puertollano R, Aguilar RC, Gorshkova I, Crouch RJ, Bonifacino JS. 2001. *Science* 292:1712–16
121. Doray B, Ghosh P, Griffith J, Geuze HJ, Kornfeld S. 2002. *Science.* 6:1700–3
122. Puertollano R, van der Wel NN, Green LE, Eisenberg E, Peters PJ, Bonifacino JS. 2003. *Mol. Biol. Cell* 14:1545–57
123. Schweizer A, Kornfeld S, Rohrer J. 1996. *J. Cell Biol.*132:577–84
124. Höning S, Sosa M, Hille-Rehfeld A, von Figura K. 1997. *J. Biol. Chem.* 272: 19884–90
125. Zhu Y, Doray B, Poussu A, Lehto VP, Kornfeld S. 2001. *Science* 292:1716–18
126. Nielsen MS, Madsen P, Christensen EI, Nykjaer A, Gliemann J, et al. 2001. *EMBO J.* 20:2180–90
127. Takatsu H, Katoh Y, Shiba Y, Nakayama K. 2001. *J. Biol. Chem.* 276:28541–45
128. Doray B, Bruns K, Ghosh P, Kornfeld S. 2002. *J. Biol. Chem.* 277:18477–82
129. Jacobsen L, Madsen P, Nielsen MS, Geraerts WP, Gliemann J, et al. 2002. *FEBS Lett.* 511:155–58
130. He X, Chang WP, Koelsch G, Tang J. 2002. *FEBS Lett.* 524:183–87
131. Boman AL, Zhang CJ, Zhu X, Kahn RA. 2000. *Mol. Biol. Cell.* 11:1241–55
132. Dell'Angelica EC, Puertollano R, Mullins C, Aguilar RC, Vargas JD, et al. 2000. *J. Cell Biol.* 149:81–94
133. Hirst J, Lui WW, Bright NA, Totty N, Seaman MN, Robinson MS. 2000. *J. Cell Biol.* 149:67–80
134. Poussu A, Lohi O, Lehto VP. 2000. *J. Biol. Chem.* 275:7176–83
135. Misra S, Puertollano R, Kato Y, Bonifacino JS, Hurley JH. 2002. *Nature.* 415: 933–37
136. Kato Y, Misra S, Puertollano R, Hurley JH, Bonifacino JS. 2002. *Nat. Struct. Biol.* 9:532–36
137. Zhdankina O, Strand NL, Redmond JM, Boman AL. 2001. *Yeast* 18:1–18
138. Costaguta G, Stefan CJ, Bensen ES, Emr SD, Payne GS. 2001. *Mol. Biol. Cell* 12:1885–96
139. Mullins C, Bonifacino JS. 2001. *Mol. Cell. Biol.* 21:7981–94
140. Shiba T, Takatsu H, Nogi T, Matsugaki N, Kawasaki M, et al. 2002. *Nature* 415: 937–41
141. Meresse S, Ludwig T, Frank R, Hoflack B. 1990. *J. Biol. Chem.* 265:18833–42
142. Meresse S, Hoflack B. 1993. *J. Cell Biol.* 120:67–75
143. Rosorius O, Mieskes G, Issinger OG, Korner C, Schmidt B, et al. 1993. *Biochem. J.* 292:833–38
144. Korner C, Herzog A, Weber B, Rosorius O, Hemer F, et al. 1994. *J. Biol. Chem.* 269:16529–32
145. Le Borgne R, Schmidt A, Mauxion F, Griffiths G, Hoflack B. 1993. *J. Biol. Chem.* 268:22552–56
146. Doray B, Bruns K, Ghosh P, Kornfeld SA. 2002. *Proc. Natl. Acad. Sci. USA* 99:8072–77
147. Jones BG, Thomas L, Molloy SS, Thulin CD, Fry MD, et al. 1995. *EMBO J.* 14:5869–83

148. Wan L, Molloy SS, Thomas L, Liu GP, Xiang Y, et al. 1998. *Cell* 94:205–16
149. Crump CM, Xiang Y, Thomas L, Gu F, Austin C, et al. 2001. *EMBO J.* 20:2191–201
150. Olson LJ, Zhang J, Dahms NM, Kim JJ. 2002. *J. Biol. Chem.* 277:10156–61
151. Borden LA, Einstein R, Gabel CA, Maxfield FR. 1990. *J. Biol. Chem.* 265:8497–504
152. Griffiths G, Hoflack B, Simons K, Mellman I, Kornfeld S. 1988. *Cell* 52: 329–41
153. Rohrer J, Schweizer A, Johnson KF, Kornfeld S. 1995. *J. Cell Biol.* 130: 1297–306
154. Schweizer A, Kornfeld S, Rohrer J. 1997. *Proc. Natl. Acad. Sci. USA* 94:14471–76
155. Schweizer A, Stahl PD, Rohrer J. 2000. *J. Biol. Chem.* 275:29694–700
156. Diaz E, Pfeffer SR. 1998. *Cell* 93:433–43
157. Krise JP, Sincock PM, Orsel JG, Pfeffer SR. 2000. *J. Biol. Chem.* 275:25188–93
158. Carroll KS, Hanna J, Simon I, Krise J, Barbero P, Pfeffer SR. 2001. *Science* 292:1373–76
159. Miura S, Gan JW, Brzostowski J, Parisi MJ, Schultz CJ, et al. 2002. *J. Biol. Chem.* 277:32253–57
160. Tan PK, Howard JP, Payne GS. 1996. *J. Cell Biol.* 135:1789–800
161. Dunn R, Hicke L. 2001. *J. Biol. Chem.* 276:25974–81
162. Howard JP, Hutton JL, Olson JM, Payne GS. 2002. *J. Cell Biol.* 157:315–26
163. Tang HY, Xu J, Cai M. 2000. *Mol. Cell. Biol.* 20:12–25
164. Munn AL. 2001. *Biochim. Biophys. Acta* 1535:236–57
165. Pickart CM. 2001. *Annu. Rev. Biochem.* 70:503–33
166. Thrower JS, Hoffman L, Rechsteiner M, Pickart CM. 2000. *EMBO J.* 19:94–102
167. Hicke L, Riezman H. 1996. *Cell* 84:277–87
168. Roth AF, Davis NG. 1996. *J. Cell Biol.* 134:661–74
169. Bonifacino JS, Weissman AM. 1998. *Annu. Rev. Cell Dev. Biol.* 14:19–57
170. Hicke L. 1999. *Trends Cell Biol.* 9:107–12
171. Rotin D, Staub O, Haguenauer-Tsapis R. 2000. *J. Membr. Biol.* 176:1–17
172. Strous GJ, van Kerkhof P, Govers R, Ciechanover A, Schwartz AL. 1996. *EMBO J.* 15:3806–12
173. Strous GJ, van Kerkhof P, Govers R, Rotwein P, Schwartz AL. 1997. *J. Biol. Chem.* 272:40–43
174. van Kerkhof P, Govers R, dos Santos CMA, Strous GJ. 2000. *J. Biol. Chem.* 275:1575–80
175. van Kerkhof P, dos Santos CMA, Sachse M, Klumperman J, Bu GJ, Strous GJ. 2001. *Mol. Biol. Cell* 12:2556–66
176. van Kerkhof P, Sachse M, Klumperman J, Strous GJ. 2001. *J. Biol. Chem.* 276: 3778–84
177. Levkowitz G, Waterman H, Ettenberg SA, Katz M, Tsygankov AY, et al. 1999. *Mol. Cell* 4:1029–40
178. Joazeiro CA, Wing SS, Huang H, Leverson JD, Hunter T, Liu YC. 1999. *Science* 286:309–12
179. Waterman H, Katz M, Rubin C, Shtiegman K, Lavi S, et al. 2002. *EMBO J.* 21:303–13
180. Hammond DE, Urbe S, Vande Woude GF, Clague MJ. 2001. *Oncogene* 20:2761–70
181. Lee PS, Wang Y, Dominguez MG, Yeung YG, Murphy MA, et al. 1999. *EMBO J.* 18:3616–28
182. Wang Y, Yeung YG, Stanley ER. 1999. *J. Cell. Biochem.* 72:119–34
183. Booth JW, Kim MK, Jankowski A, Schreiber AD, Grinstein S. 2002. *EMBO J.* 21:251–58
184. Deblandre GA, Lai EC, Kintner C. 2001. *Dev. Cell* 1:795–806
185. Lai EC, Deblandre GA, Kintner C, Rubin GM. 2001. *Dev. Cell* 1:783–94
186. Yeh E, Dermer M, Commisso C, Zhou L, McGlade CJ, Boulianne GL. 2001. *Curr. Biol.* 11:1675–79

187. Buttner C, Sadtler S, Leyendecker A, Laube B, Griffon N, et al. 2001. *J. Biol. Chem.* 276:42978–85
188. Fujita Y, Krause G, Scheffner M, Zechner D, Leddy HE, et al. 2002. *Nat. Cell Biol.* 4:222–31
189. Panigada M, Porcellini S, Barbier E, Hoeflinger S, Cazenave PA, et al. 2002. *J. Exp. Med.* 195:1585–97
190. Burbea M, Dreier L, Dittman JS, Grunwald ME, Kaplan JM. 2002. *Neuron* 35:107–20
191. Hewitt EW, Duncan L, Mufti D, Baker J, Stevenson PG, Lehner PJ. 2002. *EMBO J.* 21:2418–29
192. Lorenzo ME, Jung JU, Ploegh HL. 2002. *J. Virol.* 76:5522–31
193. Wilkinson KD. 2000. *Semin. Cell Dev. Biol.* 11:141–48
194. Amerik AY, Nowak J, Swaminathan S, Hochstrasser M. 2000. *Mol. Biol. Cell* 11:3365–80
195. Losko S, Kopp F, Kranz A, Kolling R. 2001. *Mol. Biol. Cell* 12:1047–59
196. Reggiori F, Pelham HR. 2001. *EMBO J.* 20:5176–86
197. Terrell J, Shih S, Dunn R, Hicke L. 1998. *Mol. Cell.* 1:193–202
198. Dupre S, Haguenauer-Tsapis R. 2001. *Mol. Cell. Biol.* 21:4482–94
199. Horak J, Wolf DH. 2001. *J. Bacteriol.* 183:3083–88
200. Katzmann DJ, Babst M, Emr SD. 2001. *Cell* 106:145–55
201. Urbanowski JL, Piper RC. 2001. *Traffic* 2:622–30
202. Felder S, Miller K, Moehren G, Ullrich A, Schlessinger J, Hopkins CR. 1990. *Cell* 61:623–34
203. Maeda K, Kato Y, Sugiyama Y. 2002. *J. Control Release* 82:71–82
204. Longva KE, Blystad FD, Stang E, Larsen AM, Johannessen LE, Madshus IH. 2002. *J. Cell Biol.* 156:843–54
205. Bishop N, Horman A, Woodman P. 2002. *J. Cell Biol.* 157:91–101
206. Marchese A, Benovic JL. 2001. *J. Biol. Chem.* 276:45509–12
207. Rocca A, Lamaze C, Subtil A, Dautry-Varsat A. 2001. *Mol. Biol. Cell* 12:1293–301
208. Yu A, Malek TR. 2001. *J. Biol. Chem.* 276:381–85
209. Roberg KJ, Rowley N, Kaiser CA. 1997. *J. Cell Biol.* 137:1469–82
210. Beck T, Schmidt A, Hall MN. 1999. *J. Cell Biol.* 146:1227–38
211. Helliwell SB, Losko S, Kaiser CA. 2001. *J. Cell Biol.* 153:649–62
212. Soetens O, De Craene JO, Andre B. 2001. *J. Biol. Chem.* 276:43949–57
213. Roth AF, Davis NG. 2000. *J. Biol. Chem.* 275:8143–53
214. Shih SC, Sloper-Mould KE, Hicke L. 2000. *EMBO J.* 19:187–98
215. Nakatsu F, Sakuma M, Matsuo Y, Arase H, Yamasaki S, et al. 2000. *J. Biol. Chem.* 275:26213–19
216. Hicke L. 2001. *Nat. Rev. Mol. Cell. Biol.* 2:195–201
217. Marchal C, Haguenauer-Tsapis R, Urban-Grimal D. 2000. *J. Biol. Chem.* 275:23608–14
218. Galan JM, Haguenauer-Tsapis R. 1997. *EMBO J.* 16:5847–54
219. Beal R, Deveraux Q, Xia G, Rechsteiner M, Pickart C. 1996. *Proc. Natl. Acad. Sci. USA* 93:861–66
220. Govers R, ten Broeke T, van Kerkhof P, Schwartz AL, Strous GJ. 1999. *EMBO J.* 18:28–36
221. Petrelli A, Gilestro GF, Lanzardo S, Comoglio PM, Migone N, Giordano S. 2002. *Nature* 416:187–90
222. Soubeyran P, Kowanetz K, Szymkiewicz I, Langdon WY, Dikic I. 2002. *Nature* 416:183–87
223. Jackson PK, Eldridge AG, Freed E, Furstenthal L, Hsu JY, et al. 2000. *Trends Cell Biol.* 10:429–39
224. Kay BK, Williamson MP, Sudol M. 2000. *FASEB J.* 14:231–41
225. Dunn R, Hicke L. 2001. *Mol. Biol. Cell* 12:421–35
226. Gajewska B, Kaminska J, Jesionowska

A, Martin NC, Hopper AK, Zoladek T. 2001. *Genetics* 157:91–101
227. Kölling R, Losko S. 1997. *EMBO J.* 16:2251–61
228. Miller WE, Lefkowitz RJ. 2001. *Curr. Opin. Cell Biol.* 13:139–45
229. Shenoy SK, McDonald PH, Kohout TA, Lefkowitz RJ. 2001. *Science* 294:1307–13
230. Alarcon-Vargas D, Ronai Z. 2002. *Carcinogenesis* 23:541–47
231. Staub O, Abriel H, Plant P, Ishikawa T, Kanelis V, et al. 2000. *Kidney Int.* 57:809–15
232. Asher C, Chigaev A, Garty H. 2001. *Biochem. Biophys. Res. Commun.* 286: 1228–31
233. Lott JS, Coddington-Lawson SJ, Teesdale-Spittle PH, McDonald FJ. 2002. *Biochem. J.* 361:481–88
234. Dinudom A, Harvey KF, Komwatana P, Jolliffe CN, Young JA, et al. 2001. *J. Biol. Chem.* 276:13744–49
235. Cook DI, Dinudom A, Komwatana P, Kumar S, Young JA. 2002. *Cell Biochem. Biophys.* 36:105–13
236. Shimkets RA, Warnock DG, Bositis CM, Nelson-Williams C, Hansson JH, et al. 1994. *Cell* 79:407–14
237. Hansson JH, Nelson-Williams C, Suzuki H, Schild L, Shimkets R, et al. 1995. *Nat. Genet.* 11:76–82
238. Shi HK, Asher C, Chigaev A, Yung Y, Reuveny E, et al. 2002. *J. Biol. Chem.* 277:13539–47
239. Myat A, Henry P, McCabe V, Flintoft L, Rotin D, Tear G. 2002. *Neuron* 35: 447–59
240. Schwake M, Friedrich T, Jentsch TJ. 2001. *J. Biol. Chem.* 276:12049–54
241. Qiu L, Joazeiro C, Fang N, Wang HY, Elly C, et al. 2000. *J. Biol. Chem.* 275: 35734–37
242. De Wit R, Makkinje M, Boonstra J, Verkleij AJ, Post JA. 2001. *FASEB J.* 15:306–8
243. Stang E, Johannessen LE, Knardal SL, Madshus IH. 2000. *J. Biol. Chem.* 275: 13940–47
244. Burke P, Schooler K, Wiley HS. 2001. *Mol. Biol. Cell* 12:1897–910
245. de Melker AA, van der Horst G, Calafat J, Jansen H, Borst J. 2001. *J. Cell Sci.* 114:2167–78
246. Miyake S, Mullane-Robinson KP, Lill NL, Douillard P, Band H. 1999. *J. Biol. Chem.* 274:16619–28
247. Jehn BM, Dittert I, Beyer S, von der Mark K, Bielke W. 2002. *J. Biol. Chem.* 277:8033–40
248. Wang G, McCaffery JM, Wendland B, Dupre S, Haguenauer-Tsapis R, Huibregtse JM. 2001. *Mol. Cell. Biol.* 21:3564–75
249. Reggiori F, Pelham HR. 2002. *Nat. Cell Biol.* 4:117–23
250. Reggiori F, Black MW, Pelham HR. 2000. *Mol. Biol. Cell* 11:3737–49
251. Young P, Deveraux Q, Beal RE, Pickart CM, Rechsteiner M. 1998. *J. Biol. Chem.* 273:5461–67
252. Hofmann K, Falquet L. 2001. *Trends Biochem. Sci.* 26:347–50
253. Bilodeau PS, Urbanowski JL, Winistorfer SC, Piper RC. 2002. *Nat. Cell Biol.* 4:534–39
254. Lloyd TE, Atkinson R, Wu MN, Zhou Y, Pennetta G, Bellen HJ. 2002. *Cell* 108: 261–69
255. Polo S, Sigismund S, Faretta M, Guidi M, Capua MR, et al. 2002. *Nature* 416: 451–55
256. Shih SC, Katzmann DJ, Schnell JD, Sutanto M, Emr SD, Hicke L. 2002. *Nat. Cell Biol.* 4:389–93
257. Brett TJ, Traub LM, Fremont DH. 2002. *Structure* 10:797–809
258. Kalthoff C, Alves J, Urbanke C, Knorr R, Ungewickell EJ. 2001. *J. Biol. Chem.* 277:8209–16
259. Wilkinson CR, Seeger M, Hartmann-Petersen R, Stone M, Wallace M, et al. 2001. *Nat. Cell Biol.* 3:939–43
260. Futter CE, Collinson LM, Backer JM, Hopkins CR. 2001. *J. Cell Biol.* 155: 1251–64

261. Raiborg C, Bache KG, Mehlum A, Stang E, Stenmark H. 2001. *EMBO J.* 20:5008–21
262. Chin LS, Raynor MC, Wei X, Chen HQ, Li L. 2001. *J. Biol. Chem.* 276:7069–78
263. Komada M, Soriano P. 1999. *Genes Dev.* 13:1475–85
264. Garrus JE, von Schwedler UK, Pornillos OW, Morham SG, Zavitz KH, et al. 2001. *Cell* 107:55–65
265. Babst M, Odorizzi G, Estepa EJ, Emr SD. 2000. *Traffic* 1:248–58
266. Patnaik A, Chau V, Wills JW. 2000. *Proc. Natl. Acad. Sci. USA* 97:13069–74
267. Schubert U, Ott DE, Chertova EN, Welker R, Tessmer U, et al. 2000. *Proc. Natl. Acad. Sci. USA* 97:13057–62
268. Yasuda J, Hunter E, Nakao M, Shida H. 2002. *EMBO Rep.* 3:636–40
269. Strack B, Calistri A, Gottlinger HG. 2002. *J. Virol.* 76:5472–79
270. Pornillos O, Alam SL, Rich RL, Myszka DG, Davis DR, Sundquist WI. 2002. *EMBO J.* 21:2397–406
271. Klapisz E, Sorokina I, Lemeer S, Pijnenburg M, Verkleij AJ, van Bergen en Henegouwen PMP. 2002. *J. Biol. Chem.* 277: 30746–53
272. Oldham CE, Mohney RP, Miller SL, Hanes RN, O'Bryan JP. 2002. *Curr. Biol.* 12:1112–16
273. Katz M, Shtiegman K, Tal-Or P, Yakir L, Mosesson Y, et al. 2002. *Traffic* 3:740–51
274. Babst M, Katzmann D, Snyder W, Wendland B, Emr S. 2002. *Dev. Cell* 3:283–89
275. Bean AJ, Davanger S, Chou MF, Gerhardt B, Tsujimoto S, Chang Y. 2000. *J. Biol. Chem.* 275:15271–78
276. Cadavid AL, Ginzel A, Fischer JA. 2000. *Development* 127:1727–36
277. Chen X, Zhang B, Fischer JA. 2002. *Genes Dev.* 16:289–94

Annu. Rev. Biochem. 2003. 72:449–79
doi: 10.1146/annurev.biochem.72.121801.161520
Copyright © 2003 by Annual Reviews. All rights reserved
First published online as a Review in Advance on March 19, 2003

THE RNA POLYMERASE II CORE PROMOTER

Stephen T. Smale[1] and James T. Kadonaga[2]

[1]Howard Hughes Medical Institute and Department of Microbiology, Immunology, and Molecular Genetics, University of California, Los Angeles, California 90095-1662; email: smale@mednet.ucla.edu

[2]Section of Molecular Biology, University of California, San Diego, La Jolla, California 92093-0347; email: jkadonaga@ucsd.edu

Key Words gene regulation, transcription

■ **Abstract** The events leading to transcription of eukaryotic protein-coding genes culminate in the positioning of RNA polymerase II at the correct initiation site. The core promoter, which can extend ~35 bp upstream and/or downstream of this site, plays a central role in regulating initiation. Specific DNA elements within the core promoter bind the factors that nucleate the assembly of a functional preinitiation complex and integrate stimulatory and repressive signals from factors bound at distal sites. Although core promoter structure was originally thought to be invariant, a remarkable degree of diversity has become apparent. This article reviews the structural and functional diversity of the RNA polymerase II core promoter.

CONTENTS

INTRODUCTION

Transcription of a eukaryotic protein-coding gene is preceded by multiple events; these include decondensation of the locus, nucleosome remodeling, histone modifications, binding of transcriptional activators and coactivators to enhancers and promoters, and recruitment of the basal transcription machinery to the core promoter. The core promoter includes DNA elements that can extend ~35 bp upstream and/or downstream of the transcription initiation site. Most core promoter elements appear to interact directly with components of the basal transcription machinery. The basal machinery can be defined as the factors, including RNA polymerase II itself, that are minimally essential for transcription in vitro from an isolated core promoter. The vast majority of studies of the basal machinery have been performed with promoters containing a TATA box as an essential core element. A stable preinitiation complex can form in vitro on TATA-dependent core promoters by association of the basal factors in the following order: TFIID/TFIIA, TFIIB, RNA polymerase II/TFIIF, TFIIE, and then TFIIH. The properties of the basal factors and the mechanisms by which they stimulate transcription initiation from TATA-dependent promoters have been the subject of several recent reviews (1–8). The mechanisms by which sequence-specific transcription factors and coregulators influence the frequency of transcription initiation have also been reviewed (4, 9–13).

Although core promoters for RNA polymerase II were originally thought to be invariant, they have been found to possess considerable structural and functional diversity (14, 15). Furthermore, it appears that core promoter diversity makes an important contribution to the combinatorial regulation of gene expression (15, 16). In this article, we review the basic properties of the most common core elements and our current knowledge of the strategies by which sequence specific motifs in the core promoter participate in combinatorial regulation.

PROPERTIES OF RNA POLYMERASE II CORE PROMOTER ELEMENTS

TATA Box

The TATA box (also named the Goldberg-Hogness box after its discoverers) was the first core promoter element identified in eukaryotic protein-coding genes. The discovery of the TATA box in 1979 emerged from a comparison of the 5′ flanking sequences in a number of *Drosophila*, mammalian, and viral protein-coding genes (17, 18). In virtually every RNA polymerase II-transcribed gene examined, the sequence TATAAA was present 25 to 30 bp upstream of the transcription start site. The development of transfection and in vitro transcription assays made it possible to demonstrate that mutations in the TATA box usually reduced or abolished the activity of cellular and viral promoters (18–22). If transcription initiation from the mutant promoter remained detectable, the initiation sites were often displaced from the correct location. In *Saccharomyces cerevisiae*, TATA boxes were also found to be critical for transcription initiation; but in this organism, the element was located 40–120 bp from the start site [reviewed in (23)].

PREVALENCE Following the early studies, it was speculated that the TATA box might be strictly conserved and essential for transcription initiation from all protein-coding genes from yeast to man. However, as the promoters for more and more genes were sequenced and characterized, the prevalence of the TATA box diminished. Recent database analyses of *Drosophila* genes revealed that the TATAAA consensus sequence, or a sequence with one mismatch from the consensus, was present in 43% of 205 core promoters (24) or, in another study, in 33% of 1941 potential promoters (25). A database analysis of human genes revealed that TATA boxes were present in 32% of 1031 potential core promoters (26).

TATA RECOGNITION Studies from the Roeder and Parker labs provided the first evidence that a protein binds specifically to the TATA sequence and is responsible for TATA activity (27, 28). Roeder and colleagues identified a biochemical activity, transcription factor IID (TFIID), that elutes from a phosphocellulose column between 0.6 and 1 M NaCl (29). This activity was essential for the activity of TATA-containing core promoters and was capable of binding the core promoter from the adenovirus major late promoter in a DNase I footprinting assay (28, 29). However, further purification and cloning of TFIID proved to be unusually difficult because, in *Drosophila* and man, it appeared to be a heterogeneous, multiprotein complex (30).

The initial cloning of the TATA-binding protein (TBP) gene was facilitated by its discovery in *Saccharomyces cerevisiae* and by the demonstration that the *S. cerevisiae* protein could be purified as a single polypeptide rather than a large multiprotein complex (31). Peptide sequences obtained from the purified yeast protein led to the isolation of TBP cDNA clones from several eukaryotes [reviewed in (32)]. The *Drosophila* and human homologues of yeast TBP were found to be the TATA-binding subunits of the multisubunit TFIID complex (33–35). TBP was also found to be a component of distinct multisubunit complexes that contribute to transcription initiation by RNA polymerases I and III [reviewed in (36)]. TBP-associated factors (TAFs), which are components of the TFIID complex, have been identified and their genes cloned. The biochemical activities contributed by specific TAFs include core promoter recognition (see below), an acetyltransferase activity that uses histones and other proteins as substrates, a kinase activity, ubiquitin activating and conjugating activities, and coactivator functions conferred by protein-protein interactions with gene-specific transcription factors. An in-depth discussion of TAF structure and function is beyond the scope of this article, but it has been the topic of recent reviews (37–41).

Analyses of TBP-TATA cocrystals revealed a novel mechanism of DNA binding (32, 42–45). The DNA-binding region of TBP folds into a structure that resembles a saddle. This molecular saddle consists of two quasi-symmetrical domains, each containing 89–90 amino acids. The N-terminal domain contacts the 3′ half of a consensus TATA box, and the C-terminal domain contacts the 5′ half. Each half of the large concave surface of the saddle consists of a 5-stranded antiparallel β-sheet. Eight of the 10 β-strands contact the minor groove of the duplex DNA. TBP binding to the minor groove relies on extensive hydrophobic interactions. TBP also induces kinks in the DNA at both the 5′ and 3′ ends of the TATA box and partially unwinds the duplex due to the insertion of phenylalanine residues. The distorted DNA structure is restricted to the region that is directly contacted by TBP, as the flanking DNA duplex is largely unperturbed.

ROLE IN TRANSCRIPTION DIRECTIONALITY In the TBP-DNA cocrystals, TBP is bound in a polar manner to the asymmetrical TATA sequences, TATAAAAG and TATATAAA (42, 43). The polar binding of TBP can, in theory, lead to the assembly of a properly oriented preinitiation complex (containing RNA polymerase II and other general factors) and, therefore, can influence the direction of transcription. Indeed, the orientation of an asymmetrical TATA box has been shown to influence the direction of transcription in vitro from simple synthetic core promoters (46, 47). However, multiple lines of evidence suggest that the contribution of the TATA box to directionality in the context of native promoters may be minimal. First, although TBP binds a consensus TATA box in one orientation in the TBP-DNA cocrystals, it can bind in both orientations in solution with only a modest preference toward the correct orientation (48–50). Furthermore, in the context of more complex synthetic promoters or native promoters, the main determinant of directionality appears to be the relative locations of the activator binding sites, TATA box, and other core promoter elements (46, 47). When the orientation of a consensus, asymmetric TATA box was reversed in a promoter containing distal activator binding sites, the direction of transcription was not reversed; the strength of the promoter was merely reduced, due to the lower affinity of TBP binding in the reverse (nonpreferred) orientation. Recent studies have demonstrated that distal activators can indeed enhance the polarity of TBP binding, which may be a dominant mechanism in determining the direction of transcription (51).

TATA CONSENSUS SEQUENCE Consensus sequences for TATA function and TBP binding have been difficult to define. A binding site selection analysis identified the sequence 5′-TATATAAG-3′ as the optimal TBP recognition sequence (52). However, several other studies revealed that a wide variety of A/T-rich sequences can function as TATA boxes and can interact with TBP (53–57). TBP-DNA cocrystals have been prepared with 10 different TATA sequences to

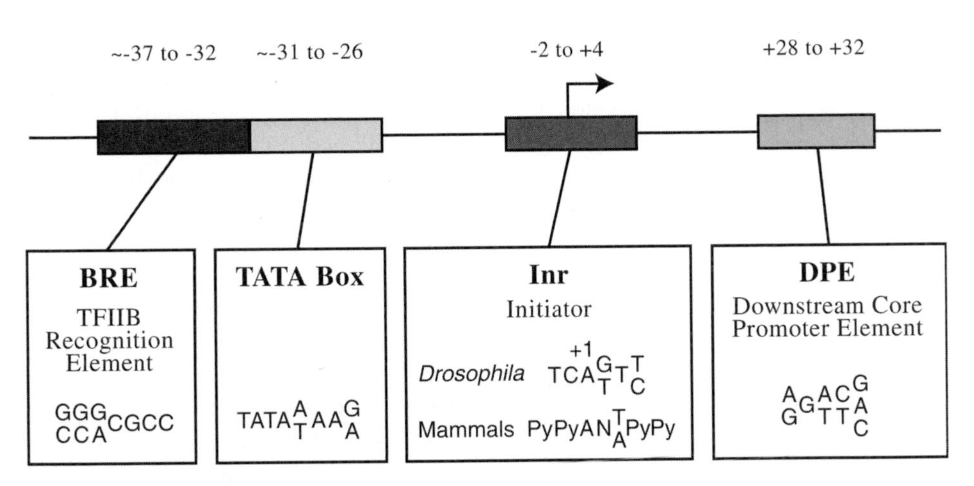

Figure 1 Core promoter motifs. This diagram depicts some of the sequence elements that can contribute to basal transcription from a core promoter. Each of these sequence motifs is found in only a subset of core promoters. A particular core promoter many contain some, all, or none of these elements. The TATA box can function in the absence of BRE, Inr, and DPE motifs. In contrast, the DPE motif requires the presence of an Inr. The BRE is located immediately upstream of a subset of TATA box motifs. The DPE consensus was determined with *Drosophila* core promoters. The Inr consensus is shown for both mammals and *Drosophila*.

examine the structural basis of the promiscuous binding of TBP (57). The results revealed that TBP can induce a similar conformational change in each of the TATA sequences examined. The structural results, combined with statistical data (58), revealed that C:G or G:C base pairs can be accommodated at all but three positions of the TATA box (positions 2, 4, and 5 of the sequence 5′-TATA-AAAG-3′) (57). The TATA definition that resulted from these structural studies was: $T \gg c > a \approx g/A \gg t/T \gg a \approx c/A \gg t/T \gg a/A \gg g > c \approx t/A \approx T > g > c/G \approx A > c \approx t$ (see Figure 1 for a more simplified version of this consensus). It is important to note, however, that this definition only predicts how well each nucleotide can be tolerated at each position, not whether TBP can bind a particular nucleotide sequence. This definition also does not take into account the possibility that TBP may be able to function without forming the stable, kinked structure that is observed in the TBP-TATA cocrystals (59, 60). This possibility may be particularly relevant in promoters that contain other strong, core elements (e.g., an Inr) (56, 61) (see below).

EVOLUTION OF THE TATA BOX Although most studies of TATA boxes have been performed in yeast, *Drosophila*, and man, analogous elements have been found in more ancient eukaryotes as well as in the archaea. In the promoters of several archaeal species, an 8-bp AT-rich sequence is located ~24 bp upstream of the transcription start site [reviewed in (62, 63)]. This sequence, originally called Box

A, is now known to interact with the archaeal homologue of TBP (64, 65). An X-ray crystal structure of TBP from an archaeal hyperthermophile, *Pyrococcus woesei*, revealed a saddle structure similar to that found in eukaryotic TBP (66). Interestingly, the DNA-binding activity of *P. woesei* TBP is optimal at high temperatures, consistent with the fact that this organism grows at 105°C (66). Another notable difference between the archaeal and eukaryotic TBPs is that the archaeal protein exhibits greater symmetry, both in its primary sequence and electrostatic charge distribution [reviewed in (49, 62)]. As discussed below (see BRE section), this increased symmetry decreases the protein's ability to bind TATA boxes in a polar manner.

TATA boxes, or AT-rich sequences located at a fixed distance upstream of the transcription start site, have been identified in essentially all animals, plants, and fungi that have been examined. In addition, TATA-like sequences are found in a number of the more recently evolved protists. For example, promoters in the protozoan parasite, *Entamoeba histolytica*, contain a sequence at −30 that matches the consensus GTATTTAAA(G/C) (67). Like its higher eukaryotic counterpart, this TATA-like element contributes to both promoter strength and start-site selection (68). However, despite the existence of TATA boxes and TBP in the archaea, TATA-like sequences are not apparent in many deep-branching eukaryotes, such as the most ancient parasitic protists [reviewed in (69)]. The absence of TATA boxes in these organisms is likely to reflect the divergence that occurred after these organisms branched from the main line of eukaryotic evolution (69). This divergence is also apparent in some protists that contain TATA boxes, such as *E. histolytica*. In addition to TATA-like and Inr-like elements, core promoters in this organism contain an unusual sequence matching the consensus GAACT (67). This element is found at variable locations within *E. histolytica* core promoters, yet it plays an important role in both promoter strength and start-site placement (68). Thus, although the studies of archaeal transcription originally suggested that core promoter structure would be highly conserved throughout the eukaryotic lineages, tremendous diversity is now apparent.

START-SITE SELECTION IN TATA-CONTAINING PROMOTERS The mechanism that determines the distance from the TATA box to the transcription start site has been the subject of a number of studies [reviewed in (70)]. A key finding emerged from an analysis of the basal factors responsible for the unusually long distance from the TATA box to the transcription start site in *S. cerevisiae* (generally 40–120 bp). By swapping basal factors purified from *S. cerevisiae* and *Schizosaccharomyces pombe*, TFIIB and RNA polymerase II were found to dictate this distance (71). That is, when *S. cerevisiae* TFIIB and RNA polymerase II were combined with the other *S. pombe* basal factors, transcription initiated 40–120 bp downstream of the TATA box. When *S. pombe* TFIIB and RNA polymerase II were combined with the other *S. cerevisiae* basal factors, transcription initiated 25–30 bp downstream of TATA. Studies of archaeal factors confirmed that its RNA polymerase and TFIIB homolog, TFB, measure the distance from the TATA box to the start site (72). TFIIB was further implicated in this process by the observation that mutations in its N-terminal charged cluster domain can shift the location of the start sites by a few nucleotides in yeast and mammalian promoters (73–75). RNA polymerase II mutations that alter the transcription start site have also been identified (76).

The precise mechanism by which TFIIB and RNA polymerase measure the TATA to start site distance is not known. Most of the TFIIB mutations that alter the start site have no effect on the interaction between TFIIB and the other basal factors with which it interacts, which include TBP, RNA polymerase II, and TFIIF (70, 77, 78). Instead, a recent study found that these mutations alter the conformation of TFIIB (70). One possibility is that this conformational change shifts the location of the RNA polymerase II catalytic center on the promoter. An alternative hypothesis is that the TFIIB mutation induces a conformational change in RNA polymerase II, which alters its position on the promoter (79).

Interestingly, electron crystallography of a yeast RNA polymerase II/TFIIB/TFIIE complex revealed that the distance between the TATA box and the active center of the polymerase is ~30 bp (80). Similarly, the melting of yeast promoter DNA has been found to begin ~20 bp downstream of the TATA box, similar to the distance observed in metazoans (81). These results suggest that, in yeast, the unusually large distance between the TATA box and start site may not be determined by the distance between TBP and the RNA polymerase II catalytic center in a stable transcription preinitiation complex. Rather, a scanning mechanism has been proposed in which the catalytic center is translocated further downstream after initially melting the DNA 20 bp downstream of the TATA box (81).

TBP-RELATED FACTORS Although TBP appears to be the major TATA-binding protein, multicellular animals from nematodes to humans express at least one additional TBP-related factor (TRF) or TBP-like factor (TLF) [reviewed in (82, 83)]. The first TRF, TRF1, was identified in *Drosophila* as a tissue-restricted regulatory protein (84). This protein has the potential to bind consensus TATA boxes, interact with TFIIA and TFIIB (the two basal factors that interact directly with TBP), and substitute for TBP in an in vitro transcription assay (85). In *Drosophila*, TRF1 is also the major component of the RNA polymerase III transcription factor, TFIIIB, whereas TBP is a component of this complex in other eukaryotes (86). Most recently, TRF1 was shown to bind preferentially the sequence TTTTCT, referred to as the TC box, within the core promoter of the *Drosophila tudor* gene (87). Microarray and chromatin immunoprecipitation assays provided further evidence that *tudor* is a direct target of TRF1 (87). TRF1 assembles into a multiprotein complex with associated factors that are distinct from those found in TFIID (85). Together, these findings suggest that TRF1 may be functionally similar to TBP and contribute to the activation of a specific subset of protein-coding genes.

Although TRF1 has been identified only in *Drosophila*, a second TRF, TRF2 (also known as TLF, TLP, and TRP), has been identified in several multicellular animals, but not in plants or fungi [reviewed in (82, 83)]. TRF2 is more distantly related to TBP than is TRF1, but it is likely to fold into a similar saddle structure. TFIIA and TFIIB interactions are retained in TRF2, but the phenylalanines, which (in TBP) are responsible for unwinding and kinking the DNA helix, are missing (82, 83). Consistent with the divergence in the DNA-binding surface, TRF2 cannot bind consensus TATA boxes.

To analyze the function of TRF2 in *Caenorhabditis elegans* early development, RNA interference was used to disrupt its expression (88, 89). The results revealed that TRF2 is essential for development and for the expression of a specific subset of genes. A 300-bp fragment of the *C. elegans* pes-10 promoter has been identified by expression and in situ studies as a direct target of TRF2 (88). In the absence of TRF2, a number of genes were also found to be aberrantly upregulated, which led to the proposal that TRF2 may be a negative regulator of transcription (89). More recently, an analysis of a purified TRF2 complex from *Drosophila* revealed that the transcription factor DREF (DNA replication-related element binding factor) is a TRF2-associated factor (90). A previously described DREF target gene, the PCNA gene, was found to require the DREF-TRF2 complex for the activity of one of its two promoters. Interestingly, an analysis of conserved motifs in 1941 *Drosophila* core promoters revealed that the DREF consensus site, the DRE, was among the most prevalent elements identified (25). However, unlike the TATA, Inr, and downstream promoter element (DPE) motifs, the DREs were not confined to a particular location relative to the transcription start site. These results suggest that DREF may target TRF2 to a subset of core promoters. Because TRF2 can interact with TFIIA and TFIIB, it may then promote the assembly of a productive preinitiation complex.

Initiator Element

The early comparisons of promoter sequences from efficiently transcribed protein-coding genes revealed that most contained an adenosine at the transcription start site (+1), a cytosine at the −1 position, and a few pyrimidines surrounding these nucleotides (91). A large deletion including this initiator region from a sea urchin histone H2A gene revealed that the efficiency of transcription was reduced and that the start-site locations became more heterogeneous (19). Subsequent studies of other metazoan TATA-containing promoters revealed similar results (91–95). In *S. cerevisiae*, where transcription does not initiate at a strictly defined distance from the TATA box, disruption of sequences in the vicinity of the transcription start site resulted in the use of alternative initiation sites (96–99). Collectively, these studies demonstrated that sequences in close proximity to the transcription start site contribute to accurate initiation and the strength of TATA-containing promoters.

The initiator element (Inr) was defined as a discrete core promoter element that is functionally similar to the TATA box and can function independently of a TATA box in an analysis of the lymphocyte-specific terminal transferase (TdT) promoter (100, 101). Transcription from this promoter initiates at a single start site, yet the region between −25 and −30 is G/C-rich and is unimportant for promoter activity. An extensive mutant analysis revealed that the sequence between −3 and +5 is necessary and sufficient for accurate transcription in vitro and in vivo (100, 102). This region matched the start site consensus sequence observed in the early studies (91). By itself, the TdT Inr supports a very low level of specific initiation by RNA polymerase II. In nuclear extracts, its activity is comparable to that of an isolated TATA box lacking an Inr at the start site (100, 103). The two elements function synergistically with one another when separated by 25 bp. Most importantly, when an Inr is inserted into a synthetic promoter downstream of six binding sites for transcription factor Sp1 (in the absence of a TATA box), the Inr supports high levels of transcription that initiate at a specific start site within the Inr. When the Inr is inserted at a different location relative to Sp1 sites, RNA synthesis consistently begins at the nucleotide dictated by the Inr. In the absence of the Inr, transcription begins from heterogeneous start sites at much lower frequencies.

INR CONSENSUS SEQUENCE AND PREVALENCE Analyses of randomly generated and specifically targeted Inr mutants by in vitro transcription and transient transfection using human cell lines led to the functional consensus sequence Py Py A(+1) N T/A Py Py (102, 104) (Figure 1). The functional consensus defined in *Drosophila* is virtually identical (104). Only a subset of the pyrimidines at the −2, +4, and +5 positions appears to be essential for Inr activity, but the activity increases with increasing numbers of pyrimidines in these positions (102, 104). This functional consensus is similar, but not identical, to the mammalian and *Drosophila* Inr consensus sequences determined by database analysis: Py C A(+1) N T Py Py in mammals (58, 91) and T C A(+1) G/T T Py in *Drosophila* (24, 25, 105, 106) (Figure 1). The *Drosophila* consensus, or a sequence containing one mismatch, was present in 69% of 205 core promoters or, in a separate study, in 69% of 1941 core promoters (24, 25). To our knowledge, the prevalence of the Inr in mammalian promoters has not been determined using rigorous methods similar to those used to calculate its prevalence in *Drosophila*.

Interestingly, transcription does not need to begin at the +1 nucleotide for the Inr to function. RNA polymerase II has been redirected to alternative start sites by reducing ATP concentrations within a nuclear extract, by altering the spacing between the TATA and Inr in a promoter containing both elements, and by dinucleotide initiation strategies (47, 107, 108). In all of these studies, the Inr continued to increase the efficiency of transcription initiation from the alternative sites.

TATA-INR SPACING AND TRANSCRIPTION DIRECTIONALITY Studies of TATA-Inr spacing have shown that the two elements act synergistically when separated by 25–30 bp but act independently when separated by more than 30 bp (47). When separated by 15 or 20 bp, synergy is retained, but the location of the start site is dictated by the location of the TATA box rather than the location of the Inr (i.e., initiation occurs 25 bp downstream of TATA) (47). Although the Inr element is not symmetrical and therefore has the potential to dictate the direction of transcription from simple synthetic promoters, its contribution to directionality from native promoters and more complex synthetic promoters appears to be minimal (47). Similar to the results obtained with the TATA box, reversal of an Inr in the context of a synthetic promoter containing distal activator binding sites reduced promoter strength, but it did not reverse the direction of transcription.

INR RECOGNITION Most studies of Inr recognition have focused on recognition by TFIID (109). The potential for TFIID recognition of the Inr was suggested by the finding that the TFIID complex is essential for Inr activity (i.e., TBP is insufficient) and that TFIID footprints on some TATA-containing promoters, such as the AdML and *Drosophila hsp70* promoters, extend downstream of the TATA box to approximately +40 (27, 28, 35, 103, 110–113). Although the contacts at the start-site and downstream regions of the AdML promoter were subsequently found to be independent of the Inr (114), DNase I footprinting studies with highly purified human TFIID and simple synthetic promoters containing only TATA and Inr elements revealed weak TFIID interactions at the Inr (115). Analysis of probes containing Inr mutations revealed a close correlation between the Inr contacts and the nucleotides required for Inr function. In the presence of a strong activator bound to distal sites, a much stronger DNase I footprint extending from −35 to +30 was detected on synthetic promoters containing both TATA and Inr elements (114). When the Inr was disrupted by a point mutation, a much weaker footprint was detected that was confined to the TATA box. Disruption of the TATA box eliminated the entire footprint, with the exception of an enhanced DNase I cleavage site at the Inr. Cooperative binding of TFIID was also disrupted by increasing or decreasing the TATA-Inr spacing by 5 bp (114). An independent study used a binding site selection assay to define the nucleotides near the transcription start site of the *Drosophila hsp70* promoter that are required for optimal TFIID binding (116). The selected sequences matched the functional Inr consensus sequence, which provided strong evidence that the Inr consensus sequence is a TFIID recognition site.

Several studies have confirmed that TFIID specifically interacts with the Inr (117–120). Notably, an analysis of TFIID binding to core promoters containing both Inr and DPE elements (see below) revealed that the Inr is essential for stable TFIID binding (120). Although stable binding of the intact TFIID complex to an isolated Inr lacking an upstream TATA box or downstream DPE has not been reported, Verrijzer and colleagues have detected stable Inr binding by a complex consisting of two TAFs, $TAF_{II}250$ and $TAF_{II}150$ (121) [TAFs 1 and 2 in the new TAF nomenclature (122)]. In this study, the Inr consensus was identified in a binding site selection analysis as the DNA sequence preferred by the $TAF_{II}150$-$TAF_{II}250$ complex.

The domains of $TAF_{II}150$ and $TAF_{II}250$ that are responsible for Inr recognition have not been determined. However, their involvement is consistent with functional studies that have shown that a trimeric TBP-$TAF_{II}250$-$TAF_{II}150$ complex is sufficient for Inr activity in reconstituted transcription assays (118). $TAF_{II}150$ was also implicated in Inr activity by the finding that *Drosophila* $TAF_{II}150$ possesses a core promoter-binding activity (123) and that human $TAF_{II}150$ corresponds to a biochemical activity in HeLa cell extracts that is required for Inr function (124, 125).

The synergistic function of TATA and Inr elements was found to correlate with the cooperative binding of TFIID to the two elements (114). The general transcription factor TFIIA was found to be critical for the cooperative binding of TFIID to the Inr element (114). This observation is consistent with earlier evidence that TFIIA can induce a conformational change in the TFIID complex, which alters its contacts with DNA in the vicinity and downstream of the transcription start site (126–128). In a separate study, crosslinking of $TAF_{II}250$ to the Inr of the AdML promoter was greatly enhanced in the presence of TFIIA, presumably due to a TFIID conformational change; this provided further support for the importance of both $TAF_{II}250$ and TFIIA for Inr function (129).

TBP BINDING TO TATA-LESS PROMOTERS Given that the TFIID complex contains subunits that recognize both TATA and Inr elements, a key question is whether the TBP subunit of TFIID must contact the −30 region of TATA-less promoters that contain only an Inr. One study of synthetic core promoters suggested that a TBP contact is necessary because the strength of synthetic Inr-containing promoters was found to be roughly proportional to the A/T content of the −30 region (56). That is, A/T nucleotides at the −30 region had a profound influence on promoter strength even if they had little resemblance to the TATA consensus sequence and were unable to interact stably with TBP. A more definitive analysis of this issue made use of a TFIID complex containing a mutant TBP subunit that cannot bind DNA (61). Two promoters that contain Inr elements but lack TATA boxes (the β-polymerase and TdT promoters) were tested. The mutant TFIID protein was inactive on the β-polymerase promoter, but it remained active on the TdT promoter, which suggested that at least some promoters can function in the absence of a TBP-DNA contact.

The implication of the studies described above, that TBP can contact sequences that have minimal similarity to the TATA consensus sequence, seemed surprising in light of the sequence requirements for the formation of the stable TBP-TATA saddle structure (57). However, a potential explanation was provided by studies showing that stable TBP binding occurs in two steps (59, 60). Possibly, at Inr-containing promoters that require a TBP interaction with a TATA-less −30 sequence, TBP only forms the unstable unbent protein-DNA complex.

OTHER INR-BINDING PROTEINS In addition to TFIID, three other proteins have been reported to recognize Inr elements: RNA polymerase II, TFII-I, and YY-1. Purified RNA polymerase II initiates transcription inefficiently from Inr elements in the absence of other basal transcription factors (130). Although RNA polymerase II is generally thought to recognize DNA nonspecifically, these results suggest that it possesses a weak, intrinsic preference for Inr-like sequences. In the absence of TAFs, RNA polymerase II, TBP, TFIIB, and TFIIF can form a stable complex on TATA-less promoters that contain Inr elements (130, 131). A subset of the nucleotides within the Inr consensus sequence is required for polymerase recognition and complex formation, but it is not known whether the preferred recognition sequence matches the Inr consensus (131). One hypothesis consistent with the recognition of Inr elements by both TFIID and RNA polymerase II is that TFIID initially recognizes the Inr during preinitiation complex formation. When the polymerase is recruited to the promoter, its intrinsic preference for the Inr may contribute to its proper positioning.

TFII-I was discovered as a factor capable of binding the Inr element within the AdML promoter (132, 133). The structure of TFII-I is complicated, with an unusual DNA-binding domain and six helix-loop-helix motifs that support both homomeric and heteromeric interactions (134, 135). Its ability to stimulate transcription in vitro from Inr-containing promoters and to promote the assembly of a preinitiation complex in an Inr-dependent manner supported the hypothesis that it is a general transcription factor dedicated to the recognition of Inr elements (132, 133). Subsequent evidence revealed that TFII-I is identical to BAP-135, a major phosphorylation target of Bruton's tyrosine kinase (136); SPIN, a protein that binds distal elements in the c-fos promoter and stabilizes the binding of serum response factor and Phox 1 to the c-fos promoter (137, 138); and endoplasmic reticulum stress response element binding factor, which induces the transcription of glucose-regulated protein genes by binding distal promoter elements (139). Furthermore, immunodepletion studies demonstrated that TFII-I is not required for the in vitro function of consensus Inr elements (140). Although several lines of evidence suggest that TFII-I contributes to the function of the Inr element in the T cell receptor Vβ 5.2 promoter, its complicated structure has made it difficult to determine its precise role (141–143).

YY-1 was discovered as a C2H2 zinc finger protein that binds a distal element in the adeno-associated virus (AAV) P5 promoter (144). YY-1 represses transcription through this element in the absence of the adenovirus E1A protein, but it activates transcription in the presence of E1A (144). A subsequent study provided evidence that YY-1 can activate the AAV P5 core promoter by binding its consensus Inr element (145). YY-1 interacts directly with both TFIIB and RNA polymerase II, and the three proteins were found to be sufficient for transcription from the AAV P5 Inr (146, 147). Although these data provide strong evidence that YY-1 can stimulate Inr-dependent transcription in vitro in an assay reconstituted with pure proteins, its contribution to Inr activity in vivo and in nuclear extracts is less certain. Specifically, a mutation at the +2 position of the AAV P5 Inr was found to abolish YY-1 binding, but it had no effect on the activity of the native P5 promoter or the activity of a synthetic promoter containing the P5 Inr (102, 104). In contrast, a mutation at the +3 position reduced promoter activity but had little effect on YY-1 binding (102, 104). A separate study found that YY-1 does not contribute to the function of the consensus Inr within the β-polymerase promoter, which binds YY-1 with high affinity (131). The precise roles of sequence-specific DNA-binding proteins like TFII-I and YY-1 in Inr function will require further exploration. However, from a more general perspective, it would not be surprising to find that Inr elements in specific genes are recognized by factors other than TFIID and RNA polymerase II.

EVOLUTION OF THE INR Although most studies of Inr elements have been performed in *Drosophila* and man, sequences in the vicinity of the start site contribute to start-site placement and promoter activity in many organisms. In archaeal promoters, the transcription start site usually contains a purine preceded by a pyrimidine [reviewed in (62, 63)]. All archaeal promoters that have been studied require an upstream TATA box for their activity; start-site mutations, however, often reduce promoter strength or shift the transcription start site to other nearby purines. Because archaeal TBP, TFB, and RNA polymerase are sufficient for archaeal transcription, the archaeal Inr is probably recognized by the RNA polymerase (63, 148).

In *S. cerevisiae* promoters, the precise locations of transcription start sites are determined by the DNA sequences in their immediate vicinity (96–99). A positioning sequence is necessary because, as mentioned above, transcription initiation by yeast RNA polymerase II begins 40–120 bp downstream of the TATA box. The sequences observed in the vicinity of yeast transcription start sites often match the consensus PuPuPyPuPu (149). Mutation of this sequence results in repositioning of the start site, but promoter strength is often unaffected or reduced only modestly (149). As in the archaea, it has been suggested that all yeast promoters rely on the presence of a TATA box (150). These properties are quite different from those observed in metazoans, where an Inr often functions in the absence of a TATA box and can greatly enhance promoter strength. Thus, the yeast Inr may be a relatively passive contributor to promoter activity. Rather than being a major contributor to core promoter recognition, it may represent a preferred initiation site for the RNA polymerase after it scans downstream from its initial interaction site, which is 20 bp downstream of the TATA box (81), see above. It is noteworthy, however, that one gene in *S. cerevisiae*, the *GAL80* gene, contains a functional Inr sequence, CACT, that exhibits greater similarity to the metazoan Inr consensus and appears to function in a TATA-independent manner (151, 152).

Similar to yeast and the archaea, start-site sequences that diverge considerably from the metazoan Inr have been identified in several protists, which include *E. histolytica* and *Giardia lamblia* (67, 68, 153, 154). For example, in *E. histolytica*, the Inr consensus sequence is AAAAATTCA (67). Interestingly, the start-site sequences in one of the most ancient eukaryotes, the parasitic protist *Trichomonas vaginalis*, conform closely to the metazoan Inr consensus (155, 156).

Although the *T. vaginalis* genome is AT-rich, functional TATA boxes have not been detected. Rather, promoter activity and the location of the start site are strongly dependent on the *T. vaginalis* Inr (69, 156). It is not yet known whether the apparent absence of TATA boxes in this organism reflects the loss of the TBP gene after branching from the main line of eukaryotic descent, or whether a TBP exists that contributes to transcription initiation from the TATA-less promoters of *T. vaginalis*.

Downstream Promoter Element

The DPE was identified as a downstream core promoter motif that is required for the binding of purified TFIID to a subset of TATA-less promoters [reviewed in (15, 157, 158)]. The DPE is conserved from *Drosophila* to humans and is typically but not exclusively found in TATA-less promoters (24). The DPE acts in conjunction with the Inr, and the core sequence of the DPE is located at precisely +28 to +32 relative to the A_{+1} nucleotide in the Inr motif (24). The DPE consensus sequence is shown in Figure 1.

BINDING OF TFIID TO DPE-DEPENDENT CORE PROMOTERS A typical DPE-dependent promoter contains an Inr and a DPE. Mutation of either the DPE or the Inr results in a loss of TFIID binding and basal transcription activity (120). Hence, TFIID binds cooperatively to the DPE and Inr motifs. In addition, a single nucleotide increase or decrease in the spacing between the DPE and Inr results in a several-fold decrease in TFIID binding and transcriptional activity (24). Consistent with this strict Inr-DPE spacing requirement, all of the ~18 characterized DPE-dependent promoters in *Drosophila* possess the identical spacing between the Inr and DPE sequence motifs (24, 120, 159). Thus, the DPE and Inr function together as a single core promoter unit. In this respect, the DPE differs from the TATA box, which is able to function independently of the presence of an Inr.

DNase I footprinting analysis of the binding of purified TFIID to DPE-dependent promoters revealed an extended region of protection (from about −10 to about +35) that encompasses the Inr and DPE motifs (24, 120). The pattern of DNase I protection and hypersensitivity is consistent with the close association of the DNA in a specific orientation along the surface of TFIID from the Inr to the DPE (24, 120). TBP alone does not bind to TATA-less, DPE-dependent promoters. Photocrosslinking studies with purified TFIID indicated that $TAF_{II}60$ and $TAF_{II}40$ [which are designated TAF6 and TAF9 in the new TAF nomenclature (122)] are in close proximity to the DPE (159). Genetic analyses of $TAF_{II}60$ and $TAF_{II}40$ have been carried out in *Drosophila* (160, 161). These studies revealed that $TAF_{II}60$ and $TAF_{II}40$ are encoded by essential genes. Mutations that alter the amino acid sequences of $TAF_{II}60$ and $TAF_{II}40$ have been found to increase as well as to decrease the expression of DPE- or putative-DPE-containing genes, but it is not yet known whether these mutations affect the function of TFIID at DPE-dependent promoters (160, 161).

THE DPE SEQUENCE MOTIF The DPE was initially identified in the *Drosophila Antennapedia* P2 and *jockey* core promoters, and therefore, the early attempts to determine a DPE consensus sequence were biased toward the DPE sequence motifs in those two promoters. To circumvent this bias, core promoter libraries with randomized sequences in the DPE (or in the vicinity of the DPE) were generated and subjected to in vitro transcription analysis. These studies led to the identification of a range of sequences that can function as a DPE motif (24). This information was then used to identify putative DPE-dependent core promoters in *Drosophila*. In a database of 205 core promoters, it was estimated that about 29% contain a TATA box and no DPE, 26% possess a DPE but no TATA, 14% contain both TATA and DPE motifs, and 31% do not appear to contain either a TATA or a DPE. This analysis further identified nonrandom sequences outside of the DPE core motif (+28 to +32). For instance, a G nucleotide is overrepresented at position +24, and the presence of a G at +24 results in a two- to fourfold higher level of basal transcription (24). Thus, the complete DPE appears to consist of the core motif along with other preferred nucleotides, such as G_{+24}, between the Inr and DPE core. The DPE is also present in human core promoters (159, 162). The analysis of the DPE consensus in humans suggests that it is similar but not identical to that in *Drosophila* (159, Alan K. Kutach, Scott M. Iyama, and J.T. Kadonaga, unpublished data). The frequency of occurrence of the DPE motif in the human genome remains to be determined.

THE DPE VERSUS THE TATA BOX There are similarities and differences between the DPE and TATA box. For example, both the DPE and TATA box are recognition sites for the binding of TFIID. On the other hand, the TATA box, but not the DPE, can function independently of an Inr. If a TATA-dependent promoter is inactivated by mutation of the TATA motif, then core promoter activity can be restored by the addition of a DPE at its downstream position (120).

A key difference between TATA- versus DPE-dependent transcription was revealed by the identification of an activity that stimulates DPE-dependent transcription and represses TATA-dependent transcription (163). This activity was purified and found to be mediated by NC2/Dr1-Drap1, which was initially identified as a repressor of TATA-dependent transcription (164). The observation that NC2/Dr1-Drap1 activates DPE-dependent transcription indicates that it functions differently at DPE- and TATA-dependent promoters. In addition, a mutant form of NC2/Dr1-Drap1 was found to activate DPE transcription but not to repress TATA transcription. Hence, the ability of NC2/Dr1-Drap1 to activate DPE transcription is distinct from its ability to repress TATA transcription. These findings indicate that NC2/Dr1-Drap1 is a multifunctional factor that can discriminate between DPE- and TATA-dependent core promoters.

TFIIB Recognition Element

The TFIIB recognition element (BRE) is the only well-characterized element in the core promoters of protein-coding genes that is recognized by a factor other than TFIID (or the TRF1 and TRF2 complexes). Initial evidence that a functionally significant element exists immediately upstream of some TATA boxes was provided by mutant analyses of archaeal promoters (165–168). The X-ray crystal structure of a TBP-TFIIB-TATA ternary complex subsequently revealed that TFIIB interacts with the major groove upstream of the TATA box and with the minor groove downstream of the TATA box (169). Protein-DNA crosslinking studies confirmed that TFIIB is in close proximity to the upstream sequences (170, 171).

Compelling evidence that TFIIB interacts with DNA in a sequence-specific manner emerged from two studies, one involving an analysis of the T6 promoter from the archaeal *Sulfolobus shibatae* virus (172). As in other archaeal promoters, a sequence immediately upstream of the TATA box was found to be critical for promoter activity. Archaeal TBP and TFB (the archaeal homolog of TFIIB) bound cooperatively to the promoter when both the TATA box and upstream element were present. A binding site selection analysis revealed no sequence preferences in this upstream region when TBP was analyzed alone. However, in the presence of TFB, strong preferences for purines were observed 3 and 6 bp upstream of the TATA box, with weaker nucleotide preferences at other positions. A parallel study with human TFIIB established the existence of a eukaryotic BRE that prefers a 7-bp sequence: G/C G/C G/A C G C C (173). Recognition of the BRE was found to be mediated by a helix-turn-helix motif at the C-terminus of TFIIB (169, 173, 174). Interestingly, this motif is missing in yeast and plants, which suggests that the BRE may not contribute to gene regulation in these organisms.

As mentioned above, one difference between archaeal TBP and eukaryotic TBP is that the two halves of the archaeal DNA-binding domain exhibit greater symmetry and therefore are incapable of binding most TATA boxes in a polar manner [reviewed in (49, 62)]. Although the relative locations of distal activator proteins and the TATA box appear to be the primary determinants of transcription directionality in eukaryotes, this mechanism may not function in the archaea due to the relative simplicity of the promoter. A series of elegant structural and biochemical studies has shown that archaeal TFB facilitates the polar binding of TBP to the TATA box via its interaction with the BRE (174–176).

Although the interaction between the archaeal TFB and BRE clearly enhances the assembly of a preinitiation complex and transcription initiation, the function of the human TFIIB-BRE interaction appears to be very different. This interaction was originally reported to stimulate RNA polymerase II transcription in an in vitro assay reconstituted with purified basal factors (173). However, it was also observed that the BRE is a repressor of basal transcription in vitro with crude nuclear extracts as well as in vivo in transfection assays (177). This repression and the TFIIB-BRE interaction were relieved when transcriptional activators were bound to distal sites, which resulted in an increased amplitude of transcriptional activation. These results suggest that the function of the BRE may have expanded during evolution. In the archaea, it stimulates promoter activity, but in eukaryotes, it may also repress transcription.

Proximal Sequence Element

Although this article focuses on core elements found in the promoters of eukaryotic protein-coding genes, a brief description of a well-characterized and highly conserved element within the core promoters of small nuclear RNA (snRNA) genes is pertinent. This element, the proximal sequence element (PSE), is located between −45 and −60 relative to the transcription start site of snRNA genes (178–183). The PSE is essential for basal transcription and dictates the location of the transcription start site. The PSE consensus sequence has been found to vary among organisms (178, 180). In humans, the consensus is T C A C C N T N A C/G T N A A A A G T/G (180).

One of the most intriguing features of the PSE is that, within a single organism, it supports transcription initiation by RNA polymerase II at some snRNA genes and by RNA polymerase III at a distinct subset of snRNA genes (178, 180–183). The core promoter sequences that determine which polymerase transcribes a given snRNA gene vary among organisms. In humans, the presence of a TATA box 15–20 nucleotides downstream of the PSE leads to the recruitment of RNA polymerase III to the promoter, whereas RNA polymerase II transcribes snRNA promoters containing a PSE in the absence of a TATA box (184, 185).

The PSE is recognized by a unique multiprotein complex called SNAPc, PBP, or PTF (181–183, 186–188). The human SNAPc complex is essential for the activity of both RNA polymerase II- and RNA polymerase III-transcribed snRNA promoters (189). The smallest subassembly of SNAPc that can bind the PSE in a sequence-specific manner consists of SNAP190 (residues 84–505), SNAP43 (residues 1–268), and SNAP50 (190). The SNAPc-PSE interaction is mediated, in part, by an atypical Myb domain within SNAP190 (190, 191). SNAPc-directed transcription also requires TBP, which binds DNA in snRNA promoters that contain a TATA box in addition to a PSE (181–183).

Other Core Promoter Elements

There are likely to be a variety of other DNA sequence elements that contribute to core promoter activity. In the analysis of a new putative core promoter motif, it is important to consider its effect upon the basal transcription process, such as the binding of TFIID or TFIIB to the core promoter. It is also useful to examine the frequency of occurrence of the sequence motif—that is, to determine whether the element is used in multiple core promoters. Another related consideration is the specific location of the motif in the core promoter. For instance, the BRE, TATA, Inr, and DPE are located at distinct positions relative to the +1 start site, which is consistent with their roles in the assembly of the transcription initiation complex. It is possible, however, that other core promoter elements with slightly different functions may be located at variable positions relative to the start site.

It will be important to extend the analysis of core promoter motifs. To understand the full range of mechanisms that are used in the basal transcription process, it is ultimately necessary to identify and to characterize all of the sequence elements that contribute to core promoter activity. In fact, in addition to the motifs described above, a number of core promoter sequences have been found to contribute to transcriptional activity. Some examples of these sequences are as follows (note that the downstream sequences in these promoters appear to be distinct from the DPE). First, the downstream core element (DCE) was identified in the human β-globin promoter (192). The DCE is located from +10 to +45, and it was observed to contribute to transcriptional activity and binding of TFIID. Second, in the human glial fibrillary acidic protein (*gfa*) gene, a downstream promoter element (from +11 to +50) was found to be required for TFIID binding and transcriptional activity (193, 194). Third, the multiple start site downstream element (MED-1) was identified in TATA-less promoters that have unclustered, multiple start sites (195). The MED-1 element was observed to contribute to transcriptional activity in two of three promoters tested (195–197). In the future, it will useful to study these elements further as well as to identify additional core promoter motifs.

CpG Islands

The CpG dinucleotide, a DNA methyltransferase substrate, is underrepresented in the genomes of many vertebrates because 5-methylcytosine can undergo deamination to form thymine, which is not repaired by DNA repair enzymes (198). However, 0.5–2 kbp stretches of DNA exist that possess a relatively high density of CpG dinucleotides. The human genome contains ~29,000 of these CpG islands. Most importantly, it has been estimated that, in mammals, CpG islands are associated with approximately half of the promoters for protein-coding genes (26, 199). During early mammalian development, DNA methylation decreases substantially throughout the genome, followed by de novo methylation to normal levels prior to implantation (198). CpG islands are largely excluded from this phase of de novo methylation, and most remain unmethylated in all tissues and at all stages of development.

Despite the prevalence of promoters associated with CpG islands, the elements that are responsible for their core promoter function remain poorly defined. CpG islands usually lack consensus or near-consensus TATA boxes, DPE elements, or Inr elements (14, 200). In addition, they are often characterized by the presence of multiple transcription start sites that span a region of 100 bp or more. The transcription start sites can coincide with sequences exhibiting weak homology to the Inr consensus or can be unrelated to this sequence. Mutations in the vicinity of the start site can lead to the use of alternative start sites, but promoter strength is often unaffected. In general, it has been difficult to identify core promoter elements within CpG islands that are essential for promoter function. One common feature of CpG islands is the presence of multiple binding sites for transcription factor Sp1 (200–

202). Transcription start sites are often located 40–80 bp downstream of the Sp1 sites; this suggests that Sp1 may direct the basal machinery to form a preinitiation complex within a loosely defined window (14, 103, 200). One possibility is that TFIID subunits that are capable of core promoter recognition (i.e., TBP, $TAF_{II}150$/$TAF_{II}250$, and $TAF_{II}40/60$) then interact with the sequences within that window that are most compatible with their DNA recognition motifs. According to this hypothesis, core promoter recognition within CpG islands relies on the same factors and elements as were discussed above. The key difference is that the binding of basal factors is more strongly dependent on recruitment by activator proteins bound to distal promoter elements.

CONTRIBUTIONS TO COMBINATORIAL GENE REGULATION

The precise reasons for the diversity of eukaryotic core promoters remain largely unknown. Different core promoter classes (e.g., TATA, TATA-Inr, Inr, Inr-DPE) may have evolved initially as functionally equivalent recognition sites for TFIID subunits and their evolutionary precursors. At some eukaryotic promoters, the core elements may continue to serve as functionally equivalent and interchangeable recognition sites for TFIID. However, at other promoters, dramatic differences have been identified, which suggest that core elements can make significant contributions to combinatorial gene regulation strategies. The principle underlying combinatorial regulation is that the limited number of transcription factors within an organism can support a much larger number of gene expression patterns if activation of each gene requires the concerted action of multiple factors. If transcription factors bound to promoter or enhancer elements were capable of activating transcription only when the core promoter contains a specific element or combination of elements, a larger number of gene expression patterns could be obtained. Specific examples of selective communication between core promoter and transcription factors or transcriptional control regions are discussed below, along with hypothetical benefits of selective communication.

Functionally Distinct TATA Sequences

An analysis of the *S. cerevisiae his3* promoter provided the first evidence that core elements can be functionally distinct (203, 204). This promoter contains two TATA boxes, a downstream consensus TATA box (T_R) and an upstream AT-rich sequence (T_C), which functions as a nonconsensus TATA box (205). T_C supports weak transcription of the *his3* gene under all growth conditions, whereas T_R supports strong transcription after the inducible activators GAL4 or GCN4 bind to upstream activating sequences (206). Thus, proper regulation of *his3* transcription appears to rely on the presence of functionally distinct TATA boxes. One benefit of restricting activation by GCN4 and GAL4 to the consensus TATA box (T_R) is that expression of a nearby, divergently transcribed gene, *pet56*, is unaltered when these factors are induced, because the *pet56* core promoter contains a weak, nonconsensus TATA box (204).

TATA sequences that respond differently to transcriptional activators have also been observed in mammalian cells (207–209). For example, the adenovirus E1A protein stimulates transcription from an *hsp70* promoter regulated by its own consensus TATA box, but E1A cannot stimulate transcription from an *hsp70* promoter containing a nonconsensus TATA box from the SV40 early promoter (207). Similarly, an enhancer associated with the myoglobin gene can activate transcription from the myoglobin promoter, which contains a consensus TATA box, or from an SV40 early promoter after insertion of a consensus TATA box (208). However, the same enhancer cannot activate transcription from an SV40 promoter containing its own nonconsensus TATA box. Although these mammalian studies did not address the benefits of the selective communication, potential benefits were suggested by the following examples.

Restricting the Stimulatory Capacity of Enhancers

Transcriptional enhancers are often located at a considerable distance from their relevant target promoters. Because several enhancers may be in the vicinity of a given promoter and because several promoters may be in the vicinity of a given enhancer, strategies that limit activation by an enhancer to a given promoter could be of great benefit. Indeed, selective communication between enhancers and promoters has been well documented (210, 211).

The myoglobin enhancer analysis cited above and two studies performed in *Drosophila* suggest that core promoter diversity may be critical for selective communication between enhancers and promoters. One study began with the observation that the AE1 enhancer of the *Drosophila* Hox gene cluster is located between the *fushi tarazu* (*ftz*) and *Sex combs reduced* (*Scr*) promoters, but it stimulates transcription only from the *ftz* promoter (212). The *ftz* promoter contains a TATA box, and the *Scr* promoter contains Inr and DPE elements. This observation led to the hypothesis that AE1 preferentially stimulates TATA-containing promoters. Analysis of a series of transgenic *Drosophila* lines demonstrated that AE1 prefers to stimulate transcription from promoters containing TATA boxes. The TATA preference was dependent on competition between the two promoter classes, as AE1 activated the Inr-DPE promoter when the nearby TATA promoter was compromised.

In the second study, P-element-mediated transformation and an enhancer trap strategy were used to introduce a promoter-reporter cassette into various locations of the *Drosophila* genome (213). The promoter-reporter cassette contained both TATA-Inr and Inr-DPE core promoters, with each core promoter linked to a green fluorescent protein (GFP) reporter gene. After integration into the genome, the TATA-Inr promoter-reporter or the Inr-DPE promoter-reporter was deleted by recombination with the FLP or Cre recombinase, respectively, resulting in matched sublines with the different core promoters in identical genomic locations. Out of 18 enhancers tested, 3 were specific for the DPE-dependent core promoter, whereas 1 was specific for the TATA-dependent core promoter. Moreover, primer extension analysis revealed that there was no detectable activation of the TATA-dependent promoters at the DPE-specific integration sites. The differential activities of TATA-Inr and Inr-DPE promoters when analyzed at identical genomic locations provide evidence that the two types of core promoters exist, at least in part, for the purpose of mediating selective enhancer function.

Restricting Activation by Members of Large Protein Families

In metazoans, most sequence-specific DNA binding proteins are members of large protein families. Multiple family members often recognize similar DNA sequences. Strategies are therefore needed to restrict the activity of a given family member to its relevant target promoters. Selective communication with the core promoter is one such strategy that can be envisioned. Initial support for this hypothesis has been provided by an analysis of the murine terminal transferase (TdT) promoter, which contains a consensus Inr at the transcription start site. The −25 to −30 region of this promoter is G/C-rich and unimportant for promoter function (100, 214). Nevertheless, because the TFIID complex is required for the in vitro activity of this promoter (61), it was anticipated that an engineered TATA box at −30 would effectively substitute for the Inr. Surprisingly, promoter activity was severely compromised when a consensus TATA box was inserted and the Inr disrupted (214). These results suggested that the function of the native promoter depends on the Inr, presumably due to an Inr preference of transcription factors bound to the distal promoter. A subsequent study revealed that Elf-1, which binds 60 bp upstream of the TdT start site and is a member of the large Ets family of DNA-binding proteins, possesses an intrinsic Inr preference (215). Perhaps, as implied above, only a small subset of Ets proteins will exhibit an Inr preference. Family members that can bind the Ets recognition sequence in the TdT promoter, but lack an Inr preference, would be incapable of activating TdT transcription. This hypothesis remains to be tested.

A strong preference for an Inr element has also been observed with a fusion protein between the GAL4 DNA-binding domain and the glutamine-rich activation domains of Sp1 (216). The biological significance of this preference is uncertain, however, because full-length Sp1 is a potent activator of promoters containing either a TATA box or Inr (216). A specific activation domain of c-Fos was found to exhibit a strong preference for activation of TATA-containing promoters (217); deletion of this domain resulted in comparable activation of TATA- and Inr-containing promoters. Transcriptional repression by p53 was also found to depend on a core promoter containing a TATA box (218). Although promoters containing TATA boxes were repressed by p53, comparable promoters containing an Inr instead of a TATA box were resistant to repression.

Finally, GAL4-VP16 activates transcription much more strongly when both TATA and Inr elements are present in the core promoter, whereas other proteins, such as full-length Sp1, are strong activators when the core promoter contains either a TATA box, an Inr, or both elements. (216). The bovine papillomavirus E2 transactivator exhibited a similar preference for core promoters containing both elements (219). The benefit of strengthening a promoter by combining two core elements is apparent in the *Drosophila Adh* gene, which is transcribed from two different promoters. The distal promoter is preferentially used at early stages of development because it contains both TATA and Inr elements (220).

Classes of Genes that Rely on Specific Core Elements

The existence of core promoter diversity has led to considerable interest in the possibility that specific classes of genes will contain specific core elements. The most widely discussed correlation of this type involves CpG-rich promoters, which are frequently associated with ubiquitously expressed housekeeping genes. The retrotransposons termed long interspersed nuclear elements (LINEs) can be used as a second example. In these retrotransposons, the entire promoter region is located downstream of the transcription start site, because these elements are propagated via an RNA intermediate in the absence of a long terminal repeat (LTR). The retrotransposons in *Drosophila* include the *jockey*, *Doc*, *G*, *I*, and *F* elements, all of which contain a DPE in their promoter regions. Thus, the *Drosophila* LINE promoters provide an example of a class of genes that could not function with an upstream TATA box and are entirely dependent on downstream promoter elements. A third class of genes that may be associated with a specific core promoter structure are genes expressed during the earliest stages of mammalian embryogenesis (221). TATA boxes may be nonfunctional during early development, which suggests that the expressed genes contain core promoters that can function in the absence of a TATA box.

MECHANISTIC EVENTS THAT DIFFER AMONG CORE PROMOTER CLASSES

A complete understanding of the mechanistic events that differ among core promoter classes can be realized only after the basal factors required for transcription initiation from each class have been identified. At this time, the reconstitution of basal transcription with a complete set of purified proteins has been achieved only with promoters containing consensus TATA boxes. Identification of the basal factors used at other core promoter classes will also be required for a mechanistic understanding of the selective communication between core promoters and regulatory factors bound to distal sites. Although this goal has not yet been achieved, a number of studies have provided insight into unique mechanistic events that occur at different types of core promoters.

The mechanisms resulting in selective communication between activators and core promoters can be envisioned most easily when the core promoter is recognized by factors other than the intact TFIID complex. For example, some core promoters are recognized by the TRF1 or TRF2 complexes. These complexes are likely to be the targets of defined sets of transcriptional activators and coactivators that cannot interact with TFIID.

A similar scenario can be envisioned at the *S. cerevisiae his3* promoter, which was discussed above. Although metazoans contain very little free TBP, yeast TBP is not tightly associated with the TAF complex; this allows TBP and the TAF complex to be recruited independently to target promoters (222, 223). Genetic analyses have shown that some promoters in yeast are dependent on TAFs, whereas others are TAF-independent (41). At the *his3* locus, Tc-directed transcription is TAF-dependent, whereas T_R-directed transcription is TAF-independent (224, 225). Chromatin immunoprecipitation studies have shown that TAFs associate only with the TAF-dependent T_C promoter and with other TAF-dependent promoters (222, 223). Interestingly, upstream activating sequences associated with a TAF-independent promoter are unable to recruit TAFs to a TAF-dependent promoter (226). GCN4 and GAL4 cannot recruit TAFs to either T_C or T_R, which explains why these activators cannot stimulate transcription from the TAF-dependent T_C promoter. If core promoter function relies on a TAF recognition event, transcription will occur only if the activators and coactivators are capable of recruiting the TAF complex. Although sequence-specific interactions between yeast TAFs and core promoters have not been observed, the promoters that require TAF recruitment, such as the promoters in the ribosomal protein genes, usually contain weak, nonconsensus TATA boxes (227, 228). Furthermore, core promoter sequences downstream of the TATA box have been shown to be responsible for the TAF dependence of some yeast promoters (229). Thus, the yeast core promoter sequence along with the activator bound to the upstream activating sequences appear to communicate by dictating the selective recruitment of the TAF complex, even though discrete TAF-binding elements, comparable to the Inr and DPE in metazoans, may not exist.

In metazoans, tissue-specific TAFs have been identified that are likely to contribute to combinatorial regulation by supporting selective communication between activators and the basal machinery (40). However, because the tissue-specific TAFs do not appear to recognize core promoter elements, their contribution to selective gene activation may be distinct from the contribution of core promoter diversity.

Although the mechanisms of selective communication are relatively easy to envision when the core promoters are recognized by factors other than the intact TFIID complex, it is important to note that most core promoter diversity that has been documented in metazoans involves the variable occurrence of TATA, Inr, and DPE elements. All of these elements are recognized by subunits of the same TFIID complex (103, 114, 120). If the same complex recognizes all core elements, how do these elements support selective communication with enhancers and distal activators? The mechanisms responsible are likely to involve differences in the mechanisms of basal transcription initiation from the various core promoter classes.

A few factors that are uniquely required for basal transcription from Inr and Inr-DPE core promoters have been identified. The most intriguing factors are NC2/Dr-1-DRAP1, TIC-2, and TIC–3. NC2/Dr1-DRAP1 was described above as a factor that stimulates transcription from Inr-DPE promoters in *Drosophila* but represses transcription from TATA promoters in both *Drosophila* and mammals (163, 164). Perhaps this factor is recruited to core promoters by a subset of distal activators or enhancers, which could provide a mechanism for selective communication between the activator and core promoter. TIC-2 and TIC-3 were identified as activities in HeLa cell nuclear extracts that are required for transcription from an Inr-containing core promoter (140). These factors had no effect on the activity of TATA or TATA-Inr core promoters and therefore could contribute to selective communication. Biochemical studies identified two other factors that are required for Inr activity: $TAF_{II}150$ and TIC-1 (118, 124, 125, 140). Unlike NC2/Dr1-DRAP1, TIC-2, and TIC-3, however, these factors also stimulate transcription from TATA-containing promoters, even though they are not required for TATA-directed transcription.

Although the above-mentioned selectivity factors are likely to contribute to core promoter preferences of transcriptional activators, the mechanisms responsible for the preferences do not necessarily involve selectivity factors. Activators may instead influence parameters of the transcription initiation reaction that are important for only one core promoter class. For example, the rate-limiting steps at TATA promoters may differ from those at Inr or Inr-DPE promoters (108). If an activator influences an event that is rate-limiting at only one core promoter class, it would preferentially stimulate transcription from that class. One biochemical difference between the initiation reaction at different core promoters is that transcription reinitiation appears to be more efficient at TATA-containing promoters than at TATA-less promoters (230). The conformation of the TFIID complex also appears to differ when it is bound to different core promoters; this might make it competent for activation by different subsets of transcriptional activators (231).

Finally, the affinity of the TFIID-DNA interaction may be important for activation by transcription factors such as VP16 and E2, which prefer promoters containing two core elements. As described above, TFIID binds with much higher affinity to core promoters containing two elements (i.e., TATA-Inr or Inr-DPE) (114, 120). VP16 and E2 may belong to a class of activators that stimulates transcription by a mechanism that benefits from the higher affinity of TFIID binding. Activators such as Sp1, which stimulate with equal efficiency from core promoters containing only one core element, may recruit TFIID to the core promoter, either directly or indirectly, via a mechanism that is less dependent on the affinity of TFIID for the core promoter.

CONCLUDING REMARKS

The studies cited above demonstrate that core promoters for RNA polymerase II exhibit considerable diversity, which has the potential to contribute to differential gene regulation. The recent computational studies by Ohler et al. (25) suggest that the TATA box, Inr, and DPE are the predominant core promoter elements whose locations relative to the transcription start site are restricted, at least in *Drosophila*. However, there is a high probability that additional core elements will exist whose locations relative to the start site are not conserved. Less prevalent core elements with restricted locations may also be found. Furthermore, elements within the core promoter that are more analogous to gene-specific elements will almost certainly be identified in many promoters. The detailed dissection of core promoters for new genes will be required to identify these elements.

In the study of core promoters that depend on the common TATA, Inr, and DPE elements, the most important goal for the immediate future is to identify the complete set of basal factors required for the activity of the DPE and Inr. The identification of these factors will be necessary for the long-term goal of determining precisely how different combinations of core elements contribute to the differential regulation of transcription.

ACKNOWLEDGMENTS

We thank Gerald Rubin and Robert Tjian for communicating results prior to their publication. We apologize to colleagues whose work was inadvertently omitted. Studies of RNA polymerase II core promoters in the laboratories of S.T.S. and J.T.K. were supported by the Howard Hughes Medical Institute and the NIH (GM41249), respectively.

The *Annual Review of Biochemistry* is online at http://biochem.annualreviews.org

LITERATURE CITED

1. Orphanides G, Lagrange T, Reinberg D. 1996. *Genes Dev.* 10:2657–83
2. Hampsey M. 1998. *Microbiol. Mol. Biol. Rev.* 62:465–503
3. Myer VE, Young RA. 1998. *J. Biol. Chem.* 273:27757–60
4. Roeder RG. 1998. *Cold Spring Harbor Symp. Quant. Biol.* 58:201–18
5. Lee TI, Young RA. 2000. *Annu. Rev. Genet.* 34:77–137
6. Dvir A, Conaway JW, Conaway RC. 2001. *Curr. Opin. Genet. Dev.* 11:209–14
7. Kornberg RD. 2001. *Biol. Chem.* 382: 1103–7
8. Woychik NA, Hampsey M. 2002. *Cell* 108:453–63
9. Carey M. 1998. *Cell* 92:5–8
10. Kornberg RD. 1999. *Trends Cell. Biol.* 9:M46–49
11. Myers LC, Kornberg RD. 2000. *Annu. Rev. Biochem.* 69:729–49
12. Lemon B, Tjian R. 2000. *Genes Dev.* 14:2551–69
13. Näär AM, Lemon BD, Tjian R. 2001. *Annu. Rev. Biochem.* 70:475–501
14. Smale ST. 1994. *Transcription: Mechanisms and Regulation,* ed. RC Conaway, JW Conaway, pp. 63–81. New York: Raven

15. Butler JEF, Kadonaga JT. 2002. *Genes Dev.* 16:2583–92
16. Smale ST. 2001. *Genes Dev.*15:2503–8
17. Goldberg ML. 1979. PhD Diss. Stanford Univ.
18. Breathnach R, Chambon P. 1981. *Annu. Rev. Biochem.* 50:349–83
19. Grosschedl R, Birnstiel ML. 1980. *Proc. Natl. Acad. Sci. USA* 77:1432–36
20. Wasylyk B, Derbyshire R, Guy A, Molko D, Roget A, et al. 1980. *Proc. Natl. Acad. Sci. USA* 77:7024–28
21. Grosveld GC, Shewmaker CK, Jat P, Flavell RA. 1981. *Cell* 25:215–26
22. Hu SL, Manley JL. 1981. *Proc. Natl. Acad. Sci. USA* 78:820–24
23. Struhl K. 1989. *Annu. Rev. Biochem.* 58:1051–77
24. Kutach AK, Kadonaga JT. 2000. *Mol. Cell. Biol.* 20:4754–64
25. Ohler U, Liao G, Niemann H, Rubin GM. 2002. *Genome Biol.* 3:0087.1–87.12
26. Suzuki Y, Tsunoda T, Sese J, Taira H, Mizushima-Sugano J, et al. 2001. *Genome Res.* 11:677–84
27. Parker CS, Topol J. 1984. *Cell* 36:357–69
28. Sawadogo M, Roeder RG. 1985. *Cell* 43:165–75
29. Matsui T, Segall J, Weil PA, Roeder RG. 1980. *J. Biol. Chem.* 255:11992–96
30. Nakajima N, Horikoshi M, Roeder RG. 1988. *Mol. Cell. Biol.* 8:4028–40
31. Buratowski S, Hahn S, Sharp PA, Guarente L. 1988. *Nature* 334:37–42
32. Burley SK, Roeder RG. 1996. *Annu. Rev. Biochem.* 65:769–99
33. Dynlacht BD, Hoey T, Tjian R. 1991. *Cell* 66:563–76
34. Tanese N, Pugh BF, Tjian R. 1991. *Genes Dev.* 5:2212–24
35. Zhou Q, Lieberman PM, Boyer TG, Berk AJ. 1992. *Genes Dev.* 6:1964–74
36. Goodrich JA, Tjian R. 1994. *Curr. Opin. Cell Biol.* 6:403–9
37. Davidson I, Romier C, Lavigne AC, Birck C, Mengus G, et al. 1998. *Cold Spring Harbor Symp. Quant. Biol.* 63:233–41
38. Shen WC, Apone LM, Virbasius CM, Li XY, Monsalve M, Green MR. 1998. *Cold Spring Harbor Symp. Quant. Biol.* 63:219–27
39. Bell B, Tora L. 1999. *Exp. Cell Res.* 246:11–19
40. Albright SR, Tjian R. 2000. *Gene* 242: 1–13
41. Green MR. 2000. *Trends Biochem. Sci.* 25:59–63
42. Kim JL, Nikolov DB, Burley SK. 1993. *Nature* 365:520–27
43. Kim YC, Geiger JH, Hahn S, Sigler PB. 1993. *Nature* 365:512–20
44. Juo ZS, Chiu TK, Leiberman PM, Baikalov I, Berk AJ, Dickerson RE. 1996. *J. Mol. Biol.* 261:239–54
45. Nikolov DB, Chen H, Halay ED, Hoffman A, Roeder RG, Burley SK. 1996. *Proc. Natl. Acad. Sci. USA* 93:4862–67
46. Xu L, Thali M, Schaffner W. 1991. *Nucleic Acids Res.* 19:6699–704
47. O'Shea-Greenfield A, Smale ST. 1992. *J. Biol. Chem.* 267:1391–402
48. Cox JM, Hayward MM, Sanchez JF, Gegnas LD, van der Zee S, et al. 1997. *Proc. Natl. Acad. Sci. USA* 94:13475–80
49. Tsai FT, Littlefield O, Kosa PF, Cox JM, Schepartz A, Sigler PB. 1998. *Cold Spring Harbor Symp. Quant. Biol.* 63:53–61
50. Liu Y, Schepartz A. 2001. *Biochemistry* 40:6257–66
51. Kays AR, Schepartz A. 2002. *Biochemistry* 41:3147–55
52. Wong J, Bateman E. 1994. *Nucleic Acids Res.* 22:1890–96
53. Hahn S, Buratowski S, Sharp PA, Guarente L. 1989. *Proc. Natl. Acad. Sci. USA* 86:5718–22
54. Singer VL, Wobbe CR, Struhl K. 1990. *Genes Dev.* 4:636–45
55. Wobbe CR, Struhl K. 1990. *Mol. Cell. Biol.* 10:3859–67
56. Zenzie-Gregory B, Khachi A, Garraway

IP, Smale ST. 1993. *Mol. Cell. Biol.* 13:3841–49
57. Patikoglou GA, Kim JL, Sun L, Yang SH, Kodadek T, Burley SK. 1999. *Genes Dev.* 13:3217–30
58. Bucher P. 1990. *J. Mol. Biol.* 212: 563–68
59. Hoopes BC, LeBlanc JF, Hawley DK. 1992. *J. Biol. Chem.* 267:11539–46
60. Zhao X, Herr W. 2002. *Cell* 108:615–27
61. Martinez E, Zhou Q, L'Etoile ND, Oelgeschläger T, Berk AJ, Roeder RG. 1995. *Proc. Natl. Acad. Sci. USA* 92:11864–68
62. Bell SD, Jackson SP. 1998. *Cold Spring Harbor Symp. Quant. Biol.* 63:41–51
63. Bell SD, Jackson SP. 2001. *Curr. Opin. Microbiol.* 4:208–13
64. Marsh TL, Reich CI, Whitelock RB, Olsen GJ. 1994. *Proc. Natl. Acad. Sci. USA* 91:4180–84
65. Rowlands T, Baumann P, Jackson SP. 1994. *Science* 264:1326–29
66. DeDecker BS, O'Brien R, Fleming PJ, Geiger JH, Jackson SP, Sigler PB. 1996. *J. Mol. Biol.* 264:1072–84
67. Purdy JE, Mann BJ, Pho LT, Petri WA Jr. 1994. *Proc. Natl. Acad. Sci. USA* 91:7099–103
68. Singh U, Rogers JB, Mann BJ, Petri WA Jr. 1997. *Proc. Natl. Acad. Sci. USA* 94:8812–17
69. Liston DR, Lau AO, Ortiz D, Smale ST, Johnson PJ. 2001. *Mol. Cell. Biol.* 21:7872–82
70. Fairley JA, Evans R, Hawkes NA, Roberts SGE. 2002. *Mol. Cell. Biol.* 22:6697–705
71. Li Y, Flanagan PM, Tschochner H, Kornberg RD. 1994. *Science* 263:805–7
72. Bell SD, Jackson SP. 2000. *J. Biol. Chem.* 275:12934–40
73. Pinto I, Ware DE, Hampsey M. 1992. *Cell* 68:977–88
74. Bangur CS, Pardee TS, Ponticelli AS. 1997. *Mol. Cell. Biol.* 17:6784–93
75. Hawkes NA, Roberts SGE. 1999. *J. Biol. Chem.* 274:14337–43
76. Berroteran RW, Ware DE, Hampsey M. 1994. *Mol. Cell. Biol.* 14:226–37
77. Cho E-J, Buratowski S. 1999. *J. Biol. Chem.* 274:25807–13
78. Ranish JA, Yudkovsky N, Hahn S. 1999. *Genes Dev.* 13:49–63
79. Zhang D-Y, Carson DJ, Ma J. 2002. *Nucleic Acids Res.* 30:3078–85
80. Leuther KK, Bushnell DA, Kornberg RD. 1996. *Cell* 85:773–79
81. Giardina C, Lis JT. 1993. *Science* 261: 759–62
82. Dantonel J-C, Wurtz J-M, Poch O, Moras D, Tora L. 1999. *Trends Biochem. Sci.* 285:335–39
83. Berk AJ. 2000. *Cell* 103:5–8
84. Crowley TE, Hoey T, Lui JK, Jan LN, Jan LY, Tjian R. 1993. *Nature* 361:557–61
85. Hansen SK, Takada S, Jacobson RH, Lis JT, Tjian R. 1997. *Cell* 91:71–83
86. Takada S, Lis JT, Zhou S, Tjian R. 2000. *Cell* 101:459–69
87. Holmes MC, Tjian R. 2000. *Science* 288: 867–70
88. Kaltenbach L, Horner MA, Rothman JH, Mango SE. 2000. *Mol. Cell* 6:705–13
89. Dantonel J-C, Quintin S, Lakatos L, Labouesse M, Tora L. 1999. *Mol. Cell* 6:715–22
90. Hochheimer A, Zhou S, Zheng S, Holmes MC, Tjian R. 2002. *Nature* 420: 439–45
91. Corden J, Wasylyk B, Buchwalder A, Sassone-Corsi P, Kedinger C, Chambon P. 1980. *Science* 209:1405–14
92. Talkington CA, Leder P. 1982. *Nature* 298:192–95
93. Dierks P, van Ooyen A, Cochran MD, Dobkin C, Reiser J, Weissmann C. 1983. *Cell* 32:695–706
94. Concino MF, Lee RF, Merryweather JP, Weinmann R. 1984. *Nucleic Acids Res.* 12:7423–33
95. Tokunaga K, Hirose S, Suzuki Y. 1984. *Nucleic Acids Res.* 12:1543–58
96. Chen W, Struhl K. 1985. *EMBO J.* 4:3273–80

97. Hahn S, Hoar ET, Guarente L. 1985. *Proc. Natl. Acad. Sci. USA* 82:8562–66
98. McNeil JB, Smith M. 1985. *Mol. Cell. Biol.* 5:3545–51
99. Nagawa F, Fink GR. 1985. *Proc. Natl. Acad. Sci. USA* 82:8557–61
100. Smale ST, Baltimore D. 1989. *Cell* 57:103–13
101. Smale ST, Jain A, Kaufmann J, Emami KH, Lo K, Garraway IP. 1998. *Cold Spring Harbor Symp. Quant. Biol.* 63:21–31
102. Javahery R, Khachi A, Lo K, Zenzie-Gregory B, Smale ST. 1994. *Mol. Cell. Biol.* 14:116–27
103. Smale ST, Schmidt MC, Berk AJ, Baltimore D. 1990. *Proc. Natl. Acad. Sci. USA* 87:4509–13
104. Lo K, Smale ST. 1996. *Gene* 182:13–22
105. Hultmark D, Klemenz R, Gehring WJ. 1986. *Cell* 44:429–38
106. Arkhipova IR. 1995. *Genetics* 139:1359–69
107. Kadonaga JT. 1990. *J. Biol. Chem.* 265:2624–31
108. Zenzie-Gregory B, O'Shea-Greenfield A, Smale ST. 1992. *J. Biol. Chem.* 267:2823–30
109. Smale ST. 1997. *Biochim. Biophys. Acta* 1351:73–88
110. Conaway JW, Hanley JP, Garrett KP, Conaway RC. 1991. *J. Biol. Chem.* 266:7804–11
111. Pugh BF, Tjian R. 1991. *Genes Dev.* 5:1935–45
112. Purnell BA, Gilmour DS. 1993. *Mol. Cell. Biol.* 13:2593–603
113. Martinez E, Chiang CM, Ge H, Roeder RG. 1994. *EMBO J.* 13:3115–26
114. Emami KH, Jain A, Smale ST. 1997. *Genes Dev.* 11:3007–19
115. Kaufmann J, Smale ST. 1994. *Genes Dev.* 8:821–29
116. Purnell BA, Emanuel PA, Gilmour DS. 1994. *Genes Dev.* 8:830–42
117. Wang JC, Van Dyke MW. 1993. *Biochim. Biophys. Acta* 1216:73–80
118. Verrijzer CP, Chen J-L, Yokomori K, Tjian R. 1995. *Cell* 81:1115–25
119. Bellorini M, Dantonel JC, Yoon JB, Roeder RG, Tora L, Mantovani R. 1996. *Mol. Cell. Biol.* 16:503–12
120. Burke TW, Kadonaga JT. 1996. *Genes Dev.* 10:711–24
121. Chalkley GE, Verrijzer CP. 1999. *EMBO J.* 18:4835–45
122. Tora L. 2002. *Genes Dev.* 16:673–75
123. Verrijzer CP, Yokomori K, Chen J-L, Tjian R. 1994. *Science* 264:933–41
124. Kaufmann J, Verrijzer CP, Shao J, Smale ST. 1996. *Genes Dev.* 10:873–86
125. Kaufmann J, Ahrens K, Koop R, Smale ST, Müller R. 1998. *Mol. Cell. Biol.* 18:233–39
126. Horikoshi M, Carey MF, Kakidani H, Roeder RG. 1988. *Cell* 54:665–69
127. Horikoshi M, Hai T, Lin YS, Green MR, Roeder RG. 1988. *Cell* 54:1033–42
128. Chi TH, Carey M. 1996. *Genes Dev.* 10:2540–50
129. Oelgeschläger T, Chiang C-M, Roeder RG. 1996. *Nature* 382:735–38
130. Carcamo J, Buckbinder L, Reinberg D. 1991. *Proc. Natl. Acad. Sci. USA* 88:8052–56
131. Weis L, Reinberg D. 1997. *Mol. Cell. Biol.* 17:2973–84
132. Roy AL, Meisterernst M, Pognonec P, Roeder RG. 1991. *Nature* 354:245–48
133. Roy AL, Malik S, Meisterernst M, Roeder RG. 1993. *Nature* 365:355–59
134. Roy AL, Du H, Gregor PD, Novina CD, Martinez E, Roeder RG. 1997. *EMBO J.* 16:7091–104
135. Cheriyath V, Roy AL. 2001. *J. Biol. Chem.* 276:8377–83
136. Yang W, Desiderio S. 1997. *Proc. Natl. Acad. Sci. USA* 94:604–9
137. Grueneberg DA, Henry RW, Brauer A, Novina CD, Cheriyath V, et al. 1997. *Genes Dev.*11:2482–93
138. Kim DW, Cheriyath V, Roy AL, Cochran BH. 1998. *Mol. Cell. Biol.* 18:3310–20

139. Parker R, Phan T, Baumeister P, Roy B, Cheriyath V, et al. 2001. *Mol. Cell. Biol.* 21:3220–33
140. Martinez E, Ge H, Tao Y, Yuan CX, Palhan V, Roeder RG. 1998. *Mol. Cell. Biol.* 18:6571–83
141. Manzano-Winkler B, Novina CD, Roy AL. 1996. *J. Biol. Chem.* 271:12076–81
142. Cheriyath V, Novina CD, Roy AL. 1998. *Mol. Cell. Biol.* 18:4444–54
143. Roy AL. 2001. *Gene* 274:1–13
144. Shi Y, Seto E, Chang LS, Shenk T. 1991. *Cell* 67:377–88
145. Seto E, Shi Y, Shenk T. 1991. *Nature* 354:241–45
146. Usheva A, Shenk T. 1994. *Cell* 76:1115–21
147. Usheva A, Shenk T. 1996. *Proc. Natl. Acad. Sci. USA* 93:13571–76
148. Qureshi SA, Bell SD, Jackson SP. 1997. *EMBO J.* 16:2927–36
149. Mosch HU, Graf R, Braus GH. 1992. *EMBO J.* 11:4583–90
150. Brandl CJ, Martens JA, Liaw PC, Furlanetto AM, Wobbe CR. 1992. *J. Biol. Chem.* 267:20943–52
151. Sakurai H, Ohishi T, Fukasawa T. 1994. *Mol. Cell. Biol.* 14:6819–28
152. Ohishi-Shofuda T, Suzuki Y, Yano K, Sakurai H, Fukasawa T. 1999. *Biochem. Biophys. Res. Commun.* 255:157–63
153. Sun CH, Tai JH. 1999. *J. Biol. Chem.* 274:19699–706
154. Yee J, Mowatt MR, Dennis PP, Nash TE. 2000. *J. Biol. Chem.* 275:11432–39
155. Quon DV, Delgadillo MG, Khachi A, Smale ST, Johnson PJ. 1994. *Proc. Natl. Acad. Sci. USA* 91:4579–83
156. Liston DR, Johnson PJ. 1999. *Mol. Cell. Biol.* 19:2380–88
157. Burke TW, Willy PJ, Kutach AK, Butler JEF, Kadonaga JT. 1998. *Cold Spring Harbor Symp. Quant. Biol.* 63:75–82
158. Kadonaga JT. 2002. *Exp. Mol. Med.* 34:259–64
159. Burke TW, Kadonaga JT. 1997. *Genes Dev.*11:3020–31
160. Soldatov A, Nabirochkina E, Georgieva S, Belenkaja T, Georgiev P. 1999. *Mol. Cell. Biol.*19:3769–78
161. Aoyagi N, Wassarman DA. 2001. *Mol. Cell. Biol.* 21:6808–19
162. Zhou T, Chiang C-M. 2001. *J. Biol. Chem.* 276:25503–11
163. Willy PJ, Kobayashi R, Kadonaga JT. 2000. *Science* 290:982–84
164. Maldonado E, Hampsey M, Reinberg D. 1999. *Cell* 99:455–58
165. Reiter WD, Hudepohl U, Zillig W. 1990. *Proc. Natl. Acad. Sci. USA* 87:9509–13
166. Hain J, Reiter WD, Hudepohl U, Zillig W. 1992. *Nucleic Acids Res.* 20:5423–28
167. Hausner W, Frey G, Thomm M. 1991. *J. Mol. Biol.* 222:495–508
168. Palmer JR, Daniels CJ. 1995. *J. Bacteriol.* 177:1844–49
169. Nikolov DB, Chen H, Halay ED, Usheva AA, Hisatake K, et al. 1995. *Nature* 377:119–28
170. Lee S, Hahn S. 1995. *Nature* 376:609–12
171. Lagrange T, Kim T-K, Orphanides G, Ebright Y, Ebright R, Reinberg D. 1996. *Proc. Natl. Acad. Sci. USA* 93:10620–25
172. Qureshi SA, Jackson SP. 1998. *Mol. Cell* 1:389–400
173. Lagrange T, Kapanidis AN, Tang H, Reinberg D, Ebright RH. 1998. *Genes Dev.* 12:34–44
174. Tsai FTF, Sigler PB. 2000. *EMBO J.* 19:25–36
175. Bell SD, Kosa PL, Sigler PB, Jackson SP. 1999. *Proc. Natl. Acad. Sci. USA* 96:13662–67
176. Littlefield O, Korkhin Y, Sigler PB. 1999. *Proc. Natl. Acad. Sci. USA* 96:13668–73
177. Evans R, Fairley JA, Roberts SGE. 2001. *Genes Dev.* 15:2945–49
178. Dahlberg JE, Lund E. 1988. In *Structure and Function of the Major and Minor Small Nuclear Ribonucleoprotein Particles*, ed. ML Birnstiel, pp. 38–70. Berlin: Springer-Verlag

179. Skuzeski JM, Lund E, Murphy JT, Steinberg TH, Burgess RR, Dahlberg JE. 1984. *J. Biol. Chem.* 259:8345–52
180. Lobo SM, Hernandez NT. 1994. See Ref. 14, pp. 127–59
181. Henry RW, Ford E, Mital R, Mittal V, Hernandez N. 1998. *Cold Spring Harbor Symp. Quant. Biol.* 63:111–20
182. Hernandez N. 2001. *J. Biol. Chem.* 276: 26733–36
183. Schramm L, Hernandez N. 2002. *Genes Dev.* 16:2593–620
184. Mattaj IW, Dathan NA, Parry HW, Carbon P, Krol A. 1988. *Cell* 55:435–42
185. Lobo SM, Hernandez N. 1989. *Cell* 58:55–67
186. Waldschmidt R, Wanandi I, Seifart KH. 1991. *EMBO J.* 10:2595–603
187. Murphy S, Yoon J-B, Gerster T, Roeder RG. 1992. *Mol. Cell. Biol.* 12:3247–61
188. Sadowski CL, Henry RW, Lobo SM, Hernandez N. 1993. *Genes Dev.* 7:1535–48
189. Henry RW, Mittal V, Ma B, Kobayashi R, Hernandez N. 1998. *Genes Dev.* 12:2664–72
190. Ma B, Hernandez N. 2001. *J. Biol. Chem.* 276:5027–35
191. Mittal V, Ma B, Hernandez N. 1999. *Genes Dev.* 13:1807–21
192. Lewis BA, Kim TK, Orkin SH. 2000. *Proc. Natl. Acad. Sci. USA* 97:7172–77
193. Nakatani Y, Brenner M, Freese E. 1990. *Proc. Natl. Acad. Sci. USA* 87:4289–93
194. Nakatani Y, Horikoshi M, Brenner M, Yamamoto T, Besnard F, et al. 1990. *Nature* 348:86–88
195. Ince TA, Scotto KW. 1995. *J. Biol. Chem.* 270:30249–52
196. Benson LQ, Coon MR, Krueger LM, Han GC, Sarnaik AA, Wechsler DS. 1999. *J. Biol. Chem.* 274:28794–802
197. Rudge TL, Johnson LF. 1999. *J. Cell. Biochem.* 73:90–96
198. Bird A. 2002. *Genes Dev.* 16:6–21
199. Antequera F, Bird A. 1995. *Proc. Natl. Acad. Sci. USA* 90:11995–99
200. Blake MC, Jambou RC, Swick AG, Kahn JW, Azizkhan JC. 1990. *Mol. Cell. Biol.* 10:6632–41
201. Brandeis M, Frank D, Keshet I, Siegfried Z, Mendelsohn M, et al. 1994. *Nature* 371:435–38
202. Macleod D, Charlton J, Mullins J, Bird AP. 1994. *Genes Dev.* 8:2282–92
203. Struhl K. 1986. *Mol. Cell. Biol.* 6:3847–53
204. Struhl K. 1987. *Cell* 49:295–97
205. Mahadevan S, Struhl K. 1990. *Mol. Cell. Biol.* 10:4447–55
206. Iyer V, Struhl K. 1995. *Mol. Cell. Biol.* 15:7059–66
207. Simon MC, Fisch TM, Benecke BJ, Nevins JR, Heintz N. 1988. *Cell* 52:723–29
208. Wefald FC, Devlin BH, Williams RS. 1990. *Nature* 344:260–62
209. Taylor ICA, Kingston RE. 1990. *Mol. Cell. Biol.* 10:165–75
210. Li X, Noll M. 1994. *EMBO J.* 13:400–6
211. Merli C, Bergstrom DE, Cygan JA, Blackman RK. 1996. *Genes Dev.* 10:1260–70
212. Ohtsuki S, Levine M, Cai HN. 1998. *Genes Dev.* 12:547–56
213. Butler JEF, Kadonaga JT. 2001. *Genes Dev.* 15:2515–19
214. Garraway IP, Semple K, Smale ST. 1996. *Proc. Natl. Acad. Sci. USA* 93:4336–41
215. Ernst P, Hahm K, Trinh L, Davis JN, Roussel MF, et al. 1996. *Mol. Cell. Biol.* 16:6121–31
216. Emami KH, Navarre WW, Smale ST. 1995. *Mol. Cell. Biol.* 15:5906–16
217. Metz R, Bannister AJ, Sutherland JA, Hagemeier C, O'Rourke EC, et al. 1994. *Mol. Cell. Biol.* 14:6021–29
218. Mack DH, Vartikar J, Pipas JM, Laimins LA. 1993. *Nature* 363:281–83
219. Ham J, Steger G, Yaniv M. 1994. *EMBO J.* 13:147–57
220. Hansen SK, Tjian R. 1995. *Cell* 82:565–75
221. Majumder S, DePamphilis ML. 1994. *Mol. Cell. Biol.* 14:4258–68

222. Li X-Y, Bhaumik SR, Green MR. 2000. *Science* 288:1242–44
223. Kuras L, Kosa P, Mencia M, Struhl K. 2000. *Science* 288:1244–48
224. Moqtaderi Z, Bai Y, Poon D, Weil PA, Struhl K. 1996. *Nature* 383:188–91
225. Moqtaderi Z, Keaveney M, Struhl K. 1988. *Mol. Cell* 2:675–82
226. Li X-Y, Bhaumik SR, Zhu X, Li L, Shen W-C, et al. 2002. *Curr. Biol.* 12:1240–44
227. Tsukihashi Y, Miyake T, Kawaichi M, Kokubo T. 2000. *Mol. Cell. Biol.* 20:2385–99
228. Mencía M, Moqtaderi Z, Geisberg JV, Kuras L, Struhl K. 2002. *Mol. Cell* 9:823–33
229. Shen WC, Green MR. 1997. *Cell* 90:615–24
230. Yean D, Gralla J. 1997. *Mol. Cell. Biol.* 17:3809–16
231. Knutson A, Castaño E, Oelgeschläger T, Roeder RG, Westin G. 2000. *J. Biol. Chem.* 275:14190–97

Annu. Rev. Biochem. 2003. 72:481–516
doi: 10.1146/annurev.biochem.72.121801.161547
Copyright © 2003 by Annual Reviews. All rights reserved
First published online as a Review in Advance on March 27, 2003

THE ESTABLISHMENT, INHERITANCE, AND FUNCTION OF SILENCED CHROMATIN IN *SACCHAROMYCES CEREVISIAE*

Laura N. Rusche,[1] Ann L. Kirchmaier,[2] and Jasper Rine[1]
[1]Department of Molecular and Cell Biology, University of California Berkeley, Berkeley, California 94720-3202; email: lrusche@uclink4.berkeley.edu; jrine@uclink4.berkeley.edu
[2]Department of Biochemistry, Purdue University, West Lafayette, Indiana 47907, email: kirchmaier@purdue.edu

Key Words silencing, heterochromatin, Sir, epigenetic, nucleosome

■ **Abstract** Genomes are organized into active regions known as euchromatin and inactive regions known as heterochromatin, or silenced chromatin. This review describes contemporary knowledge and models for how silenced chromatin in *Saccharomyces cerevisiae* forms, functions, and is inherited. In *S. cerevisiae,* Sir proteins are the key structural components of silenced chromatin. Sir proteins interact first with silencers, which dictate which regions are silenced, and then with histone tails in nucleosomes as the Sir proteins spread from silencers along chromosomes. Importantly, the spreading of silenced chromatin requires the histone deacetylase activity of Sir2p. This requirement leads to a general model for the spreading and inheritance of silenced chromatin or other special chromatin states. Such chromatin domains are marked by modifications of the nucleosomes or DNA, and this mark is able to recruit an enzyme that makes further marks. Thus, among different organisms, multiple forms of repressive chromatin can be formed using similar strategies but completely different proteins. We also describe emerging evidence that mutations that cause global changes in the modification of histones can alter the balance between euchromatin and silenced chromatin within a cell.

CONTENTS

OVERVIEW OF SILENCING IN *SACCHAROMYCES CEREVISIAE*

Silencing involves the formation of a specialized chromatin structure that blocks expression of most genes within the silenced domain, irrespective of which activator or even which RNA polymerase is utilized. Silencing was originally defined in *S. cerevisiae* as a process that represses transcription of secondary copies of the mating type genes, and studies of silencing in *S. cerevisiae* have been at the forefront of molecular insights into the mechanism of silencing ever since.

Silencing differs from gene-specific repression, in which a single promoter is specifically targeted for repression. Silencing is mediated by regulatory sites, known as silencers, that act at some distance from their target genes. In contrast, gene-specific repression is mediated by operators at, or near, the site of transcription initiation. However, in the literature of the past decade, the terms repression and silencing have had no consistent distinction. This literary sleight of hand has presaged a growing recognition of common mechanisms in the two processes, such as requirements for similar nucleosome-modifying enzymes.

Here we describe emerging mechanistic insights into the silencing of regions of the genome, with particular emphasis on the roles of nucleosome modifications. We also discuss potential mechanistic links between regional silencing and gene-specific silencing. The historical antecedents to these studies are described more fully elsewhere (1).

The Nature of Silenced Chromatin

Silenced chromatin in *S. cerevisiae* is akin to heterochromatin in organisms such as maize, flies, and mammals. Heterochromatin was originally defined cytologically as blocks of the genome in which the structure of chromatin is highly condensed throughout the cell cycle (2). Such regions are largely inactive for gene expression. The discovery that the N-terminal tail of histone H4 is required for silencing (3) revealed that silencing in *S. cerevisiae* also involves a specialized structure of chromatin. The N-terminal tails of histones H3 and H4 are hypoacetylated in silenced chromatin (4, 5).

The structural proteins of silenced chromatin in *S. cerevisiae* are known as silent information regulator, or Sir, proteins. Silenced domains consist of continuous distributions of Sir proteins (6–10) that, together with the hypoacetylated nucleosomes, are thought to form an ordered, compact structure that is restrictive to transcription. Silenced regions are less accessible to restriction nucleases and DNA methylases (11–13), display ordered, regularly spaced nucleosomes (14, 15), and have an altered topology (16, 17). The ability of silenced chromatin to block restriction enzymes and methylases indicates that it may also prevent the transcription machinery from gaining access to promoters. However, silenced chromatin may affect gene expression at other molecular levels as well.

Silenced Regions of the Genome

Genetic studies have identified several regions of the *S. cerevisiae* genome that are silenced, or are at least subject to silencing. These regions are the two silent mating type loci, the telomeres, and the rRNA-encoding DNA (the rDNA).

In *S. cerevisiae,* haploid cells are of one of two mating types, **a** or α. The mating type is determined by the allele of the mating type locus *MAT*. *MAT**a*** and *MAT*α encode regulatory proteins that, by affecting the expression of other genes, are responsible for the difference between the two mating types. In addition to *MAT*, all *S. cerevisiae* strains have unexpressed copies of mating type genes at two other loci, *HML* and *HMR*, located near the two telomeres on the same chromosome as *MAT* (Figure 1, top and center). In most strains, *HML* contains a cryptic copy of the *MAT*α gene and *HMR* contains a cryptic copy of the *MAT**a*** gene. Some *S. cerevisiae* strains, known as homothallic strains, can interconvert their mating types from **a** to α or from α to **a** as frequently as once per generation. This interconversion is catalyzed by a site-specific endonuclease, encoded by the *HO* gene, that cleaves the mating type locus. The repair of the cleavage occurs by a mechanism akin to gene conversion, in which a copy of the genes at either *HML* or *HMR* is transferred to *MAT* (18). Although most laboratory strains do not contain an active gene for the endonuclease (*ho*), and

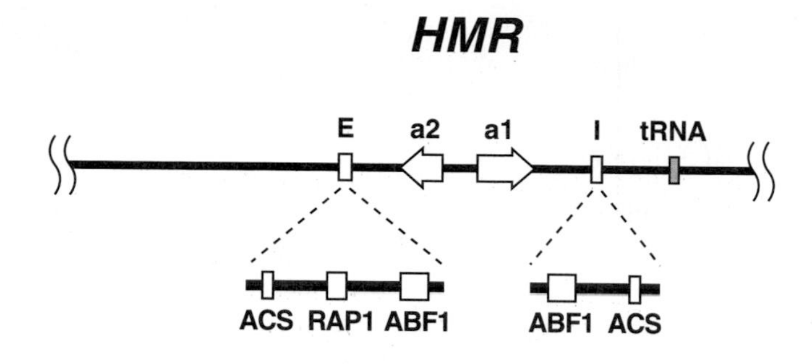

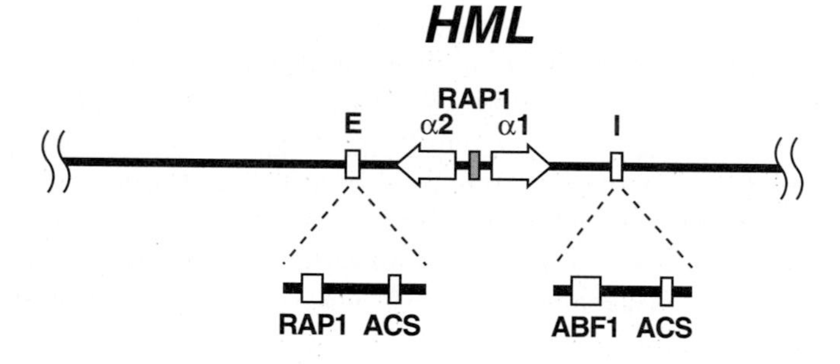

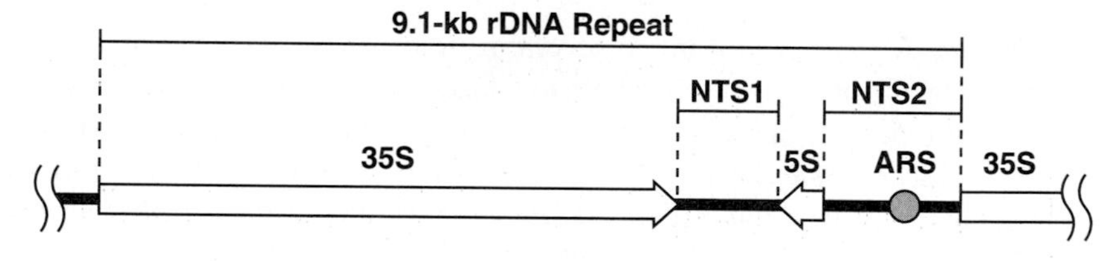

Figure 1 The *HM* loci and the rDNA array. The *HM* loci *(top, center)*. The E silencers (E), I silencers (I), binding sites for ORC (ACS), Rap1p (RAP1), and Abf1p (ABF1), open reading frames of the **a**1, **a**2, α1, α2 mating type genes, and the tRNA gene flanking *HMR* involved in boundary formation are indicated. The rDNA array *(bottom)*. The 9.1-kb repeat unit contains the 35S and 5S genes separated by the nontranscribed spacer regions, NTS1 and NTS2. The origin of replication is at ARS.

hence do not switch mating types, they all contain cryptic mating type genes at *HML* and *HMR*.

Reporter genes that are inserted adjacent to telomeres of *S. cerevisiae* chromosomes are silenced by a process termed telomere position effect (19). Telomere position effect requires most of the genes required for silencing of the

HM loci, with the exception of *SIR1* and the *ORC* genes (described below) (20). However, most studies of telomere position effect are conducted using artificial telomeres lacking subtelomeric repeat sequences, and at some natural telomeres, *SIR1* is also important for silencing (21). Thus, to a first approximation, silencing at telomeres shares many similarities with silencing at the *HM* loci. Consequently, we treat silencing at *HM* loci and telomeres as related phenomena, and unless otherwise stated, "silenced chromatin" refers to both classes of silenced loci.

The genes encoding rRNA in *S. cerevisiae* are highly expressed and arranged in a tandem repeat of 100 to 200 copies (Figure 1, bottom) (22). However, when marker genes such as those used to monitor telomeric silencing are inserted into the rDNA array, those genes are expressed at a lower level than they are at their normal chromosomal loci (23, 24). Silencing of genes in rDNA requires *SIR2*, but does not require any of the other *SIR* genes (23). This observation reflects a fundamental difference between rDNA silencing and the other two forms of silencing. The mechanism of rDNA silencing is explored in a separate section of this review.

Genome-wide expression and chromatin immunoprecipitation studies (9, 25, 26) indicate that the rest of the *S. cerevisiae* genome is largely free of Sir protein–mediated silencing. There have been hints of a few other loci that are either silenced or bound by silencing proteins. However, silencing has not yet been confirmed at any of these other loci by more direct investigations.

The Components of Silenced Chromatin

SILENCERS AND SILENCER BINDING PROTEINS The *HM* loci are each flanked by silencers, known as *E* and *I* (Figure 1, top and center). The genetic properties of silencers were originally defined by deletion studies on plasmids (27, 28). In the chromosomal context, deletion of *HMR-E* leads to the complete derepression of *HMR*, whereas deletion of *HMR-I* has no apparent effect in the absence of other mutations that weaken silencing (29). In contrast, either silencer at *HML* is able to achieve silencing in the absence of the other (30). All four silencers act as autonomously replicating sequences on plasmids (27, 28), but only *HMR-E* and *HMR-I* are bona fide origins of replication in the chromosome (31–33).

HMR-E is the most thoroughly studied of the four silencers and provides a high-resolution view of what is required to make a functional silencer. The *HMR-E* silencer is ~140 bp in length and has binding sites for three essential factors: ORC, Rap1p, and Abf1p. Rap1p and Abf1p are two of the most common transcription activators in *S. cerevisiae;* each is individually responsible for activating the expression of hundreds of different genes (34). ORC is the origin recognition complex, conserved throughout eukaryotes, that is essential for initiating DNA replication (35). A 138-base-pair synthetic sequence containing the binding sites for all of these factors in the same position and orientation as in the natural silencer is capable of functionally replacing the natural silencer (36), indicating that no additional elements are required for silencer function. This does not exclude the possibility that other elements contribute to silencer function, nor does this preclude some overlapping contribution of these proteins to silencer function. Indeed, in the context of the natural silencer, mutations in any two of the three binding sites are required to achieve complete, or near complete, loss of silencing (37). Mutations in the genes encoding ORC, Rap1p, and Abf1p proteins that disrupt silencer function have been found (38–41). Thus, the proteins that bind these sites in vitro are the same proteins required for silencer function in vivo.

One interesting issue is how the binding sites for three factors with independent roles in the cell can create a silencer. The current view is that a silencer is a site with an emergent property caused by the close juxtaposition of these three factors, each of which has some affinity for one or more Sir proteins. No single binding site for any one of these factors by itself can create a sufficiently high local concentration of Sir proteins to sustain silencing, but in combination these factors can. Consistent with this model, arrays of multiple Rap1p binding sites recruit Sir proteins to telomeres (6, 42) and can create artificial silencers (43).

THE SIR PROTEINS The four *SIR* genes were identified by mutations that activated the silent mating type genes (44, 45). Sir2p, Sir3p, and Sir4p are essential for silencing because of their role as structural components of silenced chromatin. Sir2p and Sir4p form a soluble complex, sometimes referred to as TEL (46–48). In contrast, Sir3p does not copurify with other proteins in stoichiometric amounts (46–48). Once all three Sir proteins are recruited to a silencer, cooperative interactions enable them to spread throughout the silent locus, with Sir3p and Sir4p binding to the deacetylated tails of histones H3 and H4 (6, 49, 50). Sir2p is an NAD^+-dependent histone deacetylase (51–53) that modifies the tails of histones H3 and H4 to create high-affinity binding sites for Sir3p and Sir4p. As described in more detail below, Sir2p has an additional role in rDNA silencing.

In contrast to the other Sir proteins, Sir1p contributes to, but is not essential for, silencing. Sir1p acts in the establishment of silencing by facilitating the assembly of the other Sir proteins at the silencer (8, 10, 54).

Establishment, Maintenance, and Inheritance: Definition of Terms

The present understanding of the formation and stability of silenced chromatin borrows concepts first elucidated in the study of the repression of bacteriophage lambda (55). These early studies demonstrated that a protein may contribute to the initial repression of a gene but not be required for the gene to remain repressed. Such a protein is said to contribute to the establishment of repression, but not to its maintenance. In contrast, a protein that is required continuously for a repressed target gene to remain repressed is said to be required for the maintenance of that repressed state. If the repressed state persists for more than one cell cycle, a mechanism must exist for the repressed state to be inherited as well as maintained.

In *S. cerevisiae,* the concepts of establishment, maintenance, and inheritance of a silenced state were first illuminated by the unusual phenotype of *sir1* mutants. In a population of genetically identical *MAT**a** sir1* cells, ~20% of the cells are in a silenced state at *HML*, whereas *HML* is expressed in the other 80% of cells (54). Interestingly, both the silenced and the expressed states are stably heritable for 10 generations or more (54). Several insights follow from this unusual behavior. First, because two different expression states are stable in genetically identical cells, the two states are said to be epigenetic. Second, because 20% of *sir1* cells and their immediate descendants retain the silenced state, Sir1p is not required to maintain or inherit the silenced state. Finally, because *HML* is expressed in 80% of *sir1* cells, Sir1p must promote the establishment of the silenced state.

Histones and Silencing

The understanding of silencing in *S. cerevisiae* has been dramatically increased by our rapidly expanding knowledge of the nature and function of histone modifications. These modifications are thought to represent a code indicating the expression state of the underlying chromatin (56–58). Indeed studies of silencing have contributed substantially to the emergence of the so-called histone code hypothesis. For example, active regions are associated with acetylated histones, whereas silenced regions are associated with unacetylated histones. This code is translated by effector proteins that bind specifically to particular modifications on the histones and, in turn, lead to further chromatin modifications that ultimately dictate particular expression states. Histone modifications also reinforce one another, with one modification affecting the likelihood of other modifications. This histone code hypothesis provides a framework for understanding the role of histone deacetylation in the processes of establishment, maintenance, and inheritance of silenced chromatin. Furthermore, several proteins known to modulate silencing modify histones, revealing how global changes in histone modifications can affect silenced chromatin. These issues are discussed more fully below.

BIOCHEMISTRY OF SIR PROTEINS AND SILENCED CHROMATIN

Sir2p

Sir2 ENZYMOLOGY Sir2p is a member of a large protein family that is conserved from bacteria to humans. In addition to Sir2p, *S. cerevisiae* encodes four other family members—Hst1p, Hst2p, Hst3p, and Hst4p (59). The recent discovery that members of the Sir2p family are NAD^+-dependent deacetylases (52, 53, 60) has greatly enhanced our understanding of how silencing is achieved. Sir2p family members couple a deacetylation reaction to the hydrolysis of NAD^+ (Figure 2) (51–53, 60–62). In this reaction, one molecule of NAD^+ is cleaved

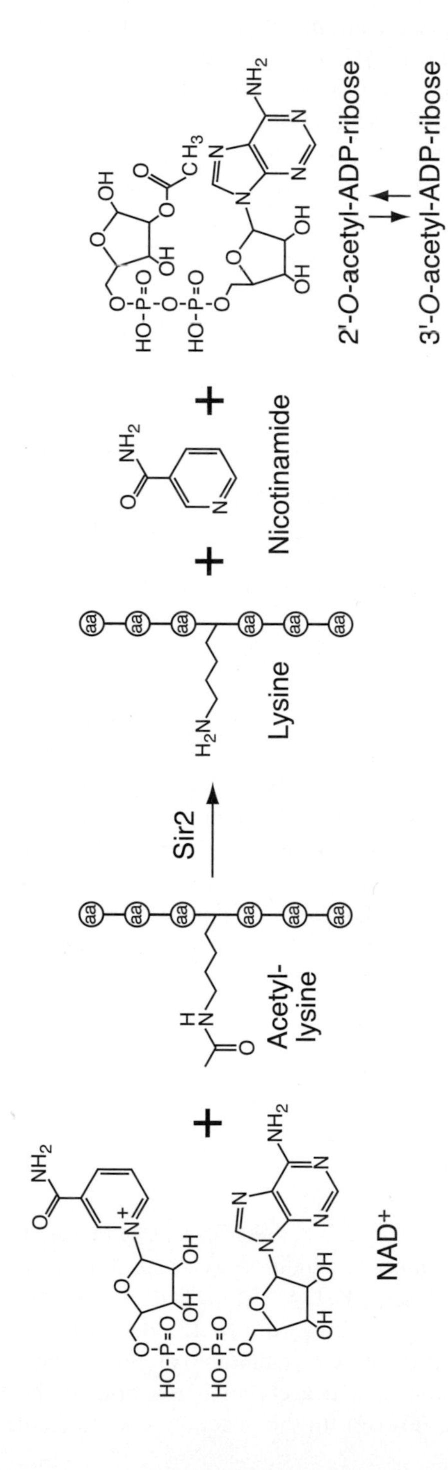

Figure 2 The NAD^+-dependent deacetylation reaction catalyzed by Sir2p.

into ADP-ribose and nicotinamide per acetyl group removed (51, 61, 62). The acetyl group is transferred to ADP-ribose, forming the product 2′-*O*-acetyl-ADP-ribose, which equilibrates to a combination of 2′-*O*-acetyl-ADP ribose and 3′-*O*-acetyl-ADP ribose (63, 64). There is much speculation but little data on what, if any, role the *O*-acetyl-ADP ribose products may have.

Interestingly, in the NAD^+-dependent deacetylation reaction, the free energy of hydrolysis of the glycosidic bond between nicotinamide and ribose is −8.2 kcal/mol (65, 66), comparable to the free energy of hydrolysis of ATP. Thus, considerable free energy is released in silencing a locus such as *HMR*, which spans 12 ordered nucleosomes (14). This energy may be required to induce conformational changes in proteins and DNA that are necessary for the formation of silenced chromatin.

In addition to their primary NAD^+-dependent deacetylase activity, Sir2p family members have ADP-ribosylation activity in vitro (53, 62, 67, 68). However, this ADP-ribosylation activity is much less efficient than the deacetylase activity and it is not known whether this reaction is biologically significant.

Several additional aspects of Sir2p biochemistry need resolution, including what dictates the activity level and the substrate specificity of the Sir2p family members. Relative to Hst2p, the histone deacetylase activity of purified Sir2p is weak (60). This difference between the activity of Sir2p and its paralog may reflect a genuine difference in the efficiency of catalysis, a difference in substrate specificity, or a requirement for Sir2p to function in a complex with other Sir proteins for full activity. Although other substrates for mammalian Sir2 orthologs have been identified (69–71), whether Sir2p in *S. cerevisiae* has substrates other than histone tails is not known.

Another unresolved issue is that Sir2p specifically targets lysines 9 and 14 of histone H3 and lysine 16 of histone H4 in vitro (53, 61), whereas in vivo all acetylatable lysines of histones H3 and H4 are fully deacetylated at silenced loci (5). Other deacetylases may act on the remaining lysines, or the substrate specificity of Sir2p may be altered in vivo such that it deacetylates these lysines as well.

INHIBITORS OF Sir2p Several strategies have been employed both in vitro and in vivo to inhibit the enzymatic activity of Sir2p family members. The inhibitors, summarized below, should be powerful tools for dissecting the biochemistry, substrate specificity, and biological function of Sir2p family members in *S. cerevisiae* and in organisms with less facile genetics.

Enzymes that use NAD^+ as an electron acceptor or as an allosteric regulator can utilize a nonhydrolyzable analog of NAD^+ in their reactions (72, 73). In contrast, the nonhydrolyzable analogs of NAD^+, carba-NAD^+ and pseudocarba-NAD^+, are competitive inhibitors of Hst2p in vitro (51). Thus, unlike most NAD^+-dependent enzymes which use NAD^+ as an electron acceptor in coupled redox reactions, members of the Sir2 family hydrolyze NAD^+ as part of their mechanism.

Nicotinamide, a product of NAD^+-dependent deacetylation, also inhibits Hst2p (51). After hydrolysis of NAD^+, nicotinamide may be retained within a deep pocket of the catalytic domain and remain there until the release of the ADP-ribose molecule (74). Importantly, certain mutations in the NAD^+ biosynthetic pathway that affect Sir2p activity (75, 76) could act by increasing the cellular concentration of nicotinamide, as described below.

Several other compounds inhibit Sir2p family members in vitro. Coumermycin A1, an inhibitor of mono-ADP-ribosyltransferases (77), blocks both NAD^+ hydrolysis and ADP ribosylation by Sir2p (47, 53). In addition, *o*-phenanthroline, a zinc chelator, blocks Hst2p activity at high concentrations, likely by chelating a zinc ion in a conserved zinc finger (74, 78). Unlike other histone deacetylases, the Sir2p family members are not inhibited by trichostatin A (79, 80).

Recently, small molecule inhibitors of *S. cerevisiae* Sir2p that act in vivo have been identified (79, 81). These inhibitors also block human SIRT2 deacetylase activity in vitro and likely inhibit *S. cerevisiae* Hst1p activity in vivo as well (79, 81). Identified inhibitors include A3 (8,9-dihydroxy-6*H*-(1)benzofuro[3,2-c]chromen-6-one), M15 (1-[(4-methoxy-2-nitro-phenylimino)-methyl]-naphthalene-2-ol), sirtinol (2-[(2-hydroxy-naphthalen-1-ylmethylene)-amino]-*N*-(1-phenyl-ethyl)-benzamide), and splitomicin (1,2,dihydro-3*H*-naphtho[2,1-b]pyran-3-one). This approach holds promise for identifying inhibitors that specifically block individual Sir2p family members.

Sir4p

Unlike its partner Sir2p, Sir4p has no known enzymatic activities. Rather, Sir4p plays a structural role in silenced chromatin, linking various proteins together. Sir4p associates with silencer-binding proteins and, in turn, is required for the association of Sir2p and Sir3p with the silencer or with telomeres. (Individual binding partners are described below.) In addition, Sir4p facilitates the association of other cellular factors with silenced chromatin. For example, Sir4p may associate with an immobile, peripheral component of the nucleus, thereby restricting the movement of silenced domains (82). Sir4p also interacts with Dis1p/Ris1p, a Swi/Snf family member that makes silenced chromatin available for recombination during mating type switching (83), and with the targeting domain of the integrase of the Ty5 transposon, explaining the preference of Ty5 for inserting into silenced chromatin (84).

One of the most intriguing features of Sir4p is its interaction with two ubiquitin hydrolases, Dot4p and Ubp3p (85, 86). Loss of Dot4p activity results in decreased telomeric silencing (85), consistent with the idea that the removal of ubiquitin from a target is important for silencing. Although the ubiquitinated target of Dot4p is not known, the level of Sir4p is lower in *dot4Δ* cells (85), leading to the proposal that the stability or function of Sir4p could be regulated by ubiquitin. However, ubiquitinated Sir4p has not been reported. In contrast to Dot4p, loss of Ubp3p slightly improves silencing at telomeres and *HM* loci (86), implying that these two ubiquitin hydrolases act on different targets. An alternative model for Dot4p action is discussed below.

Sir3p

Like Sir4p, Sir3p has no known enzymatic activity but rather plays a structural role in the assembly of silenced chromatin. Sir3p is recruited to the silencer separately from the Sir2p-Sir4p complex, through interactions described below. Recombinant Sir3p binds both nucleosomes and DNA in vitro (47, 87). Thus, by itself, Sir3p has rather promiscuous binding properties. In one study, recombinant Sir3p protected linker DNA in a dinucleosome from cleavage, supporting a role for Sir3p in binding linker DNA in silenced chromatin (47). However, in a separate study, Sir3p did not appear to protect the linker DNA of a longer nucleosome array (87).

The first 214 amino acids of Sir3p share 50% identity with the amino terminus of Orc1p (88). This region includes a bromo-adjacent homology (BAH) domain, which is thought to participate in protein-protein interactions. The N-terminal domain of Orc1p is not required for replication but is required for silencing (88) by binding Sir1p (89, 90). The Sir1p binding domain of Orc1p lies in a small domain that is positioned by the BAH domain but is not well conserved with Sir3p, explaining why Sir3p does not bind Sir1p (10). A binding partner for the Sir3p N-terminal domain has not been reported.

Another region of Sir3p with similarity to Orc1p (27% identity; 43% similarity) includes an AAA domain. The AAA superfamily members couple ATP hydrolysis to conformational changes that drive the assembly and disassembly of protein complexes (91). Cdc6p, Orc1p, Orc4p, Orc5p, and Sir3p comprise a subfamily of the AAA superfamily (91). The AAA domains of the other subfamily members bind ATP, but Sir3p lacks critical residues required to bind ATP. Perhaps Sir3p binds 2′-*O*-acetyl-ADP ribose, a product of the Sir2p deacetylation reaction, coupling deacetylation to conformational changes that drive silenced chromatin assembly (92).

Both Sir3p and Sir4p are phosphorylated, but it is not known whether this phosphorylation regulates their action. Sir3p is hyperphosphorylated in response to mating pheromone, heat shock, and starvation, and these changes in phosphorylation correlate with changes in silencing (93, 94). In another set of experiments in which Sir3p phosphorylation was not examined, silencing of *MAT* flanked by silencers did not occur on nonfermentable carbon sources or when adenylyl cyclase was disrupted (95). Together, these observations suggest that the extent of silencing could be sensitive to environmental conditions, perhaps as a result of the action of kinase signaling pathways targeting Sir proteins.

Silenced Chromatin

The structure of silenced chromatin is determined by Sir proteins and histones. Chromatin immunoprecipitation studies reveal that Sir proteins spread inward from telomeres (6, 9) and are distributed throughout *HMR* and *HML* (8–10). At the silenced loci, the tails of histones H3 and H4 are unacetylated at all positions tested (4, 5, 96). Lysine 12 of histone H4 was thought to be acetylated (96), but this observation was apparently an artifact of the antibody used (5). At *HMR*, the distributions of Sir proteins and hypoacetylated histones coincide (8), implying that these two features form a distinct chromatin structure. Other modifications of histones at silenced loci have not been reported. For example, methylation of histone H3 at lysines 4 and 79 appears to be characteristic of active regions of the genome and is probably underrepresented at silenced loci (97, 98), and methylation of lysine 9 of histone H3, which is characteristic of silenced chromatin in other organisms, has not been reported in *S. cerevisiae*.

The assembly of the Sir proteins into silenced chromatin creates a highly ordered structure, as revealed by micrococcal nuclease protection studies. *HMR* is composed of 12 ordered nucleosomes, arranged into six pairs of closely spaced nucleosomes separated from one another by longer linkers (14). *HML* is organized into at least 23 nucleosomes, most of which are precisely positioned and also arranged in pairs (15). Silenced loci are also more negatively supercoiled in wild-type cells compared to *sir* cells (16, 17). This topological change could result from an additional nucleosome or an alteration in the path of the linker DNA or the DNA around the nucleosomes.

The structure formed by Sir proteins and histones is restrictive to various enzymes (11–13), probably because of the occlusion of the DNA by the Sir proteins, yet this structure is dynamic. Studies of silenced chromatin excised from its chromosomal context reveal that silencers, or silencer-like elements, are required continuously to stabilize the silenced structure (99). Furthermore, in noncycling cells, newly expressed Sir3p associates with silenced chromatin (99). Consistent with these observations, the deacetylase activity of Sir2p is required continuously to maintain silenced chromatin in noncycling cells (81).

The Binding of Sir Proteins to Histones

The structural role of Sir proteins in silenced chromatin is mediated by binding to histones. Sir3p and Sir4p bind residues 1–25 of histone H3 and 15–34 of histone H4 in vitro (49). Importantly, the regions of histones H3 and H4 that bind to Sir3p and Sir4p overlap with silencing domains defined by mutational analysis—specifically, residues 4–20 of histone H3 (100) and 16–29 of histone H4 (101, 102). Furthermore, mutations that abolish interactions between Sir proteins and histones in vitro also abolish silencing in vivo (49). Another region of the nucleosome is predicted to contact Sir proteins based on mutations in the core domains of histones H3 and H4 that alter silencing and map to a patch of the nucleosome (103, 104). Moreover, methylation of lysine 79 of histone H3, which lies in this same region, is thought to prevent the binding of Sir proteins (97). Sir proteins appear not to bind the N-terminal tails of histones H2A or H2B (49).

The presence of hypoacetylated lysines in silenced chromatin implies that Sir3p and Sir4p might bind preferentially to unacetylated histones. Indeed, acetylation of lysine residues in a peptide that corresponds to the tail of histone H4 reduces the binding of Sir3p in a cumulative manner, and the effects are quantitatively similar at any lysine (50). However, examination of the effects of mutations of individual lysines in vivo reveals a more complex picture. In vivo, a single mutation of histone H4 at lysine 16 to glutamine or alanine completely disrupts silencing at *HML*. In contrast, mutating lysines 5, 8, and 12 of histone H4 in combination with lysines 9, 14, 18, and 23 of histone H3 results in minimal loss of silencing (102). Thus, lysine 16 of histone H4 has a special role in silencing. In keeping with this special role, Sir2p specifically deacetylates lysine 16 of histone H4 in vitro (53, 61), and Sas2p, an acetyltransferase that modulates silencing, probably acetylates this same lysine residue (105, 106, 106a).

Although the ability of histones to interact with Sir proteins is clearly very important for silencing, histones have additional functions in silenced chromatin, as evidenced by the existence of mutations in histones that disrupt silencing but do not affect the binding of Sir3p or Sir4p (49).

ESTABLISHMENT OF SILENCED CHROMATIN

Silenced chromatin formation occurs in discrete steps. First, the Sir proteins are recruited to the silencers or the telomeres. Then, Sir proteins spread throughout the target locus.

Assembly of Sir Proteins at the Silencer

Sir proteins are recruited to silencers through a series of protein-protein interactions. Sir1p binds directly to Orc1p (10, 89, 90) and enhances the probability of recruiting the other Sir proteins to the silencer (8, 10). Sir1p localizes to silencers but does not spread throughout silenced regions (8, 10), as expected given the role of Sir1p in establishing silencing. Interestingly, the silencing function of the silencer binding proteins ORC or Rap1p can be bypassed by recruiting Sir1p to a silencer through a heterologous DNA binding domain (90, 107, 108), indicating that ORC and Rap1p are not required for any subsequent event in silenced chromatin formation.

The assembly of Sir proteins at silencers or telomeres exhibits a hierarchy of recruitment. Sir1p can be recruited to a silencer in the absence of any other individual Sir protein (8). Sir4p is recruited to the silencer through its interactions with Sir1p (89) and Rap1p (109, 110). The recruitment of Sir4p to the silencer does not require Sir2p or Sir3p (8, 48, 111). Sir4p likely brings Sir2p to the silencer as a member of a Sir2p-Sir4p complex (47, 48). Sir4p is also required to recruit Sir3p to the silencer (8, 48). Sir3p binds both Rap1p (109, 110) and Sir4p (46) and likely Abf1p as well. Sir2p and Sir3p can each associate with the silencer in the absence of the other (8, 48). The catalytic activity of Sir2p is not required for its association, or the association of any other Sir protein, with the silencer (8, 48). A similar hierarchy exists at the telomere, with the difference being that Sir4p is recruited through Rap1p and Hdf1p (yKu70p) (112–114).

Spreading of Sir Proteins

After Sir proteins have assembled at a silencer, they spread from the silencer to the gene that is to be silenced. Although Sir2p and Sir3p are not required for the other Sir proteins to associate with a silencer, they are required for the Sir complex to spread stably from the silencer (8, 48). The spreading of the Sir complex at telomeres has similar requirements (111). The ability of the Sir proteins to bind to the tails of histones H3 and H4 in nucleosomes enables the Sir proteins to spread across the chromosome. As described above, Sir3p and presumably Sir4p bind more efficiently to hypoacetylated histone tails than to fully acetylated tails (50).

The NAD^+-dependent histone deacetylase activity of Sir2p is required for spreading of the Sir proteins (8, 48). This observation supports a mechanism for the spreading of Sir proteins that involves the sequential deacetylation of histone tails in nucleosomes along the chromosome. In this model, upon the recruitment of the Sir proteins to the silencers, Sir2p is brought into the proximity of its substrate, the acetylated lysines on the tails of histones H3 and H4 of a nearby nucleosome. The subsequent deacetylation of these tails by Sir2p creates a high-affinity site for binding of Sir3p and Sir4p (and its Sir2p partner), thus enabling the recruitment of additional Sir proteins to the nucleosomes flanking the silencers. This process positions the newly loaded Sir2p next to the acetylated tails of H3 and H4 on the next nucleosome. The sequential process of deacetylating neighboring nucleosomes and loading additional Sir proteins allows the Sir proteins to spread over several kilobase pairs of DNA.

The sequential deacetylation model predicts that silenced chromatin can spread continuously from points of initiation, as observed at one telomere (115). A further prediction is that catalytically inactive Sir2p should disrupt silencing even in the presence of wild-type Sir2p because its incorporation into the silenced structure should terminate the chain of deacetylated nucleosomes. In fact, many mutant alleles of *SIR2,* some of which have been demonstrated to lack deacetylase activity, are dominant negative mutations (68, 78, 116, 117). However, under other conditions some of these same alleles do not exhibit dominant negative behavior (117), implying that there is more to be learned about the spreading process.

In most cases, silenced chromatin is continuous throughout a domain. However, in some cases spreading appears to be discontinuous and to "jump" beyond sequences that, in other contexts, would behave as boundary elements (21, 118–120). These situations occur in the vicinity of additional silencer-like elements, termed protosilencers, which are binding sites for ORC or Rap1p. One explanation for these discontinuous domains of silencing is that the Sir proteins at the silencer also associate with Rap1p or ORC at the protosilencer, bringing the silencer and protosilencer into close proximity. Histones near the protosilencer would consequently be targets for Sir2p-mediated deacetylation, thereby generating a secondary site from which the Sir proteins could spread. The mechanism and regulation of this jumping await further investigation.

A paradox in our present understanding is that, in vitro, Sir2p preferentially targets lysine 16 of histone H4 (53, 61), yet deacetylation of all four lysines would seem to be required for the highest affinity binding by Sir proteins (50). Interestingly, in *S. cerevisiae* there is a higher level of acetylation at lysine 16 than at the other lysines of histone H4 (103, 121). Therefore, deacetylation of lysine 16 may be sufficient for the spreading of silenced chromatin to occur.

Finally, it is not clear why two proteins that bind histone tails, Sir3p and Sir4p, are needed. In principle, only Sir4p, which associates with Sir2p, should be necessary for spreading. Perhaps both proteins are needed to stabilize binding to nucleosomes. Alternatively, Sir3p, through its AAA domain, might drive a conformational change of either a structural protein or the chromosome itself that is required for silencing.

S Phase and Silencing

In *S. cerevisiae,* the establishment of silencing requires passage through S phase of the cell cycle (122). This S-phase requirement was widely interpreted as reflecting a role for DNA replication in silencing. However, a role for replication in silencing has been discounted by the observations that neither the initiation of replication at a silencer (108, 123) nor the passage of a replication fork through *HMR* (124–125a) is required to establish silencing. These observations motivate a reconsideration of the nature of the essential contribution that the cell cycle makes to silencing. For example, the loading of any or all of the Sir proteins at the silencers, the spreading of Sir proteins across the chromosome, or the deacetylase activity of Sir2p might be regulated by the cell cycle. In addition, other structures built during S phase, such as cohesion complexes between sister chromatids, might play a role in establishing silencing.

METABOLISM OF SILENCED CHROMATIN

The Boundaries of Silenced Chromatin

Silenced chromatin spreads efficiently from silencers, but its spreading must be limited by boundaries between silenced and expressed chromatin. At *HMR,* silencing is most robust between the *HMR-E* and *HMR-I* silencers, implying that the presence of two flanking silencers may stabilize the structure, although the silenced domain extends beyond both silencers (11, 120). The telomere-proximal boundary at *HMR* maps to a tRNA gene and depends upon the promoter of that gene being functional (126). The centromere-proximal boundary is less well mapped, and it is not clear whether there is a single boundary element or whether the endpoints of silenced chromatin vary from cell to cell in a population (8, 126).

As at *HMR*, silencing at *HML* is most robust between the silencers. On the centromere side of *HML*, no boundary element has been identified. Rather, the *HML-I* silencer may be directional (127). On the telomere side, the 1.5-kb region immediately flanking the *HML-E* silencer does not have boundary activity (127), even though silencing of a reporter gene extends only to about 1 kb (128). These observations imply that discrete boundary elements at *HML* may not exist. A genome-wide analysis detected Sir2p and Sir4p continuously from *HML* to the telomere (9), but reporter genes inserted throughout this region were not silenced (128). This discrepancy may result from Sir proteins being present in these regions, but only at low levels. Importantly, the method of analysis used in the genome-wide study did not determine the relative level of Sir protein occupancy.

The silenced domains of many natural telomeres, as determined by reporter gene silencing, are quite limited (21), and at least two telomeres (III-R and IV-L) are not associated with silenced chromatin (9, 21). These results imply either that silenced chromatin is not stable over long distances or that sequences near telomeres limit the spread of silenced chromatin. Although sequences known as STAR elements, found in X and Y' subtelomeric repeats, have been proposed to be boundaries (119), they are unlikely candidates because reporter genes are silenced when integrated on the centromere side of these STAR elements (21).

Models for boundary function start from two key premises: first, that silenced chromatin spreads continuously from points of initiation; and second, that the association of Sir3p and Sir4p with deacetylated histone tails is essential to the spreading process. Boundary elements could alter or displace nucleosomes such that they can no longer be bound by Sir proteins, thereby terminating the propagation of silenced chromatin (8, 96, 129–132). One way in which nucleosomes could be refractory to binding of Sir proteins is to be actively acetylated beyond the capacity of Sir2p to deacetylate them. In fact, some sequences shown to have boundary function are promoter elements that likely recruit acetyltransferase activities (119, 126, 133). In further support of localized acetyltransferases acting as boundaries, tethered acetyltransferases (126) or transcriptional activators (134) can act as barriers to the propagation of silenced chromatin. Likewise, mutations in the histone acetyltransferases Sas2p or Gcn5p disrupt boundary function (126), although it is not known whether these acetyltransferases act directly at boundaries. Consistent with this model, Sir3p spreads further from a telomere in *sas2*Δ cells compared to wild-type cells (106, 106a).

The propagation of silenced chromatin could also be blocked by the physical displacement or occlusion of nucleosomes such that Sir2p bound to one nucleosome no longer has an adjacent target nucleosome on which to act. Such a mechanism may act at *HMR*, where the cohesin protein Scc1p/Mcd1p, and presumably the rest of the cohesin complex, is associated with DNA sequences flanking *HMR* and overlapping with the known boundaries of the silenced domain (135). Furthermore, mutations in two cohesin components, Smc1p and Smc3p, disrupt the function of the telomere-proximal boundary element at *HMR* (120). Whether the tRNA gene coincident with this boundary is associated with a cohesin complex is not known. Alternatively, cohesins might limit the silenced domain by constraining the DNA path, thereby preventing silenced chromatin from jumping around boundary elements.

Silencing and Global Changes in Histone Modification

Genetic screens using sensitized strains have identified mutations in chromatin-modifying enzymes that perturb or enhance silencing. These mutations have generally been interpreted as acting directly at silenced loci, altering the chromatin structure in subtle ways that affect silencing. Alternatively, these mutations could act indirectly by affecting the expression of a gene required for silencing. However, recent studies suggest a third explanation for some of these mutations. In *S. cerevisiae,* acetylation and methylation of histones are associated with active chromatin, and Sir proteins may bind preferentially to unmodified histones. Consequently, the global loss of these nucleosome modifications could cause normally active chromatin to have enhanced affinity for Sir proteins. Thus, mutations in chromatin-modifying enzymes could result in promiscuous binding of Sir proteins throughout the genome, thereby diluting the effective concentration of Sir proteins and weakening silencing at normally silenced loci. Similarly, overproduction of these enzymes may result in modification of normally silenced loci, thereby preventing Sir proteins from associating. This Sir protein redistribution mechanism is consistent with observations that pools of Sir proteins are limiting and that enhancing silencing at one locus can reduce silencing at another (7, 136, 137).

The possibility that mutations in enzymes that modify histones cause the redistribution of Sir proteins was first revealed in studies of Dot1p, which methylates histone H3 at lysine 79 (97, 138, 139). Strains in which Dot1p is absent or overproduced have telomeric silencing defects (140), which result from the redistribution of Sir proteins away from telomeres (97, 139, 141). Curiously, 90% of histone H3 in a cell is methylated at lysine 79 (97), an observation that is inconsistent with this methylation being restricted to the silenced loci. Instead, methylation at lysine 79, which is located in a region of the nucleosome that probably contacts Sir proteins (103, 104), may actually disrupt the association of Sir proteins with nucleosomes (97).

Set1p, another histone methyltransferase associated with silencing (142–144), may act in a manner similar to that of Dot1p. Set1p methylates lysine 4 of histone H3 (142, 145–147). However, the requirement for methylation of lysine 4 for full silencing has been puzzling because in other organisms this methylation is generally associated with active regions of the genome (148, 149). In fact, in *S. cerevisiae,* coding regions of active genes are highly methylated, whereas telomeric regions, the *HM* loci, and rDNA are hypomethylated at lysine 4 (98). These results indicate that *set1*Δ likely influences silencing either through the misregulation of a gene required for silencing or, like *dot1*Δ, through the mislocalization of Sir proteins to unmethylated histones throughout the genome.

The ubiquitin-conjugating enzyme, Rad6p, is also important for silencing (150, 151). Rad6p ubiquitinates histone H2B at lysine 123 (152). This ubiquitination event is required for methylation of histone H3 at lysine 4 by Set1p (153, 154) and for methylation of histone H3 at lysine 79 by Dot1p (155, 156). Therefore, the effects of *rad6Δ* on silencing may be exerted through the loss of methylation at lysines 4 and 79 on histone H3. These results suggest an alternative model for the action of the ubiquitin hydrolase Dot4p that interacts with Sir4p. Dot4p may remove ubiquitin from histone H2B in nucleosomes associated with Sir4p, thereby preventing methylation of histone H3 by Dot1p and Set1p.

Given that silenced chromatin is marked by hypoacetylated histones and that histones throughout the rest of the genome are highly acetylated in *S. cerevisiae* (157), it is likely that global changes in histone acetylation also affect silencing. Indeed, mutations in the acetyltransferase enzymes Sas2p and Sas3p do modulate silencing (158, 159). Sas2p is of particular interest because genetic evidence suggests it acetylates lysine 16 of histone H4, the residue that Sir2p deacetylates. Curiously, mutations in Sas2p can both reduce and enhance silencing, depending on the situation (158, 159).

How might mutation of *SAS2* modulate silencing? Sas2p, as part of the SAS-I complex, is likely to act globally since it interacts with the chromatin assembly proteins Cac1p and Asf1p (105, 160). Thus, *sas2* mutants might be deficient in the acetylation of lysine 16 throughout the genome. By analogy with the *dot1* mutation, this reduction in acetylation could lead to Sir proteins spreading beyond the normal boundaries of silenced chromatin (106) and even to the promiscuous association of Sir proteins with nucleosomes throughout the genome, thereby reducing their effective concentration and weakening silencing. The one situation in which *sas2Δ* enhances silencing involves a mutant silencer lacking Rap1p and Abf1p binding sites (159) that is consequently unable to recruit Sir proteins (8). Perhaps in this case, the presence of underacetylated histone H4 adjacent to this weakened silencer in *sas2Δ* cells allows the Sir proteins to assemble by using the nucleosome as an anchor. The recruitment of SAS-I to chromatin, at least in part through Cac1p and Asf1p, implies that the observed loss of silencing in *cac asf1* cells (161) may also be due to a global decrease in acetylation of histones.

Importantly, the histone-modifying enzymes described here could also play a role in determining the boundaries of silenced chromatin by acting on nucleosomes at the endpoints of silenced chromatin domains.

Maintenance of Silenced Chromatin

Maintenance of silenced chromatin refers to the preservation of the silenced state within a cell cycle, whereas inheritance refers to perpetuation of that state upon DNA replication and cell division. Defects in the maintenance of silenced chromatin are characterized by rapid switching between expression states. Such maintenance defects have been detected using an assay in which silencing is measured in single cells. ***MATa*** cells exposed to alpha factor will arrest in G1 only if *HMLα* is completely silenced, and thus, cells with defects in maintenance transiently break out of the arrest. A subset of these cells reestablish silencing in subsequent cell cycles, resulting in a cluster of cells in which many cells are arrested in G1 (162).

Factors affecting the maintenance of silenced chromatin fall into three categories—structural proteins, silencers and silencer binding proteins, and chromatin assembly factors. By definition, structural components of silenced chromatin would be required for the maintenance of silencing. In fact, shifting cells bearing a temperature-sensitive allele of *SIR3* to the restrictive temperature results in the loss of silencing (122, 163). Interestingly, some mutations of structural components allow the establishment of silenced chromatin but this chromatin is not stably maintained. That is, chromatin structures with suboptimal stability can be built, but they fall apart easily. Such mutations include a *sir3* allele, reduced levels of Sir2p or Sir3p, and some histone tail mutations (162, 164). These observations are consistent with the dynamic properties of silenced chromatin described above.

Silencers, being the sites at which Sir proteins first assemble, are logically linked to the establishment of silencing. However, they also play a role in the maintenance of silenced chromatin. Excised DNA circles bearing *HMR* but no silencers become derepressed in G1-arrested cells and are less stable in vitro than circles bearing silencers (99, 165). Thus, silencers help to stabilize silenced chromatin. In contrast to *HMR, HML* retains the silenced state in the absence of silencers (99, 163) as long as a Rap1p binding site in the middle of *HML* is present (99). Individual Rap1p binding sites do not act as silencers but have been called protosilencers because of their ability to stabilize silenced chromatin (118), probably through interactions between DNA, Rap1p, and Sir proteins. Given that a Rap1p binding site is sufficient to maintain silenced chromatin in the absence of silencers, the full function of a silencer is probably not required for maintenance of silenced chromatin.

Chromatin assembly factors also contribute to the maintenance of silenced chromatin. Mutations in the chromatin assembly factor (CAF-I) subunit Cac1p result in transient loss of silencing (162) and increased frequency of switching of the expression state of telomeric reporter genes (166). The defect in maintenance of silencing in *cac1* mutants may result from silenced chromatin being assembled on a weak foundation of improperly assembled or modified nucleosomes (162). Alternatively, *cac1* mutants may affect maintenance indirectly. As outlined above, *cac* mutants might be partially deficient in the global acetylation of histone H4, leading to the promiscuous association of Sir proteins with nucleosomes throughout the genome.

Interactions between CAF-I and proliferating cell nuclear antigen (PCNA) (161, 167, 168) imply that mutations in PCNA might also have defects in maintenance of silenced chromatin. In fact, some alleles of the essential gene *POL30,* which encodes PCNA, result in intermediate levels of expression of an

ADE2 reporter in silenced chromatin (169). However, other mutations result in two semistable populations of silenced and expressed cells, a characteristic indicative of a defect in establishment. The different behavior of alleles of *pol30* is explained by these mutations disrupting interactions of PCNA with different binding partners (161).

Epigenetic Inheritance of Silenced Chromatin

Epigenetic inheritance refers to the heritability of two distinctly different expression states in otherwise identical cells. This heritability reflects the ability of a specialized chromatin structure to template its own reformation. Inheritance occurs shortly after DNA replication. This duplication is all the more remarkable considering that the act of DNA replication is at least partially disruptive to existing higher-order chromatin structures, which would then need to be rebuilt. Logistically, whether a given expression state has been duplicated and inherited is determined by examining that state in the two daughter cells following cell division.

Conceptually, the processes of establishment and inheritance of a chromatin state differ substantially. To establish a chromatin state, all the components of the new structure must be recruited to the locus de novo. In contrast, when a structure is inherited, both sister chromatids could inherit a partial structure consisting of correctly modified histones and structural proteins. Given the right affinities and circumstances, these partial structures could then seed the formation of the complete structure through both cooperative interactions and modifications of newly assembled nucleosomes. In principle, inheritance of a state should be more probable than its initial establishment. This distinction is well illustrated by mutations in certain silencing components that result in a mixed population of completely repressed and completely derepressed cells (54, 170, 171). The expression state of any given cell within these populations is stably inherited over several generations, indicating that once the silenced state is established, it is stably inherited.

The mechanism by which a chromatin state templates its own reformation involves a self-reinforcing cycle of chromatin modification (172). Specifically, the inheritance of a chromatin state requires a mark that is inherited on both DNA duplexes following replication and the ability of that mark to recruit an enzyme that makes an additional mark (8, 173, 174). This concept is rooted in Holliday's model for the inheritance of DNA methylation patterns (175). In that model, DNA methylation on both strands of a palindromic sequence provides the mark, and DNA methyltransferases that recognize the hemi-methylated state formed immediately after replication recreate the mark on the newly synthesized strands. In addition to methylated DNA, other types of marks can be inherited on both sister chromatids, such as modified histones or histone variants. Upon passage of the replication fork, the two H2A-H2B dimers disassociate from both the H3-H4 tetramer and from DNA. The H3-H4 tetramers remain associated with DNA (176) and are randomly distributed to the two sister molecules (177, 178). Thus, any epigenetic mark associated with histones would logically be on H3 and/or H4.

In Sir-mediated silencing, the hypoacetylated H3-H4 tetramer appears to be the mark, providing a high-affinity binding site for the Sir complex. It is not clear whether Sir proteins remain bound to the H3-H4 tetramer during replication or rebind immediately afterward. In either case, the Sir complex bound to the old hypoacetylated nucleosome could deacetylate the H3 and H4 tails of newly deposited nucleosomes, which are acetylated, thus creating new marks (8, 96). In addition, cooperative interactions among the nucleosome-bound Sir proteins, silencer binding proteins, and unbound Sir proteins may help direct Sir3p and Sir4p to those newly deacetylated histones in silenced chromatin rather than deacetylated histones elsewhere in the genome. Consistent with this hypothesis, silencers are required for the inheritance of silenced chromatin (17, 163). Silencers probably function in inheritance in much the same way as they do in maintenance but are more crucial during this most disruptive phase of the cell cycle.

This model for inheritance of Sir-silenced chromatin raises the issue of which lysine residues Sir2p deacetylates when silenced chromatin is duplicated. Newly synthesized histones are acetylated at lysine 9 of histone H3 and lysines 5 and 12 of histone H4 (179, 180). In vitro, Sir2p does deacetylate lysine 9 of histone H3, but lysines 5 and 12 of histone H4 are deacetylated only very inefficiently (53). However, mutational analysis of histone H4 suggests that deacetylation of lysines 5 and 12 may not be essential for functional silenced chromatin (101, 181). Sir2p may also be required to deacetylate lysine 16 of histone H4 if Sas2p is active on newly replicated DNA as proposed (105).

In addition to self-reinforcing marks, another way in which chromatin states could be inherited is through the coupling of chromatin assembly and nucleosome modification to DNA replication. For example, SAS-I could be recruited to newly replicated DNA through Cac1p, Asf1p, and PCNA. Such a link would lead to newly synthesized DNA having a particular chromatin state characterized by acetylated nucleosomes. However, it is not clear whether the same replication-coupled chromatin assembly occurs throughout the genome, creating a uniform default state of acetylation, or whether a mechanism exists by which the particular region being replicated is recognized and the correct chromatin state for that region is assembled. The multitude of different chromatin assembly factors and DNA polymerases and variations in organization of origins of replication certainly allow for more complexity than has yet been appreciated.

RELATIONSHIP BETWEEN SILENCING AND REPRESSION

The mechanisms by which Sir-silenced chromatin spreads and is restricted to particular regions of the genome apply to other types of repression in *S. cerevisiae* as well as in other organisms. As with Sir silencing, other specialized chromatin states require a mark and the ability of that mark to recruit an enzyme or complex that makes an additional mark. Interestingly, *S. cerevisiae* appears to use a unique marking system for silenced chromatin. In most other organisms studied, silenced chromatin is marked by methylation of lysine 9 on histone H3, a mark which is recognized and bound by chromodomain-containing proteins (173). In turn, these chromodomain-containing proteins recruit histone methyltransferases of the SuVar3-9 family, which methylate lysine 9 of histone H3, thereby making additional marks. Thus, silenced chromatin in other organisms is marked and maintained using strategies similar to those used in *S. cerevisiae* but with completely different proteins.

Studies of the *SUM1-1* mutation highlight an unexpected link between gene-specific repression, in which spreading is, at most, limited, and the creation of more extensive silenced domains of chromatin. The wild-type Sum1p is a repressor that binds to operator sequences at promoters of genes involved in meiosis, leading to their repression in mitotic cells (182). In contrast, Sum1-1p, which differs from Sum1p by a single critical amino acid, restores *HM* silencing in strains lacking Sir2p, Sir3p, and Sir4p (183, 184). Sum1-1p acts directly at the *HM* loci where, in association with Hst1p, a paralog of Sir2p, it creates an alternate form of heterochromatin (185, 186). In the absence of Hst1p, Sum1-1p associates with the silencer but does not spread (8), indicating that the deacetylase Hst1p is required for the propagation of Sum1-1p-containing silenced chromatin. Thus, Sum1-1p-mediated silencing occurs through a mechanism that is remarkably similar to Sir-mediated silencing and yet uses different proteins. Additionally, the minor sequence change in Sum1p that is required to convert a promoter-specific repressor to a silencing protein highlights the narrow margin separating repression from silencing and underscores the similarities in these two modes of repression.

Tup1-Ssn6 is a gene-specific repressor in *S. cerevisiae* whose mechanism also has similarities to Sir protein–mediated silencing. The Tup1-Ssn6 corepressor complex is recruited to various promoters through DNA-binding repressor proteins, much as Sir proteins are recruited to silenced loci by silencer-binding proteins. Tup1-Ssn6, in turn, brings a histone deacetylase to the promoter (187, 188). The resultant hypoacetylated histone tails can be bound by Tup1p (189), which then spreads to a limited extent (187, 190, 191). Thus, the mark created by the Tup1-Ssn6 repressor is also a hypoacetylated nucleosome. The factors that limit the spreading of Tup1-Ssn6 are not known. In principle, the inability of Tup1-Ssn6 to spread extensively could be due either to some property of the complex itself or to the presence of boundaries that block its spreading. Tup1-Ssn6 also represses transcription by contacting the mediator complex directly (192, 193).

HOW DOES SILENCED CHROMATIN INHIBIT TRANSCRIPTION?

The function of silenced chromatin at the *HM* loci is to prevent transcription, but surprisingly little is known about the exact mechanism by which silenced chromatin blocks transcription. Most early models proposed that the compact structure of Sir chromatin inhibited the association of RNA polymerase II with the DNA by steric mechanisms. However, the TATA binding protein (TBP) and pol II are both present at the *HMR* **a**1 promoter even when that gene is silenced (194). Furthermore, the promoter sequence of *HMLα* is actually more accessible to micrococcal nuclease than is the promoter at *MATα* (15), implying that this region could be accessible to transcription factors as well. These results are consistent with Sir proteins blocking a step after formation of the preinitiation complex. A comparable mechanism for repression is seen for heat shock promoters, which contain stalled RNA polymerases bound to the promoters in non-heat-shocked condition (195, 196). Similarly, components of the preinitiation complex are associated with the repressed *CYC1* promoter in *S. cerevisiae* (197) and with promoters of *Drosophila* genes repressed by polycomb group proteins (198).

Several steps in the transcription process are regulatable, including preinitiation complex assembly, open complex formation, promoter clearance, and elongation. Thus, Sir proteins could inhibit transcription in several ways. First, Sir2p might deacetylate a transcription factor, thereby reducing its activity. Consistent with this model, mammalian Sir2p homologs deacetylate p53 (70, 71) and TAF_I68 (69). Alternatively, Sir proteins might inhibit the transcription machinery through direct interactions independent of deacetylation. In keeping with this model, two *Drosophila* proteins, HP1 and the PRC1 polycomb complex, both of which are associated with repressive chromatin, interact directly with TAFs (199, 200). Finally, Sir proteins might alter the properties of nucleosomes in such a way as to block either promoter clearance or elongation. The binding of Sir proteins to nucleosomes could make their structure more rigid and less able to undergo conformational changes necessary for transcription, including subtle structural rearrangements, displacement of H2A-H2B dimers (176, 201), or displacement of whole nucleosomes (202). Clearly, a more detailed analysis of pol II at the *HMR* **a**1 promoter is required to delineate at which step(s) transcription is blocked.

rDNA SILENCING

When genes transcribed by RNA pol II are inserted into the rDNA array, their expression, as measured in a population of cells, is reduced but not completely eliminated (23, 24). This "rDNA silencing" requires Sir2p but not Sir3p or Sir4p (23). Presumably, the Sir2p deacetylase activity is involved in the propagation of a specialized chromatin state at the rDNA, much as at the *HM* loci or telomeres, but using adaptor proteins other than Sir3p and Sir4p. In fact, histone H4 is hypoacetylated at the rDNA array, and rDNA silencing is affected by mutations in all four histones and multiple genes involved with DNA replication or chromatin modulation (24, 48, 203). The mechanism of rDNA silencing remains unresolved.

The *S. cerevisiae* rDNA array consists of a 9.1-kilobase-pair sequence that is tandemly repeated 100 to 200 times. Each rDNA repeat encodes the 35S rRNA, which is the precursor to the 25S, 18S, and 5.8S rRNAs and is transcribed by pol I, and the 5S rRNA, which is transcribed by pol III. These two genes are separated by nontranscribed spacers, NTS1 and NTS2 (Figure 1, *bottom*). Only a fraction of the 35S genes are transcribed by pol I at a given time. Psoralen cross-linking and electron microscopy imaging studies reveal that the coding regions of the active repeats are largely devoid of nucleosomes, whereas the inactive repeats are associated with nucleosomes (204). The spacer regions of both active and inactive repeats are packaged in regular nucleosomal arrays, and the pol I promoter is associated with transcription factors (204, 205). Taking these observations into account, a specialized chromatin structure that is repressive to pol II is likely to be discontinuous with respect to the entire array and centered on the spacer regions. In fact, Sir2p is more highly associated with the spacer regions than with the 35S coding region (48, 206). Presumably a sequence within the spacers is responsible for recruiting Sir2p and associated proteins. However, no silencer sequence for rDNA silencing has yet been identified.

Based upon the partnership of Sir2p with Sir4p in *HM* and telomeric silencing, Sir2p would be expected to have a protein partner with high affinity for deacetylated histone tails to enable it to spread at the rDNA array. Sir2p is a component of the nucleolar RENT complex (207), and the members of this complex are currently the only proteins involved in rDNA silencing that are candidates to be structural components of chromatin at the rDNA. In addition to Sir2p, RENT contains Net1p, which is required for association of Sir2p with the rDNA, and Cdc14, a phosphatase required for mitotic exit (207). Clearly, Cdc14p and Net1p should be tested for their histone and nucleosome binding properties.

The silencing of pol II genes inserted at the rDNA is an artificial situation, yet it must reflect some aspect of rDNA structure that is incompatible with transcription by pol II. An early model proposed that Sir2p regulates the transition of rDNA repeats from the open to closed conformations, thereby regulating transcription of both the 35S coding sequence and the pol II genes that happen to reside in the rDNA array. However, the ratio of closed to open repeats does not change in *sir2*Δ cells (208). Instead, silencing may be related to a mechanism that favors transcription by pol I through the inhibition of pol II. Within the pol I promoter region is a cryptic pol II promoter (209). The pol I transcription factor UAF specifically inhibits transcription by pol II from this promoter (210, 211) and could also inhibit the expression of pol II–transcribed reporter genes inserted into the rDNA array as well as Ty1 elements that frequently integrate near the pol III transcribed 5S gene. How Sir2p might participate in this process is not clear.

SILENCING AS ENTRÉE TO OTHER ASPECTS OF CHROMATIN METABOLISM

Sir2p, rDNA Stability, and NAD^+ Metabolism

S. cerevisiae cells divide by budding asymmetrically to generate a larger cell, the mother, and a smaller cell, the daughter. Life span is measured as the number of daughters that a mother cell can produce. As *S. cerevisiae* age, mother cells divide more slowly, increase in volume, and have decreased fertility due to loss of silencing at the *HM* loci (212). Daughters from old mothers have short life spans, but those daughters' daughters have normal life spans (213). These observations indicate that an aging factor accumulates in, and is preferentially inherited by, mother cells over time.

One aging factor in *S. cerevisiae* appears to be extrachromosomal rDNA circles, or ERCs, which are formed by recombination within the rDNA repeats (214, 215). Because each rDNA repeat contains an origin of replication, the excised circles replicate each S phase as a plasmid. However, because these circles lack a centromere, they are not efficiently segregated to both cells upon division. Rather, such plasmids are preferentially retained by the larger mother cells. As a result, aging mother cells accumulate vast amounts of these circular DNA molecules (214, 215). The causal link between the accumulation of plasmids in mother cells and aging was established by the experimental creation of extrachromosomal circles lacking any rDNA sequences that still caused accelerated aging of mother cells (214). Extrachromosomal circles could affect cells by competing with chromosomal loci for essential components of the replication and transcription machinery.

Sir2p affects the rate of formation of ERCs. Recombination at the rDNA array increases in cells lacking Sir2p (215, 216) and results in a more rapid accumulation of ERCs and a decreased life span (215). In contrast, life span is increased in cells overproducing Sir2p (215). Consistent with Sir2p deacetylase activity having a role in determining life span, cells containing catalytically inactive mutants have shortened life spans relative to wild-type cells, even though such mutants of Sir2p are recruited to the rDNA (53, 116). Although the accumulation of ERCs correlates with aging, several mutants that lead to shortened life spans do not accumulate these circles (217–219).

Aging is associated with a curious redistribution of Sir3 protein. In young cells, Sir2p localizes to the nucleolus (206), and the bulk of Sir3p and Sir4p are found in telomere-containing foci at the nuclear periphery (137, 206). However, Sir3p relocalizes to the nucleolus in old cells (220), and this relocalization requires catalytically active Sir2p but not Sir4p (48, 206). Sir3p could be recruited to the rDNA locus through histones deacetylated by Sir2p, by binding Sir2p directly, or through an unidentified protein whose function is modulated by Sir2p. Although the localization of Sir4p in old cells has not been determined, the cellular redistribution of at least Sir3p late in life presumably causes the derepression of *HM* loci, which in turn results in the mating defect of old mother cells (221). It is unlikely that the relocalization of Sir3p to the nucleolus promotes longevity because *sir3* cells display only a slight reduction in life span, which is attributed to the **a**/α state (215).

The most common influence on aging in a wide range of organisms is the ability of caloric restriction to increase life span. Consequently, interest in the role of Sir2p in aging has focused on the relationship between the nutritional status of cells and the ability of Sir2p to modulate life span. The requirement of Sir2p for NAD^+ in its deacetylation activity provides one potential link (222). Like other organisms, caloric restriction in *S. cerevisiae,* through limiting glucose in media or by mimicking caloric restriction genetically, leads to an increased life span (223, 224). This prolonged life span correlates with enhanced rDNA silencing and is partially dependent on Sir2p (223, 224). Caloric restriction has been hypothesized to enhance Sir2p activity by increasing the cellular concentration of NAD^+. Presumably, the shift from fermentation to respiration upon caloric restriction leads to the oxidation of more NADH, thus increasing the ratio of NAD^+ to NADH, and to the enhancement of the activity of Sir2p (223). Although this model regarding NAD^+/NADH levels is attractive, the relationship between NAD^+ levels, Sir2p activity, and aging are likely more complex.

To appreciate this complexity, a brief review of NAD^+ biosynthetic pathways will provide the framework for discussing several recent results. In *S. cerevisiae,* NAD^+ biosynthesis occurs by both a de novo and a salvage pathway (Figure 3). The salvage pathway regenerates NAD^+ from nicotinamide, which can be produced by the reaction catalyzed by the Sir2 family of deacetylases. Nicotinamide is converted into nicotinic acid by the enzyme Pnc1p. Nicotinic acid is converted to nicotinic acid mononucleotide by the enzyme Npt1p. At this point, the salvage pathway merges with the de novo pathway to ultimately generate NAD^+.

Removal of the enzymes Qpt1p and Bna1p in the de novo pathway has no effect on cellular concentrations of NAD^+, presumably because of the import and subsequent processing of nicotinic acid by the salvage pathway (75). These mutations also have no effect on Sir2p functions under normal growth conditions (75, 224). In contrast, loss-of-function alleles of the salvage pathway enzyme (*npt1*) result in a reduction in silencing as well as a decrease in NAD^+ concentration (52, 75). Interestingly, other mutations of enzymes in the salvage pathway, including *pnc1*Δ and overproduction of Npt1p, do affect Sir2p-mediated processes but do not produce changes in NAD^+ levels (75, 76). Importantly, these mutations could alter the concentration of nicotinamide, a known inhibitor of Sir2p catalytic activity (51). For example, nicotinamide may accumulate in cells lacking Pnc1p, resulting in the inhibition of Sir2p. Also, overproduction of Sir2p can rescue silencing in *npt1*Δ mutants (75), perhaps by producing enough Sir2 enzyme to overcome the inhibitory effect of nicotinamide. Clearly, perturbations in the biosynthesis of NAD^+ could affect Sir2 function in more than one way.

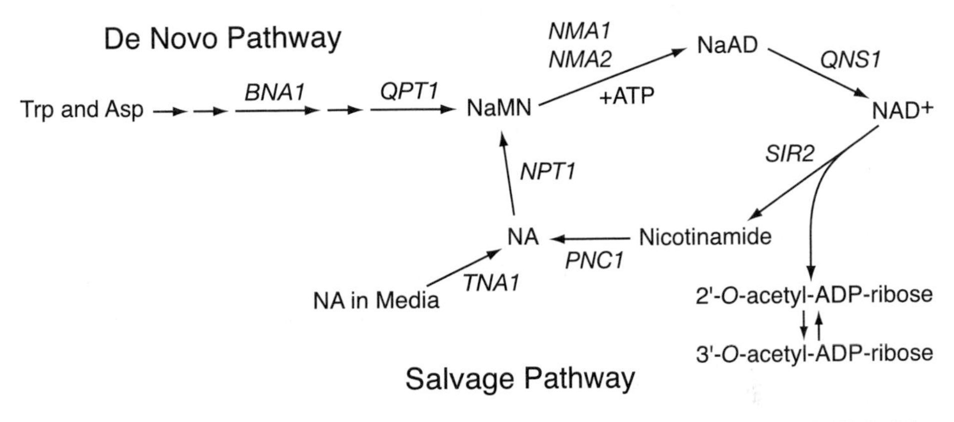

Figure 3 NAD^+ biosynthetic pathways. Trp, tryptophan; Asp, aspartate; *BNA1*, 3-hydroxyanthranilate 3,4-dioxygenase gene; *QPT1*, nicotinate-nucleotide pyrophosphorylase (carboxylating) gene; NaMN, nicotinic acid mononucleotide; *NMA1*, *NMA2*, nicotinamide mononucleotide adenyltransferase genes; NaAD, desamido NAD^+; *QNS1*, NAD^+ synthase gene; *PNC1*, nicotinamidase gene; NA, nicotinic acid; *TNA1*, nicotinic acid plasma membrane permease gene; *NPT1*, nicotinate phosphoribosyltransferase gene.

Paradoxically, aging mother cells have increased steady-state levels of NAD^+ (225), which would be expected to enhance the activity of Sir2p and counter aging by slowing the formation of extrachromosomal circles. The mechanism of this response is unknown but could represent an anti-aging defense.

Sir Proteins and Checkpoint Arrest

Recent studies identify a role for the Sir proteins in the meiotic pachytene checkpoint. On completion of DNA replication, meiotic cells undergo recombination between homologous chromosomes, and the synaptonemal complex forms along the lengths of the paired homologous chromosomes. Pachytene is the meiotic stage at which recombination is completed, the synaptonemal complex is disassembled, and the cell commits to undergoing meiotic divisions. The pachytene checkpoint is activated by defects in recombination or synaptonemal complex formation (226). Two proteins that are required to maintain the pachytene arrest, Pch2p and Dot1p, reveal a role for Sir2p in this process (141, 227).

Pch2p is a meiosis-specific protein, most of which localizes to the nucleolus (227). Pch2p is delocalized in *sir2*Δ cells, which are also defective in maintaining pachytene arrest (227). Similarly, Pch2p is delocalized in *dot1*Δ cells (141). As described earlier, Dot1p methylates lysine 79 on histone H3 in most nucleosomes throughout the genome, and the loss of this methylation is hypothesized to cause the delocalization of Sir proteins (97, 138, 139, 228). Therefore, the mislocalization of Pch2p in *dot1Δ* cells could be due to the mislocalization of Sir2p. Thus, the presence of Pch2p and Sir2p in the nucleolus correlates with checkpoint function.

The role of Sir2p in the pachytene arrest may parallel its role in the exit from mitosis. Cdc14p, a phosphatase required for exit from mitosis, is localized to the nucleolus as a component of the RENT complex, which also contains Sir2p and Net1p (207, 229). At the point in the cell cycle when Cdc14p is required to act, it is released from the nucleolus to the nucleoplasm, where it gains access to its substrates (229, 230). Perhaps Pch2p, together with Sir2p, sequesters in the nucleolus a protein that is required for exit from pachytene (226).

Sir Proteins and DNA Repair

Recent studies highlight the possibility that Sir proteins are part of the cellular response to double-strand breaks in the chromosomes. The first link between Sir proteins and DNA repair was made by a two-hybrid interaction between Sir4p and Hdf1p, a subunit of the *S. cerevisiae* Ku complex (112). *S. cerevisiae* Ku is a heterodimer composed of Hdf1p (yKu70p) and Hdf2p (yKu80p) that binds the ends of double-stranded DNA breaks to facilitate nonhomologous end joining (231). In addition, Ku binds to telomeres, where it plays a role in telomere length maintenance and facilitates the localization of telomeres in clusters at the nuclear periphery. The telomeric localization of Hdf1p and Hdf2p is Sir independent (232), indicating that these proteins have affinity for telomeres that does not require interactions with Sir proteins.

Sir proteins localize to telomeres through protein-protein interactions with the Ku complex as well as with Rap1p and histones H3 and H4 (113, 114). The importance of the Ku complex for stabilizing the association of the Sir proteins with telomeres is underscored by observations that both *hdf1* and *hdf2* mutations reduce telomere position effect (232–234) and that tethered Hdf1p can nucleate silencing (113, 114).

The interaction between Sir4p and the Ku complex suggested that Sir proteins might also be involved in DNA repair. Evidence for such a role comes from observations that DNA damage causes Ku proteins and Sir proteins to be released from telomeres and associate with double-strand breaks (114, 235, 236). Interestingly, Sir3p localization to double-strand breaks has slower kinetics than Hdf1p (114, 235), indicating that Sir proteins are not part of the initial response to damage.

The key issue raised by these observations is whether the Sir proteins actually have a role in DNA repair or merely appear at the sites of double-strand breaks by virtue of their interaction with Hdf1p and Hdf2p. Cells lacking Sir proteins do have defects in double-strand break repair through nonhomologous end joining (112, 233, 237). However, this defect is an indirect effect primarily due to the expression of both mating type genes in *sir* mutants (237–240) and the subsequent repression of the nonhomologous end-joining repair genes *NEJ1* and *LIF1* by the **a**1/α2 heterodimer (241, 242).

Another situation in which Sir proteins might create an altered chromatin structure in response to stressed DNA is on dicentric chromosomes. In cells containing dicentric chromosomes, the region between the two centromeres becomes stretched in *hdf1, hdf2,* and *sir2* mutants but not in their wild-type counterpart (243). These observations imply that together Ku and Sir proteins prevent the DNA from decompacting under stress. The role of Sir2p in this process appears to be direct since Sir2p localizes to the region between the two centromeres on dicentric but not monocentric chromosomes (243). Perhaps Sir2p and Ku associate with this DNA when it is under stress.

CONCLUSION

Our understanding of silencing in *S. cerevisiae* has increased dramatically in recent years. Two contributions in particular have revolutionized our view of silenced chromatin: the discovery of the NAD^+-dependent deacetylase activity of Sir2p and the elaboration of the histone code hypothesis. The deacetylase activity of Sir2p, together with the preferences of Sir3p and Sir4p to bind deacetylated histone tails, provides a molecular model for the formation and perpetuation of silenced chromatin. The histone code hypothesis informs our thinking about the interdependencies of histone modifications, proteins with high affinity for those modifications, and enzymes that create those modifications, making it clear that common mechanisms mediate silencing as well as gene-specific repression and even other types of specialized chromatin states. The major unresolved issues in understanding silenced chromatin are mechanistic. Although genes that have roles in silencing continue to be identified, the biggest advances in the coming years will involve in vitro assays to establish mechanisms for spreading, maintenance, and inheritance of silenced chromatin and to test hypotheses based on genetic studies.

ACKNOWLEDGMENTS

We thank Harry Charbonneau, Dan Gottschling, Paul Kaufman, Michael Kobor, and Jeffrey Smith for their helpful comments and suggestions on this manuscript. We are also grateful to many colleagues for preprints and unpublished manuscripts. This work is supported by postdoctoral fellowships from the American Cancer Society to LNR (PF-01-116-01-GMC) and ALK (PF-01-126-01-MBC), a Hatch grant from the U.S. Department of Agriculture (IND053072) to ALK, and a grant from the National Institutes of Health (GM31105) to JR. This paper is number 16919 from the Purdue University Agricultural Experiment Station.

The *Annual Review of Biochemistry* is online at http://biochem.annualreviews.org

LITERATURE CITED

1. Loo S, Rine J. 1995. *Annu. Rev. Cell Dev. Biol.* 11:519–48
2. Schultz J. 1936. *Proc. Natl. Acad. Sci. USA* 22:27–33
3. Kayne PS, Kim U-J, Han M, Mullen JR, Yoshizaki F, Grunstein M. 1988. *Cell* 55:27–39
4. Braunstein M, Rose AB, Holmes SG, Allis CD, Broach JR. 1993. *Genes Dev.* 7:592–604
5. Suka N, Suka Y, Carmen AA, Wu J, Grunstein M. 2001. *Mol. Cell* 8:473–79
6. Hecht A, Strahl-Bolsinger S, Grunstein M. 1996. *Nature* 383:92–96
7. Strahl-Bolsinger S, Hecht A, Luo K, Grunstein M. 1997. *Genes Dev.* 11:83–93
8. Rusche LN, Kirchmaier AL, Rine J. 2002. *Mol. Biol. Cell* 13:2207–22
9. Lieb JD, Liu X, Botstein D, Brown PO. 2001. *Nat. Genet.* 28:327–34
10. Zhang Z, Hayashi MK, Merkel O, Stillman B, Xu RM. 2002. *EMBO J.* 21:4600–11
11. Loo S, Rine J. 1994. *Science* 264: 1768–71
12. Gottschling DE. 1992. *Proc. Natl. Acad. Sci. USA* 89:4062–65
13. Singh J, Klar AJ. 1992. *Genes Dev.* 6:186–96
14. Ravindra A, Weiss K, Simpson RT. 1999. *Mol. Cell. Biol.* 19:7944–50
15. Weiss K, Simpson RT. 1998. *Mol. Cell. Biol.* 18:5392–403
16. Cheng T-H, Li Y-C, Gartenberg MR. 1998. *Proc. Natl. Acad. Sci. USA* 95:5521–26
17. Bi X, Broach JR. 1997. *Mol. Cell. Biol.* 17:7077–87
18. Haber JE. 1998. *Annu. Rev. Genet.* 32:561–99
19. Gottschling DE, Aparicio OM, Billington BL, Zakian VA. 1990. *Cell* 63:751–62
20. Aparicio OM, Billington BL, Gottschling DE. 1991. *Cell* 66: 1279–87
21. Pryde FE, Louis EJ. 1999. *EMBO J.* 18:2538–50
22. Petes TD, Botstein D. 1977. *Proc. Natl. Acad. Sci. USA* 74:5091–95
23. Smith JS, Boeke JD. 1997. *Genes Dev.* 11:241–54
24. Bryk M, Banerjee M, Murphy M, Knudsen KE, Garfinkel DJ, Curcio MJ. 1997. *Genes Dev.* 11:255–69
25. Bernstein BE, Tong JK, Schreiber SL. 2000. *Proc. Natl. Acad. Sci. USA* 97:13708–13
26. Wyrick JJ, Holstege FC, Jennings EG, Causton HC, Shore D, et al. 1999. *Nature* 402:418–21
27. Abraham J, Nasmyth KA, Strathern JN, Klar AJS, Hicks JB. 1984. *J. Mol. Biol.* 176:307–31
28. Feldman JB, Hicks JB, Broach JR. 1984. *J. Mol. Biol.* 178:815–34
29. Brand AH, Breeden L, Abraham J, Sternglanz R, Nasmyth K. 1985. *Cell* 41:41–48
30. Mahoney DJ, Broach JR. 1989. *Mol. Cell. Biol.* 9:4621–30
31. Dubey DD, Davis LR, Greenfeder SA, Ong LY, Zhu J, et al. 1991. *Mol. Cell. Biol.* 11:5346–55
32. Rivier DH, Rine J. 1992. *Science* 256: 659–63
33. Rivier DH, Ekena JL, Rine J. 1999. *Genetics* 151:521–29
34. Planta RJ, Goncalves PM, Mager WH. 1995. *Biochem. Cell Biol.* 73:825–34
35. Bell SP, Dutta A. 2002. *Annu. Rev. Biochem.* 71:333–74
36. McNally FJ, Rine J. 1991. *Mol. Cell. Biol.* 11:5648–59
37. Brand AH, Micklem G, Nasmyth K. 1987. *Cell* 51:709–19
38. Foss M, McNally FJ, Laurenson P, Rine J. 1993. *Science* 262:1838–44
39. Loo S, Fox CA, Rine J, Kobayashi R,

Stillman B, Bell S. 1995. *Mol. Biol. Cell* 6:741–56
40. Sussel L, Shore D. 1991. *Proc. Natl. Acad. Sci. USA* 88:7749–53
41. Loo S, Laurenson P, Foss M, Dillin A, Rine J. 1995. *Genetics* 141:889–902
42. Cockell M, Palladino F, Laroche T, Kyrion G, Liu C, et al. 1995. *J. Cell Biol.* 129:909–24
43. Stavenhagen JB, Zakian VA. 1994. *Genes Dev.* 8:1411–22
44. Rine J, Herskowitz I. 1987. *Genetics* 116:9–22
45. Ivy JM, Klar AJS, Hicks JB. 1986. *Mol. Cell. Biol.* 6:688–702
46. Moazed D, Kistler A, Axelrod A, Rine J, Johnson AD. 1997. *Proc. Natl. Acad. Sci. USA* 94:2186–91
47. Ghidelli S, Donze D, Dhillon N, Kamakaka RT. 2001. *EMBO J.* 20: 4522–35
48. Hoppe GJ, Tanny JC, Rudner AD, Gerber SA, Danaie S, et al. 2002. *Mol. Cell. Biol.* 22:4167–80
49. Hecht A, Laroche T, Strahl-Bolsinger S, Gasser SM, Grunstein M. 1995. *Cell* 80:583–92
50. Carmen AA, Milne L, Grunstein M. 2002. *J. Biol. Chem.* 277:4778–81
51. Landry J, Slama JT, Sternglanz R. 2000. *Biochem. Biophys. Res. Commun.* 278:685–90
52. Smith JS, Brachmann CB, Celic I, Kenna MA, Muhammad S, et al. 2000. *Proc. Natl. Acad. Sci. USA* 97:6658–63
53. Imai S, Armstrong CM, Kaeberlein M, Guarente L. 2000. *Nature* 403:795–800
54. Pillus L, Rine J. 1989. *Cell* 59:637–47
55. Ptashne M. 1992. *A Genetic Switch.* Cambridge, MA: Blackwell Sci.
56. Turner BM. 2000. *BioEssays* 22:836–45
57. Strahl BD, Allis CD. 2000. *Nature* 403: 41–45
58. Jenuwein T, Allis CD. 2001. *Science* 293:1074–80
59. Brachmann CB, Sherman JM, Devine SE, Cameron EE, Pillus L, Boeke JD. 1995. *Genes Dev.* 9:2888–902
60. Landry J, Sutton A, Tafrov ST, Heller RC, Stebbins J, et al. 2000. *Proc. Natl. Acad. Sci. USA* 97:5807–11
61. Tanny JC, Moazed D. 2001. *Proc. Natl. Acad. Sci. USA* 98:415–20
62. Tanner KG, Landry J, Sternglanz R, Denu JM. 2000. *Proc. Natl. Acad. Sci. USA* 97:14178–82
63. Jackson MD, Denu JM. 2002. *J. Biol. Chem.* 277:18535–44
64. Sauve AA, Celic I, Avalos J, Deng H, Boeke JD, Schramm VL. 2001. *Biochemistry* 40:15456–63
65. Zatman L. 1953. *J. Biol. Chem.* 200: 197–212
66. Rowen J. 1951. *J. Biol. Chem.* 193: 497–507
67. Frye RA. 1999. *Biochem. Biophys. Res. Commun.* 260:273–79
68. Tanny JC, Dowd GJ, Huang J, Hilz H, Moazed D. 1999. *Cell* 99:735–45
69. Muth V, Nadaud S, Grummt I, Voit R. 2001. *EMBO J.* 20:1353–62
70. Luo JY, Nikolaev AY, Imai S, Chen DL, Su F, et al. 2001. *Cell* 107:137–48
71. Vaziri H, Dessain SK, Eaton EN, Imai SI, Frye RA, et al. 2001. *Cell* 107: 149–59
72. Slama JT, Simmons AM. 1988. *Biochemistry* 27:183–93
73. Slama JT, Simmons AM. 1989. *Biochemistry* 28:7688–94
74. Min J, Landry J, Sternglanz R, Xu RM. 2001. *Cell* 105:269–79
75. Sandmeier JJ, Celic I, Boeke JD, Smith JS. 2002. *Genetics* 160:877–89
76. Anderson RM, Bitterman KJ, Wood JG, Medvedik O, Cohen H, et al. 2002. *J. Biol. Chem.* 277:18881–90
77. Banasik M, Ueda K. 1994. *Mol. Cell. Biochem.* 138:185–97
78. Sherman JM, Stone EM, Freeman-Cook LL, Brachmann CB, Boeke JD, Pillus L. 1999. *Mol. Biol. Cell* 10:3045–59
79. Grozinger CM, Chao ED, Blackwell

HE, Moazed D, Schreiber SL. 2001. *J. Biol. Chem.* 276:38837–43
80. Yoshida M, Kijima M, Akita M, Beppu T. 1990. *J. Biol. Chem.* 265:17174–79
81. Bedalov A, Gatbonton T, Irvine WP, Gottschling DE, Simon JA. 2001. *Proc. Natl. Acad. Sci. USA* 98:15113–18
82. Ansari A, Gartenberg MR. 1997. *Mol. Cell. Biol.* 17:7061–68
83. Zhang Z, Buchman AR. 1997. *Mol. Cell. Biol.* 17:5461–72
84. Xie WW, Gai XW, Zhu YX, Zappulla DC, Sternglanz R, Voytas DF. 2001. *Mol. Cell. Biol.* 21:6606–14
85. Kahana A, Gottschling DE. 1999. *Mol. Cell. Biol.* 19:6608–20
86. Moazed D, Johnson AD. 1996. *Cell* 86:667–77
87. Georgel PT, DeBeer MAP, Pietz G, Fox CA, Hansen JC. 2001. *Proc. Natl. Acad. Sci. USA* 98:8584–89
88. Bell SP, Mitchell J, Leber J, Kobayashi R, Stillman B. 1995. *Cell* 83:563–68
89. Triolo T, Sternglanz R. 1996. *Nature* 381:251–53
90. Gardner KA, Rine J, Fox CA. 1999. *Genetics* 151:31–44
91. Neuwald AF, Aravind L, Spouge JL, Koonin EV. 1999. *Genome Res.* 9:27–43
92. Gasser SM, Cockell MM. 2001. *Gene* 279:1–16
93. Stone EM, Pillus L. 1996. *J. Cell Biol.* 135:571–83
94. Roy N, Runge KW. 2000. *Curr. Biol.* 10:111–14
95. Shei G-J, Broach JR. 1995. *Mol. Cell. Biol.* 15:3496–506
96. Braunstein M, Sobel RE, Allis CD, Turner BM, Broach JR. 1996. *Mol. Cell. Biol.* 16:4349–56
97. van Leeuwen F, Gafken PR, Gottschling DE. 2002. *Cell* 109:745–56
98. Bernstein BE, Humphrey EL, Erlich RL, Schneider R, Bouman P, et al. 2002. *Proc. Natl. Acad. Sci. USA* 99:8695–700
99. Cheng T-H, Gartenberg MR. 2000. *Genes Dev.* 14:452–63
100. Thompson JS, Ling X, Grunstein M. 1994. *Nature* 369:245–47
101. Johnson LM, Kayne PS, Kahn ES, Grunstein M. 1990. *Proc. Natl. Acad. Sci. USA* 87:6286–90
102. Johnson LM, Fisher-Adams G, Grunstein M. 1992. *EMBO J.* 11:2201–9
103. Smith CM, Haimberger ZW, Johnson CO, Wolf AJ, Gafken PR, et al. 2002. *Proc. Natl. Acad. Sci. USA* 99(Suppl. 4):16454–61
104. Park JH, Cosgrove MS, Youngman E, Wolberger C, Boeke JD. 2002. *Nat. Genet.* 32:273–79
105. Meijsing SH, Ehrenhofer-Murray AE. 2001. *Genes Dev.* 15:3169–82
106. Suka N, Grunstein M. 2002. *Nat. Genet.* 32:378–83
106a. Kimura A, Umehara T, Horikoshi M. 2002. *Nat. Genet.* 32:370–77
107. Chien CT, Buck S, Sternglanz R, Shore D. 1993. *Cell* 75:531–41
108. Fox CA, Ehrenhofer-Murray AE, Loo S, Rine J. 1997. *Science* 276:1547–51
109. Moretti P, Freeman K, Coodly L, Shore D. 1994. *Genes Dev.* 8:2257–69
110. Moretti P, Shore D. 2001. *Mol. Cell. Biol.* 21:8082–94
111. Luo K, Vega-Palas MA, Grunstein M. 2002. *Genes Dev.* 16:1528–39
112. Tsukamoto Y, Kato J, Ikeda H. 1997. *Nature* 388:900–3
113. Mishra K, Shore D. 1999. *Curr. Biol.* 9:1123–26
114. Martin SG, Laroche T, Suka N, Grunstein M, Gasser SM. 1999. *Cell* 97:621–33
115. Renauld H, Aparicio OM, Zierath PD, Billington BL, Chhablani SK, Gottschling DE. 1993. *Genes Dev.* 7:1133–45
116. Armstrong CM, Kaeberlein M, Imai SI, Guarente L. 2002. *Mol. Biol. Cell* 13:1427–38
117. Cuperus G, Shafaatian R, Shore D. 2000. *EMBO J.* 19:2641–51

118. Boscheron C, Maillet L, Marcand S, Tsai-Pflugfelder M, Gasser SM, Gilson E. 1996. *EMBO J.* 15:2184–95
119. Fourel G, Revardel E, Koering CE, Gilson E. 1999. *EMBO J.* 18:2522–37
120. Donze D, Adams CR, Rine J, Kamakaka RT. 1999. *Genes Dev.* 13:698–708
121. Clarke DJ, O'Neill LP, Turner BM. 1993. *Biochem. J.* 294:557–61
122. Miller AM, Nasmyth KA. 1984. *Nature* 312:247–51
123. Fox CA, Rine J. 1996. *Curr. Opin. Cell Biol.* 8:354–57
124. Kirchmaier AL, Rine J. 2001. *Science* 291:646–50
125. Li YC, Cheng TH, Gartenberg M. 2001. *Science* 291:650–53
125a. Lau A, Blitzbau H, Bell SP. 2002. *Genes Dev.* 16:2935–45
126. Donze D, Kamakaka RT. 2001. *EMBO J.* 20:520–31
127. Bi X, Braunstein M, Shei GJ, Broach JR. 1999. *Proc. Natl. Acad. Sci. USA* 96:11934–39
128. Bi X. 2002. *Genetics* 160:1401–7
129. Donze D, Kamakaka RT. 2002. *BioEssays* 24:344–49
130. Bi X, Broach JR. 2001. *Curr. Opin. Genet. Dev.* 11:199–204
131. Sun FL, Elgin SC. 1999. *Cell* 99:459–62
132. West AG, Gaszner M, Felsenfeld G. 2002. *Genes Dev.* 16:271–88
133. Bi X, Broach JR. 1999. *Genes Dev.* 13:1089–101
134. Fourel G, Boscheron C, Revardel E, Lebrun E, Hu YF, et al. 2001. *EMBO Rep.* 2:124–32
135. Laloraya S, Guacci V, Koshland D. 2000. *J. Cell Biol.* 151:1047–56
136. Buck WW, Shore D. 1995. *Genes Dev.* 9:370–84
137. Smith JS, Brachmann CB, Pillus L, Boeke JD. 1998. *Genetics* 149:1205–19
138. Lacoste N, Utley RT, Hunter J, Poirier GG, Cote J. 2002. *J. Biol. Chem.* 277:30421–24
139. Ng HH, Feng Q, Wang H, Erdjument-Bromage H, Tempst P, et al. 2002. *Genes Dev.* 16:1518–27
140. Singer MS, Kahana A, Wolf AJ, Meisinger LL, Peterson SE, et al. 1998. *Genetics* 150:613–32
141. San-Segundo PA, Roeder GS. 2000. *Mol. Biol. Cell* 11:3601–15
142. Briggs SD, Bryk M, Strahl BD, Cheung WL, Davie JK, et al. 2001. *Genes Dev.* 15:3286–95
143. Bryk M, Briggs SD, Strahl BD, Curcio MJ, Allis CD, Winston F. 2002. *Curr. Biol.* 12:165–70
144. Nislow C, Ray E, Pillus L. 1997. *Mol. Biol. Cell* 8:2421–36
145. Roguev A, Schaft D, Shevchenko A, Pijnappel WW, Wilm M, et al. 2001. *EMBO J.* 20:7137–48
146. Nagy PL, Griesenbeck J, Kornberg RD, Cleary ML. 2002. *Proc. Natl. Acad. Sci. USA* 99:90–94
147. Krogan NJ, Dover J, Khorrami S, Greenblatt JF, Schneider J, et al. 2002. *J. Biol. Chem.* 277:10753–55
148. Strahl BD, Ohba R, Cook RG, Allis CD. 1999. *Proc. Natl. Acad. Sci. USA* 96:14967–72
149. Noma K, Allis CD, Grewal SI. 2001. *Science* 293:1150–55
150. Sun ZW, Hampsey M. 1999. *Genetics* 152:921–32
151. Huang H, Kahana A, Gottschling DE, Prakash L. 1997. *Mol. Cell. Biol.* 17:6693–99
152. Robzyk K, Recht J, Osley MA. 2000. *Science* 287:501–4
153. Sun ZW, Allis CD. 2002. *Nature* 418:104–8
154. Dover J, Schneider J, Tawiah-Boateng MA, Wood A, Dean K, et al. 2002. *J. Biol. Chem.* 277:28368–71
155. Briggs SD, Xiao T, Sun ZW, Caldwell JA, Shabanowitz J, et al. 2002. *Nature* 418:498
156. Ng HH, Xu RM, Zhang Y, Struhl K. 2002. *J. Biol. Chem.* 277:34655–57

157. Waterborg JH. 2000. *J. Biol. Chem.* 275:13007–11
158. Reifsnyder C, Lowell J, Clarke A, Pillus L. 1996. *Nat. Genet.* 14:42–49
159. Ehrenhofer-Murray AE, Rivier DH, Rine J. 1997. *Genetics* 145:923–34
160. Osada S, Sutton A, Muster N, Brown CE, Yates JR 3rd, et al. 2001. *Genes Dev.* 15:3155–68
161. Sharp JA, Fouts ET, Krawitz DC, Kaufman PD. 2001. *Curr. Biol.* 11:463–73
162. Enomoto S, Berman J. 1998. *Genes Dev.* 12:219–32
163. Holmes SG, Broach JR. 1996. *Genes Dev.* 10:1021–32
164. Enomoto S, Johnston SD, Berman J. 2000. *Genetics* 155:523–38
165. Ansari A, Gartenberg MR. 1999. *Proc. Natl. Acad. Sci. USA* 96:343–48
166. Monson EK, de Bruin D, Zakian VA. 1997. *Proc. Natl. Acad. Sci. USA* 94:13081–86
167. Shibahara K, Stillman B. 1999. *Cell* 96:575–85
168. Moggs JG, Grandi P, Quivy JP, Jonsson ZO, Hubscher U, et al. 2000. *Mol. Cell. Biol.* 20:1206–18
169. Zhang Z, Shibahara K, Stillman B. 2000. *Nature* 408:221–25
170. Sussel L, Vannier D, Shore D. 1993. *Mol. Cell. Biol.* 13:3919–28
171. Mahoney DJ, Marquardt R, Shei G-J, Rose AB, Broach JR. 1991. *Genes Dev.* 5:605–15
172. Richards EJ, Elgin SC. 2002. *Cell* 108: 489–500
173. Grewal SI, Elgin SC. 2002. *Curr. Opin. Genet. Dev.* 12:178–87
174. Jenuwein T. 2001. *Trends Cell Biol.* 11:266–73
175. Holliday R, Pugh JE. 1975. *Science* 187:226–32
176. Kimura H, Cook PR. 2001. *J. Cell Biol.* 153:1341–53
177. Sogo JM, Stahl H, Koller T, Knippers R. 1986. *J. Mol. Biol.* 189:189–204
178. Jackson V, Chalkley R. 1985. *Biochemistry* 24:6930–38
179. Kuo MH, Brownell JE, Sobel RE, Ranalli TA, Cook RG, et al. 1996. *Nature* 383:269–72
180. Sobel RE, Cook RG, Perry CA, Annunziato AT, Allis CD. 1995. *Proc. Natl. Acad. Sci. USA* 92:1237–41
181. Park E-C, Szostak JW. 1990. *Mol. Cell. Biol.* 10:4932–34
182. Xie JX, Pierce M, Gailus-Durner V, Wagner M, Winter E, Vershon AK. 1999. *EMBO J.* 18:6448–54
183. Klar AJ, Kakar SN, Ivy JM, Hicks JB, Livi GP, Miglio LM. 1985. *Genetics* 111:745–58
184. Laurenson P, Rine J. 1991. *Genetics* 129:685–96
185. Sutton A, Heller RC, Landry J, Choy JS, Sirko A, Sternglanz R. 2001. *Mol. Cell. Biol.* 21:3514–22
186. Rusche LN, Rine J. 2001. *Genes Dev.* 15:955–67
187. Wu J, Suka N, Carlson M, Grunstein M. 2001. *Mol. Cell* 7:117–26
188. Watson AD, Edmondson DG, Bone JR, Mukai Y, Yu Y, et al. 2000. *Genes Dev.* 14:2737–44
189. Edmondson DG, Smith MM, Roth SY. 1996. *Genes Dev.* 10:1247–59
190. Davie JK, Trumbly RJ, Dent SY. 2002. *Mol. Cell. Biol.* 22:693–703
191. Ducker CE, Simpson RT. 2000. *EMBO J.* 19:400–9
192. Papamichos-Chronakis M, Conlan RS, Gounalaki N, Copf T, Tzamarias D. 2000. *J. Biol. Chem.* 275:8397–403
193. Gromoller A, Lehming N. 2000. *EMBO J.* 19:6845–52
194. Sekinger EA, Gross DS. 2001. *Cell* 105:403–14
195. Rougvie AE, Lis JT. 1988. *Cell* 54:795–804
196. Rasmussen EB, Lis JT. 1993. *Proc. Natl. Acad. Sci. USA* 90:7923–27
197. Martens C, Krett B, Laybourn PJ. 2001. *Mol. Microbiol.* 40:1009–19
198. Breiling A, Turner BM, Bianchi ME, Orlando V. 2001. *Nature* 412:651–55
199. Saurin AJ, Shao Z, Erdjument-

Bromage H, Tempst P, Kingston RE. 2001. *Nature* 412:655–60

200. Vassallo MF, Tanese N. 2002. *Proc. Natl. Acad. Sci. USA* 99:5919–24
201. Kireeva ML, Walter W, Tchernajenko V, Bondarenko V, Kashlev M, Studitsky VM. 2002. *Mol. Cell* 9: 541–52
202. Ahmad K, Henikoff S. 2002. *Mol. Cell* 9:1191–200
203. Smith JS, Caputo E, Boeke JD. 1999. *Mol. Cell. Biol.* 19:3184–97
204. Dammann R, Lucchini R, Koller T, Sogo JM. 1993. *Nucleic Acids Res.* 21: 2331–38
205. Vogelauer M, Cioci F, Camilloni G. 1998. *J. Mol. Biol.* 275:197–209
206. Gotta M, Strahl-Bolsinger S, Renauld H, Laroche T, Kennedy BK, et al. 1997. *EMBO J.* 16:3243–55
207. Straight AF, Shou W, Dowd GJ, Turck CW, Deshaies RJ, et al. 1999. *Cell* 97:245–56
208. Sandmeier JJ, French S, Osheim Y, Cheung WL, Gallo CM, et al. 2002. *EMBO J.* 21:4959–68
209. Conrad-Webb H, Butow RA. 1995. *Mol. Cell. Biol.* 15:2420–28
210. Oakes M, Siddiqi I, Vu L, Aris J, Nomura M. 1999. *Mol. Cell. Biol.* 19:8559–69
211. Siddiqi IN, Dodd JA, Vu L, Eliason K, Oakes ML, et al. 2001. *EMBO J.* 20:4512–21
212. Sinclair D, Mills K, Guarente L. 1998. *Annu. Rev. Microbiol.* 52:533–60
213. Kennedy BK, Austriaco NR Jr, Guarente L. 1994. *J. Cell Biol.* 127:1985–93
214. Sinclair DA, Guarente L. 1997. *Cell* 91:1033–42
215. Kaeberlein M, McVey M, Guarente L. 1999. *Genes Dev.* 13:2570–80
216. Gottlieb S, Esposito RE. 1989. *Cell* 56:771–76
217. Hoopes LL, Budd M, Choe W, Weitao T, Campbell JL. 2002. *Mol. Cell. Biol.* 22:4136–46
218. Heo SJ, Tatebayashi K, Ohsugi I, Shimamoto A, Furuichi Y, Ikeda H. 1999. *Genes Cells* 4:619–25
219. Merker RJ, Klein HL. 2002. *Mol. Cell. Biol.* 22:421–29
220. Kennedy BK, Gotta M, Sinclair DA, Mills K, McNabb DS, et al. 1997. *Cell* 89:381–91
221. Smeal T, Claus J, Kennedy B, Cole F, Guarente L. 1996. *Cell* 84:633–42
222. Guarente L. 2000. *Genes Dev.* 14: 1021–26
223. Lin SJ, Kaeberlein M, Andalis AA, Sturtz LA, Defossez PA, et al. 2002. *Nature* 418:344–48
224. Lin SJ, Defossez PA, Guarente L. 2000. *Science* 289:2126–28
225. Ashrafi K, Lin SS, Manchester JK, Gordon JI. 2000. *Genes Dev.* 14:1872–85
226. Roeder GS, Bailis JM. 2000. *Trends Genet.* 16:395–403
227. San-Segundo PA, Roeder GS. 1999. *Cell* 97:313–24
228. Feng Q, Wang H, Ng HH, Erdjument-Bromage H, Tempst P, et al. 2002. *Curr. Biol.* 12:1052–58
229. Shou W, Seol JH, Shevchenko A, Baskerville C, Moazed D, et al. 1999. *Cell* 97:233–44
230. Traverso EE, Baskerville C, Liu Y, Shou W, James P, et al. 2001. *J. Biol. Chem.* 276:21924–31
231. Milne GT, Jin S, Shannon KB, Weaver DT. 1996. *Mol. Cell. Biol.* 16: 4189–98
232. Laroche T, Martin SG, Gotta M, Gorham HC, Pryde FE, et al. 1998. *Curr. Biol.* 8:653–56
233. Boulton SJ, Jackson SP. 1998. *EMBO J.* 17:1819–28
234. Nugent CI, Bosco G, Ross LO, Evans SK, Salinger AP, et al. 1998. *Curr. Biol.* 8:657–60
235. Mills KD, Sinclair DA, Guarente L. 1999. *Cell* 97:609–20
236. McAinsh AD, Scott-Drew S, Murray JA, Jackson SP. 1999. *Curr. Biol.* 9:963–66

237. Hegde V, Klein H. 2000. *Nucleic Acids Res.* 28:2779–83
238. Astrom SU, Okamura SM, Rine J. 1999. *Nature* 397:310
239. Heude M, Fabre F. 1993. *Genetics* 133: 489–98
240. Lee SE, Paques F, Sylvan J, Haber JE. 1999. *Curr. Biol.* 9:767–70
241. Kegel A, Sjostrand JO, Astrom SU. 2001. *Curr. Biol.* 11:1611–17
242. Valencia M, Bentele M, Vaze MB, Herrmann G, Kraus E, et al. 2001. *Nature* 414:666–69
243. Thrower DA, Bloom K. 2001. *Mol. Biol. Cell* 12:2800–12
244. Buck, SW, Sandmeier JJ, Smith JS. 2002. *Cell* 111:1003–14

Note Added in Proof

A recent important breakthrough indicates that RNA polymerase I propagates the unidirectional spreading of silencing in rDNA (244).

Annu. Rev. Biochem. 2003. 72:517–571
doi: 10.1146/annurev.biochem.72.121801.161617
Copyright © 2003 by Annual Reviews. All rights reserved
First published online as a Review in Advance on April 10, 2003.

CHALLENGES IN ENZYME MECHANISM AND ENERGETICS

Daniel A. Kraut, Kate S. Carroll, and Daniel Herschlag
Department of Biochemistry, B400 Beckman Center, 279 Campus Drive, Stanford University, Stanford, California 94305-5307; email: dkraut@stanford.edu; katecarroll@mac.com; herschla@cmgm.stanford.edu

Key Words catalysis, thermodynamics, cooperativity, protein engineering, evolution, site-directed mutagenesis, dynamics

■ **Abstract** Since the discovery of enzymes as biological catalysts, study of their enormous catalytic power and exquisite specificity has been central to biochemistry. Nevertheless, there is no universally accepted comprehensive description. Rather, numerous proposals have been presented over the past half century. The difficulty in developing a comprehensive description for the catalytic power of enzymes derives from the highly cooperative nature of their energetics, which renders impossible a simple division of mechanistic features and an absolute partitioning of catalytic contributions into independent and energetically additive components. Site-directed mutagenesis has emerged as an enormously powerful approach to probe enzymatic catalysis, illuminating many basic features of enzyme function and behavior. The emphasis of site-directed mutagenesis on the role of individual residues has also, inadvertently, limited experimental and conceptual attention to the fundamentally cooperative nature of enzyme function and energetics. The first part of this review highlights the structural and functional interconnectivity central to enzymatic catalysis. In the second part we ask: What are the features of enzymes that distinguish them from simple chemical catalysts? The answers are presented in conceptual models that, while simplified, help illustrate the vast amount known about how enzymes achieve catalysis. In the last section, we highlight the molecular and energetic questions that remain for future investigation and describe experimental approaches that will be necessary to answer these questions. The promise of advancing and integrating cutting edge conceptual, experimental, and computational tools brings mechanistic enzymology to a new era, one poised for novel fundamental insights into biological catalysis.

CONTENTS

INTRODUCTION

Much of the focus of biochemical investigations throughout the last half of the twentieth century was on the mechanism by which enzymes achieve their enormous rate enhancements and exquisite specificity. Following the identification of proteins as the primary catalysts in biology by Sumner in 1926 (1), progress unraveling the chemical pathways underlying enzyme action was rapid and extensive. Enzymatic cofactors and coenzymes were identified, their chemical properties uncovered, and by a combination of nonenzymatic and enzymatic studies, their roles in facilitating distinct classes of reactions were elucidated (2–6). Although fascinating mysteries remain concerning the chemical mechanism of numerous enzymes, especially those involving oxidation-reduction and free radical chemistry, a reasonably sophisticated student confronted with an unfamiliar enzymatic transformation can, in most cases, identify what coenzymes or cofactors are likely to be involved, determine whether energy input such as ATP hydrolysis is utilized, and write a likely chemical reaction mechanism.

But enzymes are considerably better catalysts than isolated cofactors, general acids and bases, and other simple, small molecule catalysts. Enzymatic rate enhancements of $10^{10}-10^{23}$, relative to the uncatalyzed transformations in aqueous solution, are common, as is exquisite specificity (7–10). And enzymes accomplish these enormous rate accelerations using amino acid side chains and cofactors that have limited intrinsic reactivity, relative to catalysts employed in organic synthesis. Beyond determination of the chemical mechanisms by which these side chains and cofactors operate, much attention has been paid to the energetic properties of enzymes that lead to this enhanced catalysis and to ways to describe these features (2, 3, 11–42). In this case, however, the central lessons are less clear from a casual inspection of the literature.

Why does the origin of enzymatic catalysis remain unsettled? Part of the answer is that enzymes use multiple mechanisms for catalysis. For example, some active sites take advantage of charge accumulation in the transition state to give strengthened electrostatic interactions, whereas others take advantage of charge dispersal and stabilize the transition state relative to the ground state within a relatively nonpolar pocket (43–45); some use general acids and bases, and others use metal ions. Furthermore, each enzyme uses a combination of strategies to achieve its prodigious catalysis (46–53).

But appreciation of the multiplicity of catalytic strategies is not sufficient to understand the difficulty in comprehending and describing enzymatic catalysis. It is necessary to recognize and appreciate the complexity of enzyme energetics. Catalytic mechanisms and contributions cannot be separated and summed to provide a quantitative accounting of catalysis. This is not a limitation of our experimental abilities, but rather, energetic nonadditivity is a fundamental property of enzymes.

Site-directed mutagenesis has emerged as a powerful tool to probe individual amino acids within an enzyme. The ability to change a specific amino acid and thereby modulate catalysis has been invaluable in determining which groups are directly involved in a reaction. Further, site-directed mutagenesis has allowed the consequences from a wide array of side chain substitutions to be assessed and has been instrumental, in conjunction with other techniques, in unraveling energetic, functional, structural, and dynamic properties of the protein matrix. Nevertheless, site-directed mutagenesis focuses attention on individual residues, which tempts us to ignore the interconnectivity and nonadditivity inherent to enzymatic energetics.

First, we describe why a quantitative breakdown of catalysis into independent and energetically additive factors is not possible and how this complicates the standard scientific reductionist tendency to understand via a divide and conquer approach. We then describe a series of conceptual models that address the question: What are the features of enzymes that distinguish them from simple chemical catalysts? Finally, we formulate questions and describe experimental approaches that are key in bringing us to the next level of understanding of enzyme catalysis.

THE COMPLEX ENERGETICS OF ENZYMATIC CATALYSIS

As scientists, we search for underlying patterns in Nature. This leads to the reductionist pursuit to find simple principles and commonalities that provide satisfying explanations for complex and seemingly disparate behaviors. Follow-

$$E_{WT} + S \xrightarrow{\Delta G^{\ddagger}_{WT}} E_{WT} + P$$

$$E_{Mut} + S \xrightarrow{\Delta G^{\ddagger}_{Mut}} E_{Mut} + P$$

$$\Delta\Delta G^{\ddagger}_{(WT \rightarrow Mut)} = \Delta G^{\ddagger}_{Mut} - \Delta G^{\ddagger}_{WT}$$

Scheme 1.

ing a reductionist path, one might want to interrogate each enzymatic residue, especially those in the active site, by site-directed mutagenesis to quantitatively determine its contribution to binding and catalysis. One might also want to identify catalytic strategies and determine how much of the rate enhancement arises from general base catalysis, general acid catalysis, electrostatic interactions with a substrate group that has an increased charge in the transition state, or other mechanisms.

Unfortunately, the fully reductionist approaches outlined above for enzymatic catalysis are incomplete and even misleading. By understanding these approaches and their flaws, we can appropriately evaluate specific experimental data and conclusions, develop a more general description of enzymatic catalysis, and, most importantly, define approaches that will substantially advance our appreciation for how enzymes are able to achieve their enormous rate enhancements and exquisite specificity. In this section, the limits of reductionism applied to enzyme catalysis are described and the interconnectivity of enzyme energetics is highlighted.

Assigning Specific Energetic Contributions to Catalysis: The Limits of Energetic Additivity

It is commonly stated, following a site-directed mutagenesis experiment, that a particular residue or hydrogen bond contributes a certain amount of free energy to binding or catalysis (or stability in the case of protein folding). There are two problems with such statements. The first is that the energetic value is derived from $\Delta\Delta G$, not ΔG (Scheme 1). The reaction of a mutant enzyme is compared to that of the wild type: Each is characterized by a free energy of activation ($\Delta G^{\ddagger}$), which represents the free energy difference between each ground state and transition state; thus, the difference between the mutant and wild-type reactions is a four way comparison—a difference of differences, or a $\Delta\Delta G$ value. As four different states are being compared, a single ΔG value that represents the contribution of one residue to catalysis in the wild-type enzyme cannot be extracted. Nor is it possible to devise some other scheme to do this—all such values used to assess the contribution of a residue rely on some comparison state, whether explicitly stated or not, and are thus inherently relative. [The relative nature of thermodynamic values is introduced in general terms in physical chemistry texts (54).] The second problem deals with energetic nonadditivity, as discussed below.

If it were possible to quantitatively assign an energetic value that describes the contribution from one residue, then one ought to be able to do this for each residue and, ultimately, sum the energetic contributions to obtain a quantitative description of catalysis. Stated another way, implicit in assignments of specific energetic values is an assumption that the groups involved are independent of one another; this renders their energetic effects additive. Energetic additivity has been observed in many site-directed mutagenesis experiments probing more than one mutation simultaneously (55). There is, however, no fundamental expectation of energetic additivity in chemical systems; additivity holds as an approximation only in special cases in which local factors dominate (55–59). Below, basic experimental and conceptual examples are reviewed to illustrate the limitations of energetic additivity. Recognizing the energetic and functional interconnectedness of chemical systems is a key step in developing a deeper understanding of enzyme catalysis.

The most common example of energetic additivity, taught in introductory college chemistry courses, pertains to enthalpies of formation of organic molecules. In the 1950s and 1960s, Benson and colleagues derived group additivity principles, which have proven remarkably powerful for predicting heats of formation (ΔH_f) for organic molecules (60, 61). These rules work well because local factors dominate bond enthalpies and are hardly perturbed by the remainder of the molecule. The classic exception to simple group additivity rules for ΔH_f is benzene, which is $\sim$30 kcal mol^{-1} more stable than predicted based on adding together the single and double bond energies of "cyclohexatriene" (62). Consider the thought experiment of Figure 1. Building benzene from hexane one bond at a time, a six-membered ring can first be formed to give cyclohexane, and then the double bonds can be added. (Figure 1, path *a*); the final double bond (DB_3) contributes this extra $\sim$30 kcal mol^{-1} of energy. However, if instead the three double bonds are first added to give hexatriene, and then the ring is closed with the addition of a single bond (SB_6), it is this final single bond that provides the extra energy of $\sim$30 kcal mol^{-1} (Figure 1, path *b*). The final bond formed appears to be extraordinarily stable, although it is a different bond in each path. This paradox arises because the enthalpy is not a local property of the new bond alone, but rather a property of the system, and this distributed property is not introduced until the system is fully formed—until the last bond is in place. Aromaticity and resonance stabilization provide ad hoc explanations for the unexpected stability of benzene and conjugated compounds, i.e., the observed nonadditivity. We now know that this extra stability arises from electron delocalization throughout the benzene ring or conjugated system. The properties,

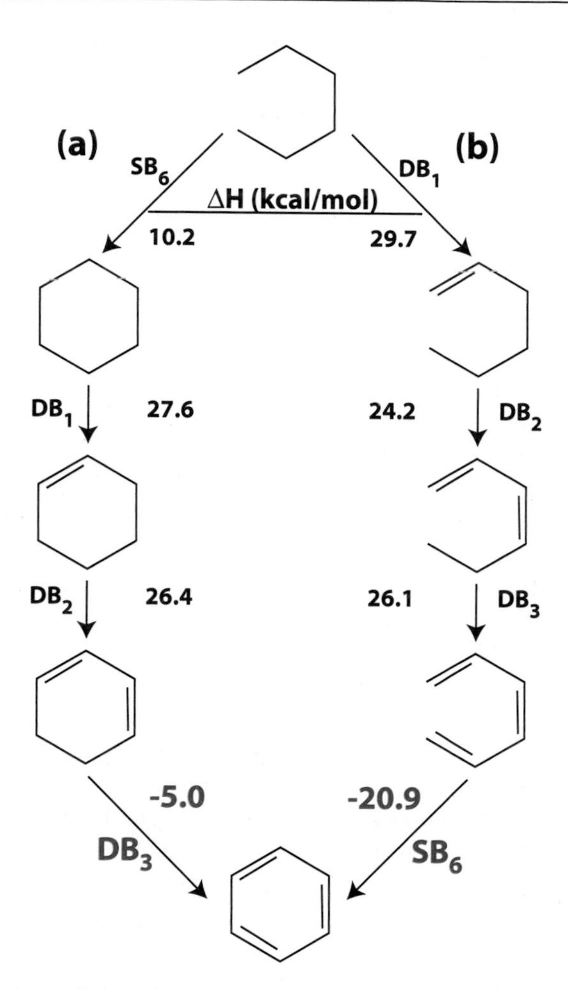

Figure 1 The bonds in benzene do not make independent, additive contributions to the molecule's stability. In pathway *a*, benzene is constructed from hexane by first forming a sixth carbon-carbon single bond (SB_6) to close the ring (with concomitant breakage of two C-H bonds and formation of H_2 gas; this occurs in each step but is omitted for clarity), followed by formation of three carbon-carbon double bonds (DB_1, DB_2, DB_3). Although the first two double bonds cost approximately the same amount of energy, the formation of the final double bond (DB_3) is more favorable by ~30 kcal mol^{-1} (-5.0 versus 27.6 and 26.4 kcal mol^{-1}). In pathway *b*, the double bonds are first added to hexane, followed by the single bond closure of the hexatriene ring. Now the three double bonds are all of about the same energy (29.7, 24.2, and 26.1 kcal mol^{-1}), while the formation of the single bond is more favorable by ~30 kcal mol^{-1} (SB_6 is 10.2 and -20.9 kcal mol^{-1} in pathway *a* and *b*, respectively) (62a). The 30 kcal mol^{-1} of resonance energy present in benzene can be expressed in a single bond or a double bond, depending on how the molecule is constructed, which indicates that the bond energies depend on one another.

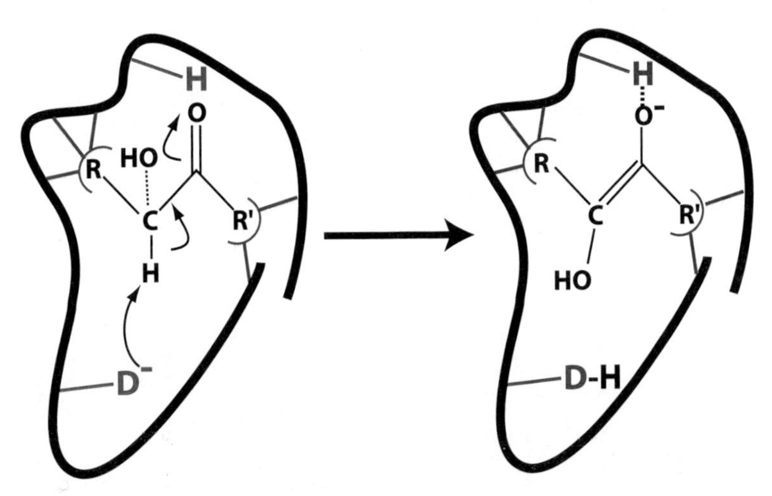

Figure 2 A hypothetical ketose isomerase enzyme. The first reaction step is shown, in which glutamate (D) is used as a general base to remove the proton alpha to the substrate's carbonyl group and histidine (H) donates a hydrogen bond to stabilize developing negative charge on the carbonyl oxygen atom. The enzyme loses essentially all catalytic activity if the His and Glu residues (shown by the magenta H and D) are both mutated to alanine, leading to their designation as "catalytic residues." "Binding residues" (i.e., other residues contacting the substrate) are depicted as blue lines, and the remaining structural residues are depicted by the black outline.

and thus energetics, of benzene are not simply the sum of nearly independent local bonding interactions.

An analogous thought experiment conducted on an enzyme demonstrates that nonlocal factors are also critical for enzyme function, so enzymes cannot be considered additive systems. We start with a wild-type enzyme that catalyzes a ketose isomerization (similar to the classic enzyme triosephosphate isomerase) and contains as "catalytic residues" a base to remove a proton (glutamate) and a hydrogen bonding group that stabilizes the negative charge that develops on a carbonyl oxygen atom (histidine) (Figure 2). When we replace these "catalytic residues" with alanine, the enzyme loses all catalytic activity, as all of the other residues are considered binding or structural residues in this model. For the purposes of illustration, imagine that we continue replacing residues until the result is an unstructured poly-alanine of the same length as the starting enzyme. We now add back the wild-type residues one at a time (Figure 3). Three paths are considered. If we first add back the residues required for structure, then the "binding residues," and only at the end add back the "catalytic residues," the addition of the last residues will cause a large increase in catalytic activity, as expected for "catalytic residues" (Figure 3, pathway *a*). However, if after restoring the structural residues the "catalytic residues" are added next, there will be little to no catalytic activity. Without binding interactions to hold the substrate

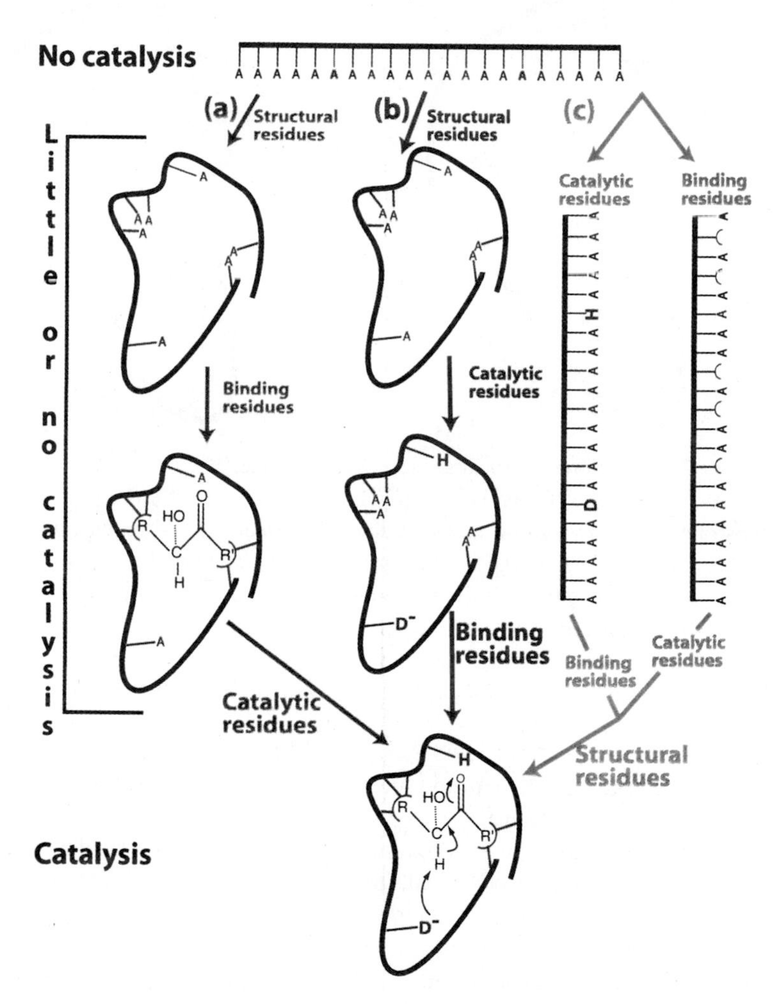

Figure 3 The interdependence of so-called catalytic, binding, and structural residues. The enzyme from Figure 2 has been mutated to polyalanine, and three different pathways for conversion back to the functional enzyme are explored (*a*, *b*, and *c*). The pathway taken determines which residues appear to be important for catalysis, demonstrating that the functions of the individual residues are interdependent. The "catalytic" histidine and glutamate are shown in magenta [either as alanine (A) before mutation or as H and D after mutation]. "Binding residues" are shown as blue alanines that are converted to blue lines upon mutagenesis to their wild-type identity, and upon formation of a binding site, the substrate is shown bound. "Structural residues" are shown either as black alanines or by the black outline. (Because enzymes typically have >100 residues, not all residues are depicted.)

in place, the "catalytic residues" cannot perform. (The relationship between binding and catalysis is discussed below.) Now addition of the "binding residues" brings the enzyme across the threshold to catalytic activity (pathway *b*). Finally, if the "catalytic residues" and the "binding residues" are restored in either order, no catalysis is realized in the poly-alanine background (pathway *c*). It is only upon addition of sufficient "structural residues" to stabilize the overall fold and position of the "binding" and "catalytic" residues that function is restored. Thus each of the structural residues will exhibit the phenotype of a catalytic residue when it is the one that tips the balance to allow formation of the active structure.

The thought experiment of Figure 3 demonstrates that binding and catalytic residues do not act in isolation—they are not independent of the other residues. Rather, all of the residues contribute to binding and catalysis by the definition typically applied in simple site-directed mutagenesis experiments, i.e., which residue when removed causes a loss in the particular function of interest. The only difference in Figure 3 is that the enzyme system is probed more deeply by carrying out more extensive mutagenesis than is typical (or practical). The resulting distributive assignment of function is equivalent to stating that the enzyme is a cooperative system, a statement we are perfectly comfortable with in other contexts. Thus, independent energetic contributions to catalysis cannot be assigned on a residue-by-residue basis. Similarly for benzene, the C-H groups together contribute to the extraordinary stability.

EXPERIMENTAL EXAMPLES AND CONCEPTUAL ANALOGIES TO FURTHER ELUCIDATE THE LIMITS OF ADDITIVITY The conserved sequences of the self-cleaving RNAs referred to as the hammerhead and hairpin ribozymes are depicted in Figure 4. Both ribozymes catalyze strand scission to give a 5′-hydroxyl and a 2′,3′-cyclic phosphate, and each self-cleaving RNA has been converted into a multiple turnover ribozyme by separating a catalytic core (outlined letters) from a substrate strand (63–66). The hammerhead ribozyme was subjected to a systematic subtractive mutagenesis approach, akin to alanine scanning for protein enzymes, in which each of the thirteen conserved bases was individually replaced with a hydrogen atom (to give an abasic residue). Despite the modest catalysis by this ribozyme of $\sim 10^8$-fold, nearly all of these residues gave enormous rate decreases, typically $10^4 - 10^6$-fold (67, 68). These large and widespread effects are distinct from observations with protein enzymes in which mutation of only a few residues typically give large effects on catalysis (57, 68).

It is of course highly unlikely that each of these ribozyme residues plays a direct role in catalysis; nor do the modifications significantly affect substrate binding (68, 69). What then is the origin of the large and distributed mutational effects in the hammerhead core? There is evidence that the resting state of the hammerhead ribozyme is a noncatalytic conformation, so that the core must assemble with each catalytic event (70, 71); this is akin to the situation with the hypothetical enzyme when the "structural residues" added last gave energetic signatures of "catalytic residues" (Figure 3, path *c*). Indeed, one form of the

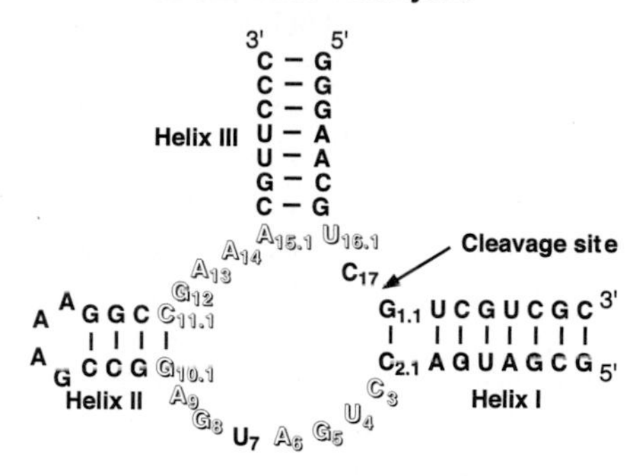

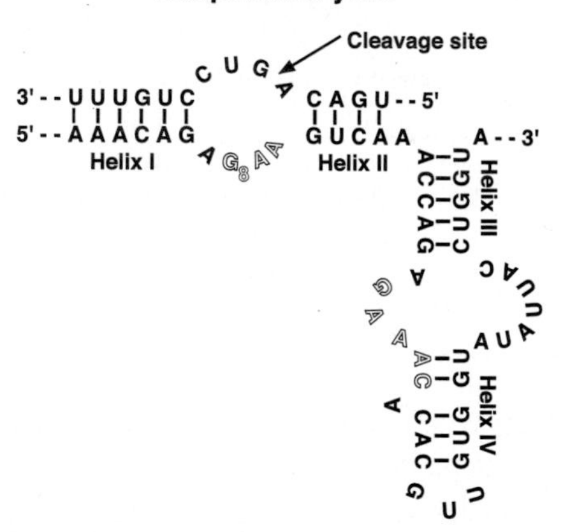

Figure 4 The secondary structure of a hammerhead (*top*) and a hairpin (*bottom*) ribozyme with bound substrates. The conserved catalytic core residues are shown in outline, and the cleavage site in the oligonucleotide substrate is shown with an arrow (63–66).

hairpin ribozyme behaves similarly to the hammerhead, with energetic signatures from mutation of many core residues, whereas addition of a remote structural element to aid proper folding removes the large effects of all but one of these conserved residues (64, 72, 73). The remaining susceptible residue presumably plays a more direct role in the chemical process. We emphasize that the discovery of a less mutationally sensitive form of the hairpin ribozyme does not mean that the other residues are unimportant—their importance is merely masked in experiments when residues are mutated individually.

Consider, by analogy, the two houses shown in Figure 5. The one on the left is a minimal unit, and the one on the right is well-built, with many reinforcing beams. The primitive house, lacking structural redundancy, represents the hammerhead ribozyme. In this primitive house removal of any board, i.e., any

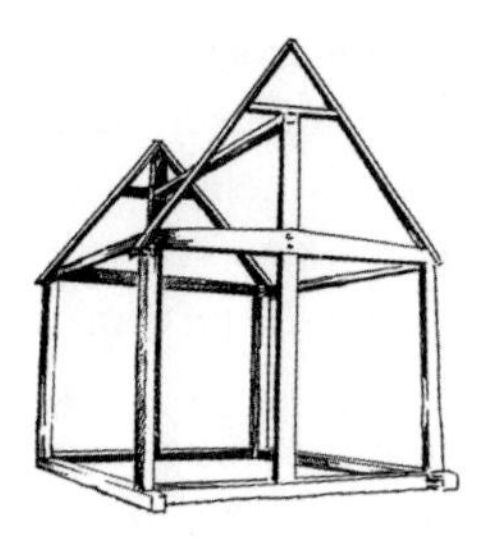

Figure 5 Two houses as metaphors for the role of structure in function and the ability to ascertain function from site-directed mutagenesis. The house on the left represents a primitive enzyme, and the house on the right represents a highly evolved enzyme with structural redundancy. For the house on the left, removal of any beam will cause collapse. In contrast, removing any individual beam will leave the house on the right standing, just as removing any individual structural residue will leave the evolved enzyme functioning. Thus, depending on context, i.e., what other beams are present in a hosue or what other residues are present in an enzyme, site-directed mutagenesis may or may not reveal components involved in overall function. From drawing by A. Peracchi.

mutation, can lead to structural collapse, thereby obliterating overall function. The well-built house is meant to represent a typical protein enzyme or the stabilized hairpin ribozyme. Removal of individual boards still leaves the structure standing in its functional form. [See (68) for a more extensive discussion of structural redundancy.]

In site-directed mutagenesis, a residue is typically defined as important (or not) based on the response of the structure or function to removal of that residue's side chain—or a particular plank in the houses of Figure 5. Residues that give a catastrophic effect in response to a single change are deemed important. Nevertheless, the rest of the enzyme is still important—as elucidated in the thought experiment of Figure 3 and the house analogy of Figure 5: Although removal of any single vertical support beam in a house will leave it standing and still strong, one would not conclude that vertical support beams are unimportant, and few would acquiesce to their removal!

Can Quantitative Energetic Contributions Be Assigned to Specific Catalytic Strategies?

The preceding section demonstrates that additivity does not underlie the energetics of complex, cooperative systems such as enzymes, so a numerical energetic contribution cannot be assigned on a residue by residue basis. As part of a reductionist approach, might one still be able to determine which residues comprise a catalytic strategy (e.g., general acid or base catalysis, electrostatic complementarity to the transition state, and ground state destabilization) and thereby quantitate the energetic contribution from this strategy? For numerous enzymes, it is known, for example, which residues donate or abstract a proton in general acid or base catalysis. There are nevertheless limitations in our ability to assign catalytic function to specific residues, and our ability to assign energetic contributions to specific catalytic strategies is even more limited. These limitations are illustrated in the following examples.

Experiments with a PI-specific phospholipase C are instructive with respect to functional interconnections between residues (74). Tsai and coworkers examined the catalytic histidines (H32, the general base, and H82, the general acid) as well as two aspartate residues, D274 and D33, thought to hydrogen bond with the histidines. Mutation of either "catalytic" histidine to alanine led to a rate decrease of 10^5, which lends support to the idea that the histidines are the catalytic residues. However, mutating D274 to alanine caused a 10^4-fold drop in rate, and the D33A mutant had a 10^3-fold drop in rate. These data indicate that mutation of the residues adjacent to the general acid and general base abrogate general acid and base catalysis, even though these are not the residues directly involved in proton donation to or removal from the substrate. The catalysis associated with proton removal or donation is not just a function of the residue that accepts or donates the proton; it is also connected to properties of the surrounding enzyme environment, i.e., the residues around the proton donor and acceptor that determine positioning, electrostatic potentials, and solvation. Although a catalytic value to acid/base catalysis or to a specific residue cannot be assigned from the above experiments, they, along with many other experiments, demonstrate the interconnectedness of the active site and the power of site-directed mutagenesis in uncovering these connections.[1,2]

Because multiple interactions influence a given catalytic strategy, it is impossible to separate the contribution of a given residue in the strategy. Furthermore, enzymes use multiple sets of catalytic strategies, and these strategies are also interconnected, preventing assignment of stabilization energies to a specific strategy. This is illustrated by a hypothetical serine esterase. The active site contains a general base to remove the proton from the attacking serine residue, an oxyanion hole that stabilizes the development of negative charge on the incipient oxyanion, and several groups that bind and position the attacking serine with respect to the ester carbon. It would be desirable to determine the amount

---

[1]In some cases there may be significant amounts of shared covalent character in hydrogen bonds, and additional proton rearrangements may accompany proton abstraction from or donation to substrates. These possibilities further illustrate the limitations in discrete assignment of catalytic function. Elucidation of the nature of bonding in these situations as the functional, structural, and energetic origins and consequences of the bonding represents an exciting challenge (75–77).

[2]Other elegant studies have combined site-directed mutagenesis with isotope effects and variation in the identity of the leaving group to reveal roles of residues in general acid catalysis (78–81).

of catalysis provided by each strategy, but a problem arises. The hydrogen bond donors in the oxyanion hole help position the ester carbon with respect to the serine nucleophile in addition to stabilizing charge buildup on the transition state oxyanion. Similarly, the residue that acts as a general base, because of its placement in the active site, helps position the incipient oxyanion with respect to the residues that make up the oxyanion hole, aiding this catalytic function in addition to directly facilitating proton removal. Thus, the catalytic strategies are interconnected: Mutating a group involved in one type of catalysis can adversely affect another catalytic strategy as well. The energetic contributions of each catalytic strategy are not cleanly separable.

In summary, while it is often possible to assign direct chemical participation in catalysis to a particular residue, the residue's capability to act depends on its neighbors and surroundings. Thus, responsibility for a catalytic strategy cannot be assigned to a single residue. Furthermore, the catalytic strategies that an enzyme uses to facilitate reaction are not independent of one another. This functional and energetic interdependency prevents a quantitative dissection of enzymatic catalysis into types of stabilization.

Assigning the Signature of a Residue: Dissecting Binding and Catalytic Contributions

The above examples demonstrate the limitations in assigning energetic contributions to individual residues and to individual catalytic strategies. One might also like to know which reaction step or steps a particular residue facilitates: Does the residue contribute to the binding or chemical step, or both? Such a partitioning of function into neat categories would be highly appealing from a reductionist standpoint, which distinguishes the residues (or functionalities) responsible for getting the substrate localized to the active site from those that actually carry out the chemical transformation. Indeed, in the literature this sort of assignment often follows site-directed mutagenesis experiments; residues that, when mutated, give increases in K_M are typically ascribed roles in binding, and residues that give decreases in k_{cat} are typically ascribed roles in the chemical step. (We assume for simplicity that $K_M = K_d$ and $k_{cat} = k_{chemical\ step}$.)

Albery & Knowles, following their classic determination of the triose phosphate isomerase (TIM) kinetic mechanism in the late 1970s, formalized this classification and suggested potential evolutionary ramifications (Figure 6) (31, 32). They noted that the addition of a residue in the course of evolution could stabilize the ground state and transition state equally (Figure 6*a*, uniform binding) or could stabilize the transition state without affecting the ground state (Figure 6*b*, specific transition state stabilization); a residue could also give a mixed effect, stabilizing both ground and transition states but providing more stabilization to the transition state (Figure 6*c*, differential binding).

As noted above, these categories correspond to what are commonly considered binding and catalytic residues. This mechanistic distinction, however, was at odds with the perspective articulated by Jencks. In 1973 Jencks stated "on closer

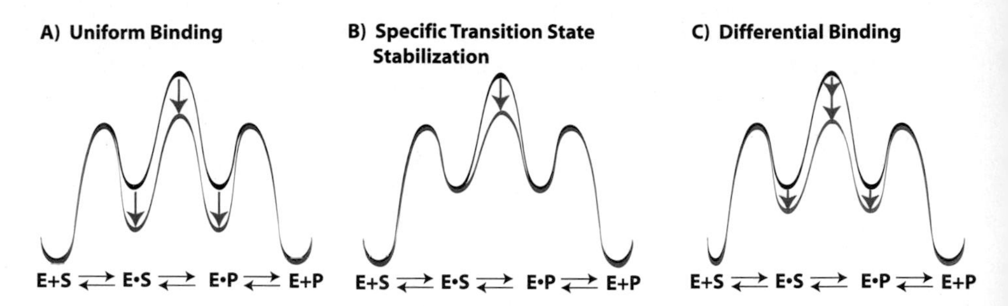

Figure 6 Free energy reaction profiles demonstrating potential catalytic effects of mutants, as described by Albery & Knowles (31, 32). The profile for a primitive enzyme is shown in black, and possible results from introduction of potentially advantageous mutations are shown in magenta. In typical site-directed mutagenesis experiments, the effects would be in the opposite direction, starting with the magenta profile and going to the black one. (*A*) Uniform binding. All enzyme-bound species are stabilized equally by the mutation, accelerating a reaction with subsaturating concentrations of substrate, but not with saturating substrate (see also Figure 9). (*B*) Specific transition state stabilization. The mutation causes stabilization of the transition state without stabilization of the ground state. This transition state interaction leads directly to enhanced reaction of the bound substrate, and it also increases the rate of reaction of unbound substrate. (*C*) Differential binding. There is a continuum of possible energetic effects between the extremes shown in (*A*) and (*B*) in which mutations can stabilize both transition state and the ground state but provide greater stabilization to the transition state.

examination. . . the classical separation of considerations of enzymatic catalysis into the specific binding of substrates and chemical catalysis breaks down completely" (15).

Although site-directed mutagenesis had not yet been applied to enzymes, experiments with substrate analogs had clearly established the interconnection of binding and catalysis. For example, addition of amino acid residues remote from the site of chemical transformation for elastase substrates increased the rate of transformation of the already bound substrate (k_{cat}); similarly, the binding of transition state analogs but not substrates was enhanced (82, 83). The most basic point emphasized by Jencks was that binding interactions (through the "intrinsic binding energy" provided) can facilitate the chemical transformation (16). The binding interactions can help by positioning substrates with respect to one another, by positioning substrates with respect to functional groups on the enzyme, and by enforcing electrostatically or sterically destabilizing ground state interactions that are relieved as the substrate undergoes electronic and geometric rearrangement in the transition state (16).

DIRECT DEMONSTRATION OF THE USE OF BINDING INTERACTIONS FOR CATALYSIS More recent experiments have directly demonstrated this interconnection between binding interactions and the chemical transformation of bound sub-

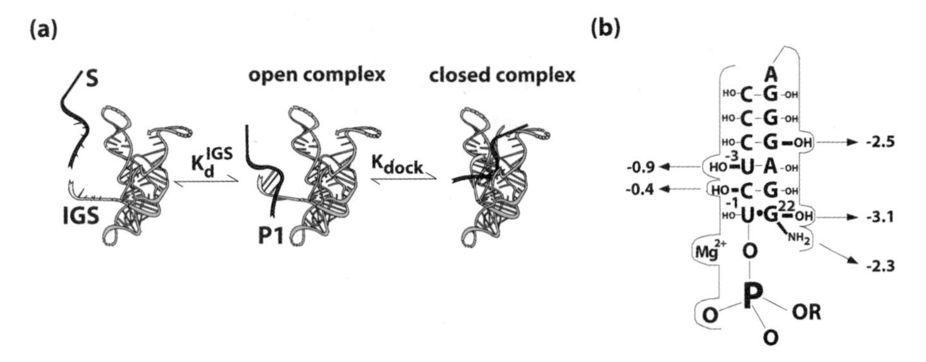

Figure 7 Substrate binding to the *Tetrahymena* group I RNA enzyme. (*a*) Binding occurs in two steps. First the oligonucleotide substrate (S) binds to the internal guide sequence (IGS) of the ribozyme via base pairing to form the open complex (K_d^{IGS}). The resulting helix then docks into the active site of the enzyme to form the closed complex (K_{dock}). (*b*) Schematic diagram showing functional groups in the substrate helix that make tertiary interactions with the active site. Only the groups varied in the works referred to in the text are shown, including the -3 (three residues 5′ of the substrate's cleavage site) 2′-hydroxyl group and the exocyclic amino group of G22 on the enzyme strand (84). The energetic cost (in kcal mol^{-1}) of replacement of each group by a hydrogen atom is also shown [(84) and references therein].

strates (84). The RNA enzyme derived from the *Tetrahymena thermophila* group I intron catalyzes a reaction analogous to the first step in intron self-splicing (Equation 1). The catalytic mechanism of this RNA enzyme has been studied extensively, using presteady state kinetics to establish a complete kinetic and thermodynamic framework for individual reaction steps and a plethora of substrate analogs with single functional group substitutions to probe interactions with the RNA enzyme and their energetic consequences (85–88).

$$\underset{(S)}{\text{CCCUCU}_{\text{P}}\text{AAAAA}} + \text{G}_{\text{OH}} \rightarrow \underset{(P)}{\text{CCCUCU}_{\text{OH}}} + \text{GpAAAAA} \qquad 1.$$

Binding of an oligonucleotide 5′-splice site analog (S) occurs in two distinct steps (Figure 7*a*). In the first step, S forms a duplex with a complementary sequence of the RNA enzyme to form the open complex. In the second step, this duplex docks into the core of the enzyme and makes tertiary interactions; Figure 7*b* shows the functional groups of the duplex involved in these binding interactions, which include several 2′-hydroxyl groups, and the exocyclic amino group of G22, which forms a G·U wobble pair to specify the cleavage site. The two binding states are key to the analysis described below, because the bound oligonucleotide substrate is either positioned for reaction (closed complex) or localized to the enzyme only by base pairing and thus bound but not positioned for reaction (open complex).

In the course of probing the roles of individual functional groups (and attempting to assign discrete roles in binding and catalysis), it was discovered that addition of a single functional group to the substrate, the 2′-hydroxyl group of U(-3), could give either uniform binding or specific transition state stabilization, depending on the context, i.e., depending on the other groups and interactions present (Figure 8*a*). [U(-3) refers to the substrate position three residues 5′ of the cleavage site (Figure 7*b*).] With the wild-type enzyme, addition of this 2′-hydroxyl gave uniform binding, as witnessed by a decreased dissociation constant without any change in the rate of the chemical step (Figure 8*a*, wild type). In a mutant with the 2′-hydroxyl and exocyclic amino group of G22 absent to give deoxyinosine (dI22), addition of the 2′-hydroxyl at U(-3) gave the same total energetic contribution of 1 kcal mol^{-1} as the wild type, but the energy was expressed only in increasing the rate of the chemical step, thereby giving specific transition state stabilization (Figure 8*a*, dI22). A group remote from the site of chemical transformation giving uniform binding in the context of the wild-type enzyme, and thus typically thought of as providing binding interactions, can instead specifically stabilize the transition state in a mutant context.

The explanation for these results provides insight into the interconnection between binding and catalysis. Recall the two binding states of S, open and closed (Figure 7*a*). In the otherwise wild-type context, the duplex with S forms enough tertiary interactions such that it docks and remains in the closed complex whether or not the 2′-hydroxyl at U(-3) is present (Figure 8*b*). Thus, even without the U(-3) 2′-hydroxyl group, the substrate is positioned for reaction. The docked substrate makes the same remote interactions in the transition state as in the ground state, so addition of the U(-3) 2′-hydroxyl group gives the same 1 kcal mol^{-1} stabilization to both states and uniform binding is observed. In the mutant context, removal of two additional tertiary contacts from the docked complex has tipped the energetic balance so that now the open or unpositioned complex is the stable ground state conformation (Figure 8*c*). Addition of the 2′-hydroxyl at U(-3) has no effect on binding—it is not sufficient to tip the balance back to the closed complex where an interaction with the enzyme core can be made. Nevertheless, the reaction must occur through the docked complex. Thus, the

Figure 8 The same functional group in different enzyme contexts can provide uniform binding or specific transition state stabilization. Free energy profiles (*a*) show that adding the same 2′-hydroxyl group (going from black to red) at the -3 position of the oligonucleotide substrate (Figure 7) of the *Tetrahymena* RNA enzyme results in uniform binding in the wild-type enzyme and specific transition state stabilization in the dI22 mutant (missing the 2′-hydroxyl group and the exocyclic amino group at position 22). Models of the significantly populated bound states of the substrate for the wild type (*b*) and mutant dI22 (*c*) that explain the respective uniform binding and specific transition state stabilization phenotypes upon addition of the 2′-hydroxyl group at position -3 (shown in red) (84).

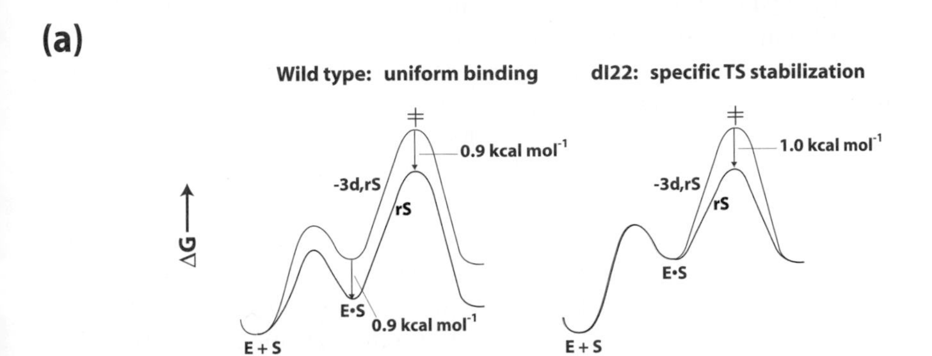

(b) Uniform binding

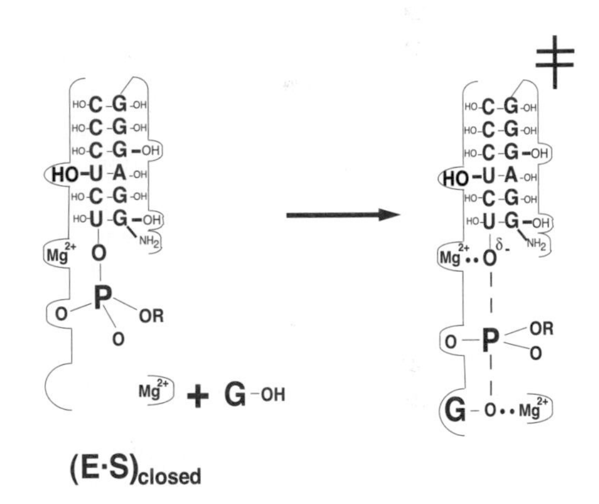

(c) Specific transition state stabilization

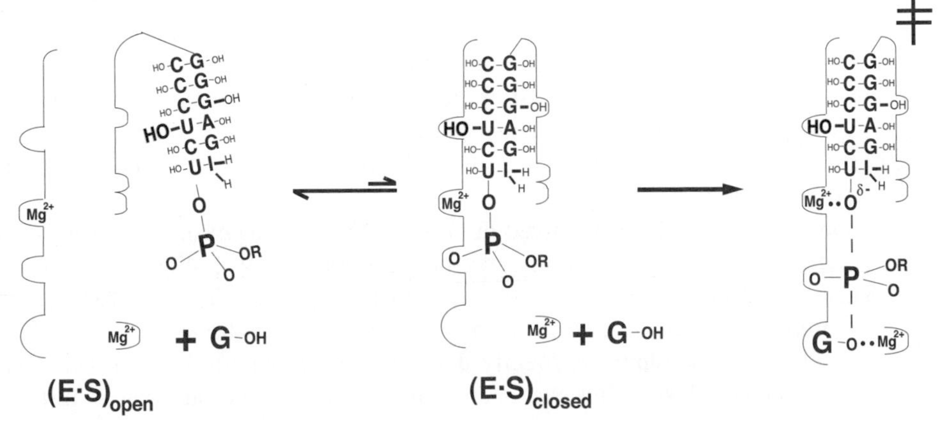

U(-3) 2′-hydroxyl interaction is made in the transition state, thereby providing specific transition state stabilization.

Indeed it was shown that each of the functional groups contributing binding energy to docking can contribute to binding or catalysis; this results in a uniform binding or a specific transition state stabilization phenotype dependent only on the total binding energy available from the other functional groups present. If few are present, the duplex is undocked and specific transition state stabilization is observed; if sufficient groups are already present to favor the docked complex, then uniform binding is observed. Functional groups that make binding interactions remote from the site of chemical transformation can contribute specific transition state stabilization instead of uniform binding.

The fundamental contribution of binding interactions to catalysis are masked by the presence of other binding and positioning interactions. Only when a sufficient number of these are removed (and an energetic threshold is crossed such that the unpositioned state is more stable than the positioned state) is the underlying contribution to catalysis with bound substrate revealed; what appears to be a uniform binding interaction is shown to provide transition state stabilization in a different context.

As Albery & Knowles noted, a uniform binding contribution such as that seen in the wild-type ribozyme does not provide catalysis for the enzyme/substrate complex, because the barrier for the chemical step is not lowered (Figure 9) (31). Nevertheless, uniform binding contributions can increase reaction from free enzyme and substrate (k_{cat}/K_M), and a "binding" residue that gives a uniform binding phenotype when mutated in the context of a modern day enzyme could have been selected early in evolution on the basis of a contribution to transition state stabilization via substrate positioning. The later selection for residues that provide additional binding interactions would mask this early and continuing role in catalysis.

These possible changes in phenotype over the course of evolution reveal a basic limitation of site-directed mutagenesis. Removal of one or two residues may be insufficient to unmask the catalytic contributions of "binding" residues, and removal of more residues almost invariably leads to rearrangement of bound complexes to a variety of nonproductive and partially productive complexes that obscure straightforward, energetic interpretation [e.g., (89, 90)].

In summary, residues involved in positioning, as in the example above, play critical roles in catalysis. Their importance, however, is not readily uncovered by site-directed mutagenesis. The *Tetrahymena* RNA enzyme has properties, such as simple two-state binding in a positioned or an unpositioned complex, that provided an opportunity to directly demonstrate the inextricable link between binding and catalysis, a link that is at the heart of enzymatic catalysis, as further elaborated in the following section.

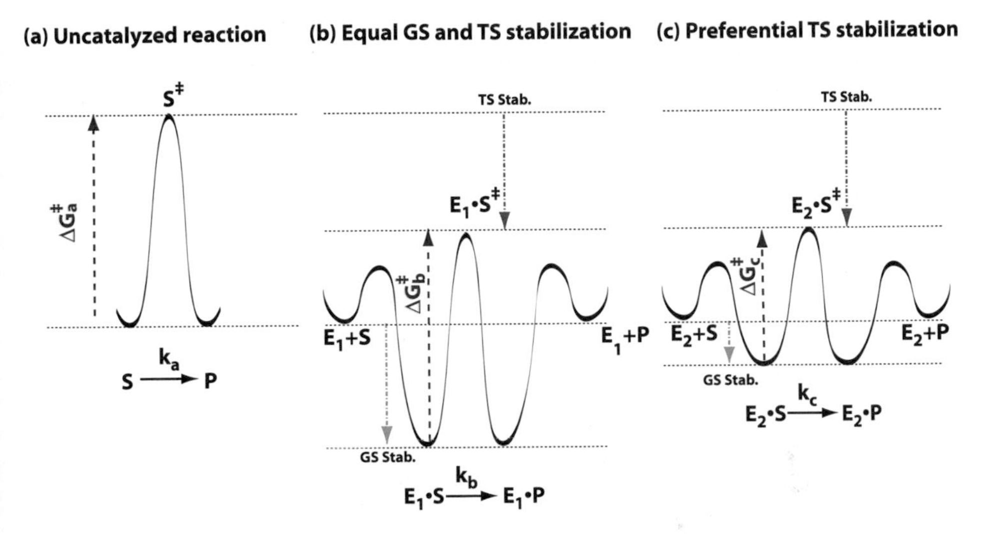

Figure 9 Free energy reaction profiles demonstrating the requirement for preferential transition state stabilization relative to ground state stabilization for catalysis. For simplicity, assume that the chemical step is rate limiting in this figure and throughout the review. (*a*) The uncatalyzed reaction of S, with an activation barrier of $\Delta G_a^\ddagger$ and a rate constant of k_a. (*b*) Enzyme 1 (E_1) stabilizes the ground state (GS) and transition state (TS) equally, leaving the reaction barrier $\Delta G_b^\ddagger$ equal to the reaction barrier $\Delta G_a^\ddagger$ for the uncatalyzed reaction; this enzyme is not a catalyst as $k_b = k_a$. (*c*) Enzyme 2 (E_2) stabilizes the transition state more than it stabilizes the ground state such that $\Delta G_c^\ddagger < \Delta G_a^\ddagger$ so that the rate constant k_c is larger than k_a. Thus, this enzyme is a catalyst.

Summary

We have considered three ways to subdivide enzyme function: assigning quantitative energetic contributions to individual residues, assigning quantitative energetic contributions to the catalytic strategies used by the enzyme, and assigning energetic contributions to residues in binding versus chemical reaction steps. Each is not possible.

Correspondingly, limitations of site-directed mutagenesis have been revealed. This approach, while providing many important insights, cannot provide a unique energetic signature for a residue or a catalytic strategy. Site-directed mutagenesis can reveal the importance of an active site residue for catalysis; for example, replacement of a Glu residue acting as a general base catalyst by Ala will greatly compromise catalysis. However, roles of "binding" residues in catalysis can be masked by the presence of multiple positioning interactions, and structural rearrangements upon removal of the interacting groups can obscure or amplify their underlying contributions.

In all cases the context, i.e., the properties of the surrounding residues, matters quantitatively and/or qualitatively and prevents a unique breakdown of enzyme function into independent, energetically additive components. This complexity arises from the cooperative nature of enzyme structure and function. As described in the final section of this review, multiple comparisons of the energetics and physical properties of wild-type and mutant enzymes with cognate and modified substrates are essential tools in revealing the behavior and properties of these cooperative systems.

THE DISTINGUISHING PROPERTIES OF ENZYMES: COMPARISON TO SMALL MOLECULE CHEMICAL CATALYSTS

The task of understanding enzyme catalysis is rendered more difficult by the absence of a single answer attainable by quantitative dissection into individual contributions. In the absence of an absolute answer, understanding must be sought through relativistic analysis. Multiple comparisons and multiple types of comparisons are needed to provide different perspectives and insights. In this section, we define a general comparison with the question: What distinguishes enzymes from simple, small molecule chemical catalysts? To address this question, we consider a series of model enzymes that successively build in features, that are associated with enzymatic catalysis, and that ask for each model: Can it account for the known properties of enzyme catalysis? The models highlight catalytic properties of enzymes that are understood and also illuminate areas that remain to be elucidated. This sets the stage for the final section of this review, where the questions, comparisons, and approaches required to further our understanding of enzymatic catalysis are considered.

A Hypothetical Enzyme and Reaction for Comparison

The enzyme models are used to recreate the hypothetical enzyme shown schematically in Figure 10. This enzyme catalyzes an enolization reaction, analogous to triose phosphate isomerase, ketosteroid isomerase, and many other well-studied enzymes (6, 47, 91). Structural work on this hypothetical enzyme identified a His residue at the active site acting as a general base to deprotonate the methylene group and a Gln residue hydrogen bonding to and thereby stabilizing the incipient anionic enolate oxygen atom. In addition, site-directed mutagenesis of these so-called "catalytic" residues to Ala gave dramatic reductions in catalysis, 10^6-fold for His and 10^4-fold for Gln upon side chain removal. As the overall rate enhancement observed by this enzyme is 10^{10}, it is tempting to conclude that all of the catalytic power of this enzyme has been identified. There is a basic flaw with this logic, however. The catalysis does not come solely from what are referred to as the "catalytic groups." This point was emphasized in the previous section, and its fundamental relationship to enzymatic catalysis is highlighted in the models below.

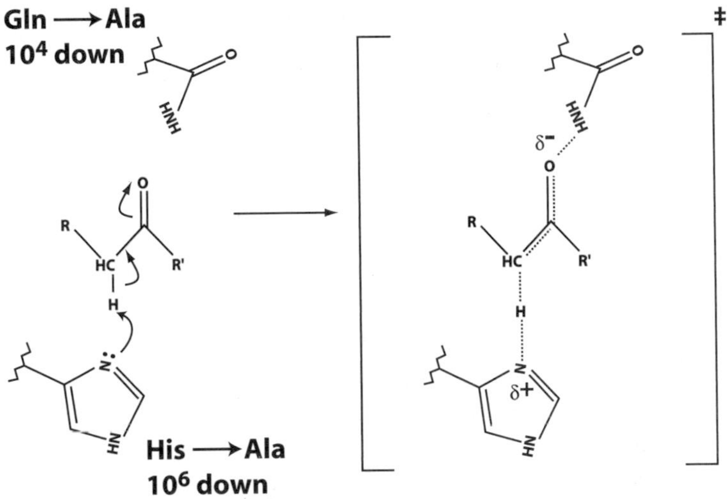

Gln ⟶ Ala
10^4 down

His ⟶ Ala
10^6 down

Figure 10 Hypothetical enzyme used as a starting point to develop models explaining enzymatic activity. The enzyme catalyzes an enolization reaction using histidine as a general base to remove the proton alpha to the carbonyl group and glutamine as a hydrogen bond donor to stabilize developing negative charge on the carbonyl oxygen atom. Removal of these two residues leads to complete loss of catalysis.

Model I. "Catalytic" Residues

Taking literally the notion that His and Gln are the catalytic residues of the hypothetical enzyme (Figure 10), it follows that adding both histidine and glutamine to solution should provide catalysis that matches that of the enzyme (Figure 11). Of course this is not the case. The "catalytic" residues only allow enormous rate enhancements when placed within the context of the folded enzyme, as shown also by the Ala mutagenesis example above (Figure 3). This realization leads to a second model.

Model II. Positioned "Catalytic" Residues

Clearly the "catalytic" groups of enzymes need to be positioned with respect to one another. This is accomplished in Model II (Figure 12). The second model enzyme has a hypothetical framework with covalent interconnections that give precise positioning of the His and Gln residues without the large size necessary to form tertiary structure in a protein. This allows us to separately address the potential special effects of the protein framework later and thus to conceptually distinguish these effects. Imagine that advances in synthetic chemistry allow construction of a nano-scaffold that provides precise positioning of the "catalytic" residues in this model.

How does the Model II enzyme rate as a catalyst? Although the His and Gln residues are properly positioned, there is nothing to position the substrate with

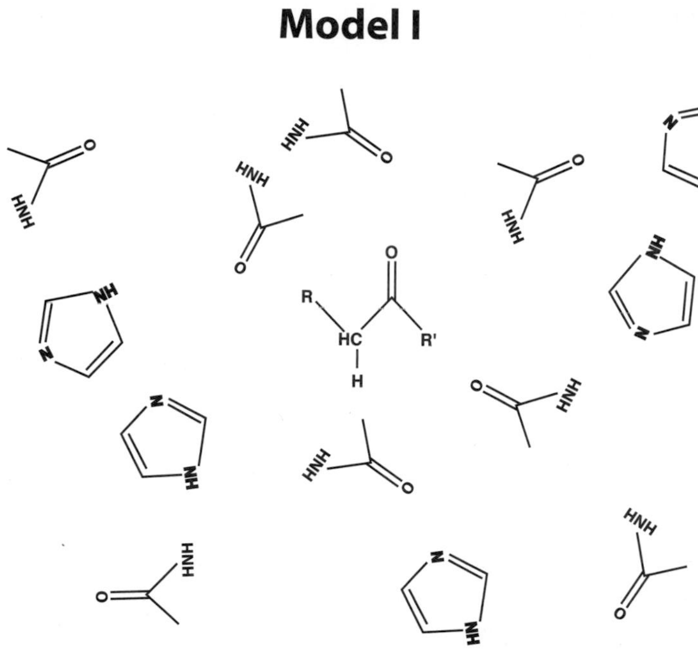

Model I

Figure 11 Model I: "catalytic residues" in solution. Histidine and glutamine (only their side chains are shown) are added to a solution containing the substrate.

respect to these catalytic groups. While reaction may be very efficient when the substrate is appropriately positioned within the active site, such positioning is highly improbable for the Model II enzyme. This problem is corrected in Model III.

Model III. Positioned "Binding" and "Catalytic" Residues

Model III maintains the His and Gln positioning from Figure 12 but introduces binding interactions with the groups that flank the enolization site (Figure 13). Two successive versions of Model III are shown. In the first (Figure 13*a*), these binding interactions are sufficient to localize the substrate to the active site under the hypothetical physiological conditions. However, even though the substrate is in the active site most of the time, it is not ordinarily positioned for reaction, as represented by its misalignment and mobility in the ground state (Figure 13*a*). This model illustrates that optimal catalysis requires positioning beyond the loss of translational entropy, as has been described previously in several different ways [e.g., (22, 92, 93–95)].

The situation of Model IIIa is precisely analogous to the *Tetrahymena* RNA enzyme example above: In a mutant context the substrate was localized to the enzyme in the open complex but not positioned for reaction (Figure 8). As for the

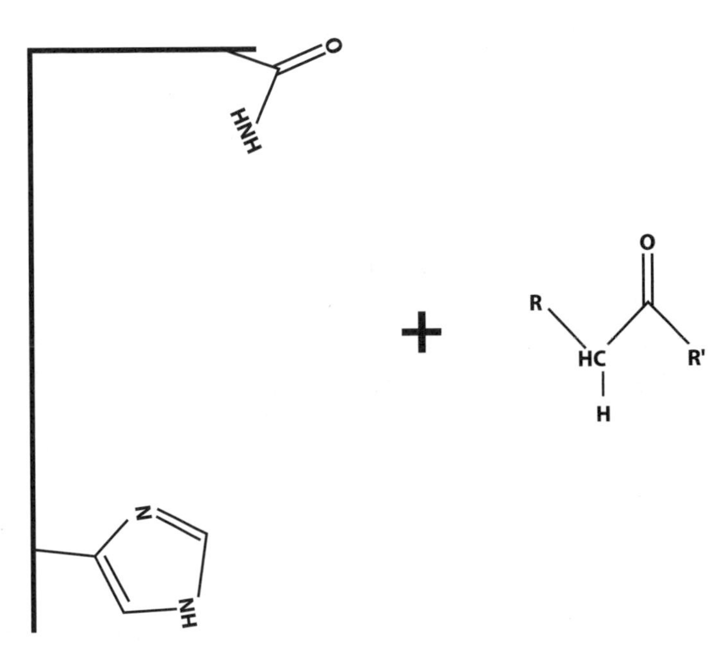

Figure 12 Model II: positioned "catalytic residues." The histidine and glutamine residues are positioned on a rigid low molecular weight framework such that they are correctly oriented for reaction. However, the model cannot position the substrate relative to the so-called catalytic groups.

RNA enzyme, the solution is to add more binding interactions (i.e., interactions remote from the site of chemical transformation) to provide the necessary fixation of the substrate within the active site (Figure 13*b*). These additional binding interactions, illustrated by the green lines in Model IIIb, position the bound substrate correctly for reaction and thereby speed the chemical transformation of the bound substrate.[3]

In Model III the groups that contribute to binding and reaction of the bound complex are interconnected energetically and functionally. The "binding" residues contribute to catalysis, and the so-called "catalytic" Gln helps in positioning by providing an additional attachment point for the substrate. This functional

[3]Binding interactions and binding energy in actual enzymes can be added not only by introducing residues that directly contact the substrate but also by adding more distant residues that help position the directly interacting residues for binding and catalysis. This later indirect mechanism was elegantly demonstrated by sequence and structural changes of a catalytic antibody through its maturation (96, 97).

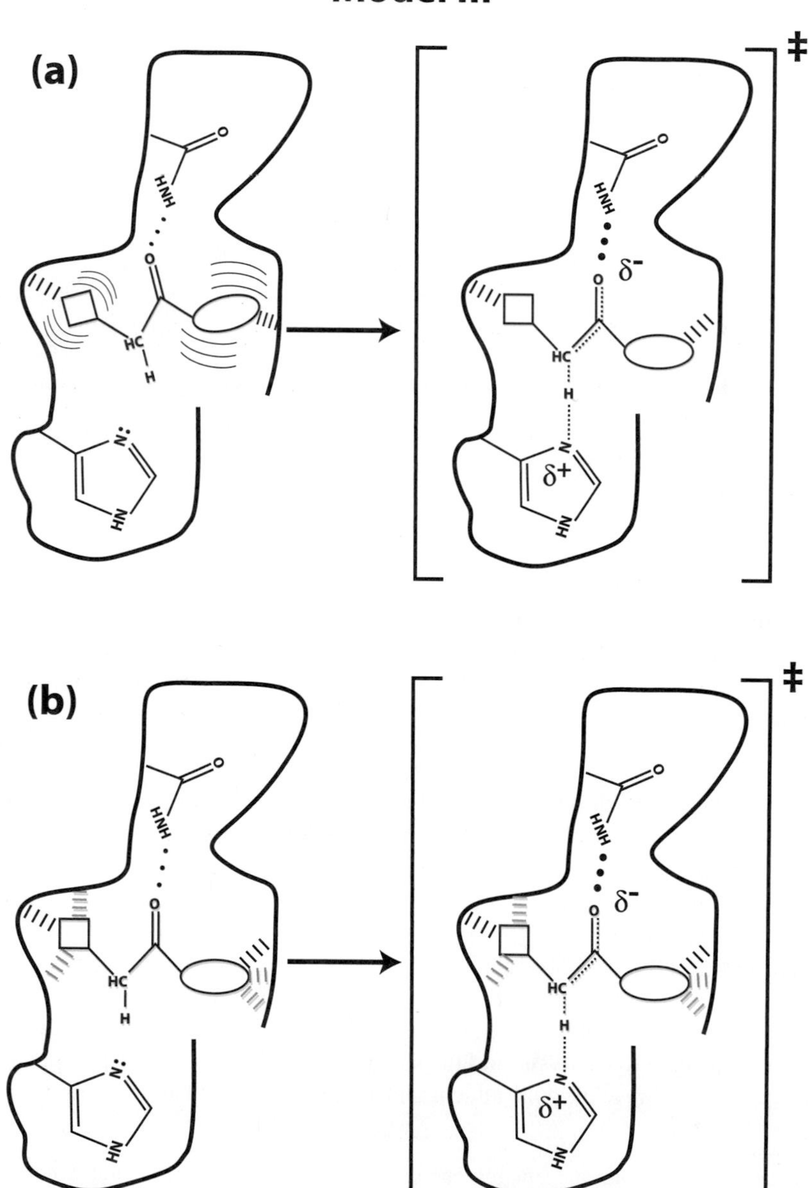
Model III
(a)
O
HNH
O
HC
H
N:
HN
‡
HNH
δ-
O
HC
H
N
δ+
HN
(b)
O
HNH
O
HC
H
N:
HN
‡
HNH
δ-
O
HC
H
N
δ+
HN

interconnectedness would not be revealed by single site-directed mutants, but it is nevertheless at the heart of enzyme function.

Model IV: Tuning Interactions and Binding Energy

Would the last model perform catalysis as efficiently as an actual enzyme? There are two likely inadequacies, both related to the strength of binding interactions, addressed and corrected below. (Additional features lacking in these models may or may not prove to be important for catalysis; these features at the frontier of scientific understanding are introduced and discussed in the last section of this review.)

Enzymes need to catalyze reactions relative to reactions in aqueous solution, and preferential stabilization of the transition state over the ground state is required to accomplish this (Figure 9). In solution, reactions for amide and ester hydrolysis and proton abstraction α to carbonyl groups (as in Figure 14) occur with oxyanion-like transition states. There is likely to be hydrogen bond donation from at least two water molecules to the oxyanion; presumably a penalty in electrostatic interaction energy would be paid, relative to aqueous solution, if only a single hydrogen bond were donated by the enzyme. Indeed, in active sites carbonyl oxygen atoms that develop oxyanion character typically accept two hydrogen bonds from the enzyme (98, 99). Therefore a second hydrogen bond donor to the oxyanion is added in Model IVa (Figure 15*a*).[4] Related issues of electrostatic stabilization within the active site are treated in greater depth in the final section.

The second inadequacy derives, paradoxically, from binding interactions being too strong. Speaking anthropomorphically, an enzyme wants to maximize binding interactions with its substrates to maximize the energy potentially

[4]In the case of ketosteroid isomerase, mutation of the catalytic base (Asp38) and a hydrogen bonding group (Tyr14) seemed to account for all of the catalytic enhancement of the enzyme, but it was later found that another group (Asp99) also donated a hydrogen bond to the enolate oxygen atom and that mutation of this residue was detrimental to catalysis (100, 101).

←

Figure 13 Model III: substrate positioning and catalysis. (*a*) This framework provides a limited number of interactions with nonreactive portions of the substrate. These interactions are sufficient to localize the substrate to the active site, but considerable motion remains (depicted by motion lines). Thus, the substrate will be properly aligned for catalysis only a fraction of the time. (*b*) Groups making additional substrate interactions are introduced. This allows the substrate to be bound in a conformation that is aligned for reaction, with the proton that will be removed positioned with respect to the histidine lone pair and the carbonyl oxygen atom positioned with respect to the active site glutamine (compare to Figure 8).

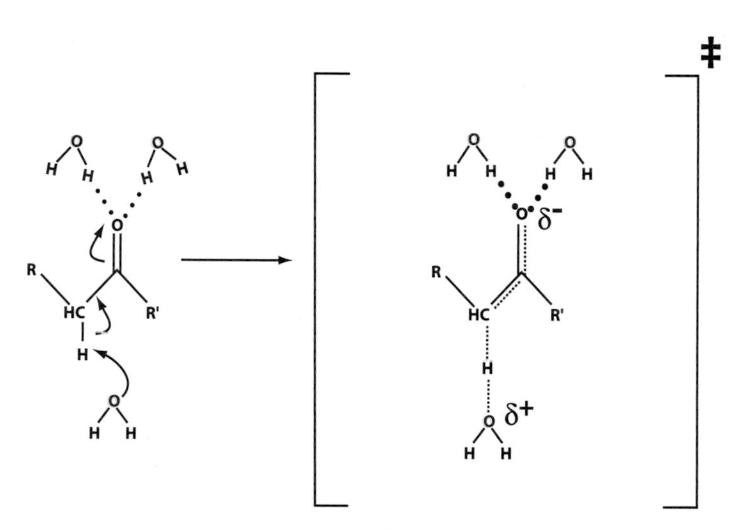

Figure 14 Multiple hydrogen bonds to anionic oxygen atoms. A solution enolization reaction comparable to that used to develop the enzyme-like models (Figure 10). Hydrogen bonds to the carbonyl oxygen atom in the ground state get stronger in the transition state (as depicted by the larger dots).

available for catalysis and specificity (16, 29). In Model IIIb binding interactions were strengthened to position the substrate in a reactive conformation. Superficially one would expect increased precision in the alignment of active site residues to provide precise complementarity for the desired substrate and an ability to provide maximal preferential stabilization of the transition state relative to the ground state, thereby increasing specificity and catalysis. However, the highly precise positioning of the enzyme models creates problems, paradoxically due to their limited flexibility.

Maximal binding interactions will require, in general, interactions of the enzyme on all sides of the substrate. Enzyme flexibility is then needed to allow substrate ingress and product egress. Indeed, enzymatic flap closures and hinge motions to accomplish this are extraordinarily common (102–107). A joint is therefore introduced in the otherwise rigid enzyme to give open and closed states that allow both ready access to the binding pocket and recognition of all elements of the substrate (Figure 15*b*, Model IVb).

The precise positioning of binding groups in the new model enzyme now seems optimal to position the substrate for reaction in the closed conformation. Nevertheless, problems derived from binding interactions persist for both catalytic turnover and specificity. Precise positioning can cause binding to be too strong, because the modest structural rearrangements that mitigate against very strong binding in real situations are eliminated in the more rigid model enzyme. With very strong binding, product dissociation is slowed and the maximal turnover rate can be lowered [e.g., (85, 108, 109)]. Indeed, it has been suggested

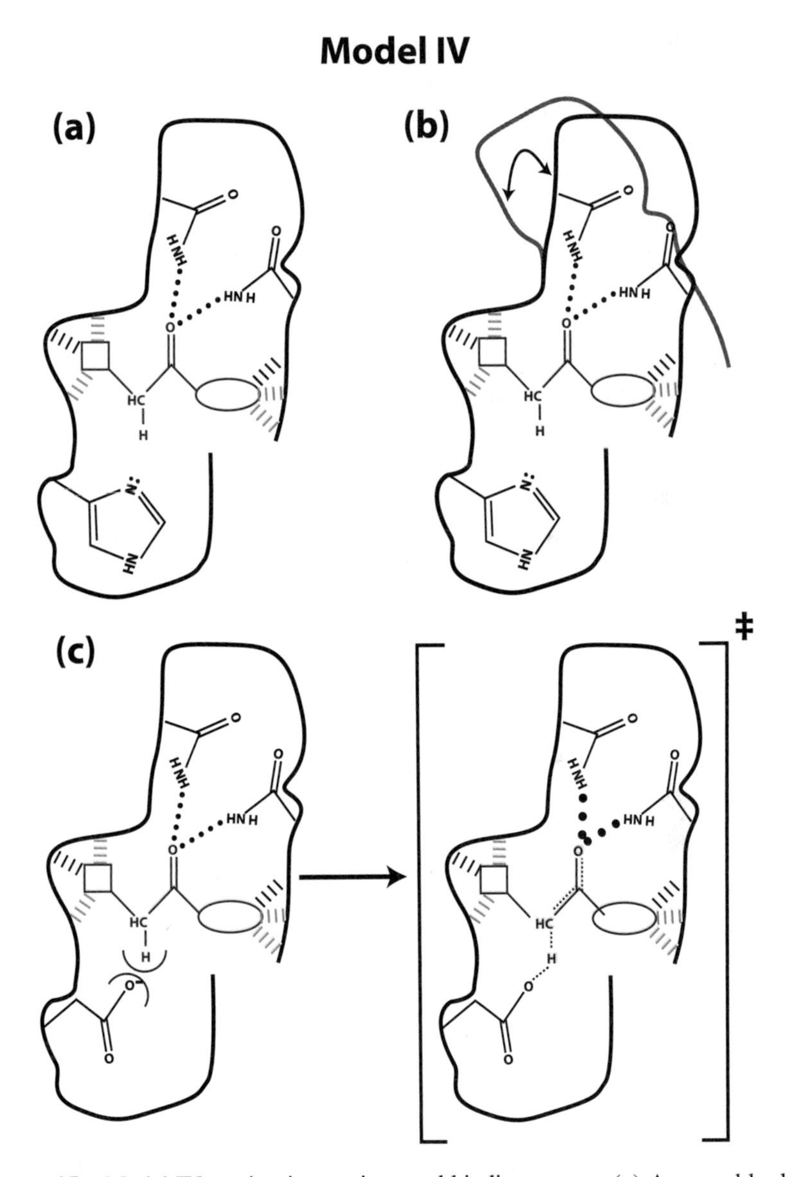

Figure 15 Model IV: tuning interactions and binding energy. (*a*) A second hydrogen bonding interaction is added to the negative charge forming on the carbonyl oxygen atom. (*b*) A hinge point is added (arrow); this allows the enzyme to open so that substrate and product can enter and exit the active site while maintaining the specific interactions made on all sides of the substrate. (*c*) The general base is changed to aspartate, which has a full negative charge. Burial of this negative charge in the ground state, upon substrate binding, is destabilizing relative to a neutral group or to interactions in solution; this is represented by the purple lines. This interaction preferentially weakens ground state binding, which allows for rapid turnover and high specificity.

that limitations of antibodies as catalysts may arise from a too-rigid structural template that leads to slow product dissociation [(110); see also (111)]. Too-strong binding also decreases specificity, because even substrates that only make a subset of the binding interactions will react faster than they dissociate, and both desirable and undesirable substrates can therefore react at the rate of diffusional binding (104, 112). To remedy these problems, additional motion is introduced into the binding residues in Model IVb, thereby weakening binding sufficiently to allow undesirable substrates to dissociate in preference to reacting (104, 113).

Another way to weaken binding interactions without compromising stabilization of the transition state is to replace the His general base with a negatively charged Asp residue (Figure 15*c*, Model IVc). Ground state binding is weakened because the negatively charged Asp is positioned adjacent to a proton of the hydrophobic substrate methylene group by the combination of the model's structural scaffold and its binding interactions. In the transition state, this destabilizing interaction is relieved as proton transfer from the methylene group neutralizes the Asp residue.[5] Ground-state destabilization has been proposed in numerous enzyme systems [e.g., (114–116)] and can contribute to an enzyme's ability to provide preferential stabilization of a reaction's transition state relative to its ground state (Figure 9) [see (16) for a more detailed discussion of the relevant energetics].

Summary

A progression of enzyme models has been presented to help identify and understand the features of enzyme catalysts that are special, or distinct, relative to simple chemical catalysts. Most fundamental is the energetic and functional interconnection of binding and catalysis. This property of enzymes intimately links specificity and catalysis, because only the correct substrates make the interactions that lead to efficient catalysis. The inextricable linkage between binding and catalysis was noted early on by Polanyi, Pauling, and Jencks (16, 27, 28, 117) and has been highlighted by many subsequent enzymologists [e.g., (13, 29, 35, 93)].

Returning to the question of whether our most advanced enzyme models would rival natural enzymes in specificity and catalysis, the answer is not known. Instead, a different question is posed below: What features of actual enzymes are not accounted for by the simplified enzyme models? This comparison helps identify and frame central questions that remain about enzymatic catalysis and the properties and behavior of enzymes.

---

[5]Assessment of ground state destabilization requires comparison to some standard condition (e.g., aqueous solution or gas phase), and one can imagine making different comparisons in the same system, some of which give ground state destabilization and others of which do not. Further, whether or not ground state destabilization occurs in any specific case will depend upon what other interactions are made with the Asp residue and upon the environment surrounding the active site.

MEETING THE CHALLENGES OF UNDERSTANDING ENZYME MECHANISM: A MODERN PERSPECTIVE

The face of mechanistic enzymology has changed dramatically over the past two decades. Researchers have gone from speculation and indirect assays for residues involved in the chemical catalysis to direct tests via site-directed mutagenesis [e.g., (33, 118, 119)]. Mechanistic schemes have progressed from cartoon representations of enzymes to atomic resolution supported by X-ray crystallography. Presteady state kinetics has supplanted steady state kinetics for in-depth enzymological investigations, allowing reactions to be defined in terms of microscopic steps. These advances leave us poised to attain an understanding of enzyme mechanism at unprecedented depth.

There are, nevertheless, critical limitations of even these contemporary approaches. Site-directed mutagenesis allows a residue to be removed and effects on function to be probed, but this response is read out in energetic terms (from rate and equilibrium effects), which must then be given molecular interpretations. The structures of wild-type and mutant enzymes can be obtained, but these are static pictures that do not capture the catalytic process.

How can we do this better? A bridge between structure and energetics must be created through both an understanding of dynamics and an understanding of the physical properties of enzymes that goes beyond a description of the differences between mutants and the wild type (Scheme 1). In this section, questions at the boundaries of the understanding of enzyme mechanism and energetics are raised. These questions are framed both by the intellectual history of mechanistic enzymology and also by the enzyme-like models developed in the last section (Figures 11, 12, 13, 15). The primary differences between actual enzymes and these enzyme-like models lie in what surrounds the groups directly interacting with substrates and the rigidity or flexibility of the interconnections between these groups. Thus, these comparisons may be particularly adept at revealing the functional and behavioral consequences from the enzyme's molecular architecture beyond the active site.

In addition to formulating the right questions to provide insights into the function and behavior of enzymes, a deep understanding of enzyme mechanism and energetics will require developing existing and new experimental and computational tools and then bringing these tools to bear on these and other questions. Below, we first describe tools that are sufficiently developed to recognize their power in this pursuit and then turn to the pressing questions in mechanistic enzymology that will require our attention, imagination, and inventiveness in the coming years.

Twenty-First Century Technology for Enzymology

X-RAY CRYSTALLOGRAPHY As noted above, high resolution X-ray crystallographic structures of enzymes, often with substrate or transition state analogs bound, are now common. More structures of individual enzymes that are under intense mechanistic scrutiny are needed: structures of one enzyme with substrates, products, and a variety of analogs bound. No transition state analog is perfect, so rearrangements of the enzyme structure often occur to accommodate these imperfections.[6] Thus, structures with multiple analogs should be sought to help unravel the types of rearrangements likely to occur within the enzyme [e.g., (120)]. And these structures are needed at ultrahigh resolution, resolution higher than 1 Å [e.g., (121, 122)]. Because many of the interactions of interest are hydrogen bonds and most enzymatic reactions involve transfer of one or more protons, the inability to directly observe the hydrogen's electron density in lower resolution structures limits mechanistic interpretation. At 0.9 Å resolution, about half of the hydrogens have assignable electron density (G.A. Petsko, personal communication); also, neutron diffraction, while technically challenging, allows observation of the hydrogen atoms (123). Another value of ultrahigh resolution X-ray structures is a reduction of the error in measuring interatomic distances, as precise distances can be of great value in understanding hydrogen bonding interactions and the positioning of groups for bond formation (G.A. Petsko, personal communication; 76, 77, 124).

ENZYME DYNAMICS AND NMR Energy cannot be discerned even from the highest resolution structure. One fundamental reason for this is that structure is not static, and free energy is a function of enthalpy and entropy, i.e., the number of substates accessible and the energy of each of these states. It is clear that enzymes are highly dynamic entities, held together by noncovalent interactions that are weak compared to covalent bonds. Thus, to understand the basic functioning of enzymes in terms of energetics and the conformational states utilized in function, the dynamic behavior of enzymes must be interrogated. Currently the most powerful approach to study dynamics is NMR, which provides (in principle) the ability to probe motions on timescales ranging from picosecond to millisecond and greater for each residue (125–129). Nevertheless, bringing such extensive data together into dynamic molecular models is a formidable challenge (see Computation of Enzyme Behavior and Function, below).

SITE-DIRECTED MUTAGENESIS WITH UNNATURAL AMINO ACIDS Limitations of site-directed mutagenesis have been highlighted throughout this review. Site-directed mutagenesis cannot provide a reductionist dissection and understanding of enzymatic catalysis (as is sometimes claimed or implied), and the functional importance of certain enzymatic features are masked from discovery by site-

---

[6]Rearrangements of noncovalent connections are typically easier than rearrangement of covalent bonds. Thus rearrangements of the packing and positioning within the enzyme is more likely than substantial rearrangement of the covalent bonds of imperfect transition state analogs.

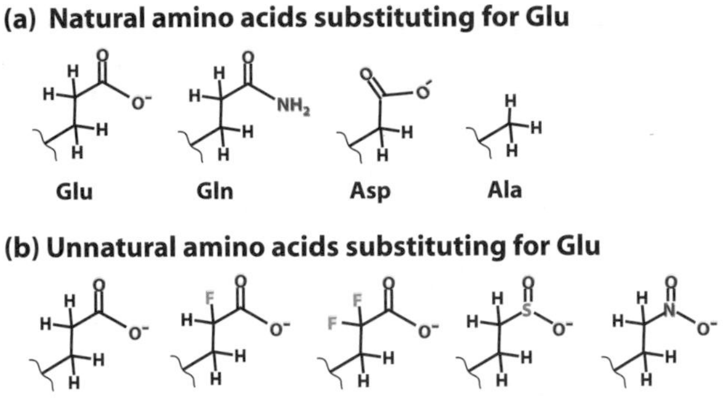

(a) Natural amino acids substituting for Glu

Glu Gln Asp Ala

(b) Unnatural amino acids substituting for Glu

Scheme 2.

directed mutagenesis. Nevertheless, site-directed mutagenesis has proven enormously powerful in identifying residues with direct roles in catalysis. Further, assessing the functional, energetic, and structural response to a vast number of mutations has provided vital information about the behavior of the protein scaffold [e.g., (130–132)].

But despite the usefulness of generalities from site-directed mutagenesis, the scale of the changes introduced often renders interpretation difficult. For example, a Glu residue thought to act as a general base is often replaced by Ala. In this mutant the ability of the group to accept a proton is not altered; rather, it is obliterated. Further, residues around the active site can rearrange in response to removal of the two hydrogen bond-accepting oxygens of Glu (Scheme 2*a*). Even if an apparently more conservative change is made of Glu to Gln, two (or three) hydrogen bond acceptors are lost, and two hydrogen bond donors are added; changes that are also likely to lead to substantial destabilization and local rearrangement. If the carboxylate functionality is instead maintained by mutating to Asp, the situation is no better. Now a conformational rearrangement or intervening solvent is required to allow the Asp to act as a general base. Indeed, there is ample evidence for active site rearrangement in response to mutation, and remaining active site functional groups have even been known to rearrange to take the place of the removed group (133).

What we would like to be able to do is to alter, in an incisive and systematic fashion, the properties of a particular enzymatic functional group. Instead of removing the general base, its electronic properties could be varied without varying [or with minimal variation in (134)] its size and geometry (Scheme 2*b*). The response to this variation can be compared to that observed in related solution reactions. The behavior of this series of analogs can also be compared with different substrates and with enzyme variants at other positions. These comparisons can help reveal the interplay of general base catalysis with other catalytic and structural features of the enzyme and substrate. This approach mirrors that of physical organic chemistry, which has uncovered much about solution reactions through careful and systematic variation of reactants and conditions (62, 135).

Accomplishing these goals will require ready access to a series of unnatural amino acids (and even nonamino acids incorporated into the peptide backbone). And large quantities of protein will be needed so that structural and spectroscopic work can be carried out alongside functional assays. Fortunately, solid phase synthesis and intein ligation methods are emerging that promise to provide ready access to the large amounts of unnaturally substituted proteins that will be required for this next level of mechanistic analysis [(136–139); see also (140)].

VIBRATIONAL PROBES: SPECTROSCOPY AND ISOTOPE EFFECTS Changes in NMR chemical shifts for residues extending well beyond the site of mutation suggest that mutagenesis changes interactions throughout the protein, even in cases in which no structural perturbation is evident from crystallography [e.g., (141)]. And even the most conservative unnatural amino acid substitutions still leave one comparing properties of a wild-type and mutant enzyme (Scheme 1).

Vibrational probes have the special quality of allowing direct measurement of bonding properties without recourse to mutation. Because bond vibrations are highly sensitive to the molecular surroundings, information is presented about the environment felt within the enzyme. This may allow one of the most vexing questions in enzymology to be addressed: What is the solvation behavior in an enzyme active site? This question is considered below.

PRESTEADY STATE KINETICS AND SINGLE MOLECULE MEASUREMENTS In the past decades, presteady state methods, which allow the examination of individual reaction steps, have continued to improve in terms of equipment availability, background literature, lower sample requirements, and data fitting approaches (142, 143). Most recently single molecule fluorescence approaches have begun to emerge in enzymology (144–147). Single molecule approaches provide a potential window to intermediate and transient states that do not accumulate and allow dissection of population averages into the behavior of the individual constituents.

COMPUTATION OF ENZYME BEHAVIOR AND FUNCTION Enzymes are complex cooperative systems. Measured rate and equilibrium constants and even thermodynamic parameters (ΔH, ΔS, ΔC_p) provide miniscule information relative to this complexity. Although NMR relaxation has the potential to assay the dynamics of every residue, it does not provide information about the correlation of these motions. Computation will be essential to integrate information from multiple experimental approaches and to provide a thorough and detailed characterization of enzyme behavior and function. Computation may also provide suggestions for new experiments and predictions of their outcomes. Indeed, computation can in principle probe the response of enzymes to systematic controlled variations with higher precision than even the unnatural amino acid approach outlined above.

The charge or size of any group can be varied in silico and the effects assessed [e.g., (148)].

It is encouraging that there has been a considerable increase in the number of investigators applying computational approaches to problems in enzymology over the past few years. Nevertheless, enormous limitations remain: Simulations cannot be carried out on the timescales of enzymatic reactions; the potentials used are limited, typically lacking group polarizability; only part of the protein can be treated quantum mechanically; and not all configurations can be sampled. Computations typically start from structural coordinates determined crystallographically. But several ensemble kinetic studies and more recent single molecule studies have suggested that the activity of individual enzyme molecules can vary on the millisecond and longer timescales (149–153); i.e., slow conformational transitions may have profound effects on catalysis. The resulting uncertainties about the catalytic competence of a given structure, if widespread, will limit current computational approaches, but these uncertanties may also provide opportunities to more incisively test computational models and, subsequently, predict functional behaviors.

Computational power will certainly continue to grow in the coming years as technical challenges are met. But this alone will not render computation a powerful tool in enzymology. Theoreticians and experimentalists will need to join forces to develop an extensive interface that provides experimental constraints for computational development and experimental tests of detailed predictions from computational models. This is a critical challenge for both communities; without this partnership, computation will remain ungrounded by experiment, and experimental observations will not be placed in the most advantageous context of comprehensive and predictive models.

Questions at the Frontier of Enzymology

As described in the first section of this review, there is no absolute or universal breakdown of energy and function into component parts. Instead, comparisons are needed to understand enzymes and their catalysis. Most basically, the enzyme catalyzed is compared to the uncatalyzed reaction in solution. But even here numerous comparisons are possible, and different comparisons can lead to different insights. The enzymatic reaction can be compared to reaction in the gas phase, to general acid, to general base, or to metal ion catalyzed reactions in aqueous or other solutions, for example. And for each comparison a standard state must be defined: What concentrations will be used to compare reactions that have different molecularities (154)?

There has been considerable discussion in the enzymological literature about the choice of standard state. One idea is to use physiological concentrations of substrates. Physiological concentrations are often near enzymes' K_M values, an observation that can be rationalized in terms of an evolutionary driving force to increase catalytic throughput. Throughput can be increased by increasing K_M while maintaining k_{cat}/K_M at a constant value; this frees enzymes that would otherwise be sequestered in enzyme/substrate, intermediate and product complexes and thus inaccessible for reaction with new substrate. Once k_{cat}/K_M is near the diffusional limit and K_M is near the physiological substrate concentration, there is little further advantage from increased K_M and k_{cat} values; most of the enzyme is free to react with substrate, and this reaction occurs at a rate near the diffusional limit (155).

But a set of comparisons that provides insight into the evolution of enzyme function may not be the most powerful tool to derive mechanistic and energetic insights. Indeed, the same interaction can help, not affect, or hinder catalysis, depending on the standard state used. Consider a residue that when added to an evolving enzyme contributes uniform binding (Figure 6*a*). With subsaturating substrate, this residue speeds the enzymatic reaction by increasing the probability that the substrate will be bound at the enzyme's active site. With saturating substrate, there is no effect for the enzyme because the ground and transition states are stabilized to the same extent; this leaves an identical barrier to overcome. For an enzyme (say more evolutionarily advanced) that already binds substrates (and products) strongly, addition of this same residue can hinder catalysis by slowing rate-limiting product release. Combining these comparisons provides a richer understanding of the enzyme than any individual comparison, but the comparisons must be defined explicitly. Often disagreements in the literature arise because two groups are asking different questions and making different comparisons, and therefore getting different answers. Spelling out models, assumptions, and comparisons as explicitly as possible fosters communication of ideas and sharing of perspectives within and beyond the enzymological community.

Comparing our small model enzymes with real enzymes, we start with two basic questions stemming from their differences: Why are most enzymes proteins, and why are enzymes so big? Several sections of questions follow; each deals with different aspects of the consequences of the large size and proteinaceous nature of enzymes. Throughout, comparisons highlight the limitations imposed by and opportunities created by the molecular properties of enzymes.

WHY ARE MOST ENZYMES PROTEINS? Modern-day biological catalysts arose from accidents of evolution, tempered by the capabilities and limitations of their molecular constituents. Currently the vast majority of biological catalysts are proteins, but several RNA enzymes remain from what many believe to have been an RNA world early in evolution (156–158).

Why did the RNAs, like the dinosaurs, lose their preeminence? The simplest explanation is that RNA side chains lack the functional groups that tend to be directly involved in chemical transformations within protein enzyme active sites, such as the carboxylate group of Asp and Glu, the amine cation of Lys, the imidazole of His, and the sulfhydryl of Cys (159). This then begs the question: Why did these side chains evolve for proteins but not RNA? There are many possible (and not mutually exclusive) answers. Perhaps the metabolic machinery could not readily explore and adapt to the synthesis of nucleic acid side chains with a wide array of functional groups, or perhaps RNA was constrained evolutionarily by simultaneous requirements to act as an information store and a functional macromolecule. A newly emerging macromolecule such as a peptide might have been freer to evolve, akin to gene duplication freeing one copy of a required gene to explore sequence space and new functions. There may be an additional constraint for RNA that is more fundamental to its molecular character: a limited ability to pack tightly, exclude water, and thereby manipulate the active site environment [see (159)].

No single experiment or approach could explain why proteins are the dominant biological catalyst, given the complexity of this question and its intimate connection with evolution. But an ability to assess the different classes of macromolecules as catalysts, deconvolved from their evolutionary history, would speak to their catalytic potential. Early mechanistic work explored the catalytic potential of lipid micelles and carbohydrate templates (160, 161), and developed and developing in vitro selection methods will provide more powerful approaches to assess catalytic potential. In vitro selection of RNA (and DNA) enzymes, the emerging ability to carry out in vitro selections with proteins, and future tools to allow encoded selection of unnatural polymers will allow mechanistic comparisons of enzymes that carry out the same molecular transformation but are made from different building blocks (159, 162–164). Protein-like side chains and functionality can be introduced into RNA, and conversely, the number and type of protein side chains can be limited (165, 166). It will also be interesting to explore whether protein or other templates can effectively use the more reactive functionalities exploited in organic and inorganic synthesis. In general, these approaches should allow isolation of catalytic consequences from different aspects of the structure of each polymer family.

WHY ARE ENZYMES SO BIG? Only a small fraction of the residues of an enzyme directly participate in the chemical transformation or contact the substrates. It is generally recognized that the remaining residues are involved in forming the overall structure necessary to position the residues that make these direct interactions [e.g., (167)]. The enzyme-like models of the last section (Figures 11, 12, 13, 15) use covalent bonds to arrange the groups directly contacting the substrates, whereas noncovalent forces accomplish this positioning for proteins. Noncovalent forces are weaker than covalent bonds, and the hydrophobic effect requires gathering a quorum of nonpolar groups. Thus, a polypeptide of substantial length is required to allow folding to a functional structure, whereas the models with covalent architecture can be quite small.

It would be wonderful to separate the catalytic consequences from the large size of protein enzymes by creating low molecular weight catalysts akin to the models of the previous section. Interestingly, several smaller structured peptides, especially those acting as neurotoxins, contain multiple disulfide bridges (168). Could disulfide fortification be used to obtain smaller polypeptide catalysts? Not all attempts to engineer disulfides into preexisting protein templates have increased stability, and many have been destabilizing [e.g., (169–171)]. Advances in computational prediction of structure and stability will be required to successfully design such a potential catalyst.

Even if much smaller polypeptides could be designed to rival enzymes in catalytic efficiency, it does not necessarily follow that Nature would have done so. In some cases, the larger size is used to recognize larger substrates or to carry out complex reactions (like the passage of DNA strands by topoisomerase). But there are likely additional causes of this variation [see (159)]. It is possible that the evolutionary driving force to conserve energy by making smaller proteins is not strong or not strong enough relative to the difficulty of this task through natural selection. From a functional standpoint, enzymes of course do more than catalyze reactions; many bind other proteins and nonsubstrate ligands as part of control mechanisms and the integration of cellular signals and responses.

CONSEQUENCES OF THE PROTEINACEOUS NATURE OF ENZYMES The low molecular weight enzyme-like models (Figures 11, 12, 13, 15) were designed to have the same groups as actual enzymes directly interacting with the substrates and directly participating in the chemical transformation. The difference lies in what surrounds these interacting groups and the rigidity or flexibility of the interconnections between these groups. These molecular properties impose limitations on what enzymes can do and also create opportunities. For example, the less rigid but closely packed protein scaffold allows residues directly interacting with substrates to be energetically and structurally interconnected with the residues that appear to constitute the enzyme's framework for the binding and catalytic residues. This linkage in turn allows communication of binding and chemical events at the active site to other sites on the enzyme. The coupling that results is the raw material for allostery, signaling, and complex coupled reactions such as the proton pumping and ATP synthesis carried out by the F_1/F_0 ATPase. In the discussions that follow, we emphasize the consequences for basic catalytic events rather than these more complex conformational events.

How does the solvation environment differ between enzymes and solution? The most obvious difference between the reaction at an enzyme active site versus our low molecular weight model is that there is a lot more "stuff" that surrounds the interacting residues in the protein enzyme. The proteinaceous material excludes solvent, thereby introducing both difficulties and opportunities for the protein enzyme.

Difficulties arise because most transition states have greater charge separation than ground states. Imagine that there is only grease (hydrophobic groups) in the area surrounding the substrate. Even with direct polar and ionic interactions at the active site, the enzyme-bound transition state may be less stable than that in aqueous solution because of the large energetic cost of full desolvation of the inner and outer water layers; this unfavorable effect could then cancel other

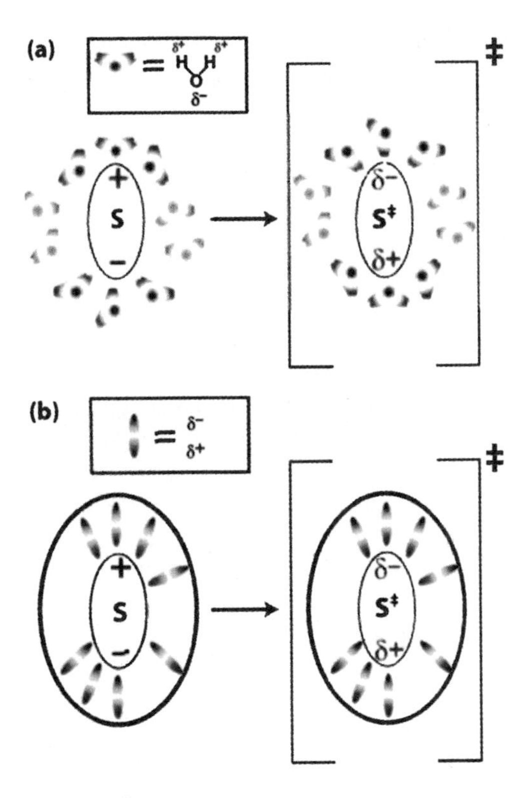

Scheme 3.

catalytic contributions, thereby rendering the enzyme ineffectual [(172); K.S. Carroll & D. Herschlag, unpublished calculations]. But enzymes do have dipoles, minimally in each backbone unit, and dipolar and charged side chains can be introduced that are not in direct contact with the substrate. The structural task would then be to align these more remote groups to stabilize the transition state via longer-range electrostatic interactions in a manner somewhat akin to solvation (Scheme 3*b*, $S^{\ddagger}$). Indeed, several examples of helix dipoles oriented to make favorable transition state interactions have been noted (173–176). It is possible that enzymes transform the potential liability of solvent exclusion into an asset. This could be accomplished by having more numerous oriented dipoles than water, by prepaying the cost to orient the stabilizing dipoles during folding of the enzyme, and by increasing the electrostatic discrimination between transition and ground states, as described below.

It remains to be determined, experimentally and computationally, to what extent such alignments exist and what the energetic and catalytic consequences of such alignment are. Computational methods are improving in this area but cannot yet be considered reliable or predictive, and thus these methods must be bolstered by recourse to experiment. Vibrational spectroscopy provides a powerful opportunity to assess the electrostatic environment at different places within an enzyme, because bond vibrations are highly sensitive to the environment (177–182).

Any interaction that can discriminate between the ground state and transition state can be beneficial to catalysis. The solvent-excluding protein matrix may be able to maximize the difference in electrostatic energy between the ground and transition states. In bulk solution, for a given change in charge (Δq_1), the energetic response in the presence of second charge (q_2) is determined by the dielectric constant (ϵ) of the medium (Equation 2). The concept of a dielectric constant loses meaning on the molecular scale,[7] but the response of the local environment to changes in charge distribution must still be considered. This is because electrostatic changes occur along a reaction coordinate, in going from a ground state to a transition state. Our model catalyst, by virtue of its small size, does not prevent water from closely approaching the bound reactants. Thus, water's strong dipole, small size, and ability to readily rearrange to stabilize either the ground state or the transition state can mitigate against large electrostatic discrimination between these states. In contrast, the larger size of protein enzymes and their tight packing could increase the electrostatic discrimination by excluding water and by placing the enzyme dipoles and charges within a structural matrix that minimizes their ability to rearrange in response to changes in the charge distribution at the active site (Scheme 3) (36, 37, 185–187).

$$\Delta E \propto \frac{\Delta q_1 q_2}{\varepsilon r} \qquad 2.$$

We are aware of two experiments that provide initial tests of whether protein enzymes do indeed provide environments that allow greater electrostatic discrimination. One utilized a series of ligands with varying hydrogen bond accepting ability and compared the change in binding affinity observed on the enzyme with changes in hydrogen bonding equilibria in solution (188, 189); the other used unnatural amino acids, a series of fluorotyrosines, and determined folding stability (190).[8] One of these experiments suggested hydrogen bonding

[7]The response to a change in charge within a molecular environment can vary depending on which individual site is substituted and can even vary for different substitutions at the same site. In addition, actual experiments typically involve more than a simple change in charge, e.g., pK_a measurements have been used to assess the electrostatic environment of an enzyme, but the observed pK_a value will be affected by the ability to rearrange to accommodate the change in size and change in hydrogen bond donors and acceptors present as a result of protonation. In other words, the observed response is a function of the local surroundings and their properties, not simply some global property of the protein. Thus, a dielectric constant or even a single effective dielectric constant cannot be defined for a protein interior (178, 183, 184).

[8]It is important to recognize that these hydrogen bonding experiments give results in terms of affinities or stabilities, not individual bond strengths, and so rely critically on an ability

energetics similar to what is observed in aqueous solution, whereas the other suggested a substantially greater sensitivity in the enzyme environment (185). Much additional work will be required to determine whether different enzyme sites give different behaviors and to more completely characterize these interactions. For example, it will be of great value to combine these energetic analyses with controls to monitor possible structural effects and vibrational spectroscopy to provide complementary information about the hydrogen bonds and their environment.

Can enzymes distort substrate conformations to facilitate reaction? As noted above, catalysis relies on preferential stabilization of a transition state relative to the corresponding ground state (Figure 9). Enzymes can pick out certain substrate conformers, and this can favor the desired reaction and also disfavor undesirable side reactions. For example, enzyme-bound Schiff base conjugates to pyridoxal phosphate position the hydrogen destined for removal perpendicular to the pyridoxal ring, favoring removal of that proton and disfavoring removal of other protons (191, 192); triose phosphate isomerase binds its substrate in such a way as to disfavor elimination of the phosphate from the enediolate intermediate (193–195).

Enzymes can select conformers from solution, but can enzymes actually distort substrates to the desired conformer, akin to the older idea that enzymes might act like the ancient torture device referred to as the rack, which stretched victims to uncomfortable and painful lengths (39)? Do enzymes cause analogous discomfort and pain for chemical bonds? On the one hand, the enzyme structure is enforced by noncovalent interactions, which tend to be weaker than the covalent bonds of substrates; thus distortions of these covalent bonds might be difficult to accomplish at an enzyme active site. Steel bars can distort rubber bands, but rubber bands are unlikely to bend steel! It is easier to imagine our model enzyme, with its rigid template enforced by a network of covalent bonds, distorting substrates. On the other hand, the cooperative formation and tight packing of protein native structure might enhance rigidity along the coordinate needed for bond distortion, and the enzyme's electrostatic environment might also promote distortion. The ability to distort substrates and the extent of this distortion will be a function of the sensitivity of a particular substrate motion to applied physical and electrostatic forces and the magnitude of these forces in the active site.

X-ray structures have suggested that glycosidases can distort sugar conformations toward the half chair, the conformation adopted in the oxycarbonium-like transition state (196–198). Distortions of bound substrates of serine

---

to vary the property of a functional group without removing it and with minimal geometrical side effects; traditional site-directed mutagenesis cannot effectively address this question.

proteases and other enzymes have been detected by changes in vibrational states, vibrational spectroscopy, and isotope effects on binding [e.g., (199, 200)].

How can these distortions be understood? Most substrate distortions may arise simply from placing substrates in an environment designed to stabilize a different species, the transition state, rather than a design by the enzyme to physically distort the ground state (178). This suggestion is in keeping with the idea that enzymes evolved active sites to complement transition states and not ground states, as discussed previously (27, 28, 117).

Knowledge of the physical environment of enzyme active sites and how this environment can affect bound ligands and transition states is in its early stages (see below). Higher-resolution structures with bound ligands will help address this question and so will additional vibrational spectroscopy and binding isotope effect data. These complementary approaches will be most powerful carried out with extensive series of substrates, substrate and transition state analogs, and subtly and systematically modified enzymes supplemented with kinetic and thermodynamic analysis.

Do enzymes change the nature of transition states from those observed in solution? This question parallels that concerning ground state distortions, but here the focus is on the transition state. In strict energetic terms, the least amount of transition state stabilization is required if the enzyme stabilizes the most stable solution transition state. However it may be easier for an enzyme to provide more stabilization along an alternative reaction path or with a modified transition state.

Modest changes in transition state structures have been surmised for glycosyl transfer and hydrolase enzymes based on extensive isotope effect analysis and on calculated vibrational modes for series of potential transition state geometries; nevertheless, all of the transition states obtained from these analyses retain considerable oxycarbonium ion character, as observed in solution (201). There have been many suggestions of altered transition states in enzymatic reactions, most prevalently for phosphoryl transfer reactions. These suggestions have typically been based on inspection of structures with bound substrate, substrate analogs, or transition state analogs, whereas functional data that can assess aspects of transition state structure—isotope effects and linear free energy relationships—do not support these proposals [e.g., (202–204)]. In general, transition state analogs cannot be assumed to faithfully mimic transition states, and resting ground state structures do not neccesarily represent reactive conformations. Further, the extent of transition state distortion will be a function of the relative energies of the different potential transition state structures, the electronic and geometrical properties of these potential transition states, and the forces at the enzyme active site, precisely analogous to the situation for ground state distortions. Experimentally the challenge is greater for transition state distortions, because transition states can only be probed indirectly. Characterization of enzymatic (and solution) transition states will remain a major challenge, one that will require a synthesis of energetic, structural, spectroscopic, and computational approaches.

DISTINGUISHING BETWEEN THE TRANSITION STATE AND GROUND STATE TO PROVIDE CATALYSIS As demonstrated schematically in Figure 9, catalysis requires preferential stabilization of a transition state relative to a ground state. The term "transition state stabilization" is often used in the literature to refer to specific electrostatic interactions that are strengthened in the transition state. However, any effect that leads to greater catalysis gives, by definition, transition state stabilization; that is stabilization of the transition state relative to the ground state is not a mechanism of catalysis but rather an energetic truism that follows from transition state theory (16, 21, 29). The imprecision in the usage of this term has led to unnecessary confusion and disagreement in the literature. Transition state stabilization does not provide an explanation for catalysis. Instead, it specifies the rate advantage relative to a comparison reaction, and thus it encompasses all possible catalytic factors, which range from general acid and base catalysis to direct electrostatic interactions, to indirect effects from the protein matrix that affect electrostatic interaction energies (185), and to changes in the number of accessible states.

Preferential stabilization of the transition state is a remarkably difficult task. Most of a substrate such as ATP or NADH is unchanged in going to the transition state, and the enzyme must focus its attention on the region(s) of bond making and breaking. The differences between ground and transition states can be divided, at least conceptually, into three areas: electrostatic changes, geometrical changes, and changes in the freedom of motion of substrates. Each is dealt with in the following questions. The rigidity versus flexibility of the active site and surrounding protein is key in all cases for determining the capabilities and limitations of enzymes to take advantage of each of these differences. The issue of enzyme dynamics is addressed in a separate question at the end of this section, because this basic property of enzymes has relevance for nearly all of the other questions posed and arguably represents the largest challenge to deepening our understanding of enzyme function and energetics.

How do enzymes discriminate between the electrostatic properties of transition and ground states, and how well do they do this? This topic has been covered in previous sections of this review, so here only a brief recap is provided. Most generally, charge build-up in transition states is stabilized by well-positioned hydrogen bonds and metal ions and by the presence of general acid and base residues to donate and accept protons. Indeed, the majority of biological reactions involves some shuttling of protons; for in-depth discussions of this topic, see (2, 205, 206). Hydrogen bond donors and acceptors are also common, and an example was emphasized in the models of Figures 11, 12, 13, and 15. Of course,

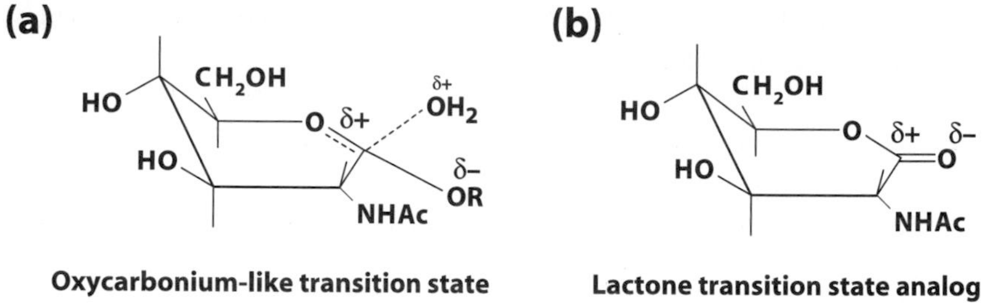

(a) Oxycarbonium-like transition state

(b) Lactone transition state analog

Scheme 4.

there are also hydrogen bonds to and from water for the corresponding solution reactions. Enzymes can provide catalysis via these interactions by (*a*) using stronger hydrogen bond donors and acceptors than water (185, 207, 208); (*b*) prepositioning the donors and acceptors to avoid a reorganization penalty from solution; and (*c*) creating an enzyme environment that allows a greater change in electrostatic stabilization energy than in solution (Equation 2; Scheme 3; Figure 15*c*). For reactions such as decarboxylations in which charge is more delocalized in the transition state than in the ground state, the active site may provide electrostatic interactions that are weaker or more repulsive than in solution (44). Nevertheless, some electrostatic interactions between the enzyme and the substrate will typically be required to position substrates for reaction (209).

Can enzymes discriminate between ground and transition states on the basis of geometrical differences? This question is highly related to those above concerning the ability of an enzyme to distort or change ground state and transition states. To underscore this, the same reaction classes, glycosidases and proteases, are used. Geometrical changes typically occur in the process of bond formation and breaking, such as planar ground states going to tetrahedral-like transition states (e.g., sp^2 carbons going to sp^3).

It has long been suggested that glycosidases can bind preferentially to compounds in the half chair geometry, the geometry adopted in the oxycarbonium ion-like transition state that has sp^2 character at C1 (198) (Scheme 4a). However, the compounds used also have electrostatic differences. The carbonyl group, introduced in the transition state analog of Scheme 4b to give an sp^2 center at C1 carbon and to favor the half chair conformation, also introduces a dipole. We cannot distinguish whether this analog binds strongly to the enzyme because of the partial positive charge (δ+) at C1 (which can interact with a neighboring active site Asp residue), because the half chair geometry fits better sterically into the active site, or because of a combination of these factors.

Similarly, tetrahedral transition state analogs of proteases often bind strongly. But the charge density on the oxygen atom of these analogs is greater than on the carbonyl oxygen atom of substrates so that stronger electrostatic interactions with the protease's oxyanion hole could be responsible for the observed affinity increase instead of geometrical factors.

The motion of atoms at the reaction center, such as the carbonyl oxygen atom in a peptide bond or the nonbridging phosphoryl oxygen atoms in a phosphate ester, will be on the scale of ~0.1–0.2 Å, very small changes [see also (210, 211)]. On the other hand, the distances between groups forming new single bonds will change on the order of an angstrom. Alternatively or in addition, the geometrical changes at the reaction center might be propagated to regions of the substrate that are more distal to give larger changes, somewhat akin to allosteric changes in enzyme structure. Such a lever arm effect would require limited conformational flexibility in the substrate, which could be intrinsic to the substrate or could be imposed by interactions with the enzyme.

The length scale that enzymes can sense is not known, and it will be necessary to understand much more about the flexibility of proteins to address this question at its most general level. For example, it has been suggested that enzymes precisely position the nucleophile and leaving group in phosphoryl and glycosyl transfer reactions, such that the predominant reaction coordinate motion is that of the phosphorus atom of the transferred phosphoryl group or the C1′ carbon of the transferred sugar as it migrates from the leaving group to the nucleophile (W.W. Cleland, personal communication; 124, 212, 213).

Testing possibilities for geometrical discrimination will require development of substrate and amino acid series that allow electrostatic and geometrical effects to be distinguished. Indeed, the desired variations to test for geometrical versus electrostatic discrimination can, in principle, best be carried out computationally, where features can be varied individually and systematically. This underscores the need to use experiment to guide the development of more powerful computational approaches.

How well can enzymes position substrates and catalytic groups for reaction? The role of positioning of substrates and catalytic groups with respect to one another has received an enormous amount of attention over the years and may have been described with more different terms than any other physical phenomena [e.g., (35, 40, 42, 95, 214–217)]. Here the intellectual history of this subject is briefly reviewed, with the idea of laying out basic concepts and emphasizing the important challenges that remain. Additional perspectives can be found in a number of treatments in the literature [e.g., (218–223)].

Early on it was thought that the largest rate enhancement obtainable from positioning was ~55 M (2, 19). The units of molarity come from comparing a unimolecular process with a bimolecular process (Equation 3), and the enhancement is often referred to as an effective concentration or effective molarity (154). The value 55 M is the concentration of pure water and represents an approximate limit for physically attainable concentrations. However, comparisons of intra-

and intermolecular nonenzymatic reactions gave effective concentrations far exceeding 55 M. For example, rate enhancements corresponding to effective concentrations of up to $\sim 10^8$ M for intramolecular formation of anhydrides were observed by Bruice & Pandit, and many other examples with large effective concentrations were also observed (224, 225). Storm & Koshland recognized that 55 M should not be a limit because even if two substrates are next to one another, they can be situated in a nonreactive orientation (42). He then surmised that enzymes and model systems could do better by orienting or steering the orbitals that needed to react. This insight allowed the limit of 55 M to be surpassed conceptually and remains the most intuitive teaching tool. While the quantitative theory of orbital steering that grew out of this insight is inconsistent with determinations of orientational restrictions in actual chemical bonds (94, 226, 227), the underlying insight is correct and instructive.

$$A + B \xrightarrow{k_2\ (M^{-1}s^{-1})} A\text{-}B \qquad A + B \xrightarrow{k_1\ (s^{-1})} A\text{-}B \tag{3.}$$

$$\frac{k_1}{k_2} = \frac{s^{-1}}{M^{-1}s^{-1}} = M$$

Page & Jencks considered this problem from a thermodynamic perspective (94). They recognized that the process of bond formation is accompanied by an entropic transformation. Basically, two molecules that together have six degrees of translational entropy and six degrees of rotational entropy are converted into one molecule with three degrees of translational entropy, three degrees of rotational entropy, and six extra degrees of vibrational entropy. Because there is typically much greater entropy in translational and (free) rotational modes than in vibrational modes, there can be a large loss in entropy upon formation of a new bond (or in the transition state for new bond formation, as this has partial covalent bonds). If an enzyme positions substrates such that this entropy has already been lost on binding, the entropy loss can result in a rate enhancement.

Entropic values for this process can be estimated reasonably well in the gas phase. Page & Jencks then used simplifying assumptions to convert estimated gas phase values to solution. The value they obtained of $\sim 10^8$-10^{10} M represents an estimated maximum rate enhancement for a bimolecular reaction from such positioning effects.

In more recent computational work, Lightstone & Bruice revisited a series of compounds that gave very large effective concentrations for intramolecular anhydride formation (228). The results suggest that as the effective concentration increases the enthalpic rather than the entropic barrier to anhydride formation decreases. Apparently, cyclization of the unconstrained parent compounds results in the formation of unfavorable eclipsed geometry of methylene groups; constraining the starting material in ring structures allows this strain energy to be paid by the introduced covalent bonds. The results are consistent with the treatment of Page & Jencks, because the rate enhancements observed in this series are larger than predicted from simply freezing the rotational entropy of the molecule (94) and thus imply that an additional factor is involved in the rate enhancement. Although the intramolecular series investigated by Bruice and colleagues does not have direct biological relevance, as there are a limited number of biological cyclization reactions and thus few requirements for directly analogous eclipsed geometries, the result makes the general and important point that constraining molecules can overcome entropic and enthalpic barriers to reaction.

Effective molarities on the order of $10^4 - 10^5$ M and larger have been obtained for enzyme ligands (229–232). These observations and the well-ordered binding sites and well-positioned ligands observed in X-ray crystallography and NMR dynamics are consistent with an ability of enzymes to effectively minimize substrate entropy and position groups for reaction [e.g., (127, 129, 233)]. But effective molarities, as noted above, can contain nonentropic contributions, and disorder and relaxation values from crystallography and NMR cannot be reliably transformed into quantitative measures of entropy.

The obvious solution would then seem to be direct measurements of entropies of binding and activation. Unfortunately, such measures are generally uninformative. The discussion above has focused on the configurational entropy of the reactants in the ground and transition states. However, thermodynamic parameters such as entropy changes are properties of the entire system. Indeed, most of the degrees of freedom are associated with solvent molecules, and many solvent molecules are displaced when a substrate binds to an enzyme. And even if one could focus only on the enzyme-ligand complex, most of these vibrational modes predominantly involve the enzyme. Thus, there is a disconnect between what we think of conceptually as entropy (related to the number of accessible configurations of the substrates) and the actual thermodynamic observable.

In principle, with sufficiently accurate potentials and advanced computational skills, one could model the system and fully describe its thermodynamics. But even if this were possible, it is not clear what would be gained conceptually. The entropy change is one number, and it represents an enormously complex sum of terms. Although these brute force approaches would certainly be useful, there is a more immediate need of conceptual advances to solve this central but vexing problem. Is it possible to understand configurational entropy and/or a positioning term that describes how preorganization of the various enzymatic components provide a rate advantage? Can the mixing of thermodynamic terms in enthalpy/entropy compensation that occurs in practice be meaningfully deconvoluted? Careful comparative work focusing on revealing thermodynamic trends may provide the best route to such advances. For example, obtaining $\Delta\Delta S$ and $\Delta\Delta H$ values for a series of enzymes and ligands varied systematically at a particular location, in conjunction with structural and dynamic data, may lead to new energetic insights [e.g., (234)].

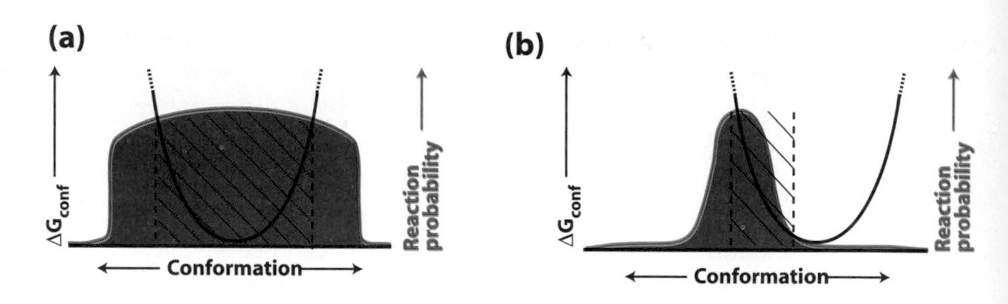

Figure 16 Potential interplay between enzymatic conformational states and reactivity. The possible conformations of an enzyme are represented by the black energy well (ΔG_{conf}); the probability of reacting from a given conformation is shown by the blue line, which is accentuated by the red under the curve. The product of these two probabilities in which 95% of the reaction will occur is shown with the hatch marks. Two possibilities are shown. (*a*) Reaction occurs from the most stable enzyme state, and fluctuations are not important for reactivity. (*b*) Most reaction occurs from enzyme conformations away from the most stable states.

WHAT IS THE DYNAMIC BEHAVIOR OF ENZYMES, AND WHAT ARE THE ENERGETIC AND FUNCTIONAL CONSEQUENCES OF THESE MOTIONS? As noted above in the section "Consequences of the Proteinaceous Nature of Enzymes," the noncovalent forces holding proteins together result in dynamic behavior, and this behavior can be important for functions such as product release, complex multistep reactions, allostery, and signal transduction. Here we discuss the role of dynamics in the catalytic event and the importance of understanding dynamics to understand this event. Two extreme views of the importance of dynamics can be considered. Both extremes can be ruled out but are instructive.

The first view is that dynamics are essentially inconsequential to catalysis—that we need only consider the average or most stable enzyme structure to account for catalysis. This model is depicted schematically in Figure 16*a*; the reaction rate or probability of reaction is a shallow function of enzyme motion, which drops off only for extreme conformational fluctuations. Indepth studies of isotope effects for enzymes that react via hydrogen tunneling events pioneered by Klinman and coworkers have provided an unexpected window into dynamics (235–238). Although the details of these systems are beyond the scope of this review, several features of these reactions appear only to be explained by models in which the majority of reaction occurs from an enzyme state that represents a conformational excursion from the most stable structure (represented schematically in Figure 16*b*). An oversimplified but conceptually useful view of this model comes from recognizing that tunneling events are extremely sensitive to barrier widths. It is therefore easy to imagine rare conformational excursions that bring the hydrogen donor and acceptor close to one another providing such a large increase in reaction probability that most of the reaction occurs through these rare states. More generally, dynamics, or conformational excursions, can lead to distance changes between reacting groups and to rearrangements of enzyme dipoles that facilitate reaction.

The opposite extreme view to the illustration in Figure 16*a* maintains that the overall dynamics of the enzyme are exquisitely tuned to facilitate the reaction. There are at least two types of arguments against this extreme view. The first comes from recognition that the barrier crossing for breaking of a covalent bond occurs on the subpicosecond timescale. Most motions and rearrangements of the protein matrix, because of its large size, are expected to occur on much slower timescales. Thus models invoking overall enzyme vibrational modes that directly correlate with and guide the bond breaking process have been strongly discounted (12, 149). Even motions of water molecules in solution reactions do not appear to be directly coupling to or in resonance with subpicosecond bond making and breaking events (239). Second, if the overall vibrations of the enzyme were exquisitely tuned, one might expect modifications that alter these vibrational modes to have major catalytic consequences. This experiment has been done many times, although not with this question in mind. Many enzymes have been extensively mutated in their cores for structural and folding studies with little effect on activity [see references in (68)].[9]

In summary, fully choreographed enzyme dynamics that mirror bond formation and cleavage events appear to be highly unlikely, but the opposite extreme, reaction probabilities that are insensitive to dynamic fluctuations (Figure 16*a*), is, at a minimum, not universal. The questions then arise: Is reaction through conformational excursions a special property of enzymes with large tunneling contributions or is it a more general phenomenon? Are the motions that lead to reaction favored on the enzyme's energy landscape, having been tuned through evolution, or are these dynamics random events that are inherent properties of the protein scaffold? For example, it is easier to imagine evolutionarily choreographed motions for enzymes that recognize substrates with high specificity, as opposed to enzymes that work more broadly on classes of substrates.

We do not yet have the tools to describe or dissect the dynamic behavior of enzymes. As noted above, NMR can provide dynamic information on a variety of timescales, but the trajectories and correlations of motions are not revealed.

[9]Similar results have been obtained from solvent isotope effect measurements. The large number of solvent isotope experiments carried out provides no indication of larger than expected rate effects that could arise if the exchange of deuterium for protium in backbone amide bonds throughout the enzyme disrupted vibrational coupling, and primary isotope effect experiments with unique substitutions at nonexchangeable sites provide no indication of smaller than expected rate effects that could be attributed to a large reduced mass from coupled motions that involve portions of the protein (13).

Computational approaches for these problems are in their infancy and will certainly be required. But even if the dynamics of an enzyme could be completely described, how would these motions be understood and related to function? If computational power were sufficiently strong, it might be possible to predict what motions are integral to catalysis and what substitutions would disrupt those motions. But as always, functional disruptions must be interpreted with extreme care: Residues that are candidates for important dynamic processes may also (or instead) contribute to catalysis in other ways.

Presumably it will still be necessary to use comparisons in an effort to identify and, if possible, isolate features of the enzyme dynamics linked to function. But what comparisons are likely to be fruitful? One idea is to compare the dynamics and catalysis of a series of homologous enzymes that catalyze the same reaction but are derived from organisms that grow at different temperatures and to carry out these comparisons as a function of temperature [e.g., (240–242)]. An additional step would be to carry out in vivo selections using this series of enzymes to replace the wild-type enzyme in organisms that grow over a range of temperatures and then to analyze the dynamics and function of the resultant enzymes, again as a function of temperature.

In summary, protein dynamics are clearly integral to some biological mechanisms. One often sees claims that a specific set of dynamic behaviors probed in a particular protein are then important to that protein's function. But consider the following analogies. A weight lifter is straining under a heavy weight, with her arms and even her entire body shaking; the shaking is a consequence of the strain on her muscles, not the causative feature allowing her to lift the weight! A magician waves a wand and would have the audience believe that these movements are the cause of the astounding acts he performs; these dynamics are not causative and, indeed, are meant to distract and mislead the audience. Although we assume no analogous intent on the part of enzymes, the same danger nevertheless lurks. There are many different timescales, directions, distances, and degrees of coupling of enzyme motions; it would be astounding if all of these dynamic events were finely choreographed by evolution to accomplish function. There are simply too many degrees of freedom to expect any but a miniscule subset of motions to be directly linked to function. But, these few motions may be critical to a complete description of catalysis.

Finally, understanding enzyme dynamics is critical for understanding catalysis of individual reaction steps, because a complete energetic description of an enzyme/substrate system requires an accounting of the number of accessible states as well as the energy (or enthalpy) of those states. Although we often refer to the transition state for a reaction and use two-dimensional reaction coordinates and transition state theory in discussing nonenzymatic and enzymatic reactions, a single transition state structure, a single reaction coordinate, and an exclusively classical view of the transition state will be inadequate to fully describe and understand enzymatic reactions, their mechanisms, and their energetics (12, 236,

237, 243, 244). These represent daunting, yet fascinating challenges in the quest to unravel the secrets of enzyme action.

PERSPECTIVE

The enormous rate accelerations and exquisite specificity of enzymes have captured the imaginations of generations of scientists. These features, rate acceleration and specificity, are inextricably linked, and in this linkage lies a key to enzyme function. Over the past decades, the level of atomic detail, kinetic description, and energetic insight obtained has revolutionized our understanding, and clarified our view, of the most basic behavior and function of biological macromolecules.

Enzymology now lies poised at the interstices of chemistry, physics, and biology. From this new perspective and with tools of unprecedented power available, we have the capacity to ask still more probing questions and to apply novel and more powerful approaches. But even with this enormous power at our hands, we cannot expect breakthroughs to come easily. The systems are complex and irreducible to independent component parts; new experimental and computational tools are needed; and conceptual insights will be required to phrase questions and to parlay these questions into understanding. We look forward to these enormous challenges in enzymology and to the next era of mechanistic understanding that lies ahead.

ACKNOWLEDGMENTS

We would like to thank John Brauman, Mo Cleland, Pehr Harbury, Judith Klinman, Geeta Narlikar, Greg Petsko, Shu-ou Shan, Vern Schramm, Peter Tonge, Dick Wolfenden, and members of the Herschlag lab for stimulating discussions and/or comments on the manuscript and Alessio Peracchi for artwork. DH would like to thank the many enzymologists, in addition to those listed above, who have shared their perspectives and ideas over the years, in conversation and in print, and Peter and Patricia Walter for acting as surrogate hosts while portions of this review were written. In most cases references represent examples rather than inclusive lists; we apologize for the numerous omissions of important citations required because of space constraints. This work was supported by grants from the National Institutes of Health and the Packard Foundation. DAK is an HHMI predoctoral fellow.

The *Annual Review of Biochemistry* is online at http://biochem.annualreviews.org

LITERATURE CITED

1. Sumner JB. 1926. *J. Biol. Chem.* 69:435–41
2. Jencks WP. 1969. *Catalysis in Chemistry and Enzymology*. New York: McGraw-Hill. 644 pp.
3. Bruice TC, Benkovic SJ. 1966. *Bioor-*

ganic Mechanisms, Vols. 1, 2. New York: Benjamin. 362 pp. 419 pp.
4. Dolphin D, Poulson R, Avramovic O, eds. 1970. *Coenzymes and Cofactors*, Vols. 1, 2, 3. New York: Wiley
5. Walsh C. 1979. *Enzymatic Reaction Mechanism*. San Francisco: Freeman. 978 pp.
6. Silverman RB. 2002. *The Organic Chemistry of Enzyme-Catalyzed Reactions*. San Diego: Academic. 717 pp.
7. Radzicka A, Wolfenden R. 1995. *Science* 267:90–93
8. Radzicka A, Wolfenden R. 1996. *J. Am. Chem. Soc.* 118:6105–9
9. Snider MJ, Wolfenden R. 2000. *J. Am. Chem. Soc.* 122:11507–8
10. Wolfenden R, Ridgeway C, Young G. 1998. *J. Am. Chem. Soc.* 120:833–34
11. Bruice TC, Benkovic SJ. 2000. *Biochemistry* 39:6267–74
12. Cannon WR, Singleton SF, Benkovic SJ. 1996. *Nat. Struct. Biol.* 3:821–33
13. Cannon WR, Benkovic SJ. 1998. *J. Biol. Chem.* 273:26257–60
14. Dewar MJ. 1986. *Enzyme* 36:8–20
15. Jencks WP. 1973. *PAABS Rev.* 2:235–43
16. Jencks WP. 1975. *Adv. Enzymol. Relat. Areas Mol. Biol.* 43:219–410
17. Kirsch JF. 1973. *Annu. Rev. Biochem.* 42:205–34
18. Koshland DE Jr. 1958. *Proc. Natl. Acad. Sci. USA* 44:98–104
19. Koshland DE Jr. 1962. *J. Theor. Biol.* 2:75–86
20. Levit M. 1974. In *Peptides, Polypeptides and Proteins*, ed. ER Blout, FA Bowery, M Goodman, N Lotan, pp. 99–113. New York: Wiley
21. Lienhard GE. 1973. *Science* 180:149–54
22. Menger FM. 1992. *Biochemistry* 31: 5368–73
23. Murphy DJ. 1995. *Biochemistry* 34: 4507–10
24. Northrop DB. 1999. *Adv. Enzymol. Relat. Areas Mol. Biol.* 73:25–55
25. Page MI. 1979. *Int. J. Biochem.* 10:471–76
26. Page MI. 1987. In *Enzyme Mechanisms*, ed. MI Page, A Williams, pp. 1–13. London: R. Soc. Chem.
27. Pauling L. 1946. *Chem. Eng. News* 24:1375–77
28. Pauling L. 1948. *Am. Sci.* 36:51–58
29. Wolfenden R. 1976. *Annu. Rev. Biophys. Bioeng.* 5:271–306
30. Schowen RL. 1978. In *Transition States of Biochemical Processes*, pp. 77–114. New York: Plenum
31. Albery WJ, Knowles JR. 1976. *Biochemistry* 15:5631–40
32. Albery WJ, Knowles JR. 1977. *Angew. Chem. Int. Ed.* 16:285–93
33. Fersht AR. 1999. *Structure and Mechanism in Protein Science*. New York: Freeman
34. Kraut J. 1988. *Science* 242:533–40
35. Mildvan AS. 1974. *Annu. Rev. Biochem.* 43:357–99
36. Warshel A. 1978. *Proc. Natl. Acad. Sci. USA* 75:5250–54
37. Warshel A. 1998. *J. Biol. Chem.* 273: 27035–38
38. Blow D. 2000. *Struct. Fold. Des.* 8(R): 77–81
39. Lumry R. 1959. *Enzymes* 1:157
40. Westheimer FH. 1962. *Adv. Enzymol.* 24:441–82
41. Mader MM, Bartlett PA. 1997. *Chem. Rev.* 97:1281–302
42. Storm DR, Koshland DE Jr. 1970. *Proc. Natl. Acad. Sci. USA* 66:445–52
43. Crosby J, Stone R, Lienhard GE. 1970. *J. Am. Chem. Soc.* 92:2891–900
44. Alston TA, Abeles RH. 1987. *Biochemistry* 26:4082–85
45. Lightstone FC, Zheng YJ, Maulitz AH, Bruice TC. 1997. *Proc. Natl. Acad. Sci. USA* 94:8417–20
46. Whitty A, Fierke CA, Jencks WP. 1995. *Biochemistry* 34:11678–89
47. Knowles JR. 1991. *Nature* 350:121–24
48. Fersht AR. 1986. *Trends Biochem. Sci.* 11:321–25
49. Davis JP, Zhou MM, Van Etten RL. 1994. *J. Biol. Chem.* 269:8734–40

50. Admiraal SJ, Schneider B, Meyer P, Janin J, Veron M, et al. 1999. *Biochemistry* 38:4701–11
51. Ray WJ, Long JW. 1976. *Biochemistry* 15:3993–4006
52. Carter P, Wells JA. 1990. *Proteins* 7:335–42
53. Rye CS, Withers SG. 2000. *Curr. Opin. Chem. Biol.* 4:573–80
54. Levine IN. 2002. *Physical Chemistry*. New York: McGraw-Hill. 986 pp.
55. Wells JA. 1990. *Biochemistry* 29: 8509–17
56. Mildvan AS, Weber DJ, Kuliopulos A. 1992. *Arch. Biochem. Biophys.* 294:327–40
57. Plapp BV. 1995. *Methods Enzymol.* 249: 91–119
57a. Dill KA. 1997. *J. Biol. Chem.* 272: 701–04
58. LiCata VJ, Ackers GK. 1995. *Biochemistry* 34:3133–39
59. Mark AE, van Gunsteren WF. 1994. *J. Mol. Biol.* 240:167–76
60. Benson SW, Cruickshank FR, Golden DM, Haugen GR, O'Neal HE, et al. 1969. *Chem. Rev.* 69:279–324
61. Benson SW, Buss JH. 1958. *J. Chem. Phys.* 29:546–72
62. Carey FA, Sundberg RJ. 1990. *Advanced Organic Chemistry: Structure and Mechanisms*. New York: Plenum
62a. Natl. Instit. Stand. Technol. 2001. *NIST Standard Reference Database Number 69*. http://webbook.nist.gov/chemistry/
63. Buzayan JM, Hampel A, Bruening G. 1986. *Nucleic Acids Res.* 14:9729–43
64. Berzal-Herranz A, Joseph S, Chowrira BM, Butcher SE, Burke JM. 1993. *EMBO J.* 12:2567–73
65. Symons RH. 1992. *Annu. Rev. Biochem.* 61:641–71
66. Hertel KJ, Pardi A, Uhlenbeck OC, Koizumi M, Ohtsuka E, et al. 1992. *Nucleic Acids Res.* 20:3252
67. Peracchi A, Beigelman L, Usman N, Herschlag D. 1996. *Proc. Natl. Acad. Sci. USA* 93:11522–27
68. Peracchi A, Karpeisky A, Maloney L, Beigelman L, Herschlag D. 1998. *Biochemistry* 37:14765–75
69. Peracchi A, Matulic-Adamic J, Wang S, Beigelman L, Herschlag D. 1998. *RNA* 4:1332–46
70. McKay DB. 1996. *RNA* 2:395–403
71. Wang S, Karbstein K, Peracchi A, Beigelman L, Herschlag D. 1999. *Biochemistry* 38:14363–78
72. Esteban JA, Walter NG, Kotzorek G, Heckman JE, Burke JM. 1998. *Proc. Natl. Acad. Sci. USA* 95:6091–96
73. Walter NG, Burke JM, Millar DP. 1999. *Nat. Struct. Biol.* 6:544–49
74. Hondal RJ, Zhao Z, Kravchuk AV, Liao H, Riddle SR, et al. 1998. *Biochemistry* 37:4568–80
75. Gerlt JA, Kreevoy MM, Cleland WW, Frey PA. 1997. *Chem. Biol.* 4:259–67
76. Cleland WW, Frey PA, Gerlt JA. 1998. *J. Biol. Chem.* 273:25529–32
77. Harris TK, Mildvan AS. 1999. *Proteins* 35:275–82
78. Thompson JE, Raines RT. 1994. *J. Am. Chem. Soc.* 116:5467–68
79. Hengge AC, Sowa GA, Wu L, Zhang ZY. 1995. *Biochemistry* 34:13982–87
80. McCain DF, Catrina IE, Hengge AC, Zhang ZY. 2002. *J. Biol. Chem.* 277: 11190–200
81. MacLeod AM, Tull D, Rupitz K, Warren RA, Withers SG. 1996. *Biochemistry* 35:13165–72
82. Thompson RC, Blout ER. 1973. *Biochemistry* 12:57–65
83. Thompson RC, Blout ER. 1973. *Biochemistry* 12:66–71
84. Narlikar GJ, Herschlag D. 1998. *Biochemistry* 37:9902–11
85. Herschlag D, Cech TR. 1990. *Biochemistry* 29:10159–71
86. Cech TR, Herschlag D, Piccirilli JA, Pyle AM. 1992. *J. Biol. Chem.* 267: 17479–82
87. Cech TR, Herschlag D. 1996. In *Nucleic Acids and Molecular Biology*, ed. F Eck-

stein, DMJ Lilley, pp. 1–17. Berlin: Springer-Verlag
88. Karbstein K, Carroll KS, Herschlag D. 2002. *Biochemistry* 41:11171–83
89. Huang Z, Wagner CR, Benkovic SJ. 1994. *Biochemistry* 33:11576–85
90. Wagner CR, Huang Z, Singleton SF, Benkovic SJ. 1995. *Biochemistry* 34: 15671–80
91. Ha NC, Choi G, Choi KY, Oh BH. 2001. *Curr. Opin. Struct. Biol.* 11:674–78
92. Koshland DE Jr, Neet KE. 1968. *Annu. Rev. Biochem.* 37:359–410
93. Bruice TC. 1976. *Annu. Rev. Biochem.* 45:331–73
94. Page MI, Jencks WP. 1971. *Proc. Natl. Acad. Sci. USA* 68:1678–83
95. Bruice TC, Lightstone FC. 1999. *Acc. Chem. Res.* 32:127–36
96. Boder ET, Midelfort KS, Wittrup KD. 2000. *Proc. Natl. Acad. Sci. USA* 97:10701–5
97. Wedemayer GJ, Patten PA, Wang LH, Schultz PG, Stevens RC. 1997. *Science* 276:1665–69
98. Menard R, Storer AC. 1992. *Biol. Chem. Hoppe-Seyler* 373:393–400
99. Holden HM, Benning MM, Haller T, Gerlt JA. 2001. *Acc. Chem. Res.* 34:145–57
100. Kuliopulos A, Talalay P, Mildvan AS. 1990. *Biochemistry* 29:10271–80
101. Wu ZR, Ebrahimian S, Zawrotny ME, Thornburg LD, Perez-Alvarado GC, et al. 1997. *Science* 276:415–18
102. Citri N. 1973. *Adv. Enzymol. Relat. Areas Mol. Biol.* 37:397–648
103. Anderson CM, Zucker FH, Steitz TA. 1979. *Science* 204:375–80
104. Herschlag D. 1988. *Bioorg. Chem.* 16:62–96
105. Kempner ES. 1993. *FEBS Lett.* 326:4–10
106. Gerstein M, Lesk AM, Chothia C. 1994. *Biochemistry* 33:6739–49
107. Gerstein M, Krebs W. 1998. *Nucleic Acids Res.* 26:4280–90
108. Young B, Herschlag D, Cech TR. 1991. *Cell* 67:1007–19
109. Gallagher DT, Mayhew M, Holden MJ, Howard A, Kim KJ, Vilker VL. 2001. *Proteins* 44:304–11
110. Benkovic SJ, Adams JA, Borders CL Jr, Janda KD, Lerner RA. 1990. *Science* 250:1135–39
111. Hilvert D. 2000. *Annu. Rev. Biochem.* 69:751–93
112. Herschlag D. 1991. *Proc. Natl. Acad. Sci. USA* 88:6921–25
113. Alber T. 1981. *Structural origins of the catalytic power of triose phosphate isomerase*. PhD thesis. MIT, Cambridge
114. Makinen MW. 1998. *Spectrochim. Acta A* 54(A):2269–81
115. Kelemen BR, Schultz LW, Sweeney RY, Raines RT. 2000. *Biochemistry* 39:14487–94
116. Narlikar GJ, Gopalakrishnan V, McConnell TS, Usman N, Herschlag D. 1995. *Proc. Natl. Acad. Sci. USA* 92:3668–72
117. Polanyi M. 1921. *Z. Elektrochem.* 27:142–50
118. Wells JA, Estell DA. 1988. *Trends Biochem. Sci.* 13:291–97
119. Fersht AR, Winter G. 1992. *Trends Biochem. Sci.* 17:292–94
120. Kicska GA, Tyler PC, Evans GB, Furneaux RH, Shi W, et al. 2002. *Biochemistry.* 41:14489–98
121. Teeter MM, Roe SM, Heo NH. 1993. *J. Mol. Biol.* 230:292–311
122. Kuhn P, Knapp M, Soltis SM, Ganshaw G, Thoene M, Bott R. 1998. *Biochemistry* 37:13446–52
123. Niimura N. 1999. *Curr. Opin. Struct. Biol.* 9:602–8
124. Schramm VL, Shi W. 2001. *Curr. Opin. Struct. Biol.* 11:657–65
125. Palmer AG III. 1993. *Curr. Opin. Biotechnol.* 4:385–91
126. Wand AJ. 2001. *Nat. Struct. Biol.* 8:926–31
127. Ishima R, Torchia DA. 2000. *Nat. Struct. Biol.* 7:740–43

128. Kay LE. 1998. *Biochem. Cell Biol.* 76:145–52
129. Stone MJ. 2001. *Acc. Chem. Res.* 34:379–88
130. Matthews BW. 1996. *FASEB J.* 10:35–41
131. Chen J, Stites WE. 2001. *Biochemistry* 40:15280–89
132. Silverman JA, Balakrishnan R, Harbury PB. 2001. *Proc. Natl. Acad. Sci. USA* 98:3092–97
133. Peracchi A. 2001. *Trends Biochem. Sci.* 26:497–503
134. Xiao G, Parsons JF, Tesh K, Armstrong RN, Gilliland GL. 1998. *J. Mol. Biol.* 281:323–39
135. Lowry TH, Richardson KS. 1997. *Mechanism and Theory in Organic Chemistry.* Baltimore: Addison-Wesley. 1090 pp.
136. Miranda LP, Alewood PF. 2000. *Biopolymers* 55:217–26
137. Kochendoerfer GG, Kent SB. 1999. *Curr. Opin. Chem. Biol.* 3:665–71
138. Dawson PE, Kent SB. 2000. *Annu. Rev. Biochem.* 69:923–60
139. Blaschke UK, Silberstein J, Muir TW. 2000. *Methods Enzymol.* 328:478–96
140. Wang L, Schultz PG. 2002. *Chem. Commun.* 1:1–11
141. Spudich G, Lorenz S, Marqusee S. 2002. *Protein Sci.* 11:522–28
142. Johnson KA. 1995. *Methods Enzymol.* 249:38–61
143. Fierke CA, Hammes GG. 1995. *Methods Enzymol.* 249:3–37
144. Ha T. 2001. *Methods* 25:78–86
145. Kagawa Y. 1999. *Adv. Biophys.* 36:1–25
146. Xie XS, Lu HP. 1999. *J. Biol. Chem.* 274:15967–70
147. Zhuang XW, Kim H, Pereira MJB, Babcock HP, Walter NG, Chu S. 2002. *Science* 296:1473–76
148. Karplus M, Evanseck JD, Joseph D, Bash PA, Field MJ. 1992. *Faraday Discuss.* 93:239–48
149. Hammes GG. 2002. *Biochemistry* 41:8221–28
150. Dunn SM, Batchelor JG, King RW. 1978. *Biochemistry* 17:2356–64
151. Lu HP, Xun LY, Xie XS. 1998. *Science* 282:1877–82
152. Beebe JA, Fierke CA. 1994. *Biochemistry* 33:10294–304
153. Matthews CR. 1993. *Annu. Rev. Biochem.* 62:653–83
154. Jencks WP. 1981. *Proc. Natl. Acad. Sci. USA* 78:4046–50
155. Fersht AR. 1974. *Proc. R. Soc. London Ser. B* 187:397–407
156. Joyce GF. 2002. *Nature* 418:214–21
157. Cech TR. 1993. *Gene* 135:33–36
158. Orgel LE. 1998. *Trends Biochem. Sci.* 23:491–95
159. Narlikar GJ, Herschlag D. 1997. *Annu. Rev. Biochem.* 66:19–59
160. Kunitake T, Shinkai S. 1980. *Adv. Phys. Org. Chem.* 17:435–87
161. Breslow R, Dong SD. 1998. *Chem. Rev.* 98:1997–2012
162. Wilson DS, Szostak JW. 1999. *Annu. Rev. Biochem.* 68:611–47
163. Roberts RW, Ja WW. 1999. *Curr. Opin. Struct. Biol.* 9:521–29
164. Gartner ZJ, Kanan MW, Liu DR. 2002. *J. Am. Chem. Soc.* 124:10304–6
165. Vaish NK, Fraley AW, Szostak JW, McLaughlin LW. 2000. *Nucleic Acids Res.* 28:3316–22
166. Roth A, Breaker RR. 1998. *Proc. Natl. Acad. Sci. USA* 95:6027–31
167. Shoichet BK, Baase WA, Kuroki R, Matthews BW. 1995. *Proc. Natl. Acad. Sci. USA* 92:452–56
168. Dutton JL, Craik DJ. 2001. *Curr. Med. Chem.* 8:327–44
169. Matsumura M, Becktel WJ, Levitt M, Matthews BW. 1989. *Proc. Natl. Acad. Sci. USA* 86:6562–66
170. Mitchinson C, Wells JA. 1989. *Biochemistry* 28:4807–15
171. Burton RE, Hunt JA, Fierke CA, Oas TG. 2000. *Protein Sci.* 9:776–85
172. Warshel A, Åqvist J, Creighton S. 1989. *Proc. Natl. Acad. Sci. USA* 86:5820–24

173. Lockhart DJ, Kim PS. 1992. *Science* 257:947–51
174. Doran JD, Carey PR. 1996. *Biochemistry* 35:12495–502
175. Davies C, Heath RJ, White SW, Rock CO. 2000. *Struct. Fold. Des.* 8:185–95
176. Sitkoff D, Lockhart DJ, Sharp KA, Honig B. 1994. *Biophys. J.* 67:2251–60
177. Bublitz GU, Boxer SG. 1997. *Annu. Rev. Phys. Chem.* 48:213–42
178. Cheng H, Nikolic-Hughes I, Wang JH, Deng H, O'Brien PJ, et al. 2002. *J. Am. Chem. Soc.* 124:11295–306
179. Tonge PJ, Carey PR. 1992. *Biochemistry* 31:9122–25
180. Tonge PJ, Carey PR. 1995. *Acc. Chem. Res.* 28:8–13
181. Gilson HS, Honig BH, Croteau A, Zarrilli G, Nakanishi K. 1988. *Biophys. J.* 53:261–69
182. Callender R, Deng H. 1994. *Annu. Rev. Biophys. Biomol. Struct.* 23:215–45
183. Schutz CN, Warshel A. 2001. *Proteins* 44:400–17
184. Sham YY, Muegge I, Warshel A. 1998. *Biophys. J.* 74:1744–53
185. Shan S, Herschlag D. 1999. *Methods Enzymol.* 308:246–76
186. Honig B, Nicholls A. 1995. *Science* 268:1144–49
187. Chong LT, Dempster SE, Hendsch ZS, Lee LP, Tidor B. 1998. *Protein Sci.* 7:206–10
188. Petrounia IP, Blotny G, Pollack RM. 2000. *Biochemistry* 39:110–16
189. Petrounia IP, Pollack RM. 1998. *Biochemistry* 37:700–5
190. Thorson JS, Chapman E, Schultz PG. 1995. *J. Am. Chem. Soc.* 117:9361–62
191. Palcic MM, Floss HG. 1986. In *Vitamin B6 Pyridoxal Phosphate*, ed. D Dolphin, R Poulson, O Avramovic, pp. 25–68. New York: Wiley
192. Kirsch JF, Eichele G, Ford GC, Vincent MG, Jansonius JN, et al. 1984. *J. Mol. Biol.* 174:497–525
193. Alber T, Banner DW, Bloomer AC, Petsko GA, Phillips D, et al. 1981. *Philos. Trans. R. Soc. London Ser. B* 293:159–71
194. Banner DW, Bloomer AC, Petsko GA, Phillips DC, Pogson CI, et al. 1975. *Nature* 255:609–14
195. Pompliano DL, Peyman A, Knowles JR. 1990. *Biochemistry* 29:3186–94
196. Kuroki R, Weaver LH, Matthews BW. 1993. *Science* 262:2030–33
197. Sidhu G, Withers SG, Nguyen NT, McIntosh LP, Ziser L, Brayer GD. 1999. *Biochemistry* 38:5346–54
198. Zechel DL, Withers SG. 1999. *Acc. Chem. Res.* 33:11–18
199. LaReau RD, Wan W, Anderson VE. 1989. *Biochemistry* 28:3619–24
200. Tonge PJ, Carey PR. 1990. *Biochemistry* 29:10723–27
201. Schramm VL. 1998. *Annu. Rev. Biochem.* 67:693–720
202. O'Brien PJ, Herschlag D. 1999. *J. Am. Chem. Soc.* 121:11022–23
203. O'Brien PJ, Herschlag D. 2002. *Biochemistry* 41:3207–25
204. Hengge AC. 1998. In *Comprehensive Biological Catalysis*, ed. ML Sinnott, pp. 517–42. London: Academic
205. Jencks WP. 1972. *J. Am. Chem. Soc.* 94:4731–32
206. Jencks WP. 1976. *Acc. Chem. Res.* 9:425–32
207. Hine J. 1972. *J. Am. Chem. Soc.* 94:5766–71
208. Stahl N, Jencks WP. 1986. *J. Am. Chem. Soc.* 108:4196–205
209. Maegley KA, Admiraal SJ, Herschlag D. 1996. *Proc. Natl. Acad. Sci. USA* 93:8160–66
210. Zhang M, Zhou M, Van Etten RL, Stauffacher CV. 1997. *Biochemistry* 36:15–23
211. Alhambra C, Wu L, Zhang ZY, Gao J. 1998. *J. Am. Chem. Soc.* 120:3858–66
212. Cleland WW, Hengge AC. 1995. *FASEB J.* 9:1585–94
213. Fedorov A, Shi W, Kicska G, Fedorov E, Tyler PC, et al. 2001. *Biochemistry* 40:853–60
214. Bruice TC. 1970. In *The Enzymes*, ed.

PD Boyer, pp. 217–79. New York: Academic
215. Winstein S, Lindegren CR, Marshall H, Ingraham LL. 1953. *J. Am. Chem. Soc.* 75:147–55
216. Page MI, Jencks WP. 1987. *Gazz. Chim. Ital.* 117:455–60
217. Milstien S, Cohen LA. 1970. *Proc. Natl. Acad. Sci. USA* 67:1143–47
218. DeLisi C, Crothers DM. 1973. *Biopolymers* 12:1689–704
219. Murphy KP, Xie D, Thompson KS, Amzel LM, Freire E. 1994. *Proteins* 18:63–67
220. Finkelstein AV, Janin J. 1989. *Protein Eng.* 3:1–3
221. Tidor B, Karplus M. 1994. *J. Mol. Biol.* 238:405–14
222. Mammen M, Shakhnovich EI, Deutch JM, Whitesides GM. 1998. *J. Org. Chem.* 63:3821–30
223. Dunitz JD. 1995. *Chem. Biol.* 2:709–12
224. Bruice TC, Pandit UK. 1960. *Proc. Natl. Acad. Sci. USA* 46:402–4
225. Kirby AJ. 1980. *Adv. Phys. Org. Chem.* 17:183–278
226. Jencks WP, Page MI. 1974. *Biochem. Biophys. Res. Commun.* 57:887–92
227. Bruice TC, Brown A, Harris DO. 1971. *Proc. Natl. Acad. Sci. USA* 68:658–61
228. Lightstone FC, Bruice TC. 1996. *J. Am. Chem. Soc.* 118:2595–605
229. Miller BG, Snider MJ, Short SA, Wolfenden R. 2000. *Biochemistry* 39:8113–18
230. Carlow D, Wolfenden R. 1998. *Biochemistry* 37:11873–78
231. Nakamura CE, Abeles RH. 1985. *Biochemistry* 24:1364–76
232. Bode W. 1979. *J. Mol. Biol.* 127:357–74
233. Sanders CR II, Tian GC, Tsai MD. 1989. *Biochemistry* 28:9028–43
234. Jen-Jacobson L, Engler LE, Jacobson LA. 2000. *Struct. Fold. Des.* 8:1015–23
235. Klinman JP. 1989. *Trends Biochem. Sci.* 14:368–73
236. Sutcliffe MJ, Scrutton NS. 2002. *Eur. J. Biochem.* 269:3096–102
237. Knapp MJ, Klinman JP. 2002. *Eur. J. Biochem.* 269:3113–21
238. Kohen A, Klinman JP. 1999. *Chem. Biol.* 6(R):191–98
239. Gertner BJ, Whitnell RM, Wilson KR, Hynes JT. 1991. *J. Am. Chem. Soc.* 113:74–87
240. Kohen A, Cannio R, Bartolucci S, Klinman JP. 1999. *Nature* 399:496–99
241. Tsai S, Klinman JP. 2001. *Biochemistry* 40:2303–11
242. Kohen A, Klinman JP. 2001. *J. Am. Chem. Soc.* 122:10738–39
243. Kramers HA. 1940. *Physica* 7:284–304
244. Borgis DC, Lee S, Hynes JT. 1989. *Chem. Phys. Lett.* 162:19–26

Annu. Rev. Biochem. 2003. 72:573–608
doi: 10.1146/annurev.biochem.72.121801.161724
Copyright © 2003 by Annual Reviews. All rights reserved

THE DYNAMICS OF CHROMOSOME ORGANIZATION AND GENE REGULATION

David L. Spector
Cold Spring Harbor Laboratory, One Bungtown Road, Cold Spring Harbor, New York 11724; email: spector@cshl.org

Key Words gene expression, transcription, nuclear structure, nuclear dynamics

■ **Abstract** With the sequence of the human genome now complete, studies must focus on how the genome is functionally organized within the confines of the cell nucleus and the dynamic interplay between the genome and its regulatory factors to effectively control gene expression and silencing. In this review I describe our current state of knowledge with regard to the organization of chromosomes within the nucleus and the positioning of active versus inactive genes. In addition, I discuss studies on the dynamics of chromosomes and specific genetic loci within living cells and its relationship to gene activity and the cell cycle. Furthermore, our current understanding of the distribution and dynamics of RNA polymerase II transcription factors is discussed in relation to chromosomal loci and other nuclear domains.

CONTENTS

INTRODUCTION

The completion of the human genome sequence has thus far indicated an estimate of 30,000–75,000 genes, distributed among 3.2 billion basepairs of DNA that is packaged into a higher-order chromatin structure [reviewed in (1–4)]. These genes are arranged in the human interphase nucleus among the 46 chromosome territories such that they are readily accessible to transcriptional regulators that will mediate their expression or repression. All of this regulation takes place within the confines of the cell nucleus having an average volume of 600–1500 μm^3. In addition, for cells exhibiting an open mitosis (i.e., mammalian, *Drosophila*), this organization is disassembled and then reassembled during each cell cycle. Determining how nuclear functions are organized and coordinated, both spatially and temporally, is central to understanding the proper workings of the cell and the alterations that are associated with various diseases. Recent advances in the areas of probe development [reviewed in (5, 6)] as well as microscopy have given us an unprecedented opportunity to visualize aspects of gene expression and dynamics at high resolution and/or in the context of the living cell. Such approaches that capitalize on previous biochemical and molecular advances allow one to delve into the inner workings of the nucleus in ways that could not have been anticipated a decade ago. In this review, I concentrate on the dynamic organization of the genome and its associated regulatory factors within the cell nucleus, focusing on the interplay between nuclear organization and RNA polymerase II transcription. For a more detailed analysis of additional aspects of nuclear structure/function the reader is directed to other reviews (7–13).

ORGANIZATION OF THE GENOME IN THE INTERPHASE NUCLEUS

Chromosome Territories

DNA is packaged into a higher-order chromatin structure, via its associations with histones and other nonhistone proteins, that directly impacts on gene expression [reviewed in (11)]. Early on, studies by Rabl and Boveri suggested that chromatin was not randomly organized within the interphase nucleus but occupied distinct territories. Rabl suggested that chromosomes in plant cells occupy discrete domains throughout interphase that reflect their mitotic orientation (14). Boveri confirmed these studies by showing that chromosomes maintained relatively fixed positions in the nuclei of *Ascaris* eggs (15). Furthermore, these studies suggested that telomeres were attached to the nuclear envelope on one side of the nucleus and centromeres were attached on the opposite nuclear side; this became known as a Rabl configuration. One of the most dramatic cases for interphase chromosome organization, and how chromosomes contribute to the assembly of a nuclear domain, comes from studies of nucleoli [reviewed in (13)]. In a classic study reported in 1934, Barbara McClintock identified a densely stained chromosomal region that she named the nucleolar-organizing body or element, and that we today refer to as the nucleolar-organizing region (NORs) (17). She went on to relate this chromosomal region to both the number and the type of nucleoli present in mutant strains of maize. In human cells, rDNA genes are clustered in the NORs of five pairs of acrocentric chromosomes (chromosomes 13, 14, 15, 21, and 22). At the end of mitosis, in HeLa cells, 6 of the 10 NORs become transcriptionally active and subsequently both the active and inactive NORs fuse to form the nucleoli.

More recently, numerous studies have readdressed the initial questions asked by Rabl and Boveri in a variety of systems at significantly higher resolution. In a groundbreaking study, Cremer et al. (18) showed that laser UV microirradiation of specific interphase nuclear areas in Chinese hamster cells damaged discrete chromosomal regions, suggesting that the genome is organized during interphase and provided some of the earliest insight into the concept of interphase chromosome territories. Early studies in *Drosophila*, examining polytene chromosomes differentially stained with vital dyes in conjunction with optical sectioning methods, have revealed these chromosomes to be closely associated with the inner surface of the nuclear membrane and to contact the membrane at specific sites (19–21). Furthermore, chromosomes were shown to occupy distinct territories within diploid and polytene nuclei and to spiral with the same handedness through the nucleus (21–23). In an elegant study, Sedat and co-workers (24) presented evidence for specific positioning of euchromatic loci within interphase nuclei of *Drosophila* embryos. These investigators used fluorescence in situ hybridization (FISH) to map the three-dimensional position of 42 DNA probes on a single interphase chromosome (24). Fourteen of 32 probes to euchromatic loci showed a nonrandom peripheral localization. Of six heterochromatic loci probed, only two, the AATAC satellite and the rDNA locus, were nuclear envelope associated, whereas four other loci were not associated with the nuclear periphery. Further analysis of the nuclear positions of these loci showed that the interphase nucleus in *Drosophila* is strongly polarized in a Rabl configuration that, taken together with specific targeting to the nuclear envelope or to the nuclear interior, results in each locus occupying a specific and reproducible position within the nucleus (24). Based upon the mapping of nuclear envelope contacts, it was estimated that there are on the order of 15 nuclear envelope interaction sites per chromosome arm, or a total of 150 nuclear envelope association sites per diploid *Drosophila* nucleus. These nuclear envelope association sites would be spaced on average 1–2 Mb apart and could thus define the boundaries of large loop domains tethered to the nuclear envelope in interphase. The nuclear envelope association sites were not found to strictly

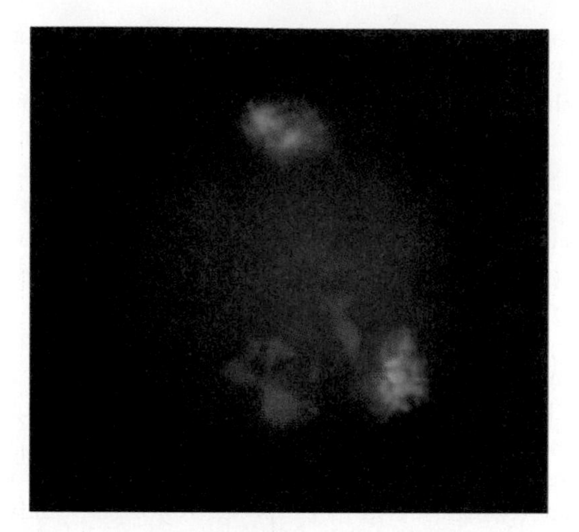

Figure 1 Fluorescence in situ hybridization to a human primary peripheral blood lymphocyte showing the peripheral localization of gene-poor chromosomes 18 (green) and the more internal localization of gene-rich chromosomes 19 (red). Note the more decondensed state of the gene-rich chromosomes. Total DNA is stained with DAPI. Photo courtesy of Wendy Bickmore, MRC Human Genetics Unit, United Kingdom.

correlate with scaffold-attachment regions, heterochromatin, or binding sites of known chromatin proteins (24).

Although the Rabl configuration of chromosomes is generally not observed in human cell nuclei, several groups have examined the organization of human chromosomes in interphase nuclei with the goal of identifying other organizational principles. Using probes specific for individual human chromosomes, and fluorescence in situ hybridization, it was shown early on that each chromosome occupies an individual interphase domain referred to as a chromosome territory (25–29). These territories occupy discrete and nonoverlapping nuclear regions (30), and in mammalian cells, homologous chromosome territories are usually not adjacent. Although repeatedly examined, the idea of a precise ordered positioning of human chromosomes is still somewhat controversial (31–33). However, in an extensive study, Bickmore and colleagues analyzed the nuclear organization of every human chromosome in diploid lymphoblasts and primary fibroblasts (34, 35). They found that most gene-rich chromosomes concentrate at more internal nuclear regions, whereas the more gene-poor chromosomes are located toward the nuclear periphery (Figure 1). These arrangements of chromosomes are established early in the cell cycle and are maintained throughout interphase (34). However, no statistically significant relationship between physical chromosome size (base pairs) and nuclear position was found in this study. Furthermore, chromosome subnuclear localization does not appear to be determined by the centromeres, as, remarkably, the distinctive localization observed is retained by regions of the chromosome arms that are translocated to chromosomes associated with reciprocal nuclear regions (34). Interestingly, in quiescent or senescent cells, gene-poor human chromosome 18 was shown to move from a location at the nuclear periphery to a more internal site in the nucleus (36). Intriguingly, the chromosome moves back toward the nuclear periphery during a 2–4 h period of time as quiescent cells enter the G1 phase of the cell cycle. Based upon these findings, the authors suggest that the spatial organization of the genome is plastic, potentially leading to the ability of quiescent nuclei to be more amenable to reprogramming (36, 37). In a parallel study, Cremer and colleagues performed a quantitative comparison of chromosome arrangements in flat-ellipsoid nuclei of human amniotic fluid cells and fibroblasts and in spherical B and T lymphocytes (38). Similar to that observed in spherically shaped lymphocyte nuclei by the Bickmore laboratory, a preferential positioning of small gene-dense chromosome territories was observed in the three-dimensional nuclear interior, whereas the gene-poor small chromosomes were peripherally localized. However, in contrast to that observed by the Bickmore group, large chromosomes were also preferentially located toward the nuclear periphery (38). The chicken karyotype represents an interesting example of chromosome organization. Chicken cells contain 9 pairs of gene-poor macrochromosomes and 30 pairs of gene-rich microchromosomes. Although the microchromosomes represent only 23% of the chicken genome, they contain 50% of its genes (39). In chicken fibroblasts and neurons, the gene-rich chromosomes are concentrated in the nuclear interior, whereas gene-poor chromosomes are located in peripheral nuclear regions (39). Gene density–correlated radial chromosome arrangements have been conserved during evolution of the higher-primate genome dating back an estimated 30–40 million years (40). In the ellipsoid nuclei of amniotic fluid cells and fibroblasts, all tested chromosome territories showed association with the upper and/or lower part of the nuclear envelope (38). Small chromosomes were located toward the center of the nuclear projection, whereas large chromosomes were located toward the nuclear periphery (38). These differences observed on the basis of chromosome positioning may reflect the different approaches used in sample preparation and/or may be indicative of the degree of dynamics of chromosome territories among different cell populations.

The specific interactions that mediate the nuclear position of a particular chromosome territory have not yet been identified. However, the nuclear envelope and nuclear lamina stand out as possible candidates based on convincing evidence that they participate in other aspects of nuclear organization [reviewed in (41)]. Numerous pathologies have recently surfaced with regard to mutations in protein constituents of the nuclear envelope/lamina (42). In one such disease, X-linked Emery-Dreifuss muscular dystrophy (X-EDMD), mutations have been identified in the emerin gene (43). Emerin is a type II integral membrane protein localized to the inner nuclear envelope and most X-EDMD-associated mutations result in a loss of emerin protein at the nuclear membrane [reviewed in (44)]. One disease model has suggested the inability of emerin-null cells to sequester inactive chromatin at the nuclear periphery, thereby leading to altered regulation of gene expression [reviewed in (44)]. To examine this possibility, Boyle et al. determined chromosomal positions in lymphoblast cells from an individual with a null mutation in emerin (35). However, the spatial positioning of chromosome territories in such cells was not significantly different from that of their normal counterparts. Therefore, although emerin has thus far not been implicated in chromosomal positioning, other components of the nuclear envelope/lamina have yet to be tested.

Aside from physical associations, changes in chromatin structure, as reflected by protein modifications and/or associations, may also play a role in the positioning of chromosomes. In a recent study, Almouzni and colleagues examined the long-term effects of deacetylase inhibitors on the maintenance and nuclear compartmentalization of pericentric heterochromatin in mouse cells (45). Incubation of cells with the histone deacetylase inhibitor, trichostatin A (TSA), for five days selectively induced large-scale movements of centromeric and pericentric heterochromatin to the nuclear periphery without any significant effect on either its methylation status or telomere position. In addition, these regions lost their association with heterochromatin 1 (HP1) proteins, which became distributed throughout the nucleoplasm (45). Interestingly, upon drug removal (20 h) these chromatin regions resumed their typical nuclear positions and their association with HP1. Based upon the necessary lengthy incubation with TSA, prior to observing the effects, the authors posit that several cell divisions may be required to destabilize the HP1 population. As each round of DNA replication would dilute by half the parental nonacetylated histones with newly incorporated acetylated histones, eventually a complete loss of HP1 would be achieved (45). The observed reversibility of the TSA effect in mammalian cells was not found to be the case in *Schizosaccharomyces pombe*, where TSA treatment resulted in a delocalization of the HP1 homologue and severe defects in centromeric regions and chromosome segregation during mitosis (46, 47). Almouzni and colleagues suggest that this difference may relate to a lack of methylation of pericentric DNA in *S. pombe*, as DNA methylation has been shown to induce histone deacetylation in mammalian cells and could thus result in the observed recovery after TSA removal (45).

In addition to pursuing the associations that mediate chromosome territory position, several groups have assessed the parameters involved in the organization of individual chromosome territories. Upon increasing concentrations of salt, up to 1.8 M, gene-poor chromosome 18 was released from the nuclear remnants, whereas gene-rich chromosome 19 remained in the center of the nuclei, suggesting a differential association with the nuclear substructure based upon gene density (34). Furthermore, gene-rich chromosome 19 assumed a more compact structure in the absence of transcription, although its overall position in the nucleus was not altered (34). In a separate study, Berezney and colleagues found chromosome territory organization to be maintained despite the extraction of greater than 90% of the histones and other soluble nuclear proteins in DNA-rich nuclear matrix preparations. Interestingly, upon complete extraction of internal nuclear matrix components with RNase treatment followed by 2 M NaCl, a disruption of higher-order chromosome territory architecture was achieved, suggesting a role for RNA/RNP in chromosome organization (48). In conjunction with the observed chromosome disruption, a small set of 40–90-kD acidic proteins, distinct from the major nuclear matrix proteins, was released (48). These proteins are being pursued as potential mediators of chromosome territory organization. Based upon the effect of RNase treatment, it is tempting to speculate that stable nuclear RNAs and/or RNPs may play a role in the maintenance of interphase chromosome organization. Interestingly, Almouzni and colleagues have recently identified the involvement of RNA in mediating the interaction of heterochromatin protein 1 (HP1) and methylation of H3 lysine-9 with pericentric heterochromatin (49).

Positioning of Active Versus Inactive Genes Within Chromosome Territories

An extensive effort has focused on examining the position of individual genes and/or chromosomal regions within chromosome territories to assess the relationship of gene activity to chromosomal position. Several studies have suggested that inactive genes may be located in interior regions of chromosomal territories and active genes may concentrate along the periphery (30, 50–52), adjacent to nonchromosomal nucleoplasmic space, termed the interchromosome domain compartment (ICD) (53, 54). However, other studies indicate that this type of organization may not be a general trend (55, 56). In addition, an overall analysis of gene-rich (GC-rich) chromosomal regions indicates that they are distributed throughout chromosomal territories with no preference for their periphery (57). Interestingly, the gene-rich major histocompatibility complex (MHC), at human chromosome 6p21.3, was localized external to the chromosome 6 territory or to large chromosomal loops that extend from the surface of the bulk chromosome 6 territory (52). Transcriptional upregulation of the MHC class II genes by interferon-gamma resulted in an increase in the frequency with which this gene cluster appeared on external chromosomal loops (52). A similar organization was observed for the human epidermal differentiation complex at 1q21, which contains genes that are involved in keratinocyte differentiation (58). This region appears to extend outside of the chromosome 1 territory in keratinocytes where the genes are highly expressed, but not in lymphoblasts where they are silent (58). The amplified and highly expressed ERBB2 genes have also been observed to extend from the chromosome 17 territories in a breast cancer cell line (59). While the position of a number of active genes have been mapped with respect to their corresponding chromosome territories (50, 51, 60), the most extensive analysis has been done on a ~1-Mb region of human chromosome 11p13 and the syntenic region in the mouse (61). This region of chromosome 11 contains ubiquitously expressed genes as well as genes whose expression is tissue type specific (62–64). In addition, large intergenic stretches of DNA are present within this region (62, 63). All the 11p13 loci studied, including expressed genes, were located within the chromosome territory, compared with a locus from 11p15 that localized to the territory edge (61). Similar observations were made with its region of conserved synteny in the mouse. Furthermore, tissue-restricted genes were not relocated to the periphery of the respective chromosome territory in expressing cells (61). Based upon these findings, it was concluded that gross remodeling of chromosome territories does not occur to accommodate relatively small changes in gene expression in mammalian cells (61). Instead, local changes in chromatin fiber conformation are likely to be associated with gene expression (65, 66). The respective factors involved in this process are likely to have accessibility to neighborhoods throughout the chromosome territory, with local changes in protein modification and chromatin structure regulating the binding affinities and subsequent gene expression profiles.

CHROMOSOME DYNAMICS

Although individual chromosome territories have been localized in fixed cells to discrete nuclear regions, several early studies provided initial evidence to support the concept that these territories are dynamic in the interphase nucleus and that their positions may be cell cycle dependent. Using a composite probe to chromosome 8, Ferguson & Ward showed by in situ hybridization that in G1 cells chromosome 8 centromeres localized adjacent to the nuclear periphery and the chromosomal arms extended toward the nuclear interior (67). However, in G2 cells the chromosome reoriented itself; the centromeres were internal, and the chromosomal arms extended toward the nuclear periphery. A similar redistribution was observed in brain tumor cells where centromeres were dispersed during G1 and S-phase and became clustered toward the nuclear interior during G2 (68). Examination of chromosome positioning in CNS nuclei in larval *Drosophila* indicated large chromosomal movements in diploid interphase nuclei (69). At the onset of S-phase, an increased separation was seen between proximal and distal positions of a long chromosome arm (69). However, a study of centromeres in living human cells using a CENP-B-GFP fusion protein found that centromeres in HeLa cells were predominantly stationary, although movement of individual or small groups of centromeres was occasionally observed at very slow rates of 7–10 μm/h (70). Therefore, different paradigms are likely to exist for chromosome movement in different cell types within an organism, as well as in similar cell types among different organisms, depending upon the overall chromatin organization within the respective cell type. One of the most provocative studies demonstrating a correlation between chromosome position and cell physiology comes from work on human epileptic foci (27). In normal male cortical neurons, the X chromosome was localized to the nuclear periphery. However, when cells in an electrophysiologically defined seizure focus were observed, there was a dramatic increase from ~7% to 45% in the number of cells exhibiting internal nuclear localization of the X chromosome (27). A similar observation was previously reported in neurons after 8 h of electrical stimulation (71). However, more recent live cell studies in tissue culture cells, examining the dynamics of fluorescently labeled chromosomes, have concluded that interphase chromosome territories in general undergo only a limited large-scale translational motion (72–75).

Dynamics of Chromosomal Regions and Genetic Loci in Living Cells

In a significant breakthrough, Belmont and coworkers developed an approach, based upon the use of the *lac* operator/repressor system, to directly visualize chromatin organization and dynamics in living cells (76). By introducing a *lac* operator array into Chinese hamster ovary cells with the dihydrofolate reductase gene and amplifying it through methotrexate selection (76), a stable cell line was selected containing a ~90-Mbp chromosomal array that can be visualized using a GFP-*lac* repressor fusion protein. The array formed a late-replicating homogeneously staining region (HSR) (77). Cell cycle analysis indicated that the integration site was peripherally localized throughout most of interphase. However, during several hours in mid- to late S-phase, the HSR decondensed and moved toward the nuclear interior correlated with its DNA replication (77). Using this system the Belmont group has been able to directly visualize activator binding in living cells, and they have found that chromatin decondensation occurs upon activator binding and in the absence of transcription (78). However, studies using a tandem array (≥2 Mbp) of the mouse mammary tumor virus (MMTV) promoter driving a *ras* reporter have shown that this array does require transcription for chromatin decondensation (79). Therefore, different loci may respond to different signals for chromatin decondensation. In a separate study, using a cell line with a smaller integration (150–300 Kbp), VP16 targeting to the locus was also shown to induce its movement from the nuclear periphery to a more internal nuclear region (80), suggesting that internal nuclear regions may be more amenable to potentially active loci. Furthermore, the recruitment of several histone acetyltransferases, including GCN5, P/CAF, and p300/CBP, and hyperacetylation of all core histones was observed (78). Examination of the extended chromosome fibers by light and electron microscopy supports the existence of a folded chromonema model based upon ~100-nm chromonema fibers formed by compaction of 10-nm and 30-nm chromatin fibers (76, 78).

The *lac* operator/repressor system has provided an unprecedented approach to visualizing chromatin organization and remodeling in real time. An elaboration of this approach has allowed for the development of a cell system that offers direct read-out of gene expression in living cells. Using the *lac* operator/repressor system and two color variants of GFP, Spector and colleagues have developed a system to visualize a genetic locus and its protein product directly in living cells (66). Dynamic morphological changes in chromatin structure, from a condensed

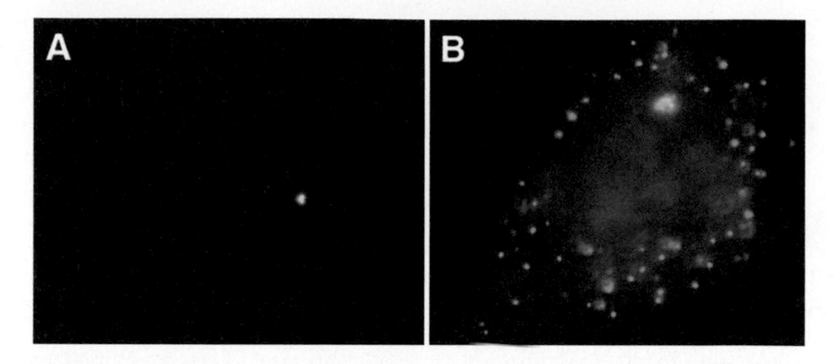

Figure 2 Localization of a stably integrated regulatable genetic locus in the "off" (*A*) and "on" (*B*) states using the *lac* operator/repressor system (66). In the "off" position the locus is visualized via CFP-*lac* repressor protein as a single dot in each interphase nucleus (*A*). Upon transcriptional activation the locus decondenses and the protein product of the transcription unit is localized to cytoplasmic peroxisomes (*B*). The protein product is visualized via a CFP-peroxisome targeting signal fusion protein.

to an open structure, were observed during gene activation, and targeting of a cyan fluorescent protein reporter to the peroxisomes was directly visualized in living cells (Figure 2) (66).

In order to monitor the movements of individual genetic loci on different human chromosomes, Bickmore and co-workers randomly integrated a *lac* operator array into the human genome and selected clones that contained a single integration site of the array (128-mer array) at different chromosomal positions (81). In general, chromatin associated with the nucleolus or nuclear periphery was more restricted in its movement than chromatin associated with other nucleoplasmic regions, indicating that these structures may act as anchoring sites (81). For example, 13q22, a region associated with the nuclear periphery, exhibited a maximum range of movement of 0.9 μm, whereas 5p14, a nucleoplasmic locus, exhibited a maximum movement of 1.5 μm. Interestingly, the 5p14 and 3q26.2 loci exhibited similar diffusion coefficients (1.25×10^{-4} $\mu m^2/s$), although their gene density and replication timing were determined to be different. The 5p14 locus resides in a G band (gene poor, late replicating), whereas 3q26.2 is in an R band (gene dense, early replicating). The diffusion coefficient for human loci (81) was estimated to be fourfold lower than that estimated for budding yeast centromeres, which move considerably less than coding regions (82, 83). This constrained motion may relate to associations with nuclear structures, with the relative concentration of DNA relative to the nuclear volume, and/or the significantly larger amount of heterochromatin present in human cells that may act as anchoring sites.

Several studies examining chromatin dynamics in *Drosophila* have also identified constrained chromatin movements. An earlier study in *Drosophila* embryos, examining a topoisomerase II-enriched 359-base pair repeat block of heterochromatin on the X chromosome (82), revealed that this region of chromatin undergoes significant diffusive Brownian motion with a diffusion constant of 2.0 X $\times$ 10^{-7} $\mu m^2/s$ within a restricted radius of ~0.9 μm (82). Interestingly, this chromatin region is a scaffold-associated region (SAR) in *Drosophila*. More recently, a study to track chromosome motion in *Drosophila* spermatocytes (73) using a *lac* operator array revealed multiple levels of constrained motion. Over short time intervals of a few seconds, chromatin movement was restricted to submicron-sized regions (0.3 $\mu m/s$) of the nucleus. Over time periods of an hour or more, loci were found to be considerably more mobile, with a range of several microns. Interestingly, this long-range motion was restricted to early G2 spermatocytes and was not observed later in G2, as cells approached meiotic prophase. The overall chromatin movements were consistent with a random walk as no directed movements were observed (73). The average diffusion coefficient of single sites in early nuclei was ~1.0 X 10^{-3} $\mu m^2/s$, about one order of magnitude greater than that observed for a centromeric domain in yeast (82). Although specific chromosomal regions demonstrate various levels of confined movement, the overall position of chromosomes was constrained to specific nuclear territories.

A similar approach to examine chromatin dynamics in *Saccharomyces cerevisiae* evaluated specific chromosomal sites that were tagged with a *lac* operator array at the LEU2 locus near the centromere of chromosome III (82). Using single particle tracking methods, the chromatin was found to undergo diffusive Brownian motion with a diffusion constant of 5.0 X 10^{-8} $\mu m^2/s$ within a restricted radius of less than 0.3 μm, a region that corresponds to ~1% of the nuclear volume (82). Interestingly, cells treated with a microtubule depolymerizing drug, nocodazole, resulted in less-confined chromatin diffusion (82). However, the motion of noncentromeric sites was not affected by such drugs (83). In a more recent study, examining G1 nuclei in *S. cerevisiae*, early and late origins of replication exhibited large rapid movements (>0.5 μm in 10 s) that surprisingly were ATP dependent, whereas smaller saltatory movements (<0.2 μm), similar in range to those observed in *Drosophila* (73), were observed throughout interphase (83). Given that the yeast nucleus measures ~2.0 μm in diameter, these 0.5 μm movements are extremely significant and demonstrate that chromosomal regions can move within large nuclear areas over short time periods. The large movements were proposed to reflect the action of ATP-dependent enzymes involved in transcription and chromatin remodeling (83). These rapid movements became constrained in S-phase, and they dropped fourfold in G1 cells, as cell density increased to ~2 X 10^7 cells/ml, or just before the diauxic shift from fermentative to oxidative metabolism (83). However, telomeres and centromeres were found to provide replication-independent constraint on chromatin movement in both G1 and S-phase cells, supporting the concept that periodic sites along a chromosome may tether it and thereby confine its position within a restricted nuclear region (83). The large rapid movements observed for regions of yeast chromosomes have been suggested to indicate that chromosome territories are loosely arranged. Given that yeast contain little heterochromatin, their chromosomes may contain fewer anchoring sites. The dynamics of chromatin in vivo may in part reflect nuclear volume: DNA length, as reduction in chromatin movement correlates with reduced nuclear volume in several instances (73, 81, 83).

An alternative live cell approach, which allows the visualization of individual or a few different chromatid territories in living cells, has made use of microinjection of the fluorescent thymidine analog Cy3-AP3-dUTP into the nuclei of cultured human cells (84). The analog incorporates into replicating DNA in S-phase, and after growth for several cell cycles random segregation of labeled and unlabeled chromatids into daughter nuclei results in nuclei exhibiting individual labeled chromatid territories that can be studied in living cells. However, as compared to the use of the *lac* operator/repressor system described above, the chromatid(s) that is labeled is randomly selected and not stable. Such studies have indicated that chromatid territories are composed of subcompartments with diameters ranging between 400–800 nm, referred to as subchromosomal foci. The foci are composed of either early or late-replicating chromatin; similar size replication foci have been reported previously during S-phase (85). Time-lapse imaging has indicated changes in the shape and positioning of individual chromatid territories, repositioning of subchromosomal foci within stable territories, and changes in patterns of folding and extension of the foci over time (84).

DISTRIBUTION OF RNA POLYMERASE II TRANSCRIPTION SITES

Several approaches have been used to examine the organization of transcription sites within the nuclei of mammalian cells. In a series of now classic studies, Fakan and colleagues used ^{3}H-uridine incorporation, after short pulses, combined with electron microscopic autoradiography and observed non-nucleolar transcription sites to be distributed throughout the nucleoplasm of mammalian cells with a preference for the borders of condensed chromatin (86–88). In addition, transcription sites are also observed on the periphery of interchromatin granule clusters and at other nucleoplasmic regions (89, 90). Transcription sites are specifically associated with nuclear structures termed perichromatin fibrils, which are thought to represent nascent transcripts based upon the pulse labeling experiments as well as antibody labeling for transcription and pre-mRNA processing factors [reviewed in (7, 91)]. Perichromatin fibrils are RNase sensitive and their appearance is inhibited by pretreatment with actinomycin D or α-amanitin [reviewed in (92)]. More recent studies have used Br-UTP incorporation (93, 94) to examine the localization of transcription sites both at the immunofluorescence and electron microscopic levels (Figure 3). Using this approach, Pombo et

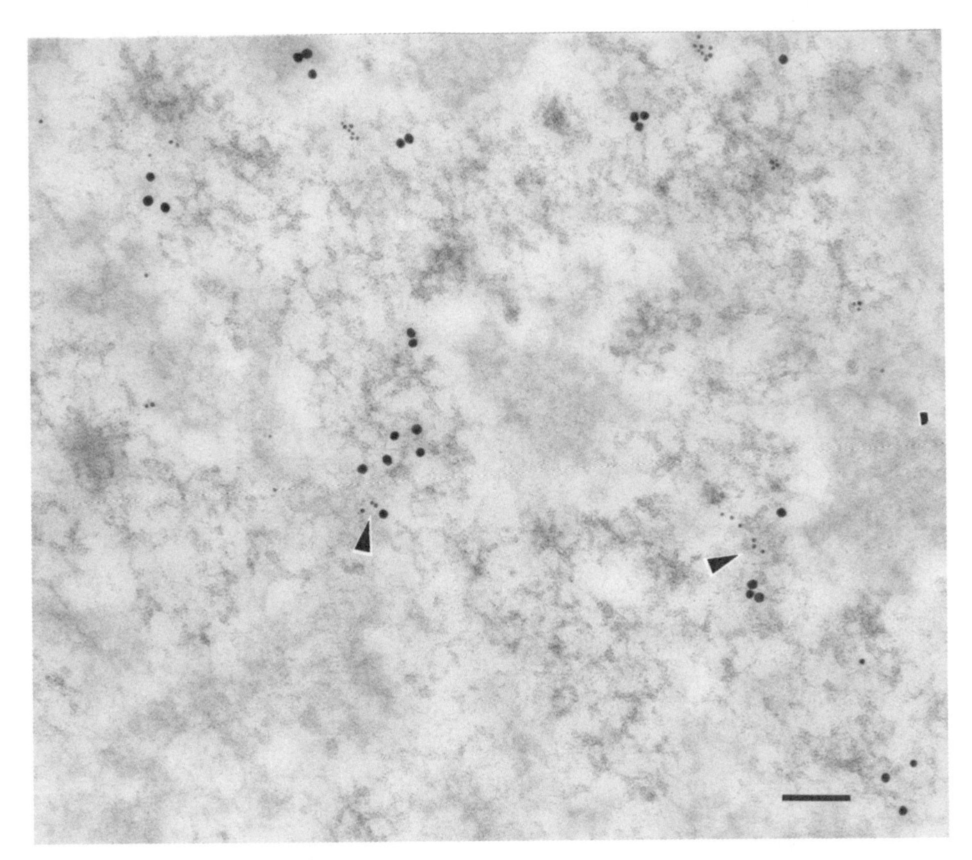

Figure 3 Human bladder carcinoma T24 cells were microinjected with Br-UTP and incubated for 12 min to label transcription sites. Double-labeling with mouse anti-BrdU antibody (6-nm colloidal gold particles) and chicken anti-RNA polymerase II antibody (15-nm colloidal gold particles) shows that both localize at perichromatin fibrils (*arrowheads*). Bar equals 100 nm. Photo courtesy of D. Cmarko & S. Fakan, University of Lausanne, Switzerland.

al. (95) identified ~10,000 non-nucleolar transcription sites in HeLa cell nuclei. Of these ~8000 are thought to represent RNA pol II transcription sites and ~2000 represent RNA pol III transcription sites. Each of these sites measures ~50 nm in diameter and has been termed a transcription factory (96), as the authors propose a model in which RNA polymerases are associated with the nuclear skeleton–forming factories, and templates surround the factories. Transcripts are extruded from the factories upon passage of templates through the positionally fixed polymerases (97).

Numerous studies have indicated that active genes as well as many transcription factors are associated with the nuclear matrix or nucleoskeleton [reviewed in (98–102)]. Most interestingly, the nuclear lamins, an indisputable structural element of the nuclear periphery as well as internal nuclear regions [reviewed in (41)], have recently been shown to be necessary for RNA polymerase II transcription both in mammalian cells as well as in nuclei from *Xenopus laevis* (103). When lamins were disrupted in BHK cells or in nuclei from early *Xenopus* gastrulae by addition of a ΔN-terminal human lamin A (ΔNLA), a dominant-negative lamin mutant, the lamins formed abnormal nucleoplasmic aggregates (103). Br-UTP incorporation was dramatically decreased in these nuclei and TATA binding protein (TBP) was found in the nucleoplasmic aggregates. However, the distribution of Sp1, a RNA pol II gene-specific transcription factor, was not affected by addition of the ΔNLA nor was RNA pol I or pol III transcription. Based upon these data, the authors proposed that RNA pol II transcription and the distribution of TBP depend upon the maintenance of normal nuclear lamin organization (103).

ORGANIZATION OF TRANSCRIPTION SITES WITHIN CHROMOSOME TERRITORIES

To examine the overall relationship of transcription to the organization of chromosomal territories, Verschure et al. (56) combined chromosome painting, to identify a specific territory, with Br-UTP incorporation, to identify transcription sites. Confocal image stacks of hybridized chromosome paint probes indicated that chromosome territories are not compact structures, as had been previously implied; instead they appeared as open structures consisting of clusters of 300–450-nm domains that were sometimes interconnected, forming a thread-like, folded structure (56). The subdomains were surrounded by less intensely labeled areas. When such analysis was combined with Br-UTP labeling to localize nascent RNA, transcription sites were observed to extend throughout the territories of gene-rich chromosome 19 and the active X chromosome (Figure 4). Interestingly, newly synthesized RNA localized in those regions that labeled less intensely with the chromosome paint probes. Based upon these data, the authors propose that the interchromosomal domain space (53, 54, 104) extends to regions within chromosome territories (56). A similar localization of transcription sites throughout chromosome territories in rapidly dividing wheat cell nuclei has been observed (55).

TRANSCRIPTION FACTOR LOCALIZATION

Dynamic Interactions with the Genome

Although many models of transcription based upon in vitro experiments implicate stable complexes that are bound to chromatin for relatively long periods of time [reviewed in (105)], recent data from live cell experiments indicate just the

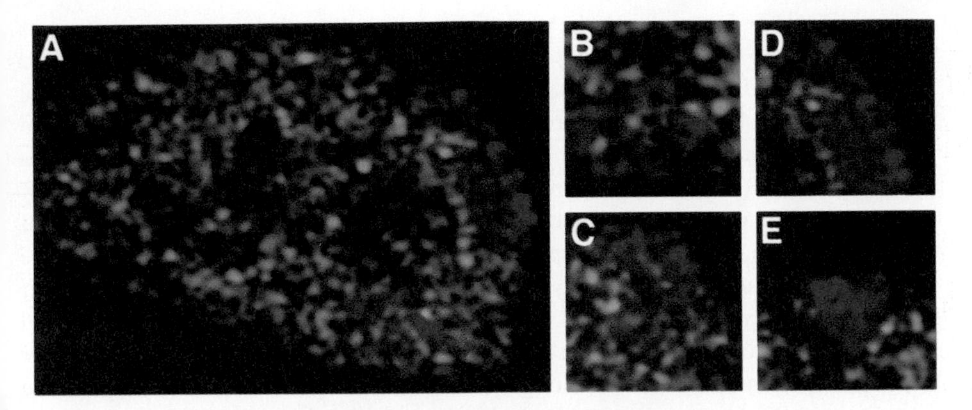

Figure 4 Confocal optical sections showing the distribution of transcription sites in relation to the two X-chromosome territories in human female primary fibroblasts. Nascent RNA was labeled by incorporation of Br-UTP (green) and the X-chromosome territories were labeled by FISH (red). Transcription sites occur as defined spots throughout one of the two X-chromosome territories (most likely the active X, *panels A, B*), whereas almost no transcription sites are observed in the other X-chromosome territory (most likely the inactive X, *panels A, D*). An additional example of an active X territory (*panel C*) and inactive X territory (*panel E*) from a different nucleus are also shown. Chromosome territories have a distinct substructure, showing strongly labeled structures that are surrounded by less intensely labeled structures. The intensely labeled structures have a diameter in the range of 300–450 nm. Nascent RNA preferentially accumulates between the intensely labeled structures, in the areas with little to no DNA FISH labeling. Reproduced from Verschure et al. *The Journal of Cell Biology*, 1999, 147:13–24, by copyright permission of The Rockefeller University Press.

contrary—highly dynamic interactions of factors with the chromatin substrate [reviewed in (106)]. Live cell imaging using GFP fused to several different transcription factors (107–109), as well as to histone H1 and the HMG 17, 14 proteins [reviewed in (12)], in combination with fluorescence recovery after photobleaching (FRAP) techniques, has shown that these nuclear proteins are highly dynamic, exhibiting rapid exchange rates, with occupancy times of minutes to a few seconds at their target substrates. In contrast, such studies have indicated that the core histones remain immobilized on chromatin for hours (110). In a groundbreaking study, the Hager group demonstrated that the glucocorticoid receptor undergoes rapid exchange, with a $t_{1/2}$ of 5 s, between chromatin regulatory elements and the nucleoplasm in the presence of ligand (107). These data support a "hit-and-run" model of factor association with the response element, and ATP-dependent chromatin remodeling activities have been suggested to be involved in this rapid exchange of proteins (111). Mancini and colleagues found the unliganded estrogen receptor exhibited a rapid exchange, $t_{1/2} > 1$ s, whereas agonist or partial agonist slowed recovery ($t_{1/2}$ 5–6

s) (108, 113). Interestingly, the dynamics of the unliganded estrogen receptor was shown to be ATP and proteasome dependent (108). A similar degree of rapid dynamics has been observed between a receptor coactivator, SRC-1, or a general coregulator (CBP), and an estrogen-receptor-*lac* repressor fusion protein bound to a *lac* operator array (113). Furthermore, the identification of members of a chaperone complex at certain promoters has resulted in a model whereby molecular chaperones may be essential for promoting continuous disassembly of transcriptional regulatory complexes, resulting in the observed turnover and the ability to respond to signaling changes (114). For example, the p23 molecular chaperone can disrupt thyroid hormone receptor DNA complexes in vitro and appears to compete with a coactivator for association with the thyroid receptor ligand binding domain, resulting in opposing effects on the stability of the receptor-DNA interaction (115). Intriguingly, regardless of this rapid dynamics of factors, protein complexes can be visualized at specific promoters and within an increasing number of different nuclear organelles. The basis for maintaining the highly local concentrations of these factors such that steady-state levels form functional complexes and are easily observed in intact cells remains to be elucidated.

Nuclear Domains and Transcription Factor Localization

The localization of the large subunit of RNA pol II as well as a number of general transcription factors (GTFs) and promoter-specific transcription factors have been examined by immunofluorescence microscopy. Although, as expected, these factors localize at transcription sites that are distributed throughout the nucleoplasm in a fine punctate localization as described above, many of them have also been found concentrated in a number of different nuclear domains in which transcription has thus far not been reproducibly detected, except in the case of the OPT domain (116) (Figure 5). As compared to the localization of RNA pol II LS, the distribution of GTFs is somewhat more variable. Grande et al. (117) examined the localization of a number of GTFs including TFIIH (p62) and TFIIF (RAP74) and BRG1, a human homologue of the yeast SWI/SNF chromatin remodeling complex. All of these proteins are distributed in a fine punctate distribution throughout the nucleoplasm. Other promoter-specific transcription factors such as E2F-1 (117), GATA-1 (118), Oct1 (116), Pit-1 (119), Sp1 (103), Sp100 (120), CBFα/AML-related factors (121), and the glucocorticoid receptor (107) have also been localized throughout the nucleoplasm in a punctate distribution. Surprisingly, except for TFIIH, little correlation was observed between the localization of other GTFs and the localization of RNA pol II LS and transcription sites (117). In addition to this broad distribution, many transcription-related proteins also localize to other nuclear domains, such as nuclear speckles, the OPT domain (116), Cajal bodies (122, 123), promyelocytic leukemia (PML) nuclear bodies [reviewed in (9, 124)], and heat shock factor 1 (HSF1) granules (125, 126), as is discussed below.

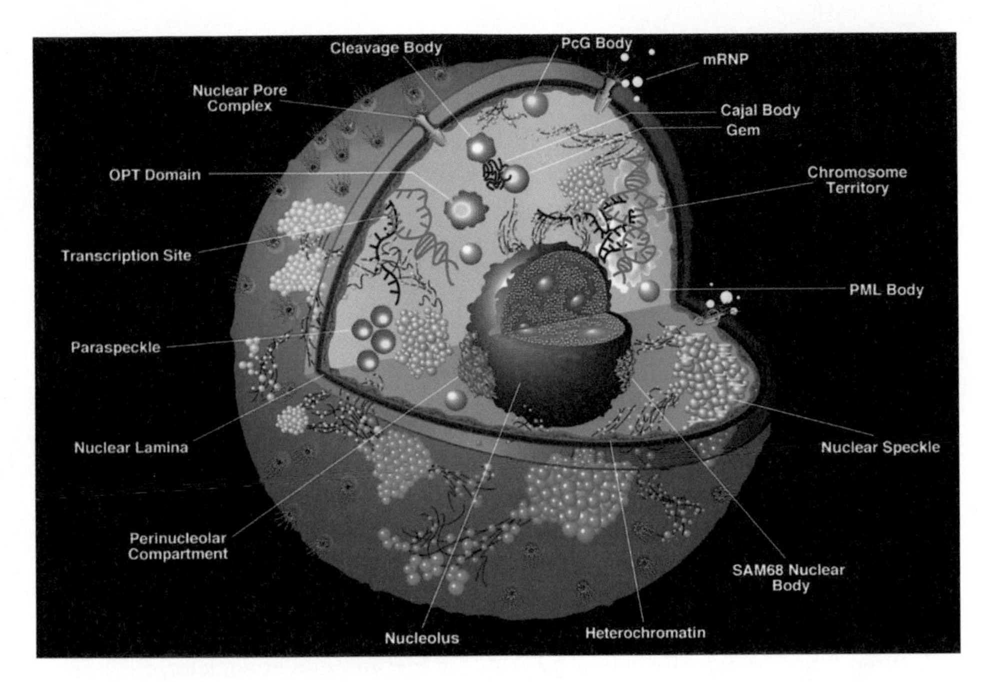

Figure 5 Cartoon of a mammalian cell nucleus showing the large number of nuclear domains that have been identified. Those involved in aspects of transcription or associated with transcription factors are discussed in the text.

THE SPECKLE CONNECTION The unphosphorylated or the serine-5 phosphorylated form of the large subunit of RNA pol II (RNA pol II LS) that is involved in transcriptional initiation is localized in a diffuse nuclear distribution that has considerable, but not complete, overlap with sites of transcription (Figure 3) (117, 127). The serine-2 phosphorylated form of RNA pol II LS, which is involved in elongation, is localized similarly, but in addition it is localized to nuclear speckles (Figure 5) when observed by immunofluorescence microscopy (127, 128). However, some studies have not observed an enrichment of RNA pol II LS in speckles [for example (117)]. These nuclear regions correspond to interchromatin granule clusters (IGCs), 0.5–1.0 μm nuclear regions composed of granules each measuring 20–25 nm in diameter. IGCs have been implicated in the modification and/or assembly of premRNA splicing factors and possibly a subset of transcription factors (129–131). As transcription does not occur in IGCs and because this localization pattern of RNA pol II LS is primarily observed by labeling with one antibody, H5 (127), it has been suggested that this epitope may be present both on the elongating form of RNA pol II as well as on a form that is stored and/or in the process of being modified in IGCs prior to its recruitment to transcription sites. Biochemical characterization of the IGC proteome has identified several subunits of RNA pol II (132; N. Saitoh, P.

Sacco-Bubulya & D.L. Spector, in preparation), supporting the localization of at least a population of RNA pol II in these nuclear domains.

The Cdk9-cyclin T complex, also known as TAK/P-TEFb, is thought to be involved in transcriptional elongation via phosphorylation of the RNA pol II LS [reviewed in (133)]. This complex was found diffusely distributed throughout the nucleoplasm, with the exception of the nucleoli (134). In addition, a significant overlap between cyclin T1 and nuclear speckles was observed. A region in the central portion of the cyclin T1 protein was found to be important for this subnuclear targeting (134). However, Cdk9 was present in the vicinity of nuclear speckles but a direct overlap was limited (134). Surprisingly, Cdk9 and cyclin T1 also showed only a limited colocalization with RNA pol II at sites on polytene chromosomes in *Drosophila* (135). This lack of colocalization has been suggested to possibly be due to a dynamic and short-lived interaction of these proteins at transcription sites in vivo (135). Other members of the Cdk family, Cdk7 and Cdk8, which are involved in transcriptional regulation, are not found in nuclear speckles (136). Cdk7 was localized to Cajal bodies (Figure 5) in addition to being diffusely distributed throughout the nucleoplasm (136).

FBI-1 is a cellular POZ-domain-containing protein that binds to the HIV-1 LTR and associates with the HIV-1 transactivator protein Tat (137). FBI-1 has been found to partially colocalize with Tat and its cellular cofactor, P-TEFb, at nuclear speckles (138). In addition, a less soluble population of FBI-1 is distributed in a peripheral-speckle pattern that is dependent upon the FBI-1 DNA binding domain and active transcription (138). This distribution may be associated with active transcription sites (perichromatin fibrils) that are found on the periphery of IGCs (139, 140). The nucleosome binding protein HMG-17, which can alter the structure of chromatin and enhance transcription, has been localized in a pattern similar to that of FBI-1 (141).

OPT DOMAIN In addition to its diffuse distribution, in ~30% of the HeLa cells examined, Oct1 occurs in a highly localized domain often located close to a nucleolus and measuring 1.0–1.5 μm in diameter (117). However, in normal human skin fibroblasts, Oct1 was found in four to six nuclear domains smaller than those observed in HeLa cells. Based upon the localization of Oct 1, PTF [PSE binding transcription factor, also known as SNAPc (snRNA-activating protein complex)], a complex involved in activating transcription of snRNA genes [reviewed in (142)], and the finding that transcription could be detected within this nuclear domain, the region was named the OPT domain (Figure 5) (116). RNA pol II as well as TBP (TATA binding protein) and Sp1 were also found within the OPT domain (116). These domains were found to contain transcripts generated by both RNA pol II and pol III (116). OPT domains assemble during G1 and disappear early in S-phase (116), and they coincide with a nuclear domain identified earlier as the polymorphic interphase karyosomal association (PIKA) domain (143). Interestingly, OPT domains are often associated with chromosome 6p21 and chromosome 7, although they were not found associated with various Oct1- and PTF-dependent genes (i.e., U1 and U2 snRNAs, 7SK, hY RNA, histones) (116).

CAJAL BODIES Cajal bodies are nuclear organelles first identified as nucleolar accessory bodies by Santiago Ramón y Cajal (144), and they are thought to function in snRNP biogenesis (123). Populations of several transcription factors including TFIIF and TFIIH have been localized to Cajal bodies in HeLa cells (117). Cajal bodies are associated with histone genes as well as gene clusters encoding the U1, U2, and U3 snRNAs (145, 146), although the organization of genes in arrays is not necessary for Cajal body association, as several single-copy snRNA loci also have a statistical preference to localize with Cajal bodies (147). Interestingly, Matera and coworkers showed that artificial tandem arrays of U2 genes colocalized with Cajal bodies and the frequency of the colocalization was directly dependent upon the transcriptional activity of the array. The association was lost upon transcriptional inhibition or by promoter mutations (147). Importantly, the U2 coding region was required for the association, suggesting that the association may be mediated through the snRNA nascent transcripts or a polymerase complex (148). Interestingly, a recent study has shown these bodies to associate with chromatin in the presence of transcription via an ATP-dependent mechanism; upon ATP depletion the bodies exhibited increased mobility that was described by anomalous diffusion (149).

PML BODIES PML nuclear bodies were first identified as nuclear domain 10 (ND10) by autoantibodies recognizing the Sp100 protein (150, 151). However, interest in these bodies was significantly increased upon finding that a fusion protein resulting from a t(15;17) translocation between the PML protein and the retinoic acid receptor-α, in acute promyelocytic leukemia, resulted in the disruption of these bodies (152–154). More interestingly, treatment of these individuals with retinoic acid or arsenic trioxide allowed them to go into remission and concomitantly the PML bodies were reformed (152–154). A clear function for these bodies has not yet been established, although roles in transcriptional regulation, as storage sites regulating the levels of active proteins within the nucleus or as sites of active proteolysis, have been pursued [reviewed in (124, 155–157)]. Earlier studies have reported transcription to occur in (158) or on the periphery (159) of these bodies. In addition, the transcriptional coactivator CREB-binding protein (CBP) has been localized to PML bodies in certain cell types (Hep-2, SKN-SH, COS-1, and CHO) (160) and the coactivator p300 and RNA polymerase II have been localized to a subset of PML bodies in some cell types (Hep-2, HeLa, MCF-7, and T24) (158, 161). Of particular interest here is the recent finding of a nonrandom association of the gene-rich MHC on chromosome 6 with PML bodies (120). At least one homologue was associated or overlapping with a PML body in ~42% of the MRC-5 cells (normal human fibroblasts) examined (120). This association was independent of active transcription. Interestingly, when several copies of a YAC containing a large proportion of the MHC class II region was stably integrated into the long arm of chromosome 18 in a B-lymphoblastoid cell line, monosomic for the short arm of chromosome 6 and containing a large deletion of the MHC class II region, PML bodies were found to be associated with the MHC class II YAC integration site on chromosome 18, demonstrating the specificity of the association (120). The localization of the transcription factor Sp100, which is involved in the constitutive expression of several genes within the MHC region, to PML bodies has been suggested to support a model in which PML bodies function in transcriptional regulation, perhaps by regulating the soluble pool of transcription factors/coactivators/corepressors (120). A recent study reexamining Br-UTP incorporation indicated that ~9% of PML nuclear bodies were associated with transcription sites and this association increased to 51% upon interferon-α treatment (162). As interferon treatment also increases the number of PML bodies, the direct relationship of these bodies to transcription remains unclear. Counter to an active role in transcription, a subset of PML bodies have been shown to contain HP1 (163, 164), a protein associated with silenced chromatin [reviewed in (165)]. In addition, the proteosome inhibitor MG132 induces a PML body/centromere association in a significant number of cells in the G2 phase of the cell cycle (163). Therefore, although it is unclear what role, if any, PML bodies has in transcriptional regulation, their role(s) may be dynamic as cells traverse the cell cycle, and/or different PML bodies within the same nucleus may be involved in different activities. In this regard, a recent study has indicated that a subset of these bodies exhibits ATP-dependent dynamics (166).

HSF1 GRANULES Upon activation of the mammalian heat shock transcription factor 1 (HSF1) by stress, heat shock genes are turned on and HSF1 relocalizes to nuclear bodies termed HSF1 granules (125) or stress-induced Src-activated during mitosis nuclear bodies (SNBs) (126). Although numerous RNA binding proteins have been localized to these granules, transcription does not occur within these nuclear regions. Recently, the granules were shown to bind to a nucleosome-containing subclass of satellite III repeats at the human 9q11-q12 pericentromeric heterochromatin region (125), as well as to the centromeric regions of chromosomes 12 and 15 (126). It is unclear why these specific chromosomal regions are the targets for HSF1 granules and what role these granules have in nuclear metabolism.

POLYCOMB GROUP GRANULES Polycomb group (PcG) protein complexes localize at specific sites in *Drosophila* called Polycomb response elements (PREs) and in doing so repress genes in *cis* [reviewed in (167)]. Such elements have not yet been identified in mammalian cells, although homologues of PcG genes have been identified and they appear to function in silencing as do their *Drosophila* counterparts. Interestingly, human polycomb group (PcG) complex proteins are localized in a fine granular distribution throughout the nucleoplasm and are also localized to a varying number of larger domains ranging in size from 0.2 to 1.5 μm in diameter; these domains have been termed PcG granules [reviewed in (167, 168)]. These granules associate with pericentromeric heterochromatin regions on human chromosome 1 (1q12) and with related sequences on other chromosomes (167, 168), and the association is maintained throughout mitosis. However, recently Fakan and co-workers have examined the localization of four members (HPC2, HPH1, BMI1, and RING1) of the human PcG protein family in two cell types and tissue sections by immunoelectron microscopy (169). Although the BMI1 and HPC2 proteins localized to PcG granules, the four proteins were found highly concentrated on the border of condensed chromatin domains in perichromatin nuclear regions. In addition, the PcG proteins were found in the interchromatin space, a region devoid of chromatin, although at a five- to tenfold lower concentration (169). However, these proteins were nearly absent from regions of condensed chromatin. Based on this localization, the authors proposed that loci silenced by PcG proteins are spatially interspersed among transcriptionally active genes (169). These findings argue against a model in which silencing of genes by association with PcG proteins gives rise to the repositioning of silenced loci inside compact chromatin domains. However, they are in agreement with similar studies in *Drosophila* in which PcG proteins were found to occupy different positions on polytene chromosomes than HP1, a marker for constituitive heterochromatin (170, 171). Interestingly, several recent studies have identified a role for the Extra Sex Combs (ESC) and Enhancer of Zeste E(z) PcG protein complex in the trimethylation of lysine-9 and the methylation of lysine-27 of histone H3 (172–174), resulting in the recruitment of other Polycomb proteins to the histone H3 amino-terminal tail and mediated silencing.

SEQUESTRATION IN THE CYTOPLASM A significant number of transcription factors are localized in the cytoplasm and only upon activation by a signaling molecule are they transported into the nucleus, where they bind DNA to activate transcription [reviewed in (175, 176)]. Among such factors are members of a broad array of transcription factor families including STATs (signal transducers and activators of transcription), NF-κBs (nuclear factors of Igκ B cells), NFATs (nuclear factors of activated T cells), and the glucocorticoid receptor. Of this class of factors, perhaps the best-studied at the cell biological level is the glucocorticoid receptor (GR) (79, 107, 177). Hager and colleagues have utilized a cell line containing a tandem array of ~200 copies of a mouse mammary tumor virus promoter-driven *ha-v-ras* gene (178) to examine the dynamics of transcription and the association of transcription factors with this endogenous template (79, 107). After steroid hormone treatment, a GFP-GR fusion was shown to enter the nuclei and associate with the arrays. The arrays decondensed within 3 h of hormone treatment and then recondensed over the next 6 h (178).

TRANSLATION IN THE NUCLEUS

The nuclear envelope was thought to functionally demarcate the nuclear compartment where transcription and RNA processing occurs from the cytoplasmic compartment where translation occurs. However, a recent study indicated that a low level of translation occurs within the nucleus (179). When biotin-lysyl-tRNA^{lys} or BODIPY-lysyl-tRNA^{lys} were used to label newly made proteins in vivo, using conditions where nascent polypeptides are extended by ~15 residues, discrete sites were identified within nucleoli and the nucleoplasm in addition to the prevalent cytoplasmic labeling (179). The nuclear signal represented 15% of the total incorporation. A similar degree of labeling was observed in isolated nuclei lacking >95% of the cytoplasmic ribosomes. Treatment with cycloheximide reduced the nuclear labeling to 4%, and an inhibition of nuclear import did not affect incorporation. In support of these incorporation studies, immunofluorescence analysis has shown that populations of a number of different translation factors including eIF2α, eIF3, eIF4γ, eIF4E as well as ribosomal proteins L7 and QM were present in nuclei (179). A portion of the observed nuclear translation was dependent upon concurrent transcription by RNA pol II, and immunogold labeling at the electron microscopic level showed a colocalization of some nascent RNA and nascent polypeptides (179). Based upon these data, the authors suggested that nuclear translation may function in nonsense-mediated decay, a quality control mechanism in which mRNAs are surveyed for the presence of nonsense codons. Support for this possibility comes from studies showing that nonsense mutations can affect pre-mRNA 3′ end processing (180) and result in an accumulation of pre-mRNAs at transcription sites (181). To determine whether ribosomes are recruited to nascent transcripts, Brogna et al. (182) used antibodies to 20 different ribosomal proteins and probes complementary to 18S and 28S rRNA to demonstrate that the translation apparatus is present at transcription sites on the giant polytene chromosomes of *Drosophila melanogaster* salivary glands. The presence of both 18S as well as 28S rRNA suggests the presence of assembled ribosomes (182). Furthermore, these components were concentrated at interbands and were particularly apparent at major puffs. The kinetics of ribosomal protein recruitment to two different ecdysone-inducible loci indicated that the association occurs cotranscriptionally and prior to the completion of pre-mRNA splicing (182).

GENE SILENCING AND NUCLEAR POSITIONING

Equally important to gene activation is gene silencing, a process by which the inactivity of a gene or set of genes is important to the overall viability of the cell and/or organism or to its differentiation. Two of the best-studied examples of mammalian cell silencing relating to subnuclear positioning are the dosage compensation achieved by X-chromosome inactivation in female mammals [reviewed in (183)] and the positioning of inactive genes adjacent to centromeric heterochromatin [reviewed in (184)]. Here, I focus on the relationship of nuclear organization to gene silencing; for an examination of changes in chromatin structure and modifications relating to silencing the reader is referred to other recent reviews (165, 185–187).

The Inactive X Chromosome

As discussed earlier, gene-rich chromosomes are generally located in more internal nuclear regions, whereas gene-poor chromosomes are located in more peripheral areas of the nucleus (34). However, one of the clearest examples of the peripheral localization of a chromosome territory is provided by the inactive X chromosome in female mammals (71, 188). The transcriptionally inactive X chromosome consists largely of heterochromatic regions and forms a dense nuclear domain, the Barr body, which is often found associated with the nuclear periphery. X-chromosome inactivation results in the transcriptional silencing of several thousand genes and ensures dosage compensation for X-linked gene products between XX females and XY males (189). However, about 15% of the genes on the human X chromosome escape X inactivation (190). X inactivation is controlled by the X-inactivation center (Xic), which is essential for the developmentally regulated initiation and spread of inactivation along the X chromosome [reviewed in (183)]. A key player in the *cis*-acting function of the *Xic* is the *Xist* gene. *Xist* produces a nuclear, untranslated RNA that is expressed uniquely from the inactive X chromosome, coating it in *cis* (191, 192). In addition, the noncoding antisense transcript Tsix represses *Xist* expression and is thought to regulate X chromosome choice at the onset of inactivation (193, 194). Other features associated with the inactive X chromosome are its late replication timing, the methylation of CpG islands, the hypoacetylation of histones H3 and H4 (195), methylation of H3 lysine-9 as well as hypoacetylation of H3 lysine-9 and hypomethylation of H3 lysine-4 (186, 196), and its enrichment in a histone H2A variant, macro H2A.1 (197). Although these and other differences in the shape, size, and surface properties of the inactive X chromosome have been reported (198, 199), the mechanism behind its commonly observed peripheral localization has not been elucidated.

Repositioning to Pericentric Heterochromatin

In a pair of now classic studies, Sedat and coworkers and Csink & Henikoff demonstrated that the brown gene in *Drosophila* was silenced by contact with centromeric heterochromatin (200, 201). The brown ^{Dominant} (bw^D) allele, a null mutation caused by the insertion of a block of heterochromatin within the coding sequences of the brown gene, causes the bw^D gene to be misdirected and to associate with centromeric heterochromatin. Most interestingly, in bw^+/bw^D larva, as a consequence of homologous chromosome pairing, the wild-type allele also localized to the centromere, providing an explanation for the variegated inactivation of the normal *bw* gene (200). Subsequently, this association of silenced genes with centromeric heterochromatin was found to also occur naturally in other cell types. In cycling but not quiescent B cells, transcriptionally active genes are generally positioned away from centromeric heterochromatin, whereas many inactive genes are localized adjacent to heterochromatin domains. For example, the expressed alleles of the IgH and Igκ immunoglobulin heavy and light chain genes in lymphocytes occupy nuclear positions away from heterochromatin as opposed to their heterochromatin-associated nonexpressed alleles (202). A similar situation occurs in developing immature thymocytes, where the Rag-1 and TdT genes are downregulated and upon transcriptional termination the genes are repositioned to centromeric regions (203). As the observed repositioning occurs after transcriptional shutdown, locus movement appears to be a result, rather than a cause, of shutdown (203). This repositioning is heritable and stably transmitted through cell division (203). The heritable silencing of genes has also been found in T lymphocytes where CD4 T cells can be induced to develop into Th1 or Th2 cells (T-helper subsets). In this case, the IL-4 (in Th1) and γ–interferon genes (in Th2) are positioned near centromeric domains (204). Based upon these studies, Fisher and co-workers have proposed a model in which an active target gene localized in a transcriptionally permissive environment is sequestered, upon transcriptional shut-down, to a repressive environment by a hypothetical "recruiter" (205). The recruitment is proposed to be initiated by increased binding of factors to motifs in the target genes, or alternatively, centromere-bound recruiters may access and interact with target genes when motifs previously occupied by transcriptional activators are vacated (205). Of particular interest has been the search for proteins that might mediate the repositioning and/or silencing of these genes. An interesting candidate with regard to the lymphoid lineage is the Ikaros protein. Ikaros is required for normal hemopoiesis, and it shares homology with the Hunchback protein in *Drosophila* (206, 207), which has been implicated in Polycomb recruitment and the establishment of silencing complexes in *Drosophila* (208, 209). Interestingly, Ikaros has been shown to interact with chromatin remodeling proteins, to colocalize with many inactive genes in lymphocyte nuclei, and to bind to lymphoid-specific promoter sequences as well as to centromere-associated repetitive DNA [reviewed in (184)]. These characteristics are consistent with either a direct role for the Ikaros protein in transcriptional repression or a role as a mediator of silent chromatin. Ikaros is expressed at very low levels and localizes predominantly within the cytoplasm of noncycling B lymphocytes (203). However, following mitogenic activation, Ikaros levels increase and at about 3 days after stimulation Ikaros is highly localized to centromeric clusters, as are transcriptionally repressed genes (203). Smale and co-workers (210) have found that Ikaros proteins compete with the ELF-1 activator for binding to the D′ region of the *TdT* promoter, resulting in TdT downregulation in T-cell lines. Similar findings have been made for a role of the Ikaros protein in silencing the λ5 promoter of transgenes in mature B cells (211).

In an interesting extension of these studies, Francastel et al. (212) assessed the ability of the β-globin 5′HS2 enhancer to influence silencing and nuclear location of a transgene in mammalian cells. At genomic integration sites where stable expression does not require the presence of an enhancer, transgenes localized away from centromeric heterochromatin regardless of their transcriptional activity. However, at sites where stable expression requires an intact enhancer, active transgenes localized away from centromeric heterochromatin when linked to a functional enhancer (212). Enhancer mutations that impair the ability of the enhancer to suppress silencing also resulted in the transgene remaining close to centromeric heterochromatin, even before the transgene was silenced. Therefore, the same enhancer motifs are required for both suppression of transgene silencing and localization of the transgene away from centromeric heterochromatin (212). Based upon these data, functional enhancers were proposed to antagonize gene silencing by preventing the localization of a gene near centromeric heterochromatin or by recruiting a gene to a nuclear region that is transcriptionally favorable and stably heritable (212). However, changes in chromatin structure are likely also to be involved in gene positioning as localization of the human β-globin locus away from heterochromatin in mouse erythroleukemia cells correlates with hyperacetylation in the promoter region, even in the absence of transcription (213).

To further assess the interplay between gene silencing and transcriptional activation in heterochromatic nuclear regions, Dillon and co-workers targeted a pre-B-cell specific λ5 transgene directly into pericentric heterochromatin in mice (214). The integration resulted in strongly variegated expression in pre-B cells. Analysis of the stability of the expression patterns indicated that the λ5 transgene undergoes a reversible switching between the active and repressed states, resembling the telomeric silencing observed in *S. cerevisiae* (215). In both expressing and nonexpressing clones, the transgene remained closely associated with the periphery of the centromeric complex, indicating that activation of the λ5 transgene does not require movement away from the heterochromatic region (214). In fibroblasts, a DNase I hypersensitive site (HS1) was shown to be responsible for locating the transgene to the outside of the pericentric heterochromatin complex in the absence of transcription. Deletion of the HS1 site resulted in the λ5 transgene being embedded within the pericentric heterochromatin, demonstrating that changes in chromatin structure are directly involved in genome organization (214). However, in pre-B cells, location of the transgene to the outside of the heterochromatin complex was not linked to the HS1 site but instead was related to the dosage of early B cell factor (EBF), as was the ability of the transgene to be transcriptionally activated (214). Therefore, the level of EBF affects both the higher-order chromatin structure and the transcriptional activity of the heterochromatic λ5 transgene. Importantly, relocation of this gene away from heterochromatin was not required for transcriptional activation, and transcription of this heterochromatin-associated gene was maintained through multiple cell divisions. This study clearly demonstrated that gene regulation is a balance between factor levels/accessibility and chromatin condensation state (214).

Interestingly, just as active genes can be juxtaposed to heterochromatin, inactivation of transcription does not always occur through an association with heterochromatin. In a study examining the expression and nuclear positioning of the α- and β-globin genes in hemopoietic cells, Brown et al. identified differences in how the nuclear position of these two loci corresponds to their transcriptional activity (216). In primary erythroblasts where both the human α- and β-globin loci are active, they were both positioned away from centromeric heterochromatin (216). In human primary T lymphocytes and immortalized B cells, where these genes are not expressed, the β-globin alleles were localized in close proximity to centromeric DNA, although the α-globin locus did not localize close to centromeric DNA in cycling lymphocytes where these genes are inactive. Even when integrated into different chromosomal sites, the α-globin genes neither remained discrete from centromeric heterochromatin, nor did they localize with other regions of heterochromatin as demarcated by the localization of HP1 (216). As the α-globin locus exhibits characteristics generally attributed to ubiquitously expressed genes, such as a lack of methylation, early replication timing, and DNAse I sensitivity in all tissues, these attributes may influence the chromatin associations and nuclear position of this locus within the nucleus (216). However, in mammalian cells there is no genetic evidence that nuclear position affects the repressed state, and it is not clear whether position is a result or a cause of repression.

Telomeres and the Nuclear Periphery

In the yeast *S. cerevisiae,* telomeres as well as proteins associated with repressed chromatin (Sir3 and Sir4) have been localized in clusters adjacent to the nuclear envelope [reviewed in (217)]. A fascinating study of chromatin domains in live yeast cells by Gasser and colleagues has shown that telomeres are highly dynamic and move rapidly along the inner nuclear membrane in G1 and S phase cells (83, 218). To test whether genes artificially drawn to the yeast nuclear envelope could become transcriptionally silenced, a prior study by Sternglantz and co-workers had shown that a HMR locus with a defective silencer can be silenced by anchoring the locus to the nuclear envelope via the fusion of an integral membrane protein to the GAL4 DNA binding domain (219). As the concentration of Sir proteins is greater at the yeast nuclear periphery, silencing was thought to be favored by artificially placing the locus in a nuclear environment that would be more favorable to silencing. However, since telomeres have been shown to remain peripherally anchored in the absence of Sir proteins, other factors are likely to play a role in this process (220, 221). In this regard, several groups have reported that the Ku heterodimer is bound to telomeres and ~50% of the telomeres are displaced from the nuclear periphery in *ku*-deficient strains, leading to a reduction in the repression of telomeres as well as silent mating-type (HM) genes (222). As was found for the Sir proteins, anchoring and repression can occur, in certain cases, in the absence of Ku and therefore there is more to the story. Recently, Feuerbach et al. have found that upon disruption of myosin-like protein 2 (Mlp2p), a Ku-binding factor associated with the nuclear periphery, a 20% increase in the number of telomeric foci was observed, suggesting telomere dispersion (223). Furthermore, these authors have shown that a membrane-anchored reporter to a subtelomeric region or to the silent mating-type locus is sensitive to mutations in Mlps and the nuclear pore protein Nup60. In contrast, repression of natural mating-type loci, which are also associated with peripheral telomeric foci, is not affected by mutations in either Mlps or nuclear pore proteins. Therefore, although it is clear that the peripheral localization of telomeres in yeast coincides with transcriptional repression, multiple mechanisms may be in play to regulate the association of telomeres with the nuclear periphery (218).

The Nuclear Periphery and Gene Activity

Although strong connections exist between the nuclear periphery and silencing, recently a genetic screen in *S. cerevisiae* to identify chromatin boundary activities (BAs) has identified a class of BAs that mediate their epigenetic function by specific physical tethering to the NPC (224). Among the BAs identified were the exportins Cse1p, Mex67p, and Los1p. These transport proteins were shown to block spreading of heterochromatin by physical tethering of the *S. cerevisiae* silent mating-type locus HML/boundary trap reporter ADE2 to the Nup2p receptor of the NPC, the major docking site for the exportins on the NPC basket (224). Genetic deletion of Nup2p abolished the BA of the transport proteins directly implicating the NPC in chromatin regulation. Based upon these findings, the authors posit that transportins block the propagation of heterochromatin by tethering of the *cis*-acting boundary elements to the Nup2p receptor of the NPC basket, thereby initiating a series of chromatin remodeling events (224). Although this reporter assay provides an interesting scenario, it remains to be determined whether genomic loci physically interact in a similar way with the NPC.

REESTABLISHING THE GENE EXPRESSION MACHINERY AFTER MITOSIS

In higher eukaryotes, mitosis is accompanied by dramatic transformations in the structural organization of both the cytoplasm and nucleus. The onset of mitosis is accompanied by chromatin condensation, breakdown of the nuclear envelope (225), and cessation of bulk cellular transcription (226–228). The constituents of many nuclear domains, such as nuclear speckles and other nuclear bodies, become diffusely distributed throughout the cytoplasm (89, 229–231). However, the gene expression machinery must be rapidly reactivated when cells exit from

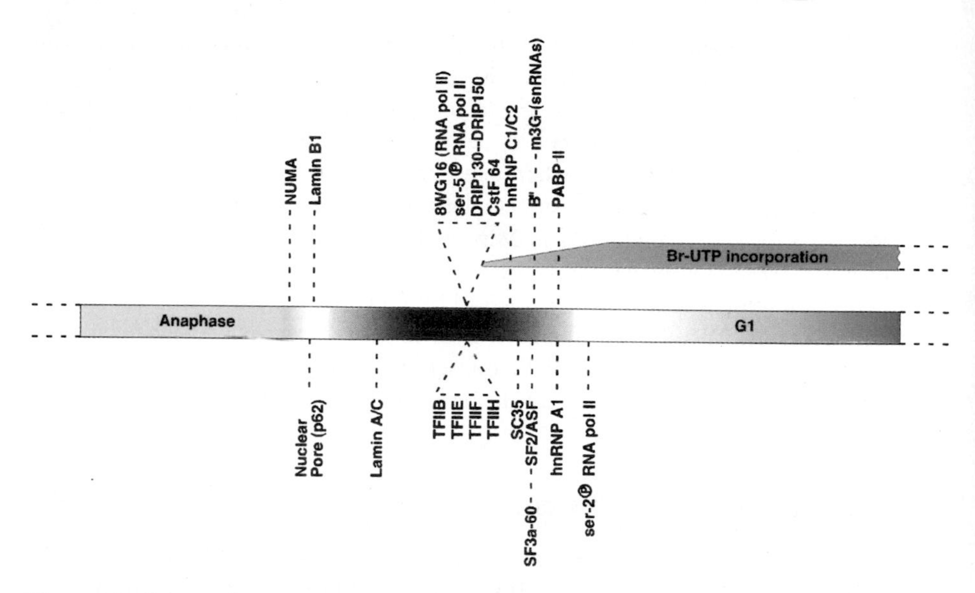

Figure 6 Schematic representation of the order of recruitment of the gene expression machinery into daughter nuclei at the end of mitosis (232).

mitosis. A problem confronted by the mitotic cell is the establishment of the gene expression machinery in daughter nuclei so that these cells become competent to undergo transcription/RNA processing immediately as they exit from mitosis. A recent study has investigated whether components of the gene expression machinery enter postmitotic nuclei individually or as a unitary complex (232). Interestingly, localization studies of numerous RNA pol II transcription and pre-mRNA processing factors revealed a nonrandom and sequential entry of these factors into daughter nuclei after nuclear envelope/lamina formation (Figure 6). The initiation-competent form of RNA pol II and general transcription factors appeared in the daughter nuclei simultaneously, but prior to premRNA processing factors, whereas the elongation competent form of RNA pol II was detected even later (232). The differential entry of these factors rules out the possibility that they are transported as a unitary complex. Furthermore, the differential entry is unlikely to be due to a nuclear retention mechanism based upon substrate binding, as previous studies have shown that pre-mRNA splicing factors (229) as well as RNA pol II (K.V. Prasanth and D.L. Spector, unpublished data not shown) enter daughter nuclei and are maintained in the absence of RNA pol II transcription. Although the mechanism that regulates the sequential recruitment has not been elucidated, it may involve the activation of different importin-β family members. Telophase nuclei were competent for transcription and pre-mRNA splicing concomitant with the initial entry of the respective factors. In addition, a low turnover rate of transcription and pre-mRNA splicing factors was found during mitosis, demonstrating that these factors were recycled into daughter nuclei as the cells progress through telophase (232). Based upon these data, the authors present a model in which the entry of the RNA pol II gene expression machinery into newly forming daughter nuclei is a staged and ordered process (232). This ordered entry of transcription factors prior to pre-mRNA splicing factors appears to be a general phenomenon as similar results were observed in transformed cells (HeLa) as well as in cells of defined passage number (IMR90). The observed sequential recruitment of proteins into daughter nuclei may establish favorable cues for transcription initiation within the context of the decondensing chromosomes.

PERSPECTIVES

Progress in the field of chromosome organization and gene regulation has been substantial over the past few years. The ability to visualize individual chromosome territories within the cell has allowed an investigation of the organization of chromosomes within the three-dimensional structure of the cell nucleus. Such studies have indicated that while there is probably not a functionally obligate position for each chromosome within the nucleus, gene complexity and/or chromosome size may influence chromosome position. The overall organization of individual chromosome territories is such that active genes are located throughout the territory and are not restricted to the periphery, as initially posited. In fact, in some cases, large chromosomal loops containing active genes extend outside of the respective chromosome territories (52, 58, 59). This finding raises concern regarding the interpretation of interphase chromosome paint studies. Based upon our current rate of advancement, it is not unrealistic to anticipate studies that will examine the organization of clusters of adjacent genes within the same chromosome territory in living cells, providing higher-resolution insight into the three-dimensional organization of the genome. Are genes repositioned upon their inactivation? Numerous studies have indicated the movement of inactive genes to centromeres, resulting in heritable silencing. However, association with centromeres or the nuclear periphery in mammalian cells in and of itself does not appear sufficient to result in silencing. Increased telomeric repression in yeast has been correlated with anchoring, and silent chromatin was shown to be tethered to the nuclear envelope in a Sir-dependent manner during S-phase (218).

Whereas chromosome territories generally remain in their nuclear neighborhood, particular genes or gene clusters appear to be dynamic and can move via constrained Brownian motion. However, overall movements seem to follow a random walk as no directed movements have been observed (73, 82). In some cases, such movements seem to correlate with entry into S-phase or transcriptional activation. Chromosome constraint may be imparted through centromeres and telomeres and/or associations with the nucleolus and nuclear envelope/lamina. The technology is currently in place for future studies to examine these associations directly in living cells through the cell cycle, and we can expect significant insight over the next few years. Surprisingly, diffusion is not the only means of chromatin movement. A recent elegant study in yeast indicated that ATP-dependent movements of chromosomal regions also occur (83). It will be important to determine how widespread such movements are and whether they occur in other cell systems containing different levels of heterochromatin.

RNA polymerase II transcription has been reported to occur at nearly 8000 sites scattered throughout the nucleoplasm (95). Contrary to earlier indications, chromosome territories are accessible to RNA polymerase and transcription factors that are freely diffusing throughout the nuclear space (56). Intriguingly, transcription factors studied to date show rapid exchange rates at transcription sites, with residency times ranging from seconds to minutes, supporting a "hit-and-run" model (107). Various transcription factors are also localized to different nuclear bodies. Although many of these bodies are preferentially associated with specific chromosomal regions, we await future studies to determine the functional relevance of these associations. Given our current ability to visualize genetic loci and factors in living cells, we are on the verge of being able to perform in-depth studies of gene expression with the ability to simultaneously visualize DNA, RNA, and protein in real time. In summary, while there have been many advances in our understanding of the organization and dynamics of the genome and its regulatory factors, one point that stands out is the balance between organization and a degree of plasticity, which may represent an inherent mechanism to ensure proper gene expression.

ACKNOWLEDGMENTS

I thank Wendy Bickmore, Susan Gasser, Prasanth Kumar, Paula Sacco-Bubulya, and Mona Spector for helpful suggestions on the manuscript. I also thank the members of my laboratory and my wife for bearing with me while I was confined to my office preparing this review. In addition, I thank Wendy Bickmore, Stan Fakan, Roel van Driel, and Pernette Verschure for providing some of the wonderful figures in the manuscript. The nucleus model is a long-standing and continuously evolving project designed in collaboration with Jim Duffy. Research in my laboratory is supported by NIH/NIGMS (GM42694).

The *Annual Review of Biochemistry* is online at http://biochem.annualreviews.org

LITERATURE CITED

1. Hogenesch JB, Ching KA, Batalov S, Su AI, Walker JR, et al. 2001. *Cell* 106: 413–15
2. Wright FA, Lemon WJ, Zhao WD, Sears R, Zhuo D, et al. 2001. *Genome Biol.* 2:0025.1–0025.18
3. Lander ES, Linton LM, Birren B, Nusbaum C, Zody MC, et al. 2001. *Nature* 409:860–921
4. Venter JC, Adams MD, Myers EW, Li PW, Mural RJ, et al. 2001. *Science* 291: 1304–51

5. Tsien RY. 1998. *Annu. Rev. Biochem.* 67:509–44
6. van Roessel P, Brand AH. 2002. *Nat. Cell Biol.* 4:E15–20
7. Spector DL. 1993. *Annu. Rev. Cell Biol.* 9:265–315
8. Lamond AI, Earnshaw WC. 1998. *Science* 280:547–53
9. Matera AG. 1999. *Trends Cell Biol.* 9:302–9
10. Spector DL. 2001. *J. Cell Sci.* 114: 2891–93
11. Jenuwein T, Allis CD. 2001. *Science* 293:1074–80
12. Misteli T. 2001. *Science* 291:843–47
13. Hernandez-Verdun D, Roussel P, Gebrane-Younes J. 2002. *J. Cell. Sci.* 115:2265–70
14. Rabl C. 1885. *Morphol. Jahrb.* 10: 214–330
15. Boveri T. 1909. *Arch. Exp. Zellforsch.* 3:181–268
16. Deleted in proof
17. McClintock B. 1934. *Z. Zellforsch. Mikroskop. Anat.* 21:294–328
18. Cremer T, Cremer C, Baumann H, Luedtke EK, Sperling K, et al. 1982. *Hum. Genet.* 60:46–56
19. Agard DA, Sedat JW. 1983. *Nature* 302: 676–81
20. Hochstrasser M, Sedat JW. 1987. *J. Cell Biol.* 104:1455–70
21. Mathog D, Hochstrasser M, Gruenbaum Y, Saumweber H, Sedat JW. 1984. *Nature* 308:414–21
22. Hilliker AJ. 1985. *Genet. Res.* 47:13–18
23. Hochstrasser M, Mathog D, Gruenbaum Y, Saumweber H, Sedat JW. 1986. *J. Cell Biol.* 102:112–23
24. Marshall WF, Dernburg AF, Harmon B, Agard DA, Sedat JW. 1996. *Mol. Biol. Cell* 7:825–42
25. Trask B, van den Engh G, Pinkel D, Mullikin J, Waldman J, et al. 1988. *Hum. Genet.* 78:251–59
26. Cremer T, Lichter P, Borden J, Ward DC, Manuelidis L. 1988. *Hum. Genet.* 80:235–46
27. Borden J, Manuelidis L. 1988. *Science* 242:1687–91
28. Pinkel D, Gray JW, Trask B, van den Engh G, Fuscoe J, et al. 1986. *Cold Spring Harbor Symp. Quant. Biol.* 51:151–57
29. Pinkel D, Landegent J, Collins C, Fuscoe J, Segraves R, et al. 1988. *Proc. Natl. Acad. Sci. USA* 85:9138–42
30. Cremer T, Cremer C. 2001. *Nat. Rev. Genet.* 2:292–301
31. Nagele R, Freeman T, McMorrow L, Lee H-Y. 1995. *Science* 270:1831–35
32. Nagele RG, Freeman T, Fazekas J, Lee KM, Thomson Z, Lee HY. 1998. *Chromosoma* 107:330–38
33. Allison DC, Nestor AL. 1999. *J. Cell Biol.* 145:1–14
34. Croft JA, Bridger JM, Boyle S, Perry P, Teague P, Bickmore WA. 1999. *J. Cell Biol.* 145:1119–31
35. Boyle S, Gilchrist S, Bridger JM, Mahy NL, Ellis JA, Bickmore WA. 2001. *Hum. Mol. Genet.* 10:211–19
36. Bridger JM, Boyle S, Kill IR, Bickmore WA. 2000. *Curr. Biol.* 10:149–529
37. Wilmut I, Campbell KH. 1998. *Science* 281:1611
38. Cremer M, von Hase J, Volm T, Brero A, Kreth G, et al. 2001. *Chromosome Res.* 9:541–67
39. Habermann FA, Cremer M, Walter J, Kreth G, von Hase J, et al. 2001. *Chromosome Res.* 9:569–84
40. Tanabe H, Muller S, Neusser M, von Hase J, Calcagno E, et al. 2002. *Proc. Natl. Acad. Sci. USA* 99:4424–29
41. Goldman RD, Gruenbaum Y, Moir RD, Shumaker DK, Spann TP. 2002. *Genes Dev.* 16:533–47
42. Hutchison CJ, Alvarez-Reyes M, Vaughan OA. 2001. *J. Cell Sci.* 114: 9–19
43. Wulff K, Parrish JE, Herrmann FH, Wehnert M. 1997. *Hum. Mutat.* 9: 526–30
44. Wilson KL. 2000. *Trends Cell Biol.* 10:125–29

45. Taddei A, Maison C, Roche D, Almouzni G. 2001. *Nat. Cell Biol.* 3:114–20
46. Ekwall K, Olsson T, Turner BM, Cranston G, Allshire RC. 1997. *Cell* 91:1021–32
47. Grewal SIS, Bonaduce MJ, Klar AJS. 1998. *Genetics* 150:563–76
48. Ma H, Siegel AJ, Berezney R. 1999. *J. Cell Biol.* 146:531–42
49. Maison C, Bailly D, Peters AH, Quivy JP, Roche D, et al. 2002. *Nat. Genet.* 30:329–34
50. Kurz A, Lampel S, Nickolenko JE, Bradl J, Benner A, et al. 1996. *J. Cell Biol.* 135:1195–205
51. Dietzel S, Schiebel K, Little G, Edelmann P, Rappold GA, et al. 1999. *Exp. Cell Res.* 252:363–75
52. Volpi EV, Chevret E, Jones T, Vatcheva R, Williamson J, et al. 2000. *J. Cell Sci.* 113(Part 9):1565–76
53. Zirbel RM, Mathieu UR, Kurz A, Cremer T, Lichter P. 1993. *Chromosome Res.* 1:93–106
54. Cremer T, Kurz A, Zirbel R, Dietzel S, Rinke B, et al. 1993. *Cold Spring Harbor Symp. Quant. Biol.* 58:777–92
55. Abranches R, Beven AF, Aragon-Alcaide L, Shaw PJ. 1998. *J. Cell Biol.* 143:5–12
56. Verschure PJ, van Der Kraan I, Manders EM, van Driel R. 1999. *J. Cell Biol.* 147:13–24
57. Tajbakhsh J, Luz H, Bornfleth H, Lampel S, Cremer C, Lichter P. 2000. *Exp. Cell Res.* 255:229–37
58. Williams RR, Broad S, Sheer D, Ragoussis J. 2002. *Exp. Cell Res.* 272:163–75
59. Park PC, De Boni U. 1998. *Chromosoma* 107:87–95
60. Nogami M, Kohda A, Taguchi H, Nakao M, Ikemura T, Okumura K. 2000. *J. Cell Sci.* 113(Part 12):2157–65
61. Mahy NL, Perry PE, Gilchrist S, Baldock RA, Bickmore WA. 2002. *J. Cell Biol.* 157:579–89
62. Kent J, Lee M, Schedl A, Boyle S, Fantes J, et al. 1997. *Genomics* 42:260–67
63. Gawin B, Niederfuhr A, Schumacher N, Hummerich H, Little PF, Gessler M. 1999. *Genome Res.* 9:1074–86
64. Kleinjan DA, Seawright A, Elgar G, van Heyningen V. 2002. *Mamm. Genome* 13:102–7
65. Belmont AS, Dietzel S, Nye AC, Strukov YG, Tumbar T. 1999. *Curr. Opin. Cell Biol.* 11:307–11
66. Tsukamoto T, Hashiguchi N, Janicki SM, Tumbar T, Belmont AS, Spector DL. 2000. *Nat. Cell Biol.* 2:871–78
67. Ferguson M, Ward DC. 1992. *Chromosoma* 101:557–65
68. Manuelidis L. 1985. *Ann. NY Acad. Sci.* 450:205–21
69. Csink AK, Henikoff S. 1998. *J. Cell Biol.* 143:13–22
70. Shelby RD, Hahn KM, Sullivan KF. 1996. *J. Cell Biol.* 135:545–57
71. Barr ML, Bertram EG. 1949. *Nature* 163:676–77
72. Abney JR, Cutler B, Fillbach ML, Axelrod D, Scalettar BA. 1997. *J. Cell Biol.* 137:1459–68
73. Vazquez J, Belmont AS, Sedat JW. 2001. *Curr. Biol.* 11:1227–39
74. Zink D, Cremer T. 1998. *Curr. Biol.* 8:321–24
75. Manders EM, Kimura H, Cook PR. 1999. *J. Cell Biol.* 144:813–21
76. Robinett CC, Straight A, Li G, Willhelm C, Sudlow G, et al. 1996. *J. Cell Biol.* 135:1685–700
77. Li G, Sudlow G, Belmont AS. 1998. *J. Cell Biol.* 140:975–89
78. Tumbar T, Sudlow G, Belmont AS. 1999. *J. Cell Biol.* 145:1341–54
79. Muller WG, Walker D, Hager GL, McNally JG. 2001. *J. Cell Biol.* 154:33–48
80. Tumbar T, Belmont AS. 2001. *Nat. Cell Biol.* 3:134–39
81. Chubb JR, Boyle S, Perry P, Bickmore WA. 2002. *Curr. Biol.* 12:439–45

82. Marshall WF, Straight A, Marko JF, Swedlow J, Dernburg A, et al. 1997. *Curr. Biol.* 7:930–39
83. Heun P, Laroche T, Shimada K, Furrer P, Gasser SM. 2001. *Science* 294: 2181–86
84. Zink D, Cremer T, Saffrich R, Fischer R, Trendelenburg MF, et al. 1998. *Hum. Genet.* 102:241–51
85. Nakayasu H, Berezney R. 1989. *J. Cell Biol.* 108:1–11
86. Fakan S, Bernhard W. 1971. *Exp. Cell Res.* 67:129–41
87. Fakan S, Nobis P. 1978. *Exp. Cell Res.* 113:327–37
88. Fakan S, Puvion E, Spohr G. 1976. *Exp. Cell Res.* 99:155–64
89. Spector DL, Fu X-D, Maniatis T. 1991. *EMBO J.* 10:3467–81
90. Cmarko D, Verschure PJ, Martin TE, Dahmus ME, Krause S, et al. 1999. *Mol. Biol. Cell* 10:211–23
91. Fakan S. 1994. *Trends Cell Biol.* 4:86–90
92. Fakan S, Puvion E. 1980. *Int. Rev. Cytol.* 65:255–99
93. Jackson DA, Hassan AB, Errington RJ, Cook PR. 1993. *EMBO J.* 12:1059–65
94. Wansink DG, Schul W, van der Kraan I, van Steensel B, van Driel R, de Jong L. 1993. *J. Cell Biol.* 122:283–93
95. Pombo A, Jackson DA, Hollinshead M, Wang Z, Roeder RG, Cook PR. 1999. *EMBO J.* 18:2241–53
96. Iborra FJ, Pombo A, Jackson DA, Cook PR. 1996. *J. Cell Sci.* 109:1427–36
97. Jackson DA, Iborra FJ, Manders EM, Cook PR. 1998. *Mol. Biol. Cell* 9:1523–36
98. Davie JR. 1995. *Int. Rev. Cytol.* 162A: 191–250
99. Nakayasu H, Berezney R. 1991. *Proc. Natl. Acad. Sci. USA* 88:10312–16
100. Berezney R. 2002. *Adv. Enzyme Regul.* 42:39–52
101. Berezney R, Mortillaro MJ, Ma H, Wei XY, Samarabandu J. 1995. *Int. Rev. Cytol.* 162A:1–65
102. Stein GS, van Wijnen AJ, Stein J, Lian JB, Montecino M. 1995. *Int. Rev. Cytol.* 162A:251–78
103. Spann TP, Goldman AE, Wang C, Huang S, Goldman RD. 2002. *J. Cell Biol.* 156:603–8
104. Cremer T, Kreth G, Koester H, Fink RH, Heintzmann R, et al. 2000. *Crit. Rev. Eukaryot. Gene Expr.* 10:179–212
105. Lemon B, Tjian R. 2000. *Genes Dev.* 14:2551–69
106. Hager GL, Elbi C, Becker M. 2002. *Curr. Opin. Genet. Dev.* 12:137–41
107. McNally JG, Muller WG, Walker D, Wolford R, Hager GL. 2000. *Science* 287:1262–65
108. Stenoien DL, Patel K, Mancini MG, Dutertre M, Smith CL, et al. 2001. *Nat. Cell Biol.* 3:15–23
109. Chen D, Hinkley CS, Henry RW, Huang S. 2002. *Mol. Biol. Cell* 13:276–84
110. Kimura H, Cook PR. 2001. *J. Cell Biol.* 153:1341–53
111. Fletcher TM, Ryu BW, Baumann CT, Warren BS, Fragoso G, et al. 2000. *Mol. Cell Biol.* 20:6466–75
112. Deleted in proof
113. Stenoien DL, Nye AC, Mancini MG, Patel K, Dutertre M, et al. 2001. *Mol. Cell Biol.* 21:4404–12
114. Freeman BC, Yamamoto KR. 2002. *Science* 296:2232–35
115. Freeman BC, Felts SJ, Toft DO, Yamamoto KR. 2000. *Genes Dev.* 14:422–34
116. Pombo A, Cuello P, Schul W, Yoon J-B, Roeder RG, et al. 1998. *EMBO J.* 17:1768–78
117. Grande MA, van der Kraan I, de Jong L, van Driel R. 1997. *J. Cell Sci.* 110: 1781–91
118. Elefanty A, Antoniou M, Custodio N, Carmo-Fonseca M, Grosveld F. 1996. *EMBO J.* 15:319–33
119. Mancini MG, Liu B, Sharp ZD, Mancini MA. 1999. *J. Cell. Biochem.* 72:322–38
120. Shiels C, Islam SA, Vatcheva R, Sasieni

P, Sternberg MJ, et al. 2001. *J. Cell Sci.* 114:3705–16
121. Stein GS, van Wijnen AJ, Stein JL, Lian JB, Pockwinse S, McNeil S. 1998. *J. Cell. Biochem.* 70:200–12
122. Gall JG. 2000. *Annu. Rev. Cell Dev. Biol.* 16:273–300
123. Sleeman JE, Lamond AI. 1999. *Curr. Biol.* 9:1065–74
124. Zhong S, Salomoni P, Pandolfi PP. 2000. *Nat. Cell Biol.* 2:E85–90
125. Jolly C, Konecny L, Grady DL, Kutskova YA, Cotto JJ, et al. 2002. *J. Cell Biol.* 156:775–81
126. Denegri M, Moralli D, Rocchi M, Biggiogera M, Raimondi E, et al. 2002. *Mol. Biol. Cell* 13:2069–79
127. Bregman DB, Du L, van der Zee S, Warren SL. 1995. *J. Cell Biol.* 129:287–98
128. Mortillaro MJ, Blencowe BJ, Wei X, Nakayasu H, Du L, et al. 1996. *Proc. Natl. Acad. Sci. USA* 93:8253–57
129. Misteli T, Spector DL. 1998. *Curr. Opin. Cell Biol.* 10:323–31
130. Huang S, Spector DL. 1996. *J. Cell Biol.* 131:719–32
131. Jiménez-García LF, Spector DL. 1993. *Cell* 73:47–59
132. Mintz PJ, Patterson SD, Neuwald AF, Spahr CS, Spector DL. 1999. *EMBO J.* 18:4308–20
133. Price DH. 2000. *Mol. Cell Biol.* 20:2629–34
134. Herrmann CH, Mancini MA. 2001. *J. Cell Sci.* 114:1491–503
135. Lis JT, Mason P, Peng J, Price DH, Werner J. 2000. *Genes Dev.* 14:792–803
136. Jordan P, Cunha C, Carmo-Fonseca M. 1997. *Mol. Biol. Cell* 8:1207–17
137. Pessler F, Pendergrast PS, Hernandez N. 1997. *Mol. Cell Biol.* 17:3786–98
138. Pendergrast PS, Wang C, Hernandez N, Huang S. 2002. *Mol. Biol. Cell* 13:915–29
139. Spector DL. 1990. *Proc. Natl. Acad. Sci. USA* 87:147–51
140. Hendzel MJ, Kruhlak MJ, Bazett-Jones DP. 1998. *Mol. Biol. Cell* 9:2491–507
141. Hock R, Wilde F, Scheer U, Bustin M. 1998. *EMBO J.* 17:6992–7001
142. Lobo SM, Hernandez N. 1994. In *Transcription Mechanisms and Regulation*, ed. RC Conaway, JW Conaway, pp. 127–59. New York: Raven
143. Saunders WS, Cooke CA, Earnshaw WC. 1991. *J. Cell Biol.* 115:919–31
144. Ramón y Cajal S. 1903. *Trab. Lab. Invest. Biol.* 2:129–221
145. Matera AG. 1998. *J. Cell. Biochem.* 70:181–92
146. Frey MR, Matera AG. 1995. *Proc. Natl. Acad. Sci. USA* 92:5915–19
147. Jacobs EY, Frey MR, Wu W, Ingledue TC, Gebuhr TC, et al. 1999. *Mol. Biol. Cell* 10:1653–63
148. Frey MR, Bailey AD, Weiner AM, Matera AG. 1999. *Curr. Biol.* 9:126–35
149. Platani M, Goldberg I, Lamond AI, Swedlow JR. 2002. *Nat. Cell Biol.* 4:502–8
150. Ascoli CA, Maul GG. 1991. *J. Cell Biol.* 112:785–95
151. Szostecki C, Guldner HH, Netter HJ, Will H. 1990. *J. Immunol.* 145:4338–47
152. Dyck JA, Maul GG, Miller WH Jr, Chen JD, Kakizuka A, Evans RM. 1994. *Cell* 76:333–43
153. Koken MH, Puvion-Dutilleul F, Guillemin MC, Viron A, Linares-Cruz G, et al. 1994. *EMBO J.* 13:1073–83
154. Weis K, Rambaud S, Lavau C, Jansen J, Carvalho T, et al. 1994. *Cell* 76:345–56
155. Borden KL. 2002. *Mol. Cell Biol.* 22:5259–69
156. Negorev D, Maul GG. 2001. *Oncogene* 20:7234–42
157. Lallemand-Breitenbach V, Zhu J, Puvion F, Koken M, Honore N, et al. 2001. *J. Exp. Med.* 193:1361–71
158. LaMorte VJ, Dyck JA, Ochs RL, Evans RM. 1998. *Proc. Natl. Acad. Sci. USA* 95:4991–96
159. Boisvert FM, Hendzel MJ, Bazett-Jones DP. 2000. *J. Cell Biol.* 148:283–92

160. Boisvert FM, Kruhlak MJ, Box AK, Hendzel MJ, Bazett-Jones DP. 2001. *J. Cell Biol.* 152:1099–106
161. von Mikecz A, Zhang S, Montminy M, Tan EM, Hemmerich P. 2000. *J. Cell Biol.* 150:265–73
162. Fuchsova B, Novak P, Kafkova J, Hozak P. 2002. *J. Cell Biol.* 158:463–73
163. Everett RD, Earnshaw WC, Pluta AF, Sternsdorf T, Ainsztein AM, et al. 1999. *J. Cell Sci.* 112:3443–54
164. Seeler JS, Marchio A, Sitterlin D, Transy C, Dejean A. 1998. *Proc. Natl. Acad. Sci. USA* 95:7316–21
165. Eissenberg JC, Elgin SC. 2000. *Curr. Opin. Genet. Dev.* 10:204–10
166. Muratani M, Gerlich D, Janicki SM, Gebhard M, Eils R, Spector DL. 2002. *Nat. Cell Biol.* 4:106–10
167. Brock HW, van Lohuizen M. 2001. *Curr. Opin. Genet. Dev.* 11:175–81
168. Saurin AJ, Shiels C, Williamson J, Satijn DP, Otte AP, et al. 1998. *J. Cell Biol.* 142:887–98
169. Cmarko D, Verschure PJ, Otte AR, van Driel R, Fakan S. 2003. *J. Cell Sci.* 116:335–43
170. Sinclair DA, Milne TA, Hodgson JW, Shellard J, Salinas CA, et al. 1998. *Development* 125:1207–16
171. Buchenau P, Hodgson J, Strutt H, Arndt-Jovin DJ. 1998. *J. Cell Biol.* 141:469–81
172. Czermin B, Melfi R, McCabe D, Seitz V, Imhof A, Pirrotta V. 2002. *Cell* 111: 185–96
173. Muller J, Hart CM, Francis NJ, Vargas ML, Sengupta A, et al. 2002. *Cell* 111: 197–208
174. Cao R, Wang L, Wang HY, Xia L, Erdjument-Bromage H, et al. 2002. *Science* 298:1039–43
175. Kaffman A, O'Shea EK. 1999. *Annu. Rev. Cell Dev. Biol.* 15:291–339
176. DeFranco DB. 1999. *Cell Biochem. Biophys.* 30:1–24
177. Hager GL, Fletcher TM, Xiao N, Baumann CT, Muller WG, McNally JG. 2000. *Biochem. Soc. Trans.* 28:405–10
178. Kramer PR, Fragoso G, Pennie W, Htun H, Hager GL, Sinden RR. 1999. *J. Biol. Chem.* 274:28590–97
179. Iborra FJ, Jackson DA, Cook PR. 2001. *Science* 293:1139–42
180. Brogna S. 1999. *RNA* 5:562–73
181. Muhlemann O, Mock-Casagrande CS, Wang J, Li S, Custodio N, et al. 2001. *Mol. Cell* 8:33–43
182. Brogna S, Sato TA, Rosbash M. 2002. *Mol. Cell* 10:93–104
183. Avner P, Heard E. 2001. *Nat. Rev. Genet.* 2:59–67
184. Fisher AG, Merkenschlager M. 2002. *Curr. Opin. Genet. Dev.* 12:193–97
185. Hwang KK, Eissenberg JC, Worman HJ. 2001. *Proc. Natl. Acad. Sci. USA* 98:11423–27
186. Heard E, Rougeulle C, Arnaud D, Avner P, Allis CD, Spector DL. 2001. *Cell* 107:727–38
187. Strahl BD, Allis CD. 2000. *Nature* 403: 41–45
188. Barr ML, Carr DH. 1962. *Acta Cytol.* 6:34–45
189. Lyon MF. 1961. *Nature* 190:372–73
190. Carrel L, Cottle AA, Goglin KC, Willard HF. 1999. *Proc. Natl. Acad. Sci. USA* 96:14440–44
191. Brockdorff N, Ashworth A, Kay GF, Cooper P, Smith S, et al. 1991. *Nature* 351:329–31
192. Brown CJ, Hendrich BD, Rupert JL, Lafreniere RG, Xing Y, et al. 1992. *Cell* 71:527–42
193. Debrand E, Chureau C, Arnaud D, Avner P, Heard E. 1999. *Mol. Cell Biol.* 19:8513–25
194. Lee JT, Lu N. 1999. *Cell* 99:47–57
195. Jeppesen P, Turner BM. 1993. *Cell* 74:281–89
196. Boggs BA, Cheung P, Heard E, Spector DL, Chinault AC, Allis CD. 2002. *Nat. Genet.* 30:73–76
197. Costanzi C, Pehrson JR. 1998. *Nature* 393:599–601
198. Clemson CM, McNeill JA, Willard HF,

Lawrence JB. 1996. *J. Cell Biol.* 132: 259–75
199. Eils R, Dietzel S, Bertin E, Schröck E, Speicher MR, et al. 1996. *J. Cell Biol.* 135:1427–40
200. Dernburg AF, Broman KW, Fung JC, Marshall WF, Philips J, et al. 1996. *Cell* 85:745–59
201. Csink AK, Henikoff S. 1996. *Nature* 381:529–31
202. Skok JA, Brown KE, Azuara V, Caparros ML, Baxter J, et al. 2001. *Nat. Immunol.* 2:848–54
203. Brown KE, Baxter J, Graf D, Merkenschlager M, Fisher AG. 1999. *Mol. Cell* 3:207–17
204. Grogan JL, Mohrs M, Harmon B, Lacy DA, Sedat JW, Locksley RM. 2001. *Immunity* 14:205–15
205. Brown KE, Guest SS, Smale ST, Hahm K, Merkenschlager M, Fisher AG. 1997. *Cell* 91:845–54
206. Georgopoulos K, Moore DD, Derfler B. 1992. *Science* 258:808–12
207. Hahm K, Ernst P, Lo K, Kim GS, Turck C, Smale ST. 1994. *Mol. Cell Biol.* 14:7111–23
208. Zhang CC, Bienz M. 1992. *Proc. Natl. Acad. Sci. USA* 89:7511–15
209. Poux S, Kostic C, Pirrotta V. 1996. *EMBO J.* 15:4713–22
210. Trinh LA, Ferrini R, Cobb BS, Weinmann AS, Hahm K, et al. 2001. *Genes Dev.* 15:1817–32
211. Sabbattini P, Lundgren M, Georgiou A, Chow C, Warnes G, Dillon N. 2001. *EMBO J.* 20:2812–22
212. Francastel C, Walters MC, Groudine M, Martin DI. 1999. *Cell* 99:259–69
213. Schubeler D, Francastel C, Cimbora DM, Reik A, Martin DI, Groudine M. 2000. *Genes Dev.* 14:940–50
214. Lundgren M, Chow CM, Sabbattini P, Georgiou A, Minaee S, Dillon N. 2000. *Cell* 103:733–43
215. Gottschling DE, Aparicio OM, Billington BL, Zakian VA. 1990. *Cell* 63:751–62
216. Brown KE, Amoils S, Horn JM, Buckle VJ, Higgs DR, et al. 2001. *Nat. Cell Biol.* 3:602–6
217. Hediger F, Gasser SM. 2002. *Nat. Cell Biol.* 4:E53–55
218. Hediger F, Neumann FR, Van Houwe G, Dubrana K, Gasser SM. 2002. *Curr. Biol.* 12:2076–89
219. Andrulis ED, Neiman AM, Zappulla DC, Sternglanz R. 1998. *Nature* 394:592–95
220. Gotta M, Laroche T, Formenton A, Maillet L, Scherthan H, Gasser SM. 1996. *J. Cell Biol.* 134:1349–63
221. Tham WH, Wyithe JS, Ferrigno PK, Silver PA, Zakian VA. 2001. *Mol. Cell* 8:189–99
222. Laroche T, Martin SG, Gotta M, Gorham HC, Pryde FE, et al. 1998. *Curr. Biol.* 8:653–56
223. Feuerbach F, Galy V, Trelles-Sticken E, Fromont-Racine M, Jacquier A, et al. 2002. *Nat. Cell Biol.* 4:214–21
224. Ishii K, Arib G, Lin C, Van Houwe G, Laemmli UK. 2002. *Cell* 109:551–62
225. John S, Workman JL. 1998. *BioEssays* 20:275–79
226. Gottesfeld JM, Forbes DJ. 1997. *Trends Biochem. Sci.* 22:197–202
227. Prescott DM, Bender MA. 1962. *Exp. Cell Res.* 26:260–68
228. Johnson LH, Holland JJ. 1965. *J. Cell Biol.* 27:565–74
229. Ferreira JA, Carmo-Fonseca M, Lamond AI. 1994. *J. Cell Biol.* 126:11–23
230. Reuter R, Appel B, Rinke J, Lührmann R. 1985. *Exp. Cell Res.* 159:63–79
231. Spector DL, Smith HC. 1986. *Exp. Cell Res.* 163:87–94
232. Prasanth KV, Sacco-Bubulya P, Prasanth SG, Spector DL. 2003. *Mol. Biol. Cell www.molbiolcell.org/cgi/doi/10.1091/mbc.E02-10-0669*

Annu. Rev. Biochem. 2003. 72:609–642
doi: 10.1146/annurev.biochem.72.121801.161629
First published online as a Review in Advance on March 27, 2003

TRK RECEPTORS: ROLES IN NEURONAL SIGNAL TRANSDUCTION*

Eric J. Huang[1] and Louis F. Reichardt[2]

[1]*Department of Pathology, University of California Veterans Administration Medical Center, San Francisco, California 94143; email: ejhuang@itsa.ucsf.edu*

[2]*Department of Physiology, Howard Hughes Medical Institute at the University of California, San Francisco, California 94143; email: lfr@cgl.ucsf.edu*

Key Words tyrosine kinase, neurotrophin, apoptosis, signaling, differentiation

■ **Abstract** Trk receptors are a family of three receptor tyrosine kinases, each of which can be activated by one or more of four neurotrophins—nerve growth factor (NGF), brain-derived neurotrophic factor (BDNF), and neurotrophins 3 and 4 (NT3 and NT4). Neurotrophin signaling through these receptors regulates cell survival, proliferation, the fate of neural precursors, axon and dendrite growth and patterning, and the expression and activity of functionally important proteins, such as ion channels and neurotransmitter receptors. In the adult nervous system, the Trk receptors regulate synaptic strength and plasticity. The cytoplasmic domains of Trk receptors contain several sites of tyrosine phosphorylation that recruit intermediates in intracellular signaling cascades. As a result, Trk receptor signaling activates several small G proteins, including Ras, Rap-1, and the Cdc-42-Rac-Rho family, as well as pathways regulated by MAP kinase, PI 3-kinase and phospholipase-C-γ (PLC-γ). Trk receptor activation has different consequences in different cells, and the specificity of downstream Trk receptor-mediated signaling is controlled through expression of intermediates in these signaling pathways and membrane trafficking that regulates localization of different signaling constituents. Perhaps the most fascinating aspect of Trk receptor-mediated signaling is its interplay with signaling promoted by the pan-neurotrophin receptor $p75^{NTR}$. $p75^{NTR}$ activates a distinct set of signaling pathways within cells that are in some instances synergistic and in other instances antagonistic to those activated by Trk receptors. Several of these are proapoptotic but are suppressed by Trk receptor-initiated signaling. $p75^{NTR}$ also influences the conformations of Trk receptors; this modifies ligand-binding specificity and affinity with important developmental consequences.

*The U.S. Government has the right to retain a nonexclusive, royalty-free license in and to any copyright covering this paper.

CONTENTS

INTRODUCTION

The Trk family of receptor tyrosine kinases derives its name from the oncogene that resulted in its discovery (1). This oncogene was isolated in gene transfer assays from a carcinoma and, when cloned, was found to consist of the first seven of eight exons of nonmuscle tropomyosin fused to the transmembrane and cytoplasmic domains of a novel tyrosine kinase. Consequently, the proto-oncogene was named tropomyosin-related kinase (trk) and is now commonly referred to as trkA. The *trkB and trkC* genes were identified because of their high homology to *trkA*. Comparisons of their sequences to those of other transmembrane tyrosine kinases indicated that they constitute a novel family of cell surface receptor tyrosine kinases. Specific patterns of expression within the nervous system suggested roles in neuronal development and function, but the Trk receptors were only a small percentage of the large number of orphan tyrosine kinases with high expression in the nervous system. In 1991, though, two groups independently presented convincing evidence that nerve growth factor bound to and activated the tyrosine kinase activity of TrkA (2, 3). Subsequently, TrkB and TrkC were shown to be receptors for other members of the neurotrophin family: BDNF and NT4 activated TrkB, and NT3 activated TrkC (4, 5). Subsequent work has shown that NT3 is able to activate each of the Trk receptors in some cell types.

The discovery of the first neurotrophin, NGF, preceded by several decades the identification of Trk receptors and was a seminal advance in developmental neurobiology (6). Ablation and transplantation studies had previously indicated that targets of innervation secrete limiting amounts of survival factors that ensure a balance between target tissue size and innervation. NGF was the first protein identified that fulfilled this role. The availability of NGF made it possible to identify mechanisms of intercellular communication (7). For example, NGF was shown to be internalized by a receptor-dependent process and to be transported for long distances in small vesicles within axons by an energy- and microtubule-dependent process. NGF was shown to have both local and nuclear actions, which regulate, respectively, growth cone motility and expression of genes encoding the biosynthetic enzymes for neurotransmitters. Without receptors in hand, though, it was not possible to understand the molecular bases for these actions. The discovery of the Trk receptors had a revolutionary impact on this field, because it provided essential tools for pursuing the signaling pathways controlled by neurotrophins. In addition, the literature on other tyrosine kinases suggested that neurotrophins might have much more extensive roles in the nervous system and implicated a number of tyrosine kinase-regulated pathways, such as those activated by Ras, PI 3-kinase, and the Cdc-42-Rac-Rho family, that might mediate these functions. More recent studies on the signaling mechanisms and functions of tyrosine kinases in other systems continue to provide guidance cues for neuroscientists.

The pathways regulated by neurotrophin-mediated activation of Trk receptors include proliferation and survival; axonal and dendritic growth and remodeling; assembly and remodeling of the cytoskeleton; membrane trafficking and fusion; and synapse formation, function, and plasticity (Figure 1). Because of space constraints, comparatively little of the biology of neurotrophins and their receptors themselves are critically examined in this review, which focuses instead on the molecular interactions and pathways regulated by the Trk receptors. Interested readers are referred to many excellent reviews on the biological actions of neurotrophins (4, 5, 7–14).

CONTROL OF NEUROTROPHIN RESPONSIVENESS BY TRK

In general, the repertoire of endogenous Trk receptors expressed by a neuron predicts the set of neurotrophins able to promote a neuron's survival and differentiation. The presence of TrkA, TrkB, or TrkC confers responsiveness, respectively, to NGF, BDNF, and NT4, or to NT3. Ectopic expression of a Trk receptor has been shown to confer responsiveness to the neurotrophins able to activate that receptor in most but not all neurons (15). Differential splicing of the mRNAs encoding each of the Trk receptors, however, makes this generalization an oversimplification (Figure 2). The presence or absence of short sequences of

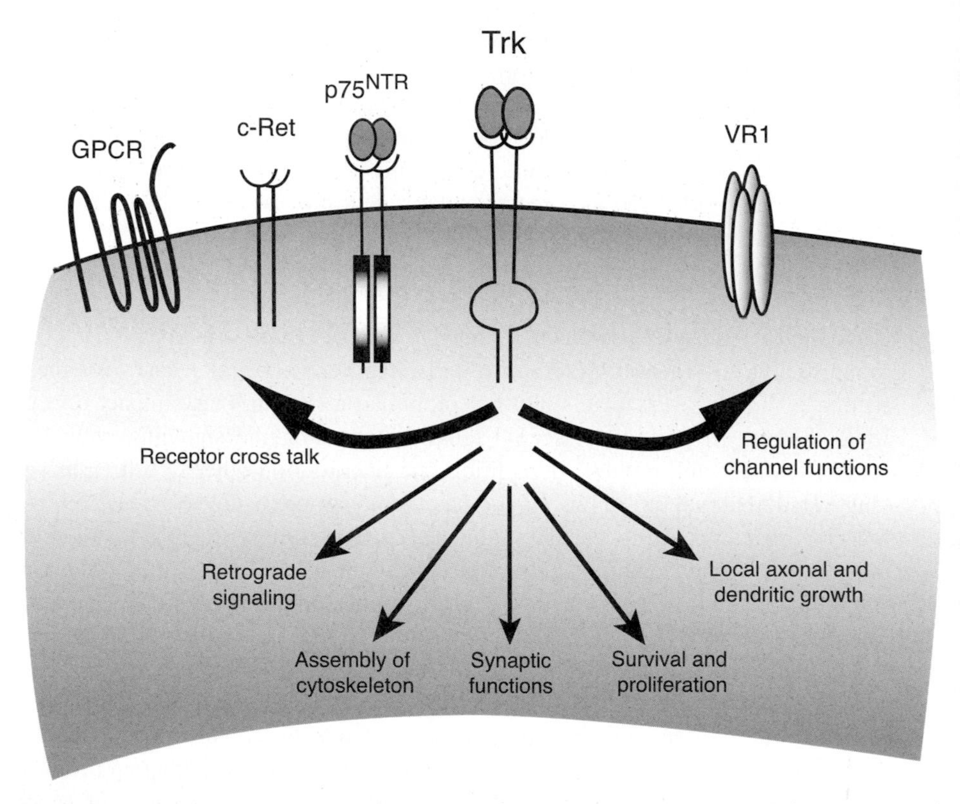

Figure 1 Major functions of Trk receptors. Neurotrophin-mediated activation of Trk receptors leads to a variety of biological responses, which include proliferation and survival, axonal and dendritic growth and remodeling, assembly and remodeling of cytoskeleton, membrane trafficking, and modifications of synaptic functions. In addition, cross talk has been reported between Trk receptors and other membrane receptors such as $p75^{NTR}$, G protein-coupled receptors (GPCRs), vannilloid receptor (VR1), and c-Ret.

amino acids in the juxtamembrane region of each Trk receptor has been shown to regulate the specificity of Trk receptor responsiveness. For example, an isoform of TrkA lacking this short insert is activated efficiently only by NGF, but the presence of this insert increases activation of TrkA by NT3 without affecting its activation by NGF (16). A TrkB isoform lacking a similar short insert can be activated only by BDNF, whereas TrkB containing the insert is also activatable by NT3 and NT4 (17, 18). TrkB with and without the insert is expressed differentially in subpopulations of sensory neurons, which suggests that regulation of splicing of the exon encoding this insert is important for normal neuronal development or function in vivo (18). The short polypeptide sequence in TrkA is immediately adjacent to the major ligand-binding domain in the membrane proximal Ig-like domain (19, 20). These residues were not included in the

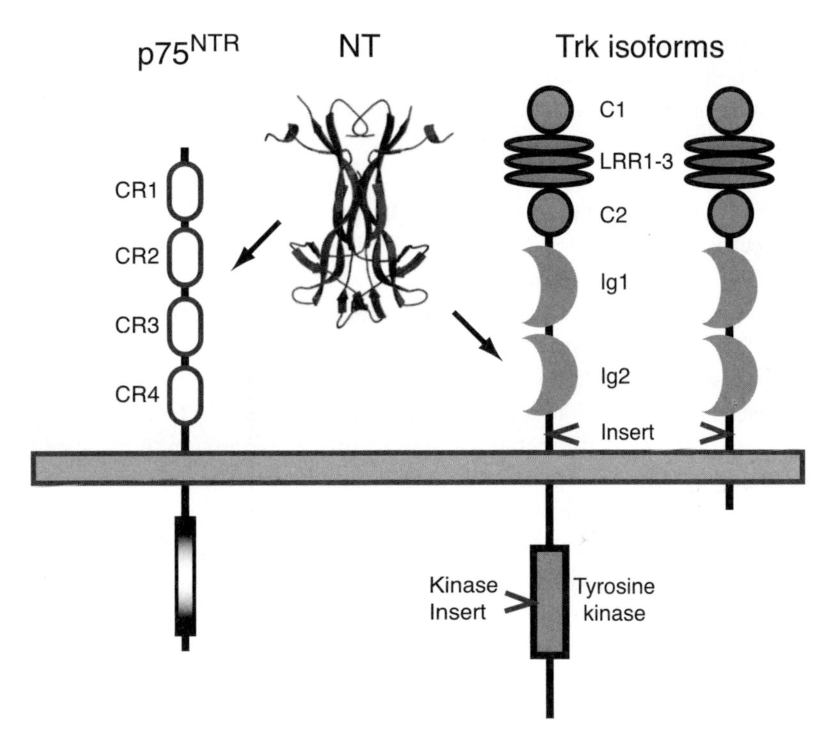

Figure 2 Interactions of neurotrophins (NT) with Trk and $p75^{NTR}$ receptors. The domain structures of the Trk receptors and of $p75^{NTR}$ are schematized in this figure. The locations in the Trk receptor extracellular domains of cysteine clusters (C), leucine-rich repeats (LRR), and immunoglobin-like domains (Ig) are depicted. The locations of the alternatively spliced juxtamembrane inserts that affect the ligand-binding specificity of Trk receptors are also depicted. Each Trk has a single transmembrane domain and a single cytoplasmic tyrosine kinase domain. An isoform of TrkC has been identified with a kinase insert domain. In addition, several truncated isoforms of TrkB and TrkC that lack the tyrosine kinase domain have been identified, and one of these is depicted on the far right. The second immunoglobin-like domain (Ig2) of TrkA and TrkB is the major ligand-binding interface. The domain structure of $p75^{NTR}$ is depicted on the far left. The extracellular domain of $p75^{NTR}$ consists of four cysteine-repeat domains (CR). Both CR2 and CR3 have been implicated in neurotrophin-binding interactions. $p75^{NTR}$ has single transmembrane and cytoplasmic domains. The latter contains a "death domain" similar to those identified in TNF receptors.

NGF-TrkA ligand binding domain complex that was solved at atomic resolution, but the organization of the interface between NGF and TrkA was considered compatible with an additional contribution to binding by these amino acids (19). Organization of the interface between NT4 and TrkB is very similar to that between NGF and TrkA and also appears compatible with a direct contribution to binding by the amino acids in the differentially spliced TrkB insert (21). In addition to the structural data, mutational analysis of the TrkB insert indicates that there are interactions between negatively-charged amino acids in this insert with positively charged residues in NT3 (18).

The pan-neurotrophin receptor, $p75^{NTR}$, also regulates the responsiveness of Trk receptors to neurotrophins. In the presence of $p75^{NTR}$, NT3 is much less effective at activating TrkA, and NT3 and NT4 are much less effective at activating TrkB. In other words, presence of $p75^{NTR}$ enhances the specificity of TrkA and TrkB for their primary ligands, NGF and BDNF, respectively (16, 22–26). Thus, the specificity of neuronal responses to neurotrophins can be modulated by the type of receptor, differential splicing, and the absence or presence of $p75^{NTR}$. Not all isoforms of TrkB and TrkC contain their tyrosine kinase domains. Differential splicing results in expression of truncated receptors lacking the kinase domain (27–29). The functions of these truncated isoforms of TrkB and TrkC are poorly understood. Despite some evidence suggesting that truncated receptors alone can affect intracellular signaling directly (30, 31), tyrosine kinase activity is essential for the vast majority of Trk receptor-mediated responses to neurotrophins. When expressed in *trans*, e.g., on nonneural cells, it has been suggested that the truncated receptors can raise the local effective neurotrophin concentration by capturing and presenting neurotrophins to neurons expressing full-length Trk receptors. When expressed in *cis*, e.g., on the same neuron as a full-length Trk receptor, experiments indicate that truncated receptors inhibit activation of Trk kinases by forming nonproductive heterodimers (32). Consistent with this model, transgenic mice overexpressing a truncated TrkC receptor show neuronal loss in sensory ganglia and cardiac defects similar to those observed in mice lacking NT3 (33). Recent evidence also indicates that truncated isoforms help regulate the surface expression of full-length TrkB (34). Finally, another isoform of TrkC contains an amino acid insert within the kinase domain. The presence of this insert clearly modifies the substrate specificity of the TrkC kinase and alters cellular responses to its activation (35–37).

For a neuron to be responsive to a neurotrophin requires that a Trk receptor be expressed on the surface of the cell. In some cultured CNS neurons, Trk receptors are localized to intracellular vesicles in the absence of signals. Electrical activity, cAMP, and Ca^{2+} stimulate Trk insertion into the cell surface by exocytosis of cytoplasmic membrane vesicles containing Trk (38, 39). Thus, interactions with neighboring cells clearly affect the ability of these neurons to respond to neurotrophins. In summary, most survival-promoting and differentiation-promoting responses to neurotrophins require the presence of a Trk receptor on a neuron, but the competence of a Trk receptor to convey appropriate signals to the interior of the cell is regulated by additional factors, which include the proportions of truncated or insert-containing receptors produced by differential splicing, the presence or absence of $p75^{NTR}$, and second messengers that promote vesicle-mediated receptor insertion into the plasma membrane.

TRK RECEPTOR STRUCTURE AND LIGAND INTERACTIONS

Trk receptors share a common structural organization of their extracellular domains that clearly distinguishes them from other receptor tyrosine kinases (Figure 2) (40). Immediately following the cleaved signal sequence is an array of three leucine-rich 24-residue motifs flanked by two cysteine clusters. Two C2-type immunoglobulin-like domains are adjacent to these structures, which are followed by a single transmembrane domain and a cytoplasmic domain that contains a tyrosine kinase domain plus several tyrosine-containing motifs similar to those present in other receptor tyrosine kinases. Like other receptor tyrosine kinases, phosphorylation of cytoplasmic tyrosines in Trk receptors regulates tyrosine kinase activity and provides phosphorylation-dependent recruitment sites for adaptor molecules and enzymes that mediate initiation of intracellular signaling cascades (4, 5, 9–11).

The unique structural organization of the extracellular domain of Trk receptors suggests that they might mediate adhesive interactions in addition to neurotrophin signaling, but subsequent work has provided no support for this proposal. The major ligand-binding domains of the Trk receptors have been localized to the membrane-proximal Ig-C2-like domain (Ig2) (41). Structures of each of these domains have been solved (42), as have those of several neurotrophins (43–45). In addition, as mentioned earlier, the structures of NGF bound to the TrkA Ig-C2 domain and of NT4 associated with the TrkB Ig-C2 domain have been determined to high resolution. Comparisons between these structures have provided detailed descriptions of the interactions that regulate the specificity and strength of ligand binding to Trk receptors (19, 21). Ligand binding is promoted partly through a set of relatively conserved contacts, shared among the receptor family, while a second set of contacts is responsible for further promoting binding to cognate ligands and for selecting against binding to inappropriate neurotrophins. The N termini of neurotrophins are important in controlling specificity, and the structure of this region is reorganized upon binding to a Trk receptor. Interactions with Trk receptors also alter neurotrophin structures in other regions. This deformability appears important for permitting some neurotrophins to activate more than one type of Trk receptor. To achieve a definitive understanding of this flexibility will require solutions of structures of each of the Trk receptor ligand-binding domains with NT3, the most promiscuous of the neurotrophins.

Although the membrane proximal Ig domain (Ig-2) is the major interface for neurotrophin binding by Trk receptors, other regions in the Trk extracellular domains are also important for ligand binding, either contributing to binding directly or indirectly through effects on conformation of the ligand binding site. For example, mutation of a conserved cysteine in the membrane distal Ig-1 domain of TrkA abolishes NGF binding, suggesting that this domain may also be involved in ligand engagement (46). Studies using a series of TrkA-TrkB chimeras also indicate that the Ig-1 domain of TrkB is required in addition to the TrkB Ig-2 domain for BDNF or NT3-dependent receptor activation (47). Several studies indicate that the cysteine-rich and leucine-rich domains of Trk receptors also participate in ligand binding, at least in some circumstances. For example, high-affinity binding of NT3 to TrkA is not observed in cells expressing a truncated TrkA lacking these regions (48). Mutation of the linker region between the leucine repeats and the Ig-1 domain actually increases the binding affinity of TrkA for NGF, which also implicates this region (46). A series of analyses of TrkA-TrkB chimeras has revealed that the TrkA Ig-2 domain is essential for NGF binding and receptor activation; but in the additional presence of $p75^{NTR}$, this Ig domain is no longer essential to observe ligand-mediated receptor activation (47). The presence of the first cysteine-rich domain of TrkA in a TrkA-TrkB chimera is sufficient to permit effective NGF-dependent activation of a chimeric receptor in the presence of $p75^{NTR}$.

In addition to regulating ligand binding, the different regions in the extracellular domains of Trk receptors also control ligand-independent Trk receptor dimerization. Deletion of Ig-1, Ig-2, or both domains of TrkA increases spontaneous receptor dimerization and activation, which suggests that these domains inhibit spontaneous dimerization in the absence of ligand (49). In contrast, analysis of a mutation in the leucine-rich repeat indicates that it may promote dimerization of ligand-engaged receptor without increasing the affinity of NGF binding (46, 49).

The picture that emerges from these studies suggests that each of the extracellular domains of Trk receptors helps to modulate ligand binding, either by directly interacting with neurotrophins or by modulating conformational changes in the ligand-binding Ig-2 domains of these receptors. Each of the extracellular subdomains also modulates receptor dimerization through interactions that are poorly understood. Despite the impressive three-dimensional crystal structures of neurotrophins complexed to Trk Ig-2 domains, much additional effort will be necessary to characterize the allosteric conformations and ligand-binding interactions of these receptors.

TRK RECEPTOR ACTIVATION MECHANISMS

Binding by neurotrophins provides the primary mechanism for activation of Trk receptors, but the affinity and specificity of Trk receptor activation by neurotrophins is regulated by the pan-neurotrophin receptor $p75^{NTR}$ (Figure 2). The presence of $p75^{NTR}$ is required to observe high-affinity binding of NGF to TrkA (16, 23, 24, 50, 51). Kinetic characterization of NGF interactions with TrkA and with $p75^{NTR}$ demonstrated that, although dissociation constants for each receptor are very similar, the kinetics are quite different. NGF associates with and dissociates from $p75^{NTR}$ much more rapidly than from TrkA, and the presence of $p75^{NTR}$ increases the rate of NGF association with TrkA (52). Recent data have shown that mutations of the cytoplasmic or transmembrane domains of either TrkA or $p75^{NTR}$ prevent the formation of high-affinity binding sites on TrkA (53). Surprisingly, an intact ligand-binding site in $p75^{NTR}$ is not essential for it to promote high-affinity binding. Thus these data argue persuasively that the presence of $p75^{NTR}$ favors a conformation of TrkA that has a high-affinity binding site for NGF. In the presence but not in the absence of $p75^{NTR}$, the presence of only the first cysteine-rich domain of TrkA in an TrkA-TrkB chimera is able to mediate NGF-dependent receptor activation (47), which also supports an allosteric model. Inhibition by $p75^{NTR}$ of TrkA interactions with NT3 also appears to be caused by induction of a conformational change in TrkA (25). Although the extracellular domain of $p75^{NTR}$ must be present to observe inhibition, NT3 binding to $p75^{NTR}$ is not required.

Exciting recent data demonstrate that, similar to the EGF receptor (54, 55), Trk receptors can also be activated by at least two G protein–coupled receptors, the adenosineA_{2a} and PAC-1 receptors (56, 57) (Figure 1). Trk receptor activation through these G protein-coupled receptors has very different kinetics and requirements than activation of the same receptors by neurotrophins. G protein receptor-mediated activation occurs with exceedingly slow kinetics (hours, not minutes) and is prevented by chelators of internal Ca^{2+} and by an inhibitor of Src-family tyrosine kinases, but not by inhibitors of protein kinase C or protein kinase A (57). Compared to neurotrophin action, TrkA activation by pituitary adenylate cyclase-activating peptide (PACAP) results in preferential stimulation of Akt versus Erk1 and Erk2 (58). Potentially facilitating G protein-coupled receptor-mediated activation of Trk receptors, TrkA has been shown to be associated via a PDZ-containing linker protein named GIPC with GAIP (59). GAIP is a protein that contains a regulator of G protein signaling (RGS) domain.

Other receptor tyrosine kinases, such as the EGF receptor, can be activated by "lateral propagation" initiated by local receptor activation or by cell adhesion (60, 61). While propagated Trk receptor activation is a particularly attractive means to explain the rapid kinetics of retrograde signaling by Trk receptors from the axonal tip to the cell body in neurons (12, 62, 63), no experiments directly demonstrate that Trk receptors are activated through this mechanism in vitro or in vivo. There is also no evidence that adhesive interactions activate Trk receptors, although there are almost certainly synergistic interactions with integrins (64).

TRK RECEPTOR INTERACTIONS WITH CYTOPLASMIC ADAPTOR PROTEINS

Trk receptor activation results in phosphorylation of several of ten evolutionarily conserved tyrosines present in the cytoplasmic domains of each receptor (65, 66) (Figure 3). Three of these—Y670, Y674, and Y675 in human TrkA—are in the activation loop of the kinase domain. Phosphorylation of these residues poten-

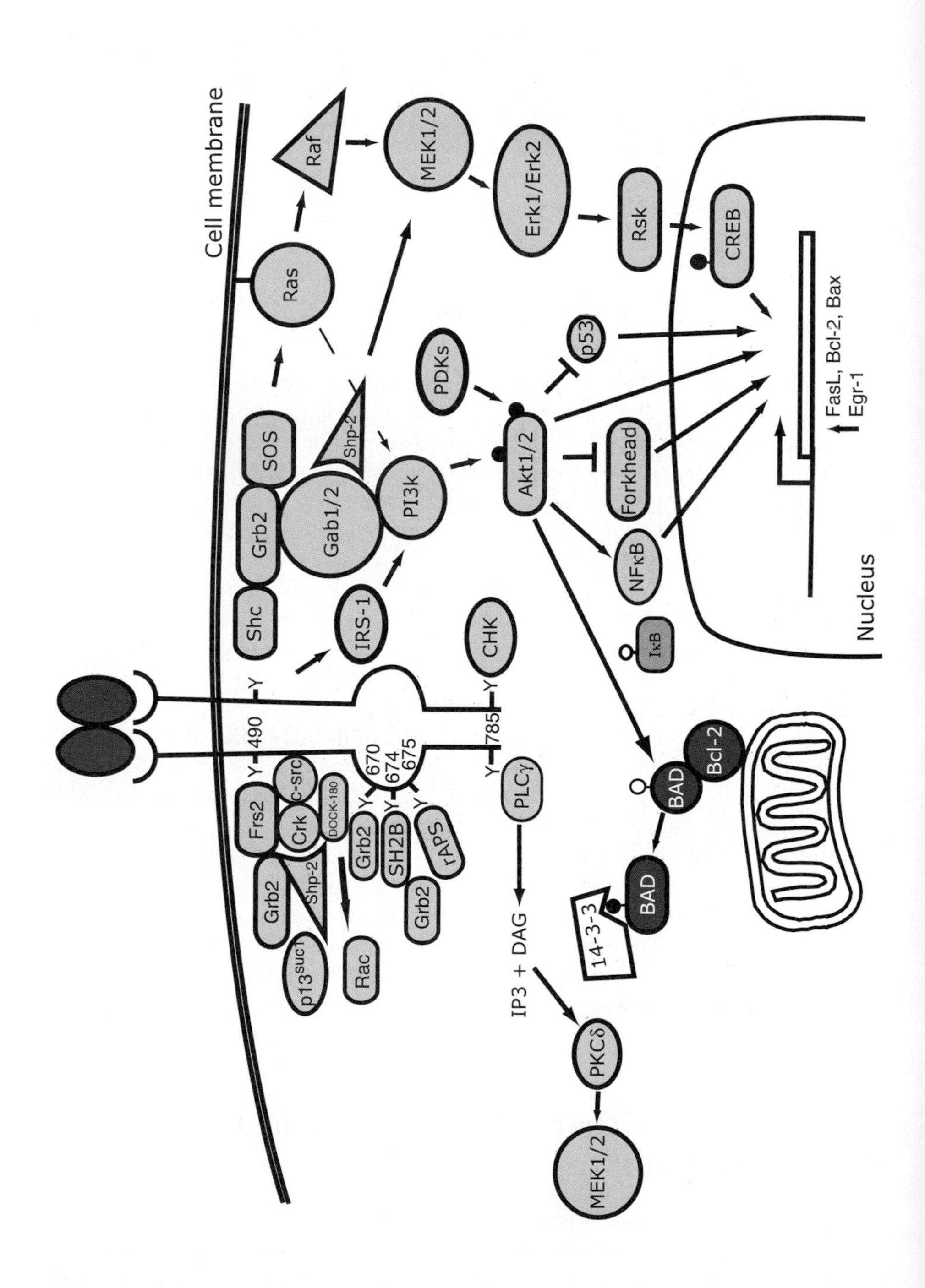

Cell membrane
Raf
MEK1/2
Erk1/Erk2
Rsk
CREB
Ras
SOS
Grb2
Shc
Gab1/2
Shp-2
PI3k
IRS-1
PDKs
Akt1/2
p53
Forkhead
NFκB
IκB
CHK
FasL, Bcl-2, Bax
Egr-1
Nucleus
490
785
670
674
675
Y
Frs2
c-src
Crk
DOCK-180
Grb2
Shp-2
p13suc1
Rac
Grb2
SH2B
rAPS
Grb2
PLCγ
IP3 + DAG
PKCδ
MEK1/2
BAD
Bcl-2
BAD
14-3-3

tiates tyrosine kinase activity by pairing these negatively charged phosphotyrosine residues with basic residues in their vicinity (67). Phosphorylation of additional tyrosines creates docking sites for proteins containing PTB or SH2 domains. Intracellular signaling events activated by these adaptor proteins include Ras-Raf-Erk, PI3 kinase-Akt, PLC-γ-Ca^{2+}, NFκB, and atypical protein kinase C pathways (11, 68, 69). Research has focused on interactions mediated by two phosphorylated tyrosines. Y490 and Y785 are the major phosphorylated tyrosine residues that are not in the kinase activation domain (65). Phosphotyrosine-490 interacts with Shc, fibroblast growth factor receptor substrate 2 (Frs2), and other adaptors, which provide mechanisms for activation of ras and PI3 kinase. Phosphotyrosine-785 recruits PLC-γ1. It should be noted that single mutations of three of the remaining five tyrosines have been shown to inhibit NGF-dependent neurite outgrowth by PC12 cells (66). Thus at least eight tyrosines in the cytoplasmic domain contribute to Trk-mediated signaling, but the details of how several of them contribute are not very clear.

Although most work has focused on pathways controlled by Shc, Frs2, and PLC-γ1, several additional adaptors and signaling complexes have been identified that interact with activated Trk receptors, some of which depend on transport of Trk receptors to intracellular membrane compartments (70–74). In addition to interactions with cytoplasmic adaptors, Trk receptors interact with a number of other membrane proteins, including $p75^{NTR}$, either directly or indirectly (75, 76). Association of $p75^{NTR}$ and Trk undoubtedly facilitates the synergistic and antagonistic interactions between them.

Almost all models that describe the details of Trk receptor signaling focus on the activation of survival and differentiation pathways activated by the adaptors, Frs-2 and Shc, that interact with phospho-Y490 in TrkA and similarly positioned phosphotyrosine residues in TrkB and TrkC. Nothing illustrates the limitations of these models more clearly than the surprisingly modest phenotypes of mice homozygous for a Y-to-F mutation at this site in TrkB or TrkC (77, 78). In each case, major populations of neurons survive that are lost in mice homozygous for a kinase-deletion mutant in TrkB or TrkC. Thus, phospho-Y490-independent interactions must be capable of promoting neuronal survival.

It seems likely that phospho-Y490-independent neuronal differentiation and survival reflect the presence of additional adaptors that have not been incorporated into current models. For example, two additional adaptors, rAPS and

←

Figure 3 Schematic diagram of Trk receptor-mediated signal transduction pathways. Binding of neurotrophins to Trk receptors leads to the recruitment of proteins that interact with specific phosphotyrosine residues in the cytoplasmic domains of Trk receptors These interactions trigger the activation of signaling pathways, such as the Ras, PI 3-kinase, and PLCγ pathways and ultimately result in activation of gene expression, neuronal survival, and neurite outgrowth. The nomenclature for tyrosine residues in the cytoplasmic domains of Trk receptors is based on the human sequence for TrkA.

SH2-B, are similar to PH- and SH2-domain-containing proteins with tyrosines that are phosphorylated following Trk activation (73, 74, 79). They are recruited by phosphorylated tyrosines in the activation loops of each Trk receptor and may also interact with additional phosphorylated tyrosines elsewhere in the cytoplasmic domains of these receptors. Both rAPS and SH2-B form both homo- and heterodimers and, after Trk activation, associate with Grb2, which provides a potential link through SOS to Ras signaling and through both Ras and the Gab adaptor proteins to PI3-kinase. A dominant-interfering construct of SH2-B and antibodies to SH2-B interfere with NGF-dependent survival, Erk activation, and axonal outgrowth by sympathetic neurons (74). Insulin receptor substrate-1 and -2 are adaptors that are tyrosine phosphorylated, following which they promote sustained association with and activation of PI3-kinase in cortical neurons after BDNF application (80). Ligand engagement of TrkA in PC12 cells does not result in insulin receptor substrate-1 or 2 phosphorylation. This indicates that an adaptor protein must be missing in this cell line. CHK, a homolog of the cytoplasmic tyrosine kinase CSK (control of Src kinase), interacts with activated TrkA in PC12 cells and enhances MAP kinase signaling and neurite outgrowth by these cells (81). The mechanism by which it acts is not understood. Finally, a transmembrane protein with three extracellular Ig domains and four tyrosines in the cytoplasmic domain is phosphorylated after BDNF application to cortical neurons and provides a docking site for the protein tyrosine phosphatase, Src-homology phosphatase (Shp2) (82). Overexpression of this protein enhances BDNF-dependent activation of PI3-kinase and survival of cortical neurons.

Providing additional potential complexity, the presence of more than one gene and differential splicing result in expression of more cytoplasmic adaptor and signaling proteins than are considered in simplified signaling models. For example, Shc is expressed at much higher levels in PC12 cells than in neurons, where expression of two closely related homologs, ShcB and ShcC, predominates. Similar to Shc, ShcB and ShcC interact with and are activated by neurotrophin binding to Trk receptors, but they then interact differentially with the repertoire of signaling proteins and exhibit differences in signal transmission (83, 84). Absence of both ShcB and ShcC results in apoptosis of sympathetic and sensory neuron populations whose survival is dependent upon Trk receptor-mediated signaling (85). Despite their potential importance, ShcB and ShcC have been much less intensely studied than Shc itself.

Evidence also indicates that not all interactions with Trk receptors depend upon phosphorylation of tyrosine. c-Abl, a cytoplasmic tyrosine kinase, interacts with the juxtamembrane domain of TrkA, whether or not tyrosines in this region are phosphorylated (86). Deletion of five conserved amino acids in this region blocks differentiation of PC12 cells without preventing phosphorylation of Shc or Frs2 (87). Because c-Abl is involved in many aspects of differentiation, it may well prove to have a role in Trk receptor-mediated signaling that is prevented by this deletion. Recent experiments also indicate that Trk receptors interact with components of the dynein motor complex, which suggests a role for these interactions in Trk receptor trafficking (88). Later in this review, we review compelling evidence implicating membrane trafficking in regulation of Trk receptor-mediated signaling.

In summary, there is far more complexity to Trk receptor-mediated interactions with cytoplasmic adaptor proteins than is currently either appreciated or understood. Different adaptors almost certainly compete with each other for binding to activated Trk receptors. Different neurons clearly may have different subsets of these adaptors. Many adaptors have been identified in studies of other neuronal tyrosine kinases that may also prove to function in Trk receptor-mediated signaling (89). In the future, it will be important to characterize signaling interactions using discrete, well-identified populations of neurons. Almost certainly, this will require technological advances to permit both spatial and temporal examination of protein-protein interactions in single neurons or small populations of identical neurons purified from the complex mixture of neurons present in the nervous system (90).

TRK RECEPTOR EFFECTOR MECHANISMS

PLC-γ1 Signaling

When phosphorylated, Y785 on TrkA and analogous sites on TrkB and TrkC bind PLC-γ1, which is then activated through phosphorylation by the Trk receptor kinase (11) (Figure 3). Activated PLC-γ1 hydrolyses PtdIns(4,5)P_2 to generate inositol tris-phosphate (IP3) and diacylglycerol (DAG). IP3 promotes release of Ca^{2+} from internal stores, which results in activation of enzymes, such as Ca^{2+}-regulated isoforms of protein kinase C and Ca^{2+}-calmodulin-regulated protein kinases. DAG stimulates DAG-regulated protein kinase C isoforms. In PC12 cells, one of these, PKCδ, is required for NGF-promoted neurite outgrowth and for activation of Erk1 and Erk2 (91). PKCδ appears to act between Raf and MEK in this cascade because inhibition of PKCδ reduces activation of MEK, but not of c-Raf.

Not surprisingly, the signaling pathways activated in neuronal cells by Trk-mediated activation of PLC-γ1 extend to the nucleus. Of particular interest, a brief pulse of NGF has been shown to activate a sequence of transcriptional events that results in long-term induction of a sodium channel gene (92). Recently, use of site-specific phosphotyrosine antibodies demonstrated that a brief exposure of PC12 cells to NGF resulted in prolonged phosphorylation of Y785 lasting for several hours (93). Resistance of this site to protein phosphatases provides a likely explanation for the unexpectedly long duration of sodium channel gene induction. This observation illustrates the critical role that protein tyrosine phosphatases play in controlling Trk receptor signaling. It will be exceedingly useful to identify and characterize the expression patterns of the phosphatases responsible for dephosphorylation of each of the cytoplasmic phosphotyrosine residues in Trk receptors.

The physiological functions of TrkB-mediated PLCγ signaling pathways have been tested in vivo by mutating the recruitment site, Y816, to phenylalanine (94). Mice homozygous for the Y816F mutation (*trkB*$^{PLC-/PLC-}$) have a normal life span but are hyperactive compared with control littermates. Electrophysiological experiments show that the *trkB*$^{PLC-/PLC-}$ mutants have significant deficiencies in the induction of both the early and late phases of hippocampal CA1 long-term potentiation (LTP). Results are similar to those observed in animals in which a floxed *trkB* allele was deleted in the postnatal forebrain using Cre recombinase expressed under control of the Ca^{2+}-CaMKII promoter. Surprisingly, while the Erk-MAPK pathways have been implicated in late phase hippocampal LTP (95, 96), BDNF-dependent phosphorylation of Erk and the distribution of phospho-Erk appear to be unaffected in cortical neurons of *trkB*$^{PLC-/PLC-}$ mutants. In contrast, phosphorylation of CREB, CaMKII, and CaMKIV are severely impaired in these neurons. Interestingly, expression of the zinc finger transcription factor Egr-1 (Krox24, Zif268), which is a downstream target of both Ras-Erk and CREB signaling and has been shown to be important for hippocampal LTP, is markedly reduced in the hippocampus of *trkB*$^{PLC-/PLC-}$ mutants. Taken together, these data indicate that signaling initiated at the PLCγ1 docking site on TrkB is important for the initiation and maintenance of hippocampal LTP. Future analyses of *trkB*$^{PLC-/PLC-}$ mutants in other parts of the nervous system will probably reveal additional roles for PLCγ1-initiated signaling. In the future, it will be important to determine whether the entire phenotype of this site mutation is caused by deficiencies in PLCγ1 signaling or also reflects contributions from other proteins that are normally recruited to this same site, such as the cytoplasmic tyrosine kinase CHK (81).

Ras-MAP Kinase Signaling

Activation of the Ras-MAPK/Erk signaling cascade is essential for neurotrophin-promoted differentiation of neurons and PC12 cells. In the response of PC12 cells to neurotrophins, transient versus prolonged activation of Erk signaling is closely but not absolutely associated, respectively, with mitogenic-promoting and differentiation-promoting outcomes (97, 97a).

Several pathways lead from Trk receptors to activation of Ras; most of these appear to involve phosphorylation at Y490. For example, phosphorylated Y490 provides a recruitment site for binding of the PTB domain of the adaptor protein, Shc. After its own phosphorylation, Shc recruits the adaptor protein, Grb2, complexed with SOS, an exchange factor for Ras (and Rac) (98). The presence of activated Ras stimulates signaling through several downstream pathways, which include those mediated by Class I PI3-kinases, Raf, and p38MAP kinase (99, 99a). Activation of Erk1 and Erk2 requires sequential phosphorylation by Raf of MEK1 and/or MEK2 and then phosphorylation of Erk1 and Erk2 by MEK1 or MEK2 (100) (Figure 3). Ras-GTP probably activates p38MAP kinase through a pathway involving sequential activation of RalGDS, Ral, and Src

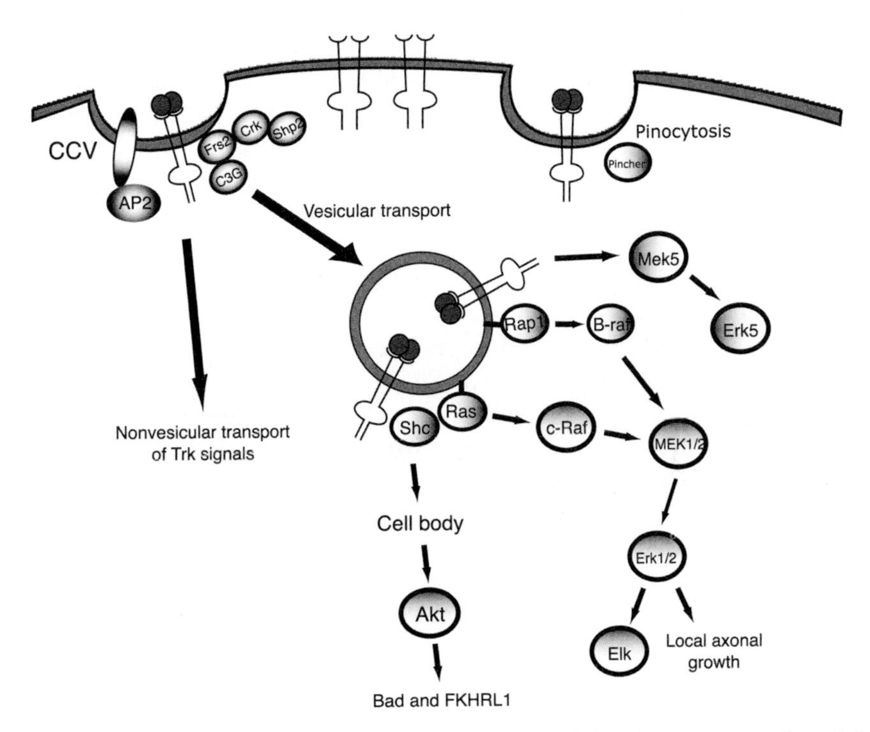

Figure 4 Signaling of Trk receptors by vesicle-mediated transport and vesicle-independent propagation of signals. This diagram illustrates the two distinct pathways that have been shown to propagate signals transmitted upon the activation of Trk receptors. Vesicular transport can be mediated through the clathrin-coated endosomal vesicles and/or by pinocytosis, which can be enhanced by Pincher, a novel protein induced by NGF in PC12 cells.

(99b). p38MAP kinase activates MAP kinase-activated protein kinase-2 (99). Ras activation also triggers a signaling cascade independent of Raf that results in activation of Erk5 through sequential activation of MEKK3 and MEK5 (101–103) (Figure 4). Trk receptor-mediated stimulation of Ras through Shc and Grb2/SOS promotes transient but not prolonged activation of Erk signaling (104). Termination of signaling through this pathway appears to be caused by Erk- and Rsk-mediated phosphorylation of SOS, which results in dissociation of the SOS-Grb2 complex (105).

In contrast to the transient activation of MAP kinase signaling promoted through Shc, Grb2/SOS, and Ras, prolonged Erk activation depends on a distinct signaling pathway involving the adapter protein Crk, the guanine nucleotide exchange factor C3G, the small G protein Rap1, the protein tyrosine phosphatase Shp2, and the serine threonine kinase B-Raf (72, 104, 105) (Figure 4). Trk signaling appears to initiate activation of this pathway by recruitment of an adaptor fibroblast growth factor receptor substrate-2 (Frs2) to phosphorylated Y490 (106, 107). Trk activation results in Frs2 phosphorylation (Figure 3). Frs2 provides binding sites for numerous additional signaling proteins, including the adaptors Grb2 and Crk, the enzymes c-Src and Shp2, and the cyclin-dependent kinase substrate p13^{suc1} (106). Trk signaling increases the association of Crk with Frs2 (105). Association with Crk results in activation of the guanyl nucleotide exchange factor C3G for Rap1(108). Rap-1-GTP stimulates B-Raf, which initiates the Erk cascade (Figures 3 and 4). In addition, Frs2 also contains several phosphorylation-dependent recruitment sites for Grb2; Grb2 provides a mechanism independent of Shc for activation of Ras through the Grb2/SOS exchange factor complex. As expected, overexpression of intermediates in this pathway promotes differentiation of PC12 cells (106, 109–111). Although associated with prolonged MAP kinase signaling, differentiation-promoting effects of Frs2 and Crk are almost certainly not mediated solely through this pathway. The activity of Shp2, which is associated with Frs2, is essential for NGF-dependent activation of the MAP kinase pathway and probably acts by inactivation of an inhibitor, such as Ras-GAP or MAP kinase phosphatase (112) (Figure 3). The recruitment of c-Src to Frs2 may well promote differentiation that synergizes with, but is not dependent upon Frs2-mediated promotion of MAP kinase signaling through B-raf, because activation of endogenous c-Src has been shown to promote neurite outgrowth without altering the kinetics of MAP kinase signaling in PC12 cells (97). Crk contains a binding site for DOCK-180, an exchange factor for Rac (113). Rac activation is also likely to promote differentiative responses through Jun kinase, the Arp2/3 complex, and other effectors that are not completely dependent upon MAP kinase signaling. Finally, despite a consensus on the importance of Crk-associated signaling proteins in mediating activation of MAP kinases by Trk receptors, different groups have reached quite different conclusions on the constituents and assembly kinetics of Crk-associated signaling complexes (72, 105). In particular, these groups have reached different conclusions about the presence of Frs2 in these complexes.

The various MAP kinases activated through Ras and Rap1 have different downstream targets that synergize with each other to mediate gene transcription and cell differentiation. Thus, Erk1, Erk2, and Erk5 substrates include the Rsk family of protein kinases (Figure 3). Rsks and MAP kinase-activated protein kinase-2 can each phosphorylate CREB. CREB has been shown to regulate genes whose products are essential for normal differentiation and prolonged survival of neurons in vitro and in vivo (114, 115). However, different MAPKs also have specific transcription factor targets. For example, Erk5, but not Erk1 or Erk2, activates MEF2 directly; whereas Erk1and Erk2, but not Erk5, activate Elk-1 (116). Recent results indicate that NGF signaling can activate Erk1- and Erk2-mediated signaling locally, but not at a distance (102), which may further contribute to specificity. NGF that is internalized at the growth cone and transported to the cell body appears to activate only Erk5 and not Erk1 or Erk2.

The transcriptional targets of NGF-regulated transcription factors are diverse. Of particular interest, recent work implicates the transcription factor Egr-1, which is the product of an immediate early response gene, in the signaling pathway leading to cell cycle withdrawal and neurite outgrowth in PC12 cells (117–119). NGF application induces transcription of *egr-1*, which in turn induces transcription of the *p35* gene, whose protein product activates Cdk5. Inhibition of either Egr-1 or Cdk5 suppresses NGF-stimulated but not cAMP-stimulated neurite outgrowth. Egr-1 acts, at least in part, through interaction with and activation of c-Jun and not by direct binding to DNA (118). Forced expression of a constitutively active MEK is reported to activate Erk, which results in transcription of the *egr-1* and *p35* genes and neurite outgrowth (119). Unfortunately, there is disagreement on whether MEK inhibitors block NGF-mediated induction of this interesting transcriptional and signaling network (117, 119). In any event, this work illustrates what will undoubtedly become a major effort in this field, namely to identify the intermediates between proximal Trk-induced signaling and the various differentiated phenotypes of neurons ultimately evoked by neurotrophins.

PI3-Kinase Signaling

Production of P3-phosphorylated phosphoinositides promotes survival of many populations of neurons. Phosphatidylinositides are generated by PI3-kinase and activate phosphatidylinositide-dependent protein kinase (PDK-1). Together with these 3-phosphoinositides, PDK-1 activates the protein kinase Akt (also known as PKB), which then phosphorylates several proteins important in promoting cell survival (Figure 3). Class I PI3-kinases are activated through Ras-dependent and independent pathways (99a, 120, 121). Direct activation by Ras of PI3-kinase is a major pathway through which Trk signaling promotes survival in many, but not all, neurons (122, 123). In addition, phosphorylated Grb2 recruits the adaptor proteins Gab1 and Gab2 (122–124). Subsequently, phosphorylated Gab proteins recruit and facilitate the activation of Class I PI3-kinases. In many neurons, but not in PC12 cells, Trk signaling has been shown to result in phosphorylation of insulin receptor substrate-1, which in turn recruits and activates Class I PI3-kinases (80).

In response to growth factor signaling, Gab proteins also nucleate formation of complexes that include the tyrosine phosphatase Shp2 (124). As discussed earlier, Shp2 enhances activation of MAP kinase signaling.

Substrates of Akt include proteins involved in several steps of cell death pathways (120, 121). One of these, Bad, is a Bcl-2 family member that promotes apoptosis through sequestration of Bcl-XL, which otherwise would inhibit Bax, a proapoptotic protein (125). 14-3-3 proteins bind to phosphorylated Bad; this interaction prevents Bad from promoting apoptosis. In addition, Akt-mediated phosphorylation of Bad at S136 has been shown to increase the accessibility of a conserved phosphorylation site (S155) in the BH3 domain of Bad to protein kinases; these include protein kinase A (126). Phosphorylation of S155 directly prevents binding of Bad to Bcl-XL. 14-3-3 proteins, binding to phospho-S136, function as essential cofactors that facilitate phosphorylation of S155. Phosphorylation of both S136 and S155 is necessary to inhibit completely the proapoptotic activity of Bad. Activation of MAP kinase signaling also results in phosphorylation, probably by a RSK, of Bad at S112 (126a, 126b). Phospho-S112 may also promote binding to 14-3-3 proteins and increase the accessibility of S155 to survival-promoting kinases (126). Many additional proteins in the apoptotic pathway, including Bcl-2, Apaf-1, caspase inhibitors, and caspases, also have consensus sites for Akt-mediated phosphorylation, but they have not been shown to actually be substrates of this kinase (121).

The activity of glycogen synthase kinase 3-β (GSK3β) is also negatively regulated by Akt-mediated phosphorylation (127). In cultured neurons, elevated GSK3β promotes apoptosis, so inhibition through phosphorylation undoubtedly contributes to the prosurvival effects of Trk activation. The inhibitory binding partner for NFκB, IκB, is another substrate of Akt. Akt-mediated phosphorylation of IκB promotes its degradation, which results in liberation of active NFκB. NFκB-promoted gene transcription has been shown to promote neuronal survival. Activation of the Trk and $p75^{NTR}$ receptors both promote activation of NFκB (68, 76). Finally, the forkhead transcription factor, FKHRL1, is another substrate of Akt (128, 129). Phosphorylation of FKHRL1 within the nucleus is followed by formation of a complex of FKHRL1 and 14-3-3 (128). The complex is exported from the nucleus and sequestered in the cytoplasm. Because FKHRL1 promotes expression of several proapoptotic proteins, including the FasL (Figure 3), preventing its nuclear entry promotes survival.

PI3 kinase-mediated signaling does not simply promote cell survival. Akt has been shown to activate effectors not involved directly in the survival response, e.g., p70 and p85 S6 kinases, which are important for promoting translation of a subset of mRNAs (for example, those encoding some of the cyclins) (130). In addition, the 3-phosphoinositides generated by PI3 kinase recruit many signaling proteins to the membrane; these include regulators of the Cdc-42-Rac-Rho family of G proteins that are activated through ligand engagement of Trk receptors and control the behavior of the F-actin cytoskeleton (130a). Localized activation of PI3 kinase has been shown to result in localized activation of these G proteins, which permits directed cell motility (131, 132). As mentioned earlier, activated Ras has also been shown to activate an exchange factor activity of SOS for Rac through a mechanism dependent upon PI 3-kinase (98). Localized Trk-promoted activation of Ras and PI 3-kinase almost certainly accounts for the ability of neurotrophin gradients to steer growth cones (133, 134). Activation of Akt by Trk receptors also appears to have effects on axon diameter and branching that are distinct from those observed after MAP kinase activation (135). In the future, it will be very interesting to dissect the pathways responsible for this intriguing phenotype.

REGULATION OF SIGNALING THROUGH MEMBRANE TRANSPORT OF TRK RECEPTORS

Neurotrophin sources are frequently localized in tissues at substantial distances from the cell bodies of innervating neurons. Although neurotrophins have local effects on signaling at axon terminals that affect growth cone motility and exocytosis, it is well established that signaling in the cell soma and nucleus is essential to promote neuronal survival and differentiation (6, 7). During the past few years, there have been intensive studies aimed at identifying mechanisms of retrograde signal transduction, but many issues remain to be clarified.

On the cell surface, Trk receptors are enriched in caveoli-like areas of the plasma membrane (136). Ligand engagement stimulates internalization of Trk receptors through clathrin-coated pits and by macropinocytosis in cell surface ruffles (137, 138) (Figure 4). After internalization, neurotrophins are localized with Trk receptors in endosomes that also contain activated signaling intermediates, such as Shc and PLC-γ1 (139). Retrograde transport of these endosomes provides a conceptually attractive mechanism through which neurotrophin-mediated signals can be conveyed to the cell soma and nucleus.

While there is unambiguous evidence that TrkA is transported together with NGF to the cell soma and that TrkA activity within the soma is required for transmission of a neurotrophin-initiated signal (115), a recent report has demonstrated that neuronal survival can be supported by bead-coupled NGF application to distal axons with little or no internalization and transport of NGF (62). These results suggest the existence of a mechanism for propagation of neurotrophin signaling that does not require internalization and transport of the neurotrophin. Lateral propagation of Trk receptor activation, similar to the activation of ErbB1 observed after focal application of bead-attached EGF (90), provides a potential mechanism through which unliganded Trk receptors might be activated, internalized, and retrogradely transported. Alternatively, there may be retrograde transport of some of the signaling intermediates activated by Trk receptor action. It is unclear how the activation state of these receptors or signaling intermediates would be maintained in the absence of NGF, but it is conceivable that, after concentration in endocytotic vesicles, the basal activity of Trk receptors is sufficiently high to maintain activity. When overexpressed, Trk receptors have been shown to have relatively high levels of basal activation (63).

A second topic of contemporary interest is the role of endocytosis and subsequent membrane trafficking of Trk receptors in controlling Trk receptor-initiated signaling cascades. Recent evidence indicates that membrane sorting events help determine which pathways are activated by Trk receptors, probably because signaling intermediates are preferentially localized in different membrane compartments. For example, unusually rapid internalization of TrkA is induced by exposure to a complex of NGF and a monoclonal antibody (mAb) that does not interfere with receptor binding (70). This NGF-mAb complex promotes Shc phosphorylation, transient Erk activation, and survival of PC12 cells, but it does not promote phosphorylation of Frs2 or differentiation of PC12 cells. Frs2, which is N-myristoylated, is preferentially concentrated in lipid rafts on the cell surface (140). Perhaps because of this localization, it is only recruited by TrkA with comparatively slow kinetics, so that rapid internalization of TrkA prevents its recruitment. Future experiments will be essential to test this hypothesis. Counterintuitively, a thermosensitive dynamin that inhibits clathrin-mediated endocytosis at high temperature has an effect similar to that of the NGF-mAb complex, which acts to inhibit differentiation-promoting but not survival-promoting effects of Trk receptor signaling (141). As work described earlier has indicated that prolonged MAP kinase activation through a complex containing Crk, C3G, and Rap1 is required for differentiation-promoting responses of PC12 cells, the data suggest that Frs2 must be recruited by Trk on the cell surface, but Trk and Frs2-containing signaling complexes must then be internalized to efficiently activate Rap1. Examination of the differential distributions of Ras and Rap1 provide an attractive explanation for the data (71, 72). Ras is prominently expressed on the cell surface where it can be activated without the necessity of TrkA internalization. In contrast, Rap1 appears to be localized almost exclusively in small intracellular vesicles. Thus the data suggest that Trk must be internalized into intracellular vesicles that then fuse with endosomal vesicles containing Rap1 to permit sustained activation of the Erk pathway and normal differentiation of PC12 cells.

Two provocative recent papers have suggested that Trk signaling is regulated by additional, poorly understood mechanisms. In one contribution, local application of NGF to either the distal axons or cell bodies of dorsal root ganglion sensory neurons was shown to activate both the Erk1-Erk2 and Erk5 signaling cascades (102). Surprisingly, application of NGF to the distal axons resulted at the cell somas in activation of Erk5 in the cell soma, but not of Erk1 or Erk2. Activation of Erk5 at the cell somas required receptor internalization because it was inhibited by a thermosensitive dynamin. Local application of K252a, a relatively specific Trk inhibitor, at either the distal axon or cell body also prevented activation of Erk5 in the cell soma by NGF applied at the distal axon. Thus, the results strongly support a model in which active Trk receptors are internalized into signaling endosomes, but the mechanism through which distance affects the specificity of signaling is very uncertain.

One possibility, supported by some recent data, is that Trk receptor activation generates more than one population of signaling endosomes that differ in their potential for retrograde transport. In addition to clathrin-dependent endocytosis through coated pits, it has recently been shown that Trk receptors can also be internalized by pinocytosis (138). A novel protein, Pincher, has been identified that promotes pinocytosis, but not endocytosis, of TrkA together with NGF (138). In PC12 cells, overexpression of Pincher mutated in an ATP-binding site inhibits pinocytosis and prevents the intracellular accumulation of phospho-Erk5 without affecting that of phospho-Erk1 and -Erk2. This result suggests that Clathrin and Pincher may promote formation of endosomal vesicles that differ in their signaling potential. If only the contents of the latter vesicles are transported efficiently in the retrograde direction, Erk5 might be preferentially activated through retrograde transport. To achieve a molecular understanding of these observations, it will be essential to confirm these results and to compare the molecular compositions of the vesicle populations formed by endocytosis and pinocytosis.

In conclusion, Trk receptor-mediated signaling is regulated by both the kinetics and specificity of membrane transport. Regulation of membrane transport provides many additional potential steps at which specificity can be imposed upon signaling.

ACTIVATION OF ION CHANNELS, RECEPTORS AND OTHER RECEPTOR TYROSINE KINASES

Activation of Trk receptors regulates the expression and activities of ion channels, neurotransmitter receptors, and other receptor tyrosine kinases. Trk receptor activation also modulates exocytosis and endocytosis of synaptic vesicles. Some of these effects are observed in seconds to minutes and clearly do not require protein synthesis, while other actions do involve regulation of gene expression through control of transcription factors, some of which have been described above. Work to date implicates regulation of activity through lipid metabolism and protein phosphorylation, localization of proteins and organelles, and local regulation of protein translation in addition to control of gene expression. Each of these mechanisms affects function of the synapse, and these mechanisms are attractive candidates to mediate the important roles that neurotrophins have in regulating synaptic plasticity in the hippocampus and elsewhere. Only a few of the most recent examples of short-term regulation can be cited here. This area is more extensively reviewed elsewhere (4, 14).

It has only recently been appreciated that Trk receptors control membrane properties on a time scale similar to the actions of classical neurotransmitters, and studies on these phenomena are consequently of intense current interest (75, 142–144). For example, low concentrations of BDNF and NT4 have been shown to activate a sodium ion conductance as rapidly as glutamate in slices of CA1 hippocampus, cortex, and cerebellum (144). Neurotrophin action is reversibly blocked by K252a, a reasonably selective Trk receptor inhibitor, which indicates that signaling occurs through TrkB. Until the Na^+ channel regulated so directly by Trk activation is identified, it will be difficult to pursue mechanistic studies, but the extremely rapid response time argues that this channel is very likely to physically associate with TrkB in these neurons. BDNF and NT4 also have been shown to activate postsynaptic Ca^{2+} currents in dentate granule cells. Activation of the Ca^{2+} channels may well be secondary to membrane depolarization resulting from the activation of the Na^+ channel described above. Interestingly, dendritic spines appear to be the exclusive site of rapid activation by BDNF of these Ca^{2+} channels, which synergize with NMDA receptors to induce robust LTP. Not surprisingly, several groups studying different preparations have observed that neurotrophins also enhance intracellular Ca^{2+} levels and potentiate synaptic transmission through PLCγ1-mediated generation of IP3 that results in release of intracellular Ca^{2+} stores (145).

In addition to synaptic actions on Na^+ and Ca^{2+} currents, Trk receptor activation has also been shown to activate several members of the TRP family of cation channels. Many members of this family had previously been shown to be activated through PLC. In the CNS, TRPC3 is broadly expressed and is activated through TrkB-dependent activation of PLC (146). In nociceptive sensory neurons, activity of the heat-activated TRP channel, VR1 is potentiated by NGF signaling through TrkA resulting in hypersensitization to thermal and mechanical stimuli (75). In this instance, the channel is activated by depletion of phosphatidylinositol-4,5-bisphosphate (PtdIns(4,5)P_2) following hydrolysis by PLC or antibody sequestration. TrkA and VR1 can be coimmunoprecipitated, which indicates that they exist in a macromolecular complex in the plasma membrane. It seems quite likely that relief from PtdIns(4,5)P_2-mediated repression will provide an explanation for the PLC-dependent activation of other TRP channels. It will also be interesting to determine whether other members of this family form complexes with Trk receptors. While it seems important to examine in the TrkB-PLCγ1 docking site mutant (94), the response to BDNF of the rapidly activated CNS Na^+ channel described in the previous paragraph (144), this channel appears to be activated far more rapidly than would be possible through recruitment and activation of PLCγ1. Trk receptors almost certainly activate this channel by a more direct mechanism.

Trk receptor signaling also controls the activity and localization of neurotransmitter receptors through protein phosphorylation. For example, BDNF activation of TrkB promotes the phosphorylation and dephosphorylation of the NMDA receptor subunit NR2B with phosphorylation increasing the open probability of the NMDA receptor ion channel and thereby rapidly enhancing synaptic transmission (147, 148). Trk signaling has been shown to be essential to maintain the integrity of the neuromuscular junction in vivo and clustering of neuronal acetylcholine receptors in vitro by signaling mechanisms that have not been further dissected (149, 150). TrkB-mediated signaling also promotes formation of synapses by inhibitory interneurons in the mouse cerebellum and by the axons of retinal ganglion cells in the optic tectum of *Xenopus laevis* (151, 152).

Rapid potentiation of synaptic transmission has been seen in many systems, including developing *Xenopus* neuromuscular synapses, hippocampal cultures, and brain synaptosomes. In the latter, potentiation by BDNF of transmitter release is mediated by an Erk-mediated phosphorylation of synapsin I (153). Potentiation is not observed in synaptosomes isolated from synapsin I-deficient mice and is prevented by inhibitors of the Erk cascade. Synapsins immobilize synaptic vesicles through attachment to the F-actin cytoskeleton and phosphorylation by Erk inhibits synapsin interactions with F-actin. Thus Erk signaling in this system probably potentiates synaptic transmission by releasing synaptic vesicles from the cytoskeleton. In addition to these acute effects of Trk activation, additional deficits are seen in mice lacking normal TrkB-mediated signaling. These include reductions in vesicles docked at release sites and reduced expression of synaptic proteins (154).

While its functional significance has not yet been fully explored, BDNF has recently been shown to regulate local translation of a GFP reporter mRNA in dendrites (155). This reporter contained the 5′ and 3′ untranslated regions of the *Ca^{2+}-CAMKII-alpha* mRNA that confer dendritic localization and BDNF-controlled translational regulation. In a variety of previous studies, protein synthesis-dependent synaptic potentiation has been observed in response to neurotrophins (14). The mechanisms revealed through use of this GFP reporter seem almost certain to be contributors to these phenomena. In the future it will be exceedingly interesting to identify the repertoire of mRNAs whose translation is controlled locally by neurotrophins.

Finally, in postnatal sympathetic neurons, Trk receptors can also activate at least one additional receptor tyrosine kinase, the long-tailed Ret51 isoform of c-Ret (156). TrkA-mediated phosphorylation of Ret51 increases with postnatal age and does not involve the ligand-binding coreceptors required for c-Ret activation by GDNF and related trophic factors. Activation occurs with unusually extended kinetics and does not depend upon PI3 kinase or MAP kinase signaling. The presence of c-Ret is essential to observe full trophic responses of sympathetic neurons to NGF in vivo. Signaling mechanisms are not known but should be vigorously pursued in the future.

INTERACTIONS WITH P75^{NTR}- AND P75^{NTR}-REGULATED SIGNALING PATHWAYS

Each of the neurotrophins also binds to the pan-neurotrophin receptor p75^{NTR}, a member of the TNF receptor superfamily (5, 8). p75^{NTR} binds each of the neurotrophins with approximately equal affinity. Recent work has also shown that p75^{NTR} interacts with the Nogo receptor as a signal-transmitting subunit that mediates inhibitory effects on axon growth of three myelin-associated glycoproteins—Nogo, MAG, and Omgp (156a, 156b). Neurotrophin binding to p75^{NTR} promotes survival of some cells and apoptosis of others as well as affects axon outgrowth both in vivo and in vitro. p75^{NTR} exerts these diverse actions through a set of signaling pathways largely distinct from those activated by Trk receptors that can only be summarized in this review because of space constraints (8) (Figure 5). Prosurvival pathways activated by p75^{NTR} include NFκB and Akt (76, 157). Ligand binding to p75^{NTR} also stimulates several proapototic pathways, which include the Jun kinase signaling cascade, sphingolipid turnover, and association with several adaptors (e.g., NRAGE and NADE) that directly promote

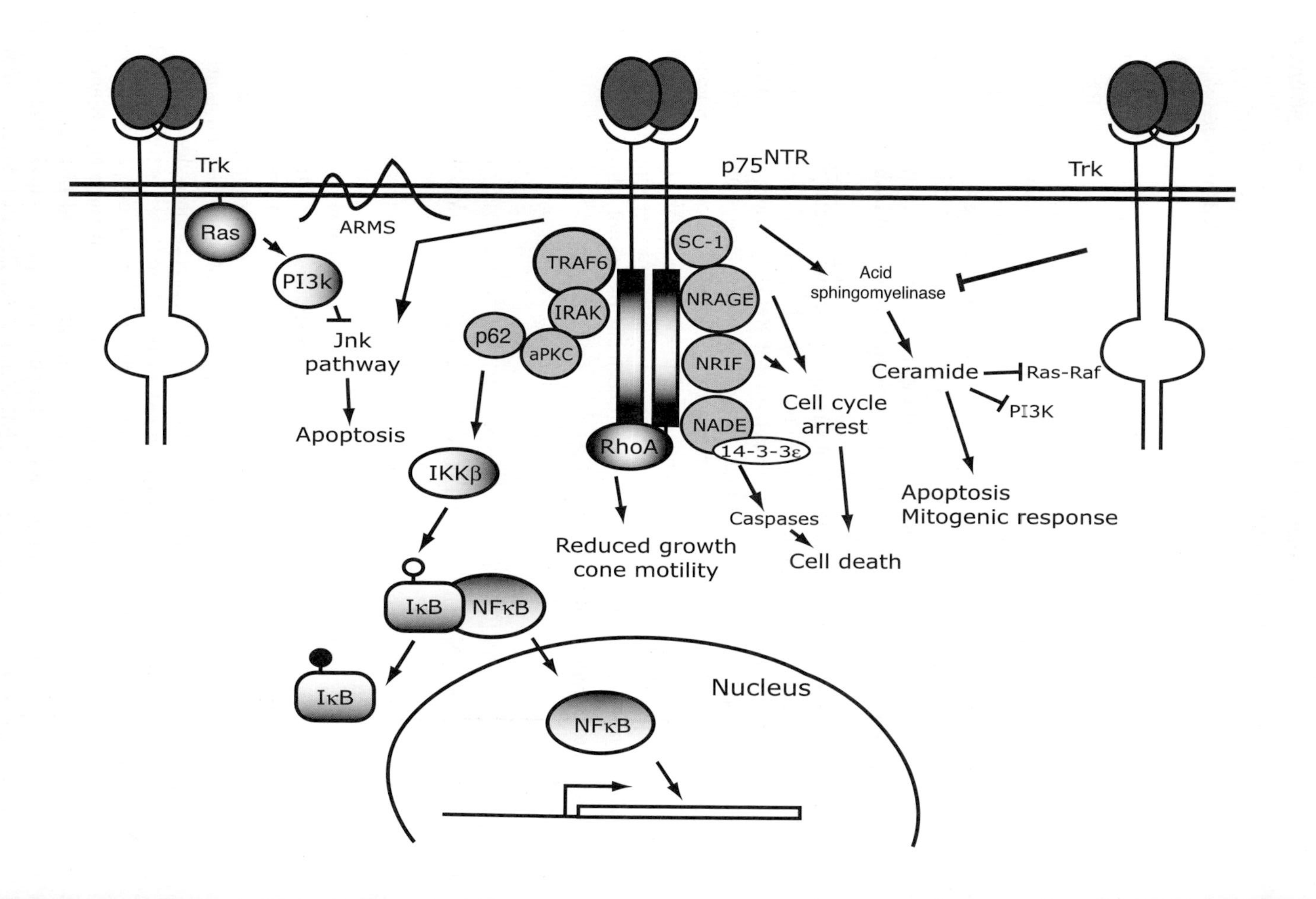
Trk
ARMS
p75^{NTR}
Trk
Ras
PI3k
Jnk
pathway
Apoptosis
TRAF6
IRAK
p62
aPKC
SC-1
NRAGE
NRIF
NADE
14-3-3ε
RhoA
Acid
sphingomyelinase
Ceramide
Ras-Raf
PI3K
Cell cycle
arrest
IKKβ
Caspases
Apoptosis
Mitogenic response
Reduced growth
cone motility
Cell death
IκB
NFκB
IκB
Nucleus
NFκB

cell cycle arrest and apoptosis (158–161). $p75^{NTR}$ also activates the small G proteins Rac and Rho that directly affect growth cone motility (162). Signaling by Trk receptors modulates signaling through many of these $p75^{NTR}$-mediated pathways, thereby altering the nature of the signals conveyed by neurotrophins to neurons. An important consequence is that in the absence of Trk receptor activation neurotrophins are much more effective at inducing apoptosis through $p75^{NTR}$.

Recent experiments have identified adaptors that appear able to nucleate formation of multiprotein complexes including both $p75^{NTR}$ and Trk, which provide scaffolds for the reciprocal interactions between these two receptors. One of these, named ankyrin repeat-rich membrane spanning adaptors (ARMS), has been shown to interact with both receptors and to be phosphorylated following either Trk or Eph receptor activation (163) (Figure 5). Caveolin provides another platform that may mediate interplay between Trk and $p75^{NTR}$ signaling because it has been shown to interact with both of these receptors (164). A third platform is a multiprotein complex required for NFκB activation (76). Neurotrophins have been shown to promote NFκB activation and NFκB-promoted survival through $p75^{NTR}$ (165). NGF promotes the association of the adaptor protein TRAF6 with the juxtamembrane region of $p75^{NTR}$ (166). Interleukin 1 receptor-associated kinase (IRAK) is recruited to this complex leading to formation of a complex of TRAF6 and IRAK with atypical protein kinase C-ι (aPKC-ι) and the aPKC-interacting protein, p62 (167, 76). IκB kinase-β (IKKβ), a known substrate of aPKC, is recruited to the complex; this results in phosphorylation and degradation of IκB and leads to activation of NFκB. Activation of NFκB depends upon p62 and the kinase activity of IRAK. In addition to mediating NFκB activation, the p62-TRAF6-IRAK complex also can function as a scaffold for association of TrkA and $p75^{NTR}$ with TRAF6 binding to $p75^{NTR}$ and p62 to TrkA (76). NGF mediates assembly of this complex, which stimulates association of TRAF6 with both $p75^{NTR}$ and p62 prior to the recruitment of p62 to TrkA. Intriguingly, even though TrkA-mediated signaling in the absence of $p75^{NTR}$ stimulation results in activation of NFκB (68), probably through phosphorylation of IκB by Akt (121), activation of Trk in the presence of $p75^{NTR}$ stimulation actually suppresses NFκB activation and retards the degradation of IκB in PC12 cells (168). An attractive but unproven possibility is that interactions of TrkA with p62 prevent p62 from functioning as a scaffold to facilitate activation of IRAK, aPKC, or IKKβ. In neurons, at least, suppression can not be complete, because in the

←

Figure 5 Schematic diagram of $p75^{NTR}$-mediated signal transduction pathways. $p75^{NTR}$ interacts with proteins (TRAF6, RhoA, NRAGE, SC-1, and NRIF) and regulates gene expression, the cell cycle, apoptosis, mitogenic responses, and growth cone motility. Binding of neurotrophins to $p75^{NTR}$ has also been shown to activate the Jnk pathway, which can be inhibited by activation of the Ras-PI 3-kinase pathway by Trk receptors. See text for detailed descriptions.

presence of Trk signaling, activation of the NFκB cascade by p75^{NTR} has been shown to make a synergistic contribution to survival (165).

Trk activation also affects p75^{NTR}-mediated signaling through other mechanisms. Activation of a Trk receptor completely suppresses the activation by p75^{NTR} of acidic sphingomyelinase through association of activated PI3-kinase with acidic sphingomyelinase in caveoli-related domains (169, 170). Acidic sphingomyelinase activation results in generation of ceramide that promotes apoptosis and mitogenic responses in different cell types through control of many signaling pathways, which include Erk, PI3-kinase, and aPKC. For example, ceramide binds to Raf and may induce formation of inactive Ras-Raf complexes, thereby inhibiting Erk signaling (171). Ceramide also inhibits the activity of PI3-kinase, either by modifying the association of receptor tyrosine kinases and PI3-kinase with caveolin in lipid rafts or by directly inhibiting PI3-kinase activity (172, 173). Thus Trk receptor-mediated suppression of these p75^{NTR}-mediated signaling pathways prevents interference by p75^{NTR} with the survival and differentiation-promoting actions of neurotrophins.

Trk receptor activation also suppresses activation of the Jnk cascade by p75^{NTR}-mediated signaling. In the absence of Trk receptor activation, the activated Jnk cascade promotes apoptosis through elevation of p53 (13). Both Akt and c-Raf have recently been shown to interact with and inactivate the protein kinase ASK-1, an upstream activator of the Jnk pathway (174, 175). Akt also inhibits Jnk signaling by binding to a Jnk-associated scaffolding protein Jip1 (176).

Although Trk receptors suppress p75^{NTR}-mediated signaling, Trk receptors are not always completely efficient at preventing p75^{NTR}-mediated apoptosis. NGF, for example, increases apoptosis of cultured motor neurons from wild-type, but not from p75$^{NTR-/-}$ embryos (177). In PC12 cells, p75^{NTR} activation is reported to reduce NGF-dependent TrkA autophosphorylation (178).

The studies cited above indicate that proapoptotic signaling by p75^{NTR} is strongly inhibited by Trk-mediated signaling through Raf and Akt. As described earlier, p75^{NTR} also appears to refine the ligand-binding specificities of Trk receptors. This may result in apoptosis of neurons not exposed to their preferred trophic factor environment. Consistent with this role for p75^{NTR}, *p75$^{NTR-/-}$* embryos exhibit less apoptosis in both the retina and spinal cord (179). Recent experiments indicate that in the embryonic chick retina coordinated signaling by TrkA and p75^{NTR} functions to regulate the number of mature retinal ganglion cells (180). Differentiated retinal ganglion cells have been shown to express TrkA, p75^{NTR}, and NGF, while migrating precursors express only p75^{NTR}. Survival of the migrating precursors is impaired by the presence of the differentiated retinal ganglion cells and by NGF; this suggests that NGF secreted by mature retinal ganglion cells functions to promote apoptosis only in the immature cells that express p75^{NTR}, but not TrkA (180).

In contrast to these observations, the absence of p75^{NTR} reduces the number of surviving sensory neurons (181). The presence of p75^{NTR} has been shown to increase retrograde transport of neurotrophins (182, 183). In addition, sensory axon outgrowth is slower than normal during development in the absence of $p75^{NTR}$, which possibly reflects the recent demonstration that $p75^{NTR}$ activates Rho in the absence but not in the presence of NGF (184). As a result, axons may not be exposed to adequate levels of neurotrophins in $p75^{NTR-/-}$ animals. The recent demonstration that $p75^{NTR}$ interacts with the NOGO receptor to mediate the inhibitory effects of several myelin-associated glycoproteins on growth cone motility also raises that possibility that absence of $p75^{NTR}$ disrupts signaling pathways important in sensory neuron survival not related to neurotrophin function (156a, 156b). Determination of the mechanisms underlying the neuronal deficits observed in $p75^{NTR-/-}$ animals is an important issue for future investigation. It will also be interesting to examine the effects of the presence of myelin-associated glycoproteins on the signaling events initiated by neurotrophin-binding to Trk and $p75^{NTR}$.

SPECIFICITY IN TRK RECEPTOR-MEDIATED SIGNALING

Although the proximal signaling mechanisms of Trk receptors appear very similar, the effects of signaling through these different receptors are occasionally quite different, and it is a major challenge to understand how these differences arise at the molecular level. As the best example, gradients of neurotrophins have long been known to steer growth cones in vitro (185). In these assays, each of the neurotrophins can be either a chemoattractant or chemorepellant, depending upon the second messenger levels within the neurons (186). When assayed using *Xenopus* motor neurons expressing an appropriate Trk receptor, for example, NGF and BDNF normally attract growth cones, but in the presence of cAMP inhibitors, these factors instead repel growth cones. Similarly, NT3 is a chemoattractant for these growth cones, but is a chemorepellant for the same growth cones when cultured in the presence of cGMP. Inhibitors of cAMP-mediated signaling have no effect on NT3-mediated chemotropism, and inhibitors of cGMP-mediated signaling have no effect on the chemotropic responses to NGF and BDNF. In addition, a uniformly high concentration of netrin, a guidance protein whose actions are also regulated by cAMP, extinguishes chemotropic responses to NGF and BDNF but does not affect responses to NT3. These neurotrophins have also been shown to differ in their effects on synaptic transmission at the neuromuscular junction where BDNF-mediated potentiation of neurotransmitter release requires extracellular Ca^{2+} influx, whereas NT3-mediated potentiation is independent of extracellular Ca^{2+} but requires Ca^{2+} stores and activation of Ca^{2+}-calmodulin-dependent kinase II and PI 3-kinase (14, 187). NT3 and BDNF also have strikingly different effects on the physiological properties and ion channel composition of spiral ganglion neurons (14, 188). Thus, despite their similarities in signaling, NGF and BDNF appear to regulate axon guidance, synaptic function, and neuronal differentiation through signaling pathways that sometimes differ fundamentally from those utilized by NT3. Achieving a satisfactory understanding of the molecular underpinnings of these differences is one of the most interesting challenges facing workers in this field.

In a preliminary effort to examine differences between TrkB- and TrkC-mediated signaling, mice have been generated with Shc-site mutations in TrkB and TrkC (equivalent to Y490F in TrkA) (77, 78). As described earlier, phospho-Y490 is a major recruitment site for Shc and Frs2. In comparisons of mice with Shc-site mutations in TrkB and TrkC, surviving sensory neurons lost target contact in the former but not the latter animals (78). In biochemical assays using the Shc-site mutants, autophosphorylation of TrkB, but not of TrkC, was reduced compared to autophosphorylation of wild-type receptors. Thus these data represent a first attempt to understand specificity in Trk receptor-mediated signaling. It is uncertain whether these differences observed in biochemical assays are relevant for understanding the fundamental differences in the pathways utilized by BDNF and NT3 in axon guidance.

Intriguingly, the Shc-site mutation in TrkB resulted in losses of neurons dependent upon NT4, but with only modest affects on BDNF-dependent neurons (77). In culture, NT4-stimulated Erk activation was more severely compromised than BDNF-stimulated Erk activation in neurons homozygous for the Shc-site mutant. NT4-dependent neuronal survival was similarly impaired in vitro. Thus, these data argue that specificity in signaling through a receptor can be controlled by interactions with different ligands.

CONCLUSION

Neurotrophins are now known to have surprisingly diverse roles in development and function of the nervous system, considering that they were discovered as target-derived trophic factors that ensure during development a match between the number of neurons and the requirement for their function. Neurotrophins have been shown to regulate cell fate, axon growth and guidance, dendrite structure and pruning, synaptic function, and synaptic plasticity. The discovery of Trk receptors was a tremendous advance for these studies and made it possible to initiate studies of the signaling pathways involved in each of these phenotypes. Analyses of Trk receptor structures and signaling properties have been very productive but are still incomplete. While progress in mechanistic studies has been rapid using model cells, such as PC12, much less is known about the pathways utilized by subpopulations of neurons, and it has become clear that different neurons exhibit very different responses to the same ligand. A major challenge for the future will be to devise methodologies appropriate for studying these cell populations that differentiate and function in the context of an exceedingly complex nervous system.

ACKNOWLEDGMENTS

We thank our many colleagues who shared papers and preprints with us. We apologize to many of these same colleagues for our inability to discuss all work in this field because of space and time constraints. We thank Mike Greenberg and David Kaplan for help in interpreting the literature and acknowledge with appreciation an extremely thorough editing job by Jeremy Thorner. Work from our laboratories has been supported by the National Institutes of Health, the Veterans Administration, and the Howard Hughes Medical Institute.

The *Annual Review of Biochemistry* is online at http://biochem.annualreviews.org

LITERATURE CITED

1. Barbacid M, Lamballe F, Pulido D, Klein R. 1991. *Biochim. Biophys. Acta* 1072:115–27
2. Kaplan DR, Hempstead BL, Martin-Zanca D, Chao MV, Parada LF. 1991. *Science* 252:554–58
3. Klein R, Jing SQ, Nanduri V, O'Rourke E, Barbacid M. 1991. *Cell* 65:189–97
4. Huang EJ, Reichardt LF. 2001. *Annu Rev. Neurosci.* 24:677–736
5. Bibel M, Barde YA. 2000. *Genes Dev.* 14:2919–37
6. Levi-Montalcini R. 1987. *Science* 237: 1154–62
7. Shooter EM. 2001. *Annu Rev. Neurosci.* 24:601–29
8. Hempstead BL. 2002. *Curr. Opin. Neurobiol.* 12:260–67
9. Patapoutian A, Reichardt LF. 2001. *Curr. Opin. Neurobiol.* 11:272–80
10. Sofroniew MV, Howe CL, Mobley WC. 2001. *Annu Rev. Neurosci.* 24:1217–81
11. Kaplan DR, Miller FD. 2000. *Curr. Opin. Neurobiol.* 10:381–91
12. Miller FD, Kaplan DR. 2002. *Science* 295:1471–73
13. Miller FD, Kaplan DR. 2001. *Cell. Mol. Life. Sci.* 58:1045–53
14. Poo MM. 2001. *Nat. Rev. Neurosci.* 2:24–32
15. Allsopp TE, Robinson M, Wyatt S, Davies AM. 1994. *Gene Ther.* 1(Suppl.1):S59
16. Clary DO, Reichardt LF. 1994. *Proc. Natl. Acad. Sci. USA* 91:11133–37
17. Strohmaier C, Carter BD, Urfer R, Barde YA, Dechant G. 1996. *EMBO J.* 15:3332–37
18. Boeshore KL, Luckey CN, Zigmond RE, Large TH. 1999. *J. Neurosci.* 19:4739–47
19. Wiesmann C, Ultsch MH, Bass SH, de Vos AM. 1999. *Nature* 401:184–88
20. Robertson AG, Banfield MJ, Allen SJ, Dando JA, Mason GG, et al. 2001. *Biochem. Biophys. Res. Commun.* 282: 131–41
21. Banfield MJ, Naylor RL, Robertson AG, Allen SJ, Dawbarn D, Brady RL. 2001. *Structure* 9:1191–99
22. Lee KF, Davies AM, Jaenisch R. 1994. *Development* 120:1027–33
23. Bibel M, Hoppe E, Barde YA. 1999. *EMBO J.* 18:616–22
24. Benedetti M, Levi A, Chao MV. 1993. *Proc. Natl. Acad. Sci. USA* 90:7859–63
25. Mischel PS, Smith SG, Vining ER, Valletta JS, Mobley WC, Reichardt LF. 2001. *J. Biol. Chem.* 276:11294–301
26. Brennan C, Rivas-Plata K, Landis SC. 1999. *Nat. Neurosci.* 2:699–705
27. Barbacid M. 1994. *J. Neurobiol.* 25: 1386–403
28. Shelton DL, Sutherland J, Gripp J,

Camerato T, Armanini MP, et al. 1995. *J. Neurosci.* 15:477–91
29. Stoilov P, Castren E, Stamm S. 2002. *Biochem. Biophys. Res. Commun.* 290: 1054–65
30. Baxter GT, Radeke MJ, Kuo RC, Makrides V, Hinkle B, et al. 1997. *J. Neurosci.* 17:2683–90
31. Hapner SJ, Boeshore KL, Large TH, Lefcort F. 1998. *Dev. Biol.* 201:90–100
32. Eide FF, Vining ER, Eide BL, Zang K, Wang XY, Reichardt LF. 1996. *J. Neurosci.* 16:3123–29
33. Palko ME, Coppola V, Tessarollo L. 1999. *J. Neurosci.* 19:775–82
34. Haapasalo A, Sipola I, Larsson K, Akerman KE, Stoilov P, et al. 2002. *J. Biol. Chem.* 277:43160–67
35. Meakin SO, Gryz EA, MacDonald JI. 1997. *J. Neurochem.* 69:954–67
36. Guiton M, Gunn-Moore FJ, Glass DJ, Geis DR, Yancopoulos GD, Tavare JM. 1995. *J. Biol. Chem.* 270:20384–90
37. Tsoulfas P, Stephens RM, Kaplan DR, Parada LF. 1996. *J. Biol. Chem.* 271: 5691–97
38. Du J, Feng LY, Yang F, Lu B. 2000. *J. Cell Biol.* 150:1423–33
39. Meyer-Franke A, Wilkinson GA, Kruttgen A, Hu M, Munro E, et al. 1998. *Neuron* 21:681–93
40. Schneider R, Schweiger M. 1991. *Oncogene* 6:1807–11
41. Pérez P, Coll PM, Hempstead BL, Martín-Zanca D, Chao MV. 1995. *Mol. Cell. Neurosci.* 6:97–105
42. Ultsch MH, Wiesmann C, Simmons LC, Henrich J, Yang M, et al. 1999. *J. Mol. Biol.* 290:149–59
43. McDonald NQ, Lapatto R, Murray-Rust J, Gunning J, Wlodawer A, Blundell TL. 1991. *Nature* 354:411–14
44. Butte MJ, Hwang PK, Mobley WC, Fletterick RJ. 1998. *Biochemistry* 37: 16846–52
45. Robinson RC, Radziejewski C, Spraggon G, Greenwald J, Kostura MR, et al. 1999. *Protein Sci.* 8:2589–97
46. Arevalo JC, Conde B, Hempstead BI, Chao MV, Martin-Zanca D, Perez P. 2001. *Oncogene* 20:1229–34
47. Zaccaro MC, Ivanisevic L, Perez P, Meakin SO, Saragovi HU. 2001. *J. Biol. Chem.* 276:31023–29
48. MacDonald JI, Meakin SO. 1996. *Mol. Cell. Neurosci.* 7:371–90
49. Arevalo JC, Conde B, Hempstead BL, Chao MV, Martin-Zanca D, Perez P. 2000. *Mol. Cell. Biol.* 20:5908–16
50. Hempstead BL, Martin-Zanca D, Kaplan DR, Parada LF, Chao MV. 1991. *Nature* 350:678–83
51. Davies AM, Lee KF, Jaenisch R. 1993. *Neuron* 11:565–74
52. Mahadeo D, Kaplan L, Chao MV, Hempstead BL. 1994. *J. Biol. Chem.* 269:6884–91
53. Esposito D, Patel P, Stephens RM, Perez P, Chao MV, et al. 2001. *J. Biol. Chem.* 276:32687–95
54. Daub H, Wallasch C, Lankenau A, Herrlich A, Ullrich A. 1997. *EMBO J.* 16:7032–44
55. Luttrell LM, Della Rocca GJ, van Biesen T, Luttrell DK, Lefkowitz RJ. 1997. *J. Biol. Chem.* 272:4637–44
56. Lee FS, Chao MV. 2001. *Proc. Natl. Acad. Sci. USA* 98:3555–60
57. Lee FS, Rajagopal R, Kim AH, Chang PC, Chao MV. 2002. *J. Biol. Chem.* 277:9096–102
58. Lee FS, Rajagopal R, Chao MV. 2002. *Cytokine Growth Factor Rev.* 13:11–17
59. Lou XJ, Yano H, Lee F, Chao MV, Farquhar MG. 2001. *Mol. Biol. Cell* 12:615–27
60. Sawano A, Takayama S, Matsuda M, Miyawaki A. 2002. *Dev. Cell* 3:245
61. Moro L, Dolce L, Cabodi S, Bergatto E, Erba EB, et al. 2002. *J. Biol. Chem.* 277:9405–14
62. MacInnis BL, Campenot RB. 2002. *Science* 295:1536–39
63. Miller FD, Kaplan DR. 2001. *Neuron* 32:767–70

64. Lindsay RM, Thoenen H, Barde YA. 1985. *Dev. Biol.* 112:319–28
65. Stephens RM, Loeb DM, Copeland TD, Pawson T, Greene LA, Kaplan DR. 1994. *Neuron* 12:691–705
66. Inagaki N, Thoenen H, Lindholm D. 1995. *Eur. J. Neurosci.* 7:1125–33
67. Cunningham ME, Greene LA. 1998. *EMBO J.* 17:7282–93
68. Foehr ED, Lin X, O'Mahony A, Geleziunas R, Bradshaw RA, Greene WC. 2000. *J. Neurosci.* 20:7556–63
69. Wooten MW, Seibenhener ML, Zhou G, Vandenplas ML, Tan TH. 1999. *Cell Death Differ.* 6:753–64
70. Saragovi HU, Zheng W, Maliartchouk S, DiGugliemo GM, Mawal YR, et al. 1998. *J. Biol. Chem.* 273:34933–40
71. York RD, Molliver DC, Grewal SS, Stenberg PE, McCleskey EW, Stork PJ. 2000. *Mol. Cell. Biol.* 20:8069–83
72. Wu C, Lai CF, Mobley WC. 2001. *J. Neurosci.* 21:5406–16
73. Qian X, Ginty DD. 2001. *Mol. Cell. Biol.* 21:1613–20
74. Qian X, Riccio A, Zhang Y, Ginty DD. 1998. *Neuron* 21:1017–29
75. Chuang HH, Prescott ED, Kong H, Shields S, Jordt SE, et al. 2001. *Nature* 411:957–62
76. Wooten MW, Seibenhener ML, Mamidipudi V, Diaz-Meco MT, Barker PA, Moscat J. 2001. *J. Biol. Chem.* 276: 7709–12
77. Minichiello L, Casagranda F, Tatche RS, Stucky CL, Postigo A, et al. 1998. *Neuron* 21:335–45
78. Postigo A, Calella AM, Fritzsch B, Knipper M, Katz D, et al. 2002. *Genes Dev.* 16:633–45
79. Rui LY, Herrington J, Carter-Su C. 1999. *J. Biol. Chem.* 274:10590–94
80. Yamada M, Ohnishi H, Sano S, Nakatani A, Ikeuchi T, Hatanaka H. 1997. *J. Biol. Chem.* 272:30334–39
81. Yamashita H, Avraham S, Jiang S, Dikic I, Avraham H. 1999. *J. Biol. Chem.* 274:15059–65
82. Takai S, Yamada M, Araki T, Koshimizu H, Nawa H, Hatanaka H. 2002. *J. Neurochem.* 82:353–64
83. Nakamura T, Komiya M, Gotoh N, Koizumi S, Shibuya M, Mori N. 2002. *Oncogene* 21:22–31
84. Liu HY, Meakin SO. 2002. *J. Biol. Chem.* 277:26046–56
85. Sakai R, Henderson JT, O'Bryan JP, Elia AJ, Saxton TM, Pawson T. 2000. *Neuron* 28:819–33
86. Yano H, Cong F, Birge RB, Goff SP, Chao MV. 2000. *J. Neurosci. Res.* 59:356–64
87. Meakin SO, MacDonald JI. 1998. *J. Neurochem.* 71:1875–88
88. Yano H, Lee FS, Kong H, Chuang JZ, Arevalo JC, et al. 2001. *J. Neurosci.* 21:RC125
89. Grimm J, Sachs M, Britsch S, Di Cesare S, Schwarz-Romond T, et al. 2001. *J. Cell Biol.* 154:345–54
90. Verveer PJ, Wouters FS, Reynolds AR, Bastiaens PI. 2000. *Science* 290: 1567–70
91. Corbit KC, Foster DA, Rosner MR. 1999. *Mol. Cell. Biol.* 19:4209–18
92. Toledo-Aral JJ, Brehm P, Halegoua S, Mandel G. 1995. *Neuron* 14:607–11
93. Choi DY, Toledo-Aral JJ, Segal R, Halegoua S. 2001. *Mol. Cell. Biol.* 21: 2695–705
94. Minichiello L, Calella AM, Medina DL, Bonhoeffer T, Klein R, Korte M. 2002. *Neuron* 36:121–37
95. Impey S, Obrietan K, Wong ST, Poser S, Yano S, et al. 1998. *Neuron* 21:869–83
96. English JD, Sweatt JD. 1997. *J. Biol. Chem.* 272:19103–6
97. Yang LT, Alexandropoulos K, Sap J. 2002. *J. Biol. Chem.* 277:17406–14
97a. Marshall CJ. 1995. *Cell* 80:179–85
98. Nimnual AS, Yatsula BA, Bar-Sagi D. 1998. *Science* 279:560–63
99. Xing J, Kornhauser JM, Xia Z, Thiele EA, Greenberg ME. 1998. *Mol. Cell. Biol.* 18:1946–55

99a. Vanhaesebroeck B, Leevers SJ, Ahmadi K, Timms J, Katso R, et al. 2001. *Annu. Rev. Biochem.* 70:535–602
99b. Ouwens DM, de Ruiter ND, van der Zon GC, Carter AP, Schouten J, et al. 2002. *EMBO J.* 21:3782–93
100. English J, Pearson G, Wilsbacher J, Swantek J, Karandikar M, et al. 1999. *Exp. Cell Res.* 253:255–70
101. Esparis-Ogando A, Diaz-Rodriguez E, Montero JC, Yuste L, Crespo P, Pandiella A. 2002. *Mol. Cell. Biol.* 22:270–85
102. Watson FL, Heerssen HM, Bhattacharyya A, Klesse L, Lin MZ, Segal RA. 2001. *Nat. Neurosci.* 4:981–88
103. Sun W, Kesavan K, Schaefer BC, Garrington TP, Ware M, et al. 2001. *J. Biol. Chem.* 276:5093–100
104. Grewal SS, York RD, Stork PJ. 1999. *Curr. Opin. Neurobiol.* 9:544–53
105. Kao S, Jaiswal RK, Kolch W, Landreth GE. 2001. *J. Biol. Chem.* 276:18169–77
106. Meakin SO, MacDonald JI, Gryz EA, Kubu CJ, Verdi JM. 1999. *J. Biol. Chem.* 274:9861–70
107. Yan KS, Kuti M, Yan S, Mujtaba S, Farooq A, et al. 2002. *J. Biol. Chem.* 277:17088–94
108. Nosaka Y, Arai A, Miyasaka N, Miura O. 1999. *J. Biol. Chem.* 274:30154–62
109. Tanaka S, Hattori S, Kurata T, Nagashima K, Fukui Y, et al. 1993. *Mol. Cell. Biol.* 13:4409–15
110. Matsuda M, Hashimoto Y, Muroya K, Hasegawa H, Kurata T, et al. 1994. *Mol. Cell. Biol.* 14:5495–500
111. Hempstead BL, Birge RB, Fajardo JE, Glassman R, Mahadeo D, et al. 1994. *Mol. Cell. Biol.* 14:1964–71
112. Wright JH, Drueckes P, Bartoe J, Zhao Z, Shen SH, Krebs EG. 1997. *Mol. Biol. Cell* 8:1575–85
113. Kiyokawa E, Hashimoto Y, Kobayashi S, Sugimura H, Kurata T, Matsuda M. 1998. *Genes Dev.* 12:3331–36
114. Lonze BE, Riccio A, Cohen S, Ginty DD. 2002. *Neuron* 34:371–85
115. Riccio A, Ahn S, Davenport CM, Blendy JA, Ginty DD. 1999. *Science* 286:2358–61
116. Pearson G, Robinson F, Gibson TB, Xu BE, Karandikar M, et al. 2001. *Endocr. Rev.* 22:153–83
117. Levkovitz Y, O'Donovan KJ, Baraban JM. 2001. *J. Neurosci.* 21:45–52
118. Levkovitz Y, Baraban JM. 2002. *J. Neurosci.* 22:3845–54
119. Harada T, Morooka T, Ogawa S, Nishida E. 2001. *Nat. Cell Biol.* 3:453–59
120. Yuan J, Yankner BA. 2000. *Nature* 407:802–9
121. Datta SR, Brunet A, Greenberg ME. 1999. *Genes Dev.* 13:2905–27
122. Vaillant AR, Mazzoni I, Tudan C, Boudreau M, Kaplan DR, Miller FD. 1999. *J. Cell Biol.* 146:955–66
123. Holgado-Madruga M, Moscatello DK, Emlet DR, Dieterich R, Wong AJ. 1997. *Proc. Natl. Acad. Sci. USA* 94:12419–24
124. Liu Y, Rohrschneider LR. 2002. *FEBS Lett.* 515:1–7
125. Datta SR, Dudek H, Tao X, Masters S, Fu HA, et al. 1997. *Cell* 91:231–41
126. Datta SR, Katsov A, Hu L, Petros A, Fesik SW, et al. 2000. *Mol. Cell* 6:41–51
126a. Fang X, Yu S, Eder A, Mao M, Bast RC Jr, et al. 1999. *Oncogene* 18:6635–40
126b. Scheid MP, Schubert KM, Duronio V. 1999. *J. Biol. Chem.* 274:31108–13
127. Hetman M, Cavanaugh JE, Kimelman D, Xia ZG. 2000. *J. Neurosci.* 20:2567–74
128. Brunet A, Kanai F, Stehn J, Xu J, Sarbassova D, et al. 2002. *J. Cell Biol.* 156:817–28
129. Brunet A, Bonni A, Zigmond MJ, Lin MZ, Juo P, et al. 1999. *Cell* 96:857–68
130. Kimball SR, Farrell PA, Jefferson LS. 2002. *J. Appl Physiol* 93:1168–80

130a. Yuan XB, Jin M, Xu X, Song YQ, Wu CP, et al. 2003. *Nat. Cell Biol.* 5:38–45
131. Weiner OD, Neilsen PO, Prestwich GD, Kirschner MW, Cantley LC, Bourne HR. 2002. *Nat. Cell Biol.* 4:509–12
132. Wang F, Herzmark P, Weiner OD, Srinivasan S, Servant G, Bourne HR. 2002. *Nat. Cell Biol.* 4:513–18
133. Ming GL, Wong ST, Henley J, Yuan XB, Song HJ, et al. 2002. *Nature* 417: 411–18
134. Song HJ, Ming GL, Poo MM. 1997. *Nature* 388:275–79
135. Markus A, Zhong J, Snider WD. 2002. *Neuron* 35:65–76
136. Huang CS, Zhou J, Feng AK, Lynch CC, Klumperman J, et al. 1999. *J. Biol. Chem.* 274:36707–14
137. Beattie EC, Howe CL, Wilde A, Brodsky FM, Mobley WC. 2000. *J. Neurosci.* 20:7325–33
138. Shao Y, Akmentin W, Toledo-Aral JJ, Rosenbaum J, Valdez G, et al. 2002. *J. Cell Biol.* 157:679–91
139. Howe CL, Valletta JS, Rusnak AS, Mobley WC. 2001. *Neuron* 32:801–14
140. Paratcha G, Ledda F, Baars L, Coulpier M, Besset V, et al. 2001. *Neuron* 29:171–84
141. Zhang Y, Moheban DB, Conway BR, Bhattacharyya A, Segal RA. 2000. *J. Neurosci.* 20:5671–78
142. Kafitz KW, Rose CR, Konnerth A. 2000. *Prog. Brain Res.* 128:243–49
143. Kovalchuk Y, Hanse E, Kafitz KW, Konnerth A. 2002. *Science* 295: 1729–34
144. Kafitz KW, Rose CR, Thoenen H, Konnerth A. 1999. *Nature* 401:918–21
145. Canossa M, Gartner A, Campana G, Inagaki N, Thoenen H. 2001. *EMBO J.* 20:1640–50
146. Li HS, Xu XZ, Montell C. 1999. *Neuron* 24:261–73
147. Levine ES, Crozier RA, Black IB, Plummer MR. 1998. *Proc. Natl. Acad. Sci. USA* 95:10235–39
148. Lin SY, Wu K, Len GW, Xu JL, Levine ES, et al. 1999. *Mol. Brain Res.* 70:18–25
149. Gonzalez M, Ruggiero FP, Chang Q, Shi YJ, Rich MM, et al. 1999. *Neuron* 24:567–83
150. Kawai H, Zago W, Berg DK. 2002. *J. Neurosci.* 22:7903–12
151. Rico B, Xu BJ, Reichardt LF. 2002. *Nat. Neurosci.* 5:225–33
152. Alsina B, Vu T, Cohen-Cory S. 2001. *Nat. Neurosci.* 4:1093–101
153. Jovanovic JN, Benfenati F, Siow YL, Sihra TS, Sanghera JS, et al. 1996. *Proc. Natl. Acad. Sci. USA* 93:3679–83
154. Pozzo-Miller LD, Gottschalk W, Zhang L, McDermott K, Du J, et al. 1999. *J. Neurosci.* 19:4972–83
155. Aakalu G, Smith WB, Nguyen N, Jiang C, Schuman EM. 2001. *Neuron* 30: 489–502
156. Tsui-Pierchala BA, Milbrandt J, Johnson EM Jr. 2002. *Neuron* 33:261–73
156a. Wang KC, Kim JA, Sivasankaran R, Segal R, He Z. 2002. *Nature* 420:74–78
156b. Wong ST, Henley JR, Kanning KC, Huang KH, Bothwell M, Poo MM. 2002. *Nat. Neurosci.* 5:1302–8
157. Roux PP, Bhakar AL, Kennedy TE, Barker PA. 2001. *J. Biol. Chem.* 276: 23097–104
158. Mukai J, Hachiya T, Shoji-Hoshino S, Kimura MT, Nadano D, et al. 2000. *J. Biol. Chem.* 275:17566–70
159. Salehi AH, Roux PP, Kubu CJ, Zeindler C, Bhakar A, et al. 2000. *Neuron* 27:279–88
160. Jordan BW, Dinev D, LeMellay V, Troppmair J, Gotz R, et al. 2001. *J. Biol. Chem.* 276:39985–89
161. Whitfield J, Neame SJ, Paquet L, Bernard O, Ham J. 2001. *Neuron* 29:629–43
162. Harrington AW, Kim JY, Yoon SO. 2002. *J. Neurosci.* 22:156–66
163. Kong H, Boulter J, Weber JL, Lai C, Chao MV. 2001. *J. Neurosci.* 21:176–85

164. Bilderback TR, Gazula VR, Lisanti MP, Dobrowsky RT. 1999. *J. Biol. Chem.* 274:257–63
165. Hamanoue M, Middleton G, Wyatt S, Jaffray E, Hay RT, Davies AM. 1999. *Mol. Cell. Neurosci.* 14:28–40
166. Khursigara G, Orlinick JR, Chao MV. 1999. *J. Biol. Chem.* 274:2597–600
167. Vandenplas ML, Mamidipudi V, Seibenhener ML, Wooten MW. 2002. *Cell Signal* 14:359–63
168. Mamidipudi V, Li X, Wooten MW. 2002. *J. Biol. Chem.* 277:28010–18
169. Dobrowsky RT, Jenkins GM, Hannun YA. 1995. *J. Biol. Chem.* 270:22135–42
170. Bilderback TR, Gazula VR, Dobrowsky RT. 2001. *J. Neurochem.* 76: 1540–51
171. Muller G, Storz P, Bourteele S, Doppler H, Pfizenmaier K, et al. 1998. *EMBO J.* 17:732–42
172. Zhou H, Summers SA, Birnbaum MJ, Pittman RN. 1998. *J. Biol. Chem.* 273: 16568–75
173. Zundel W, Swiersz LM, Giaccia A. 2000. *Mol. Cell. Biol.* 20:1507–14
174. Chen J, Fujii K, Zhang L, Roberts T, Fu H. 2001. *Proc. Natl. Acad. Sci. USA* 98:7783–88
175. Kim AH, Khursigara G, Sun X, Franke TF, Chao MV. 2001. *Mol. Cell. Biol.* 21:893–901
176. Kim AH, Yano H, Cho H, Meyer D, Monks B, et al. 2002. *Neuron* 35: 697–709
177. Wiese S, Metzger F, Holtmann B, Sendtner M. 1999. *Eur. J. Neurosci.* 11:1668–76
178. MacPhee IJ, Barker PA. 1997. *J. Biol. Chem.* 272:23547–51
179. Frade JM, Barde YA. 1999. *Development* 126:683–90
180. Gonzalez-Hoyuela M, Barbas JA, Rodriguez-Tebar A. 2001. *Development* 128:117–24
181. Stucky CL, Koltzenburg M. 1997. *J. Neurosci.* 17:4398–405
182. Harrison SM, Jones ME, Uecker S, Albers KM, Kudrycki KE, Davis BM. 2000. *J. Comp. Neurol.* 424:99–110
183. Curtis R, Adryan KM, Stark JL, Park JS, Compton DL, et al. 1995. *Neuron* 14:1201–11
184. Yamashita T, Tucker KL, Barde YA. 1999. *Neuron* 24:585–93
185. Gundersen RW, Barrett JN. 1979. *Science* 206:1079–80
186. Song HJ, Ming GL, He ZG, Lehmann M, McKerracher L, et al. 1998. *Science* 281:1515–18
187. Yang F, He XP, Feng LY, Mizuno K, Liu XW, et al. 2001. *Nat. Neurosci.* 4:19–28
188. Adamson CL, Reid MA, Davis RL. 2002. *J. Neurosci.* 22:1385–96

Annu. Rev. Biochem. 2003. 72:643–691
doi: 10.1146/annurev.biochem.72.121801.161809
Copyright © 2003 by Annual Reviews. All rights reserved
First published online as a Review in Advance on March 27, 2003

A Genetic Approach to Mammalian Glycan Function

John B. Lowe[1] and Jamey D. Marth[2]

[1]Department of Pathology and Howard Hughes Medical Institute, University of Michigan, Ann Arbor, Michigan 48109; email: johnlowe@umich.edu
[2]Department of Cellular and Molecular Medicine, Howard Hughes Medical Institute, University of California San Diego, La Jolla, California 92093; email: jmarth@ucsd.edu

Key Words glycosylation, physiology, disease, gene-targeting, transgenics

■ **Abstract** The four essential building blocks of cells are proteins, nucleic acids, lipids, and glycans. Also referred to as carbohydrates, glycans are composed of saccharides that are typically linked to lipids and proteins in the secretory pathway. Glycans are highly abundant and diverse biopolymers, yet their functions have remained relatively obscure. This is changing with the advent of genetic reagents and techniques that in the past decade have uncovered many essential roles of specific glycan linkages in living organisms. Glycans appear to modulate biological processes in the development and function of multiple physiologic systems, in part by regulating protein-protein and cell-cell interactions. Moreover, dysregulation of glycan synthesis represents the etiology for a growing number of human genetic diseases. The study of glycans, known as glycobiology, has entered an era of renaissance that coincides with the acquisition of complete genome sequences for multiple organisms and an increased focus upon how posttranslational modifications to protein contribute to the complexity of events mediating normal and disease physiology. Glycan production and modification comprise an estimated 1% of genes in the mammalian genome. Many of these genes encode enzymes termed glycosyltransferases and glycosidases that reside in the Golgi apparatus where they play the major role in constructing the glycan repertoire that is found at the cell surface and among extracellular compartments. We present a review of the recently established functions of glycan structures in the context of mammalian genetic studies focused upon the mouse and human species.

Nothing tends so much to the advancement of knowledge as the application of a new instrument. The native intellectual powers of men in different times are not so much the causes of the different success of their labours, as the peculiar nature of the means and artificial resources in their possession.

T. Hager: *Force of Nature* (1)

CONTENTS

INTRODUCTION

The study of glycans encompasses a long and robust history and yet today appears as a frontier area of biological research. Although glycans are one of the four major components of living cells, their functions have remained more enigmatic by comparison with our understanding of DNA, protein, and lipids. Nevertheless, glycans appear as the most abundant and structurally diverse biopolymers formed in nature. The numerous pathways and enzymatic activities involved in glycan biosynthesis (also termed glycosylation) generate this diversity in the secretory pathway. Subsequently, most glycans are destined for extracellular compartments. Glycan synthesis begins in the endoplasmic reticulum and continues in the Golgi apparatus where most of the structural variation is produced (1a–3). A remarkable foundation of biochemical knowledge was formed prior to the availability of genetic approaches involving genes that encode the enzymes responsible for glycan biosynthesis. The term glycobiology came to describe the study of glycans, and more recently, the term *glycomics* has emerged in reference to the repertoire of glycan structures present among organisms and cell types.

Unlike many other fields of biological research, the use of cultured cells in glycobiology has rarely led to physiologic insights into glycan function. The same in vitro and mutant cell systems that so significantly contributed to the understanding of mammalian glycan biosynthesis were found to metabolize and proliferate in the absence of one or more glycan linkages. These observations were in some ways discouraging to a field struggling to enter the genomic era and assert the relevance of glycan variation in biological processes. Numerous observations and correlative findings nevertheless suggested a large number of biological roles for glycans (4). A notable high point was the finding that a missing N-glycan structure is responsible for Inclusion Cell Disease, wherein the absence of mannose-6-phosphate results in failure to sort specific glycoproteins to the lysosome (5). Efforts persisted in many laboratories to purify and sequence the low-abundance enzymes that participated in glycan formation. In the late 1980s, the first glycan-synthetic gene sequences were identified from mammalian sources (6–10). Within a few years, their numbers increased dramatically. All were remarkably conserved among mammalian species, often with 90% or more amino acid sequence identity. It now seemed possible to answer questions glycobiologists had posed for years. For example, could a mammal develop to term and survive if its N-glycans failed to diversify into hybrid and complex forms, like those found in fungi and yeast? However, such a defect due to mutation of the *Mgat1* gene was found to be innocuous in somatic cell cultures (11–13). With newly developed mammalian transgenetic technologies, this and other N-glycan deficiencies could be engineered in vivo. The consequence in the case of *Mgat1* deficiency was embryonic lethality with defects in the vasculature, neural tube, and in the determination of heart loop asymmetry (14, 15). Such findings have pointed to the need for an in vivo genetic approach to decipher the nature and extent of information contained in the mammalian glycan repertoire.

In the emerging postgenomic era there has been a call for closer evaluation of structural modifications to molecules as a major mechanistic theme in the control of organismal development, physiology, and disease. Much of the increased interest in glycobiology is further due to the advent of genetic tools and the application of genetic technologies that have been used to increase our knowledge of glycan function. Identical approaches to determine the function of nucleic acids, lipids, and proteins have been applied to other enzyme families, for example the kinases and phosphatases, resulting in an appreciably increased knowledge of metabolic regulation and the etiology of various diseases. Parallels exist among studies of phosphorylation and glycosylation that are useful in formulating a constructive view of glycobiology. Both include the study of posttranslational modifications to protein, but with phosphorylation functioning inside the cell while glycosylation functions in the secretory pathway and among extracellular compartments. However, both processes involve discrete pathways such that eliminating specific steps yields information on the function of downstream steps in these pathways. In glycobiology, the regulated expression of glycosynthetic and glycolytic genes among cell types is primarily responsible for the dynamic complexity of glycan structures.

A common misconception is that cellular glycan structures do not change and are under little, if any, regulatory control. Glycan structures actually change in association with variations in cellular metabolism. Such glycan structural diversification is highly regulated by signals that control cell differentiation, normal physiology, and even neoplastic transformation (16–19). Glycosynthetic and glycolytic gene families with overlapping activities but unique expression patterns are common, and glycan structures can be rapidly altered by glycosidase activity. Rapid modification of glycan structures in vivo can also occur by increases in glycoprotein turnover rates along with alterations in glycan biosynthesis. The glycan repertoire, or glycome, is currently an assemblage of structures that change in normal and disease metabolism, yet are incompletely defined and, for the most part, bear an unknown amount of biologic information. Nevertheless, from a compilation of enzymatic steps and isozymes families in a previous assessment (20), and recent genome sequence analyses, it is estimated that 1% of mammalian genes are devoted to glycan production and modification.

Molecules that bind to carbohydrates are termed lectins. Mammalian lectins appear to be involved in recognizing structural diversity among the glycans and thereby modulate glycan-dependent molecular associations that can alter cellular physiology (21, 22). Lectins exist in most organisms, and those characterized among plants and invertebrates have been of tremendous value in establishing the presence or absence of specific glycan linkages in cells and organisms. Like the glycosynthetic and glycolytic genes, mammalian lectins continue to be discovered, although currently their specific functions are mostly unresolved. These molecules are not discussed further in this review, nor are studies of specific substrate molecules (lipids or proteins) that carry common or unique glycan structures. Neither does this review discuss the lysosomal storage disorders wherein defects in glycosidases attenuate glycan catabolism. Although physiologic abnormalities have been observed from augmenting the formation of specific glycan linkages in transgenic systems, we focus this review upon findings involving the induced and inheritable deficiency of endogenous glycosynthetic or glycolytic genes in the context of an otherwise normal genomic, ontogenic, and physiologic background (Table 1 and Table 2).

OVERVIEW OF MAMMALIAN GLYCOSYLATION IN THE SECRETORY PATHWAY

There are significant differences in glycosylation among organisms in the major evolutionary branches, including the vertebrates, invertebrates, multicellular fungi, yeast, and the prokaryotes. This review is focused upon mammalian glycosylation, which includes pathways often also present among nonmammalian vertebrates and some invertebrates. [For information on evolutionary aspects of glycan formation and structures found among other organisms, readers are referred to (23–25)].

The six major classes of extracellular glycans include three variations among one class (Figure 1). There is in addition some diversity in specific glycan linkages among mammalian species, including the ABO blood group structures

(26), the α1–3Gal "xenotransplantation" antigen (27), and the lack of N-glycolylneuraminic acid in humans (28, 29). These and other glycan variations have been linked to mutations in one or more genes involved in glycan synthesis. Although some of this variation may have little functional consequence, a portion could be selected for in evolution by pathogen-host interactions. Pathogen infection typically begins by glycan recognition, e.g., the requirement of influenza virus binding to host sialic acid linkages for infection (30, 31).

A hallmark of glycosyltransferase and glycosidase enzymes is their remarkable degree of substrate specificity. The substrate is typically the previous step in the pathway, and therefore absence of a single enzyme can restrict further diversification of the repertoire as downstream steps cannot occur. The stepwise process of glycan formation in the Golgi incorporates their transit through the Golgi cisternae from the *cis* to *medial* and then *trans* compartments. Because the cast of glycosynthetic enzymes may change, steps in the pathway are not necessarily intermediates but can represent a "mature" glycan structure to be exported. Sometimes, such variation may occur on the same glycoprotein, and the variable use of peptide sequences as glycosylation sites has been noted. The term "microheterogeneity" refers to this phenomenon, and has been considered as either a stochastic event or a mechanism in glycan formation and function under stringent regulatory control. The mechanisms by which glycosylation is regulated appear diverse and need further study to resolve.

FUNCTIONS OF CORE AND PROXIMAL GLYCAN STRUCTURES

Core glycan structures are shared among particular classes of glycans. Due to their proximal position in biosynthesis, they remain present regardless of further steps in glycan linkage formation or subsequent hydrolysis. Because a significant amount of glycan structure is typically missing when core linkage formation is altered, it is expected that multiple systems and cell types would be affected. This is essentially correct; however, the unexpected presence of isozymes that contribute to core and proximal steps in glycan formation has also yielded evidence for specificity in function among cell types and physiologic systems.

N-Glycans

Asparagine (N)-linked modification of proteins by oligosaccharides is universally observed in eukaryotic cells. N-glycans are assembled in a step-wise manner by the sequential actions of glycosyltransferases and glycosidases (Figure 2) [reviewed in (1a)]. Initial steps in this synthetic pathway are characterized by the assembly of a dolichol-linked oligosaccharide precursor that is subsequently transferred en bloc to a nascent polypeptide chain (1a). Synthesis of this oligosaccharide precursor initiates with the enzyme UDP-GlcNAc: dolichol

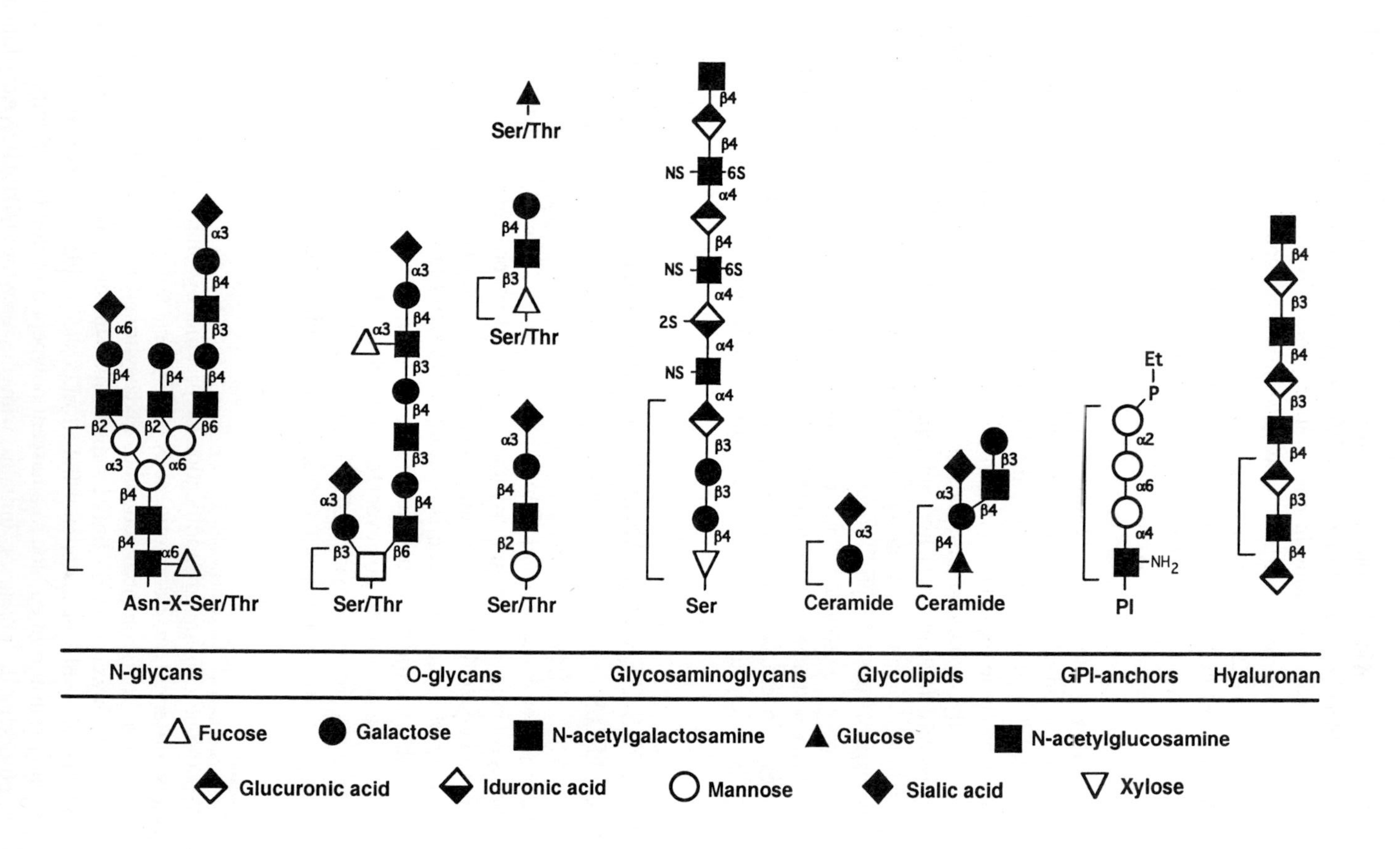
Asn-X-Ser/Thr
Ser/Thr
Ser/Thr
Ser/Thr
Ser/Thr
Ser
Ceramide
Ceramide
PI
Et
P
NH2
NS
2S
6S
N-glycans
O-glycans
Glycosaminoglycans
Glycolipids
GPI-anchors
Hyaluronan
Fucose
Galactose
N-acetylgalactosamine
Glucose
N-acetylglucosamine
Glucuronic acid
Iduronic acid
Mannose
Sialic acid
Xylose

phosphate N-acetylglucosamine-1-phosphate transferase (GPT). The addition of this oligosaccharide precursor to a nascent polypeptide chain is in turn followed by glucosidase- and mannosidase-dependent processing events that remodel the N-glycan.

These initial processing events are an integral part of a system of quality control in protein folding, characterized by calnexin- and calreticulin-dependent recognition of the terminal glucose moieties displayed by the asparagine-linked glycan (32, 33). The pivotal nature of GPT in N-glycan assembly and the role of these glycans in assisting protein folding imply an essential requirement for this enzyme in the elaboration of many, if not all, membrane-associated and secreted polypeptides, and perhaps a requirement in cellular homeostasis. Indeed, studies with tunicamycin, a GlcNAc analogue that is a competitive inhibitor of GPT, indicate that tunicamycin-dependent inhibition of GPT in cells or in animals is associated with severe cellular dysfunction and death (34–36). The role of GPT inhibition per se by tunicamycin in phenotype formation has been validated and made more precise by the construction and characterization of mice with a targeted deletion of GPT (37, 38). Homozygous GPT-null embryos become inviable at or before E5.5 of gestation, in association with degeneration of embryonic and extraembryonic cells. Nonetheless, such embryos can develop normally to the morula and blastocyst stages in vitro and implant on the uterine epithelium, presumably reflecting a maternal source for GPT early in development. These observations provide evidence that N-glycans are required for cellular viability and organismal function likely through essential processes occurring in the ER.

Studies with somatic cell glycosylation mutants demonstrate that genetic defects in Golgi-based glycosylation processes do not lead to cell intrinsic defects that preclude cell viability (11–13). However, such studies do not address the contribution of these processes to muticellular organismal viability and

Figure 1 The six different classes of mammalian glycans. Representative examples of each distinct class of mammalian glycans are schematically illustrated. In mammals, many alternative structures, both simpler and more complex, exist for N-glycans, O-glycans, glycolipids, and the glycosaminoglycans. Symbols are used to denote each component monosaccharide within each glycan and are identified in the legend. The anomeric nature of each glycosidic linkage is identified by an α or β symbol, which is adjacent to a number that identifies the carbon atom of the acceptor monosaccharide that is involved in the glycosidic linkage. N-sulfation is indicated by NS. O-sulfation is indicated by S, with a number that identifies the carbon atom of the O-sulfated monosaccharide to which the sulfate moiety is linked. A bracket adjacent to each representative glycan identifies the part that is generally common to that particular glycan class. Et-P denotes the ethanolamine phosphate moiety linked to the COOH-terminus (not shown) of GPI-modified proteins. PI denotes the phosphatidylinositol moiety and its fatty acyl group (not shown) that together promote membrane anchoring by insertion into the plasma membrane lipid bilayer.

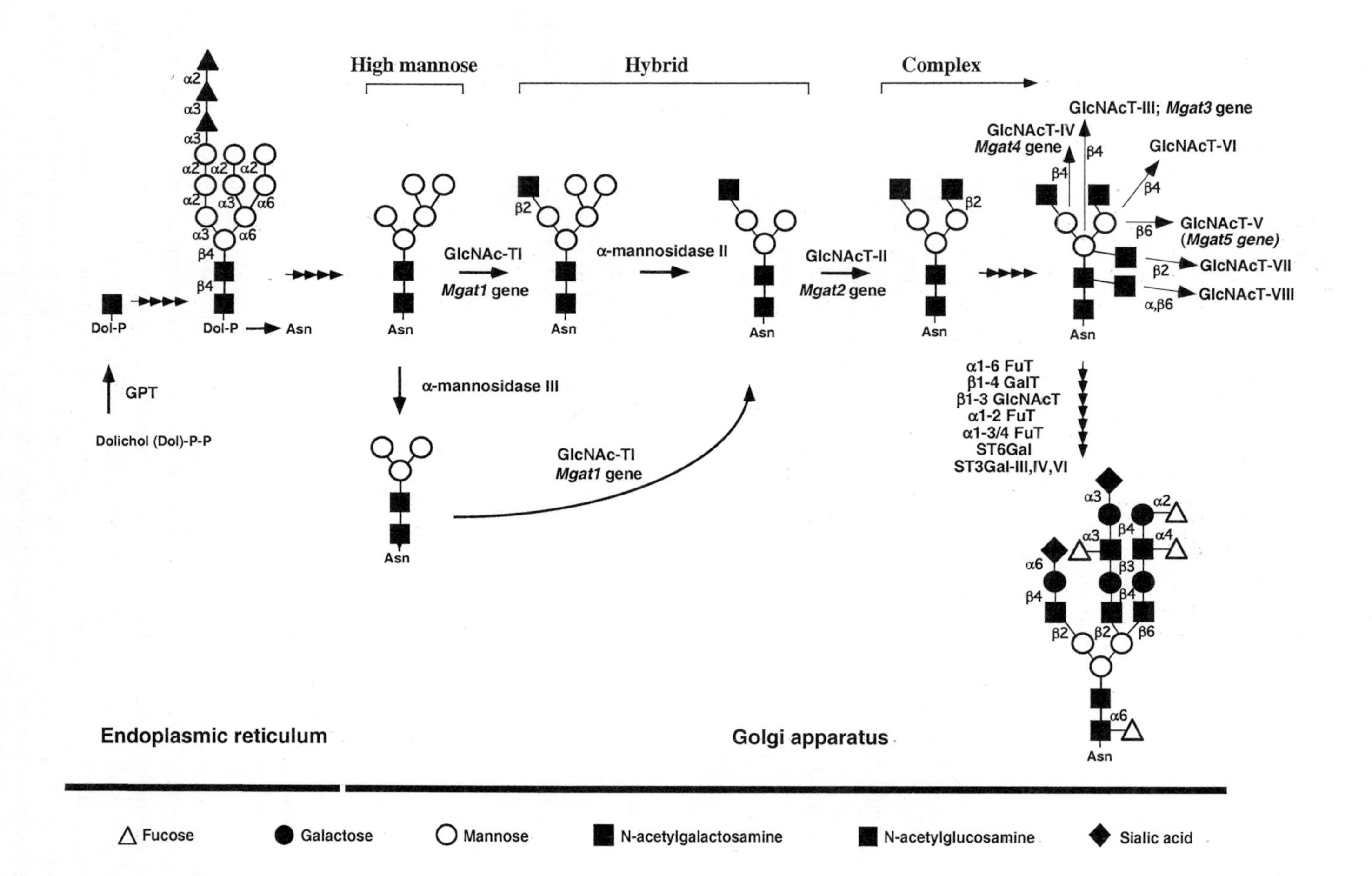
High mannose
Hybrid
Complex
Dol-P
Dol-P
Asn
GPT
Dolichol (Dol)-P-P
GlcNAc-TI
Mgat1 gene
α-mannosidase II
GlcNAcT-II
Mgat2 gene
α-mannosidase III
GlcNAc-TI
Mgat1 gene
GlcNAcT-III; Mgat3 gene
GlcNAcT-IV
Mgat4 gene
GlcNAcT-VI
GlcNAcT-V
(Mgat5 gene)
GlcNAcT-VII
GlcNAcT-VIII
α1-6 FuT
β1-4 GalT
β1-3 GlcNAcT
α1-2 FuT
α1-3/4 FuT
ST6Gal
ST3Gal-III,IV,VI
Endoplasmic reticulum
Golgi apparatus
Fucose
Galactose
Mannose
N-acetylgalactosamine
N-acetylglucosamine
Sialic acid

development. This issue has been addressed, in part, through the creation and analysis of mice that are deficient in the *Mgat1*-encoded GlcNAc-TI glycosyltransferase. This enzyme catalyzes an essential step in the conversion of high mannose to hybrid and complex N-glycans (Figure 1) (39, 40). *Mgat1*$^{+/-}$ embryos develop normally, and adult *Mgat1*$^{+/-}$ mice are indistinguishable from wild-type littermates. By contrast, *Mgat1*-null embryos become inviable at midgestation, evident by E9.5–E10.5, with morphogenic abnormalities including defects in neural tube formation and vascularization, as well as a variable frequency of situs inversus of the heart (41) becoming becoming evident by E9.5–E10.5 (14, 15, 42). *Mgat1*-deficient embryos also express complex N-glycans at E3.5 (preimplantation), although expression declines to an undetectable level by E7.5, coincident with the stage in embryogenesis where *Mgat1* transcripts begin to accumulate. Characterization of the glycans and *Mgat1* transcript levels in early embryogenesis discloses that maternally determined complex N-glycans, consequent at least in part to maternally derived *Mgat1* mRNA in preimplantation embryos, may sustain the viability of *Mgat1*-null embryos through preimplantation development, and that subsequent viability is compromised by embryonic deficiency of *Mgat1* and the corresponding deficiency of hybrid and complex N-glycans (14, 15, 42, 43). Roles for *Mgat1* in organogenesis have been further explored through the creation and analysis of chimeric embryos assembled with *Mgat1*-null embryonic stem cells (44). In chimeras analyzed at E10.5–E16.5, *Mgat1*$^{-/-}$ cells were observed to contribute to several different tissues, but not to the bronchial epithelium. These observations indicate an essential requirement for complex N-glycans in the establishment of this

---

Figure 2 Mammalian N-glycan synthesis. The schematic includes biosynthetic steps that correspond to mammalian mutants discussed in the text. The general scheme for mammalian N-glycan synthesis starts with dolichol pyrophosphate (Dol-P-P) (*left*), and ends with a final multiantennary structure (*bottom right*). This product is only one representative example of the numerous distinct N-glycans that exist in mammals. Potential branching patterns are represented by multiple arrows without labels and are made possible by the actions of several distinct GlcNAc transferases (GlcNAcT-III through -VIII) indicated in the structure (*top right*), together with the name of the responsible enzyme and genetic locus if defined. Further synthesis to form structures like the representative product shown is accomplished by multiple glycosylation events denoted by enzymes identified adjacent to the multiple thick arrows immediately preceding the product. The general category of each class of N-glycan (high mannose, hybrid, or complex) is identified at top. The general subcellular compartmental location of each segment of the N-glycan synthetic scheme (endoplasmic reticulum, Golgi) is identified at bottom. Glycosidic linkages are indicated as in Figure 1; these are shown fully in the Dol-P-linked structure (*left*), whereas in subsequent structures, only the linkage created in the preceding step is shown. Symbols used to denote each component monosaccharide are identified in the legend at bottom.

epithelial cell layer, although mechanisms to account for this phenotype remain to be identified.

Efforts to further refine the contributions of Golgi N-glycan structures to cellular and tissue function have included the construction and characterization of mice with a deficiency in α-mannosidase II (αM-II). This enzyme is a Golgi-localized processing glycosidase once thought to be uniquely responsible for trimming the hybrid N-glycan product of GlcNAcT-I prior to the formation of GlcNAcT-II-dependent synthesis of complex N-glycans (Figure 2) (45). αM-II-deficient mice are viable, fertile, exhibit normal behavior, and are morphologically indistinguishable from wild-type littermates (46). However, analysis of the cellular elements of the blood disclosed a dyserythropoietic anemia accompanied by splenomegaly. Surprisingly, even though all tissues in the *αM-II*-null mice were devoid of detectable αM-II activity, among the tissues analyzed, complex N-glycan expression was deleted only in the erythroid compartment, unveiling a previously undetected alternative pathway for the elaboration of GlcNAcT-II substrate catalyzed by another α-mannosidase activity termed αM-III (Figure 2). The alternative pathway does not, however, fully complement αM-II deficiency. As these mice age, a significant fraction of them begin to suffer from kidney dysfunction and a few die prematurely. This syndrome is characterized by immune complex glomerulonephritis with autoantibody formation, features that are reminiscent of human systemic lupus erythrematosus (47).

The alternate pathway in complex N-glycan formation may involve a previously identified alpha-mannoidase termed $\alpha-$mannosidase-IIx [αMX (48)]. The contribution of αMX in the control of N-glycan synthesis and function has been explored through the construction and characterization of αMX-deficient mice (49). These mice survive to term, although breeding experiments implied a deficiency in the function or production of spermatogenic cells. Females were fully fertile, whereas subfertility in αMX-null male mice was associated with testicular hypoplasia, impaired spermatogenesis, and failure of spermatogenic cells to adhere to Sertoli cells. The likelihood that αMX provides the αM-III activity disclosed in αM-II-deficient mice is increased by the overlapping phenotypes observed in comparisons with mice lacking the next step in N-glycan formation.

The enzyme termed GlcNAc-TII catalyzes the formation of complex N-glycans in the subsequent step following GlcNAcT-I and αM-II/αMX (Figure 2). Deletion of this gene [*Mgat2* (50)] in mice (51) models in many respects the human disease known as Congenital Disorders of Glycosylation type IIa (CDG-IIa), caused by mutational inactivation of the human *MGAT2* locus [reviewed in (52, 53)]. Biochemical analyses of these mice indicated a deficiency of GlcNAcT-II activity in all tissues (excepting possibly the intestine), and loss of complex type N-glycans associated with, in the kidney, a novel N-glycan structure bearing a Lewis x-modified branch on the bisected GlcNAc (the position modified by *Mgat3*; Figure 2). In a C57BL/6 genetic background, a modest degree of embryonic lethality is observed between E9 and E15, most live-born *Mgat2*-null animals perish during the first week of postembryonic life, and all die within 4 weeks of birth. By contrast, in an "outbred" ICR strain, mice lacking *Mgat2* function can survive with similar yet less severe disease signs, implying that one or more genetic modifiers exist. Overall, the *Mgat2*-null mouse largely recapitulates the phenotype of human CDG-IIa and may provide an opportunity to preview symptoms in aged CDG-IIa patients and explore therapeutic approaches to this disease.

N-glycan formation characterized by GlcNAc addition to the tri-mannosyl core can occur on both hybrid and complex types by *Mgat3*-encoded GlcNAcT-III (54), which provides the "bisecting" β1–4 GlcNAc linkage to the β-linked mannose of the tri-mannosyl core (Figure 2). Deletion of the mouse *Mgat3* locus yielded no reported phenotype (55). By contrast, a different *Mgat3* germline mutation in mice by gene insertion mutagenesis resulted in a neurological defect (56). This latter *Mgat3*-null allele has been termed $Mgat3^{t37}$ because it encodes an inactive GlcNAc-TIII of ~37 kDa. Analysis of $Mgat3^{\Delta}/Mgat3^{t37}$ compound heterozygotes disclosed that some aspects of the neurological deficits associated with homozygosity for the $Mgat3^{t37}$ allele, and absent from $Mgat3^{\Delta}/Mgat3^{\Delta}$ mice, are likely due to a combination of expression of the inactive truncated GlcNAcT-III protein in the brain, in conjunction with absence of wild-type GlcNAcT-III. Homozygosity for either allele is associated with diminished growth of diethylnitrosamine-induced hepatic neoplasms via mechanisms that remain to be defined (57, 58).

The tri-mannosyl core can also be modified by the addition of GlcNAc in β1–4 linkage to the α1–3-linked mannose through the action of at least two GlcNAcT-IV isoenzymes [GlcNAcT-IVa (59); GlcNAcT-IVb (60)]. The major role of these enzymes is to elaborate tri- and tetra-antennary N-glycans. The function of the glycan products of these two enzymes is not yet clear; *Mgat4a*-null mice are viable and unremarkable at 6 weeks of age (K. Ohtsubo & J.D. Marth, unpublished data), whereas *Mgat4b*-null mice have yet to be studied.

The *Mgat5*-encoded GlcNAcT-V glycosyltransferase generates a β1–6-linked GlcNAc branch from the α1–3-linked mannose (Figure 2) (61). GlcNAcT-V and its corresponding branched glycans have been studied extensively in the context of malignant transformation. N-glycans bearing a β1–6-linked GlcNAc branch are consistently elevated in concert with increased expression of GlcNAcT-V, and numerous direct correlations have been made between the presence of this N-glycan branch and enhanced tumor growth and metastasis [reviewed in (62)]. Adult *Mgat5*-null mice are devoid of detectable GlcNAcT-V activity and its glycan product and exhibit reduced leukocyte recruitment in inflammation, as well as defects in maternal nurturing behavior (63). To further define the contributions to tumor biology made by GlcNAcT-V and its β1–6-linked GlcNAc branched N-glycans, these mice were crossed with a strain of transgenic mice that express the Polyomavirus middle T antigen (PyMT) in mammary gland epithelium. In these mice, PyMT activates Ras- and Akt-dependent signaling pathways that transform breast epithelium and promote the metastasis of the resulting tumors. In this model, *Mgat5*-null mice experienced a significant delay in the onset of tumor formation, their tumors grew more slowly in vivo, and the incidence of lung metastasis was significantly decreased (63). Mechanisms that account for these observations are not yet understood.

Mgat5-null mice are also characterized by enhanced clustering of the TCR when presented with TCR agonist-coated beads and glomerulonephritis, suggesting a possible autoimmune disease (64). Enhanced TCR clustering is associated with increased TCR internalization, reorganization of the lymphocyte microfilaments, and enhanced intracellular signal transduction events. Exploration of this phenotype further suggested that in wild-type lymphocytes, *Mgat5*-dependent β1–6 GlcNAc branching on the TCR attracts members of the galectin family of carbohydrate-binding proteins. These proteins have relatively specific affinities for lactosamine chains that typically decorate β1–6-branched N-glycans, and may form a "multivalent galectin-*Mgat5*-dependent glycoprotein lattice" on the surface of a normal T cell. This lattice is hypothesized to actively restrain the mobility of the TCR within the plane of the membrane, thus restricting the threshold for TCR complex clustering and for TCR-initiated signal transduction events to situations where TCR-peptide-MHC interactions are appropriate (64).

Glycan structural analyses of chicken glycoproteins have identified a β1–4-linked branch on the α1–6-linked mannose of the tri-mannosyl core (65), and a corresponding GlcNAc transferase activity that creates the corresponding pentasaccharide was reported. The chicken enzyme, termed GlcNAcT-VI, has been purified and characterized recently and exhibits a requirement for N-glycans bearing the *Mgat5*-dependent branch (66). There is evidence for a corresponding pentasaccharide from analysis of human glycoproteins [discussed in (66)], but a mammalian GlcNAcT-VI activity has not yet been described. Recent biochemical analyses of lectin-resistant mutant CHO cell lines have identified GlcNAc transferases activities termed GlcNAcT-VII and GlcNAcT-VIII, with corresponding glycan structures (67) (Figure 2). Genes encoding these three enzymes are not yet identified, and the functions of their corresponding glycans remain undefined.

Considered together, studies involving genetic modulation of N-glycan structures in mice identify a hierarchical correlation between the breadth of biological contributions by specific types of N-glycans during development and the stage of their synthesis and diversification. Specifically, termination of N-glycan synthesis at a stage prior to the formation of hybrid N-glycans is incompatible with embryonic development, indicating that these molecules and their products are essential in ontogeny. By contrast, embryogenesis is authorized in genetic circumstances that permit synthesis of high-mannose and hybrid N-glycans while blocking formation of complex N-glycans. These circumstances are nevertheless incompatible with some major aspects of development and normal function in the adult, engender the CDG-IIa syndrome accompanied by a high frequency of postnatal lethality, and unveil evidence that glycan-dependent disturbances in tissue morphology or function can be modulated by variation at other genetic loci. Finally, deletion of single N-glycan branching moieties distal to the transition to complex N-glycan formation (*Mgat3*, *Mgat4a*, *Mgat5*), or deletion of a class of N-glycans in specific tissues (*αM-II*, *αMX*), yields phenotypes that range from undetected to moderate.

These observations and accompanying experimental evidence indicate that the correspondingly limited subset of N-glycans dictated by these enzymes has evolved to effect specific functions in certain cells or tissues. Novel or atypical glycan structures have been found in association with instances where N-glycan structure has been perturbed, indicating biochemical compensation by the remaining glycan synthetic processes, which has in some instances revealed structures present normally at very low levels. Similar considerations apply when other classes of mammalian glycans are genetically perturbed, as is discussed below.

O-Glycans

Some serine and threonine residues in proteins are subject to covalent modification by mono- and/or oligosaccharide chains. Classically, these so-called O-linked glycans initiate with GalNAc modification of serine or threonine (68). However, recent research indicates that some mammalian glycoproteins contain serine and threonine that are modified by fucose, glucose, mannose, and N-acetylglucosamine. This section focuses first on genetic experiments where the classical O-glycans have been experimentally ablated in mice. Additional coverage of mutations takes into account other classes of O-linked glycans, including O-linked xylose, which lies within the province of a subsequent section on glycosaminoglycans.

Modification of specific serine and threonine residues by GalNAc in α linkage represents the initiating step in classical O-glycan synthesis (Figure 3). Genes encoding more than nine functional mammalian polypeptide GalNAc transferases have been identified (ppGalNAcT-1 through ppGalNAcT-9), and the enzymes that they encode have been characterized to variable degrees [reviewed in (69–71)]. Despite some degree of tissue-specific expression of these genes in embryos and in the adult mammal, in general, each tissue or cell type expresses more than one such gene (72). Members of this family of enzymes exhibit distinct substrate specificities dictated by the nature of the polypeptide sequence context of the serines and threonines that are modified, and by an interesting hierarchical mode of operation dictated by the presence of other neighboring O-glycans (69, 73–76). The overlapping substrate specificities, and partially redundant expression patterns that characterize this family, likely account for the absence of any detectable phenotype and for the retention of essentially wild-type levels of ppGalNAc-T activity and cell surface O-glycan levels in mice with homozygous deletion of the murine ppGalNAcT-8 locus, either in T cells or in all somatic tissues (77). More recently, however, deletion of ppGalNAcT-8 locus assigns this enzyme an essential role in Tn antigen expression in early postnatal brain development (78). The significant degree of overlap in expression patterns and substrate specificities among the members of this family present a formidable

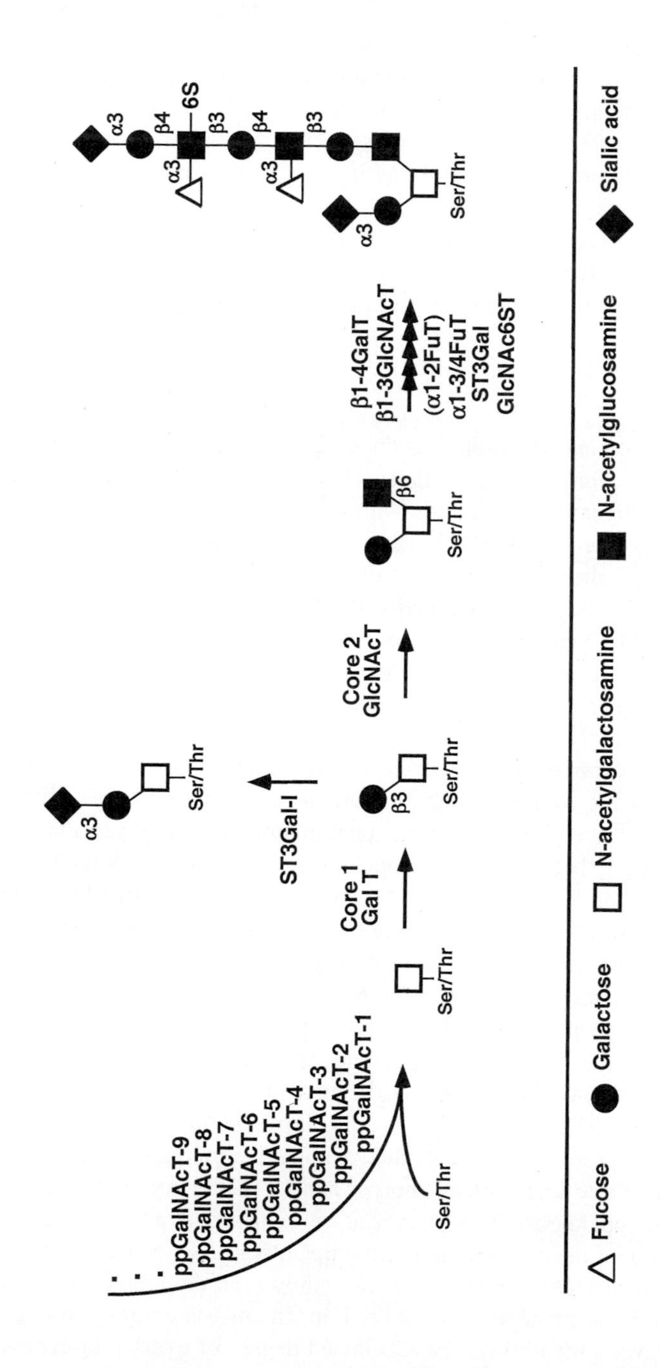
ppGalNAcT-1
ppGalNAcT-2
ppGalNAcT-3
ppGalNAcT-4
ppGalNAcT-5
ppGalNAcT-6
ppGalNAcT-7
ppGalNAcT-8
ppGalNAcT-9
Ser/Thr
Ser/Thr
Core 1
Gal T
β3
Ser/Thr
ST3Gal-I
α3
Ser/Thr
Core 2
GlcNAcT
β6
Ser/Thr
β1-4GalT
β1-3GlcNAcT
(α1-2FuT)
α1-3/4FuT
ST3Gal
GlcNAc6ST
α3
β4
6S
α3
β3
β4
α3
β3
α3
Ser/Thr
Fucose
Galactose
N-acetylgalactosamine
N-acetylglucosamine
Sialic acid

challenge to define the function of individual and multiple ppGalNAc-T genes.

O-glycan chain initation by the ppGalNAc-Ts is followed by modification involving distinct branch-specific glycosyltransferases. Glycosidases are not known to be required for O-glycan synthesis, as they are in N-glycan synthesis. A subset of mammalian O-glycans, termed Core 2 O-glycans, are highly abundant. These have been studied extensively because of their regulated expression in association with lymphocyte activation [reviewed in (79)], and because they contribute to the activities of counter-receptors for the selectin family of cell adhesion molecules [reviewed in (80)]. Core 2 O-glycan synthesis is controlled by three different Core 2 β1–6GlcNAc transferases, termed Core 2 GlcNAcT-I, -II, and -III (81–83). The biology of the Core 2 O-glycans elaborated by Core 2 GlcNAcT-I has been explored in mice with targeted deletion of this locus (84–87). Mice that are homozygous null for the *Core 2 GlcNAcT-I* locus exhibit no overt abnormalities, whereas splenocytes are devoid of both Core 2 GlcNAcT-I activity and Core 2 O-glycans (84). The manifestation of a leukocytosis focused further characterization primarily on control of E-, P-, and L-selectin ligand formation.

In Core 2 GlcNAcT-I-null mice, neutrophil P-selectin ligand activity and P-selectin-dependent cell adhesion are virtually absent (84–86), consistent with the role for Core 2 GlcNAcT activity in the reconstitution of P-selectin counter-receptor activity in cultured cell lines [reviewed in (80)]. L-selectin ligands that may mediate neutrophil-dependent neutrophil tethering under shear (80) are also essentially inactive on Core 2 GlcNAcT-I-null neutrophils. By contrast, Core 2 GlcNAcT-I-type O-glycans make less of a contribution to E-selectin ligands on mouse neutrophils. In vitro, E-selectin ligand defects range from subtle to considerable, probably reflecting methodological differences in the in vitro adhesion assays used in these studies (84, 85). A third study in which intravital

Figure 3 Mammalian O-glycan synthesis. The scheme identifies specific steps in that correspond to mammalian mutants discussed in the text. These involve Core 1 and Core 2 subtypes as illustrated, and they start with the addition of GalNAc to some serines and threonines that is common to GalNAc-initiated O-glycan synthesis. GalNAc addition is catalyzed by one of several distinct polypeptide GalNAc transferases (ppGalNAc-Ts). Subsequent synthetic steps are illustrated here only by way of representative examples of many other possible pathways not shown. Core 1 glycans can be directly sialylated, as shown, or can be elongated and then sialylated (not shown). Core 1 O-glycans can also be modified to become Core 2 O-glycans, which may then be further modified by the actions of other glycosyltransferases and sulfotransferases. The final synthetic structure shown is only one representative example of the numerous distinct O-glycans that exist in mammals. Structures and synthetic pathways corresponding to serine and threonine-linked glycans that initiate with mannose, fucose, glucose, or N-acetylglucosamine are not illustrated. Glycosidic linkages are indicated as in Figures 1and 2. Symbols used to denote each component monosaccharide are identified in the legend at bottom.

microscopy was used to examine E-selectin-dependent adhesion in vivo (86) assigns a modest contribution to E-selectin-dependent leukocyte rolling. These studies indicate that Core 2 GlcNAcT-I is virtually essential to neutrophil P-selectin ligand formation in the mouse neutrophil but makes a less prominent contribution to E-selectin ligands. Intravital microscopy also identifies a significant role for Core 2 GlcNAcT-I in the elaboration of as yet unidentified L-selectin ligands that are displayed by inflamed venules (87).

Unexpectedly, Core 2 GlcNAc-T-null mice exhibited only a slight reduction in L-selectin counter-receptor activity on the high endothelial venules (HEV) of their peripheral nodes (84). These mice also retain essentially normal L-selectin-dependent lymphocyte homing efficiencies, and maintain a nearly normal number of lymphocytes in peripheral nodes, mesenteric nodes, and Peyer's patches, but Core 2 O-glycans are absent from their HEV (88). Glycan structural analyses disclose that retention of L-selectin ligands in these mice may be accounted for by compensatory elaboration of extended Core 1-type O-linked glycans bearing 6-sulfo sialyl Lewis x capping groups (88). These studies identified the 6-sulfated, extended core 1 glycan Galβ1–4[6-sulfo]GlcNAcβ1–3Galβ1–3GalNAcα-Ser/Thr as the chemical structure of the glycan epitope recognized by the MECA-79 monoclonal antibody, a widely used reagent that binds to HEV (80).

Forms of O-glycosylation distinct from those determined by the ppGalNAcTs are initiated on serine- and threonine-linked fucose, glucose, and mannose. Fucose-initiated O-linked chains are catalyzed by an enzyme encoded by a recently cloned GDP-fucose protein O-fucosyltransferase locus (*POFUT1*) and best characterized in the context of serines and threonines within epidermal growth factor-like repeat domains (EGF repeats) [see (51, 89) and references therein]. EGF domains within the Notch family of transmembrane signal transduction proteins contain O-linked fucose moieties that are elongated by the fringe family of β1–3GlcNAc transferases (90–93). EGF domains on Notch ligands are also O-fucosylated and can be elongated by fringe (94). Each of the three fringe enzymes (Radical fringe, Rfng; Lunatic fringe, Lfng; and Manic fringe, Mfng) requires O-linked fucose as a substrate and forms a disaccharide that is further elongated by at least one specific β1–4 galactosyltransferase (95) and then by the subsequent actions of other glycosyltransferases.

Fringe-dependent elongation of fucosylated EGF domains on Notch can modulate the signal transduction properties of Notch when Notch interacts with its ligands (91–93). Corresponding biological consequences can also be strongly modulated [for an example in mammals, see (96); for a review of fringe-dependent control of morphogenesis in *Drosophila*, see (97)], through mechanisms that may involve glycan-dependent alterations on the strength of interactions between Notch and its ligands (90; reviewed in 98) or that may modulate the *trans*-endocytic processes that are apparently required for Notch signaling in some contexts [reviewed in (99)]. Targeted inactivation of the *Lunatic fringe* locus in the mouse is associated with perinatal death; *Lfng*-null embryos display defective somite segmentation, which is presumed to lead to the distortions of the axial skeleton characteristic of these mice (100, 101). By contrast, abnormalities have not been observed in mice with an inactivation of the *Rfng* locus (102), and mice deficient in the *Mfng* locus have not been described. The *Brainiac* locus also regulates some Notch-dependent biological events in *Drosophila* and shares sequence similarity with glycosyltransferase loci. Mammalian homologues have been identified; some exhibit β1–3-N-acetylglucosaminyltransferase activity (103, 104). Deletion of one of these homologues in the mouse [a murine enzyme termed β3-GalT-III, now known to be a β1–3GlcNAc transferase (104); encoded by a locus termed *Brainiac 1*, or *Mbrn 1* in (105)] is associated with embryonic lethality, as null blastocyts are unable to implant. It remains to be determined if this enzyme, or other members of this family, elongates glycans associated with Notch or its ligands, and if so, whether such *Braniac*-mediated glycosylation modulates Notch signaling, as is observed with the fringe family members.

Glucose is also found as a covalent modification of serine and threonine residues of some proteins, including EGF domains of Notch, and may be elongated by subsequent glycosylation (106, 107). A corresponding protein O-glucosyltransferase activity has been reported (107), but biological functions for this modification are not yet identified. O-mannosylation of serine and threonine residues is present if not abundant among mammals (108). However, an elongated O-mannose-linked glycan (Siaα2–3Galβ1–4GlcNAcβ1–2Man-Ser/Thr) is enriched in muscle tissue and on α-dystroglycan, a component of the dystrophin-glycoprotein-complex that is altered in human muscular dystrophies (109). Subsequent to mannosylation, POMGnT1 is a glycosyltransferase that catalyzes the linkage of N-acetylglucosamine to O-mannosyl-Ser/Thr. Recent work identifies inactivating point mutations in the human *POMGnT1* locus as causal in human muscle eye brain disease (MEB) (110), an autosomal recessive disorder characterized by congenital muscular dystrophy, ocular abnormalities, and lissencephaly. Deletion of a putative glycosyltransferase locus (*LARGE*) in the myodystrophy mouse (111), mutations in a putative glycosyltransferase locus (*fukutin*) in Fukuyama congenital muscular dystrophy (112), and mutations in the fukutin-related protein locus (*FKRP*) in congenital muscular dystrophy MDC1C (113) are also implicated in the pathogenesis of these muscular dystrophies. Underglycosylation of α-dystroglycan is characteristic of each of these disorders; underglycosylated α-dystroglycan is incapable of engaging in productive adhesive interactions with its extraceullular matrix ligands laminin, neurexin, and agrin (110, 114), and accounts for the dystrophic muscle and neuronal migration phenotypes that typify these syndromes.

Glycolipids

Glycosphingolipids or glycolipids are relatively abundant components of mammalian cell plasma membranes. Glycolipids are characterized by a membrane-embedded ceramide moiety that anchors the associated, extracellular glycan component of the glycolipid to the cell. As with N- and O-linked glycans, the

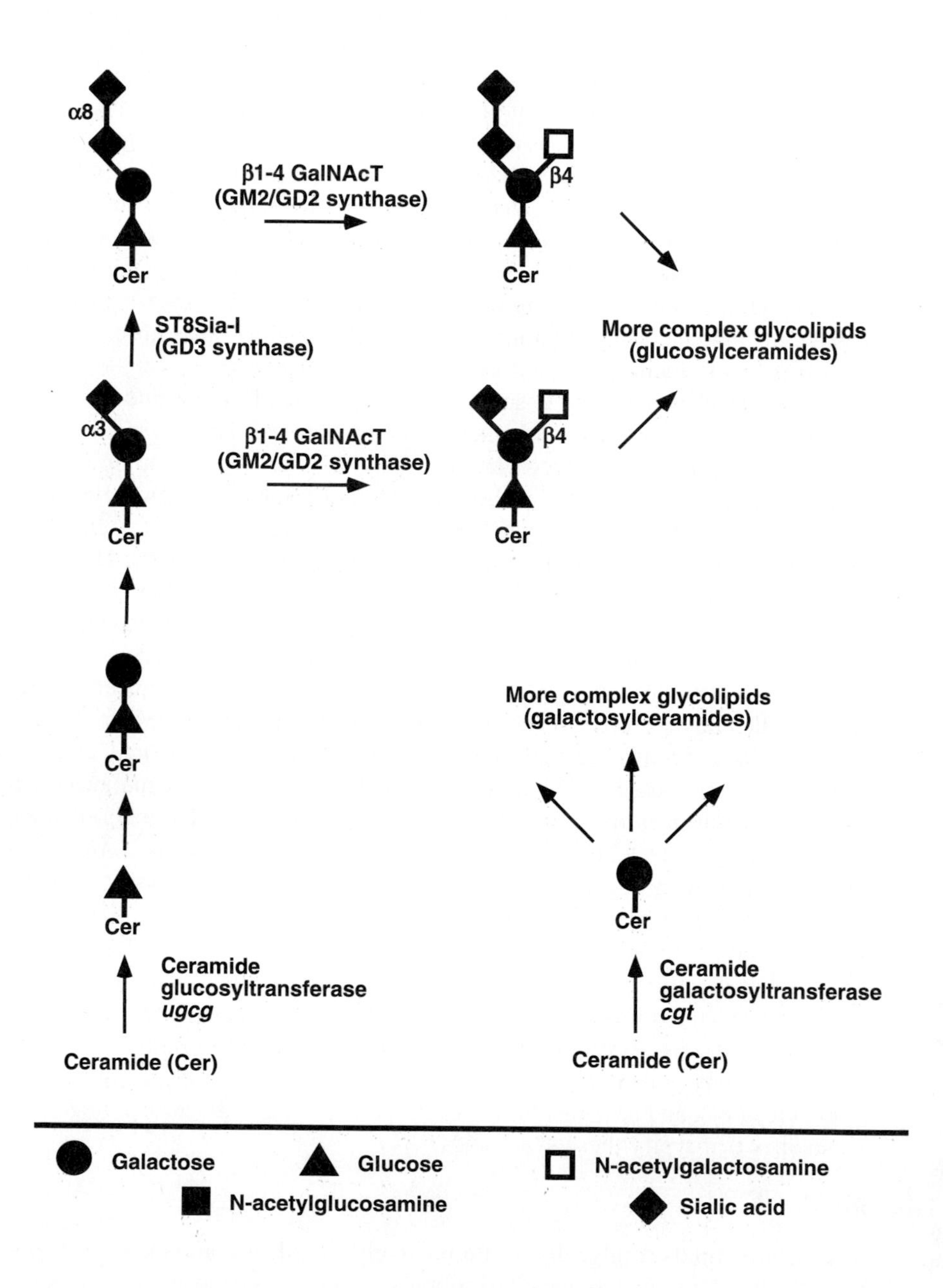
α8
β1-4 GalNAcT
(GM2/GD2 synthase)
β4
Cer
Cer
ST8Sia-I
(GD3 synthase)
More complex glycolipids
(glucosylceramides)
α3
β1-4 GalNAcT
(GM2/GD2 synthase)
β4
Cer
Cer
More complex glycolipids
(galactosylceramides)
Cer
Cer
Cer
Ceramide
glucosyltransferase
ugcg
Ceramide
galactosyltransferase
cgt
Ceramide (Cer)
Ceramide (Cer)
Galactose
Glucose
N-acetylgalactosamine
N-acetylglucosamine
Sialic acid

sequential actions of distinct glycosyltransferases elaborate structurally diverse, linear, and branched chain glycans that characterize the glycolipids (Figure 4) [reviewed in (115)]. Neutral and sialylated glycolipids exist in mammals; the sialylated forms, often termed gangliosides, are relatively enriched in neuronal tissues. As can be expected for a class of abundant, structurally diverse membrane-associated molecules, the glycolipids have been implicated in a diverse array of cell biological processes, including signal transduction and cell-adhesion events essential to the function of cells, tissues, and intact organisms (4).

Glucosylceramide comprises the core of most glycolipids in mammals. Glucosylceramide is synthesized in the cytosol by glucosylceramide synthase (116), which transfers the glucose moiety from UDP-glucose to ceramide. Glucosylceramide is subsequently translocated to the Golgi lumen, where it is elongated by Golgi-localized glycosyltransferases. In mice, homozygosity for a null allele otherwise encoding the glucosylceramide synthase locus (*Ugcg* locus) leads to embryonic lethality at gastrulation (between E6.5 and E7.5), and is associated with an increased apopotic rate in cells of the embryonic ectoderm (117). Defective differentiative events were observed in vitro and in vivo in cells homozygous for the mutant *Ugcg* locus, together indicating that meaningful differentiation is impaired in the absence of glucosylceramide and subsequent glycolipid synthesis.

Galactosylceramide synthase [UDP-galactose:ceramide galactosyltransferase, CGT (118)] initiates the elaboration of the less abundant glycolipids, which include galactosylceramide and its sulfated derivative, sulfatide. Inactivation of the *Cgt* locus in mice disables the synthesis of galactosylceramide and its downstream product, sulfatide. *Cgt*-null mice are grossly normal until postnatal days 12–16, followed by the onset of tremor, and progressive dysfunction of the hind limbs until death between 18 and 90 days (119–121). Nerve conduction is disrupted in these animals, and is associated with localized myelin defects, superimposed on a relatively normal myelin histology. Regional defects in myelin integrity occur in these mice, presumably consequent to the deficiency of

Figure 4 Mammalian glycolipid synthesis. The two basic schemes for mammalian glycolipid synthesis are illustrated, involving the addition of either glucose or galactose to ceramide. The scheme identifies specific steps in glycolipid biosynthesis that correspond to mammalian mutants discussed in the text. Glucose or galactose addition to ceramide is catalyzed, respectively, by the actions of glucosylceramide synthase (ceramide glucosyltransferase encoded by the *ugcg* locus) or galactosylceramide synthase (ceramide galactosyltransferase encoded by the *cgt* locus). The products of these two reactions may then be modified by subsequent sequential actions of numerous other glycosyltransferases. The figure illustrates synthesis of three sialylated forms of glucosylceramide-based glycolipids (gangliosides) that are discussed in the text. Glycosidic linkages are indicated as in Figures 1 and 2. Symbols used to denote each component monosaccharide are identified in the legend at bottom.

galactosylceramide and sulfatide. Defective sulfatide synthesis in the testes of these mice is associated with arrest of spermatogenesis and apopototic degeneration of spermatogenic cells (122). The enzyme GM2/GD2 synthetase (123) occupies a pivotal position in the complex ganglioside synthesis pathway (Figure 4). Targeted inactivation of this locus in mice leads to a deficiency of all complex gangliosides (124), but initial evaluations, made in mice at 10 weeks of age, found no histological defects and identified only subtle nerve conduction deficits. Subsequent analyses of an independently derived strain of GM2/GD2 synthetase-null mice, focusing on mice at 12–16 weeks of age, identified decreased central myelination, axonal degeneration in the peripheral and central nervous systems, and decreased myelination in peripheral nerves (125). These defects progress with age, eventually leading to serious motor and behavioral deficits (126), and are presumed to be consequent to disruptions of the interactions between complex gangliosides and myelin-associated glycoprotein. GM2/GD2 synthetase-null mice also exhibit a defect in the transport of testosterone from the Leydig cells to the blood and seminiferous tubules (127).

Considered together, these observations identify essential roles for glucosylceramide-based glycolipids in development and in cellular and tissue function. Mechanistic insights into how each of the many molecules in this arm of the glycolipid synthetic pathway contributes to these processes are not yet available. By contrast, the complex gangliosides and galactosylceramide-based glycans generally appear to make contributions to cellular and tissue physiology that are subtle, or that may be subsumed by compensatory molecules, including those of the glucosylceramide-based pathway. The physiological impact of compensatory interactions that occur between these two glycolipid pathways remains uncertain.

Glycosaminoglycans

Glycosaminoglycans are also often termed proteoglycans; however, this latter term actually refers to a protein that is modified by one or more glycosaminoglycan chains. Glycosaminoglycans are comprised of four distinct types of sulfated, linear oligosaccharide polymers (Figure 5). Glycosaminoglycan synthesis may be generally characterized by chain initiation, polymerization, and modification, via the actions of different glycosyltransferases, sulfotransferases, and epimerase and deacetylase activities. Glycosaminoglycans are implicated in growth factor binding, presentation, and internalization, as well as in cell adhesion and the maintenance of extracellular matrix integrity [reviewed in (128–130)]. Mutational inactivation of genes required for glycosaminoglycan synthesis currently involves those relevant to heparan sulfate synthesis. EXT1 and EXT2 are GlcNAc and GlcA glycosyltransferases that contribute to the biosynthesis of heparan sulfate (HS) glycosaminoglycans (131–133). Haploinsufficency for either locus in humans causes a disease known as hereditary multiple exostoses, or HME. These patients are typically of short stature, and suffer from multiple bony exostoses in childhood. Mice with an induced null mutation in the *EXT1* locus (Figure 5) (134) have been constructed and characterized (135). In contrast to $EXT1^{+/-}$ humans with HME, $EXT1^{+/-}$ mice do not manifest exostoses, despite a reduced ability to synthesize heparan sulfate (135). Homozygosity for the *EXT1* null allele in mice deletes the synthesis of heparan sulfate and leads to embryonic lethality at gastrulation; extraembryonic tissues are absent, and the mesoderm in these mice is severely disrupted. The roles of heparan sulfation have also been explored in mice by targeted inactivation of genes encoding two of the four known glucosaminyl N-deacetylase/N-sulfotransferases (NDSTs). These enzymes deacetylate N-acetylglucosamine residues and then modify them with sulfate, and their activities are required for many subsequent steps in heparan sulfate biosynthesis (130). In *NDST-2*-null mice, only mast cells appear to manifest a phenotype (136, 137). Sulfated heparin is absent from these cells, and apparently participates in the control of mast cell granule formation since *NDST-2*-null mast cells have few granules, and the content of granule-associated proteases and histamine is decreased. By contrast, N-sulfated heparan sulfate is largely normal elsewhere in these mice. *NDST-2* appears to be the only *NDST* locus expressed in mast cells, thus accounting for the observed deficits in these cells. By contrast, although *NDST-2* is widely expressed elsewhere in wild-type mice, retention of essentially normal heparan sulfate synthesis is observed in other tissues in NDST-2-null mice, presumably the result of one or more of the other NDSTs. N-sulfation of heparan sulfate meanwhile is severely reduced in virtually all tissues of *NDST-1*-null mice (138, 139). Reduced N-sulfation is accompanied only by an inability of the type II epithelial cells in the lung to elaborate surfactant, yielding a postnatal respiratory distress-like syndrome and death shortly after birth. Mice with deficiencies in both NDST-1 and NDST-2 apparently die early in embryogenesis [unpublished observations cited in (129)], identifying a requirement for N-sulfated heparan modification in the murine developmental program and functional overlap by the two enzymes.

Glucuronic acid moieties within the products of the NDSTs are epimerized to iduronic acid; some of these iduronic acid residues, and some remaining glucuronic acid moieties, are sulfated by a 2-O-sulfotransferase activity. A random gene trap event was discovered to have inactivated the only *2-O-sulfotransferase* locus (*Hs2st*) identified thus far in the mouse. Mice that are homozygous for a null allele exhibit renal agenesis due to failed ureteric bud branching and mesenchymal condensation (140). Variable frequencies and/or severity of limb and eye formation are also observed, and the mice die in the perinatal period. The structure of heparan sulfate chains in these mice is not yet known. As with the other heparan sulfation-defective mice, mechanisms to account for the phenotypes in the *2-O-sulfotransferase*-null mice remain to be discovered. Some aspects of the phenotypes may possibly result from disruption of heparan sulfate-growth factor complexes, typified by HS-FGF2 and HS-HGF interactions, that are postulated to be necessary for physiologic dimerization and activation of their receptors [reviewed in (141)].

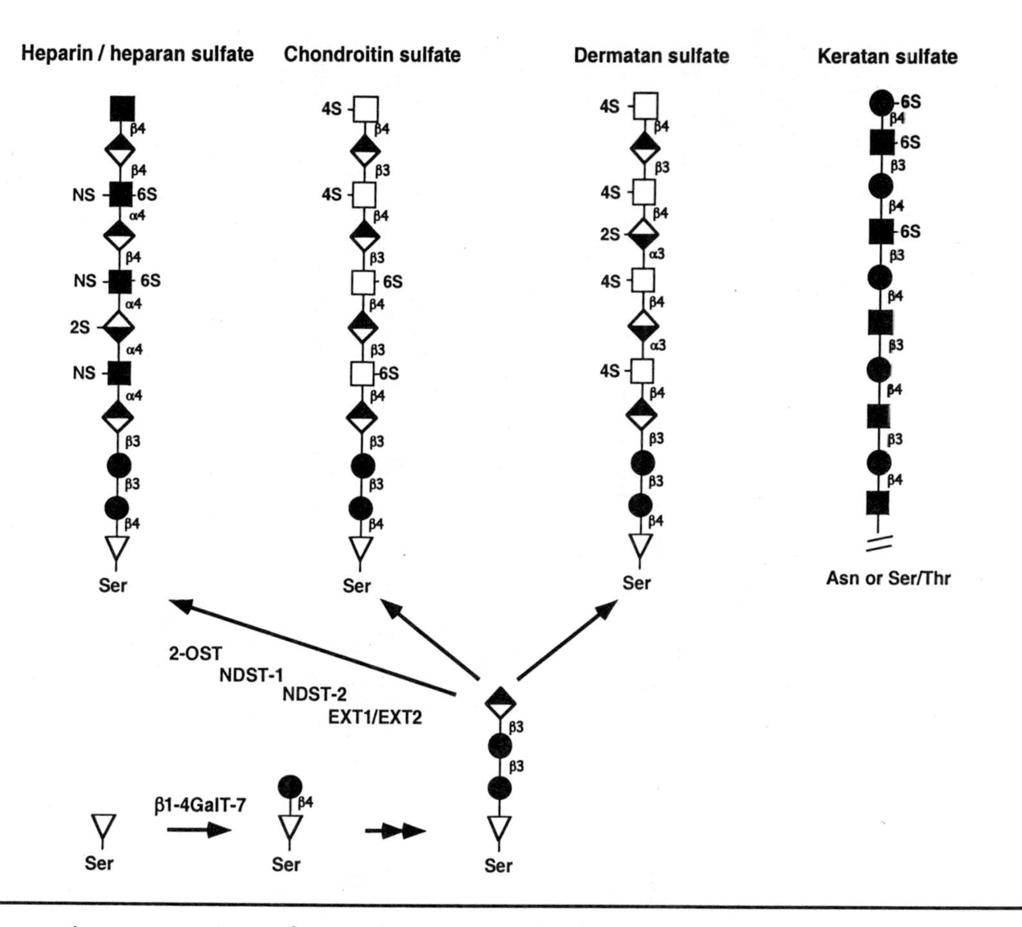

Heparin / heparan sulfate
Chondroitin sulfate
Dermatan sulfate
Keratan sulfate
NS
2S
6S
4S
Ser
Asn or Ser/Thr
2-OST
NDST-1
NDST-2
EXT1/EXT2
β1-4GalT-7

Galactose
Glucuronic acid
Iduronic acid
N-acetylgalactosamine
N-acetylglucosamine
Xylose

Glycophosphatidylinositol Linkages

Glycophosphatidylinositol (GPI)-linked proteins are anchored to the cell surface by an inositol phosphate attached to a glycan that is, in turn, linked to a phosphatidylinositol-linked fatty acyl chain residing in the outer leaflet of the plasma membrane [reviewed in (142, 143)]. GPI-linked proteins are widely expressed in animal tissues. As is the case with other glycan-containing polymers, the glycan linkages comprising the glycan moiety are elaborated by the sequential actions of specific enzymes, including several glycosyltransferases.

The human disease known as paroxysmal nocturnal hemoglobinuria (PNH) is caused by a somatic mutation in the X-linked *PIG-A* locus [reviewed in (144)], which encodes the catalytic subunit of the enzyme complex that forms GlcNAc-phosphatidylinositol, the first intermediate in GPI anchor formation (145). *PIG-A* mutation and deficiency lead to the absence of GPI-linked complement-regulating molecules CD59 and decay accelerating factor, and increased susceptibility

Figure 5 Mammalian glycosaminoglycan synthesis. An abbreviated schematic outline of mammalian glycosaminoglycan synthesis is illustrated [reviewed in (128, 130)]. The scheme identifies specific steps in heparan sulfate/heparin biosynthesis that correspond to mammalian mutants discussed in the text. Xylosyltransferase modification of specific serines on proteoglycan core proteins initiates glycosaminoglycan synthesis. Xylosyl residues are elongated by the sequential actions of β1–4galactosyltransferase (β1–4GalT-7, also known as GalTI), β1–3galactosyltransferase (also known as GalTII), and β1–3glucuronosyltransferase (also known as GlcATI) to form the core protein linkage tetrasaccharide common to heparin, heparan sulfate, chondroitin sulfate, and dermatan sulfate. Sulfation of the linkage tetrasaccharide is not shown. Phosphorylation of this tetrasaccharide has also been reported. Modification by β1–4-linked GalNAc addition initiates assembly of chondroitin sulfate and dermatan sulfate, which are not discussed further. Modification of the core tetrasaccharide by α1–4-linked GlcNAc addition initiates assembly of heparan sulfate. In heparan sulfate assembly, α1–4-linked GlcNAc addition to the core tetrasaccharide (by one or more EXT-like enzymes, or EXTLs) is followed by alternating addition of β1–4-linked GlcA and α1–4-linked GlcNAc residues, catalyzed by EXT1 and EXT2. Polymerization is associated with chain modification reactions characterized by N-deacetylation/N-sulfation (catalyzed by NDST-1 and NDST-2), epimerization that forms iduronic acid, and several types of O-sulfation, including nonstoichiometric 2-O sulfation of iduronic acid catalyzed by 2-OST. The single arrow underlining these enzymes serves to emphasize that these reactions are likely to take place more or less simultaneously. Keratan sulfate, representing variably sulfated N- and/or O-linked polymers of the polylactosamine disaccharide (Galβ1–4GlcNAcβ1–3) is elaborated by synthetic processes largely distinct from those relevant to heparin, heparan sulfate, chondroitin sulfate, and dermatan sulfate, and are shown only for structural comparison. Glycosidic linkages are indicated as in Figures 1 and 2. Symbols used to denote each component monosaccharide are identified in the legend at bottom.

to complement-mediated red cell lysis. Mouse ES cells with an inactivated *Pig-A* locus are apparently unable to contribute in a robust way to chimericism in mice. However, mice that are mosaic for *Pig-A* deficiency maintain significant numbers of circulating GPI-deficient erythrocytes. These cells have an increased sensitivity to complement-mediated lysis and shortened life span, but the erythroid lineage does not exhibit clonal expansion as occurs in human PNH. GPI-deficient cells within the lymphoid lineage are without apparent abnormality in such experiments, whereas keratinocyte-specific deletion of the *Pig-A* locus is associated with aberrant skin histopathology. These studies [(146–149); reviewed in (150)] imply that clonal expansion of *Pig-A*-deficient cells in human PNH may not result from GPI-dependent mechanisms, and indicate tolerance for loss of GPI-anchoring in the development of some, but not all, complex tissues. Mutational studies of other loci in the GPI synthetic pathway have not yet been reported.

Hyaluronan

Hyaluronan (HA) is a high-molecular-weight glycosaminoglycan composed of the repeating disaccharide glucuronic acid β1–3-N-acetylglucosamine β1–4. HA is a ligand for CD44 and RHAMM, two cell adhesion molecules that also participate in signal transduction events. HA synthesis is determined by at least three distinct HA synthase genes (*Has1*, *Has2*, *Has3*) in humans and mice [reviewed in (151)]. Deletion of the Has-2 locus in mice leads to loss of HA during embryogenesis, and is associated with embryonic lethality at E9.5–10. Lethality is characterized by absence of cardiac endothelial transformation to mesenchyme and deficiency in formation of the AV cushions that precede morphogenesis of the AV canal and other major cardiac structures (152). Defects in HA- and Ras-dependent signal transduction processes likely contribute to these defects (152). By contrast, normal embryonic development apparently proceeds in mice deleted for either the *Has1* or *Has3* loci, or both [unpublished data cited in (152)]. Functions for the *Has2* locus at later stages of development remain to be identified.

Catabolism of HA is thought to be effected by the actions of one or more of six hyaluronidase-like genes that exist in both human and mouse genomes [reviewed in (153)]. Humans with an inactivating mutation in the *HYAL1* locus suffer from Mucopolysaccharidosis IX, but no other mammalian mutants have yet been reported.

Galactosylation

Galactose may be considered a component of core or proximal glycan linkages because it is virtually always present in β1–3 or β1–4 anomeric linkage to an underlying GlcNAc moiety and is often itself "masked" by the actions of glycosyltransferases that use it as a glycan acceptor. A number of endogenous lectins exist in mammals that can bind terminal GlcNAc and mannose-terminated glycans, and some function in facilitating the innate immune response to microbial infection. A large family of galactosyltransferases has been identified and characterized [reviewed in (154)].

Only one of these genes, encoding the β1–4 galactosyltransferase-1 (GalT-1) (7), has been mutated in the mouse (155–159), in part to help confirm its suspected major role in the elaboration of so-called type II β1–4 galactosylated N-glycans (Galβ1–4GlcNAc-) and to explore nonenzymatic roles in cell adhesion assigned to an isoform of this protein (155, 156). GalT-1-null mice survive to term at a normal frequency, and although male and female adults are fertile, they exhibit growth retardation. Most die prematurely, owing to incompletely characterized defects in growth and differentiation of epithelia, and exhibit signs of endocrine insufficiency. Structural analysis of the glycans in these mice indicates that on glycoproteins from red cells, hepatocytes, and plasma proteins and the liver, the normally predominating type II chains are dramatically reduced in concert with an apparently compensatory increase in type I chains (Galβ1–3GlcNAc-). The sialylation pattern is also altered, with repression of the normally dominant α2–6-linked sialylation characteristic of type II chains, and marked increase in α2–3-linked sialylation characteristic of type I chains. It remains to be determined if or how such changes contribute to the phenotypes in these mice.

FUNCTIONS OF TERMINAL GLYCAN LINKAGES AND MODIFICATIONS

Terminal glycan linkages and modifications to terminal linkages may be distinguished from core linkages by virtue of their most distal positions in biosynthesis and because these linkages are often expressed in a lineage-specific manner, in contrast to core linkages, which are generally more likely to be expressed in many cell lineages and tissue types. There is obviously overlap in these characteristics. For example, some core glycans are elaborated by isoenzymes with tissue-specific expression patterns, and their glycan products may be "exposed" without typical terminal linkages. A strict distinction between terminal and core glycan linkages is not therefore always possible. Nevertheless, sialylation and fucosylation represent prominent and definitive examples of terminal linkages. Occasionally, galactose and GalNAc modifications may also be found at the termini of glycan chains. This section also focuses on experiments in which these structures have been genetically ablated in vivo to expose physiologic functions.

Sialylation

Sialylation of glycans is characteristic of all tissues in mammals and is represented by sialic acid in α2–3, α2–6, or α2–8 linkage to a variety of underlying glycan precursors. Among all mammals studied, humans appear unique in the inability to form the N-glycolyl type of sialic acid (NeuGc). This is due to germline mutations in the gene encoding CMP-N-acetylneuraminic acid hydroxylase (28, 29). There is no known effect of this mutation. A large family of sialyltransferases is responsible for glycan sialylation; 20 are presently known, and these exhibit unique substrate specificities as well as distinct and dynamically regulated expression patterns.

Some N-glycans are characterized by terminal α2–6-sialylation of subterminal β1–4-linked galactose moieties (6). The sialyltransferase ST6Gal-I is responsible for this specific type of α2–6-linked sialylation. This sialic acid linkage contributes essentially to counter-receptor activities for CD22, a member of the SigLec family that down-modulates B lymphocyte function. A role for terminal α2–6-linked sialic acid in immune system function has been identified through the creation and analysis of mice with a deletion of the *ST6Gal-I* locus (160). Mice lacking this sialyltransferase are overtly normal in all respects; however, B lymphocytes in these mice do not bind CD22. Unlike CD22-deficient mice [reviewed in (161)], loss of ST6Gal-I results in a profound immunosuppression characterized by attenuated humoral immune responses of B lymphocytes, defective tyrosine phosphorylation following IgM cross-linking, and an apparent inability to elaborate T-dependent and T-independent antibody responses. The altered phosphotyrosine patterns in the B cells of this animal imply that α2–6-linked sialylation of type II N-glycan chains modulates signal transduction events required for B cell function.

Sialylation of the Core 1 O-glycan Galβ1–3GalNAc-α-Ser/Thr on thymocytes during the cortical to medullary transition of these cells is a widely studied example of a precisely regulated glycosylation event. The nonsialylated glycan is recognized by peanut agglutinin (PNA) (as are cortical thymocytes), whereas the sialylated product (Siaα2–3Galβ1–3GalNAc-α-Ser/Thr) is not, and medullary thymocytes do not bind PNA. The sialyltransferase ST3Gal-I has been implicated in the control of this process, as it sialylates the Core 1 O-linked glycan Galβ1–3GalNAc in vitro, and the expression pattern of the ST3Gal-I in the thymus varies inversely with PNA binding (162). Functions for the sialylated glycan elaborated by ST3Gal-I have been explored in mice lacking a functional ST3Gal-I enzyme (163). *ST3Gal-I*-null mice are grossly normal, as is thymic histology; however, the medullary thymocytes in these mice bind PNA, indicating loss of the terminal α2–3 linked sialic acid on Core 1 O-glycans that normally masks PNA binding in wild-type medullary thymocytes. These mice exhibit a marked decrement in the number of $CD8^+$ T lymphocytes in peripheral compartments, which is associated with increased rates of apoptosis and reduced cytotoxic activity upon tumor challenge. The phenotype is intrinsic to the lymphoid lineage and is accompanied by a compensatory increase in the expression of Core 2 O-glycans but without an increase in Core 2 GlcNAcT activity, revealing a synthetic block in Core 2 O-glycan biosynthesis by expression of ST3Gal-I. The proapopotic phenotype of the ST3Gal-I-null CD8 cells can be recapitulated in vitro by O-glycan cross-linking, and is suppressed by caspase inhibition as well as prior immune activation.

Considered together, these observations have invoked a model in which regulated desialylation of Core 1 O-glycans in the postactivation phase of an immune response induces apoptosis by default in the absence of further stimuli or immune memory cell differentiative signals, thereby resetting the number of $CD8^+$ T cells to a homeostatic level during the resolution of an immune response (163).

Regulated expression of ST3Gal-I has also been implicated in the control of thymocyte differentiation involving peptide-MHC class I presentation (164). These studies disclose that ST3Gal-I-dependent sialylation of Core 1 O-glycans reduces the avidity by which MHC class I molecules interact with the CD8$\alpha\beta$ coreceptor. Cortical thymocytes lack sialylated Core 1 glycans on the CD8 beta chain since ST3Gal-I is not yet expressed at this stage of thymocyte development. By contrast, transition to the medulla with T cell differentiation is associated with induction of ST3Gal-I, expression of α2–3 sialylated Core 1 O-glycans and a reduction in MHC class I binding avidity by CD8 heterodimers. Desialylation of medullary thymocytes, as well as a genetic deficiency of ST3Gal-I, enhances CD8$\alpha\beta$ binding avidity to MHC class I. These and related observations indicate that ST3Gal-I-regulated Core 1 O-glycan sialylation affects T cell receptor-dependent thymic selection and may diminish the likelihood of autoreactive T lymphocyte formation.

ST3Gal-I is one member of a family of at least six distinct ST3Gal sialyltransferases (ST3Gal-I through -VI). Mice deficient in ST3Gal-II, ST3Gal-III, and ST3Gal-IV have been constructed and are the subject of ongoing analyses. ST3Gal-II exhibits a preference for type III chains (Galβ1–3GalNAc-), with a notable preference for glycolipid substrates (165). ST3Gal-III can utilize type II chains, but is most active with type I chains (166). Mice deficient in either enzyme are born at normal rates and have not yet been reported to display abnormal phenotypes (167). ST3Gal-IV utilizes type I and type II glycan substrates without a marked preference for either (168, 169). Mice with a deletion of the *ST3Gal-IV* locus are also produced in normal Mendelian ratios from heterozygous intercrosses, are fertile, and grossly normal (167, 170).

Two notable phenotypes have been reported in *ST3Gal-IV* null mice. One is characterized by a bleeding disorder evident in both heterozygotes and homozygotes and associated with markedly decreased circulating levels of von Willebrand factor (vWF) akin to low vWF levels observed in human von Willebrand disease (171). The autosomal nature of the vWF deficiency in the absence of ST3Gal-IV is accounted for by a role in modifying the glycans of vWF. ST3Gal-IV normally sialylates murine vWF, and loss of ST3Gal-IV is associated with increased exposure of terminal galactose moieties, which are recognized by the asialoglycoprotein receptor (172, 173). Unexpectedly, loss of ST6Gal-I also reduced vWF sialylation and exposed galactose residues, albeit without a decrease in vWF levels or increased clearance by asialoglycoprotein receptors.

The low vWF levels, its altered glycosylation, and the associated bleeding disorder are consequent in these mice to increased asialoglycoprotein-dependent clearance of circulating vWF. This syndrome appears to be mimicked in some human patients with abnormally low levels of vWF (170).

ST3Gal-IV null mice also exhibit a leukocyte adhesion defect characterzed by diminished P- and E-selectin ligand activities on neutrophils analyzed by cytometry or in vitro rolling on CHO cells bearing either P- or E-selectins (167), again in contrast to other sialyltransferase-deficient mice studied thus far. Interestingly, in vivo analysis of leukocyte rolling on endothelium in a TNF-α induced inflammation model revealed a defect in E-selectin-mediated rolling, but indicated essentially normal levels of P-selectin ligands. Flow cytometric studies indicated further reductions in P- and E-selectin ligand activities following sialidase treatment of ST3Gal-IV-null neutrophils, implying that other ST3Gal loci, likely involving ST3Gal-VI, are responsible for a significant fraction of selectin ligand synthesis.

ST8Sia-I is one of six sialyltransferases presently identified in the mammalian genome that add sialic acid in α2–8 linkage to glycan chains of glycolipids or glycoproteins. In particular, ST8Sia-I, also termed GD3 synthase, specifically contributes to a key step in glycolipid formation (Figure 4) (174). In the absence of ST8Sia-I, the formation of more complex glycolipid structures requires the presence of β1–4 GalNAcT (GM2/GD2 synthase) (175). Mice deficient in ST8Sia-I were found to exist in normal Mendelian frequencies among littermates in the absence of detectable GD3 synthase activity (175). The animals were viable and fertile, without observable behavorial defects, and did not show indications of demyelination in the brain. By breeding mice lacking ST8Sia-I into the β1–4GalNAcT (GM2/GD2 synthase)-deficient background, it was possible to ablate the formation of all types of complex glycolipids, terminating glycolipid synthesis subsequent to the formation of of GM3 (Figure 4). Such mice recapitulated the GM2/GD2-deficient phenotype (as described above in Glycolipids), but in addition had a high mortality rate. These double-null mice were found to be sensitive to audiogenic and handling-induced seizures in which few recovered and death typically ensued (175). Histologic analysis of the central nervous system did not find notable defects, and the cause of the seizures is not known, although aberrant modulation of various signal transduction systems due to the reduced diversification of glycolipids was postulated.

The presence of α2–8-linked sialic acid was initially described on the neural cell adhesion molecule (NCAM) as an abundant polymer of α2–8-linked sialic acid repeated 50 or more times, and comprising a significant fraction of molecular weight and negative charge associated with NCAM at specific developmental stages. This modification has been termed polysialic acid (PSA) and has been observed on a subset of neuronal glycoproteins. PSA appears to reduce the homotypic binding of NCAM, yet PSA can also enhance adhesion involving other membrane-bound counter-receptors. Experiments involving NCAM bearing various PSA levels suggested a role for PSA in modulating axonal guidance during innervation and in the migration of neuronal cells during neural development (176–178). Efforts to further explore PSA function were subsequently facilitated by the cloning of the gene encoding ST8Sia-IV, a polysialic acid transferase also termed PST (179–181).

The production and characterization of mice bearing a homozygous mutation in the gene encoding ST8Sia-IV revealed a significant role for PSA in synaptic plasticity (182). These mice bore normal anatomic features, including a typical olfactory bulb that was found poorly developed in NCAM-deficient mice (183). Olfactory precursor cells continued to express PSA and migrate normally. However, *ST8Sia-IV*-null mice exhibited deficits in long-term potentiation (LTP) and long-term depression (LTD) among the Schaffer collateral-CA1 synapses of the hippocampus concident with a decrease of PSA. No effect was noted among hippocampal mossy fiber-CA3 synapses, and no evidence of delamination of these fibers was found, unlike results obtained among NCAM-deficient mice (183, 184). These findings provided discrimination between the roles of NCAM and PSA in neuronal processes, revealing a key function of PSA in LTP and LDP, which are processes involved in the formation of memory, and indicated the continued presence of PSA in some brain regions and developmental periods, likely due to the presence of other ST8Sia isozymes that exhibit PSA synthesis activity in vitro.

Fucosylation

In mammals fucose is found as a "Core" O-glycan on serine and threonine residues, as discussed above. Fucose also modifies the chitobiose core of N-glycans, in α1–6 linkage (Figure 2). This modification is constructed by fucosyltransferase VIII (FUT8); however, functions for this glycan are not known, and publications describing *Fut8*-deficient mice are not yet available (185). α1–2-, α1–3-, and α1–4-linked fucose also decorates terminal and subterminal substitutents of N-, O-, and lipid-linked glycans. Some α1–2-, α1–3-, and α1–4 fucosylation events are polymorphic in humans, where they contribute to the H, Secretor, Lewis, and ABO blood groups [reviewed in (186)].

Functions for α1–2-fucosylated glycans are being sought in mice with targeted deletions of two α1–2 fucosyltransferase loci (*Fut1* and *Fut2*) (187). These mice are born at normal rates and exhibit no gross physical or behavioral abnormalities. In both strains, males and females are fully fertile, eliminating a postulated requirement for α1–2-fucosylated glycans in implantation and sperm maturation. It remains to be determined if these mice will manifest abnormalities in other organ systems where α1–2-fucosylated glycans are expressed and inferred to have function.

Sialylated α1–3-fucosylated glycans have been implicated as essential components of the counter-receptors for E-, P-, and L-selectins [reviewed in (80)]. Two human α1–3-fucosyltransferases, termed FUT4 and FUT7 (also termed FucT-IV and FucT-VII; 188–190), are conserved among higher mammals, are expressed in cell types that elaborate selectin ligands, and have thus been implicated in the synthesis of the α1–3-linkedfucose associated with selectin counter-receptors. This has been explored in mice with targeted inactivation of the *Fut4* and *Fut7* loci, and in mice deficient in both genes (188–190). All three strains of mice are born at normal rates and are without gross postnatal behavioral abnormalities. *Fut7*-null mice exhibit a moderately severe deficiency of E- and P-selectin ligand activity on neutrophils, monocytes, Th1 and Tc1 lymphocytes, and L-selectin ligand activity on HEV. Residual selectin ligand activities in these mice are elaborated by Fut4 since the doubly deficient mice are virtually devoid of all E-, P-, and L-selectin ligand activities. Contributions by Fut4 to selectin ligand activities are largely superseded by Fut7, since selectin ligand activities are only subtly deficient in *Fut4*-null mice. E- and P-selectin ligand deficiencies in these mice are accompanied by correspondingly depressed leukocyte recruitment in inflammation. Deficiencies in HEV-borne L-selectin ligand activities, and thus in homing of T and B cells to peripheral lymph nodes, are predicted to compromise adaptive immunity in these mice. Roles for the α1–3-fucosylated glycans elaborated by these two enzymes may extend to the control of myelopoiesis (191) through mechanisms that remain to be defined.

Two human disorders have been reported that are associated with altered expression of sialyl Lewis x-type glycans. In one such instance, heterozygosity and homozygosity for an inactivating missense mutation in the *FUT7* locus was associated with diminished expression of the sialyl Lewis x epitope on neutrophils (192). Residual sialyl Lewis x activity was ascribed to FUT4-dependent synthesis and may account for apparent lack of a leukocyte adhesion phenotype in such individuals. In another instance, macrothrombocytopenia and neutropenia with a complete lack of sialyl Lewis-x antigen on leukocytes has been described (193). The molecular defect to account for this phenotype has not yet been defined.

Biological roles for the murine α1–3-fucosyltransferase Fut9 (194) remain to be defined, and the phenotype of Fut9-deficient mice has not yet been reported. A pair of mammalian proteins with primary sequence similarity to the α1–3-fucosyltransferases have been identified in a genome database search and have been termed FUT10 and FUT11 (195). Catalytic specificities for this pair of "orphan" glycosyltransferases, and functions for any cognate glycans, will be the subject of further study.

Sulfation

Glycan sulfation is characteristic of the glycosaminoglycans, as noted above, but is also represented among other classes of glycans. Glycan sulfation is elaborated by at least 30 sulfotransferases with distinct and overlapping catalytic specificities, and idiosyncratic tissue-specific expression patterns that predict substantial diversity in glycan sulfation [reviewed in (196, 197)]. Glycan sulfotransferases are distinct from tyrosylprotein sulfotransferases that sulfate some tyrosine residues (198). All sulfotransferases utilize 3′-phosphoadenosine 5′-phosphosulfate (PAPS) as a sulfate donor. In mammals, PAPS is synthesized by bifunctional enzymes known as PAPS synthetases. Mice with brachymorphism have a disabling mutation in the *PAPS synthase-2* locus (*SK2*) (199). These mice have a normal life span, but exhibit postnatal growth retardation associated with aberrant morphology of epiphyseal growth plates and undersulfated cartilage glycosaminoglycans. Other glycan-independent, sulfation-dependent phenotypes are also observed. Redundancy afforded by a multiplicity of *PAPS synthetase* loci likely allows PAPS synthesis to proceed in most tissues in the absence of SK2 activity.

There are four examples currently where the functions of glycan sulfation have been explored by genetic deletion of sulfotransferase loci. Three of these enzymes [NDST-1 (135), NDST-2 (138, 139), heparan-2-O-sulfotransferase (140)] have been described above in the context of the glycosaminoglycans. The fourth concerns one member of a family of galactose/N-acetylgalactosamine/N-acetylglucosamine 6-O-sulfotransferases (GSTs), now termed GST-3 (197) [once known as LSST (200) and as HEC-GlcNAc6ST (201)]. The catalytic and expression pattern characteristics of GST-3 had assigned to it a possible role in the elaboration of HEV-borne ligands for L-selectin, which require sulfation for optimal activity (200, 201). An in vivo role for GST-3 in the control of L-selectin ligand activity and in other functions has been explored by creating and characterizing *GST-3*-null mice (202). *GST-3*-null mice are born at normal rates, and adults are fertile and grossly normal. In peripheral lymph nodes, GST-3 deficiency is characterized by a reduction in the activity of L-selectin counter-receptor activity on the lumenal face of HEV, with retention of activity on the ablumenal aspect. Loss of lumenal activity is associated with a prominent reduction in homing of lymphocytes to the peripheral nodes in these mice and a corresponding deficit in the nodal lymphocyte content. The prominent contribution by GST-3 to L-selectin counter-receptor activity disclosed by these studies corresponds to its role in the synthesis of the 6-sulfo sialyl Lewis x moiety, a sulfated O-glycan moiety that decorates L-selectin counter-receptors CD34, GlyCAM-1, and other such glycoproteins. A significant level of GST-3-independent L-selectin counter-receptor activity is evident in these mice, however. The nature of the glycans that contribute to this activity and whether and how these may be sulfated remain to be explored. Other functions for GST-3-dependent sulfation have also not yet been identified.

In humans, defects in a sulfate anion transporter (diastrophic dysplasia sulfate transporter or DTDST) and consequent undersulfation of GAG sulfation account for dwarfism, spinal deformation, and joint abnormalities in chondrodysplasias termed diastrophic dysplasia (DTD) (203), atelosteogenesis type II (AO2) (204), multiple epiphyseal dysplasia (MED) (204), and achondrogenesis type 1B (ACG1B) (205, 206). Marked variation in clinical severity is observed in these disorders, reflecting the degree of residual sulfate transport activity afforded by each mutant transporter allele.

Macular corneal dystrophy (MCD), an autosomal recessive human disease characterized by progressive corneal opacity, has been associated with sequence polymorphism in, and deletions of, the *N-acetylglucosamine-6-sulfotransferase* locus termed *CHST6* (207). It is not yet known how the observed sequence polymorphisms in this locus may modulate the activity of this enzyme and how the predicted undersulfation of keratan sulfate may contribute to the phenotype in this disease.

Other Terminal Structures

The α-Gal epitope (Galα1–3-Galβ1–4GlcNAc-R, where R can be a glycolipid or glycoprotein) is a common carbohydrate structure expressed in many tissue in most in mammals [reviewed in (208)]. It is absent from cells and tissues of Old World monkeys, apes, and humans. Absence of the α-gal epitope in these species is accounted for by inactivation of the *α1–3GalT* locus that remains active in other mammals (27). Species that lack a functional *α1–3GalT* locus maintain circulating levels of naturally occurring IgM α-Gal antibody. The α-Gal determinants expressed by animal organs are the target of antibody-mediated hyperacute graft rejection in xenotransplantation. Functions for this glycan have been sought by analysis of mice with targeted mutation of the *α1–3Gal-T* locus (9) responsible for α-Gal synthesis. These mice are distinguished from wild-type mice only by the absence of α-Gal epitopes on their cells, by the variable presence of anti-α-Gal in their serum, and by cataracts (209, 210). Mechanisms to account for the cataracts in these mice and functions for the α-Gal structure are not yet evident.

The glycan structure (GalNAcβ1–4[NeuAcα2–3]Galβ1–4GlcNAc-R) is polymorphic in human owing to the presence or absence of the terminal β-linked GalNac residue and defines the Sda blood group locus (211). In mice, this determinant is known as the CT1 or CT2 determinant, defined by monoclonal antibodies that recognize this glycan on cytotoxic T lymphocytes (CTL) and that block CTL-dependent killing of target cells. Synthesis of the CT1/CT2 epitope is controlled by tissue-specific, regulated expression of a β1–4GalNAc transferase (Galgt2) (211). This enzyme decorates Siaα2–3Galβ1–4GlcNAc-R structures common to many N- and O-linked glycans. Among some strains of mice, decreased levels of vWF in the circulation are associated with aberrant expression of this locus in endothelial cells where vWF is synthesized. In these strains, this leads to modification of vWF-associated glycans with the CT1/CT2 epitope and rapid glycan-dependent clearance of vWF from the circulation (212). Functions for this epitope in the immune system and other contexts are under exploration with *Galgt2*-null mice (J.B. Lowe, unpublished data).

CONSEQUENCES OF SUGAR NUCLEOTIDE DEFECTS

Mice have been constructed with a targeted deletion of the *FX* locus, which encodes an enzyme required for the constitutive pathway for GDP-fucose synthesis (191). These mice exhibit a conditional defect in the formation of

GDP-fucose, the required substrate for GDP-fucose protein O-fucosyltransferase that intiates fringe-modified, Notch-associated glycan modification. The defect in GDP-fucose synthesis in these mice can be circumvented by fucose supplementation via a salvage pathway for GDP-fucose synthesis. These mice exhibit a strain-dependent embryonic lethal phenotype of as yet undetermined cause (191). *FX*-null newborns develop into essentially normal adults when their diets are supplemented with fucose. By contrast, in the absence of exogenous fucose, the glycans in these mice are deficient in fucose (with consequent deficiencies in E-, P-, and L-selectin ligand activities), and they suffer from a colitis of the large bowel. These mice are predicted to be deficient in O-linked fucose on Notch (and its ligands), but this remains to be confirmed experimentally. Also remaining to be determined is how such fucose-deficiency and the loss of fringe-dependent elongation processes may contribute to the phenotypes in these mice.

Every mammalian sialyltransferase utilizes the substrate CMP-sialic acid, whose synthesis is elaborated in the cytosol from UDP-N-acetylglucosamine by four enzymatic steps. The first of these steps is catalyzed by the bifunctional enzyme UDP-GlcNAc 2-epimerase. A requirement for sialylation in mouse development has been demonstrated by the observation that homozygosity for an inactivating mutation in the mouse *UDP-GlcNAc 2-epimerase* locus is associated with embryonic lethality evident between E8.5 and E9.5 of gestation (213). A viable homozygous null ES cell line isolated in these experiments retains expression of small amounts of cell surface sialic acid, thought to be derived via catabolism of endocytosed sialylated glycans in the culture media.

ESSENTIAL ROLE OF NUCLEAR AND CYTOPLASMIC GLYCOSYLATION

Several glycan modifications have been reported to exist in the nucleus and cytoplasm of mammalian cells, including O-mannose, O-fucose, and nuclear glycosaminoglycans (214). Little is known of where these modifications are initially produced, or whether they have a role in cell metabolism. However, a significant amount of information has been obtained regarding a highly expressed nuclear and cytoplasmic form of glycosylation involving N-acetylglucosamine linkage to serine and threonine residues of proteins, termed O-GlcNAc [reviewed in (215, 216)].

Protein O-GlcNAc is abundant in the cytoplasm and nucleus of mammalian cells, and this modification is also found among invertebrates, although not in yeast. Studies by Hart and colleagues, who discovered this modification, have disclosed its close relationship to protein phosphorylation. Sites of O-GlcNAc addition are frequently also sites of phosphate modification, suggesting a role for O-GlcNAc in modulating signal transduction systems within cells. Like phosphate addition, O-GlcNAc is a reversible modification, and recent studies have reported the isolation of genes encoding both the O-GlcNAc transferase (OGT)

TABLE 1 Genetic defects in mouse glycan formation and physiologic consequence

Enzyme	Phenotype associated with enzyme deficiency	Reference(s)
GlcNAc-1-phosphotransferase	Embryonic lethality (E4.5)	(38)
GlcNAcT-I	Embryonic lethality (E9.5) with defects in vascularization, neural tube formation, and situs inversus of the heart	(42–44)
α-mannosidase-II	Dyserythropoiesis and SLE-like autoimmune disease	(46, 47)
α-mannosidase-IIx	Spermatogenic failure, male sterility	(49)
GlcNAcT-II	Frequent postnatal lethality with defects in multiple physiologic systems Survivors phenocopy human CDG-IIa disease	(51)
GlcNAcT-III	Neurological deficit and tumor inhibition reported upon gene truncation No defect yet reported upon complete gene deletion	(55, 56)
GlcNAc-IVa	Viable, under further study	(K. Ohtsubo & J.D. Marth, unpublished data)
GlcNAcT-V	Immune dysfunction with hyperactive T cells and intestinal hyperplasia	(63, 64)
Polypeptide GalNAcT-1	Viable/under further study	(77)
Polypeptide GalNAcT-8	Viable/under further study	(78)
Core 2 GlcNAcT-I	Myeloid leukocytosis and inflammation deficit	(84–88)
Polypeptide-O-fucosyltransferase	Not reported	(51, 89)
Lunatic fringe β1-3GlcNAcT	Defective somite formation, axial skeleton deformities, perinatal death	(100, 101)
Radical fringe β1-3GlcNAcT	Viable, no reported phenotype	(102)
Manic fringe β1-3GlcNAcT	Not reported	(100–102)
Braniac 1 β1-3GlNAcT	Defective implantation with early embryonic lethality	(105)
Myd gene (putative glycosyltransferase POMGNT1)	Neuro-muscular disease similar to human muscular dystrophy	(111)
Ceramide glucosyltransferase	Embryonic lethality at gastrulation (E6.5 to E7.5)	(117)
Ceramide galactosyltransferase	Myelin abnormalities, paralysis, and postnatal death; arrested spermatogenesis	(119–122)
β1-4GalNAcT (GM2/GD2 synthase)	Myelin and axonal degeneration, male sterility with defective testosterone transport	(125–127)
EXT1 α4GlcNAc & β4GlcA transferase	Embryonic lethal, gastrulation defect	(135)
N-deactylase/N-sulfotransferase-1 (NDST-1)	Defective pulmonary surfactant formation, postnatal respiratory distress and lethality	(138, 139)
N-deactylase/N-sulfotransferase-1 (NDST-2)	Defects in mast cell heparin synthesis and granule formation	(136)
Heparan sulfate 2-sulfotransferase (HS-2OST)	Renal agenesis with neonatal lethality	(140)
GPI synthesis: X-linked *Pig-A*	Lethality if mutation is germline or prevalent in somatic tissue	(146–149)
Hyaluronan synthase-2 (Has-2)	Embryonic lethality (E9.5); defects in cardiac morphogenesis and mesenchyme formation	(152)
β1-4 GalT-1	Multiple defects including epithelial and endocrine abnormalities, frequent postnatal lethality; CDG-IId defect in humans	(155–159)
ST6Gal-I	Immunodeficiency with attenuated B cell function	(160)
ST3Gal-I	Cytotoxic T cell apoptosis, enhanced CD8ab-MHC class I binding	(163, 164)
ST3Gal-II	Viable/under further study	(167, 170)
ST3Gal-III	Viable/under further study	(167, 170)
ST3Gal-IV	Inflammation response deficit, vWF and platelet deficiencies	(167, 170)
ST8Sia-I (GD3 synthase)	Viable alone. Lethal audiogenic seizures in collaboration with GM2/GD2 synthase deficiency	(175)
ST8Sia-IV (PST)	Abnormal neuronal development and LTP deficit	(182)
ST8Sia-II (STX)	Viable, under further study	(J.D. Marth, unpublished data)

TABLE 1 Continued

Enzyme	Phenotype associated with enzyme deficiency	Reference(s)
FUT1	Viable, under further study	(187)
FUT2	Viable, under further study	(187)
FUT4	Partial neutrophil adhesion deficit, subtle selectin ligand defects	(188–190)
FUT7	Defective selectin ligands, neutrophil adhesion, and lymphoid homing	(188–190)
FUT8	Not yet reported	
FUT9	Not yet reported	
GST-3 (LSST, HEC-GlcNAc6ST)	Moderate defect in lymphocyte homing and HEV-borne L-selectin ligands	(202)
β1-3 GalT	Cataracts, model for hyperacute acute xenograft rejection	(209, 210)
Galgt2 β1-4GalNAcT	Viable, under further study	(212, J.B. Lowe, unpublished data)
GDP-4-keto,6-deoxymannose epimerase/reductase (FX)	Strain-dependent embryonic lethality; block in GDP-fucose synthesis conditionally defective fucosylation in all tissues	(191)
UDP-GlcNAc 2-epimerase	Embryonic lethality; bock to CMP-sialic acid synthesis with sialic acid deficiency	(213)
O-GlcNAc transferase: X-linked *Ogt*	Lethality unless mutation is paternally acquired in females	(221)

and the O-GlcNAcase (OGA) (217–219). The OGT enzyme contains tetratricopeptide repeats and is tyrosine phosphorylated, suggesting involvement in protein-protein association and signal transduction events modulated by tyrosine phosphorylation.

Numerous functions for O-GlcNAc have been proposed with various degrees of accompanying evidence, including regulation of transcription (220), nuclear importation, glucose homeostasis, cytoskeletal function, and neurologic disease etiology. To gain further insight into the physiology of O-GlcNAc, a mouse model of OGT deficiency was generated. Characterization during breeding, followed by karyotypic analyses, revealed that the gene encoding OGT was on the X chromosome (Xq13 in humans and syntenic region D in mouse), and its function was found to be essential for the viability of mouse ES cells and during mouse embryogenesis (221). Hence, the information that could be gathered in this animal model was highly restricted without additional somatic chimeric approaches, or through other genetic studies including those that may link *OGT* mutations to cases of X-linked disease syndromes mapping to the q13 region in humans.

HUMAN DEFICITS IN GLYCAN FORMATION

Examples of genetic alterations in glycan structure and expression are numerous in humans, and are increasingly helping to inform us about glycan function (Table 2). The most prominently studied examples of genetic variation of human

TABLE 2 Human genetic defects in glycan formation and clinical outcome

Enzyme (*Locus*)	Phenotype associated with enzyme deficiency	References
Blood group glycosyltransferases	ABO, H, Se, P, and Lewis blood group polymorphisms; function(s) not known	Reviewed in (222)
UDP-N-acetylglucosamine:lysosomal enzyme N-acetylglucosamine 1-phosphotransferase (two loci encode this multi-subunit enzyme) (*GNPTA*)	Inclusion cell disease (Mucolipidosis II and Mucolipidosis III)	(5, 223)
Phosphomannomutase 2 (*PMM2*)	CDG-Ia	(228, 229)
Phosphomannose isomerase (*PMI*)	CDG-Ib	(230, 231)
Dolichyl-P-Glc:Man9GlcNAc2-PP-dolichyl α-1-3-glucosyltransferase (*ALG6*)	CDG-Ic	(232, 233)
Dolichyl-P-Man:Man5GlcNAc2-PP-dolichyl α-1-3-mannosyltransferase (*ALG3*)	CDG-Id	(234)
Dolichol-P-Man synthase 1 (*DPM1*)	CDG-Ie	(235, 236)
Dolichol-P-Man utilization defect 1 (*MPDU1*)	CDG-If	(237, 238)
Dolichyl-P-Man:Man7GlcNAc2-PP-dolichyl α-1-6-mannosyltransferase (*ALG12*)	CDG-Ig	(224–226)
GnT II (*MGAT2*)	CDG-IIa (phenocopy of *Mgat2*-null mouse)	(239)
α-1-2-glucosidase I (*GCS1*)	CDG-IIb	(23)
GDP-fucose transporter (*FUCT1*)	CDG-IIc (leukocyte adhesion deficiency (LAD)-II)	(241–243)
β1-4 GalT-I galactosyltransferase (*β4GalTI*)	CDG-IId	(227)
GlcNAc-phosphatidylinositol synthase (*PIG-A*)	Paroxysmal nocturnal hemoglobinuria (PNH)	Reviewed in (144)
β1-4GlcA-T/α1-4GlcNAcT heparan sulfate copolymerase (*EXT1*)	Inherited multiple exostoses	Reviewed in (133)
β1-4GlcA-T/α1-4GlcNAcT heparan sulfate copolymerase (*EXT2*)	Inherited multiple exostoses	Reviewed in (133)
Protein O-mannose β-1-2-N-acetyl-glucosaminyl-transferase (*POMGNT1*)	Muscle-eye brain disease	(110)
Fukutin, a putative glycosyltransferase (*FCMD*)	Fukuyama congenital muscular dystrophy	(112)
Fukutin-related protein, a putative glycosyltransferase (*FKRP*)	Congenital muscular dystrophy type MDC1C	(113)
Sulfate anion transporter (*DTDST*)	Chondrodysplasias (DTD, AO2, MED, ACG1B)	(203–206)
β1-4 GalT-7 xylose galactosyltransferase (*XGPT*)	Ehlers-Danlos syndrome (progerioid form)	(250–252)
Fucosyltransferase VII (*FUT7*)	Modest defect in leukocyte selectin ligand activity	(192)
Galactose-1-phosphate uridyl transferase (*GALT*)	Galactosemia (classical galactosemia)	(249)
UDP-galactose 4′ epimerase (*GALE*)	Galactosemia	(247)
Galactokinase (*GALK*)	Galactosemia	(248)
α1-3GalT pseudogenes	Absent from Old World monkeys, apes and humans; xenotransplantation antigen	Reviewed in (208)
CMP-N-acetylneuraminic acid hydroxylase	Humans lack NeuGc form of sialic acid, no known phenotype	(28, 29)

glycan structure come from the field of transfusion medicine, which was spawned by the consequence of polymorphism in the structure of red cell glycans among different humans. In general, individuals whose red cells lack a particular glycan epitope elaborate natural antibody or alloantibody responses to the missing determinant. Such antibodies can lead to the lysis of transfused red cells taken from a donor whose red cells do make the determinant(s) missing from the recipient, leading to the signs and symptoms of transfusion reaction. Polymorphism exists in the ABO blood group locus (encoding α1–3Gal-T/α1–3Gal-T), in the H and Secretor blood group loci (α1–2fucosyltransferases), in the Lewis blood group locus (α1–3/1–4fucosyltransferases), and in the P blood group systems (several Gal-Ts) [reviewed in (223)]. Although there are associations between various disease states and some such polymorphisms, these are generally weak, without clear mechanism-based cause and effect relationships (223).

By contrast, there are now numerous human diseases with a clear pathogenic relationship to a genetic defect in glycan structure (Table 2). Some of these have been discussed above in the context of murine gene targeting experiments; the remainder are addressed in this section.

Inclusion Cell Disease is the first recognized human congenital defect in glycan formation to have pathogenic consequences (5). In individuals with this disease, a genetic defect in N-acetylglucosamine-1-phosphotransferase (223) abrogates modification of lysosomal enzymes by the mannose-6-phosphate moiety required for trafficking of lysosomal enzymes to the lysosome. Individuals with Inclusion Cell Disease thus suffer from a severe lysosomal storage disease syndrome.

Defects in distinct genetic loci involved in N-glycan synthesis have been causally associated with a class of human disorders termed Congenital Disorders of Glycosylation (CDG). Nine distinct CDGs had been identified through the end of 2001 [CDG-Ia–f, and CDG-IIa–c (53)]. Two additional CDGs [CDG-IIg (224–226) and CDG-IId (227)] have since been reported. These disorders and the relevant literature are tabulated in Table 2 and have been reviewed recently (52, 53).

CDGS are generally characterized by varying degrees of mental and psychomotor retardation, coagulopathies, and gastrointestinal signs and symptoms, with the exception of CDGIb, where CNS disease is not noted. The full spectrum of signs, symptoms, and laboratory abnormalities in the CDGs is wide and variable within patients with the same CDG, and among the various CDGs (52, 53). An estimated 20% of patients with signs and symptoms consistent with CDG and with abnormal serum transferrin structure suffer as a consequence of as yet uncharacterized mutations in one or more of the ~50 genes known to be required for N-glycan synthesis (53).

In the 11 well-characterized CDGs, alterations of glycosylation loci generally represent missense mutations that leave residual activity of the corresponding enzyme or sugar nucleotide transporter. Low levels of glycan formation are thus retained, apparently sustaining viability but engendering pathophysiology [reviewed in (52, 53)]. Diagnosis of CDG has historically included isoelectric focusing analysis of the N-glycan changes of serum transferrin (244, 245). This approach is now supplemented with other diagnostic measures due to the poor specificity and sensitivity of this method [as in trisomy 7 mosaicism, see (246); reviewed in (53)]. In general, there are no satisfactory therapies for most CDGs. Exceptions include mannose therapy in CDG-Ib (a defect in phosphomannose isomerase and loss of mannose-6-phosphate synthesis), and fucose therapy for CDG-IIc.

Galactosemia is a disease caused by genetic deficiency in one of three enzymes (GALK, galactokinase; GALT, galactose-1-phosphate uridyl transferase; GALE; UDP-galactose-4′-epimerase) in a pathway that converts galactose into glucose-6-phosphate, and that also contributes to the synthesis of UDP-galactose, the nucleotide sugar substrate for all galactosyltransferases. Defects in GALE (247) and GALK (248) apparently do not lead to impaired UDP-galactose synthesis or accumulation. Inherited deficiency of GALT causes classical galactosemia, is variably associated with modest UDP-Gal deficiency, and is characterized by failure to thrive in infancy, hepatomegaly, and cataracts (249). These signs are ameliorated by a lactose-free diet, which decreases pathogenic accumulation of pathway intermediates. Cognitive defects, ataxia, and ovarian dysfunction in this disease are unresponsive to lactose restriction. Hypogalactosylation of glycoproteins is observed in some GALT-deficient patients, but it is not clear whether this is a direct consequence of a relative UDP-Gal deficiency or if this is pathogenic.

The progeroid form of Ehlers-Danlos syndrome is characterized by connective tissue abnormalities, and defects in glycosaminoglycan modification of the proteoglycan known as decorin (250–252). Defective glycosaminoglycan chain formation is associated with diminished activity of GalT-7, the enzyme that utilizes the O-linked xylose moiety to establish glycosaminoglycan elongation in the proximal part of the proteoglycan linkage region. Mutations in the GalT-7 locus have been associated with diminished GalT-7 activity, suggesting causality in this disease. However, some of these mutants retain modest levels of GalT-7 activity and can direct normal HS synthesis in GalT-7-null CHO cell lines. In some instances, a substantial fraction of glycosaminoglycan chain synthesis in patients' cells is retained, and diminished activity of a second Gal-T activity has been observed. These considerations leave open the possibility that simple mutational events at the GalT-7 locus may not fully account for the phenotype in this disorder.

Tn syndrome is an acquired disorder apparently caused by transcriptional repression of Core 1 O-glycan β1–3 GalT (253) in a fraction of these patients' early hematopoietic progenitor cells. Loss of expression of this enzyme leads to absence of the Core 1 glycan (Galβ1–3GalNAcα1-Ser/Thr) and widespread expression of the GalNAcα1-Ser/Thr moiety (the Tn antigen). These patients may have thrombocytopenia, leukopenia, and evidence of an autoimmune hemolytic anemia caused by auto-anti-Tn antibodies.

Hereditary erythroblastic multinuclearity with a positive acidified-serum lysis test (HEMPAS) is an autosomal recessive disorder of glycosylation also known as congenital dyserythropoietic anemia type II. The term HEMPAS is derived from the observation that an antibody in some normal serum binds to HEMPAS red cells and will lyse these cells when the serum is acidified. HEMPAS red cell progenitors are prone to lysis, accounting for ineffective erythropoiesis, hyperplasia of erythroid precursors, multinucleated erythroblasts, and anemia. Red cell glycoproteins bands 3 and 4.5 in HEMPAS are devoid of polylactosamine chains normally present on bi-antennary complex N-linked glycans. Instead, HEMPAS red cells express abnormally large amounts of polylactosamine chains on glycolipids.

In some HEMPAS patients, the structure of the abnormal N-glycans suggests a decrease or absence of GlcNAcT-II, and substantial reductions in GlcNAcT-II activity have been reported in some HEMPAS patients (254). However, GlcNAcT-II deficiency in CDGS type IIa is causal of a much more severe disease, involves multiple organ systems, and band 3 polylactosamine chains are reduced by only ~50%, making it unlikely that systemic loss of GlcNAcT-II activity accounts for HEMPAS in most instances (255). In other HEMPAS patients, hybrid-type N-glycans that are substrates for αM-II accumulate, suggesting a causal defect in expression or activity of αM-II. In at least one HEMPAS patient, αM-II activity is nearly absent, and αM-II transcripts do not accumulate to normal levels (256). The mutational defect for these observations is not yet defined, however, so it is not known if the defect lies in the *αM-II* locus or in a locus that controls its expression. Further complicating matters is the observation that genetic linkage studies in some HEMPAS pedigrees exclude defects in the *αM-II* locus, the *α-mannosidase-IIx* (*MX*) locus, and the *GlcNAcT-II* locus (257). Heterogeneity in glycan structures between different HEMPAS pedigrees implies genetic heterogeneity, but a final molecular definition remains to be defined for any patients with this disorder.

In rheumatoid arthritis, a correlation has been observed between disease severity and the presence of galactose-deficient N-glycans on serum IgG (258). It has been suggested that the "uncovered" GlcNAc moieties of agalactosyl glycoforms of IgG enable IgG recognition by mannose binding lectin, promoting complexes that fix complement and initiating or perpetuating synovial inflammation (259). Cause and effect relationships between hypogalactosylation and disease status in rheumatoid arthritis and mechanisms accounting for hypogalactosylation of IgG in this context remain to be further extended. Similar considerations apply to the O-glycan abnormalities on IgA that are associated with IgA nephropathy (260, 261).

CONCLUSIONS AND PERSPECTIVE

The relatively recent application of genetic tools and approaches to perturb glycan structure in the context of developing and mature mammals has led to a more precise and widespread understanding of glycan function. The majority of mammalian glycan biosynthetic mutations characterized to date do not preclude embryonic development or adult reproduction; however, informative physiologic abnormalities are typically observed. More than three dozen glycan synthetic deficiencies in the mouse have been produced, of which a third result in lethality or preclude development. Phenotypes are commonly relegated to distinct cell and tissue types, indicating a high degree of specificity in glycan function. Where mechanism is defined, glycan-dependent phenotypes represent a remarkably wide range of molecular processes, including cell-cell adhesion, cell-matrix adhesion, extracellular receptor-ligand interactions, regulation of intracellular signal transduction processes, and modulation of transcriptional activity. These studies have informed us about new regulatory aspects of various fundamental biological processes, including cell migration, cell fate determination, morphogenesis, and mechanisms that modulate innate and adaptive immunity.

Significant progress in the discovery of inherited human disorders of glycosylation has paralleled directed mutational perturbation of glycan structure in the mouse, with significant potential for enhancement by these latter efforts. This is in part attributable to the highly conserved nature of the enzymes and corresponding glycosyation pathways in humans and mice and comparable physiology of the two species. To date, however, there is modest, although promising, overlap between naturally occurring disorders of glycosylation in humans and phenotypes associated with induced mutations in glycosylation loci in mice, which significantly limits the information from this interspecies comparison. Incomplete overlap between these two sets of mutational data reflects, in part, the small size of the set of targeted inactivation of murine glycosylation pathways, relative to the much larger set of glycosylation "targets" that exist in both species [1% of a mammalian genome, by estimates, corresponds to loci devoted to glycosylation (53)]. Incomplete overlap among murine and human cases of glycan defects at present is also a function of the fact that some human disorders of glycosylation are characterized by hypomorphic missense mutations in glycosylation loci that permit viability while engendering pathophysiology. By contrast, and in virtually all instances, targeted mutations of glycosylation loci in mice are characterized by the creation of null alleles that may lead to embryonic or early postnatal lethality. Early lethality obviously precludes the emergence of a phenotype in an adult mouse that might be informative for human diseases that become evident postnatally. In the future, targeted replacement of wild-type glycosylation loci with hypomorphic alleles in murine experiments may be a useful approach to generate murine phenotypes that might assign cause to a human syndrome of undefined mechanistic basis.

Diagnostics for glycan alterations in mammals are poorly developed outside of a few research laboratories, and this further contributes to insufficient information on the prevalence of human glycan defects. Nevertheless, glycan linkage-specific lectins and antibodies are abundant and inexpensive to employ. This irony reflects the more recent appearance of glycan-based diseases and the difficulty in establishing these syndromes as more mainstream clinical possibilities in undiagnosed and multigenic disorders. As a result, the majority of human glycosylation defects remain unrecognized and upon diagnosis and inspection would likely provide informative genetic and physiologic correlates to many of the known mouse mutants.

In virtually all instances where a mechanistic understanding of glycan function has been derived from a phenotypic perturbation due to a defect in glycan formation, biochemical characterization of the corresponding mutant and wild-type glycan structures has been essential. In the future, progress in understanding of glycan function will continue to rely heavily on glycan structural analyses in the context of mutational analysis, and it is likely to be enhanced by emerging approaches that can rapidly, accurately, and thoroughly define glycan structure with amounts of experimental material that are often modest. The identification of physiologically relevant glycan linkages among specific proteins and lipids will also be needed to determine, for any such glycan linkage, which protein- or lipid-specific glycan modifications actually play critical functional roles and which are neither essential nor selected for during organism evolution. Similar considerations have applied to intracellular phosphorylation cascades, where there is the potential for the existence of nonessential bystander modifications. Genetic perturbation of glycan function in the laboratory mouse and efforts to characterize new inherited defects of glycosylation in humans, aided by characterization of glycan structure in these contexts, will continue to significantly expand the known range of glycan functions and the mechanisms by which they modulate development, differentiation, physiology, and disease.

ACKNOWLEDGMENTS

The authors apologize for the inability to cite all related and important literature due to space limitations. We thank Thierry Hennet for helpful discussions on the human CDG syndromes and Bob Haltiwanger for sharing data ahead of publication. J.D.M. is supported by grants from the NIH (DK48247, P01-HL57345), and as an Investigator of the Howard Hughes Medical Institute. J.B.L. is supported by grants from the NIH (GM62116, 1P01-CA71932), and as an Investigator of the Howard Hughes Medical Institute.

The *Annual Review of Biochemistry* is online at http://biochem.annualreviews.org

LITERATURE CITED

1. Hager T. 1995. *Force of Nature,* p. 86. New York: Simon & Schuster
1a. Kornfeld R, Kornfeld S. 1985. *Annu. Rev. Biochem.* 54:631–64
2. Rademacher TW, Parekh RB, Dwek RA. 1988. *Annu. Rev. Biochem.* 57:785–38
3. Schachter H. 1991. *Glycobiology* 1:453–61
4. Varki A. 1993. *Glycobiology* 3:97–30
5. Reitman ML, Varki A, Kornfeld S. 1981. *J. Clin. Invest.* 67:1574–79
6. Weinstein J, Lee EU, McEntee K, Lai

PH, Paulson JC. 1987. *J. Biol. Chem.* 262:17735–43
7. Shaper NL, Hollis GF, Douglas JK, Kirsch IR, Shaper JH. 1988. *J. Biol. Chem.* 263:10420–28
8. Joziasse DH, Shaper JH, Van den Eijnden DH, Van Tunen AJ, Shaper NL. 1989. *J. Biol. Chem.* 264:14290–97
9. Larsen RD, Rajan VP, Ruff MM, Kukowska-Latallo J, Cummings RD, Lowe JB. 1989. *Proc. Natl. Acad. Sci. USA* 86:8227–31
10. Rajan VP, Larsen RD, Ajmera S, Ernst LK, Lowe JB. 1989. *J. Biol. Chem.* 264:11158–67
11. Gottlieb C, Baenziger J, Kornfeld S. 1975. *J. Biol. Chem.* 250:3303–9
12. Stanley P, Narasimhan S, Siminovitch L, Schachter H. 1975. *Proc. Natl. Acad. Sci. USA* 72:3323–27
13. Kumar R, Stanley P. 1989. *Mol. Cell. Biol.* 9:5713–17
14. Ioffe E, Stanley P. 1994. *Proc. Natl. Acad. Sci. USA* 91:728–32
15. Metzler M, Gertz A, Sarkar M, Schachter H, Schrader JW, Marth JD. 1994. *EMBO J.* 13:2056–65
16. Buck CA, Glick MC, Warren L. 1971. *Science* 172:169–71
17. Feizi T. 1985. *Nature* 314:53–57
18. Dennis JW, Laferte E, Waghorne C, Breitman ML, Kerbel RS. 1987. *Science* 236:582–85
19. Marth JD. 1999. In *Essentials of Glycobiology*, ed. A Varki, R Cummings, J Esko, H Freeze, G Hart, J Marth, pp. 515–36. New York: Cold Spring Harbor Press
20. Varki A, Marth JD. 1995. *Semin. Dev. Biol.* 6:127–38
21. Dodd RB, Drickamer K. 2001. *Glycobiology* 11:R71–79
22. Sharon N, Goldstein IJ. 1998. *Science* 282:1049
23. Kukuruzinska MA, Bergh MLE, Jackson BJ. 1987. *Annu. Rev. Biochem.* 56: 915–44
24. Tanner W, Lehle L. 1987. *Biochim. Biophys. Acta* 906:81–99
25. Varki A. 1999. See Ref. 19, pp. 31–39
26. Koda Y, Tachida H, Soejima M, Takenaka O, Kimura H. 2000. *J. Mol. Evol.* 50:243–48
27. Galili U, Swanson K. 1991. *Proc. Natl. Acad. Sci. USA* 88:7401–4
28. Chou HH, Takematsu H, Diaz S, Iber J, Nickerson E, et al. 1998. *Proc. Natl. Acad. Sci. USA* 95:11751–56
29. Irie A, Suzuki A. 1998. *Biochem. Biophys. Res. Commun.* 248:330–33
30. Gagneux P, Varki A. 1999. *Glycobiology* 9:747–55
31. Suzuki Y, Ito T, Suzuki T, Holland RE Jr, Chambers TM, et al. 2000. *J. Virol.* 74:11825–31
32. Ellgaard L, Molinari M, Helenius A. 1999. *Science* 286:1882–88
33. Trombetta ES, Parodi AJ. 2001. *Adv. Protein Chem.* 59:303–44
34. Takasuki A, Tamura G. 1971. *J. Antibiot.* 24:785–94
35. Surani MA. 1979. *Cell* 18:217–27
36. Atienza-Samols SB, Pine PR, Sherman MI. 1980. *Dev. Biol.* 79:19–32
37. Rajput B, Muniappa N, Vijay IK. 1994. *J. Biol. Chem.* 269:16054–61
38. Marek KW, Vijay I, Marth JD. 1999. *Glycobiology* 9:1263–71
39. Kumar R, Yang J, Larsen RD, Stanley P. 1990. *Proc. Natl. Acad. Sci. USA* 87:9948–52
40. Sarkar M, Hull E, Nishikawa Y, Simpson RJ, Moritz RL, et al. 1991. *Proc. Natl. Acad. Sci. USA* 88:234–38
41. Layton WM. 1976. *J. Hered.* 67: 336–38
42. Campbell RM, Metzler M, Granovsky M, Dennis JW, Marth JD. 1995. *Glycobiology* 5:535–43
43. Ioffe E, Liu Y, Stanley P. 1997. *Glycobiology* 7:913–19
44. Ioffe E, Liu Y, Stanley P. 1996. *Proc. Natl. Acad. Sci. USA* 93:11041–46
45. Moremen KW, Robbins PW. 1991. *J. Cell Biol.* 115:1521–34

46. Chui D, Oh-Eda M, Liao YF, Panneerselvam K, Lal A, et al. 1997. *Cell* 90:157–67
47. Chui D, Sellakumar G, Green RS, Sutton-Smith M, McQuistan T, et al. 2001. *Proc. Natl. Acad. Sci. USA* 98:1142–47
48. Misago M, Liao YF, Kudo S, Eto S, Mattei MG, et al. 1995. *Proc. Natl. Acad. Sci. USA* 92:11766–70
49. Akama TO, Nakagawa H, Sugihara K, Narisawa S, Ohyama C, et al. 2002. *Science* 295:124–27
50. D'Agostaro GA, Zingoni A, Moritz RL, Simpson RJ, Schachter H, Bendiak B. 1995. *J. Biol. Chem.* 270:15211–21
51. Wang Y, Tan J, Sutton-Smith M, Ditto D, Panico M, et al. 2001. *Glycobiology* 11:1051–70
52. Jaeken J, Matthijs G, Carchon H, Van Schaftingen E. 2001. In *The Metabolic and Molecular Bases of Inherited Disease*, ed. CR Scriver, AL Beaudet, WS Sly, D Valle, pp. 1600–22. New York: McGraw-Hill
53. Freeze HH. 2001. *Glycobiology* 11: R129–143
54. Bhaumik M, Seldin MF, Stanley P. 1995. *Gene* 164:295–300
55. Priatel JJ, Sarkar M, Schachter H, Marth JD. 1997. *Glycobiology* 7:45–56
56. Bhattacharyya R, Bhaumik M, Raju TS, Stanley P. 2002. *J. Biol. Chem.* 277: 26300–9
57. Bhaumik M, Harris T, Sundaram S, Johnson L, Guttenplan J, et al. 1998. *Cancer Res.* 58:2881–87
58. Yang X, Bhaumik M, Bhattacharyya R, Gong S, Rogler CE, Stanley P. 2000. *Cancer Res.* 60:33313–19
59. Minowa MT, Oguri S, Yoshida A, Hara T, Iwamatsu A, et al. 1998. *J. Biol. Chem.* 273:11556–62
60. Yoshida A, Minowa MT, Takamatsu S, Hara T, Ikenaga H, Takeuchi M. 1998. *Glycoconj. J.* 15:1115–23
61. Shoreibah M, Perng GS, Adler B, Weinstein J, Basu R, et al. 1993. *J. Biol. Chem.* 268:15381–85
62. Dennis JW, Granovsky M, Warren CE. 1999. *Biochim. Biophys. Acta* 1473: 21–34
63. Granovsky M, Fata J, Pawling J, Muller WJ, Khokha R, Dennis JW. 2000. *Nat. Med.* 6:306–12
64. Demetriou M, Granovsky M, Quaggin S, Dennis JW. 2001. *Nature* 409:733–39
65. Brockhausen I, Hull E, Hindsgaul O, Schachter H, Shah RN, et al. 1989. *J. Biol. Chem.* 264:11211–21
66. Taguchi T, Ogawa T, Inoue S, Inoue Y, Sakamoto Y. 2000. *J. Biol. Chem.* 275: 32598–602
67. Raju TS, Stanley P. 1998. *J. Biol. Chem.* 273:14090–108
68. Schachter H, Brockhausen I. 1989. *Symp. Soc. Exp. Biol.* 43:1–26
69. Ten Hagen KG, Bedi GS, Tetaert D, Kingsley PD, Hagen FK, et al. 2001. *J. Biol. Chem.* 276:17395–104
70. Clausen H, Bennett EP. 1996. *Glycobiology* 6:635–46
71. Marth JD. 1996. *J. Clin. Invest.* 97: 1999–102
72. Kingsley PD, Hagen KG, Maltby KM, Zara J, Tabak LA. 2000. *Glycobiology* 12:1317–23
73. Brockhausen I, Toki D, Brockhausen J, Peters S, Bielfeldt T, et al. 1996. *Glycoconj. J.* 13:849–56
74. Nehrke K, Hagen FK, Tabak LA. 1996. *J. Biol. Chem.* 271:7061–65
75. Ten Hagen KG, Tetaert D, Hagen FK, Richet C, Beres TM, et al. 1999. *J. Biol. Chem.* 274:27867–74
76. Hanisch FG, Muller S, Hassan H, Clausen H, Zachara N, et al. 1999. *J. Biol. Chem.* 274:9946–54
77. Hennet T, Hagen FK, Tabak LA, Marth JD. 1995. *Proc. Natl. Acad. Sci. USA* 92:12070–74
78. Zhang Y, Iwasaki H, Wang H, Guo J, Kudo T, Kalka TB, et al. 2003. *J. Biol. Chem.* 278:573–84
79. Tsuboi S, Fukuda M. 2001. *BioEssays* 23:46–53

80. Lowe JB. 2002. *Immunol. Rev.* 186: 19–36
81. Bierhuizen MF, Fukuda M. 1992. *Proc. Natl. Acad. Sci. USA* 81:9326–30
82. Yeh JC, Ong E, Fukuda M. 1999. *J. Biol. Chem.* 274:3215–21
83. Schwientek T, Yeh JC, Levery SB, Keck B, Merkx G. 2000. *J. Biol. Chem.* 275:11106–13
84. Ellies LG, Tsuboi S, Petryniak B, Lowe JB, Fukuda M, Marth JD. 1998. *Immunity* 9:881–90
85. Snapp KR, Heitzig CE, Ellies LG, Marth JD, Kansas GS. 2001. *Blood* 97:3806–11
86. Sperandio M, Thatte A, Foy D, Ellies LG, Marth JD, Ley K. 2001. *Blood* 97:3812–19
87. Sperandio M, Forlow SB, Thatte J, Ellies LG, Marth JD, Ley K. 2001. *J. Immunol.* 167:2268–74
88. Yeh JC, Hiraoka N, Petryniak B, Nakayama J, Ellies LG, et al. 2001. *Cell* 105:957–69
89. Wang Y, Shao L, Shi S, Harris RJ, Spellman MW, et al. 2001. *J. Biol. Chem.* 276:40338–45
90. Bruckner K, Perez L, Clausen H, Cohen S. 2000. *Nature* 406:411–15. Erratum. 2000. *Nature* 407:654
91. Moloney DJ, Shair LH, Lu FM, Xia J, Locke R, et al. 2000. *J. Biol. Chem.* 275:9604–11
92. Moloney DJ, Panin VM, Johnston SH, Chen J, Shao L. 2000. *Nature* 406: 369–75
93. Hicks C, Johnston SH, diSibio G, Collazo A, Vogt TF, Weinmaster G. 2000. *Nat. Cell Biol.* 2:515–20
94. Panin VM, Shao L, Lei L, Moloney DJ, Irvine KD, Haltiwanger RS. 2002. *J. Biol. Chem.* 277:29945–52
95. Chen J, Moloney DJ, Stanley P. 2001. *Proc. Natl. Acad. Sci. USA* 98:13716–21
96. Koch U, Lacombe TA, Holland D, Bowman JL, Cohen BL, et al. 2001. *Immunity* 15:225–36
97. Irvine KD. 1999. *Curr. Opin. Genet. Dev.* 9:434–41
98. Haltiwanger RS, Stanley P. 2002. *Biochim. Biophys. Acta* 1573:328–35
99. Kramer H. 2001. *Dev. Cell* 1:725–26
100. Evrard YA, Lun Y, Aulehla A, Gan L, Johnson RL. 1998. *Nature* 394:377–81
101. Zhang N, Grindley T. 1998. *Nature* 394:374–77
102. Moran JL, Levorse JM, Vogt TF. 1999. *Nature* 399:742–43
103. Egan S, Cohen B, Sarkar M, Ying Y, Cohen S, et al. 2000. *Glycoconj. J.* 17:867–75
104. Hennet T, Dinter A, Kuhnert P, Mattu TS, Rudd PM, Berger EG. 1998. *J. Biol. Chem.* 273:58–65
105. Vollrath B, Fitzgerald KJ, Leder P. 2001. *Mol. Cell. Biol.* 16:5688–97
106. Hase S, Kawabata S, Nishimura H, Takeya H, Sueyoshi T, et al. 1988. *J. Biochem.* 104:867–68
107. Shao L, Luo Y, Moloney DJ, Haltiwanger R. 2003. *Glycobiology.* In press
108. Strahl-Bolsinger S, Gentzsch M, Tanner W. 1999. *Biochim. Biophys. Acta* 1426:297–307
109. Chiba A, Matsumura K, Yamada H, Inazu T, Shimizu T, et al. 1997. *J. Biol. Chem.* 272:2156–62
110. Yoshida A, Kobayashi K, Manya H, Taniguchi K, Kano H, et al. 2001. *Dev. Cell* 1:717–24
111. Grewal PK, Holzfeind PJ, Bittner RE, Hewitt JE. 2001. *Nat. Genet.* 28:151–54
112. Kobayashi K, Nakahori Y, Miyake M, Matsumura K, Kondo-Iida E, et al. 1998. *Nature* 394:388–92
113. Brockington M, Blake DJ, Brown SC, Muntoni F. 2002. *Neuromuscul. Disord.* 12:233–34
114. Michele DE, Barresi R, Kanagawa M, Saito F, Cohn RD, et al. 2002. *Nature* 418:417–21
115. Varki A. 1999. See Ref. 19, pp. 115–29
116. Ichikawa S, Ozawa K, Hirabayashi Y.

1998. *Biochem. Biophys. Res. Commun.* 253:707–11
117. Yamashita T, Wada R, Sasaki T, Deng C, Bierfreund U, et al. 1999. *Proc. Natl. Acad. Sci. USA* 96:9142–47
118. Coetzee T, Li X, Fujita N, Marcus J, Suzuki K, et al. 1996. *Genomics* 35:215–22
119. Coetzee T, Fujita N, Dupree J, Shi R, Blight A, et al. 1996. *Cell* 86:209–19
120. Bosio A, Binczek E, Stoffel W. 1996. *Proc. Natl. Acad. Sci. USA* 93: 13280–85
121. Dupree JL, Coetzee T, Blight A, Suzuki K, Popko B. 1998. *J. Neurosci.* 18: 1642–49
122. Fujimoto H, Tadano-Aritomi K, Tokumasu A, Ito K, Hikita T, et al. 2000. *J. Biol. Chem.* 275:22623–26
123. Takamiya L, Yamamoto A, Yamashiro S, Furukawa K, Haraguchi M, et al. 1995. *FEBS Lett.* 358:79–83
124. Takamiya K, Yamamoto A, Furukawa K, Yamashiro S, Shin M, et al. 1996. *Proc. Natl. Acad. Sci. USA* 93: 10662–67
125. Sheikh KA, Sun J, Liu Y, Kawai H, Crawford TO, et al. 1999. *Proc. Natl. Acad. Sci. USA* 96:7532–37
126. Chiavegatto S, Sun J, Nelson RJ, Schnaar RL. 2000. *Exp. Neurol.* 166: 227–34
127. Takamiya K, Yamamoto A, Furukawa K, Zhao J, Fukumoto S, et al. 1998. *Proc. Natl. Acad. Sci. USA* 95:12147–52
128. Esko JD. 1999. See Ref. 19, pp. 145–59
129. Forsberg E, Kjellen L. 2001. *J. Clin. Invest.* 108:175–80
130. Esko JD, Selleck SB. 2002. *Annu. Rev. Biochem.* 71:435–71
131. Kim BT, Kitagawa H, Tamura J, Saito T, Kusche-Gullberg M, et al. 2001. *Proc. Natl. Acad. Sci. USA* 98:7176–81
132. Kitagawa H, Shimakawa H, Sugahara K. 1999. *J. Biol. Chem.* 274:13933–37
133. Duncan G, McCormick C, Tufaro F. 2001. *J. Clin. Invest.* 108:511–16
134. Lohmann DR, Buiting K, Ludecke HJ, Horsthemke B. 1997. *Cytogenet. Cell Genet.* 76:164–66
135. Lin X, Wei G, Shi Z, Dryer L, Esko JD, et al. 2000. *Dev. Biol.* 224:299–11
136. Forsberg E, Pejler G, Ringvall M, Lunderius C, Tomasini-Johansson B, et al. 1999. *Nature* 400:773–76
137. Humphries DE, Lanciotti J, Karlinsky JB. 1998. *Biochem. J.* 332:303–7
138. Ringvall M, Ledin J, Holmborn K, van Kuppevelt T, Ellin F, et al. 2000. *J. Biol. Chem.* 275:25926–30
139. Fan GP, Xiao L, Cheng L, Wang XH, Sun B, Hu GX. 2000. *FEBS Lett.* 467: 7–11
140. Merry CL, Bullock SL, Swan DC, Backen AC, Lyon M, et al. 2001. *J. Biol. Chem.* 276:35429–34
141. Gallagher JT. 2001. *J. Clin. Invest.* 108: 357–61
142. Ferguson MA. 1999. *J. Cell Sci.* 112: 2799–109
143. Hart GW. 1999. See Ref. 19, pp. 131–43
144. Bessler M, Schaefer A, Keller P. 2001. *Transfus. Med. Rev.* 15:255–67
145. Miyata T, Takeda J, Lida Y, Yamada N, Inoue N, et al. 1993. *Science* 259: 1318–20
146. Kawagoe K, Kitamura D, Okabe M, Taniuchi I, Ikawa M, et al. 1996. *Blood* 87:3600–6
147. Tarutani M, Itami S, Okabe M, Ikawa M, Tezuka T, et al. 1997. *Proc. Natl. Acad. Sci. USA* 94:7400–5
148. Takahama Y, Ohishi K, Tokoro Y, Sugawara T, Sugawara T, et al. 1998. *Eur. J. Immunol.* 28:2159–66
149. Nozaki M, Ohishi K, Yamada N, Kinoshita T, Nagy A, Takeda J. 1999. *Lab. Invest.* 79:293–99
150. Rosti V. 2002. *Ann. NY Acad. Sci.* 963: 290–96
151. Spicer AP, McDonald JA. 1998. *J. Biol. Chem.* 273:1923–32
152. Camenisch TD, Spicer AP, Brehm-Gibson T, Biesterfeldt J, Augustine ML, et al. 2000. *J. Clin. Invest.* 106:349–60

153. Csoka AB, Frost GI, Stern R. 2001. *Matrix Biol.* 20:499–508
154. Amado M, Almeida R, Schwientek T, Clausen H. 1999. *Biochim. Biophys. Acta* 1473:35–53
155. Lu Q, Hasty P, Shur BD. 1997. *Dev. Biol.* 181:257–67
156. Lu Q, Shur BD. 1997. *Development* 124:4121–31
157. Asano M, Furukawa K, Kido M, Matsumoto S, Umesaki Y, et al. 1997. *EMBO J.* 16:1850–57
158. Kotani N, Asano M, Iwakura Y, Takasaki S. 1999. *Biochem. Biophys. Res. Commun.* 260:94–108
159. Kotani N, Asano M, Iwakura Y, Takasaki S. 2001. *Biochem. J.* 357: 827–34
160. Hennet T, Chui D, Paulson JC, Marth JD. 1998. *Proc. Natl. Acad. Sci. USA* 95:4504–9
161. Poe JC, Hasegawa M, Tedder TF. 2001. *Int. Rev. Immunol.* 20:739–62
162. Gillespie W, Kelm S, Paulson JC. 1992. *J. Biol. Chem.* 267:21004–10
163. Priatel JJ, Chui D, Hiraoka N, Simmons CJT, Richardson KB, et al. 2000. *Immunity* 12:273–83
164. Moody AM, Chui D, Reche PA, Priatel JJ, Marth JD, Reinherz EL. 2001. *Cell* 107:501–12
165. Lee YC, Kojima N, Wada E, Kurosawa N, Nakaoka T, et al. 1994. *J. Biol. Chem.* 269:10028–33
166. Wen DX, Livingston BD, Medzihradszky KF, Kelm S, Burlingame AL, Paulson JC. 1992. *J. Biol. Chem.* 267: 21011–19
167. Ellies LG, Sperandio M, Underhill GH, Yousif J, Smith M, et al. 2002. *Blood* 100:3618–25
168. Kitagawa H, Paulson JC. 1993. *Biochem. Biophys. Res. Commun.* 194:375–82
169. Sasaki K, Watanabe E, Kawashima K, Sekine S, Dohi T. 1993. *J. Biol. Chem.* 268:22782–87
170. Ellies LG, Ditto D, Levy GG, Wahrenbrock M, Ginsburg D, et al. 2002. *Proc. Natl. Acad. Sci. USA* 99:10042–47
171. Nichols WC, Ginsburg D. 1997. *Medicine* 76:1–20
172. Stockert RJ, Morell AG, Ashwell G. 1991. *Targeted Diagn. Ther.* 4:41–64
173. Rice KG, Lee YC. 1993. *Adv. Enzymol. Relat. Areas Mol. Biol.* 66:41–83
174. Matsuda Y, Nara K, Watanabe Y, Saito T, Sanai Y. 1996. *Genomics* 32:137–39
175. Kawai H, Allende ML, Wada R, Kono M, Sango K, et al. 2001. *J. Biol. Chem.* 276:6885–88
176. Doherty P, Cohen J, Walsh FS. 1990. *Neuron* 5:209–19
177. Landmesser I, Dahm L, Tang J, Rutishauser U. 1990. *Neuron* 4:655–67
178. Rutishauser U. 1996. *Curr. Opin. Cell Biol.* 8:679–84
179. Eckhardt M, Muhlenhoff M, Bethe A, Koopman J, Frosch M, et al. 1995. *Nature* 373:715–18
180. Yoshida Y, Kojima N, Tsuji S. 1995. *J. Biochem.* 118:658–64
181. Nakayama J, Fukuda MN, Fredette B, Ranscht B, Fukuda M. 1995. *Proc. Natl. Acad. Sci. USA* 92:7031–35
182. Eckhardt M, Bukalo O, Chazal G, Wang L, Goridis C, et al. 2000. *J. Neurosci.* 20:5234–44
183. Cremer H, Lange R, Christoph A, Plomann M, Vopper G, et al. 1994. *Nature* 367:455–59
184. Muller D, Wang C, Skibo G, Toni G, Cremer H, et al. 1996. *Neuron* 17:413–22
185. Miyoshi E, Noda K, Yamaguchi Y, Inoue S, Ikeda Y, et al. 1999. *Biochim. Biophys. Acta* 1473:9–20
186. Becker DJ, Lowe JB. 2003. *Glycobiology.* In press
187. Domino SE, Zhang L, Gillespie PJ, Saunders TL, Lowe JB. 2001. *Mol. Cell. Biol.* 21:8336–45
188. Maly P, Thall AD, Petryniak B, Rogers GE, Smith PL, et al. 1996. *Cell* 86: 643–53
189. Homeister JW, Thall AD, Petryniak B,

Maly P, Rogers GE, et al. 2001. *Immunity* 15:115–26
190. Smithson G, Rogers GE, Smith PL, Scheidegger EP, Petryniak B, et al. 2001. *J. Exp. Med.* 194:601–14
191. Smith PL, Myers JT, Rogers GE, Zhou L, Petryniak B, et al. 2002. *J. Cell Biol.* 158:801–15
192. Bengtson P, Larson C, Lundblad A, Larson G, Pahlsson P. 2001. *J. Biol. Chem.* 276:31575–82
193. Willig TB, Breton-Gorius J, Elbim C, Mignotte V, Kaplan C, et al. 2001. *Blood* 97:826–28
194. Kudo T, Ikehara Y, Togayachi A, Kaneko M, Hiraga T, et al. 1998. *J. Biol. Chem.* 273:26729–38
195. Roos C, Kolmer M, Mattila P, Renkonen R. 2002. *J. Biol. Chem.* 277:3168–75
196. Fukuda M, Hiraoka N, Akama TO, Fukuda MN. 2001. *J. Biol. Chem.* 276:47747–50
197. Hemmerich S, Lee JK, Bhakta S, Bistrup A, Ruddle NR, Rosen SD. 2001. *Glycobiology* 11:75–87
198. Ouyang YB, Crawley JT, Aston CE, Moore KL. 2002. *J. Biol. Chem.* 277:23781–87
199. Kurima K, Warman ML, Krishnan S, Domowicz M, Krueger RC Jr, et al. 1998. *Proc. Natl. Acad. Sci. USA* 95:8681–85
200. Hiraoka N, Tsuboi S, Suzuki M, Petryniak B, Nakayama J, et al. 1999. *Immunity* 11:79–89
201. Bistrup A, Bhakta S, Lee JK, Belov YY, Gunn MD, et al. 1999. *J. Cell Biol.* 145:899–910
202. Hemmerich S, Bistrup A, Singer MS, van Zante A, Lee JK, et al. 2001. *Immunity* 15:237–47
203. Hästbacka J, de la Chapelle A, Mahtani MM, Clines G, Reeve-Daly MP. 1994. *Cell* 78:1073–87
204. Hästbacka J, Superti-Furga A, Wilcox WR, Rimoin DL, Cohn DH, Lander ES. 1996. *Am. J. Hum. Genet.* 58:255–62
205. Superti–Furga A, Hästbacka J, Wilcox WR, Cohn DH, van der Harten HJ, et al. 1996. *Nat. Genet.* 12:100–2
206. Rossi A, Bonaventure J, Delezoide AL, Cetta G, Superti-Furga A. 1996. *J. Biol. Chem.* 271:18456–64
207. Akama TO, Nishida K, Nakayama J, Watanabe H, Ozaki K, et al. 2000. *Nat. Genet.* 26:237–41
208. Galili U. 2001. *Biochimie* 83:557–63
209. Thall AD, Maly P, Lowe JB. 1995. *J. Biol. Chem.* 270:21437–40
210. Tearle RG, Tange MJ, Zannettino ZL, Katerelos M, Shinkel TA, et al. 1996. *Transplantation* 61:13–19
211. Smith PL, Lowe JB. 1994. *J. Biol. Chem.* 269:15162–71
212. Mohlke KL, Purkayastha AA, Westrick RJ, Smith PL, Petryniak B, et al. 1999. *Cell* 96:111–20
213. Scharzkopf M, Knobeloch KP, Rohde E, Hinderlich S, Wiechens N, et al. 2002. *Proc. Natl. Acad. Sci. USA* 99:5267–70
214. Hart GW. 1999. See Ref. 19, pp. 171–82
215. Wells L, Vossler K, Hart GW. 2001. *Science* 291:2376–78
216. Hart GW. 1999. See Ref. 19, pp. 183–94
217. Kreppel LK, Blomberg MA, Hart GW. 1997. *J. Biol. Chem.* 272:9308–15
218. Lubas WA, Frank DW, Krause M, Hanover JA. 1997. *J. Biol. Chem.* 272:9316–24
219. Gao Y, Wells L, Comer FI, Parker GJ, Hart GW. 2001. *J. Biol. Chem.* 276:9838–45
220. Yang XY, Zhang FX, Kudlow JE. 2002. *Cell* 110:69–80
221. Shafi R, Iyler SPN, O'Donnell N, Ellies LG, Marek KW, et al. 2000. *Proc. Natl. Acad. Sci. USA* 97:5735–39
222. Lowe JB. 2001. In *The Molecular Basis of Blood Diseases,* ed. G Stamatoyannopoulos, AW Nienhuis, PW Majerus, H Varmus, pp. 314–61. Orlando, FL: Saunders. 3rd ed.

223. Raas-Rothschild A, Cormier-Daire V, Bao M, Genin E, Salomon R, et al. 2000. *J. Clin. Invest.* 105:673–81
224. Chantret I, Dupre T, Delenda C, Bucher S, Dancourt J, et al. 2002. *J. Biol. Chem.* 277:25815–22
225. Grubenmann CE, Frank CG, Kjaergaard S, Berger EG, Aebi M, Hennet T. 2002. *Hum. Mol. Genet.* 11:2331–39
226. Thiel C, Schwarz M, Hasilik M, Grieben U, Hanefeld F, et al. 2002. *Biochem. J.* 367:195–201
227. Hansske B, Thiel C, Lubke T, Hasilik M, Honing S, et al. 2002. *J. Clin. Invest.* 109:725–33
228. Van Schaftingen E, Jaeken J. 1995. *FEBS Lett.* 377:318–20
229. Matthijs G, Schollen E, Pardon E, Veiga-Da-Cunha M, Jaeken J, et al. 1997. *Nat. Genet.* 16:88–92
230. Jaeken J, Matthijs G, Saudubray JM, Dionisi-Vici C, Bertini E, et al. 1998. *Am. J. Hum. Genet.* 62:1535–39
231. Niehues R, Hasilik M, Alton G, Korner C, Schiebe-Sukumar M, et al. 1998. *J. Clin. Invest.* 101:1414–20
232. Imbach T, Burda P, Kuhnert P, Wevers RA, Aebi M, et al. 1999. *Proc. Natl. Acad. Sci. USA* 96:6982–87
233. Körner C, Knauer R, Holzbach U, Hanefeld F, Lehle L, von Figura K. 1998. *Proc. Natl. Sci. USA* 95:13200–5
234. Körner C, Knauer R, Stephani U, Marquardt T, Lehle L, von Figura K. 1999. *EMBO J.* 18:6818–22
235. Imbach T, Schenk B, Schollen E, Burda P, Stutz A, et al. 2000. *J. Clin. Invest.* 105:233–39
236. Kim S, Westphal V, Srikrishna G, Mehta DP, Peterson S, et al. 2000. *J. Clin. Invest.* 105:191–98
237. Kranz C, Denecke J, Lehrman MA, Ray S, Kienz P, et al. 2001. *J. Clin. Invest.* 108:1613–19
238. Schenk B, Imbach T, Frank CG, Grubenmann CE, Raymond GV, et al. 2001. *J. Clin. Invest.* 108:1687–95
239. Tan J, Dunn J, Jaeken J, Schachter H. 1996. *Am. J. Hum. Genet.* 59:810–17
240. Deleted in proof
241. Luhn K, Wild MK, Eckhardt M, Gerardy-Schahn R, Vestweber D. 2001. *Nat. Genet.* 28:69–72
242. Lubke T, Marquardt T, Etzioni A, Hartmann E, von Figura K, Körner C. 2001. *Nat. Genet.* 28:73–76
243. Marquardt T, Luhn K, Srikrishna G, Freeze HH, Harms E, Vestweber D. 1999. *Blood* 94:3976–85
244. Wada Y, Nikishawa A, Okamoto N, Inui K, Tsukamoto H, et al. 1992. *Biochem. Biophys. Res. Commun.* 189:832–36
245. Yamashita K, Hiroko I, Ohkura T, Fukushima K, Yuasa I, et al. 1993. *J. Biol. Chem.* 268:5783–89
246. Knopf C, Rod R, Jaeken J, Berant M, van Schaftingen E, et al. 2000. *J. Inherit. Metab. Dis.* 23:399–403
247. Quimby BB, Alano A, Almashanu S, DeSandro AM, Cowan TM, Fridovich-Keil JL. 1997. *Am. J. Hum. Genet.* 61:590–98
248. Stambolian D, Ai Y, Sidjanin D, Nesburn K, Sathe G, et al. 1995. *Nat. Genet.* 10:307–12
249. Reichardt JK, Woo SL. 1991. *Proc. Natl. Acad. Sci. USA* 88:2633–37
250. Quentin E, Gladen A, Roden L, Kresse H. 1990. *Proc. Natl. Acad. Sci. USA* 87:1342–46
251. Okajima T, Fukumoto S, Furukawa K, Urano T. 1999. *J. Biol. Chem.* 274: 28841–44
252. Almeida R, Levery SB, Mandel U, Kresse H, Schwientek T, et al. 1999. *J. Biol. Chem.* 247:26165–71
253. Berger EG. 1999. *Biochim. Biophys. Acta* 1455:255–68
254. Fukuda MN, Dell A, Scartezzini P. 1987. *J. Biol. Chem.* 262:7195–206
255. Charuk JH, Tan J, Bernardini M, Haddad S, Reithmeier RA, et al. 1995. *Eur. J. Biochem.* 230:797–805
256. Fukuda MN, Masri KA, Dell A,

Luzzatto L, Moremen KW. 1990. *Proc. Natl. Acad. Sci. USA* 87: 7443–47
257. Iolascon A, delGiudice EM, Perrotta S, Granatiero M, Zelante L, Gasparini P. 1997. *Blood* 90:4197–200
258. Parekh RB, Dwek RA, Sutton BJ, Fernandes DL, Leung A, et al. 1985. *Nature* 316:452–57
259. Malhotra R, Wormald MR, Rudd PM, Fischer PB, Dwek RA, Sim RB. 1995. *Nat. Med.* 1:237–43
260. Baharaki D, Dueymes M, Perrichot R, Basset C, Le Corre R, et al. 1996. *Glyconconj. J.* 13:505–11
261. Hiki Y, Kokubo T, Iwase H, Masaki Y, Sano T, et al. 1999. *J. Am. Soc. Nephrol.* 10:760–69

Annu. Rev. Biochem. 2003. 72:693–715
doi: 10.1146/annurev.biochem.72.121801.161551
Copyright © 2003 by Annual Reviews. All rights reserved
First published online as a Review in Advance on March 27, 2003

THE RNA POLYMERASE II ELONGATION COMPLEX

Ali Shilatifard[1], Ronald C. Conaway[2], and Joan Weliky Conaway[2]

[1]*Edward A. Doisey Department of Biochemistry, St. Louis University School of Medicine, St. Louis, Missouri 63104; email: shilatia@slu.edu*

[2]*Stowers Institute for Medical Research, Kansas City, Missouri 64110; email: rcc@stowers-institute.org, jlc@stowers-institute.org*

Key Words chromatin, ELL, Elongin, DSIF, Elongator, FACT, NELF, P-TEFb, RNA polymerase II, SII, THO complex

■ **Abstract** Synthesis of eukaryotic mRNA by RNA polymerase II is an elaborate biochemical process that requires the concerted action of a large set of transcription factors. RNA polymerase II transcription proceeds through multiple stages designated preinitiation, initiation, and elongation. Historically, studies of the elongation stage of eukaryotic mRNA synthesis have lagged behind studies of the preinitiation and initiation stages; however, in recent years, efforts to elucidate the mechanisms governing elongation have led to the discovery of a diverse collection of transcription factors that directly regulate the activity of elongating RNA polymerase II. Moreover, these studies have revealed unanticipated roles for the RNA polymerase II elongation complex in such processes as DNA repair and recombination and the proper processing and nucleocytoplasmic transport of mRNA. Below we describe these recent advances, which highlight the important role of the RNA polymerase II elongation complex in regulation of eukaryotic gene expression.

CONTENTS

INTRODUCTION

Over the past few years, advances in several lines of research on the elongation stage of eukaryotic mRNA synthesis have provided significant new insights into the mechanisms governing transcription elongation by RNA polymerase II. First and perhaps most notable among these advances is the recent determination of the high resolution structures of free and elongating forms of RNA polymerase II by Kornberg and coworkers (1–3). Since the discovery and purification of RNA polymerase II nearly 30 years ago, a seemingly insurmountable obstacle opposing efforts to understand RNA polymerase II transcription has been the size and complexity of the polymerase itself. While biochemical studies of purified RNA polymerase II have revealed many of its key catalytic properties (such as its extraordinary processivity and inherent tendencies to suffer transcriptional pause and arrest) (4, 5), a lack of high resolution information on the enzyme's structure has thwarted efforts to understand detailed mechanisms of RNA polymerase II transcription at a molecular level. Indeed, for years the prospects of obtaining high resolution structural information on a macromolecular assembly as complex as RNA polymerase II seemed bleak. The enzyme is composed of 12 distinct subunits with an aggregrate molecular mass of more than 500 kDa. It is present in cells in low abundance and, perhaps as a result of its sheer size, is often labile and difficult to handle. Thus, the recent report by Kornberg and coworkers of the high resolution crystal structures of both free and transcribing RNA polymerase II is a scientific landmark and a *tour de force* of X-ray crystallography that has brought to light for the first time RNA polymerase II at atomic resolution (1–3). As discussed in more detail below, the polymerase structures are providing new insights into (*a*) the enzyme's catalytic mechanism and the coupling of phosphodiester bond synthesis with the enzyme's translocation along DNA, (*b*) the nature of the enzyme's interactions with its DNA template and nascent transcript, and (*c*) the mechanisms underlying the enzyme's propensity to pause and arrest.

Second, advances in ongoing studies of the cadre of known RNA polymerase II elongation factors, which include SII, ELL, Elongin, the P-TEFb/NELF/DSIF system, Elongator, and FACT, are illuminating their biochemical mechanisms of action and their unique roles in cells. Finally, advances in studies outside the realm of traditional transcription research have brought new and unanticipated insights into the repertoire of functions of the RNA polymerase II elongation complex. These studies have solidified the notion that the polymerase is not simply an RNA synthetic machine, but it serves a more fundamental role as a critical platform for recruiting and coordinating the actions of a host of additional enzymes and proteins responsible for such diverse cellular processes as DNA recombination and repair and the proper processing and subcellular trafficking of the polymerase's primary product, mRNA (6, 7). Below we discuss these recent advances, which are beginning to provide a detailed picture of the intricacies of the elongation stage of eukaryotic mRNA synthesis.

THE HIGH RESOLUTION STRUCTURE OF ELONGATING RNA POLYMERASE II

With their reports of the 2.8 Å resolution structure of a free, catalytically active 10 subunit form of *Saccharomyces cerevisiae* RNA polymerase II (1) and of a 3.3 Å resolution structure of an elongating form of the same enzyme (2), Kornberg and coworkers have taken great strides toward settling longstanding controversies over the structure of the RNA polymerase II elongation complex and its catalytic mechanism. Their structures provide for the first time unequivocal evidence that the RNA polymerase II elongation complex includes (*a*) an ~9 basepair RNA-DNA hybrid that holds the newly synthesized transcript in register with the DNA template, (*b*) a maximum of ~4 basepairs of unwound template DNA ahead of the 3′-OH terminus of the growing transcript, and (*c*) a discrete RNA exit groove that is located near the base of the largely unstructured carboxy-terminal domain (CTD) of the largest polymerase subunit and accepts the nascent transcript as it is separated from the coding strand of the DNA template ~10 nucleotides upstream of the transcript's 3′-end (Figure 1).

In addition to elucidating these basic features of the RNA polymerase II elongation complex, the crystal structures have furnished in-depth information on the overall architecture of the complex and significant insights into its mechanism of action. The structures show that the enzyme's two largest Rpb1 and Rpb2 subunits interface along an extended surface to create an ~20 Å cleft or channel that accomodates the downstream region of the DNA template and extends from the leading edge of the elongation complex to the primary catalytic Mg^{+2} ion that lies buried deep in the interior of the enzyme. The Rpb1 and Rpb2 subunits are situated at the center of the RNA polymerase II elongation complex with the enzyme's smaller subunits arrayed on its surface. Although the enzyme's surface

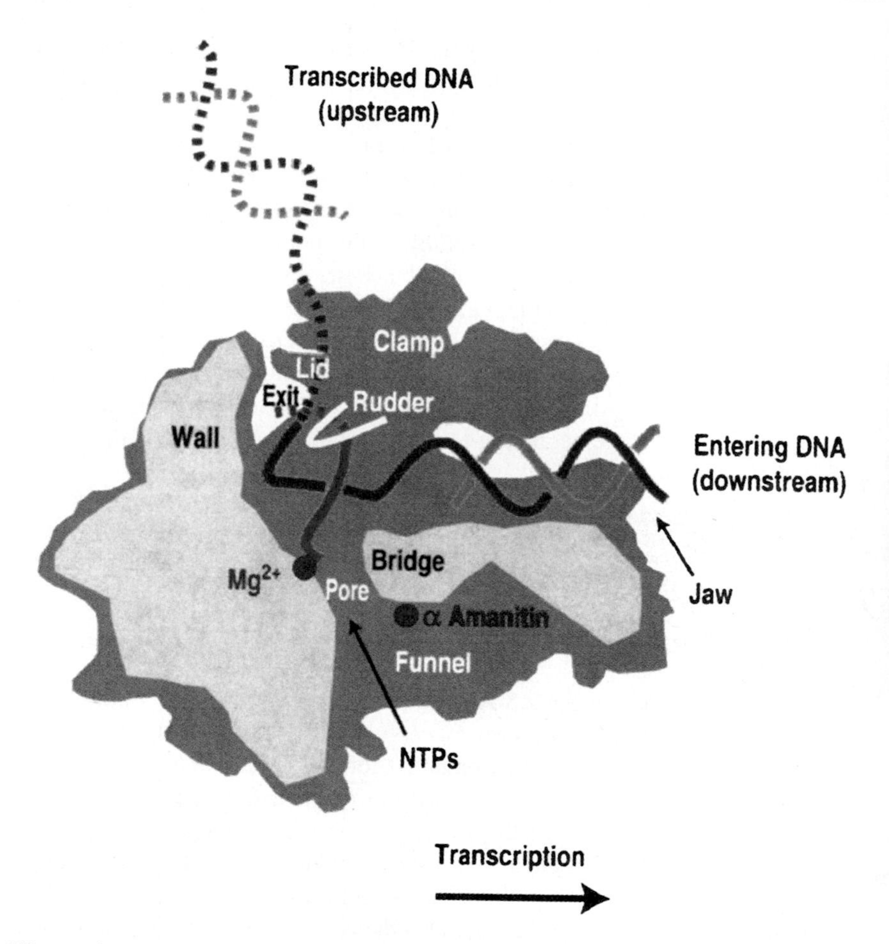

Figure 1 Cutaway view of the RNA polymerase II elongation complex. Original figure from D.A. Bushnell, P. Cramer, and R.D. Kornberg in the Proceedings of the National Academy of Sciences, Volume 99, pages 1218–22. Copyright 2002 National Academy of Sciences, U.S.A.

is largely negatively charged, the DNA binding cleft is lined with a uniform positive charge, consistent with its role in DNA binding.

Perhaps the most notable structural difference between free and transcribing RNA polymerase II is the position of a large, ~50 kDa hinged region referred to as the "clamp," which in free polymerase is in an open conformation that would allow easy entry of the DNA template into the catalytic site, but closes tightly over the DNA and nascent transcript in the elongating enzyme. The clamp is composed of portions of the Rpb1, Rpb2, and Rpb3 subunits. Because of its location and its extensive series of contacts with the DNA template and nascent transcript, the clamp is likely to be largely responsible for the extreme stability

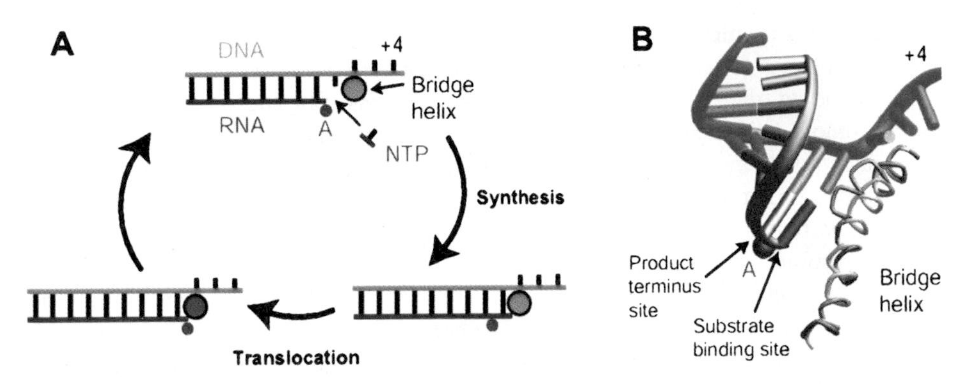

Figure 2 Model for translocation by RNA polymerase II.The straight conformation of the bridging helix found in yeast RNA polymerase II is shown in gray, and the bent conformation found in bacterial RNA polymerase is shown in purple. The nascent transcript is red, and the DNA template is light blue. The pink dot (A) represents the position of the active site magnesium. Computer-generated image courtesy of R.D. Kornberg and D.A. Bushnell.

of the elongation complex. At the base of the clamp are five loops or "switches," which appear unfolded in free polymerase. Upon transition of RNA polymerase II from the free to the transcribing state, the switches undergo dramatic conformational changes to well-folded structures, in a process that is likely induced by their extensive contacts with the RNA-DNA hybrid near the catalytic site. Thus, ordering of the switches may drive closure of the clamp once a short transcript has been synthesized and the RNA-DNA hybrid is present.

Three loops, referred to as the "rudder," "lid," and "zipper," emanate from the clamp. Based on their positions with respect to the the RNA-DNA hybrid in both RNA polymerase II and bacterial RNA polymerase and on evidence from biochemical cross-linking studies probing the bacterial RNA polymerase elongation complex, it was suggested that these loops might be involved in separation of the nascent transcript from the DNA template at the upstream end of the RNA-DNA hybrid (2, 8); however, deletion of the rudder from bacterial RNA polymerase was recently shown not to affect separation of the nascent transcript from the DNA template, but rather to destabilize the ternary elongation complex (9). Thus, a major function of the rudder, and perhaps the lid and zipper, may be to help lock the DNA template and nascent transcript in position in the enzyme's active site.

Among the most intriguing structural features of RNA polymerase II is a long "bridging helix" that emanates from the Rpb1 subunit and crosses the cleft between Rpb1 and Rpb2 near the catalytic Mg^{+2} ion (Figure 2). In the crystal structure of the RNA polymerase II elongation complex, amino acid side chains of the bridging helix make direct contact with the coding base in the DNA template. Notably, the corresponding helix in the bacterial RNA polymerase crystal structure is bent by about 3 Å (8). Similar bending of the RNA polymerase II bridging helix could cause the helix to push on the basepair at the 3′-end of the RNA-DNA hybrid; this suggests that the bridging helix may promote translocation of RNA polymerase II along the DNA by acting as a ratchet that moves the RNA-DNA hybrid through the active site by switching from straight to bent conformations at each step of nucleotide addition. Consistent with this model, the recent report by Kornberg and coworkers (10) of the 2.8 Å resolution crystal structure of RNA polymerase II bound to α-amanitin, a potent inhibitor of RNA chain elongation, revealed that α-amanitin binds specifically to residues in the bridging helix in a manner that should decrease its conformational flexibility, thereby preventing the enzyme's translocation along the DNA.

In addition to the clamp domain and bridging helix, RNA polymerase II possesses several additional notable structural features. First, at the distal end of the DNA binding cleft near the leading edge of the elongation complex are two large "jaws" that surround the cleft: an upper jaw, formed by portions of Rpb1, Rpb2, and Rpb9, and a lower jaw, formed by Rpb5 and portions of Rpb1 and Rpb6. The jaws are perfectly positioned to grab the incoming DNA template; thus, they serve not only to lend additional stability to the elongation complex, but also to guide the DNA into the catalytic site. Direct contacts between the incoming DNA and amino acid side chains in the Rpb5 subunit are likely to contribute significantly to the interaction of the jaws with the DNA template. Second, a narrow pore, which lies directly beneath the catalytic Mg^{+2} ion and opens into a wider-diameter, solvent-accessible funnel, is likely to provide incoming ribonucleoside triphosphates with their only access to the catalytic site. Finally, the nascent transcript exits RNA polymerase II through a groove near the base of the CTD of its largest subunit. In light of evidence that the CTD recruits a diverse collection of mRNA processing enzymes to the site of transcription (6, 7), the positioning of the RNA exit groove near the base of the CTD would ensure convenient coupling of transcription with subsequent steps of mRNA maturation and export from the nucleus.

THE SII FAMILY OF RNA POLYMERASE II ELONGATION FACTORS

SII (also referred to as TFIIS) was the first RNA polymerase II transcription factor to be purified (11). In addition to the major, ubiquitously expressed form of SII, there are multiple, tissue- or developmental stage-specific SII family members (12–14). SII family members are all ~30 kDa proteins and are found in eukaryotes from yeast to man. SII was initially identified and purified by its ability to promote efficient synthesis of long transcripts in vitro (11). Biochemical studies subsequently revealed that SII promotes efficient RNA polymerase II elongation by preventing premature arrest of polymerase at discrete sites within transcribed DNA sequences (15).

Although the exact mechanism of transcriptional arrest has not been unequivocally established, a plausible model for this process has come from studies of both bacterial and eukaryotic RNA polymerases. Elongating RNA polymerases appear to oscillate between inactive and active conformations at each step of nucleotide addition (16–23). As a consequence, the overall RNA polymerase elongation rate is not determined solely by the rate of phosphodiester bond formation (K_{cat}), but by the fraction of time polymerase spends in paused, inactive conformations. Moreover, at some sites along the DNA template, paused polymerase falls into an arrested state that can be reversed only by the action of elongation factors like SII or its bacterial counterparts GreA and GreB (24).

Both RNA polymerase pausing and arrest are proposed to result from backward movement or back-tracking of the enzyme and its associated transcription bubble along the DNA template and RNA transcript (23, 25, 26). During back-tracking, the 8–9 base pair RNA-DNA hybrid is maintained with (*a*) reannealing of a portion of the nascent transcript on the 5′-side of the transcription bubble into the RNA-DNA hybrid, (*b*) unwinding of the 3′-end of the transcript from the RNA-DNA hybrid, and (*c*) displacement of the now single-stranded 3′-OH terminus of the transcript from its proper position in alignment with the enzyme's catalytic Mg^{+2} ion.

RNA polymerase pausing is believed to result from reversible back-tracking of the enzyme by ~2 to 4 nucleotides; this leaves the transcript's misaligned 3′-OH terminus unable to serve as an acceptor for the incoming ribonucleoside triphosphate in synthesis of the next phosphodiester bond of the transcript. After a variable length of time, the enzyme spontaneously slides forward again, and the transcript's 3′-OH terminus is realigned with the enzyme's catalytic Mg^{+2} ion, which allows RNA synthesis to continue. In contrast, RNA polymerase arrest is believed to result from irreversible backsliding of the enzyme by ~7 to 14 nucleotides (25, 27, 28). Based on analysis of the high resolution structures of free and elongating RNA polymerase II, it has been suggested that, during arrest, the single-stranded 3′-end of the displaced transcript is extruded from the polymerase through the small pore near the catalytic Mg^{+2} ion, because the pore is the only potential site of egress of the displaced transcript from the enzyme's active site (1–3). Thus, arrest may result from irreversible trapping of the 3′-end of the displaced transcript in the pore.

Results of elegant biochemical studies have shown that SII reactivates arrested RNA polymerase II by interacting directly with the enzyme and inducing endonucleolytic cleavage of the nascent transcript ~7 to 14 nucleotides upstream of the transcript's 3′-end, in a reaction that is catalyzed by the polymerase active site and results in creation of a new transcript 3′-OH terminus that is correctly basepaired to the DNA template and can be reextended (29–33). SII can also promote nascent transcript cleavage by paused RNA polymerase II; in this case, cleavage of the nascent transcript typically occurs 2 or 3 nucleotides upstream of the transcript's 3′-end.

Although the exact mechanism of SII action in reactivating arrested RNA polymerase II has not been established, it has been suggested that the SII domain responsible for this process could reach deeply into the pore and induce endonucleolytic cleavage of the extruded portion of the displaced transcript through direct interactions with both the transcript and polymerase active site residues (1–3). Consistent with this hypothesis, SII is known to interact with RNA polymerase II at sites near the pore (34, 35), and SII can be cross-linked to the 3′-end of transcripts present in the arrested RNA polymerase II elongation complex (36). In addition, parallel studies of the mechanism of arrest by *Escherichia coli* RNA polymerase have found that the 3′-end of transcripts present in the arrested elongation complex can be cross-linked to residues within the pore, and the bacterial GreA and GreB proteins, which function similarly to SII, can be cross-linked to sites near the pore of the bacterial enzyme (37).

The Role of SII In Vivo

A major impediment of efforts to understand what roles SII and other RNA polymerase II elongation factors play in vivo has been a lack of suitable assays for measuring their activities in cells. Unlike DNA binding transcriptional activators, SII does not appear to function exclusively through sequence-specific interactions with DNA regulatory elements of genes. As a consequence, the commonly used strategy of employing DNA-sequence-based reporter assays for investigating transcription factor function has not proven particularly useful in studies of the roles of SII and other RNA polymerase II elongation factors in cells. Nevertheless, evidence supporting a role for SII in transcriptional activation in yeast has come from a variety of studies.

An important tool in these studies has been the drug 6-azauracil (6-AU), which is believed to impede RNA polymerase II elongation by reducing intracellular GTP or UTP levels through inhibition of their biosynthetic enzymes IMPDH and orotidylate decarboxylase (38). Although the SII gene is not essential for yeast viability (39), yeast lacking the SII gene exhibit transcriptional defects when grown in the presence of 6-AU; these transcriptional defects include reductions in both the magnitude and rate of transcriptional activation of such genes as *GAL1*, which is induced by galactose (40), *ENA1*, which is induced by osmotic stress (40), and *PUR5* and *SSM1*, which are both induced by 6-AU (41, 42).

Several lines of evidence suggest that SII exerts its effects on transcriptional activation in cells primarily at the elongation stage of mRNA synthesis. First, the transcriptional defects of yeast SII mutants are syngeristically enhanced when SII mutations are present in combination with mutations in genes encoding RNA polymerase II subunits (34, 43–45) and other proteins, such as DSIF/NELF (46), Elongator (47), and FACT (48), which are also implicated in elongation. Second, results of chromatin immunoprecipitation experiments suggest that SII is associated with both the promoter-regulatory and coding regions of multiple genes in yeast, whereas transcription initiation factors, such as TBP, TFIIB, and Mediator subunits, are found only at promoters (49). Finally, although attempts to detect SII-dependent transcription in cells through a site in the H3.3 gene that functions as a very strong transcriptional arrest site in vitro have not been successful (40), mutations in SII have recently been shown to affect transcription in yeast through an artificial arrest site inserted into the coding sequence of a β-galactosidase reporter gene (50). This artificial arrest site should provide an excellent model for further studies investigating the mechanism of SII action in cells.

A ROLE FOR PHOSPHORYLATION OF THE RNA POLYMERASE II CTD IN REGULATION OF ELONGATION

In addition to its well-known tendency to suffer SII-sensitive arrest, elongating RNA polymerase II will also arrest transcription in cells and in vitro by a different mechanism, in a process that is exacerbated by protein kinase inhibitors such as the nucleotide analog 5,6-dichloro-1-β-D-ribofuranosylbenzimidazole (DRB) (51–54). Biochemical fractionation of DRB-sensitive transcription systems led to the resolution and purification of two negatively acting factors, DRB sensitivity inducing factor (DSIF) (55) and negative elongation factor (NELF) (56) (which function together to promote RNA polymerase II arrest), and a positively acting DRB-sensitive factor, positive transcription elongation factor b (P-TEFb) (57, 58) (which prevents DSIF- and NELF-dependent RNA polymerase II arrest).

P-TEFb is a DRB-sensitive cyclin-dependent protein kinase that is capable of phosphorylating the heptapeptide repeats in the CTD of the largest RNA polymerase II subunit (59). The RNA polymerase II CTD is composed of tandemly repeated copies of a heptapeptide with consensus $[Y_1S_2P_3T_4S_5P_6S_7]$. The CTD heptapeptide is repeated ~26 times in the largest Rpb1 subunit of *S. cerevisiae* RNA polymerase II and ~32, ~45, and ~52 times in the largest subunits of the enzymes from *C. elegans*, *Drosophila melanogaster*, and mammals (60). Genetic studies have shown that the CTD is essential for the function of RNA polymerase II because deletions that remove only a fraction of the heptapeptide repeats can be lethal (61, 62).

A requirement for phosphorylation of the CTD in elongation by RNA polymerase II was initially proposed on the basis of evidence (*a*) that the CTDs of actively elongating polymerases are highly phosphorylated and (*b*) that polymerases containing hypophosphorylated CTDs preferentially enter the preinitiation complex, where they are subsequently phosphorylated during or shortly after initiation (63–69). Phosphorylation of the CTD occurs on both serine 2 and serine 5. Serine 5 phosphorylation by the TFIIH-associated Cdk7/cyclin H kinase occurs at or near the promoter, while serine 2 phosphorylation is seen primarily on polymerase molecules that have moved away from the promoter region and are engaged in transcript elongation (70).

The P-TEFb CTD Kinase

The DRB-sensitive cyclin-dependent kinase P-TEFb is composed of the ~43 kDa Cdk9 kinase (also referred to as PITALRE) and either cyclin T1, cyclin T2a, cyclin T2b, or cyclin K. In addition to phosphorylating the RNA polymerase CTD (58), the P-TEFb kinase is capable of phosphorylating the Spt 5 subunit of DSIF (71, 72). Although the precise mechanism by which P-TEFb relieves DSIF- and NELF-dependent inhibition of elongation is not yet known, this P-TEFb activity appears to depend on CTD phosphorylation, because P-TEFb cannot prevent arrest by RNA polymerase II lacking the CTD (58). In addition, (*a*) overexpression of Cdk9, but not Cdk9 lacking kinase activity, will activate expression of reporter genes in mammalian cell transfection experiments (73, 74), and (*b*) direct recruitment of a GAL4-Cdk9 fusion protein, but not GAL4-Cdk9 lacking kinase activity, to the *D. melanogaster* heat-shock activatable HSP70 promoter is sufficient to activate transcription even in the absence of a heat shock signal (75). Finally, results of RNA interference experiments reveal that Cdk9 is essential for phosphorylation of serine 2 in the heptapeptide repeats of the CTD in *C. elegans* embryos (76).

The first evidence implicating P-TEFb in regulation of RNA polymerase II transcription elongation in vivo came with discovery that P-TEFb is required for efficient synthesis of human immunodeficiency virus-1 (HIV-1) transcripts (59, 74, 77–80). These studies demonstrated that P-TEFb is recruited to early RNA polymerase II elongation complexes through interaction of its cyclin T subunit with the HIV-1-encoded Tat protein, which binds to the transactivation response (TAR) element in the nascent HIV-1 transcript.

Evidence for a more general role for P-TEFb in regulation of RNA polymerase II transcription in vivo has come from studies performed in *D. melanogaster* and *C. elegans*. P-TEFb is broadly distributed at sites of transcription on *D. melanogaster* polytene chromosomes and is rapidly redistributed to heat shock genes when they are activated by temperature shift (75). Inhibition of Cdk9 expression in *C. elegans* blocks transcription from heat shock promoters and from a large group of promoters expressed during early embryogenesis (76).

The Negative Factors DSIF and NELF

DSIF and NELF were identified and purified from HeLa cells by Handa and coworkers (55, 56). DSIF is a heterodimer composed of the ~14 kDa Spt4 and ~160 kDa Spt5 proteins, which are expressed ubiquitously in eukaryotes from yeast to man. NELF is an ~300 kDa multiprotein complex composed of five polypeptides of ~66, 61, 59, 58, and 46 kDa, which are all required for NELF activity in inhibition of RNA polymerase II elongation. Notable among the NELF subunits are NELF-A, which is encoded by a gene potentially involved in

Wolf-Hirschhorn syndrome, and NELF-E, a putative RNA binding protein with arginine-aspartic acid (RD) dipeptide repeats. Although the exact mechanisms of action of DSIF and NELF in promoting RNA polymerase II arrest are not known, evidence indicating (*a*) that DSIF interacts directly with transcribing polymerase, (*b*) that NELF binds directly to the DSIF-RNA polymerase binary complex, and (*c*) that NELF RNA binding activity is required for arrest has led to the proposal that DSIF and NELF act together to induce polymerase arrest by a mechanism that involves their direct interactions with polymerase and the nascent transcript (81).

Evidence in support of the model that DSIF and P-TEFb function together in cells has come from the demonstration that P-TEFb and the DSIF Spt4 and Spt5 proteins colocalize at a large number of transcriptionally active chromosomal sites on *D. melanogaster* polytene chromosomes (82, 83). More importantly, although inhibiting expression of the Cdk9 subunit of P-TEFb prevents activation of heat shock gene expression in *C. elegans* embryos, heat shock gene activation does not require Cdk9 when Spt4 and Spt5 expression is inhibited (76). Notably, expression of a number of other genes in *C. elegans* lacking functional P-TEFb cannot be rescued by inhibition of Spt4 and Spt5; this raises the possibility that P-TEFb may also function through Spt4- and Spt5-independent pathways in vivo (76).

THE TFIIF, ELL, AND ELONGIN FAMILIES OF ELONGATION FACTORS

mRNA is synthesized in cells, on chromatin templates, at rates approaching 2000 nucleotides per minute (84–87). In contrast, purified RNA polymerase II is capable of elongating RNA chains in vitro, even on naked DNA templates, at rates of only 100 to 300 nucleotides per minute (88). These observations prompted a search for activities that would boost the elongation rate of the purified enzyme on both chromatin templates (discussed in more detail below) and on naked DNA templates.

Efforts to identify activities that stimulate the rate of elongation by RNA polymerase II on naked DNA templates led to the demonstration that a number of transcription factors, which include TFIIF and members of the ELL and Elongin families, can interact directly with elongating polymerase and increase its elongation rate by suppressing transient pausing by the enzyme at all or most steps of nucleotide addition (5, 89, 90). The exact mechanism by which TFIIF, ELL, and Elongin suppress RNA polymerase II pausing is not known. Nevertheless, several lines of evidence suggest that they may all function similarly to promote RNA polymerase II elongation in vitro by helping to maintain the 3′-OH terminus of the nascent transcript in proper alignment with the polymerase catalytic site, thereby preventing backtracking by the enzyme. First, all three elongation factors can stimulate both the binding of RNA polymerase II to the

3′-OH termini of DNA primers and the enzyme's subsequent DNA template-directed addition of ribonucleotides to those primers using a mechanism that may mimic transcription by the RNA polymerase II elongation complex (91). Second, interaction of TFIIF and Elongin with paused RNA polymerase II decreases the likelihood that the enzyme will suffer arrest (20), which, as discussed above, occurs when the 3′-end of the nascent transcript becomes misaligned with the polymerase active site due to backtracking of the enzyme. Third, TFIIF, ELL, and Elongin have all been shown to inhibit SII-induced cleavage of nascent transcripts by paused RNA polymerase II (92), consistent with the idea that they promote proper alignment of the 3′-end of nascent transcripts with the polymerase catalytic site, which reduces their susceptibility to SII-induced cleavage. Finally, TFIIF, ELL, and Elongin can all stimulate the rate at which nascent transcripts undergo cleavage by pyrophosphorolysis, a reaction that is the reverse of nucleotide addition by RNA polymerase II (32, 92).

The similarities between the transcription elongation activities of TFIIF, Elongin, and ELL in vitro raise the possibility that they may play overlapping or redundant roles in the control of elongation in vivo. As discussed below, however, increasing evidence suggests that they may serve quite different functions in cells.

A Role for TFIIF Elongation Activity in Transcription Initiation and Promoter Escape

TFIIF is ubiquitously expressed in eukaryotes from yeast to man. In higher eukaryotes,TFIIF is a heterodimer composed of ~30 kDa RAP30 and ~70 kDa RAP70 subunits. In *S. cerevisiae*, TFIIF is a heterotrimer composed of ~105 kDa Tfg1, ~54 kDa Tfg2, and ~30 kDa Tfg3 subunits (93). TFIIF was initially identified and purified as an essential RNA polymerase II initiation factor that is required for stable binding of polymerase to TFIID and TFIIB at the promoter and for entry of TFIIE and TFIIH into the preinitiation complex (94). Greenleaf and coworkers subsequently demonstrated that *Drosophila* TFIIF is capable of potently stimulating the rate of elongation by RNA polymerase II (95). Subsequent studies confirmed that mammalian TFIIF has a similar activity (88, 96, 97).

Although the exact role of TFIIF elongation activity in transcription has been unclear, Yan et al. (98) recently exploited TFIIF mutants that selectively lack transcription initiation and elongation activities to obtain evidence suggesting that TFIIF elongation activity is required for efficient transcription initiation by RNA polymerase II from promoters. Their findings argued that TFIIF increases the efficiency of transcription initiation by significantly reducing the frequency at which RNA polymerase II aborts transcription during synthesis of the first few phosphodiester bonds of nascent transcripts. TFIIF elongation activity could expedite transcription initiation by increasing the rate of synthesis of the first few phosphodiester bonds of nascent transcripts, ensuring that growing transcripts reach a sufficient length (>6 to 8 nucleotides) to be relatively resistant to abortion before abortion occurs. In light of evidence (*a*) that TFIIF dissociates from early RNA polymerase II elongation complexes shortly after polymerase has synthesized 9 or 10 nucleotide long transcripts (99, 100) and (*b*) that TFIIF apparently localizes in cells only near promoter regions, as judged by results of chromatin immunoprecipitation assays of *S. cerevisiae* TFIIF (49, 101), it is possible that TFIIF elongation activity plays an important role in RNA polymerase II transcription initiation and promoter escape, but that it does not play a major role in elongation once polymerase has escaped the promoter.

Evidence for an Essential Role for ELL in Elongation

ELL, the founding member of the ELL family, was initially purified to homogeneity from rat liver nuclei by its ability to stimulate the rate of elongation by RNA polymerase II in vitro (102). To date, two additional ELL family members, ELL2 (103) and ELL3 (104), have been identified in mammalian cells, and a single ELL homolog has been identified in *D. melanogaster* (105, 106). The ELL proteins range in size from ~50 kDa to ~120 kDa and contain a conserved N-terminal elongation activation domain and C-terminal Occludin-like domain. All of the ELLs function similarly to stimulate the overall rate of elongation by RNA polymerase II in vitro.

The human *ELL* gene was initially identified as a gene that undergoes translocations with the *MLL* gene in acute myeloid leukemia (107, 108). Chromosomal translocations involving the *MLL* and *ELL* genes generate a chimeric *MLL-ELL* gene that encodes a fusion protein that contains the entire ELL elongation activation domain and the C-terminal Occludin-like domain, but that lacks the first 45 N-terminal amino acids of the ELL protein.

Results of structure-function analyses of the MLL-ELL chimera indicate (*a*) that the ELL portion of the chimera is essential for transformation of hematopoietic precursors in vitro and for tumor formation in nude mice (109, 110), (*b*) that the ELL C-terminal Occludin-like domain, which is not required for ELL elongation activity in vitro (103), is sufficient for transformation of hematopoietic precursors (109, 110), but (*c*) that both the ELL elongation activation domain near the N-terminus of the protein and the ELL CTD have roles in tumor formation in nude mice. Thus, although the ELL Occludin-like CTD is sufficient for immortalization of hematopoietic precursors in vitro, both the ELL elongation activation domain and Occludin-like C-terminus appear to contribute to oncogenesis induced by the MLL-ELL fusion protein in mice.

Even though ELL has not yet been shown unequivocally to function in transcription elongation in vivo, it exhibits properties expected of an RNA polymerase II elongation factor. First, like the Cdk9 subunit of P-TEFb and the Spt5 subunit of DSIF, *D. melanogaster* ELL is found at a large number of transcriptionally active sites on polytene chromosomes and is rapidly relocalized with RNA polymerase II to heat shock genes when embryos are subjected to elevated temperatures. Second, *D. melanogaster* ELL has recently been identified as the product of the *Su(Tpl)* (*Suppressor of triplo-lethal*) gene (111). *D. melanogaster* ELL is essential for early embryonic development (106), and ELL mutations suppress the triplo-lethality of *Tpl*, a dosage-effect gene that is lethal in flies when present in three copies or one copy (111). As would be expected of a transcription elongation factor, ELL mutations preferentially affect expression of large genes: ELL mutations decrease expression of the 65 kb *Cut* gene, the 35 kb *Notch* gene, and the 23 kb *Sex combs reduced* gene, but they have little effect on expression of the shorter *white* and *rudimentary* genes (111). These results also argue strongly that ELL has functions distinct from the biochemically similar transcription factors TFIIF and Elongin, because the phenotypic consequences of *Su(Tpl)* mutations can be seen in flies expressing wild-type TFIIF and Elongin.

A Potential Role for Elongin as an E3 Ubiquitin Ligase

Elongin was originally purified to homogeneity from rat liver nuclei by its ability to stimulate the rate of elongation by RNA polymerase II in vitro (112, 113). Elongin is a heterotrimeric protein composed of a transcriptionally active ~100 kDa Elongin A subunit and two smaller ~18 kDa Elongin B and ~15 kDa Elongin C subunits, which can form an isolable Elongin BC subcomplex (114–116). Two additional Elongin A family members, Elongin A2 and Elongin A3, have been identified in mammalian cells (117, 118), and a single Elongin A homolog has been identified in *C. elegans* (119). All four Elongin A proteins function similarly to stimulate the overall rate of elongation by RNA polymerase II in vitro, and all four proteins have been shown to bind to the Elongin BC complex through a short, degenerate sequence motif referred to as the BC-box with consensus [(A,P,S,T)LXXXCXXX(V,I,L)].

Recently, Elongin A was found to belong to a larger family of Elongin BC-box proteins that can all be linked via Elongins B and C to a heterodimeric module composed of Cullin family proteins Cul2 or Cul5 and the RING-H2 finger protein Rbx1 (also referred to as ROC1 or Hrt1) to form multisubunit complexes that can function as E3 ubiquitin ligases (120). The best characterized of these Elongin BC-based ubiquitin ligases is the von Hippel-Lindau (VHL) tumor suppressor complex, where the VHL protein serves as a substrate recognition subunit that recruits target proteins for ubiquitylation and the Cullin/Rbx1 module functions to activate ubiquitylation of target proteins by an E2 ubiquitin conjugating enzyme. Accordingly, some, if not all, BC-box proteins are also likely to function as substrate recognition subunits of ubiquitin ligases.

Although it is not yet known whether Elongin A assembles into a ubiquitin ligase complex in cells and, if so, what its substrates might be, it is tempting to speculate that one function of Elongin A may be to recruit a Cullin/Rbx1 module directly to transcribing RNA polymerase II to target ubiquitylation of polymerase or other components of the transcription apparatus. It is noteworthy that RNA polymerase II is a known target for ubiquitylation, even though the regulatory significance of this modification remains unclear. A significant fraction of the largest RNA polymerase II subunit is rapidly ubiquitylated and degraded by the proteasome in cells exposed to UV light or treated with the genotoxic agent cisplatin (121, 122) in a process that appears to depend on the transcription repair coupling factors Rad26 and Def1 in yeast (123) and CSA and CSB in mammalian cells (121, 122). In addition, ubiquitylation of RNA polymerase II in vitro has been shown to be coupled to transcription elongation and to be strongly stimulated by DNA damage (124). In yeast, the ubiquitin ligase Rsp5 appears to be responsible for all or most ubiquitylation of RNA polymerase II (125, 126); however, the ubiquitin ligases responsible for ubiquitylation of polymerase in higher eukaryotes have not yet been defined. Because elongation by RNA polymerase II is blocked at many sites of DNA damage, it is has been suggested that ubiquitin-dependent degradation of polymerases stalled at these sites would provide a convenient mechanism for removing them from the DNA to allow the DNA repair machinery access to the damage. Whether ubiquitylation of RNA polymerase II actually plays an essential role in this process has been controversial, however, because it has been reported that removal, repression, or overexpression of Rsp5 in yeast has no affect on transcription-coupled DNA repair in yeast (127).

ELONGATOR AND FACT PROMOTE RNA POLYMERASE II ELONGATION THROUGH CHROMATIN

Unlike the RNA polymerase II elongation factors discussed thus far, Elongator and FACT do not appear to affect polymerase activity directly; instead, Elongator and FACT appear to promote RNA polymerase II elongation by modifying nucleosomes in ways that facilitate passage of polymerase through chromatin. Based on substantial evidence that nucleosomes and other chromatin proteins can severely impede RNA polymerase II elongation, Elongator and FACT may function in cells to help overcome the negative effects of chromatin on efficient mRNA synthesis.

Elongator

Elongator is a multiprotein complex composed of two subassemblies, one containing three subunits designated Elp1–3 (47) and the other containing three subunits, Elp 4–6 (128–130). Elongator was initially identified through its tight association with the hyperphosphorylated, elongating form of RNA polymerase II and was therefore proposed to function during transcription elongation (47). Consistent with this idea, yeast cells lacking genes encoding Elongator subunits exhibit phenotypes that are similar to those observed upon mutation of genes encoding other proteins implicated in elongation, which include 6-AU sensitivity and a decreased rate and magnitude of induction of some genes. In addition, the phenotypes of *elp* mutants are synergistically enhanced when combined with mutations in SII (47). On the other hand, in contrast to SII and the DSIF subunits,

Elongator subunits have not yet been found in association with the transcribed regions of genes in chromatin immunoprecipitation experiments in yeast (49, 101).

Elongator is highly conserved from yeast to man (47, 128–130). Notable among the Elongator subunits are Elp3, a histone acetyltransferase (HAT) of the GCN5-related N-acetyltransferase (GNAT) family, and Elp1, a protein encoded by the human *IKAP* gene, which is mutated in the neurodegenerative disease familial dysautonomia. The HAT activity associated with Elongator can acetylate both core and nucleosomal histones, predominantly on lysine 14 of histone H3 and lysinc 8 of histonc H4. Although there have been no reports that Elongator stimulates elongation by purified RNA polymerase II on naked DNA templates, addition of purified Elongator to nuclear extracts that have been immunodepleted with antibodies against Elp1 and Elp3 strongly stimulates transcription in a reaction that depends on acetyl-CoA (131).

FACT

FACT was initially identified and purified by its ability to promote RNA polymerase II elongation on chromatin DNA templates in vitro (132). FACT is a heterodimer composed of the Cdc68/Spt16 and HMG-like SSRP1 proteins (48). FACT is evolutionarily conserved from yeast to man. Mechanistic studies suggest that FACT stimulates RNA polymerase II elongation through nucleosomes by binding them and promoting removal of nucleosomal histones H2A and H2B. In addition, genetic evidence suggests that FACT interacts with DSIF in cells (48), and biochemical studies have shown that FACT can function together with P-TEFb to relieve DSIF- and NELF-mediated RNA polymerase II arrest in vitro (133), which supports a functional link between FACT and the P-TEFb/DSIF/NELF systems.

THE RNA POLYMERASE II ELONGATION COMPLEX FUNCTIONS AS A PLATFORM THAT COORDINATES mRNA PROCESSING AND EXPORT

Finally, advances in studies emanating from outside the realm of traditional transcription research are providing new and unanticipated insights into the RNA polymerase II elongation complex. These studies have solidified the notion that the polymerase is not simply an RNA synthetic machine, but it serves a more fundamental role as a critical platform for recruiting and coordinating the actions of enzymes and proteins responsible for mRNA maturation and export to the cytoplasm.

A Role for the RNA Polymerase CTD in Pre-mRNA Capping, Splicing, and Polyadenylation

As discussed above, the CTD of the largest subunit of RNA polymerase II is composed of tandemly repeated copies of a heptapeptide with consensus $[Y_1S_2P_3T_4S_5P_6S_7]$. The CTD is a target in vitro and in cells for extensive phosphorylation by a variety of cyclin-dependent kinases, which include P-TEFb and its potential *S. cerevisiae* homologue Ctk1, the TFIIH associated Cdk7(Kin28)/cyclin H kinase (134–138), the Mediator-associated Cdk8(Srb10)/cyclin C (Srb11) kinase (139, 140), and the *S. cerevisiae* Bur1 kinase (141, 142). During or shortly after initiation, RNA polymerase II undergoes TFIIH-dependent CTD phosphorylation predominantly on serine 5 of the heptapeptide repeats. Entry of RNA polymerase II into the elongation stage of mRNA synthesis is accompanied by loss of serine 5 phosphorylation and extensive serine 2 phosphorylation (70), regulated in yeast at least in part by the opposing action of the Ctk1 kinase and Fcp1 CTD phosphatase (143).

A mechanistic link between the RNA polymerase II elongation complex and the mRNA capping machinery was brought to light by the observation that mRNAs transcribed in cells by RNA polymerase II mutants that lack an intact CTD are not efficiently capped (144, 145). Capping of RNA polymerase II transcripts occurs when they reach a length of 25 to 30 nucleotides and requires the concerted action of three enzymes: an RNA triphosphatase, which removes the γ-phosphate at the 5′-end of the nascent transcript; a guanylyltransferase, which transfers a GMP cap to the resulting diphosphate terminus of the nascent transcript; and a methyltransferase, which adds a methyl group to the N7 position of the GMP cap to form the transcript's mature m7G(5′)ppp(5′)N cap structure (6).

Biochemical studies have shown that both mammalian and yeast guanylyltransferase are recruited to transcribing RNA polymerase II through specific interactions with the phosphorylated CTD (144, 146–150). Furthermore, the purified mammalian guanylyltransferase was shown to bind tightly and specifically to synthetic CTD peptides that contain as few as two copies of the heptapeptide repeats phosphorylated on either serine 2 or 5 (148). Guanylyltransferase activity is stimulated by CTD peptides phosphorylated on serine 5, but not on serine 2, which suggests that CTD phosphorylation by the TFIIH CTD kinase may provide a signal that links capping to the synthesis of nascent transcripts (148). Consistent with this model, genetic evidence indicates that the TFIIH CTD kinase is indeed required for recruitment of capping enzymes (guanylyltransferase in particular) to RNA polymerase II initiation and/or elongation complexes (151). In addition, results of chromatin immunoprecipitation assays indicate that, in cells, the guanylyltransferase is transiently associated with early RNA polymerase II elongation complexes near the promoter, whereas the guanine methyltransferase appears to associate with the enzyme continuously during transcription of a gene (152). Interestingly, the Spt5 component of DSIF also appears to stimulate capping; this provides an additional linkage between early RNA polymerase II elongation and capping (153).

Besides its role in capping of nascent transcripts, the RNA polymerase II CTD also links the transcript elongation complex to polyadenylation of pre-mRNAs (145, 154–156). Polyadenylation requires the action of a complex set of proteins and occurs in two steps: endonucleolytic cleavage of the mRNA precursor at a site downstream of the critical AAUAAA signal in the transcript and addition of 200 to 300 nucleotides of poly(A) to the newly generated transcript 3′-OH terminus by poly (A) polymerase (6, 7). Biochemical studies have revealed that polyadenylation of pre-mRNAs requires multiple proteins including cleavage/polyadenylation specificity factor (CPSF), cleavage stimulation factor (CstF), cleavage factors CFI and CFII, and poly(A) polymerase (6, 7).

Biochemical experiments investigating the mechanism of polyadenylation in vitro have revealed that the RNA polymerase CTD is required for efficient 3′ cleavage of transcripts prior to poly(A) addition by poly(A) polymerase (155, 157). RNA polymerase II containing either the hyperphosphorylated or hypophosphorylated forms of the CTD stimulate 3′ cleavage, which argues that, unlike the capping reaction, polyadenylation does not depend on phosphorylation of the CTD.

Although it is not yet known exactly how the CTD promotes 3′ cleavage of pre-mRNAs, it is noteworthy (*a*) that both CPSF and CstF have been shown to be capable of associating with both RNA polymerase II CTD and a GST-CTD fusion protein and (*b*) that the 50 kDa subunit of CstF appears to interact directly with the CTD (144, 145, 154, 158). In addition, evidence suggests that CPSF may be recruited to the CTD in the preinitiation complex through interactions with TFIID (159).

Another role for the CTD in linking transcription with the RNA processing machinery was suggested by the observation that cotranscriptional splicing in cells is inhibited by truncation of the RNA polymerase II CTD (145). Further strengthening the link between the CTD and pre-mRNA splicing, purified RNA polymerase II containing the hyperphosphorylated but not hypophosphorylated CTD strongly stimulates pre-mRNA splicing in vitro (6). Although the mechanisms underlying CTD action in splicing are not known, it is likely that the CTD recruits components of the splicing machinery to the RNA polymerase II elongation complex. SR and snRNP proteins have been shown to interact with phosphorylated RNA polymerase II and phosphorylated CTD peptides (160–162). Thus, it is reasonable that the CTD could stimulate splicing at least in part by promoting assembly of early splicing intermediates, such as the U1 snRNP-5′ splice site complex or the U2 snRNP-branch site complex (6).

The THO Complex Links the RNA Polymerase II Elongation Complex to the mRNA Export Machinery

The THO complex is best characterized in *S. cerevisiae*, where it is a heterotetrameric complex composed of the ~160 kDa Tho2, ~90 kDa Hpr1, ~60 kDa

Mft1, and ~40 kDa Thp2 proteins (163). Even though it has not been shown directly to regulate the activity of the RNA polymerase II elongation complex in vitro, components of the THO complex interact with free (164) and transcribing polymerase (165), and yeast strains containing mutations in subunits of the THO complex exhibit apparent defects in elongation by RNA polymerase II of both long and G+C-rich transcripts (166).

A link between the RNA polymerase II elongation complex and the mRNA export machinery was recently brought to light by findings indicating that yeast strains containing mutations in subunits of the THO complex are defective in nucleocytoplasmic transport of properly processed mRNAs (165, 167). The THO complex interacts with the mRNA export factors Sub2 and Yra1, as well as with a novel protein designated TREX component 1 (Tex-1) (165) to form a discrete and isolable complex called the transcription/export complex, or TREX. Taken together, these findings raise the intriguing possibility that the THO complex may recruit the mRNA export machinery directly to transcribing RNA polymerase II to facilitate expeditious transport of mRNA from the nucleus to their sites of translation on the ribosomes in the cytoplasm.

ACKNOWLEDGMENTS

We thank F. Barnett for assistance in preparation of the manuscript. We thank R. Kornberg for Figures 1 and 2. A. Shilatifard was supported by American Cancer Society Grant RP69921801, National Institutes of Health Grant R01CA089455, and a Mallinckrodt Foundation Award. R.C.Conaway and J.W.Conaway were supported by National Institutes of Health Grant R37 GM41628.

The *Annual Review of Biochemistry* is online at http://biochem.annualreviews.org

LITERATURE CITED

1. Cramer P, Bushnell DA, Kornberg RD. 2001. *Science* 292:1863–75
2. Gnatt AL, Cramer P, Fu J, Bushnell DA, Kornberg RD. 2001. *Science* 292: 1876–82
3. Cramer P, Bushnell DA, Fu J, Gnatt AL, Maier-Davis B, et al. 2000. *Science* 288: 640–49
4. Kerppola TK, Kane CM. 1991. *FASEB J.* 5:2833–41
5. Uptain SM, Kane CM, Chamberlin MJ. 1997. *Annu. Rev. Biochem.* 66:117–72
6. Hirose Y, Manley JL. 2000. *Genes Dev.* 14:1415–29
7. Barabino SML, Keller W. 1999. *Cell* 99:9–11
8. Zhang G, Campbell EA, Minakhin L, Richter C, Severinov K, Darst SA. 1999. *Cell* 98:811–24
9. Kuznedelov K, Kozheva N, Mustaev A, Severinov K. 2002. *EMBO J.* 21:1369–78
10. Bushnell DA, Cramer P, Kornberg RD. 2002. *Proc. Natl. Acad. Sci. USA* 99:1218–22
11. Sekimizu K, Kobayashi N, Mizuno D, Natori S. 1976. *Biochemistry* 15:5064–70
12. Kanai A, Kuzuhara T, Sekimizu K, Natori S. 1991. *J. Biochem.* 109:674–77

13. Xu Q, Nakanishi T, Sekimizu K, Natori S. 1994. *J. Biol. Chem.* 269:3100–3
14. Taira Y, Kubo T, Natori S. 1998. *Genes Cells* 3:289–96
15. Reines D, Chamberlin MJ, Kane CM. 1989. *J. Biol. Chem.* 264:10799–809
16. Rhodes D, Chamberlin MJ. 1974. *J. Biol. Chem.* 249:6675–83
17. Krummel B, Chamberlin MJ. 1992. *J. Mol. Biol.* 225:221–37
18. Matsuzaki H, Kassavetis GA, Geiduschek EP. 1994. *J. Mol. Biol.* 235:1173–92
19. Erie DA, Hajiseyedjavadi O, Young MC, von Hippel PH. 1993. *Science* 262:867–73
20. Gu W, Reines D. 1995. *J. Biol. Chem.* 270:11238–44
21. Davenport R, Wuite GJL, Landick R, Bustamante C. 2000. *Science* 287:2497–500
22. Adelman K, La Porta A, Santagelo TJ, Lis JT, Roberts JW, Wang MD. 2002. *Proc. Natl. Acad. Sci. USA* 99:13538–43
23. Nudler E. 1999. *J. Mol. Biol.* 288:1–12
24. Wind M, Reines D. 2000. *BioEssays* 22:327–36
25. Komissarova N, Kashlev M. 1997. *J. Biol. Chem.* 272:15329–38
26. Palangat M, Landick R. 2001. *J. Mol. Biol.* 311:265–82
27. Gu W, Reines D. 1995. *J. Biol. Chem.* 270:30441–47
28. Izban MG, Luse DS. 1993. *J. Biol. Chem.* 268:12874–85
29. Reines D. 1992. *J. Biol. Chem.* 267:3795–800
30. Reines D, Ghanouni P, Li Q, Mote J. 1992. *J. Biol. Chem.* 267:15516–22
31. Izban MG, Luse DS. 1992. *Genes Dev.* 6:1342–56
32. Wang DG, Hawley DK. 1993. *Proc. Natl. Acad. Sci. USA* 90:843–47
33. Rudd MD, Izban MG, Luse DS. 1994. *Proc. Natl. Acad. Sci. USA* 91:8057–61
34. Archambault J, Lacroute F, Ruet A, Friesen JD. 1992. *Mol. Cell. Biol.* 12:4142–52
35. Wu J, Awrey DE, Edwards AM, Archambault J, Friesen JD. 1996. *Proc. Natl. Acad. Sci. USA* 93:11552–57
36. Powell W, Bartholomew B, Reines D. 1996. *J. Biol. Chem.* 271:22301–4
37. Korzheva N, Mustaev A, Kozlov M, Malhotra A, Nikiforov V, et al. 2000. *Science* 289:619–25
38. Exinger G, Lacroute F. 1992. *Curr. Genet.* 142:749–59
39. Nakanishi T, Nakano A, Nomura K, Sekimizu K, Natori S. 1992. *J. Biol. Chem.* 267:13200–4
40. Wind-Rotolo M, Reines D. 2001. *J. Biol. Chem.* 276:11531–38
41. Shaw RJ, Reines D. 2000. *Mol. Cell. Biol.* 20:7427–37
42. Shimoaraiso M, Nakanishi T, Kubo T, Natori S. 2000. *J. Biol. Chem.* 275:29623–27
43. Lennon JC, Wind M, Saunders L, Hock MB, Reines D. 1998. *Mol. Cell. Biol.* 18:5771–79
44. Hemming SA, Jansma DB, Macgregor PF, Goryachev A, Friesen JD, Edwards AM. 2000. *J. Biol. Chem.* 275:35506–11
45. Lindstrom DL, Hartzog GA. 2001. *Genetics* 159:487–97
46. Hartzog GA, Wada T, Handa H, Winston F. 1998. *Genes Dev.* 12:357–69
47. Otero G, Fellows J, Li Y, de Bizemont T, Dirac A, et al. 1999. *Mol. Cell* 3:125–29
48. Orphanides G, Wu WH, Lane WS, Hampsey M, Reinberg D. 1999. *Nature* 400:284–88
49. Pokholok DK, Hannett NM, Young RA. 2002. *Mol. Cell* 9:799–809
50. Kulish D, Struhl K. 2001. *Mol. Cell. Biol.* 21:4162–68
51. Fraser NW, Sehgal PB, Darnell JE. 1978. *Nature* 272:590–93
52. Zandomeni R, Zandomeni MC, Shugar D, Weinmann R. 1986. *J. Biol. Chem.* 261:3414–19
53. Chodosh LA, Fire A, Samuels M, Sharp PA. 1989. *J. Biol. Chem.* 264:2250–57
54. Marshall NF, Price DH. 1992. *Mol. Cell. Biol.* 12:2078–90
55. Wada T, Takagi T, Yamaguchi Y, Fer-

dous A, Imai T, et al. 1998. *Genes Dev.* 12:343–56
56. Yamaguchi Y, Takagi T, Wada T, Yano K, Furuya A, et al. 1999. *Cell* 97:41–51
57. Marshall NF, Price DH. 1995. *J. Biol. Chem.* 270:12335–38
58. Marshall NF, Peng J, Xie Z, Price DH. 1996. *J. Biol. Chem.* 271:27176–83
59. Zhu YR, Pe'ery T, Peng TM, Ramanathan Y, Marshall N, et al. 1997. *Genes Dev.* 11:2622–32
60. Young RA. 1991. *Annu. Rev. Biochem.* 60:689–715
61. Nonet M, Sweetser D, Young RA. 1987. *Cell* 50:909–15
62. Bartolomei MS, Halden NF, Cullen CR, Corden JL. 1988. *Mol. Cell. Biol.* 8:330–39
63. Bartholomew B, Dahmus ME, Meares CF. 1986. *J. Biol. Chem.* 261:14226–31
64. Cadena DL, Dahmus ME. 1987. *J. Biol. Chem.* 262:12468–74
65. Payne JM, Laybourn PJ, Dahmus ME. 1989. *J. Biol. Chem.* 264:19621–29
66. Laybourn PJ, Dahmus ME. 1990. *J. Biol. Chem.* 265:13165–73
67. O'Brien T, Hardin S, Greenleaf A, Lis JT. 1994. *Nature* 370:75–77
68. Lu H, Flores O, Weinmann R, Reinberg D. 1991. *Proc. Natl. Acad. Sci. USA* 88:10004–8
69. Serizawa H, Conaway JW, Conaway RC. 1993. *Nature* 363:371–74
70. Komarnitsky PB, Cho E-J, Buratowski S. 2000. *Genes Dev.* 14:2452–60
71. Kim JB, Sharp PA. 2001. *J. Biol. Chem.* 276:12317–23
72. Ivanov D, Kwak YT, Guo J, Gaynor RB. 2000. *Mol. Cell. Biol.* 20:2970–83
73. Peng J, Zhu Y, Milto JT, Price DH. 1998. *Genes Dev.* 12:755–62
74. Mancebo HSY, Lee G, Flygare J, Tomassini J, Luu P, et al. 1998. *Genes Dev.* 11:2633–44
75. Lis JT, Mason P, Peng J, Price DH, Werner J. 2000. *Genes Dev.* 14:792–803
76. Shim EY, Walker AK, Shi Y, Blackwell TK. 2002. *Genes Dev.* 16:2135–46
77. Herrmann CH, Rice AP. 1995. *J. Virol.* 69:1612–20
78. Yang XZ, Gold MO, Tang DN, Lewis DE, Aguilar-Cordova E, et al. 1997. *Proc. Natl. Acad. Sci. USA* 94:12331–36
79. Gold MO, Yang XZ, Herrmann CH, Rice AP. 1998. *J. Virol.* 72:4448–53
80. Wei P, Garber ME, Fang SM, Fischer WH, Jones KA. 1998. *Cell* 92:451–62
81. Yamaguchi Y, Inukai N, Narita T, Wada T, Handa H. 2002. *Mol. Cell. Biol.* 22:2918–27
82. Kaplan CD, Morris JR, Wu C, Winston F. 2000. *Genes Dev.* 14:2623–34
83. Andrulis ED, Guzman E, Doring P, Werner J, Lis JT. 2000. *Genes Dev.* 14:2635–49
84. Ucker DS, Yamamoto KR. 1984. *J. Biol. Chem.* 259:7416–20
85. Thummel CS, Burtis KC, Hogness DS. 1990. *Cell* 61:101–11
86. Shermoen AW, O'Farrell PH. 1991. *Cell* 67:303–10
87. Tennyson CN, Klamut HJ, Worton RG. 1995. *Nat. Genet.* 9:184–90
88. Izban MG, Luse DS. 1992. *J. Biol. Chem.* 267:13647–55
89. Shilatifard A, Conaway JW, Conaway RC. 1997. *Curr. Opin. Genet. Dev.* 7:199–204
90. Shilatifard A. 1998. *FASEB J.* 12: 1437–46
91. Takagi Y, Conaway JW, Conaway RC. 1995. *J. Biol. Chem.* 270:24300–5
92. Elmendorf BJ, Shilatifard A, Yan Q, Conaway JW, Conaway RC. 2001. *J. Biol. Chem.* 276:23109–14
93. Conaway RC, Conaway JW. 1993. *Annu. Rev. Biochem.* 62:161–90
94. Roeder RG. 1996. *Trends Biochem. Sci.* 21:327–35
95. Price DH, Sluder AE, Greenleaf AL. 1989. *Mol. Cell. Biol.* 9:1465–75
96. Flores O, Maldonado E, Reinberg D. 1989. *J. Biol. Chem.* 264:8913–21
97. Tan S, Aso T, Conaway RC, Conaway JW. 1994. *J. Biol. Chem.* 269:25684–91
98. Yan Q, Moreland RJ, Conaway JW,

Conaway RC. 1999. *J. Biol. Chem.* 274: 35668–75
99. Zawel L, Kumar KP, Reinberg D. 1995. *Genes Dev.* 9:1479–90
100. Moreland RJ, Hanas JS, Conaway JW, Conaway RC. 1998. *J. Biol. Chem.* 273: 26610–17
101. Krogan NJ, Kim M, Ahn SH, Zhong G, Kobor MS, et al. 2002. *Mol. Cell. Biol.* 22:6979–92
102. Shilatifard A, Lane WS, Jackson KW, Conaway RC, Conaway JW. 1996. *Science* 271:1873–76
103. Shilatifard A, Haque D, Conaway RC, Conaway JW. 1997. *J. Biol. Chem.* 272: 22355–63
104. Miller T, Williams K, Johnstone RW, Shilatifard A. 2001. *J. Biol. Chem.* 275: 32052–56
105. Gerber M, Ma J, Dean K, Eissenberg JC, Shilatifard A. 2001. *EMBO J.* 20: 6104–14
106. Khattak S, Im H, Park T, Ahnn J, Spoerel NA. 2002. *Cell Biochem. Funct.* 20: 119–27
107. Thirman MJ, Levitan DA, Kobayashi H, Simon MC, Rowley JD. 1994. *Proc. Natl. Acad. Sci. USA* 91:12110–14
108. Mitani K, Kanda Y, Ogawa S, Tanaka T, Inazawa J, et al. 1995. *Blood* 85:2017–24
109. DiMartino JF, Miller T, Ayton PM, Landewe T, Hess JL, et al. 2000. *Blood* 96:3887–93
110. Luo RT, Lavau C, Du C, Simone F, Polak PE, et al. 2001. *Mol. Cell. Biol.* 21:5678–87
111. Eissenberg JC, Ma JY, Gerber MA, Christensen A, Kennison JA, Shilatifard A. 2002. *Proc. Natl. Acad. Sci. USA* 99:9894–99
112. Bradsher JN, Jackson KW, Conaway RC, Conaway JW. 1993. *J. Biol. Chem.* 268:25587–93
113. Bradsher JN, Tan S, McLaury H-J, Conaway JW, Conaway RC. 1993. *J. Biol. Chem.* 268:25594–603
114. Aso T, Lane WS, Conaway JW, Conaway RC. 1995. *Science* 269:1439–43
115. Garrett KP, Tan S, Bradsher JN, Lane WS, Conaway JW, Conaway RC. 1994. *Proc. Natl. Acad. Sci. USA* 91:5237–41
116. Garrett KP, Aso T, Bradsher JN, Foundling SI, Lane WS, et al. 1995. *Proc. Natl. Acad. Sci. USA* 92:7172–76
117. Aso T, Yamazaki K, Amimoto K, Kuroiwa A, Higashi H, et al. 2000. *J. Biol. Chem.* 275:6546–52
118. Yamazaki K, Guo L, Sugahara K, Zhang C, Enzan H, et al. 2002. *J. Biol. Chem.* 277:26444–51
119. Aso T, Haque D, Barstead RJ, Conaway RC, Conaway JW. 1996. *EMBO J.* 15:5557–66
120. Kamura T, Burian D, Yan Q, Schmidt SL, Lane WS, et al. 2001. *J. Biol. Chem.* 276:29748–53
121. Bregman DB, Halaban R, van Gool AJ, Henning KA, Friedberg EC, Warren SL. 1996. *Proc. Natl. Acad. Sci. USA* 93:11586–90
122. Ratner JN, Balasubramanian B, Corden J, Warren SL, Bregman DB. 1998. *J. Biol. Chem.* 273:5184–89
123. Woudstra EC, Gilbert C, Fellows J, Jensen L, Brouwer J, et al. 2002. *Nature* 415:929–33
124. Lee KB, Wang D, Lippard SJ, Sharp PA. 2002. *Proc. Natl. Acad. Sci. USA* 99:4239–44
125. Huibregtse JM, Yang JC, Beaudenon SL. 1997. *Proc. Natl. Acad. Sci. USA* 94:3656–61
126. Beaudenon SL, Huacani MR, Wang G, McDonnell DP, Huibregtse JM. 2002. *Mol. Cell. Biol.* 19:6972–79
127. Lommel L, Bucheli ME, Sweder KS. 2000. *Proc. Natl. Acad. Sci. USA* 97:9088–92
128. Winkler GS, Petrakis TG, Ethelberg S, Tokunaga M, Erdjument-Bromage H, et al. 2001. *J. Biol. Chem.* 276:32743–49
129. Krogan NJ, Greenblatt JF. 2001. *Mol. Cell. Biol.* 21:8203–12
130. Li Y, Takagi Y, Jiang Y, Tokunaga M, Erdjument-Bromage H, et al. 2001. *J. Biol. Chem.* 276:29628–31

131. Kim JH, Lane WS, Reinberg D. 2002. *Proc. Natl. Acad. Sci. USA* 99:1241–46
132. Short NJ. 1987. *Nature* 326:740–41
133. Wada T, Orphanides G, Hasegawa J, Kim DK, Yamaguchi Y, et al. 2000. *Mol. Cell* 5(6):1067–72
134. Feaver WJ, Gileadi O, Li Y, Kornberg RD. 1991. *Cell* 67:1223–30
135. Feaver WJ, Svejstrup JQ, Henry NL, Kornberg RD. 1994. *Cell* 79:1103–9
136. Roy R, Adamczewski JP, Seroz T, Vermeulen W, Tassan JP, et al. 1994. *Cell* 79:1093–101
137. Serizawa H, Makela TP, Conaway JW, Conaway RC, Weinberg RA, Young RA. 1995. *Nature* 374:280–82
138. Shiekhattar R, Mermelstein F, Fisher RP, Drapkin R, Dynlacht B, et al. 1995. *Nature* 374:283–87
139. Liao SM, Zhang J, Jeffery DA, Koleske AJ, Thompson CM, et al. 1995. *Nature* 374:193–96
140. Hengartner CJ, Myer VE, Liao SM, Wilson CJ, Koh SS, Young RA. 1998. *Mol. Cell* 2:43–53
141. Murray S, Udupa R, Yao S, Hartzog G, Prelich G. 2001. *Mol. Cell. Biol.* 21:4089–96
142. Yao S, Neiman A, Prelich G. 2000. *Mol. Cell. Biol.* 20:7080–87
143. Cho E-J, Kobor MS, Kim M, Greenblatt J, Buratowski S. 2001. *Genes Dev.* 15:3319–29
144. Cho E-J, Takagi T, Moore CR, Buratowski S. 1997. *Genes Dev.* 11:3319–26
145. McCracken S, Fong N, Yankulov K, Ballantyne S, Pan G, et al. 1997. *Nature* 385:357–61
146. McCracken S, Fong N, Rosonina E, Yankulov K, Brothers G, et al. 1997. *Genes Dev.* 11:3306–18
147. Cho E-J, Rodriguez CR, Takagi T, Buratowski S. 1998. *Genes Dev.* 12:3482–87
148. Ho CK, Shuman S. 1999. *Mol. Cell* 3:405–11
149. Yue ZY, Maldonado E, Pillutla R, Cho H, Reinberg D, Shatkin AJ. 1997. *Proc. Natl. Acad. Sci. USA* 94:12898–903
150. Pillutla RC, Yue Z, Maldonado E, Shatkin AJ. 1998. *J. Biol. Chem.* 273: 21443–46
151. Rodriguez CR, Cho E-J, Keogh M-C, Moore CL, Greenleaf AL, Buratowski S. 2000. *Mol. Cell. Biol.* 20:104–12
152. Takase Y, Takagi T, Komarnitsky PB, Buratowski S. 2000. *Mol. Cell. Biol.* 20:9307–16
153. Wen Y, Shatkin AJ. 1999. *Genes Dev.* 13:1774–79
154. Licatalosi DD, Geiger G, Minet M, Schroeder S, Cilli K, et al. 2002. *Mol. Cell* 9:1101–11
155. Hirose Y, Manley JL. 1998. *Nature* 395: 93–96
156. Proudfoot NJ, Furger A, Dye MJ. 2002. *Cell* 108:501–12
157. Ryan K, Murthy KGK, Kaneko S, Manley JL. 2002. *Mol. Cell. Biol.* 22: 1684–92
158. Fong N, Bentley DL. 2001. *Genes Dev.* 15:1783–95
159. Dantonel J-C, Murthy KGK, Manley JL, Tora L. 1997. *Nature* 389:399–402
160. Chabot B, Bisotto S, Vincent M. 1995. *Nucleic Acids Res.* 23:3206–13
161. Mortillaro MJ, Blencowe BJ, Wei S, Nakayasu H, Du L, et al. 1996. *Proc. Natl. Acad. Sci. USA* 93:8253–57
162. Yuryev A, Patturajan M, Litingtung Y, Joshi RV, Gentile C, et al. 1996. *Proc. Natl. Acad. Sci. USA* 93:6975–80
163. Chavez S, Beilharz T, Rondon AG, Erdjument-Bromage H, Tempst P, et al. 2000. *EMBO J.* 19:5824–34
164. Chang M, French-Cornay D, Fan H, Klein H, Denis C, Jaehning JA. 1999. *Mol. Cell. Biol.* 19:1056–67
165. Straber K, Masuda S, Mason P, Pfannstiel J, Oppizzi M, et al. 2002. *Nature* 746:1–4
166. Chavez S, Garcia-Rubio M, Prado F, Aguilera A. 2001. *Mol. Cell. Biol.* 21:7054–64
167. Jimeno S, Rondon AG, Luna R, Aguilera A. 2002. *EMBO J.* 21:3526–35

Annu. Rev. Biochem. 2003. 72:717–742
doi: 10.1146/annurev.biochem.72.121801.161625
Copyright © 2003 by Annual Reviews. All rights reserved

DYNAMICS OF CELL SURFACE MOLECULES DURING T CELL RECOGNITION

Mark M. Davis,[1] Michelle Krogsgaard,[1] Johannes B. Huppa,[1] Cenk Sumen,[1] Marco A. Purbhoo,[1] Darrell J. Irvine,[2] Lawren C. Wu,[3] and Lauren Ehrlich[4]

[1]Howard Hughes Medical Institute and the Department of Microbiology and Immunology, Stanford University, Stanford, California 94305-5323; email: mdavis@cmgm.stanford.edu, mkrogsgd@cmgm.stanford.edu, jhuppa@cmgm.stanford.edu, csumen@stanford.edu, purbhoo@cmgm.stanford.edu
[2]Department of Materials Science and Engineering, Massachusetts Institute of Technology, Cambridge, Massachusetts 02139; email: djirvine@mit.edu
[3]Genentech, Incorporated, South San Francisco, California 94080-4990; email: lawren@gene.com
[4]Department of Microbiology and Immunology, University of California, San Francisco, California 94143; email: lehr5027@itsa.ucsf.edu

Key Words T lymphocytes, cell:cell recognition, cell surface molecules, T cell receptors, immunological synapse

■ **Abstract** Recognition of foreign antigens by T lymphocytes is a very important component of vertebrate immunity—vital to the clearance of pathogenic organisms and particular viruses and necessary, indirectly, for the production of high affinity antibodies. T cell recognition is mediated by the systematic scanning of cell surfaces by T cells, which collectively express many antigen receptors. When the appropriate antigenic peptide bound to a molecule of the major histocompatibility complex is found—even in minute quantities—a series of elaborate cell-surface molecule and internal rearrangements take place. The sequence of events and the development of techniques required to observe these events have significantly enhanced our understanding of T cell recognition and may find application in other systems of transient cell:cell interactions as well.

CONTENTS

INTRODUCTION

Cell-cell interactions are fundamental to the development and day-to-day functions of multicellular organisms. These interactions may be long lasting or transient, but in most cases, they are mediated by cell surface glycoproteins. While a great deal is known about the identity and structure of these molecules in some systems, what has been lacking is an understanding of how they function in situ and particularly how multiple cell surface molecules on one cell act together to recognize and dock with another cell. This situation has persisted because cell-cell interactions can only be studied in the context of intact membranes and thus do not easily lend themselves to the classical "purify and reconstitute" strategy that has worked so well with soluble molecules and pathways.

Recent advances in labeling and imaging technologies have dramatically changed this situation, however, and we are beginning to get a glimpse of how at least some cell-cell interactions are accomplished. These analyses depend on labeling cell surface molecules either intrinsically [e.g., as fusions with Green Fluorescence Protein (GFP)] or extrinsically and observing their dynamics in either intact cells or in artificial membranes. Although these methods can be applied to virtually any system where cells interact with each other, this review will focus on the special case of T lymphocyte recognition of antigen bearing cells.

While this type of interaction has a number of specialized requirements that are not shared with other types of cell-cell recognition, there has been a great deal of activity in this area recently, and it seems likely that at least some of the principles and surprising complexity that are emerging will be generalizable, at least for transient interactions. T lymphocytes readily lend themselves to such studies because of the wealth of information about molecules involved in recognition (both on the cell surface and internally), the extensive work on the signaling pathways, and the fact that the recognition process can be studied conveniently in vitro on a time scale of minutes to hours.

COMPONENTS OF T CELL RECOGNITION

T lymphocytes are an essential component of the adaptive immune response and have been documented in almost all chordates (1). Two distinct lineages of T cells have been identified based on whether they express an $\alpha\beta$ or $\gamma\delta$ type of T cell receptor for antigen. These are disulfide linked heterodimeric molecules, which are equivalent to surface antibodies on B lymphocytes with many of the same properties of rearranging gene segments to create a large repertoire of different binding sites in the V region domains (see below) (2). The $\alpha\beta$ type of T cell predominates in most mammals and is geared toward recognizing fragments of antigens (usually peptides) bound to molecules of the major histocompatibility complex (MHC). $\gamma\delta$ T cells are not as well understood, but in at least some cases they recognize the surfaces of intact molecules and in others are able to detect small phosphate containing molecules (likely presented by unknown molecules on cell surfaces) (3, 4).

Here we deal exclusively with $\alpha\beta$ T cell receptor for antigen (TCR) bearing T cells, as that is where the most information is available. Cells of this type can be subdivided into helper and cytotoxic T cells. The T helper cells express the CD4 molecule and secrete cytokines and other factors when they recognize antigens displayed on class II MHC molecules. They are particularly important in inducing B cells to proliferate and mature into antibody secreting cells and probably perform a similar function for other cells in the immune system. They have also been implicated in some forms of autoimmunity, and it is, perhaps, to avoid this happening that T helper interactions with B cells and other cells normally go on for many hours. The other major subset, the cytotoxic or killer T cells typically express CD8 dimers and induce apoptosis in cells on which they recognize foreign antigens bound to class I MHC molecules.

A further distinction is that class I MHC molecules generally gather and display internal antigens (5), whereas class II MHC present serum antigens taken up through the endosomal pathway (6). Thus cytotoxic T cells will normally only kill cells that express the offending antigen (such as a viral protein). Cytotoxic T cell recognition and reaction also occur on a much faster time scale than do T helper cells, having preloaded granules that can release their contents within minutes of engagement.

Self-reactive $\alpha\beta$ T cells are largely removed during differentiation in the thymus, hence reactivity among those that remain is generally against foreign entities.

IMPORTANT T CELL SURFACE MOLECULES AND THEIR LIGANDS

There is a considerable body of work on the biochemistry and biological properties of many T cell surface molecules and their ligands, as shown in Figure 1 and Table 1 (from References 7–17, 19–32). An important general principle is that essentially all membrane proteins that bind to proteins in another cell's membrane are of very low affinity—in the micromolar range (17, 33). Yet these interactions are highly specific and sensitive to even small changes in key contact residues. There are at least two reasons for this. One is that these receptor-ligand pairs operate in an essentially two dimensional environment (34, 35), and the second is that these are transient interactions (34, 35). This last point is particularly important as Leckband and colleagues (35) showed that affinities $<10^5$ M make it difficult to detach two membranes without disrupting their structure.

TCR and CD3

TCR is an antibody-like heterodimeric molecule with each chain having a variable and a single constant region. As with antibody genes, the variable domain is encoded in separate pieces that rearrange differently in each T cell to create a potential repertoire of $\sim 10^{13}$ possible species (36). For $CD4^+$ T cells (helper cells) TCRs recognize short [10–25 amino acids (aa)] peptides bound to class II MHC molecules (6), and for $CD8^+$ T cells (cytotoxic) 8–10 aa peptides are recognized when bound to class I MHCs (5). Signaling functions are carried out by the associated CD3 polypeptides (γ, δ, ϵ, ζ,) whose cytoplasmic domains are phosphorylated by Src kinases in the early stages of activation. In addition to the binding properties of TCRs with peptide-MHC (pMHC) ligands summarized in Table 1, recent structural, thermodynamic, and mutagenesis studies have suggested properties of TCRs that may be uniquely suited for scanning the surfaces of other cells and other properties. These analyses will be discussed in the last section.

LFA-1

This α_1 β_2 integrin molecule serves both an adhesive function and as a costimulatory receptor (meaning it synergizes with TCR/CD3 signaling as discussed in more detail below). Its ligand is ICAM-1 in the mouse and ICAM-1, 2 and 3 in humans. The intriguing feature of LFA-1 is that it increases its affinity ($\sim$100x) (37) for ligand early in the activation process, and this markedly enhances adhesiveness.

TABLE 1 Binding characteristics of T cell surface molecules with their ligands

Receptor	Ligand	Affinity K_D (μM)	K_{on} ($M^{-1}s^{-1}$)	k_{off} (s^{-1})	Temp. (°C)	ΔG[a] (kcal/mol)	TΔS (kcal/mol)	References
TCR	Peptide/MHC[b,c]	0.1–57	850–210000	0.011–0.08	25	5.9–9.7	−6 to −15.9	(7–17, 19–22)
LFA-1[d]	ICAM-1	0.13–0.36	224000	0.0298	25	8.9–9.5	–	(23, 24)
LFA-1	ICAM-1	91–100	266–367	0.032–0.033	37	5.5–5.6	–	(24)
CD4	Class II MHC	199	–	–	25	5.1	–	(25)
CD8$\alpha\alpha$	Class I MHC	200	≥100000	≥18	37	5.1	–	(26)
CD28	B7-1	4	≥660000	≥1.6	37	7.5	–	(27, 28)
CD28	B7-2	20	≥1400000	≥28	37	6.5	–	(27, 28)
CTLA-4	B7-1	0.2–1.4	≥2150000	≥0.43	37	8.1–9.3	–	(27, 28)
CTLA-4	B7-2	2.6–22.4	1960000	5.1	37	6.4–7.7	–	(27, 28)
CD2	CD58	9–22	≥400000	≥4	37	6.4–7	–	(29)
CD2	CD48	≥90	≥100000	≥6	37	≤5.6	–	(29–32)
CD45	Unknown							

[a]Calculated from $\Delta G = RT\ln K_a$, where RT equals 0.6.

[b]Only numbers for agonist peptide ligands derived from the original antigen are shown.

[c]If more than one value can be found in the literature, a range of measured values is indicated.

[d]Activated form of LFA-1.

CD4

This molecule binds weakly to class II MHC molecules, is associated with TCR/CD3 on activated T cells, and binds the Src kinase lck on its cytoplasmic tail. These features facilitate the delivery of lck to the TCR/CD3 molecules early in activation and seem a key part of the triggering mechanism. Absence or blockage of CD4 results in a requirement for much more peptide-MHC (10–100x) (38).

CD8$\alpha\beta$, $\alpha\alpha$

This dimeric molecule binds to class I MHC molecules (Table 1) and potentiates the reactivities of $CD8^+$ T cells, also (as with CD4) by delivering lck to TCR/CD3 (38).

CD28

This is the principal costimulatory receptor that colocalizes with the TCR in the central region of the synapse (39) and also activates a distinct signaling pathway, which (*a*) triggers the active transport of cell surface molecules into the synapse (see below), (*b*) influences nuclear transcription directly (40), and (*c*) blocks the egress of NFAT from the nucleus—thereby enhancing the effect of this family of transcription factors (41). It has at least two ligands, B7-1 and B7-2.

CTLA-4

This molecule also binds to B7-1 and B7-2, but it comes to the surface some minutes after activation begins and has a negative regulatory effect—that is, it inhibits T cell proliferation (42, 43).

CD2

This costimulatory molecule also binds to CD48 and CD58. The exact circumstances under which this is critical are not clear (44).

CD45

This large surface molecule has no known ligand, but its phosphatase activity seems crucial for both lck function and later events in activation (45, 46). At least late in synapse formation, it is on the periphery of the interface, which may serve to promote src kinase activity in the central regions.

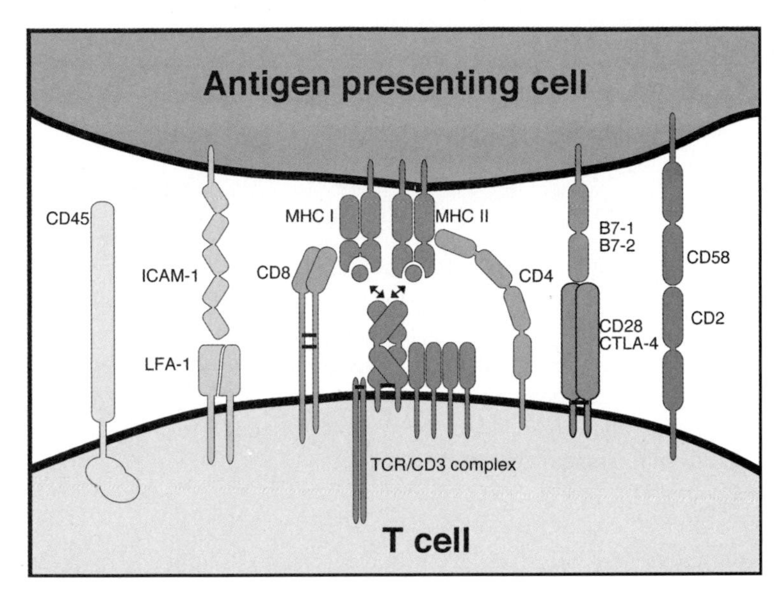

Figure 1 Cell surface molecules involved in T cell recognition. Molecules thought to play key roles in T cell recognition, as discussed in the text.

OVERVIEW OF T CELL RECOGNITION

A cartoon of the T cell molecules discussed above and their ligands is shown in Figure 1, with their approximate dimensions. The arrows between the TCR and its possible peptide-MHC ligands reflect the uncertainty as to whether the TCR binds to the same MHC as its associated CD4/CD8 molecules (on most mature T cells, either one or the other is expressed). Figure 2 shows the cellular behavior associated with recognition (47, 48) in which a T cell first probes another cell with pseudopodial extensions (the scanning phase) and then, finding the correct ligand, becomes activated with a sustained elevation of calcium (achieved through both the release of stored material and imported through specific channels). At this point, it begins to form a synapse characterized by a discrete pattern of central TCR/CD3 accumulation surrounded by a ring of integrin (LFA-1) (49–53) (see later sections). Following this synapse phase, the T cell will secrete either cytokines to stimulate the cell it is recognizing (if it is a T helper cell) or factors to induce cell death (for a T cytotoxic cell).

INITIATION OF T CELL ACTIVATION

It has been known for some time that T cells are quite sensitive to antigen, with estimates ranging from 1 to 400 peptide-MHC complexes on a cell necessary to achieve complete T cell responses (54–59). These studies were

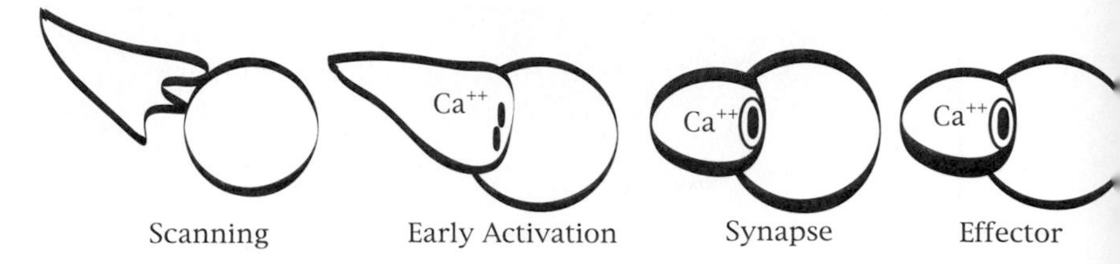

Figure 2 Overview of T cell recognition. Cell biology of T cell recognition, from first probing of potential target cells for specific peptide-MHC complexes to early activation, synapse formation, and the effector phase in which either cytokines (from T helper cells) or cell death promoting molecules (from cytotoxic T cells) are directed toward the cell being recognized. For T helper cells the time scale prior to this last phase is many hours (6–12) and for T cytotoxic cells it can be minutes to hours [adapted from references (47, 48)].

indirect, however, because they relied on average values of peptide loading on a given MHC and because there is no very firm idea of the average surface area surveyed by a T cell (58). Very recently, we developed a new approach to this problem, which gives more precise information about T cell sensitivity and the nature of T cell responsiveness to different numbers of ligands (60).

In these experiments, B cells are pulsed with varying concentrations of a model antigenic peptide, which is modified to have a 15 aa extension from the MHC binding core and which terminates in a biotin. After washing away free peptide, a streptavidin-phycoerythrin conjugate is introduced and saturates the exposed biotin groups on the peptide-MHC complexes. Phycoerythrin is a large (~240 kDa) seven chain protein with ~34 flurophores (61). Because of its consequent brightness, it is possible to visualize single phycoerythrin molecules with a standard cooled CCD camera (60). In this way, we can count the exact number of ligands that a T cell encounters on another cell and then follow the consequences of that interaction with respect to the elevation and the behaviors of different GFP labeled proteins. Figure 3 shows some of this data, which demonstrates that these $CD4^+$ T cells can respond to even a single ligand by stopping and weakly fluxing calcium. Two ligands in the interface produce a more sustained rise in calcium, and ten or more produce a maximal response (not shown). This dose-response relationship can be seen more clearly in the data of Figure 4, which shows the total calcium elevation over background in the fifteen minutes after ligand encounter. Here each additional ligand produces more calcium release/import up to saturation.

It is also at ~10 ligands that an immunological synapse forms; this suggests that some important threshold has been reached and that activation will proceed to full cytokine release. Interestingly, parallel experiments using a cytotoxic T cell model give a very similar dose response curve (M. A. Purbhoo et al., in preparation). Thus we can see a very similar sensitivity in two different CD4 T cells and now two different cytotoxic T cell models. At least three of these can detect even one molecule of ligand, but all require ~10 to fully elevate and maintain calcium levels. Extensive work on T cell signaling has shown that calcium elevation is a key indicator of activation and is responsible for inducing the nuclear localization of the transcription factor NFAT, which seems responsible for many of the gene expression changes associated with activated T helper cells (62).

Another interesting finding from this study was that CD4 blockade dramatically affected the ability of the T helper cells to detect one or a few ligands. Without CD4, it took 25–30 peptide-MHCs in the interface to induce T cells to stop and flux calcium, which they did at normal levels (60). These data indicate that CD4 is intimately involved in detecting antigens at very low surface densities, and we have suggested a pseudodimer model in which CD4 cross-links two TCRs, one that is binding to an agonist peptide-MHC and a second that is binding very weakly to an endogenous peptide-MHC. Alternatively, it may be that engagement of even one ligand, together with the delivery of lck to the site by CD4 or CD8, is sufficient to initiate a signaling cascade (63), although most data initiating T cell activation with soluble peptide-MHC complexes show a requirement for dimers or higher order multimers (64–66).

In the absence of CD4, we suggest that T cells must depend on seeing dimers of agonist complexes, which must be numerous in order for some to dimerize during a T cell scan. In the future, we hope to critically test this model as well as survey the response of other key molecules during T cell activation by defined numbers of ligands. It is also important to follow the chain of events to a full fledged effector function (e.g., cell killing or cytokine release), although sustained calcium elevation and synapse formation correlate very well with these results.

FORMATION OF AN IMMUNOLOGICAL SYNAPSE: ROLE OF COSTIMULATION

After T cell activation has begun, various cell surface molecules on the T cell and their corresponding ligands on the cell, which are recognized, segregate into discrete regions of the interface where TCR/CD3, CD28, and (briefly) CD4 or CD8 occupy the central region with an outer ring with LFA-1 and CD45. On the recognized cell, the relevant MHC forms a core opposite the T cell core, and the outer ring has ICAM-1 (49–53, 67, 68). Figure 5 shows a time lapse view of this reorganization from the work of Dustin and colleagues (53) who have studied the interaction of specific CD4 T cells with an artifical bilayer containing fluorescently labeled peptide-MHC (green) and ICAM-1 (red). Accompanying this reorganization of the cell membrane is significant interior activity as well, where lck (briefly), PKC θ, and actin congregate underneath the synapse core, and the

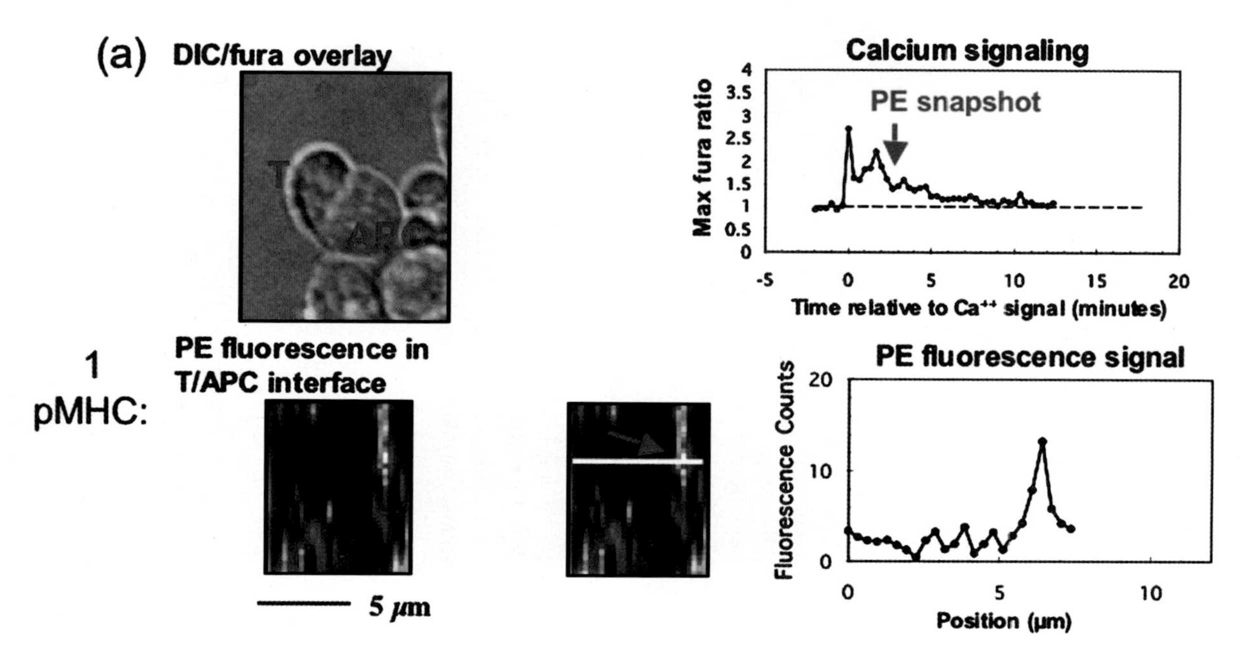

Figure 3 T cell activation can be initiated by even a single ligand. (*a*) Individual peptides loaded onto MHC molecules on the surface of B cells can be labeled with streptavidin-phycoerythrin conjugated such that the precise number of ligands that a T cell contacts can be seen and counted. This panel shows a T cell:B cell conjugate (*upper left*) in which imaging through the interface (*lower left*) shows a fluorescence signal corresponding to a single phycoerythrin molecule, which denotes a peptide at that position. The *lower right* panel shows that a trace through this fluorescence is well above background, and the *upper right* panel shows significant but transient calcium elevation in the T cell. This indicates partial activation. (*b*) Here the same analysis shows two phycoerythrin molecules close together at the interface and a more robust calcium elevation, although well short of full activation (60).

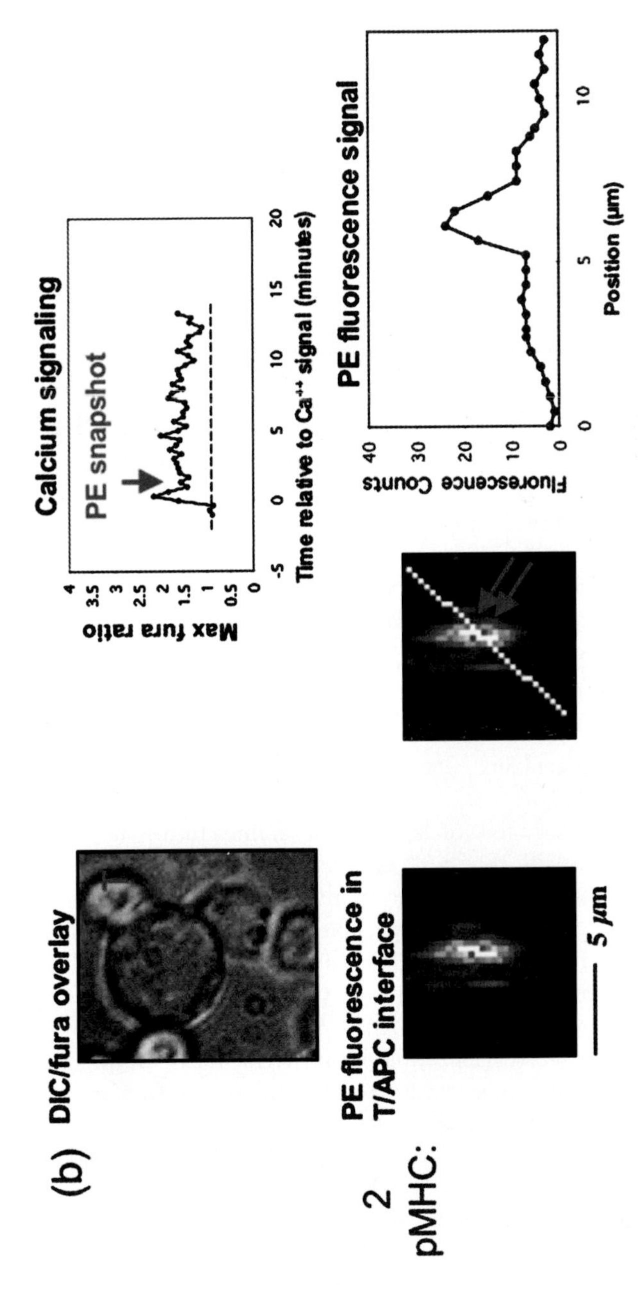

Figure 3 *Continued.*

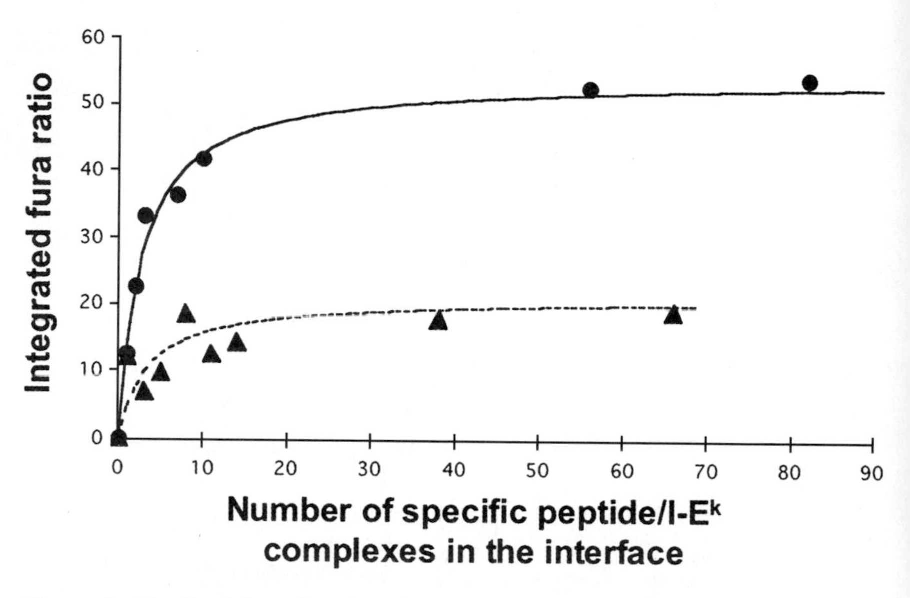

Figure 4 T cell calcium elevation plateaus at ~ten ligands. Here, in a dose response curve, the total calcium elevation produced by a given number of peptide-MHC ligands is plotted for two different T helper cells lines (derived from TCR transgenic animals). In each case, a peak calcium response is reached at ~10 ligands. Similar results were obtained with two other cytotoxic T cell lines specific for other peptide-MHC (60; M. Purbhoo, in preparation).

microtubule organizing center (MTOC) reorients between the synapse and the nucleus (69–71).

While the exact role of the synapse architecture is not known, its presence correlates well with robust T cell activation, and it is dependent on continuous stimulation through the TCR, even many hours after it has formed (J. Huppa, in preparation). Although initial speculation suggested a signaling function because it involves a concentration (2–3x) of important signaling molecules (50, 53, 68), others have proposed that its primary role is in subsequent effector cytokines/cytotoxic granule release (72, 73).

How does this reorganization occur? Evidence of an important role for the T cell cytoskeleton in this process was first shown by Kupfer and colleagues, who documented a distinct polarization in the T cell toward the cell it was recognizing. This involves the movement of the MTOC to a position in between the nucleus and the interface area (69–71). This is accompanied by an accumulation of the cystoskeletal protein Talin (69) and the mobilization of PKC θ to the region as well (49). Later studies, by Wülfing & Davis (74) and Viola et al. (75), showed that a combination of TCR/CD3 signals and costimulatory signals (through either CD28 and LFA-2 in the first study and CD28 in the second) triggered a movement of cell surface material toward the interface. In the

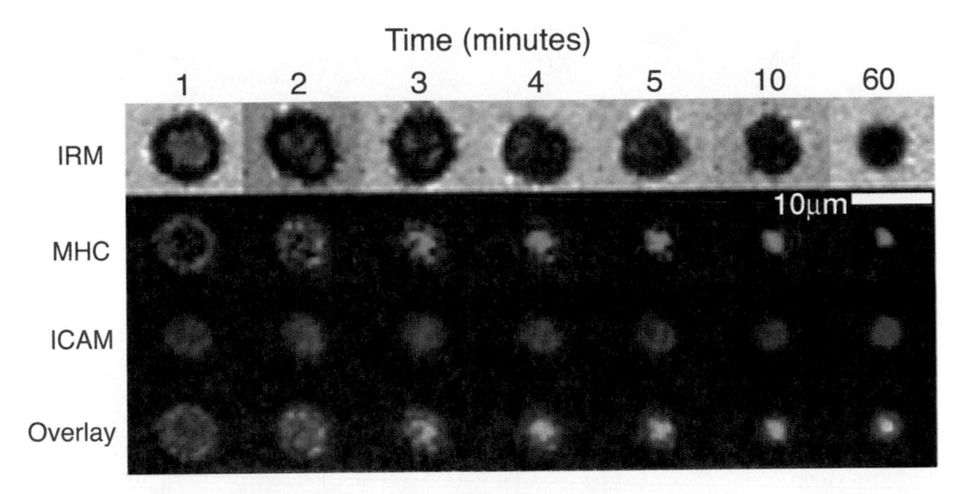

Figure 5 Synapse formation. As shown in Grakoui et al. (53), artificial bilayers loaded with labeled peptide-MHC (green) and ICAM-1 (red) in mobile form are rearranged into an immunological synapse by specific T cells within minutes of contact. Time from first contact is given at the top of the figure. The outline of the T cell is shown in the interference contrast (IRM) panel.

Wülfing study, this was directly linked to cytoskeletal transport as the myosin motor inhibitor BDM blocked this movement. In the Viola et al. study, a marker for lipid rafts was also shown to congregate at this location, but there is some controversy about this (76). Further work by Wülfing et al. (77) has shown that the accumulation of MHC molecules into a tight cluster in the center of the bull's eye is dependent on costimulation, and analyses by Moss et al. (78) showed that TCR/CD3 from all over the T cell surface is transported to the synapse. Both computer reconstruction and tracking small beads attached to TCRs have shown that they achieve a velocity of $\sim$0.25 μM/sec (78).

Experiments using T cell ligands on artificial membranes also indicate that unless costimulation is provided through ICAM-1, an ordered synapse does not form (C. Wülfing, C. Sumen, M.M. Davis, and M. Dustin, unpublished information). Previously, it had been thought that the clustering of surface molecules at the interface of a T cell and another cell was the result of passive diffusion and trapping of molecules due to ligand binding (79), but this study and other recent data show quite clearly that an active transport mechanism on the T cell side is involved and that this requires costimulatory signals and the cytoskeleton (74, 77, 78). Although the data are incomplete, it seems likely that myosin motor proteins transport TCR/CD3 and perhaps other membrane proteins to the synapse. This process is schematized in Figure 6, which shows TCRs being transported toward the synapse upon receiving signals through the TCR and costimulatory receptors (Figure 6*A*, *B*, *C*).

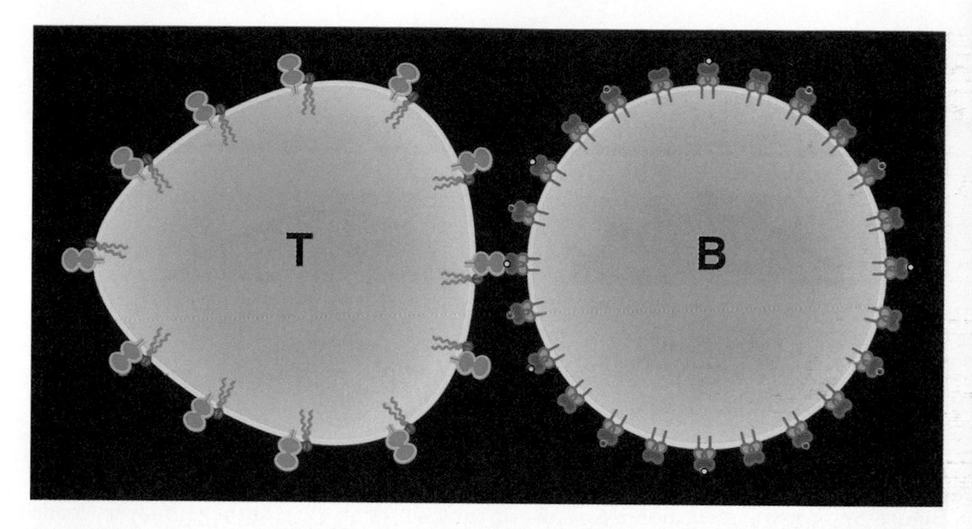

Figure 6A Costimulation mediates the active transport of TCRs into the synapse. The synergy between TCR/CD3 signaling and costimulation through CD28/LFA-1 triggers the active transport of TCR molecules from all over the T cell surface toward the interface. Following initial recognition (*a*), TCRs are mobilized through linkage to the cytoskeleton to move toward the center of the cell:cell interaction. As these TCRs move past the B cell membrane, they gather compatible peptide-MHC molecules (*b*) and bring them into the center of the synapse (*c*).

As TCRs come in contact with membrane molecules on the cell being recognized, they can bind peptide-MHC complexes and pull these into the synapse as well (Figure 6*B*, *C*). This would explain how MHC on B cells treated with cytochalasin D (74), which inhibits actin polymerization, or even artificial membranes (53) can be gathered into a synapse structure. There are a number of interesting aspects to this recruitment of MHC molecules into the synapse. One is the observation of Grakoui et al. (53) that the density of MHC in these structures is proportional to the half-life of monomeric TCR binding to peptide-MHC in solution. Previously it has been shown that TCR half-life for a ligand is generally a good predictor of T cell activation (15, 80, 81) [although a number of exceptions suggest that some other variable(s) is important as well (6, 11, 17, 83, 84)]. While this finding suggests a simple model of binding, transport, and release of different MHC ligands, recent experiments show that there is a massive recruitment of poor peptide-MHC ligands into the synapse as long as there are some TCR peptide-MHC interactions of high quality (77). This suggests a rheostat model in which the strength (likely the velocity) of TCR transport to the synapse is determined by a small number of the best ligands. With respect to those that are recruited, Wülfing et al. (77) showed that these could include a peptide in which several key T cell recognition residues have been replaced by those with smaller side chains but that these exclude an antigen peptide analog

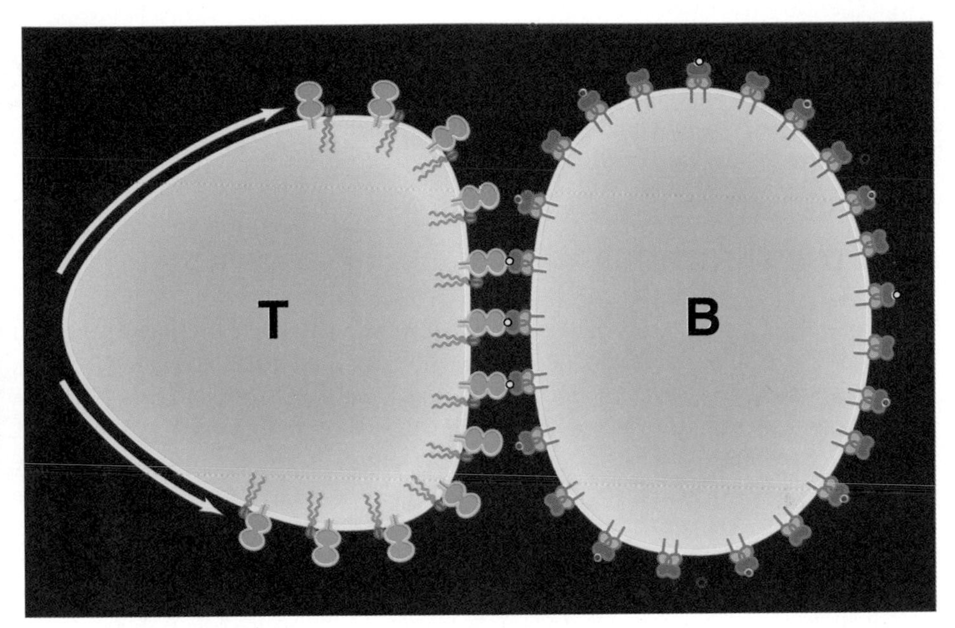

Figure 6B *Continued.*

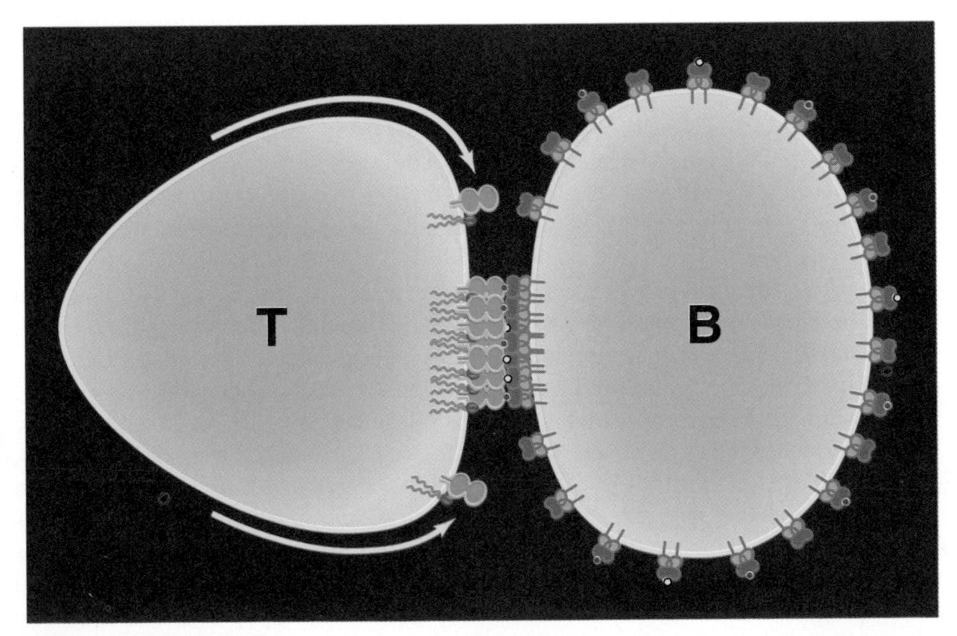

Figure 6C *Continued.*

in which a central lysine recognized by the TCR has been replaced by a glutamic acid—presumably interfering with TCR binding. Also interesting is the fact that this recruitment seems only dependent on TCR and not CD4 (77); this suggests that the latter cannot bind class II MHC on other cells in this context or does so very selectively (see previous section).

EFFECTOR FUNCTIONS

For helper T cells, a cell:cell couple can last for many hours (~10–12 h) with at least 2–3 h being necessary for an irreversible commitment to activation (85, 86). This typically involves continuously high calcium levels, which keeps NFAT in the nucleus and correlates with the maintenance of an immunological synapse. RNA encoding cytokines is first detectable within several hours, and after a few hours, secretory vesicles become loaded with protein cargo that is released in the direction of the cell being recognized (79). In this way, the T cell can deliver signals specifically to the target cell. Cell division is also underway in the T cell, and it is probably this activity that breaks up the conjugate. In a cytotoxic response, everything happens on an accelerated time scale, with a much shorter period of elevated calcium (~2–20 min), and because granules are preloaded with cytotoxic agents, killing can occur very quickly, in 5 min or less, or in some cases in several hours for reasons that relate to cell physiology or perhaps the density of antigen (87).

OTHER T CELL SYNAPSES

As predicted by Dustin and colleagues (88), other varieties of T cell synapses have recently come to light. These include both the interactions immature T cells make with thymic stromal cells during thymic selection (next section) and the interactions of mature T cells with dendritic cells and antigen presenting cells of various types.

From the examples that we have so far, it seems very likely that T cells at different developmental stages will all have unique variations of the type of synapse described in the previous sections. A particularly interesting example is that of the T cell:dendritic cell interaction as characterized by the work of Trautman, Ploegh, and Mellman and colleagues (89–92). In contrast to the passive B cell models of antigen presenting cell, discussed earlier, dendritic cells, which are the most potent of antigen presenting cells with respect to eliciting a T cell response, have the interesting property of becoming stimulated themselves early in T cell recognition and thus establishing a cell:cell dialogue of sorts. This involves both an occasional calcium flux in a dendritic cell that has formed a synapse with one or more T cells (89) and also, as very recently described, the ability to direct recently processed antigen in the direction of an antigen specific

T cell forming the synapse (91, 92). Presumably this serves to reinforce the initial antigenic stimulus seen by the T helper cell and push it toward complete activation.

Another important area of investigation involves the activation of naive T cells; these are cells that have had no previous encounter with antigen. Although most studies have focused on previously activated T cells (T cell blasts) or lines that have been repeatedly stimulated, the activation of naive cells is a very important topic because it is more difficult to accomplish and likely to be crucial to the initiation of autoimmunity (where an inappropriate activation takes place). On this topic, two groups have provided some clues as to what sort of immune synapses are formed with naive cells (73, 93). Normally, naive T cells will not form synapses with B cells pulsed with antigen. Yet, as Dustin and colleagues have shown, the addition of a chemokine, which promotes T cell activation, can induce naive T cells to form conjugates and synapses (93). It will be interesting to learn more about this system.

Another series of experiments involving naive T cells has been described by Shaw and colleagues (73) in which cells of a given specificity are lightly centrifuged onto splenic monocytes pulsed with peptide. In this case, TCR/CD3 only accumulates at the periphery of the interface in the first thirty min, and the active form of lck (which phophorylates motifs on the cytoplasmic domains of CD3 molecules) localizes there also. Later a more central synapse forms, but by then the lck activity has died down. Shaw et al. argue that the bulk of signaling has occurred before a mature synapse has formed and conclude that this structure may not fulfill a signaling function (73). However, we have observed T cells in which much of the cell surface TCR has been internalized in the intermediate (synapse) phase of recognition, but when the specific pMHC is blocked with an antibody (even six h after cell:cell contact), the synapse dissolves, calcium levels decline, and the cells detach. Thus continuous signaling through the TCR seems necessary for the maintenance of the synapse structure (J. Huppa, M. Gleimer, unpublished observations).

IMAGING LYMPHOCYTES IN VIVO

In an organism, T lymphocyte precursors arise in the bone marrow (or fetal liver) and then migrate to the thymus, where they sequentially rearrange their TCR genes while going through a series of well-defined stages and selection (94). One type of selection is referred to as negative selection because it induces programmed cell death in most strongly self-reactive T cells, based on their TCR reactivity. Another more mysterious process is referred to as positive selection, as it permits the survival of T cells that are weakly reactive to self peptide-MHCs displayed in the thymus. The 1% to 5% of thymocytes that survive these selections (95) then leave the thymus and circulate in the body, migrating through lymph nodes and the spleen (where antigen presenting cells and antigens tend to collect). These secondary lymphoid organs are where T cells encounter antigens as displayed by B cells, macrophages, and dendritic cells (the so-called professional antigen presenting cells), which are the most potent T cell stimulators (because they express costimulatory ligands among other attributes). Given this biology, it is of great interest to use imaging technology to observe T cell interactions with other cells in the thymus and in lymph nodes, as the standard tissue culture assay used in the experiments described thus far is only a crude approximation of these complex environments.

In the case of thymic differentiation and selection, an approximation of the thymic microenvironment can be made by mixing fetal thymic stromal cells with thymocytes under conditions that promote the growth of a multicellular reaggregate entity in which thymic positive and negative selection occurs readily and in an appropriate time scale. Using such a system, Richie et al. (96) transduced immature thymocytes with retroviral constructs containing CD3ζ-GFP or lck-GFP and characterized the synapses that these cells formed with thymic stromal cells when the appropriate negatively selecting peptide was introduced into the culture. It was found that cell conjugates formed very rapidly upon peptide introduction and that, in contrast to mature T cells, recognizing professional antigen presenting cells (APCs), CD3ζ-GFP did not form stable, central accumulations but instead tended to accumulate on the periphery of the synapse—as did lck (but that molecule behaved the same way in mature T cells) (97). This different synapse architecture is at least partly due to a lack of costimulation but also appears to reflect a difference in the capabilities of the immature T cells (as judged by exposing these thymocytes to professional APCs). In any event, characterizing these synapses further and those that may result in positive selection as well should give important insights into these processes (88).

Another system for the study of thymic development with great potential for the future is the use of 2-photon microscopy, which has recently been utilized by Bousso et al. (98) to image live thymocytes within a fetal thymus. Thus far it has not been possible to detect GFP-labeled proteins by this technique, but advances in sensitivity and/or improved labeling procedures may make that possible in the near future.

2-photon microscopy has also been applied to the movements of T cells within a lymph node. Here an important issue has been raised by Gunzer et al. (99) who found that in a collagen matrix cell culture system naive T cells had only very transient interactions with dendritic cells presenting antigen, yet they still managed to proliferate at levels approaching those achieved in static cultures, which suggests that a stable synapse is not the only way to activate T cells.

Nonetheless in this same system, naive T cells do form tight conjugates with B cells (although they continue to migrate), and mature T cell blasts (e.g., antigen experienced) frequently form stable interactions with dendritic cells bearing the correct antigen. These data and those of others (M.D. Cahalan, personal communication) raise the issue of how well any of these ex vivo systems mimic the in vivo reality. It also supports the argument that the synapse structure's primary function is not signaling, but is a delivery vehicle for effector molecules (72, 73).

Recent data with lymph node cultures and 2-photon microscopy give conflicting results with respect to the issue. In one study with nodes that are perfused with 100% O_2, T cells move quite rapidly and form few stable contacts (100). In another system, the tissue is not oxygenated; the T cells form very stable contacts (101). In a third, the node is perfused with blood from a living mouse; there are both stable and unstable contacts (M. Dustin, personal communication).

RELATIONSHIP OF CELL SURFACE MOVEMENTS TO T CELL SIGNALING CASCADES

Another very interesting area of investigation is to link up our emerging knowledge about the cell surface movements discussed previously and how these trigger the various signaling pathways that have been studied extensively in T cells. Previously, it has been shown that one of the earliest events, which occurs within fifteen seconds of TCR engagement, is the phosphorylation of tyrosines within the ITAM motifs of CD3 cytoplasmic regions, particularly CD3ζ, and that lck kinase is responsible for this (102). Recent work has also indicated another change in the first few seconds; it is a conformational change in the CD3ϵ cytoplasmic region that allows the adapter protein Nck to bind to it (103). How dimerization or clustering affects these events is not clear. Results from Freiberg et al. (104) show that active lck appears in the early stages of synapse formation in the central region in T cell blasts, but then it dissipates to the periphery (104) [as does CD4 in a T cell line (97) although not in a hybridoma model (105)]. Consistent with this finding, Ehrlich et al. found that total lck translocates to the center and then to the periphery of the synapse in T cells undergoing activation (106). As discussed earlier, Shaw and colleagues found active lck at the periphery of an early synapse in naive T cells, which then becomes undetectable as the synapse takes on a mature form (73). This is also the site of lck and CD3 in immature thymocytes undergoing negative selection, so it seems likely that lck is not required to be at either location in order to signal. One surprising finding with lck is that of Ehrlich et al. (106) who found that an intracellular bolus of lck translocates across the cell to be delivered at or near the underside of an active synapse in mature T cells ten min or so after the initial activation; it suggests that additional lck is needed to sustain or potentiate signaling.

In any event, we can look forward to a great deal of work on the dynamics of signaling molecules vis-à-vis those on the cell surface in the near future, with colocalization studies and other analysis adding to our understanding of cause and effect.

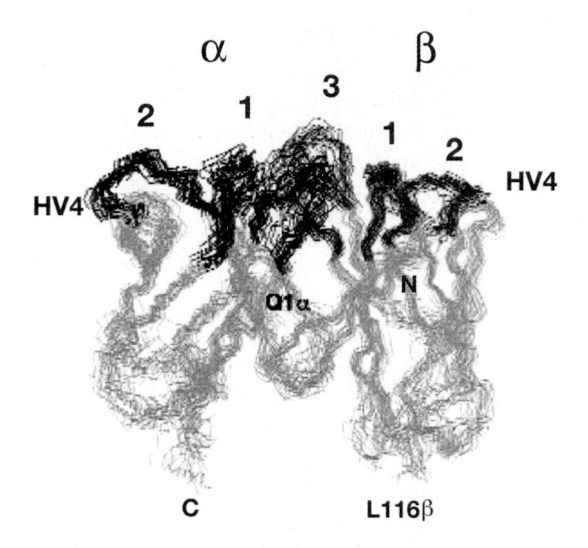

Figure 7 CDR3 mobility. The 2D NMR studies of Hare et al. (114) show that the central CDR3 loops of V_{α} and V_{β} are more mobile than the rest of the binding surface.

BINDING CHARACTERISTICS OF T CELL RECEPTORS THAT FACILITATE SCANNING AND CROSS-REACTIVITY

Recent thermodynamic analyses of T cell receptor binding have shown that in most cases there is a significant reduction in entropy and a conformational change in the binding site upon engagement with peptide-MHC ligands, characteristics of an induced-fit type of interaction (20–22). As pointed out by Boniface et al. (21), there is an interesting parallel between TCRs and DNA recognition proteins, which also exhibit an induced fit mode of binding, because both molecules need to scan a large number of very similar entities efficiently to find the correct fit. This is in marked contrast to other cell surface receptors that have been characterized, which exhibit a lock and key type of binding (110, 111). We are beginning to get a better idea about how this might work with TCRs from both binding analyses and structural work. In the case of the later, it is becoming increasingly apparent from structures of bound versus free TCRs that there is a significant movement of some of the complementary determining loops (CDRs) that form the binding site. In particular, large movements in CD3β have been observed now in two different TCRs (112, 113); a remarkable 13 Å was traversed in one case (112). In addition, a 2D NMR structure of a TCR shows that the CDR3 regions are much more flexible than the rest of the binding site, as shown in Figure 7 (114). Previous work has shown that the CDR3 regions of TCRs (in the center of the binding site) are the primary determinants of peptide specificity (115, 116).

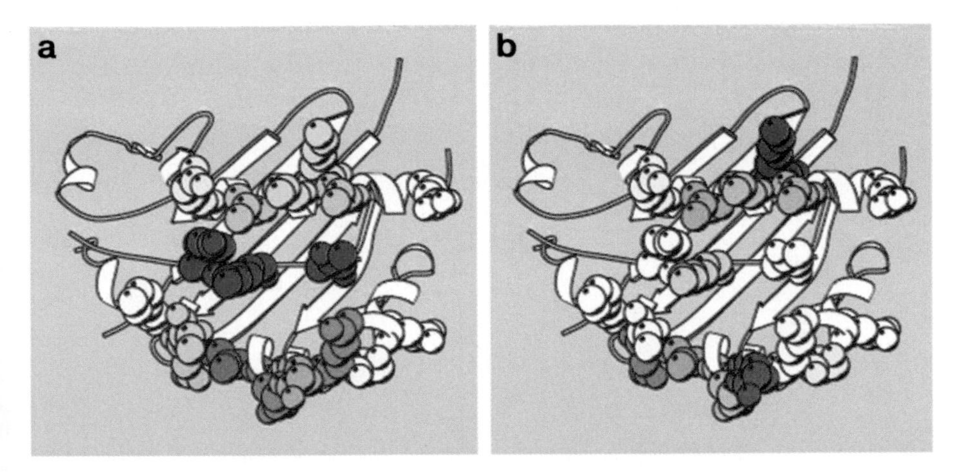

Figure 8 Dichotomy between peptide and MHC residues affecting TCR binding. Mutagenesis of MHC or peptide determinants that affect TCR binding reveals that while peptide residues exert a great deal of influence on stability (*a*) they have very little to do with the initial contact (*b*). This is in contrast to MHC residues that exhibit the reverse properties (see text). Here a color scale represents residues having the least (white) to the most (red) effect (116).

Another piece of the puzzle can be found in the work of Wu et al. (19) who analyzed the relative contribution of MHC versus peptide to both the initial interaction between TCR and pMHC in the transition state, using the phi analysis system of Fersht and colleagues (117, 118), as well as the contribution to stability ($\Delta\Delta G$). These data, summarized in Figure 8, reveal that peptide-TCR contacts had very little to do with the transition state, yet they were very important in the stability for MHC mutants, in contrast to MHC residues, which are important in the initial contact but not of much consequence to stability.

Although this data is from only one example of TCR-ligand binding, partial data from a second case (as discussed in Reference 19) and now a third complete analysis (M. Krogsgaard, unpublished data) supports the generality of this finding. A straightforward conclusion from these data is that TCR binding is a two-stage process wherein the first consists of TCR contact with the MHC helices, followed by folding of the CDR3 loops around the peptide to achieve the final, stable state, as shown in Figure 9. Thus scanning would consist of an initial contact with the MHC that would orient the TCR such that it could quickly determine whether the peptide occupying the binding groove is appropriate. It is important to note that we find no evidence for distinct kinetic intermediates for these two stages (M. Krogsgaard, personal communication); thus, we believe that the initial interaction of TCR with the MHC can be considered a nucleation process, with the later CDR3-peptide contacts forming very quickly thereafter. Because the CDR3 loops could adopt their final conformation in a variety of ways—perhaps thousands of different possibilities [as first suggested by the

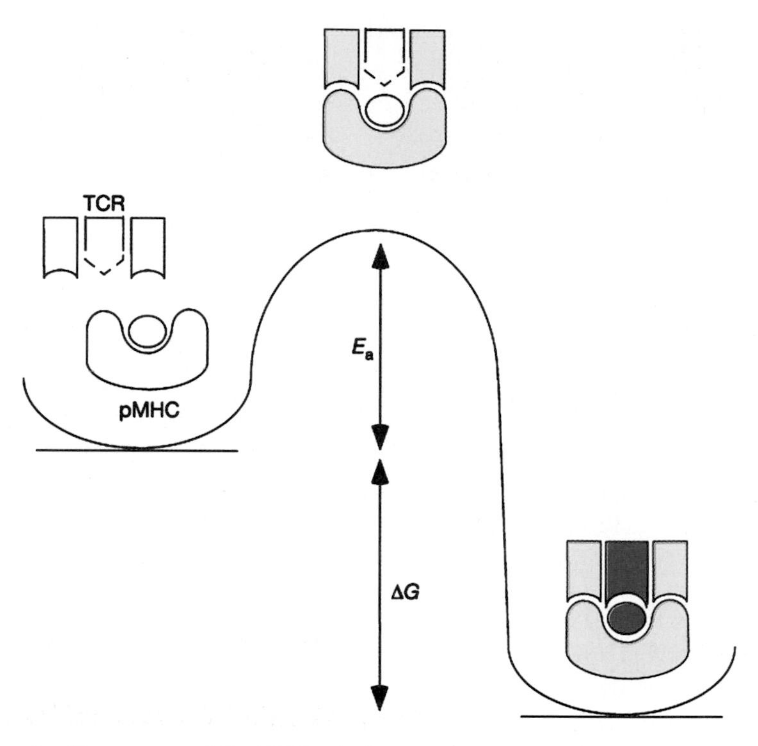

Figure 9 Two step model of TCR binding. As shown by Wu et al. (19), mutational analysis of TCR-peptide/MHC binding indicates that the TCR first contacts MHC residues (in the transition state) with the peptide having very little influence. Subsequently, however, the peptide residues contribute greatly to the stability of the complex. Thus we have proposed that the transition state largely involves TCR-MHC contact followed by stabilization of mobile CDR3 residues into a stable state, which usually involves significant conformational change and loss of entropy.

structural studies of Garcia and colleagues (112)], this model would also account for the inherent cross-reactivity of TCRs—required by positive selection in the thymus and thought to be a major factor in autoimmunity.

CONCLUSION

It is our hope that the experiments related here demonstrate the very rapid progress that is being made in piecing together the early steps required for T cell recognition. They also are of general interest because they illustrate how the traditionally intractable problem of discerning the biochemical principles behind cell:cell recognition are at last yielding to new methodologies that allow the visualization of cell surface molecules in intact cells or in artificial membranes. Application of these new approaches should make accessible some of the many instances of cell:cell recognition that are vital to the development and function of higher organisms. It is also instructive that these analyses of T cell recognition are revealing novel mechanisms of cell surface molecule movement, such as synapse formation and costimulation-driven active transport, that were not predicted by early attempts to model such interactions. Thus we might look forward to the discovery of more new phenomenon in this and other systems.

ACKNOWLEDGMENTS

We thank the Howard Hughes Medical Institute and the National Institutes of Health for financial support. M.K. was a fellow of the Alfred Benzon Foundation, J.H. was supported by the Cancer Research Institute, L.W. by the Damon-Runyon Cancer Foundation, and D.I. by the Damon Runyon Cancer Foundation.

The *Annual Review of Biochemistry* is online at http://biochem.annualreviews.org

LITERATURE CITED

1. Litman GW, Anderson MK, Rast JP. 1999. *Annu. Rev. Immunol.* 17:109–47
2. Davis MM, Boniface JJ, Reich Z, Lyons D, Hampl J, et al. 1998. *Annu. Rev. Immunol.* 16:523–44
3. Chien YH, Hampl J. 2000. *Springer Semin. Immunopathol.* 22:239–50
4. Hayday AC. 2000. *Annu. Rev. Immunol.* 18:975–1026
5. Cresswell P. 1994. *Annu. Rev. Immunol.* 12:259–93
6. Wolf PR, Ploegh HL. 1995. *Annu. Rev. Cell Dev. Biol.* 11:267–306
7. Alam SM, Travers PJ, Wung JL, Nasholds W, Redpath S, et al. 1996. *Nature* 381:616–20
8. Alam SM, Davies GM, Lin CM, Zal T, Nasholds W, et al. 1999. *Immunity* 10:227–37
9. Degano M, Garcia KC, Apostolopoulos V, Rudolph MG, Teyton L, Wilson IA. 2000. *Immunity* 12:251–61
10. Garcia KC, Tallquist MD, Pease LR, Brunmark A, Scott CA, et al. 1997. *Proc. Natl. Acad. Sci. USA* 94:13838–43
11. Baker BM, Gagnon SJ, Biddison WE, Wiley DC. 2000. *Immunity* 13:475–84
12. Ding YH, Baker BM, Garboczi DN, Biddison WE, Wiley DC. 1999. *Immunity* 11:45–56
13. Sykulev Y, Brunmark A, Jackson M, Cohen RJ, Peterson PA, et al. 1994. *Immunity* 1:15–22
14. Sykulev Y, Vugmeyster Y, Brunmark A, Ploegh HL, Eisen HN. 1998. *Immunity* 9:475–83
15. Kersh GJ, Kersh EN, Fremont DH, Allen P. 1998. *Immunity* 9:817–26
16. Margulies DH. 1997. *Curr. Opin. Immunol.* 9:390–95
17. Slansky JE, Rattis FM, Boyd LF, Fahmy T, Jafee EM, et al. 2002. *Immunity* 13:529–38
18. Deleted in proof
19. Wu LC, Tuot DS, Lyons DS, Garcia KC, Davis MM. 2002. *Nature* 418:552–56
20. Willcox BE, Gao GF, Wyer JR, Ladbury JE, Bell JI, et al. 1999. *Immunity* 10:357–65
21. Boniface JJ, Reich Z, Lyons DS, Davis

MM. 1999. *Proc. Natl. Acad. Sci. USA* 96:11446–51
22. Garcia KC, Radu CG, Ho J, Ober RJ, Ward S. 2001. *Proc. Natl. Acad. Sci. USA* 98:6818–23
23. Labadia ME, Jeanfavre DD, Caviness GO, Morelock MM. 1998. *J. Immunol.* 161:836–42
24. Lollo BA, Chan KWH, Hanson EM, Moy VT, Brian AA. 1993. *J. Biol. Chem.* 268:21693–700
25. Xiong Y, Kern P, Chang H, Reinherz EL. 2001. *J. Biol. Chem.* 276:5659–67
26. Wyer JR, Willcox BE, Gao GF, Gerth UC, Davis SJ, et al. 1999. *Immunity* 10:219–25
27. van der Merwe PA, Bodian DL, Daenke S, Linsley P, Davis SJ. 1997. *J. Exp. Med.* 185:393–403
28. Collins AV, Brodie DW, Gilbert RJC, Iaboni A, Manso-Sancho R, et al. 2002. *Immunity* 17:201–10
29. van der Merwe PA, Barclay AN, Mason DW, Davies EA, Morgan BP, et al. 1994. *Biochemistry* 33:10149–60
30. van der Merwe PA, Brown MH, Davis SJ, Barclay AN. 1993. *EMBO J.* 12:4945–54
31. Pierres A, Benoliel AM, Bongrand P, van der Merwe PA. 1996. *Proc. Natl. Acad. Sci. USA* 93:15114–18
32. Davis SJ, Davies EA, Tucknott MG, Jones EY, van der Merwe PA. 1998. *Proc. Natl. Acad. Sci. USA* 95:5490–94
33. van der Merwe PA, Barclay AN. 1994. *Trends Biochem. Sci.* 19:354–58
34. Dustin ML, Ferguson LM, Chan PY, Springer TA, Golan D. 1996. *J. Cell. Biol.* 132:465–74
35. Leckband DE, Israelachvili JN, Schmitt FJ, Knoll W. 1992. *Science* 255:1419–21
36. Davis MM, Bjorkman PJ. 1988. *Nature* 334:395–402
37. Dustin ML, Springer TA. 1991. *Annu. Rev. Immunol.* 9:27–66
38. Janeway CA Jr. 1992. *Annu. Rev. Immunol.* 10:645–74
39. Chambers CA, Allison JP. 1997. *Curr. Opin. Immunol.* 9:396–404
40. Kane LP, Andres PG, Howland KC, Abbas AK, Weiss A. 2001. *Nat. Immunol.* 1:37–44
41. Diehn M, Alizadeh AA, Rando OJ, Liu CL, Stankunas K, et al. 2002. *Proc. Natl. Acad. Sci. USA* 99:11796–801
42. Chambers CA, Kuhns MS, Egen JG, Allison JP. 2001. *Annu. Rev. Immunol.* 19:565–94
43. Egen JG, Allison JP. 2002. *Immunity* 16:23–35
44. van der Merwe PA. 1999. *J. Exp. Med.* 190:1371–74
45. Trowbridge IS, Thomas ML. 1994. *Annu. Rev. Immunol.* 12:85–116
46. Pradhan D, Morrow JS. 2002. *Immunity* 17:303–15
47. Donnadieu E, Bismuth G, Trautmann A. 1994. *Curr. Biol.* 4:584
48. Negulescu PA, Krasieva TB, Khan A, Kerschbaum HH, Cahalan MD. 1996. *Immunity* 4:421
49. Monks CRF, Kupfer H, Tamir I, Barlow A, Kupfer A. 1997. *Nature* 385:83–86
50. Monks CRF, Freiberg BA, Kupfer H, Sciaky N, Kupfer A. 1998. *Nature* 395:82–86
51. Wülfing C, Sjaastad MD, Davis MM. 1998. *Proc. Natl. Acad. Sci. USA* 95:6302–7
52. Dustin ML, Olszowy MW, Holdorf AD, Li J, Bromley S, et al. 1998. *Cell* 94:667–77
53. Grakoui A, Bromley SK, Sumen C, Davis MM, Shaw AS, et al. 1999. *Science* 285:221–27
54. Harding CV, Unanue ER. 1990. *Nature* 346:574–76
55. Demotz S, Grey HM, Sette A. 1990. *Science* 249:1028–30
56. Christinck ER, Luscher MA, Barber BH, Williams DB. 1991. *Nature* 352:67–70
57. Sykulev Y, Joo M, Vturina I, Tsomides TJ, Eisen HN. 1996. *Immunity* 4:565–71
58. Brower RC, England R, Takeshita T,

Kozlowski S, Margulies DH, et al. 1994. *Mol. Immunol.* 31:1285–93
59. Reay PA, Matsui K, Haase K, Wülfing C, Chien YH, Davis MM. 2002. *J. Immunol.* 164:5626–34
60. Irvine DJ, Purbhoo MA, Krogsgaard M, Davis MM. 2002. *Nature* 419:845–49
61. Jiang T, Zhang JP, Liang DC. 1999. *Proteins* 34:224–31
62. Crabtree GR, Olson EN. 2002. *Cell* 109(Suppl. S):S67–79
63. Delon J, Gregoire C, Mallissen B, Darche S, Lamaitre F, et al. 1998. *Immunity* 9:467–73
64. Boniface JJ, Rabinowitz JD, Wülfing C, Hampl J, Reich Z, et al. 1998. *Immunity* 9:459–66
65. Cochran JR, Cameron TO, Stern LJ. 2000. *Immunity* 12:241–50
66. Ge O, Stone JD, Thompson MT, Cochran JR, Rushe M, et al. 2002. *Proc. Natl. Acad. Sci. USA* 99:13729–34
67. Potter TA, Grebe K, Freiberg B, Kupfer A. 2001. *Proc. Natl. Acad. Sci. USA* 98:12624–29
68. Krummel MF, Sjaastad M, Wülfing C, Davis MM. 2002. *Science* 289:1349–52
69. Kupfer A, Singer SJ, Dennert G. 1986. *J. Exp. Med.* 163:489–98
70. Kupfer A, Mosmann TR, Kupfer A. 1991. *Proc. Natl. Acad. Sci. USA* 88:775–79
71. Kupfer A, Swain SL, Singer SJ. 1987. *J. Exp. Med.* 165:1565–80
72. Davis SJ, van der Merwe PA. 2001. *Curr. Biol.* 11:R289–90
73. Lee KH, Holdorf AD, Dustin ML, Chan AC, Allen PM, Shaw AS. 2002. *Science* 295:1539–42
74. Wülfing C, Davis MM. 1998. *Science* 282:2266–69
75. Viola A, Schroeder S, Sakakibara Y, Lanzavecchia A. 1991. *Science* 283: 680–82
76. Burack WR, Lee KH, Holdorf AD, Dustin ML, Shaw AS. 2002. *J. Immunol.* 169:2837–41
77. Wülfing C, Sumen C, Sjaastad MD, Wu LC, Dustin ML, Davis MM. 2002. *Nat. Immunol.* 3:42–47
78. Moss WC, Irvine DJ, Davis MM, Krummel MF. 2002. *Proc. Natl. Acad. Sci. USA.* 99:15024–29
79. Kupfer A, Singer SJ. 1989. *Annu. Rev. Immunol.* 7:309–37
80. Matsui K, Boniface JJ, Steffner P, Reay PA, Davis MM. 1994. *Proc. Natl. Acad. Sci. USA* 91:12862–66
81. Lyons DS, Lieberman SA, Hampl J, Boniface JJ, Reay PA, et al. 1996. *Immunity* 5:53–61
82. Deleted in proof
83. Al-Ramadi BK, Jelonek MT, Boyd LF, Margulies DH, Bothwell ALM. 1995. *J. Immunol.* 155:662–73
84. van der Merwe PA. 2001. *Immunity* 14:665–68
85. Karttunen J, Shastri N. 1991. *Proc. Natl. Acad. Sci. USA* 88:3972–76
86. Iezzi G, Karjalainen K, Lanzavecchia A. 1998. *Immunity* 8:89–95
87. Stinchcombe JC. 2001. *Immunity* 15: 751–61
88. Dustin ML, Bromley SK, Davis MM, Zhu C. 2001. *Annu. Rev. Cell Dev. Biol.* 17:133–57
89. Delon J, Bercovici N, Raposo G, Liblau R, Trautmann A. 1998. *J. Exp. Med.* 188:1473–84
90. Revy P, Sospedra M, Barbour B, Trautmann A. 2001. *Nat. Immunol.* 2:925–31
91. Boes M, Cerny J, Massol R, den Brouw MO, Kirchhausen T, et al. 2002. *Nature* 418:983–88
92. Chow A, Toomre D, Garrett W, Mellman I. 2002. *Nature* 418:988–94
93. Bromley SK, Dustin ML. 2002. *Immunology* 106:289–98
94. Borowski C, Martin C, Gounari F, Haughn L, Aifantis L, et al. 2002. *Curr. Opin. Immunol.* 14:200–6
95. Scollay RG, Butcher EC, Weissman IL. 1980. *Eur. J. Immunol.* 10:210–18
96. Richie LI, Ebert PJR, Wu LC, Owen JTT, Davis MM. 2002. *Immunity* 16:595–606

97. Hailman E, Burack WR, Shaw AS, Dustin ML, Allen PM. 2002. *Immunity* 16:839–48
98. Bousso P, Bhakta NR, Lewis RS, Robey E. 2002. *Science* 296:1876–80
99. Gunzer M, Schafer A, Borgmann S, Grabbe S, Zanker KS, et al. 2000. *Immunity* 13:323–32
100. Miller MJ, Wei SH, Parker I, Cahalan MD. 2002. *Science* 296:1869–73
101. Stoll S, Delon J, Brotz TM, Germain RN. 2002. *Science* 296:1873–76
102. Kane LP, Lin J, Weiss A. 2000. *Curr. Opin. Immunol.* 12:242–49
103. Gil D, Schamel WW, Montoya M, Sanchez-Madrid F, Alarcon B. 2002. *Cell* 109:901–12
104. Freiberg BA, Kupfer H, Maslanik W, Delli J, Kappler J, et al. 2002. *Nat. Immunol.* 3:911–17
105. Zal T, Zal MA, Gascoigne NRJ. 2002. *Immunity* 16:521–34
106. Ehrlich LI, Ebert PJR, Krummel MF, Weiss A, Davis MM. 2002. *Immunity* 17:809–22
107. Deleted in proof
108. Deleted in proof
109. Deleted in proof
110. Maenaka K, Juji T, Nakayama T, Wyer JR, Gao GF, et al. 1999. *J. Biol. Chem.* 274:28329–34
111. Maenaka K, van der Merwe PA, Stuart DI, Jones EY, Sondermann P. 2001. *J. Biol. Chem.* 276:44898–904
112. Garcia KC, Degano M, Pease LR, Huang M, Peterson PA, et al. 1998. *Science* 279:1166–72
113. Reiser JB, Gregoire C, Darnault C, Mosser T, Guimezanes A, et al. 2002. *Immunity* 16:345–54
114. Hare BJ, Wyss DF, Osburne MS, Kern PS, Reinherz EL, Wagner G. 1999. *Nat. Struct. Biol.* 6:574–81
115. Jorgensen JL, Esser U, Fazekas de St Groth B, Reay PA, Davis MM. 1992. *Nature* 355(6357):224–30
116. Sant'Angelo DB, Waterbury G, Preston-Hurlburt P, Medzhituv R, et al. 1996. *Immunity* 4:367–76
117. Fersht AR. 1997. *Curr. Opin. Struct. Biol.* 7:3–9
118. Schreiber G, Fersht AR. 1996. *Nat. Struct. Biol.* 3:427–31

Annu. Rev. Biochem. 2003. 72:743–81
doi: 10.1146/annurev.biochem.72.121801.161742
Copyright © 2003 by Annual Reviews. All rights reserved
First published online as a Review in Advance on March 27, 2003

BIOLOGY OF THE P21-ACTIVATED KINASES

Gary M. Bokoch
Departments of Immunology and Cell Biology, The Scripps Research Institute, 10550 North Torrey Pines Road, La Jolla, California 92037; email: bokoch@scripps.edu

Key Words Rho GTPases, Rac, Cdc42, cytoskeleton

■ **Abstract** The p21-activated kinases (PAKs) 1–3 are serine/threonine protein kinases whose activity is stimulated by the binding of active Rac and Cdc42 GTPases. Our understanding of the regulation and biology of these important signaling proteins has increased tremendously since their discovery in the mid-1990s. PAKs 1–3 are activated by a variety of GTPase-dependent and -independent mechanisms. This complexity reflects the contributions of PAK function in many cellular signaling pathways and the need to carefully control PAK action in a highly localized manner. PAKs serve as important regulators of cytoskeletal dynamics and cell motility, transcription through MAP kinase cascades, death and survival signaling, and cell-cycle progression. Consequently, PAKs have also been implicated in a number of pathological conditions and in cell transformation. We propose here a key role for PAK action in coordinating the dynamics of the actin and microtubule cytoskeletons during directional motility of cells, as well as in other functions requiring cytoskeletal polarization.

CONTENTS

0066-4154/03/0707-0743$14.00

INTRODUCTION TO p21-ACTIVATED KINASES (PAKS)

The Rho-related GTPases, Rac and Cdc42, regulate a wide variety of cellular activities involved in both normal and pathological cell function (1–3). These GTPases are able to regulate diverse processes through their ability to interact with multiple targets. Each of these effectors in turn modulates multiple specific downstream signaling pathways or events. One class of important mediators of Rac and Cdc42 GTPase activity is the p21-activated kinases (PAKs). Additionally, the PAKs have been implicated in the action of a number of other intracellular signaling processes.

The p21-activated kinases were initially identified in a screen for specific Rho GTPase (p21) binding partners in rat brain cytosol (4). [γ-^{32}P]GTP-labeled Rac or Cdc42, but not RhoA, were shown to interact in a gel overlay assay with proteins of ~68, 65, and 62 kDa. These species did not interact with GDP-bound Rac or Cdc42. These three Rac and Cdc42 targets turned out to be the three major isoforms of p21-activated kinases (Figure 1), subsequently termed PAK1 (or α-PAK), PAK3 (or β-PAK), and PAK2 (or γ-PAK). The phosphotransferase activity of PAKs was stimulated dramatically in vitro by the GTP-bound, but not GDP-bound, forms of Rac1 and Cdc42, suggesting that PAKs are effectors for these GTPases. Soon after the initial report by Manser et al. (4), a number of other studies verified the existence of PAKs 1–3 as Rac and Cdc42 targets (5–7).

PAKs 1–3 are serine/threonine protein kinases with significant sequence homology in their catalytic domains to Ste20, a protein kinase from budding yeast (8–10). It has subsequently been determined that the genomes of all eukaryotes examined contain a large family of Ste20-related protein kinases (10). Most closely related in their catalytic domains to the mammalian PAKs 1–3 are the more recently discovered PAKs 4–6. Although the latter have also been termed PAKs, PAKs 4–6 differ significantly in their structural organization and regulation from PAKs 1–3 [see (11) for review]. For the purposes of this review, we focus on PAKs 1–3, describing their regulation, molecular interactions, and biology.

→

Figure 1 Amino acid sequence alignment of the human PAK1, PAK2, and PAK3 isoforms. Sections of 10 amino acid residues (aa) each are indicated by the single dots above the sequences, and identical (*) and conserved (. or :) amino acids are indicated below the sequences. Proline-rich regions containing putative SH3-binding PXXP motifs are boxed in black, and the noncanonical PIX binding site is indicated by the open box. Areas boxed in gray indicate highly charged basic or acidic tracts. The homodimerization domain (aa 78–87), CRIB motif (aa 75–90), p21-binding domain or PBD (aa 67–113), and the autoinhibitory switch domain (aa 83–149) for PAK 1 are indicated by the dark overhead lines. Diagnostic kinase motifs in the catalytic domain are boxed and numbered per convention.

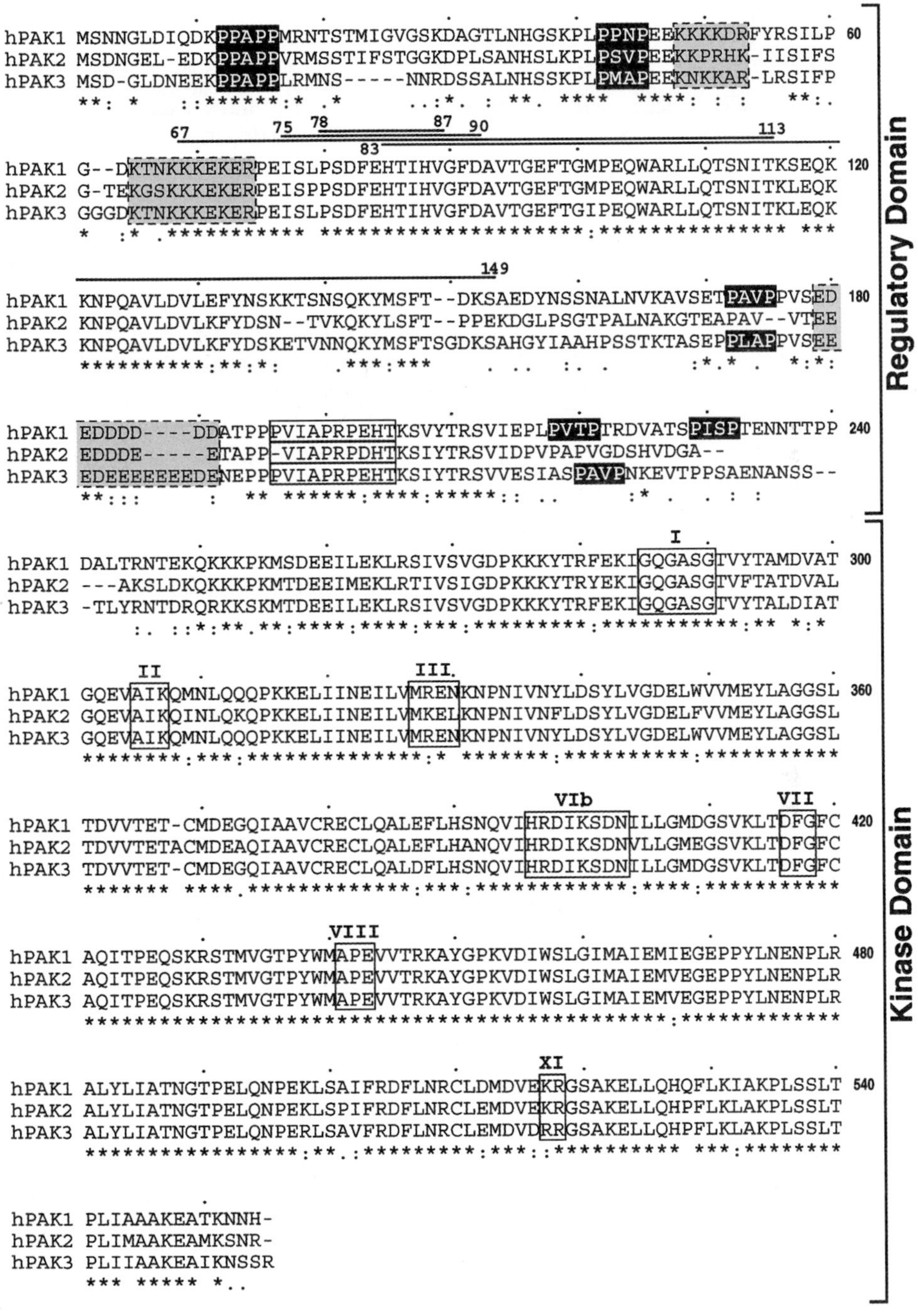

```
hPAK1 MSNNGLDIQDKPPAPPMRNTSTMIGVGSKDAGTLNHGSKPLPPNPEEKKKKDRFYRSILP  60
hPAK2 MSDNGEL-EDKPPAPPVRMSSTIFSTGGKDPLSANHSLKPLPSVPEEKKPRHK-IISIFS
hPAK3 MSD-GLDNEEKPPAPPLRMNS-----NNRDSSALNHSSKPLPMAPEEKNKKAR-LRSIFP
      **: *   ::******:* .*      ..:*. : **. ****  ****: : :   **:.

           67       75 78      87 90                          113
                              83
hPAK1 G--DKTNKKKEKERPEISLPSDFEHTIHVGFDAVTGEFTGMPEQWARLLQTSNITKSEQK 120
hPAK2 G-TEKGSKKKEKERPEISPPSDFEHTIHVGFDAVTGEFTGMPEQWARLLQTSNITKLEQK
hPAK3 GGGDKTNKKKEKERPEISLPSDFEHTIHVGFDAVTGEFTGIPEQWARLLQTSNITKLEQK
      *  :* .*********** ********************:*************** ***

                                     149
hPAK1 KNPQAVLDVLEFYNSKKTSNSQKYMSFT--DKSAEDYNSSNALNVKAVSETPAVPPVSED 180
hPAK2 KNPQAVLDVLKFYDSN--TVKQKYLSFT--PPEKDGLPSGTPALNAKGTEAPAV--VTEE
hPAK3 KNPQAVLDVLKFYDSKETVNNQKYMSFTSGDKSAHGYIAAHPSSTKTASEPPLAPPVSEE
      **********:**:*:      .***:***     . ..  :. .       :*.*  .  *:*:

hPAK1 EDDDD----DDATPPPVIAPRPEHTKSVYTRSVIEPLPVTPTRDVATSPISPTENNTTPP 240
hPAK2 EDDDE-----ETAPP-VIAPRPDHTKSIYTRSVIDPVPAPVGDSHVDGA--
hPAK3 EDEEEEEEEDENEPPPVIAPRPEHTKSIYTRSVVESIASPAVPNKEVTPPSAENANSS--
      **:::       :  ** ******:****:*****::  ..       :*  .  . : :

                                                   I
hPAK1 DALTRNTEKQKKKPKMSDEEILEKLRSIVSVGDPKKKYTRFEKIGQGASGTVYTAMDVAT 300
hPAK2 ---AKSLDKQKKKPKMTDEEIMEKLRTIVSIGDPKKKYTRYEKIGQGASGTVFTATDVAL
hPAK3 -TLYRNTDRQRKKSKMTDEEILEKLRSIVSVGDPKKKYTRFEKIGQGASGTVYTALDIAT
         :.  ::*:**.**:****:****:***:*********:***********:**  *:*

         II                III
hPAK1 GQEVAIKQMNLQQQPKKELIINEILVMRENKNPNIVNYLDSYLVGDELWVVMEYLAGGSL 360
hPAK2 GQEVAIKQINLQKQPKKELIINEILVMKELKNPNIVNFLDSYLVGDELFVVMEYLAGGSL
hPAK3 GQEVAIKQMNLQQQPKKELIINEILVMRENKNPNIVNYLDSYLVGDELWVVMEYLAGGSL
      *******:***:**************:*  ******:**********:**********

                                        VIb               VII
hPAK1 TDVVTET-CMDEGQIAAVCRECLQALEFLHSNQVIHRDIKSDNILLGMDGSVKLTDFGFC 420
hPAK2 TDVVTETACMDEAQIAAVCRECLQALEFLHANQVIHRDIKSDNVLLGMEGSVKLTDFGFC
hPAK3 TDVVTET-CMDEGQIAAVCRECLQALDFLHSNQVIHRDIKSDNILLGMDGSVKLTDFGFC
      ******* ****.*************:***:************:****:**********

                          VIII
hPAK1 AQITPEQSKRSTMVGTPYWMAPEVVTRKAYGPKVDIWSLGIMAIEMIEGEPPYLNENPLR 480
hPAK2 AQITPEQSKRSTMVGTPYWMAPEVVTRKAYGPKVDIWSLGIMAIEMVEGEPPYLNENPLR
hPAK3 AQITPEQSKRSTMVGTPYWMAPEVVTRKAYGPKVDIWSLGIMAIEMVEGEPPYLNENPLR
      *********************************************:************

                                          XI
hPAK1 ALYLIATNGTPELQNPEKLSAIFRDFLNRCLDMDVEKRGSAKELLQHQFLKIAKPLSSLT 540
hPAK2 ALYLIATNGTPELQNPEKLSPIFRDFLNRCLEMDVEKRGSAKELLQHPFLKLAKPLSSLT
hPAK3 ALYLIATNGTPELQNPERLSAVFRDFLNRCLEMDVDRRGSAKELLQHPFLKLAKPLSSLT
      ****************:**.:*********:***::**********  ***:********

hPAK1 PLIAAAKEATKNNH-
hPAK2 PLIMAAKEAMKSNR-
hPAK3 PLIIAAKEAIKNSSR
      *** ***** *..
```

Regulatory Domain

Kinase Domain

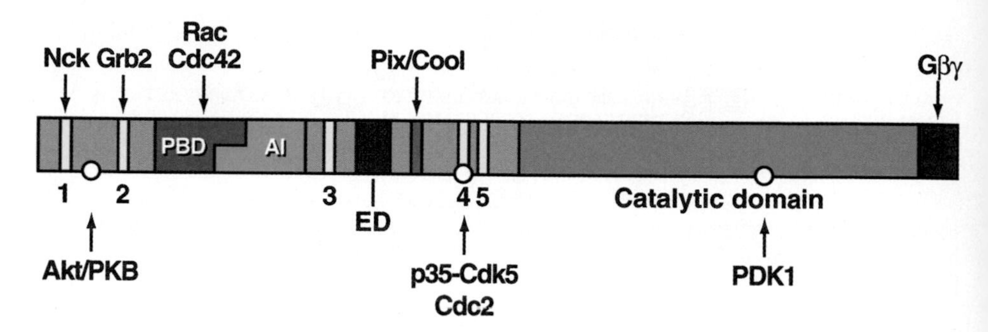

Figure 2 Schematic diagram indicating features of PAK1 structure. The PAK1 backbone is shown in orange and the catalytic domain (aa 255–529) in bright blue. The p21 (Rac/Cdc42)-binding domain or PBD is shown in purple overlapping with the pale green autoinhibitory (AI) domain. Yellow boxes are the five PXXP putative SH3-binding motifs, the green box is the noncanonical prorich Pix/Cool SH-3 binding motif, and the red box is the ED-rich region of unknown significance. The white circles represent identified sites of phosphorylation by Akt (Ser21), Cdc2/Cdk5 (Thr212), and PDK1 (Thr423).

PAK STRUCTURE AND REGULATION OF ACTIVITY

Structural Features

PAKs 1–3 contain an N-terminal extension that is a regulatory domain and a highly conserved C-terminal catalytic domain [amino acid residues (aa) 255–529, aa 235–509, and aa 254–528 in PAKs 1, 2, and 3 respectively]. Distinguishing features of the regulatory domain are the presence of five (PAK1), two (PAK2), or four (PAK3) canonical PXXP SH3-binding motifs and one nonclassical (PXP) SH3 binding site (Figure 1). Interaction of the first site in PAK1 with the SH3-containing adapter protein Nck (12, 13), the second site with the adapter protein Grb2 (14), and the noncanonical site with the PIX family of proteins (15) have been described (Figure 2). Rac and Cdc42 minimally bind to the so-called CRIB (for Cdc42 and Rac interactive binding) domain (aa 75–90 in PAK1). The more inclusive p21-binding domain (PBD) (aa 67–113 in PAK1) contributes to overall binding affinity (8, 9, 16). The PBD overlaps, but is not coincident with an autoinhibitory segment (aa 83–149) that forms part of an "inhibitory switch" (16) that controls the basal kinase activity of PAKs 1–3. Interposed between the N terminus and the kinase domain is an acidic (ED) residue–rich region of unknown significance. Additionally, a conserved binding site for the $G_{\beta\gamma}$ subunit complex of heterotrimeric G proteins exists at the extreme C terminus (17, 18).

The crystal structure of PAK1 in an autoinhibited conformation has been determined to 2.3-Å resolution (16). PAK1 exists as a homodimer in solution and in cells (16, 19), with the protein in a *trans*-inhibited conformation where the N-terminal regulatory domain of one PAK1 molecule binds and inhibits the

C-terminal catalytic domain of the other (Figure 3). The dimerization interface overlaps the PBD/CRIB and inhibitory switch domains. The inhibitory switch consists of a bundle of three α helices with a short N-terminal β hairpin. Helices Iα2 and Iα3 pack against the catalytic domain in a high-affinity ($K_d \approx 90$ nM) interaction and prevent kinase activation. Consistent with earlier mutational data, PAK1 residues Leu107, Glu116, and Asp126 critically contribute to this inhibitory interface (16).

Activation Mechanism

The structural data (16, 20–23), as well as genetic and biochemical studies (17, 24–31), support a model in which GTPase binding disrupts dimerization and leads to a series of conformational changes that destabilize the folded structure of the inhibitory switch domain, inducing its dissociation from the catalytic domain, and that rearrange the kinase active site into a catalytically competent state [(Figure 3); see (16) for a detailed mechanistic proposal for PAK activation]. Phosphorylation at Thr 423 in the activation loop of the PAK1 catalytic domain is important both for maintaining relief from autoinhibition and for full catalytic function toward exogenous substrates (26–28). Although Thr423 will autophosphorylate when PAK1 is activated in solution, there is evidence that this does not readily occur and that phosphorylation of this site by an exogenous kinase, such as PDK1, may be required for effective PAK activation in vivo (32). Exogenous phosphorylations and/or autophosphorylation of PAK at additional site(s) also contribute to kinase activation and/or maintenance of kinase activity, including Ser144 (PAK1) (27, 30). Interestingly, Ser144 autophosphorylation was reported to not be required for PAK activation by sphingosine (see following section), in contrast to activation by GTPases (30). In combination with the observation that PAK1 activated by sphingosine was more sensitive to PDK1-mediated phosphorylation and activation than the GTPase-activated enzyme (32), these data suggest differences in the conformational changes in PAK evoked by lipid versus those induced by GTPase binding.

PAK ACTIVATION BY GTPases AND SPHINGOLIPIDS PAK1 binds and is activated by Rac1, Rac2, and Rac3 (4, 33, 34), Cdc42 (4) and also by CHP (35), TC10 (36), and Wrch-1 (37), but not by Rho A-G or by other Ras superfamily members. On the basis of in vitro binding studies, it has been suggested that PAK2 may be selective for Cdc42 (versus Rac1) (20, 38, 39). However, the basis for the differential binding reported (38) is not clear, since residues within PAK1 conferring selective binding and activation by Rac (versus Cdc42) are conserved in PAK2. Reeder et al. (40) have mapped residues conferring GTPase selectivity to sites primarily within the overall PBD domain. Knaus et al. (33) have reported that a short Lys-rich tract (aa 66–68 of PAK1) just upstream of the CRIB domain is required for effective Rac GTPase binding, as well as for optimal stimulation of the kinase activity by bound Rac GTPase.

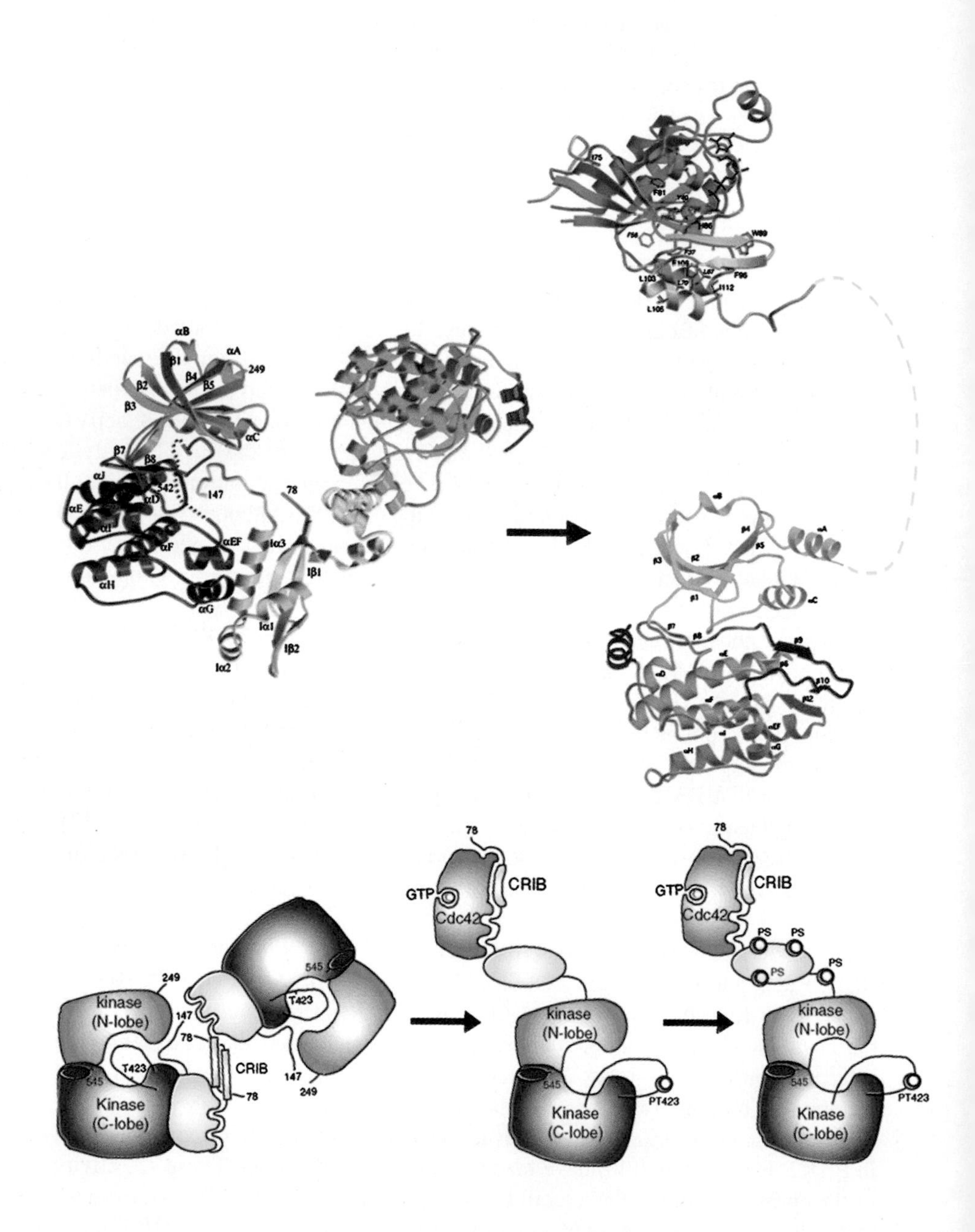
αB
αA
249
β1
β4
β2
β5
β3
αC
β7
β8
αJ
542
147
78
αD
αE
αI
αEF
αF
Iα3
Iβ1
αH
αG
Iα1
Iβ2
Iα2
78
GTP
CRIB
Cdc42
78
GTP
CRIB
Cdc42
PS
PS
PS
PS
249
kinase
(N-lobe)
545
T423
147
78
CRIB
T423
Kinase
(C-lobe)
78
147
249
kinase
(N-lobe)
545
Kinase
(C-lobe)
PT423
kinase
(N-lobe)
545
Kinase
(C-lobe)
PT423

For the GTP-bound forms of the GTP-hydrolysis deficient Rac1 (Q61L) and Cdc42 (Q61L) mutants, the binding constants to peptides encompassing PAK1 residues 70–118 and 75–132 have been measured as 20–50 nM and 10–30 nM, respectively (20). [Note: The Q61L mutation increases the affinity for effector binding over that seen with the wild-type protein loaded with a nonhydrolyzable GTP analog (20, 41).] It has been argued that the binding of active GTPase does more than just relieve inhibitory constraints imposed on the catalytic domain by the inhibitory switch and may also act by an allosteric mechanism to promote autophosphorylation events necessary for full kinase activation (28, 30, 31). Additionally, since Rac and Cdc42 are membrane localized via their prenylated C terminus, the association of PAK with a GTPase positions it at the plasma membrane where sphingolipids and PDK1 reside, potentially stimulating activation.

Once PAKs are activated by the binding of Rac or Cdc42 in vitro, the continued binding of the GTPase is not necessary for PAK kinase activity, and it has been reported that active GTPase dissociates from the PAK3 complex once activation has occurred (4). Release of GTPase is unlikely to readily occur from PAK1, since the dissociation constant for Rac1- or Cdc42-GTP from PAK1 is so low (31). Binding of Rac and/or Cdc42 to PAK inhibits their intrinsic- and GAP-stimulated GTP hydrolysis (4); this property has made the PAK PBD a useful affinity reagent for assaying the state of Rac and Cdc42 activation (42).

GTPase-INDEPENDENT PAK ACTIVATION While PAKs are usually considered a downstream target for the Rac and Cdc42 GTPases, a number of GTPase-independent activation mechanisms have been identified (Figure 4). Although their identity was unknown at the time, PAKs were first studied as kinases whose autophosphorylation and activity could be stimulated by limited protease digestion (24, 39). A physiological correlate for protease-mediated activation came

Figure 3 Schematic diagram of PAK1 autoregulation and activation by Cdc42. PAK1 in its "off" state is a dimer (*lower left*). The autoregulatory segment of the polypeptide chain (yellow) associates with the kinase domain (blue) and blocks its active site (16). The interaction promotes dimerization because the autoregulatory segment of one partner contacts the kinase domain of the other (19). Cdc42 or Rac (green), in their GTP-bound states, bind the CRIB sequence within the autoregulatory region, thereby dissociating the dimer and unblocking both kinase domains (*center*). Phosphorylation of the activation loop (purple disk) fully activates the enzyme. Further phosphorylation of sites within the regulatory region (*lower right*) would prevent reversal of these steps, even if the GTPase were to dissociate. Ribbon representations of the dimer (*upper left*) and of the activated monomer (*upper right*) are shown. The diagram of the active kinase is based on a structure of a kinase domain with mutations that lock the activation loop (purple) in its ordered, active conformation (M. Lei and S. Harrison, personal communication). The C-terminal helices (red) may be targets of other signaling inputs.

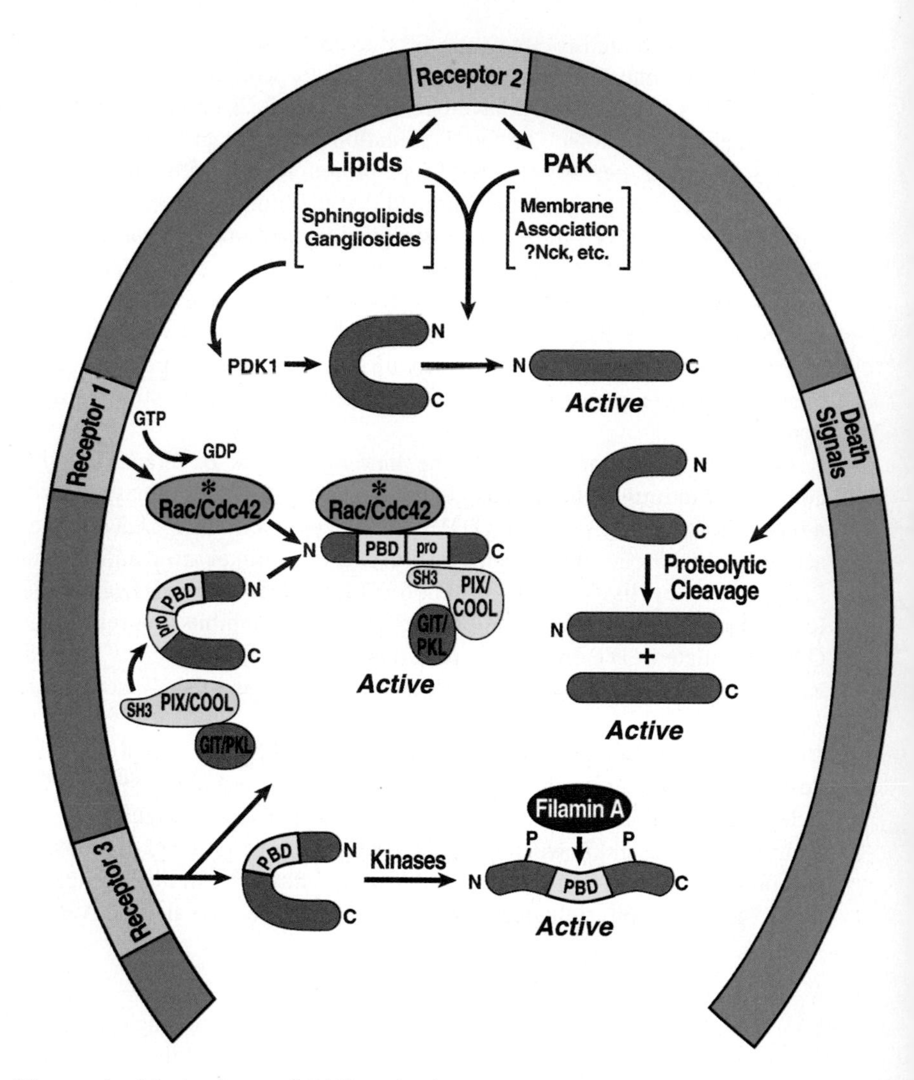

Figure 4 Mechanisms of PAK activation. Multiple mechanisms for activation of PAKs are indicated, as described in more detail in the text. PBD, p21-binding domain; pro, proline-rich binding site. [Figure modified from original in the *International Journal of Biochemistry and Cell Biology*, Vol. 30 (8): Knaus and Bokoch, "The p21 Rac/Cdc42 activated protein kinases," pp. 857–862, 1998, with permission from Elsevier Science].

with the discovery of PAK2 as a substrate for effector caspases during the apoptotic response (43, 44). PAK2 is cleaved by caspase 3 at Asp212, resulting in the formation of 28-kDa N-terminal and 34-kDa C-terminal fragments. The cleaved PAK2 appears to become catalytically activated, probably as a result of a reduction in autoinhibitory constraints normally imposed by the N-terminal inhibitory switch.

PAKs can interact via their N-terminal PXXP motifs with SH3-containing adapter proteins, such as Nck and Grb2. These adapters can act to recruit PAKs to activated tyrosine kinase receptors at the plasma membrane [e.g., (12–14, 45–47)]. Several studies have shown that membrane recruitment of PAK1 via adapter proteins, or via the addition of membrane-targeting sequences, results in the stimulation of kinase activity (48, 49). Although the molecular basis for PAK activation induced by membrane association remains ill defined (50), there is evidence that such activation is GTPase independent (51). Membrane recruitment of PAK via adapter proteins provides a GTPase-independent means to localize PAK in proximity to PDK1 and in an environment that promotes phosphorylation at the critical Thr423 residue. Indeed, many growth factor receptors stimulate the formation and metabolism of sphingolipids. PAK has been shown to be directly stimulated by sphingosine and sphingosine-derived lipids (51). Activation occurs at levels similar to that induced by GTPases and may occur through a similar relief from autoinhibition (28), although there may be mechanistic differences, as suggested by the lack of the requirement for autophosphorylation at Ser 144 for activation by sphingosine (30). Also active in stimulating PAK activity were gangliosides and phosphatidic acids, both of which may have signaling functions in certain biological contexts (51). Indeed, the potential contribution of PAK activation to the myriad biological effects reported for sphingolipids and gangliosides remains to be adequately investigated.

As indicated above, Thr423 is a critical phosphorylation site in the PAK activation process. Autophosphorylation at this site does not readily occur (32), suggesting the possibility that either *trans*-phosphorylation by PAK itself and/or phosphorylation by an exogenous kinase is a prerequisite to PAK activation. PDK1 may be one such kinase. PDK1 is a 3-phosphoinositide-dependent Ser/Thr kinase shown to control the activation of a number of kinases by phosphorylating a conserved critical threonine residue in the activation loop (52). King et al. (32) showed that PAK1 is a substrate for PDK1, which phosphorylates PAK primarily at the critical Thr423 residue and promotes activation. PDK1 also phosphorylated PAK1 at Thr423 in vivo after platelet-derived growth factor (PDGF) stimulation of NIH-3T3 cells. The effect of PDK1 on PAK activation appears to be sphingosine-dependent and may be independent of 3-phosphoinositides, as it is not blocked in the presence of wortmannin and occurs with a PDK1 mutant deficient in 3-phosphoinositide binding (32). The relative contributions of Rac and Cdc42 versus sphingosine in regulating localized PAK activation in response to growth factor stimulation remains to be defined.

ACTIVATION OF PAK BY PHOSPHORYLATION AND PROTEIN BINDING A number of kinases have been reported to phosphorylate PAK and regulate its function, both by direct effects on kinase activity and through regulation of the binding of

TABLE 1 Reported proteins that interact with PAKs

Category	Molecular partners	Reference(s)	PAK isoform
Cytoskeletal	Paxillin	(213)	3
Kinase	Akt	(53)	1
	PDK1	(32)	1
	PI 3-kinase/p85	(214)	1
	Cdk5	(57)	1
	Cdc2	(58)	1
	Src kinases	(62)	2
	Abl	(59)	2
	PKA (protein kinase A)	(215)	
Phosphatase	PP2A	(63)	1, 3
	POPX1, 2	(64)	1, * via PIX
Adapter	Nck	(12, 13)	1
			1
	Grb2	(14)	1
	NESH	(113)	1, 2
Other	CIB	(216)	1
	Rac/Cdc42	(4–6)	1, 2
			1, 2, 3
			1, 2
	Sphingolipids	(30, 51)	1
			1, 2
	Gβγ	(17, 18)	Ste20
			1
	PIX/COOL	(15, 65)	1, 2, 3
			3
	GIT/PKL	(71, 115)	* via PIX/COOL

various PAK-interacting proteins (see Table 1). Akt phosphorylates PAK1 at Ser21, and this modification both decreases binding of Nck to the PAK1 N terminus and stimulates PAK1 activity in a GTPase-independent manner in vivo (53; G. Zhou, Y. Zhou, C.C. King, B. Fryer, G.M. Bokoch, and J. Field, unpublished observations). The p35-bound form of Cdk5, a neuron-specific protein kinase, associates with and phosphorylates PAK1 at Thr212, a site also targeted by cyclin B-bound Cdc2 in a cell-cycle-dependent manner (55–58). Although phosphorylation by Cdk5 was initially reported to inhibit PAK activity, it now appears that Thr212 phosphorylation has no direct effect on PAK activity (56).

Tyrosine phosphorylation of PAK has been described in several recent studies. The tyrosine kinase, Abl, was reported to associate with PAK2 in vivo and to decrease PAK2 kinase activity concomitantly with tyrosine phosphorylation of PAK2 on multiple sites (59). A multiprotein complex containing tyrosine-phosphorylated and highly active PAK1 was identified in constitutively activated v-ErbB receptor–transformed cells. Formation of this complex appeared to be Rho dependent (60). The non-receptor tyrosine kinase Etk/Bmx, a Tec family member, was reported to directly bind and phosphorylate PAK1 (61). Renkema et al. (62) observed that coexpression of Src family tyrosine kinases with PAK2 enhanced activation by Rac1 or Cdc42 concomitantly with phosphorylation of Tyr130 in a small subpopulation of cellular PAK2. Src-mediated tyrosine phosphorylation of PAK2 appeared to be dependent upon conformational changes in PAK2 induced by GTPase binding. The modulation of PAK activity and/or molecular partners resulting from phosphorylation of various sites by other Ser/Thr and Tyr kinases may provide important mechanisms for enabling cells to respond appropriately to different external stimuli (e.g., 30, 60).

The importance of phosphorylation in PAK activation suggests that dephosphorylation reactions will be equally important in shutting off PAK activity. Although an interaction of PAK with PP2A has been reported, its physiological significance is unclear (63). Koh et al. (64) described the isolation of two closely related human protein phosphatases that efficiently dephosphorylate PAK1, including Thr423. The two phosphatases, termed POPX1 and POPX2 (for partner of Pix-1 and Pix-2), bind to various forms of PIX and form multimeric cellular complexes containing PAK. These PP2C-related phosphatases are either ubiquitously expressed (POPX2) or enriched in brain and testis (POPX1), similar to PAK1. Overexpression of either enzyme antagonizes the cellular effects of active PAK1; however, the physiological roles and regulation of the POPX phosphatases in the context of PAK signaling remain to be defined.

PAKs bind to a family of Rac- and Cdc42-specific guanine nucleotide exchange factors, termed PIX [for PAK-interacting exchangers (15)] or COOL [for cloned out of a library (65)]. The interaction of αPIX with PAK1 was shown to induce PAK1 activation that was distinct from, and synergistic with, Rac or Cdc42 binding (66, 67). The molecular mechanism for the stimulation of PAK activity by PIX binding is unclear. This effect is independent from the guanine nucleotide exchange activity of αPIX. There is evidence suggesting that the regulation of PAK activity by PIX may contribute to both growth factor (68–70) and focal adhesion signaling (71, 72) (see following sections), as well as apoptosis induced by the carcinogen, benzo[α]pyrene (73). In contrast to the stimulatory effect of αPIX, a 50-kDa form of βPIX (also termed $p50^{Cool-1}$) inhibits Rac- and Cdc42-stimulated PAK activity (65, 67). This inhibitory effect has been localized to an 18-amino-acid T1 domain present in the βPIX proteins (67). The exact mechanism of, and the physiological significance of, such negative regulation of PAK by βPIX has not yet been elucidated. Similarly, the mammalian homolog (hPIP1) of a WD repeat–containing regulator, Skb15, from fission yeast exists that binds and inhibits the ability of PAK to be activated by Rac or Cdc42 both in vitro and in vivo (74). Finally, in contrast to the stimulation of yeast Ste20 activity by $G_{\beta\gamma}$ binding (17), $G_{\beta\gamma}$ binding to mammalian PAK1 was found to inhibit kinase activity (18).

THE BIOLOGY OF PAKS 1–3

It has become increasingly evident that the PAK family of kinases plays significant roles in modulating a range of biological activities. These pleiotropic actions are reflected in the plethora of interacting proteins and phosphorylation substrates for PAKs 1–3, as well as in the many stimuli that regulate PAK activity (see Tables 1, 2, and 3). In the following sections, we discuss the major biological areas in which PAK function plays important regulatory roles, according to current information. The biochemical properties of various dominant active and dominant negative PAK mutants and their use in elucidating the multiple cellular roles of PAK in cell biological studies are described in detail in (75).

Regulation of Cytoskeletal Dynamics

One of the first indications that PAK1 might be involved in cytoskeletal regulation by Rac or Cdc42 came from localization studies. Immunofluorescence analysis using a PAK1-specific, affinity-purified antibody showed that PAK1 redistributed from the cytosol into cortical actin structures after cell stimulation by PDGF, insulin, wounding of monolayers, or transformation by v-Src (76). The cortical actin structures to which PAK1 localized included lamellae at the leading edge of polarized cells, circular dorsal ruffles, and some, but not all, peripheral membrane ruffles. A fraction of PAK1 was also localized to a vesicular fraction that remains incompletely characterized, although colocalization with a fluid-phase marker and a subsequent functional study suggested these may be macropinocytotic vesicles (77). Additionally, several studies have shown that PAK1 localizes to focal adhesions (78, 79, 81). Sells et al. (82) established that constitutively active PAK1 (activated by several different types of N-terminal mutations) induced the formation of filipodia, the accumulation of large, polarized membrane ruffles, and retraction of the trailing edge of the cell in Swiss 3T3 and REF52 cells. These changes are accompanied by the formation of vinculin-containing focal complexes within these structures. Highly kinase-active versions of PAK1 caused the loss of actin stress fibers and increased focal adhesion turnover (82). Interestingly, the cytoskeletal and morphological effects induced by PAK1 exhibited both kinase-dependent and kinase-independent components, as verified in subsequent studies (e.g., 49, 79). Manser et al. (78) also reported PAK1-dependent loss of actin stress fibers, as well as dissolution of focal adhesions. Use of an autoinhibitory domain fragment from PAK1 as a means to inhibit endogenous PAK activity showed that PAK1 activity was required for formation of actin microspikes and loss of focal adhesions and stress fibers promoted by active Rac or Cdc42 (25, 79). It should be noted that the cytoskeletal effects of PAK1 observed in the above studies were largely insensitive to dominant negative versions of Rac or Cdc42, suggesting that PAK was acting downstream of and/or independently from these Rho GTPases.

Initially noted in mammalian cell studies, the ability of PAK and its Ste20 relatives to modulate the actin cytoskeleton has been supported by genetic studies in other organisms. The *Caenorhabditis elegans* PAK homolog localizes with CeRac and CeCdc42 to hypodermal cell boundaries during embryonic body elongation, which involves dramatic cytoskeletal reorganization (83). The *mbt* (*mushroom bodies tiny*) gene of *Drosophila melanogaster* encodes a PAK-related kinase that is involved with neurogenesis in the adult fly nervous system (84). *Drosophila* PAK has been genetically linked to the signal transduction pathway from axonal guidance receptors to the actin cytoskeleton in photoreceptor (R cell) growth cones (85). Several PAK homologs have been identified in *Dictyostelium discoideum*, and PAKa colocalizes with myosin II to the cleavage furrow of dividing cells and to the posterior of polarized cells undergoing chemotaxis (86). Genetic manipulation of PAKa disrupts both cytokinesis and the ability of the cells to maintain directionality during chemotaxis. A function in suppressing lateral pseudopod formation and in proper tail retraction during chemotaxis is indicated. Finally, in both *Saccharomyces cerevisiae* and *Schizosaccharomyces pombe,* roles for the multiple existing homologs of PAK in proper actin cytoskeletal assembly, cytokinesis, and polarized growth via budding have been identified (87–93). Hyphal formation during filamentous growth of both *S. cerevisiae* and *Candida albicans* requires PAK function (94).

The observations in these multiple organisms, including genetic data, strongly support the hypothesis that PAKs and related kinases play multiple and critical physiological roles in controlling cytoskeletal organization and regulation. Indeed, mammalian studies of cell motility, neurogenesis, and angiogenesis indicate that PAKs are critical components of these complex processes as well (see below).

CELL MOTILITY PAK1 is recruited to the leading edge and lamellae of motile cells, including polarized fibroblasts (76) and human leukocytes (95). Analysis using an antiphosphopeptide antibody directed against phospho-T423 in PAK1 showed that PAK1 becomes activated in a spatial and temporal pattern consistent with a role for PAK1 in ongoing cortical actin remodeling (81). PAK1 and PAK2 are also strongly activated by the chemoattractant fMLP in human neutrophils (6, 96) and by the chemokine CXCL1 in a model cell line (97). Chemotaxis to CXCL1 was blocked by expression of a dominant negative PAK1 COOH terminal construct (aa 232–544, K298A). Studies in inducible, stable NIH-3T3 cell lines (98) and in transfected human microvascular endothelial (HMEC) cells (99) showed that expression of either constitutively activated or inhibitory PAK1 mutants could markedly alter cell motility. NIH-3T3 cells expressing active versions of PAK1 had large leading edge lamellipodia and were more motile than their normal counterparts (98). In contrast, expression of kinase-inactive PAK1 induced the formation of random, multiple lamellipodia, which decreased the ability of these cells to move in a directed manner. These results differed from expression of PAK1 constructs in HMEC, where motility was decreased by either active or inactive versions of PAK1 (99). This inhibition of motility was accompanied by effects on focal contact dynamics and stability. Both studies noted that expression of active PAK1 enhanced overall cell contractility associated with increased phosphorylation of myosin II regulatory light chain.

NEUROGENESIS As indicated above, roles for PAK function in neurogenesis and in axon pathfinding have been established by genetic studies in model metazoan organisms. Role(s) for PAK(s) in neuronal guidance in humans may be reflected in a disease, nonsyndromic X-linked mental retardation, which is caused by point mutations in PAK3, the brain-specific PAK isoform (100). Interestingly, one such PAK3 mutation, R67C, found in these patients (101) lies in the conserved polybasic tract reported by Knaus et al. (33) to be critical for GTPase binding and PAK activation.

Expression of membrane-targeted versions of PAK1 in PC12 cells in culture induced formation of growth cone–tipped neurites (49). Interestingly, this effect did not appear to be kinase dependent, and expression of pieces of the N terminal regulatory domain of PAK1 containing its first and second SH3-binding motifs blocked nerve growth factor–induced neurogenesis. To some degree, these effects, including induction of membrane ruffling, may result from interactions of PAK with the Rac and Cdc42 guanine nucleotide exchange factor, PIX (102). PAK5, which is abundant in brain tissue and has similarity to *Drosophila* MBT (see above), has also been shown to induce neurite outgrowth in N1E-115 neuroblastoma cells (103). This effect correlated with kinase activity and, unlike PAK1, did not require membrane targeting of the kinase. PAK1 is phosphorylated by neuronal p35-bound Cdk5 on Thr212, apparently inhibiting PAK1 activity indirectly and, somewhat paradoxically, promoting neurite outgrowth (55). This seeming paradox can be explained, however, by the fact that phosphorylation of PAK1 on Thr212 appears to be required to maintain normal cytoskeletal and microtubule morphology in the neuron (56, 57) (see following sections).

ANGIOGENESIS Blood vessel formation is akin to the formation and branching of neurons in the brain. Endothelial cells must respond to external signals (e.g., binding of angiogenic growth factors to their receptors) with cell proliferation and cytoskeletal rearrangements. These changes lead to increased endothelial cell motility, alterations of cell morphology, and the formation of tubes that evolve into blood vessels. Not surprisingly, PAK plays a role in this process.

Normal PAK function was shown to be required for endothelial cell motility (99). In addition, the overexpression of kinase-dead PAK1 and/or the autoinhibitory domain of PAK1 blocks the formation of branched capillary networks in an in vitro tube formation assay using human vascular endothelial cells (HUVEC) suspended in Matrigel (104). In a separate study, introduction (via HIV Tat-mediated entry) of a portion of the N-terminal Nck-binding domain of PAK1 inhibited endothelial cell migration and tube formation in vitro (105). Strikingly, the same peptide blocked angiogenesis in the chick chorioallantoic membrane assay. These results suggest an important role for the Nck adapter in coupling PAK to angiogenic growth factor signaling. Indeed, Nck appears to couple PAK to angiogenic signaling by the basic FGF receptor and is required for basic FGF activation of the MAP kinase, ERK, known to be required for angiogenesis (J. Hood and D.A. Cheresh, submitted for publication). Additional supporting evidence for the function of PAK in angiogenesis comes from studies in which the recruitment and activation of PAK1 via Nck to the Tek/Tie-2 angiopoietin receptors potentiated the ability of these angiogenic ligands to stimulate endothelial cell motility (106). Similarly, cytoskeletal rearrangements necessary to maintain endothelial cell barrier integrity induced by sphingosine 1-phosphate required PAK-dependent actin remodeling (107). Endothelial cell remodeling induced by sheer stress is also reported to require PAK function (108), although the PAK dominant negative construct used to make this conclusion was not adequately described. PAK1, activated in breast cancer epithelial cells by heregulin-β1, can also enhance angiogenic responses by inducing the up-regulation of VEGF expression (109).

CANCER METASTASIS PAK(s) may play important roles in modulating the ability of cancer cells to move and metastasize. A number of human breast cancer lines exhibit constitutively elevated PAK1 and PAK2 activity, in some cases associated with the presence of an activated Rac GTPase (34). Heregulin, a potent stimulator of breast cancer cell growth, motility, and progression acting through the HER2/erb2 receptor, stimulates PAK1 activity (110). Heregulin induces a redistribution of PAK1 into the leading edge of motile cells, and cytoskeletal reorganization can be blocked by a dominant inhibitory PAK1 construct. In subsequent studies, a kinase-dead mutant, PAK1 (K299R), inhibited motility and invasiveness of the highly invasive MDA-MB435 cell line (111), whereas active versions of PAK1 stimulated MCF-7 cell motility (112). A correlation between the status of PAK1 activity and basal invasiveness of human breast cancer–derived cells and breast tumor grades was suggested (112). This correlation may be overly simplistic, however, as our laboratory has observed that certain breast cancer lines expressing high basal PAK activity have poor motility (M. Stofega and G.M. Bokoch, unpublished observations). High basal PAK activity is associated with the abnormal distribution of active PAK into very large focal adhesion–like structures. It is of interest, though, that a number of other modulators of tumorigenicity activate PAK [e.g., Etk/Bmx (61)] or inhibit its activity [e.g., NESH (113)].

Mechanisms of Cytoskeletal Regulation by PAK

There are a number of PAK substrates and/or interacting proteins identified that can account for some, but probably not all, of the effects of PAK to modulate cell morphology, actin/microtubule dynamics, and cell motility (Table 1, Table 2). While a preferred sequence for phosphorylation by PAK2, -KRES, has been derived from peptide studies in vitro (114), it is clear from Table 2 that not all PAK substrates identified conform to this motif, suggesting other factors (e.g., protein secondary structure, surrounding sequences, and affinity of substrate binding to PAK) may influence substrate selection. Several of the best-described targets of PAK action are discussed here.

PAK Targets Acting as Downstream Effectors

PIX/COOL, GIT/CAT/PKL As described above, a fraction of cellular PAKs are constitutively associated with the PIX/COOL guanine nucleotide exchange factors. PIX/COOL binds to a nonconventional proline-enriched motif in PAK (Figures 1 and 2) (15, 65). Complexes of PAK and PIX were found associated with a family of serine- and tyrosine-phosphorylated 90- to 95-kDa proteins, termed PKL [for paxillin kinase linker (71)] or CAT1 (COOL-associated, tyrosine-phosphorylated) and CAT2 (115). These proteins had been previously identified as GIT1 [G protein–coupled receptor kinase-interactor 1 (116, 117)], an Arf6 GAP domain–containing protein. The GIT/CAT/PKL proteins are phosphorylated when immunoprecipitated complexes with PAK and PIX are incubated in vitro with [32P]ATP (15) and can be directly phosphorylated by PAK in vitro (30). Tyrosine phosphorylation is observed in vivo and is stimulated by coexpression of FAK and by members of the Src family of nonreceptor tyrosine kinases (115). GIT/CAT/PKL family proteins also contain multiple ankyrin-binding repeats, Ca^{2+}- and calmodulin-binding motifs, two paxillin-binding subdomains, and a yeast Spa2 homology domain that is present in a region that binds focal adhesion kinase (FAK) and PIX (72). The latter domains appear to be important for mediating the localization of GIT/CAT/PKL to focal adhesions via interaction with paxillin and/or FAK. GIT/CAT/PKL proteins can induce the disassembly of focal adhesions accompanied by loss of paxillin, possibly under regulation by the PIX interaction (68, 118–121). Recruitment of PAK to focal adhesions appears to occur primarily via the PIX/COOL-GIT/CAT/PKL connection, although Nck may contribute in some cases (121). Complexes of PAK-PIX-GIT have also been found to distribute between focal adhesions, a cytoplasmic vesicular compartment, and the leading edge of motile cells (120,

TABLE 2 Reported PAK Substrates

Category	Substrate	Phosphorylation site	Reference(s)	PAK isoform
Cytoskeletal	Myosin light chain kinase (MLCK)	RPKS^{439} SLP	(137)	1
		GRLS^{941} SMA	(138)	2
	Regulatory Myosin light chain (R-MLC)	RATS^{19} NVF	(133, 134)	2
				2
	Myosin I heavy chain	—	(141, 205)	3, Ste20, Cla4
	Myosin II heavy chain	—	(136, 141, 206)	
	Myosin VI	—	(143)	3
			(L. Aschenbrenner & T. Hasson, unpublished observations)	1, 2
	Caldesmon	NIKS^{657} MWE	(207, 208)	3
		GVSS^{687} RIN		3
	Desmin		(209)	
	Op18/stathmin	KRAS^{16} GQA	(151, 152)	1
	Merlin	KRLS^{518} MEI	(177, 178)	1
				2
	Filamin A	RAPS^{2152} VAN	(156)	1
Kinase	LIM kinase (1 and 2)	KRYT^{508} VVG	(129, 130)	1
				4
	Raf-1	QRDS^{338} SYY	(171, 172)	2, 3
				1
	MEK-1	RPLS^{298} SYG	(165)	1
Other	p47*phox*	RRNS^{328} VRF	(6)	1
	Bad	RHSS^{112} YP	(187)	1, 2
		RSRS^{136} AP	(53, 185)	1
				2
	Estrogen receptor	KKNS^{305} LAL	(210)	1
	GEF-H1	RRRS^{885} LPA	(161)	1
	Gαz	ARRS^{16} RRI	(18)	1
	NET1	—	(A. Alberts, personal communication)	
	Phosphoglycerate mutase-B (PGAM-B)	NRFS^{23} GWY	(203)	1
		WRRS^{118} YDV		
	RhoGDI	KKQS^{101} FVL	(C. DerMardirossian & G. M. Bokoch, in preparation)	1
		ARGS^{174} YSI		
	Prolactin	RRDS^{177} HK	(212)	2

122). Trafficking of components between these various compartments appears to be regulated in a complex fashion, and the existing data remain conflicting (118–123). Similarly, the contributions of Arf6 regulation via the GIT/CAT/PKL Arf6 GAP domain (124) to these biological effects remains to be clarified. It appears likely, though, that the activities of PAK mediated via the PIX/COOL-GIT/CAT/PKL complex are important components of the dynamic processes that contribute to localized cytoskeletal regulation during changes in cell shape and motility. Also intriguing is the possible regulation of vesicular/membrane trafficking related to motile responses by vesicle-associated GIT (122) and PAK1 (77).

LIM *kinase* LIM kinases-1 and -2 are serine kinases implicated in the regulation of actin cytoskeletal dynamics through their ability to specifically phosphorylate members of the cofilin/actin depolymerizing factor (ADF) family. The ~70-kDa LIM kinases are characterized by the presence of two N-terminal LIM domains followed by a PDZ domain and a C-terminal kinase domain. Cofilin/ADF proteins are phosphorylated solely at Ser3 by LIM kinases and are the only known substrates for this kinase family. When phosphorylated at Ser3, cofilin/ADF can no longer bind effectively to F-actin, and the ability of these proteins to catalyze both F-actin depolymerization and severing is thus inhibited. These functions appear to be critical for normal actin dynamics, and the action of LIM kinases and cofilin/ADF in a variety of processes requiring rapid actin rearrangements has been documented [reviewed in (125, 126)]. LIM kinases are abundant in neuronal tissues and have been implicated in axonal motility. Of note, deletion of the LIM kinase gene is linked to Williams syndrome, a human visuospatial cognitive disorder associated with mild mental retardation and vascular disease (126).

LIM kinase-1 activity was originally reported to be activated by receptors that stimulate Rac GTPase (127, 128). However, Rac was unable to directly bind to or regulate LIM kinase-1 activity, which suggested the existence of an additional mediator acting downstream of Rac and upstream of LIM kinase. Edwards et al. (129) identified PAK1 as the missing link in this signaling pathway (Figure 5). Active PAK1 bound LIM kinase-1 and phosphorylated it primarily at Thr508 within the activation loop, resulting in a 10-fold increase in LIM kinase-1 activity toward cofilin/ADF in vitro. Inhibition of PAK activity in vivo by overexpression of its autoinhibitory domain blocked LIM kinase-1-mediated modulation of actin assembly, whereas expression of dominant inhibitory LIM kinase-1 blocked the ability of overexpressed PAK1 to induce dorsal ruffle formation in BHK cells. These data suggest that LIM kinase activity is one important downstream mediator of PAK1 function. We have observed regulation of both LIM kinase-1 and -2 in vitro by PAK1, PAK2, and PAK4 (G.M. Bokoch, unpublished observations), and cytoskeletal changes induced by PAK4 in C2C12 cells were LIM kinase-1 dependent (130). Interestingly, it was subsequently found that the critical Thr508 regulatory site in LIM kinase-1 (as well as the equivalent Thr505

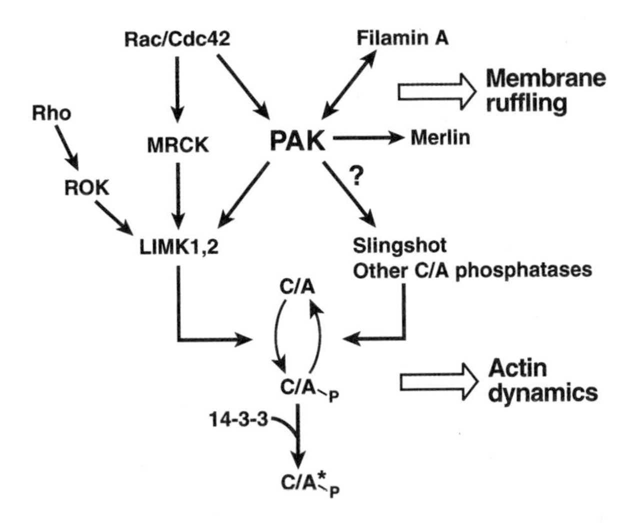

Figure 5 Mechanisms of actin cytoskeletal regulation by PAK1. This figure emphasizes the ability of PAK1 to modulate actin dynamics through regulation of the actin severing/depolymerizing function of cofilin and ADF (C/A). Activation of PAK1 by Rac/Cdc42 and/or filamin A regulates the activity of LIM kinases 1 and 2 to phosphorylate Ser3 of C/A. Similarly, the ability of Cdc42, acting via myotonic dystrophy kinase-related Cdc42-binding kinase (MRCK), and Rho, acting via Rho kinase (ROK), to stimulate LIM kinase activity is indicated. A role for PAK in modulating the action of C/A phosphatases (e.g., Slingshot) is speculatively (?) indicated, although there is no evidence for this as yet. 14–3-3 protein has recently been shown to bind and stabilize (*) Ser3-phosphorylated C/A (204). PAK1 may also modulate membrane ruffling and cortical actin assembly through effects on filamin A and/or Merlin.

site in LIM kinase-2) can also be phosphorylated by the myotonic dystrophy-related Cdc42-binding kinase (MRCK) (131) and by Rho kinase (132), see Figure 5. These findings suggest that the ability to control actin severing and/or depolymerization may be an important step in cytoskeletal regulation by all Rho GTPase family members.

Regulatory myosin light chain and myosin light chain kinase Myosins are actin-activated Mg-ATPases that convert the energy of ATP hydrolysis into force between actin and myosin filaments, leading to either contraction or tension. "Conventional" myosins that form filaments (e.g., myosin II) consist of two heavy chain (MHC) actin-activated ATPases and two light chains, one essential and one regulatory. The phosphorylation of the regulatory myosin light chain (R-MLC) at Ser19 (and Thr18) by Ca^{2+}-dependent myosin light chain kinase (MLCK) has been shown to be a critical regulatory step for physiological modulation of myosin contractility (Figure 6).

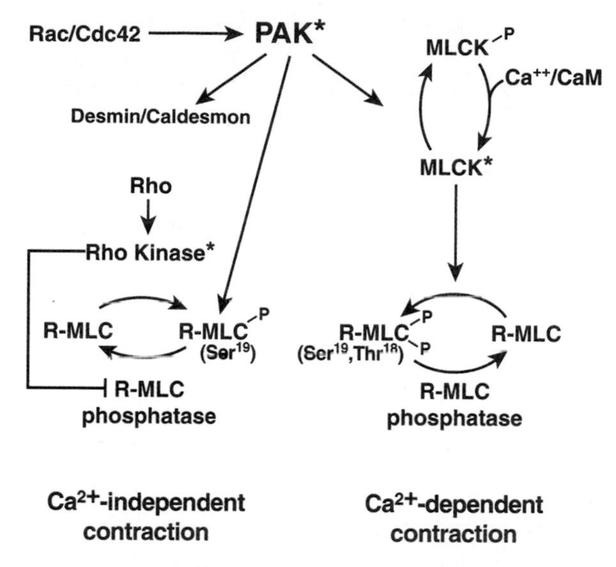

Figure 6 Pathways for regulation of myosin II–dependent contractility. Activated PAK (PAK*) mediates the effects of Rac and Cdc42 to promote cellular relaxation by phosphorylating myosin light chain kinase (MLCK), thereby decreasing its ability to phosphorylate regulatory myosin light chain (R-MLC) on Thr18 and Ser19. In contrast, PAK can also promote contractility by direct phosphorylation of R-MLC on Ser19 and/or by phosphorylation of the regulatory proteins desmin and caldesmon. Rho can also directly phosphorylate R-MLC on Ser19 and decrease MLC phosphatase activity via Rho kinase activation (*).

A number of studies have reported that PAK1 and/or PAK2 are able to directly phosphorylate R-MLC at the critical Ser19 site, with a resulting increase in contractility (133–135). It has been reported that both intact myosin and the isolated R-MLC are phosphorylated effectively by PAK (134, 135). However, some laboratories have had difficulty in observing phosphorylation of intact myosin, raising the question of whether this reaction indeed occurs under physiological conditions (e.g., G.M. Bokoch & P. de Lanerolle, unpublished observations). In contrast, it has also been suggested that PAK may phosphorylate MHC, based on the observation that dominant inhibitory PAK1 (H83L,H86L,K299R) blocked myosin II heavy chain phosphorylation induced in a Rac-dependent manner by bradykinin in PC12 cells (136). However, expression of either wild-type or constitutively active PAK1 was unable to increase MHC phosphorylation. In *Dictyostelium,* a PAK-related kinase increases the assembly of myosin II in response to chemoattractant stimulation but does not appear to directly phosphorylate myosin II (86). Instead, this effect may be mediated through PAK-dependent inhibition of an MHC kinase (see below).

PAK1 also modulates R-MLC function via direct phosphorylation and inhibition of MLCK (137). Phosphorylation of MLCK by PAK1 in vitro decreased MLCK activity toward R-MLC by more than 50%. Expression of activated PAK1 decreased R-MLC phosphorylation at Ser19 and inhibited cell spreading. It was also shown that PAK2 had similar effects, blunting the development of calcium-induced isometric tension in saponin-permeabilized endothelial monolayers by 75% (138). Constitutively active PAK2 phosphorylated MLCK to a stoichiometry of 1.71 $\pm$ 0.21 mol PO_4/mol MLCK, and two sites of phosphorylation, Ser439 and Ser991, were identified. Interestingly, the binding of calmodulin to MLCK inhibited phosphorylation primarily at the Ser991 site, which may be the more important site for inhibition of catalytic activity by PAKs, although this has not been directly demonstrated (138). These data suggest that the regulation of myosin function by PAK may be both stimulus dependent and complex and that there may be substantial interplay between regulation by PAK and other physiological regulators of MLCK (e.g., Ca^{2+}/calmodulin) and/or R-MLC (e.g., Rho kinase) (Figure 5). Rho kinase has been shown to enhance R-MLC phosphorylation at Ser19 by both direct phosphorylation of this site and through inhibition of MLC phosphatase [reviewed in (139)]. The ability of Rac- and Cdc42-signaling to antagonize the functional responses of cells to Rho is well documented, and at least part of this effect appears to be via opposing myosin II–mediated contraction through PAK-imposed inhibition of MLCK. The action of PAK to inhibit myosin II–based contractility via inhibition of MLCK may assist in disassembly of actin stress fibers and focal adhesions by PAK (140).

The regulation of multiple myosins is likely to be an important component of PAK-mediated cytoskeletal signaling. Both *S. cerevisiae* Ste20-related kinases and mammalian PAK1 phosphorylate the motor domain of the unconventional myosin I (141), and this phosphorylation appears to positively regulate the ability of myosin I to promote actin assembly via the Arp2/3 complex (142).

Myosin VI is a unique unconventional myosin isoform that translocates along actin filaments toward the pointed ends; all other myosins move toward the barbed ends. Myosin VI is phosphorylated by PAK3 on Ser406 in vitro, significantly enhancing the actin-translocating activity of myosin VI (143). Myosin VI has been implicated in hair cell development and movement of stereocilia in the inner ear (L. Aschenbrenner & T. Hasson, unpublished observations). Expression of active PAK1 and PAK2 in sensory hair cells of the chick cochlea induced myosin VI phosphorylation in the motor domain in an actin-sensitive manner, as well as in the cargo binding tail domain. PAKs 1 and 2 colocalized with areas enriched in myosin VI during the development of the hair cells, suggesting that PAK1 and PAK2 may be involved in the physiological regulation of myosin VI.

STATHMIN/Op18; GEF-H1 During cell division, a series of checkpoints ensure that the process proceeds normally. These include monitors of both the microtubule and actin cytoskeletons, which ensure that the structure of the dividing cell is not compromised. Recently, several lines of evidence have implicated PAKs as possible regulators of microtubule dynamics in addition to modulating the actin cytoskeleton and potentially coordinating events necessary for cell division. Studies of S. cerevisiae and S. pombe PAK homologs suggested effects of these kinases on cell cycle and cell division/cytokinesis (87–93). Injection of active PAK2 into frog embryos inhibited blastomere division (36), and introduction of Xenopus PAK into oocytes blocked progesterone-induced maturation, perhaps linked to mitotic spindle pole migration (145, 146). The S. pombe PAK, Shk1, is necessary for normal interphase and mitotic microtubule organization (147). Consistent with a role in microtubule dynamics, Shk1 localized to interphase microtubules and mitotic spindles. A subset of the Xenopus PAK-related kinase, termed X-PAK5, also appeared to localize to microtubule networks (148), and X-PAK5 interfered with microtubule dynamics by promoting formation of nonactively growing, stabilized microtubules.

PAK1 is phosphorylated at specific phases of the cell cycle on Thr212, and this phosphorylation can be mediated via cyclin B1/Cdc2 (57, 58). The same site is phosphorylated by the related neuronal kinase, p35-bound Cdk5 (57). Interestingly, the Thr212 site is not conserved in PAK2 or PAK3, indicating this may reflect a PAK1-specific function. Phosphorylation at Thr212 was not associated with changes in PAK1 activity but rather modulated the interaction of PAK1 with several unidentified proteins (58). In mitotic cells, PAK1 phosphorylated at T212 accumulated at the microtubule organizing centers (MTOC), and introduction of a synthetically T212-phosphorylated PAK1 peptide into cells by coupling it to a penetratin peptide (capable of mediating translocation across the cell membrane) resulted in localization along spindle microtubules. Expression of the same PAK1 T212-PO_4 peptide induced microtubule abnormalities upon reformation after nocodazole washout in Swiss 3T3 fibroblasts. Cells approaching mitosis exhibited markedly increased numbers and length of astral microtubules extending from the MTOCs. Since the T212-PO_4 peptide acted as a dominant negative to displace endogenous PAK bound to microtubules, as also determined after detergent extraction, these data suggest that when PAK1 is phosphorylated at T212, microtubules are less stable. Activated versions of PAK1 induced centrosomal abnormalities in ~10% of breast epithelial MCF-7 cells (102), and a portion of the active PAK1 was localized to centrosomes during metaphase and on the contractile ring during cytokinesis (149).

A possible direct link between PAK and microtubule dynamics is via the protein stathmin/Op18. Op18 binds $\alpha\beta$ tubulin dimers, inhibits tubulin polymerization, and promotes microtubule catastrophe [reviewed in (150)]. Phosphorylation of Op18 at several sites, particularly Ser16, inhibits the binding of Op18 to microtubules and prevents its destabilizing activity. Phosphorylation of Op18 at Ser16 is stimulated by the epidermal growth factor receptor in a Rac/Cdc42-dependent manner, and this effect was blocked by expression of the PAK1 autoinhibitory domain (151). A subsequent study demonstrated that phosphorylation of Op18 by PAK1 at Ser16 directly inhibited the microtubule destabilizing function of Op18, both in vitro and in vivo in an interphase cell (152). Microtubule function is associated with formation and maintenance of the leading edge of the cell [reviewed in (153)]. PAK activation promoted stabilization and growth of microtubules into the leading edge. This role for PAK is further discussed below.

FILAMIN A Filamin A is a 280,000-Da actin-binding protein that induces high-angle cross-linking of actin filaments [reviewed in (154)]. Cells lacking filamin A exhibit reduced cytoplasmic viscoelasticity and unstable membranes that constitutively bleb, and the cells are unable to support effective locomotion (154). There is increasing evidence that filamin A and the Arp2/3 complex act in complementary ways to maintain actin structure and mechanics at the leading edge of cells (155). Filamin A was identified as a PAK1 binding partner in a two-hybrid screen, and a physical interaction was confirmed in an in vitro GST-pulldown assay and in vivo by coimmunoprecipitation (156). Interestingly, filamin A binds to a region that overlaps the CRIB/PBD domain of PAK1 (aa 52–132), and this binding appears to be sufficient to relieve autoinhibition, thereby enhancing intrinsic PAK1 activity. Filamin A contains a series of 96-amino acid tandem "repeats" forming antiparallel sheet domains that overlap so as to generate a rod (154). PAK1 binds to tandem repeat 23 in the C terminus of filamin A and phosphorylates Ser2152 in filamin A. Filamin A appears to be necessary for the ability of PAK-mediated signaling to induce formation of dorsal membrane ruffles (Figure 5) (156). Thus, ruffle formation induced by PAK-activating stimuli (heregulin, sphingosine) or by expression of a constitutively active PAK mutant, T423E, was almost totally blocked in the filamin A deficient M2 cell line, but it was normal in the A7 cell line in which filamin A expression had been restored. PAK1 activated by filamin A also stimulated cofilin phosphorylation, presumably through the PAK1 substrate, LIM kinase. Because filamin A also appeared to be required for full and sustained PAK activation in response to heregulin, it was suggested that filamin A may serve both as a direct PAK activator and/or as a platform for GTPase-mediated activation, because filamin A can itself directly bind Rac and Cdc42 (154). Additional investigation of how PAK might be regulated by interactions with filamin A, and how filamin A contributes to PAK-mediated cytoskeletal regulation, seems warranted.

A Model for PAK Action in Cell Polarity and Motility

Based on our current knowledge of PAK function in animal cells, in combination with genetic data obtained on the functions of PAK homologs in lower organisms, a working model for the integrated regulation of the actin and microtubule cytoskeletons during processes such as cell motility can be proposed (Figure 7).

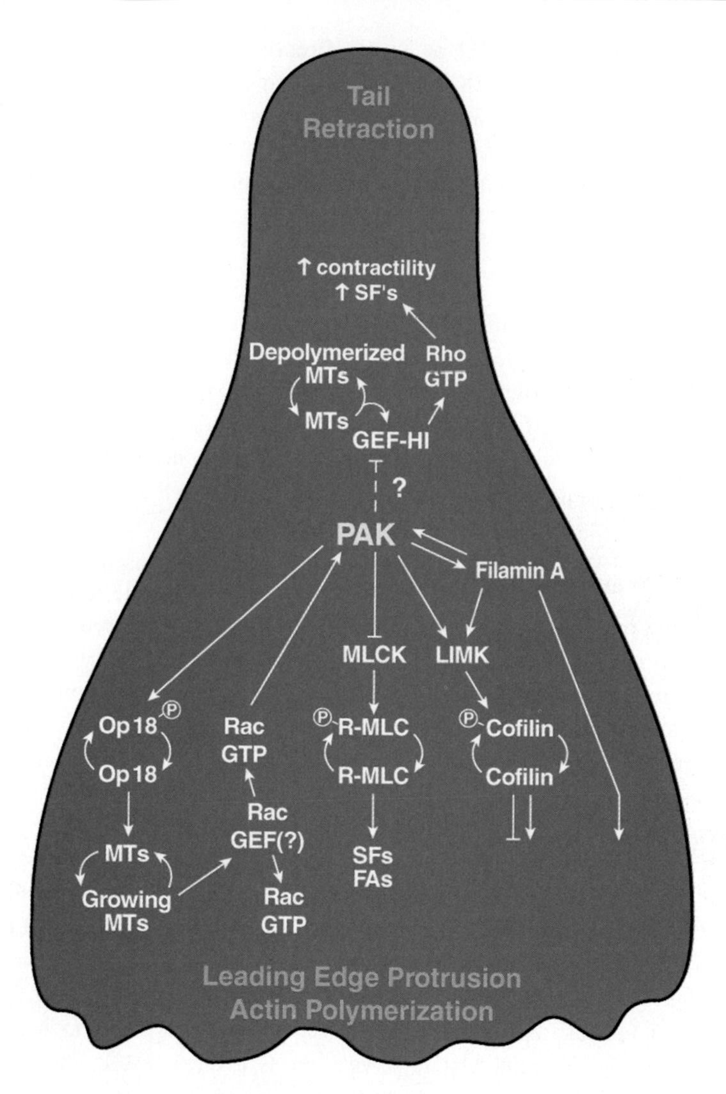

Figure 7 Hypothetical model of PAK action as a critical controlling element in directed cell migration. See text for details of the proposed model. FAs, focal adhesions; MTs, microtubules; SFs, stress fibers.

As described above, activated PAK becomes localized to the leading edge of polarized fibroblasts, leukocytes, and other motile cells. The ability of PAK to inhibit myosin light chain kinase activity would act to suppress R-MLC phosphorylation and contractility in the leading lamellipodium. This action would serve to inhibit contraction at the leading edge back into the main body of the cell, whereas the lack of active PAK at the sides and rear of the cell would permit retraction of pseudopods in these regions, thus aiding in maintenance of directionality. Inhibition of contractility would also promote the loss of established actin stress fibers and focal adhesions at the leading edge (140, 157), thereby readying this area of the cell for movement. PAK activation at the front of the cell would also enhance the phosphorylation and inactivation of cofilin/ADF (via LIM kinase), thereby suppressing actin filament turnover, thereby promoting stabilization of the leading edge. Through its interaction with filamin A, PAK would promote the formation of membrane ruffles and the leading edge. This effect could potentially be independent of PAK activity (see above). Alternatively, or concomitantly, PAK may recruit the Nck adapter protein as a means to interact with and activate the Arp2/3 complex to stimulate actin polymerization at select sites (158).

How might actin-based events be coordinated with microtubule dynamics? Microtubules exhibit a distinctly polarized arrangement in motile cells, with actively growing microtubules extending into sites at the leading edge, often associated with focal adhesions. In contrast, microtubules within the cell body are actively breaking down. An extensive literature, reviewed in (153), indicates that the active microtubule growth occurring at the leading edge of polarized, motile cells is required for maintenance of the motile phenotype. For example, disruption of microtubules with the drug nocodazole also prevents the establishment and/or maintenance of cell polarity and directed movement. Waterman-Storer et al. (159) showed that associated with the growth of microtubules is an activation of the Rac GTPase by an unknown mechanism, which promotes actin polymerization and lamellipod formation. Conversely, the loss of microtubules at the cell rear is associated with activation of Rho GTPase, which promotes retraction of the cell body and rear to enable forward movement. This activation of Rho was recently shown to be mediated, at least in part, via release of the microtubule-inhibited Rho-specific guanine nucleotide exchange factor, GEF-H1 (160).

Activation of PAK at the leading edge, promoted either by Rac activation elicited by specific stimuli of motility (i.e., chemoattractants) or by the Rac activation that occurs upon microtubule growth (159), would lead to stathmin/Op18 phosphorylation, further increasing microtubule stability. These events would act as a positive feedback loop to further increase Rac-GTP levels and PAK activity, in turn promoting more actin assembly and leading edge stability, thus leading to a self-sustaining reinforcement of cell polarity. In support of this positive feedback mechanism, we have shown that Rac activation promotes microtubule growth rates by three- to fourfold in Ptk1 epithelial cells (152). This effect required the inactivation of stathmin/Op18 through a PAK1-dependent mechanism. Mutant stathmin/Op18 not phosphorylatable at Ser16 antagonized Rac-dependent leading edge formation and cell motility. Similarly, expression of the PAK1 autoinhibitory domain blocked leading edge formation. PAK may also modulate the action of GEF-H1, as we have recently shown that GEF-H1 is phosphorylated by PAK1 and that this phosphorylation results in the binding of

14-3-3 protein(s) to GEF-H1 (F.T. Zenke, M. Krendel, G.M. Bokoch, submitted for publication).

Of course, this working model is likely to be oversimplified. There are almost certainly other PAK substrates and/or interacting proteins remaining to be identified that might influence the motile phenotype and cell polarization. However, it is evident that PAK plays a critical role as an effector and orchestrator of cytoskeletal dynamics in processes such as motility that require polarized cell behavior.

Cell Growth Signaling and Transformation

Several studies have implicated PAKs in the signaling pathways involved in growth and/or transformation of cells. An early study by Field and colleagues demonstrated that the expression of kinase-deficient PAK1 mutant prevented the transformation of Rat-1 fibroblasts by the Ras oncogene (161). This inhibitory effect was not evident in NIH-3T3 cells but was subsequently observed in Ras-dependent Schwann cell and neurofibrosarcoma cell transformation (162). Active versions of PAK1 were not themselves sufficient to induce transformation (161), although another study observed that PAK1 could synergize with Ras signaling to enhance transformation (163–165). The effect of PAK1 to enhance transformation could be correlated with effects of PAK1 on signaling through the ERK MAP kinase pathway and was dissociable from effects on the JNK or p38 MAP kinase pathways. Finally, Ras activation effectively induced PAK1 activation through a pathway involving PI 3-kinase, Rac and Cdc42, and Akt, although other factors may also contribute (53). PAK4 has been reported to also have transforming potential; constitutively active PAK4 induces anchorage-independent growth of NIH-3T3 cells (166). The contributions of PAK2-dependent modulation of Abl tyrosine kinase signaling (59) in cell transformation remain to be explored. Mira et al. (34) have shown that endogenous, hyperactive Rac3 and hyperactive PAKs 1 and 2 were present in highly proliferative human breast cancer–derived cell lines. The possibility that PAK contributes to both enhanced proliferative capacity of certain cancers (102) and their resistance to apoptotic signals (see below) requires additional investigation.

The ability of PAK to regulate MAP kinase pathway signaling was originally examined because of the analogy suggested by the yeast PAK homolog, Ste20, which acts as a MAP kinase kinase kinase kinase (MEKK kinase) in the yeast alpha-mating factor signaling pathway (10). PAKs 1, 2, and 3 have been shown in several studies to stimulate activation of the JNK and p38 MAP kinases (167, 168), whereas dominant-negative versions of PAK blocked activation in response to a number of upstream stimuli [e.g., (169)]. Interestingly, the activation of JNK and p38 mediated by PAK appears to occur in only certain cell types, and how PAK stimulates JNK and p38 activities is still not clear. Gallagher et al. (170) have recently reported that PAK1 phosphorylates MEKK1 on Ser67, thereby inhibiting the binding (and activation) of JNK by MEKK1. Modest effects of this phosphorylation on the ability of MEKK1 to phosphorylate kinase-dead MEK4 substrate were observed.

In the studies cited above, no stimulatory effect of PAK on the ERK MAP kinase pathway was detected. One of the first reports to observe enhanced ERK activation induced by PAK was by Lu et al. (48), who showed that PAK1 activation by membrane targeting resulted in stimulation of coexpressed p38, JNK, and ERK1. Frost et al. (164, 165) and Tang et al. (163) also reported that the observed synergy between Rho GTPases and Ras in activating ERK required active PAK. This requirement appeared to be due to the ability of PAK1 to phosphorylate MEK1 on Ser298, a site important for the binding of the upstream regulatory kinase, Raf-1 (165). It is worth noting that this site, like several reported PAK substrates, does not conform to the preferred KRES motif favored by the catalytic domain of PAK2. In a search for kinases that might phosphorylate critical sites in Raf-1, Marshall and colleagues identified PAK2 (originally reported erroneously to be PAK3) as a positive regulator of Raf-1 activity through phosphorylation of Ser338, an essential regulatory site for Ras-dependent Raf-1 activation (171). It has subsequently been established that phosphorylation at the essential Ser338 site by PAKs 1, 2, or 3 is critical for Raf-1 activation and ERK stimulation by a variety of growth factors and integrins (172, 173), although a dissenting study suggests this effect may not be physiological (174). The direct binding of PAK1 via the C-terminal portion of its catalytic domain to Raf-1 was regulated by the Raf-1 N-terminal regulatory domain, suggesting that this interaction would require upstream activation of both proteins (173). The significance of these interactions is evidenced by studies showing the role of PAK-dependent MAP kinase signaling in T cell receptor stimulation of serum response factor (SRF) (175) and in ERK activation by the basic FGF receptor during angiogenesis (J. Hood and D.A. Cheresh, submitted for publication). Based on these observations, as well as others, blockade of PAK activity has been suggested as a potential means to treat Ras-induced cancers (176).

PAK activity has also been linked to tumor suppression via regulatory effects on the action of the neurofibromatosis type 2 (NF2) gene product, Merlin (177, 178). Loss of heterozygosity and mutations in Merlin are associated with the development of inherited forms of Schwann cell tumors, as well as other spontaneous tumors that are highly malignant (179). Loss of Merlin function is correlated with high metastatic potential in an animal model (180), whereas overexpression inhibits growth and impairs cell motility, adhesion, and spreading (181, 182). Merlin is homologous to the members of the ERM (ezrin, radixin, moesin) family of cell membrane–cytoskeletal linker proteins and localizes to cortical actin structures. It is phosphorylated at low cell density and is growth permissive, whereas the hypophosphorylated form is growth inhibitory (183). The activity of Merlin is suppressed by Rac and Cdc42 through phosphorylation at Ser518 mediated by PAK, perhaps specifically by PAK2 (177, 178). Dominant inhibitory forms of PAK block Merlin phosphorylation and function, whereas active versions of PAK directly phosphorylate Merlin on Ser518 and increase its co-localization with cortical actin structures. Since Merlin itself appears to inhibit Rac-induced signaling (by an unknown mechanism), this situation suggests the possibility of a feed-forward loop that may enhance growth and metastasis under the regulation of PAK (178).

Cell Death and Survival Signaling

Apoptosis—programmed cell death—plays a central role both in development and in homeostasis in most organisms [see (184) for review]. Improper regulation of apoptosis contributes to pathological conditions: For example, a reduction of apoptosis leads to cancer and autoimmune diseases, and an increase in apoptotic events results in neurodegenerative disorders. The cellular death program can be divided into several phases. In the initiation phase, cells receive both pro- and antiapoptotic (survival) signals whose balance ultimately determines whether the cell will undergo cell death. The commitment phase is the point at which death signals become irreversible. A common biochemical feature of this phase is the activation of cysteine proteases termed caspases, which activate one another in a proteolytic cascade (amplification phase). Activation of the terminal effector caspases in this cascade leads to cleavage of important regulatory proteins (often separating regulatory and effector domains), activation of other degradative enzymes (e.g., endonucleases), and disassembly and breakdown of cell structures, resulting in cell death and dissolution. Cellular events characteristic of the destruction phase include DNA fragmentation, chromatin condensation, membrane blebbing, cell shrinkage, and disassembly into membrane-enclosed apoptotic bodies.

One major apoptotic pathway is initiated by the activation of cell-surface death receptors, such as Fas (also known as CD95 or APO-1) (184). Binding of "death ligands" leads to intracellular recruitment of the adapter protein FADD (Fas-associated death domain). FADD interacts via its death effector domain with the zymogen form of caspase 8, resulting in the activation of caspase 8 through self-cleavage. Caspase 8 then activates, either directly or indirectly, effector caspases, including caspases 3, 6, or 7. A second apoptotic pathway is induced by either increased cellular stress or the withdrawal of trophic (survival) factors, thereby changing the balance between prosurvival and proapoptotic proteins of the Bcl-2 family (184). This leads to the translocation of proapoptotic Bcl-2 family members, such as Bad, from the cytosol to the outer membrane of the mitochondria, where it interacts with Bcl-2/Bcl-xL to release cytochrome *c*. Cytochrome *c* release induces the formation of a multiprotein "apoptosome" complex (comprising Apaf1, cytochrome *c*, dATP or ATP, and procaspase 9) in which procaspase 9 is cleaved to active caspase 9, which in turn proteolyzes and activates downstream effector caspases.

It has been demonstrated that different PAK family members participate in these two apoptotic pathways to promote both apoptotic and antiapoptotic events. Simplistically, this dichotomy is reflected in the activation of different PAK isoforms (see below): PAK2, which promotes apoptosis, is activated in response to stimuli that result in cytostasis or cell death, and by other types of cellular stress, e.g., hyperosmolarity (39). PAK1, which stimulates prosurvival pathways, has been shown to be activated by growth factors (EGF and PDGF), cytokines, and cell adhesion (see Table 3). However, the possibility that the early activation of PAK2 by cell survival signals also regulates antiapoptotic pathways in this context has not been adequately examined [see (185)].

PAK2 ACTIVATION DURING APOPTOSIS Several groups have shown that after activation of various effector caspases, PAK2 is cleaved into a 28-kDa N-terminal and a 34-kDa C-terminal fragment (43, 44, 186), and this event occurs in response to multiple stimuli that induce apoptosis. Caspase 3 cleaves PAK2 adjacent to Asp212, which is absent in PAK3 and inaccessible in PAK1. The 34-kDa cleavage product of PAK2 contains the complete catalytic kinase domain, which appears to be constitutively active after proteolytic removal of the regulatory domain. The proteolytic activation of PAK2 provides a GTPase-independent mechanism of PAK activation that may contribute to the regulation of the morphological and biochemical alterations necessary for apoptosis. However, binding of Rac or Cdc42 may still be necessary for dissociation of the cleaved N-terminal fragment from the catalytic domain [see (28, 31)]. Microinjection or transfection into cell lines of the active C-terminal PAK2 catalytic fragment induced striking alterations in cellular and nuclear morphology that promoted apoptosis (44). On the other hand, expression of a dominant negative PAK2 (kinase dead) mutant delayed apoptosis and decreased apoptotic body formation (43). PAK2 activity is required for activation of the JNK pathway by Fas (44, 169), but inhibition of JNK activity is not sufficient to block cell death induced by Fas (169).

PAK ACTIVATION AND ACTION IN CELL SURVIVAL PATHWAYS PAK2 cleavage and activation is a relatively late event; it takes place well after cells have committed to the apoptotic program in response to proapoptotic signals. PAK1, however, is activated as an early event by certain receptors that promote cell survival. A consequence of PAK1 activation is phosphorylation of the death-promoting Bcl-2 family member, Bad, on Ser112 and 136 (53, 187). In the absence of survival factors, Bad is not phosphorylated and interacts with the Bcl-2 family members, Bcl-2 or Bcl-xL, via its BH3 domain to inhibit their antiapoptotic activities [reviewed in (188, 189)]. Exposure to factors promoting cell survival, for example IL-3, induces phosphorylation of Bad on Ser112, Ser136, and Ser155 (187, 190). Phosphorylation of any of these sites causes dissociation of Bad from Bcl-2 or Bcl-xL, and the association of Bad with the cytosolic adapter protein, 14-3-3, freeing Bcl-2 and Bcl-xL to exert their antiapoptotic actions.

PAK1 was activated by IL-3 in FL5.12 cells and phosphorylated Bad on Ser112 and Ser136, both in vitro and in vivo (187). The time course of PAK activation paralleled the time course of Bad phosphorylation. As a result of this

TABLE 3 Cellular stimuli reported to activate PAK

Stimulus	PAK isoform activated	Reference
Angiopoietin-1	1	(106)
Angiotensin II (Ang II)	1	(217)
Apoptotic stimuli (Fas, TNFα, etc.)	2	(43)
CXCL1	1	(97)
DNA-damaging agents (AraC, cisplatin, etc.)	1, 2	(218)
Epidermal growth factor (EGF)	1	(12)
Extracellular matrices (via integrins), e.g., fibronectin, collagen	1	(219, 220)
Heat shock	1	(221)
Helicobacter pylori	1	(222)
Hepatocyte growth factor (HGF)	1	(223)
Heregulin (HRG)	1	(110)
Hyperosmotic shock	1	(224)
Immune complexes (via FcγR)	1	(225)
Interleukin-1	1	(168)
Interleukin-3 (IL-3)	1	(187)
Interleukin-8 (IL-8)	1, 2	(226)
Ionizing, UV radiation	1, 2	(218)
Lipopolysaccharide (LPS)	1	(192)
Lysophosphatidic acid (LPA)	1	(227)
Macrophage scavenger receptor ligands (e.g., fucoidin)	?	(228)
N-formylmethionyl phenylalanine (fMLP)	1, 2	(6)
Opioids (e.g., EKC; α_{S_1} casomorphin)	1	(214)
Opsonized zymosan (via CR_3)	1	(95)
Platelet activating factor (PAF)	1, 2	(226)
Platelet-derived growth factor (PDGF)	1	(76)
RANTES	1, 2	(226)
Salmonella typhimurium	?	(229)
Thrombin	3	(230)

phosphorylation, Bad dissociated from Bcl-2/Bcl-xL and bound to cytosolic 14-3-3 protein. Overexpression of a constitutively active mutant, PAK1(T423E), promoted cell survival of FL5.12 cells (in the absence of IL-3) and NIH-3T3 cells, whereas overexpression of the PAK1 autoinhibitory domain enhanced apoptotic responsiveness. PAK1 is thus implicated in survival pathways through direct phosphorylation of Bad. A recent paper has described PAK1-mediated Bad phosphorylation as a critical event in survival signaling induced by the HIV viral Nef protein (191). The pathways regulating cell survival signaling and transformation via the 3-phosphoinositide-regulated protein kinase Akt and PAK (see above sections) are intimately connected, and additional studies will be necessary to sort out the complexities of their regulation.

PAK has been reported to be required for upstream stimulus-dependent activation of the NFκB transcription factor, and constitutively active PAK stimulates NFκB activity (192). When *Helicobacter pylori* infects gastric epithelial cells, NFκB becomes activated through a PAK1-dependent mechanism involving activation of the upstream regulatory kinase, NFκB-inducing kinase (NIK) (193). Since NFκB has been shown to promote cell survival (194), some of the antiapoptotic effects of PAK may be mediated through this mechanism.

PAK in HIV/Nef Signaling

The HIV accessory protein, Nef, is a membrane-associated protein required for the development of AIDS [reviewed in (195)]. Three major functions of Nef include: (*a*) enhancing viral particle infectivity; (*b*) reduction of cell surface expression of the CD4 and MHC I molecules; and (*c*) induction of cell signaling pathways leading to cytoskeletal rearrangements, JNK activation, and enhanced or decreased apoptotic responses. A Nef-binding kinase was identified as PAK1 and/or PAK2 in various studies (196–200), and the subpopulation of PAK associated with Nef is highly active (199). Nef-associated PAK appears to mediate the effects of Rac and Cdc42 to promote viral replication and pathogenesis of HIV, and PAK activity is required for survival of infected cells, JNK activation, cytoskeletal rearrangements, and the production of viral particles (191, 195–200). Thus, the HIV virus has clearly subverted PAK function for its own purposes.

The NADPH Oxidase of Phagocytic Cells

PAKs are abundant proteins in human leukocytes, and human neutrophils were one of the first cell types in which PAK was identified (5, 6). As a primary component of the innate immune response, human neutrophils internalize invading microorganisms through the process of phagocytosis. PAK1 has been localized to phagocytic cups during phagocytosis, where it is thought to participate in the cytoskeletal rearrangements and/or contractile processes necessary for particle uptake (95, 201). The engulfed bacteria can be killed by mechanisms utilizing the formation of reactive oxygen species via a membrane-associated

NADPH oxidase. This multiprotein oxidase consists of a membrane-localized cytochrome b_{558}, the cytosolic factors p47^{phox} and p67^{phox}, and the Rac2 GTPase [reviewed in (202)]. Assembly of the active oxidase occurs at the plasma membrane after the cytosolic components are induced to translocate during phagocytic activation.

In intact leukocytes, translocation of the p47^{phox} and p67^{phox} proteins requires multiple phosphorylation events on p47^{phox} that alter its conformation and expose protein-protein (e.g., SH3, PXXP) interaction motifs that drive NADPH oxidase membrane assembly. The first putative PAK substrate identified was p47^{phox} (6), suggesting PAK might play a role in coordinating assembly of the cytosolic oxidase components with Rac GTPase activation. The recent demonstration that PAK autoinhibitory domain is able to substantially inhibit NADPH oxidase assembly supports this hypothesis (M.-J. Kim & U.G. Knaus, personal communication). PAK also appears to enhance NADPH oxidase activity by regulating NADPH availability through the phosphorylation of phosphoglycerate mutase-B (PGAM-B) (203). Phosphorylation of serine residues 23 and 118 on PGAM-B by PAK1 inhibits the activity of this glycolytic pathway enzyme, resulting in the switching off of glycolysis and enhanced generation of NADPH via the hexose monophosphate shunt. PAK thus acts to redirect energy production to the pathway that supports optimal generation of reactive oxidant species. It will be interesting to determine if the ability of PAK to serve as a metabolic switch is operative in other metabolic pathways that require the generation of reducing equivalents via NADPH.

PERSPECTIVE

Many basic questions remain to be addressed with regard to the biology of PAKs 1–3. Certainly, the question of their substrate specificity in vivo needs to be addressed. Although we have observed that PAKs 1–3 all phosphorylate a number of identified PAK substrates to a similar extent when tested in vitro (G.M. Bokoch, unpublished observations), there are suggestions that the isoforms might exhibit more specificity under physiological circumstances (e.g., PAK2 may specifically phosphorylate Merlin, or Raf-1). Whether this reflects differing subcellular localization of PAK isoforms, distinct molecular partners in vivo, or other contributing factors is unknown. Comparisons of the physiological consequences of knocking out PAK isoforms in mice or in specific cell types by using short interfering RNA (siRNA) should prove informative. The basis for regulation and coordination of cytoskeletal dynamics by PAKs needs to be defined in greater biochemical detail, as do the contributions of kinase-dependent effects versus those resulting from the scaffolding function of PAKs. A better understanding of the mechanisms that regulate localized activation and modulation of PAK activity is also necessary. How the action of PAKs is modulated through homodimerization, the action of lipids and membrane localization (versus GTPases), phosphorylation by other kinases, and interaction with regulatory proteins (PIX, POPX, etc.) are all important questions to be investigated.

Since their initial discovery as targets for Rac and Cdc42 GTPases, PAK family members have been shown to modulate many aspects of cellular biology. Emerging themes, such as regulation of cell division and the cell cycle, promise to expand the theater of PAK action even further. The continuing identification of novel PAK 1–3 phosphorylation targets, and the recognition that PAKs can interact with a wide variety of proteins that influence their activity, protein-protein interactions, and subcellular localization, indicate that the complexities of PAK biology will continue to grow. Understanding how PAKs integrate these many regulatory activities in response to specific stimuli will provide insights into both normal and pathological cell function, perhaps leading to new avenues for therapeutic intervention. Studies in the second decade of PAK investigation should yield considerable new insight into our understanding of PAK and Rho GTPase biology.

ACKNOWLEDGMENTS

I would like to thank Mary Stofega for assistance in table preparation, Jon Chernoff for comments on the manuscript, and Lia Marshall for excellent secretarial assistance. I am especially grateful to Stephen Harrison (Harvard University) for providing Figure 3 to illustrate PAK structure and activation mechanism. This work was supported by the National Institutes of Health (GM39434, GM44428).

The *Annual Review of Biochemistry* is online at http://biochem.annualreviews.org

LITERATURE CITED

1. Ridley AJ. 1996. *Curr. Biol.* 10: 1256–64
2. Bishop AL, Hall A. 2000. *Biochem. J.* 348:241–55
3. Schmidt A, Hall A. 2002. *Genes Dev.* 16:1587–609
4. Manser E, Leung T, Salihuddin H, Zhao Z-S, Lim L. 1994. *Nature* 367:40–46
5. Martin GA, Bollag G, McCormick F, Abo A. 1995. *EMBO J.* 14:1970–78
6. Knaus UG, Morris S, Dong H-J, Chernoff J, Bokoch GM. 1995. *Science* 269:221–23
7. Bagrodia S, Taylor SJ, Creasy CL, Chernoff J, Cerione RA. 1995. *J. Biol. Chem.* 270:22731–37
8. Sells MA, Chernoff J. 1997. *Trends Cell Biol.* 7:162–67
9. Knaus UG, Bokoch GM. 1998. *Int. J. Biochem. Cell Biol.* 30:857–62
10. Dan I, Watanabe NM, Kusumi A. 2001. *Trends Cell Biol.* 11(5):220–30
11. Jaffer ZM, Chernoff J. 2002. *Int. J. Biochem. Cell Biol.* 34:713–17
12. Galisteo ML, Chernoff J, Su Y-C, Skolnik EY, Schlessinger J. 1996. *J. Biol. Chem.* 271(35):20997–1000
13. Bokoch GM, Wang Y, Bohl BP, Sells MA, Quilliam LA, Knaus UG. 1996. *J. Biol. Chem.* 271(42):25746–49
14. Puto L, Pestonjamasp K, King CC,

Bokoch GM. 2003. *J. Biol. Chem.* 278:(11):9388–93; 10.1074/jbc.M208414200
15. Manser E, Loo T-H, Koh C-G, Zhao Z-S, Chen X-Q, et al. 1998. *Mol. Cell* 1:183–92
16. Lei M, Lu W, Meng W, Parrini M-C, Eck M-J, et al. 2000. *Cell* 102:387–97
17. Leeuw T, Wu C, Schrag JD, Whiteway M, Thomas DY, Leberer E. 1998. *Nature* 391:191–95
18. Wang J, Frost JA, Cobb MH, Ross EM. 1999. *J. Biol. Chem.* 274:31641–47
19. Parrini MC, Lei M, Harrison SC, Mayer BJ. 2002. *Mol. Cell* 9:73–83
20. Thompson G, Owen D, Chalk PA, Lowe PN. 1998. *Biochemistry* 37:7885–91
21. Morreale A, Venkatesan M, Mott HR, Owen D, Nietlispach D, et al. 2000. *Nat. Struct. Biol.* 7(5):384–88
22. Gizachew D, Guo W, Chohan KK, Sutcliffe MJ, Oswald RE. 2000. *Biochemistry* 39:3963–71
23. Hoffman GR, Cerione RA. 2000. *Cell* 102:403–6
24. Benner GE, Dennis PB, Masaracchia RA. 1995. *J. Biol. Chem.* 270:21121–28
25. Zhao ZS, Manser E, Chen X-Q, Chong C, Leung T, Lim L. 1998. *Mol. Cell. Biol.* 18(4):2153–63
26. Yu J-S, Chen W-J, Ni M-H, Chan W-H, Yang S-D. 1998. *Biochem. J.* 334:121–31
27. Gatti A, Huang Z, Tuazon PT, Traugh JA. 1999. *J. Biol. Chem.* 274:8022–28
28. Zenke FT, King CC, Bohl BP, Bokoch GM. 1999. *J. Biol. Chem.* 274(46):32565–73
29. Tu H, Wigler M. 1999. *Mol. Cell. Biol.* 19:602–11
30. Chong C, Tan L, Lim L, Manser E. 2001. *J. Biol. Chem.* 276(20):17347–53
31. Buchwald G, Hostinova E, Rudolph MG, Kraemer A, Sickmann A, et al. 2001. *Mol. Cell. Biol.* 21(15):5179–89
32. King CC, Gardiner EM, Zenke FT, Bohl BP, Newton AC, et al. 2000. *J. Biol. Chem.* 275:41201–9
33. Knaus UG, Wang Y, Reilly AM, Warnock D, Jackson JH. 1998. *J. Biol. Chem.* 273:21512–18
34. Mira JP, Bernard V, Groffen J, Sanders LC, Knaus UG. 2000. *Proc. Natl. Acad. Sci. USA* 97:185–89
35. Aronheim A, Broder YC, Cohen A, Fritsch A, Belisle B, Abo A. 1998. *Curr. Biol.* 8(20):1125–28
36. Neudauer CL, Joberty G, Tatsis N, Macara IG. 1998. *Curr. Biol.* 8:1151–60
37. Tao W, Pennica D, Xu L, Kalejta RF, Levine AJ. 2001. *Genes Dev.* 15:1796–807
38. Jakobi R, Chen C-J, Tuazon PT, Traugh JA. 1996. *J. Biol. Chem.* 271(11):6206–11
39. Roig J, Traugh JA. 2001. *Vitam. Horm.* 62:167–98
40. Reeder MK, Serebriiskii IG, Golemis EA, Chernoff J. 2001. *J. Biol. Chem.* 276(44):40606–13
41. Xu X, Barry DC, Settleman J, Schwartz MA, Bokoch GM. 1994. *J. Biol. Chem.* 269:23569–74
42. Benard V, Bohl BP, Bokoch GM. 1999. *J. Biol. Chem.* 274:13198–204
43. Rudel T, Bokoch GM. 1997. *Science* 276:1571–74
44. Lee N, MacDonald H, Reinhard C, Halenbeck R, Roulston A, et al. 1997. *Proc. Natl. Acad. Sci. USA* 94:13642–47
45. Yablonski D, Kane LP, Qian DP, Weiss A. 1998. *EMBO J.* 17(19):5647–57
46. Wardenburg JB, Pappu R, Bu J-Y, Mayer B, Chernoff J, et al. 1998. *Immunity* 9:607–16
47. Izadi KD, Erdreich-Epstein A, Liu Y, Durdern DL. 1998. *Exp. Cell. Res.* 245:330–42
48. Lu W, Katz S, Gupta R, Mayer BJ. 1997. *Curr. Biol.* 7:85–94
49. Daniels RH, Hall P, Bokoch GM. 1998. *EMBO J.* 17:754–64
50. Lu W, Mayer BJ. 1999. *Oncogene* 18:797–806
51. Bokoch GM, Reilly AM, Daniels RH, King CC, Olivera A, et al. 1998. *J. Biol. Chem.* 273(14):8137–44

52. Toker A, Newton AC. 2000. *Cell* 103: 185–88
53. Tang Y, Zhou H, Chen A, Pittman RN, Field J. 2000. *J. Biol. Chem.* 275(13): 9106–9
54. Deleted in proof
55. Nikolic M, Chou MM, Lu W, Mayer BJ, Tsai LH. 1998. *Nature* 395:194–98
56. Rashid T, Banerjee M, Nikolic M. 2001. *J. Biol. Chem.* 176(52):49043–52
57. Banerjee M, Worth D, Prowse DM, Nikolic M. 2002. *Curr. Biol.* 12:1233–39
58. Thiel DA, Reeder MK, Pfaff A, Coleman TR, Sells MA, Chernoff J. 2002. *Curr. Biol.* 12:1227–32
59. Roig J, Tuazon PT, Zipfel PA, Pendergast AM, Traugh J. 2000. *Proc. Natl. Acad. Sci. USA* 97(26):14346–51
60. McManus MJ, Boerner JL, Danielsen AJ, Wang Z, Matsumura F, Maihle NJ. 2000. *J. Biol. Chem.* 275(45):35328–34
61. Bagheri-Yarmand R, Mandal M, Taludker AH, Wang R-A, Vadlamudi RK, et al. 2001. *J. Biol. Chem.* 276(31): 29403–9
62. Renkema GH, Pulkkinen K, Saksela K. 2002. *Mol. Cell. Biol.* 22(19):6719–25
63. Westphal RS, Coffee RL, Marotta A, Pelech SL, Wadzinski BE. 1999. *J. Biol. Chem.* 274(2):687–92
64. Koh C-G, Tan E-J, Manser E, Lim L. 2002. *Curr. Biol.* 12:317–21
65. Bagrodia S, Taylor SJ, Jordon KA, Van Aelst L, Cerione RA. 1998. *J. Biol. Chem.* 273:23633–36
66. Daniels RH, Zenke FT, Bokoch GM. 1999. *J. Biol. Chem.* 274:6047–50
67. Feng Q, Albeck JG, Cerione RA, Yang W. 2002. *J. Biol. Chem.* 277(7):5644–50
68. Obermeier A, Ahmed S, Manser E, Yen SC, Hall C, Lim L. 1998. *EMBO J.* 17:4328–39
69. Lee S-H, Eom M, Lee SD, Kim S, Park H-J, Park D. 2001. *J. Biol. Chem.* 276(27):25066–72
70. Shin EY, Shin KS, Lee CS, Woo KN, Quan S-H, et al. 2002. *J. Biol. Chem.* 277(46):44417–30
71. Turner CE, Brown MC, Perrotta JA, Riedy MC, Nikolopoulos SN, et al. 1999. *J. Cell. Biol.* 145(4):851–63
72. Zhao Z-S, Manser E, Loo T-H, Lim L. 2000. *Mol. Cell. Biol.* 20(17):6354–63
73. Yoshii S, Tanaka M, Otsuki Y, Fujiyama T, Kataoka H, et al. 2001. *Mol. Cell. Biol.* 21(20):6796–807
74. Xia C, Ma W, Stafford LJ, Marcus S, Xiong W-C, Liu M. 2001. *Proc. Natl. Acad. Sci. USA* 98(11):6174–79
75. King CC, Sanders LC, Bokoch GM. 2000. *Methods Enzymol.* 325:315–27
76. Dharmawardhane S, Sanders LC, Martin SS, Daniels RH, Bokoch GM. 1997. *J. Cell Biol.* 138:1265–78
77. Dharmawardhane S, Schurmann A, Sells MA, Chernoff J, Schmid SL, Bokoch GM. 2000. *Mol. Biol. Cell* 11(10): 3341–52
78. Manser E, Huang HY, Loo TH, Chen XQ, Dong JM, et al. 1997. *Mol. Cell. Biol.* 17:1129–43
79. Frost JA, Khokhlatchev A, Stippec S, White MA, Cobb MH. 1998. *J. Biol. Chem.* 273:29191–98
80. Deleted in proof
81. Sells MA, Pfaff A, Chernoff J. 2000. *J. Cell. Biol.* 151:1449–57
82. Sells MA, Knaus UG, Bagrodia S, Ambrose DM, Bokoch GM, Chernoff J. 1997. *Curr. Biol.* 7:202–10
83. Chen W, Chen S, Yap SF, Lim L. 1996. *J. Biol. Chem.* 271:26362–68
84. Melzig J, Rein K-H, Schafer U, Pfister H, Jackle H, et al. 1998. *Curr. Biol.* 8:1223–26
85. Hing H, Xiao J, Harden N, Lim L, Zipursky SL. 1999. *Cell* 97:853–63
86. Chung CY, Firtel RA. 1999. *J. Cell. Biol.* 147(3):559–75
87. Cvrckova F, De Virgilio C, Manser E, Pringle JR, Nasmyth K. 1995. *Genes Dev.* 9:1817–30
88. Leberer E, Thomas DY, Whiteway M. 1997. *Curr. Opin. Genet. Dev.* 7:59–66
89. Eby JJ, Holly SP, van Drogen F, Grishin

AV, Peter M, et al. 1998. *Curr. Biol.* 8:967–70

90. Holly SP, Blumer KJ. 1999. *J. Curr. Biol.* 147(4):845–56
91. Weiss EL, Bishop AC, Shokat KM, Drubin DG. 2000. *Nat. Cell Biol.* 2:677–85
92. Sawin KE, Hajibagheri MAN, Nurse P. 1999. *Curr. Biol.* 9:1335–38
93. Ottilie SJ, Miller P, Johnson DI, Creasy CL, Sells MA, et al. 1995. *EMBO J.* 14(23):5908 19
94. Leberer E, Ziegelbauer K, Schmidt A, Harcus D, Dignard D, et al. 1997. *Curr. Biol.* 7:539–46
95. Dharmawardhane S, Brownson D, Lennartz M, Bokoch GM. 1999. *J. Leukocyte Biol.* 66:521–27
96. Ding JB, Knaus UG, Lian JP, Bokoch GM, Badwey JA. 1996. *J. Biol. Chem.* 271:24869–73
97. Wang D, Sai J, Carter G, Sachpatzidis A, Lolis E, Richmond A. 2002. *Biochemistry* 41:7100–7
98. Sells MA, Boyd JT, Chernoff J. 1999. *J. Cell. Biol.* 145:837–49
99. Kiosses WB, Daniels RH, Otey RH, Bokoch GM, Schwartz MA. 1999. *J. Biol. Chem.* 147:831–44
100. Allen KM, Gleeson JG, Bagrodia S, Partington MW, MacMillan JC, et al. 1998. *Nat. Genet.* 20:25–30
101. Bienvenu T, des Portes V, McDonell N, Carrie A, Zemni R, et al. 2000. *Am. J. Med. Genet.* 93:294–98
102. Obermeier A, Ahmed S, Manser E, Yen SC, Hall C, Lim L. 1998. *EMBO J.* 17(15):4328–39
103. Dan C, Nath N, Liberto M, Minden A. 2002. *Mol. Cell. Biol.* 22(2):567–77
104. Connolly JO, Simpson N, Hewlett L, Hall A. 2002. *Mol. Biol. Cell.* 13: 2474–85
105. Kiosses WB, Hood J, Yang S, Gerritsen ME, Cheresh DA, et al. 2002. *Circ. Res.* 90:697–702
106. Master Z, Jones N, Tran J, Jones J, Kerbel RS, et al. 2001. *EMBO J.* 20(21): 5919–28
107. Garcia JGN, Lui F, Verin AD, Birukova A, Dechert MA, et al. 2001. *J. Clin. Invest.* 108:689–701
108. Birukov KG, Birukova AA, Dudek SM, Verin AD, Crow MT, et al. 2002. *Am. J. Respir. Cell Mol. Biol.* 26:453–64
109. Bagheri-Yarmand R, Vadlamudi RK, Wang R-A, Mendelsohn J, Kumar R. 2000. *J. Biol. Chem.* 275:39451–57
110. Adam L, Vadlamudi R, Kondapaka SB, Chernoff J, Mendelsohn J, Kumar R. 1998. *J. Biol. Chem.* 273:28238–46
111. Adam L, Vadlamudi R, Mandal M, Chernoff J, Kumar R. 2000. *J. Biol. Chem.* 275(16):12041–50
112. Vadlamudi R, Adam L, Wang R-A, Mandal M, Nguyen D, et al. 2000. *J. Biol. Chem.* 275(46):36238–44
113. Ichigotani Y, Yokozaki S, Fukuda Y, Hamaguchi M, Matsuda S. 2002. *Cancer Res.* 62:2215–19
114. Tuazon PT, Spanos WC, Gump EL, Monnig CA, Traugh JA. 1997. *Biochemistry* 36:16059–64
115. Bagrodia S, Bailey D, Lenard Z, Hart M, Guan JL, et al. 1999. *J. Biol. Chem.* 274(32):22393–400
116. Premont RT, Claing A, Vitale N, Freeman JLR, Pitcher JA, et al. 1998. *Proc. Natl. Acad. Sci. USA* 95:14082–87
117. Premont RT, Claing A, Vitale N, Perry SJ, Lefkowitz RJ. 2000. *J. Biol. Chem.* 275(29):22373–80
118. Kondo A, Hashimoto S, Yano H, Nagayama K, Mazaki Y, Sabe H. 2000. *Mol. Biol. Cell.* 11:1315–27
119. Mazaki Y, Hashimoto S, Katsuya O, Tsbouchi A, Nakamura K, et al. 2001. *Mol. Biol. Cell.* 12:645–62
120. Manabe R-I, Kovalenko M, Webb DJ, Horowitz AR. 2002. *J. Cell. Sci.* 115: 1497–510
121. Brown MC, West KA, Turner CE. 2002. *Mol. Biol. Cell.* 13:1550–65
122. Di Cesare A, Paris S, Albertinazzi C,

Dariozzi S, Andersen J, et al. 2000. *Nat. Cell. Biol.* 2:521–30
123. Turner CE. 2000. *Nat. Cell. Biol.* 2:E231–36
124. Vitale N, Patton WA, Moss J, Vaughan M, Lefkowitz RJ, Premont RT. 2000. *J. Biol. Chem.* 275(18):13901–9
125. Bamburg JR. 1999. *Annu. Rev. Cell Dev. Biol.* 15:185–230
126. Stanyon CA, Bernard O. 1999. *Int. J. Biochem. Cell. Biol.* 31:389–94
127. Arber S, Barbayannis FA, Hanser H, Schneider C, Stanyon CA, et al. 1998. *Nature* 393:805–9
128. Yang N, Higuchi O, Ohashi K, Nagata K, Wada A. 1998. *Nature* 393:809–12
129. Edwards DC, Sanders LC, Bokoch GM, Gill GN. 1999. *Nat. Cell Biol.* 1:253–59
130. Dan C, Kelly A, Bernard O, Minden A. 2001. *J. Biol. Chem.* 276(34):32115–21
131. Sumi T, Matsumoto K, Shibuya A, Nakamura T. 2001. *J. Biol. Chem.* 276(25):23092–96
132. Maekawa M, Ishizaki T, Boku S, Watanabe N, Fujita A, et al. 1999. *Science* 285:895–98
133. Ramos E, Wysolmerski RB, Masaracchia RA. 1997. *Recept. Signal Transduct.* 7:99–110
134. Chew TL, Masaracchia RA, Goeckeler ZM, Wysolmerski RB. 1998. *J. Muscle Res. Cell Motil.* 19:839–54
135. Zeng Q, Lagunoff D, Masaracchia R, Goeckeler Z, Cote G, Wysolmerski R. 2000. *J. Cell. Sci.* 113:471–82
136. Van Leeuwen FN, van Delft S, Kain HE, van der Kammen RA, Collard JG. 1999. *Nat. Cell Biol.* 1:242–48
137. Sanders LC, Matusmura F, Bokoch GM, de Lanerolle P. 1999. *Science* 283(5410):2083–85
138. Goeckeler ZM, Masaracchia RA, Zeng Q, Chew TL, Gallagher P, Wysolmerski RB. 2000. *J. Biol. Chem.* 275:18366–74
139. Kaibuchi K, Kuroda S, Amano M. 1999. *Annu. Rev. Biochem.* 68:459–86
140. Burridge K. 1999. *Science* 283:2028–29
141. Wu C, Lee S-F, Furmaniak-Kazmierczak E, Cotes GP, Thomas DY, Leberer E. 1996. *J. Biol. Chem.* 271(50):31787–90
142. Lechler T, Shevchenko A, Shevchenko A, Li R. 2000. *J. Cell. Biol.* 148(2): 363–73
143. Yoshimura M, Homma K, Siato J, Inoue A, Ikebe R, Ikebe M. 2001. *J. Biol. Chem.* 276(43):39600–7
144. Deleted in proof.
145. Faure S, Vigneron S, Doree M, Morin NA. 1997. *EMBO J.* 16:5550–61
146. Cau J, Faure S, Vigneron S, Labbé JC, Delsert C, Morin N. 2000. *J. Biol. Chem.* 275(4):2367–75
147. Qyang Y, Yang P, Du H, Lao H, Kim HW, Marcus S. 2002. *Mol. Microbiol.* 44(2):325–34
148. Cau J, Faure S, Comps M, Delsert C, Morin N. 2001. *J. Biol. ·Chem.* 155(6): 1029–42
149. Li F, Adam L, Vadlamudi R, Zhou H, Sen S, et al. 2002. *EMBO Rep.* 3(8): 767–73
150. Cassimersis L. 2002. *Curr. Opin. Cell Biol.* 14:18–24
151. Daub H, Gevaert K, Vandekerchove J, Sobel A, Hall A. 2001. *J. Biol. Chem.* 276:1677–80
152. Wittman T, Bokoch GM, Waterman-Storer CM. 2003. *J. Cell. Biol.* In press
153. Wittman T, Waterman-Storer CM. 2001. *J. Cell. Sci.* 114:3795–803
154. Stossel TP, Condeelis J, Cooley L, Hartwig JH, et al. 2001. *Nat. Rev. Mol. Cell. Biol.* 2:138–45
155. Nakamura F, Osborn E, Janmey PA, Stossel TP. 2002. *J. Biol. Chem.* 277(11):9148–54
156. Vadlamudi RK, Li F, Adam L, Nguyen D, Ohta Y, et al. 2002. *Nat. Cell Biol.* 4:681–90
157. Chrzanowska-Wodnicka M, Burridge K. 1996. *J. Cell Biol.* 133(6):1403–15
158. Gruenheid S, DeVinney R, Bladt F, Goosney D, Gelkop S, et al. 2002. *Nat. Cell Biol.* 3(9):856–59
159. Waterman-Storer CM, Worthylake RA,

Liu B, Burridge K, Salmon ED. 1999. *Nat. Cell Biol.* 1:45–50
160. Krendel M, Zenke FT, Bokoch GM. 2002. *Nat. Cell Biol.* 4:294–301
161. Tang Y, Chen Z, Ambrose D, Liu J, Gibbs JB, et al. 1997. *Mol. Cell. Biol.* 17(8):4454–64
162. Tang Y, Marwaha S, Rutkowski JL, Tennekoon GI, Phillips PC, Field J. 1998. *Proc. Natl. Acad. Sci. USA* 95:5139–44
163. Tang Y, Yu J, Field J. 1999. *Mol. Cell. Biol.* 19(3):1881–91
164. Frost JA, Xu S, Hutchison MR, Marcus S, Cobb MH. 1996. *Mol. Cell. Biol.* 16:3707–13
165. Frost JA, Steen H, Shapiro P, Lewis T, Ahn N, et al. 1997. *EMBO J.* 16:6426–38
166. Callow MG, Clairvoyant F, Zhu S, Schryver B, Whyte DB, et al. 2002. *J. Biol. Chem.* 277(1):550–58
167. Bagrodia S, Derijard B, Davis RJ, Cerione RA. 1995. *J. Biol. Chem.* 270:27995–98
168. Zhang S, Han J, Sells MA, Chernoff J, Knaus UG, et al. 1995. *J. Biol. Chem.* 270:23934–36
169. Rudel T, Zenke FT, Chuang T-H, Bokoch GM. 1998. *J. Immunol.* 160: 7–11
170. Gallagher ED, Xu S, Moomaw C, Slaughter CA, Cobb MH. 2002. *J. Biol. Chem.* 277:45785–92
171. King AJ, Sun H, Diaz B, Barnard D, Miao W, et al. 1998. *Nature* 396:180–83
172. Chaudhary A, King WG, Mattaliano MD, Frost JA, Diaz B, et al. 2000. *Curr. Biol.* 10:551–54
173. Zang MW, Hayne C, Luo ZJ. 2002. *J. Biol. Chem.* 277(6):4395–405
174. Chiloeches A, Mason CS, Marais R. 2001. *Mol. Cell. Biol.* 21(7):2423–34
175. Charvet C, Auberger P, Tartare-Deckert S, Bernard A, Deckert M. 2002. *J. Biol. Chem.* 277(18):15376–84
176. He H, Hirokawa Y, Manser E, Lim L, Levitzki A, Maruta H. 2001. *Cancer J.* 7:191–202
177. Xiao G-H, Beeser A, Chernoff J, Testa JR. 2001. *J. Biol. Chem.* 277(2):883–86
178. Kissil JL, Johnson KC, Eckman MS, Jacks T. 2002. *J. Biol. Chem.* 277(12): 10394–99
179. Gusella JF, Ramesh V, MacCollin M, Jacoby LB. 1999. *Biochem. Biophys. Acta* 1423:M29–36
180. McClatchey AI, Saotome I, Mercer K, Crowley D, Gusella JF, et al. 1998. *Genes Dev.* 12:1121–33
181. Sherman L, Xu HM, Geist RT, Saporito-Irwin S, Howells N, et al. 1997. *Oncogene* 15:2505–9
182. Gutmann DH, Sherman L, Seftor L, Haipek C, Lu KH, Hendrix M. 1999. *Hum. Mol. Genet.* 8:267–75
183. Morrison H, Sherman LS, Legg H, Banine F, Isacke G, et al. 2001. *Genes Dev.* 15:968–80
184. Strasser A, O'Connor L, Dixit VM. 2000. *Annu. Rev. Biochem.* 69:217–45
185. Jakobi R, Moertl E, Koeppel MA. 2001. *J. Biol. Chem.* 276(20):16624–34
186. Walter BN, Huang Z, Jakobi R, Tuazon PT, Alnemri ES, et al. 1998. *J. Biol. Chem.* 273:28733–39
187. Schurmann A, Mooney AF, Sanders LC, Sells MA, Wang HG, et al. 2000. *Mol. Cell. Biol.* 20(2):453–61
188. Gajewski TF, Thompson CB. 1996. *Cell* 87:589–92
189. Downward J. 1999. *Nat. Cell. Biol.* 1:E33–35
190. Zhou X-M, Liu Y, Payne G, Lutz RJ, Chittenden T. 2000. *J. Biol. Chem.* 275(32):25046–51
191. Wolf D, Witte V, Laffert B, Blume K, Stromer E, et al. 2001. *Nat. Med.* 7(11): 1217–24
192. Frost JA, Swantek JL, Stippec S, Yin MJ, Gaynor R, Cobb MH. 2000. *J. Biol. Chem.* 275:19693–99
193. Foryst-Ludwig A, Naumann M. 2000. *J. Biol. Chem.* 275(50):39779–85
194. Karin M, Lin A. 2002. *Nat. Immunol.* 3(3):221–27

195. Geyer M, Fackler OT, Peterlin BM. 2001. *EMBO Rep.* 2(7):580–85
196. Sawai ET, Khan IH, Montbraind PM, Peterlin BM, Cheng-Mayer C, Luciw PA. 1996. *Curr. Biol.* 6(11):1519–27
197. Fackler OT, Lu X, Frost JA, Geyer M, Jiang B, et al. 2000. *Mol. Cell. Biol.* 20(7):2619–27
198. Renkema GH, Manninen A, Mann DA, Harris M, Saksela K. 1999. *Curr. Biol.* 9:1407–10
199. Renkema GH, Manninen A, Saksela KJ. 2001. *J. Virol.* 75(5):2154–60
200. Linnemann T, Zheng Y-H, Mandic R, Peterlin BM. 2002. *Virology* 294: 246–55
201. Diakonova A, Bokoch GM, Swanson JA. 2002. *Mol. Biol. Cell.* 13:402–11
202. Bokoch GM, Diebold BA. 2002. *Blood* 100(8):2692–96
203. Shalom-Barak T, Knaus UG. 2002. *J. Biol. Chem.* 277:40659–65
204. Gohla A, Bokoch GM. 2002. *Curr. Biol.* 12:1704–10
205. Brzeska H, Korn ED. 1996. *J. Biol. Chem.* 271:16983–86
206. Lee S-F, Egelhoff TT, Mahasneh A, Cote GP. 1996. *J. Biol. Chem.* 271: 27044–48
207. Van Eyk JE, Arrell DK, Foster DB, Strauss JD, Heinonen TYK, et al. 1998. *J. Biol. Chem.* 273:23433–39
208. Foster DB, Shen LH, Kelly J, Thibault D, Van Eyk JE, Mak AS. 2000. *J. Biol. Chem.* 275:1959–65
209. Ohtakara K, Inada H, Goto H, Taki W, Manser E, et al. 2000. *Biochem. Biophys. Res. Commun.* 272:712–16
210. Wang R-A, Mazumdar A, Vadlamudi RK, Kumar R. 2002. *EMBO J.* 21(20): 5437–47
211. Deleted in proof
212. Tuazon PT, Lorenson MY, Walker AM, Traugh JA. 2002. *FEBS Lett.* 515:84–88
213. Hashimoto S, Tsubouchi A, Mazaki Y, Sabe H. 2001. *J. Biol. Chem.* 276: 6037–45
214. Papakonstanti EA, Stournaras C. 2002. *Mol. Biol. Cell* 13:2946–62
215. Howe AK, Juliano RL. 2000. *Nat. Cell Biol.* 2:593–600
216. Leisner TM, Saunders R, Chernoff J, Parise LV. 2001. *Mol. Biol. Cell. Suppl.* 12:1682a
217. Schmitz U, Ishida T, Ishida M, Surapisitchat J, Hashan MI, et al. 1998. *Circ. Res.* 82:1272–78
218. Roig J, Traugh JA. 1999. *J. Biol. Chem.* 274:31119–22
219. Price LS, Leng J, Schwartz MA, Bokoch GM. 1998. *Mol. Biol. Cell* 9:1863–71
220. Suzuki-Inoue K, Yatomi Y, Asazuma N, Kainoh M, Tanaka T, et al. 2001. *Blood* 98:3708–16
221. Chan W-H, Yu JS, Yang SD. 1998. *J. Protein Chem.* 17:485–94
222. Churin Y, Kardalinou E, Meyer TF, Naumann M. 2001. *Mol. Microbiol.* 40:815–23
223. Royal I, Lamarche-Vane N, Lamorte L, Kaibuchi K, Park M. 2000. *Mol. Biol. Cell* 11:1709–25
224. Chan WH, Yu JS, Yang SD. 1999. *J. Cell. Physiol.* 178:397–408
225. Jones SL, Knaus UG, Bokoch GM, Brown EJ. 1998. *J. Biol. Chem.* 273: 10556–66
226. Huang RY, Lian JP, Robinson D, Badwey JA. 1998. *Mol. Cell. Biol.* 18:7130–38
227. Schmitz U, Thommes K, Beier I, Vetter H. 2002. *Biochem. Biophys. Res. Commun.* 291:687–91
228. Hsu H-Y, Chin SL, Wen MH, Chen KY, Hua KF. 2001. *J. Biol. Chem.* 276: 28719–30
229. Chen L-M, Bagrodia S, Cerione RA, Galan JE. 1999. *J. Exp. Med.* 189:1479–88
230. Malcolm KC, Chambard JC, Grall D, Pouyssegur J, van Obberghen-Schilling E. 2000. *J. Cell. Physiol.* 185:235–43

Annu. Rev. Biochem. 2003. 72:783–812
doi: 10.1146/annurev.biochem.72.121801.161511
Copyright © 2003 by Annual Reviews. All rights reserved

PROTEOMICS

Heng Zhu,[1] Metin Bilgin,[3] and Michael Snyder[1,2]
[1]*Department of Molecular, Cellular, and Developmental Biology, Yale University, New Haven, Connecticut 06520; email: heng.zhu@yale.edu*
[2]*Department of Molecular Biophysics and Biochemistry, Yale University, New Haven, Connecticut 06520; email: michael.snyder@yale.edu*
[3]*Biological Sciences and Bioengineering, Sabanc University, Orhanli Tuzla Istanbul, 81474 Turkey; email: mbilgin@sabanciuniv.edu*

Key Words 2-D/MS, yeast two-hybrid, protein localization, proteome microarray, data integration

■ **Abstract** Fueled by ever-growing DNA sequence information, proteomics–the large scale analysis of proteins–has become one of the most important disciplines for characterizing gene function, for building functional linkages between protein molecules, and for providing insight into the mechanisms of biological processes in a high-throughput mode. It is now possible to examine the expression of more than 1000 proteins using mass spectrometry technology coupled with various separation methods. High-throughput yeast two-hybrid approaches and analysis of protein complexes using affinity tag purification have yielded valuable protein-protein interaction maps. Large-scale protein tagging and subcellular localization projects have provided considerable information about protein function. Finally, recent developments in protein microarray technology provide a versatile tool to study protein-protein, protein–nucleic acid, protein-lipid, enzyme-substrate, and protein-drug interactions. Other types of microarrays, though not fully developed, also show great potential in diagnostics, protein profiling, and drug identification and validation. This review discusses high-throughput technologies for proteome analysis and their applications. Also discussed are the approaches used for the integrated analysis of the voluminous sets of data generated by proteome analysis conducted on a global scale.

CONTENTS

INTRODUCTION

With the DNA sequences of more than 90 genomes completed, as well as a draft sequence of the human genome, a major challenge in modern biology is to understand the expression, function, and regulation of the entire set of proteins encoded by an organism—the aims of the new field of proteomics. This information will be invaluable for understanding how complex biological processes occur at a molecular level, how they differ in various cell types, and how they are altered in disease states.

A rapidly emerging set of key technologies is making it possible to identify large numbers of proteins in a mixture or complex, to map their interactions in a cellular context, and to analyze their biological activities (1). Mass spectrometry has evolved into a versatile tool for examining the simultaneous expression of more than 1000 proteins and the identification and mapping of posttranslational modifications (2, 3). High-throughput methods performed in an array format have enabled large-scale projects for the characterization of protein localization, protein-protein interactions, and the biochemical analysis of protein function (4, 5). Finally, the plethora of data generated in the last few years has led to approaches for the integration of diverse data sets that greatly enhance our understanding of both individual protein function and elaborate biological processes (6).

In this review, we discuss recent developments in various technologies for characterizing protein function at the level of the entire proteome of a given organism. Much of this work was initially established in microorganisms such as yeast but is currently being applied to multicellular organisms.

PROTEIN PROFILING

The spectrum of proteins expressed in a cell type provides that cell with its unique identity. Elucidating how the protein complement changes in a cell type during development in response to environmental stimuli and in disease states is crucial for understanding how these processes occur at a molecular level. Recent years have witnessed a revolution in the development of new approaches for identifying large numbers of proteins expressed in cells and also for globally detecting the differences in levels of proteins in different cell states. In this section, we discuss these newly emerged technologies for profiling the proteins expressed in different cell types.

Two-Dimensional Gels and Mass Spectroscopy

Traditionally, two-dimensional (2-D) gel electrophoresis has been the primary tool for obtaining a global picture of the expression levels of a proteome under various conditions. In this method, proteins are first separated in one direction by isoelectric focusing usually in a tube gel and then in the orthogonal direction by molecular mass using electrophoresis in a slab gel containing sodium dodecyl sulfate (SDS) (7). Using this approach, several thousand protein species can be resolved in a single slab gel. However, 2-D gels are cumbersome to run, have a poor dynamic range, and are biased toward abundant and soluble proteins. Also 2-D gel analysis alone cannot provide the identity of the proteins that have been resolved.

In recent years, protein separation methods coupled with various mass spectrometry (MS) technologies have evolved as the dominant tools in the field of protein identification and protein complex deconvolution (8). The key developments were the invention of the time-of-flight (TOF) MS and relatively nondestructive methods to convert proteins into volatile ions. Two "soft ionization" methods, namely matrix-assisted laser desorption ionization (MALDI) and electrospray ionization (ESI), have made it possible to analyze large biomolecules, such as peptides and proteins (9–11).

In initial studies, protein mixtures were first separated using 2-D gel electrophoresis followed by the excision of protein spots from the gel. In those and more recent studies with other protein separation methods, the next step is digestion using a sequence-specific protease such as trypsin, and then the resulting peptides are analyzed by MS. When MALDI is used, the samples of interest are solidified within an acidified matrix, which absorbs energy in a specific UV range and dissipates the energy thermally. This rapidly transferred energy generates a vaporized plume of matrix and thereby simultaneously ejects the analytes into the gas phase where they acquire charge. A strong electrical field between the MALDI plate and the entrance of the MS tube forces the charged analytes to rapidly reach the entrance at different speeds based on their mass-to-charge (*m/z*) ratios. A significant advantage of MALDI-TOF is that it is relatively easy to perform protein or peptide identification with moderate throughput (96 samples at a time). MALDI-MS provides a rapid way to identify proteins when a fully decoded genome is available because the deduced masses of the resolved analytes can be compared to those calculated for the predicted products of all of the genes in the genomes of an organism

The ESI method is also widely used to introduce mixtures of biomolecules into the MS instrument. The unique feature of ESI is that at atmospheric pressure it allows the rapid transfer of analytes from the liquid phase to the gas phase (8).

The spray device creates droplets, which once in the MS go through a repetitive process of solvent evaporation until the solvent has disappeared and charged analytes are left in the gas phase. Normally, ESI is coupled with either a triple quadrupole, ion trap, or hybrid TOF MS. Compared with MALDI, ESI has a significant advantage in the ease of coupling to separation techniques such as liquid chromatography (LC) and high-pressure LC (HPLC), allowing high-throughput and on-line analysis of peptide or protein mixtures (12, 13). Typically, a mixture of proteins is first separated by LC followed by tandem MS (MS/MS). In this procedure, a mixture of charged peptides is separated in the first MS according to their *m/z* ratios to create a list of the most intense peptide peaks. In the second MS analysis, the instrument is adjusted so that only a specific *m/z* species is directed into a collision cell to generate "daughter" ions derived from the "parent" species (Figure 1). Using the appropriate collision energy, fragmentation occurs predominantly at the peptide bonds such that a ladder of fragments, each of which differs by the mass of a single amino acid, is generated. The daughter fragments are separated according to their *m/z,* and the sequence of the peptide can then be deduced from the resulting fragments (8, 10). By comparison with predicted sequences in the databases, the identity of the peptide is revealed.

The coupling of liquid chromatography (LC) with MS has had a great impact on small molecule and protein profiling, and has proven to be an important alternative method to 2-D gels (14). Typically, proteins in a complex mixture are separated by ionic or reverse phase column chromatography and subjected to MS analysis. Of the various ionization methods developed for coupling liquid chromatography to MS, including thermospray (15), continuous-flow fast atom bombardment (16), and particle beam (17) techniques, ESI is the most widely used interface technique (18).

LC-MS has been applied to large-scale protein characterization and identification. The Yates group (19) was able to resolve and identify 1484 proteins from yeast in a single experiment. Unlike the 2-D/MS approaches, the authors demonstrated that even low-abundance proteins could be clearly identified, such as certain protein kinases. In addition, 131 of the proteins identified have three or more predicted transmembrane domains, suggesting that this approach was able to readily detect membrane proteins. In addition to its role in protein profiling, LC-MS is perhaps the most powerful technique for the monitoring, characterization, and identification of impurities in pharmaceuticals.

An instrument that combines the benefits of high mass accuracy with highly sensitive detection is the Fourier transform ion cyclotron resonance mass spectrometer (FTICR-MS). FTICR-MS has recently been applied to identify low-abundance compounds or proteins in complex mixtures and to resolve species of closely related *m/z* ratios (20). Coupled with HPLC and ESI, FTICR-MS is able to characterize single compounds (up to 500 Da) from large combinatorial chemistry libraries and to accurately detect the masses of peptides in a complex protein sample in a high-throughput mode. For example, Nawrocki et al. studied the diversity and degeneracy of a small-peptide combinatorial library containing

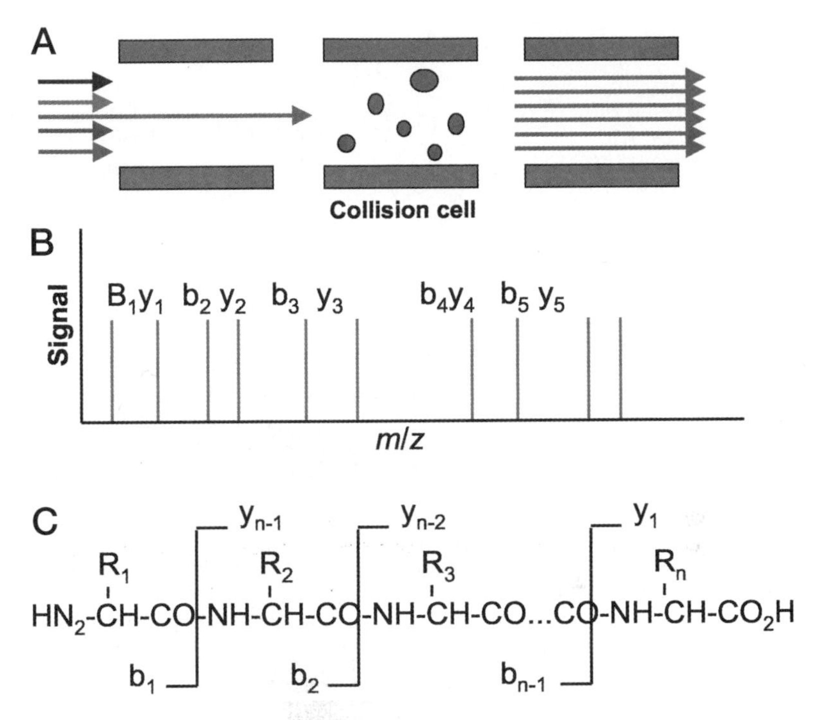

Figure 1 MS/MS analysis of peptide sequences. A protein mixture is first separated by LC followed by ESI ionization to generate fragment patterns (MS/MS spectra). In the first pass, a mixture of charged peptides, indicated as *arrows*, are separated according to their *m/z* ratios to create a list of the most intense peptide peaks. (*A*) The instrument is adjusted so that only a specific *m/z* species (indicated as the *longer arrow*) is directed into a collision cell to generate "daughter" ions derived from the "parent" species. (*B*) The newly generated fragments are separated according to their *m/z* ratio, creating the MS/MS spectrum. Using appropriate collision energy, fragmentation occurs predominantly at the peptide bonds such that a ladder of fragments, each of which differs by the mass of one amino acid, is generated. (*C*) The sequence of the peptide can then be deduced by a ladder-walk. [Adapted from (8, 100).]

up to 10^4 compounds using FTICR-MS (21). Lipton et al. (22) developed a high-throughput and LC-coupled FTICR-MS approach to characterize the proteome of a radiation-resistant bacterium, *Deinococcus radiodurans*. The authors combined global enzymatic digestion of the whole cell lysates, high-resolution LC separation, and analysis by FTICR-MS to resolve 6997 peptides [termed accurate mass tags (AMT)] with high confidence. The 6997 AMTs corresponded to 1910 predicted open reading frames (ORFs), which covered 61% of the *D. radiodurans* proteome. Others have used a similar strategy to characterize proteins in human body fluids (23).

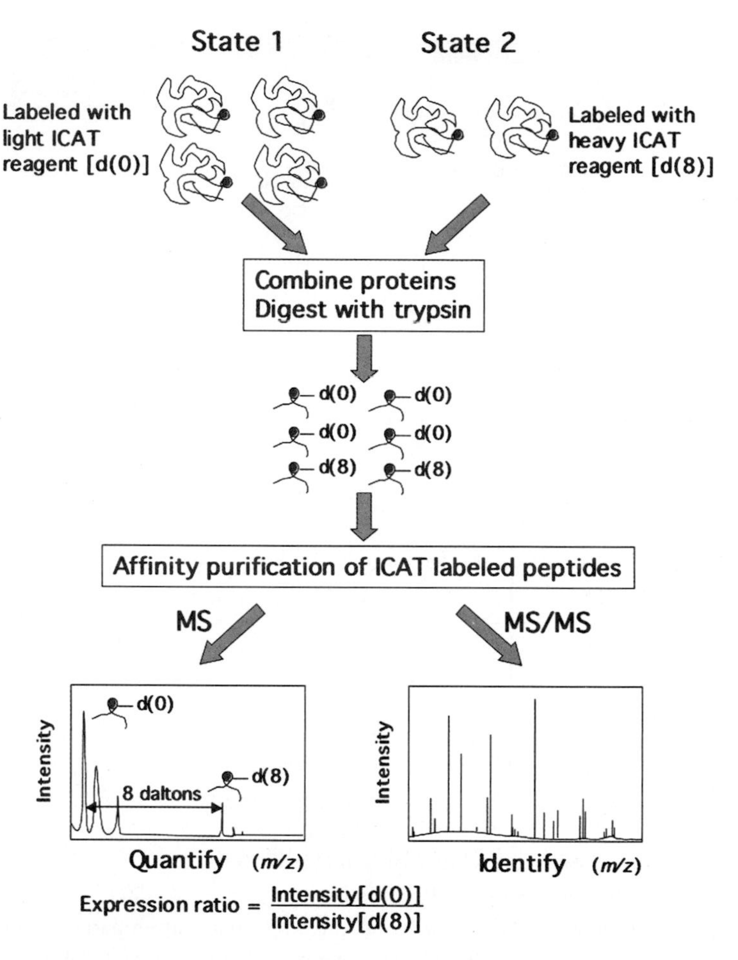

Figure 2 Schematic illustration of ICAT technology. Equal amounts of proteins extracted from two different biological states are separately labeled with heavy [d(8)] and light [d(0)] ICAT reagents. The samples are combined, digested with protease, and separated with multidimensional chromatography, and then analyzed by MS and MS/MS for quantification and identification, respectively. The relative abundances of labeled peptides are determined by comparison of peak intensities between the light and heavy forms of the peptides, which are separated by 8 Da.

Because of the complexity of any given proteome and the separation limits of both 2-D gel electrophoresis and liquid chromatography, only a fraction of that proteome can be analyzed. An alternative approach is to reduce the complexity prior to protein separation and characterization. The Aebersold group (24) designed a pair of isotope-coded affinity tag (ICAT) reagents to differentially label protein samples on their cysteine residues (Figure 2). The ICAT reagent contains a biotin moiety and a linker chain with either eight deuterium or eight hydrogen atoms. Two samples, each labeled with the ICAT reagent carrying one of the two different isotopes, were mixed and subjected to site-specific protease digestion. The labeled peptides containing Cys can be highly enriched by binding the biotin tags to streptavidin, resulting in a greatly simplified peptide mixture. Characterization of the peptide mixture was carried out by the LC-MS approach as described above. Quantitation of differential protein expression level can be achieved by comparing the areas under the doublet peaks that are separated by eight mass units. The authors demonstrated that they could follow the differential expression of more than 1400 different proteins in yeast. When dealing with the human proteome, Han et al. (25) further simplified the protein mixture by focusing on microsomal proteins that were isolated from cells untreated or treated with a differentiation-inducing stimulus and then labeled the proteins with ICAT reagents. They were able to detect the abundance ratios of 149 proteins in the microsomal fraction of human myeloid leukemia HL-60 cells. Thus, the ICAT method works well for the differential analysis of many proteins in a complex mixture. The obvious limitation of the ICAT labeling approach is that a protein has to contain at least one cysteine residue to be detected.

Antibody Microarrays

Although mass spectroscopy has demonstrated considerable promise for examining simultaneously the expression of large numbers of proteins in a complex mixture, such as a cell lysate, antibody microarrays hold potential promise for the high-throughput profiling of a smaller number of proteins (Figure 3). Briefly, antibodies (or other affinity reagents directed against defined proteins) are spotted onto a surface such as a glass slide; a complex mixture, such as a cell lysate or serum, is passed over the surface to allow the antigens present to bind to their cognate antibodies (or targeted reagents). The bound antigen is detected either by using lysates containing fluorescently tagged or radioactively labeled proteins, or by using a secondary antibodies against each antigen of interest. Low-density antibody arrays have been constructed that measure the levels of several proteins in blood (26) and sera (27, 28). In high-density arrays constructed recently, Sreekumar et al. (29) spotted 146 distinct antibodies on glass to monitor the changes in quantity of a number of antigens expressed in LoVo colon carcinoma cells. They found that radiation treatment of the cells up-regulated the levels of many interesting proteins, including p53, DNA fragmentation factors 40 and 45, and tumor necrosis factor–related ligand, and down-regulated the levels of other proteins.

The biggest problem with antibody arrays is antibody specificity. Haab and colleagues (30) analyzed the reactivity of 115 antibodies with their respective antigens. Protein microarrays containing either immobilized antigen or immobilized antibody were probed with antibodies or antigens, respectively. Only 30% of the antibody/antigen pairs showed the linear relationships expected for specific

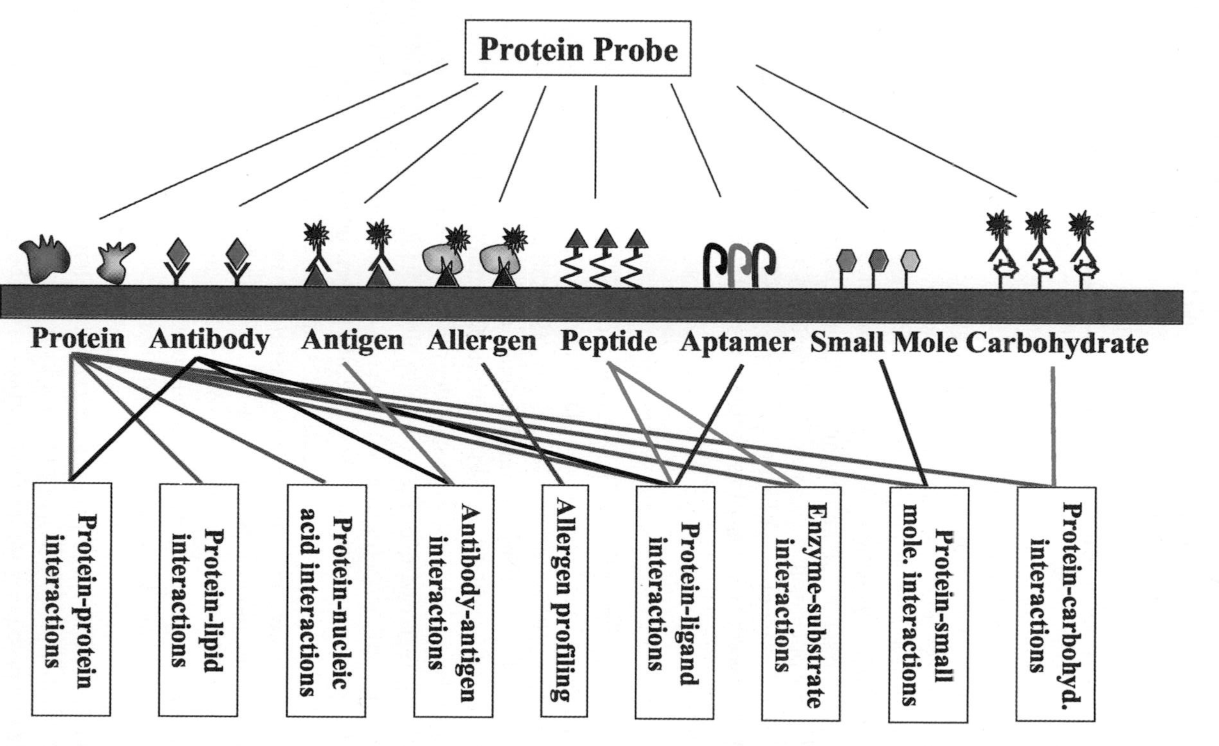

Figure 3 Protein microarrays and their applications. Ligands, such as proteins, peptides, antibodies, antigens, allergens, and small molecules, are immobilized in high density on modified surfaces to form functional and analytical protein microarrays. These protein microarrays can also be used for various kinds of biochemical analysis.

binding, indicating that most antibodies are not suitable for quantitative detection. Nonetheless, for those antibodies that are specific, quantitative detection of antigen abundance in a complex mixture could be determined. In a follow-up report, they showed that antibody microarrays could be applied to obtain serum profiles (31).

To profile the biological activities of a living cell or tissue, however, the capture molecules immobilized on the surface are not restricted to antibodies or antibody mimics. They can be short peptides, aptamers (DNA, RNA, or protein molecules selected for their ability to bind nucleic acid, proteins, small organic compounds, or cells), polysaccharides, allergens, or synthetic small molecules (2–4, 41). To profile antibodies in human sera, Robinson et al. (32) fabricated microarrays of autoantigen by arraying hundreds of such autoantigens, including proteins, peptides, and other biomolecules. These arrays were incubated with sera from patients to study the specificity and pathogenesis of autoantibody responses. In a similar approach, Hiller et al. (33) robotically arrayed 94 purified allergens on glass to monitor the IgE activity profiles of allergy patients. Using serum samples of minute amounts, they could profile an allergic patient's IgE reactivity in a single measurement. By comparing the reactivity to controls, specific IgE profiles could then be related to a large number of disease-causing allergens. Some of the findings from the allergen microarrays were further validated by classical skin tests. As another example, Joos and colleagues (34) used 18 diagnostic markers for autoimmune diseases to form a microarray of autoantigens and used it to monitor antigen-antibody interactions. Thus, protein microarrays can be used to profile the presence of a limited number of proteins and to analyze the antibody reactivity profile of individuals.

PROTEIN POSTTRANSLATIONAL MODIFICATION

Covalent modifications to protein structures, which occur either co- or posttranslationally, play a pivotal role in regulating protein activity. Identification of the type of modification and its location often provide crucial information for understanding the function or regulation of a given protein in biological pathways. So far, more than 200 different modifications have been reported, many of which are known to control signaling pathways and cellular processes (35). Many strategies have been developed to analyze protein modifications, however, most of them focus on only a specific type of modification, such as protein phosphorylation. For example, one strategy to identify phosphoproteins is to enrich the phosphorylated peptides using either immunoprecipitation with phosphopeptide-specific antibodies or by metal-chelate affinity chromatography (36). The latter uses resins with chelated trivalent metal ions such as Fe(III) and Ge(III) to bind the phosphopeptides or phosphoproteins (37–40). The enriched proteins are then subjected to trypsin digestion, and the resulting fragments are identified using MS techniques. This approach can provide important information on the sites of phosphorylation in proteins. This method was recently applied on a large scale to proteins of a yeast lysate (42). Phosphopeptides were purified using metal-chelated columns and subjected to MS/MS; 383 sites of phosphorylation were identified on 216 peptides.

To identify multiple types of modifications in a single experiment, MacCoss et al. (43) also employed a so-called shotgun MS approach. This approach used multidimensional liquid chromatography (LC/LC), tandem mass spectrometry (MS/MS), and database-searching algorithms. In brief, the protein mixture was first digested with one site-specific and two nonspecific proteases; the resulting peptides were separated by multidimensional liquid chromatography; and finally their identities were revealed by MS/MS. The digestion with multiple proteases produced overlapping peptides of a given protein thereby providing thorough coverage of the protein and increasing the chance of pinpointing a modification on a specific amino acid residue. Using this strategy to analyze protein samples from the human lens tissue, the researchers identified 270 proteins. Further analysis of a family of 11 lens crystallins proteins revealed a total of 73 modifications including phosphorylation, methylation, oxidation, and acetylation in these proteins. Thus, although lens tissue is not extremely complex, this method has demonstrated its great potential for revealing a more comprehensive picture of protein modifications in a complex sample.

In summary, MS has played an important role in identifying posttranslational protein modifications. Protein microarrays, described below, have played an important role in identifying the enzymes responsible for many modifications and the substrate specificity of the modifying enzymes.

PROTEIN LOCALIZATION

Protein localization data provide valuable information in elucidating eukaryotic protein function. To monitor the relative levels of protein expression and obtain a snapshot of protein localization in yeast, our laboratory has developed a random transposon tagging strategy to generate a library of expressed ORFs fused to coding sequences for an epitope tag (Figure 4) (44). Briefly, a transposon containing an *Escherichia coli lacZ* gene lacking its ATG translation initiation codon and promoter lies adjacent to a *lox* site at one end of the transposon; a coding sequence for three copies of a hemagluttinin epitope tag lies adjacent to another *lox* site at the other end of the transposon (Figure 4) (44). The transposition is mediated in *E. coli* and the mutagenized DNA is then shuttled into yeast. When the *lacZ* cassette is inserted in-frame in an ORF in yeast, its transcription and translation can be detected. A large portion of the inserted cassette can be looped out in yeast via recombination between the *lox* sites (mediated by the phage *Cre* recombinase), leaving behind the ORF with a short in-frame epitope tag. Using high-throughput immunostaining, the subcellular locations of 2744 yeast proteins have been determined over the years (44–47).

Using this approach as well as an approach in which 2000 ORFs were fused directly to an epitope tag, Kumar et al. (48) localized approximately 55% of the proteome and described the first "localizome"—the subcellular localization of most proteins of an organism. They showed that 47% of yeast proteins were cytoplasmic, 13% mitochondrial, 13% exocytic, and 27% nuclear/nucleolar. A subset of nuclear proteins was further analyzed by using surface-spread preparations of meiotic chromosomes, and 38% were found associated with chromosomal DNA. The major shortcoming of the transposon approach is that the library is not complete yet—it contains roughly 60% of the 6300 annotated genes (48). Furthermore, because the tagged proteins are expressed from their native promoters, the localization information is biased toward abundant proteins.

Because the transposon tagging approach visualizes proteins via indirect immunostaining on fixed cells, the dynamics of protein localization and transportation cannot be analyzed. To develop a real-time detection method, Ding et al. (49) attempted to tag *Schizosaccharomyces pombe* proteins using green fluorescent protein (GFP) in a genomic library. The tagged plasmid library was transformed into the *S. pombe* cells, and 6954 transformants exhibiting GFP fluorescence were obtained, 728 of which showed distinct localization patterns. By recovering plasmids from these strains, 250 unique genes were confirmed to have GFP tags in-frame. For mammalian cells, systematic GFP tagging of complementary DNA (cDNA) clones has been accomplished. Simpson et al. (50) cloned 107 novel human cDNAs to produce both N- and C-terminal fusions to GFP; ~100 proteins showed a clear pattern of intracellular localization. On the basis of sequence homology, they were able to predict the locations of 47% of these novel cDNAs; these predictions were in good agreement with the experimental results. Although considerable effort is needed to cover the entire proteome of any organism, these studies indicate that it is feasible to analyze protein localization globally in microbes and multicellular organisms.

INTERACTION PROTEOMICS AND PATHWAY BUILDING

It is widely acknowledged that proteins rarely act as single isolated species when performing their functions in vivo (1). The analysis of proteins with known functions indicates that proteins involved in the same cellular processes often interact with each other (6). Following this observation, one valuable approach for elucidating the function of an unknown protein is to identify other proteins with which it interacts, some of which may have known activities. On a large scale, mapping protein-protein interactions has not only provided insight into protein function but facilitated the modeling of functional pathways to elucidate the molecular mechanisms of cellular processes.

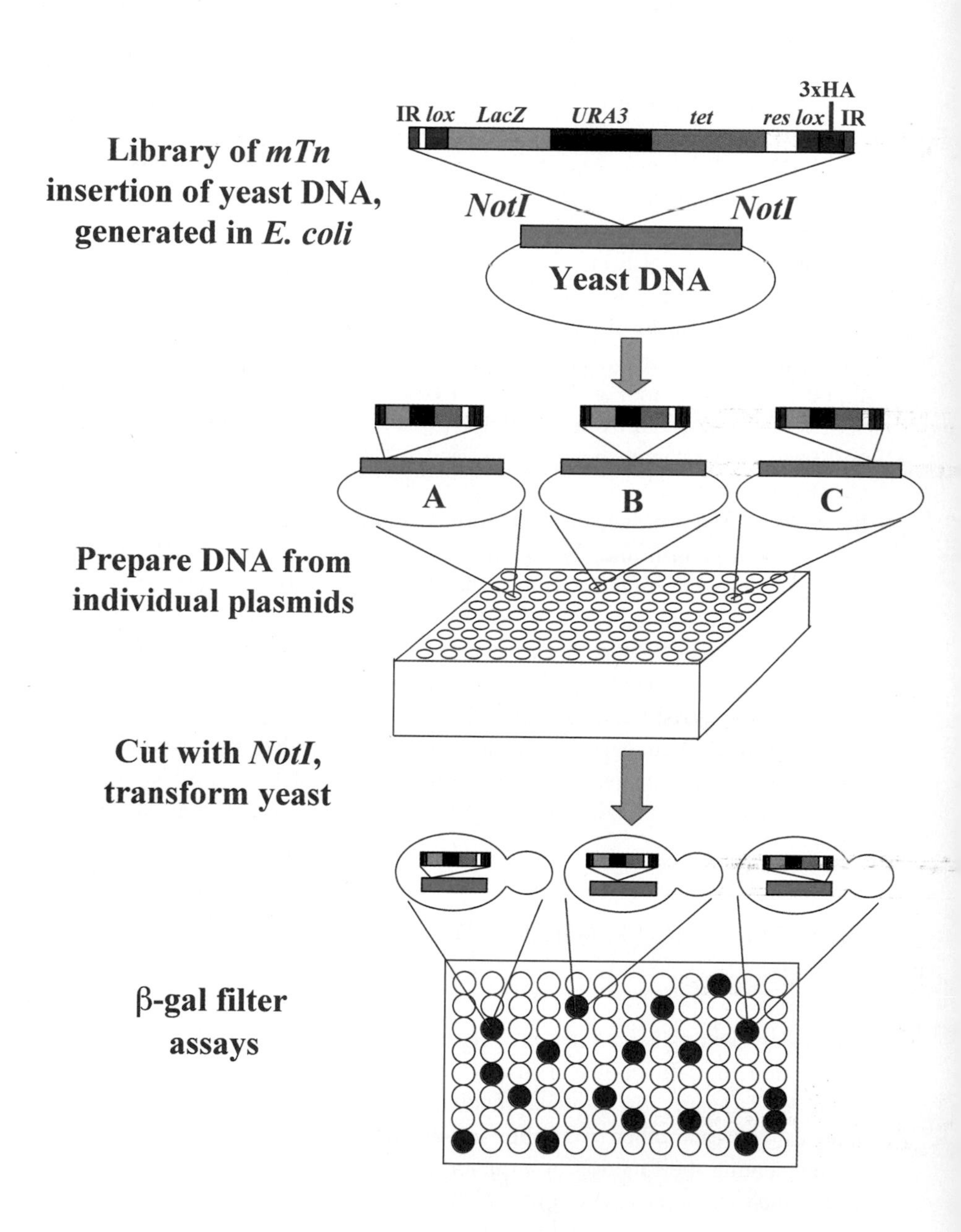

Library of *mTn* insertion of yeast DNA, generated in *E. coli*
3xHA
IR *lox* *LacZ* *URA3* *tet* *res lox* IR
NotI
NotI
Yeast DNA
A
B
C
Prepare DNA from individual plasmids
Cut with *NotI*, transform yeast
β-gal filter assays

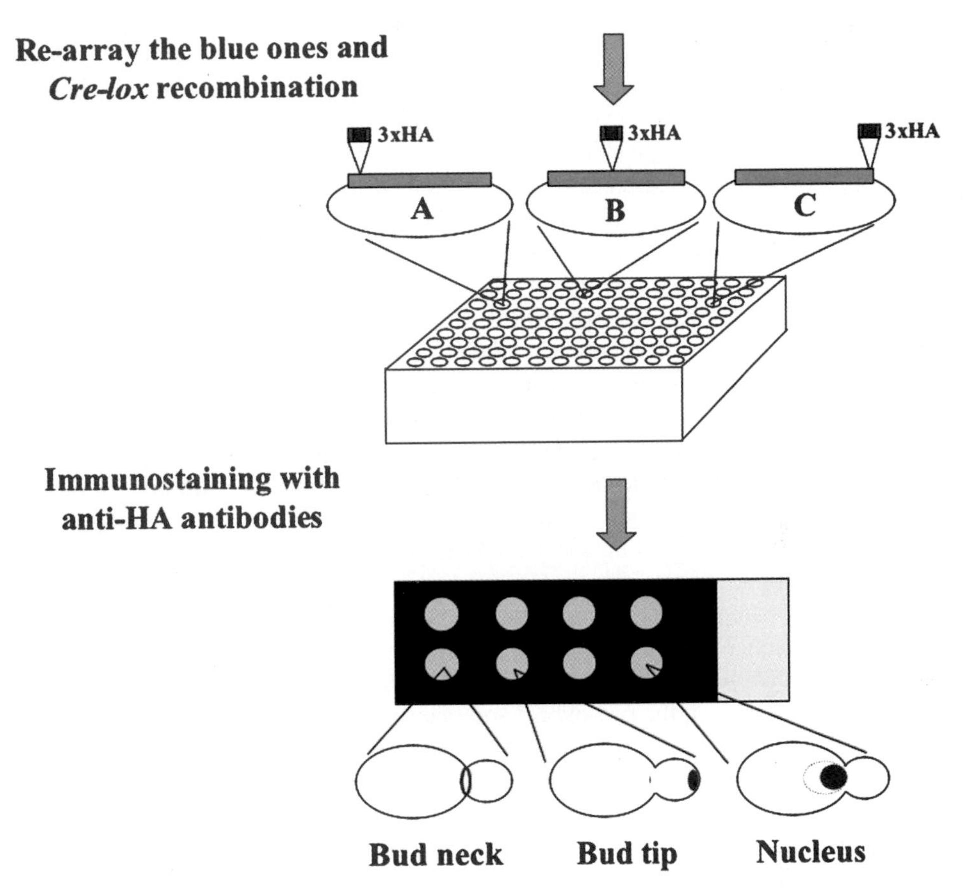

Figure 4 *Continued.*

←

Figure 4 Transposon tagging strategy for protein localization. To monitor the relative levels of protein expression, a transposon tagging strategy using a mini-transposon (mTn) was used to generate a library of random insertions in yeast DNA in an *E. coli* plasmid. The mTn contains a promoterless and 5′-truncated *lacZ* gene near one end of the transposon and coding sequence for three copies of an epitope tag at the other end. The mutagenized yeast DNA is prepared in a 96 well format, cleaved with the restriction endonuclease NotI to free the yeast DNA for the plasmid, and individually transformed into a diploid yeast strain. The insertion allele replaces one of the chromosomal copies. When the *lacZ* gene is inserted in-frame within a yeast ORF, its transcription and translation can be visualized in yeast cells using assays for β-galactosidase; such clonies will turn blue. A large portion of the inserted cassette is then excised at the *lox* sites via *Cre*-mediated recombination, leaving behind the ORF with a short in-frame epitope tag coding sequence. The subcellular location of the protein is then determined by indirect immunofluorescence. [Adapted from (44).]

Two-Hybrid Studies

One of the best-established in vivo approaches to map protein-protein interactions is the yeast two-hybrid method (51). In this technique, a component of interest (bait) is typically fused to a DNA-binding domain. Other proteins (preys), which are fused to a transcription-activating domain, are screened for physical interactions with the bait protein using the activation of a transcription reporter construct as the detection method (read out). This approach is scalable and can be fully automated.

Recently, systematic two hybrid projects have been undertaken to analyze protein-protein interactions at a global level in budding yeast, the nematode, *Cuenorhabditis elegans*, and human gastric bacterial pathogen, *Helicobacter pylori* (51–55). More than 4500 interactions have been identified in yeast, unraveling a host of unexpected and interesting interactions (51, 52). Likewise, Walhout et al. (53) and Boulton et al. (54) were able to map protein-protein interaction networks involved in *C. elegans* vulval development and DNA-damaging response components, respectively. Finally, Rain et al. (55) have recently built a large-scale protein-protein interaction map of the human gastric bacterial pathogen *Helicobacter pylori,* whose genome encodes 1590 predicted coding sequences (56). A total of 261 bait proteins were used to identify 1280 interactions, resulting in a protein interaction map covering much of the proteome. In general, the two-hybrid approach is especially powerful when analyzing smaller genomes because most of genes are easily known from the genome sequence and because the method has the potential to uncover all the possible interaction combinations quickly and relatively easily.

However, there are several drawbacks to the two-hybrid studies. First, they are not comprehensive. Based on the observations that most well-characterized proteins interact with 5–7 other proteins, estimates of the number of interactions expected in yeast are ~30,000–40,000 (59), significantly higher than the 4500 identified thus far. This is likely due to the fact that most of the studies performed thus far have been carried out in pools of yeast strains; this probably fails to detect all possible individual interactions. Direct tests of all possible individual interactions have not yet been performed, and thus the screens are not saturated.

A second disadvantage of two-hybrid methods is that they identify a large number of false positives, presumably through spurious interactions between proteins that do not normally occur in vivo (6). Approximately 50% of the interactions are estimated to be false positives.

A third drawback of the conventional two-hybrid is that interaction occurs in the nucleus and uses a transcriptional readout. Consequently, the interaction of many membrane proteins and transcription factors cannot be measured. To circumvent this problem, other two-hybrid methods have been developed. One promising method is the "split ubiquitin system," which appears to be especially useful for detecting interactions among membrane proteins (57). The split ubiquitin system involves bringing together two halfs of ubiquitin; the N terminus ubiquitin fragment is fused to one protein (e.g., the bait) and the C-terminal fragment, which is fused to a transcription factor, is also fused to a potential interaction protein. When the proteins interact, the ubiquitin fragments interact, and the transcription factor is released by cleavage from the C-terminal fragment and activates a reporter construct. This system has been used successfully to detect interaction among a variety of test proteins, including yeast oligosaccharyl transferases (57), sucrose transporters (57a), proteins involved in viral replication in *Arabidopsis* (57b), and transmembrane proteins normally present in the endoplasmic reticulum (57c, 57d).

In spite of the drawbacks of two-hybrid studies, the data have proven to be exteremely valuable, and even more so when integrated with protein-protein interaction data from other sources. Schwikowski et al. (58) compiled a list of 2709 published yeast protein interactions (including two hybrid) and found that 1548 of them could be mapped in a single network containing 2358 interactions. Based on this network, a putative function category can often be assigned to many novel proteins. Thus, although the data are incomplete, they have enormous utility.

Affinity Tagging and Mass Spectroscopy

Affinity purification has long been used to identify protein-protein interactions or validate their existance in cell extracts on a one-protein-at-a-time basis. However, the throughput has been dramatically improved by two recent reports on the systematic characterization of protein complexes in yeast (60, 61). In one study, endogenous protein coding genes were fused to the coding sequences of a tandem affinity purification (TAP) tag. The tagged proteins were then purified under gentle conditions along with their associated partners, and subsequently separated by gel electrophoresis. The copurifying proteins were identified by MS. Starting with a set of more than 500 chromosomal tagged genes, Gavin et al. (60) were able to purify and subsequently resolve 232 protein complexes encompassing 1440 distinct proteins in yeast. In another study, Ho et al. (61) constructed 725 inducible FLAG epitope–tagged fusions, which they overexpressed and purified along with their associated proteins. This study identified more than 3000 protein-protein interactions involving 1578 individual yeast proteins. The differences in the results likely reflect the fact that one study used endogenous protein levels whereas the latter used overexpressed proteins. The overexpressed proteins are likely to interact with more proteins, but they may also associate with proteins that they do not normally bind and thereby yield false positives. By combining both data sets, much greater accuracy is achieved (see below).

ANALYSIS OF PROTEIN BIOCHEMICAL ACTIVITIES

One of the most direct methods for elucidating protein function and regulation is to determine its biochemical activity. The past few years have brought several powerful approaches for the large-scale analysis of biochemical activities in yeast (summarized in Table 1). These approaches involve overexpressing the protein-coding genes from the organism of interest and then screening the expressed proteins for biochemical activities of interest. Much of this work is possible because of the development of efficient methods to clone sets of open reading frames (ORFs) into plasmids for expression in an appropriate host. As a result, large-scale biochemical analysis of proteins has been initiated.

Analysis of Biochemical Activities Using Pooling Strategies

In a pioneering work, the Phizicky group (62) applied a recombination-based cloning strategy to fuse >85% of the yeast ORFs to glutathione-*S*-transferase (GST) on a plasmid under the control of an inducible promoter. Yeast strains containing these fusion plasmids were grouped into 64 pools, each containing 96 clones, and induced to produce fusion proteins. GST fusion proteins from each pool were purified and screened for biochemical activities. Individual strains from positive pools were screened again to identify the specific clone expressing the activity of interest. A number of ORFs carrying new biochemical activities were identified, including tRNA ligase, 2′-phosphotransferase, phosphodiesterase, and cytochrome *c* methyltransferase. Compared to conventional approaches, this method allows a rapid and sensitive assignment of catalytic function to ORFs and is generally applicable for detection of virtually any type of activity. The disadvantages of this approach are that it requires several steps and that prevalent enzymes or activities, such as kinases and phosphatases, must usually be screened in smaller pools.

Functional Protein Microarrays for the Analysis of Biochemical Activities

A more direct approach for the global identification of biochemical activities of interest is using functional protein microarrays. In this technique, sets of proteins of interest or an entire proteome is overexpressed, purified, distributed in an addressable array format, and then assayed (Figure 3). These arrays can be used not only to screen for biochemical activities of interest but also to examine posttranslational modifications and to detect binding to small molecules, proteins, antibodies, and drugs. The latter feature has potentially powerful applications in the discovery and development of pharmaceuticals.

There are several types of functional protein array formats [reviewed in (63)]: nanowells, which are miniature wells; solid surface supports such as glass slides; and thick absorbent surfaces, such as hydrogels. The latter have not been used extensively and thus are not discussed here.

Nanowell arrays typically contain wells 1 mm or less in diameter; they can be made of a plastic such as polydimethylsiloxane (PDMS) by using a mold; alternatively, wells can be etched in glass. The wells compartmentalize samples and reduce evaporation. Using this format, Zhu et al. (64) analyzed the kinase-substrate specificity of almost all (119 of 122) yeast kinases using 17 different

TABLE 1 Comparison of different technologies for interaction proteomics

Approach	Application	Advantage	Disadvantage
Yeast two-hybrid	Protein-protein interactions, protein-DNA interactions	High-throughput and systematic to reveal protein interactions	No control over interaction condition; interactions are usually in the nucleus
Affinity tagging/MS	Dissecting protein complexes	In vivo interactions that involve multiple partners	May miss transient or weak interactions, hard to identify false positives
Antibody array	Protein profiling, protein detection, clinical diagnostics	Very sensitive and low sample consumption, great potential in biomarker and drug development	Highly restricted by the quantity and quality of available antibodies; semiquantitative protein detection
Functional protein array	Diverse, e.g., protein-protein, protein-lipid, protein–small molecule, enzyme-substrate interactions as well as drug discovery and posttranslational modifications	Great potentials for analyzing biochemical activities of proteins and high-throughput drug and drug target screening	In vitro assays
Peptide array	Enzyme-substrate interaction and drug discovery	Sensitive and straightforward way to identify epitopes	Expensive to fabricate; in vitro assays
Carbohydrate array	Carbohydrate-mediated molecular recognition and anti-infection response	A new and sensitive way to study carbohydrate-mediated molecular events	In vitro arrays; tough to acquire carbohydrate molecules in pure forms
Small molecule array	Protein–small molecule interaction, drug discovery, enzyme specificity profiling	Minimum small molecule consumption and high sensitivity	In vitro assays; necessary to improve throughput to cover 10^6 molecules in a normal combinatorial chemistry library

substrates. The substrates were first covalently immobilized on the surface of individual nanowells, and individual protein kinases in kinase buffer with [^{33}P]ATP were incubated with the substrates. After washing away the kinases and unincorporated ATP, the nanowell chips were analyzed for phosphorylated substrates using a Phosphoimager (Molecular Dynamics, Inc.). Not only were known kinase-substrate interactions identified but many novel activities were revealed. These studies showed that approximately one fourth of yeast protein kinases are capable of phosphorylating tyrosine residues on an artificial substrate (poly Glutamine-Tyrosine), even though by sequence they are all members of the Ser-Thr family of protein kinases. Thus, many kinases are capable of phosphorylating tyrosine in vivo. Thus far, at least two of the in vitro phosphorylating enzymes have been shown to phosphorylate their substrates on tyrosine in vivo, suggesting that many of them are also tyrosine kinases in vivo (64a; M. Snyder, unpublished).

The more common approach for functional protein microarrays is to use glass microscope slides, as these are compatible with many commercial scanners. Proteins are attached to the surface using either direct covalent methods, linkers, or affinity tags (63, 65, 66). The bound proteins are then assayed for binding or enzymatic activities. MacBeath & Schreiber (65) used this format to demonstrate that they could detect antibody-antigen interactions, protein kinase activities, and protein interaction with small molecules using several test systems.

The major limitation in functional protein microarrays has been the preparation of proteins to analyze. This requires high-quality and comprehenisve expression libraries and methods that yield a large number of functional active proteins. This problem was recently surmounted. For example, as mentioned above, it was possible to produce in functional forms nearly all yeast protein kinases. More recently, the first eukaryotic proteome chip was prepared. This microarray is composed of >5800 individually cloned, overexpressed, and purified proteins (66). A high-throughput protein purification protocol was developed to purify 80% of the yeast proteome as full-length proteins (Figure 5). In initial studies, the proteome chips were used to identify protein-protein interactions by screening for binding targets of calmodulin (Figure 6). Calmodulin is a highly conserved protein that regulates signaling pathways and other cellular processes. The proteome chip was probed with biotinylated calmodulin and washed stringently; the bound calmodulin was detected by binding of streptavidin, which was labeled with a cyanine dye, Cy3. The identities of the calmodulin-interacting proteins on the proteome chips were deconvoluted using a laser scanner and the known addresses on the array. In addition to six known targets, 33 potential new binding partners of calmodulin were identified. Sequence comparison revealed that 14 of the 39 calmodulin-binding proteins shared a common motif, which is similar to a previously known calmodulin-binding motif, called the IQ sequence.

To explore the possibility of using proteome chips to identify the binding targets of secondary messengers, the chips were probed with phosphatidylinosi-

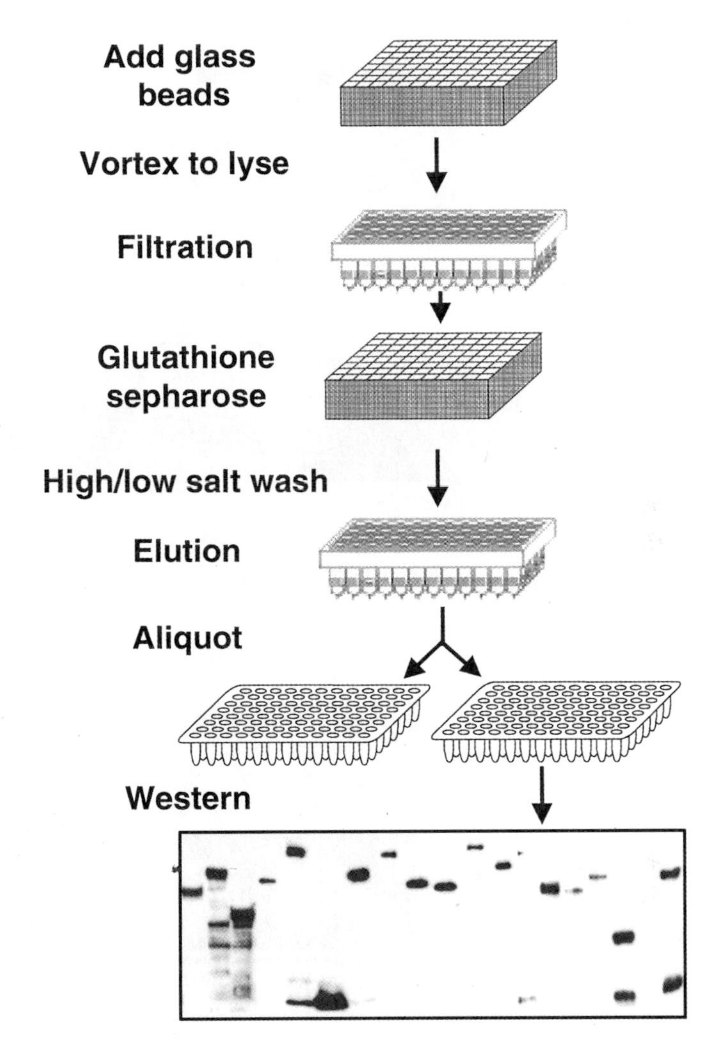

Figure 5 High-throughput procedure for purification of protein from yeast cells. Yeast strains each carrying a different fusion protein expression vector were grown and purified in a 96-well format. Western analysis revealed that 80% of the purified proteins were full length.

tides incorporated in liposomes. These liposomes should represent the most relevant physiological binding environment and also contain 1% biotinylated lipid as a detection tag. A total of 150 proteins were identified as binding either phosphotidylcholine or phosphotidylcholine vesicles containing one of five different phosphatidylinositides. Of the 98 annotated proteins, 45 are membrane associated and 8 more are involved in lipid metabolism. One interesting result from this study is that many protein kinases bind liposomes containing specific phosphatidylinositides.

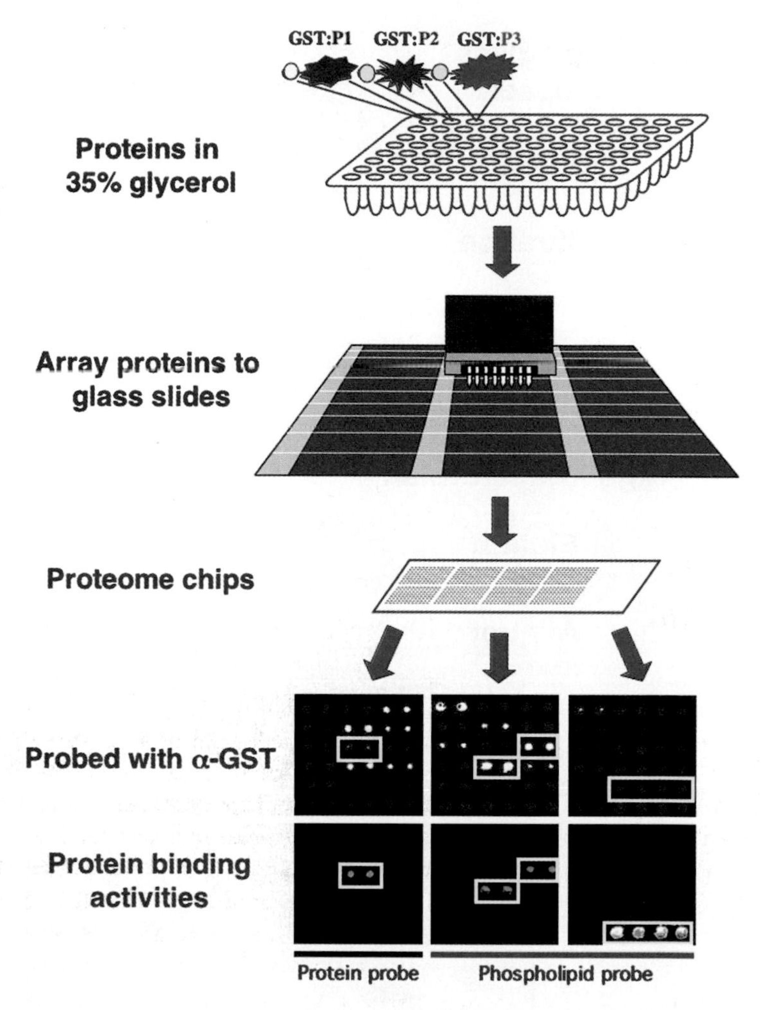

Figure 6 Protein activity analysis using the yeast proteome microarrays. GST-purified proteins were arrayed in duplicate onto nickel-coated glass slides at high density using a commercially available microarrayer. These chips were subsequently probed with biotinylated calmodulin and several phospholipid-containing liposomes. The binding activities were then detected using fluorescently labeled streptavidins. Positive signals are indicated in green and encapsulated in a yellow box. For comparison, the corresponding section of an array from the same printing was probed with anti-GST antibodies. All of the proteins present on the array react with this antibody (positives shown in red).

The proteome chips can also be used to study other binding activities, such as protein–nucleic acid, protein–small molecule, and protein-drug interactions. Although these are all in vitro binding assays, the advantage is that the experimental conditions can be well controlled. For example, different cofactors or inhibitors can be included in the binding assays to adjust the strignency of the binding activities. Another advantage is that these highly parallel assays are not biased toward abundant proteins. In addition, with proper detection methods, proteome chips can be used to identify the downstream targets of various enzymes such as protein kinases, phosphatases, methyl transferases, and proteases (4, 63). Finally, protein microarrays can be used to identify in vivo posttranslational modifications by probing for specific modifications, such as glycosylation or phosphorylation, using lectins or antibodies, respectively (M. Bilgin, H. Zhu, & M. Snyder, unpublished observations). This potentially allows the identification of all of the proteins that carry those modifications in a single experiment.

Protein microarrays also have the potential advantage of permitting analysis of the kinetics of protein-protein interactions via real-time detection methods. Surface plasmon resonance (SPR) has matured as a versatile detection tool to analyze the kinetics of protein-ligand interactions over a wide range of molecular weights, affinities, and binding rates (67–69). Although the commercially available SPR chips are not yet high throughput, Myszka & Rich (70) recently described a sensor surface with 64 individual immobilization sites in a single flow cell. Alternatively, Sapsford et al. (71) used a planar waveguide as the detection method to develop an antibody array biosensor and studied the kinetics of antigen-antibody interactions. More importantly, they demonstrated that significant signal intensity could be achieved from spots as small as 200 μm in diameter. However, to date, most of these alternative approaches have been applied successfully on a small scale using only a handful of samples. Some of these techniques may prove robust enough to be applied in a fully automated fashion.

A modified version of protein microarrays is peptide microarrays, which can be used as substrates for enzymatic activites and as potential ligands when probing with proteins or other molecules. In one recent example, Houseman et al. (72) characterized the substrate specificity of a nonreceptor tyrosine kinase, the *c-Src* product, using immobilized 9-mer peptide substrates arrayed at high density on a gold-coated glass surface. They characterized the kinase–peptide substrate interactions using SPR and the phosphorylation events using ATP derivatives, fluorescent labeling, and phosphoimaging methods. They could also quantitatively evaluate the effect of three known inhibitors of the kinase. Although the study was still primitive, the authors demonstrated the potential of coupling peptide chips with various detection methods to quantitatively study the dynamics of enzyme-substrate interactions, with obvious applications in drug discovery. In another example, Lizcano et al. (73) studied the molecular basis for the substrate specificity of a human protein kinase Nek6 using peptide microarrays harboring >1000 peptide species. They observed that a Leu located three residues N-terminal to the phosphorylation site in a substrate was important for the phosphorylation of this enzyme. However, genetic analysis indicated that selectivity for a Leu-X-X-Ser/Thr site might not occur in cells, which raises the challenge of using consensus sequences determined in vitro as a means to identify physiologically relevant substrates for Nek6 or any other kinase.

Finally, in addition to protein and peptide arrays, carbohydrate arrays have been introduced recently. Wang et al. (74) used carbohydrate-based microarrays to analyze the different types of anticarbohydrate antibodies in human and mammalian sera. An array of 48 different carbohydrate macromolecules was prepared and probed with sera from 20 normal individuals. A variety of different reactivities were observed. Interestingly, many of the carbohydrates that react with the sera are normally present in pathogen microbes, suggesting that the individuals may have acquired these antibodies during a microbial infection. Carbohydrate arrays can also be used to profile other types of binding activities, as demonstrated by Houseman & Mrksich (75). The authors conjugated and self-assembled a monolayer of 10 monosaccharides on a glass surface. The arrays were then used to profile the binding specificities to several lectins. Both SPR and competition experiments demonstrated that the carbohydrate-protein interactions were highly specific. Although the density of the chips was not high, it is scalable for large-scale and high-throughput analysis in the future.

In summary, protein microarrays can be used to globally analyze the activities of proteins including their binding to proteins, nucleic acids, lipids, carbohydrates, and small molecules. Because of its miniaturized and versatile nature, microarray technology is expected to flourish in the field of proteomics.

PROTEIN ENGINEERING

Another way to probe protein function is through protein engineering. The Shokat group (76) developed an elegant approach for analyzing kinase activities in a variety of organisms. Protein kinases were engineered to have a "hole" in the ATP-binding pocket. A modified ATP analogue containing an extra side chain (i.e., a "bump") is used to selectively inhibit that kinase in vivo. This approach was used to analyze the role of three yeast protein kinases, Cdc28p, Cla4p, and Pho85p. The inhibitor was added to cells containing the engineered protein kinase, and the resulting phenotype examined (76–78). In each case, in addition to their known function, a new role of each affected protein kinase could be deduced.

A modified version of this approach has been used to identify substrates of protein kinases in vitro (79). In this case, a different ATP analogue that can be used only by the engineered kinases is employed. The ATP analogue is added to a cell extract containing the engineered protein kinase, and the resulting phosphorylated proteins are likely to be the substrates of the engineered kinase. This technology can theoretically be applied to analyze other classes of enzymes for which detailed structural information about active site geometry and substrate recognition determinants are available.

PROTEOMICS AND DRUG DISCOVERY

In recent years, a number of approaches have been developed to identify molecules that bind or inhibit protein function. Molecules that bind a protein of interest can be used to develop diagnostic tests for that protein, or they can be used to probe that protein's function. Molecules that inhibit a protein of interest have the potential to be used directly for therapeutic applications. There are two basic types of protein binding agents: macromolecules such as proteins or nucleic acids and small molecules.

Macromolecular Inhibitors

Antibodies, both monoclonal and polyclonal have been used for many years as affinity reagents to probe and inhibit protein function. Recently, alternative methods, such as phage antibody-display, ribosome display, and mRNA display, have been developed to expedite the process of drug discovery (2, 80). All these approaches involve the construction of large repertoires of folded domains with potential binding activity, which are then selected by multiple rounds of affinity purification. The binding affinity of the resulting candidate clones can be further improved using subsequent mutagenesis and selection strategies.

Oligonucleotides can also be selected to bind proteins. Using a protocol for in vitro evolution, called systematic evolution of ligands by exponential enrichment (SELEX), nucleic acids that bind a wide variety of molecules have been selected. Using this method, many powerful antagonists of proteins have been found with K_d (equilibrium dissociation constant) values in the range of 1 pM to 1 nM. For example, Biesecker et al. (81) used the SELEX procedure to identify specific aptamer inhibitors of the human complement C5 component. In an initial round of selection, seven aptamers were isolated; these formed a closely related family based on sequence homology with a K_d values of 20–40 nM. The binding affinity was improved by a second round of SELEX, which produced an aptamer with a K_d of 2–5 nM.

Small Molecules

For several decades, natural and synthetic small molecules have provided powerful and invaluable means to dissect protein functions and regulatory mechanisms (82). Overwhelmed by the ever-growing sequence information, a more efficient and systematic approach to identify these small molecules is needed. Recently, several groups have devised methods for identifying small molecules using a microarray format.

MacBeath et al. (83) immobilized small organic compounds from a combinatorial chemistry library to form a high-density small molecule microarray. These microarrays were probed with fluorescently labeled target proteins to identify new ligands for these proteins. Similarly, Winssinger et al. (84) constructed a library of small molecules by tethering them to a peptide nucleic acid

(PNA) tag. These PNA tags provide the address for the structure of the corresponding small molecules as well as immobilize them at specific sites on the chip. The study further proved that the immobilized small molecule could withstand stringent washing conditions. As a test case, these arrays were used to identify a small molecule inhibitor of a capase.

As a third example, Kuruvilla et al. (85) constructed a small molecule microarray containing a collection of 3780 structurally complex 1,3-dioxane compounds to dissect the function of a yeast protein, Ure2. Ure2 is a central regulator of the nitrogen metabolic pathway. The library of small molecules was synthesized with a technology platform based on one bead–one stock solution and parsed out to form the small molecule microarray (86, 87). The microarray then probed with fluorescently labeled Ure2. One compound (uretupamine) was identified as specifically inhibiting Ure2 function in a subsequent cellular reporter assay. An analysis of gene expression profiles determined that uretupamine inhibits a particular function of Ure2 without affecting all of the functions of the protein globally. This approach uses a structurally complex small molecule library that is unbiased toward any particular protein targets (82).

The microarray format used in these approaches allows high throughput and highly parallel screening with minimum consumption of small molecules from a combinatorial chemistry library, and yet the signal-to-noise ratio is still high. This strategy is therefore expected to generate several tailored probes for every protein of interest in an entire proteome, which may greatly facilitate the development of pharmaceutical agents.

INTEGRATION OF DIVERSE DATA SETS

High-throughput methods and proteomics projects have exploded during the past few years. These projects have generated overwhelming amounts of data that help identify and characterize components of biological pathways as well as elucidate the response of these pathways (5). However, one important conclusion is that no single set of any high-throughput data is definitive. Instead, integration of multiple sets of data and verification using alternative methods is always required before drawing any firm conclusion. For example, Jansen et al. (88) investigated the relationship of protein-protein interactions with the expression level of mRNAs encoding the components in clearly defined protein complexes. The mRNA levels of the subunits in the same protein complexes showed significant coexpression patterns over a time course. By contrast, protein interactions identified by the yeast two-hybrid approach applied genome-wide had only a weak relationship with gene expression. Therefore, protein-protein interactions identified by the yeast two-hybrid approach should be independently confirmed using other methods such as protein chip and/or affinity tagging/MS approaches (54, 60, 61, 66), or by more traditional methods (e.g., coimmuno-precipitation).

To provide a systematic approach for analyzing multiple sets of data generated by gene expression profile studies, Hughes et al. (89) constructed a reference database of expression profiles corresponding to 300 diverse mutations and chemical treatments in *Saccharomyces cerevisiae.* By pattern matching to the database, even subtle differences in profiles can be revealed. Using this approach, the authors could predict the function of eight uncharacterized genes and confirm the predictions experimentally. In addition, they showed that it can be applied to characterize pharmacological perturbations, which means that this in silico (computed) approach has great potential in drug discovery and drug target identifications.

Large sets of data generated from different types of high-throughput methods can also be integrated to obtain a more comprehensive view of a biological process. Ideker et al. (90), for example, analyzed the galactose metabolic pathway in yeast using an integrated approach that combined expression profiles, MS analysis, and known protein-protein interaction information in the database. They were able to build, test, and refine the existing model, suggesting new hypotheses about the regulation of galactose utilization and physical interactions between this and a variety of other metabolic pathways.

Global sets of data generated with high-throughout approaches can be integrated to evaluate the quality and improve the confidence of individual data sets. In a recent report, Kemmeren et al. (91) applied a collection of expression profiles to assess the quality of several high-throughput protein interaction data sets. Out of 5342 putative two-hybrid interactions, the confidence levels of 973 interactions have been dramatically increased. In addition, the integration of the expression profiles and two-hybrid interactions has functionally annotated more than 300 previously uncharacterized genes. Furthermore, such in silico predictions were validated experimentally.

The Bork group (6) has taken such in silico evaluation of data one step further by examining protein interactions revealed by high-throughput yeast two-hybrid experiments, affinity protein complex purification analyses, correlated mRNA expression profiles, genetic interactions, and in silico prediction. Many interesting and intriguing results were discovered. Of the 80,000 purported interactions between yeast proteins, only ~2400 interactions are supported by more than one method. Each technique produced a unique coverage of interactions in terms of gene categories, which suggests that these methods have their specific strengthes and weaknesses. For example, most protein interaction data sets are heavily biased toward proteins of high abundance and toward particular cellular localizations of interacting proteins. In addition, the degree of evolutionary novelty of proteins plays a role in causing biased interaction coverage. To assess the accuracy and coverage of large interaction data, the authors compared the data with a reference set of trusted interactions. Not surprisingly, the highest accuracy was achieved for interactions supported by more than one method. Therefore, to increase coverage and accuracy for protein interactions, as many complementary methods as possible should be used; however, the ever-growing body of high-throughput data remains a challenge to integrate using data storage and analysis approaches (92–95).

In summary, the integration of multiple sets of high-throughput data has shown its power in evaluating and improving the quality of the data. The strategy also provides insight into hidden properties to better understand the molecular mechanisms of various biological pathways and processes.

CONCLUSIONS

The rapid progress in the field of large-scale biology has provided us opportunities to understand the function of biological networks as a whole. Genomics decodes sequence information of an organism and provides the "parts catalog," while proteomics attempts to elucidate the functions and relationships of the individual "parts" and predict the outcomes of the modules they form on a higher level (97). Since protein is closer to biological function than DNA, much emphasis has been devoted to the development of new tools for proteomics. The recent advances discussed in this review have demonstrated that new technologies are powering proteomics research.

Among these new methodologies, microarray technology has served as a versatile tool to analyze protein activities and holds great promise in the identification of drugs and drug targets, as well as in clinical diagnostics. As discussed above, protein microarray technologies have shown their great potential in basic proteomics research such as determining protein-protein, protein-lipid, protein-ligand, and enzyme-substrate interactions. The reduced sample consumption in the microarray format is important in both basic proteomics research and diagnostics, where only minimal amounts of samples are available (3, 63). It is expected that real-time patient monitoring during disease treatment and therapy will be developed based on this emerging technology.

With the massive amount of data produced by high-throughput assays, it has become obvious that the integration of different sets of data provided by different systematic methods will greatly enhance the understanding of biological systems. For example, efforts have been devoted to combine all the protein information to generate a uniform numerical ranking system, which provides a more comprehensive picture of the biological context of each protein (96). Davidson et al. (98) borrowed the concepts used in integrated circuit design to dissect DNA-based regulatory gene networks using the sea urchin embryo as a model system. This network was derived from the combination of large-scale perturbation analyses, *cis*-regulatory analysis, molecular embryology, and computational algorithms (99). In such networks, the relationships between the transcriptional factors and their targets are represented by "and," "or," or "non" logic. Such networks could reveal the hidden interactions between pathways and deepen our understanding of biological processes. Thus, it is increasingly important to properly integrate multiple sets of data to form an interaction network composed of all kinds of biological components, which will help elucidate the molecular mechanisms of life.

ACKNOWLEDGMENTS

We thank Jeremy Thorner and Jesslyn Holombo for critical comments on the manuscript. H. Zhu is supported by a postdoctoral fellowship from the Damon Runyon–Walter Winchell Cancer Research Foundation. Research in the Snyder lab is supported by grants from the National Institutes of Health.

The *Annual Review of Biochemistry* is online at http://biochem.annualreviews.org

LITERATURE CITED

1. Yanagida M. 2002. *J. Chromatogr. B* 771:89–106
2. Templin MF, Stoll D, Schrenk M, Traub PC, Vohringer CF, et al. 2002. *Trends Biotechnol.* 20:160–66
3. Stoll D, Templin MF, Schrenk M, Traub PC, Vohringer CF, Joos TO. 2002. *Front. Biosci.* 7:C13–32
4. Zhu H, Snyder M. 2001. *Curr. Drug Disc.* Sept:31–34
5. Zhu H, Snyder M. 2002. *Curr. Opin. Cell Biol.* 14:173–79
6. von Mering C, Krause R, Snel B, Cornell M, Oliver SG, et al. 2002. *Nature* 417: 399–403
7. O'Farrell PZ, Goodman HM, O'Farrell PH. 1977. *Cell* 12:1133–41
8. Figeys D, Linda D, McBroom LD, Moran MF. 2001. *Methods* 24:230–39
9. Yates JR 3rd. 1998. *J. Mass Spectrom.* 33:1–19
10. Godovac-Zimmermann J, Brown LR. 2001. *Mass Spectrom. Rev.* 20:1–57
11. Mann M, Hendrickson RC, Pandey A. 2001. *Annu. Rev. Biochem.* 70:437–73
12. Ducret A, Van Oostveen I, Eng JK, Yates JR 3rd, Aebersold R. 1998. *Protein Sci.* 7:706–19
13. Gatlin CL, Kleemann GR, Hays LG, Link AJ, Yates JR 3rd. 1998. *Anal. Biochem.* 263:93–101
14. Ermer J, Vogel M. 2000. *Biomed. Chromatogr.* 14:373–83
15. Vestal ML. 1984. *Science* 226:275–81
16. Caprioli RM. 1990. *Anal. Chem.* 62: A477–85
17. Baczynskyj L. 1991. *J. Chromatogr.* 562: 13–29
18. Covey TR, Huang EC, Henion JD. 1991. *Anal. Chem.* 63:1193–200
19. Washburn MP, Wolters D, Yates JR 3rd. 2001. *Nat. Biotechnol.* 19:242–47
20. Schmid DG, Grosche P, Bandel H, Jung G. 2000. *Biotechnol. Bioeng.* 71:149–61
21. Nawrocki JP, Wigger M, Watson CH, Hayes TW, Senko MW, et al. 1996. *Rapid Commun. Mass Spectrom.* 10: 1860–64
22. Lipton MS, Pasa-Tolic' L, Anderson GA, Anderson DJ, Auberry DL, et al. 2002. *Proc. Natl. Acad. Sci. USA* 99:11049–54
23. Bergquist J, Palmblad M, Wetterhall M, Hakansson P, Markides KE. 2002. *Mass Spectrom. Rev.* 21:2–15
24. Gygi SP, Rist B, Gerber SA, Turecek F, Gelb MH, et al. 1999. *Nat. Biotechnol.* 17:994–99

25. Han DK, Eng J, Zhou H, Aebersold R. 2001. *Nat. Biotechnol.* 19:946–51
26. Wiese R, Belosludtsev Y, Powdrill T, Thompson P, Hogan M. 2001. *Clin. Chem.* 47:1451–57
27. Moody MD, Van Arsdell SW, Murphy KP, Orencole SF, Burns C. 2001. *BioTechniques* 31:186–90
28. Huang RP, Huang R, Fan Y, Lin Y. 2001. *Anal. Biochem.* 294:55–62
29. Sreekumar A, Nyati MK, Varambally S, Barrette TR, Ghosh D, et al. 2001. *Cancer Res.* 61:7585–93
30. Haab BB, Dunham MJ, Brown PO. 2001. *Genome Biol.* 2:RESEARCH0004
31. Miller JC, Butler EB, Teh BS, Haab BB. 2001. *Dis. Markers* 17:225–34
32. Robinson WH, DiGennaro C, Hueber W, Haab BB, Kamachi M, et al. 2002. *Nat. Med.* 8:295–301
33. Hiller R, Laffer S, Harwanegg C, Huber M, Schmidt WM, et al. 2002. *FASEB J.* 16:414–16
34. Joos TO, Schrenk M, Hopfl P, Kroger K, Chowdhury U, et al. 2000. *Electrophoresis* 21:2641–50
35. Krishna RG, Wold F. 1993. *Adv. Enzymol. Relat. Areas Mol. Biol.* 67:265–98
36. Griffin TJ, Aebersold R. 2001. *J. Biol. Chem.* 276:45497–500
37. Andersson L, Porath J. 1986. *Anal. Biochem.* 154:250–54
38. Neville DC, Rozanas CR, Price EM, Gruis DB, Verkman AS, et al. 1997. *Protein Sci.* 6:2436–45
39. Posewitz MC, Tempst P. 1999. *Anal. Chem.* 71:2883–92
40. Stensballe A, Andersen S, Jensen ON. 2001. *Proteomics* 1:207–22
41. Seetharaman S, Zivarts M, Sudarsan N, Breaker RR. 2001. *Nat. Biotechnol.* 19:336–41
42. Ficarro SB, McCleland ML, Stukenberg PT, Burke DJ, Ross MM, et al. 2002. *Nat. Biotechnol.* 20:301–5
43. MacCoss MJ, McDonald WH, Saraf A, Sadygov R, Clark JM, et al. 2002. *Proc. Natl. Acad. Sci. USA* 99:7900–5
44. Ross-Macdonald P, Sheehan A, Roeder GS, Snyder M. 1997. *Proc. Natl. Acad. Sci. USA* 94:190–95
45. Ross-Macdonald P, Coelho PS, Roemer T, Agarwal S, Kumar A, et al. 1999. *Nature* 402:413–18
46. Ross-Macdonald P, Sheehan A, Friddle C, Roeder GS, Snyder M. 1999. *Methods Enzymol.* 303:512–32
47. Kumar A, Agarwal S, Heyman JA, Matson S, Heidtman M, et al. 2002. *Genes Dev.* 16:707–19
48. Kumar A, Cheung KH, Tosches N, Masiar P, Liu Y, et al. 2002. *Nucleic Acids Res.* 30:73–75
49. Ding DQ, Tomita Y, Yamamoto A, Chikashige Y, Haraguchi T, et al. 2000. *Genes Cells* 5:169–90
50. Simpson JC, Wellenreuther R, Poustka A, Pepperkok R, Wiemann S. 2000. *EMBO Rep.* 1:287–92
51. Uetz P, Giot L, Cagney G, Mansfield TA, Judson RS, et al. 2000. *Nature* 403:623–27
52. Ito T, Chiba T, Ozawa R, Yoshida M, Hattori M, et al. 2001. *Proc. Natl. Acad. Sci. USA* 98:4569–74
53. Walhout AJ, Sordella R, Lu X, Hartley JL, Temple GF, et al. 2000. *Science* 287:116–22
54. Boulton SJ, Gartner A, Reboul J, Vaglio P, Dyson N, et al. 2002. *Science* 295:127–31
55. Rain JC, Selig L, De Reuse H, Battaglia V, Reverdy C, et al. 2001. *Nature* 409:211–15
56. Tomb JF, White O, Kerlavage AR, Clayton RA, Sutton GG, et al. 1997. *Nature* 388:539–47
57. Stagljar I, Korostensky C, Johnsson N, te Heesen S. 1998. *Proc. Natl. Acad. Sci. USA* 95:5187–92
57a. Reinders A, Schulze W, Kuhn C, Barker L, Schulz A, et al. 2002. *Plant Cell* 14:1567–77
57b. Tsujimoto Y, Numaga T, Ohshima K, Yano MA, Ohsawa R, et al. 2003. *EMBO J.* 22:335–43

57c. Wang B, Nguyen M, Breckenridge DG, Stojanovic M, Clemons PA, et al. 2003. *J. Biol. Chem.* In press.
57d. Wittke S, Dunnwald M, Albertsen M, Johnsson N. 2002. *Mol. Biol. Cell.* 13:2223–32
58. Schwikowski B, Uetz P, Fields S. 2000. *Nat. Biotechnol.* 18:1257–61
59. Snyder M, Kumar A. 2002. *Funct. Integr. Genomics* 2:135–37
60. Gavin AC, Bosche M, Krause R, Grandi P, Marzioch M, et al. 2002. *Nature* 415: 141–47
61. Ho Y, Gruhler A, Heilbut A, Bader GD, Moore L, et al. 2002. *Nature* 415:180–83
62. Martzen MR, McCraith SM, Spinelli SL, Torres FM, Fields S, et al. 1999. *Science* 286:1153–55
63. Zhu H, Snyder M. 2001. *Curr. Opin. Chem. Biol.* 5:40–45
64. Zhu H, Klemic JF, Chang S, Bertone P, Casamayor A, et al. 2000. *Nat. Genet.* 26:283–89
64a. Malathi K, Xiao Y, Mitchell AP. 1999. *Genetics* 153:1145–52
65. MacBeath G, Schreiber SL. 2000. *Science* 289:1760–63
66. Zhu H, Bilgin M, Bangham R, Hall D, Casamayor A, et al. 2001. *Science* 293: 2101–5
67. McDonnell JM. 2001. *Curr. Opin. Chem. Biol.* 5:572–77
68. Nieba L, Nieba-Axmann SE, Persson A, Hamalainen M, Edebratt F, et al. 1997. *Anal. Biochem.* 252:217–28
69. Salamon Z, Brown MF, Tollin G. 1999. *Trends Biochem. Sci.* 24:213–19
70. Myszka DG, Rich RL. 2000. *Pharm. Sci. Technol. Today* 3:310–17
71. Sapsford KE, Liron Z, Shubin YS, Ligler FS. 2001. *Anal. Chem.* 73:5518–24
72. Houseman BT, Huh JH, Kron SJ, Mrksich M. 2002. *Nat. Biotechnol.* 20:270–74
73. Lizcano JM, Deak M, Morrice N, Kieloch A, Hastie CJ, et al. 2002. *J. Biol. Chem.* 277:27839–49
74. Wang D, Liu S, Trummer BJ, Deng C, Wang A. 2002. *Nat. Biotechnol.* 20:275–81
75. Houseman BT, Mrksich M. 2002. *Chem. Biol.* 9:443–54
76. Bishop AC, Ubersax JA, Petsch DT, Matheos DP, Gray NS, et al. 2000. *Nature* 407:395–401
77. Weiss EL, Bishop AC, Shokat KM, Drubin DG. 2000. *Nat. Cell Biol.* 2:677–85
78. Carroll AS, Bishop AC, DeRisi JL, Shokat KM, O'Shea EK. 2001. *Proc. Natl. Acad. Sci. USA* 98:12578–83
79. Habelhah H, Shah K, Huang L, Burlingame AL, Shokat KM, et al. 2001. *J. Biol. Chem.* 276:18090–95
80. Haab BB. 2001. *Curr. Opin. Drug Discov. Dev.* 4:116–23
81. Biesecker G, Dihel L, Enney K, Bendele RA. 1999. *Immunopharmacology* 42: 219–30
82. Chen J. 2002. *Chem. Biol.* 9:543–44
83. MacBeath G, Koehler AN, Schreiber SL. 1999. *J. Am. Chem. Soc.* 121:7967–68
84. Winssinger N, Ficarro S, Schultz PG, Harris JL. 2002. *Proc. Natl. Acad. Sci. USA* 99:11139–44
85. Kuruvilla FG, Shamji AF, Sternson SM, Hergenrother PJ, Schreiber SL. 2002. *Nature* 416:653–57
86. Blackwell HE, Perez L, Stavenger RA, Tallarico JA, Eatough EC, et al. 2001. *Chem. Biol.* 8:1167–82
87. Clemons PA, Koehler AN, Wagner BK, Sprigings TG, Spring DR, et al. 2001. *Chem. Biol.* 8:1183–95
88. Jansen R, Greenbaum D, Gerstein M. 2002. *Genome Res.* 12:37–46
89. Hughes TR, Marton MJ, Jones AR, Roberts CJ, Stoughton R, et al. 2000. *Cell* 102:109–26
90. Ideker T, Thorsson V, Ranish JA, Christmas R, Buhler J, et al. 2001. *Science* 292:929–34
91. Kemmeren P, van Berkum NL, Vilo J, Bijma T, Donders R, et al. 2002. *Mol. Cell* 9:1133–43
92. Xenarios I, Salwinski L, Duan XJ,

Higney P, Kim SM, et al. 2002. *Nucleic Acids Res.* 30:303–5
93. Mellor JC, Yanai I, Clodfelter KH, Mintseris J, DeLisi C. 2002. *Nucleic Acids Res.* 30:306–9
94. Bader GD, Hogue CW. 2000. *Bioinformatics* 16:465–77
95. Paton NW, Khan SA, Hayes A, Moussouni F, Brass A, et al. 2000. *Bioinformatics* 16:548–57
96. Qian J, Stenger B, Wilson CA, Lin J, Jansen R, et al. 2001. *Nucleic Acids Res.* 29:1750–64
97. Noble D. 2002. *Science* 295:1678–82
98. Davidson EH, Rast JP, Oliveri P, Ransick A, Calestani C, et al. 2002. *Dev. Biol.* 246:162–90
99. Davidson EH, Rast JP, Oliveri P, Ransick A, Calestani C, et al. 2002. *Science* 295:1669–78
100. Goodlett DR, Yi EC. 2002. *Funct. Integr. Genomics* 2:138–53

Annu. Rev. Biochem. 2003. 72:813–850
doi: 10.1146/annurev.biochem.71.110601.135450
Copyright © 2003 by Annual Reviews. All rights reserved

The Structural Basis of Large Ribosomal Subunit Function

Peter B. Moore [1, 2] and Thomas A. Steitz [1, 2, 3]
[1]*Departments of Molecular Biophysics and Biochemistry, and* [2]*Chemistry, Yale University, and* [3]*Howard Hughes Medical Institute, New Haven, Connecticut 06520; email: moore@proton.chem.yale.edu, eatherton@csb.yale.edu*

Key Words ribosomes, crystallography, peptidyl transferase, antibiotics

■ **Abstract** The ribosome crystal structures published in the past two years have revolutionized our understanding of ribonucleoprotein structure, and more specifically, the structural basis of the peptide bonding forming activity of the ribosome. This review concentrates on the crystallographic developments that made it possible to solve these structures. It also discusses the information obtained from these structures about the three-dimensional architecture of the large ribosomal subunit, the mechanism by which it facilitates peptide bond formation, and the way antibiotics inhibit large subunit function. The work reviewed, taken as a whole, proves beyond doubt that the ribosome is an RNA enzyme, as had long been surmised on the basis of less conclusive evidence.

CONTENTS

0066-4154/03/0707-0813$14.00

INTRODUCTION

Background

The last step in the gene expression pathway, protein synthesis, is executed by a formidable cellular apparatus. In addition to messenger RNAs for every protein currently in production, each cell contains transfer RNAs for (almost) every amino acid it uses to make protein, aminoacyl tRNA synthetases to charge those tRNAs, an array of facilitating protein factors, and finally ribosomes, the massive enzymes that catalyze mRNA-directed protein synthesis.

Elucidation of the chemical basis of protein synthesis has been a goal of biochemists for half a century. In its pursuit, crystallographers have determined the structures of many of the macromolecules involved. The crystal structures obtained for yeast phenylalanine tRNA in the 1970s were the first fruit of this enterprise (1, 2), and over the years many other important structures have been obtained, e.g., all but one of the aminoacyl tRNA synthetases and many of the protein synthesis factors. Long sought, but missing until 2000, were atomic resolution structures for the ribosome and its subunits. The crystal structures of the ribosomal subunits that have recently been determined are the subject of this review.

THE ROLE OF THE RIBOSOME IN PROTEIN SYNTHESIS Functionally, ribosomes are polymerases. Like DNA polymerase and RNA polymerase, they catalyze the synthesis of biopolymers of a single chemical class, and the sequence of the specific member of that class a ribosome makes is determined by its interaction with a nucleic acid template. The substrates ribosomes consume are aminoacyl tRNAs, their products are proteins, and their templates are messenger RNAs.

There are three binding sites for tRNA on the ribosome: an A site, to which aminoacyl tRNAs are delivered in an mRNA-directed fashion, a P site where peptidyl tRNAs reside, and an E site through which deacylated tRNAs pass as they are released from the ribosome (3). Nascent polypeptides are elongated by a cyclic process that starts with a molecule of mRNA bound to the ribosome, a deacylated tRNA in the E site, a peptidyl tRNA in the P site, and a vacant A site. In the first step, an aminoacylated tRNA whose anticodon is complementary to the mRNA codon presented in the A site is delivered to the ribosome by a protein factor that in prokaryotes is called EF-Tu, and the deacylated tRNA in the E site leaves the ribosome. Once delivery has occurred, the peptide esterified to the 3′-terminal A of the P-site bound tRNA is transferred to the amino group of the amino acid esterified to the 3′ terminal A of the tRNA in the A site, which elongates the nascent peptide chain by one amino acid. During the final step in the cycle, translocation, the deacylated tRNA in the P site moves to the E site, the peptidyl tRNA in the A site migrates to the P site, and the ribosome moves down the mRNA, in the 3′ direction, by 1 codon. In prokaryotes, translocation is catalyzed by a protein factor called EF-G. Once translocation has occurred, the ribosome can accept the next aminoacyl tRNA.

The ribosomes of all species are a 1:1 complex of two subunits of unequal size, the larger being about twice the mol wt of the smaller. During the initiation phase of protein synthesis, subunits are recruited from the pool of dissociated subunits in the cell with the aid of initiation factors. The product of these interactions is a ribosome with a mRNA bound so that the first codon in the message and the corresponding aminoacyl tRNA occupy the P site. Termination occurs when an elongating ribosome encounters a stop codon, which is a codon for which no cognate tRNA exists. Ribosomes stalled at stop codons are recognized by termination factors, which promote the hydrolysis of the ester bond linking the now completed polypeptide to its tRNA, the release of the bound mRNA, and the return of subunits to the cellular pool. Unlike the elongation cycle discussed above, which is effectively the same in all organisms, the mechanism of both initiation and termination differs substantially among kingdom to kingdom.

RIBOSOME STRUCTURE AT LOW RESOLUTION The two subunits of the ribosome perform distinctly different functions during protein synthesis. The small ribosomal subunit programs protein synthesis; it binds mRNA and mediates the interaction between mRNA codons and tRNA anticodons. The large subunit takes care of production; it contains the peptidyl transferase site, the site at which peptide bonds are formed. Consistent with their functions, the small subunit interacts with the anticodon-containing ends of tRNAs, and the large subunit interacts primarily with their CCA termini. There is an A site, a P site,and an E site on both subunits. Both subunits interact with the protein factors that facilitate ribosome function, and intersubunit interactions are important in all phases of protein synthesis.

The mol wt of the ribosome ranges from about 2.5 $x10^6$ in prokaryotes to about 4.5 $x10^6$ in higher eukaryotes, and the typical ribosome is roughly two-thirds RNA and one-third protein. The large subunit of the prokaryotic ribosome sediments at about 50S, and it contains one large RNA (23S rRNA), one small RNAS (5S rRNA), and about 35 different proteins, most of them in single copy.

The overall shape of the ribosome was determined by electron microscopy over 25 years ago (4) [see also (5)]. At ~40 Å resolution, the large subunit is (crudely) a hemisphere about 250 Å in diameter with three projections protruding radially from the edge of its flat face, a large one in the middle (the central protuberance), and two smaller ones at roughly 2 o'clock and 10 o'clock relative to the central protuberance. The central protuberance includes 5S rRNA and its associated proteins (6). Looking at the flat face of the subunit with its central protuberance up, the so-called crown view, the right protuberance includes ribosomal proteins L7/L12, and the left protuberance contains ribosomal protein L1 (7, 8).

Around 1995, the resolution of electron microscopic density maps of the ribosome improved dramatically as a consequence of the perfection of single particle image reconstruction techniques that could be applied to micrographs of ribosomes embedded in vitreous ice (9, 10). In addition to better delineating the morphology of the ribosome, these improved images proved that the large subunit contains a tunnel that runs from roughly the middle of its flat face to the side away from the small subunit in the 70S ribosome, as earlier electron microscopic studies had suggested (11–13, 13a). They also demonstrated that the peptidyl transferase site is located at the end of the tunnel closest to the small subunit in the ribosome (14–17).

The Scope of This Review

The first papers describing atomic structures of ribosomes derived from high resolution X-ray studies that appeared in August 2000, and since then many more have been published. The currently known structures include a 2.4 Å resolution structure of the large subunit from *Haloarcula marismortui* (18), a 3.0 Å resolution structure (19), and a 3.2 Å resolution structure (20) of the small subunit from *Thermus thermophilus*, and a 3.1 Å resolution structure of the large subunit from *Deinococcus radiodurans* (21). In addition, there are numerous structures of complexes of low mol wt substrate, inhibitor and antibiotic ligands bound to these ribosomal subunits, as well as a model of the 70S ribosome from *T. thermophilus* with mRNA, tRNA, and tRNA analogs bound derived from a 5.5 Å resolution electron density map (22).

This chapter begins with an account of how the crystal structures of ribosomes were determined, and it concludes with a review of the information that has emerged that is limited to the large subunit because it is impossible to do justice to both subunits and the 70S ribosome in a single review of this format. Additional information can be found in reviews that have appeared elsewhere in the last two years, e.g., (23–28).

RIBOSOME CRYSTALLOGRAPHY

The year the structure of the large ribosomal subunit from *H. marismortui* was published, it was the largest structure of an asymmetric assembly determined at atomic resolution by about a factor of 4. Ribosomes were, and still remain, challenging targets for crystallographic investigation.

Crystals

Crystals that diffract to resolutions of ~3.5 Å, or better, are the *sine qua non* for macromolecular crystal structure determination. The first three-dimensional crystals of ribosome subunits were grown in the laboratory of H.G. Wittman in 1980 by Yonath et al. from the 50S subunits of *Bacillus stearothermophillus* (29). Although they diffracted only to low resolution, their existence suggested that X-ray studies of the ribosome might be possible and motivated the search for better crystals. The crystals of 30S subunits (and also 70S ribosomes) from *T. thermophilus*, first prepared in Pushchino in 1987 (30) and in Berlin at about the same time (31), were the progenitors of those that yielded structures, but they too diffracted only to 10–12 Å resolution (32, 33). Likewise, the crystals of the large ribosomal subunit of the archeon, *H. marismortui*, diffracted poorly when first grown in Berlin in 1987 (34) but were gradually improved, and by 1991 crystals that diffracted to 3 Å resolution had been obtained (35). The existence of these crystals showed that an atomic structure was in principle possible, but not until a method could be found for the phasing of diffraction patterns.

Although the *H. marismortui* 50S crystals diffracted well, they had some serious flaws. A number of their defects have been described by Yonath and coworkers (36). They include "... severe nonisomorphism, high radiation sensitivity, ... nonuniform mosaic spread, uneven reflection shape, and high fragility." Later Harms et al. (37) pointed out other problems, "... the unfavorable crystal habit (plates, made of sliding layers, reaching typically up to 0.5 mm^2 with an average thickness of a few microns in the direction of the *c* axis), and variations in the *c* axis length (567–570 Å) as a function of irradiation."

The Yale group discovered yet another problem, crystal twinning (38). Depending on how these crystals are handled after they have grown, a 10 Å shift in the relative positions of subunits may occur that changes their space group symmetry from orthorhombic, $C222_1$, to monoclinic, $P2_1$, with almost no alteration in unit cell dimensions, or crystal cracking, and, surprisingly, no alteration in the symmetry of the crystal's diffraction pattern. The reason for the latter is that the direction of the packing shift varies randomly from one mosaic block to the next within a crystal, and hence crystals in which it has occurred are twinned.

These crystal pathologies were all eventually overcome. Crystal to crystal isomorphism was improved, twinning was suppressed using an appropriate stabilizing solution, and crystal thickness was increased to between 0.1 and 0.2 mm using a reverse extraction procedure for crystal growth that also increased crystal strength and singularity (18). The result was extension of the maximum resolution of the diffraction patterns of these crystals to 2.2 Å.

The reports of the two groups that solved the structure of the 30S subunit from *T. thermophilis* indicated that the crystallization issues they overcame were less dramatic, albeit no less essential for success (39, 40). A full account of how the 70S crystals were grown that were used to obtain the 5.5 Å resolution structure referred to above has yet to be published. Interestingly, the paper that reports the crystal structure of the *D. radiodurans* large subunit structure is also the first report of the existence of the corresponding crystals (21).

Crystallography

Although crystallographic technology was perhaps not up to the task of ribosome structure determination when the first crystals were grown in 1980, 10 years later, when crystals that diffract to high resolution had been obtained, all the X-ray and computational technologies required were in place: synchrotron radiation, crystal freezing, phosphorimaging detection, and the requiste computers and programs. The strategy for solving the structure of the large ribosomal subunit from *H. marismortui*, in brief, was to begin at extremely low resolution using data from 100 to 16 Å resolution, where the scattering power of heavy atom cluster compounds is enhanced by up to 100-fold, and to check the validity of heavy atom positioning with phases derived from a 20 Å resolution cryo-electron microscopy (EM) reconstruction. This led to the first ribosome electron density map, at 9 Å resolution, that showed density corresponding to recognizable macromolecular features (46). This map provided the means to proceed to progressively higher resolution. A more complete description of the trials and tribulations that led from the first crystals to the first atomic structure follows.

X-RAY SOURCES AND COMPUTERS While it is certainly the case that the ribosome crystal structures could not have been solved without the use of synchrotron X-ray sources, progress was never impeded by the lack of synchrotron beam lines of appropriate quality. The X-ray data used in the Yale 5 Å resolution map of the *H. marismortui* 50S subunit, for example, were measured on a bending magnet beamline at Brookhaven National Lab, X12B, which produces what is by today's standards a quite weak X-ray beam, and data for a 3 Å resolution map that allowed complete interpretation of the electron density corresponding to 23S rRNA were measured at beamline X-25 using a phosphorimaging plate detector.

Similarly, ribosome crystallography was not hampered by the lack of computational tools any more than virus crystallography had been. The supercomputers available in the early 1990s would have sufficed to do all the computation needed to solve ribosome crystal structures, and by the late 1990s all of the computation required could be executed in a few days on a high-end workstation.

DETECTORS Because the unit cell dimensions of the virus crystals are comparable to those of ribosome crystals, and because the first virus structures were solved in the late 1970s and 80s, it could be argued that the determination of ribosome crystal structures was technically feasible decades before it was accomplished. However, the low quality of the data obtainable using film-based data collection methods, which were limited by the high background and the narrow range of linearity, would have made it extremely difficult, since very small isomorphous and anomalous diffraction intensity differences had to be measured with high accuracy. The limitations of film were far less important for the determination of virus structures because the internal symmetry of viruses makes it possible to use powerful averaging methods for improving electron density maps.

Both charge coupled device (CCD) and phosphorimaging plate X-ray detectors, which came into use in the 1990s, can measure diffraction data with the requisite accuracy, though the CCD detectors are far faster. The disadvantage of CCD detectors, for ribosome crystallography in the 1990s derived from their small size. Even the CCD detector available at ID19 at the Advanced Photon Source (APS) at Argonne National Laboratory, which was exceptionally large, limited data collection from the *H. marismortui* large subunit crystals by the Yale group to under 2.4 Å resolution. The speed of data collection was stunning, however. More than 6 million reflections were measured in a few hours, which is a data collection rate about 10^5 times faster than could be achieved in the 1960s using state of the art laboratory equipment and crystals of molecules 50 times smaller.

CRYSTAL FREEZING The development of ways to freeze crystals and to collect data from them at -140°C was one of the most important developments in macromolecular crystallography of the last decade or so because it enabled the use of the high intensity synchrotron X-ray sources that were necessary for solving the structure of the ribosome and for the rapid solving of other protein and nucleic acid structures. At room temperature, the crystals of all biological macromolecules wither due to radiation damage when exposed to the intense and tunable X-ray beams that are necessary for crystallographic studies of large structures and for MAD phasing. The in-beam lifetime of crystals is greatly enhanced if they are flash frozen before exposure and remain frozen during data collection.

The first successful X-ray diffraction experiments done with frozen crystals were executed in 1970 by Rossmann and colleagues using crystals of lactate dehydrogenase that had been immersed in sucrose solutions (40a). Because the benefits of freezing were not commensurate with the technical difficulties involved at the time, the method did not come into general use. Nearly two decades later, after the development of synchrotron X-ray sources had brought the issue of radiation damage to the fore, Hope devised a method for flash-freezing protein crystals in liquid nitrogen (40b) and then, in collaboration with Yonath, demonstrated its efficacy using ribosome crystals (41). Ribosome crystals are so sensitive to radiation damage that data cannot be collected from any of them effectively unless they are frozen, and even when frozen, some are still significantly damaged by short exposures to synchrotron radiation.

PHASING STRATEGIES Once crystallographically suitable crystals have been obtained, the major barrier to solving any crystal structure is determining the phases that are associated with its X-ray diffraction amplitudes. Although the ribosome is not larger than the viruses whose structures were solved in the late 1970s and early 1980s, the lack of internal symmetry in the ribosome precluded the use of averaging methods for phase (map) improvement that have been so essential for the determination of virus structures. The only phasing method that could be used for the ribosome was heavy atom-multiple isomorphous replacement appropriately combined with anomalous scattering.

To obtain measurable diffraction intensity changes from heavy atom derivatized crystals of such a large asymmetric assembly, either a large number (50 to 100) of single heavy atoms or a smaller number (1 to 10) of heavy atom cluster compounds need to be bound and their positions in the crystal located. The crucial step in the process of obtaining an interpretable electron density map is determining the positions of the bound heavy atoms, and when the number of heavy atom sites is large, it is difficult to accomplish using standard difference Patterson methods.

The strategy used to solve the phasing problem for ribosome crystals involved multiple steps (18, 38, 46). The objective of the initial step was to correctly determine the positions of heavy atom cluster compounds at very low resolution (20Å), which were then used to produce reliable phases to 5 Å resolution. These low resolution phases were then used to locate the positions of heavy atoms in derivatized crystals containing a large number of single-atom heavy atom molecules. The difference Fourier methods enabled by low resolution phases are effective for locating large numbers (~100) of heavy atom positions. Once this was achieved, phasing could then be extended to high resolution. Phases to 20 Å resolution were obtained by two methods: isomorphous replacement using heavy atom cluster compounds and molecular replacement using electron microscopic images. All of the de novo ribosome structure determinations reported so far have used similar phasing strategies.

Heavy atom cluster compounds are particularly powerful at very low resolutions (out to 20 Å) because they produce extraordinarily large diffraction intensity changes. One of the cluster compounds used to prepare isomorphous derivatives of *H. marismortui* large subunit crystals, for example, contained 18 tungsten atoms. At 20 Å resolution the intensity change, $|F|^2$, produced by one W18 molecule bound per asymmetric unit approximates what would be observed if 324 (= 18^2) tungsten atoms were bound at random positions. It scatters like a super heavy atom containing 18x74 electrons. Not only is the intensity change produced by one to five such cluster compounds easy to observe, but the low resolution difference Patterson maps that emerge are easy to interpret. A derivative containing one to five single heavy atoms would generally not produce measurable intensity changes with the ribosome. It should be noted, however, that the intensity changes produced by heavy atom cluster compounds fall off very rapidly with resolution due to interference effects, and at resolutions approaching that of radius of the cluster compound and beyond (~ 10 Å), the changes will be less than that of the same number of heavy atoms bound at random sites. Even if cluster compound data do not yield high resolution phases, the low resolution phase information provided can be used to obtain the positions of multiple site, single heavy atom compounds by difference Fourier methods, which will work with even more than 100 sites.

Electron density maps derived by electron microscopy can also be used to obtain very low resolution phases for X-ray diffraction patterns and are very useful for validating the positions of heavy atom clusters that have been determined by other means. In the 1970s, Harrison and coworkers used the three-dimensional images of viruses derived from electron micrographs of negatively stained particles to phase viral X-ray diffraction patterns at low resolution by molecular replacement and used these phases to locate heavy atoms (42). The three-dimensional cyro-EM images of ribosomes produced by the mid-1990s (see above) were of a substantially higher quality and resolution and were useful in producing phases to validate heavy atom positions (46).

SOLVING THE STRUCTURE OF THE *H. MARISMORTUI* LARGE SUBUNIT The first attempts to phase ribosome diffraction patterns were made by the Yonath group using heavy atom cluster compounds, but the electron density maps that resulted did not show the continuous density features characteristic of RNA and protein (36, 37, 43, 44), and it appears that the subunit packings inferred from solvent flatterned maps presented for both the *H. marismortui* large subunit and the *T. thermophilus* small subunit (45) were incorrect because they are not the same as those deduced for the same crystals subsequently (18, 19, 46).

The first heavy atom-derivatized ribosome crystal in which heavy atom positions were successfully determined was obtained by soaking a cluster compound containing 18 tungsten atoms into crystals of the *H. marismortui* large ribosomal subunit (46). Fortuitously, this compound binds to those crystals predominantly at a single location, which simplified determination of its position. It was located initially by difference Patterson methods, and its location was confirmed later by a difference electron density map computed using low resolution X-ray diffraction amplitudes and phases obtained by molecular replacement. The model used for molecular replacement was a 20 Å resolution cryo-electron microscopic reconstruction of the *H. marismortui* 50S subunit provided by Frank and colleagues. By far the largest feature in the difference map that emerged was a 13 σ peak at the position predicted from difference Patterson maps that was 6 σ higher than the next highest peak, which turned out to be that of another site with lower occupancy (46). Molecular replacement was also used to understand the twinning problem discussed above.

The first electron density map published of any ribosome or ribosomal subunit that showed features clearly interpretable in molecular terms was a 9 Å resolution map of the *H. marismortui* large ribosmal subunit (Figure 1*a*), which included rods of density of the appropriate size that show the right handed but irregular twist characteristic of RNA duplex (46).

Because the heavy atom cluster compound derivatives obtained for *H. marismortui* large subunit crystals had little phasing power beyond about 5 Å resolution, experimental phase extension depended on derivatives prepared using single-atom, heavy-atom compounds such as osmium pentamine and iridium hexamine. The positions of the more than 100 osmium and iridium atoms bound

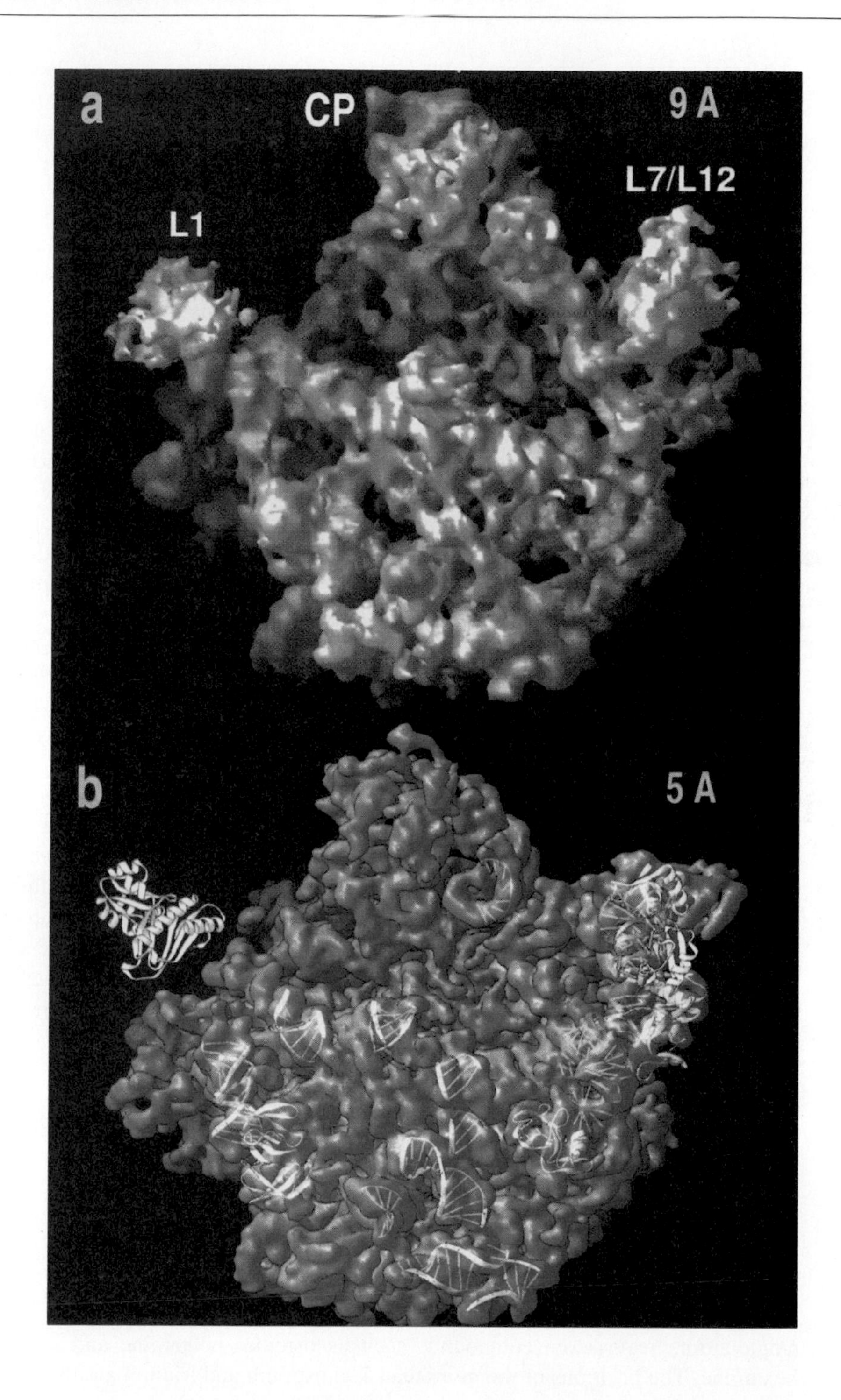
a
CP
9 A
L7/L12
L1
b
5 A

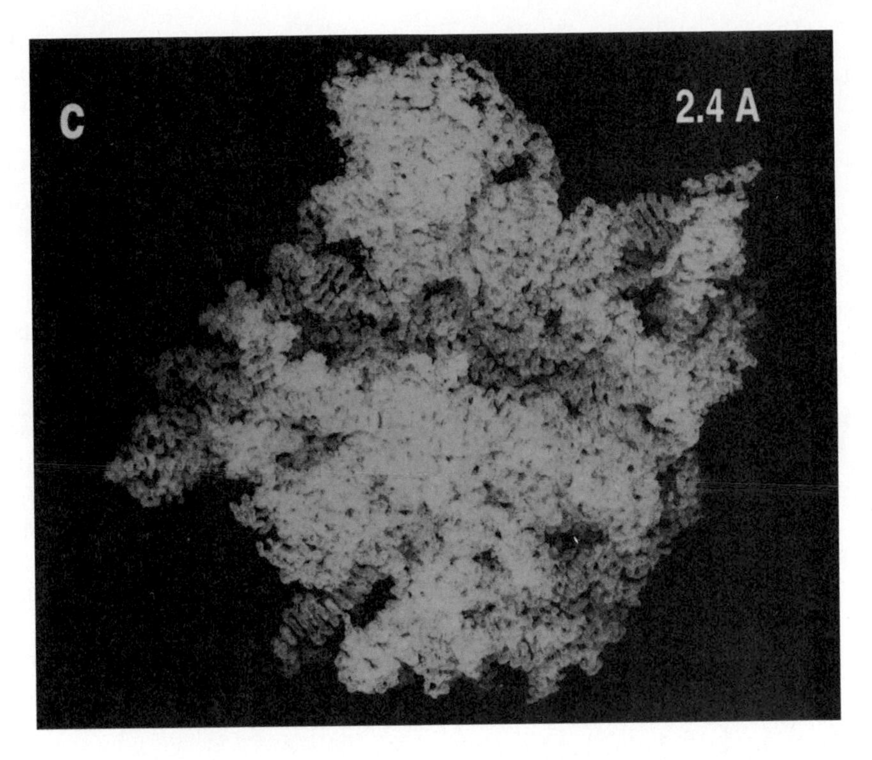

Figure 1 Continued

in the unit cell were located using difference Fourier maps computed employing the 5 Å resolution phases previously determined. Only at this stage in the investigation, when the structure was already solved in principle, did it become necessary to collect data using high intensity beam lines. The high resolution data produced by two trips to high brilliance beam lines resulted finally in a 2.4 Å resolution electron density map of the *H. marismortui* large ribosomal subunit.

The unusually high quality of the final, experimentally phased electron density map of the *H. marismortui* large subunit is due to several factors. It is attributable in part to the well-collimated, high intensity, 80 μ diameter X-ray beam used to collect data at beam line ID19 and the small pixel size and relatively large size of the CCD detector used, which made it possible to collect very accurate data.

Figure 1 The appearance of the large ribosomal subunit from *H. marismortui* in electron density maps at different resolutions. The subunit is shown in the crown view at (*a*) 9 Å resolution (46), (*b*) 5 Å resolution (38), and (*c*) 2.4 Å resolution (18). CP designates the central protuberance. The L1 stalk, which is visible at low resolution, disappears as resolution improves. [Reproduced with permission from (46a).]

The solvent flipping density modification (46b) program as implemented in the Crystallography and NMR System (CNS) program package proved remarkably effective at improving both the quality and resolution of the heavy atom phased maps (Figure 1*c*).

SOLVING THE STRUCTURE OF THE SMALL SUBUNIT The first clear indication that a structure would soon be obtained for the small subunit was the partially interpreted 5.5 Å resolution electron density map of the 30S subunit from *T. thermophilus* presented by Ramakrishnan and colleagues at a meeting in the summer of 1999 (47) and published soon thereafter (48). A year later, the same group published a fully interpreted structure for the 30S subunit at 3.05 Å resolution (19). Their structure appeared shortly after a paper from the Yonath group describing a less accurate and less completely interpreted 3.3 Å resolution electron density map of the small subunit (40). The differences between the two structures have been resolved in favor of the Ramakrishnan structure (20). Initial low-resolution phases for both of these 30S subunit structures were obtained using heavy atom cluster compounds (39, 40). The extension of the phasing of the Ramakrishnan 30S map to 3 Å resolution depended on data from multiple site, single, heavy atom derivatives (39) including good anomalous scattering measurements. Remarkably, one of the tungsten cluster compounds employed by the Yonath group in their studies of the small subunit studies contained one well ordered site that produced useful phasing extending to 3.3 Å resolution (40).

OTHER RIBOSOMAL STRUCTURES The crystal structure of the 70S ribosome derives from a phasing process that also began at very low resolution with molecular replacement using EM-derived ribosome images to assist in heavy atom locations in derivatives that ultimately extended the resolution to 5.5 Å (22, 50). As pointed out earlier, a resolution of 5.5 Å is well below that needed for the independent interpretation of electron density maps. Nevertheless, most of the sequence of 16S rRNA was fit into the electron density of the small subunit part of the map, albeit less accurately, without reference to other structures, and the large subunit electron density was interpreted using the Yale *H. marismortui* large subunit structure as a guide. At this resolution, the positions of bases cannot be seen, and 50S proteins of unknown structure could not be solved. However, the resulting 70S structure provides important information about the conformational differences between free subunits and subunits in 70S couples, the structures of the connections between the two subunits, and the placement of mRNA and tRNAs on the intact ribosome.

In the last three years, the Yonath group has obtained structures for two different large ribosomal subunits. The first was a 3.6 Å resolution structure of the large subunit from *H. marismortui* derived from crystals similar to those used by the Yale group, but it was stabilized in a buffer containing more K^+ and less Na^+ (52). This structure, which was obtained from a map calculated using a phasing strategy that combined molecular replacement using the Yale structure and anomalous scattering from isomorphous derivatives, is reported to differ from the Yale structure in several ways. First, the Yale structure is reported to be systematically somewhat larger. Second, the Yonath subunit includes structures for the 23S rRNA sequences omitted from the Yale structure because of their low visibility. Third, the conformations found for some of the proteins are different. Because a full description of this structure has not appeared and its coordinates have not been released, it is impossible to comment on it further. However, since the *D. radiodurans* 50S subunit and the large subunit of the 70S structure are the same size as the Yale *H. marismortui* structure and homologous proteins have similar conformations, it is unclear why the *H. marismortui* 50S structure reported by the Yonath group is so different.

The Yonath group has also solved the structure of the 50S subunit from *D. radiodurans* (21). Phases were obtained initially by molecular replacement using the Yale *H. marismortui* structure and then improved using heavy atom isomorphous replacement and anomalous scattering data. The latest version of this structure (PDB # 1LNR) has been refined to a free R of 27.9%. However, for the purposes of refinement, the atoms of each nucleotide were divided into 2 groups, a ribose group and a base + phosphorus group, which were assigned independent B-factors. The B-factor variations in 1LNR between groups of atoms within single nucleotides, and from one nucleotide to the next are too large to be physically meaningful, and their existence indicates that the structure is not fully refined. Refinement limitations may be the root source of the controversy that has arisen about the structures of some of the proteins in earlier releases of this structure (53, 54).

THE ARCHITECTURE OF THE LARGE RIBOSOMAL SUBUNIT

Below we discuss the structure and function of the large subunit, using the Yale *H. marismortui* large subunit crystal structure (PDB # 1JJ2) as the standard. The other structures of the large subunit that have been determined are the large subunit part of the *T. thermophilus* 70S ribosome (22), the *D. radiodurans* large ribosomal subunit (21), and the *H. marismortui* large ribosomal subunit examined at higher K^+ concentrations (52). These structures are similar but not identical in every detail, and unfortunately, it is hard to interpret the sources of the differences. The refinement problem evident in the *D. radiodurans* structure raises questions about its coordinate accuracy. The 70S-derived large subunit structure is not fully independent because it was derived originally from the standard *H. marismortui* structure, and the 5.5 Å resolution of the map is not sufficient to enable a reliable modeling of RNA differences or interpretation of the uniquely eubacterial proteins, which are not fully interpreted. In addition, because *H. marismortui* is archaeal, and the other two species are eubacterial, there are substantial sequence differences in both RNA and protein. Finally, both of the eubacterial structures were determined at moderate ionic strength, although the standard structure, which derives from an extreme halophile, was obtained from crystals in which the ionic strength is about 3 M.

One difference between the standard *H. marismortui* structure and the others is that all the latter include many of the parts of the large ribosomal subunit omitted from the standard structure because of local disorder. The visibility of these components in the high-K^+ *H. marismortui* structure is particularly puzzling. The Yale group has determined the structure of *H. marismortui* large subunit crystals in which the predominant cation is K^+, not Na^+, and finds the visibility of these regions unaffected (J.L. Hansen, unpublished results). In thinking about this problem, it should not be forgotten that the lower the resolution of an electron density map, the easier it is to visualize its less well ordered regions. The features missing from the Yale structure at 2.4 Å resolution were visible in its 9 Å resolution electron density maps (46), and the high-K^+ *H. marismortui* structure was determined at 3.6 Å resolution.

All of these structures support two important conclusions. First, the shape of the large subunit, like that of the small ribosomal subunit, is determined by its RNA; pictures of the large subunit drawn with its proteins omitted look just like the pictures obtained when they are included. Second, the large subunit is monolithic. Except for its two lateral protuberances, which are flexibly attached to the rest of the subunit by single RNA stems, the object has no obvious morphological subdivisions.

RNA Structure in the Large Subunit

The secondary structure of 23S rRNAs has been investigated by comparative sequence analysis since the early 1980s (55, 56). The crystal structures of the large subunit prove that the secondary structure model obtained by sequence comparisons was remarkably accurate. Better still, the overwhelming majority of its errors proved to be pairings that exist but were omitted from the model because they had not been detected, rather than pairings included in the model that do not exist.

DOMAINS IN 23S rRNA Like all large RNAs, 23S rRNA secondary structure is discussed in terms of domains, a concept that also works well in three dimensions for 16S rRNA and many other RNAs, but it is less useful here. RNA secondary structure can be represented as a succession of stem/loops; a domain is a stem/loop that has a loop so large it contains stem/loops of its own. Implied in the use of the word "domain" to describe such structures is the notion that they will fold properly in isolation. There are 11 independent stem/loops in the secondary structure of 23S rRNA, 6 of which meet the domain definition. The remaining five are stems capped by loops too small to contain stem/loops of their own. In addition, as is the case with many other large RNAs, the 5′ and 3′ terminal sequences of 23S rRNA form a helix that, formally, makes the entire molecule a single domain. Nevertheless, 23S rRNAs are deemed to consist of six domains, one for each of the stems that has a hypertrophied loop, and its five simple stem/loops are considered to be components of the complex stem/loops they adjoin (55). The assignment of the simple stem/loops to domains that resulted did not work out in every case. For example, helix 25 of 23S rRNA, which is one of the simple stem/loops, has been considered to be part of domain 1, but its interactions in the three-dimensional structure of the ribosome clearly indicate that it belongs to domain 2 (18). 5S rRNA is effectively the large subunit's seventh RNA domain.

There is some evidence that the domains of 23S rRNA can fold autonomously, but there is also reason to believe the significance of these observations is not as deep as it is for the domains of 16S rRNA. Oligonucleotides having sequences corresponding to domains I, IV, and VI of 23S rRNA appear to assume ribosome-like secondary structures independently, and there are indications that their tertiary structures are also native-like in isolation (57–59). In addition, in isolation, all six domains form complexes in vitro with at least some of the proteins that bind to them in the ribosome (60). Nevertheless, as already noted, the large ribosomal subunit is not divided into morphologically distinct domains. The secondary structure domains in 23S rRNA interact extensively with each other in the intact ribosome, and the intimacy of their interdomain interactions are indistinguishable from that of their intradomain interactions.

SECONDARY STRUCTURE MOTIFS IN rRNAs An RNA secondary structure motif is a specific sequence or family of sequences or set of non-Watson-Crick base-base juxtapositions that occurs with an appreciable frequency in RNAs generally and that has a distinctive conformation, independent of context. Prior to 2000, fewer than 10 RNA motifs were known, e.g., the GNRA tetraloop and the bulged G motif (61, 62). Examples of all of the previously known motifs can be found in rRNA.

It seemed possible in 2000 that the enormous increase in the amount of RNA structure known at high resolution embodied in the ribosome crystal structures would lead to the discovery of many new secondary structure motifs. However, only one entirely new motif has emerged so far, the kink-turn (63). A motif found previously in tRNA, the T-loop (64) is now seen to be more generally observed, and a secondary structure feature, the hook-turn (65), has been recognized. The small size of the harvest suggests that our knowledge of RNA secondary structure is close to complete.

Kink-turns, or K-turns, are asymmetric internal loops embedded in RNA double helices (63) (Figure 2). Each asymmetric loop is flanked by C-G base pairs on one side and sheared G-A base pairs on the other, with an A-minor interaction between these two helical stems (Figure 2). A consensus sequence and secondary structure derived from the 9 K-turns in the ribosome have a 3-nucleotide loop and 10 consensus nucleotides out of 15 that predict its presence in at least 5 other RNAs. In three dimensions, the most striking feature of a K-turn is the sharp bend in the phosphodiester backbone of the 3-nucleotide

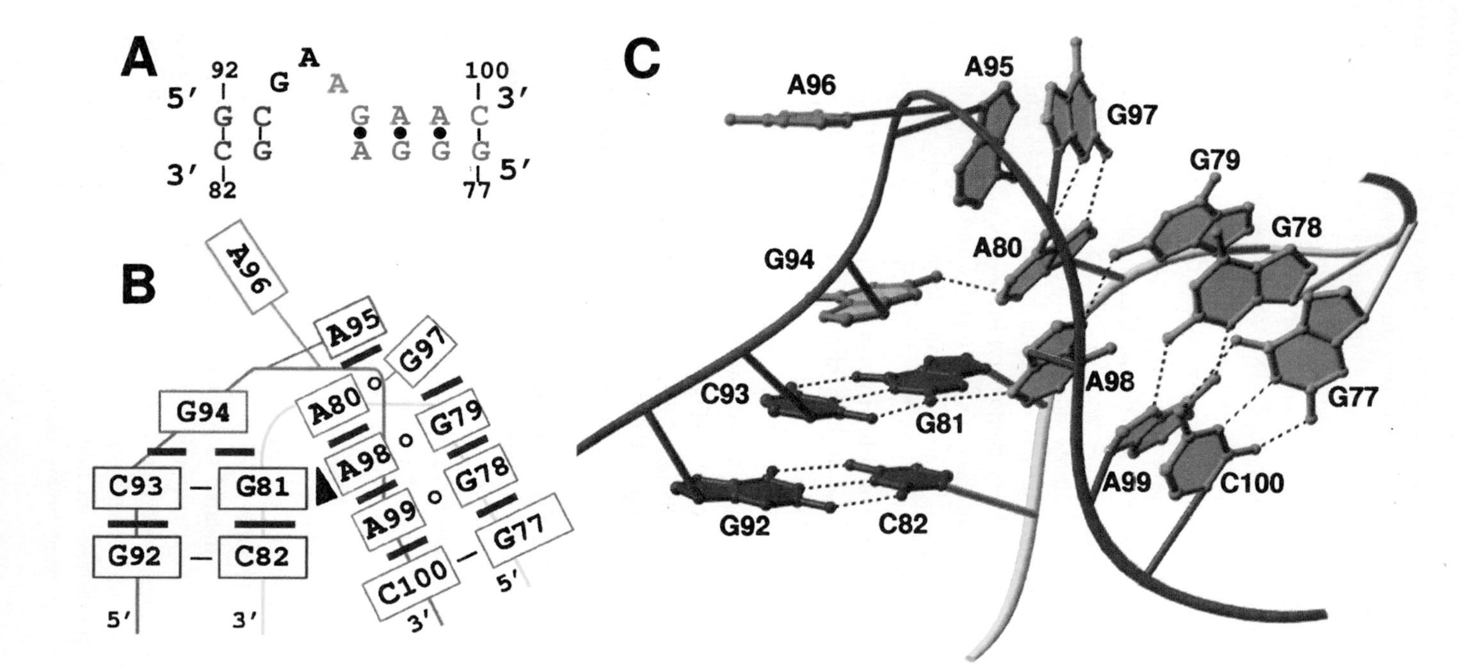

Figure 2 The structure of kink-turn 7 in the 23s rRNA of the *H. marismortui* ribosome. (*a*) The secondary structure of KT-7. The C-stem is red, the NC-stem blue, and the bulged nucleotide is green. (*b*) Base pairing and stacking interactions in KT-7. The black triangle identifies an A-minor interaction. (*c*) KT-7 in three dimensions. The backbone of the kinked strand is orange, and that of the unkinked strand is yellow. Dashed lines indicate hydrogen bonds. [Reproduced with permission from (63).]

bulge that leaves its third base pointing out into solution and results in an ~60° difference in the orientation of the axes of the two flanking helices. Five of the 6 K-turns in the *H. marismortui* large subunit interact with 9 of the 27 proteins visible in that structure, but they interact with proteins of unrelated structures in entirely different ways.

The T-loop motif is a terminal loop structure that has been known for years (64). The TψC-loop of every tRNA adopts that conformation. What had not been previously realized is that T-loops occur in other contexts, e.g., in rRNAs. There is some variation among T-loops, but in RNA secondary structure diagrams, they are usually seven-base loops. The 5′ base of the loop, which is always a U, forms a Hoogsteen pair with the A at the fifth position. That pair stacks on the Watson-Crick pair that terminates the stem, leaving the bases 3′ to that A bulged out. A base from another part of the molecule inserts between the same A and the base at the fourth position.

Hook-turns are found at places where the two strands of a helix are separating to interact with other sequences (65). Their conformations are not determined entirely by local interactions, and their consensus is short enough to make them difficult to identify in secondary structure diagrams so that their status may not quite rise to the level of a bona fide motif. Nevertheless, they tend to follow helices that end with a GC base pair and an AG juxtaposition (65), and in the space of two nucleotides, the direction of the backbone of the strand 3′ to the G changes about 180°. The first nucleotide in the sequence that forms the bend is often an A.

LONG-RANGE INTERACTIONS IN 23S rRNA Prior to the determination of the structure of the ribosome, little was known about long-range RNA-RNA interactions because so few of the RNA structures known were large enough to have tertiary structures. It was understood, of course, that base pairing between remote single stranded bases stabilizes the tertiary and quaternary structure of RNAs, and two tertiary/quaternary structure motifs had been identified: the tetraloop-tetraloop receptor motif (66) and the ribose zipper (67). These interactions all occur in the ribosome.

So far, only one new long-range RNA interaction motif has been discovered: the A-minor motif (68) (Figure 3). This motif was uncovered by examining the reason for the large number of highly conserved A residues in 23S rRNA, particularly in single stranded regions. This motif is termed "A-minor" because it involves the insertion of the smooth minor groove (C2-N3) edges of adenines into the minor grooves of neighboring helices, preferentially at C-G base pairs where they form hydrogen bonds with one or both of the 2′ OHs of those pairs. In the most abundant and presumably most stable type I interaction, the inserted A forms a triple base pair with the G-C having its N3 hydrogen bond to the N3 of the G and its 2′ OH hydrogen bond to the O2 of the C (Figure 3). Less intimate and less abundant A-minor interactions termed types O, II, and III also occur,

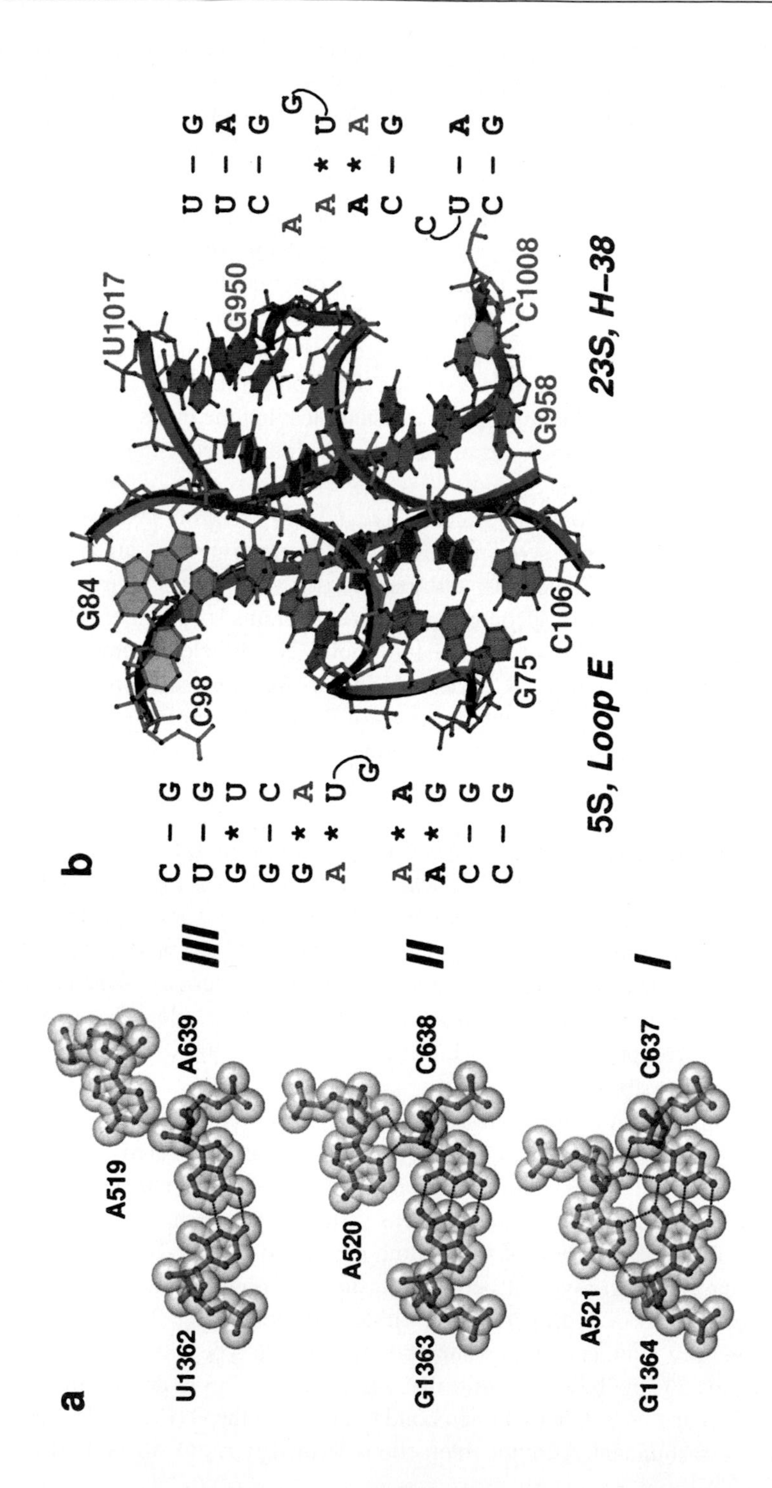
a
b
III
II
I
U1362
A519
A639
G1363
A520
C638
G1364
A521
C637
G84
C98
G75
C106
U1017
G950
G958
C1008
5S, Loop E
23S, H–38

usually in sequential association with type I interaction as short runs of As (up to 3), which stack on each other to form what is termed an "A-patch" (68).

These A-minor interactions often involve As in single-stranded regions, but the As in a shared GA pair that is followed by a reverse Hoogsteen UA may also participate because even these paired As are accessible for A-minor interactions (68). The A-minor motif is by far the most abundant tertiary structure interaction in the large ribosomal subunit; 186 adenines in 23S and 5S rRNA are involved compared with about 100 long-range base pairs. As vital structural elements, they are highly conserved; 99 of the 170 A-minor interactions in 23S RNA are greater than 90% conserved across all kingdoms.

In addition to stabilizing specific tertiary structure, A-minor interactions appear to play functionally significant roles in the ribosome. tRNAs are involved in functionally significant A-minor interactions in both the large and small subunits. The 3′-terminal A76 of tRNA molecules bound to both the A site and the P site make type I interactions with 23S rRNA (68, 70). In the small subunits, type I A-minor interactions made by A1492 and A1493 play a critical role in messenger decoding by assuring Watson-Crick pairing between the first and second bases of the codon and the appropriate anticodon (71).

METAL IONS Specifically bound metal ions play an important role in stabilizing the compact tertiary structure of the large 23S and 5S RNA polyanions; these account for the well known capacity of Mg^{2+} ions to stabilize RNA structure. The Yale *H. marismortui* large subunit structure is the only one determined at high enough resolution to allow identification of metal ion interactions with a high degree of confidence. The current structure [PDB# 1JJ2; (63)] includes 88 monovalent cations, 117 Mg^{2+} ions, 22 Cl^- ions, and 5 Cd^{2+} ions. The Cd^{2+} ions included in crystallization buffers replace the Zn^{2+} in the several zinc-finger proteins contained in the *H. marismortui* large subunit (18). Mg^{2+} ions are seen in this structure in regions where the local density of phosphate groups is unusually high, and there they help stabilize structure by neutralizing charge. (D. Klein, T.M. Schmeing, P.B. Moore, and T.A. Steitz, in preparation).

Proteins of the Large Subunit

Following the initial reports on the structures of proteins and their interactions with each other and with rRNA in the large (18, 21) and small (19) subunits,

Figure 3 The A-minor motif. (*a*) Examples of the three most important kinds of A-minor motifs from the 23S rRNA of *H. marismortui*. Types I and II are A-specific. Type III interactions involving other base types are seen, but A is preferred. (*b*) The interaction between helix 38 of 23S rRNA and 5S rRNA in *H. marismortui*. The only direct contact between these two molecules includes six A-minor interactions, involving three As in 23S rRNA and 3 As in 5S rRNA that are symmetrically disposed. Secondary structure diagrams are provided for the interacting sequences. [Reproduced with permission from (68).]

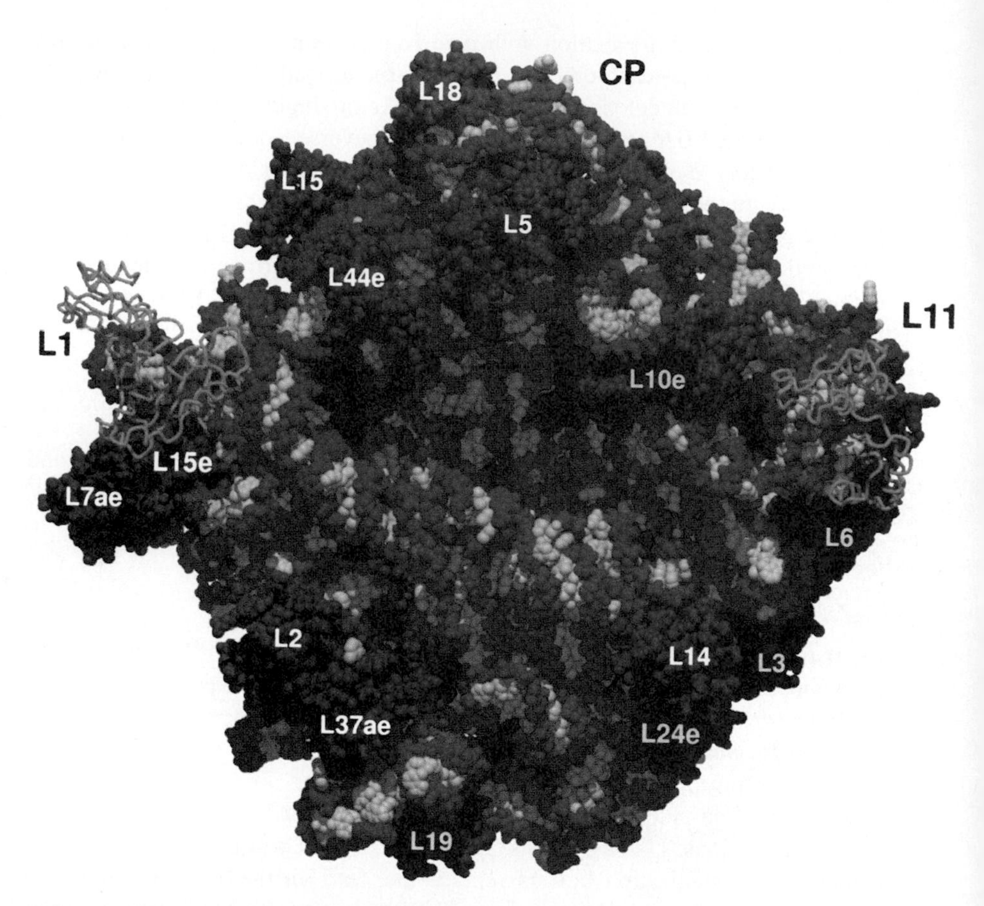

Figure 4 A space filling model of the large ribosomal subunit from *H. marismortui* with a transition state analog bound viewed down the active site cleft. Bases are white, the sugar-phosphate backbone is orange, and the substrate analog (in the center) is red. Proteins whose structures are defined by the 2.4 Å resolution map are blue. Cyan ribbon represents proteins whose structures are independently known that have been positioned approximately using lower resolution electron density maps. Identification numbers are provided for all proteins, and CP designates the central protuberance. [Reproduced with permission from (70).]

more complete analyses have been achieved (72; D. Klein, T.M. Schmeing, P.B. Moore, and T.A. Steitz, in preparation). The major message conveyed is that the primary function of ribosomal proteins is stabilization of rRNA structure, though some are also involved in functionally important interactions with protein factors.

Except for the four proteins that form the tips of the two lateral protuberances, the proteins of the large subunit do not extend significantly beyond the envelope defined by the RNA (Figure 4) (18). Their globular domains are on the exterior of the particle, often nestled in the gaps and crevices formed by the RNA, thus acting like mortar filling the gaps and cracks between RNA bricks. The distribution of proteins on the subunit surface is nearly uniform, except for the active site cleft and the flat surface that interacts with the 30S subunit, places where they are notably absent.

All of the proteins in the particle, except L12, interact directly with RNA, and all but 7 of the remaining 30 proteins interact with 2 rRNA domains or more (18). Protein L22 interacts with RNA sequences belonging to all six domains of 23S rRNA. About one third of the average protein surface area becomes buried when the subunit forms from its isolated but fully structured components, an average of 3000 Å^2 of protein surface area buried per protein, which is comparable to the 2,700 Å^2 of glutaminyl-tRNA synthetase that interacts with $tRNA^{Gln}$. Presumably, this enormous contact surface area is essential to energetically compensate for the considerable conformational entropy associated with both a flexible rRNA and the protein tails, which is lost upon subunit assembly.

Unlike proteins that bind to specific DNA sequences, the majority of ribosomal proteins bind to specific locations by recognizing unique RNA shapes through interactions largely with the sugar phosphate backbone rather than through interactions with bases via the major or minor grooves. Proteins that share sequence and structural homology, such as L15 and L18e, interact with their RNA targets in very similar ways. However, unrelated proteins that contain similar structural motifs, such as the RRM-like folds or SH3-like barrels, interact with rRNA in unrelated ways (D. Klein, T.M. Schmeing, P.B. Moore, and T.A. Steitz, in preparation).

PROTEIN TAILS Of all the interactions between ribosomal proteins and the rRNA, the interactions between extended protein tails that penetrate into the interior of the ribosome and the forest of RNA helices are the most unprecedented and the biggest surprise in the *H. marismortui* large subunit structure. At the time the first atomic resolution crystal structures appeared for ribosomes, the structures of a large number of ribosomal proteins in isolation had been established (73); it was anticipated that the conformations of those proteins in the ribosome would be similar to their conformations in isolation, which turned out to be true. What almost no one anticipated (72a) was that the ribosomal proteins that had resisted structure determination contain sequences whose conformations are determined entirely by their interactions with ribosomal RNA, i.e., tails. The *H. marismortui* large subunit includes 12 such proteins (18). In the intact subunit, their globular domains, like the globular domains that constitute the totality of nontailed ribosomal proteins, are found on the surface of the particle, and their tails insert into the ribosome's interior making numerous, idiosyncratic interactions with the surrounding RNA. While being only 18% of the total large subunit protein, they constitute 44% of the total rRNA interaction surface (D. Klein, T.M. Schmeing, P.B. Moore, and T. A. Steitz, in preparation).

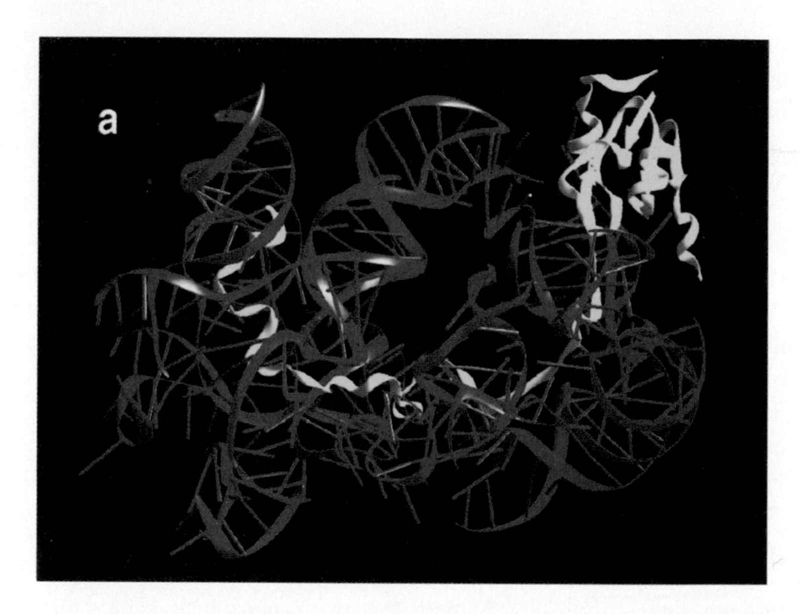

Figure 5 Ribosomal proteins in the *H. marismortui* large ribosomal subunit. (*a*) A ribbon representation of L15 (yellow) and the RNA sequences with which it interacts (red). The globular domain of the protein is exposed to solvent on the surface of the ribosome, but its extended tail penetrates deeply into the subunit. (*b*, *c*) A space-filling representation of proteins in the *H. marismortui* large ribosomal subunit, with the RNA removed, color-coded to display electrostatic charge potential. Negative regions are red, positive regions are blue, and neutral regions are white. The ribosome is seen in the crown view in (*b*), and rotated 180° about its vertical axis in (*c*). The surface of the globular domains of those proteins that face the exterior are acidic, but the surfaces that face the interior, including their tails, are basic (D. Klein, T.M. Schmeing, P.B. Moore, and T.A. Steitz, in preparation).

In many respects, the tails of the ribosomal proteins are reminiscent of those highly basic termini of histone proteins, which also interact with nucleic acid (73a). In all species, the tails of ribosomal proteins are highly basic being about one quarter arginine plus lysine, which makes them suitable for stabilizing the structure of a polyanion like RNA (D. Klein, T.M. Schmeing, P.B. Moore, and T.A. Steitz, in preparation). However, they are not just high mol wt counterions. They also contain an abundance of glycines and prolines, and their sequences are more conserved (20% in archaea and eukaryotes) than those of the globular domains (10%) to which they are attached. Further, the interactions they make with the RNAs surrounding them are highly specific (D. Klein, T.M. Schmeing, P.B. Moore, and T.A. Steitz, in preparation). Because the globular domains of these proteins are acidic, there is a striking segregation of protein charge in the large subunit from *H. marismortui* (Figure 5*b*) with the surface facing the outside of the particle being negative and that in the interior being positive.

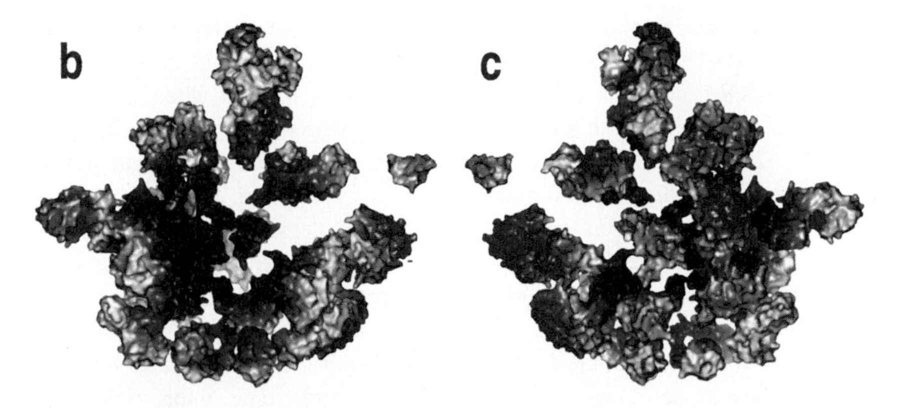

Figure 5 Continued

Examination of the pathway of these protein extensions in and among the spaces between RNA helices suggests that the assembly of the ribosomal RNA and proteins must be a complex, ordered, and cooperative process (Figure 5*a*). Clearly, these tails do not thread their way into the RNA after it has reached its final conformation, but rather must cofold with the RNA. One attractive hypothesis is that the globular domains first interact with a contiguous piece of 23S rRNA, burying more than 1000 $Å^2$ of surface, and then other more remote RNAs cofold around the tails burying significantly smaller surface areas (D. Klein, T.M. Schmeing, P.B. Moore, and T.A. Steitz, in preparation). Indeed, each of the tailed proteins has a single large surface on its globular domain that interacts with a contiguous rRNA structure.

LARGE SUBUNIT ASSEMBLY In the 1960s and 1970s, it was discovered that both prokaryotic ribosomal subunits can be reconstituted in vitro from their isolated components (74, 75). The studies of reconstitution that followed produced most of what is known about the details of ribosome assembly. The general conclusion that emerges when reconstitution data are compared with subunit structures is that the primary function of small subunit proteins is stabilization of 16S rRNA domains, whereas the primary function of large subunit proteins is stabilization of interdomain interactions.

One way to document this difference is to count the number of rRNA domains with which each ribosomal protein makes more than incidental contact. The average for the 30S proteins in *T. thermophilus* is 1.2 per protein (72), while the corresponding average for the 50S protein in *H. marismortui* is 2.3 per protein (18). The large subunit average is skewed because it includes the contributions of L1, L10, L11, and L12, which are components of its two lateral protuberances and do not function in stabilization of the structure. Leaving them out, the average for the large subunit rises to 2.6 different domains contacted per protein.

The interdomain character of large subunit protein/RNA interactions becomes manifest also when large subunit proteins are allowed to interact with isolated 23S rRNA domains (60). The nucleoproteins formed when the 6 23S rRNA domains interact with total large subunit protein one at a time include only 17 of the 32 proteins in the *Escherichia coli* large ribosomal subunit, but another 5 bind when RNAs that include pairs of domains are tested. In contrast, 16 of the 20 small subunit proteins tested associate with isolated 16S rRNA domains (77–79).

It is hard to make detailed comparisons between the experimental reconstitution assembly data available for the large subunit and the structure of the large ribosomal subunit from *H. marismortui* because the assembly data were obtained using *E. coli* ribosomes, and only 20 *H. marismortui* large ribosomal subunit proteins have *E. coli* equivalents. There are some interesting correlations, however. The reconstitution process of Nierhaus and coworkers generates two intermediate ribonucleoproteins, the earlier one is called $RI_{50}(1)$ and the later one called $RI_{50}(2)$ (80). Proteins in *E. coli* that have *H. marismortui* homologues tend to add to assembling subunits early. Of the 21 proteins found in the $RI_{50}(1)$ intermediate, 16 have homologues in *H. marismortui*, but only 6 of the 10 proteins found in the $RI_{50}(2)$ (in addition to the 21 that are already there) have *H. marismortui* homologues. In addition, 5 of the 6 proteins identified as essential for $RI_{50}(1)$ formation have homologues in *H. marismortui*. These observations suggest that the proteins that have homologues in all kingdoms are somehow more important than those that do not.

It is also the case that tailed large subunit proteins are likely to be involved in the early stages of assembly. Eight of the 20 proteins in the homologous set have tails, and all of them are found in the $RI_{50}(1)$ complex. Moreover, of the subset of the proteins essential for the formation of the $RI_{50}(1)$ intermediate that have *H. marismortui* equivalents, 4 out of 5 have tails. This correlation makes structural sense because it is very difficult to understand how protein tails could get inserted into the interior of the RNA core of the subunit once it attains its fully folded form. Tailed proteins ought to join the large subunit early.

THE STRUCTURAL BASIS OF LARGE SUBUNIT ACTIVITY

The most significant result to emerge from the crystal structures of the large ribosomal subunit and its substrate complexes is the finding that the ribosome is a ribozyme. All of the components of the ribosome involved in orienting both the A-site α-amino group and the P-site bound carbonyl carbon it is to attack are made of RNA, as is the rest of the peptidyl transferase center (70). Although proteins do contact ribosome-bound tRNAs and undoubtedly help orient them, the catalytic business of the ribosome is conducted entirely by RNA.

Investigating the Enzymatic Activity of the Peptidyl Transferase Center

The process catalyzed by the peptidyl transferase center of the ribosome is the aminolysis of an ester bond. The nucleophilic α-amino group of an aminoacyl-tRNA bound to the A site of the peptidyl transferase center attacks the carbonyl carbon of the ester bond linking the peptide moiety of a peptidyl-tRNA bound to the P site of the peptidyl transferase center. The anionic, tetrahedral intermediate that results breaks down to yield a deacylated tRNA in the P site and an A-site bound tRNA joined by an ester bond to a peptide that has been extended by one amino acid.

At present, only the structures of 50S ribosomal subunits have been established at sufficiently high resolution to enable a detailed analysis of the peptide bond forming process. The 5.5 Å resolution maps of the 70S ribosome with A- and P-site tRNAs bound do not allow an atomic interpretation of the catalytic site. Although full tRNA molecules are too large to bind in 50S subunit crystals, the 50S subunit will catalyze the formation of a peptide bond using substrate analogues that are fragments of the CCA acceptor ends of tRNA (Figure 6) in what is termed "the fragment reaction" (81, 82), and these fragments do bind to crystals. Puromycin, which is a di-methylated adenosine aminoacylated with a hydroxymethyl tyrosine, is an acceptable A-site substrate, though either C-puromycin or C-C-puromycin is better; molecules as small a CCA aminoacylated with N-formyl methionine will serve as P-site substrates.

All of the published structures of large subunits with substrate, intermediate, and product analogues bound have been obtained using crystals of *H. marismortui* subunits. Not only do fragment substrates diffuse into these crystals and bind, but these subunits will make peptide bonds *in crystalo* (83). Separate structures have been obtained of the large subunit complexed with an A-site fragment substrate, a P-site fragment substrate, and with products bound together in one complex (Figure 7), as well as a complex with an intermediate analogue (70, 83, 84). These structures have been combined to produce a movie showing how the peptide bond forming process could occur on the large subunit (Figure 8) (84). A model of the peptidyl transferase center with both its A site and its P site occupied by substrate analogues (Figure 8*a*) was generated by superimposing the structure of the large subunit with a P-site substrate bound on the independently determined structure of the large subunit with the A site occupied. Because these crystals are enzymatically active and because the activity of the *H. marismortui* subunit in solution is independent of salt concentration and cation identity (83), these structures are likely to be relevant to understanding peptide bond formation by the large subunit.

THE CATALYSIS OF PEPTIDE BOND FORMATION BY THE LARGE RIBOSOMAL SUBUNIT

The structures of these complexes as a group not only prove that the peptidyl transferase site is entirely composed of 23S rRNA, they also provide insight into

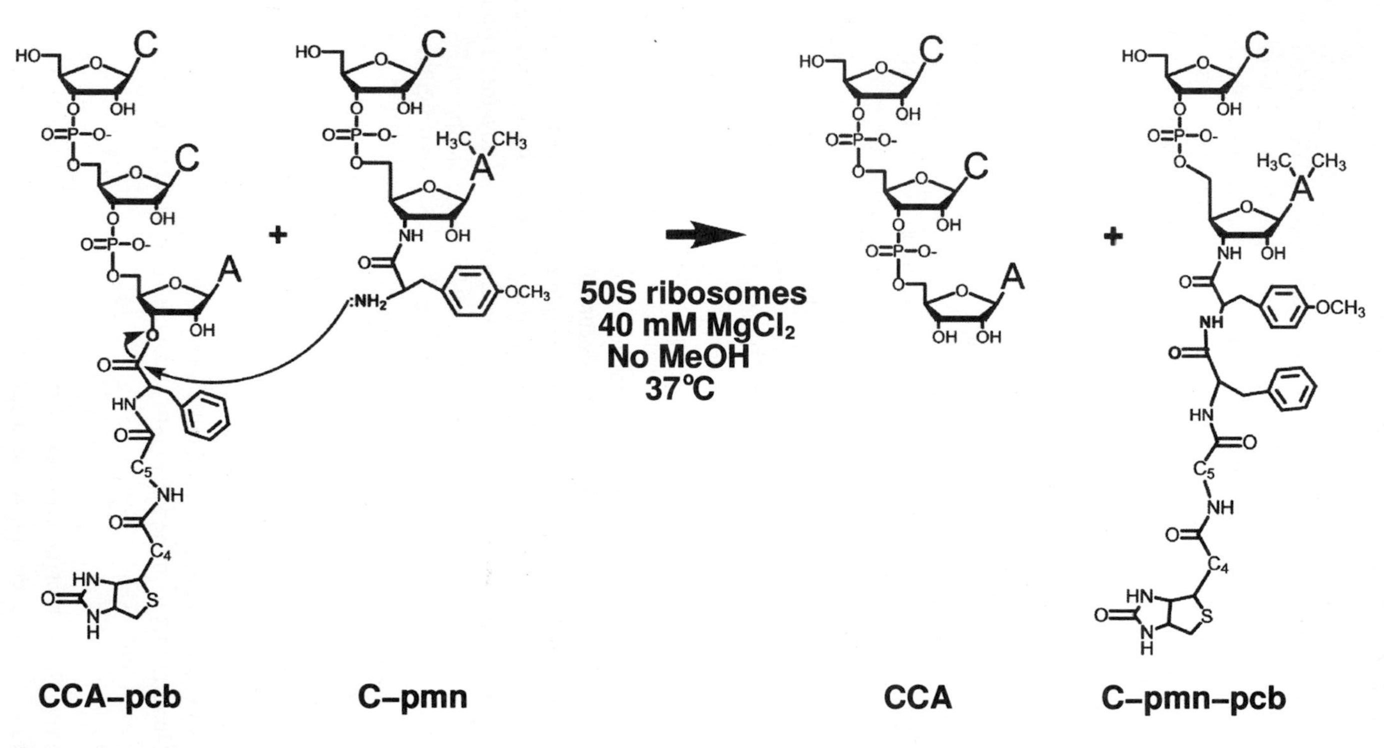

Figure 6 A ribosome-catalyzed peptide bond forming reaction involving low mol wt substrates. The reaction of CCA-phenylalanine-caproic acid-biotin (CCA-pcb) and C-puromycin (C-pmn) that yields CCA and C-puromycin-phenylalanine-caproic acid-biotin (C-pmn-pcb) is catalyzed by large ribosomal subunits. Reactions of this type are analogous to the peptidyl transferase reaction, which occurs on intact ribosomes in vivo and is referred to as the "fragment reaction," because its substrates resemble the 3′ termini of aminoacyl and peptidyl tRNAs. [Reproduced with permission from (83).]

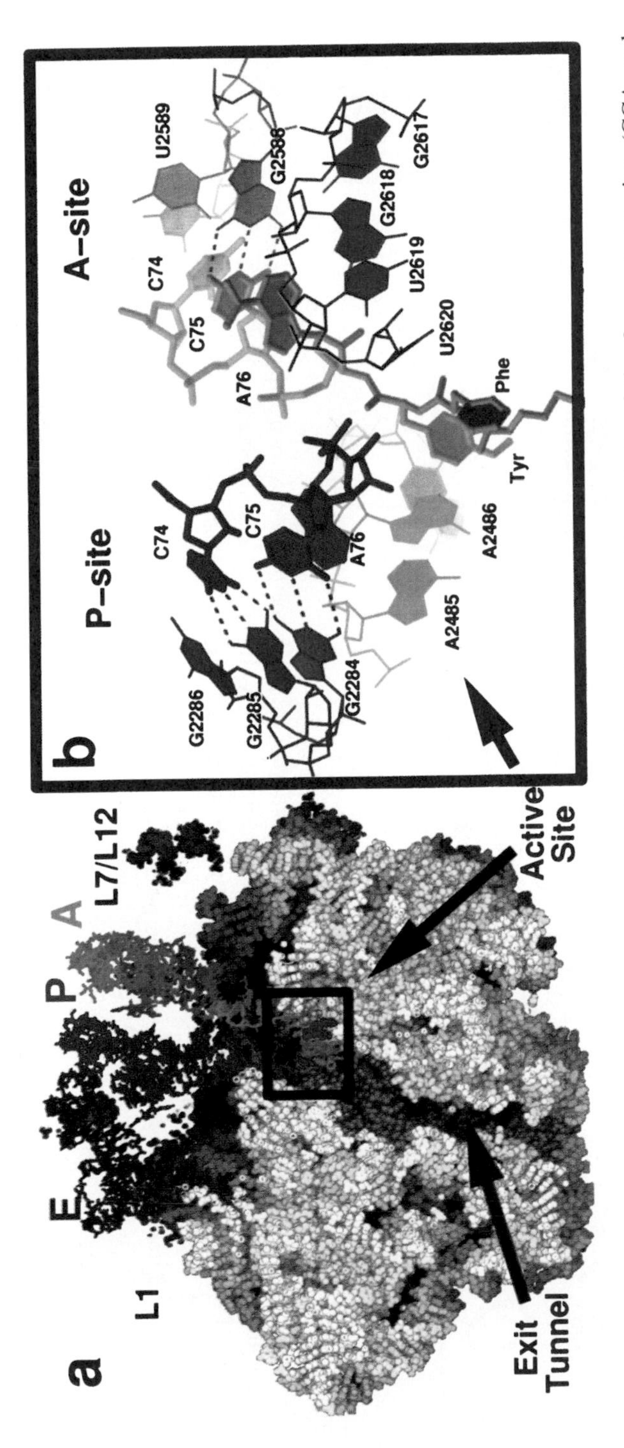

Figure 7 The structure of the large ribosomal subunit from *H. marismortui* with products of the fragment reaction (CCA and C-pmn-pcb: see Figure 6) bound in the peptidyl transferase center. (*a*) A space-filling representation of the complex with three intact tRNAs added in the positions tRNAs assume when bound to the A, P, and E sites of the 70S ribosome (22). rRNA is white, and ribosomal proteins are yellow. The subunit, which is oriented in the crown view, has been cut in half along a plane that passes through the peptide exit tunnel, and the front of the structure has been removed to expose the tunnel lumen. The active site area is in a box. (*b*) A close-up of the active site showing the peptidyl product (CC-pmn-pcb; green) bound to the A-loop (orange), and the deacylated product (CCA; violet) bound to the P-loop (dark blue). The N3 of A2486 (A2451 in *E. coli*) (light blue) is close to the 3′ OH of the CCA, and the base of U2620 (U2585) (red) has moved close to the new peptide bond and the 3′ OH of A76. [Reproduced with permission from (83).]

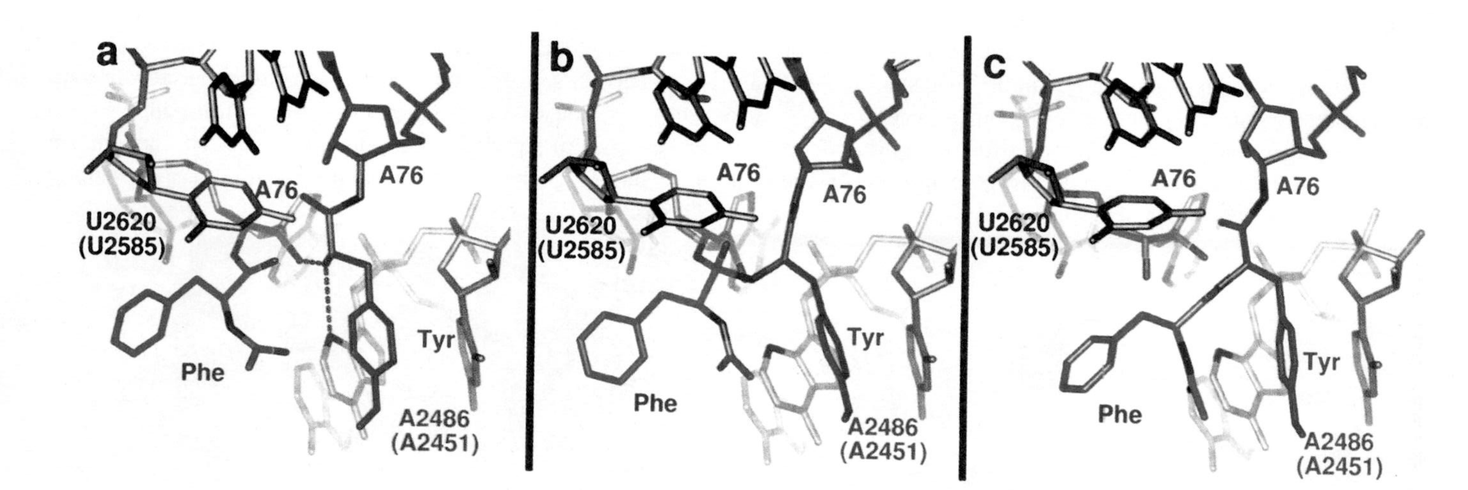

Figure 8 Steps in the peptidyl transferase reaction pathway. (*a*) The superposition of two independently determined cocrystal structures suggests that the α-amino group of an A-site substrate is positioned for a pro-R attack on the carbonyl carbon of the ester bond of the P-site substrate (green). (*b*) A model for the tetrahedral intermediate that would result if the reaction were to occur in the manner suggested by Figure 8*a*. Note that the oxyanion points away from A2486 (2451). (*c*) The structure of the products of the the peptidyl transferase reaction bound to the peptidyl transferase center. [Reproduced with permission from (84).]

the ways the large subunit facilitates peptide bond formation. Of primary importance for the catalytic activity of the peptidyl transferase center is the proper alignment of substrates, as is the case for all enzymes (70, 85, 86). It is clear that the peptidyl transferase center, including the A- and P-loops and A2486 (2451), is organized to bring the nucleophilic α-amino group of an A-site aminoacyl-tRNA into contact with the carbonyl carbon with which it reacts during peptide bond formation (Figure 8*a*).

We still do not know the extent to which, or how, the peptidyl transferase center also chemically enhances the rate of peptide bond formation. There is no evidence for metal ion involvement, although until a structure is obtained with substrates bound to both the A site and the P site simultaneously, metal ion involvement will be difficult to exclude. In the present model for A-site and P-site substrates bound to the center there are, however, three rRNA groups close enough to the reaction site to form hydrogen bonds with the reactive α-amino group: (1) the 2′ OH of A76 of the tRNA in the P site, (2) the N3 of A2486 (A2451 in *E. coli*5) of 23S rRNA, and (3) the 2′ OH of A2486 (2451) (84). These hydrogen bonds must help align the α-amino group, and there is reason to believe that hydrogen bond formation alone may enhance reactivity by several orders of magnitude beyond what it contributes to alignment (86a). If any of these groups were in addition to have an elevated pK_a, its hydrogen bonding interactions could further accelerate the rate of peptide bond formation by enhancing the nucleophilicity of the α-amino group.

Nothing is known about the role of the 2′ OH of A2486 (2451) in protein synthesis. However, tRNAs that terminate with a 2′ deoxyadenosine can be aminoacylated and are A-site substrates, but 2′deoxy-A76 peptidyl tRNAs are inactive as P-site substrates (87). The 2′OHs of A76 of tRNAs bound in the P site are thus likely to be important for A-site substrate alignment, but a more active role for them cannot be ruled out.

Many experiments have been done to test the proposal that the N3 of A2486 (2451) enhances the rate of peptide bond formation chemically by general acid/base catalysis and intermediate stabilization (70, 88). Two important points about that hypothesis have now been clarified. First, present structural data indicate that A2486 is unlikely to help stabilize the oxyanion intermediate (84) because the anionic oxygen of the tetrahedral intermediate constructed from the substrate complexes appears to point away from the N3 of A2486 (2451) (Figure 8*b*). Second, contrary to earlier reports (89, 91–93), mutation of A2486 (2451) to U reduces the rate of the chemical step of peptide bond formation in 70S ribosomes by about 100-fold at basic pHs. Furthermore, the A2486 (2451)-associated rate effect titrates with a pK_a of about 7.5 (94). These observations are compatible with the hypothesis that the N3 of A2486 has a pK_a of 7.5 (but see 90), and activates the α-amino group as a general base (70). A more indirect hypothesis that the peptidyl transferase center is activated by a pH-dependent conformational change that for some unknown reason depends on the identity of the base at position 2486 (2451) is also consistent with these kinetic data (94). Further experimentation will be required to settle these issues.

THE LARGE SUBUNIT-CATALYZED PEPTIDE BOND FORMATION AS A MODEL FOR PEPTIDE BOND FORMATION Most of the statements about the chemistry of peptide bond formation found in textbooks today derive from studies of large subunit catalyzed fragment reactions like those just described. There are reasons for believing the fragment reaction is biologically relevant. First, both the 70S ribosome and the 50S subunit promote the reaction of the same low mol wt substrates. Second, the substrates and products of the fragment reaction are obviously related to those of the normal protein synthesizing system. Third, the fragment reaction and the normal peptidyl transferase reaction occur at the same site in the large ribosomal subunit. Fourth, both reactions are sensitive to many of the same inhibitors.

However, it appears that the full peptidyl transferase reaction catalyzed by the 70S ribosome may have additional catalytically important features, since its rate is about 10^4 times faster. At 37°C, bacteria synthesize protein at a rate of about 20 amino acids per second (95), and the rate of the chemical, bond-forming step in that sequence of reactions is probably $>100\ s^{-1}$ (96). Consistent with this estimate, under saturating substrate conditions, puromycin will react with peptidyl tRNA bound to mRNA-programmed 70S ribosomes at a rate of $\sim 70\ s^{-1}$ (94), which is 3 to 4 orders of magnitude faster than 50S subunits catalyze the reaction of puromycin with low mol wt P-site substrates (97). This rate difference may mean that the peptidyl transferase center in reaction-ready 70S ribosomes has a configuration that differs from that seen in the large ribosomal subunit, although the structural differences are likely to be modest. Alignment of the *H. marismortui* 50S structure and that from the 70S shows that the A- and P-site tRNAs bound to the 70S ribosome can connect smoothly with the fragment substrates bound to the 50S subunit (84). However, the low resolution of the 70S ribosome study (22) cannot exclude (or support) a small, but catalytically important, alteration in the presentation of the nucleophilic 5 α-amino group to the carbonyl group it attacks or in its relationship to surrounding ribosomal groups.

The Polypeptide Exit Tunnel

The polypeptide exit tunnel of the large subunit is a passage about 100 Å long that begins immediately below the peptidyl transferase center, and ends at the back side of the large subunit (Figure 7*a*). Most nascent proteins pass though the tunnel as they are synthesized, but there is evidence that at least some can leave the ribosome by other routes [reviewed in (98)]. The signal recognition particle (SRP), which recognizes the signal sequences of proteins destined to be inserted into membranes or secreted through them, interacts with those sequences when they reach the distal end of the tunnel. The SRP also interacts with proteins that

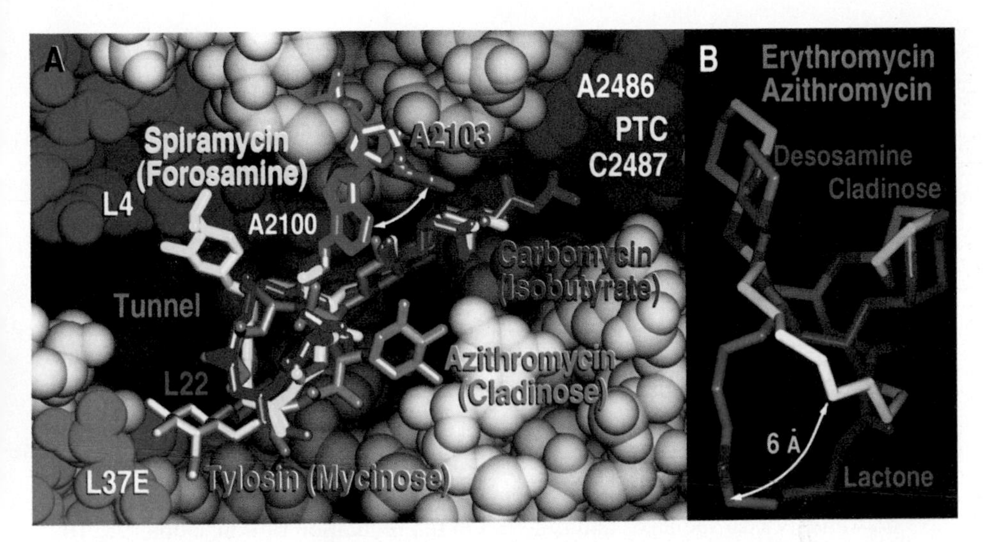

Figure 9 The interaction of macrolide antibiotics with the large ribosomal subunit of *H. marismortui*. (*a*) A superposition of several macrolides/large ribosomal subunit complex structures. Structures containing carbomycin (red), tylosin (orange), spiramycin (yellow), and azithromycin (blue) have been superimposed by aligning corresponding 23S rRNA atoms in four independently determined crystal structures. The macrolide rings of the four antibiotics bind to virtually the same site in the proximal portion of the peptide exit tunnel. In the case of the 16-membered macrolides examined (tylosin, carbomycin, and spiramycin), A2103 (2062) swings down so that its N6 can form a covalent bond with their aldehyde substituents. The differences between these drugs are due primarily to the substitutents on their macrolide rings, which differ in chemical nature, bulk, and placement. Some extend into the peptidyl transferase center. (*b*) The position assumed by erythromycin (white) when bound to the large ribosomal subunit from *D. radiodurans* (104) compared to that adopted by azithromycin (blue) bound to the large ribosomal subunit from *H. marismortui*. [Reproduced with permission from (105).]

surround the end of the tunnel (100). By a mechanism that is not fully understood but may involve interaction of the SRP with the factor binding site, the interaction of a ribosome engaged in the synthesis of such a protein with a SRP prevents it from completing the synthesis of that protein until it has become membrane bound in association with a translocon, the pore structure through which the protein must pass (101).

The wall of the tunnel is formed entirely by RNA for the first third of its length extending from the PTC to a constriction that is formed by portions of ribosomal proteins L4 and L22, which are highly conserved components of the ribosome. Below the constriction, the bore of the tunnel becomes large enough to accommodate an α-helix, and below the point where L39e contributes sequences to its wall, it becomes even wider. Electron microscopists claim that the tunnel bifurcates in its lower reaches, and hence that there are two ways for nascent polypeptides to exit from the ribosome (10). This feature of the tunnel is not confirmed by the atomic resolution crystal structures (70). The appearance of a side tunnel in the EM may be a resolution artifact, which results from Fourier series termination effects; a feature in the crystal structure that may appear as a side tunnel in the EM is in fact completely occluded by protein. There is only one passage big enough to accept a polypeptide that passes through the large subunit.

The tunnel, of course, has to allow the passage of nascent proteins of all sequences, and hence its wall ought to possess a nonstick character. Indeed, consistent with this expectation, its wall is a mosaic of small hydrophobic and hydrophilic patches, which for most of the tunnel appear to be too small to promote strong interactions with nascent peptides (18). A possible exception is the upper part of the tunnel near the PTC where two bases splay apart at two locations and form binding sites for antibiotics (105). Furthermore, there are some short hydrophobic peptide sequences that appear to stick to the tunnel, presumably near the PTC, and function in the regulation of translation (98, 102). It is possible that these regulatory peptide sequences and some antibiotics exploit the same binding opportunities. Sequences that stick in the tunnel have probably been selected against for the vast majority of proteins.

One of the most intriguing unanswered questions about the tunnel is how nascent polypeptides proceed down its lumen: Is it a passive diffusion process or are there shape changes that occur in the tunnel during protein synthesis that actively promote the passage of nascent polypeptides? The conformation of the tunnel is indistinguishable in the two 50S structures and the 70S structure as well as all of the substrate analogue and antibiotic complexes established thus far, which implies that the process may be passive. However, Frank and coworkers (103, 103a) have proposed that the tunnel functions like a peristaltic pump to facilitate the passage of peptides. They report large differences in the tunnel diameter between wild-type ribosomes and ribosomes possessing mutations in either protein L4 or L22 that render these ribosomes resistant to macrolides. It will be of interest to see in an atomic structure how such shape changes can be

---

Figure 10 Antibiotic structures and antibiotic interactions with the *H. marismortui* large ribosomal subunit. (*a*) The structures of two macrolide antibiotics. (*b*) The interactions of sparsomycin (green), puromycin (red), blasticidin S (magenta), chloramphenicol (light blue), carbomycin (dark blue), and streptogramin A (blue) with the large ribosomal subunit. The ribosome has been split open as in Figure 7*a* to reveal the lumen of the exit tunnel and adjacent regions of the peptidyl transferase site. Ribosomal components are depicted as a continuous surface. Seven independently determined cocrystal structures have been aligned by superimposing the 23S rRNA in each complex. The sites to which these antibiotics bind are all different, but there is extensive overlap (J. Hansen, P.B. Moore, and T.A. Steitz, manuscript in preparation).

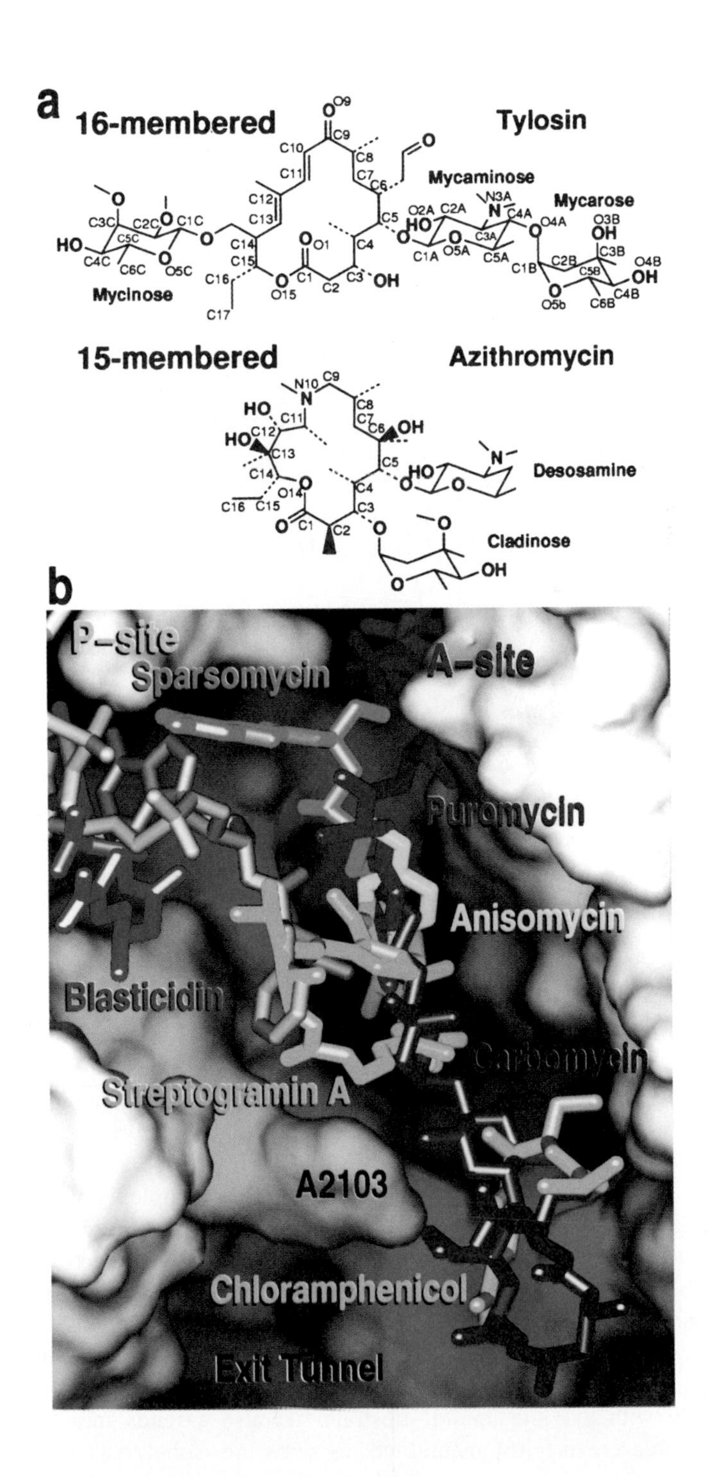
a
16-membered
Tylosin
Mycaminose
Mycarose
Mycinose
15-membered
Azithromycin
Desosamine
Cladinose
b
P-site
Sparsomycin
A-site
Puromycin
Anisomycin
Blasticidin
Carbomycin
Streptogramin A
A2103
Chloramphenicol
Exit Tunnel

achieved and whether or not they might be related normal functions of the native ribosome.

Antibiotics Targeting the Large Subunit

For the biochemical community interested in antibiotics that block ribosome function, the last two years have been one long field day in which a wealth of information on the structures of antibiotic complexes of both the large and small subunit has been published. The structures of five different antibiotic complexes with the *D. radiodurans* large subunit were established: chloramphenicol, clindamycin, and the 14-membered macrolides (erythromycin, clarithromycin, and roxithromycin) (104). Complexes of the *H. marismortui* large subunit with the 16-membered macrolides (tylosin, carbomycin, and spiramycin), as well as the 15-membered macrolide, azithromycin, (105), and complexes of the *H. marismortui* large subunit with the nonmacrolide antibiotics (anisomysin, blasticidin, sparsomycin, virginiamycin M, and chloramphenicol) bound to a second site have all been determined (J.L. Hansen, P.B. Moore, and T.A. Steitz, unpublished observations).

All of the macrolide antibiotics bind in largely the same place in the proximal part of polypeptide exit tunnel adjacent to the peptidyl transferase center and before the constriction formed by L4 and L22 (Figure 9). They all appear to inhibit protein syntheses by blocking the passage of nascent peptides down the tunnel or, as in the case of carbomycin A, by inhibiting A-site substrate binding. The saccharide branch attached to C5 of the lactone rings extends toward the PTC, and the isobuterate extension of the carbomycin A dissaccharide overlaps the A site. In the case of the 16-membered macrolides, a reversible covalent bond forms between the ethyladehyde substituent at the C6 position and the N6 of A2063 (A2062, *E. coli*). The orientation of the 14-membered lactose rings of the macrolides bound to the eubacteria *D. radiodurans* large subunit differs significantly from that of the 15- and 16-membered lactose rings bound to the archaeal *H. marismortui* large subunit. Whether this is due to a difference between the binding of 14-membered and the 15- or 16-membered macrolides or due to a difference between eubacteria and archaea, such as occurs at position 2058 in 23S RNA, is not yet established. Interactions of all seven macrolides with the large subunit seen in the structures are consistent with prior biochemical and genetic data.

The structures of seven nonmacrolide antibiotics complexed with either the *H. marismortui* or the *D. radiodurans* large subunit show that these compounds bind to sites that overlap those of either peptidyl tRNA or aminoacyl-tRNA, consistent with their functioning as competitive inhibitors of peptide bond formation (104; J.L. Hansen, P.B. Moore, and T.A. Steitz, in preparation). One hydrophobic crevice in the PTC forms the binding site for a tyrosine side chain of A-site analogues as well as for anisomycin and chloramphenicol. Sparsomycin interacts primarily with a P-site bound substrate but also extends into the active site hydrophobic crevice just mentioned, as does the isobuterate extension of the carbomycin A dissaccharide. Virginiamycin M occupies portions of both the A site and P site, and it induces a conformational change in A2103, while blasticidin S exploits another strategy by mimicking C74 and C75 of a P-site bound tRNA and base pairing with the P-loop.

Many of these antibiotics bind in nearby but, in some cases, nonoverlapping sites (Figure 10), which suggests a strategy for the design of new antibiotics that may overcome many known antibiotic resistance mutations. One could imagine synthesizing hybrid antibiotics by combining portions of two antibiotics that bind adjacent sites. In any case, these structures provide important leads for the design and synthesis of new antibiotics that target the large ribosomal subunit. Thus, it is now possible to use the same structure based drug design approach that was so successful with HIV protease to design novel antibiotics targeting the large ribosomal subunit, in spite of its being 100 times larger than the protease.

ACKNOWLEDGMENTS

We thank Nenad Ban, Jeff Hansen, Joe Ippolito, Dan Klein, Poul Nissen, and Martin Schmeing. This research was supported by NIH grant GM22778 and an Agouron Institute grant to P.B.M. and T.A.S.

The *Annual Review of Biochemistry* is online at http://biochem.annualreviews.org

LITERATURE CITED

1. Robertus JD, Ladner JE, Finch JT, Rhodes D, Brown RS, et al. 1974. *Nature* 250:546–51
2. Kim SH, Suddath FL, Quigley G, McPherson A, Sussman JL, et al. 1974. *Science* 185:435–40
3. Rheinberger H-J, Geigenmuller U, Gnirke A, Hausner T-P, Remmer J, et al. 1990. In *The Ribosome*, ed. WE Hill, AE Dahlberg, RA Garrett, PB Moore, D Schlessigner, JR Warner, pp. 318–30. Washington, DC: Am. Soc. Microbiol.
4. Lake JA. 1976. *J. Mol. Biol.* 105:131–59
5. Stoeffler G, Stoeffler-Meilicke M. 1984. *Annu. Rev. Biophys. Bioeng.* 13:303–30
6. Shatsky IN, Evstafieva AG, Bystrova TF, Bogdanov AA, Vasiliev VD. 1980. *FEBS Lett.* 121:97–100
7. Oakes M, Henderson E, Scheinman A, Clark M, Lake JA. 1986. See Ref. 106, pp. 47–67
8. Stoeffler G, Stoeffler-Meilicke M. 1986. See Ref. 106, pp. 28–46
9. Stark H, Mueller F, Orlova EV, Schatz M, Dube P, et al. 1995. *Structure* 3:815–21
10. Frank J, Zhu J, Penczek P, Li YH, Srivastava S, et al. 1995. *Nature* 376: 441–44
11. Bernabeu C, Lake JA. 1982. *Proc. Natl. Acad. Sci. USA* 79:3111–15
12. Bernabeu C, Tobin EM, Fowler A, Zabin I, Lake JA. 1983. *J. Cell Biol.* 96:1471–74
13. Milligan RA, Unwin PNT. 1986. *Nature* 319:693–95

13a. Yonath A, Leonard KR, Wittmann HG. 1987. *Science* 236:813–16

14. Stark H, Orlova EV, Rinke-Appel J, Junke N, Mueller F, et al. 1997. *Cell* 88:19–28

15. Stark H, Rodnina MV, Rinke-Appel J, Brimacombe R, Wintermeyer W, van Heel M. 1997. *Nature* 389:403–6
16. Agrawal RK, Penczek P, Grassucci RA, Li YH, Leith A, et al. 1996. *Science* 271:1000–2
17. Agrawal RK, Penczek P, Grassucci RA, Frank J. 1998. *Proc. Natl. Acad. Sci. USA* 95:6134–38
18. Ban N, Nissen P, Hansen J, Moore PB, Steitz TA. 2000. *Science* 289:905–20
19. Wimberly BT, Brodersen DE, Clemons WM, Morgan-Warren RJ, Carter AP, et al. 2000. *Nature* 407:327–39
20. Pioletti M, Schlunzen F, Harms J, Zarivach R, Glühmann M, et al. 2001. *EMBO J.* 20:1829–39
21. Harms J, Schluenzen F, Zarivach R, Bashan A, Gat S, et al. 2001. *Cell* 107:679–88
22. Yusupov MM, Yusupova GZ, Baucom A, Lieberman K, Earnest TN, et al. 2001. *Science* 292:883–96
23. Yonath A. 2002. *Annu. Rev. Biophys. Biomol. Struct.* 31:257–73
24. Ramakrishnan V. 2002. *Cell* 108:557–72
25. Ramakrishnan V, Moore PB. 2001. *Curr. Opin. Struct. Biol.* 11:144–54
26. Moore PB. 2001. *Biochemistry* 40: 3241–50
27. Moore PB, Steitz TA. 2003. *RNA* 9:155–59
28. Moore PB, Steitz TA. 2002. *Nature* 418:229–35
29. Yonath A, Mussig J, Tesche B, Lorenz S, Erdmann VA, Wittmann HG. 1980. *Biochem. Int.* 1:428–35
30. Trakanov SD, Yusupov MM, Agalarov SC, Garber MB, Ryazantsev SN, et al. 1987. *FEBS Lett.* 220:319–22
31. Glotz C, Mussig J, Gewitz HS, Makowski I, Arad T, et al. 1987. *Biochem. Int.* 15:953–60
32. Yusupov MM, Tichenko SV, Trakanov SD, Ryazantsev SN, Garber MB. 1988. *FEBS Lett.* 238:113–15
33. Yonath A, Glotz C, Gewitz HS, Bartels H, von Bohlen K, et al. 1988. *J. Mol. Biol.* 203:831–34
34. Shoham M, Wittmann HG, Yonath A. 1987. *J. Mol. Biol.* 193:819–22
35. von Bohlen K, Makowski I, Hansen HAS, Bartels H, Berkovitch-Yellin Z, et al. 1991. *J. Mol. Biol.* 222:11–15
36. Yonath A, Franceschi F. 1998. *Structure* 6:679–84
37. Harms J, Tocilj A, Levin I, Agmon I, Stark H, et al. 1999. *Structure* 7:931–41
38. Ban N, Nissen P, Hansen J, Capel M, Moore PB, Steitz TA. 1999. *Nature* 400:841–47
39. Clemons WMJ, Brodersen DE, McCutcheon JP, May JLC, Carter AP, et al. 2001. *J. Mol. Biol.* 310:827–43
40. Schluenzen F, Tocilj A, Zarivach R, Harms J, Gluehmann M, et al. 2000. *Cell* 102:615–23
40a. Haas DJ, Rossmann MG. 1970. *Acta Crystallogr. B* 26:998–1004
40b. Hope H. 1988. *Acta Crystallogr. B* 44:22–26
41. Hope H, Frolow F, von Bohlen K, Makowski I, Kratky C, et al. 1989. *Acta Crystallogr. B* 45:190–99
42. Jack A, Harrison SC, Crowther RA. 1975. *J. Mol. Biol.* 97:163–72
43. Schlunzen F, Hansen HAS, Thygesen J, Bennett WS, Volkmann N, et al. 1995. *Biochem. Cell Biol.* 73:739–49
44. Yonath A, Harms J, Hansen HAS, Bashan A, Schlunzen F, et al. 1998. *Acta Crystallogr. A* 54:945–55
45. Bashan A, Pioletti M, Bartles H, Janell D, Schlunzen F, et al. 2000. In *The Ribosome*, ed. RA Garrett, S Douthwaite, A Liljas, AT Matheson, PB Moore, HF Noller, pp. 21–33. Washington, DC: Am. Soc. Microbiol. Press
46. Ban N, Freeborn B, Nissen P, Penczek P, Grassucci RA, et al. 1998. *Cell* 93:1105–15
46a. Hansen JL, Schmeing TM, Klein DJ, Ippolito JA, Ban N, et al. 2001. *Cold*

Spring Harbor Symp. Quant. Biol. 66:33–42
46b. Abrahams JP, Leslie AGW. 1996. *Acta Crystallogr. D* 52:30–42
47. Ramakrishnan V, Capel MS, Clemons WM, May JLC, Wimberly BT. 2000. See Ref. 45, pp. 1–10
48. Clemons WM Jr, May JLC, Wimberly BT, McCutcheon JP, Capel MS, Ramakrishnan V. 1999. *Nature* 400: 833–40
49. Deleted in proof
50. Cate JH, Yusupov MM, Yusupova GZ, Earnest TN, Noller HF. 1999. *Science* 285:2095–104
51. Deleted in proof
52. Bashan A, Agmon I, Zarivach R, Schluenzen F, Harms J, et al. 2001. *Cold Spring Harbor Symp. Quant Biol.* 64:43–56
53. Bujnicki JM, Feder M, Rychlewski L, Fischer D. 2002. *FEBS Lett.* 525:174–75
54. Harms J, Schluenzen F, Zarivach R, Bashan A, Bartels H, et al. 2002. *FEBS Lett.* 525:176–78
55. Noller HF, Kop J, Wheaton V, Brosius J, Gutell RR, et al. 1981. *Nucleic Acids Res.* 9:6167–89
56. Cannone JJ, Subramanian S, Schnare MN, Collett JR, D'Souza LM, et al. 2002. *BioMed. Cent. Bioinf.* 3:2
57. Andersen A, Larsen N, Leffers H, Kjems J, Garrett RA. 1986. In *Structure and Dynamics of RNA*, ed. PH Knippenberg, CW Hilbers, pp. 221–37. New York: Plenum
58. Egebjerg J, Leffers H, Christensen A, Andersen H, Garrett RA. 1987. *J. Mol. Biol.* 196:125–36
59. Leffers H, Egebjerg J, Andersen A, Christensen T, Garrett RA. 1988. *J. Mol. Biol.* 204:507–22
60. Ostergaard P, Phan H, Johansen LB, Egebjerg J, Ostergaard L, et al. 1998. *J. Mol. Biol.* 284:227–40
61. Moore PB. 1999. *Annu. Rev. Biochem.* 67:287–300
62. Westhof E, Fritsch V. 2000. *Struct. Fold. Des.* 8:R55–65
63. Klein DJ, Schmeing TM, Moore PB, Steitz TA. 2001. *EMBO J.* 20:4214–21
64. Nagaswamy U, Fox GE. 2002. *RNA* 8:1112–19
65. Szep S, Wang J, Moore PB. 2003. *RNA* 9:44–51
66. Michel F, Westhof E. 1990. *J. Mol. Biol.* 216:585–610
67. Cate J, Gooding AR, Podell E, Zhou K, Golden BL, et al. 1996. *Science* 273: 1678–85
68. Nissen P, Ippolito JA, Ban N, Moore PB, Steitz TA. 2001. *Proc. Natl. Acad. Sci. USA* 98:4899–903
69. Deleted in proof
70. Nissen P, Hansen J, Ban N, Moore PB, Steitz TA. 2000. *Science* 289:920–30
71. Ogle JM, Brodersen DE, Clemons WM, Tarry MJ, Carter AP, Ramakrishnan V. 2001. *Science* 292:897–902
72. Brodersen DE, Clemons WM, Carter AP, Wimberly BT, Ramakrishnan V. 2002. *J. Mol. Biol.* 316:725–68
72a. Liljas A. 1991. *Int. Rev.Cytol.* 124: 103–36
73. Ramakrishnan V, White SW. 1998. *Trends Biochem. Sci.* 23:208–12
73a. Luger K, Mader AW, Richmond RK, Sargent DF, Richmond TJ. 1997. *Nature* 389:251–60
74. Traub P, Nomura M. 1968. *Proc. Natl. Acad. Sci. USA* 59:777–84
75. Nierhaus KH, Dohme F. 1974. *Proc. Natl. Acad. Sci. USA* 71:4713–17
76. Deleted in proof
77. Weitzmann CJ, Cunningham PR, Nurse K, Ofengand J. 1993. *FASEB J.* 7: 177–80
78. Agalarov SC, Selivanova OM, Zheleznyakova EN, Zheleznaya IA, Matvienko NI, Spirin AS. 1999. *Eur. J. Biochem.* 266:533–37
79. Samaha RR, O'Brien B, O'Brien TW, Noller HF. 1994. *Proc. Natl. Acad. Sci. USA* 91:7884–88

80. Herold M, Nierhaus KH. 1987. *J. Biol. Chem.* 262:8826–33
81. Traut RR, Monro RE. 1964. *J. Mol. Biol.* 10:63–72
82. Monro RE. 1967. *J. Mol. Biol.* 26:147–51
83. Schmeing TM, Seila AC, Hansen JL, Freeborn B, Soukup JK, et al. 2002. *Nat. Struct. Biol.* 9:225–30
84. Hansen JL, Schmeing TM, Moore PB, Steitz TA. 2002. *Proc. Natl. Acad. Sci. USA* 99:11670–75
85. Page MI, Jencks WP. 1971. *Proc. Natl. Acad. Sci. USA* 68:1678–83
86. Nierhaus KH, Schulze H, Cooperman BS. 1980. *Biochem. Int.* 1:185–92
86a. Chamberlin SI, Weeks KM. 2002. *Proc. Natl. Acad. Sci. USA* 99:14688–93
87. Quiggle K, Kumar G, Ott TW, Ryu EK, Chladek S. 1981. *Biochemistry* 20:3480–85
88. Muth GW, Ortoleva-Donnelly L, Strobel SA. 2000. *Science* 289:947–50
89. Bayfield MA, Dahlberg AE, Schulmeister U, Dorner S, Barta A. 2001. *Proc. Natl. Acad. Sci. USA* 98:10096–101
90. Muth GW, Chen L, Kosek A, Strobel S. 2001. *RNA* 7:1403–15
91. Xiong L, Polacek N, Sander P, Boettger EG, Mankin AS. 2001. *RNA* 7:1365–69
92. Polacek N, Gaynor M, Yassin A, Mankin AS. 2001. *Nature* 411:498–501
93. Thompson J, Kim DF, O'Connor M, Lieberman KR, Bayfield MA, et al. 2001. *Proc. Natl. Acad. Sci. USA* 98:9002–7
94. Katunin VI, Muth GW, Strobel S, Wintermeyer W, Rodnina MV. 2002. *Mol. Cell* 10:339–46
95. Kjeldgaard NO, Gaussing K. 1974. In *Ribosomes*, ed. M Nomura, A Tisseres, P Lengyel, pp. 369–92. Cold Spring Harbor, NY: Cold Spring Harbor Press
96. Rodnina MV, Pape T, Savelsbergh A, Mohr D, Matassova NB, Wintermeyer W. 2000. See Ref. 45, pp. 301–17
97. Maden BEH, Traut RR, Monro RE. 1968. *J. Mol. Biol.* 35:333–45
98. Tenson T, Ehrenberg M. 2002. *Cell* 108:591–94
99. Deleted in proof
100. Pool MR, Stumm J, Fulga TA, Sinning I, Dobberstein B. 2002. *Science* 297:1345–48
101. Keenan RJ, Freymann DM, Stroud RM, Walter P. 2001. *Annu. Rev. Biochem.* 70:755–75
102. Gong F, Yanofsky C. 2002. *Science* 297:1864–67
103. Frank J, Agrawal RK. 2000. *Nature* 406:318–22
103a. Gabasvili IS, Gregory ST, Valle M, Grassucci R, Worbs M, et al. 2001. *Mol. Cell* 8:181–88
104. Schluenzen F, Zarivach R, Harms J, Bashan A, Tocilj A, et al. 2001. *Nature* 413:814–21
105. Hansen JL, Ban N, Nissen P, Moore PB, Steitz TA. 2002. *Mol. Cell* 10:117–26
106. Hardesty B, Kramer G, eds. 1986. *Structure, Function, and Genetics of Ribosomes.* New York: Springer-Verlag

AUTHOR INDEX

D

E

F

H

K

M

P

S

Z

Subject Index

D

F

I

T

U

Cumulative Indexes

CONTRIBUTING AUTHORS, VOLUMES 68–72

CHAPTER TITLES, VOLUMES 68–72

PREFATORY

DNA

DNA Chemistry and Structure

Methodology

Replication

ENZYMOLOGY

CARBOHYDRATES

OTHER BIOMOLECULES

BIOENERGETICS

SIGNAL TRANSDUCTION